Atomic Masses of the [Elements]
Based on the 1995 IUPAC Table[s]

Name	Symbol	Atomic Number	Atomic Mass	Name	Symbol	Atomic Number	Atomic Mass
Actinium*	Ac	89	227.0277	Mercury	Hg	80	200.59
Aluminum	Al	13	26.981538	Molybdenum	Mo	42	95.94
Americium*	Am	95	243.0614	Neodymium	Nd	60	144.24
Antimony	Sb	51	121.760	Neon	Ne	10	20.1797
Argon	Ar	18	39.948	Neptunium*	Np	93	237.0482
Arsenic	As	33	74.92160	Nickel	Ni	28	58.6934
Astatine*	At	85	209.9871	Niobium	Nb	41	92.90638
Barium	Ba	56	137.327	Nitrogen	N	7	14.00674
Berkelium*	Bk	97	247.0703	Nobelium*	No	102	259.1011
Beryllium	Be	4	9.012182	Osmium	Os	76	190.23
Bismuth	Bi	83	208.98038	Oxygen	O	8	15.9994
Bohrium	Bh	107	—	Palladium	Pd	46	106.42
Boron	B	5	10.811	Phosphorus	P	15	30.973762
Bromine	Br	35	79.904	Platinum	Pt	78	195.078
Cadmium	Cd	48	112.411	Plutonium*	Pu	94	244.0642
Calcium	Ca	20	40.078	Polonium*	Po	84	208.9824
Californium*	Cf	98	251.0796	Potassium	K	19	39.0983
Carbon	C	6	12.0107	Praseodymium	Pr	59	140.90765
Cerium	Ce	58	140.116	Promethium*	Pm	61	144.9127
Cesium	Cs	55	132.90545	Protactinium*	Pa	91	231.03588
Chlorine	Cl	17	35.4527	Radium*	Ra	88	226.0254
Chromium	Cr	24	51.9961	Radon*	Rn	86	222.0176
Cobalt	Co	27	58.933200	Rhenium	Re	75	186.207
Copper	Cu	29	63.546	Rhodium	Rh	45	102.90550
Curium*	Cm	96	247.0703	Rubidium	Rb	37	85.4678
Dubnium	Db	105	—	Ruthenium	Ru	44	101.07
Dysprosium	Dy	66	162.50	Rutherfordium	Rf	104	261.1089
Einsteinium*	Es	99	252.0830	Samarium	Sm	62	150.36
Erbium	Er	68	167.26	Scandium	Sc	21	44.955910
Europium	Eu	63	151.964	Seaborgium	Sg	106	—
Fermium*	Fm	100	257.0951	Selenium	Se	34	78.96
Fluorine	F	9	18.9984032	Silicon	Si	14	28.0855
Francium*	Fr	87	233.0197	Silver	Ag	47	107.8682
Gadolinium	Gd	64	157.25	Sodium	Na	11	22.989770
Gallium	Ga	31	69.723	Strontium	Sr	38	87.62
Germanium	Ge	32	72.61	Sulfur	S	16	32.066
Gold	Au	79	196.96655	Tantalum	Ta	73	180.9479
Hafnium	Hf	72	178.49	Technetium*	Tc	43	97.9072
Hassium	Hs	108	—	Tellurium	Te	52	127.60
Helium	He	2	4.002602	Terbium	Tb	65	158.92534
Holmium	Ho	67	164.93032	Thallium	Tl	81	204.3833
Hydrogen	H	1	1.00794	Thorium*	Th	90	232.0381
Indium	In	49	114.818	Thulium	Tm	69	168.93421
Iodine	I	53	126.90447	Tin	Sn	50	118.710
Iridium	Ir	77	192.217	Titanium	Ti	22	47.867
Iron	Fe	26	55.845	Tungsten	W	74	183.84
Krypton	Kr	36	83.80	Ununnilium	Uun	110	—
Lanthanum	La	57	138.9055	Unununium	Uuu	111	—
Lawrencium*	Lr	103	262.110	Ununbium	Uub	112	—
Lead	Pb	82	207.2	Uranium*	U	92	238.0289
Lithium	Li	3	6.941	Vanadium	V	23	50.9415
Lutetium	Lu	71	174.967	Xenon	Xe	54	131.29
Magnesium	Mg	12	24.3050	Ytterbium	Yb	70	173.04
Manganese	Mn	25	54.938049	Yttrium	Y	39	88.90585
Meitnerium	Mt	109	—	Zinc	Zn	30	65.39
Mendelevium*	Md	101	258.0984	Zirconium	Zr	40	91.224

*This element has no stable isotopes. The atomic mass given is that of the isotope with the longest known half-life.

INTRODUCTION TO GENERAL, ORGANIC, AND BIOCHEMISTRY

INTRODUCTION TO GENERAL, ORGANIC, AND BIOCHEMISTRY

SEVENTH EDITION

Morris Hein
Mount San Antonio College

Leo R. Best
Mount San Antonio College

Scott Pattison
Ball State University

Susan Arena
University of Illinois, Urbana–Champaign

BROOKS/COLE

THOMSON LEARNING

Australia ▪ Canada ▪ Mexico ▪ Singapore ▪ Spain ▪ United Kingdom ▪ United States

BROOKS/COLE

THOMSON LEARNING

Sponsoring Editor: *Jennifer Huber*
Marketing Team: *Carla Martin-Falcone*
Editorial Assistant: *Faith Riley*
Production Editor: *Tom Novack*
Production Service: *Ex Libris/Julie Kranhold*
Manuscript Editor: *Luana Richards*
Permissions Editor: *Sue Ewing*
Interior Design: *Nancy Benedict Breuer*
Cover Design: *Vernon T. Boes*
Cover Illustration: *Internal structure of hemoglobin by Kenneth Eward*

Interior Illustration: *Lotus Art, Pat Rogondino*
Photo Researcher: *Julie Kranhold*
Print Buyer: *Vena Dyer*
Typesetting: *The Clarinda Company*
Cover Printing: *R. R. Donnelley & Sons Co.—Willard*
Printing and Binding: *R. R. Donnelley & Sons Co.—Willard*

For more information about this or any other Brooks/Cole products, contact:

BROOKS/COLE
511 Forest Lodge Road
Pacific Grove, CA 93950 USA
www.brookscole.com
1-800-423-0563 (Thomson Learning Academic Resource Center)

For permission to use material from this work, contact us by
www.thomsonrights.com
fax: 1-800-730-2215
phone: 1-800-730-2214

Printed in United States of America

10 9 8 7 6 5 4 3 2 1

Library of Congress Cataloging-in-Publication Data
Introduction to general, organic, and biochemistry / Morris Hein . . . [et al.].—7th ed.
    p. cm.
    Includes index.
    ISBN 0-534-37998-2
    1. Chemistry. I. Hein, Morris.
QD31.2.I568 2001
540—dc21   00-043625

Chemistry is the central science, drawing on the basic principles of physics while enabling biological phenomena to be understood at a molecular level. From this unique position, chemistry pervades virtually every aspect of modern life, influencing the quality of the water that sustains our lives, the pharmaceuticals that save our lives, and the advanced materials that enrich our lives. It would seem reasonable that a truly informed person in the modern world should have at least a rudimentary scientific literacy in the molecular sciences (as well as being able to read and write). . . .

Donald Weaver
FELLOW OF THE CANADIAN INSTITUTE OF CHEMISTS
PROFESSOR OF MEDICINE AND CHEMISTRY AT QUEEN'S UNIVERSITY
KINGSTON, ONTARIO, CANADA

Source: _Chemistry International,_ 2000, Vol. 22, No. 1.

PREFACE

Introduction to General, Organic, and Biochemistry was originally intended for students who had never taken a chemistry class but have a limited mathematical background, or those who had a significant interruption in their studies and are now returning to school to pursue various career objectives. Our goal for this edition is to continue to present chemistry in a clear, engaging manner that will stimulate students to further their scientific knowledge as they prepare for health sciences, nursing, and other careers. The central focus is the same as it has been from the first edition: How can we make the wide variety of chemical processes, occurring both within our bodies and in our surroundings, accessible to students and teach them the problem-solving skills they will need?

In preparing the Seventh Edition we considered the comments and suggestions of students and instructors to design a revision that builds on the strengths of previous editions and presents chemistry as a modern, vital subject. We have especially tried to relate chemistry to the real lives of our students as we develop the principles that form the foundation for the study of general, organic, and biochemistry.

Development of Problem-Solving Skills

We all want our students to develop solid skills in solving problems. We believe that a key to the success of this text is the fact that our problem-solving approach works for students. It is a step-by-step process that teaches the use of units and shows the change from one unit to the next. Students learn concepts most easily in a step-by-step process. In this edition we continue to incorporate fundamental mathematical skills, scientific notation, and significant figures in our examples. Painstaking care has been taken to show each step in the problem-solving process *(see pp. 108, 140–141)*. We continue to use four significant figures for atomic and molar masses for consistency and for rounding off answers appropriately. We have been meticulous in providing answers, correctly rounded, for students who have difficulty with mathematics.

Fostering Student Skills   *Attitude* plays a critical role in problem solving. We encourage students to learn that a systematic approach to solving problems is better than simple memorization. Throughout the book we encourage students to begin by writing down the facts or data given in a problem *(see pp. 134 and 180)* and to think their way through the problem to an answer, which is then checked to see if it makes sense. Once we have laid the foundations of concepts, we highlight the steps in blue so students can locate them easily. Important rules and equations are highlighted for emphasis and ready reference.

Student Practice   Practice problems follow the examples in the text, with answers provided at the end of the chapter. The end of each chapter begins with *Questions,* which help students review key terms and concepts, as well as material presented in tables and figures. This is followed by *Paired Exercises,* covering concepts and numerical exercises, where two similar exercises are presented side by side. The final section, *Additional Exercises,* presents questions and problems that are much more challenging to the student. Answers for *all* even-numbered questions and exercises appear in Appendix VI.

Organization

We continue to emphasize the less theoretical aspects of chemistry early in the book, leaving the more abstract theory for later. This sequence seems especially appropriate in a course where students are encountering chemistry for the very first time. Atoms, molecules, and reactions are all an integral part of the chemical nature of matter. A sound understanding of these topics allows the student to develop a basic understanding of chemical properties and vocabulary.

We build toward a basic knowledge of organic and biochemistry for the health science student. Thus, we stress the nomenclature, structure, and reactivity of each major organic functional group. In turn, the basic biochemical concepts rest on this foundation. We encourage the students to apply their understanding to examples drawn from medicine, nutrition, agriculture, and so on.

Chapter 1 gives students an understanding of the scientific method by introducing the course with a narrative account of the discovery of nitinol, often called memory metal. Also included in this chapter is material on the benefits and risks of science in our high tech world.

Chapters 2 and 3 present the basic mathematics and the language of chemistry, including an explanation of the metric system and significant figures. In Chapter 4 we present chemical properties—the ability of a substance to form new substances. Then, in Chapter 5 students encounter the history and language of basic atomic theory.

We continue to present new material at a level appropriate for the beginning student by emphasizing nomenclature, composition of compounds, and reactions in Chapters 6 through 9 before moving into the details of modern atomic theory. The Periodic Table is introduced and discussed in Chapters 10 and 11. Students gain confidence in their own ability to identify and work with chemicals in the laboratory before tackling the molecular models of matter. As practicing chemists we have little difficulty connecting molecular models and chemical properties. Students, especially those with no prior chemistry background, may not share this ability to connect the molecular models and the macroscopic properties of matter. Those instructors who feel it is essential to teach atomic theory and bonding early in the course can cover Chapters 10 and 11 immediately following Chapter 5.

In Chapters 19 through 26 we introduce organic chemistry. We have reviewed and carefully selected organic reactions to illustrate the reactivities of each important functional group. Three general categories of organic reactions—substitution, elimination, and addition—are introduced and, where possible, subsequent chapters present reactions within this conceptual framework.

IUPAC nomenclature is emphasized in this edition, but we have also considered how organics are named in everyday usage. Thus, we present a common name if it continues to be widely used.

Finally, we examine the principles of biochemistry in Chapters 27 through 35 and again strive to illustrate current, relevant applications such as the Human

Genome Project, gene therapy, and sources for, and uses of, industrial enzymes. Because biochemistry is increasingly a visual science, molecular models have been incorporated to help students better comprehend biochemical functions. As such, Chapter 29, "Amino Acids, Polypeptides, and Proteins," features a three-dimensional structure-to-function approach and Chapter 30, "Enzymes," stresses a qualitative and visual approach to enzymes. Finally, the entire text has been reexamined and the prose updated and rewritten to improve its clarity.

Learning Aids

To help the beginning student gain the confidence necessary to master technical material, we have refined and enhanced a series of learning aids:

- Important **terms** are set off in bold type where they are defined, and are printed in blue in the margin. These terms are printed in bold type in the index.
- Worked **examples** show students the how of problem solving before they are asked to tackle problems on their own.
- **Practice problems** give immediate reinforcement of a skill shown in the example problems. Answers are provided at the end of the chapter to encourage students to check their problem solving immediately.
- **Marginal notations** remind students of basic concepts and problem-solving techniques. These are printed in magenta ink to clearly distinguish them from text and vocabulary terms.
- **Each chapter opens** with a *color photograph* and a *chapter outline* assists students in viewing the topics.
- **Chemistry in Action** essays show the impact of chemistry in a variety of applications. These essays cover such topics as Hydrogen: A Fuel of the Future, Fast Food or Fast Fat, The Double Helix Revisited, and Designing Low-Calorie Fats.
- **Steps for solving problems** are printed in blue for easy reference.
- **Important statements,** equations, and laws are highlighted for emphasis.
- A list of **Concepts in Review** at the end of each chapter guides students in determining the most important concepts in the chapter.
- **Key terms** are listed alphabetically at the end of each chapter with section references to assist students in review of new vocabulary. Terms are also printed in bold type in the index.
- **End-of-chapter exercises** provide practice and review of the chapter material. *Paired exercises* present two parallel exercises, side by side, so the student can use the same problem-solving skills with two sets of similar information.
- **Additional exercises** at the end of each chapter provide students with more challenging questions and problems.
- **Answers** to even-numbered questions and exercises are given in Appendix VI.
- A **Review of Mathematics** is provided in Appendix I.
- Units of measurement are shown in table format in Appendix IV and in the endpapers.

New to This Edition

A number of changes have been made in this edition, but the level of the material remains the same. The entire text was reviewed and reworked to reflect the latest in chemistry education and to provide students with greater assistance in solving problems. Specific new features and changes include the following:

- Chapter 14, "Solutions," has been revised. Material on saturated solutions is now included in the discussion of solubility. Normality has been deleted from the chapter to allow students to focus on the role of the mole in concentrations of solutions.
- In Chapter 15, "Acids, Bases, and Salts," the discussion of colloids has been streamlined to focus attention on properties and applications.
- Chapter 16, "Chemical Equilibrium," has been significantly revised to focus student attention on the changes occurring as a system approaches equilibrium.
- A fresh, new design includes a modern art and photo program; illustrations for molecules and chemical reactions have been enhanced to improve students' ability to visualize.
- Nine new **Chemistry in Action** essays have been added, including topics that address current environmental concerns such as household pollution and fuel-efficient cars. Chemistry in Action features an essay by James Watson, "The Double Helix Revisited."
- **Putting It Together,** a new review section, summarizes every 2 to 4 chapters and includes additional conceptually oriented exercises for effective self-review.
- New illustrations and examples from recent biochemical, medical, and nutritional literature have been included.
- Chapter 21, "Polymers and Macromolecules," now focuses on addition polymers and has been moved to follow Chapter 20, "Unsaturated Hydrocarbons." Condensation polymerization (e.g., polyesters and polyamides) is now covered in those chapters that contain the relevant functional groups.
- New in-chapter practice problems have been added to both the organic and biochemistry chapters.
- Practice problems in the organic chapters now involve biochemical applications.
- More detailed instruction is provided for naming organic compounds.
- Additional color-coding has been used in chemical equations and formulas to aid student comprehension.
- Appendix II, "Using a Scientific Calculator," instructs students on how to perform the basic calculations used in this text (addition and subtraction, multiplication and division, exponents, logarithms and antilogs).
- **Online and Multimedia Resources.** Web icons encourage students to explore chemistry topics on the World Wide Web. The icons are placed at section headings to indicate related Web material.

A Complete Ancillary Package

The following comprehensive teaching and learning package accompanies this book.

For the student:

Media Resources

Brooks/Cole Chemistry Resource Center. This is Brooks/Cole's website for chemistry, which contains a homepage for *Introduction to General, Organic, and Biochemistry.* All information is arranged according to the text's Table of Contents. Students can access practice quizzes for every chapter, and hyperlinks that relate to each chapter's contents.

Other Study Aids

Solutions Manual by Morris Hein, Scott Pattison, and Susan Arena, includes answers and solutions to all end-of-chapter questions and exercises.

A Basic Math Approach to Concepts of Chemistry, Sixth Edition, by Leo Michels, is a self-paced paperbound workbook that has proven itself an excellent resource for students needing help with mathematical aspects of chemistry. Evaluation tests are provided for each unit and the test answers are given in the back of the book. A glossary is also included.

For the instructor:

For the Laboratory

Introduction to General, Organic, and Biochemistry in the Laboratory, Seventh Edition, by Morris Hein, Leo R. Best, Robert L. Miner, and Scott Ritchey, with contributions from Judith N. Peisen, includes 42 experiments for a laboratory program that may accompany the lecture course. Featuring updated information on waste disposal and emphasizing safe laboratory procedures, the lab manual also includes study aids and exercises. Six new class-tested experiments have been added. The lab manual includes:

- Directions for waste disposal for each experiment
- Directions for the preparation of all reagents
- Six study aids including Reading and Preparing a Graph and How to Use a Scientific Calculator
- Twenty-six exercises covering a host of subjects from writing chemical names and formulas to molecular models

Instructor's Manual for Introduction to General, Organic, and Biochemistry in the Laboratory, Seventh Edition, includes information on the management of the lab, evaluation of experiments, notes for individual experiments, a list of reagents and amounts needed for each experiment, and answer keys to each experiment's report form and all exercises.

Assessment Tools and Materials

Instructor's Manual with Test Items for Introduction to General, Organic, and Biochemistry, Seventh Edition, includes Chapter Tests, Test Questions, Answers to Chapter Tests, Answers to Test Questions, and the answers to all *Putting It Together* cumulative reviews. A copy of all test questions is provided electronically in *Brooks/Cole Assessment.*

Brooks/Cole Assessment is a browser-based testing, tutorial, and course management system. With a balance of efficiency and high performance, simplicity and versatility, *Brooks/Cole Assessment* gives you the power to transform the learning and teaching experience. This new-generation tool enables the instructor to invest less time on drills of basic skills and more time on higher-level learning activities (i.e., concepts and applications). Students can engage in unlimited practice in questions and problems, thus building their own confidence and skills. Results flow automatically to a grade book for tracking so that instructors can better assess students' understanding of the material, even before a class or an actual test. Instructors can also communicate with individual students, sections, or entire courses.

Presentation Tools and Online Resources

Transparencies in full color include illustrations from the text, enlarged for use in the classroom and lecture halls.

CNN Chemistry Video, produced by Turner Learning, can stimulate and engage your students by launching a lecture, sparking a discussion, or demonstrating an application. Each chemistry-related segment from recent CNN broadcasts clearly demonstrates the relevancy of chemistry to everyday life.

Acknowledgments

Books are the result of a collaborative effort of many talented and dedicated people. Among the friends and colleagues who have helped us, we appreciate the enthusiasm and energy of Connie Grosse and Ralph Martinez, who revised test items, and Rachael Henriques-Porter, who helped develop the Putting It Together sections and created the online quizzes for this edition. We are grateful for the many helpful comments from colleagues and students who, over the years, have made this book possible. We hope they will continue to share their ideas for change with us, either directly or through our publisher.

We are especially thankful for the support, friendship, and constant encouragement of our spouses, Edna, Joan, and Steve, who have endured many lost weekends and been patient and understanding through the long hours of this process. Their optimism and good humor have given us a sense of balance and emotional stability.

No textbook can be completed without the untiring efforts of many publishing professionals. Special thanks to the talented staff at Brooks/Cole, especially Tom Novack, Production Editor; Jennifer Huber, Chemistry Editor; and Stephanie Schmidt, Assistant Editor; who worked very hard to control the many aspects of this project. Much credit goes to Julie Kranhold of Ex Libris for her unfailing attention to detail and persistence in moving the book through the numerous stages of production. Julie is amazing in her ability to track each change and translate our ideas for art and photos into the reality of this edition. Our special thanks also to Nancy Benedict for the colorful and appealing design of the book. Our heartfelt thanks also to a great group of Brooks/Cole Thomson Learning book "reps" for their interest and enthusiasm as they talk with instructors across the country.

Our sincere appreciation goes to the following reviewers who were kind enough to read and give their professional comments: Kathleen Ashworth, Yakima Valley Community College; Gabriele Backes, Portland Community College; Aaron Brown, Los Angeles Pierce College; Thomas Everton, Ventura College; Michael Fields, Louisiana State University at Alexandria; Donna G. Friedman, St. Louis Community College at Florissant Valley; Robert Fremland, San Diego Mesa College; Allan A. Gahr, Gordon College; Robert Howell, University of Cincinnati—Raymond Walters College; E. Jerome Maas, Oakton Community College; Peggy J. McClure, Midlands Technical College; Kathy Mitchell, St. Petersburg Junior College; Tracy Lynn Moore, Louisiana State University at Eunice; Judith N. Peisen, Hagerstown Community College; Jeffrey A. Schneider, State University of New York College at Oswego; Tony Taylor, Santa Rosa Junior College; and Carolyn Yoder, Millersville University.

Morris Hein, Leo Best, Scott Pattison, and Susan Arena

About the Authors

Morris Hein is professor emeritus of chemistry at Mt. San Antonio College, where he regularly taught the preparatory chemistry course. His name has become synonymous with clarity, meticulous accuracy, and a step-by-step approach that students can follow. Over the years, more than three million students have learned chemistry using a text by Morris Hein. In addition to *Foundations of College Chemistry, Tenth Edition,* he is co-author of *Introduction to General, Organic, and Biochemistry, Seventh Edition,* and *Introduction to Organic and Biochemistry.* He is also co-author of *Foundations of Chemistry in the Laboratory, Tenth Edition,* and *Introduction to General, Organic, and Biochemistry in the Laboratory, Seventh Edition.*

Leo R. Best taught chemistry for twenty-four years at Mt. San Antonio College. He has been a collaborator and co-author with Morris Hein since the first edition of *Introduction to General, Organic, and Biochemistry.* He is also co-author of *Foundations of Chemistry in the Laboratory* and *Introduction to General, Organic, and Biochemistry in the Laboratory.*

Scott Pattison lives in Muncie, Indiana, and is a dedicated teacher of general, organic, and biochemistry at Ball State University. He has been a co-author with Morris Hein since the third edition of *Introduction to General, Organic, and Biochemistry.*

Susan Arena currently teaches general chemistry and is director of the Merit Program for Emerging Scholars at the University of Illinois, Urbana-Champaign. She is co-author with Morris Hein on *Foundations of College Chemistry, Tenth Edition,* and is also co-author of *Introduction to General, Organic, and Biochemistry, Seventh Edition,* and *Introduction to Organic and Biochemistry.*

BRIEF CONTENTS

CONTENTS

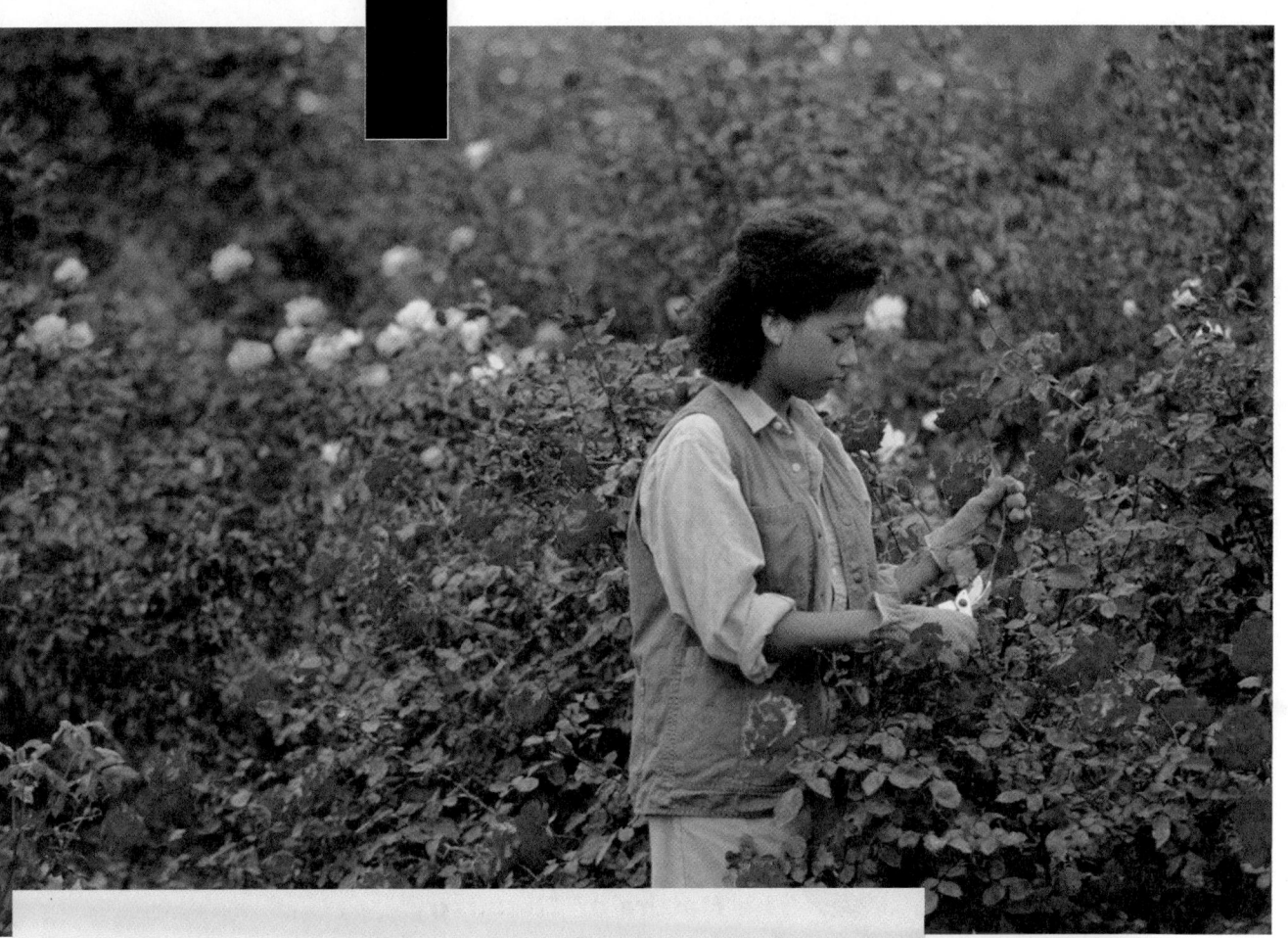

CHAPTER

1

An Introduction to Chemistry

CHAPTER 1 / OUTLINE

▲
The colors, fragrances, and textures of a rose garden are all the results of chemical reactions—examples of chemistry in our lives.

CHAPTER

1

Have you ever wondered what causes a rose to be fragrant? Perhaps you have been mesmerized by the flames in your fire on a romantic evening as they change color and form. And think of your relief when you dropped a container and found that it was plastic, not glass. These phenomena are the result of chemistry that occurs all around us, all the time. Chemical changes bring us beautiful colors, warmth, light, and products to make our lives function more smoothly. Understanding, explaining, and using the diversity of materials we find around us is what chemistry is all about.

1.1   Why Study Chemistry?

Chemistry is fascinating to many people. Learning about the composition of the world around us can lead to interesting and useful inventions and new technology. More than likely you are taking this chemistry course because someone has decided that it is an important part of your career goals. Chemistry is central to understanding many fields including agriculture, astronomy, animal science, geology, medicine, applied health technology, fire science, biology, molecular biology, and materials science. Even if you are not planning to work in any of these fields, chemistry is used by each of us every day in our struggle to cope with our technological world. Learning about the benefits and risks associated with chemicals will help you to be an informed citizen, able to make intelligent choices concerning the world around you. Studying chemistry teaches you to solve problems and communicate with others in an organized and logical manner. These skills will be helpful in college and throughout your career.

1.2   The Nature of Chemistry

chemistry

Key words are highlighted in bold and color in the margin to alert you to new terms defined in the text.

What is chemistry? One dictionary gives this definition: "**Chemistry** is the science of the composition, structure, properties, and reactions of matter, especially of atomic and molecular systems." Another, somewhat simpler definition is "Chemistry is the science dealing with the composition of *matter* and the changes in composition that matter undergoes." Neither of these definitions is entirely adequate. Chemistry, and physics, form a fundamental branch of knowledge. Chemistry is also closely related to biology, not only because living organisms are made of material substances but also because life itself is essentially a complicated system of interrelated chemical processes.

The scope of chemistry is extremely broad. It includes the whole universe and everything, animate and inanimate, in it. Chemistry is concerned with the composition and changes in the composition of matter, and also with the energy and energy changes associated with matter. Through chemistry we seek to learn and to understand the general principles that govern the behavior of all matter.

The chemist, like other scientists, observes nature and attempts to understand its secrets: What makes a rose red? Why is sugar sweet? What is occurring when iron rusts? Why is carbon monoxide poisonous? Problems such as these—some of which have been solved, some of which are still to be solved—are all part of what we call chemistry.

A chemist may interpret natural phenomena; devise experiments that reveal the composition and structure of complex substances; study methods for improving natural processes; or synthesize substances. Ultimately, the efforts of successful chemists advance the frontiers of knowledge and at the same time contribute to the well-being of humanity.

1.3 The Process of Chemistry

How does a chemist discover a new compound and study its behavior? A combination of hard work, accident, and luck produced "memory metal," an advanced substance now used to make toys, braces for teeth, and to treat problems such as blood clots and reattachment of tendons. However, none of these applications was the target of the original search for a new material. They all developed from the investigation of the properties of memory metal and the creativity of people associated with the work.

The "memory metal" discovery story began in Maryland at the Naval Ordnance Laboratory (NOL) where William J. Buehler had been given the task of finding a suitable substance for the nose cone of underwater missile launchers. Buehler was experiencing personal difficulties at this particular time in his life and so he became particularly immersed in his work. He eventually eliminated all but 12 compounds. He then began measuring their impact resistance. His test was simple but effective: He made buttons of the compound to be tested and hit them with a hammer. One substance, a 50-50 mixture of nickel and titanium, demonstrated greater impact resistance, elasticity, malleability, and resistance to fatigue than the others. Buehler named it *nitinol,* from *Ni*ckel *Ti*tanium *N*aval *O*rdnance *L*aboratory.

As Buehler and his staff tested nitinol, they varied the percentages of nickel and titanium to determine the effect of composition on the properties of the compound. They made a series of bars in a furnace and, after cooling them, smoothed them on a shop grinder. Buehler accidentally dropped one of the bars and noticed that it made a dull thud—much like that of a bar of lead. This aroused his curiosity and he began dropping other bars. To his surprise, he found that the cooled bars made a dull sound whereas the warm ones made a bell-like tone. Fascinated, he then began reheating and cooling the bars, discovering that the sound changed consistently back and forth from dull (cool) to bell-like (warm). Why was this so intriguing? Because these variances indicated a change in the atomic structure of the metal!

During a project review at NOL, Buehler demonstrated the fatigue resistance of nitinol by bending a long strip of wire into accordion folds. He passed it around, letting the directors flex it and straighten it. One of them wanted to see what would happen when it was heated and held the pleated nitinol over a flame. To everyone's amazement it stretched out into a straight wire. Buehler recognized that this behavior related to the different sounds made when nitinol was heated or cooled.

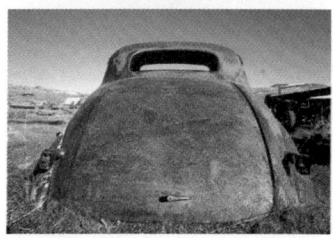

▲
The fragrance of this beautiful rose and the rust on this car are two examples of the chemical reactions.

Buehler recruited Frederick Wang, a crystallographer, to define the "memory" properties of the metal. Wang determined the structural changes that give nitinol its unique characteristic of memory—changes that involve the rearrangement of the position of particles within the solid. Changes between solids and liquids or liquids and gases are well known (e.g., boiling water or melting ice), but the same sort of changes can also occur between two solids. In the case of nitinol, a *parent shape* (the return shape) has to be defined. To fix the parent shape of nitinol it must be heated. No changes are apparent, but when it cools, the atoms in the solid rearrange into a slightly different structure. Thus, whenever the nitinol is heated, the atoms rearrange themselves back to the structure that produces the parent shape.

Buehler and Wang then proceeded to move this "memory" metal from the experimental world of the laboratory to the commercial world of applications. By the late sixties, nitinol was being used in pipe couplings in the aircraft industry.

In 1968, George Andreason, a dentist, began experimenting with nitinol in his metalworking shop where he made jewelry as a hobby. He developed a fine wire that could be molded to fit a patient's mouth (parent shape), which when cooled could be bent to fit the misaligned teeth. When warmed to body temperature, it would exert a gentle constant pressure on the teeth. A major breakthrough in orthodontics, nitinol braces cut the treatment time in half from that of steel braces.

Dr. Wang left NOL in 1980 to become a supplier of nitinol to the numerous manufacturers who wanted to use this amazing substance for such things as eyeglass frames that can take extreme abuse (e.g., bending them, sitting on them, or twisting them) and return to their original shape; antiscalding devices for showerheads and faucets that automatically shut off if the water approaches 120°F; and toys such as blinking movie posters and tail-swishing dinosaurs.

The uses of nitinol are varied and growing. Today it has many applications in medicine, engineering, and safety; in housewares; and even in the lingerie business in underwire brassieres. Nitinol was the first of the "intelligent" materials—ones that respond to changes in the environment.

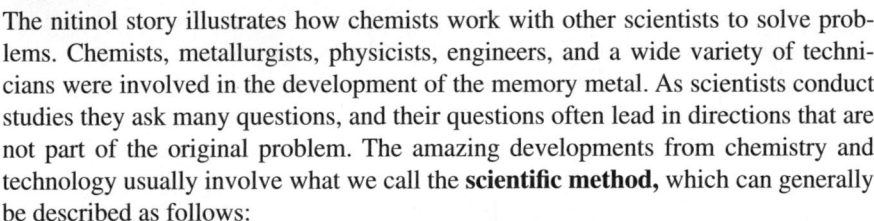

◀ FIGURE 1.1
The scientific method.

▲
Scientists employ the scientific method every day in their laboratory work.

scientific method

1.4   The Scientific Method

The nitinol story illustrates how chemists work with other scientists to solve problems. Chemists, metallurgists, physicists, engineers, and a wide variety of technicians were involved in the development of the memory metal. As scientists conduct studies they ask many questions, and their questions often lead in directions that are not part of the original problem. The amazing developments from chemistry and technology usually involve what we call the **scientific method,** which can generally be described as follows:

1. **Collect the facts or data** that are relevant to the problem or question at hand. This is usually done by planned experimentation. The data are then analyzed to find trends or regularities that are pertinent to the problem. In the nitinol story, Buehler compared the properties of various substances until he selected nitinol. He then investigated the properties of nitinol, including its ability to produce different sounds when warm and cool, and its unique ability to return to its original shape.
2. **Formulate a hypothesis** that will account for the data and that can be tested by further experimentation. Wang and Buehler proposed a hypothesis regarding the two phases of nitinol. Further tests supported their model.
3. **Plan and do additional experiments to test the hypothesis.** The nitinol group continued its testing process until the mechanism was well understood.
4. **Modify the hypothesis** as necessary so that it is compatible with all the pertinent data.

Confusion sometimes arises regarding the exact meanings of the words *hypothesis, theory,* and *law.* A **hypothesis** is a tentative explanation of certain facts that provides a basis for further experimentation. A well-established hypothesis is often called a **theory** or model. Thus a theory is an explanation of the general principles of certain phenomena with considerable evidence or facts to support it. Hypotheses and theories explain natural phenomena, whereas **scientific laws** are simple statements of natural phenomena to which no exceptions are known under the given conditions.

hypothesis

theory

scientific laws

These four steps are a broad outline of the general procedure that is followed in most scientific work, but they are not a "recipe" for doing chemistry or any other science (Figure 1.1). Chemistry is an experimental science, however, and much of its progress has been due to application of the scientific method through systematic research.

We study many theories and laws in chemistry; this makes our task as students easier because theories and laws summarize important aspects of the sciences. Certain theories and models advanced by great scientists in the past have since been substantially altered and modified. Such changes do not mean that the discoveries of the past are any less significant. Modification of existing theories and models in the light of new experimental evidence is essential to the growth and evolution of scientific knowledge. Science is dynamic.

1.5   Relationship of Chemistry to Other Sciences and Industry

Besides being a science in its own right, chemistry is the servant of other sciences and industry. Chemical principles contribute to the study of physics, biology, agriculture, engineering, medicine, space research, oceanography, and many other sciences. Chemistry and physics are overlapping sciences, because both are based on the properties and behavior of matter. Biological processes are chemical in nature. The metabolism of food to provide energy to living organisms is a chemical process. Knowledge of the molecular structure of proteins, hormones, enzymes, and the nucleic acids assists biologists in their investigations of the composition, development, and reproduction of living cells.

Chemistry plays an important role in our quest to alleviate the growing shortage of food in the world. Agricultural production has been increased with the use of chemical fertilizers, pesticides, and improved varieties of seeds. Chemical refrigerants make possible the frozen food industry, which preserves large amounts of food that might otherwise spoil. Chemistry is also producing synthetic nutrients, but much remains to be done as the world population increases relative to the land available for cultivation. Expanding energy needs have brought about difficult environmental problems in the form of air and water pollution. Chemists and other scientists are working diligently to alleviate these problems.

Advances in medicine and chemotherapy, through the development of new drugs, have prolonged life and relieved human suffering. More than 90% of the drugs and pharmaceuticals being used in the United States today have been developed commercially within the past 50 years. The plastics and polymer industry, unknown 60 years ago, has revolutionized the packaging and textile industries and continues to produce durable and useful construction materials. Energy derived from chemical processes is used for heating, lighting, and transportation. Virtually every industry is dependent on chemistry.

People outside of science usually have the perception that science is an intensely logical field. However, scientific discoveries are often the result of trial and error. In the nitinol story its memory nature was discovered only because someone wanted to see what would happen if a flame was applied to compressed nitinol. Buehler's creativity and insight enabled him to make a key connection. This is what we mean by serendipity in science. Buehler's search led him in a different direction than his assigned task because the results of his experiments were quite unexpected. The Navy found a new material for nose cones and at the same time discovered a material that has applications in medicine, engineering, and everyday life.

Household Pollution— A Health Hazard??

How often are you exposed to potentially harmful substances? Scientists have developed highly sensitive instruments and portable monitors to judge where, when, and how people come in contact with potentially dangerous chemicals. Since 1980, Wayne Ott from Stanford University and Lance Wallace from the Environmental Protection Agency (EPA) have been studying our everyday exposure to toxic substances. They have analyzed exposure rates in 15 states and parts of Canada. To look for volatile organic compounds and pesticides they often have people carry monitors during their daily activities.

The results are surprising: Most people have the greatest contact with potentially toxic chemicals in places we generally consider to be free of pollution—our homes, offices, and cars. Places that are usually targets of environmental laws (such as factories in urban areas) are relatively small contributors to our toxic exposure. Does this mean our homes pose a greater health risk than industrial pollution? It appears that this is exactly the case. Let's consider three types of indoor pollution: (1) volatile organic compounds, (2) fine dust particles, and (3) pesticides.

We are regularly exposed to low levels of volatile organic chemicals in our daily routines. Freshly dry-cleaned clothes contain traces of tetrachloroethylene ("perc"), known to cause cancer in lab animals. Benzene, which is known to cause leukemia in people exposed to high concentrations, is found in gasoline and tobacco smoke. In 1985, Wallace determined that the majority of benzene exposure comes from homes and offices. We can most easily lower our exposure to these and other volatile organic compounds by avoiding products containing these pollutants and by improving ventilation.

Fine dust particles in the air contribute to such health problems as asthma and allergies. A study conducted in Riverside, California, found that exposure to these tiny particles is 60% greater indoors than outdoors. As we move through our homes and offices we constantly stir up our own "dust cloud." When this is combined with particles from combustion (cooking, candles, and fires), we find levels of dust that would be called hazardous if found outdoors. Carpets are effective reservoirs for these dust particles (as well as for bacteria and allergens). Even regularly vacuumed carpets contain high levels of these pollutants. What can you do to lower these levels? Use floor coverings that are smooth and easy to clean (such as wood, tile, or linoleum) and wipe your feet on a doormat before entering your home.

In the late 1980s, studies were completed on pesticide levels in indoor air. In both Florida and Massachusetts, indoor air contained a minimum of 5 times the pesticide concentration of outdoor air. Pesticide residues enter our homes in many ways. Some are directly applied, while others are tracked in on the bottom of shoes. Pesticides can last for years in carpets protected from sunlight. Simply removing your shoes before entering your home can lower indoor levels of pesticides by about 90%.

Although environmental laws governing air quality have helped lower levels of pollution outdoors, the same air pollutants are found at much higher levels indoors. You can reduce your exposure to these chemicals by making small changes in your daily routine and by carefully examining the products you use. Knowledge of chemistry gives you the ability to make choices that lead to a less polluted and healthier lifestyle.

▲ **Indoor plants and less carpeting can help curb the pollution in our homes.**

The nitinol story also illustrates the fact that scientists do not work alone. Buehler, an engineer, was joined by Wang, a crystallographer, and many other chemists, engineers, dentists, and physicians in developing the applications of nitinol. Each of them had a contribution to make to the body of knowledge surrounding the substance nitinol. In chemistry teamwork and cooperation are vital.

1.6   Risks and Benefits

Many problems that rely on science for an answer confront us today. Virtually every day we read or hear about stories on

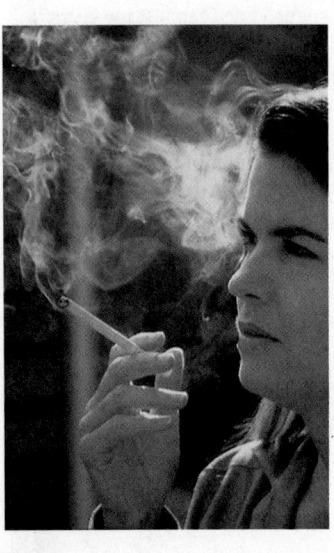

▲
Chemists continue to explore the effects of nicotine on the smoker in order to help those who want to quit smoking.

- developing an AIDS vaccine
- banning the use of herbicides and pesticides
- analyzing DNA to determine genetic disease, biological parents, or to place a criminal at the scene of a crime
- removing asbestos from public buildings
- removing lead from drinking water
- the danger of radon in our homes
- global warming
- the hole in the ozone layer
- health risks associated with coffee, alcohol, margarine, saturated fats, and other foods
- burning of tropical rain forests and the effect on global ecology
- health risks from tobacco use

Which of these risks present true danger to us and which are less threatening? These problems will be around for many years, and new ones will be continually added to the list. Wherever we live and whatever our occupation, each of us is exposed to chemicals and chemical hazards every day. The question we must address is: Do the risks outweigh the benefits?

Risk assessment is a process that brings together professionals from the fields of chemistry, biology, toxicology, and statistics in order to determine the risk associated with exposure to a certain chemical. Assessment of risk involves determining both the probability of exposure and the severity of that exposure. Once this is done, an estimate of the overall risk is made. Studies have shown how people perceive various risks. The perception of risk depends on some rather interesting factors. Voluntary risks, such as smoking or flying, are much more easily accepted than involuntary ones, such as herbicides on apples or asbestos in buildings. People also often conclude that anything "synthetic" is bad, while anything "organic" is good. Risk assessment may provide information on the degree of risk but not on whether the chemical is "safe." Safety is a qualitative judgment based on many personal factors including beliefs, preferences, benefits, and costs.

Once a risk has been assessed, the next step is to manage it. This involves ethics, economics, and equity as well as government and politics. For example, some things perceived as low risk by scientists (such as asbestos in buildings) are classified as high risk by the general public. This inconsistency may result in the expenditure of millions of dollars to rid us of a perceived threat that may be much lower than the general public believes.

Risk management involves value judgments that integrate social, economic, and political issues. These risks must then be weighed against the benefits of new tech-

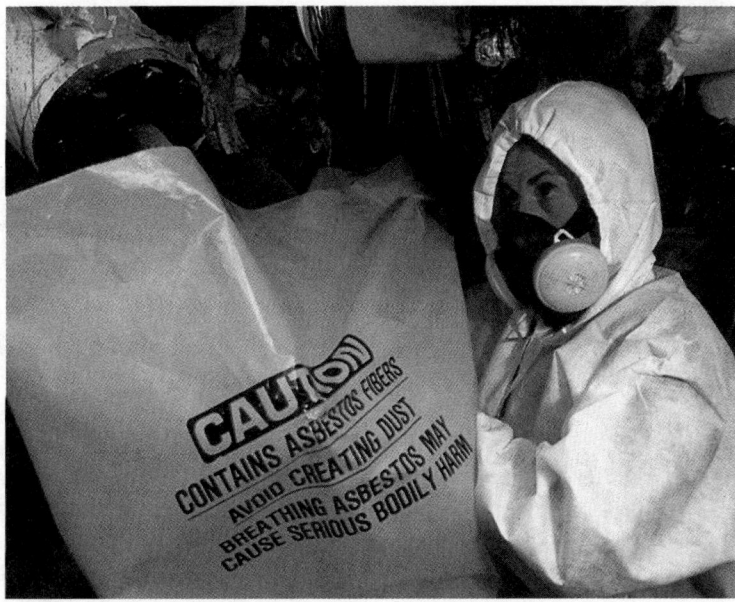

◀ Asbestos, once a widely used building material, was banned by the EPA in 1986 because of its health hazards. Today, the removal of asbestos from our homes and our public buildings has become a big concern.

nologies and products that will replace the old problem. We use both risk assessment and risk management to decide whether to buy a certain product (such as a pesticide), take a certain drug (such as a pain reliever), or eat certain foods (such as hot dogs). We must realize that all risks can never be completely eliminated. Our goal should be to minimize unnecessary risks and to make responsible decisions regarding necessary risks.

The theories and models used in risk assessment are based on assumptions and therefore contain uncertainties. Improve your understanding of the concepts of chemistry and you will be better able to understand the capabilities and limitations of science. You can then intelligently question the process of risk assessment and make decisions that will lead to a better understanding of our world and our responsibilities to each other.

Key Terms

The terms listed here are defined in the chapter. Section numbers are given in parentheses. More detailed definitions are given in the Glossary.

chemistry (1.2)   scientific method (1.4)
hypothesis (1.4)   theory (1.4)
scientific laws (1.4)

CHAPTER

2

Standards for Measurement

▲ Measuring instruments come in various shapes and sizes, all necessary to allow components to fit together, and permit us to quantify our world.

CHAPTER 2 / OUTLINE

Doing an experiment in chemistry is very much like cooking a meal in the kitchen. It's important to know the ingredients *and* the amounts of each in order to have a tasty product. Working on your car requires specific tools, in exact sizes. Buying new carpeting or draperies is an exercise in precise and accurate measurement for a good fit. A small difference in the concentration or amount of medication a pharmacist gives you may have significant effects on your well-being. In all of these cases making and properly using good measurements is key to a successful outcome. In chemistry, we begin by learning the metric system and the proper units for measuring mass, length, volume, and temperature.

2.1 Mass and Weight

Chemistry is an experimental science. The results of experiments are usually determined by making measurements. In elementary experiments the quantities that are

◄ **Mark Lee and Steven Smith work in the weightless environment of space to repair the Hubble Space telescope.**

commonly measured are mass, length, volume, pressure, temperature, and time. Measurements of electrical and optical quantities may also be needed in more sophisticated experimental work.

Although we often use mass and weight interchangeably in our everyday lives, **mass** they have quite different meanings in chemistry. In science we define the **mass** of a body as the amount of matter in that body. The mass of an object is a fixed and un**weight** varying quantity that is independent of the object's location. The mass of an object can be measured on a balance by comparison with other known masses.

An example of a balance is shown in Figure 2.1. These balances were used by assayers during the gold rush to determine the mass of gold brought in by the prospectors. The gold was placed on one pan of the balance and known masses on the other side until the pans were level with each other. (Several modern balances are shown in Figure 2.4, page 27.)

Scientists define the **weight** of an object as the measure of the earth's gravitational attraction for that object. Weight is measured on an instrument called a scale, which measures force against a spring. Unlike mass, weight varies in relation to the position of an object on Earth or its distance from Earth.

Consider an astronaut of mass 70.0 kilograms (154 pounds) who is traveling to space. At the instant before blast-off the weight of the astronaut is also 70.0 kilograms. As his distance from Earth increases and the rocket turns into an orbiting course, the gravitational pull on the astronaut's body decreases until a state of weightlessness (zero weight) is attained. However, the mass of the astronaut's body has remained constant at 70.0 kilograms during the entire event.

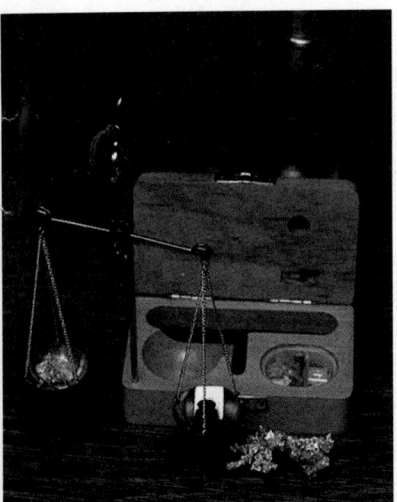

▲
FIGURE 2.1
An assayer's balance is used for weighing gold.

2.2   Measurement and Significant Figures

To understand certain aspects of chemistry it is necessary to set up and solve problems. Problem solving requires an understanding of the mathematical operations used to manipulate numbers. Numerical values or data are obtained from measurements made in an experiment. Chemists use these data to calculate the extent of the physical and chemical changes occurring in the substances that are being studied. By appropriate calculations an experiment's results can be compared with those of other experiments and summarized in ways that are meaningful.

A measurement is expressed by a numerical value together with a unit of that measurement. For example,

            ⌒ numerical value ⌝
        70.0 kilograms = 154 pounds
            ⌞_____ unit _____⌟

Numbers obtained from a measurement are never exact values. They always have some degree of uncertainty due to the limitations of the measuring instrument and the skill of the individual making the measurement. The numerical value recorded for a measurement should give some indication of its reliability (precision). To express maximum precision this number should contain all the digits that are known plus one digit that is estimated. This last estimated digit introduces some uncertainty. Because of this uncertainty every number that expresses a measurement can have only a limited number of digits. These digits, used to express a measured quantity, are known
significant figures as **significant figures.**

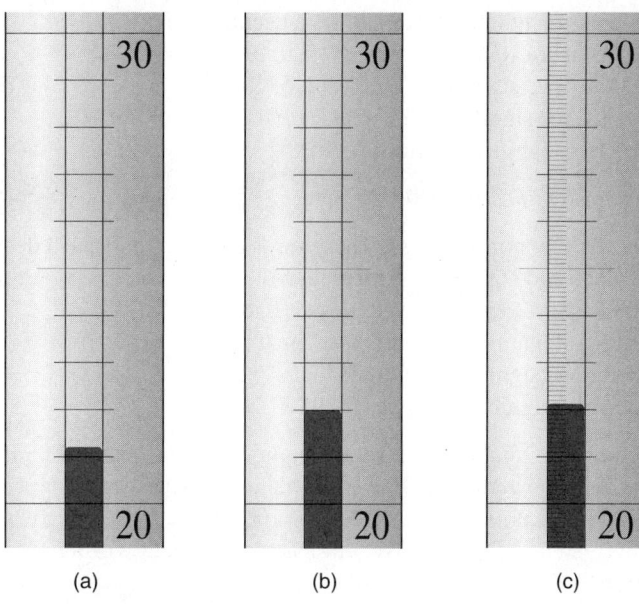

(a)   (b)   (c)

◄ **FIGURE 2.2**
Measuring temperature with various degrees of precision.

Suppose we measure temperature on a thermometer calibrated in degrees and observe that the mercury stops between 21 and 22 (see Figure 2.2a). We then know that the temperature is at least 21 degrees and is less than 22 degrees. To express the temperature with greater precision, we estimate that the mercury is about two-tenths the distance between 21 and 22. The temperature is therefore 21.2 degrees. The last digit (2) has some uncertainty because it is an estimated value. The recorded temperature, 21.2 degrees, is said to have three significant figures. If the mercury stopped exactly on the 22 (Figure 2.2b), the temperature would be recorded as 22.0 degrees. The zero is used to indicate that the temperature was estimated to a precision of one-tenth degree. Finally, look at Figure 2.2c. On this thermometer, the temperature is recorded as 22.11°C (four significant figures). Since the thermometer is calibrated to tenths of a degree, the first estimated digit is the hundredths.

Some numbers are exact and have an infinite number of significant figures. Exact numbers occur in simple counting operations; when you count 25 dollars, you have exactly 25 dollars. Defined numbers, such as 12 inches in 1 foot, 60 minutes in 1 hour, and 100 centimeters in 1 meter, are also considered to be exact numbers. Exact numbers have no uncertainty.

Evaluating Zero

In any measurement all nonzero numbers are significant. However, zeros may or may not be significant depending on their position in the number. Here are some rules for determining when zero is significant.

A zero is *significant* when it is:

1. between nonzero digits:
   205 has three significant figures
   2.05 has three significant figures
   61.09 has four significant figures

Rules for significant figures should be memorized for use throughout the text.

2. at the end of a number that includes a decimal point:
   0.500 has three significant figures (5, 0, 0)
   25.160 has five significant figures (2, 5, 1, 6, 0)
   3.00 has three significant figures (3, 0, 0)
   20. has two significant figures (2, 0)

A zero is *not significant* when it is:

1. before the first nonzero digit. These zeros are used to locate a decimal point:
   0.0025 has two significant figures (2, 5)
   0.0108 has three significant figures (1, 0, 8)
2. at the end of a number without a decimal point:
   1000 has one significant figure (1)
   590 has two significant figures (5, 9)

One way of indicating that these zeros are significant is to write the number using a decimal point and a power of 10. Thus if the value 1000 has been determined to four significant figures, it is written as 1.000×10^3. If 590 has only two significant figures, it is written as 5.9×10^2. We will learn more about writing numbers in this form in Section 2.4.

Answers to Practice Exercises are found at the end of each chapter.

Practice 2.1

How many significant figures are in each of these numbers?
(a) 4.5 inches   (e) 25.0 grams
(b) 3.025 feet   (f) 12.20 liters
(c) 125.0 meters   (g) 100,000 people
(d) 0.001 mile   (h) 205 birds

2.3 Rounding Off Numbers

When we do calculations we often obtain answers that have more digits than are justified. It is therefore necessary to drop the excess digits in order to express the answer with the proper number of significant figures. When digits are dropped from a number, the value of the last digit retained is determined by a process known as **rounding off numbers.** Two rules will be used in this book for rounding off numbers:

rounding off numbers

Not all schools use the same rules for rounding. Check with your instructor for variations in these rules.

Rule 1. When the first digit after those you want to retain is 4 or less, that digit and all others to its right are dropped. The last digit retained is not changed. The following examples are rounded off to four digits:

$74.693 = 74.69$ $1.00629 = 1.006$
    This digit is dropped.   These two digits are dropped.

Rule 2. When the first digit after those you want to retain is 5 or greater, that digit and all others to the right are dropped and the last digit retained is increased by one. These examples are rounded off to four digits:

1.026868 = 1.027

—— These three digits are dropped.
—— This digit is changed to 7.

18.02500 = 18.03

—— These three digits are dropped.
—— This digit is changed to 3.

12.899 = 12.90

—— This digit is dropped.
—— These two digits are changed to 90.

Practice 2.2

Round off these numbers to the number of significant figures indicated:
(a) 42.246 (four)   (d) 0.08965 (two)
(b) 88.015 (three) (e) 225.3 (three)
(c) 0.08965 (three) (f) 14.150 (three)

2.4 Scientific Notation of Numbers

Earth's age has been estimated to be about 4,500,000,000 (4.5 billion) years. Because this is an estimated value, say to the nearest 0.1 billion years, we are justified in using only two significant figures to express it. Thus we write it, using a power of 10, as 4.5×10^9 years.

Very large and very small numbers are often used in chemistry and can be simplified and conveniently written using a power of 10. Writing a number as a power of 10 is called **scientific notation.**

To write a number in scientific notation, move the decimal point in the original number so that it is located after the first nonzero digit. This new number is multiplied by 10 raised to the proper power (exponent). The power of 10 is equal to the number of places that the decimal point has been moved. If the decimal is moved to

scientific notation

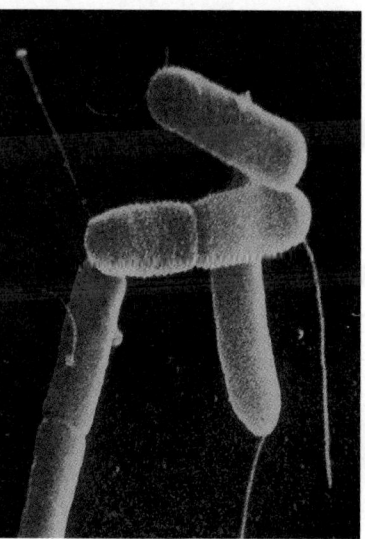

◀ Scientific notation is a useful way to write very large numbers such as the distance between Earth and the moon, or very small numbers such as the length of these E. coli bacteria (shown here as a colored scanning electron micrograph × 30,000).

the left, the power of 10 will be a positive number. If the decimal is moved to the right, the power of 10 will be a negative number.

Examples show you problem-solving techniques in a step-by-step form. Study each one and then try the Practice Exercises.

The scientific notation of a number is the number written as a factor between 1 and 10 multiplied by 10 raised to a power. For example,

$$2468 = 2.468 \times 10^3$$

number   scientific notation
                 of the number

Example 2.1   Write 5283 in scientific notation.

Solution

5283.   Place the decimal between the 5 and the 2. Since the decimal was moved
  3   three places to the left, the power of 10 will be 3 and the number 5.283
         is multiplied by 10^3.

5.283×10^3

Example 2.2   Write 4,500,000,000 in scientific notation (two significant figures).

Solution

4 500 000 000.   Place the decimal between the 4 and the 5. Since the decimal
     9   was moved nine places to the left, the power of 10 will be 9
            and the number 4.5 is multiplied by 10^9.

4.5×10^9

Example 2.3   Write 0.000123 in scientific notation.

Solution

0.000123   Place the decimal between the 1 and the 2. Since the decimal was
     4   moved four places to the right, the power of 10 will be -4 and the
         number 1.23 is multiplied by 10^{-4}.

1.23×10^{-4}

Practice 2.3

Write the following numbers in scientific notation:
(a) 1200 (four digits)   (c) 0.0468
(b) 6,600,000 (two digits)   (d) 0.00003

2.5   Significant Figures in Calculations

The results of a calculation based on measurements cannot be more precise than the least precise measurement.

Multiplication or Division

In calculations involving multiplication or division, the answer must contain the same number of significant figures as in the measurement that has the least number of significant figures. Consider the following examples:

Use your calculator to check your work in the examples. Compare your results to be sure you understand the mathematics.

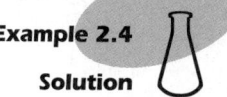

Example 2.4

Solution

$(190.6)(2.3) = 438.38$

The value 438.38 was obtained with a calculator. The answer should have two significant figures because 2.3, the number with the fewest significant figures, has only two significant figures.

Round off this digit to 4.

Drop these three digits.

438.38

Move the decimal 2 places to the left to express in scientific notation.

The correct answer is 440 or 4.4×10^2.

Example 2.5

Solution

$$\frac{(13.59)(6.3)}{12} = 7.13475$$

The value 7.13475 was obtained with a calculator. The answer should contain two significant figures because 6.3 and 12 each have only two significant figures.

Drop these four digits.

7.13475

This digit remains the same.

The correct answer is 7.1.

Practice 2.4

(a) $(134 \text{ in.})(25 \text{ in.}) = ?$

(b) $\dfrac{213 \text{ miles}}{4.20 \text{ hours}} = ?$

(c) $\dfrac{(2.2)(273)}{760} = ?$

Addition or Subtraction

The results of an addition or a subtraction must be expressed to the same precision as the least precise measurement. This means the result must be rounded to the same number of decimal places as the value with the fewest decimal places (blue line in examples).

Example 2.6   Add 125.17, 129, and 52.2.

Solution

$$
\begin{array}{r}
125.|17 \\
129.| \\
\underline{52.|2} \\
306.|37
\end{array}
$$

The number with the least precision is 129. Therefore the answer is rounded off to the nearest unit: 306.

Example 2.7   Subtract 14.1 from 132.56.

Solution

$$
\begin{array}{r}
132.5|6 \\
\underline{-14.1|} \\
118.4|6
\end{array}
$$

14.1 is the number with the least precision. Therefore the answer is rounded off to the nearest tenth: 118.5.

Example 2.8   Subtract 120 from 1587.

Solution

$$
\begin{array}{r}
158|7 \\
\underline{-12|0} \\
146|7
\end{array}
$$

120 is the number with least precision. The zero is not considered significant; therefore the answer must be rounded to the nearest ten: 1470 or 1.47×10^3.

Example 2.9   Add 5672 and 0.00063.

Solution

$$
\begin{array}{r}
5672| \\
\underline{+\ \ \ 0.|00063} \\
5672.|00063
\end{array}
$$

Note when a very small number is added to a large number the result is simply the original number.

The number with least precision is 5672. So the answer is rounded off to the nearest unit: 5672.

Example 2.10

$$
\frac{1.039 - 1.020}{1.039} = 0.018286814
$$

Solution   The value 0.018286814 was obtained with a calculator. When the subtraction in the numerator is done,

$$
1.039 - 1.020 = 0.019
$$

the number of significant figures changes from four to two. Therefore the answer should contain two significant figures after the division is carried out:

Drop these six digits.

0.018286814

This digit remains the same.

The correct answer is 0.018, or 1.8×10^{-2}.

Practice 2.5

How many significant figures should the answer in each of these calculations contain?

(a) $(14.0)(5.2)$ (e) $119.1 - 3.44$

(b) $(0.1682)(8.2)$ (f) $\dfrac{94.5}{1.2}$

(c) $\dfrac{(160)(33)}{4}$ (g) $1200 + 6.34$

(d) $8.2 + 0.125$ (h) $1.6 + 23 - 0.005$

If you need to brush up on your math skills refer to the "Mathematical Review" in Appendix I.

metric system or SI

Additional material on mathematical operations is given in Appendix I, "Mathematical Review." Study any portions that are not familiar to you. You may need to do this at various times during the course when additional knowledge of mathematical operations is required.

2.6 The Metric System

The **metric system,** or **International System** (**SI**, from *Système International*), is a decimal system of units for measurements of mass, length, time, and other physical quantities. Built around a set of standard units, the metric system uses factors of 10 to express larger or smaller numbers of these units. To express quantities that are larger or smaller than the standard units, prefixes are added to the names of the units. These prefixes represent multiples of 10, making the metric system a decimal system of measurements. Table 2.1 shows the names, symbols, and numerical values of the prefixes. Some examples of the more commonly used prefixes are

1 *kilo*meter = 1000 meters
1 *kilo*gram = 1000 grams
1 *milli*meter = 0.001 meter
1 *micro*second = 0.000001 second

The seven standard units in the International System, their abbreviations, and the quantities they measure are given in Table 2.2. Other units are derived from these units. The metric system, or International System, is currently used by most of the countries in the world, not only in scientific and technical work but also in commerce and industry.

▲
Most products today list both systems of measurement on their labels.

TABLE 2.1	Prefixes and Numerical Values for SI Units		
Prefix	Symbol	Numerical value	Power of 10 equivalent
exa	E	1,000,000,000,000,000,000	10^{18}
peta	P	1,000,000,000,000,000	10^{15}
tera	T	1,000,000,000,000	10^{12}
giga	G	1,000,000,000	10^{9}
mega	M	1,000,000	10^{6}
kilo	k	1,000	10^{3}
hecto	h	100	10^{2}
deka	da	10	10^{1}
—	—	1	10^{0}
deci	d	0.1	10^{-1}
centi	c	0.01	10^{-2}
milli	m	0.001	10^{-3}
micro	μ	0.000001	10^{-6}
nano	n	0.000000001	10^{-9}
pico	p	0.000000000001	10^{-12}
femto	f	0.000000000000001	10^{-15}
atto	a	0.000000000000000001	10^{-18}

The prefixes most commonly used in chemistry are shown in bold.

TABLE 2.2	International System's Standard Units of Measurement	
Quantity	Name of unit	Abbreviation
Length	meter	m
Mass	kilogram	kg
Temperature	kelvin	K
Time	second	s
Amount of substance	mole	mol
Electric current	ampere	A
Luminous intensity	candela	cd

 Measurement of Length

Standards for the measurement of length have a long history. The Old Testament mentions such units as the *cubit* (the distance from a man's elbow to the tip of his outstretched hand). In ancient Scotland the inch was once defined as a distance equal to the width of a man's thumb.

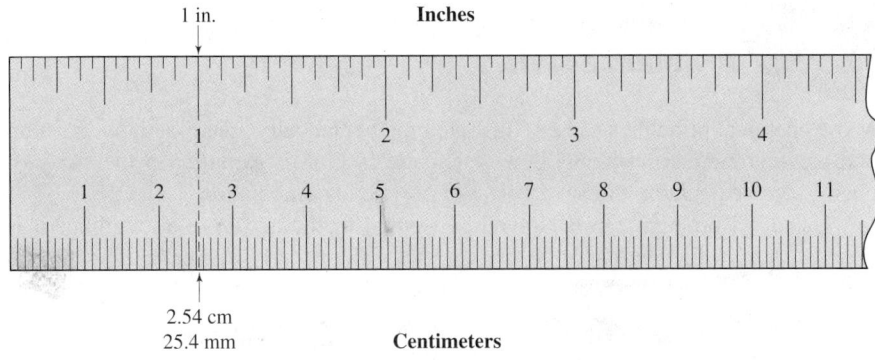

◀ **FIGURE 2.3**
**Comparison of the metric
and American systems of
length measurement:
2.54 cm = 1 in.**

Reference standards of measurements have undergone continuous improvements in precision. The standard unit of length in the metric system is the **meter (m)**. When the metric system was first introduced in the 1790s, the meter was defined as one ten-millionth of the distance from the equator to the North Pole, measured along the meridian passing through Dunkirk, France. In 1889, the meter was redefined as the distance between two engraved lines on a platinum–iridium alloy bar maintained at 0° Celsius. This international meter bar is stored in a vault at Sèvres near Paris. Duplicate meter bars have been made and are used as standards by many nations.

By the 1950s, length could be measured with such precision that a new standard was needed. Accordingly, the length of the meter was redefined in 1960 and again in 1983. The latest definition describes a meter as the distance that light travels in a vacuum during 1/299,792,458 of a second.

A meter is 39.37 inches, a little longer than 1 yard. One meter equals 10 decimeters, 100 centimeters, or 1000 millimeters (see Figure 2.3). A kilometer contains 1000 meters. Table 2.3 shows the relationships of these units.

The nanometer (10^{-9} m) is used extensively in expressing the wavelength of light as well as in atomic dimensions. See inside back cover for a complete table of common conversions.

meter (m)

Common length relationships:
$1\ m = 10^{6}\ \mu m = 10^{10}\ \text{Å}$
$= 100\ cm = 1000\ mm$
$1\ cm = 10\ mm = 0.01\ m$
$1\ in. = 2.54\ cm$
$1\ mile = 1.609\ km$

**See inside back cover for a
table of conversions.**

TABLE **2.3**	**Metric Units of Length**		
Unit	**Abbreviation**	**Meter equivalent**	**Exponential equivalent**
kilometer	km	1000 m	10^{3} m
meter	m	1 m	10^{0} m
decimeter	dm	0.1 m	10^{-1} m
centimeter	cm	0.01 m	10^{-2} m
millimeter	mm	0.001 m	10^{-3} m
micrometer	μm	0.000001 m	10^{-6} m
nanometer	nm	0.000000001 m	10^{-9} m
angstrom	Å	0.0000000001 m	10^{-10} m

2.8   Problem Solving

See Appendix II for help in using a scientific calculator.

Many chemical principles can be illustrated mathematically. Learning how to set up and solve numerical problems in a systematic fashion is *essential* in the study of chemistry. A calculator will save you a lot of time in computation.

    Usually a problem can be solved by several methods. But in all methods it is best, especially for beginners, to use a systematic, orderly approach. The *dimensional analysis method* is emphasized in this book because it

1. provides a systematic, straightforward way to set up problems.
2. gives a clear understanding of the principles involved.
3. trains you to organize and evaluate data.
4. helps to identify errors because unwanted units are not eliminated if the setup of the problem is incorrect.

Steps for solving problems are highlighted in color for easy reference.

The basic steps for solving problems are

1. Read the problem carefully. Determine what is to be solved for, and write it down.
2. Tabulate the data given in the problem. It is important to label *all* factors and measurements with the proper units.
3. Determine which principles are involved and which unit relationships are needed to solve the problem. You may need to refer to tables for needed data.
4. Set up the problem in a neat, organized, and logical fashion, making sure that unwanted units cancel. Use sample problems in the text as guides for setting up the problem.
5. Proceed with the necessary mathematical operations. Make certain that your answer contains the proper number of significant figures.
6. Check the answer to see if it is reasonable.

    A few more words about problem solving: Don't allow any formal method of problem solving to limit your use of common sense and intuition. If a problem is clear to you and its solution seems simpler by another method, by all means use it. But in the long run you should be able to solve many otherwise difficult problems by using dimensional analysis.

    Dimensional analysis converts one unit to another unit by using conversion factors.

Important equations are boxed or highlighted in color.

$$\text{unit}_1 \times \text{conversion factor} = \text{unit}_2$$

If you want to know how many millimeters are in 2.5 meters, you need to convert meters (m) to millimeters (mm). Start by writing

    m × conversion factor = mm

This conversion factor must accomplish two things: It must cancel (or eliminate) meters *and* it must introduce millimeters—the unit wanted in the answer. Such a conversion factor will be in fractional form and have meters in the denominator and millimeters in the numerator:

$$m \times \frac{mm}{m} = mm$$

We know that 1 m = 1000 mm. From this relationship we can write two conversion factors—by dividing both sides of the equation by the same quantity:

$$\frac{1 \text{ m}}{1000 \text{ mm}} = \frac{1000 \text{ mm}}{1000 \text{ mm}} \longrightarrow \frac{1 \text{ m}}{1000 \text{ mm}} = 1$$

or

$$\frac{1 \text{ m}}{1 \text{ m}} = \frac{1000 \text{ mm}}{1 \text{ m}} \longrightarrow 1 = \frac{1000 \text{ mm}}{1 \text{ m}}$$

Therefore the two conversion factors are

$$\frac{1 \text{ m}}{1000 \text{ mm}} \quad \text{and} \quad \frac{1000 \text{ mm}}{1 \text{ m}}$$

Choosing the conversion factor $\frac{1000 \text{ mm}}{1 \text{ m}}$, we can set up the calculation for the conversion of 2.5 m to millimeters:

$$(2.5 \text{ m})\left(\frac{1000 \text{ mm}}{1 \text{ m}}\right) = 2500 \text{ mm} \qquad \text{or} \qquad 2.5 \times 10^3 \text{ mm}$$

(two significant figures)

Note that, in making this calculation, units are treated as numbers; meters in the numerator are canceled by meters in the denominator.

Now suppose you need to change 215 centimeters to meters. We start with

cm × conversion factor = m

The conversion factor must have centimeters in the denominator and meters in the numerator:

$$cm \times \frac{m}{cm} = m$$

From the relationship 100 cm = 1 m, we can write a factor that will accomplish this conversion:

$$\frac{1 \text{ m}}{100 \text{ cm}}$$

Now set up the calculation using all the data given.

$$(215 \text{ cm})\left(\frac{1 \text{ m}}{100 \text{ cm}}\right) = 2.15 \text{ m}$$

Some problems require a series of conversions to reach the correct units in the answer. For example, suppose we want to know the number of seconds in 1 day. We need to convert from the unit of days to seconds in this manner:

day ↔ hours → minutes → seconds

We know that multiplying a measurement by 1 does not change its value. Since our conversion factors both equal 1 we can multiply the measurement by the appropriate one to convert units.

Units are emphasized in problems by using color and flow diagrams to help visualize the steps in the process.

This series requires three conversion factors, one for each step. We convert days to hours (hr), hours to minutes (min), and minutes to seconds (s). The conversions can be done individually or in a continuous sequence:

$$\cancel{day} \times \frac{hr}{\cancel{day}} \longrightarrow \cancel{hr} \times \frac{min}{\cancel{hr}} \longrightarrow \cancel{min} \times \frac{s}{\cancel{min}} = s$$

$$\cancel{day} \times \frac{\cancel{hr}}{\cancel{day}} \times \frac{\cancel{min}}{\cancel{hr}} \times \frac{s}{\cancel{min}} = s$$

Inserting the proper factors we calculate the number of seconds in 1 day to be

$$(1 \ \cancel{day})\left(\frac{24 \ \cancel{hr}}{1 \ \cancel{day}}\right)\left(\frac{60 \ \cancel{min}}{1 \ \cancel{hr}}\right)\left(\frac{60 \ s}{1 \ \cancel{min}}\right) = 86{,}400. \ s$$

All five digits in 86,400 are significant, since all the factors in the calculation are exact numbers.

> **Label all factors with the proper units.**

The dimensional analysis used in the preceding examples shows how unit conversion factors are derived and used in calculations. As you become more proficient, you can save steps by writing the factors directly in the calculation. The following examples show the conversion from American units to metric units.

Example 2.11   How many centimeters are in 2.00 ft?

Solution   The stepwise conversion of units from feet to centimeters may be done in this manner. Convert feet to inches; then convert inches to centimeters:

$$ft \rightarrow in. \rightarrow cm$$

The conversion factors needed are

$$\frac{12 \ in.}{1 \ ft} \quad and \quad \frac{2.54 \ cm}{1 \ in.}$$

$$(2.00 \ \cancel{ft})\left(\frac{12 \ in.}{1 \ \cancel{ft}}\right) = 24.0 \ in.$$

$$(24.0 \ \cancel{in.})\left(\frac{2.54 \ cm}{1 \ \cancel{in.}}\right) = 61.0 \ cm$$

Since 1 ft and 12 in. are exact numbers, the number of significant figures allowed in the answer is three, based on the number 2.00.

Example 2.12   How many meters are in a 100.-yd football field?

Solution   The stepwise conversion of units from yards to meters may be done in this manner, using the proper conversion factors:

$$yd \rightarrow ft \rightarrow in. \rightarrow cm \rightarrow m$$

$$(100. \text{ yd})\left(\frac{3 \text{ ft}}{1 \text{ yd}}\right) = 300. \text{ ft}$$

$$(300. \text{ ft})\left(\frac{12 \text{ in.}}{1 \text{ ft}}\right) = 3600 \text{ in.}$$

$$(3600 \text{ in.})\left(\frac{2.54 \text{ cm}}{1 \text{ in.}}\right) = 9144 \text{ cm}$$

$$(9144 \text{ cm})\left(\frac{1 \text{ m}}{100 \text{ cm}}\right) = 91.4 \text{ m} \qquad \text{(three significant figures)}$$

Examples 2.11 and 2.12 may be solved using a linear expression and writing down conversion factors in succession. This method often saves one or two calculation steps and allows numerical values to be reduced to simpler terms, leading to simpler calculations. The single linear expressions for Examples 2.11 and 2.12 are

$$(2.00 \text{ ft})\left(\frac{12 \text{ in.}}{1 \text{ ft}}\right)\left(\frac{2.54 \text{ cm}}{1 \text{ in.}}\right) = 61.0 \text{ cm}$$

$$(100. \text{ yd})\left(\frac{3 \text{ ft}}{1 \text{ yd}}\right)\left(\frac{12 \text{ in.}}{1 \text{ ft}}\right)\left(\frac{2.54 \text{ cm}}{1 \text{ in.}}\right)\left(\frac{1 \text{ m}}{100 \text{ cm}}\right) = 91.4 \text{ m}$$

If you need help in doing this calculation all at once on your calculator see Appendix II.

Using the units alone (Example 2.12), we see that the stepwise cancellation proceeds in succession until the desired unit is reached.

$$\text{yd} \times \frac{\text{ft}}{\text{yd}} \times \frac{\text{in.}}{\text{ft}} \times \frac{\text{cm}}{\text{in.}} \times \frac{\text{m}}{\text{cm}} = \text{m}$$

Practice 2.6

How many meters are in 10.5 miles?

How many cubic centimeters (cm^3) are in a box that measures 2.20 in. by 4.00 in. by 6.00 in.? **Example 2.13**

First we need to determine the volume of the box in cubic inches ($in.^3$) by multiplying the length × width × height: **Solution**

$$(2.20 \text{ in.})(4.00 \text{ in.})(6.00 \text{ in.}) = 52.8 \text{ in.}^3$$

Now we convert $in.^3$ to cm^3 by using the inches and centimeters relationship three times:

$$\text{in.}^3 \times \frac{\text{cm}}{\text{in.}} \times \frac{\text{cm}}{\text{in.}} \times \frac{\text{cm}}{\text{in.}} = \text{cm}^3$$

$$(52.8 \text{ in.}^3)\left(\frac{2.54 \text{ cm}}{1 \text{ in.}}\right)\left(\frac{2.54 \text{ cm}}{1 \text{ in.}}\right)\left(\frac{2.54 \text{ cm}}{1 \text{ in.}}\right) = 865 \text{ cm}^3$$

Busy as a Bee

Monitoring pollutants in the air is expensive and time-consuming. What if you could employ a force of millions of free workers to collect samples for you? That's exactly what Jerry Bromenshenk and his colleagues at the University of Montana, Missoula, are doing. The scientists use honeybees to collect contaminants as they go about their daily business of collecting nectar. The bees ingest water, nectar, and pollen, and all sorts of particles and pollutants from the air, soil, water, and plants stick to their bodies. When the bees return to their hive, they release the pollutants as they fan the air to regulate the temperature. The air in the hives is then analyzed to identify the chemicals in the area surrounding the hive. Studies of the Puget Sound (on the coast of Washington) measuring metal contamination found that monitoring bees gives results that correspond closely to independent results done through soil analysis.

Now researchers have learned how to measure organic contaminants as well. Electronic hives, each with 7000–10,000 bees, have been installed at a variety of sites. Small copper tubes attached to the hives pump out air without disturbing the bees. The scientists then analyze the collected air to determine the chemical components. Millions of hives exist across the United States so a huge network of pollution monitoring stations could be established with each hive serving an area of about a mile in diameter. Instead of a reputation for producing a sting, bees may be changing reputation to monitors for pollution!

▲ **Honeybee on goldenrod plant.**

Example 2.14   A driver of a car is obeying the speed limit of 55 miles per hour. How fast is the car traveling in kilometers per second?

Solution   Two conversions are needed to solve this problem:

$$mi \rightarrow km$$
$$hr \rightarrow min \rightarrow s$$

To convert mi → km,

$$\left(\frac{55 \text{ mi}}{hr}\right)\left(\frac{1.609 \text{ km}}{1 \text{ mi}}\right) = 88 \frac{km}{hr}$$

Next we must convert hr → min → s. Notice that hours is in the denominator of our quantity, so the conversion factor must have hours in the numerator:

$$\left(\frac{88 \text{ km}}{hr}\right)\left(\frac{1 \text{ hr}}{60 \text{ min}}\right)\left(\frac{1 \text{ min}}{60 \text{ s}}\right) = 0.024 \frac{km}{s}$$

Practice 2.7

How many cubic meters are in a room measuring 8 ft × 10 ft × 12 ft?

2.9   Measurement of Mass

The gram is a unit of mass measurement, but it is a tiny amount of mass; for instance, a U.S. nickel has a mass of about 5 grams. Therefore the *standard unit* of mass in the SI

(a)

(b)

(c)

FIGURE 2.4
(a) A quadruple beam balance with a precision of 0.01 g; (b) digital electronic, top-loading balance with a precision of 0.001 g (1 mg); and (c) a digital electronic analytical balance with a precision of 0.0001 g (0.1 mg).

TABLE 2.4 Metric Units of Mass			
Unit	Abbreviation	Gram equivalent	Exponential equivalent
kilogram	kg	1000 g	10^3 g
gram	g	1 g	10^0 g
decigram	dg	0.1 g	10^{-1} g
centigram	cg	0.01 g	10^{-2} g
milligram	mg	0.001 g	10^{-3} g
microgram	μg	0.000001 g	10^{-6} g

system is the **kilogram** (equal to 1000 g). The amount of mass in a kilogram is defined by international agreement as exactly equal to the mass of a platinum–iridium cylinder (international prototype kilogram) kept in a vault at Sèvres, France. Comparing this unit of mass to 1 lb (16 oz), we find that 1 kg is equal to 2.205 lb. A pound is equal to 453.6 g (0.4536 kg). The same prefixes used in length measurement are used to indicate larger and smaller gram units (see Table 2.4).

A balance is used to measure mass. Some balances can determine the mass of objects to the nearest microgram. The choice of balance depends on the precision required and the amount of material. Several balances are shown in Figure 2.4.

To change grams to milligrams, we use the conversion factor 1000 mg/g. The setup for converting 25 g to milligrams is

$$(25 \text{ g})\left(\frac{1000 \text{ mg}}{1 \text{ g}}\right) = 25,000 \text{ mg} \qquad (2.5 \times 10^4 \text{ mg})$$

Note that multiplying a number by 1000 is the same as multiplying the number by 10^3 and can be done simply by moving the decimal point three places to the right:

$$(6.428)(1000) = 6428 \qquad (6.428)$$

kilogram

Common mass relationships:
1 g = 1000 mg
1 kg = 1000 g
1 kg = 2.205 lb
1 lb = 453.6 g

To change milligrams to grams, we use the conversion factor 1 g/1000 mg. For example, to convert 155 mg to grams:

$$(155 \text{ mg})\left(\frac{1 \text{ g}}{1000 \text{ mg}}\right) = 0.155 \text{ g}$$

Mass conversions from American to metric units are shown in Examples 2.15 and 2.16.

Example 2.15   A 1.50-lb package of baking soda contains how many grams?

Solution   We are solving for the number of grams equivalent to 1.50 lb. Since 1 lb = 453.6 g, the conversion factor is 453.6 g/lb:

$$(1.50 \text{ lb})\left(\frac{453.6 \text{ g}}{1 \text{ lb}}\right) = 680. \text{ g}$$

Example 2.16   Suppose four ostrich feathers weigh 1.00 lb. Assuming that each feather is equal in mass, how many milligrams does a single feather weigh?

Solution   The unit conversion in this problem is from 1 lb/4 feathers to milligrams per feather. Since the unit *feathers* occurs in the denominator of both the starting unit and the desired unit, the unit conversions are

$$lb \rightarrow g \rightarrow mg$$

$$\left(\frac{1.00 \text{ lb}}{4 \text{ feathers}}\right)\left(\frac{453.6 \text{ g}}{1 \text{ lb}}\right)\left(\frac{1000 \text{ mg}}{1 \text{ g}}\right) = \frac{113,400 \text{ mg}}{\text{feather}} \; (1.13 \times 10^5 \text{ mg/feather})$$

Practice 2.8

You are traveling in Europe and wake up one morning to find your mass is 75.0 kg. Determine the American equivalent to see whether you need to go on a diet before you return home.

Practice 2.9

A tennis ball has a mass of 65 g. Determine the American equivalent in pounds.

2.10   Measurement of Volume

volume   **Volume,** as used here, is the amount of space occupied by matter. The SI unit of volume is the *cubic meter* (m^3). However, the liter (pronounced *leeter* and abbreviated L) and the milliliter (abbreviated mL) are the standard units of volume used in
liter   most chemical laboratories. A **liter** is usually defined as 1 cubic decimeter (1 kg) of water at 4°C.

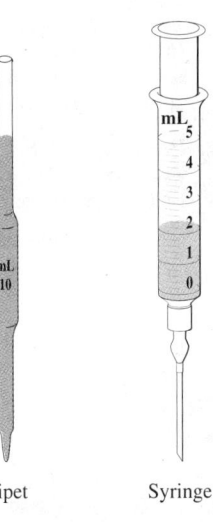

| Graduated cylinder | Volumetric flask | Buret | Pipet | Syringe |

◄ **FIGURE 2.5**
Calibrated glassware for measuring the volume of liquids.

The most common instruments or equipment for measuring liquids are the graduated cylinder, volumetric flask, buret, pipet, and syringe, which are illustrated in Figure 2.5. These pieces are usually made of glass and are available in various sizes.

The volume of a cubic or rectangular container can be determined by multiplying its length × width × height. Thus a box 10 cm on each side has a volume of $(10 \text{ cm})(10 \text{ cm})(10 \text{ cm}) = 1000 \text{ cm}^3$. Let's try some examples.

Common volume relationships:
1 L = 1000 mL = 1000 cm³
1 mL = 1 cm³
1 L = 1.057 qt
946.1 mL = 1 qt

How many milliliters are contained in 3.5 liters?

The conversion factor to change liters to milliliters is 1000 mL/L:

$$(3.5 \text{ L})\left(\frac{1000 \text{ mL}}{\text{L}}\right) = 3500 \text{ mL} \qquad (3.5 \times 10^3 \text{ mL})$$

Liters may be changed to milliliters by moving the decimal point three places to the right and changing the units to milliliters:

$$1.500 \text{ L} = 1500. \text{ mL}$$

Example 2.17

Solution

How many cubic centimeters are in a cube that is 11.1 inches on a side?

First we change inches to centimeters; our conversion factor is 2.54 cm/in.:

$$(11.1 \text{ in.})\left(\frac{2.54 \text{ cm}}{1 \text{ in.}}\right) = 28.2 \text{ cm on a side}$$

Then determine volume (length × width × height):

$$(28.2 \text{ cm})(28.2 \text{ cm})(28.2 \text{ cm}) = 22{,}426 \text{ cm}^3 \qquad (2.24 \times 10^4 \text{ cm}^3)$$

Example 2.18

Solution

When doing problems with multiple steps, you should round only at the end of the problem. We are rounding at the end of each step in example problems to illustrate the proper significant figures.

Practice 2.10

A bottle of excellent chianti holds 750. mL. What is its volume in quarts?

Practice 2.11

Milk is often purchased by the half gallon. Determine the number of liters equal to this amount.

 2.11   Measurement of Temperature

heat   **Heat** is a form of energy associated with the motion of small particles of matter. The term *heat* refers to the quantity of energy within a system or to a quantity of energy added to or taken away from a system. *System* as used here simply refers to the entity that is being heated or cooled. Depending on the amount of heat energy present, a
temperature   given system is said to be hot or cold. **Temperature** is a measure of the intensity of heat, or how hot a system is, regardless of its size. Heat always flows from a region of higher temperature to one of lower temperature. The SI unit of temperature is the kelvin. The common laboratory instrument for measuring temperature is a thermometer (see Figure 2.6).

The temperature of a system can be expressed by several different scales. Three commonly used temperature scales are Celsius (pronounced *sell-see-us*), Kelvin (absolute), and Fahrenheit. The unit of temperature on the Celsius and Fahrenheit scales is called a *degree,* but the size of the Celsius and the Fahrenheit degree is not the same. The symbol for the Celsius and Fahrenheit degrees is °, and it is placed as a superscript after the number and before the symbol for the scale. Thus 100°C means 100 *degrees Celsius.* The degree sign is not used with Kelvin temperatures.

degrees Celsius = °C
Kelvin (absolute) = K
degrees Fahrenheit = °F

On the Celsius scale the interval between the freezing and boiling temperatures of water is divided into 100 equal parts, or degrees. The freezing point of water is assigned a temperature of 0°C and the boiling point of water a temperature of 100°C. The Kelvin temperature scale is known as the absolute temperature scale, because 0 K is the lowest temperature theoretically attainable. The Kelvin zero is 273.15 kelvins below the Celsius zero. (A kelvin is equal in size to a Celsius degree.) The freezing point of water on the Kelvin scale is 273.15 K. The Fahrenheit scale has 180 degrees between the freezing and boiling temperatures of water. On this scale the freezing point of water is 32°F and the boiling point is 212°F.

$$0°C \cong 273 \text{ K} \cong 32°F$$

The three scales are compared in Figure 2.6. Although absolute zero (0 K) is the lower limit of temperature on these scales, temperature has no upper limit. (Temperatures of several million degrees are known to exist in the Sun and in other stars.)

◀ **FIGURE 2.6**
Comparison of Celsius, Kelvin, and Fahrenheit temperature scales.

By examining Figure 2.6 we can see that there are 100 Celsius degrees and 100 kelvins between the freezing and boiling points of water, but there are 180 Fahrenheit degrees between these two temperatures. Hence, the size of the Celsius degree and the kelvin are the same, but 1 Celsius degree is equal to 1.8 Fahrenheit degrees.

$$\frac{180}{100} = 1.8$$

From these data, mathematical formulas have been derived to convert a temperature on one scale to the corresponding temperature on another scale:

$$K = {}^\circ C + 273.15$$

$${}^\circ F = (1.8 \times {}^\circ C) + 32$$

$${}^\circ C = \frac{{}^\circ F - 32}{1.8}$$

The temperature at which table salt (sodium chloride) melts is 800.°C. What is this temperature on the Kelvin and Fahrenheit scales?

Example 2.19

To calculate K from °C, we use the formula

Solution

$$K = {}^\circ C + 273.15$$

$$K = 800.{}^\circ C + 273.15 = 1073 \ K$$

To calculate °F from °C, we use the formula

$$°F = (1.8 \times °C) + 32$$

$$°F = (1.8)(800.°C) + 32$$

$$°F = 1440 + 32 = 1472°F$$

Summarizing our calculations we see

$$800.°C = 1073 \text{ K} = 1472°F$$

Remember, the original measurement of 800.°C was to the units place, so the converted temperature is also to the units place.

Example 2.20   The temperature for December 1 in Honolulu, Hawaii, was 110.°F, a new record. Convert this temperature to °C.

Solution   We use the formula

$$°C = \frac{°F - 32}{1.8}$$

$$°C = \frac{110. - 32}{1.8} = \frac{78}{1.8} = 43°C$$

Example 2.21   What temperature on the Fahrenheit scale corresponds to −8.0°C? (Notice the negative sign in this problem.)

Solution $$°F = (1.8 \times °C) + 32$$

$$°F = (1.8)(-8.0) + 32 = -14 + 32$$

$$°F = 18°F$$

Temperatures used throughout this book are in degrees Celsius (°C) unless specified otherwise. The temperature after conversion should be expressed to the same precision as the original measurement.

Practice 2.12

Helium boils at 4 K. Convert this temperature to °C and then to °F.

Practice 2.13

"Normal" human body temperature is 98.6°F. Convert this to °C and K.

CHEMISTRY IN ACTION

Taking the Temperature of Old Faithful

If you have ever struggled to take the temperature of a sick child imagine the difficulty in taking the temperature of a geyser. Such are the tasks scientists set for themselves! In 1984, James A. Westphal and Susan W. Keiffer, geologists from the California Institute of Technology, measured the temperature and pressure inside Old Faithful during several eruptions in order to learn more about how a geyser functions. The measurements, taken at eight depths along the upper part of the geyser, were so varied and complicated the researchers returned to Yellowstone in 1992 to further investigate Old Faithful's structure and functioning. To see what happened between eruptions Westphal and Keiffer lowered an insulated 2-inch video camera into the geyser. Keiffer had assumed the vent (opening in the ground) was a uniform vertical tube, but this is not the case. Instead, the geyser appears to be an east–west crack in the earth that extends downward at least 14 m. In some places it is over 1.8 m wide, and in other places it narrows to less than 15 cm. The walls of the vent contain many cracks allowing water to enter at several depths. The complicated nature of the temperature data is explained by these cracks. Cool water enters the vent at depths of 5.5 m and 7.5 m. Superheated water and steam blast into the vent 14 m underground. According to Westphal temperature increases of up to 130°C at the beginning of an eruption suggest that water and steam also surge into the vent from deeper geothermal sources.

During the first 20–30 seconds of an eruption, steam and boiling water shoot through the narrowest part of the vent at near the speed of sound. The narrow tube limits the rate at which the water can shoot from the geyser. When the pressure falls below a critical value, the process slows and Old Faithful begins to quiet down again.

The frequency of Old Faithful's eruptions are not on a precise schedule but vary from 45 to 105 minutes—the average is about 79 minutes. The variations in time between eruptions depend on the amount of boiling water left in the fissure. Westphal says, "There's no real pattern except that a short eruption is always followed by a long one." Measurements of temperatures inside Old Faithful have given scientists a better understanding of what causes a geyser to erupt.

Old Faithful eruptions shoot into the air an average of 130 feet.
▼

33

Mass, 10.0 g

(a)

Volume, 1.00 cm³

(b)

▲
FIGURE 2.7
(a) Comparison of the volumes of equal masses (10.0 g) of water, sulfur, and gold. (b) Comparison of the masses of equal volumes (1.00 cm³) of water, sulfur, and gold. (Water is at 4°C; the two solids, at 20°C.)

2.12   Density

Density (d) is the ratio of the mass of a substance to the volume occupied by that mass; it is the mass per unit of volume and is given by the equation

$$d = \frac{\text{mass}}{\text{volume}}$$

Density is a physical characteristic of a substance and may be used as an aid to its identification. When the density of a solid or a liquid is given, the mass is usually expressed in grams and the volume in milliliters or cubic centimeters.

$$d = \frac{\text{mass}}{\text{volume}} = \frac{\text{g}}{\text{mL}} \quad \text{or} \quad d = \frac{\text{g}}{\text{cm}^3}$$

Since the volume of a substance (especially liquids and gases) varies with temperature, it is important to state the temperature along with the density. For example, the volume of 1.0000 g of water at 4°C is 1.0000 mL; at 20°C, it is 1.0018 mL; and at 80°C, it is 1.0290 mL. Density therefore also varies with temperature.

The density of water at 4°C is 1.0000 g/mL, but at 80°C the density of water is 0.9718 g/mL.

$$d^{4°C} = \frac{1.0000 \text{ g}}{1.0000 \text{ mL}} = 1.0000 \text{ g/mL}$$

$$d^{80°C} = \frac{1.0000 \text{ g}}{1.0290 \text{ mL}} = 0.97182 \text{ g/mL}$$

The density of iron at 20°C is 7.86 g/mL.

$$d^{20°C} = \frac{7.86 \text{ g}}{1.00 \text{ mL}} = 7.86 \text{ g/mL}$$

The densities of a variety of materials are compared in Figure 2.7.

Densities for liquids and solids are usually represented in terms of grams per milliliter (g/mL) or grams per cubic centimeter (g/cm^3). The density of gases, however, is expressed in terms of grams per liter (g/L). Unless otherwise stated, gas densities are given for 0°C and 1 atmosphere pressure (discussed further in Chapter 13). Table 2.5 lists the densities of some common materials.

Suppose that water, Karo syrup, and vegetable oil are successively poured into a graduated cylinder. The result is a layered three-liquid system. Can we predict the order of the liquid layers? Yes, by looking up the densities in Table 2.5. Karo syrup has the greatest density (1.37 g/mL), and vegetable oil has the lowest density (0.91 g/mL). Karo syrup will be the bottom layer and vegetable oil will be the top layer. Water, with a density between the other two liquids, will form the middle layer. This information can also be determined by experiment. Vegetable oil, being less dense than water, will float when added to the graduated cylinder. Figure 2.8 shows a more complex density column made in the same way.

The density of air at 0°C is approximately 1.293 g/L. Gases with densities less than this value are said to be "lighter than air." A helium-filled balloon will rise rapidly in air because the density of helium is only 0.178 g/L.

When an insoluble solid object is dropped into water, it will sink or float, depending on its density. If the object is less dense than water, it will float, displacing a *mass* of water equal to the mass of the object. If the object is more dense than water, it will sink, displacing a *volume* of water equal to the volume of the object. This information can be used to determine the volume (and density) of irregularly shaped objects.

▲
FIGURE 2.8
Relative density of liquids. When liquids are carefully poured into a cylinder the liquid with highest density will form the bottom layer (maple syrup). The remaining liquids decreasing in density are antifreeze, dish detergent, shampoo, water, and corn oil.

TABLE 2.5	Densities of Some Selected Materials			

Liquids and solids		Gases	
Substance	**Density (g/mL at 20°C)**	**Substance**	**Density (g/L at 0°C)**
Wood (Douglas fir)	0.512	Hydrogen	0.090
Ethyl alcohol	0.789	Helium	0.178
Vegetable oil	0.91	Methane	0.714
Water (4°C)	**1.000***	Ammonia	0.771
Sugar	1.59	Neon	0.90
Glycerin	1.26	Carbon monoxide	1.25
Karo syrup	1.37	Nitrogen	1.251
Magnesium	1.74	**Air**	**1.293***
Sulfuric acid	1.84	Oxygen	1.429
Sulfur	2.07	Hydrogen chloride	1.63
Salt	2.16	Argon	1.78
Aluminum	2.70	Carbon dioxide	1.963
Silver	10.5	Chlorine	3.17
Lead	11.34		
Mercury	13.55		
Gold	19.3		

*For comparing densities the density of water is the reference for solids and liquids; air is the reference for gases.

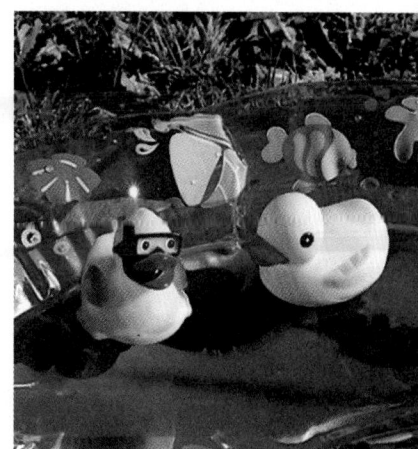

▲
These ducks float since their density is less than water.

specific gravity

The **specific gravity** (sp gr) of a substance is the ratio of the density of that substance to the density of another substance, usually water at 4°C. Specific gravity has no units because the density units cancel. The specific gravity tells us how many times as heavy a liquid, a solid, or a gas is as compared to the reference material. Since the density of water at 4°C is 1.00 g/mL, the specific gravity of a solid or liquid is the same as its density in g/mL without the units.

$$\text{sp gr} = \frac{\text{density of a liquid or solid}}{\text{density of water}}$$

Sample calculations of density problems follow.

Example 2.22   What is the density of a mineral if 427 g of the mineral occupy a volume of 35.0 mL?

Solution   We need to solve for density, so we start by writing the formula for calculating density:

$$d = \frac{\text{mass}}{\text{volume}}$$

Then we substitute the data given in the problem into the equation and solve:

$$\text{mass} = 427 \text{ g} \qquad \text{volume} = 35.0 \text{ mL}$$

$$d = \frac{\text{mass}}{\text{volume}} = \frac{427 \text{ g}}{35.0 \text{ mL}} = 12.2 \text{ g/mL}$$

Example 2.23   The density of gold is 19.3 g/mL. What is the mass of 25.0 mL of gold?

Solution   There are two ways to solve this problem: (1) Solve the density equation for mass, then substitute the density and volume data into the new equation and calculate. (2) Solve by dimensional analysis.

When alternative methods of solution are available, more than one is shown in the example. Choose the method you are most comfortable with.

Method 1. (a) Solve the density equation for mass:

$$d = \frac{\text{mass}}{\text{volume}} \qquad d \times \text{volume} = \text{mass}$$

(b) Substitute the data and calculate.

$$\text{mass} = \left(\frac{19.3 \text{ g}}{\text{mL}}\right)(25.0 \text{ mL}) = 483 \text{ g}$$

Method 2. Dimensional analysis: Use density as a conversion factor, converting

$$\text{mL} \rightarrow \text{g}$$

The conversion of units is

$$\text{mL} \times \frac{\text{g}}{\text{mL}} = \text{g}$$

$$(25.0 \text{ mL})\left(\frac{19.3 \text{ g}}{\text{mL}}\right) = 483 \text{ g}$$

Calculate the volume (in mL) of 100. g of ethyl alcohol.

Example 2.24

Solution

From Table 2.5 we see that the density of ethyl alcohol is 0.789 g/mL. This density also means that 1 mL of the alcohol has a mass of 0.789 g (1 mL/0.789 g).

   Method 1. Solve the density equation for volume and then substitute the data in the new equation.

$$d = \frac{\text{mass}}{\text{volume}}$$

$$\text{volume} = \frac{\text{mass}}{d}$$

$$\text{volume} = \frac{100.\cancel{g}}{0.789\ \cancel{g}/\text{mL}} = 127\ \text{mL}$$

   Method 2. Dimensional analysis. For a conversion factor, we can use either

$$\frac{\text{g}}{\text{mL}} \quad \text{or} \quad \frac{\text{mL}}{\text{g}}$$

In this case the conversion is from g → mL, so we use mL/g. Substituting the data,

$$100.\ \cancel{g} \times \frac{1\ \text{mL}}{0.789\ \cancel{g}} = 127\ \text{mL of ethyl alcohol}$$

The water level in a graduated cylinder stands at 20.0 mL before and at 26.2 mL after a 16.74-g metal bolt is submerged in the water. (a) What is the volume of the bolt? (b) What is the density of the bolt?

Example 2.25

(a) The bolt will displace a volume of water equal to the volume of the bolt. Thus the increase in volume is the volume of the bolt.

Solution

$$\begin{aligned} 26.2\ \text{mL} &= \text{volume of water plus bolt} \\ -20.0\ \text{mL} &= \text{volume of water} \\ \hline 6.2\ \text{mL} &= \text{volume of bolt} \end{aligned}$$

(b) $d = \dfrac{\text{mass of bolt}}{\text{volume of bolt}} = \dfrac{16.74\ \text{g}}{6.2\ \text{mL}} = 2.7\ \text{g/mL}$

Practice 2.14

Pure silver has a density of 10.5 g/mL. A ring sold as pure silver has a mass of 25.0 g. When placed in a graduated cylinder, the water level rises 2.0 mL. Determine whether the ring is actually pure silver or if the customer should see the Better Business Bureau.

> **Practice 2.15**
>
> The water level in a metric measuring cup is 0.75 L before the addition of 150. g of shortening. The water level after submerging the shortening is 0.92 L. Determine the density of the shortening.

Concepts in Review

Major concepts of the chapter are listed in this section to help you review the chapter.

1. Differentiate between mass and weight. Indicate the instruments used to measure each.

2. Know the metric units of mass, length, and volume.

3. Know the numerical equivalent for the metric prefixes *deci, centi, milli, micro, nano, kilo,* and *mega.*

4. Express any number in scientific notation.

5. Express answers to calculations to the proper number of significant figures.

6. Set up and solve problems using dimensional analysis.

7. Convert measurements of mass, length, and volume from American units to metric units, and vice versa.

8. Make temperature conversions among Fahrenheit, Celsius, and Kelvin scales.

9. Differentiate between heat and temperature.

10. Calculate the density, mass, and volume of an object.

Key Terms

The terms listed here are defined in the chapter. Section nuumbers are given in parentheses. More detailed definitions are given in the Glossary.

density (2.12)	mass (2.1)	scientific notation (2.4)	temperature (2.11)
heat (2.11)	meter (2.7)	SI (2.6)	volume (2.10)
kilogram (2.9)	metric system (SI) (2.6)	significant figures (2.2)	weight (2.1)
liter (2.10)	rounding off numbers (2.3)	specific gravity (2.12)	

Questions

Questions refer to tables, figures, and key words and concepts defined within the chapter. A particularly challenging question or exercise is indicated with an asterisk. Answers to even-numbered questions are given in Appendix VI.

1. How many centimeters make up 1 km? (Table 2.3)

2. What is the metric equivalent of 3 in.? (Figure 2.3)

3. Why is the neck of a 100-mL volumetric flask narrower than the top of a 100-mL graduated cylinder? (Figure 2.5)

4. Describe the order of the following substances (top to bottom) if these three substances were placed in a 100-mL graduated cylinder: 25 mL glycerin, 25 mL mercury, and a cube of magnesium 2.0 cm on an edge. (Table 2.5)

5. Arrange these materials in order of increasing density: salt, vegetable oil, lead, and ethyl alcohol. (Table 2.5)

6. Ice floats in vegetable oil and sinks in ethyl alcohol. The density of ice must lie between what numerical values? (Table 2.5)

7. Distinguish between heat and temperature.

8. Distinguish between density and specific gravity.

9. State the rules used in this text for rounding off numbers.

10. Compare the number of degrees between the freezing point of water and its boiling point on the Fahrenheit, Kelvin, and Celsius temperature scales. (Figure 2.6)

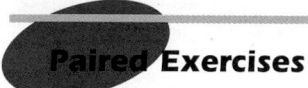

Paired Exercises

Answers to the even-numbered exercises are given in Appendix VI.

Metric Abbreviations

11. State the abbreviation for each of the following units:
  (a) gram
  (b) microgram
  (c) centimeter
  (d) micrometer
  (e) milliliter
  (f) deciliter

12. State the abbreviation for each of the following units:
  (a) milligram
  (b) kilogram
  (c) meter
  (d) nanometer
  (e) angstrom
  (f) microliter

Significant Figures, Rounding, Exponential Notation

13. For the following numbers, tell whether the zeros are significant:
  (a) 503
  (b) 0.007
  (c) 4200
  (d) 3.0030
  (e) 100.00
  (f) 8.00×10^2

14. Are the zeros significant in these numbers?
  (a) 63,000
  (b) 6.004
  (c) 0.00543
  (d) 8.3090
  (e) 60.
  (f) 5.0×10^{-4}

15. How many significant figures are in each of the following numbers?
  (a) 0.025
  (b) 22.4
  (c) 0.0404
  (d) 5.50×10^3

16. State the number of significant figures in each of the following numbers:
  (a) 40.0
  (b) 0.081
  (c) 129,042
  (d) 4.090×10^{-3}

17. Round each of the following numbers to three significant figures:
  (a) 93.246
  (b) 0.02857
  (c) 4.644
  (d) 34.250

18. Round each of the following numbers to three significant figures:
  (a) 8.8726
  (b) 21.25
  (c) 129.509
  (d) 1.995×10^6

19. Express each of the following numbers in exponential notation:
  (a) 2,900,000
  (b) 0.587
  (c) 0.00840
  (d) 0.0000055

20. Write each of the following numbers in exponential notation:
  (a) 0.0456
  (b) 4082.2
  (c) 40.30
  (d) 12,000,000

21. Solve the following problems, stating answers to the proper number of significant figures:
    (a) $12.62 + 1.5 + 0.25 = ?$
    (b) $(2.25 \times 10^3)(4.80 \times 10^4) = ?$
    (c) $\dfrac{(452)(6.2)}{14.3} = ?$
    (d) $(0.0394)(12.8) = ?$
    (e) $\dfrac{0.4278}{59.6} = ?$
    (f) $10.4 + 3.75(1.5 \times 10^4) = ?$

22. Evaluate each of the following expressions. State the answer to the proper number of significant figures:
    (a) $15.2 - 2.75 + 15.67$
    (b) $(4.68)(12.5)$
    (c) $\dfrac{182.6}{4.6}$
    (d) $1986 + 23.84 + 0.012$
    (e) $\dfrac{29.3}{(284)(415)}$
    (f) $(2.92 \times 10^{-3})(6.14 \times 10^5)$

23. Change these fractions into decimals. Express each answer to three significant figures:
    (a) $\dfrac{5}{6}$
    (b) $\dfrac{3}{7}$
    (c) $\dfrac{12}{16}$
    (d) $\dfrac{9}{18}$

24. Change each of the following decimals to fractions in lowest terms:
    (a) 0.25
    (b) 0.625
    (c) 1.67
    (d) 0.8888

25. Solve each of these equations for x:
    (a) $3.42x = 6.5$
    (b) $\dfrac{x}{12.3} = 7.05$
    (c) $\dfrac{0.525}{x} = 0.25$

26. Solve each equation for the variable:
    (a) $x = \dfrac{212 - 32}{1.8}$
    (b) $8.9\,\dfrac{g}{mL} = \dfrac{40.90\ g}{x}$
    (c) $72°F = 1.8x + 32$

Unit Conversions

27. Complete the following metric conversions using the correct number of significant figures:
    (a) 28.0 cm to m
    (b) 1000. m to km
    (c) 9.28 cm to mm
    (d) 10.68 g to mg
    (e) 6.8×10^4 mg to kg
    (f) 8.54 g to kg
    (g) 25.0 mL to L
    (h) 22.4 L to μL

28. Complete the following metric conversions using the correct number of significant figures:
    (a) 4.5 cm to Å
    (b) 12 nm to cm
    (c) 8.0 km to mm
    (d) 164 mg to g
    (e) 0.65 kg to mg
    (f) 5.5 kg to g
    (g) 0.468 L to mL
    (h) 9.0 μL to mL

29. Complete the following American/metric conversions using the correct number of significant figures:
    (a) 42.2 in. to cm
    (b) 0.64 mi to in.
    (c) 2.00 in.^{2} to cm^2
    (d) 42.8 kg to lb
    (e) 3.5 qt to mL
    (f) 20.0 gal to L

30. Make the following conversions using the correct number of significant figures:
    (a) 35.6 m to ft
    (b) 16.5 km to mi
    (c) 4.5 in.^{3} to mm^3
    (d) 95 lb to g
    (e) 20.0 gal to L
    (f) 4.5×10^4 ft^3 to m^3

31. An automobile traveling at 55 mi/hr is moving at what speed in kilometers per hour?

32. A cyclist is traveling downhill at 55 km/hr. How fast is she moving in feet per second?

33. Carl Lewis, a sprinter in the 1988 Olympic Games, ran the 100.-m dash in 9.92 s. What was his speed in feet per second?

34. Al Unser, Jr., qualified for the pole position at the 1994 Indianapolis 500 at a speed of 229 mph. What was his speed in kilometers per second?

35. When the space probe *Galileo* reached Jupiter in 1995, it was traveling at an average speed of 27,000 mi/hr. What was its speed in kilometers per second?

36. The Sun is approximately 93 million miles from the Earth. How many seconds will it take for light from the Sun to travel to the Earth if the velocity of light is 3.00×10^8 m/s?

37. How many kilograms does a 176-lb man weigh?

38. The average mass of the heart of a human baby is about 1 oz. What is its mass in milligrams?

39. A regular aspirin tablet contains 5.0 grains of aspirin. How many grams of aspirin are in one tablet? (1 grain = 1/7000. lb)

40. An adult ruby-throated hummingbird has an average mass of 3.2 g, while an adult California condor may attain a mass of 21 lb. How many hummingbirds would it take to equal the mass of one condor?

41. A bag of pretzels has a mass of 283.5 g and costs $1.49. If a bag contains 18 pretzels, what is the cost of a pound of pretzels?

42. The price of gold varies greatly and has been as high as $875 per ounce. What is the value of 250 g of gold at $350 per ounce? Gold is priced by troy ounces (14.58 troy ounces = 1 lb).

43. At 35¢/L how much will it cost to fill a 15.8-gal tank with gasoline?

44. How many liters of gasoline will be used to drive 525 miles in a car that averages 35 mi/gal?

*45. Assuming that there are 20. drops in 1.0 mL, how many drops are in 1.0 gallon?

46. How many liters of oil are in a 42-gal barrel of oil?

*47. Calculate the number of milliliters of water in a cubic foot of water.

*48. Oil spreads in a thin layer on water called an "oil slick." How much area in m^2 will 200 cm^3 of oil cover if it forms a layer 0.5 nm thick?

49. A textbook is 27 cm long, 21 cm wide, and 4.4 cm thick. What is the volume in:
(a) cubic centimeters?
(b) liters?
(c) cubic inches?

50. An aquarium measures 16 in. $\times$ 8 in. $\times$ 10 in. How many liters of water does it hold? How many gallons?

Temperature Conversions

51. Normal body temperature for humans is 98.6°F. What is this temperature on the Celsius scale?

52. Driving to the grocery store you notice the temperature is 45°C. Determine what this temperature is on the Fahrenheit scale and what season of the year it might be.

53. Make the following conversions and include an equation for each one:
(a) 162°F to °C
(b) 0.0°F to K
(c) −18°C to °F
(d) 212 K to °C

54. Make the following conversions and include an equation for each one:
(a) 32°C to °F
(b) −8.6°F to °C
(c) 273°C to K
(d) 100 K to °F

*55. At what temperature are the Fahrenheit and Celsius temperatures exactly equal?

*56. At what temperature are Fahrenheit and Celsius temperatures the same in value but opposite in sign?

Density

57. Calculate the density of a liquid if 50.00 mL of the liquid has a mass of 78.26 g.

58. A 12.8-mL sample of bromine has a mass of 39.9 g. What is the density of bromine?

59. When a 32.7-g piece of chromium metal was placed into a graduated cylinder containing 25.0 mL of water, the water level rose to 29.6 mL. Calculate the density of the chromium.

60. An empty graduated cylinder has a mass of 42.817 g. When filled with 50.0 mL of an unknown liquid it has a mass of 106.773 g. What is the density of the liquid?

61. Concentrated hydrochloric acid has a density of 1.19 g/mL. Calculate the mass of 250.0 mL of this acid.

62. What mass of mercury (density 13.6 g/mL) will occupy a volume of 25.0 mL?

Additional Exercises

These exercises are not paired or labeled by topic and provide additional practice on the concepts covered in this chapter.

63. One liter of homogenized whole milk has a mass of 1032 g. What is the density of the milk in grams per milliliter? in kilograms per liter?

64. The volume of blood plasma in adults is 3.1 L. Its density is 1.03 g/cm^3. Approximately how many pounds of blood plasma are there in your body?

*65. The dashed lane markers on an interstate highway are 2.5 ft long and 4.0 in. wide. One (1.0) qt of paint covers 43 ft^2. How many dashed lane markers can be painted with 15 gal of paint?

66. Will a hollow cube with sides of length 0.50 m hold 8.5 L of solution? Depending on your answer, how much additional solution would be required to fill the container or how many times would the container need to be filled to measure the 8.5 L?

*67. The accepted toxic dose of mercury is 300 μg/day. Dental offices sometimes contain as much as 180 μg of mercury per cubic meter of air. If a nurse working in the office ingests 2×10^4 L of air per day, is he or she at risk for mercury poisoning?

68. Which is the higher temperature, 4.5°F or −15°C?

*69. A flask containing 100. mL of alcohol ($d = 0.789$ g/mL) is placed on one pan of a two-pan balance. A larger container, with a mass of 11.0 g more than the empty flask, is placed on the other pan of the balance. What volume of turpentine ($d = 0.87$ g/mL) must be added to this container to bring the two pans into balance?

70. Suppose you have samples of two metals, A and B. Use the data below to determine which sample occupies the larger volume.

	A	B
Mass	25 g	65 g
Density	10 g/mL	4 g/mL

*71. As a solid substance is heated its volume increases but its mass remains the same. Sketch a graph of density versus temperature showing the trend you expect. Briefly explain.

72. A 35.0-mL sample of ethyl alcohol (density 0.789 g/mL) is added to a graduated cylinder that has a mass of 49.28 g. What will be the mass of the cylinder plus the alcohol?

73. You are given three cubes, A, B, and C; one is magnesium, one is aluminum, and the third is silver. All three cubes have the same mass, but cube A has a volume of 25.9 mL, cube B has a volume of 16.7 mL, and cube C has a volume of 4.29 mL. Identify cubes A, B, and C.

*74. A cube of aluminum has a mass of 500. g. What will be the mass of a cube of gold of the same dimensions?

75. A 25.0-mL sample of water at 90°C has a mass of 24.12 g. Calculate the density of water at this temperature.

76. The mass of an empty container is 88.25 g. The mass of the container when filled with a liquid ($d = 1.25$ g/mL) is 150.50 g. What is the volume of the container?

77. Which liquid will occupy the greater volume, 50 g of water or 50 g of ethyl alcohol? Explain.

78. A gold bullion dealer advertised a bar of pure gold for sale. The gold bar had a mass of 3300 g and measured 2.00 cm × 15.0 cm × 6.00 cm. Was the bar pure gold? Show evidence for your answer.

79. The largest nugget of gold on record was found in 1872 in New South Wales, Australia, and had a mass of 93.3 kg. Assuming the nugget is pure gold, what is its volume in cubic centimeters? What is it worth by today's standards if gold is $345/oz? (14.58 troy oz = 1 lb)

*80. Forgetful Freddie placed 25.0 mL of a liquid in a graduated cylinder with a mass of 89.450 g when empty. When Freddie placed a metal slug with a mass of 15.454 g into the cylinder, the volume rose to 30.7 mL. Freddie was asked to calculate the density of the liquid and of the metal slug from his data, but he forgot to obtain the mass of the liquid. He was told that if he found the mass of the cylinder containing the liquid and the slug, he would have enough data for the calculations. He did so and found its mass to be 125.934 g. Calculate the density of the liquid and of the metal slug.

Answers to Practice Exercises

2.1 (a) 2; (b) 4; (c) 4; (d) 1; (e) 3; (f) 4; (g) 1; (h) 3

2.2 (a) 42.25 (Rule 2); (b) 88.0 (Rule 1); (c) 0.0897 (Rule 2); (d) 0.090 (Rule 2); (e) 225 (Rule 1); (f) 14.2 (Rule 2)

2.3 (a) $1200 = 1.200 \times 10^3$ (left means positive exponent)
(b) $6,600,000 = 6.6 \times 10^6$ (left means positive exponent)
(c) $0.0468 = 4.68 \times 10^{-2}$ (right means negative exponent)
(d) $0.00003 = 3 \times 10^{-5}$ (right means negative exponent)

2.4 (a) 3350 in.^{2} = 3.4×10^3 in.^{2}; (b) 50.7 mi/hr; (c) 0.79

2.5 (a) 2; (b) 2; (c) 1; (d) 2; (e) 4; (f) 2; (g) 2; (h) 2

2.6 1.69×10^4 m

2.7 30 m^3 or 3×10^1 m^3

2.8 165 lb

2.9 0.14 lb

2.10 0.793 qt

2.11 1.89 L (the number of significant figures is arbitrary)

2.12 $-269°C$, $-452°F$

2.13 37.0°C, 310.2 K

2.14 The density is 13 g/mL; therefore the ring is *not* pure silver

2.15 0.88 g/mL

CHAPTER

3

Classification of Matter

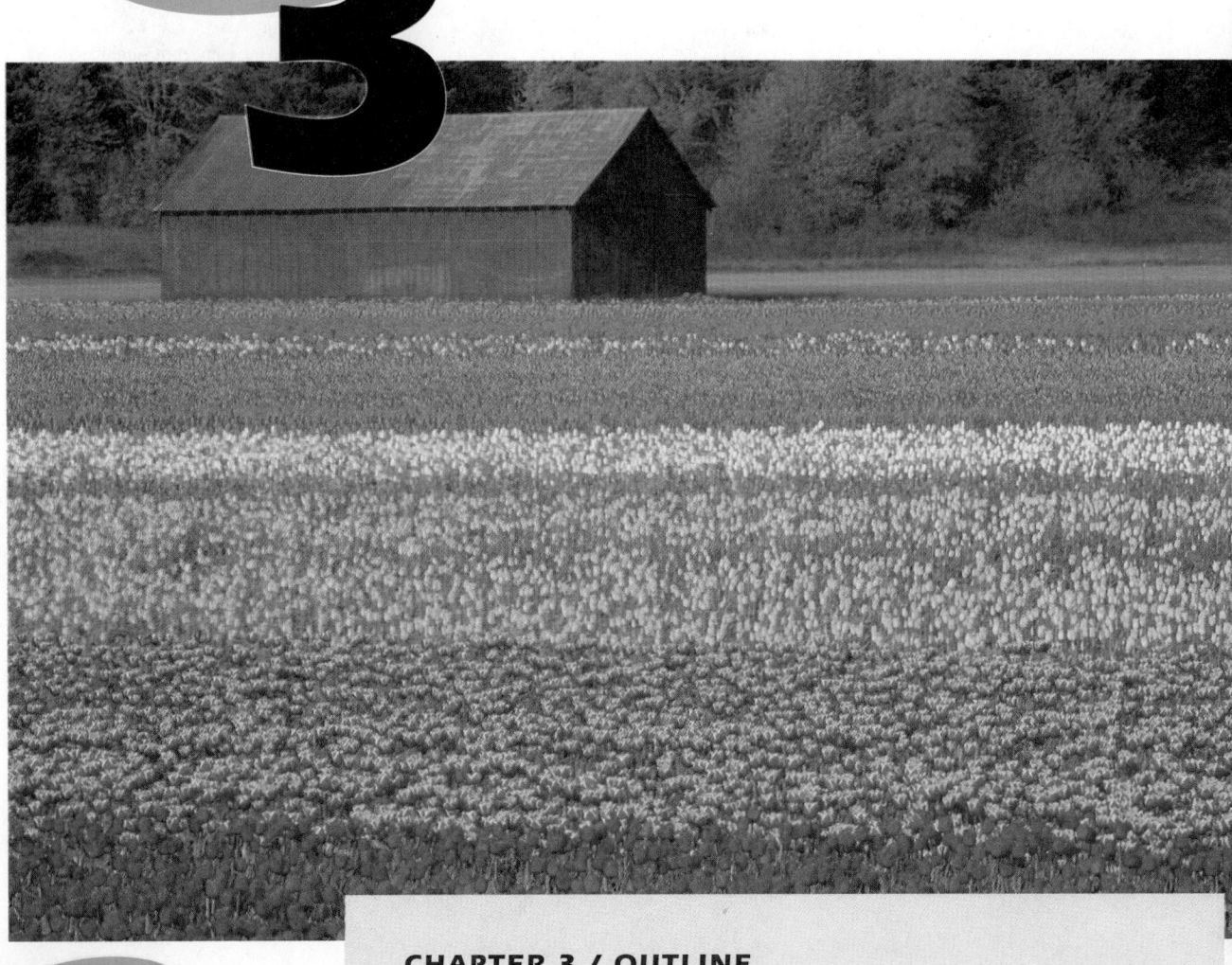

▲
Flower varieties are often classified by color in the nursery.

Throughout our lives we seek to bring order into the chaos that surrounds us. To do this we classify things according to their similarities. In the library we find books grouped according to the subject, and then by author. Department stores organize their merchandise by the size and style of clothing, as well as by the type of customer. Ballparks and theaters classify their seats by price and location. Chemists also classify things according to properties they can observe. For example, chemists classify substances as liquid (like water), solid (ice), or gas (steam). Chemists are also interested in identifying the tiniest part of a substance that still has the properties of the substance.

3.1   Matter Defined

The entire universe consists of matter and energy. Every day we come into contact with countless kinds of matter. Air, food, water, rocks, soil, glass, and this book are all different types of matter. Broadly defined, **matter** is *anything* that has mass and occupies space.

Matter may be quite invisible. For example, if an apparently empty test tube is submerged mouth downward in a beaker of water, the water rises only slightly into the tube. The water cannot rise further because the tube is filled with invisible matter: air (see Figure 3.1).

To the eye, matter appears to be continuous and unbroken. However, it is actually discontinuous and is composed of discrete, tiny particles called *atoms*. The particulate nature of matter will become evident when we study atomic structure and the properties of gases.

▲
FIGURE 3.1
An apparently empty test tube is submerged, mouth downward, in water. Only a small volume of water rises into the tube, which is actually filled with invisible matter—air.

matter

3.2   Physical States of Matter

Matter exists in three physical states: solid, liquid, and gas. A **solid** has a definite shape and volume, with particles that cohere rigidly to one another. The shape of a solid can be independent of its container. For example, a crystal of sulfur has the same shape and volume whether it is placed in a beaker or simply laid on a glass plate.

Most commonly occurring solids, such as salt, sugar, quartz, and metals, are *crystalline.* The particles that form crystalline materials exist in regular, repeating, three-dimensional, geometric patterns. Because their particles do not have any regular, internal geometric pattern, such solids as plastics, glass, and gels are called **amorphous** solids. (*Amorphous* means without shape or form.)

A **liquid** has a definite volume but not a definite shape, with particles that cohere firmly but not rigidly. Although the particles are held together by strong attractive forces and are in close contact with one another, they are able to move freely. Particle mobility gives a liquid fluidity and causes it to take the shape of the container in which it is stored.

A **gas** has indefinite volume and no fixed shape, with particles that move independently of one another. Particles in the gaseous state have gained enough energy to

solid

amorphous
liquid

gas

overcome the attractive forces that held them together as liquids or solids. A gas presses continuously in all directions on the walls of any container. Because of this quality a gas completely fills a container. The particles of a gas are relatively far apart compared with those of solids and liquids. The actual volume of the gas particles is very small compared to the volume of the space occupied by the gas. A gas therefore may be compressed into a very small volume or expanded almost indefinitely. Liquids cannot be compressed to any great extent, and solids are even less compressible than liquids.

If a bottle of ammonia solution is opened in one corner of the laboratory, we can soon smell its familiar odor in all parts of the room. The ammonia gas escaping from the solution demonstrates that gaseous particles move freely and rapidly and tend to permeate the entire area into which they are released.

Although matter is discontinuous, attractive forces exist that hold the particles together and give matter its appearance of continuity. These attractive forces are strongest in solids, giving them rigidity; they are weaker in liquids but still strong enough to hold liquids to definite volumes. In gases the attractive forces are so weak that the particles of a gas are practically independent of one another. Table 3.1 lists common materials that exist as solids, liquids, and gases. Table 3.2 compares the properties of solids, liquids, and gases.

TABLE 3.1   Common Materials in the Solid, Liquid, and Gaseous States of Matter

Solids	Liquids	Gases
Aluminum	Alcohol	Acetylene
Copper	Blood	Air
Gold	Gasoline	Butane
Polyethylene	Honey	Carbon dioxide
Salt	Mercury	Chlorine
Sand	Oil	Helium
Steel	Vinegar	Methane
Sulfur	Water	Oxygen

TABLE 3.2   Physical Properties of Solids, Liquids, and Gases

State	Shape	Volume	Particles	Compressibility
Solid	Definite	Definite	Rigidly cohering; tightly packed	Very slight
Liquid	Indefinite	Definite	Mobile; cohering	Slight
Gas	Indefinite	Indefinite	Independent of each other and relatively far apart	High

◄ **Water can exist as a solid (snow), a liquid (water), and a gas (steam) as shown here at Yellowstone National Park.**

3.3

3.3   Substances and Mixtures

The term *matter* refers to all materials that make up the universe. Many thousands of different, and distinct, kinds of matter exist. A **substance** is a particular kind of matter with a definite, fixed composition. Sometimes known as *pure substances,* substances are either elements or compounds. Familiar examples of elements are copper, gold, and oxygen. Familiar compounds are salt, sugar, and water.

  We classify a sample of matter as either *homogeneous* or *heterogeneous* by examining it. **Homogeneous** matter is uniform in appearance and has the same properties throughout. Matter consisting of two or more physically distinct phases is **heterogeneous.** A **phase** is a homogeneous part of a system separated from other parts by physical boundaries. A **system** is simply the body of matter under consideration. Whenever we have a system in which visible boundaries exist between the parts or components, that system has more than one phase and is heterogeneous. It does not matter whether these components are in the solid, liquid, or gaseous states.

  A pure substance may exist as different phases in a heterogeneous system. Ice floating in water, for example, is a two-phase system made up of solid water and liquid water. The water in each phase is homogeneous in composition, but because two phases are present, the system is heterogeneous.

  A **mixture** is a material containing two or more substances and can be either heterogeneous or homogeneous. Mixtures are variable in composition. If we add a spoonful of sugar to a glass of water, a heterogeneous mixture is formed immediately. The two phases are a solid (sugar) and a liquid (water). But upon stirring the sugar dissolves to form a homogeneous mixture or solution. Both substances are still

substance

homogeneous

heterogeneous
phase
system

mixture

Figure 3.2 ▶
Classification of matter. A pure substance is always homogeneous in composition, whereas a mixture always contains two or more substances and may be either homogeneous or heterogeneous.

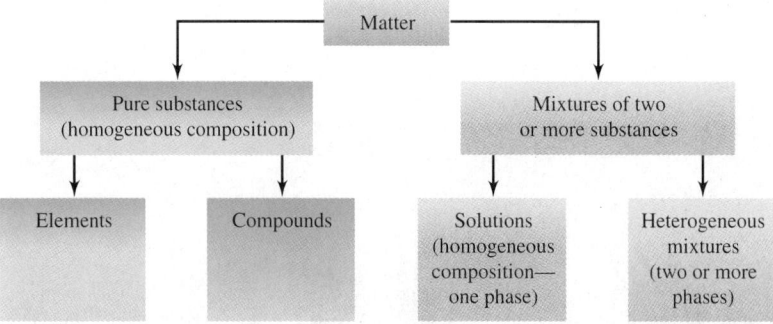

present: All parts of the solution are sweet and wet. The proportions of sugar and water can be varied simply by adding more sugar and stirring to dissolve.

Many substances do not form homogeneous mixtures. If we mix sugar and fine white sand, a heterogeneous mixture is formed. Careful examination may be needed to decide that the mixture is heterogeneous because the two phases (sugar and sand) are both white solids. Ordinary matter exists mostly as mixtures. If we examine soil, granite, iron ore, or other naturally occurring mineral deposits, we find them to be heterogeneous mixtures. Air is a homogeneous mixture (solution) of several gases. Figure 3.2 illustrates the relationships of substances and mixtures.

Flowcharts can help you to visualize the connections between concepts.

3.4   Elements

All words in English are formed from an alphabet consisting of only 26 letters. All known substances on Earth—and most probably in the universe, too—are formed from a sort of "chemical alphabet" consisting of 111 presently known elements. An **element** is a fundamental or elementary substance that cannot be broken down by chemical means to simpler substances. Elements are the building blocks of all substances. The elements are numbered in order of increasing complexity beginning with hydrogen, number 1. Of the first 92 elements, 88 are known to occur in nature. The other four—technetium (43), promethium (61), astatine (85), and francium (87)—either do not occur in nature or have only transitory existences during radioactive decay. With the exception of number 94, plutonium, elements above number 92 are not known to occur naturally but have been synthesized, usually in very small quantities, in laboratories. The discovery of trace amounts of element 94 (plutonium) in nature has been reported recently. The syntheses of elements 110 and 111 were reported in 1994. No elements other than those on the Earth have been detected on other bodies in the universe.

Most substances can be decomposed into two or more simpler substances. Water can be decomposed into hydrogen and oxygen. Sugar can be decomposed into carbon, hydrogen, and oxygen. Table salt is easily decomposed into sodium and chlorine. An element, however, cannot be decomposed into simpler substances by ordinary chemical changes.

If we could take a small piece of an element, say copper, and divide it and subdivide it into smaller and smaller particles, we would finally come to a single unit of copper that we could no longer divide and still have copper. This smallest particle of an element that can exist is called an **atom,** which is also the smallest unit of an element that can enter into a chemical reaction. Atoms are made up of still smaller

element (margin note)

atom (margin note)

subatomic particles. However, these subatomic particles (described in Chapter 5) do not have the properties of elements.

3.5 Distribution of Elements

Elements are distributed unequally in nature, as shown in Figure 3.3. At normal room temperature two of the elements, bromine and mercury, are liquids. Eleven elements, hydrogen, nitrogen, oxygen, fluorine, chlorine, helium, neon, argon, krypton, xenon, and radon, are gases. All the other elements are solids.

Ten elements make up about 99% of the mass of the Earth's crust, seawater, and atmosphere. Oxygen, the most abundant of these, constitutes about 50% of this mass. The distribution of the elements shown in Figure 3.3 includes the Earth's crust to a depth of about 10 miles, the oceans, fresh water, and the atmosphere but does not include the mantle and core of the Earth, which are believed to consist of metallic iron and nickel. Because the atmosphere contains relatively little matter, its inclusion has almost no effect on the distribution. But the inclusion of fresh and salt water does have an appreciable effect since water contains about 11.2% hydrogen. Nearly all of the 0.87% hydrogen shown in the table is from water.

The average distribution of the elements in the human body is shown in Figure 3.3. Note again the high percentage of oxygen.

▲
Mercury (left) and bromine are liquid elements at room temperature.

3.6 Names of the Elements

The names of the elements come to us from various sources. Many are derived from early Greek, Latin, or German words that describe some property of the element. For example, iodine is taken from the Greek word *iodes,* meaning violetlike, and iodine is certainly violet in the vapor state. The name of the metal bismuth originates from the German words *weisse masse,* which means white mass. Miners called it *wismat;* it was later changed to *bismat,* and finally to bismuth. Some elements are named for the location of their discovery—for example, germanium, discovered in 1886 by a German chemist. Others are named in commemoration of famous scientists, such as einsteinium and curium, named for Albert Einstein and Marie Curie, respectively.

3.7 Symbols of the Elements

We all recognize Mr., N.Y., and Ave. as abbreviations for mister, New York, and avenue. In a like manner each element also has an abbreviation; these are called **symbols** of the elements. Fourteen elements have a single letter as their symbol, and the rest have two letters. A symbol stands for the element itself, for one atom of the element, and (as we shall see later) for a particular quantity of the element.

symbol

Rules governing symbols of elements are as follows:

1. Symbols have either one or two letters.
2. If one letter is used, it is capitalized.
3. If two letters are used, only the first is capitalized.

Examples: Iodine I   Barium Ba

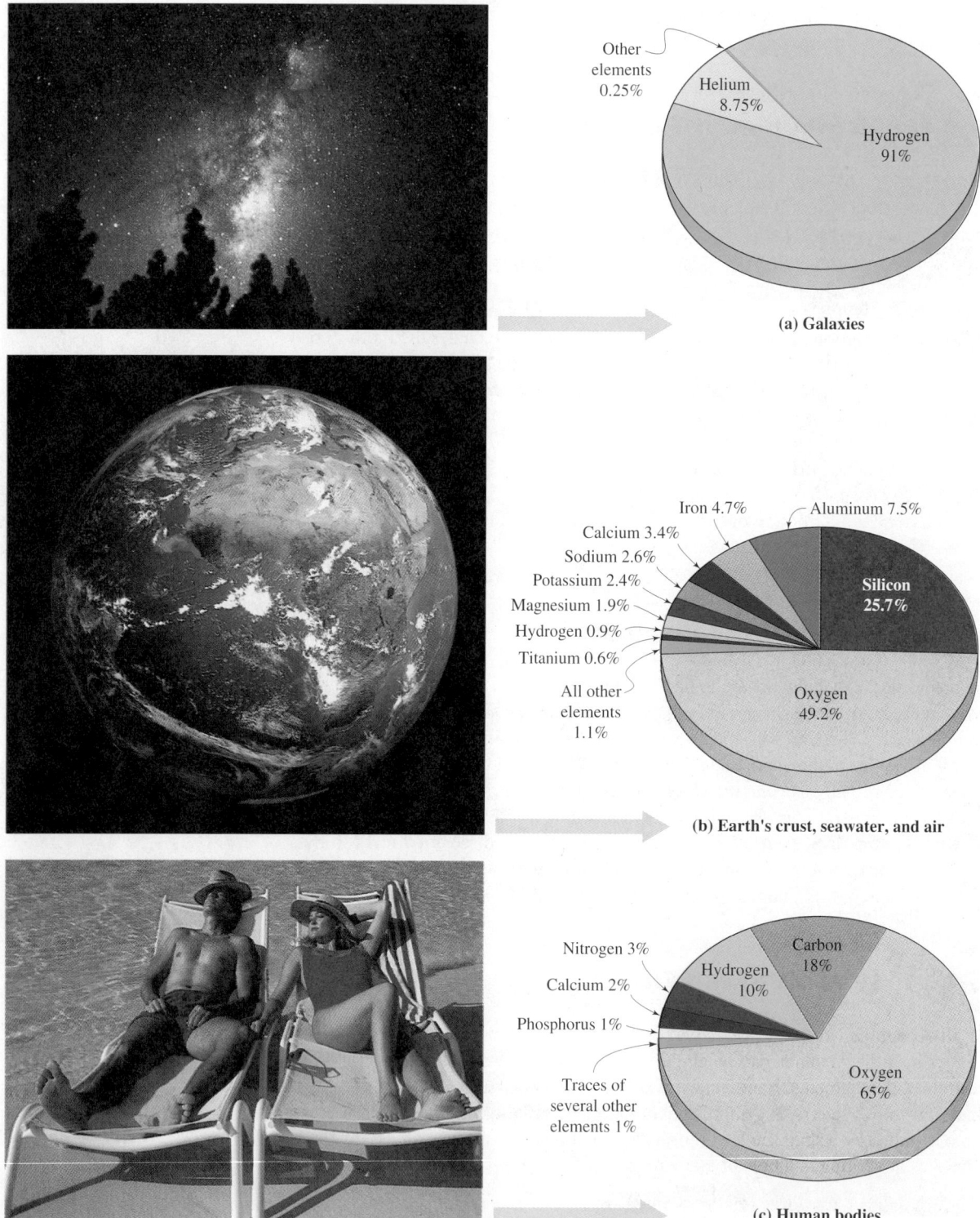

(a) Galaxies

Other elements 0.25%
Helium 8.75%
Hydrogen 91%

(b) Earth's crust, seawater, and air

Iron 4.7%
Aluminum 7.5%
Calcium 3.4%
Sodium 2.6%
Potassium 2.4%
Magnesium 1.9%
Hydrogen 0.9%
Titanium 0.6%
All other elements 1.1%
Silicon 25.7%
Oxygen 49.2%

(c) Human bodies

Nitrogen 3%
Calcium 2%
Phosphorus 1%
Traces of several other elements 1%
Carbon 18%
Hydrogen 10%
Oxygen 65%

Figure 3.3 Distribution of the common elements in nature.

The symbols and names of all the elements are given in the table on the inside front cover of this book. Table 3.3 lists the more commonly used symbols. Examine this table carefully and you will note that most of the symbols start with the same letter as the name of the element that is represented. A number of symbols, however, appear to have no connection with the names of the elements they represent (see Table 3.4). These symbols have been carried over from earlier names (usually in Latin) of the elements and are so firmly implanted in the literature that their use is continued today.

TABLE 3.3	Symbols of the Most Common Elements				
Element	**Symbol**	**Element**	**Symbol**	**Element**	**Symbol**
Aluminum	Al	Fluorine	F	Phosphorus	P
Antimony	Sb	Gold	Au	Platinum	Pt
Argon	Ar	Helium	He	Potassium	K
Arsenic	As	Hydrogen	H	Radium	Ra
Barium	Ba	Iodine	I	Silicon	Si
Bismuth	Bi	Iron	Fe	Silver	Ag
Boron	B	Lead	Pb	Sodium	Na
Bromine	Br	Lithium	Li	Strontium	Sr
Cadmium	Cd	Magnesium	Mg	Sulfur	S
Calcium	Ca	Manganese	Mn	Tin	Sn
Carbon	C	Mercury	Hg	Titanium	Ti
Chlorine	Cl	Neon	Ne	Tungsten	W
Chromium	Cr	Nickel	Ni	Uranium	U
Cobalt	Co	Nitrogen	N	Zinc	Zn
Copper	Cu	Oxygen	O		

▲ **Iodine in its elemental form is a dark purple crystal. The symbol for iodine is I.**

TABLE 3.4	Symbols of the Elements Derived from Early Names*	
Present name	**Symbol**	**Former name**
Antimony	Sb	Stibium
Copper	Cu	Cuprum
Gold	Au	Aurum
Iron	Fe	Ferrum
Lead	Pb	Plumbum
Mercury	Hg	Hydrargyrum
Potassium	K	Kalium
Silver	Ag	Argentum
Sodium	Na	Natrium
Tin	Sn	Stannum
Tungsten	W	Wolfram

*These symbols are in use today even though they do not correspond to the current name of the element.

▲ **The metal sodium is soft enough to cut with a knife. The symbol for sodium is Na.**

Special care must be taken in writing symbols. Capitalize only the first letter and use a lowercase second letter if needed. This is important. For example, consider Co, the symbol for the element cobalt. If you write CO (capital C and capital O), you will have written the two elements carbon and oxygen (the *formula* for carbon monoxide), *not* the single element cobalt. Also make sure you write the letters distinctly; otherwise Co (for cobalt) may be misread as Ca (for calcium).

Knowledge of symbols is essential for writing chemical formulas and equations, and will be needed in the remainder of this book and in any future chemistry courses you may take. One way to learn the symbols is to practice a few minutes a day by making flash cards of names and symbols and then practicing daily. Initially it is a good plan to learn the symbols of the most common elements shown in Table 3.3.

▲
The colors of these varieties of quartz are the result of the presence of different metallic elements in the sample.

3.8   Metals, Nonmetals, and Metalloids

The elements are classified as metals, nonmetals, and metalloids. Most of the elements are metals. We are familiar with them because of their widespread use in tools, construction materials, automobiles, and so on. But nonmetals are equally useful in our everyday life as major components of clothing, food, fuel, glass, plastics, and wood. Metalloids are often used in the electronics industry.

metal

The **metals** are solids at room temperature (mercury is an exception). They have high luster, are good conductors of heat and electricity, are *malleable* (can be rolled or hammered into sheets), and are *ductile* (can be drawn into wires). Most metals

Samples of various metals, ▶ including aluminum, copper, mercury, titanium, beryllium, cadmium, calcium, and nickel.

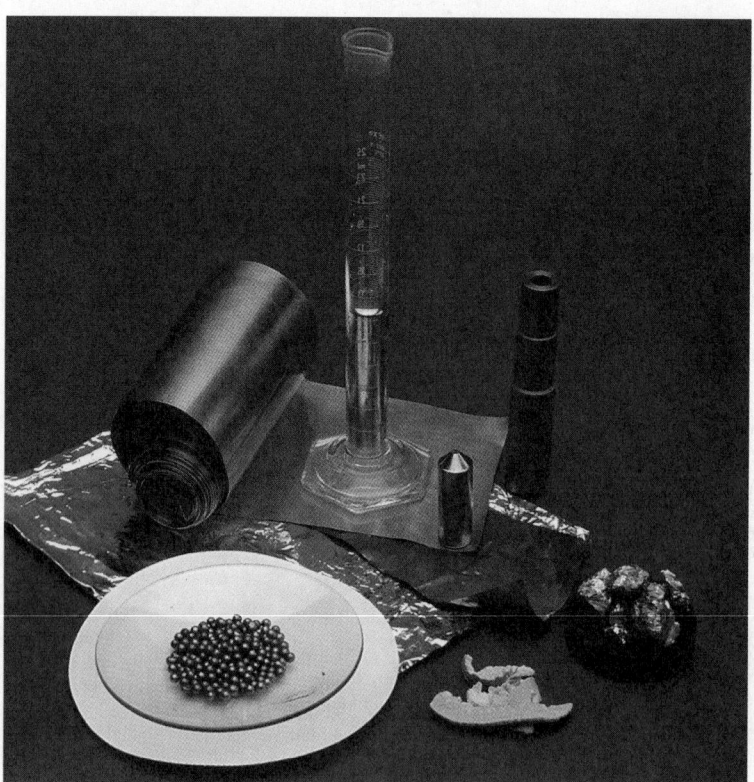

have a high melting point and high density. Familiar metals are aluminum, chromium, copper, gold, iron, lead, magnesium, mercury, nickel, platinum, silver, tin, and zinc. Less familiar but still important metals are calcium, cobalt, potassium, sodium, uranium, and titanium.

Metals have little tendency to combine with each other to form compounds. But many metals readily combine with nonmetals such as chlorine, oxygen, and sulfur to form ionic compounds such as metallic chlorides, oxides, and sulfides. In nature minerals are composed of the more reactive metals combined with other elements. A few of the less reactive metals such as copper, gold, and silver are sometimes found in a native, or free, state.

Nonmetals, unlike metals, are not lustrous, have relatively low melting points and densities, and are generally poor conductors of heat and electricity. Carbon, phosphorus, sulfur, selenium, and iodine are solids; bromine is a liquid; the rest of the nonmetals are gases. Common nonmetals found uncombined in nature are carbon (graphite and diamond), nitrogen, oxygen, sulfur, and the noble gases (helium, neon, argon, krypton, xenon, and radon).

Nonmetals combine with one another to form molecular compounds such as carbon dioxide (CO_2), methane (CH_4), butane (C_4H_{10}), and sulfur dioxide (SO_2). Fluorine, the most reactive nonmetal, combines readily with almost all other elements.

Several elements (boron, silicon, germanium, arsenic, antimony, tellurium, and polonium) are classified as **metalloids** and have properties that are intermediate between those of metals and those of nonmetals. The intermediate position of these elements is shown in Table 3.5. Certain metalloids, such as boron, silicon, and germanium, are the raw materials for the semiconductor devices that make our modern electronics industry possible.

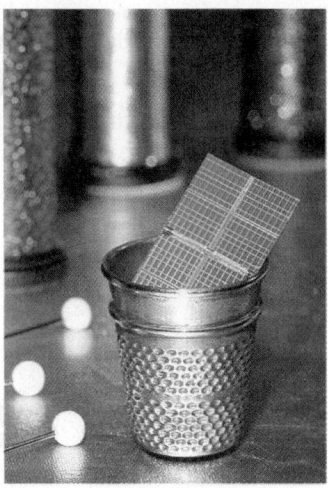

▲
This tiny and powerful computer chip is made of silicon, a metalloid.

nonmetal

metalloid

◄ **Samples of various nonmetals, including iodine (far left), bromine, oxygen, neon (in box), and sulfur.**

TABLE 3.5   Classification of the Elements into Metals, Metalloids, and Nonmetals

1 H																	2 He
3 Li	4 Be	☐ Metals										5 B	6 C	7 N	8 O	9 F	10 Ne
11 Na	12 Mg	▨ Metalloids ☐ Nonmetals										13 Al	14 Si	15 P	16 S	17 Cl	18 Ar
19 K	20 Ca	21 Sc	22 Ti	23 V	24 Cr	25 Mn	26 Fe	27 Co	28 Ni	29 Cu	30 Zn	31 Ga	32 Ge	33 As	34 Se	35 Br	36 Kr
37 Rb	38 Sr	39 Y	40 Zr	41 Nb	42 Mo	43 Tc	44 Ru	45 Rh	46 Pd	47 Ag	48 Cd	49 In	50 Sn	51 Sb	52 Te	53 I	54 Xe
55 Cs	56 Ba	57 La*	72 Hf	73 Ta	74 W	75 Re	76 Os	77 Ir	78 Pt	79 Au	80 Hg	81 Tl	82 Pb	83 Bi	84 Po	85 At	86 Rn
87 Fr	88 Ra	89 Ac†	104 Rf	105 Db	106 Sg	107 Bh	108 Hs	109 Mt	110 —	111 —	112 —						

*	58 Ce	59 Pr	60 Nd	61 Pm	62 Sm	63 Eu	64 Gd	65 Tb	66 Dy	67 Ho	68 Er	69 Tm	70 Yb	71 Lu	
†	90 Th	91 Pa	92 U	93 Np	94 Pu	95 Am	96 Cm	97 Bk	98 Cf	99 Es	100 Fm	101 Md	102 No	103 Lr	

3.9   Compounds

compound   A **compound** is a distinct substance that contains two or more elements chemically combined in definite proportions by mass. Compounds, unlike elements, can be decomposed chemically into simpler substances—that is, into simpler compounds and/or elements. Atoms of the elements in a compound are combined in whole-number ratios, never as fractional parts. Compounds fall into two general types, *molecular* and *ionic*.

molecule   A **molecule** is the smallest uncharged individual unit of a compound formed by the union of two or more atoms. Water is a typical molecular compound. If we divide a drop of water into smaller and smaller particles, we finally obtain a single molecule of water consisting of two hydrogen atoms bonded to one oxygen atom. This molecule is the ultimate particle of water; it cannot be further subdivided without destroying the water molecule and forming hydrogen and oxygen.

ion   An **ion** is a positively or negatively charged atom or group of atoms. An ionic compound is held together by attractive forces that exist between positively and negatively charged ions. A positively charged ion is called a **cation** (pronounced *cat-eye-on*); a negatively charged ion is called an **anion** (pronounced *an-eye-on*). Figure 3.4 illustrates the classification of compounds.

cation

anion

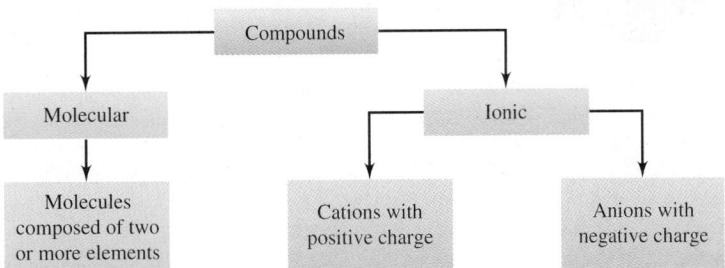

Sodium chloride is a typical ionic compound. The ultimate particles of sodium chloride are positively charged sodium ions and negatively charged chloride ions. Sodium chloride is held together in a crystalline structure by the attractive forces ex-isting between these oppositely charged ions. Although ionic compounds consist of large aggregates of cations and anions, their formulas are normally represented by the simplest possible ratio of the atoms in the compound. For example, in sodium chloride the ratio is one sodium ion to one chlorine ion, so the formula is NaCl. The two types of compounds, molecular and ionic, are illustrated in Figure 3.5.

There are more than 9 million known registered compounds, with no end in sight as to the number that will be prepared in the future. Each compound is unique and has characteristic properties. Let's consider two compounds, water and sodium chloride, in some detail. Water is a colorless, odorless, tasteless liquid that can be changed to a solid (ice) at 0°C and to a gas (steam) at 100°C. Composed of two atoms of hydrogen and one atom of oxygen per molecule, water is 11.2% hydrogen and 88.8% oxygen by mass. Water reacts chemically with sodium to produce hydro-gen gas and sodium hydroxide, with lime to produce calcium hydroxide, and with sulfur trioxide to produce sulfuric acid. No other compound has all these exact phys-ical and chemical properties; they are characteristic of water alone.

Sodium chloride is a colorless crystalline substance with a ratio of one atom of sodium to one atom of chlorine. Its composition by mass is 39.3% sodium and 60.7% chlorine. It does not conduct electricity in its solid state; it dissolves in water to produce a solution that conducts electricity. When a current is passed through molten sodium chloride, solid sodium and gaseous chlorine are produced. These spe-cific properties belong to sodium chloride and to no other substance. Thus, a com-pound may be identified and distinguished from all other compounds by its charac-teristic properties.

(a) H_2O

(b) NaCl

▲ **FIGURE 3.5**
Representation of molecular and ionic (nonmolecular) com-pounds. (a) Two hydrogen atoms combined with an oxygen atom to form a mole-cule of water. (b) A positively charged sodium ion and a negatively charged chloride ion form the compound sodium chloride.

3.10 Elements That Exist as Diatomic Molecules

Seven elements (all nonmetals) occur as **diatomic molecules.** These elements and their symbols, formulas, and brief descriptions are listed in Table 3.6. Whether found free in nature or prepared in the laboratory, the molecules of these elements always contain two atoms. The formulas of the free elements are therefore always written to show this molecular composition: H_2, N_2, O_2, F_2, Cl_2, Br_2, and I_2.

diatomic molecules

It is important to see that symbols can designate either an atom or a molecule of an element. Consider hydrogen and oxygen. Hydrogen gas is present in volcanic gases and can be prepared by many chemical reactions. Regardless of their source,

Hydrogen: Fuel of the Future

Hydrogen, the lightest element on the periodic table, could provide society with a fuel that is nearly inexhaustible, environmentally harmless, and available everywhere. The U.S. space program has used hydrogen for nearly 30 years to power rockets, provide electrical power, and produce drinking water for astronauts. How can we harness hydrogen's energy and use it in our daily lives?

Mercedes Benz has developed a prototype minivan that is powered by hydrogen fuel cells. These cells combine compressed hydrogen with oxygen to form water and power the car. Because of the difficulties in storing enough compressed hydrogen these vehicles have limited range. Hydrogen gas has a low density (requiring high pressure to avoid large, bulky tanks) and liquid hydrogen (boiling point, $-253°$ C) requires constant refrigeration and insulated containers. To improve range and eliminate storage problems engineers are currently considering ways to generate hydrogen as needed for fuel.

The U.S. Department of Energy and Arthur D. Little Co. of Cambridge, Massachusetts, have developed an alternative system that may solve the storage problem. Their gasoline-fueled system contains a processor that partially converts the gasoline to hydrogen gas, which is then sent to a fuel cell and used to generate power. Preliminary tests have produced hydrogen fast enough to generate 50 kW of electric power, plenty to run a mid-size car. The processor uses gasoline, ethanol, methanol, or natural gas to supply hydrogen to the fuel cell. A car using this system gets about twice the mileage of a traditional car.

This alternative fuel system doesn't require large batteries (like those needed in electric cars) or storage tanks for the hydrogen, so it may be a significant step toward a hydrogen-powered economy. This hybrid system also takes advantage of the gasoline distribution system already in place. You could soon drive a hydrogen car and fill up at your local gas station!

TABLE 3.6	Elements That Exist as Diatomic Molecules		
Element	**Symbol**	**Molecular formula**	**Normal state**
Hydrogen	H	H_2	Colorless gas
Nitrogen	N	N_2	Colorless gas
Oxygen	O	O_2	Colorless gas
Fluorine	F	F_2	Pale yellow gas
Chlorine	Cl	Cl_2	Yellow-green gas
Bromine	Br	Br_2	Reddish-brown liquid
Iodine	I	I_2	Bluish-black solid

all samples of free hydrogen gas consist of diatomic molecules. Free hydrogen is designated by the formula H_2, which also expresses its composition. Oxygen makes up about 21% by volume of the air that we breathe. This free oxygen is constantly being replenished by photosynthesis; it can also be prepared in the laboratory by several reactions. The majority of free oxygen is diatomic and is designated by the formula O_2. Now consider water, a compound designated by the formula H_2O (sometimes HOH). Water contains neither free hydrogen (H_2) nor free oxygen (O_2). The H_2 part of the formula H_2O simply indicates that two atoms of hydrogen are combined with one atom of oxygen to form water.

> **Symbols are used to designate elements, show the composition of molecules of elements, and give the elemental composition of compounds.**

3.11   Chemical Formulas

Chemical formulas are used as abbreviations for compounds. A **chemical formula** shows the symbols and the ratio of the atoms of the elements in a compound. Sodium chloride contains one atom of sodium per atom of chlorine; its formula is NaCl. The formula for water is H_2O; it shows that a molecule of water contains two atoms of hydrogen and one atom of oxygen.

chemical formula

The formula of a compound tells us which elements it is composed of and how many atoms of each element are present in a formula unit. For example, a unit of sulfuric acid is composed of two atoms of hydrogen, one atom of sulfur, and four atoms of oxygen. We could express this compound as HHSOOOO, but this is cumbersome, so we write H_2SO_4 instead. The formula may be expressed verbally as "H-two-S-O-four." Numbers that appear partially below the line and to the right of a symbol of an element are called **subscripts.** Thus the 2 and the 4 in H_2SO_4 are subscripts (see Figure 3.6). Characteristics of chemical formulas are

subscript

1. The formula of a compound contains the symbols of all the elements in the compound.
2. When the formula contains one atom of an element, the symbol of that element represents that one atom. The number one (1) is not used as a subscript to indicate one atom of an element.
3. When the formula contains more than one atom of an element, the number of atoms is indicated by a subscript written to the right of the symbol of that atom. For example, the two (2) in H_2O indicates two atoms of H in the formula.
4. When the formula contains more than one of a group of atoms that occurs as a unit, parentheses are placed around the group, and the number of units of the group is indicated by a subscript placed to the right of the parentheses. Consider the nitrate group, NO_3. The formula for sodium nitrate, $NaNO_3$, has only one nitrate group, so no parentheses are needed. Calcium nitrate, $Ca(NO_3)_2$, has two nitrate groups, as indicated by the use of parentheses and the subscript 2. $Ca(NO_3)_2$ has a total of nine atoms: one Ca, two N, and six O atoms. The formula $Ca(NO_3)_2$ is read as "C-A [pause] N-O-three taken twice."

Figure 3.6 ▶
Explanation of the formulas
NaCl, H₂SO₄, and Ca(NO₃)₂.

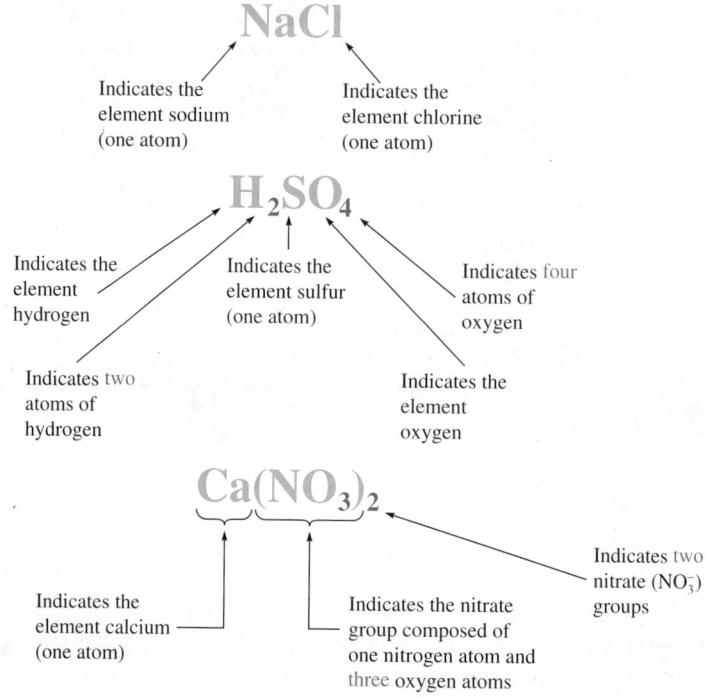

5. Formulas written as H_2O, H_2SO_4, $Ca(NO_3)_2$, and $C_{12}H_{22}O_{11}$ show only the number and kind of each atom contained in the compound; they do not show the arrangement of the atoms in the compound or how they are chemically bonded to one another.

Example 3.1   Write formulas for the following compounds; the atomic composition is given. (a) Hydrogen chloride: 1 atom hydrogen + 1 atom chlorine; (b) methane: 1 atom carbon + 4 atoms hydrogen; (c) glucose: 6 atoms carbon + 12 atoms hydrogen + 6 atoms oxygen.

Solution   (a) First write the symbols of the atoms in the formula: H Cl. Since the ratio of atoms is one to one, we bring the symbols together to give the formula for hydrogen chloride as HCl.

(b) Write the symbols of the atoms: C H. Now bring the symbols together and place a subscript 4 after the hydrogen atom. The formula is CH_4.

(c) Write the symbols of the atoms: C H O. Now write the formula, bringing together the symbols followed by the correct subscripts according to the data given (six C, twelve H, six O). The formula is $C_6H_{12}O_6$.

3.12   Mixtures

Single substances—elements or compounds—seldom occur naturally in a pure state. Air is a mixture of gases; seawater is a mixture of a variety of dissolved minerals; ordinary soil is a complex mixture of minerals and various organic materials.

(a)   (b)   (c)

▲
(a) When iron and sulfur exist as pure substances, only the iron is attracted to a magnet. (b) A mixture of iron and sulfur can be separated by using the difference in magnetic attraction. (c) The compound iron(II) sulfide cannot be separated into its elements with a magnet.

How is a mixture distinguished from a pure substance? A mixture always contains two or more substances that can be present in varying concentrations. Let's consider two examples.

Homogeneous Mixture Homogeneous mixtures (solutions) containing either 5% or 10% salt in water can be prepared simply by mixing the correct amounts of salt and water. These mixtures can be separated by boiling away the water, leaving the salt as a residue.

Heterogeneous Mixture The composition of a heterogeneous mixture of sulfur crystals and iron filings can be varied by merely blending in either more sulfur or more iron filings. This mixture can be separated physically by using a magnet to attract the iron.

Iron(II) sulfide (FeS) contains 63.5% Fe and 36.5% S by mass. If we mix iron and sulfur in this proportion, do we have iron(II) sulfide? No, it is still a mixture; the iron is still attracted by a magnet. But if this mixture is heated strongly, a chemical change (reaction) occurs in which the reacting substances, iron and sulfur, form a new substance, iron(II) sulfide. Iron(II) sulfide, FeS, is a compound of iron and sulfur and has properties that are different from those of either iron or sulfur: It is not attracted by a magnet as shown in the photo. The general characteristics of mixtures and compounds are compared in Table 3.7. The differences between the iron and sulfur *mixture* and the iron(II) sulfide *compound* are as follows:

Iron(II) sulfide is the correct name for the compound formed from iron and sulfur. We will discuss the reason for the (II) in Chapter 6 when we learn to name compounds.

	Mixture of iron and sulfur	Compound of iron and sulfur
Formula	Has no definite formula; consists of Fe and S.	FeS
Composition	Contains Fe and S in any proportion by mass.	63.5% Fe and 36.5% S by mass.
Separation	Fe and S can be separated by physical means.	Fe and S can be separated only by chemical change.

Carbon—The Chameleon

Just as a chameleon changes its color to reflect its environment, the element carbon is found in many different forms. Carbon is found in the mineral *graphite,* as *diamond,* and as *buckminsterfullerene.* The physical properties of each form of carbon are quite distinct.

Diamonds are transparent crystals that are colorless when pure—but they can range from pale blue to jet black when impurities are present. Diamond is the hardest natural substance, an excellent heat conductor, and when certain impurities are added, it becomes an electrical semiconductor. Diamonds are used in cutting and drilling tools and displayed as beautiful gems in jewelry.

Graphite consists of layered sheets of carbon atoms. Because the sheets easily slip over each other, graphite is slippery making it an excellent choice for lubricants. Mixtures of clay and graphite are molded into the "lead" used in

pencils. The higher the clay content, the harder the "lead" in the pencil (and the more difficult to use in writing). An excellent conductor of electricity, graphite is mined as massive crystals or is obtained from heating coal and pitch in very high temperature furnaces. Graphite is formed into electrodes for batteries and used as a lubricant in locks.

Charcoal, consisting of tiny crystals of graphite, can adsorb large quantities of substances onto its surfaces. For this reason it is very useful in water purification systems, in the manufacture of gas masks, and in removing color from solutions (as in refining sugar). Carbon is added to iron to form steel; carbon black (formed when natural gas is burned with an insufficient quantity of oxygen) is used in ink, shoe polish, and as an additive to rubber to make tires black.

Buckminsterfullerene is composed of clusters of carbon atoms arranged in

the shape of a soccer ball. This cagelike structure permits capture of other atoms, leading to some interesting applications. More information on buckminster-fullerene can be found in the Chemistry in Action in Chapter 11.

In addition, carbon combines with other elements to form millions of useful compounds. Hydrocarbons (molecules containing carbon and hydrogen) are found in petroleum and natural gas. So many hydrocarbons and their derivatives exist that an entire branch of chemistry, organic chemistry, is dedicated to studying them. Finally, carbon is an essential constituent of biomolecules. Found in all living organisms as part of carbohydrates, proteins, and fats, carbon is also a critical atom in both DNA and RNA, the molecules that determine the genetic composition of each organism.

Diamonds

Graphite

Petroleum production

Buckminsterfullerene

DNA component

Electrodes

TABLE 3.7	Comparison of Mixtures and Compounds	
	Mixture	**Compound**
Composition	May be composed of elements, compounds, or both in variable composition.	Composed of two or more elements in a definite, fixed proportion by mass.
Separation of components	By physical or mechanical means.	Only by chemical changes in the elements.
Identification of components	Components do not lose their identity.	A compound does not resemble the elements from which it is formed.

Concepts in Review

1. Identify the three physical states of matter.

2. Distinguish between substances and mixtures.

3. Classify common materials as elements, compounds, or mixtures.

4. Write the symbols when given the names, or write the names when given the symbols, of the common elements listed in Table 3.3.

5. Understand how symbols, including subscripts and parentheses, are used to write chemical formulas.

6. Differentiate among atoms, molecules, and ions.

7. List the characteristics of metals, nonmetals, and metalloids.

8. List the elements that occur as diatomic molecules.

Key Terms

amorphous (3.2)
anion (3.9)
atom (3.4)
cation (3.9)
chemical formula (3.11)

compound (3.9)
diatomic molecules (3.10)
element (3.4)
gas (3.2)
heterogeneous (3.3)

homogeneous (3.3)
ion (3.9)
liquid (3.2)
matter (3.1)
metal (3.8)

metalloid (3.8)
mixture (3.3)
molecule (3.9)
nonmetal (3.8)
phase (3.3)

solid (3.2)
subscripts (3.11)
substance (3.3)
symbol (3.7)
system (3.3)

Questions

1. List four different substances in each of the three states of matter.

2. In terms of the properties of the ultimate particles of a substance, explain
   (a) why a solid has a definite shape but a liquid does not
   (b) why a liquid has a definite volume but a gas does not
   (c) why a gas can be compressed rather easily but a solid cannot be compressed appreciably

3. What evidence can you find in Figure 3.1 that gases occupy space?

4. Which liquids listed in Table 3.1 are not mixtures?

5. Which of the gases listed in Table 3.1 are not pure substances?

6. When the stopper is removed from a partly filled bottle containing solid and liquid acetic acid at 16.7°C, a strong vinegarlike odor is noticeable immediately. How many acetic acid phases must be present in the bottle? Explain.

7. Is the system enclosed in the bottle in Question 6 homogeneous or heterogeneous? Explain.

8. Is a system that contains only one substance necessarily homogeneous? Explain.

9. Is a system that contains two or more substances necessarily heterogeneous? Explain.

10. Are there more atoms of silicon or hydrogen in the Earth's crust, seawater, and atmosphere? Use Figure 3.3 and the fact that the mass of a silicon atom is about 28 times that of a hydrogen atom.

11. What does the symbol of an element stand for?

12. Write down what you believe to be the symbols for the elements phosphorus, aluminum, hydrogen, potassium, magnesium, sodium, nitrogen, nickel, and silver. Check yourself by looking up the correct symbols in Table 3.3.

13. Interpret the difference in meanings for each of these pairs:
    (a) Si and SI (b) Pb and PB (c) 4 P and P_4

14. List six elements and their symbols in which the first letter of the symbol is different from that of the name. (Table 3.4)

15. Write the names and symbols for the 14 elements that have only one letter as their symbol. (See tables on inside front cover.)

16. Distinguish between an element and a compound.

17. How many metals are there? nonmetals? metalloids? (Table 3.5)

18. Of the ten most abundant elements in the Earth's crust, seawater, and atmosphere, how many are metals? nonmetals? metalloids? (Figure 3.3)

19. Of the six most abundant elements in the human body, how many are metals? nonmetals? metalloids? (Figure 3.3)

20. Why is the symbol for gold Au rather than G or Go?

21. Give the names of (a) the solid diatomic nonmetal and (b) the liquid diatomic nonmetal. (Table 3.6)

22. Distinguish between a compound and a mixture.

23. What are the two general types of compounds? How do they differ from each other?

24. What is the basis for distinguishing one compound from another?

25. How many atoms are contained in (a) one molecule of hydrogen, (b) one molecule of water, and (c) one molecule of sulfuric acid?

26. What is the major difference between a cation and an anion?

27. Write the names and formulas of the elements that exist as diatomic molecules. (Table 3.6)

28. Distinguish between homogeneous and heterogeneous mixtures.

29. Tabulate the properties that characterize metals and nonmetals.

30. Which of the following are diatomic molecules?
    (a) H_2 (c) HCl   (e) NO   (g) $MgCl_2$
    (b) SO_2 (d) H_2O (f) NO_2

Paired Exercises

31. What elements are present in each compound?
    (a) potassium iodide KI
    (b) sodium carbonate Na_2CO_3
    (c) aluminum oxide Al_2O_3
    (d) calcium bromide $CaBr_2$
    (e) acetic acid $HC_2H_3O_2$

32. What elements are present in each compound?
    (a) magnesium bromide $MgBr_2$
    (b) carbon tetrachloride CCl_4
    (c) nitric acid HNO_3
    (d) barium sulfate $BaSO_4$
    (e) aluminum phosphate $AlPO_4$

33. Write the formula for each compound (the composition is given after each name):
    (a) zinc oxide   1 atom Zn, 1 atom O
    (b) potassium chlorate 1 atom K, 1 atom Cl, 3 atoms O
    (c) sodium hydroxide   1 atom Na, 1 atom O, 1 atom H
    (d) ethyl alcohol   2 atoms C, 6 atoms H, 1 atom O

34. Write the formula for each compound (the composition is given after each name):
    (a) aluminum bromide 1 atom Al, 3 atoms Br
    (b) calcium fluoride 1 atom Ca, 2 atoms F
    (c) lead(II) chromate 1 atom Pb, 1 atom Cr, 4 atoms O
    (d) benzene 6 atoms C, 6 atoms H

35. Explain the meaning of each symbol and number in these formulas:
    (a) H_2O
    (b) Na_2SO_4
    (c) $HC_2H_3O_2$

36. Explain the meaning of each symbol and number in these formulas:
    (a) $AlBr_3$
    (b) $Ni(NO_3)_2$
    (c) $C_{12}H_{22}O_{11}$ (sucrose)

37. How many atoms are represented in each formula?
    (a) KF   (d) $NaC_2H_3O_2$
    (b) $CaCO_3$ (e) $(NH_4)_2C_2O_4$
    (c) $K_2Cr_2O_7$

38. How many atoms are represented in each formula?
    (a) NaCl   (d) CCl_2F_2 (Freon)
    (b) N_2 (e) $Al_2(SO_4)_3$
    (c) $Ba(ClO_3)_2$

39. How many atoms of oxygen are represented in each formula?
   (a) H_2O (d) $Fe(OH)_3$
   (b) $CuSO_4$ (e) $Al(ClO_3)_3$
   (c) H_2O_2

41. Classify each material as an element, compound, or mixture:
   (a) air   (c) sodium chloride
   (b) oxygen   (d) wine

43. Classify each material as an element, compound, or mixture:
   (a) paint   (c) copper
   (b) salt   (d) beer

45. Reduce the following chemical formulas to the smallest whole-number relationship among atoms. (In chemistry this is called the empirical formula. You will learn more about it later.)
   (a) $C_6H_{12}O_6$ glucose
   (b) C_8H_{18} octane
   (c) $C_{25}H_{52}$ paraffin wax

47. Is there a pattern to the location of the gaseous elements on the periodic table? If so, describe it.

49. What percent of the first 36 elements on the periodic table are metals?

40. How many atoms of hydrogen are represented in each formula?
   (a) H_2 (d) $HC_2H_3O_2$
   (b) $Ba(C_2H_3O_2)_2$ (e) $(NH_4)_2Cr_2O_7$
   (c) $C_6H_{12}O_6$

42. Classify each material as an element, compound, or mixture:
   (a) platinum   (c) iodine
   (b) sulfuric acid   (d) crude oil

44. Classify each material as an element, compound, or mixture:
   (a) hydrochloric acid   (c) milk
   (b) silver   (d) sodium hydroxide

46. Reduce the following chemical formulas to the smallest whole-number relationship among atoms. (In chemistry this is called the empirical formula. You will learn more about it later.)
   (a) H_2O_2 hydrogen peroxide
   (b) C_2H_6O ethyl alcohol
   (c) $Na_2Cr_2O_7$ sodium dichromate

48. Is there a pattern to the location of the liquid elements on the periodic table? If so, describe it.

50. What percent of the first 36 elements on the periodic table are solids at room temperature?

Additional Exercises

51. The formula for vitamin B_{12} is $C_{63}H_{88}CoN_{14}O_{14}P$.
   (a) How many atoms make up one molecule of vitamin B_{12}?
   (b) What percentage of the total atoms are carbon?
   (c) What fraction of the total atoms are metallic?

52. How many total atoms are there in seven dozen molecules of nitric acid, HNO_3?

53. These formulas look similar but represent different things.

   8 S S_8

   Compare and contrast them. How are they alike? How are they different?

54. Calcium dihydrogen phosphate is an important fertilizer. How many atoms of hydrogen are there in ten formula units of $Ca(H_2PO_4)_2$?

55. How many total atoms are there in one molecule of $C_{145}H_{293}O_{168}$?

56. Name the following:
   (a) three elements, all metals, beginning with the letter M
   (b) four elements, all solid nonmetals
   (c) five elements, all solids in the first five rows of the periodic table, whose symbols start with letters different than the element name

57. It has been estimated that there is 4×10^{-4} mg of gold per liter of seawater. At a price of $19.40/g$, what would be the value of the gold in 1 km^3 (1×10^{15} cm^3) of the ocean?

58. Make a graph using the data below. Plot the density of air in grams per liter along the x-axis and temperature along the y-axis.

Temperature (°C)	Density (g/L)
0	1.29
10	1.25
20	1.20
40	1.14
80	1.07

   (a) What is the relationship between density and temperature according to your graph?
   (b) From your plot find the density of air at these temperatures:

   5°C   25°C   70°C

CHAPTER

4

Properties of Matter

▲ Iron can be melted at very
high temperatures and then
cast into a variety of shapes.

CHAPTER 4 / OUTLINE

CHAPTER
4

The world we live in is a kaleidoscope of sights, sounds, smells, and tastes. Our senses help us to describe these objects in our lives. For example, the smell of freshly baked cinnamon rolls creates a mouthwatering desire to gobble down a sample. Just as sights, sounds, smells, and tastes form the properties of the objects around us, each substance in chemistry has its own unique properties that allow us to identify it and predict its interactions.

These interactions produce both physical and chemical changes. When you eat an apple, the ultimate metabolic result is carbon dioxideand water. These same products are achieved by burning logs. Not only does a chemical change occur in these cases, but an energy change as well. Some reactions release energy (as does the apple or the log) whereas others require energy, such as the production of steel or the melting of ice. Over 90% of our current energy comes from chemical reactions.

4.1   Properties of Substances

How do we recognize substances? Each substance has a set of **properties** that are characteristic of that substance and give it a unique identity. Properties—the "personality traits" of substances—are classified as either physical or chemical. **Physical properties** are the inherent characteristics of a substance that can be determined without altering its composition; they are associated with its physical existence. Common physical properties include color, taste, odor, state of matter (solid, liquid, or gas), density, melting point, and boiling point. **Chemical properties** describe the ability of a substance to form new substances, either by reaction with other substances or by decomposition.

Let's consider a few of the physical and chemical properties of chlorine. Physically, chlorine is a gas about 2.4 times heavier than air. It is yellowish-green in color and has a disagreeable odor. Chemically, chlorine will not burn but will support the combustion of certain other substances. It can be used as a bleaching agent, as a disinfectant for water, and in many chlorinated substances such as refrigerants and insecticides. When chlorine combines with the metal sodium, it forms a salt called sodium chloride. These properties, among others, help us characterize and identify chlorine.

Substances, then, are recognized and differentiated by their properties. Table 4.1 lists four substances and several of their common physical properties. Information about physical properties, such as that given in Table 4.1, is available in handbooks of chemistry and physics. Scientists don't pretend to know all the answers or to remember voluminous amounts of data, but it is important for them to know where to look for data in the literature and on the Internet.

properties

physical properties

chemical properties

Many chemists have reference books such as Handbook of Chemistry and Physics to use as a resource.

> **No two substances have identical physical and chemical properties.**

TABLE 4.1	Physical Properties of Chlorine, Water, Sugar, and Acetic Acid					
Substance	Color	Odor	Taste	Physical state	Melting point (°C)	Boiling point (°C)
Chlorine	Yellowish-green	Sharp, suffocating	Sharp, sour	Gas	−101.6	−34.6
Water	Colorless	Odorless	Tasteless	Liquid	0.0	100.0
Sugar	White	Odorless	Sweet	Solid	—	Decomposes 170–186
Acetic acid	Colorless	Like vinegar	Sour	Liquid	16.7	118.0

4.2   Physical Changes

physical change

Matter can undergo two types of changes, physical and chemical. **Physical changes** are changes in physical properties (such as size, shape, and density) or changes in the state of matter without an accompanying change in composition. The changing of ice into water and water into steam are physical changes from one state of matter into another (Figure 4.1). No new substances are formed in these physical changes.

When a clean platinum wire is heated in a burner flame, the appearance of the platinum changes from silvery metallic to glowing red. This change is physical because the platinum can be restored to its original metallic appearance by cooling and, more importantly, because the composition of the platinum is not changed by heating and cooling.

FIGURE 4.1 ▶
Ice melting into water or water turning into steam are physical changes from one state of matter to another.

4.3   Chemical Changes

In a **chemical change,** new substances are formed that have different properties and
composition from the original material. The new substances need not resemble the
initial material in any way.

chemical change

When a clean copper wire is heated in a burner flame, the appearance of the
copper changes from coppery metallic to glowing red. Unlike the platinum wire, the
copper wire is not restored to its original appearance by cooling but becomes instead a
black material. This black material is a new substance called copper(II) oxide. It was
formed by chemical change when copper combined with oxygen in the air during the
heating process. The unheated wire was essentially 100% copper, but the copper(II)
oxide is 79.9% copper and 20.1% oxygen. One gram of copper will yield 1.251 g of
copper(II) oxide (see Figure 4.2). The platinum is changed only physically when
heated, but the copper is changed both physically and chemically when heated.

When 1.00 g of copper reacts with oxygen to yield 1.251 g of copper(II) oxide,
the copper must have combined with 0.251 g of oxygen. The percentage of copper
and oxygen can be calculated from these data—the copper and oxygen each being a
percent of the total mass of copper(II) oxide.

$$1.00 \text{ g copper} + 0.251 \text{ g oxygen} \rightarrow 1.251 \text{ g copper(II) oxide}$$

$$\frac{1.00 \text{ g copper}}{1.251 \text{ g copper(II) oxide}} \times 100 = 79.9\% \text{ copper}$$

$$\frac{0.251 \text{ g oxygen}}{1.251 \text{ g copper(II) oxide}} \times 100 = 20.1\% \text{ oxygen}$$

Water can be decomposed chemically into hydrogen and oxygen. This is usually
accomplished by passing electricity through the water in a process called *electrolysis.* Hydrogen collects at one electrode while oxygen collects at the other (see Figure
4.3). The composition and the physical appearance of the hydrogen and the oxygen
are quite different from that of water. They are both colorless gases, but each
behaves differently when a burning splint is placed into the sample: The hydrogen

(a)   (b)   (c)

◀ **FIGURE 4.2**
**Chemical change: Forming
of copper(II) oxide from
copper and oxygen**
**(a) Before heating the wire
is 100% copper (1.00g)**
**(b) Copper and oxygen
from the air combine
chemically when the
wire is heated.**
**(c) After heating the wire is
black copper(II) oxide
(79.9% copper, 20.1%
oxygen) (1.251g)**

FIGURE 4.3 ▶
Electrolysis of water is forming hydrogen gas in the left tube and oxygen gas in the right tube.

explodes with a pop while the flame brightens considerably in the oxygen (oxygen supports and intensifies the combustion of the wood). From these observations we conclude that a chemical change has taken place.

chemical equations

Chemists have devised a shorthand method for expressing chemical changes in the form of **chemical equations.** The two previous examples of chemical changes can be represented by the following word equations:

$$\text{water} \xrightarrow{\text{electrical energy}} \text{hydrogen} + \text{oxygen} \tag{1}$$

$$\text{copper} + \text{oxygen} \xrightarrow{\Delta} \text{copper(II) oxide} \tag{2}$$

Equation (1) states that water decomposes into hydrogen and oxygen when electrolyzed. Equation (2) states that copper plus oxygen when heated produce copper(II) oxide. The arrow means "produces," and it points to the products. The Greek letter delta (Δ) represents heat. The starting substances (water, copper, and oxygen) are called the **reactants** and the substances produced (hydrogen, oxygen, and copper(II) oxide) are called the **products.** These chemical equations can be presented in still more abbreviated form by using symbols to represent the substances:

reactants
products

$$2\ H_2O \xrightarrow{\text{electrical energy}} 2\ H_2 + O_2$$

$$2\ Cu + O_2 \xrightarrow{\Delta} 2\ CuO$$

We will learn more about writing chemical equations in later chapters.

Physical change usually accompanies a chemical change. Table 4.2 lists some common physical and chemical changes; note that wherever a chemical change occurs, a physical change occurs also. However, wherever a physical change is listed, only a physical change occurs.

Process taking place	Type of change	Accompanying observations
Rusting of iron	Chemical	Shiny, bright metal changes to reddish-brown rust.
Boiling of water	Physical	Liquid changes to vapor.
Burning of sulfur in air	Chemical	Yellow, solid sulfur changes to gaseous, choking sulfur dioxide.
Boiling an egg	Chemical	Liquid white and yolk change to solids.
Combustion of gasoline	Chemical	Liquid gasoline burns to gaseous carbon monoxide, carbon dioxide, and water.
Digesting food	Chemical	Food changes to liquid nutrients and partially solid wastes.
Sawing of wood	Physical	Smaller pieces of wood and sawdust are made from a larger piece of wood.
Burning of wood	Chemical	Wood burns to ashes, gaseous carbon dioxide, and water.
Heating of glass	Physical	Solid becomes pliable during heating, and the glass may change its shape.

TABLE 4.2 Physical or Chemical Changes of Some Common Processes

4.4 Conservation of Mass

The **law of conservation of mass** states that no change is observed in the total mass   *law of conservation of mass*
of the substances involved in a chemical change. This law, tested by extensive laboratory experimentation, is the basis for the quantitative mass relationships among reactants and products.

The decomposition of water into hydrogen and oxygen illustrates this law. Thus 100.0 g of water decomposes into 11.2 g of hydrogen and 88.8 g of oxygen:

water $\longrightarrow$ hydrogen + oxygen
100.0 g   11.2 g   88.8 g

100.0 g reactant $\longrightarrow$ 100.0 g products

mass of reactants = mass of products

4.5   Energy

From the early discovery that fire can warm us and cook our food to our recent discovery that nuclear reactors can be used to produce vast amounts of controlled energy, our technical progress has been directed by our ability to produce, harness, and utilize energy. **Energy** is the capacity of matter to do work. Energy exists in many forms; some of the more familiar forms are mechanical, chemical, electrical, heat, nuclear, and radiant or light energy. Matter can have both potential and kinetic energy.

energy

potential energy
    Potential energy (PE) is stored energy, or energy that an object possesses due to its relative position. For example, a ball located 20 ft above the ground has more potential energy than when located 10 ft above the ground and will bounce higher when allowed to fall. Water backed up behind a dam represents potential energy that can be converted into useful work in the form of electrical or mechanical energy. Gasoline is a source of chemical potential energy. When gasoline burns (combines with oxygen), the heat released is associated with a decrease in potential energy. The new substances formed by burning have less chemical potential energy than the gasoline and oxygen.

kinetic energy
    Kinetic energy (KE) is energy that matter possesses due to its motion. When the water behind the dam is released and allowed to flow, its potential energy is changed into kinetic energy, which can be used to drive generators and produce electricity. Moving bodies possess kinetic energy. We all know the results when two moving vehicles collide: Their kinetic energy is expended in the crash that occurs. The pressure exerted by a confined gas is due to the kinetic energy of rapidly moving gas particles.

    Energy can be converted from one form to another form. Some kinds of energy can be converted to other forms easily and efficiently. For example, mechanical energy can be converted to electrical energy with an electric generator at better than 90% efficiency. On the other hand, solar energy has thus far been directly converted to electrical energy at an efficiency of only about 15%. In chemistry, energy is most frequently expressed as heat.

▲
The mechanical energy of falling water is converted to electrical energy at the hydro-electric plant at Niagra Falls.

4.6   Heat: Quantitative Measurement

The SI-derived unit for energy is the joule (pronounced *jool* and abbreviated J). Another unit for heat energy, which has been used for many years, is the calorie (abbreviated cal). The relationship between joules and calories is

$$4.184\ J = 1\ cal \quad (exactly)$$

joule
calorie
    To give you some idea of the magnitude of these heat units, 4.184 **joules** or 1 **calorie** is the quantity of heat energy required to change the temperature of 1 gram of water by 1°C, usually measured from 14.5°C to 15.5°C.

    Since joules and calories are rather small units, kilojoules (kJ) and kilocalories (kcal) are used to express heat energy in many chemical processes. The kilocalorie is

1 kJ = 1000 J
1 kcal = 1000 cal = 1 Cal

TABLE 4.3	Specific Heat of Selected Substances	
Substance	Specific heat J/g°C	Specific heat cal/g°C
Water	4.184	1.00
Ethyl alcohol	2.138	0.511
Ice	2.059	0.492
Aluminum	0.900	0.215
Iron	0.473	0.113
Copper	0.385	0.0921
Gold	0.131	0.0312
Lead	0.128	0.0305

also known as the nutritional or large Calorie (spelled with a capital C and abbreviated Cal). In this book heat energy will be expressed in joules.

The difference in the meanings of the terms *heat* and *temperature* can be seen by this example: Visualize two beakers, A and B. Beaker A contains 100 g of water at 20°C, and beaker B contains 200 g of water also at 20°C. The beakers are heated until the temperature of the water in each reaches 30°C. The temperature of the water in the beakers was raised by exactly the same amount, 10°C. But twice as much heat (8368 J) was required to raise the temperature of the water in beaker B as was required in beaker A (4184 J).

In the middle of the 18th century, Joseph Black, a Scottish chemist, was experimenting with the heating of elements. He heated and cooled equal masses of iron and lead through the same temperature range. Black noted that much more heat was needed for the iron than for the lead. He had discovered a fundamental property of matter—namely that every substance has a characteristic heat capacity. Heat capacities may be compared in terms of specific heats. The **specific heat** of a substance is the quantity of heat (lost or gained) required to change the temperature of 1 g of that substance by 1°C. It follows then that the specific heat of liquid water is 4.184 J/g°C (1.000 cal/g°C). The specific heat of water is high compared with that of most substances. Aluminum and copper, for example, have specific heats of 0.900 J/g°C and 0.385 J/g°C, respectively (see Table 4.3). The relation of mass, specific heat, temperature change (Δt), and quantity of heat lost or gained by a system is expressed by this general equation.

specific heat

$$\left(\begin{array}{c}\text{mass of}\\\text{substance}\end{array}\right)\left(\begin{array}{c}\text{specific heat}\\\text{of substance}\end{array}\right)\Delta t = \text{heat}$$

Thus the amount of heat needed to raise the temperature of 200. g of water by 10.0°C can be calculated as follows:

$$(200.\ g)\left(\frac{4.184\ J}{g°C}\right)(10.0°C) = 8.37 \times 10^3\ J$$

Fast Food or Fast Fat?

Fat in our diets today has become a major problem. High-fat diets have been connected with heart disease and to a variety of cancers. In the average American diet, fat supplies about 40% of the Calories (kcal). Nutritionists suggest that fat should be not more than 30% of our total daily Calories. For the average person eating about 2000 Calories per day, this is about 67 grams, 600 Calories, or 15 teaspoons of fat (e.g., a double hamburger with cheese). A small amount of fat is necessary in the diet to provide a natural source of vitamins A, D, E, and K.

Fats also supply us with energy. One gram of fat supplies 9 Calories of energy; the same amount of protein or carbohydrate supplies only 4 Calories. We use energy in a variety of ways but one of the most important is to maintain body temperature. Our bodies tend to lose heat to the surroundings since heat flows from an area of higher temperature to an area of lower temperature. Additional heat energy is used to evaporate moisture and cool our bodies as we perspire. Still more energy is demanded by our daily physical activities. The source for all this energy is the chemical oxidation of the food we eat.

The energy content of food is determined by burning it in a calorimeter and measuring the heat released. Since the initial substances and the final products of the combustion in the calorimeter are the same as those accomplished in our bodies, a calorie content can be assigned to each food. Caloric values are now found on food packages along with nutri-

- Broiled Chicken Salad
  (200 Calories; 7 grams of fat)
- Newman's Own Light Italian Dressing
  (30 Calories; 1 gram of fat)
- Vanilla shake
  (310 Calories; 7 grams of fat)

Total—540 Calories; 15 grams of fat 25% Calories from fat

- Grilled Chicken Sandwich
  (290 Calories; 7 grams of fat)
- Black coffee
  (0 Calories; 0 grams fat)
- Chocolate Frosty Dairy Dessert (medium)
  (460 Calories; 13 grams of fat)

Total—750 Calories; 20 grams of fat 24% Calories from fa

tional information regarding the contents of the food we eat.

How does fast food stack up in the nutrition department? The sample meals are shown here from fast food restaurants that get less than 30% of their calories from fat. In general, chicken and turkey sandwiches that are not fried have less fat than hamburgers or roast beef. Salads with the lowest fat have little or no cheese. The best way to reduce fat in salads is to eliminate the dressing or use a low-fat dressing.

Believe it or not, most of the milk-shakes and frozen desserts served in fast-food restaurants have less than 30% Calories from fat per serving. This is primarily because they are made with skim milk. Fast food does not necessarily mean fast fat if you make careful selections. Balancing a full day's diet is more important than worrying about each and every food eaten, although choosing foods that have less than 30% Calories from fat makes maintaining the balance easier.

Calculate the specific heat of a solid in J/g°C and cal/g°C if 1638 J raise the temperature of 125 g of the solid from 25.0°C to 52.6°C.

Example 4.1

Solution

Substitute the data into the equation (mass)(specific heat)(Δt) = heat and solve for specific heat.

$$\text{specific heat} = \frac{\text{heat}}{\text{g} \times \Delta t}$$

heat = 1638 J   mass = 125 g Δt = 52.6°C − 25.0°C = 27.6°C

$$\text{specific heat} = \frac{1638 \text{ J}}{(125 \text{ g})(27.6°\text{C})} = 0.475 \text{ J/g°C}$$

Now convert joules to calories using 1.000 cal/4.184 J:

$$\text{specific heat} = \left(\frac{0.475 \text{ J}}{\text{g°C}}\right)\left(\frac{1.000 \text{ cal}}{4.184 \text{ J}}\right) = 0.114 \text{ cal/g°C}$$

A sample of a metal with a mass of 212 g is heated to 125.0°C and then dropped into 375 g water at 24.0°C. If the final temperature of the water is 34.2°C, what is the specific heat of the metal? (Assume no heat losses to the surroundings.)

Example 4.2

Solution

When the metal enters the water, it begins to cool, losing heat to the water. At the same time the temperature of the water rises. This process continues until the temperature of the metal and the temperature of the water are equal, at which point (34.2°C) no net flow of heat occurs.

The heat lost or gained by a system is given by (mass)(specific heat)(Δt) = energy change. First calculate the heat gained by the water and then calculate the specific heat of the metal.

temperature rise of the water   (Δt) = 34.2°C − 24.0°C = 10.2°C

$$\text{heat gained by the water} = (375 \text{ g})\left(\frac{4.184 \text{ J}}{\text{g°C}}\right)(10.2°\text{C}) = 1.60 \times 10^4 \text{ J}$$

The metal dropped into the water must have a final temperature the same as the water (34.2°C):

temperature drop by the metal   (Δt) = 125.0°C − 34.2°C = 90.8°C

heat lost by the metal = heat gained by the water = 1.60 × 10⁴ J

To determine the specific heat of the metal we rearrange the equation (mass)(specific heat)(Δt) = heat to obtain

$$\text{specific heat} = \frac{\text{heat}}{(\text{mass})(\Delta t)}$$

$$\text{specific heat of the metal} = \frac{1.60 \times 10^4 \text{ J}}{(212 \text{ g})(90.8°\text{C})} = 0.831 \text{ J/g°C}$$

Practice 4.1

Calculate the quantity of energy needed to heat 8.0 g of water from 42.0°C to 45.0°C.

Practice 4.2

A 110.0-g sample of iron at 55.5°C raises the temperature of 150.0 mL of water from 23.0°C to 25.5°C. Determine the specific heat of the iron in J/g°C.

4.7   Energy in Chemical Changes

In all chemical changes, matter either absorbs or releases energy. Chemical changes can produce different forms of energy. For example, electrical energy to start automobiles is produced by chemical changes in the lead storage battery. Light energy is produced by the chemical change that occurs in a light stick. Heat and light are released from the combustion of fuels. All the energy needed for our life processes—breathing, muscle contraction, blood circulation, and so on—is produced by chemical changes occurring within the cells of our bodies.

Conversely, energy is used to cause chemical changes. For example, a chemical change occurs in the electroplating of metals when electrical energy is passed through a salt solution in which the metal is submerged. A chemical change occurs when radiant energy from the sun is used by green plants in the process of photosynthesis. And as we saw, a chemical change occurs when electricity is used to decompose water into hydrogen and oxygen. Chemical changes are often used primarily to produce energy rather than to produce new substances. The heat or thrust generated by the combustion of fuels is more important than the new substances formed.

▲
Energy from the sun is used to produce the chemical changes that occur during photosynthesis in the rain forests.

4.8   Conservation of Energy

An energy transformation occurs whenever a chemical change occurs (see Figure 4.4). If energy is absorbed during the change, the products will have more chemical potential energy than the reactants. Conversely, if energy is given off in a chemical change, the products will have less chemical potential energy than the reactants. Water can be decomposed in an electrolytic cell by absorbing electrical energy. The products, hydrogen and oxygen, have a greater chemical potential energy level than that of water (see Figure 4.4a). This potential energy is released in the form of heat and light when the hydrogen and oxygen are burned to form water again (see Figure 4.4b). Thus energy can be changed from one form to another or from one substance to another, and therefore is not lost.

The energy changes occurring in many systems have been thoroughly studied. No system has ever been found to acquire energy except at the expense of energy possessed by another system. This is the **law of conservation of energy**: Energy can be neither created nor destroyed, though it can be transformed from one form to another.

law of conservation of
energy

CHEMISTRY IN ACTION *Instant Relief!*

Have you ever used a heat pack when you have tired, aching muscles? Remember the relief you felt when you grabbed a cold pack to reduce the inflammation and swelling of your sprained ankle? How do these hot and cold packs remain ready to use and provide the relief just when we need it? Instant hot and cold packs use the properties of matter to release or absorb energy.

In a cold pack, a small sealed package of ammonium nitrate is placed in a sealed pouch containing water. As long as the substances remain separated, nothing happens. When ammonium nitrate dissolves in water energy is absorbed. If the small package is broken (as the pack is activated) the substances mix and, as the solid dissolves, the temperature of the solution falls. The cold pack absorbs heat from its surroundings and relief from pain begins.

Instant hot packs work in one of two ways. The first type depends on a spontaneous chemical reaction that releases heat energy. In one product, an inner paper bag perforated with tiny holes is contained in a plastic envelope. The inner bag contains a mixture of powdered iron, salt, activated charcoal, and dampened sawdust. The heat pack is activated by removing the inner bag, shaking to mix the chemicals, and replacing it in the outer envelope. The heat is the result of a

chemical change—that of iron rusting (oxidizing) very rapidly. These packs work slowly and cannot be reused. In the second type of heat pack a physical property of matter is responsible for producing the heat. This hot pack consists of a tough sealed plastic envelope containing a solution of sodium acetate or sodium thiosulfate. A small crystal of the chemical is formed in the solution by squeezing the corner of the pack or bending a small metal activator. Crystals then form throughout the solution and release heat to the surroundings. This type of hot pack does not overheat and is reusable. To reuse it simply pop it into your microwave and heat it until the crystals are all dissolved, then cool it carefully, and store it away. Instant relief for the weekend athlete or accident victim results directly from the properties of matter!

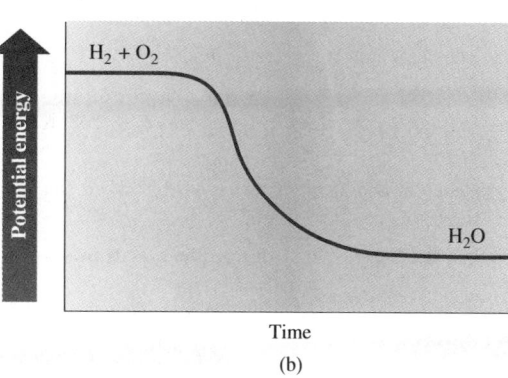

▲
FIGURE 4.4
(a) In electrolysis of water, energy is absorbed by the system so the products H_2 and O_2 have a higher potential energy. (b) When hydrogen is burned (in O_2), energy is released and the product (H_2O) has lower potential energy.

Concepts in Review

1. List the physical properties used to characterize a substance.
2. Distinguish between the physical and chemical properties of matter.
3. Classify changes undergone by matter as either physical or chemical.
4. Distinguish between kinetic and potential energy.

5. State the law of conservation of mass.
6. State the law of conservation of energy.
7. Differentiate clearly between heat and temperature.
8. Make calculations using the equation:

$$\text{heat} = (\text{mass})(\text{specific heat})(\Delta t)$$

Key Terms

calorie (4.6)
chemical change (4.3)
chemical equations (4.3)
chemical properties (4.1)
energy (4.5)

joule (4.6)
kinetic energy (4.5)
law of conservation of energy (4.8)
law of conservation of mass (4.4)

physical change (4.2)
physical properties (4.1)
potential energy (4.5)
products (4.3)

properties (4.1)
reactants (4.3)
specific heat (4.6)

Questions

1. In what physical state does acetic acid exist at 10°C? (Table 4.1)
2. In what physical state does chlorine exist at 102 K? (Table 4.1)
3. What evidence of chemical change is visible when electricity is run through water? (Figure 4.2)
4. What physical changes occur during the electrolysis of water?

5. Distinguish between chemical and physical properties.
6. What is the fundamental difference between a chemical change and a physical change?
7. Distinguish between potential and kinetic energy.
8. Calculate the boiling point of acetic acid in
   (a) Kelvins
   (b) degrees Fahrenheit (Table 4.1)

Paired Exercises

9. Classify the following as being primarily a physical or primarily a chemical change:
   (a) formation of a snowflake
   (b) freezing ice cream
   (c) boiling water
   (d) churning cream to make butter
   (e) boiling an egg
   (f) souring milk

10. Classify the following as being primarily a physical or primarily a chemical change:
    (a) lighting a candle
    (b) stirring cake batter
    (c) dissolving sugar in water
    (d) decomposition of limestone by heat
    (e) a leaf turning yellow
    (f) formation of bubbles in a pot of water long before the water boils

11. Cite the evidence that indicates that only physical changes occur when a platinum wire is heated in a Bunsen burner flame.

12. Cite the evidence that indicates that both physical and chemical changes occur when a copper wire is heated in a Bunsen burner flame.

13. Identify the reactants and products for heating a copper wire in a Bunsen burner flame.

14. Identify the reactants and products for the electrolysis of water.

15. What happens to the kinetic energy of a speeding car when the car is braked to a stop?

16. What energy transformation is responsible for the fiery reentry of the space shuttle?

17. Indicate with a plus sign (+) any of these processes that require energy, and a negative sign (−) any that release energy:
    (a) melting ice
    (b) relaxing a taut rubber band
    (c) a rocket launching
    (d) striking a match
    (e) a Slinky toy (spring) "walking" down stairs

18. Indicate with a plus sign (+) any of these processes that require energy and a negative sign (−) any that release energy:
    (a) boiling water
    (b) releasing a balloon full of air with the neck open
    (c) a race car crashing into the wall
    (d) cooking a potato in a microwave oven
    (e) ice cream freezing in an ice cream maker

19. How many joules of energy are required to raise the temperature of 75 g of water from 20.0°C to 70.0°C?

20. How many joules of energy are required to raise the temperature of 65 g of iron from 25°C to 95°C?

21. A 250.0-g metal bar requires 5.866 kJ to change its temperature from 22°C to 100.0°C. What is the specific heat of the metal?

22. A 1.00-kg sample of antimony absorbed 30.7 kJ, thus raising the temperature of the antimony from 20.0°C to its melting point (630.0°C). Calculate the specific heat of antimony.

*23. A 325-g piece of gold at 427°C is dropped into 200.0 mL of water at 22.0°C. The specific heat of gold is 0.131 J/g°C. Calculate the final temperature of the mixture. (Assume no heat loss to the surroundings.)

*24. A 500.0-g iron bar at 212°C is placed in 2.0 L of water at 24.0°C. What will be the change in temperature of the water? (Assume no heat is lost to the surroundings.)

Additional Exercises

*25. The specific heat of zinc is 0.096 cal/g°C. Determine the energy required to raise the temperature of 250.0 g of zinc from room temperature (24°C) to 150.0°C.

*26. If 40.0 kJ of energy is absorbed by 500.0 g of water at 10.0°C what is the final temperature of the water?

*27. The heat of combustion of a sample of coal is 5500 cal/g. What quantity of this coal must be burned to heat 500.0 g of water from 20.0°C to 90.0°C?

28. One gram of anthracite coal gives off 7000. cal when burned. How many joules is this? How many grams of anthracite are required to raise the temperature of 4.0 L of water from 20.0°C to 100.0°C?

29. A 100.0-g sample of copper is heated from 10.0°C to 100.0°C.
    (a) Determine the number of calories needed. (Specific heat of copper is 0.0921 cal/g°C.)
    (b) The same amount of heat is added to 100.0 g of Al at 10.0°C. (Specific heat of Al is 0.215 cal/g°C.) Which metal gets hotter, the copper or the aluminum?

30. A 500.0-g piece of iron is heated in a flame and dropped into 400.0 g of water at 10.0°C. The temperature of the water rises to 90.0°C. How hot was the iron when it was first removed from the flame? (Specific heat of iron is 0.473 J/g°C.)

*31. A 20.0-g piece of metal at 203°C is dropped into 100.0 g of water at 25.0°C. The water temperature rises to 29.0°C. Calculate the specific heat of the metal (J/g°C). Assume that all of the heat lost by the metal is transferred to the water and no heat is lost to the surroundings.

*32. Assuming no heat loss by the system, what will be the final temperature when 50.0 g of water at 10.0°C is mixed with 10.0 g of water at 50.0°C?

33. Three 500.0-g pans of iron, aluminum, and copper are each used to fry an egg. Which pan fries the egg (105°C) the quickest? Explain.

*34. At 6:00 P.M., you put a 300.0-g copper pan containing 800.0 mL of water (all at room temperature, which is 25°C) on the stove. The stove supplies 628 J/s. When will the water reach the boiling point? (Assume no heat loss.)

35. Why does blowing gently across the surface of a hot cup of coffee help to cool it? Why does inserting a spoon into the coffee do the same thing?

36. If you are boiling some potatoes in a pot of water, will they cook faster if the water is boiling vigorously than if the water is only gently boiling? Explain your reasoning.

37. Homogenized whole milk contains 4% butterfat by volume. How many milliliters of fat are there in a glass (250 mL) of milk? How many grams of butterfat ($d = 0.8$ g/mL) are in this glass of milk?

38. A 100.0-mL volume of mercury (density = 13.6 g/mL) is put into a container along with 100.0 g of sulfur. The two substances react when heated and result in 1460 g of a dark, solid matter. Is that material an element or a compound? Explain. How many grams of mercury were in the container? How does this support the law of conservation of matter?

Answers to Practice Exercises

4.1 1.0×10^2 J = 24 cal.

4.2 0.477 J/g°C.

PUTTING IT TOGETHER
Review for Chapters 1–4

Multiple Choice: *Choose the correct answer to each of the following.*

1. 1.00 cm is equal to how many meters?
   (a) 2.54   (b) 100.   (c) 10.0   (d) 0.0100

2. 1.00 cm is equal to how many inches?
   (a) 0.394   (b) 0.10   (c) 12   (d) 2.54

3. 4.50 ft is how many centimeters?
   (a) 11.4   (b) 21.3   (c) 454   (d) 137

4. The number 0.0048 contains how many significant figures?
   (a) 1   (b) 2   (c) 3   (d) 4

5. Express 0.00382 in scientific notation.
   (a) 3.82×10^3 (c) 3.82×10^{-2}
   (b) 3.8×10^{-3} (d) 3.82×10^{-3}

6. 42.0°C is equivalent to
   (a) 273 K   (b) 5.55°F   (c) 108°F   (d) 53.3°F

7. 267°F is equivalent to
   (a) 404 K   (b) 116°C   (c) 540 K   (d) 389 K

8. An object has a mass of 62 g and a volume of 4.6 mL. Its density is
   (a) 0.074 mL/g   (c) 7.4 g/mL
   (b) 285 g/mL   (d) 13 g/mL

9. The mass of a block is 9.43 g and its density is 2.35 g/mL. The volume of the block is
   (a) 4.01 mL   (c) 22.2 mL
   (b) 0.249 mL   (d) 2.49 mL

10. The density of copper is 8.92 g/mL. The mass of a piece of copper that has a volume of 9.5 mL is
    (a) 2.6 g   (b) 85 g   (c) 0.94 g   (d) 1.1 g

11. An empty graduated cylinder has a mass of 54.772 g. When filled with 50.0 mL of an unknown liquid, it has a mass of 101.074 g. The density of the liquid is
    (a) 0.926 g/mL   (c) 2.02 g/mL
    (b) 1.08 g/mL   (d) 1.85 g/mL

12. The conversion factor to change grams to milligrams is
    (a) $\dfrac{100 \text{ mg}}{1 \text{ g}}$ (c) $\dfrac{1 \text{ g}}{1000 \text{ mg}}$
    (b) $\dfrac{1 \text{ g}}{100 \text{ mg}}$ (d) $\dfrac{1000 \text{ mg}}{1 \text{ g}}$

13. What Fahrenheit temperature is twice the Celsius temperature?
    (a) 64°F   (b) 320°F   (c) 200°F   (d) 546°F

14. A gold alloy has a density of 12.41 g/mL and contains 75.0% gold by mass. The volume of this alloy that can be made from 255 g of pure gold is
    (a) 4.22×10^3 mL   (c) 27.4 mL
    (b) 2.37×10^3 mL   (d) 20.5 mL

15. A lead cylinder ($V = \pi r^2 h$) has radius 12.0 cm and length 44.0 cm and a density of 11.4 g/mL. The mass of the cylinder is
    (a) 2.27×10^5 g   (c) 1.78×10^3 g
    (b) 1.89×10^5 g   (d) 3.50×10^5 g

16. The following units can all be used for density *except*
    (a) g/cm^3 (b) kg/m^3 (c) g/L   (d) kg/m^2

17. 37.4 cm × 2.2 cm equals
    (a) 82.28 cm^2 (c) 82 cm^2
    (b) 82.3 cm^2 (d) 82.2 cm^2

18. The following elements are among the five most abundant by mass in the Earth's crust, seawater, and atmosphere *except*
    (a) oxygen   (c) silicon
    (b) hydrogen   (d) aluminum

19. Which of the following is a compound?
    (a) lead   (c) potassium
    (b) wood   (d) water

20. Which of the following is a mixture?
    (a) water   (c) wood
    (b) chromium   (d) sulfur

21. How many atoms are represented in the formula Na_2CrO_4?
    (a) 3   (b) (5)   (c) (7)   (d) 8

22. Which of the following is a characteristic of metals?
    (a) ductile   (c) extremely strong
    (b) easily shattered   (d) dull

23. Which of the following is a characteristic of nonmetals?
    (a) always a gas
    (b) poor conductor of electricity
    (c) shiny
    (d) combines only with metals

24. When a pure substance was analyzed, it was found to contain carbon and chlorine. This substance must be classified as
    (a) an element
    (b) a mixture
    (c) a compound
    (d) both a mixture and a compound

25. Chromium, fluorine, and magnesium have the symbols
    (a) Ch, F, Ma   (c) Cr, F, Mg
    (b) Cr, Fl, Mg   (d) Cr, F, Ma

26. Sodium, carbon, and sulfur have the symbols
    (a) Na, C, S   (c) Na, Ca, Su
    (b) So, C, Su   (d) So, Ca, Su

27. Coffee is an example of
    (a) an element   (c) a homogeneous mixture
    (b) a compound   (d) a heterogeneous mixture

28. The number of oxygen atoms in $Al(C_2H_3O_2)_3$ is
    (a) 2   (b) 3   (c) 5   (d) 6

29. Which of the following is a mixture?
    (a) water   (c) sugar solution
    (b) iron(II) oxide   (d) iodine

30. Which is the most compact state of matter?
    (a) solid   (c) gas
    (b) liquid   (d) amorphous

31. Which is not characteristic of a solution?
    (a) a homogeneous mixture
    (b) a heterogeneous mixture
    (c) contains two or more substances
    (d) has a variable composition

32. A chemical formula is a combination of
    (a) symbols   (c) elements
    (b) atoms   (d) compounds

33. The number of nonmetal atoms in $Al_2(SO_3)_3$ is
    (a) 5   (b) 7   (c) 12   (d) 14

34. Which of the following is not a physical property?
    (a) boiling point   (c) bleaching action
    (b) physical state   (d) color

35. Which of the following is a physical change?
    (a) A piece of sulfur is burned.
    (b) A firecracker explodes.
    (c) A rubber band is stretched.
    (d) A nail rusts.

36. Which of the following is a chemical change?
    (a) Water evaporates.
    (b) Ice melts.
    (c) Rocks are ground to sand.
    (d) A penny tarnishes.

37. When 9.44 g of calcium are heated in air, 13.22 g of calcium oxide are formed. The percent by mass of oxygen in the compound is
    (a) 28.6%   (b) 40.0%   (c) 71.4%   (d) 13.2%

38. Barium iodide, BaI_2, contains 35.1% barium by mass. An 8.50-g sample of barium iodide contains what mass of iodine?
    (a) 5.52 g   (b) 2.98 g   (c) 3.51 g   (d) 6.49 g

39. Mercury(II) sulfide, HgS, contains 86.2% mercury by mass. The mass of HgS that can be made from 30.0 g of mercury is
    (a) 2586 g   (b) 2.87 g   (c) 25.9 g   (d) 34.8 g

40. The changing of liquid water to ice is known as a
    (a) chemical change
    (b) heterogeneous change
    (c) homogeneous change
    (d) physical change

41. Which of the following does not represent a chemical change?
    (a) heating of copper in air
    (b) combustion of gasoline
    (c) cooling of red-hot iron
    (d) digestion of food

42. Heating 30·g of water from 20·°C to 50·°C requires
    (a) 30. cal   (c) 3.8×10^3 J
    (b) 50. cal   (d) 6.3×10^3 J

43. The specific heat of aluminum is 0.900 J/g°C. How many joules of energy are required to raise the temperature of 20.0 g of Al from 10.0°C to 15.0°C?
    (a) 79 J   (b) 90. J   (c) 100. J   (d) 112 J

44. A 100.-g iron ball (specific heat = 0.473 J/g°C) is heated to 125°C and is placed in a calorimeter holding 200. g of water at 25.0°C. What will be the highest temperature reached by the water?
    (a) 43.7°C   (c) 65.3°C
    (b) 30.4°C   (d) 35.4°C

45. Which has the highest specific heat?
    (a) ice   (b) lead   (c) water   (d) aluminum

46. When 20.0 g of mercury is heated from 10.0°C to 20.0°C, 27.6 J of energy are absorbed. What is the specific heat of mercury?
    (a) 0.726 J/g°C   (c) 2.76 J/g°C
    (b) 0.138 J/g°C   (d) no correct answer given

47. Changing hydrogen and oxygen into water is a
    (a) physical change
    (b) chemical change
    (c) conservation reaction
    (d) no correct answer given

Free Response Questions: *Answer each of the following. Be sure to include your work and explanations in a clear, logical form.*

1. You decide to go sailing in the tropics with some friends. Once there you listen to the marine forecast that predicts in-shore wave heights of 1.5 meters, offshore wave heights of 4 meters, and temperature of 27°C. Your friend the captain is unfamiliar with the metric system and he needs to know whether it is safe for your small boat and if it will be warm. He asks you to convert the measurements to feet and Fahrenheit respectively.

2. Jane is melting butter in a copper pot on the stove. If she knows how much heat her stove releases per minute, what other measurements does she need to take to determine how much heat the butter absorbed? She assumes that the stove does not lose any heat to the surroundings.

3. Julius decided to heat 75 g $CaCO_3$ to determine how much carbon dioxide was produced. (Note: When $CaCO_3$ is heated it produces CaO and carbon dioxide.) He collected the carbon dioxide in a balloon. Julius found the mass of the

CaO remaining was 42 g. If 44 g of carbon dioxide takes up 24 dm^3 of space, how many *liters* of gas were trapped in the balloon?

Use these pictures to answer the following questions.

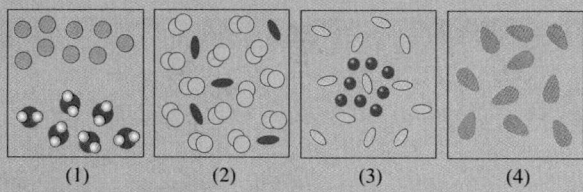

(1)　　　(2)　　　(3)　　　(4)

4. (a) Which picture best describes a homogeneous mixture?
   (b) How would you classify the contents of the other containers?
   (c) Which picture contains a compound? Explain how you made your choice.

Use these pictures to answer the following questions.

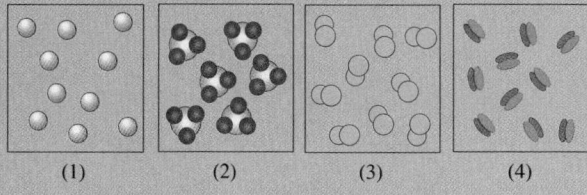

(1)　　　(2)　　　(3)　　　(4)

5. (a) Which picture best represents fluorine gas? Why?
   (b) Which other elements could that picture also represent?
   (c) Which of the pictures could represent SO_3 gas?

6. Sue and Tim each left a one-quart bowl outside one night. Sue's bowl was full of water and covered while Tim's was empty and open. The next day there was a huge snowstorm that filled Tim's bowl with snow. The temperature that night went down to 12°F.
   (a) Which bowl would require less energy to bring its contents to room temperature (25°C)? Why?
   (b) What temperature change (°C) is required to warm the bowls to 25°C?

(c) How much heat (in kJ) is required to raise the temperature of the contents of Sue's bowl to 0°C (without converting the ice to water)?
(d) Did the water in Sue's bowl undergo chemical or physical changes or both?

7. One cup of Raisin Bran provides 60.% of the U.S. recommended daily allowance (RDA) of iron.
   (a) If the cereal provides 11 mg of iron, what is the U.S. RDA for Fe?
   (b) When the iron in the cereal is extracted, it is found to be the pure element. What is the volume of iron in a cup of the cereal?

8. 16% of the U.S. RDA for Ca is 162 mg.
   (a) What mass of calcium phosphate, $Ca_3(PO_4)_2$, provides 162 mg of calcium?
   (b) Is $Ca_3(PO_4)_2$ an element, mixture, or compound?
   (c) Milk is a good source of calcium in the diet. If 120 mL of skim milk provides 13% of the U.S. RDA for Ca, how many cups of milk should you drink per day if that is your only source of calcium?

9. Absentminded Alfred put down a bottle containing silver on the table. When he went to retrieve it, he realized he had forgotten to label the bottle. Unfortunately there were two full bottles of the same size side-by-side. Alfred realized he had placed a bottle of mercury on the same table last week. State two ways Alfred can determine which bottle contains silver without opening the bottles.

10. Suppose 25 g of solid sulfur and 35 g of oxygen gas are placed in a sealed container.
    (a) Does the container hold a mixture or a compound?
    (b) After heating the contents, the container was weighed. From a comparison of the total mass before heating to the total mass after heating, can you tell whether a reaction took place? Explain.
    (c) After the container is heated, all the contents are gaseous. Has the density of the container including its contents changed?

CHAPTER

5

Early Atomic Theory and Structure

▲ Lightning occurs when electrons move to neutralize a charge difference between the clouds and the Earth.

CHAPTER 5 / OUTLINE

CHAPTER
5

Pure substances are classified as elements or compounds, but just what makes a substance possess its unique properties? How small a piece of salt will still taste salty? Carbon dioxide puts out fires, is used by plants to produce oxygen, and forms dry ice when solidified. But how small a mass of this material still behaves like carbon dioxide? Substances are in their simplest identifiable form at the atomic, ionic, or molecular level. Further division produces a loss of characteristic properties.

What particles lie within an atom or ion? How are these tiny particles alike? How do they differ? How far can we continue to divide them? Alchemists began the quest, early chemists laid the foundation, and modern chemists continue to build and expand on models of the atom.

5.1 Early Thoughts

The structure of matter has long intrigued and engaged us. The earliest models of the atom were developed by the ancient Greek philosophers. About 440 B.C. Empedocles stated that all matter was composed of four "elements"—earth, air, water, and fire. Democritus (about 470–370 B.C.) thought that all forms of matter were divisible into tiny invisible particles, which he called atoms, derived from the Greek word *atomos,* meaning indivisible. He held that atoms were in constant motion and that they combined with one another in various ways. This purely speculative hypothesis was not based on scientific observations. Shortly thereafter, Aristotle (384–322 B.C.) opposed the theory of Democritus and instead endorsed and advanced the Empedoclean theory. So strong was the influence of Aristotle that his theory dominated the thinking of scientists and philosophers until the beginning of the 17th century.

5.2 Dalton's Model of the Atom

More than 2000 years after Democritus, the English schoolmaster John Dalton (1766–1844) revived the concept of atoms and proposed an atomic model based on facts and experimental evidence (Figure 5.1). His theory, described in a series of papers published from 1803 to 1810, rested on the idea of a different kind of atom for each element. The essence of **Dalton's atomic model** may be summed up as follows:

Dalton's atomic model

1. Elements are composed of minute, indivisible particles called atoms.
2. Atoms of the same element are alike in mass and size.
3. Atoms of different elements have different masses and sizes.
4. Chemical compounds are formed by the union of two or more atoms of different elements.
5. Atoms combine to form compounds in simple numerical ratios, such as one to one, two to two, two to three, and so on.
6. Atoms of two elements may combine in different ratios to form more than one compound.

FIGURE 5.1 ▶
(a) Dalton's atoms were individual particles; the atoms of each element being alike in mass and size but different in mass and size from other elements. (b) and (c) Dalton's atoms combine in specific ratios to form compounds.

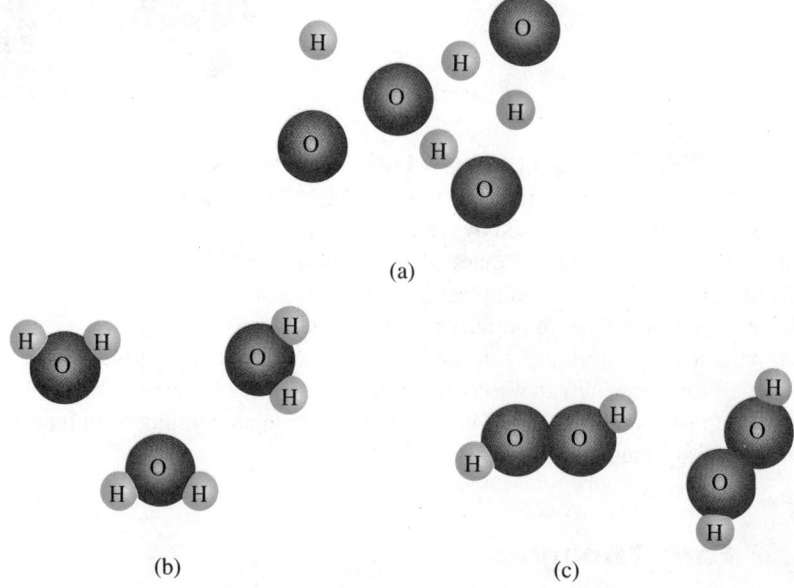

(a)

(b)   (c)

Dalton's atomic model stands as a landmark in the development of chemistry. The major premises of his model are still valid, but some of his statements must be modified or qualified because later investigations have shown that (1) atoms are composed of subatomic particles; (2) not all the atoms of a specific element have the same mass; and (3) atoms, under special circumstances, can be decomposed.

5.3   Composition of Compounds

A large number of experiments extending over a long period of time have established the fact that a particular compound always contains the same elements in the same proportions by mass. For example, water will always contain 11.2% hydrogen and 88.8% oxygen by mass. The fact that water contains hydrogen and oxygen in this particular ratio does not mean that hydrogen and oxygen cannot combine in some other ratio, but that a compound with a different ratio would not be water. In fact, hydrogen peroxide is made up of two atoms of hydrogen and two atoms of oxygen per molecule and contains 5.9% hydrogen and 94.1% oxygen by mass; its properties are markedly different from those of water.

	Water	Hydrogen peroxide
Percent H	11.2	5.9
Percent O	88.8	94.1
Atomic composition	2 H + 1 O	2 H + 2 O

The **law of definite composition** states

> **A compound always contains two or more elements combined in a definite proportion by mass.**

Let's consider two elements, oxygen and hydrogen, that form more than one compound. In water, 8.0 g of oxygen are present for each gram of hydrogen. In hydrogen peroxide, 16.0 g of oxygen are present for each gram of hydrogen. The masses of oxygen are in the ratio of small whole numbers, 16:8 or 2:1. Hydrogen peroxide has twice as much oxygen (by mass) as does water. Using Dalton's atomic model, we deduce that hydrogen peroxide has twice as many oxygen atoms per hydrogen atom as water. In fact, we now write the formulas for water as H_2O and for hydrogen peroxide as H_2O_2. See Figure 5.1b and c.

The **law of multiple proportions** states

> **Atoms of two or more elements may combine in different ratios to produce more than one compound.**

The reliability of this law and the law of definite composition is the cornerstone of the science of chemistry. In essence these laws state that (1) the composition of a particular substance will always be the same no matter what its origin or how it is formed, and (2) the composition of different compounds formed from the same elements will always be unique.

5.4 The Nature of Electric Charge

You've probably received a shock after walking across a carpeted area on a dry day. You've also experienced the static associated with combing your hair, and have had your clothing cling to you. These phenomena result from an accumulation of *electric charge*. This charge may be transferred from one object to another. The properties of electric charge are

1. Charge may be of two types, positive and negative.
2. Unlike charges attract (positive attracts negative), and like charges repel (negative repels negative and positive repels positive).
3. Charge may be transferred from one object to another, by contact or induction.
4. The less the distance between two charges, the greater the force of attraction between unlike charges (or repulsion between identical charges).

5.5 Discovery of Ions

The great English scientist Michael Faraday (1791–1867) made the discovery that certain substances when dissolved in water conduct an electric current. He also

▲
When ions are present in a solution of salt water and an electric current is passed through the solution, the light bulb glows.

noticed that certain compounds decompose into their elements by passing an electric current through the compound. Atoms of some elements are attracted to the positive electrode, while atoms of other elements are attracted to the negative electrode. Faraday concluded that these atoms are electrically charged. He called them *ions* after the Greek word meaning "wanderer."

Any moving charge is an electric current. The electrical charge must travel through a substance known as a conducting medium. The most familiar conducting media are metals formed into wires.

The Swedish scientist Svante Arrhenius (1859–1927) extended Faraday's work. Arrhenius reasoned that an ion is an atom carrying a positive or negative charge. When a compound such as sodium chloride (NaCl) is melted, it conducts electricity. Water is unnecessary. Arrhenius's explanation of this conductivity was that upon melting, the sodium chloride dissociates, or breaks up, into charged ions, Na^+ and Cl^-. The Na^+ ions move toward the negative electrode (cathode), whereas the Cl^- migrate toward the positive electrode (anode). Thus positive ions are *cations,* and negative ions are *anions.*

From Faraday's and Arrhenius's work with ions, Irish physicist G. J. Stoney (1826–1911) realized there must be some fundamental unit of electricity associated with atoms. He named this unit the *electron* in 1891. Unfortunately he had no means of supporting his idea with experimental proof. Evidence remained elusive until 1897 when English physicist Joseph Thomson (1856–1940) was able to show experimentally the existence of the electron.

5.6   Subatomic Parts of the Atom

The concept of the atom—a particle so small that until recently it could not be seen even with the most powerful microscope—and the subsequent determination of its structure stand among the greatest creative intellectual human achievements.

Any visible quantity of an element contains a vast number of identical atoms. But when we refer to an atom of an element, we isolate a single atom from the multitude in order to present the element in its simplest form. Figure 5.2 shows individual atoms as we can see them today.

FIGURE 5.2 ▶
A scanning tunneling microscope shows an array of copper atoms.

CHEMISTRY IN ACTION Triboluminescence

Some substances give off light when they are rubbed, crushed, or broken in a phenomenon called triboluminescence. Examples of substances exhibiting this property include crystals of quartz, sugar cubes, adhesive tape torn off certain surfaces, and Wintergreen Lifesavers.

Linda M. Sweeting, a chemist at Towson State University in Maryland, investigated the sparks created by a Wintergreen Lifesaver when crushed inside one's mouth in a dark room. When a sugar crystal is fractured, separate areas of positive and negative charge form on opposite sides of the crack. Electrons tend to leap the gap and neutralize the charge. When the electrons collide with nitrogen molecules in the air, small amounts of light are emitted. (Lightning is a somewhat similar phenomenon but on a much grander scale.) The addition of wintergreen molecules changes the outcome though because they absorb some of the light energy from the electron leap and reemit it as bright blue-green flashes. These are the sparks we see as we crush a Lifesaver in the dark.

▲
Triboluminescence occurs when Wintergreen Lifesavers are crushed in the mouth.

What is this tiny particle we call the atom? The diameter of a single atom ranges from 0.1 to 0.5 nanometers (1 nm = 1×10^{-9} m). Hydrogen, the smallest atom, has a diameter of about 0.1 nm. To arrive at some idea of how small an atom is, consider this dot (•), which has a diameter of about 1 mm, or 1×10^6 nm. It would take 10 million hydrogen atoms to form a line of atoms across this dot. As inconceivably small as atoms are, they contain smaller particles, the **subatomic particles,** such as electrons, protons, neutrons.

subatomic particles

The development of atomic theory was helped in large part by the invention of new instruments. For example, the Crookes tube, developed by Sir William Crookes in 1875, opened the door to the subatomic structure of the atom (Figure 5.3). The emissions generated in a Crookes tube are called *cathode rays.* Joseph Thomson demonstrated in 1897 that cathode rays (1) travel in straight lines, (2)

  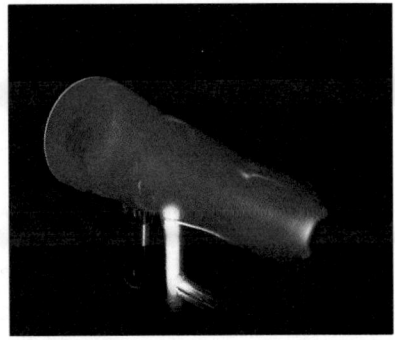

▲
FIGURE 5.3
Crookes tube. Emissions generated in Crookes tube (a) travel in straight lines and are negative in charge, (b) are deflected by a magnetic field, and (c) provide a sharp shadow of the cross in the center of the tube.

are negative in charge, (3) are deflected by electric and magnetic fields, (4) produce sharp shadows, and (5) are capable of moving a small paddle wheel. This was the experimental discovery of the fundamental unit of charge—the electron.

electron

The **electron** (e^-) is a particle with a negative electrical charge and a mass of 9.110×10^{-28} g. This mass is 1/1837 the mass of a hydrogen atom and corresponds to 0.0005486 atomic mass unit (amu) (defined in Section 5.11). One atomic mass unit has a mass of 1.6606×10^{-24} g. Although the actual charge of an electron is known, its value is too cumbersome for practical use and has therefore been assigned a relative electrical charge of -1. The size of an electron has not been determined exactly, but its diameter is believed to be less than 10^{-12} cm.

Protons were first observed by German physicist Eugen Goldstein (1850–1930) in 1886. However, it was Thomson who discovered the nature of the proton. He showed that the proton is a particle, and he calculated its mass to be about 1837 times that of an electron. The **proton** (p) is a particle with a relative mass of 1 amu and an actual mass of 1.673×10^{-24} g. Its relative charge ($+1$) is equal in magnitude, but of opposite sign, to the charge on the electron. The mass of a proton is only very slightly less than that of a hydrogen atom.

proton

Thomson had shown that atoms contained both negatively and positively charged particles. Clearly, the Dalton model of the atom was no longer acceptable. Atoms are not indivisible but are instead composed of smaller parts. Thomson proposed a new model of the atom.

Thomson model of the atom

In the **Thomson model of the atom,** the electrons are negatively charged particles embedded in the atomic sphere. Since atoms are electrically neutral, the sphere also contains an equal number of protons, or positive charges. A neutral atom could become an ion by gaining or losing electrons.

Positive ions were explained by assuming that the neutral atom loses electrons. An atom with a net charge of $+1$ (for example, Na^+ or Li^+) has lost one electron. An atom with a net charge of $+3$ (for example, Al^{3+}) has lost three electrons (Figure 5.4a).

(a)   (b)

▲
FIGURE 5.4
(a) When an electron is lost from an atom, a cation is formed. (b) When electrons are added to a neutral atom, an anion is formed.

TABLE 5.1	Electrical Charge and Relative Mass of Electrons, Protons, and Neutrons			
Particle	Symbol	Relative electrical charge	Relative mass (amu)	Actual mass (g)
Electron	e^-	-1	$\dfrac{1}{1837}$	9.110×10^{-28}
Proton	p	$+1$	1	1.673×10^{-24}
Neutron	n	0	1	1.675×10^{-24}

Negative ions were explained by assuming that extra electrons can be added to atoms. A net charge of -1 (for example, Cl^- or F^-) is produced by the addition of one electron. A net charge of -2 (for example, O^{2-} or S^{2-}) requires the addition of two electrons (Figure 5.4b).

The third major subatomic particle was discovered in 1932 by James Chadwick (1891–1974). This particle, the **neutron** (n), has neither a positive nor a negative charge and has a relative mass of about 1 amu. Its actual mass (1.675×10^{-24} g) is only very slightly greater than that of a proton. The properties of these three subatomic particles are summarized in Table 5.1.

neutron

Nearly all the ordinary chemical properties of matter can be explained in terms of atoms consisting of electrons, protons, and neutrons. The discussion of atomic structure that follows is based on the assumption that atoms contain only these principal subatomic particles. Many other subatomic particles, such as mesons, positrons, neutrinos, and antiprotons, have been discovered, but it is not yet clear whether all these particles are actually present in the atom or whether they are produced by reactions occurring within the nucleus. The fields of atomic and particle or high-energy physics have produced a long list of subatomic particles. Descriptions of the properties of many of these particles are to be found in recent physics textbooks.

The mass of a helium atom is 6.65×10^{-24} g. How many atoms are in a 4.0-g sample of helium?

Example 5.1

$$(4.0 \text{ g})\left(\frac{1 \text{ atom He}}{6.65 \times 10^{-24} \text{ g}}\right) = 6.0 \times 10^{23} \text{ atoms He}$$

Solution

Practice 5.1

The mass of an atom of hydrogen is 1.673×10^{-24} g. How many atoms are in a 10.0-g sample of hydrogen?

5.7   The Nuclear Atom

The discovery that positively charged particles are present in atoms came soon after the discovery of radioactivity by Henri Becquerel in 1896. Radioactive elements spontaneously emit alpha particles, beta particles, and gamma rays from their nuclei (see Chapter 18).

By 1907 Ernest Rutherford had established that the positively charged alpha particles emitted by certain radioactive elements are ions of the element helium. Rutherford used these alpha particles to establish the nuclear nature of atoms. In experiments performed in 1911, he directed a stream of positively charged helium ions (alpha particles) at a very thin sheet of gold foil (about 1000 atoms thick). See Figure 5.5a. He observed that most of the alpha particles passed through the foil with little or no deflection; but a few of the particles were deflected at large angles, and occasionally one even bounced back from the foil (Figure 5.5b). It was known that like charges repel each other and that an electron with a mass of 1/1837 amu could not possibly have an appreciable effect on the path of a 4-amu alpha particle, which is about 7350 times more massive than an electron. Rutherford therefore reasoned that each gold atom must contain a positively charged mass occupying a relatively tiny volume and that, when an alpha particle approaches close enough to this positive mass, it is deflected. Rutherford spoke of this positively charged mass as the *nucleus* of the atom. Because alpha particles have relatively high masses, the extent of the deflections (some actually bounced back) indicated to Rutherford that the nucleus is very heavy and dense. (The density of the nucleus of a hydrogen atom is about 10^{12} g/cm^3—about 1 trillion times the density of water.) Because most of the alpha particles passed through the thousand or so gold atoms without any apparent deflection, he further concluded that most of an atom consists of empty space.

 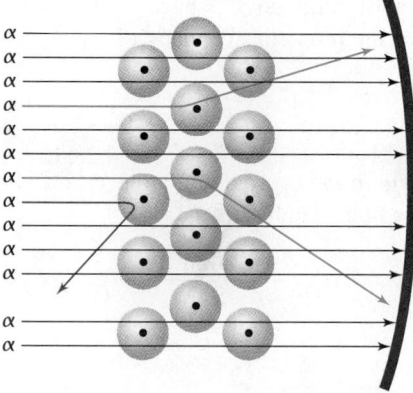

▲
FIGURE 5.5
Left: Rutherford's experiment on alpha-particle scattering, where positive alpha particles (a), emanating from a radioactive source, were directed at a thin gold foil. Right: Deflection (red) and scattering (blue) of the positive alpha particles by the positive nuclei of the gold atoms.

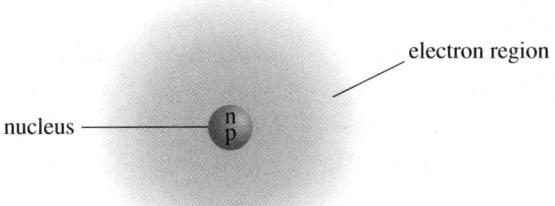

nucleus

electron region

nucleus

When we speak of the mass of an atom, we are referring primarily to the mass of the nucleus. The nucleus contains all the protons and neutrons, which represent more than 99.9% of the total mass of any atom (see Table 5.1). By way of illustration, the largest number of electrons known to exist in an atom is 112. The mass of even 112 electrons is only about 1/17 of the mass of a single proton or neutron. The mass of an atom therefore is primarily determined by the combined masses of its protons and neutrons.

5.8 General Arrangement of Subatomic Particles

The alpha-particle scattering experiments of Rutherford established that the atom contains a dense, positively charged nucleus. The later work of Chadwick demonstrated that the atom contains neutrons, which are particles with mass but no charge. Rutherford also noted that light, negatively charged electrons are present and offset the positive charges in the nucleus. Based on this experimental evidence, a model of the atom and the location of its subatomic particles was devised in which each atom consists of a **nucleus** surrounded by electrons (see Figure 5.6). The nucleus contains protons and neutrons but not electrons. In a neutral atom the positive charge of the nucleus (due to protons) is exactly offset by the negative electrons. Because the charge of an electron is equal but of opposite sign to the charge of a proton, a neutral atom must contain exactly the same number of electrons as protons. However, this model of atomic structure provides no information on the arrangement of electrons within the atom.

nucleus

A neutral atom contains the same number of protons and electrons.

5.9 Atomic Numbers of the Elements

The **atomic number** of an element is the number of protons in the nucleus of an atom of that element. The atomic number determines the identity of an atom. For

atomic number

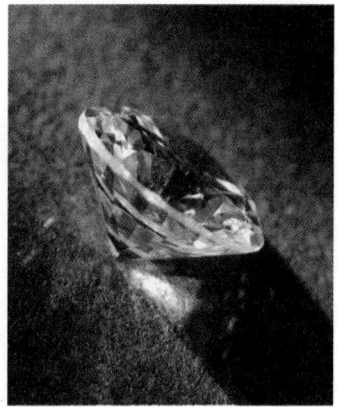

▲
Carbon, shown here as a beautiful diamond, has 6 protons and 6 electrons in each atom.

example, every atom with an atomic number of 1 is a hydrogen atom; it contains one proton in its nucleus. Every atom with an atomic number of 6 is a carbon atom; it contains 6 protons in its nucleus. Every atom with an atomic number of 92 is a uranium atom; it contains 92 protons in its nucleus. The atomic number tells us not only the number of positive charges in the nucleus, but also the number of electrons in the neutral atom.

atomic number = number of protons in the nucleus

You don't need to memorize the atomic numbers of the elements because a periodic table is commonly provided in texts, laboratories, and on examinations. The atomic numbers of all elements are shown in the periodic table on the inside front cover of this book and are also listed in the table of atomic masses on the inside front endpapers.

5.10   Isotopes of the Elements

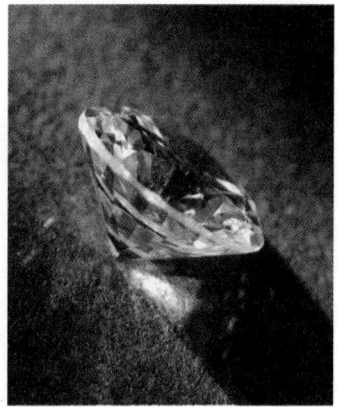

▲
A sample of uranium contains 92 protons in each nucleus.

Shortly after Rutherford's conception of the nuclear atom, experiments were performed to determine the masses of individual atoms. These experiments showed that the masses of nearly all atoms were greater than could be accounted for by simply adding up the masses of all the protons and electrons that were known to be present in an atom. This fact led to the concept of the neutron, a particle with no charge but with a mass about the same as that of a proton. Because this particle has no charge, it was very difficult to detect, and the existence of the neutron was not proven experimentally until 1932. All atomic nuclei, except that of the simplest hydrogen atom, are now believed to contain neutrons.

All atoms of a given element have the same number of protons. Experimental evidence has shown that, in most cases, all atoms of a given element do not have identical masses. This is because atoms of the same element may have different numbers of neutrons in their nuclei.

isotopes   Atoms of an element having the same atomic number but different atomic masses are called **isotopes** of that element. Atoms of the various isotopes of an element therefore have the same number of protons and electrons but different numbers of neutrons.

Three isotopes of hydrogen (atomic number 1) are known. Each has one proton in the nucleus and one electron. The first isotope (protium), without a neutron, has a mass number of 1; the second isotope (deuterium), with one neutron in the nucleus, has a mass number of 2; the third isotope (tritium), with two neutrons, has a mass number of 3 (see Figure 5.7).

The three isotopes of hydrogen may be represented by the symbols $^{1}_{1}H$, $^{2}_{1}H$, $^{3}_{1}H$, indicating an atomic number of 1 and mass numbers of 1, 2, and 3, respectively. This method of representing atoms is called *isotopic notation.* The subscript (Z) is the

mass number   atomic number; the superscript (A) is the **mass number,** which is the sum of the number of protons and the number of neutrons in the nucleus. The hydrogen isotopes may also be referred to as hydrogen-1, hydrogen-2, and hydrogen-3.

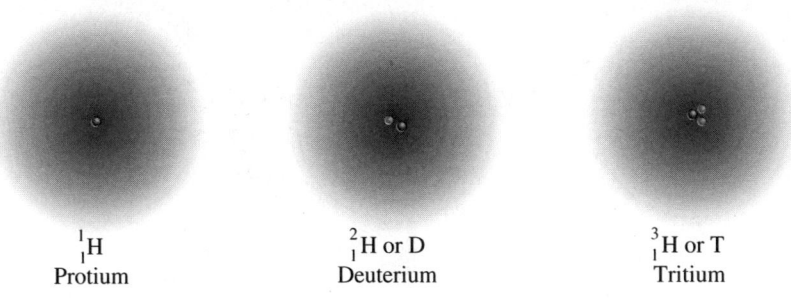

◀ **FIGURE 5.7**
**The isotopes of hydrogen.
The number of protons
(purple) and neutrons
(blue) are shown within the
nucleus. The electron (e⁻)
exists outside the nucleus.**

^1_1H
Protium

^2_1H or D
Deuterium

^3_1H or T
Tritium

**Mass number
(sum of protons and
neutrons in the nucleus)** ↘

$^A_Z E$ ⟵ **Symbol of element**

**Atomic number
(number of protons
in the nucleus)** ↗

The mass number of an
element is the sum of the
protons and neutrons in the
nucleus.

Most of the elements occur in nature as mixtures of isotopes. However, not all isotopes are stable; some are radioactive and are continuously decomposing to form other elements. For example, of the seven known isotopes of carbon, only two, carbon-12 and carbon-13, are stable. Of the seven known isotopes of oxygen, only three, $^{16}_8O$, $^{17}_8O$, and $^{18}_8O$, are stable. Of the fifteen known isotopes of arsenic, $^{75}_{33}As$ is the only one that is stable.

5.11 Atomic Mass

The mass of a single atom is far too small to measure on a balance. But fairly precise determinations of the masses of individual atoms can be made with an instrument called a *mass spectrometer* (see Figure 5.8). The mass of a single hydrogen atom is 1.673×10^{-24} g. However, it is neither convenient nor practical to compare the actual masses of atoms expressed in grams; therefore a table of relative atomic masses using *atomic mass units* was devised. (The term *atomic weight* is sometimes used instead of atomic mass.) The carbon isotope having six protons and six neutrons and designated carbon-12, or $^{12}_6C$, was chosen as the standard for atomic masses. This reference isotope was assigned a value of exactly 12 atomic mass units (amu). Thus 1 **atomic mass unit** is defined as equal to exactly 1/12 of the mass of a carbon-12 atom. The actual mass of a carbon-12 atom is 1.9927×10^{-23} g, and that of one atomic mass unit is 1.6606×10^{-24} g. In the table of atomic masses, all elements then have values that are relative to the mass assigned to the reference isotope carbon-12.

atomic mass unit

1 amu = 1.6606 × 10⁻²⁴ g

▲
These copper plates and
bracelets made in Mexico City
all contain a mixture of the
isotopes of copper.

A table of atomic masses is given on the inside front cover of this book. Hydrogen atoms, with a mass of about 1/12 that of a carbon atom, have an average atomic mass of 1.00797 amu on this relative scale. Magnesium atoms, which are about twice as heavy as carbon, have an average mass of 24.305 amu. The average atomic mass of oxygen is 15.9994 amu.

Since most elements occur as mixtures of isotopes with different masses, the atomic mass determined for an element represents the average relative mass of all the naturally occurring isotopes of that element. The atomic masses of the individual isotopes are approximately whole numbers, because the relative masses of the protons and neutrons are approximately 1.0 amu each. Yet we find that the atomic masses given for many of the elements deviate considerably from whole numbers.

For example, the atomic mass of rubidium is 85.4678 amu, that of copper is 63.546 amu, and that of magnesium is 24.305 amu. The deviation of an atomic mass from a whole number is due mainly to the unequal occurrence of the various isotopes of an element. The two principal isotopes of copper are $^{63}_{29}Cu$ and $^{65}_{29}Cu$. It is apparent that copper-63 atoms are the more abundant isotope, since the atomic mass of copper, 63.546 amu, is closer to 63 than to 65 amu (see Figure 5.9). The actual values of the copper isotopes observed by mass spectra determination are shown here:

Isotope	Isotopic mass (amu)	Abundance (%)	Average atomic mass (amu)
$^{63}_{29}Cu$	62.9298	69.09	63.55
$^{65}_{29}Cu$	64.9278	30.91	

The average atomic mass is a
weighted average of the
masses of all the isotopes
present in the sample.

The average atomic mass can be calculated by multiplying the atomic mass of each isotope by the fraction of each isotope present and adding the results. The calculation for copper is

$$(62.9298 \text{ amu})(0.6909) = 43.48 \text{ amu}$$
$$(64.9278 \text{ amu})(0.3091) = \underline{20.07 \text{ amu}}$$
$$63.55 \text{ amu}$$

The **atomic mass** of an element is the average relative mass of the isotopes of that element compared to the atomic mass of carbon-12 (exactly 12.0000 . . . amu). **atomic mass**

The relationship between mass number and atomic number is such that, if we subtract the atomic number from the mass number of a given isotope, we obtain the number of neutrons in the nucleus of an atom of that isotope. Table 5.2 shows this method of determining the number of neutrons. For example, the fluorine atom ($^{19}_{9}F$), atomic number 9, having a mass of 19 amu, contains 10 neutrons:

$$\text{mass number} - \text{atomic number} = \text{number of neutrons}$$

$$19 \quad - \quad 9 \quad = \quad 10$$

The atomic masses given in the table on the front endpapers of this book are values accepted by international agreement. You need not memorize atomic masses. In the calculations in this book, the use of atomic masses to four significant figures will give results of sufficient accuracy.

TABLE 5.2 Determination of the Number of Neutrons in an Atom by Subtracting Atomic Number from Mass Number					
	Hydrogen ($^{1}_{1}H$)	Oxygen ($^{16}_{8}O$)	Sulfur ($^{32}_{16}S$)	Fluorine ($^{19}_{9}F$)	Iron ($^{56}_{26}Fe$)
Mass number	1	16	32	19	56
Atomic number	(−)1	(−)8	(−)16	(−)9	(−)26
Number of neutrons	0	8	16	10	30

CHEMISTRY IN ACTION   Real or Imitation?

What's the difference between vanilla extract and imitation vanilla extract? Both substances taste the same because they contain the same molecules, and taste is related to molecular structure.

Vanilla extract is an alcohol and water solution of materials extracted from vanilla beans. Imitation vanilla extract is also an alcohol and water solution, but it is made from lignin, a waste product in the wood pulp industry, which is converted to vanillin—the same molecule found in vanilla extract. Why would anyone want to use wood pulp in place of vanilla beans? The main reason is cost—natural vanilla costs twice as much to produce as synthetic vanilla.

One role of a government chemist is to prevent mislabeling of products. Because the compounds in both products are chemically the same, detecting the source of the molecule is done by "inspecting" the carbon atoms in the vanilla. Two isotopes in organic compounds are carbon-12 and carbon-13. The ratio of these isotopes is slightly different for natural vanillin and for imitation vanillin. It is from those slight differences that government scientists distinguish whether your vanilla comes from the bean or from wood pulp. Chemists also work as consumer advocates—making sure you get what you pay for in your vanilla extract!

▲ The source of "real" vanilla is the vanilla bean; imitation vanilla comes from wood pulp.

Example 5.2   How many protons, neutrons, and electrons are found in an atom of $^{14}_{6}C$?

Solution   The element is carbon, atomic number 6. The number of protons or electrons equals the atomic number and is 6. The number of neutrons is determined by subtracting the atomic number from the mass number: $14 - 6 = 8$.

Practice 5.2

How many protons, neutrons, and electrons are in each element?
(a) $^{16}_{8}O$, (b) $^{80}_{35}Br$, (c) $^{235}_{92}U$, (d) $^{64}_{29}Cu$

Example 5.3

Chlorine is found in nature as two isotopes, $^{37}_{17}Cl$ (24.47%) and $^{35}_{17}Cl$ (75.53%). The atomic masses are 36.96590 and 34.96885 amu, respectively. Determine the average atomic mass of chlorine.

Solution

Multiply each mass by its percentage and add the results to find the average.

$$(0.2447)(36.96590 \text{ amu}) + (0.7553)(34.96885 \text{ amu})$$

$$= 35.4575 \text{ amu}$$

$$= 35.46 \text{ amu (4 significant figures)}$$

Practice 5.3

Silver is found in two isotopes with atomic masses 106.9041 and 108.9047 amu, respectively. The first isotope represents 51.82% and the second 48.18%. Determine the average atomic mass of silver.

Concepts in Review

1. State the major provisions of Dalton's atomic model.

2. State the law of definite composition and indicate its significance.

3. State the law of multiple proportions and indicate its significance.

4. Give the names, symbols, and relative masses of the three principal subatomic particles.

5. Describe the Thomson model of the atom.

6. Describe the atom as conceived by Ernest Rutherford after his alpha-scattering experiment.

7. Determine the atomic number, mass number, or number of neutrons of an isotope when given the values of any two of these three items.

8. Name and distinguish among the three isotopes of hydrogen.

9. Calculate the average atomic mass of an element, given the isotopic masses and the abundance of its isotopes.

10. Determine the number of protons, neutrons, and electrons from the atomic number and atomic mass of an atom.

Key Terms

atomic mass (5.11)
atomic mass unit (5.11)
atomic number (5.9)
Dalton's atomic model (5.2)

electron (5.6)
isotopes (5.10)
law of definite composition (5.3)
law of multiple proportions (5.3)

mass number (5.10)
neutron (5.6)
nucleus (5.8)

proton (5.6)
subatomic particles (5.6)
Thomson model of the atom (5.6)

Questions

1. What are the atomic numbers of (a) copper, (b) nitrogen, (c) phosphorus, (d) radium, and (e) zinc?

2. A neutron is approximately how many times heavier than an electron?

3. From the chemist's point of view, what are the essential differences among a proton, a neutron, and an electron?

4. Distinguish between an atom and an ion.

5. What letters are used to designate atomic number and mass number in isotopic notation of atoms?

6. In what ways are isotopes alike? In what ways are they different?

Paired Exercises

7. Explain why, in Rutherford's experiments, some alpha particles were scattered at large angles by the gold foil or even bounced back.

8. What experimental evidence led Rutherford to conclude the following?
   (a) The nucleus of the atom contains most of the atomic mass.
   (b) The nucleus of the atom is positively charged.
   (c) The atom consists of mostly empty space.

9. Describe the general arrangement of subatomic particles in the atom.

10. What part of the atom contains practically all its mass?

11. What contribution did these scientists make to atomic models of the atom?
    (a) Dalton
    (b) Thomson
    (c) Rutherford

12. Consider the following models of the atom: (a) Dalton, (b) Thomson, (c) Rutherford. How does the location of the electrons in an atom vary? How does the location of the atom's positive matter compare?

13. Explain why the atomic masses of elements are not whole numbers.

14. Is the isotopic mass of a given isotope ever an exact whole number? Is it always? (Consider the masses of $^{12}_{6}C$ and $^{63}_{29}Cu$.)

15. What special names are given to the isotopes of hydrogen?

16. List the similarities and differences in the three isotopes of hydrogen.

17. What is the symbol and name of the element that has an atomic number of 24 and a mass number of 52?

18. An atom of an element has a mass number of 201 and has 121 neutrons in its nucleus.
    (a) What is the electrical charge of the nucleus?
    (b) What is the symbol and name of the element?

19. What is the nuclear composition of the six naturally occurring isotopes of calcium having mass numbers of 40, 42, 43, 44, 46, and 48?

20. Which of the isotopes of calcium in the previous exercise is the most abundant isotope? Can you be sure? Explain your choice.

21. Write isotopic notation symbols for the following:
    (a) $Z = 26$, $A = 55$
    (b) $Z = 12$, $A = 26$
    (c) $Z = 3$, $A = 6$
    (d) $Z = 79$, $A = 188$

22. Give the nuclear composition and isotopic notation for
    (a) an atom containing 27 protons, 32 neutrons, and 27 electrons
    (b) an atom containing 15 protons, 16 neutrons, and 15 electrons
    (c) an atom containing 110 neutrons, 74 electrons, and 74 protons
    (d) an atom containing 92 electrons, 143 neutrons, and 92 protons

23. Naturally occurring lead exists as four stable isotopes: ^{204}Pb with a mass of 203.973 amu (1.480%); ^{206}Pb, 205.974 amu (23.60%); ^{207}Pb, 206.9759 amu (22.60%); and ^{208}Pb, 207.9766 amu (52.30%). Calculate the average atomic mass of lead.

24. Naturally occurring magnesium consists of three stable isotopes: ^{24}Mg, 23.985 amu (78.99%); ^{25}Mg, 24.986 amu (10.00%); and ^{26}Mg, 25.983 amu (11.01%). Calculate the average atomic mass of magnesium.

25. 68.9257 amu is the mass of 60.4% of the atoms of an element with only two naturally occurring isotopes. The atomic mass of the other isotope is 70.9249 amu. Determine the average atomic mass of the element. Identify the element.

26. A sample of enriched lithium contains 30.00% ^{6}Li (6.015 amu) and 70.00% ^{7}Li (7.016 amu). What is the average atomic mass of the sample?

*27. An average dimension for the radius of an atom is 1.0×10^{-8} cm, and the average radius of the nucleus is 1.0×10^{-13} cm. Determine the ratio of atomic volume to nuclear volume. Assume that the atom is spherical ($V = (4/3)\pi r^3$ for a sphere).

*28. An aluminum atom has an average diameter of about 3.0×10^{-8} cm. The nucleus has a diameter of about 2.0×10^{-13} cm. Calculate the ratio of the atom's diameter to its nucleus.

Additional Exercises

These exercises are not paired and provide additional practice on the concepts covered in this chapter.

29. What experimental evidence supports these statements?
    (a) The nucleus of an atom is small.
    (b) The atom consists of both positive and negative charges.
    (c) The nucleus of the atom is positive.

30. What is the relationship between the following two atoms:
    (a) one atom with 10 protons, 10 neutrons, and 10 electrons; and another atom with 10 protons, 11 neutrons, and 10 electrons
    (b) one atom with 10 protons, 11 neutrons, and 10 electrons; and another atom with 11 protons, 10 neutrons, and 11 electrons

31. How will an atom's nucleus differ if it loses an alpha particle?

32. The radius of a carbon atom in many compounds is 0.77×10^{-8} cm. If the radius of a Styrofoam ball used to represent the carbon atom in a molecular model is 1.5 cm, how much of an enlargement is this?

33. How is it possible for there to be more than one kind of atom of the same element?

34. Which element contains the largest number of neutrons per atom: ^{210}Bi, ^{210}Po, ^{210}At, or ^{211}At?

*35. An unknown element Q has two known isotopes: ^{60}Q and ^{63}Q. If the average atomic mass is 61.5 amu, what are the relative percentages of the isotopes?

*36. The actual mass of one atom of an unknown isotope is 2.18×10^{-22} g. Calculate the atomic mass of this isotope.

37. The mass of an atom of argon is 6.63×10^{-24} g. How many atoms are in a 40.0-g sample of argon?

38. Using the periodic table inside the front cover of the book, determine which of the first 20 elements have isotopes that you would expect to have the same number of protons, neutrons, and electrons.

39. Complete this table with the appropriate data for each isotope given:

Atomic number	Mass number	Symbol of element	Number of protons	Number of neutrons
(a) 8	16			
(b)		Ni		30
(c)	199		80	

40. Complete this table (all are neutral atoms):

Element	Symbol	Atomic number	Number of protons	Number of neutrons	Number of electrons
(a) Platinum					
(b)	^{30}P				
(c)		53			
(d)			36		
(e)				45	34
(f)	^{40}Ca				

Answers to Practice Exercises

5.1 5.98×10^{24} atoms

5.2

	protons	neutrons	electrons
(a)	8	8	8
(b)	35	45	35
(c)	92	143	92
(d)	29	35	29

5.3 107.9 amu

CHAPTER

6

Nomenclature of Inorganic Compounds

CHAPTER 6 / OUTLINE

▲
This exquisite chambered nautilus shell is formed from calcium carbonate, commonly called limestone.

As children, we begin to communicate with other people in our lives by learning the names of objects around us. As we continue to develop, we learn to speak and use language to complete a wide variety of tasks. As we enter school, we begin to learn of other languages—the languages of mathematics, of other cultures, of computers. In each case, we begin by learning the names of the building blocks, and then proceed to more abstract concepts. Chemistry has a language all its own—a whole new way of describing the objects so familiar to us in our daily lives. Only after learning the language are we able to understand the complexities of the modern model of the atom and its applications.

6.1 Common and Systematic Names

Chemical nomenclature is the system of names that chemists use to identify compounds. When a new substance is formulated, it must be named in order to distinguish it from all other substances (see Figure 6.1). In this chapter, we will restrict our discussion to the nomenclature of inorganic compounds—compounds that do not generally contain carbon.

Common names are arbitrary names that are not based on the chemical composition of compounds. Before chemistry was systematized, a substance was given a name that generally associated it with one of its outstanding physical or chemical properties. For example, *quicksilver* is a common name for mercury, and nitrous oxide (N_2O), used as an anesthetic in dentistry, has been called *laughing gas* because it induces laughter when inhaled. Water and ammonia are also common names because neither provides any information about the chemical composition of the compounds. If every substance were assigned a common name, the amount of memorization required to learn over 9 million names would be astronomical.

Common names have distinct limitations, but they remain in frequent use. Common names continue to be used in industry because the systematic name is too long or too technical for everyday use. For example, calcium oxide (CaO) is called *lime* by plasterers; photographers refer to *hypo* rather than sodium thiosulfate

Water (H_2O) and ammonia (NH_3) are almost always referred to by their common names.

FIGURE 6.1 ▶
Where to find rules for naming inorganic substances in this book.

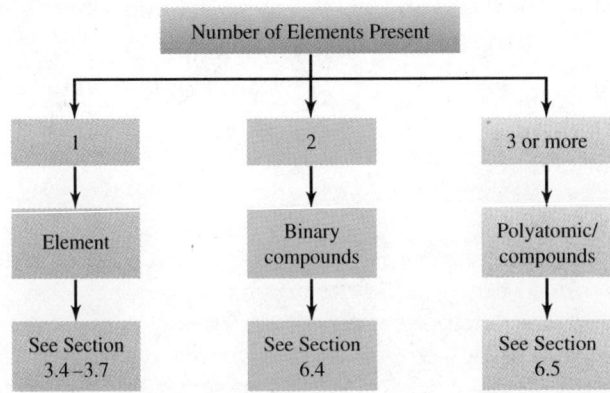

TABLE 6.1	Common Names, Formulas, and Chemical Names of Familiar Substances	
Common name	**Formula**	**Chemical name**
Acetylene	C_2H_2	ethyne
Lime	CaO	calcium oxide
Slaked lime	$Ca(OH)_2$	calcium hydroxide
Water	H_2O	water
Galena	PbS	lead(II) sulfide
Alumina	Al_2O_3	aluminum oxide
Baking soda	$NaHCO_3$	sodium hydrogen carbonate
Cane or beet sugar	$C_{12}H_{22}O_{11}$	sucrose
Borax	$Na_2B_4O_7 \cdot 10\,H_2O$	sodium tetraborate decahydrate
Brimstone	S	sulfur
Calcite, marble, limestone	$CaCO_3$	calcium carbonate
Cream of tartar	$KHC_4H_4O_6$	potassium hydrogen tartrate
Epsom salts	$MgSO_4 \cdot 7\,H_2O$	magnesium sulfate heptahydrate
Gypsum	$CaSO_4 \cdot 2\,H_2O$	calcium sulfate dihydrate
Grain alcohol	C_2H_5OH	ethanol, ethyl alcohol
Hypo	$Na_2S_2O_3$	sodium thiosulfate
Laughing gas	N_2O	dinitrogen monoxide
Lye, caustic soda	NaOH	sodium hydroxide
Milk of magnesia	$Mg(OH)_2$	magnesium hydroxide
Muriatic acid	HCl	hydrochloric acid
Plaster of Paris	$CaSO_4 \cdot \frac{1}{2}\,H_2O$	calcium sulfate hemihydrate
Potash	K_2CO_3	potassium carbonate
Pyrite (fool's gold)	FeS_2	iron disulfide
Quicksilver	Hg	mercury
Saltpeter (chile)	$NaNO_3$	sodium nitrate
Table salt	NaCl	sodium chloride
Vinegar	$HC_2H_3O_2$	acetic acid
Washing soda	$Na_2CO_3 \cdot 10\,H_2O$	sodium carbonate decahydrate
Wood alcohol	CH_3OH	methanol, methyl alcohol

($Na_2S_2O_3$); and nutritionists use the name *vitamin D$_3$*, instead of 9,10-secocholesta-5,7,10(19)-trien-3-β-ol ($C_{27}H_{44}O$). Table 6.1 lists the common names, formulas, and systematic names of some familiar substances.

Chemists prefer systematic names that precisely identify the chemical composition of chemical compounds. The system for inorganic nomenclature was devised by the International Union of Pure and Applied Chemistry (IUPAC), which was founded in 1921. The IUPAC meets regularly and constantly reviews and updates the system.

6.2 Elements and Ions

In Chapter 3, we studied the names and symbols for the elements as well as their general location on the periodic table. In Chapter 5, we investigated the composition of the atom and learned that all atoms are composed of protons, electrons, and neutrons; that a particular element is defined by the number of protons it contains; and that atoms are uncharged because they contain equal numbers of protons and electrons.

What's in a Name?

When a scientist discovered a new element in the early days of chemistry, he or she had the honor of naming it. Now researchers must submit their choices for a name to an international committee called the International Union of Pure and Applied Chemistry before they can be placed on the periodic table. In September 1997, the IUPAC decided on names for the elements from 104 through109. These six elements are now called rutherfordium (Rf), dubnium (Db), seaborgium (Sg), bohrium (Bh), hassium (Hs), and meitnerium (Mt).

The new names are a compromise among choices presented by different research teams. The Russians gained recognition for work done at a laboratory in Dubna. Americans gained recognition for Glenn Seaborg, the first living scientist to have an element named after him. The British recognized Ernest Rutherford, who discovered the atomic nucleus. The Germans won recognition both for Lise Meitner, the woman who co-discovered atomic fission, and for one of their labs in the German state of Hesse. Both the Germans and the Russians won recognition for Niels Bohr, whose model of the atom led the way toward modern ideas about atomic structure. As researchers continue to discover elements and expand the periodic table, the job of deciding on a name and symbol becomes an increasingly complex task.

Russian lab in Dubna

Niels Bohr

Ernest Rutherford

104	105	106	107	108	109
Rf	Db	Sg	Bh	Hs	Mt

Lise Meitner

Glenn Seaborg

German lab in Hesse

The formula for most elements is simply the symbol of the element. In chemical reactions or mixtures the element behaves as though it were a collection of individual particles. A small number of elements have formulas that are not single atoms at normal temperatures. Seven of the elements are *diatomic* molecules—that is, two atoms bonded together to form a molecule. These diatomic elements are hydrogen, H_2; oxygen, O_2; nitrogen, N_2; fluorine, F_2; chlorine, Cl_2; bromine, Br_2; and iodine, I_2. Two other elements that are commonly polyatomic are S_8, sulfur; and P_4, phosphorus.

Elements Occurring as Molecules					
Hydrogen	H_2	Chlorine	Cl_2	Sulfur	S_8
Oxygen	O_2	Fluorine	F_2	Phosphorus	P_4
Nitrogen	N_2	Bromine	Br_2		
		Iodine	I_2		

A charged particle, known as an *ion,* can be produced by adding or removing one or more electrons from a neutral atom. For example, potassium atoms contain 19 protons and 19 electrons. To make a potassium ion, we remove one electron leaving 19 protons and only 18 electrons. This gives an ion with a positive one $(+1)$ charge:

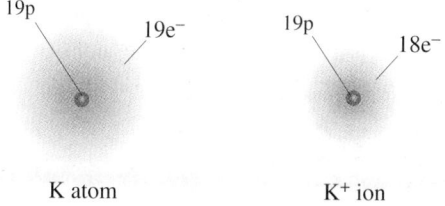

$$K \text{ atom} \qquad\qquad K^+ \text{ ion}$$

Written in the form of an equation, $K \rightarrow K^+ + e^-$. A positive ion is called a *cation.* Any neutral atom that *loses* an electron will form a cation. Sometimes an atom may lose one electron, as in the potassium example. Other atoms may lose more than one electron:

$$Mg \rightarrow Mg^{2+} + 2e^-$$

or

$$Al \rightarrow Al^{3+} + 3e^-$$

Cations are named the same as their parent atoms, as shown here:

Atom		Ion	
K	potassium	K^+	potassium ion
Mg	magnesium	Mg^{2+}	magnesium ion
Al	aluminum	Al^{3+}	aluminum ion

Ions can also be formed by adding electrons to a neutral atom. For example, the chlorine atom contains 17 protons and 17 electrons. The equal number of positive charges and negative charges results in a net charge of zero for the atom. If one electron is added to the chlorine atom, it now contains 17 protons and 18 electrons resulting in a net charge of negative one (-1) on the ion:

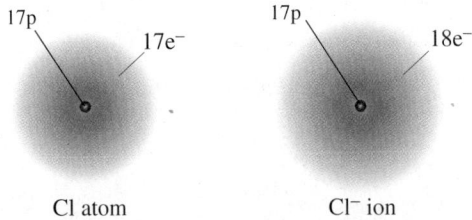

Cl atom   Cl⁻ ion

In a chemical equation this process is summarized as $Cl + e^- \rightarrow Cl^-$. A negative ion is called an *anion*. Any neutral atom that *gains* an electron will form an anion. Atoms may gain more than one electron to form anions with different charges:

$$O + 2e^- \rightarrow O^{2-}$$

$$N + 3e^- \rightarrow N^{3-}$$

Anions are named differently than cations. To name an anion consisting of only one element, use the stem of the parent element name and change the ending to *-ide*. For example, the Cl^- ion is named by using the stem *chlor-* from chlorine and adding *-ide* to form chloride ion. Here are some examples:

Symbol	Name of atom	Ion	Name of ion
F	fluorine	F^-	fluoride ion
Br	bromine	Br^-	bromide ion
Cl	chlorine	Cl^-	chloride ion
I	iodine	I^-	iodide ion
O	oxygen	O^{2-}	oxide ion
N	nitrogen	N^{3-}	nitride ion

Ions are always formed by adding or removing electrons from an atom. Atoms do not form ions on their own. Most often ions are formed when metals combine with nonmetals.

The charge on an ion can often be predicted from the position of the element on the periodic table. Figure 6.2 shows the charges of selected ions from several groups on the periodic table. Notice that all the metals in the far left column (Group IA) are ($1+$), all those in the next column (Group IIA) are ($2+$), and the metals in the next tall column (Group IIIA) form ($3+$) ions. The elements in the lower center part of the table are called *transition metals*. These elements tend to form cations with various positive charges. There is no easy way to predict the charges on these cations. All metals lose electrons to form positive ions.

IA																	
H⁺	IIA											IIIA	IVA	VA	VIA	VIIA	
Li⁺	Be²⁺													N³⁻	O²⁻	F⁻	
Na⁺	Mg²⁺											Al³⁺		P³⁻	S²⁻	Cl⁻	
K⁺	Ca²⁺															Br⁻	
Rb⁺	Sr²⁺			Transition metals												I⁻	
Cs⁺	Ba²⁺																

▲ **FIGURE 6.2**
Charges of selected ions in the periodic table.

In contrast, the nonmetals form anions by gaining electrons. On the right side of the periodic table in Figure 6.2, you can see that the atoms in Group VIIA form $(1-)$ ions. The nonmetals in Group VIA form $(2-)$ ions. It's important to learn the charges on the ions shown in Figure 6.2 and their relationship to the group number at the top of the column. We will learn more about why these ions carry their particular charges later in the course.

6.3 Writing Formulas from Names of Compounds

In Chapters 3 and 5, we learned that compounds can be composed of ions. These substances are called *ionic compounds* and will conduct electricity when dissolved in water. An excellent example of an ionic compound is ordinary table salt. It's composed of crystals of sodium chloride. When dissolved in water, sodium chloride conducts electricity very well, as shown in Figure 6.3.

A chemical compound must have a net charge of zero. If it contains ions, the charges on the ions must add up to zero in the formula for the compound. This is relatively easy in the case of sodium chloride. The sodium ion $(1+)$ and the chloride ion $(1-)$ add to zero, resulting in the formula NaCl. Now consider an ionic compound containing calcium (Ca^{2+}) and fluoride (F^-) ions. How can we write a formula with a net charge of zero? To do this we need one Ca^{2+} and two F^- ions. The correct formula is CaF_2. The subscript 2 indicates two fluoride ions are needed for each calcium ion.

FIGURE 6.3
(a) Distilled water does not conduct electricity. (b) Salt is added to the distilled water. (c) The solution of salt water contains ions and conducts electricity.
▼

(a)   (b)   (c)

Aluminum oxide is a bit more complicated because it consists of Al^{3+} and O^{2-} ions. Since 6 is the least common multiple of 3 and 2, we have $2(3+) + 3(2-) = 0$ or a formula containing $2\,Al^{3+}$ ions and $3\,O^{2-}$ ions for Al_2O_3. Here are a few more examples of formula writing for ionic compounds:

Name of compound	Ions	Lowest common multiple	Sum of charges on ions	Formula
Sodium bromide	Na^+, Br^-	1	$(+1) + (-1) = 0$	NaBr
Potassium sulfide	K^+, S^{2-}	2	$2(+1) + (-2) = 0$	K_2S
Zinc sulfate	Zn^{2+}, SO_4^{2-}	2	$(+2) + (-2) = 0$	$ZnSO_4$
Ammonium phosphate	NH_4^+, PO_4^{3-}	3	$3(+1) + (-3) = 0$	$(NH_4)_3PO_4$
Aluminum chromate	Al^{3+}, CrO_4^{2-}	6	$2(+3) + 3(-2) = 0$	$Al_2(CrO_4)_3$

Example 6.1   Write formulas for (a) calcium chloride, (b) magnesium oxide, and (c) barium phosphide.

Solution   (a) Use the following steps for calcium chloride.

Step 1. From the name we know that calcium chloride is composed of calcium and chloride ions. First write down the formulas of these ions:

$$Ca^{2+} \qquad Cl^-$$

Step 2. To write the formula of the compound, combine the smallest numbers of Ca^{2+} and Cl^- ions to give the charge sum equal to zero. In this case the lowest common multiple of the charges is 2:

$$(Ca^{2+}) + 2(Cl^-) = 0$$

$$(2+) \;\; + 2(1-) = 0$$

Therefore the formula is $CaCl_2$.

(b) Use the same procedure for magnesium oxide.

Step 1. From the name we know that magnesium oxide is composed of magnesium and oxide ions. First write down the formulas of these ions:

$$Mg^{2+} \qquad O^{2-}$$

Step 2. To write the formula of the compound, combine the smallest numbers of Mg^{2+} and O^{2-} ions to give the charge sum equal to zero:

$$(Mg^{2+}) + (O^{2-}) = 0$$

$$(2+) \;\; + (2-) = 0$$

The formula is MgO.

◀ Compounds of transition elements are typically very colorful and are useful as paint pigments.

1. Cobalt(II) chloride, $CoCl_2$
2. Sodium chloride, $NaCl$
3. Lead(II) sulfide, PbS
4. Sulfur, S
5. Zinc, Zn
6. Marble chips, $CaCO_3$
7. Logwood chips
8. Charcoal, C
9. Mercury(II) iodide, HgI_2
10. Pyrite, FeS
11. Chromium(III) oxide, Cr_2O_3
12. Iron(II) sulfate, $FeSO_4$
13. Sodium sulfite, Na_2SO_3
14. Rosin
15. Sodium thiosulfate, $Na_2S_2O_3$
16. Iron, Fe
17. Aluminum, Al
18. Potassium hexacyanoferrate, $K_3Fe(CN)_6$
19. Potassium chromium(III) sulfate, $KCr(SO_4)_2$
20. Menthol, $C_{10}H_{19}OH$
21. Potassium permanganate, $KMnO_4$
22. Ammonium nickel(II) sulfate, $(NH_4)_2Ni(SO_4)_2$
23. Copper(II) sulfate pentahydrate, $CuSO_4 \cdot 5H_2O$
24. Sodium chromate, Na_2CrO_4
25. Trilead tetraoxide, Pb_3O_4
26. Hydroquinone, $C_6H_4(OH)_2$
27. Copper, Cu

(c) Use the same procedure for barium phosphide.

Step 1. From the name we know that barium phosphide is composed of barium and phosphide ions. First write down the formulas of these ions:

Ba^{2+} P^{3-}

Step 2. To write the formula of the compound, combine the smallest numbers of Ba^{2+} and P^{3-} ions to give the charge sum equal to zero. In this case the lowest common multiple of the charges is 6:

$$3(Ba^{2+}) + 2(P^{3-}) = 0$$

$$3(2+) + 2(3-) = 0$$

The formula is Ba_3P_2.

Practice 6.1

Write formulas for compounds containing the following ions:
(a) K^+ and F^- (d) Na^+ and S^{2-}
(b) Ca^{2+} and Br^- (e) Ba^{2+} and O^{2-}
(c) Mg^{2+} and N^{3-}

Binary Compounds

6.4

Binary compounds contain only two different elements. Many binary compounds are formed when a metal combines with a nonmetal to form a *binary ionic compound.* The metal loses one or more electrons to become a cation while the nonmetal gains one or more electrons to become an anion. The cation is written first in the formula, followed by the anion.

A. Binary Ionic Compounds Containing a Metal Forming Only One Type of Cation

The chemical name is composed of the name of the metal followed by the name of the nonmetal, which has been modified to an identifying stem plus the suffix *-ide.*

For example, sodium chloride, NaCl, is composed of one atom each of sodium and chlorine. The name of the metal, sodium, is written first and is not modified. The second part of the name is derived from the nonmetal, chlorine, by using the stem *chlor-* and adding the ending *-ide*; it is named *chloride.* The compound name is *sodium chloride.*

To name these compounds:
1. **Write the name of the cation.**
2. **Write the stem for the anion with the suffix -ide.**

NaCl
Elements: Sodium (metal)
                         Chlorine (nonmetal)
                              name modified to the stem *chlor-* + *-ide*
Name of compound: Sodium chloride

Stems of the more common negative-ion-forming elements are shown in Table 6.2. Table 6.3 lists some compounds with names ending in *-ide.*

Compounds may contain more than one atom of the same element, but as long as they contain only two different elements and only one compound of these two elements exists, the name follows the rules for binary compounds:

Examples:

$CaBr_2$ Mg_3N_2 Li_2O

calcium bromide   magnesium nitride   lithium oxide

Example 6.2   Name the compound CaF_2.

Solution

Step 1. From the formula it is a two-element compound and follows the rules for binary compounds.

Step 2. The compound is composed of Ca, a metal, and F, a nonmetal. Elements in the IIA column form only one type of cation. Thus we name the positive part of the compound *calcium.*

Step 3. Modify the name of the second element to the stem *fluor-* and add the binary ending *-ide* to form the name of the negative part, *fluoride*.

Step 4. The name of the compound is therefore *calcium fluoride*.

Practice 6.2

Write formulas for these compounds:
(a) strontium chloride   (b) potassium iodide   (c) aluminum nitride
(d) calcium sulfide   (e) sodium oxide

▲
The mineral fluorite contains CaF$_2$.

B. Binary Ionic Compounds Containing a Metal That Can Form Two or More Types of Cations

The metals in the center of the periodic table (including the transition metals) often form more than one type of cation. For example, iron can form Fe^{2+} and Fe^{3+} ions, and copper can form Cu^+ and Cu^{2+} ions. This can be confusing when you are naming compounds. For example, copper chloride could be $CuCl_2$ or $CuCl$. To

TABLE 6.2   Examples of Elements Forming Anions

Symbol	Element	Stem	Anion name
Br	bromine	brom	bromide
Cl	chlorine	chlor	chloride
F	fluorine	fluor	fluoride
H	hydrogen	hydr	hydride
I	iodine	iod	iodide
N	nitrogen	nitr	nitride
O	oxygen	ox	oxide
P	phosphorus	phosph	phosphide
S	sulfur	sulf	sulfide

TABLE 6.3   Examples of Compounds with Names Ending in -ide

Formula	Name	Formula	Name
AlCl$_3$	aluminum chloride	BaS	barium sulfide
Al$_2$O$_3$	aluminum oxide	LiI	lithium iodide
CaC$_2$	calcium carbide	MgBr$_2$	magnesium bromide
HCl	hydrogen chloride	NaH	sodium hydride
HI	hydrogen iodide	Na$_2$O	sodium oxide

Stock System

resolve this difficulty the IUPAC devised a system, known as the **Stock System,** to name these compounds. This system is currently recognized as the official system to name these compounds, although another older system is sometimes used. In the Stock System, when a compound contains a metal that can form more than one type of cation, the charge on the cation of the metal is designated by a Roman numeral placed in parentheses immediately following the name of the metal. The negative element is treated in the usual manner for binary compounds.

To name these compounds:
1. **Write the name of the cation.**
2. **Write the charge on the cation as a Roman numeral in parentheses.**
3. **Write the stem of the anion with the suffix -ide.**

Cation charge	+1	+2	+3	+4	+5
Roman numeral	(I)	(II)	(III)	(IV)	(V)

Examples:	FeCl$_2$	iron(II) chloride	Fe^{2+}
	FeCl$_3$	iron(III) chloride	Fe^{3+}
	CuCl	copper(I) chloride	Cu$^+$
	CuCl$_2$	copper(II) chloride	Cu^{2+}

The fact that FeCl$_2$ has two chloride ions, each with a -1 charge, establishes that the charge of Fe is $+2$. To distinguish between the two iron chlorides, FeCl$_2$ is named iron(II) chloride and FeCl$_3$ is named iron(III) chloride.

When a metal forms only one possible cation, we need not distinguish one cation from another, so Roman numerals are not needed. Thus we do not say calcium(II) chloride for CaCl$_2$, but rather calcium chloride, since the charge of calcium is understood to be $+2$.

In classical nomenclature, when the metallic ion has only two cation types, the name of the metal (usually the Latin name) is modified with the suffixes *-ous* and *-ic* to distinguish between the two. The lower charge cation is given the *-ous* ending, and the higher one, the *-ic* ending.

Examples:	FeCl$_2$	ferrous chloride	Fe^{2+}	(lower charge cation)
	FeCl$_3$	ferric chloride	Fe^{3+}	(higher charge cation)
	CuCl	cuprous chloride	Cu$^+$	(lower charge cation)
	CuCl$_2$	cupric chloride	Cu^{2+}	(higher charge cation)

Table 6.4 lists some common metals that have more than one type of cation.

In this book we will use only the Stock System.

Notice that the *ous–ic* naming system does not give the charge of the cation of an element but merely indicates that at least two types of cations exist. The Stock System avoids any possible uncertainty by clearly stating the charge on the cation.

Example 6.3   Name the compound FeS.

Solution

Step 1. This compound follows the rules for a binary compound.

Step 2. It is a compound of Fe, a metal, and S, a nonmetal, and Fe is a transition metal that has more than one type of cation. In sulfides, the charge on

TABLE 6.4	Names and Charges of Some Common Metal Ions That Have More Than One Type of Cation	
Formula	Stock System name	Classical name
Cu^{1+}	copper(I)	cuprous
Cu^{2+}	copper(II)	cupric
Hg^{1+} $(Hg_2)^{2+}$	mercury(I)	mercurous
Hg^{2+}	mercury(II)	mercuric
Fe^{2+}	iron(II)	ferrous
Fe^{3+}	iron(III)	ferric
Sn^{2+}	tin(II)	stannous
Sn^{4+}	tin(IV)	stannic
Pb^{2+}	lead(II)	plumbous
Pb^{4+}	lead(IV)	plumbic
As^{3+}	arsenic(III)	arsenous
As^{5+}	arsenic(V)	arsenic
Ti^{3+}	titanium(III)	titanous
Ti^{4+}	titanium(IV)	titanic

the S is -2. Therefore, the charge on Fe must be $+2$, and the name of the positive part of the compound is *iron(II)*.

Step 3. We have already determined that the name of the negative part of the compound will be *sulfide*.

Step 4. The name of FeS is *iron(II) sulfide*.

Practice 6.3

Write the name for each of the following compounds using the Stock System:
(a) PbI_2 (b) SnF_4 (c) Fe_2O_3 (d) CuO

C. Binary Compounds Containing Two Nonmetals

Compounds between nonmetals are molecular, not ionic. Therefore a different system for naming them is used. In a compound formed between two nonmetals, the element that occurs first in this series is written and named first:

Si, B, P, H, C, S, I, Br, N, Cl, O, F

The name of the second element retains the *-ide* ending as though it were an anion. A Latin or Greek prefix (*mono-, di-, tri-,* and so on) is attached to the name of each element to indicate the number of atoms of that element in the molecule. The prefix *mono-* is never used for naming the first element. Some common prefixes and their numerical equivalences follow.

mono = 1	*tetra* = 4	*hepta* = 7	*nona* = 9
di = 2	*penta* = 5	*octa* = 8	*deca* = 10
tri = 3	*hexa* = 6		

Here are some examples of compounds that illustrate this system:

N_2O_3

(di) nitrogen   (tri) oxide

Indicates two   Indicates three
nitrogen atoms   oxygen atoms

To name these compounds:
1. **Write the name for the first element using a prefix if there is more than one atom of this element.**
2. **Write the stem for the second element with the suffix -ide. Use a prefix to indicate the number of atoms for the second element.**

CO	carbon monoxide	N_2O	dinitrogen monoxide
CO_2	carbon dioxide	N_2O_4	dinitrogen tetroxide
PCl_3	phosphorus trichloride	NO	nitrogen monoxide
SO_2	sulfur dioxide	N_2O_3	dinitrogen trioxide
P_2O_5	diphosphorus pentoxide	S_2Cl_2	disulfur dichloride
CCl_4	carbon tetrachloride	S_2F_{10}	disulfur decafluoride

These examples illustrates that we sometimes drop the final *o* (mono) or *a* (penta) of the prefix when the second element is oxygen. This avoids creating a name that is awkward to pronounce. For example, CO is carbon monoxide instead of carbon mon*o*oxide.

Example 6.4

Solution

Name the compound PCl_5.

Step 1. Phosphorus and chlorine are nonmetals, so the rules for naming binary compounds containing two nonmetals apply. Phosphorus is named first. Therefore the compound is a chloride.

Step 2. No prefix is needed for phosphorus because each molecule has only one atom of phosphorus. The prefix *penta-* is used with chloride to indicate the five chlorine atoms. (PCl_3 is also a known compound.)

Step 3. The name for PCl_5 is *phosphorus pentachloride.*

Practice 6.4

Name these compounds:
(a) Cl_2O (b) SO_2 (c) CBr_4 (d) N_2O_5 (e) NH_3

D. Acids Derived from Binary Compounds

Certain binary hydrogen compounds, when dissolved in water, form solutions that have *acid* properties. Because of this property, these compounds are given acid names in addition to their regular *-ide* names. For example, HCl is a gas and is called *hydrogen chloride,* but its water solution is known as *hydrochloric acid.* Binary acids are composed of hydrogen and one other nonmetallic element. However, not all binary hydrogen compounds are acids. To express the formula of a binary acid, it's customary to write the symbol of hydrogen first, followed by the symbol of the second element (e.g., HCl, HBr, H_2S). When we see formulas such as CH_4 or NH_3, we understand that these compounds are not normally considered to be acids.

TABLE 6.5	Names and Formulas of Selected Binary Acids		
Formula	Acid name	Formula	Acid name
HF	Hydrofluoric acid	HI	Hydroiodic acid
HCl	Hydrochloric acid	H_2S	Hydrosulfuric acid
HBr	Hydrobromic acid	H_2Se	Hydroselenic acid

To name a binary acid, place the prefix *hydro-* in front of, and the suffix *-ic* after, the stem of the nonmetal name. Then add the word *acid.*

HCl	H_2S

*Examples: Hydro-*chlor-*ic acid*   *Hydro-*sulfur-*ic acid*
  (hydrochloric acid)   (hydrosulfuric acid)

Acids are hydrogen-containing substances that liberate hydrogen ions when dissolved in water. The same formula is often used to express binary hydrogen compounds, such as HCl, regardless of whether or not they are dissolved in water. Table 6.5 shows several examples of binary acids.

Naming binary compounds is summarized in Figure 6.4.

To name these compounds:
1. **Write the prefix hydro- with the stem of the second element and add the suffix -ic.**
2. **Write the word acid.**

◀ FIGURE 6.4
Flow diagram for naming binary compounds.

Practice 6.5

Name these binary compounds:
(a) KBr (b) Ca_3N_2 (c) SO_3 (d) SnF_2 (e) $CuCl_2$ (f) N_2O_4

6.5   Naming Compounds Containing Polyatomic Ions

polyatomic ion

A **polyatomic ion** is an ion that contains two or more elements. Compounds containing polyatomic ions are composed of three or more elements and usually consist of one or more cations combined with a negative polyatomic ion. In general, naming compounds containing polyatomic ions is similar to naming binary compounds. The cation is named first, followed by the name for the negative polyatomic ion.

To name these compounds you must learn to recognize the common polyatomic ions (Table 6.6) and know their charges. Consider the formula $KMnO_4$. You must be able to recognize that it consists of two parts $KMnO_4$. These parts are composed of a K^+ ion and a MnO_4^- ion. The correct name for this compound is potassium permanganate. Many polyatomic ions that contain oxygen are called *oxy-anions,* and generally have the suffix *-ate* or *-ite.* Unfortunately, the suffix doesn't indicate the number of oxygen atoms present. The *-ate* form contains more oxygen atoms than the *-ite* form. Examples include sulfate (SO_4^{2-}), sulfite (SO_3^{2-}), nitrate (NO_3^-), and nitrite (NO_2^-).

Some elements form more than two different polyatomic ions containing oxygen. To name these ions, prefixes are used in addition to the suffix. To indicate more oxygen than in the *-ate* form, we add the prefix *per-,* which is a short form of *hyper-,* meaning more. The prefix *hypo-,* meaning less (oxygen in this case), is used for the ion containing less oxygen than the *-ite* form. An example of this system is shown for the polyatomic ions containing chlorine and oxygen in Table 6.7. The prefixes are also used with other similar ions, such as iodate (IO_3^-), bromate (BrO_3^-), and phosphate (PO_4^{3-}).

▲ **Potassium permanganate crystals are dark purple.**

To name these compounds:
1. Write the name of the cation.
2. Write the name of the anion.

TABLE 6.6   Names, Formulas, and Charges of Some Common Polyatomic Ions

Name	Formula	Charge	Name	Formula	Charge
Acetate	$C_2H_3O_2^-$	−1	Cyanide	CN^-	−1
Ammonium	NH_4^+	+1	Dichromate	$Cr_2O_7^{2-}$	−2
Arsenate	AsO_4^{3-}	−3	Hydroxide	OH^-	−1
Hydrogen carbonate	HCO_3^-	−1	Nitrate	NO_3^-	−1
Hydrogen sulfate	HSO_4^-	−1	Nitrite	NO_2^-	−1
Bromate	BrO_3^-	−1	Permanganate	MnO_4^-	−1
Carbonate	CO_3^{2-}	−2	Phosphate	PO_4^{3-}	−3
Chlorate	ClO_3^-	−1	Sulfate	SO_4^{2-}	−2
Chromate	CrO_4^{2-}	−2	Sulfite	SO_3^{2-}	−2

Only three of the common negatively charged polyatomic ions do not use the *ate/ite* system. These exceptions are hydroxide (OH^-), hydrogen sulfide (HS^-), and cyanide (CN^-). Care must be taken with these, as their endings can easily be confused with the *-ide* ending for binary compounds (Section 6.4).

There are two common positively charged polyatomic ions as well—the ammonium and the hydronium ions. The ammonium ion (NH_4^+) is frequently found in polyatomic compounds (Section 6.5), whereas the hydronium ion (H_3O^+) is usually seen in aqueous solutions of acids (Chapter 15).

Practice 6.6

Name these compounds:
(a) $NaNO_3$ (b) $Ca_3(PO_4)_2$ (c) KOH (d) Li_2CO_3 (e) $NaClO_3$

Inorganic compounds are also formed from more than three elements (see Table 6.8). In these cases one or more of the ions is often a polyatomic ion. Once you have

All three of these samples (sodium hydrogen carbonate, baking soda, and sodium bicarbonate) are the same chemical, named three different ways.

TABLE 6.7	Oxy-Anions and Oxy-Acids of Chlorine		
Anion formula	Anion name	Acid formula	Acid name
ClO^-	*hypo*chlor*ite*	$HClO$	*hypo*chlor*ous* acid
ClO_2^-	chlor*ite*	$HClO_2$	chlor*ous* acid
ClO_3^-	chlor*ate*	$HClO_3$	chlor*ic* acid
ClO_4^-	*per*chlor*ate*	$HClO_4$	*per*chloric acid

TABLE 6.8	Names of Selected Compounds That Contain More Than One Kind of Positive Ion
Formula	Name of compound
$KHSO_4$	potassium hydrogen sulfate
$Ca(HSO_3)_2$	calcium hydrogen sulfite
NH_4HS	ammonium hydrogen sulfide
$MgNH_4PO_4$	magnesium ammonium phosphate
NaH_2PO_4	sodium dihydrogen phosphate
Na_2HPO_4	disodium hydrogen phosphate
KHC_2O_4	potassium hydrogen oxalate
$KAl(SO_4)_2$	potassium aluminum sulfate
$Al(HCO_3)_3$	aluminum hydrogen carbonate

learned to recognize the polyatomic ions, naming these compounds follows the patterns we have already learned. First identify the ions. Name the cations in the order given, and follow them with the names of the anions. Study the following examples:

Compound	Ions	Name
$NaHCO_3$	Na^+; HCO_3^-	sodium hydrogen carbonate
NaHS	Na^+; HS^-	sodium hydrogen sulfide
$MgNH_4PO_4$	Mg^{2+}; NH_4^+; PO_4^{3-}	magnesium ammonium phosphate
$NaKSO_4$	Na^+; K^+; SO_4^{2-}	sodium potassium sulfate

6.6　Acids

While we will learn much more about acids later (see Chapter 15), it is helpful to be able to recognize and name common acids both in the laboratory and in class. The simplest way to recognize many acids is to know that acid formulas often begin with hydrogen. Naming binary acids was covered in Section 6.4D. Inorganic compounds containing hydrogen, oxygen, and one other element are called *oxy-acids*. The element other than hydrogen or oxygen in these acids is often a nonmetal, but it can also be a metal.

The first step in naming these acids is to determine that the compound in question is really an oxy-acid. The keys to identification are (1) hydrogen is the first element in the compound's formula and (2) the second part of the formula consists of a polyatomic ion containing oxygen.

Hydrogen in an oxy-acid is not specifically designated in the acid name. The presence of hydrogen in the compound is indicated by the use of the word *acid* in the name of the substance. To determine the particular type of acid, the polyatomic ion following hydrogen must be examined. The name of the polyatomic ion is modified in the following manner: (1) *-ate* changes to an *-ic* ending; (2) *-ite* changes to an *-ous* ending. (See Table 6.9.) The compound with the *-ic* ending contains more oxygen than the one with the *-ous* ending. Consider these examples:

H_2SO_4　sulf*ate* ⟶ sulfur*ic* acid (contains 4 oxygens)

H_2SO_3　sulf*ite* ⟶ sulfur*ous* acid (3 oxygens)

HNO_3　nitr*ate* ⟶ nitr*ic* acid (3 oxygens)

HNO_2　nitr*ite* ⟶ nitr*ous* acid (2 oxygens)

The complete system for naming oxy-acids is shown in Table 6.7 for the various acids containing chlorine. Examples of other oxy-acids and their names are shown in Table 6.10.

Naming polyatomic compounds is summarized in Figure 6.5 on page 120. We have now looked at ways of naming a variety of inorganic compounds—binary compounds consisting of a metal and a nonmetal and of two nonmetals, binary acids and polyatomic compounds. These compounds are just a small part of the classified chemical compounds. Most of the remaining classes are in the broad field of organic chemistry under such categories as hydrocarbons, alcohols, ethers, aldehydes, ketones, phenols, and carboxylic acids.

Practice 6.7

Name these compounds:
(a) Cu_2CO_3 (b) $Fe(ClO)_3$ (c) $Sn(C_2H_3O_2)_2$ (d) $HBrO_3$

Practice 6.8

Write formulas for
(a) lead(II) nitrate (b) potassium phosphate (c) mercury(II) cyanide (d) ammonium chromate

TABLE 6.9 Comparison of Acid and Anion Names for Selected Oxy-Acids

Acid	Anion	Acid	Anion
H_2SO_4 Sulfuric acid	SO_4^{2-} Sulfate ion	H_3PO_4 Phosphoric acid	PO_4^{3-} Phosphate ion
H_2SO_3 Sulfurous acid	SO_3^{2-} Sulfite ion	H_3PO_3 Phosphorous acid	PO_3^{3-} Phosphite ion
HNO_3 Nitric acid	NO_3^- Nitrate ion	HIO_3 Iodic acid	IO_3^- Iodate ion
HNO_2 Nitrous acid	NO_2^- Nitrite ion	$HC_2H_3O_2$ Acetic acid	$C_2H_3O_2^-$ Acetate ion
H_2CO_3 Carbonic acid	CO_3^{2-} Carbonate ion	$H_2C_2O_4$ Oxalic acid	$C_2O_4^{2-}$ Oxalate ion

TABLE 6.10 Formulas and Names of Selected Oxy-Acids

Formula	Acid name	Formula	Acid name
H_2SO_3	sulfurous acid	$HC_2H_3O_2$	acetic acid
H_2SO_4	sulfuric acid	$H_2C_2O_4$	oxalic acid
HNO_2	nitrous acid	H_2CO_3	carbonic acid
HNO_3	nitric acid	$HBrO_3$	bromic acid
H_3PO_3	phosphorous acid	HIO_3	iodic acid
H_3PO_4	phosphoric acid	H_3BO_3	boric acid

CHEMISTRY IN ACTION **Charges in Your Life**

Ions are used in living organisms to perform many important functions. For example, electrical neutrality must be maintained both inside and outside the body cells. Within the cell, neutrality is maintained by potassium ions (K^+) and hydrogen phosphate ions (HPO_4^{2-}). Outside the cell in the intercellular fluid the ions responsible for neutrality are sodium (Na^+) and chloride (Cl^-).

Another important ion within organisms is magnesium (Mg^{2+}), found in chlorophyll and used during nerve and muscle activity, as well as in conjunction with certain enzymes. Iron ions (Fe^{2+}) are incorporated within the hemoglobin molecule and are an integral part of the oxygen transport system within the body. Calcium ions (Ca^{2+}) are part of the matrix of both bones and teeth and play a significant role in the clotting of blood.

Ions also have a significant role in the detergent industry. Water is said to be "hard" when it contains relatively high concentrations of Ca^{2+} and Mg^{2+} ions. In solution, soaps combine with these ions to form an insoluble scum. This material forms the common bathtub ring and, in laundry, settles on the clothes

leaving them gray and dingy. Although soaps and detergents are similar in their cleansing actions, detergents are less likely to form this scum. For this reason, many people who live in areas with "hard" water use detergents instead of soaps.

▲
This chloroplast from a corn plant contains Mg^{2+} ions.

▲
Calcium ions (Ca^{2+}) are important in the formation of bones.

FIGURE 6.5 ▶
Flow diagram for naming polyatomic compounds.

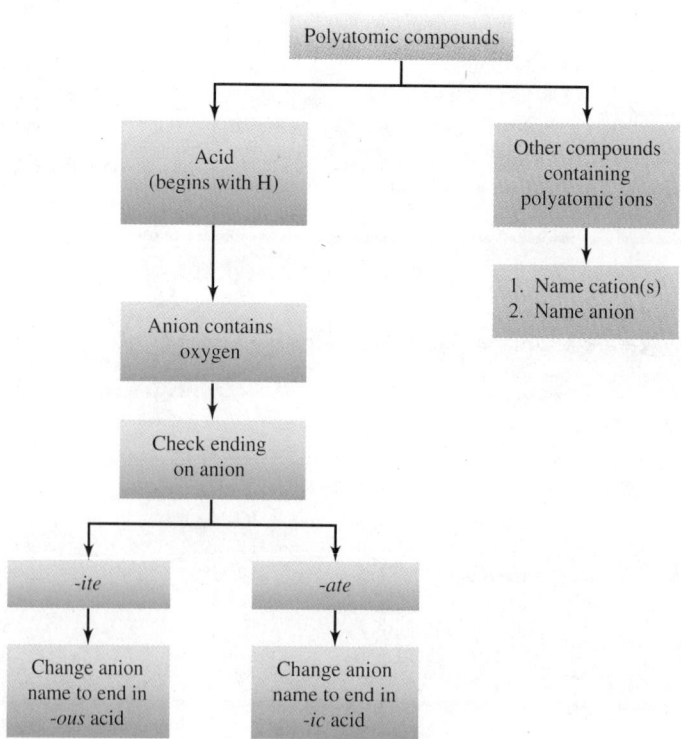

Concepts in Review

1. Write the formulas of compounds formed by combining the ions from Figure 6.2 (or from the inside front cover of this book) in the correct ratios.

2. Write the names or formulas for inorganic binary compounds in which the metal has only one type of cation.

3. Write the names or formulas for inorganic binary compounds that contain metals with multiple types of cations, using the Stock System.

4. Write the names or formulas for inorganic binary compounds that contain two nonmetals.

5. Write the names or formulas for binary acids.

6. Write the names or formulas for oxy-acids.

7. Write the names or formulas for compounds that contain polyatomic ions.

Key Terms

polyatomic ion (6.5)
Stock System (6.4)

Questions

1. Use the common ion table on the inside front cover of your text to determine the formulas for compounds composed of the following ions:
   (a) sodium and chlorate
   (b) hydrogen and sulfate
   (c) tin(II) and acetate
   (d) copper(I) and oxide
   (e) zinc and hydrogen carbonate
   (f) iron(III) and carbonate

2. Does the fact that two elements combine in a one-to-one atomic ratio mean that the charges on their ions are both 1? Explain.

3. Write the names and formulas for the four oxy-acids containing (a) bromine, (b) iodine. (Table 6.7)

4. Explain why N_2O_5 is called dinitrogen pentoxide.

5. Write the formulas for the compounds formed when a chromium(III) ion is combined with the following anions: (a) hydroxide, (b) nitrate, (c) nitrite, (d) hydrogen carbonate, (e) carbonate, (f) dichromate, (g) phosphate, (h) oxalate, (i) oxide, (j) fluoride. (See Table of Names, Formulas, and Charges of Common Ions on the inside front cover.)

6. Explain why the name for $MgCl_2$ is magnesium chloride but the name for $CuCl_2$ is copper(II) chloride.

Paired Exercises

7. Write the formula of the compound that would be formed between these elements:
   (a) Na and I   (d) K and S
   (b) Ba and F   (e) Cs and Cl
   (c) Al and O   (f) Sr and Br

9. Write formulas for the following cations (don't forget to include the charges): sodium, magnesium, aluminum, copper(II), iron(II), iron(III), lead(II), silver, cobalt(II), barium, hydrogen, mercury(II), tin(II), chromium(III), tin(IV), manganese(II), bismuth(III).

8. Write the formula of the compound that would be formed between these elements:
   (a) Ba and O   (d) Be and Br
   (b) H and S   (e) Li and Se
   (c) Al and Cl (f) Mg and P

10. Write formulas for the following anions (don't forget to include the charges): chloride, bromide, fluoride, iodide, cyanide, oxide, hydroxide, sulfide, sulfate, hydrogen sulfate, hydrogen sulfite, chromate, carbonate, hydrogen carbonate, acetate, chlorate, permanganate, oxalate.

11. Complete the table, filling in each box with the proper formula.

Anions

Cations	Br^-	O^{2-}	NO_3^-	PO_4^{3-}	CO_3^{2-}
K^+	KBr				
Mg^{2+}					
Al^{3+}					
Zn^{2+}				$Zn_3(PO_4)_2$	
H^+					

12. Complete the table, filling in each box with the proper formula.

Anions

Cations	SO_4^{2-}	Cl^-	AsO_4^{3-}	$C_2H_3O_2^-$	CrO_4^{2-}
NH_4^+			$(NH_4)_3AsO_4$		
Ca^{2+}					
Fe^{3+}	$Fe_2(SO_4)_3$				
Ag^+					
Cu^{2+}					

13. Write formulas for these binary compounds, all of which are composed of nonmetals:
    (a) carbon monoxide
    (b) sulfur trioxide
    (c) carbon tetrabromide
    (d) phosphorus trichloride
    (e) nitrogen dioxide
    (f) dinitrogen pentoxide
    (g) iodine monobromide
    (h) silicon tetrachloride
    (i) phosphorus pentaiodide
    (j) diboron trioxide

14. Name these binary compounds, all of which are composed of nonmetals:
    (a) CO_2 (f) N_2O_4
    (b) N_2O (g) P_2O_5
    (c) PCl_5 (h) OF_2
    (d) CCl_4 (i) NF_3
    (e) SO_2 (j) CS_2

15. Write formulas for these compounds:
    (a) sodium nitrate
    (b) magnesium fluoride
    (c) barium hydroxide
    (d) ammonium sulfate
    (e) silver carbonate
    (f) calcium phosphate
    (g) potassium nitrite
    (h) strontium oxide

16. Name these compounds:
    (a) K_2O (e) Na_3PO_4
    (b) NH_4Br (f) Al_2O_3
    (c) CaI_2 (g) $Zn(NO_3)_2$
    (d) $BaCO_3$ (h) Ag_2SO_4

17. Name these compounds by the Stock (IUPAC) System:
    (a) $CuCl_2$ (d) $FeCl_3$
    (b) $CuBr$ (e) SnF_2
    (c) $Fe(NO_3)_2$ (f) $HgCO_3$

18. Write formulas for these compounds:
    (a) tin(IV) bromide   (d) mercury(II) nitrite
    (b) copper(I) sulfate   (e) titanium(IV) sulfide
    (c) iron(III) carbonate   (f) iron(II) acetate

19. Write formulas for these acids:
    (a) hydrochloric acid   (d) carbonic acid
    (b) chloric acid   (e) sulfurous acid
    (c) nitric acid   (f) phosphoric acid

20. Write formulas for these acids:
    (a) acetic acid   (d) boric acid
    (b) hydrofluoric acid   (e) nitrous acid
    (c) hypochlorous acid   (f) hydrosulfuric acid

21. Name these acids:
    (a) HNO_2 (d) HBr   (g) HF
    (b) H_2SO_4 (e) H_3PO_3 (h) $HBrO_3$
    (c) $H_2C_2O_4$ (f) $HC_2H_3O_2$

22. Name these acids:
    (a) H_3PO_4 (d) HCl   (g) HI
    (b) H_2CO_3 (e) HClO   (h) $HClO_4$
    (c) HIO_3 (f) HNO_3

23. Write formulas for these compounds:
    (a) silver sulfite
    (b) cobalt(II) bromide
    (c) tin(II) hydroxide
    (d) aluminum sulfate
    (e) manganese(II) fluoride
    (f) ammonium carbonate
    (g) chromium(III) oxide
    (h) copper(II) chloride
    (i) potassium permanganate
    (j) barium nitrite
    (k) sodium peroxide
    (l) iron(II) sulfate
    (m) potassium dichromate
    (n) bismuth(III) chromate

24. Write formulas for these compounds:
    (a) sodium chromate
    (b) magnesium hydride
    (c) nickel(II) acetate
    (d) calcium chlorate
    (e) lead(II) nitrate
    (f) potassium dihydrogen phosphate
    (g) manganese(II) hydroxide
    (h) cobalt(II) hydrogen carbonate
    (i) sodium hypochlorite
    (j) arsenic(V) carbonate
    (k) chromium(III) sulfite
    (l) antimony(III) sulfate
    (m) sodium oxalate
    (n) potassium thiocyanate

25. Write the name of each compound.
  (a) $ZnSO_4$
  (b) $HgCl_2$
  (c) $CuCO_3$
  (d) $Cd(NO_3)_2$
  (e) $Al(C_2H_3O_2)_3$
  (f) CoF_2
  (g) $Cr(ClO_3)_3$
  (h) Ag_3PO_4
  (i) NiS
  (j) $BaCrO_4$

26. Write the name of each compound.
  (a) $Ca(HSO_4)_2$
  (b) $As_2(SO_3)_3$
  (c) $Sn(NO_2)_2$
  (d) $FeBr_3$
  (e) $KHCO_3$
  (f) $BiAsO_4$
  (g) $Fe(BrO_3)_2$
  (h) $(NH_4)_2HPO_4$
  (i) $NaClO$
  (j) $KMnO_4$

27. Write the chemical formula for these substances:
  (a) baking soda   **(e)** vinegar
  (b) lime   **(f)** potash
  (c) epsom salts   **(g)** lye
  (d) muriatic acid

28. Write the chemical formula for these substances:
  (a) fool's gold   **(e)** milk of magnesia
  (b) saltpeter   **(f)** washing soda
  (c) limestone   **(g)** grain alcohol
  (d) cane sugar

Additional Exercises

29. Name these compounds:
  (a) $Ba(NO_3)_2$ **(d)** $MgSO_4$ **(g)** NiS
  (b) $NaC_2H_3O_2$ **(e)** $CdCrO_4$ **(h)** $Sn(NO_3)_4$
  (c) PbI_2 **(f)** $BiCl_3$ **(i)** $Ca(OH)_2$

30. State how each of the following is used in naming inorganic compounds: *ide, ous, ic, hypo, per, ite, ate,* Roman numerals.

31. Translate the following sentences into unbalanced formula chemical equations:
  (a) Silver nitrate and sodium chloride react to form silver chloride and sodium nitrate.
  (b) Iron(III) sulfate and calcium hydroxide react to form iron(III) hydroxide and calcium sulfate.
  (c) Potassium hydroxide and sulfuric acid react to form potassium sulfate and water.

32. How many of each type of subatomic particle (protons and electrons) is in
  (a) an atom of tin?
  (b) an Sn^{2+} ion?
  (c) an Sn^{4+} ion?

33. The compound X_2Y_3 is a stable solid. What ionic charge do you expect for X and Y? Explain.

*34. The ferricyanide ion has the formula $Fe(CN)_6^{3-}$. Write the formula for the compounds that ferricyanide would form with the cations of elements 3, 13, and 30.

35. Compare and contrast the formulas of
  (a) nitride with nitrite
  (b) nitrite with nitrate
  (c) nitrous acid with nitric acid

36. In the beaker below there is a pool of ions. Write all possible formulas for ionic compounds that could form using these ions. Write the name of each compound next to its formula.

Answers to Practice Exercises

6.1 (a) KF; (b) $CaBr_2$; (c) Mg_3N_2; (d) Na_2S; (e) BaO.
6.2 (a) $SrCl_2$; (b) KI; (c) AlN; (d) CaS; (e) Na_2O.
6.3 (a) lead(II) iodide; (b) tin(IV) fluoride; (c) iron(III) oxide; (d) copper(II) oxide.
6.4 (a) dichlorine monoxide; (b) sulfur dioxide; (c) carbon tetrabromide; (d) dinitrogen pentoxide; (e) ammonia.
6.5 (a) potassium bromide; (b) calcium nitride; (c) sulfur trioxide; (d) tin(II) fluoride; (e) copper(II) chloride; (f) dinitrogen tetroxide.

6.6 (a) sodium nitrate; (b) calcium phosphate; (c) potassium hydroxide; (d) lithium carbonate; (e) sodium chlorate.
6.7 (a) copper(I) carbonate; (b) iron(III) hypochlorite; (c) tin(II) acetate; (d) bromic acid.
6.8 (a) $Pb(NO_3)_2$; (b) K_3PO_4; (c) $Hg(CN)_2$; (d) $(NH_4)_2CrO_4$.

PUTTING IT TOGETHER
Review for Chapters 5–6

Multiple Choice: *Choose the correct answer to each of the following.*

1. The concept of positive charge and a small, "heavy" nucleus surrounded by electrons was the contribution of
   (a) Dalton   (c) Thomson
   (b) Rutherford (d) Chadwick

2. The neutron was discovered in 1932 by
   (a) Dalton   (c) Thomson
   (b) Rutherford (d) Chadwick

3. An atom of atomic number 53 and mass number 127 contains how many neutrons?
   (a) 53   (c) 127
   (b) 74   (d) 180

4. How many electrons are in an atom of $^{40}_{18}Ar$?
   (a) 20   (c) 40
   (b) 22   (d) no correct answer given

5. The number of neutrons in an atom of $^{139}_{56}Ba$ is
   (a) 56   (c) 139
   (b) 83   (d) no correct answer given

6. The name of the isotope containing one proton and two neutrons is
   (a) protium   (c) deuterium
   (b) tritium   (d) helium

7. Each atom of a specific element has the same
   (a) number of protons
   (b) atomic mass
   (c) number of neutrons
   (d) no correct answer given

8. Which pair of symbols represents isotopes?
   (a) $^{23}_{11}Na$ and $^{23}_{12}Na$ (c) $^{63}_{29}Cu$ and $^{29}_{64}Cu$
   (b) $^{7}_{3}Li$ and $^{6}_{3}Li$ (d) $^{12}_{24}Mg$ and $^{12}_{26}Mg$

9. Two naturally occurring isotopes of an element have masses and abundance as follows: 54.00 amu (20.00%) and 56.00 amu (80.00%). What is the relative atomic mass of the element?
   (a) 54.20   (c) 54.80
   (b) 54.40   (d) 55.60

10. Substance X has 13 protons, 14 neutrons, and 10 electrons. Determine its identity.
    (a) ^{27}Mg (c) $^{27}Al^{3+}$
    (b) ^{27}Ne (d) ^{27}Al

11. The mass of a chlorine atom is 5.90×10^{-23} g. How many atoms are in a 42.0-g sample of chlorine?
    (a) 2.48×10^{-21} (c) 1.40×10^{-24}
    (b) 7.12×10^{23} (d) no correct answer given

12. The number of neutrons in an atom of $^{108}_{47}Ag$ is
    (a) 47   (c) 155
    (b) 108   (d) no correct answer given

13. The number of electrons in an atom of $^{27}_{13}Al$ is
    (a) 13   (c) 27
    (b) 14   (d) 40

14. The number of protons in an atom of $^{65}_{30}Zn$ is
    (a) 65   (c) 30
    (b) 35   (d) 95

15. The number of electrons in the nucleus of an atom of $^{24}_{12}Mg$ is
    (a) 12   (c) 36
    (b) 24   (d) no correct answer given

Names and Formulas: *In which of the following is the formula correct for the name given?*

1. copper(II) sulfate, $CuSO_4$
2. ammonium hydroxide, NH_4OH
3. mercury(I) carbonate, $HgCO_3$
4. phosphorus triiodide, PI_3
5. calcium acetate, $Ca(C_2H_3O_2)_2$
6. hypochlorous acid, $HClO$
7. dichlorine heptoxide, Cl_2O_7
8. magnesium iodide, MgI
9. sulfurous acid, H_2SO_3
10. potassium manganate, $KMnO_4$
11. lead(II) chromate, $PbCrO_4$
12. ammonium hydrogen carbonate, NH_4HCO_3
13. iron(II) phosphate, $FePO_4$
14. calcium hydrogen sulfate, $CaHSO_4$
15. mercury(II) sulfate, $HgSO_4$
16. dinitrogen pentoxide, N_2O_5
17. sodium hypochlorite, $NaClO$
18. sodium dichromate, $Na_2Cr_2O_7$
19. cadmium cyanide, $Cd(CN)_2$
20. bismuth(III) oxide, Bi_3O_2
21. carbonic acid, H_2CO_3
22. silver oxide, Ag_2O
23. ferric iodide, FeI_2

24. tin(II) fluoride, TiF_2

25. carbon monoxide, CO

26. phosphoric acid, H_3PO_3

27. sodium bromate, Na_2BrO_3

28. hydrosulfuric acid, H_2S

29. potassium hydroxide, POH

30. sodium carbonate, Na_2CO_3

31. zinc sulfate, $ZnSO_3$

32. sulfur trioxide, SO_3

33. tin(IV) nitrate, $Sn(NO_3)_4$

34. ferrous sulfate, $FeSO_4$

35. chloric acid, HCl

36. aluminum sulfide, Al_2S_3

37. cobalt(II) chloride, $CoCl_2$

38. acetic acid, $HC_2H_3O_2$

39. zinc oxide, ZnO_2

40. stannous fluoride, SnF_2

Free Response Questions: *Answer each of the following. Be sure to include your work and explanations in a clear, logical form.*

1. (a) What is an ion?
   (b) The average mass of a calcium atom is 40.08 amu. Why do we also use 40.08 amu as the average mass of a calcium ion (Ca^{2+})?

2. Congratulations! You discover a new element you name wyzzlebium (Wz). The average atomic mass of Wz was found to be 303.001 amu and its atomic number is 120.
   (a) If the masses of the two isotopes of wyzzlebium are 300.9326 amu and 303.9303 amu, what is the relative abundance of each isotope?
   (b) What are the isotopic notations of the two isotopes? (e.g., $_Z^A E$)
   (c) How many neutrons are in one atom of the more abundant isotope?

3. How many protons are in one molecule of dichlorine heptoxide? Is it possible to determine precisely how many electrons and neutrons are in a molecule of dichlorine heptoxide? Why or why not?

4. An unidentified metal forms an ionic compound with phosphate. The metal forms a 2+ cation. If the minimum ratio of protons in the metal to the phosphorus is 6:5, what metal is it? (*Hint:* First write the formula for the ionic compound formed with phosphate anion?)

5. For each of the following compounds, indicate what is wrong with the name and why. If possible, fix the name.
   (a) iron hydroxide
   (b) dipotassium dichromium heptoxide
   (c) sulfur oxide

6. Sulfur dioxide is a gas formed as a by-product of burning coal. Sulfur trioxide is a significant contributor to acid rain. Does the existence of these two substances violate the law of multiple proportions? Explain.

7. (a) Which subatomic particles are not in the nucleus?
   (b) What happens to the size of an atom when it becomes an anion?
   (c) What do an ion of Ca and an atom of Ar have in common?

8. An unidentified atom is found to have an atomic mass 7.18 times that of the carbon-12 isotope.
   (a) What is the mass of the unidentified atom?
   (b) What are the possible identities of this atom?
   (c) Why are you unable to positively identify the element based on the atomic mass and the periodic table?
   (d) If the element formed a compound M_2O, where M is the unidentified atom, identify M by writing the isotopic notation for the atom.

9. Scientists such as Dalton, Thomson, and Rutherford proposed important models, which were ultimately challenged by later technology. What do we know to be false in Dalton's atomic model? What was missing in Thomson's model of the atom? What was Rutherford's experiment that led to the current model of the atom?

Quantitative Composition of Compounds

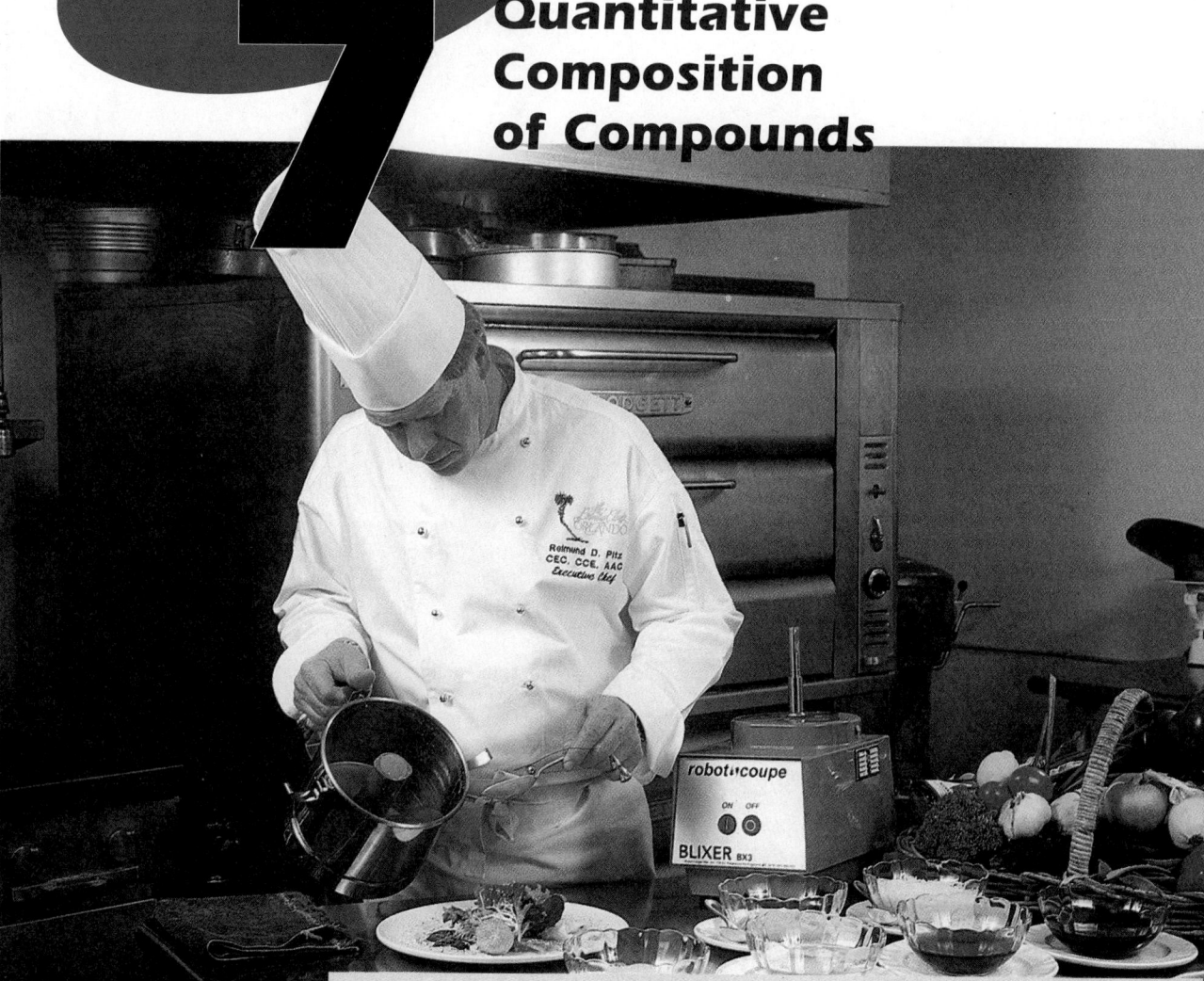

For the chef, preparing gourmet meals is an exercise in combining precise quantities of various ingredients to produce a delightful experience for the patrons.

CHAPTER 7 / OUTLINE

Cereals, cleaning products, and pain remedies all list their ingredients on the package label. The ingredients are listed in order from most to least but the amounts are rarely given. However, it is precisely these amounts that give products their desired properties and distinguish them from the competition. Understandably, manufacturers carefully regulate the amounts of ingredients to maintain quality and hopefully their customers' loyalty. In the medicines we purchase these quantities are especially important for safety reasons—for example, they determine whether a medicine is given to children or is safe only for adults.

The composition of compounds is an important concept in chemistry. Determining numerical relationships among the elements in compounds and measuring exact quantities of particles are fundamental tasks that chemists routinely perform in their daily work.

 The Mole

The atom is an incredibly tiny object. Its mass is far too small to measure on an ordinary balance. In Chapter 5 (Section 5.11), we learned to compare atoms using a table of atomic mass units. These units are valuable when we compare the masses of individual atoms (mentally), but they have no practical use in the laboratory. The mass in grams for an "average" carbon atom (atomic mass 12.00 amu) would be 2.00×10^{-23} g, which is much too tiny for the best laboratory balance.

So how can we confidently measure masses for these very tiny atoms? We increase the number of atoms in a sample until we have an amount large enough to measure on a laboratory balance. The problem then is how to count our sample of atoms.

Consider for a moment the produce in a supermarket. Frequently apples and oranges are sorted by size and then sold by weight, not by the piece of fruit. The grocer is counting by weighing. To do this he needs to know the mass of an "average" apple (235 g) and the mass of an "average" orange (186 g). Now suppose he has an order from the local college for 275 apples and 350 oranges. It would take a long time to count and package this order. The grocer can quickly count fruit by weighing—that is,

$$(275 \text{ apples})\left(\frac{235 \text{ g}}{\text{apple}}\right) = 6.46 \times 10^4 \text{ g} = 64.6 \text{ kg}$$

$$(350 \text{ oranges})\left(\frac{186 \text{ g}}{\text{orange}}\right) = 6.51 \times 10^4 \text{ g} = 65.1 \text{ kg}$$

He can now weigh 64.6 kg of apples and 65.1 kg of oranges and pack them without actually counting them. Manufacturers and suppliers often count by weighing. Other examples of counting by weighing include nuts, bolts, and candy.

Chemists also count atoms by weighing. We know the "average" masses of atoms so we can count atoms by defining a unit to represent a larger number of

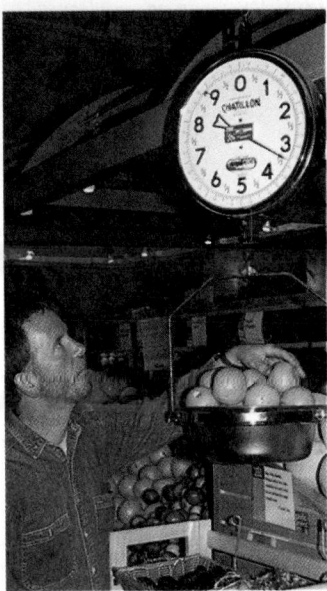

▲
Oranges can be "counted" by weighing them in the store.

Units of measurement need ▶ to be appropriate for the object being measured. (a) Eggs are measured by the dozen, (b) paper is measured by the ream (500 sheets), and (c) pencils are measured by the gross (144).

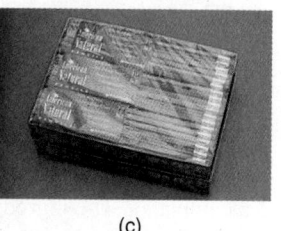

(a)   (b)   (c)

atoms. Chemists have chosen the mole (mol) as the unit for counting atoms. The mole is a unit for counting just as a dozen or a ream or a gross is used to count:

$$1 \text{ dozen} = 12 \text{ objects}$$
$$1 \text{ ream} = 500 \text{ objects}$$
$$1 \text{ gross} = 144 \text{ objects}$$
$$1 \text{ mole} = 6.022 \times 10^{23} \text{ objects}$$

Note that we use a unit only when it is appropriate. A dozen eggs is practical in our kitchen, a gross might be practical for a restaurant, but a ream of eggs would not be very practical. Chemists can't use dozens, grosses, or reams because atoms are so tiny that a dozen, gross, or ream of atoms still couldn't be measured in the laboratory.

Avogadro's number

The number represented by 1 mol, 6.022×10^{23}, is called **Avogadro's number** in honor of Amadeo Avogadro (1776–1856), who investigated several quantitative aspects in chemistry. It's difficult to imagine how large Avogadro's number really is, but this example may help: If 10,000 people started to count Avogadro's number and each counted at the rate of 100 numbers per minute each minute of the day, it would take them over 1 trillion (10^{12}) years to count the total number. Even the tiniest amount of matter contains extremely large numbers of atoms.

Remember that ^{12}C is the reference isotope for atomic masses.

Avogadro's number has been experimentally determined by several methods. How does it relate to atomic mass units? Remember that the atomic mass for an element is the average relative mass of all the isotopes for the elements. The atomic mass (expressed in grams) of 1 mole of any element contains the same number of particles (Avogadro's number) as there are in exactly 12 g of ^{12}C. Thus 1 **mole** of anything is the amount of the substance that contains the same number of items as there are atoms in exactly 12 g of ^{12}C.

mole

$$1 \text{ mole} = 6.022 \times 10^{23} \text{ items}$$

From the definition of mole, we can see that the atomic mass in grams of any element contains 1 mol of atoms. The term *mole* is so commonplace in chemistry that chemists use it as freely as the words *atom* or *molecule*. A mole of atoms, molecules, ions, or electrons represent Avogadro's number of these particles. If we can speak of a mole of atoms, we can also speak of a mole of molecules, a mole of electrons, or a mole of ions, understanding that in each case we mean 6.022×10^{23} particles.

$$1 \text{ mol of atoms} = 6.022 \times 10^{23} \text{ atoms}$$
$$1 \text{ mol of molecules} = 6.022 \times 10^{23} \text{ molecules}$$
$$1 \text{ mol of ions} = 6.022 \times 10^{23} \text{ ions}$$

The atomic mass of an element in grams contains Avogadro's number of atoms and is defined as the **molar mass** of that element. To determine molar mass, we change the units of the atomic mass (found in the periodic table) from atomic mass units to grams. For example, sulfur has an atomic mass of 32.07 amu, so 1 mol of sulfur has a molar mass of 32.07 g and contains 6.022×10^{23} atoms of sulfur. Here are some other examples:

molar mass

Element	Atomic mass	Molar mass	Number of atoms
H	1.008 amu	1.008 g	6.022×10^{23}
Mg	24.31 amu	24.31 g	6.022×10^{23}
Na	22.99 amu	22.99 g	6.022×10^{23}

To summarize:

1. The atomic mass expressed in grams is the *molar mass* of an element. It is different for each element. In this text, molar masses are expressed to four significant figures.

    1 molar mass = atomic mass of element in grams

2. One mole of any element contains Avogadro's number of atoms.

    1 mol of atoms = 6.022×10^{23} atoms

We can use these relationships to make conversions between number of atoms, mass, and moles, as shown in the following examples.

How many moles of iron does 25.0 g of iron, Fe, represent?

Example 7.1

We need to change grams of Fe to moles of Fe. The atomic mass of Fe (from either the periodic table or the table of atomic masses) is 55.85. Use the proper conversion factor to obtain moles:

Solution

$$\text{grams Fe} \longrightarrow \text{moles Fe} \qquad (\text{grams Fe})\left(\frac{1 \text{ mol Fe}}{55.85 \text{ g Fe}}\right)$$

$$(25.0 \text{ g Fe})\left(\frac{1 \text{ mol Fe}}{55.85 \text{ g Fe}}\right) = 0.448 \text{ mol Fe}$$

How many magnesium atoms are contained in 5.00 g of Mg?

Example 7.2

We need to change grams of Mg to atoms of Mg.

Solution

$$\text{grams Mg} \longrightarrow \text{atoms Mg}$$

We find the atomic mass of magnesium to be 24.31 and set up the calculation using a conversion factor between atoms and grams:

$$(\text{grams Mg})\left(\frac{6.022 \times 10^{23} \text{ atoms Mg}}{24.31 \text{ g Mg}}\right)$$

$$(5.00 \text{ g Mg})\left(\frac{6.022 \times 10^{23} \text{ atoms Mg}}{24.31 \text{ g Mg}}\right) = 1.24 \times 10^{23} \text{ atoms Mg}$$

Alternatively we could first convert grams of Mg to moles of Mg, which are then changed to atoms of Mg.

$$\text{grams Mg} \longrightarrow \text{moles Mg} \longrightarrow \text{atoms Mg}$$

Use conversion factors for each step. The calculation setup is

$$(5.00 \text{ g Mg})\left(\frac{1 \text{ mol Mg}}{24.31 \text{ g Mg}}\right)\left(\frac{6.022 \times 10^{23} \text{ atoms Mg}}{1 \text{ mol Mg}}\right) = 1.24 \times 10^{23} \text{ atoms Mg}$$

Thus 1.24×10^{23} atoms of Mg are contained in 5.00 g of Mg.

Example 7.3   What is the mass, in grams, of one atom of carbon, C?

Solution   The molar mass of C is 12.01 g. First we create a conversion factor between molar mass and atoms:

$$\text{atoms C} \longrightarrow \text{grams C}$$

$$(\text{atoms C})\left(\frac{12.01 \text{ g C}}{6.022 \times 10^{23} \text{ atoms C}}\right)$$

$$(1 \text{ atom C})\left(\frac{12.01 \text{ g C}}{6.022 \times 10^{23} \text{ atoms C}}\right) = 1.994 \times 10^{-23} \text{ g C}$$

Example 7.4   What is the mass of 3.01×10^{23} atoms of sodium, Na?

Solution   We need the molar mass of Na (22.99 g) and a conversion factor between molar mass and atoms.

$$\text{atoms Na} \longrightarrow \text{grams Na}$$

$$(\text{atoms Na})\left(\frac{22.99 \text{ g Na}}{6.022 \times 10^{23} \text{ atoms Na}}\right)$$

$$(3.01 \times 10^{23} \text{ atoms Na})\left(\frac{22.99 \text{ g Na}}{6.022 \times 10^{23} \text{ atoms Na}}\right) = 11.5 \text{ g Na}$$

Example 7.5   What is the mass of 0.252 mol of copper (Cu)?

Solution   We need the molar mass of Cu (63.55 g) and a conversion factor between molar mass and moles.

$$\text{moles Cu} \longrightarrow \text{grams Cu}$$

$$(\text{moles Cu})\left(\frac{1 \text{ molar mass Cu}}{1 \text{ mol Cu}}\right)$$

$$(0.252 \text{ mol Cu})\left(\frac{63.55 \text{ g Cu}}{1 \text{ mol Cu}}\right) = 16.0 \text{ g Cu}$$

How many oxygen atoms are present in 1.00 mol of oxygen molecules?

Example 7.6

Solution

Oxygen is a diatomic molecule with the formula O_2. Therefore a molecule of oxygen contains two oxygen atoms.

$$\frac{2 \text{ atoms O}}{1 \text{ molecule O}_2}$$

The sequence of conversions is

$$\text{moles O}_2 \longrightarrow \text{molecules O}_2 \longrightarrow \text{atoms O}$$

The two conversion factors needed are

$$\frac{6.022 \times 10^{23} \text{ molecules O}_2}{1 \text{ mol O}_2} \quad \text{and} \quad \frac{2 \text{ atoms O}}{1 \text{ molecule O}_2}$$

The calculation is

$$(1.00 \text{ mol O}_2)\left(\frac{6.022 \times 10^{23} \text{ molecules O}_2}{1 \text{ mol O}_2}\right)\left(\frac{2 \text{ atoms O}}{1 \text{ molecules O}_2}\right)$$

$$= 1.20 \times 10^{24} \text{ atoms O}$$

Practice 7.1

What is the mass of 2.50 mol of helium, He?

Practice 7.2

How many atoms are present in 0.025 mol of iron?

7.2 Molar Mass of Compounds

One mole of a compound contains Avogadro's number of *formula units* of that compound. The terms *molecular weight*, *molecular mass*, *formula weight*, and *formula mass* have been used in the past to refer to the mass of 1 mol of a compound. However, the term *molar mass* is more inclusive, because it can be used for all types of compounds.

A formula unit is indicated by the formula, e.g., Mg, MgS, H_2O, NaCl.

If the formula of a compound is known, its molar mass can be determined by adding the molar masses of all the atoms in the formula. If more than one atom of any element is present, its mass must be added as many times as it is used.

A mole of table salt (in ▶ front of a salt shaker) and a mole of water (in the film container) have different sizes but both contain Avogadro's number of formula units.

Example 7.7   The formula for water is H_2O. What is its molar mass?

Solution   First we look up the molar masses of H (1.008 g) and O (16.00 g), then we add the masses of all the atoms in the formula unit. Water contains two H atoms and one O atom. Thus

$$2\ H = 2(1.008\ g) = 2.016\ g$$
$$1\ O = 1(16.00\ g) = \underline{16.00\ \ g}$$
$$18.02\ \ g = \text{molar mass of } H_2O$$

Example 7.8   Calculate the molar mass of calcium hydroxide, $Ca(OH)_2$.

Solution   The formula of this substance contains one atom of Ca and two atoms each of O and H. We proceed as in Example 7.7:

$$1\ Ca = 1(40.08\ g) = 40.08\ \ g$$
$$2\ O = 2(16.00\ g) = 32.00\ \ g$$
$$2\ H = 2(1.008\ g) = \underline{\ 2.016\ g}$$
$$74.10\ \ g = \text{molar mass of } Ca(OH)_2$$

In this text we round all molar masses to four significant figures, although you may need to use a different number of significant figures for other work (in the lab).

Practice 7.3

Calculate the molar mass of KNO_3.

The mass of 1 mol of a compound contains Avogadro's number of formula units. Consider the compound hydrogen chloride, HCl. One atom of H combines with one atom of Cl to form HCl. When 1 mol of H (1.008 g of H or 6.022×10^{23} H atoms) combines with 1 mol of Cl (35.45 g of Cl or 6.022×10^{23} Cl atoms), 1 mol of HCl (36.46 g of HCl or 6.022×10^{23} HCl molecules) is produced. These relationships are summarized in the following table:

H	Cl	HCl
6.022×10^{23} H *atoms*	6.022×10^{23} Cl *atoms*	6.022×10^{23} HCl *molecules*
1 mol H *atoms*	1 mol Cl *atoms*	1 mol HCl *molecules*
1.008 g H	35.45 g Cl	36.46 g HCl
1 molar mass H *atoms*	1 molar mass Cl *atoms*	1 molar mass HCl

In dealing with diatomic elements (H_2, O_2, N_2, F_2, Cl_2, Br_2, and I_2), we must take special care to distinguish between a mole of atoms and a mole of molecules. For example consider 1 mol of oxygen molecules, which has a mass of 32.00 g. This quantity is equal to 2 mol of oxygen atoms. Remember that 1 mol represents Avogadro's number of the particular chemical entity that is under consideration.

$$1 \text{ mol } H_2O = 18.02 \text{ g } H_2O \ = 6.022 \times 10^{23} \text{ molecules}$$

$$1 \text{ mol NaCl} = 58.44 \text{ g NaCl} \ = 6.022 \times 10^{23} \text{ formula units}$$

$$1 \text{ mol } H_2 = 2.016 \text{ g } H_2 \ = 6.022 \times 10^{23} \text{ molecules}$$

$$1 \text{ mol } HNO_3 = 63.02 \text{ g } HNO_3 = 6.022 \times 10^{23} \text{ molecules}$$

$$1 \text{ mol } K_2SO_4 = 174.3 \text{ g } K_2SO_4 = 6.022 \times 10^{23} \text{ formula units}$$

Formula units are often used in place of molecules for substances that contain ions.

$$1 \text{ mol} = 6.022 \times 10^{23} \text{ formula units or molecules}$$
$$= 1 \text{ molar mass of a compound}$$

Create the appropriate conversion factor by placing the unit desired in the numerator and the unit to be eliminated in the denominator.

What is the mass of 1 mol of sulfuric acid, H_2SO_4?

Example 7.9

We look up the masses of hydrogen, sulfur, and oxygen and solve in a manner similar to Examples 7.7 and 7.8.

Solution

$$2 \text{ H} = 2(\ 1.008 \text{ g}) = \ 2.016 \text{ g}$$

$$1 \text{ S} = 1(32.07 \ \text{ g}) = 32.07 \ \text{ g}$$

$$4 \text{ O} = 4(16.00 \ \text{ g}) = \underline{64.00 \ \text{ g}}$$

$$98.09 \ \text{ g} = \text{mass of 1 mol of } H_2SO_4$$

Example 7.10   How many moles of sodium hydroxide, NaOH, are there in 1.00 kg of sodium hydroxide?

Solution   First we know that

$$\text{molar mass} = (\overset{Na}{22.99}\text{ g} + \overset{O}{16.00}\text{ g} + \overset{H}{1.008}\text{ g}) \text{ or } 40.00 \text{ g NaOH}$$

To convert grams to moles we use the conversion factor $\dfrac{1 \text{ mol NaOH}}{40.00 \text{ g NaOH}}$ and this conversion sequence:

$$\text{kg NaOH} \longrightarrow \text{g NaOH} \longrightarrow \text{mol NaOH}$$

The calculation is

$$(1.00 \text{ kg NaOH})\left(\frac{1000 \text{ g NaOH}}{\text{kg NaOH}}\right)\left(\frac{1 \text{ mol NaOH}}{40.00 \text{ g NaOH}}\right) = 25.0 \text{ mol NaOH}$$

$$1.00 \text{ kg NaOH} = 25.0 \text{ mol NaOH}$$

▲
Sodium hydroxide pellets gain moisture from the atmosphere, making them difficult to weigh.

Example 7.11   What is the mass of 5.00 mol of water?

Solution   First we know that

$$1 \text{ mol } H_2O = 18.02 \text{ g (Example 7.7)}$$

The conversion is

$$\text{mol } H_2O \longrightarrow \text{g } H_2O$$

To convert moles to grams, we use the conversion factor $\dfrac{18.02 \text{ g } H_2O}{1 \text{ mol } H_2O}$

The calculation is

$$(5.00 \text{ mol } H_2O)\left(\frac{18.02 \text{ g } H_2O}{1 \text{ mol } H_2O}\right) = 90.1 \text{ g } H_2O$$

Example 7.12   How many molecules of hydrogen chloride, HCl, are there in 25.0 g of hydrogen chloride?

Solution   From the formula, we find that the molar mass of HCl is 36.46 g ($\overset{H}{1.008}\text{g} + \overset{Cl}{35.45}\text{g}$). The sequence of conversions is

$$\text{g HCl} \longrightarrow \text{mol HCl} \longrightarrow \text{molecules HCl}$$

Using the conversion factors

$$\frac{1 \text{ mol HCl}}{36.46 \text{ g HCl}} \quad \text{and} \quad \frac{6.022 \times 10^{23} \text{ molecules HCl}}{1 \text{ mol HCl}}$$

$$(25.0 \text{ g HCl})\left(\frac{1 \text{ mol HCl}}{36.46 \text{ g HCl}}\right)\left(\frac{6.022 \times 10^{23} \text{ molecules HCl}}{1 \text{ mol HCl}}\right)$$

$$= 4.13 \times 10^{23} \text{ molecules HCl}$$

Practice 7.4

What is the mass of 0.150 mol of Na_2SO_4?

Practice 7.5

How many moles are there in 500.0 g of $AlCl_3$?

Percent Composition of Compounds

Percent means parts per 100 parts. Just as each piece of pie is a percent of the whole pie, each element in a compound is a percent of the whole compound. The **percent composition of a compound** is the *mass percent* of each element in the compound. The molar mass represents the total mass, or 100%, of the compound. Thus the percent composition of water, H_2O, is 11.19% H and 88.79% O by mass. According to the Law of Definite Composition, the percent composition must be the same no matter what size sample is taken.

percent composition of a compound

The percent composition of a compound can be determined (1) from knowing its formula or (2) from experimental data.

Percent Composition from Formula

If the formula is known, a two-step process is needed to determine the percent composition.

Step 1. Calculate the molar mass (Section 7.2).

Step 2. Divide the total mass of each element in the formula by the molar mass and multiply by 100. This gives the percent composition.

$$\frac{\text{total mass of the element}}{\text{molar mass}} \times 100 = \text{percent of the element}$$

Calculate the percent composition of sodium chloride, NaCl.

Example 7.13

Solution

Step 1. Calculate the molar mass of NaCl:

$$1 \text{ Na} = 1(22.99 \text{ g}) = 22.99 \text{ g}$$

$$1 \text{ Cl} = 1(35.45 \text{ g}) = \underline{35.45 \text{ g}}$$

$$58.44 \text{ g (molar mass)}$$

Step 2. Now calculate the percent composition. We know there are 22.99 g Na and 35.45 g Cl in 58.44 g NaCl.

$$\text{Na: } \left(\frac{22.99 \text{ g Na}}{58.44 \text{ g}} \right)(100) = 39.34\% \text{ Na}$$

$$\text{Cl: } \left(\frac{35.45 \text{ g Cl}}{58.44 \text{ g}} \right)(100) = \underline{60.66\% \text{ Cl}}$$

$$100.00\% \text{ total}$$

In any two-component system, if the percent of one component is known, the other is automatically defined by the difference; that is, if Na is 39.34%, then Cl is 100% − 39.34% = 60.66%. However, the calculation of the percent of each component should be carried out, since this provides a check against possible error. The percent composition data should add up to 100 ± 0.2%.

Example 7.14   Calculate the percent composition of potassium chloride, KCl.

Solution   **Step 1.** Calculate the molar mass of KCl:

$$1 \text{ K } = 1(39.10 \text{ g}) = 39.10 \text{ g}$$
$$1 \text{ Cl} = 1(35.45 \text{ g}) = \underline{35.45 \text{ g}}$$
$$74.55 \text{ g (molar mass)}$$

Step 2. Now calculate the percent composition. We know there are 39.10 g K and 35.45 g Cl in 74.55 g KCl.

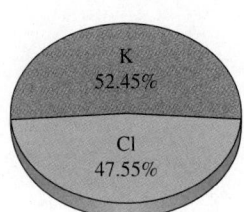

$$\text{K: } \left(\frac{39.10 \text{ g K}}{74.55 \text{ g}} \right)(100) = \quad 52.45\% \text{ K}$$

$$\text{Cl: } \left(\frac{35.45 \text{ g Cl}}{74.55 \text{ g}} \right)(100) = \quad \underline{47.55\% \text{ Cl}}$$

$$100.00\% \text{ total}$$

Comparing the results calculated for NaCl and for KCl, we see that NaCl contains a higher percentage of Cl by mass, although each compound has a one-to-one atom ratio of Cl to Na and Cl to K. The reason for this mass percent difference is that Na and K do not have the same atomic masses.

It is important to realize that, when we compare 1 mol of NaCl with 1 mol of KCl, each quantity contains the same number of Cl atoms—namely, 1 mol of Cl atoms. However, if we compare equal masses of NaCl and KCl, there will be more Cl atoms in the mass of NaCl since NaCl has a higher mass percent of Cl.

1 mol NaCl contains	100.00 g NaCl contains
1 mol Na	39.34 g Na
1 mol Cl	60.66 g Cl
	60.66%Cl

1 mol KCl contains	100.00 g KCl contains
1 mol K	52.45 g K
1 mol Cl	47.55 g Cl
	47.55%Cl

Calculate the percent composition of potassium sulfate, K_2SO_4.

Example 7.15

Solution

Step 1. Calculate the molar mass of K_2SO_4:

$$2\ K = 2(39.10\ g) = \quad 78.20\ g$$

$$1\ S = 1(32.07\ g) = \quad 32.07\ g$$

$$4\ O = 4(16.00\ g) = \underline{\quad 64.00\ g}$$

$$\qquad\qquad\qquad 174.3\quad g\ (molar\ mass)$$

Step 2. Now calculate the percent composition. We know there are 78.20 g of K, 32.07 g of S, and 64.00 g of O in 174.3 g of K_2SO_4.

$$K:\left(\frac{78.20\ g\ K}{174.3\ g}\right)(100) = 44.87\%\ K$$

$$S:\left(\frac{32.07\ g\ S}{174.3\ g}\right)(100) = 18.40\%\ S$$

$$O:\left(\frac{64.00\ g\ O}{174.3\ g}\right)(100) = \underline{36.72\%\ O}$$

$$\qquad\qquad\qquad\qquad 99.99\%\ total$$

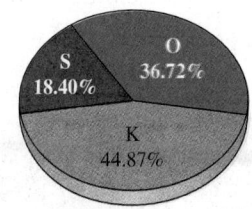

Practice 7.6

Calculate the percent composition of $Ca(NO_3)_2$.

Practice 7.7

Calculate the percent composition of K_2CrO_4.

Percent Composition from Experimental Data

The percent composition can be determined from experimental data without knowing the formula of a compound.

Step 1. Calculate the mass of the compound formed.

Step 2. Divide the mass of each element by the total mass of the compound and multiply by 100.

The Taste of Chemistry

Flavorings, seasonings, and preservatives have been added to foods since ancient times. Spices originally served as preservatives when refrigeration was unavailable. These spices contained mild antiseptics and antioxidants and were effective in prolonging the time during which food could be eaten. Now, a wide variety of food additives are used—preservatives, colorings, flavorings, antioxidants, sweeteners, and so on. Many of these substances are regarded as virtual necessities for processing foods. However, there is genuine concern that some additives, particularly those not naturally present in foods, may be detrimental to our health. This has led to the passage of federal and state laws and regulations.

The U.S. Food and Drug Administration (FDA) is the principal agency charged with enforcing federal laws concerning food and must approve the use of all food additives. The FDA divides food additives into several categories. These include a classification known as GRAS (Generally Regarded As Safe). The GRAS designation includes substances that were in use in 1958 and that met certain specifications for safety. All substances introduced after 1958 have been approved on an individual basis.

Before commercial use of a new food additive, a company must provide the FDA with satisfactory evidence that the chemical is safe for the proposed usage. Providing this evidence of safety is complex and expensive. This task requires the services of many people trained in a variety of disciplines—biochemistry, microbiology, medicine, physiology, and so on. Research and testing may have to be done in several laboratories before final FDA approval (or rejection) is obtained.

The amount of an additive considered to be safe in foods is determined by the maximum tolerable daily intake (MTDI). This is the amount of the food additive that can be eaten daily for a lifetime without adverse effects. It is calculated on the basis of body mass (mg/kg/day). If there is any doubt regarding the safety of an additive, a time limit is determined, and a conditional MTDI is issued and reviewed, with further testing at the end of the initial time period. To establish the MTDI, experiments are run on animals, increasing the quantities of the additive in successive experiments until acute and chronic toxicity occurs. A minimum of two species of animals must be tested, the most sensitive species forming the basis for the appropriate level for the additive. This quantity is then divided by 100 (for additional safety) to set the MTDI.

Once the MTDI is established, all foods in which the use of the additive is proposed must be considered. An estimate is made of the maximum amount of the additive that could be ingested. On the basis of this information, the use may then be restricted to only certain foods (as in the exclusion of sulfites from most meats) or broadened to include additional foods.

Unfortunately, it is extremely difficult to determine safe levels of additives because the majority of toxicological data is from animal testing. Results in animals are often not the same as those in humans—chemicals toxic to one species may not be so for humans.

Children add a further complication. They cannot be considered simply as small adults. A child has a much greater energy demand per kilogram than an adult, and children's chemical defense mechanisms are often much different than those of adults.

◀ **These are a few foods that contain additives.**

When heated in the air, 1.63 g of zinc, Zn, combines with 0.40 g of oxygen, O_2, to form zinc oxide. Calculate the percent composition of the compound formed.

Example 7.16

Step 1. First calculate the total mass of the compound formed.

Solution

$$\begin{array}{l} 1.63 \text{ g Zn} \\ \underline{0.40 \text{ g } O_2} \\ 2.03 \text{ g} \quad = \text{total mass of product} \end{array}$$

Step 2. Divide the mass of each element by the total mass (2.03 g) and multiply by 100.

$$\left(\frac{1.63 \text{ g Zn}}{2.03 \text{ g}}\right)(100) = 80.3\% \text{ Zn} \qquad \left(\frac{0.40 \text{ g O}}{2.03 \text{ g}}\right)(100) = 19.7\% \text{ O}$$

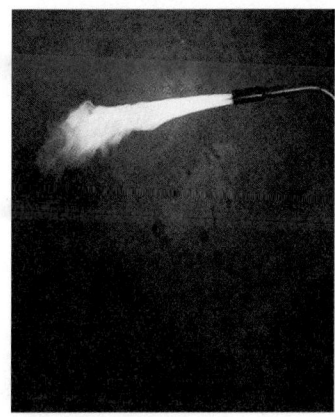

The compound formed contains 80.3% Zn and 19.7% O.

Practice 7.8

Aluminum chloride is formed by reacting 13.43 g aluminum with 53.18 g chlorine. What is the percent composition of the compound?

7.4   Empirical Formula versus Molecular Formula

The **empirical formula,** or *simplest formula,* gives the smallest whole-number ratio of atoms present in a compound. This formula gives the relative number of atoms of each element in the compound.

empirical formula

The **molecular formula** is the true formula, representing the total number of atoms of each element present in one molecule of a compound. It is entirely possible that two or more substances will have the same percent composition yet be distinctly different compounds. For example, acetylene, C_2H_2, is a common gas used in welding; benzene, C_6H_6, is an important solvent obtained from coal tar and is used in the synthesis of styrene and nylon. Both acetylene and benzene contain 92.3% C and 7.7% H. The smallest ratio of C and H corresponding to these percentages is CH (1:1). Therefore the *empirical* formula for both acetylene and benzene is CH, even though the *molecular* formulas are C_2H_2 and C_6H_6, respectively. Often the molecular formula is the same as the empirical formula. If the molecular formula is not the same, it will be an integral (whole number) multiple of the empirical formula. For example,

molecular formula

CH = empirical formula

$(CH)_2 = C_2H_2$ = acetylene   (molecular formula)

$(CH)_6 = C_6H_6$ = benzene   (molecular formula)

Table 7.1 compares the formulas of these substances. Table 7.2 shows empirical and molecular formula relationships of other compounds.

▲
Acetylene is used as a fuel in welding torches.

TABLE 7.1	Molecular Formulas of Two Compounds Having an Empirical Formula with a 1:1 Ratio of Carbon and Hydrogen Atoms		

| | Composition | | |
Formula	% C	% H	Molar mass
CH (empirical)	92.3	7.7	13.02 (empirical)
C_2H_2 (acetylene)	92.3	7.7	26.04 (2×13.02)
C_6H_6 (benzene)	92.3	7.7	78.12 (6×13.02)

TABLE 7.2	Some Empirical and Molecular Formulas					

Compound	Empirical formula	Molecular formula	Compound	Empirical formula	Molecular formula
Acetylene	CH	C_2H_2	Diborane	BH_3	B_2H_6
Benzene	CH	C_6H_6	Hydrazine	NH_2	N_2H_4
Ethylene	CH_2	C_2H_4	Hydrogen	H	H_2
Formaldehyde	CH_2O	CH_2O	Chlorine	Cl	Cl_2
Acetic acid	CH_2O	$C_2H_4O_2$	Bromine	Br	Br_2
Glucose	CH_2O	$C_6H_{12}O_6$	Oxygen	O	O_2
Hydrogen chloride	HCl	HCl	Nitrogen	N	N_2
Carbon dioxide	CO_2	CO_2			

7.5   Calculating Empirical Formulas

We can establish empirical formulas because (1) individual atoms in a compound are combined in whole-number ratios, and (2) each element has a specific atomic mass.

To calculate an empirical formula, we need to know (1) the elements that are combined, (2) their atomic masses, and (3) the ratio by mass or percentage in which they are combined. If elements A and B form a compound, we may represent the empirical formula as A_xB_y, where x and y are small whole numbers that represent the atoms of A and B. To write the empirical formula we must determine x and y:

Step 1. Assume a definite starting quantity (usually 100.0 g) of the compound, if not given, and express the mass of each element in grams.

Step 2. Convert the grams of each element into moles using each element's molar mass. This conversion gives the number of moles of atoms of each element in the quantity assumed in Step 1. At this point these numbers will usually not be whole numbers.

Step 3. Divide each value obtained in Step 2 by the smallest of these values. If the numbers obtained are whole numbers, use them as subscripts and write the empirical formula. If the numbers obtained are not whole numbers, go on to Step 4.

Step 4. Multiply the values obtained in Step 3 by the smallest number that will convert them to whole numbers. Use these whole numbers as the subscripts in the empirical formula. For example, if the ratio of A to B is 1.0:1.5, multiply both numbers by 2 to obtain a ratio of 2:3. The empirical formula then is A_2B_3.

In many of these calculations results will vary somewhat from an exact whole number; this can be due to experimental errors in obtaining the data or from rounding off numbers. Calculations that vary by no more than ± 0.1 from a whole number usually are rounded off to the nearest whole number. Deviations greater than about 0.1 unit usually mean that the calculated ratios need to be multiplied by a factor to make them all whole numbers. For example, an atom ratio of 1:1.33 should be multiplied by 3 to make the ratio 3:4.

> **Some common fractions and their decimal equivalents are**
> $\frac{1}{4} = 0.25$
> $\frac{1}{3} = 0.333\ldots$
> $\frac{2}{3} = 0.666\ldots$
> $\frac{1}{2} = 0.5$
> $\frac{3}{4} = 0.75$
> **Multiply the decimal equivalent by the number in the denominator of the fraction to get a whole number: 4(0.75) = 3.**

Calculate the empirical formula of a compound containing 11.19% hydrogen, H, and 88.79% oxygen, O.

 Example 7.17

Solution

Step 1. Express each element in grams. Assuming 100.00 g of material, the percent of each element equals the grams of each element, and the percent sign can be omitted:

H = 11.19 g

O = 88.79 g

Step 2. Convert the grams of each element to moles:

H: $(11.19\ \cancel{g\,H})\left(\dfrac{1\ \text{mol H atoms}}{1.008\ \cancel{g\,H}}\right) = 11.10\ \text{mol H atoms}$

O: $(88.79\ \cancel{g\,O})\left(\dfrac{1\ \text{mol O atoms}}{16.00\ \cancel{g\,O}}\right) = 5.549\ \text{mol O atoms}$

The formula could be expressed as $H_{11.10}O_{5.549}$. However, it's customary to use the smallest whole-number ratio of atoms. This ratio is calculated in Step 3.

Step 3. Change these numbers to whole numbers by dividing them by the smaller number.

$H = \dfrac{11.10\ \text{mol}}{5.549\ \text{mol}} = 2.000 \qquad O = \dfrac{5.549\ \text{mol}}{5.549\ \text{mol}} = 1.000$

In this step the ratio of atoms has not changed, because we divided the number of moles of each element by the same number.

The simplest ratio of H to O is 2:1.
Empirical formula = H_2O

Example 7.18   The analysis of a salt shows that it contains 56.58% potassium, K; 8.68% carbon, C; and 34.73% oxygen, O. Calculate the empirical formula for this substance.

Solution

Steps 1 and 2. After changing the percentage of each element to grams, we find the relative number of moles of each element by multiplying by the proper mole/molar mass factor:

$$K: (56.58 \text{ g K})\left(\frac{1 \text{ mol K atoms}}{39.10 \text{ g K}}\right) = 1.447 \text{ mol K atoms}$$

$$C: (8.68 \text{ g C})\left(\frac{1 \text{ mol C atoms}}{12.01 \text{ g C}}\right) = 0.723 \text{ mol C atoms}$$

$$O: (34.73 \text{ g O})\left(\frac{1 \text{ mol O atoms}}{16.00 \text{ g O}}\right) = 2.171 \text{ mol O atoms}$$

Step 3. Divide each number of moles by the smallest value:

$$K = \frac{1.447 \text{ mol}}{0.723 \text{ mol}} = 2.00$$

$$C = \frac{0.723 \text{ mol}}{0.723 \text{ mol}} = 1.00$$

$$O = \frac{2.171 \text{ mol}}{0.723 \text{ mol}} = 3.00$$

The simplest ratio of K:C:O is 2:1:3.
Empirical formula = K_2CO_3

Example 7.19   A sulfide of iron was formed by combining 2.233 g of iron, Fe, with 1.926 g of sulfur, S. What is the empirical formula of the compound?

Solution

Steps 1 and 2. The grams of each element are given, so we use them directly in our calculations. Calculate the relative number of moles of each element by multiplying the grams of each element by the proper mole/molar mass factor:

$$Fe: (2.233 \text{ g Fe})\left(\frac{1 \text{ mol Fe atoms}}{55.85 \text{ g Fe}}\right) = 0.03998 \text{ mol Fe atoms}$$

$$S: (1.926 \text{ g S})\left(\frac{1 \text{ mol S atoms}}{32.07 \text{ g S}}\right) = 0.06006 \text{ mol S atoms}$$

Step 3. Divide each number of moles by the smaller of the two numbers:

$$Fe = \frac{0.03998 \text{ mol}}{0.03998 \text{ mol}} = 1.000 \qquad S = \frac{0.06006 \text{ mol}}{0.03998 \text{ mol}} = 1.502$$

Step 4. We still have not reached a ratio that gives whole numbers in the formula so we double each value to obtain a ratio of 2.000 Fe atoms to

3.000 S atoms. Doubling both values does not change the ratio of Fe and S atoms.

Fe: $(1.000)2 = 2.000$

S: $(1.502)2 = 3.004$

Empirical formula $= Fe_2S_3$

Practice 7.9

Calculate the empirical formula of a compound containing 52.14% C, 13.12% H, and 34.73% O.

Practice 7.10

Calculate the empirical formula of a compound that contains 43.7% phosphorus and 56.3% O by mass.

Calculating the Molecular Formula from the Empirical Formula

The molecular formula can be calculated from the empirical formula if the molar mass is known. The molecular formula, as stated in Section 7.4, will be equal either to the empirical formula or some multiple of it. For example, if the empirical formula of a compound of hydrogen and fluorine is HF, the molecular formula can be expressed as $(HF)_n$, where $n = 1, 2, 3, 4, \ldots$. This n means that the molecular formula could be HF, H_2F_2, H_3F_3, H_4F_4, and so on. To determine the molecular formula, we must evaluate n.

$$n = \frac{\text{molar mass}}{\text{mass of empirical formula}} = \text{number of empirical formula units}$$

What we actually calculate is the number of units of the empirical formula contained in the molecular formula.

A compound of nitrogen and oxygen with a molar mass of 92.00 g was found to have an empirical formula of NO_2. What is its molecular formula?   **Example 7.20**

Let n be the number of NO_2 units in a molecule; then the molecular formula is   **Solution**
$(NO_2)_n$. Each NO_2 unit has a mass of $[14.01 \text{ g} + 2(16.00 \text{ g})]$ or 46.01 g. The molar mass of $(NO_2)_n$ is 92.00 g and the number of 46.01 units in 92.00 is 2:

$$n = \frac{92.00 \text{ g}}{46.01 \text{ g}} = 2 \quad \text{(empirical formula units)}$$

The molecular formula is $(NO_2)_2$ or N_2O_4.

Example 7.21   The hydrocarbon propylene has a molar mass of 42.00 g and contains 14.3% H and 85.7% C. What is its molecular formula?

Solution   First find the empirical formula:

$$\text{C:} \quad (85.7 \ \cancel{g\ C})\left(\frac{1 \ \text{mol C atoms}}{12.01 \ \cancel{g\ C}}\right) = 7.14 \ \text{mol C atoms}$$

$$\text{H:} \quad (14.3 \ \cancel{g\ H})\left(\frac{1 \ \text{mol H atoms}}{1.008 \ \cancel{g\ H}}\right) = 14.2 \ \text{mol H atoms}$$

▲
When propylene is polymerized into polypropylene it can be used for rope and other recreational products.

Divide each value by the smaller number of moles:

$$C = \frac{7.14 \ \text{mol}}{7.14 \ \text{mol}} = 1.00$$

$$H = \frac{14.2 \ \text{mol}}{7.14 \ \text{mol}} = 1.99$$

Empirical formula $= CH_2$

We determine the molecular formula from the empirical formula and the molar mass:

Molecular formula $= (CH_2)_n$

Molar mass $= 42.00$ g

Each CH_2 unit has a mass of (12.01 g + 2.016 g) or 14.03 g. The number of CH_2 units in 42.00 g is 3:

$$n = \frac{42.00 \ g}{14.03 \ g} = 3 \quad \text{(empirical formula units)}$$

The molecular formula is $(CH_2)_3$, or C_3H_6.

Practice 7.11

Calculate the empirical and molecular formulas of a compound that contains 80.0% C, 20.0% H, and has a molar mass of 30.00 g.

Concepts in Review

1. Explain the meaning of the mole.
2. Discuss the relationship between a mole and Avogadro's number.
3. Convert grams, atoms, molecules, and molar masses to moles, and vice versa.
4. Determine the molar mass of a compound from its formula.
5. Calculate the percent composition of a compound from its formula.

6. Calculate the percent composition of a compound from experimental data.
7. Explain the relationship between an empirical formula and a molecular formula.
8. Determine the empirical formula for a compound from its percent composition.
9. Calculate the molecular formula of a compound from its percent composition and molar mass.

Key Terms

Avogadro's number (7.1) molecular formula (7.4)
empirical formula (7.4) percent composition of a compound (7.3)
molar mass (7.1)
mole (7.1)

Questions

1. What is a mole?

2. Which would have a higher mass: a mole of K atoms or a mole of Au atoms?

3. Which would contain more atoms: a mole of K atoms or a mole of Au atoms?

4. Which would contain more electrons: a mole of K atoms or a mole of Au atoms?

*5. If the atomic mass scale had been defined differently, with an atom of $^{12}_{6}C$ defined as a mass of 50 amu, would this have any effect on the value of Avogadro's number? Explain.

6. What is the numerical value of Avogadro's number?

7. What is the relationship between Avogadro's number and the mole?

8. Complete these statements, supplying the proper quantity.
   (a) A mole of O atoms contains _____ atoms.
   (b) A mole of O_2 molecules contains _____ molecules.
   (c) A mole of O_2 molecules contains _____ atoms.
   (d) A mole of O atoms has a mass of _____ grams.
   (e) A mole of O_2 molecules has a mass of _____ grams.

9. How many molecules are present in 1 molar mass of sulfuric acid (H_2SO_4)? How many atoms are present?

10. In calculating the empirical formula of a compound from its percent composition, why do we choose to start with 100.0 g of the compound?

Paired Exercises

Molar Masses

11. Determine the molar masses of these compounds:
    (a) KBr   (f) Fe_3O_4
    (b) Na_2SO_4 (g) $C_{12}H_{22}O_{11}$
    (c) $Pb(NO_3)_2$ (h) $Al_2(SO_4)_3$
    (d) C_2H_5OH (i) $(NH_4)_2HPO_4$
    (e) $HC_2H_3O_2$

12. Determine the molar masses of these compounds:
    (a) NaOH   (f) C_6H_5COOH
    (b) Ag_2CO_3 (g) $C_6H_{12}O_6$
    (c) Cr_2O_3 (h) $K_4Fe(CN)_6$
    (d) $(NH_4)_2CO_3$ (i) $BaCl_2 \cdot 2 H_2O$
    (e) $Mg(HCO_3)_2$

Moles and Avogadro's Number

13. How many moles of atoms are contained in the following?
    (a) 22.5 g Zn
    (b) 0.688 g Mg
    (c) 4.5×10^{22} atoms Cu
    (d) 382 g Co
    (e) 0.055 g Sn
    (f) 8.5×10^{24} molecules N_2

14. How many moles are contained in the following?
    (a) 25.0 g NaOH
    (b) 44.0 g Br_2
    (c) 0.684 g $MgCl_2$
    (d) 14.8 g CH_3OH
    (e) 2.88 g Na_2SO_4
    (f) 4.20 lb ZnI_2

15. Calculate the number of grams in each of the following:
    (a) 0.550 mol Au
    (b) 15.8 mol H_2O
    (c) 12.5 mol Cl_2
    (d) 3.15 mol NH_4NO_3

16. Calculate the number of grams in each of the following:
    (a) 4.25×10^{-4} mol H_2SO_4
    (b) 4.5×10^{22} molecules CCl_4
    (c) 0.00255 mol Ti
    (d) 1.5×10^{16} atoms S

17. How many molecules are contained in each of the following:
    (a) 1.26 mol O_2
    (b) 0.56 mol C_6H_6
    (c) 16.0 g CH_4
    (d) 1000. g HCl

18. How many molecules are contained in each of the following:
    (a) 1.75 mol Cl_2
    (b) 0.27 mol C_2H_6O
    (c) 12.0 g CO_2
    (d) 100. g CH_4

19. Calculate the mass in grams of each of the following:
    (a) 1 atom Pb
    (b) 1 atom Ag
    (c) 1 molecule H_2O
    (d) 1 molecule $C_3H_5(NO_3)_3$

20. Calculate the mass in grams of each of the following:
    (a) 1 atom Au
    (b) 1 atom U
    (c) 1 molecule NH_3
    (d) 1 molecule $C_6H_4(NH_2)_2$

21. Make the following conversions:
    (a) 8.66 mol Cu to grams Cu
    (b) 125 mol Au to kilograms Au
    (c) 10 atoms C to moles C
    (d) 5000 molecules CO_2 to moles CO_2

22. Make the following conversions:
    (a) 28.4 g S to moles S
    (b) 2.50 kg NaCl to moles NaCl
    (c) 42.4 g Mg to atoms Mg
    (d) 485 mL Br_2 (d = 3.12 g/mL) to moles Br_2

23. Exactly 1 mol of carbon disulfide contains
    (a) how many carbon disulfide molecules?
    (b) how many carbon atoms?
    (c) how many sulfur atoms?
    (d) how many total atoms of all kinds?

24. One mole of ammonia contains
    (a) how many ammonia molecules?
    (b) how many nitrogen atoms?
    (c) how many hydrogen atoms?
    (d) how many total atoms of all kinds?

25. How many atoms of oxygen are contained in each of the following?
    (a) 16.0 g O_2
    (b) 0.622 mol MgO
    (c) 6.00×10^{22} molecules $C_6H_{12}O_6$

26. How many atoms of oxygen are contained in each of the following?
    (a) 5.0 mol MnO_2
    (b) 255 g $MgCO_3$
    (c) 5.0×10^{18} molecules H_2O

27. Calculate the number of
    (a) grams of silver in 25.0 g AgBr
    (b) grams of nitrogen in 6.34 mol $(NH_4)_3PO_4$
    (c) grams of oxygen in 8.45×10^{22} molecules SO_3

28. Calculate the number of
    (a) grams of chlorine in 5.0 g $PbCl_2$
    (b) grams of hydrogen in 4.50 mol H_2SO_4
    (c) grams of hydrogen in 5.45×10^{22} molecules NH_3

Percent Composition

29. Calculate the percent composition by mass of these compounds:
    (a) NaBr   (d) $SiCl_4$
    (b) $KHCO_3$ (e) $Al_2(SO_4)_3$
    (c) $FeCl_3$ (f) $AgNO_3$

30. Calculate the percent composition by mass of these compounds:
    (a) $ZnCl_2$ (d) $(NH_4)_2SO_4$
    (b) $NH_4C_2H_3O_2$ (e) $Fe(NO_3)_3$
    (c) MgP_2O_7 (f) ICl_3

31. Calculate the percent of iron in the following compounds:
    (a) FeO
    (b) Fe_2O_3
    (c) Fe_3O_4
    (d) $K_4Fe(CN)_6$

32. Which of the following chlorides has the highest and which has the lowest percentage of chlorine, by mass, in its formula?
    (a) KCl
    (b) $BaCl_2$
    (c) $SiCl_4$
    (d) LiCl

33. A 6.20-g sample of phosphorus was reacted with oxygen to form an oxide with a mass of 14.20 g. Calculate the percent composition of the compound.

34. A sample of ethylene chloride was analyzed to contain 6.00 g of C, 1.00 g of H, and 17.75 g of Cl. Calculate the percent composition of ethylene chloride.

35. Examine the following formulas. Which compound has the
    (a) higher percent by mass of hydrogen, H_2O or H_2O_2?
    (b) lower percent by mass of nitrogen, NO or N_2O_3?
    (c) higher percent by mass of oxygen, NO_2 or N_2O_4?
    Check your answers by calculation if you wish.

36. Examine the following formulas. Which compound has the
    (a) lower percent by mass of chlorine, $NaClO_3$ or $KClO_3$?
    (b) higher percent by mass of sulfur, $KHSO_4$ or K_2SO_4?
    (c) lower percent by mass of chromium, Na_2CrO_4 or $Na_2Cr_2O_7$?
    Check your answers by calculation if you wish.

Empirical and Molecular Formulas

37. Calculate the empirical formula of each compound from the percent compositions given:
    (a) 63.6% N, 36.4% O
    (b) 46.7% N, 53.3% O
    (c) 25.9% N, 74.1% O
    (d) 43.4% Na, 11.3% C, 45.3% O
    (e) 18.8% Na, 29.0% Cl, 52.3% O
    (f) 72.02% Mn, 27.98% O

38. Calculate the empirical formula of each compound from the percent compositions given:
    (a) 64.1% Cu, 35.9% Cl
    (b) 47.2% Cu, 52.8% Cl
    (c) 51.9% Cr, 48.1% S
    (d) 55.3% K, 14.6% P, 30.1% O
    (e) 38.9% Ba, 29.4% Cr, 31.7% O
    (f) 3.99% P, 82.3% Br, 13.7% Cl

39. A sample of tin having a mass of 3.996 g was oxidized and found to have combined with 1.077 g of oxygen. Calculate the empirical formula of this oxide of tin.

40. A 3.054-g sample of vanadium (V) combined with oxygen to form 5.454 g of product. Calculate the empirical formula for this compound.

41. Hydroquinone is an organic compound commonly used as a photographic developer. It has a molar mass of 110.1 g/mol and a composition of 65.45% C, 5.45% H, and 29.09% O. Calculate the molecular formula of hydroquinone.

42. Fructose is a very sweet natural sugar that is present in honey, fruits, and fruit juices. It has a molar mass of 180.1 g/mol and a composition of 40.0% C, 6.7% H, and 53.3% O. Calculate the molecular formula of fructose.

Additional Exercises

43. White phosphorus is one of several forms of phosphorus and exists as a waxy solid consisting of P_4 molecules. How many atoms are present in 0.350 mol of P_4?

44. How many grams of sodium contain the same number of atoms as 10.0 g of potassium?

45. One atom of an unknown element is found to have a mass of 1.79×10^{-23} g. What is the molar mass of this element?

*46. If a stack of 500 sheets of paper is 4.60 cm high, what will be the height, in meters, of a stack of Avogadro's number of sheets of paper?

47. There are about 5.0 billion (5.0×10^9) people on earth. If exactly 1 mol of dollars were distributed equally among these people, how many dollars would each person receive?

*48. If 20. drops of water equal 1.0 mL (1.0 cm^3),
    (a) how many drops of water are there in a cubic mile of water?
    (b) what would be the volume in cubic miles of a mole of drops of water?

*49. Silver has a density of 10.5 g/cm^3. If 1.00 mol of silver were shaped into a cube,
    (a) what would be the volume of the cube?
    (b) what would be the length of one side of the cube?

*50. A sulfuric acid solution contains 65.0% H_2SO_4 by mass and has a density of 1.55 g/mL. How many moles of the acid are present in 1.00 L of the solution?

*51. A nitric acid solution containing 72.0% HNO_3 by mass has a density of 1.42 g/mL. How many moles of HNO_3 are present in 100. mL of the solution?

52. Given 1.00-g samples of each of the following compounds, CO_2, O_2, H_2O, and CH_3OH,
    (a) which sample will contain the largest number of molecules?
    (b) which sample will contain the largest number of atoms?
    Show proof for your answers.

53. How many grams of Fe_2S_3 will contain a total number of atoms equal to Avogadro's number?

54. How many grams of lithium will combine with 20.0 g of sulfur to form the compound Li_2S?

55. Calculate the percentage of
(a) mercury in $HgCO_3$
(b) oxygen in $Ca(ClO_3)_2$
(c) nitrogen in $C_{10}H_{14}N_2$ (nicotine)
(d) Mg in $C_{55}H_{72}MgN_4O_5$ (chlorophyll)

*56. Zinc and sulfur react to form zinc sulfide, ZnS. If we mix 19.5 g of zinc and 9.40 g of sulfur, have we added sufficient sulfur to fully react all the zinc? Show evidence for your answer.

57. Aspirin is well known as a pain reliever (analgesic) and as a fever reducer (antipyretic). It has a molar mass of 180.2 g/mol and a composition of 60.0% C, 4.48% H, and 35.5% O. Calculate the molecular formula of aspirin.

58. How many grams of oxygen are contained in 8.50 g of $Al_2(SO_4)_3$?

59. Gallium arsenide is one of the newer materials used to make semiconductor chips for use in supercomputers. Its composition is 48.2% Ga and 51.8% As. What is the empirical formula?

60. The compositions of four different compounds of carbon and chlorine follow. Determine the empirical formula and the molecular formula for each compound.

Percent C	Percent Cl	Molar mass (g)
(a) 7.79	92.21	153.8
(b) 10.13	89.87	236.7
(c) 25.26	74.74	284.8
(d) 11.25	88.75	319.6

61. How many years is a mole of seconds?

62. A normal penny has a mass of about 2.5 g. If we assume the penny to be pure copper (which means the penny is very old since newer pennies are a mixture of copper and zinc), how many atoms of copper does it contain?

63. What would be the mass (in grams) of one thousand trillion molecules of glycerin, $C_3H_8O_3$?

64. If we assume there are 5.0 billion people on the Earth, how many moles of people is this?

65. An experimental catalyst used to make polymers has the following composition: Co, 23.3%; Mo, 25.3%; and Cl, 51.4%. What is the empirical formula for this compound?

66. If a student weighs 18 g of aluminum and needs twice as many atoms of magnesium as she has of aluminum, how many grams of Mg does she need?

*67. If 10.0 g of an unknown compound composed of carbon, hydrogen, and nitrogen contains 17.7% N and 3.8×10^{23} atoms of hydrogen, what is its empirical formula?

68. A substance whose formula is A_2O (A is a mystery element) is 60.0% A and 40.0% O. Identify the element A.

69. For the following compounds whose molecular formulas are given, indicate the empirical formula:
(a) $C_6H_{12}O_6$ glucose
(b) C_8H_{18} octane
(c) $C_3H_6O_3$ lactic acid
(d) $C_{25}H_{52}$ paraffin
(e) $C_{12}H_4Cl_4O_2$ dioxin (a powerful poison)

Answers to Practice Exercises

7.1 10.0 g helium	**7.5** 3.751 mol $AlCl_3$	**7.9** C_2H_6O
7.2 1.5×10^{22} atoms	**7.6** 24.42% Ca; 17.07% N; 58.50% O	**7.10** P_2O_5
7.3 101.1 g KNO_3	**7.7** 40.27% K; 26.78% Cr; 32.96% O	**7.11** The empirical formula is CH_3; the
7.4 21.3 g Na_2SO_4	**7.8** 20.16% Al; 79.84% Cl	molecular formula is C_2H_6.

CHAPTER

8

Chemical Equations

CHAPTER 8 / OUTLINE

▲
The thermite reaction is a reaction between elemental aluminum and iron oxide. This reaction produces so much energy the iron becomes molten. The thermite reaction is used to weld railroad rails.

149

CHAPTER

8

In the world today, we continually strive to express information in a concise, useful manner. From early childhood, we are taught to translate our ideas and desires into sentences. In mathematics, we learn to describe numerical relationships and situations through mathematical expressions and equations. Historians describe thousands of years of history in 500-page textbooks. Filmmakers translate entire events, such as the Olympics, into a few hours of entertainment.

Chemists use chemical equations to describe reactions they observe in the laboratory or in nature. Chemical equations provide us with the means to (1) summarize the reaction, (2) display the substances that are reacting, (3) show the products, and (4) indicate the amounts of all component substances in a reaction.

8.1   The Chemical Equation

Chemical reactions always involve change. Atoms, molecules, or ions rearrange to form new substances, sometimes in a spectacular manner. For example, the thermite reaction (shown in the chapter opening photo) is a reaction between aluminum metal and iron(III) oxide, which produces molten iron and aluminum oxide. The substances entering the reaction are called the *reactants* and the substances formed are called the *products*. In our example,

reactants   aluminum
             iron(III) oxide

products   iron
            aluminum oxide

During reactions chemical bonds are broken and new bonds are formed. The reactants and products may be present as solids, liquids, or gases, or in solution.

In a chemical reaction atoms are neither created nor destroyed.
All atoms present in the reactants must also be present in the products.

chemical equation

A **chemical equation** is a shorthand expression for a chemical change or reaction. A chemical equation uses the chemical symbols and formulas of the reactants and products and other symbolic terms to represent a chemical reaction. The equations are written according to this general format:

1. Reactants are separated from products by an arrow ($\longrightarrow$) that indicates the direction of the reaction. The reactants are placed to the left and the products to the right of the arrow. A plus sign ($+$) is placed between reactants and between products when needed.

$$Al + Fe_2O_3 \longrightarrow Fe + Al_2O_3$$

reactants   products

2. Coefficients (whole numbers) are placed in front of substances to balance the equation and to indicate the number of units (atoms, molecules, moles, ions)

TABLE 8.1	Symbols Commonly Used in Chemical Equations

Symbol	Meaning
+	Plus or added to (placed between substances)
$\longrightarrow$	Yields; produces (points to products)
(s)	Solid state (written after a substance)
(l)	Liquid state (written after a substance)
(g)	Gaseous state (written after a substance)
(aq)	Aqueous solution (substance dissolved in water)
Δ	Heat (written above arrow)

of each substance reacting or being produced. When no number is shown, it is understood that one unit of the substance is indicated.

$$2\,Al + Fe_2O_3 \longrightarrow 2\,Fe + Al_2O_3$$

3. Conditions required to carry out the reaction may, if desired, be placed above or below the arrow or equality sign. For example, a delta sign placed over the arrow ($\xrightarrow{\Delta}$) indicates that heat is supplied to the reaction.

$$2\,Al + Fe_2O_3 \xrightarrow{\Delta} 2\,Fe + Al_2O_3$$

4. The physical state of a substance is indicated by the following symbols: (s) for solid state; (l) for liquid state; (g) for gaseous state; and (aq) for substances in aqueous solution. States are not always given in chemical equations.

$$2\,Al(s) + Fe_2O_3(s) \xrightarrow{\Delta} Fe(l) + Al_2O_3(s)$$

Commonly used symbols are given in Table 8.1.

8.2 Writing and Balancing Equations

To represent the quantitative relationships of a reaction, the chemical equation must be balanced. A **balanced equation** contains the same number of each kind of atom on each side of the equation. The balanced equation therefore obeys the law of conservation of mass.

balanced equation

Every chemistry student must learn to *balance* equations. Many equations are balanced by trial and error, but care and attention to detail are still required. The way to balance an equation is to adjust the number of atoms of each element so that they are the same on each side of the equation, but a correct formula is never changed in order to balance an equation. The general procedure for balancing equations is:

Correct formulas are not changed to balance an equation.

Study this procedure carefully and refer to it when you work examples.

Step 1. Identify the reaction. Write a description or word equation for the reaction. For example, let's consider mercury(II) oxide decomposing into mercury and oxygen.

mercury (II) oxide $\longrightarrow$ mercury + oxygen.

Step 2. Write the unbalanced (skeleton) equation. Make sure that the formula for each substance is correct and that reactants are written to the left and products to the right of the arrow. For our example

$$HgO \longrightarrow Hg + O_2$$

The correct formulas must be known or determined from the periodic table, lists of ions, or experimental data.

Step 3. Balance the equation. Use the following process as necessary:

(a) Count and compare the number of atoms of each element on each side of the equation and determine those that must be balanced.

Hg is balanced   (1 on each side)

O needs to be balanced   (1 on reactant side, 2 on product side)

(b) Balance each element, one at a time, by placing whole numbers (coefficients) in front of the formulas containing the unbalanced element. It is usually best to balance metals first, then nonmetals, then hydrogen and oxygen. Select the smallest coefficients that will give the same number of atoms of the element on each side. A coefficient placed before a formula multiplies every atom in the formula by that number (e.g., $2 \, H_2SO_4$ means two molecules of sulfuric acid and also means four H atoms, two S atoms, and eight O atoms). Place a 2 in front of HgO to balance O.

> **Leave elements that are in two or more formulas (on the same side of the equation) unbalanced until just before balancing hydrogen and oxygen.**

$$2 \, HgO \longrightarrow Hg + O_2$$

(c) Check all other elements after each individual element is balanced to see whether, in balancing one element, other elements have become unbalanced. Make adjustments as needed. Now Hg is not balanced. To adjust this we write a 2 in front of Hg.

$$2 \, HgO \longrightarrow 2 \, Hg + O_2 \quad \text{(balanced)}$$

(d) Do a final check, making sure that each element and/or polyatomic ion is balanced and that the smallest possible set of whole-number coefficients has been used.

$$2 \, HgO \longrightarrow 2 \, Hg + O_2 \quad \text{(correct form)}$$

$$4 \, HgO \longrightarrow 4 \, Hg + 2 \, O_2 \quad \text{(incorrect form)}$$

Not all chemical equations can be balanced by the simple method of inspection just described. The following examples show *stepwise* sequences leading to balanced equations. Study each one carefully.

Example 8.1   Write the balanced equation for the reaction that takes place when magnesium metal is burned in air to produce magnesium oxide.

Solution

Step 1. *Word equation:*

$$\text{magnesium} + \text{oxygen} \longrightarrow \text{magnesium oxide}$$
$$\text{reactants (R)} \qquad\qquad\qquad \text{products (P)}$$

Step 2. *Skeleton equation:*

Mg + O$_2$ ⟶ MgO   (unbalanced)

R = reactant	R	1 Mg	2 O
P = product	P	1 Mg	1 O

Step 3. *Balance:*

(a) Mg is balanced.
Oxygen is not balanced. Two O atoms appear on the left side and one on the right side.

(b) Place the coefficient 2 in front of MgO:

Mg + O$_2$ ⟶ 2 MgO   (unbalanced)

R	1 Mg	2 O
P	2 Mg	2 O

(c) Now Mg is not balanced. One Mg atom appears on the left side and two on the right side. Place a 2 in front of Mg:

2 Mg + O$_2$ ⟶ 2 MgO   (balanced)

(d) *Check:* Each side has two Mg and two O atoms.

R	2 Mg	2 O
P	2 Mg	2 O

When methane, CH$_4$, undergoes complete combustion, it reacts with oxygen to produce carbon dioxide and water. Write the balanced equation for this reaction.

Example 8.2

Solution

Step 1. *Word equation:*

methane + oxygen ⟶ carbon dioxide + water

Step 2. *Skeleton equation:*

CH$_4$ + O$_2$ ⟶ CO$_2$ + H$_2$O   (unbalanced)

Step 3. *Balance:*

(a) Carbon is balanced.
Hydrogen and oxygen are not balanced.

R	1 C	4 H	2 O
P	1 C	2 H	3 O

(b) Balance H atoms by placing a 2 in front of H$_2$O:

CH$_4$ + O$_2$ ⟶ CO$_2$ + 2 H$_2$O   (unbalanced)

R	1 C	4 H	2 O
P	1 C	4 H	4 O

Each side of the equation has four H atoms; oxygen is still not balanced. Place a 2 in front of O$_2$ to balance the oxygen atoms:

CH$_4$ + 2 O$_2$ ⟶ CO$_2$ + 2 H$_2$O   (balanced)

R	1 C	4 H	4 O
P	1 C	4 H	4 O

(c) The other atoms remain balanced.

(d) *Check:* The equation is correctly balanced; it has one C, four O, and four H atoms on each side.

Oxygen and potassium chloride are formed by heating potassium chlorate. Write a balanced equation for this reaction.

Example 8.3

Solution

Step 1. *Word equation:*

potassium chlorate $\xrightarrow{\Delta}$ potassium chloride + oxygen

Step 2. *Skeleton equation:*

$$KClO_3 \xrightarrow{\Delta} KCl + O_2 \quad \text{(unbalanced)}$$

Step 3. *Balance:*

R | 1 K | 1 Cl | 3 O
P | 1 K | 1 Cl | 2 O

(a) Potassium and chlorine are balanced.
Oxygen is unbalanced (three O atoms on the left and two on the right side).

(b) How many oxygen atoms are needed? The subscripts of oxygen (3 and 2) in $KClO_3$ and O_2 have a least common multiple of 6. Therefore coefficients for $KClO_3$ and O_2 are needed to get six O atoms on each side. Place a 2 in front of $KClO_3$ and a 3 in front of O_2:

R | 2 K | 2 Cl | 6 O
P | 1 K | 1 Cl | 6 O

$$2\,KClO_3 \xrightarrow{\Delta} KCl + 3\,O_2 \quad \text{(unbalanced)}$$

(c) Now K and Cl are not balanced. Place a 2 in front of KCl, which balances both K and Cl at the same time:

$$2\,KClO_3 \xrightarrow{\Delta} 2\,KCl + 3\,O_2 \quad \text{(balanced)}$$

R | 2 K | 2 Cl | 6 O
P | 2 K | 2 Cl | 6 O

(d) *Check:* Each side now contains two K, two Cl, and six O atoms.

Example 8.4   Silver nitrate reacts with hydrogen sulfide to produce silver sulfide and nitric acid. Write a balanced equation for this reaction.

Solution

Step 1. *Word equation:*

silver nitrate + hydrogen sulfide $\longrightarrow$ silver sulfide + nitric acid

Step 2. *Skeleton equation:*

$$AgNO_3 + H_2S \longrightarrow Ag_2S + HNO_3 \quad \text{(unbalanced)}$$

R | 1 Ag | 2 H | 1 S | 1 NO₃
P | 2 Ag | 1 H | 1 S | 1 NO₃

Step 3. *Balance:*

(a) Ag and H are unbalanced.
(b) Place a 2 in front of $AgNO_3$ to balance Ag:

$$2\,AgNO_3 + H_2S \longrightarrow Ag_2S + HNO_3 \quad \text{(unbalanced)}$$

R | 2 Ag | 2 H | 1 S | 2 NO₃
P | 2 Ag | 1 H | 1 S | 1 NO₃

(c) H and NO_3^- are still unbalanced. Balance by placing a 2 in front of HNO_3:

$$2\,AgNO_3 + H_2S \longrightarrow Ag_2S + 2\,HNO_3 \quad \text{(balanced)}$$

In this example N and O atoms are balanced by balancing the NO_3^- ion as a unit.

R | 2 Ag | 2 H | 1 S | 2 NO₃
P | 2 Ag | 2 H | 1 S | 2 NO₃

(d) The other atoms remain balanced.
(e) *Check:* Each side has two Ag, two H, and one S atom. Also, each side has two NO_3^- ions.

When aluminum hydroxide is mixed with sulfuric acid the products are aluminum sulfate and water. Write a balanced equation for this reaction.

Example 8.5

Step 1. *Word equation:*

    aluminum hydroxide + sulfuric acid $\longrightarrow$ aluminum sulfate + water

Solution

Step 2. *Skeleton equation:*

    $Al(OH)_3 + H_2SO_4 \longrightarrow Al_2(SO_4)_3 + H_2O$ (unbalanced)

Step 3. *Balance:*

    (a) All elements are unbalanced.

    (b) Balance Al by placing a 2 in front of $Al(OH)_3$. Treat the unbalanced SO_4^{2-} ion as a unit and balance by placing a 3 in front of H_2SO_4:

R	1 Al	1 SO₄	3 O	5 H
P	2 Al	3 SO₄	1 O	2 H

        $2\ Al(OH)_3 + 3\ H_2SO_4 \longrightarrow Al_2(SO_4)_3 + H_2O$ (unbalanced)

        Balance the unbalanced H and O by placing a 6 in front of H_2O:

R	2 Al	3 SO₄	6 O	12 H
P	2 Al	3 SO₄	1 O	2 H

        $2\ Al(OH)_3 + 3\ H_2SO_4 \longrightarrow Al_2(SO_4)_3 + 6\ H_2O$ (balanced)

    (c) The other atoms remain balanced.

R	2 Al	3 SO₄	6 O	12 H
P	2 Al	3 SO₄	6 O	12 H

    (d) *Check:* Each side has two Al, twelve H, three S, and eighteen O atoms.

When the fuel in a butane gas stove undergoes complete combustion, it reacts with oxygen to form carbon dioxide and water. Write the balanced equation for this reaction.

Example 8.6

Step 1. *Word equation:*

    butane + oxygen $\longrightarrow$ carbon dioxide + water

Solution

Step 2. *Skeleton equation:*

    $C_4H_{10} + O_2 \longrightarrow CO_2 + H_2O$ (unbalanced)

Step 3. *Balance:*

    (a) All elements are unbalanced.

    (b) Balance C by placing a 4 in front of CO_2:

R	4 C	10 H	2 O
P	1 C	2 H	3 O

        $C_4H_{10} + O_2 \longrightarrow 4\ CO_2 + H_2O$ (unbalanced)

        Balance H by placing a 5 in front of H_2O:

        $C_4H_{10} + O_2 \longrightarrow 4\ CO_2 + 5\ H_2O$ (unbalanced)

R	4 C	10 H	2 O
P	4 C	10 H	13 O

        Oxygen remains unbalanced. The oxygen atoms on the right side are fixed, because 4 CO_2 and 5 H_2O are derived from the single C_4H_{10} molecule on the left. When we try to balance the O atoms, we find there is no whole number that can be placed in front of O_2 to bring about a balance. So we double the coefficients of each substance, and then balance the oxygen:

R	8 C	20 H	26 O
P	8 C	20 H	26 O

        $2\ C_4H_{10} + 13\ O_2 \longrightarrow 8\ CO_2 + 10\ H_2O$ (balanced)

(c) The other atoms remain balanced.
(d) *Check:* Each side now has eight C, twenty H, and twenty-six O atoms.

Practice 8.1

Write a balanced formula equation for:

aluminum + oxygen $\longrightarrow$ aluminum oxide

Practice 8.2

Write a balanced formula equation for:

magnesium hydroxide + phosphoric acid $\longrightarrow$

magnesium phosphate + water

8.3   What Information Does an Equation Tell Us?

Depending on the particular context in which it is used, a formula can have different meanings. A formula can refer to an individual chemical entity (atom, ion, molecule, or formula unit) or to a mole of that chemical entity. For example, the formula H_2O can mean any of the following:

1. 2 H atoms and 1 O atom
2. 1 molecule of water
3. 1 mol of water
4. 6.022×10^{23} molecules of water
5. 18.02 g of water

Formulas used in equations can represent units of individual chemical entities or moles, the latter being more commonly used. For example, in the reaction of hydrogen and oxygen to form water,

$$2 H_2 \quad + \quad O_2 \quad \longrightarrow \quad 2 H_2O$$

| 2 molecules hydrogen | 1 molecule oxygen | 2 molecules water |
| 2 mol hydrogen | 1 mol oxygen | 2 mol water |

We generally use moles in equations because molecules are so small.

As indicated earlier, a chemical equation is a shorthand description of a chemical reaction. Interpretation of a balanced equation gives us the following information:

1. What the reactants are and what the products are
2. The formulas of the reactants and products
3. The number of molecules or formula units of reactants and products in the reaction
4. The number of atoms of each element involved in the reaction
5. The number of moles of each substance

Colorful Trees

The wondrous array of color we see in the fall is the result of chemical reactions. These reactions contribute to the metabolic cycle in plants while they give us our annual autumn show of foliage. Three chemical substances play key roles in this process: chlorophylls, carotenoids, and anthocyanins.

The green pigments in leaves come from chlorophylls, which are necessary for the tree to produce food in the process called *photosynthesis.* Photosynthesis involves hundreds of chemical reactions but can be summarized by the following chemical equation:

$$6 \; CO_2 + 6 \; H_2O \xrightarrow[\text{chlorophyll}]{\text{sunlight}}$$

carbon   water   chlorophyll
dioxide

$$C_6H_{12}O_6 + 6 \; O_2$$

glucose   oxygen

There are several types of chlorophyll. All photosynthetic plants contain chlorophyll *a*, and some contain chlorophyll *b* and chlorophyll *c*, known as *accessory pigments.* The accessory pigments extend the range of colored light trees can use for photosynthesis. Chlorophyll *b* and *c* absorb light of slightly different colors and pass the energy on to chlorophyll *a*. This allows the trees to produce more food during the summer months.

The carotenoids are responsible for the yellow, orange, and some of the reds in the autumn leaves. These substances protect the plant from the destructive potential of chlorophyll. When chlorophyll absorbs energy (from light or the other chlorophylls) it produces a high energy electron that is normally passed on to other compounds in the chemical reactions of photosynthesis. Sometimes there is a shortage of these secondary compounds and the high energy electron can then combine with oxygen (in other chemical reactions) and destroy the cell. The carotenoid pigments act to save the cell from destruction by combining with the high energy oxygen and releasing it later when it can be used successfully to produce sugars. Carotenoids and anthocyanins (red and blue pigment molecules) also help the tree to attract insects for pollination and make fruits noticeable to birds and animals, which then spread seeds for reproduction of the tree. The carotenoids and anthocyanins are masked by the green of the chlorophylls during the summer and can be seen only as the chlorophyll disintegrates in the fall.

What conditions and chemical reactions are necessary to make the autumn show spectacular? The colors of autumn leaves are produced by the carotenoids and anthocyanins. Trees need lots of each molecule to produce a fall show. Trees produce the carotenoids and anthocyanins during photosynthesis. A long and moist summer provides the proper reactants and time to make plenty of sugar. When the tree signals its leaves to stop photosynthesis the chlorophylls are released from the leaf proteins. The chlorophylls break down and the protein nutrients are carried to the roots for winter storage. Now the carotenoids that have been hard at work protecting the leaf from too much chlorophyll activity can show their color! The more productive the leaf has been (more photosynthesis) the greater the concentration of carotenoids and the more beautiful the orange and yellow colors.

But what about the anthocyanins? What is their role in leaf color? As fall approaches the tree stops producing new wood from the sugar made in photosynthesis. Sugar manufactured in the leaves is now sent to the roots for storage. If the weather is chilly the sugary fluid becomes slow flowing (viscous). Some of the sugar made during the day cannot get out of the leaf to the roots during the night. The leftover sugar is changed into anthocyanins. The greater the amount of these pigment molecules the brighter the leaf color.

In the hills of Pennsylvania a parade of chemical reactions sets the stage for the brilliant colors of fall. The chlorophyll molecules begin to disintegrate, the critical biomolecules migrate to the roots, and the carotenoids begin by showing their yellow and orange colors. The sugar that remains in the leaves produces red anthocyanins and the leaves of the maples change from yellow and orange to flaming red. At last all of the pigments disintegrate in decomposition reactions and the leaves become brown, the color of the woody fibers of their skeletons, and eventually fall to the ground. If we are fortunate, we are treated to a spectacular show of color before the leaves die!

▲
Vivid colors are reflected in Tuscarora Lake in central Pennsylvania.

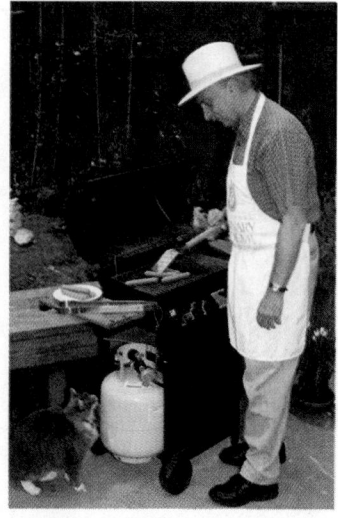

▲
Propane is burned in many outdoor grills.

Consider the reaction that occurs when propane gas (C_3H_8) is burned in air; the products are carbon dioxide, CO_2, and water, H_2O. The balanced equation and its interpretation are as follows:

Propane		Oxygen		Carbon dioxide		Water
$C_3H_8(g)$	+	$5\,O_2(g)$	→	$3\,CO_2(g)$	+	$4\,H_2O(g)$
1 molecule		5 molecules		3 molecules		4 molecules
3 atoms C 8 atoms H		10 atoms O		3 atoms C 6 atoms O		8 atoms H 4 atoms O
1 mol 44.09 g		5 mol 5 (32.00 g) (160.0 g)		3 mol 3 (44.01 g) (132.0 g)		4 mol 4 (18.02 g) (72.08 g)

Practice 8.3

Consider the reaction that occurs when hydrogen gas reacts with chlorine gas to produce gaseous hydrogen chloride.
(a) Write a word equation for this reaction.
(b) Write a balanced formula equation including state for each substance.
(c) Label each reactant and product to show relative amounts of each substance.
   1) number of molecules
   2) number of atoms
   3) number of moles
   4) mass
(d) What mass of HCl would be produced if you reacted 2 mol hydrogen gas with 2 mol chlorine gas?

The quantities involved in chemical reactions are important when working in industry or the laboratory. We will study the relationship among quantities of reactants and products in the next chapter.

8.4   Types of Chemical Equations

Chemical equations represent chemical changes or reactions. Reactions are classified into types to assist in writing equations and in predicting other reactions. Many chemical reactions fit one or another of the four principal reaction types that we discuss in the following paragraphs. Reactions are also classified as oxidation–reduction. Special methods are used to balance complex oxidation–reduction equations. (See Chapter 17.)

Combination Reaction

In a **combination reaction,** two reactants combine to give one product. The general form of the equation is

combination reaction

$$A + B \longrightarrow AB$$

in which A and B are either elements or compounds and AB is a compound. The formula of the compound in many cases can be determined from a knowledge of the ionic charges of the reactants in their combined states. Some reactions that fall into this category are given here.

1. metal + oxygen $\longrightarrow$ metal oxide

$$2 \, Mg(s) + O_2(g) \xrightarrow{\Delta} 2 \, MgO(s)$$

$$4 \, Al(s) + 3 \, O_2(g) \xrightarrow{\Delta} 2 \, Al_2O_3(s)$$

2. nonmetal + oxygen $\longrightarrow$ nonmetal oxide

$$S(s) + O_2(g) \xrightarrow{\Delta} SO_2(g)$$

$$N_2(g) + O_2(g) \xrightarrow{\Delta} 2 \, NO(g)$$

3. metal + nonmetal $\longrightarrow$ salt

$$2 \, Na(s) + Cl_2(g) \longrightarrow 2 \, NaCl(s)$$

$$2 \, Al(s) + 3 \, Br_2(l) \longrightarrow 2 \, AlBr_3(s)$$

4. metal oxide + water $\longrightarrow$ metal hydroxide

$$Na_2O(s) + H_2O(l) \longrightarrow 2 \, NaOH(aq)$$

$$CaO(s) + H_2O(l) \longrightarrow Ca(OH)_2(aq)$$

5. nonmetal oxide + water $\longrightarrow$ oxy-acid

$$SO_3(g) + H_2O(l) \longrightarrow H_2SO_4(aq)$$

$$N_2O_5(s) + H_2O(l) \longrightarrow 2 \, HNO_3(aq)$$

▲
Flames and sparks result when aluminum foil is dropped into liquid bromine.

Decomposition Reaction

In a **decomposition reaction,** a single substance is decomposed or broken down to give two or more different substances. This reaction may be considered the reverse of combination. The starting material must be a compound, and the products may be elements or compounds. The general form of the equation is

decomposition reaction

$$AB \longrightarrow A + B$$

Predicting the products of a decomposition reaction can be difficult and requires an understanding of each individual reaction. Heating oxygen-containing compounds often results in decomposition. Some reactions that fall into this category are

1. Metal oxides. Some metal oxides decompose to yield the free metal plus oxygen; others give another oxide, and some are very stable, resisting decomposition by heating:

$$2 \, HgO(s) \xrightarrow{\Delta} 2 \, Hg(l) + O_2(g)$$

$$2 \, PbO_2(s) \xrightarrow{\Delta} 2 \, PbO(s) + O_2(g)$$

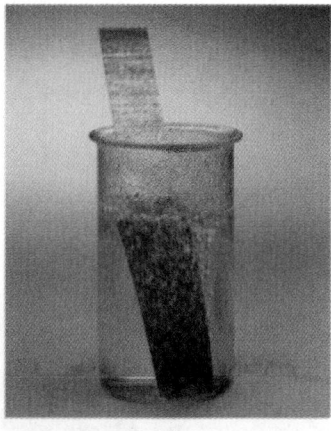

When a strip of zinc is placed in HCl, hydrogen gas bubbles form immediately.

single-displacement reaction

TABLE 8.2	Activity Series	
	Metals	**Halogens**
increasing activity ↑	K Ca Na Mg Al Zn Fe Ni Sn Pb H Cu Ag Hg Au	F_2 Cl_2 Br_2 I_2

2. Carbonates and hydrogen carbonates decompose to yield CO_2 when heated:

$$CaCO_3(s) \xrightarrow{\Delta} CaO(s) + CO_2(g)$$

$$2\,NaHCO_3(s) \xrightarrow{\Delta} Na_2CO_3(s) + H_2O(g) + CO_2(g)$$

3. Miscellaneous reactions in this category:

$$2\,KClO_3(s) \xrightarrow{\Delta} 2\,KCl(s) + 3\,O_2(g)$$

$$2\,NaNO_3(s) \xrightarrow{\Delta} 2\,NaNO_2(g) + O_2(g)$$

$$2\,H_2O_2(l) \xrightarrow{\Delta} 2\,H_2O(l) + O_2(g)$$

Single-Displacement Reaction

In a **single-displacement reaction,** one element reacts with a compound to replace one of the elements of that compound, yielding a different element and a different compound. The general form of the equation is

$$A + BC \longrightarrow B + AC \quad \text{or} \quad A + BC \longrightarrow C + BA$$

If A is a metal, A will replace B to form AC, provided A is a more reactive metal than B. If A is a halogen, it will replace C to form BA, provided A is a more reactive halogen than C.

A brief activity series of selected metals (and hydrogen) and halogens is shown in Table 8.2. This series is listed in descending order of chemical activity, with the most active metals and halogens at the top. Many chemical reactions can be predicted from an activity series because the atoms of any element in the series will replace the atoms of those elements below it. For example, zinc metal will replace hydrogen from a hydrochloric acid solution. But copper metal, which is below hydrogen on the list and thus less reactive than hydrogen, will not replace hydro-gen from a hydrochloric acid-solution. Here are some reactions that fall into this category:

1. metal + acid ⟶ hydrogen + salt

$$Zn(s) + 2\,HCl(aq) \longrightarrow H_2(g) + ZnCl_2(aq)$$

$$2\,Al(s) + 3\,H_2SO_4(aq) \longrightarrow 3\,H_2(g) + Al_2(SO_4)_3(aq)$$

2. metal + water ⟶ hydrogen + metal hydroxide or metal oxide

$$2\,Na(s) + 2\,H_2O \longrightarrow H_2(g) + 2\,NaOH(aq)$$

$$Ca(s) + 2\,H_2O \longrightarrow H_2(g) + Ca(OH)_2(aq)$$

$$\underset{\text{steam}}{3\,Fe(s) + 4\,H_2O(g)} \longrightarrow 4\,H_2(g) + Fe_3O_4(s)$$

3. metal + salt ⟶ metal + salt

$$Fe(s) + CuSO_4(aq) \longrightarrow Cu(s) + FeSO_4(aq)$$

$$Cu(s) + 2\,AgNO_3(aq) \longrightarrow 2\,Ag(s) + Cu(NO_3)_2(aq)$$

4. halogen + halide salt ⟶ halogen + halide salt

$$Cl_2(g) + 2\,NaBr(aq) \longrightarrow Br_2(l) + 2\,NaCl(aq)$$

$$Cl_2(g) + 2\,KI(aq) \longrightarrow I_2(s) + 2\,KCl(aq)$$

A common chemical reaction is the displacement of hydrogen from water or acids. This reaction is a good illustration of the relative reactivity of metals and the use of the activity series. For example,

- K, Ca, and Na displace hydrogen from cold water, steam, and acids.
- Mg, Al, Zn, and Fe displace hydrogen from steam and acids.
- Ni, Sn, and Pb displace hydrogen only from acids.
- Cu, Ag, Hg, and Au do not displace hydrogen.

Will a reaction occur between (a) nickel metal and hydrochloric acid and (b) tin metal and a solution of aluminum chloride? Write balanced equations for the reactions.

Example 8.7

(a) Nickel is more reactive than hydrogen, so it will displace hydrogen from hydrochloric acid. The products are hydrogen gas and a salt of Ni^{2+} and Cl^- ions:

Solution

$$Ni(s) + 2\ HCl(aq) \longrightarrow H_2(g) + NiCl_2(aq)$$

(b) According to the activity series, tin is less reactive than aluminum, so no reaction will occur:

$$Sn(s) + AlCl_3(aq) \longrightarrow \text{no reaction}$$

Practice 8.4

Write balanced equations for these reactions:
(a) iron metal and a solution of magnesium chloride
(b) zinc metal and a solution of lead(II) nitrate

Double-Displacement Reaction

In a **double-displacement reaction,** two compounds exchange partners with each other to produce two different compounds. The general form of the equation is

double-displacement reaction

$$AB + CD \longrightarrow AD + CB$$

This reaction can be thought of as an exchange of positive and negative groups, in which A combines with D and C combines with B. In writing formulas for the products, we must account for the charges of the combining groups.

It's also possible to write an equation in the form of a double-displacement reaction when a reaction has not occurred. For example, when solutions of sodium chloride and potassium nitrate are mixed, the following equation can be written:

$$NaCl(aq) + KNO_3(aq) \longrightarrow NaNO_3(aq) + KCl(aq)$$

When the procedure is carried out, no physical changes are observed, indicating that no chemical reaction has taken place.

A double-displacement reaction is accompanied by evidence of such reactions as the evolution of heat, the formation of an insoluble precipitate, or the production of gas bubbles. Let's look at some of these reactions more closely.

▲
A double-displacement reaction results from pouring a clear, colorless solution of $Pb(NO_3)_2$ into a clear, colorless solution of KI forming a yellow precipitate of PbI_2.

Neutralization of an acid and a base The production of a molecule of water from an H^+ and an OH^- ion is accompanied by a release of heat, which can be detected by touching the reaction container. For neutralization reactions, $H^+ + OH^- \longrightarrow H_2O$.

$$acid + base \longrightarrow salt + water$$

$$HCl(aq) + NaOH(aq) \longrightarrow NaCl(aq) + H_2O(l)$$

$$H_2SO_4(aq) + Ba(OH)_2(aq) \longrightarrow BaSO_4(s) + 2\,H_2O(l)$$

Formation of an insoluble precipitate The solubilities of the products can be determined by consulting the Solubility Table in Appendix V to see whether one or both of the products are insoluble in water. An insoluble product (precipitate) is indicated by placing an (*s*) after its formula in the equation.

$$BaCl_2(aq) + 2\,AgNO_3(aq) \longrightarrow 2\,AgCl(s) + Ba(NO_3)_2(aq)$$

$$Pb(NO_3)_2(aq) + 2\,KI(aq) \longrightarrow PbI_2(s) + 2\,KNO_3(aq)$$

Metal oxide + acid Heat is released by the production of a molecule of water.

$$metal\ oxide + acid \longrightarrow salt + water$$

$$CuO(s) + 2\,HNO_3(aq) \longrightarrow Cu(NO_3)_2(aq) + H_2O(l)$$

$$CaO(s) + 2\,HCl(aq) \longrightarrow CaCl_2(aq) + H_2O(l)$$

Formation of a gas A gas such as HCl or H_2S may be produced directly, as in these two examples:

$$H_2SO_4(l) + NaCl(s) \longrightarrow NaHSO_4(s) + HCl(g)$$

$$2\,HCl(aq) + ZnS(s) \longrightarrow ZnCl_2(aq) + H_2S(g)$$

A gas can also be produced indirectly. Some unstable compounds formed in a double-displacement reaction, such as H_2CO_3, H_2SO_3, and NH_4OH, will decompose to form water and a gas:

$$2\,HCl(aq) + Na_2CO_3(aq) \longrightarrow 2\,NaCl(aq) + H_2CO_3(aq) \longrightarrow 2\,NaCl(aq) + H_2O(l) + CO_2(g)$$

$$2\,HNO_3(aq) + K_2SO_3(aq) \longrightarrow 2\,KNO_3(aq) + H_2SO_3(aq) \longrightarrow 2\,KNO_3(aq) + H_2O(l) + SO_2(g)$$

$$NH_4Cl(aq) + NaOH(aq) \longrightarrow NaCl(aq) + NH_4OH(aq) \longrightarrow NaCl(aq) + H_2O(l) + NH_3(g)$$

Example 8.8

Write the equation for the reaction between aqueous solutions of hydrobromic acid and potassium hydroxide.

Solution

First we write the formulas for the reactants. (They are HBr and KOH.) Then we classify the type of reaction that would occur between them. Because the reactants are compounds, one an acid and the other a base, the reaction will be of the neutralization type:

$$acid + base \longrightarrow salt + water$$

Now rewrite the equation using the formulas for the known substances:

$$HBr(aq) + KOH(aq) \longrightarrow salt + H_2O$$

In this reaction, which is a double-displacement type, the H^+ from the acid combines with the OH^- from the base to form water. The ionic compound must be composed of the other two ions, K^+ and Br^-. We determine the formula of the ionic compound to be KBr from the fact that K is a +1 cation and Br is a −1 anion. The final balanced equation is

$$HBr(aq) + KOH(aq) \longrightarrow KBr(aq) + H_2O(l)$$

Example 8.9

Complete and balance the equation for the reaction between aqueous solutions of barium chloride and sodium sulfate.

Solution

First determine the formula for the reactants. (They are $BaCl_2$ and Na_2SO_4.) Then classify these substances as acids, bases, or ionic compounds. (Both substances are salts.) Since both substances are compounds, the reaction will be of the double-displacement type. Start writing the equation with the reactants:

$$BaCl_2(aq) + Na_2SO_4(aq) \longrightarrow$$

If the reaction is double-displacement, Ba^{2+} will be written combined with SO_4^{2-}, and Na^+ with Cl^- as the products. The balanced equation is

$$BaCl_2(aq) + Na_2SO_4(aq) \longrightarrow BaSO_4 + 2\,NaCl$$

The final step is to determine the nature of the products, which controls whether or not the reaction will take place. If both products are soluble, we have a mixture of all the ions in solution. But if an insoluble precipitate is formed, the reaction will definitely occur. We know from experience that NaCl is fairly soluble in water, but what about $BaSO_4$? Consulting the Solubility Table in Appendix V we see that $BaSO_4$ is insoluble in water, so it will be a precipitate in the reaction. Thus the reaction will occur, forming a precipitate. The equation is

$$BaCl_2(aq) + Na_2SO_4(aq) \longrightarrow BaSO_4(s) + 2\,NaCl(aq)$$

▲
When barium chloride is poured into a solution of sodium sulfate, a white precipitate of barium sulfate forms.

Practice 8.5

Complete and balance the equations for these reactions:
(a) potassium phosphate + barium chloride
(b) hydrochloric acid + nickel carbonate
(c) ammonium chloride + sodium nitrate

Some of the reactions you attempt may fail because the substances are not reactive or because the proper conditions for reaction are not present. For example, mercury(II) oxide does not decompose until it is heated; magnesium does not burn in air or oxygen until the temperature reaches a certain point. When silver is placed in a solution of copper(II) sulfate, no reaction occurs. When a strip of copper is placed

A copper wire is placed in ▶
a solution of silver nitrate
(left). After 24 hours crystals
of silver are seen hanging on
the copper wire and the solution has turned blue indicating copper ions are present
there.

A copper wire is placed in ▶
a solution of silver nitrate
(left). After 24 hours crystals
of silver are seen hanging on
the copper wire and the solution has turned blue indicating copper ions are present
there.

in a solution of silver nitrate, a single-displacement reaction takes place, because copper is a more reactive metal than silver.

The successful prediction of the products of a reaction is not always easy. The ability to predict products correctly comes with knowledge and experience. Although you may not be able to predict many reactions at this point, as you continue to experiment you will find that reactions can be categorized, and that prediction of the products thereby becomes easier, if not always certain.

8.5   Heat in Chemical Reactions

Energy changes always accompany chemical reactions. One reason why reactions occur is that the products attain a lower, more stable energy state than the reactants. When the reaction leads to a more stable state, energy is released to the surroundings as heat (and/or work). When a solution of a base is neutralized by the addition of an acid, the liberation of heat energy is signaled by an immediate rise in the temperature of the solution. When an automobile engine burns gasoline, heat is certainly liberated; at the same time, part of the liberated energy does the work of moving the automobile.

exothermic reaction
endothermic reaction

Reactions are either exothermic or endothermic. **Exothermic reactions** liberate heat; **endothermic reactions** absorb heat. In an exothermic reaction, heat is a product and may be written on the right side of the equation for the reaction. In an endothermic reaction, heat can be regarded as a reactant and is written on the left side of the equation. Here are two examples:

$$H_2(g) + Cl_2(g) \longrightarrow 2\,HCl(g) + 185\text{ kJ} \quad \textit{(exothermic)}$$

$$N_2(g) + O_2(g) + 181\text{ kJ} \longrightarrow 2\,NO(g) \quad \textit{(endothermic)}$$

heat of reaction

The quantity of heat produced by a reaction is known as the **heat of reaction.** The units used can be kilojoules or kilocalories. Consider the reaction represented by this equation:

$$C(s) + O_2(g) \longrightarrow CO_2(g) + 393\text{ kJ}$$

When the heat liberated is expressed as part of the equation, the substances are expressed in units of moles. Thus when 1 mol (12.01 g) of C combines with 1 mol (32.00 g) of O_2, 1 mol (44.01 g) of CO_2 is formed and 393 kJ of heat is liberated. In

◀ **This cornfield is a good example of the endothermic reactions happening through photosynthesis in plants.**

this reaction, as in many others, the heat energy is more useful than the chemical products.

Aside from relatively small amounts of energy from nuclear processes, the sun is the major provider of energy for life on Earth. The Sun maintains the temperature necessary for life and also supplies light energy for the endothermic photosynthetic reactions of green plants. In photosynthesis, carbon dioxide and water are converted to free oxygen and glucose:

$$6\ CO_2 + 6\ H_2O + 2519\ kJ \longrightarrow C_6H_{12}O_6 + 6\ O_2$$
$$\text{glucose}$$

Nearly all of the chemical energy used by living organisms is obtained from glucose or compounds derived from glucose.

The major source of energy for modern technology is fossil fuel—coal, petroleum, and natural gas. The energy is obtained from the combustion (burning) of these fuels, which are converted to carbon dioxide and water. Fossil fuels are mixtures of **hydrocarbons,** compounds containing only hydrogen and carbon.

hydrocarbon

Natural gas is primarily methane, CH_4. Petroleum is a mixture of hydrocarbons (compounds of carbon and hydrogen). Liquefied petroleum gas (LPG) is a mixture of propane (C_3H_8) and butane (C_4H_{10}).

The combustion of these fuels releases a tremendous amount of energy, but reactions won't occur to a significant extent at ordinary temperatures. A spark or a flame must be present before methane will ignite. The amount of energy that must be supplied to start a chemical reaction is called the **activation energy.** In an exothermic reaction once this activation energy is provided, enough energy is then generated to keep the reaction going.

activation energy

Here are some examples:

$$CH_4(g) + 2 O_2(g) \longrightarrow CO_2(g) + 2 H_2O(g) + 890 \text{ kJ}$$

$$C_3H_8(g) + 5 O_2(g) \longrightarrow 3 CO_2(g) + 4 H_2O(g) + 2200 \text{ kJ}$$

$$2 C_8H_{18}(l) + 25 O_2(g) \longrightarrow 16 CO_2(g) + 18 H_2O(g) + 10,900 \text{ kJ}$$

Be careful not to confuse an exothermic reaction that requires heat (activation energy) to get it *started* with an endothermic process that requires energy to keep it going. The combustion of magnesium, for example, is highly exothermic, yet magnesium must be heated to a fairly high temperature in air before combustion begins. Once started, however, the combustion reaction goes very vigorously until either the magnesium or the available supply of oxygen is exhausted. The electrolytic decomposition of water to hydrogen and oxygen is highly endothermic. If the electric current is shut off when this process is going on, the reaction stops instantly. The relative energy levels of reactants and products in exothermic and in endothermic processes are presented graphically in Figures 8.1 and 8.2.

Examples of endothermic and exothermic processes can be easily demonstrated in the laboratory. In Figure 8.1, the products are at a higher potential energy than the reactants. Energy has therefore been absorbed, and the reaction is endothermic. Here, solid $Ba(OH)_2$ and solid NH_4SCN are mixed in a beaker, which is standing in a puddle of water. The solids liquefy and absorb heat from the surroundings causing the beaker to freeze to the board. In Figure 8.2, the products are at a lower potential energy than the reactants. Energy (heat) is given off, producing an exothermic reaction. Here, potassium chlorate ($KClO_3$) and sugar are mixed and placed into a pile. A drop of concentrated sulfuric acid is added, creating a spectacular exothermic reaction.

Outside the laboratory you can experience an endothermic process when you apply a cold pack to an injury. In this case ammonium chloride (NH_4Cl) dissolves in water. For example, temperature changes from 24.5°C to 18.1°C result when 10 g of NH_4Cl are added to 100 mL of water. Energy, in the form of heat, is taken from the immediate surroundings (water) causing the salt solution to become cooler. (See Chemistry in Action, p. 75.)

FIGURE 8.1
Endothermic reaction between Ba(OH)₂ and NH₄SCN. Equal amounts of the two solids are placed in separate beakers. After placing a beaker on a wet board the two solids are mixed in the beaker, causing the board to freeze to the beaker.
▼

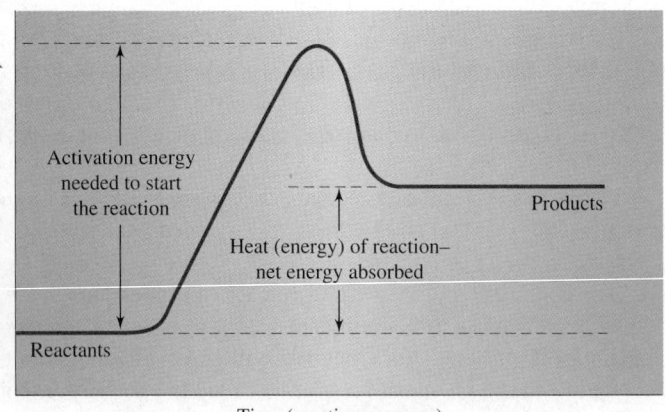

Time (reaction progress)

Endothermic reaction

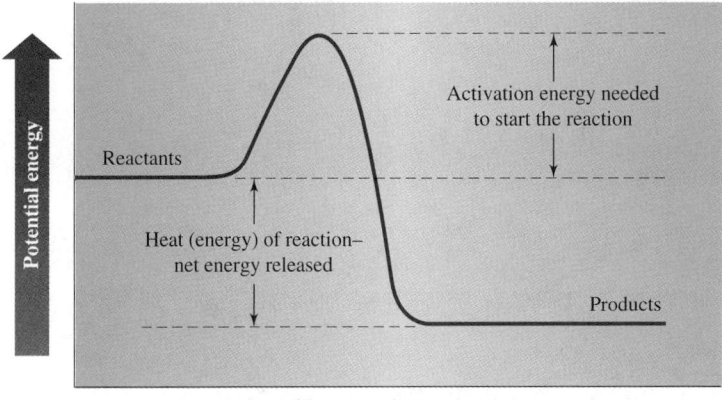

Potential energy

Reactants

Activation energy needed
to start the reaction

Heat (energy) of reaction–
net energy released

Products

Time (reaction progress)

Exothermic reaction

▲
FIGURE 8.2
Exothermic Reaction between
$KClO_3$ **and sugar. A sample of**
$KClO_3$ **and sugar are well**
mixed and placed on a
fireproof pad. Several drops
of concentrated H_2SO_4 **are**
used to ignite the mixture.

8.6   Global Warming: The Greenhouse Effect

Fossil fuels, derived from coal and petroleum, provide the energy we use to power our industries, heat and light our homes and workplaces, and run our cars. As these fuels are burned they produce carbon dioxide and water, releasing over 50 billion tons of carbon dioxide into our atmosphere each year.

The concentration of carbon dioxide has been monitored by scientists since 1958. Analysis of the air trapped in a core sample of snow from Antartica provides data on carbon dioxide levels for the past 160,000 years. The results of this study show that as the carbon dioxide increased, the global temperature increased as well. The levels of carbon dioxide remained reasonably constant from the last ice age, 100,000 years ago, until the industrial revolution. Since then the concentration of carbon dioxide in our atmosphere has risen 15% to an all-time high.

Carbon dioxide is a minor component in our atmosphere and is not usually considered to be a pollutant. The concern expressed by scientists arises from the dramatic increase occurring in the Earth's atmosphere. Without the influence of humans in the environment, the exchange of carbon dioxide between plants and animals would be relatively balanced. Our continued use of fossil fuels has led to an increase of 7.4% in carbon dioxide between 1900 and 1970 and an additional 3.5% increase during the 1980s. Continued increases are being seen in the 1990s.

Besides our growing consumption of fossil fuels, other factors contribute to increased carbon dioxide levels in the atmosphere: Rain forests are being destroyed by cutting and burning to make room for increased population and agricultural needs. Carbon dioxide is added to the atmosphere during the burning, and the loss of trees diminishes the uptake of carbon dioxide by plants.

About half of all the carbon dioxide released into our atmosphere each year remains there, thus increasing its concentration. The other half is absorbed by plants during photosynthesis or is dissolved in the ocean to form hydrogen carbonates and carbonates.

Carbon dioxide and other greenhouse gases, such as methane and water, act to warm our atmosphere by trapping heat near the surface of the Earth. Solar radiation strikes the Earth and warms the surface. The warmed surface then reradiates this

**Burning rainforests and ▶
deforestation throughout the
world are adding to the
greenhouse problem.**

**FIGURE 8.3 ▶
Elements of a global
temperature warming are
caused by the greenhouse
effect.**

energy as heat. The greenhouse gases absorb some of this heat energy from the
surface, which then warms our atmosphere. A similar principle is illustrated in a
greenhouse where sunlight comes through the glass yet heat cannot escape. The air
in the greenhouse warms, producing a climate considerably different than in nature.
In the atmosphere these greenhouse gases may produce dramatic changes in our
climate. See Figure 8.3.

Long-term effects of global warming are still a matter of speculation and debate.
One consideration is whether the polar ice caps would melt; this would cause a rise
in sea level and lead to major flooding on the coasts of our continents. Further effects
could include shifts in rainfall patterns, producing droughts and extreme seasonal
changes in such major agricultural regions as California. To reverse these trends will
require major efforts in the following areas:

- The development of new energy sources to cut our dependence on fossil fuels
- An end to deforestation worldwide
- Intense efforts to improve conservation

On an individual basis each of us can play a significant role. For example, the simple conversion of a 100-watt incandescent light bulb to a compact fluorescent bulb

can reduce the electrical consumption for that light by 20%, and the bulb can last ten times longer. Recycling, switching to more fuel-efficient cars, and energy-efficient appliances, heaters, and air conditioners all would result in decreased energy consumption and less carbon dioxide being released into our atmosphere.

Concepts in Review

1. Know the format used in setting up chemical equations.
2. Recognize the various symbols commonly used in writing chemical equations.
3. Be able to balance simple chemical equations.
4. Interpret a balanced equation in terms of the relative numbers or amounts of molecules, atoms, grams, or moles of each substance represented.
5. Classify equations as combination, decomposition, single-displacement, or double-displacement reactions.

6. Use the activity series to predict whether a single-displacement reaction will occur.
7. Complete and balance equations for simple combination, decomposition, single-displacement, and double-displacement reactions when given the reactants.
8. Distinguish between exothermic and endothermic reactions, and relate the quantity of heat to the amounts of substances involved in the reaction.
9. Identify the major sources of chemical energy and their uses.

Key Terms

activation energy (8.5)
balanced equation (8.2)
chemical equation (8.1)
combination reaction (8.4)

decomposition reaction (8.4)
double-displacement reaction (8.4)
endothermic reaction (8.5)
exothermic reaction (8.5)

heat of reaction (8.5)
hydrocarbon (8.5)
single-displacement reaction (8.4)

Questions

1. What is the purpose of balancing equations?
2. What is represented by the numbers (coefficients) that are placed in front of the formulas in a balanced equation?

3. In a balanced chemical equation:
   (a) are atoms conserved?
   (b) are molecules conserved?
   (c) are moles conserved?
   Explain yours answers briefly.
4. Explain how endothermic reactions differ from exothermic reactions.

Paired Exercises

5. Balance these equations:
   (a) $H_2 + O_2 \longrightarrow H_2O$
   (b) $C + Fe_2O_3 \longrightarrow Fe + CO$
   (c) $H_2SO_4 + NaOH \longrightarrow H_2O + Na_2SO_4$
   (d) $Al_2(CO_3)_3 \xrightarrow{\Delta} Al_2O_3 + CO_2$
   (e) $NH_4I + Cl_2 \longrightarrow NH_4Cl + I_2$

7. Classify the reactions in Exercise 5 as combination, decomposition, single-displacement, or double-displacement.

6. Balance these equations:
   (a) $H_2 + Br_2 \longrightarrow HBr$
   (b) $Al + C \xrightarrow{\Delta} Al_4C_3$
   (c) $Ba(ClO_3)_2 \xrightarrow{\Delta} BaCl_2 + O_2$
   (d) $CrCl_3 + AgNO_3 \longrightarrow Cr(NO_3)_3 + AgCl$
   (e) $H_2O_2 \longrightarrow H_2O + O_2$

8. Classify the reactions in Exercise 6 as combination, decomposition, single-displacement, or double-displacement.

9. Balance the following equations:
   (a) $MnO_2 + CO \longrightarrow Mn_2O_3 + CO_2$
   (b) $Mg_3N_2 + H_2O \longrightarrow Mg(OH)_2 + NH_3$
   (c) $C_3H_5(NO_3)_3 \longrightarrow CO_2 + H_2O + N_2 + O_2$
   (d) $FeS + O_2 \longrightarrow Fe_2O_3 + SO_2$
   (e) $Cu(NO_3)_2 \longrightarrow CuO + NO_2 + O_2$
   (f) $NO_2 + H_2O \longrightarrow HNO_3 + NO$
   (g) $Al + H_2SO_4 \longrightarrow Al_2(SO_4)_3 + H_2$
   (h) $HCN + O_2 \longrightarrow N_2 + CO_2 + H_2O$
   (i) $B_5H_9 + O_2 \longrightarrow B_2O_3 + H_2O$

10. Balance the following equations:
    (a) $SO_2 + O_2 \longrightarrow SO_3$
    (b) $Al + MnO_2 \overset{\Delta}{\longrightarrow} Mn + Al_2O_3$
    (c) $Na + H_2O \longrightarrow NaOH + H_2$
    (d) $AgNO_3 + Ni \longrightarrow Ni(NO_3)_2 + Ag$
    (e) $Bi_2S_3 + HCl \longrightarrow BiCl_3 + H_2S$
    (f) $PbO_2 \overset{\Delta}{\longrightarrow} PbO + O_2$
    (g) $LiAlH_4 \overset{\Delta}{\longrightarrow} LiH + Al + H_2$
    (h) $KI + Br_2 \longrightarrow KBr + I_2$
    (i) $K_3PO_4 + BaCl_2 \longrightarrow KCl + Ba_3(PO_4)_2$

11. Change these word equations into formula equations and balance them:
    (a) water $\longrightarrow$ hydrogen + oxygen
    (b) acetic acid + potassium hydroxide $\longrightarrow$
        potassium acetate + water
    (c) phosphorus + iodine $\longrightarrow$ phosphorus triiodide
    (d) aluminum + copper(II) sulfate $\longrightarrow$
        copper + aluminum sulfate
    (e) ammonium sulfate + barium chloride $\longrightarrow$
        ammonium chloride + barium sulfate
    (f) sulfur tetrafluoride + water $\longrightarrow$
        sulfur dioxide + hydrogen fluoride
    (g) chromium(III) carbonate $\overset{\Delta}{\longrightarrow}$
        chromium(III) oxide + carbon dioxide

12. Change these word equations into formula equations and balance them:
    (a) copper + sulfur $\overset{\Delta}{\longrightarrow}$ copper(I) sulfide
    (b) phosphoric acid + calcium hydroxide $\longrightarrow$
        calcium phosphate + water
    (c) silver oxide $\overset{\Delta}{\longrightarrow}$ silver + oxygen
    (d) iron(III) chloride + sodium hydroxide $\longrightarrow$
        iron(III) hydroxide + sodium chloride
    (e) nickel(II) phosphate + sulfuric acid $\longrightarrow$
        nickel(II) sulfate + phosphoric acid
    (f) zinc carbonate + hydrochloric acid $\longrightarrow$
        zinc chloride + water + carbon dioxide
    (g) silver nitrate + aluminum chloride $\longrightarrow$
        silver chloride + aluminum nitrate

13. Use the activity series to predict which of the following reactions will occur. Complete and balance the equations. Where no reaction will occur, write "no reaction" as the product.
    (a) $Ag(s) + H_2SO_4(aq) \longrightarrow$
    (b) $Cl_2(g) + NaBr(aq) \longrightarrow$
    (c) $Mg(s) + ZnCl_2(aq) \longrightarrow$
    (d) $Pb(s) + AgNO_3(aq) \longrightarrow$

14. Use the activity series to predict which of the following reactions will occur. Complete and balance the equations. Where no reaction will occur, write "no reaction" as the product.
    (a) $Cu(s) + FeCl_3(aq) \longrightarrow$
    (b) $H_2(g) + Al_2O_3(aq) \longrightarrow$
    (c) $Al(s) + HBr(aq) \longrightarrow$
    (d) $I_2(s) + HCl(aq) \longrightarrow$

15. Complete and balance the equations for these reactions. All reactions yield products.
    (a) $H_2 + I_2 \longrightarrow$
    (b) $CaCO_3 \overset{\Delta}{\longrightarrow}$
    (c) $Mg + H_2SO_4 \longrightarrow$
    (d) $FeCl_2 + NaOH \longrightarrow$

16. Complete and balance the equations for these reactions. All reactions yield products.
    (a) $SO_2 + H_2O \longrightarrow$
    (b) $SO_3 + H_2O \longrightarrow$
    (c) $Ca + H_2O \longrightarrow$
    (d) $Bi(NO_3)_3 + H_2S \longrightarrow$

17. Complete and balance the equations for these reactions. All reactions yield products.
    (a) $Ba + O_2 \longrightarrow$
    (b) $NaHCO_3 \overset{\Delta}{\longrightarrow} Na_2CO_3 +$
    (c) $Ni + CuSO_4 \longrightarrow$
    (d) $MgO + HCl \longrightarrow$
    (e) $H_3PO_4 + KOH \longrightarrow$

18. Complete and balance the equations for these reactions. All reactions yield products.
    (a) $C + O_2 \longrightarrow$
    (b) $Al(ClO_3)_3 \overset{\Delta}{\longrightarrow} O_2 +$
    (c) $CuBr_2 + Cl_2 \longrightarrow$
    (d) $SbCl_3 + (NH_4)_2S \longrightarrow$
    (e) $NaNO_3 \overset{\Delta}{\longrightarrow} NaNO_2 +$

19. Interpret these chemical reactions in terms of the number of moles of each reactant and product:
    (a) $MgBr_2 + 2\,AgNO_3 \longrightarrow Mg(NO_3)_2 + 2\,AgBr$
    (b) $N_2 + 3\,H_2 \longrightarrow 2\,NH_3$
    (c) $2\,C_3H_7OH + 9\,O_2 \longrightarrow 6\,CO_2 + 8\,H_2O$

20. Interpret these equations in terms of the relative number of moles of each substance involved and indicate whether the reaction is exothermic or endothermic:
    (a) $2\,Na + Cl_2 \longrightarrow 2\,NaCl + 822\ kJ$
    (b) $PCl_5 + 92.9\ kJ \longrightarrow PCl_3 + Cl_2$

21. Write balanced equations for each of these reactions, including the heat term:

(a) Lime (CaO) is converted to slaked lime $Ca(OH)_2$ by reaction with water. The reaction liberates 65.3 kJ of heat for each mole of lime reacted.

(b) The industrial production of aluminum metal from aluminum oxide is an endothermic electrolytic process requiring 1630 kJ per mole of Al_2O_3. Oxygen is also a product.

22. Write a balanced equation for these reactions. Include a heat term on the appropriate side of the equation.

(a) Powdered aluminum will react with crystals of iodine when moistened with dishwashing detergent. The reaction produces violet sparks and flaming aluminum. The major product is aluminum iodide (AlI_3). The detergent is not a reactant.

(b) Copper(II) oxide (CuO), a black powder, can be decomposed to produce pure copper by heating the powder in the presence of methane gas (CH_4). The products are copper, carbon dioxide, and water vapor.

Additional Exercises

23. Name one piece of evidence that a chemical reaction is actually taking place in each of these situations:

(a) making a piece of toast

(b) frying an egg

(c) striking a match

24. Balance this equation, using the smallest possible whole numbers. Then determine how many atoms of oxygen appear on each side of the equation:

$$P_4O_{10} + HClO_4 \longrightarrow Cl_2O_7 + H_3PO_4$$

25. Suppose that in a balanced equation the term $7 Al_2(SO_4)_3$ appears.

(a) How many atoms of aluminum are represented?

(b) How many atoms of sulfur are represented?

(c) How many atoms of oxygen are represented?

(d) How many atoms of all kinds are represented?

26. Name two pieces of information that can be obtained from a balanced chemical equation. Name two pieces of information that the reaction does not provide.

27. Make a drawing to show six molecules of ammonia gas decomposing to form hydrogen and nitrogen gases.

28. Explain briefly why this single-displacement reaction will not take place:

$$Zn + Mg(NO_3)_2 \longrightarrow \text{no reaction}$$

29. A student does an experiment to determine where titanium metal should be placed on the activity series chart. He places newly cleaned pieces of titanium into solutions of nickel(II) nitrate, lead(II) nitrate, and magnesium nitrate. He finds that the titanium reacts with the nickel(II) nitrate and lead(II) nitrate solutions, but not with the magnesium nitrate solution. From this information place titanium in the activity series in a position relative to these ions.

30. Complete and balance the equations for these combination reactions:

(a) $K + O_2 \longrightarrow$

(b) $Al + Cl_2 \longrightarrow$

(c) $CO_2 + H_2O \longrightarrow$

(d) $CaO + H_2O \longrightarrow$

31. Complete and balance the equations for these decomposition reactions:

(a) $HgO \xrightarrow{\Delta}$

(b) $NaClO_3 \xrightarrow{\Delta}$

(c) $MgCO_3 \xrightarrow{\Delta}$

(d) $PbO_2 \xrightarrow{\Delta} PbO +$

32. Complete and balance the equations for these single-displacement reactions:

(a) $Zn + H_2SO_4 \longrightarrow$

(b) $AlI_3 + Cl_2 \longrightarrow$

(c) $Mg + AgNO_3 \longrightarrow$

(d) $Al + CoSO_4 \longrightarrow$

33. Complete and balance the equations for these double-displacement reactions:

(a) $ZnCl_2 + KOH \rightarrow$

(b) $CuSO_4 + H_2S \rightarrow$

(c) $Ca(OH)_2 + H_3PO_4 \rightarrow$

(d) $(NH_4)_3PO_4 + Ni(NO_3)_2 \rightarrow$

(e) $Ba(OH)_2 + HNO_3 \rightarrow$

(f) $(NH_4)_2S + HCl \rightarrow$

34. Predict which of the following double-displacement reactions will occur. Complete and balance the equations. Where no reaction will occur, write "no reaction" as the product.

(a) $AgNO_3(aq) + KCl(aq) \longrightarrow$

(b) $Ba(NO_3)_2(aq) + MgSO_4(aq) \longrightarrow$

(c) $H_2SO_4(aq) + Mg(OH)_2(aq) \longrightarrow$

(d) $MgO(s) + H_2SO_4(aq) \longrightarrow$

(e) $Na_2CO_3(aq) + NH_4Cl(aq) \longrightarrow$

35. Write balanced equations for the combustion of these hydrocarbons:

(a) ethane, C_2H_6

(b) benzene, C_6H_6

(c) heptane, C_7H_{16}

36. List the factors that contribute to an increase in carbon dioxide in our atmosphere.

37. List three gases considered to be greenhouse gases. Explain why they are given this name.

38. How can the effects of global warming be reduced?

39. What happens to carbon dioxide released into our atmosphere?

Answers to Practice Exercises

8.1 $4 \text{Al} + 3 \text{O}_2 \longrightarrow 2 \text{Al}_2\text{O}_3$

8.2 $3 \text{Mg(OH)}_2 + 2 \text{H}_3\text{PO}_4 \longrightarrow \text{Mg}_3(\text{PO}_4)_2 + 6 \text{H}_2\text{O}$

8.3 (a) hydrogen gas + chlorine gas $\longrightarrow$ hydrogen chloride

(b) $\text{H}_2(g) + \text{Cl}_2(g) \longrightarrow 2 \text{HCl}(g)$

(c)

$\text{H}_2(g)$	+	$\text{Cl}_2(g)$	$\longrightarrow$	$2 \text{HCl}(g)$
1 molecule		1 molecule		2 molecules
2 atoms H		2 atoms Cl		2 atoms H
				2 atoms Cl
1 mol		1 mol		2 mol
2.016 g		70.90 g		2 (36.46 g)
				(72.92 g)

(d) $2 \text{ mol H}_2 + 2 \text{ mol Cl}_2 \longrightarrow 4 \text{ mol HCl}$ (145.8 g)

8.4 (a) $\text{Fe} + \text{MgCl}_2 \longrightarrow$ no reaction

(b) $\text{Zn}(s) + \text{Pb(NO}_3)_2(aq) \longrightarrow \text{Pb}(s) + \text{Zn(NO}_3)_2(aq)$

8.5 (a) $2 \text{K}_3\text{PO}_4(aq) + 3 \text{BaCl}_2(aq) \longrightarrow \text{Ba}_3(\text{PO}_4)_2(s) + 6 \text{KCl}(aq)$

(b) $2 \text{HCl}(aq) + \text{NiCO}_3(aq) \longrightarrow \text{NiCl}_2(aq) + \text{H}_2\text{O}(l) + \text{CO}_2(g)$

(c) $\text{NH}_4\text{Cl}(aq) + \text{NaNO}_3(aq) \longrightarrow$ no reaction

9

Calculations from Chemical Equations

CHAPTER 9 / OUTLINE

▲
A chemist must measure exact quantities of each reactant to produce new chemical compounds.

The old adage "waste not, want not" is appropriate in our daily life and in the laboratory. Determining correct amounts comes into play in most all professions. A seamstress determines the amount of material, lining, and trim necessary to produce a gown for her client by relying on a pattern or her own experience to guide the selection. A carpet layer determines the correct amount of carpet and padding necessary to recarpet a customer's house by calculating the floor area. The IRS determines the correct deduction for federal income taxes from your paycheck based on your expected annual income.

The chemist also finds it necessary to calculate amounts of products or reactants by using a balanced chemical equation. With these calculations, the chemist can control the amount of product by scaling the reaction up or down to fit the needs of the laboratory, and can thereby minimize waste or excess materials formed during the reaction.

9.1 A Short Review

Molar mass Molar mass is the sum of the atomic masses of all the atoms in a molecule. Molar mass also applies to the mass of a mole of any formula unit—atoms, molecules, or ions; it is the atomic mass of an atom, or the sum of the atomic masses in a molecule or an ion (in grams).

Relationship between molecule and mole A molecule is the smallest unit of a molecular substance (e.g., Br_2), and a mole is Avogadro's number (6.022×10^{23}) of molecules of that substance. A mole of bromine (Br_2) has the same number of molecules as a mole of carbon dioxide, a mole of water, or a mole of any other molecular substance. When we relate molecules to molar mass, 1 molar mass is equivalent to 1 mol, or 6.022×10^{23} molecules.

The term *mole* also refers to any chemical species. It represents a quantity (6.022×10^{23} particles) and may be applied to atoms, ions, electrons, and formula units of nonmolecular substances. In other words,

$$1 \text{ mole} = \begin{cases} 6.022 \times 10^{23} \text{ molecules} \\ 6.022 \times 10^{23} \text{ formula units} \\ 6.022 \times 10^{23} \text{ atoms} \\ 6.022 \times 10^{23} \text{ ions} \end{cases}$$

Other useful mole relationships are

$$\text{molar mass} = \frac{\text{grams of a substance}}{\text{number of moles of the substance}}$$

$$\text{molar mass} = \frac{\text{grams of a monatomic element}}{\text{number of moles of the element}}$$

$$\text{number of moles} = \frac{\text{number of molecules}}{6.022 \times 10^{23} \text{ molecules/mole}}$$

▲
A mole of water, salt and any gas all have the same number of particles (6.022×10^{23}).

Balanced equations When using chemical equations for calculations of mole–mass–volume relationships between reactants and products, the equations must be balanced. *Remember*: The number in front of a formula in a balanced chemical equation represents the number of moles of that substance in the chemical reaction.

Introduction to Stoichiometry: The Mole-Ratio Method

We often need to calculate the amount of a substance that is either produced from, or needed to react with, a given quantity of another substance. The area of chemistry that deals with quantitative relationships among reactants and products is known as **stoichiometry** *(stoy-key-ah-meh-tree)*. The *mole* or *mole-ratio* method is generally best for solving problems in stoichiometry.

A **mole ratio** is a ratio between the number of moles of any two species involved in a chemical reaction. For example, in the reaction

$$2\ H_2 + O_2 \longrightarrow 2\ H_2O$$
2 mol   1 mol   2 mol

six mole ratios can be written:

$$\frac{2\ \text{mol}\ H_2}{1\ \text{mol}\ O_2} \qquad \frac{2\ \text{mol}\ H_2}{2\ \text{mol}\ H_2O} \qquad \frac{1\ \text{mol}\ O_2}{2\ \text{mol}\ H_2}$$

$$\frac{1\ \text{mol}\ O_2}{2\ \text{mol}\ H_2O} \qquad \frac{2\ \text{mol}\ H_2O}{2\ \text{mol}\ H_2} \qquad \frac{2\ \text{mol}\ H_2O}{1\ \text{mol}\ O_2}$$

We use the mole ratio to convert the number of moles of one substance to the corresponding number of moles of another substance in a chemical reaction. For example, if we want to calculate the number of moles of H_2O that can be obtained from 4.0 mol of O_2, we use the mole ratio 2 mol H_2O/1 mol O_2:

$$(4.0\ \text{mol}\ O_2)\left(\frac{2\ \text{mol}\ H_2O}{1\ \text{mol}\ O_2}\right) = 8.0\ \text{mol}\ H_2O$$

Since you will encounter stoichiometric problems frequently, master the mole-ratio method now. There are three basic steps:

1. Convert the quantity of starting substance to moles (if it is not given in moles).
2. Convert the moles of starting substance to moles of desired substance.
3. Convert the moles of desired substance to the units specified in the problem.

Like balancing chemical equations, making stoichiometric calculations requires practice. A detailed description of the mole-ratio method and several worked examples follow. Study this material and practice on the problems at the end of this chapter.

Use a balanced equation.

Step 1. Determine the number of moles of starting substance.
Identify the starting substance from the data given in the problem state-

stoichiometry

mole ratio

Write and balance the equation before you begin the problem.

ment. Convert the quantity of the starting substance to moles, if it is not already done.

$$\text{moles} = (\text{grams}) \left(\frac{1 \text{ mole}}{\text{molar mass}} \right)$$

As in all problems with units, the desired quantity is in the numerator, and the quantity to be eliminated is in the denominator.

Step 2. Determine the mole ratio of the desired substance to the starting substance.

The number of moles of each substance in the balanced equation is indicated by the coefficient in front of each substance. Use these coefficients to set up the mole ratio:

$$\text{mole ratio} = \frac{\text{moles of } \boxed{\text{desired substance}} \text{ in the equation}}{\text{moles of } \boxed{\text{starting substance}} \text{ in the equation}}$$

Multiply the number of moles of starting substance (from Step 1) by the mole ratio to obtain the number of moles of desired substance:

Units of moles of starting substance cancel in the numerator and denominator.

$$\begin{array}{c} \text{moles of desired} \\ \text{substance} \end{array} = \left(\begin{array}{c} \overset{\text{From Step 1}}{} \\ \text{moles of starting} \\ \text{substance} \end{array} \right) \left(\frac{\begin{array}{c} \text{moles of desired substance} \\ \text{in the equation} \end{array}}{\begin{array}{c} \text{moles of starting substance} \\ \text{in the equation} \end{array}} \right)$$

Step 3. Calculate the desired substance in the units specified in the problem.

If the answer is to be in moles, the calculation is complete. If units other than moles are wanted, multiply the moles of the desired substance (from Step 2) by the appropriate factor to convert moles to the units required. For example, if grams of the desired substance are wanted,

$$\text{grams} = (\text{moles}) \left(\frac{\text{molar mass}}{1 \text{ mol}} \right)$$

If moles $\longrightarrow$ atoms, use $\dfrac{6.022 \times 10^{23} \text{ atoms}}{1 \text{ mol}}$

If moles $\longrightarrow$ molecules, use $\dfrac{6.022 \times 10^{23} \text{ molecules}}{1 \text{ mol}}$

The steps for converting the mass of a starting substance A to either the mass, atoms, or molecules of desired substance B are summarized in Figure 9.1.

FIGURE 9.1 ▶
Steps for converting starting substance A to mass, atoms, or molecules of desired substance B.

9.3 Mole–Mole Calculations

Let's use the mole-ratio method to solve stoichiometric problems for mole–mole calculations. The quantity of starting substance is given in moles and the quantity of desired substance is requested in moles.

How many moles of carbon dioxide will be produced by the complete reaction of 2.0 mol of glucose ($C_6H_{12}O_6$), according to the following reaction?

Example 9.1

$$C_6H_{12}O_6 + 6\ O_2 \longrightarrow 6\ CO_2 + 6\ H_2O$$

   1 mol   6 mol   6 mol   6 mol

The balanced equation states that 6 mol of CO_2 will be produced from 1 mol of $C_6H_{12}O_6$. Even though we can readily see that 12 mol of CO_2 will be formed from 2.0 mol of $C_6H_{12}O_6$, let's use the mole-ratio method to solve the problem.

Solution

Step 1. The number of moles of starting substance is 2.0 mol $C_6H_{12}O_6$.

Step 2. The conversion needed is

      moles $C_6H_{12}O_6 \longrightarrow$ moles CO_2

Multiply 2.0 mol of glucose (given in the problem) by this mole ratio:

$$(2.0\ \text{mol } C_6H_{12}O_6) \left(\frac{6\ \text{mol } CO_2}{1\ \text{mol } C_6H_{12}O_6} \right) = 12\ \text{mol } CO_2$$

The mole ratio (shown in color) is exact and does not affect the number of significant figures in the answer.

Note how the units work; the moles of $C_6H_{12}O_6$ cancel, leaving the answer in units of moles of CO_2.

How many moles of ammonia can be produced from 8.00 mol of hydrogen reacting with nitrogen? The balanced equation is

Example 9.2

$$3\ H_2 + N_2 \longrightarrow 2\ NH_3$$

Step 1. The starting substance is 8.00 mol of H_2.

Solution

Step 2. The conversion needed is

      moles $H_2 \longrightarrow$ moles NH_3

The balanced equation states that we get 2 mol of NH_3 for every 3 mol of H_2 that react. Set up the mole ratio of desired substance (NH_3) to starting substance (H_2):

$$\text{mole ratio} = \frac{2\ \text{mol } NH_3}{3\ \text{mol } H_2} \quad \text{(from equation)}$$

Multiply the 8.00 mol H_2 by the mole ratio:

$$(8.00\ \text{mol } H_2) \left(\frac{2\ \text{mol } NH_3}{3\ \text{mol } H_2} \right) = 5.33\ \text{mol } NH_3$$

Example 9.3   Given the balanced equation

$$K_2Cr_2O_7 + 6\ KI + 7\ H_2SO_4 \longrightarrow Cr_2(SO_4)_3 + 4\ K_2SO_4 + 3\ I_2 + 7\ H_2O$$

1 mol   6 mol   3 mol

calculate (a) the number of moles of potassium dichromate ($K_2Cr_2O_7$) that will react with 2.0 mol of potassium iodide (KI) and (b) the number of moles of iodine (I_2) that will be produced from 2.0 mol of potassium iodide.

Solution   Since the equation is balanced, we are concerned only with $K_2Cr_2O_7$, KI, and I_2, and we can ignore all the other substances. The equation states that 1 mol of $K_2Cr_2O_7$ will react with 6 mol of KI to produce 3 mol of I_2.

(a) Calculate the number of moles of $K_2Cr_2O_7$.

Step 1. The starting substance is 2.0 mol of KI.

Step 2. The conversion needed is

$$\text{moles KI} \longrightarrow \text{moles } K_2Cr_2O_7$$

Set up the mole ratio of desired substance to starting substance:

$$\text{mole ratio} = \frac{1\ \text{mol } K_2Cr_2O_7}{6\ \text{mol KI}} \quad \text{(from equation)}$$

Multiply the moles of starting material by this ratio:

$$(2.0\ \text{mol KI}) \left(\frac{1\ \text{mol } K_2Cr_2O_7}{6\ \text{mol KI}} \right) = 0.33\ \text{mol } K_2Cr_2O_7$$

(b) Calculate the number of moles of I_2.

Step 1. The moles of starting substance are 2.0 mol KI as in part (a).

Step 2. The conversion needed is

$$\text{moles KI} \longrightarrow \text{moles } I_2$$

Set up the mole ratio of desired substance to starting substance:

$$\text{mole ratio} = \frac{3\ \text{mol } I_2}{6\ \text{mol KI}} \quad \text{(from equation)}$$

Multiply the moles of starting material by this ratio:

$$(2.0\ \text{mol KI}) \left(\frac{3\ \text{mol } I_2}{6\ \text{mol KI}} \right) = 1.0\ \text{mol } I_2$$

Example 9.4   How many molecules of water can be produced by reacting 0.010 mol of oxygen with hydrogen?

Solution   The balanced equation is $2\ H_2 + O_2 \longrightarrow 2\ H_2O$.

The sequence of conversions needed in the calculation is

moles $O_2 \longrightarrow$ moles $H_2O \longrightarrow$ molecules H_2O

Step 1. The starting substance is 0.010 mol O_2.

Step 2. The conversion needed is moles $O_2 \longrightarrow$ moles H_2O. Set up the mole ratio of desired substance to starting substance:

$$\text{mole ratio} = \frac{2 \text{ mol } H_2O}{1 \text{ mol } O_2} \quad \text{(from equation)}$$

Multiply 0.010 mol O_2 by the mole ratio:

$$(0.010 \text{ mol } O_2) \left(\frac{2 \text{ mol } H_2O}{1 \text{ mol } O_2} \right) = 0.020 \text{ mol } H_2O$$

Step 3. Since the problem asks for molecules instead of moles of H_2O, we must convert moles to molecules. Use the conversion factor $(6.022 \times 10^{23}$ molecules)/mole:

$$(0.020 \text{ mol } H_2O) \left(\frac{6.022 \times 10^{23} \text{ molecules}}{1 \text{ mol}} \right) = 1.2 \times 10^{22} \text{ molecules } H_2O$$

Note that 0.020 mol is still quite a large number of water molecules.

▲
The space shuttle is powered by H_2 and O_2, which react to produce H_2O.

Practice 9.1

How many moles of aluminum oxide will be produced from 0.50 mol of oxygen?

$$4 \text{ Al} + 3 \text{ O}_2 \longrightarrow 2 \text{ Al}_2O_3$$

Practice 9.2

How many moles of aluminum hydroxide are required to produce 22.0 mol of water?

$$2 \text{ Al(OH)}_3 + 3 \text{ H}_2SO_4 \longrightarrow \text{Al}_2(SO_4)_3 + 6 \text{ H}_2O$$

9.4   Mole–Mass Calculations

The object of this type of problem is to calculate the mass of one substance that reacts with or is produced from a given number of moles of another substance in a chemical reaction. If the mass of the starting substance is given, we need to convert it to moles. We use the mole ratio to convert moles of starting substance to moles of desired substance. If required, we can then change moles of desired substance to mass. Each example is solved in two ways.

Method 1: Step-by-step.
Method 2: Continuous calculation, where the individual steps are combined in a single line.

Select the method that is easiest for you and use it to practice solving problems.

Example 9.5   What mass of hydrogen can be produced by reacting 6.0 mol of aluminum with hydrochloric acid?

Solution   The balanced equation is $2\,Al(s) + 6\,HCl(aq) \longrightarrow 2\,AlCl_3(aq) + 3\,H_2(g)$.

Method 1. Step-by-step

First calculate the moles of hydrogen produced, using the mole-ratio method; then calculate the mass of hydrogen by multiplying the moles of hydrogen by its grams per mole.

Step 1. The starting substance is 6.0 mol of aluminum.

Step 2. Calculate moles of H_2 by the mole-ratio method:

$$\text{moles Al} \longrightarrow \text{moles H}_2$$

$$(6.0\ \text{mol Al})\left(\frac{3\ \text{mol H}_2}{2\ \text{mol Al}}\right) = 9.0\ \text{mol H}_2$$

Step 3. Convert moles of H_2 to grams:

$$\text{moles H}_2 \longrightarrow \text{grams H}_2$$

$$(9.0\ \text{mol H}_2)\left(\frac{2.016\ \text{g H}_2}{1\ \text{mol H}_2}\right) = 18\ \text{g H}_2$$

We see that 18 g of H_2 can be produced by reacting 6.0 mol of Al with HCl.

Method 2. Continuous calculation

$$\text{moles Al} \longrightarrow \text{moles H}_2 \longrightarrow \text{grams H}_2$$

$$(6.0\ \text{mol Al})\left(\frac{3\ \text{mol H}_2}{2\ \text{mol Al}}\right)\left(\frac{2.016\ \text{g H}_2}{1\ \text{mol H}_2}\right) = 18\ \text{g H}_2$$

▲
Al wire in HCl.

Example 9.6   How many moles of water can be produced by burning 325 g of octane (C_8H_{18})? The balanced equation is $2\,C_8H_{18}(g) + 25\,O_2(g) \longrightarrow 16\,CO_2(g) + 18\,H_2O(g)$.

Solution   **Method 1. Step-by-step**

Step 1. The starting substance is 325 g C_8H_{18}. Convert 325 g of C_8H_{18} to moles:

$$\text{grams C}_8\text{H}_{18} \longrightarrow \text{moles C}_8\text{H}_{18}$$

$$(325\ \text{g C}_8\text{H}_{18})\left(\frac{1\ \text{mol C}_8\text{H}_{18}}{114.2\ \text{g C}_8\text{H}_{18}}\right) = 2.85\ \text{mol C}_8\text{H}_{18}$$

Step 2. Calculate the moles of water by the mole-ratio method:

$$\text{moles } C_8H_{18} \longrightarrow \text{moles } H_2O$$

$$(2.85 \text{ mol } C_8H_{18}) \left(\frac{18 \text{ mol } H_2O}{2 \text{ mol } C_8H_{18}} \right) = 25.7 \text{ mol } H_2O$$

Method 2. Continuous calculation

$$\text{grams } C_8H_{18} \longrightarrow \text{moles } C_8H_{18} \longrightarrow \text{moles } H_2O$$

$$(325 \text{ g } C_8H_{18}) \left(\frac{1 \text{ mol } C_8H_{18}}{114.2 \text{ g } C_8H_{18}} \right) \left(\frac{18 \text{ mol } H_2O}{2 \text{ mol } C_8H_{18}} \right) = 25.6 \text{ mol } H_2O$$

The answers for the different methods vary in the last digit. This results from rounding off at different times in the calculation. Check with your instructor to find the appropriate rules for your course.

Practice 9.3

How many moles of potassium chloride can be produced from 100.0 g of potassium chlorate?

$$2 \text{ KClO}_3 \longrightarrow 2 \text{ KCl} + 3 \text{ O}_2$$

Practice 9.4

How many grams of silver nitrate are required to produce 0.25 mol of silver sulfide?

$$2 \text{ AgNO}_3 + H_2S \longrightarrow Ag_2S + 2 \text{ HNO}_3$$

9.5 Mass–Mass Calculations

Solving mass–mass stoichiometry problems requires all the steps of the mole-ratio method. The mass of starting substance is converted to moles. The mole ratio is then used to determine moles of desired substance, which in turn is converted to mass.

Use either the step-by-step method or the continuous calculation method to solve these problems.

What mass of carbon dioxide is produced by the complete combustion of 100. g of the hydrocarbon pentane, C_5H_{12}? The balanced equation is

$$C_5H_{12} + 8 \text{ O}_2 \qquad 5 \text{ CO}_2 + 6 \text{ H}_2O$$

Method 1. Step-by-step

Step 1. The starting substance is 100. g of C_5H_{12}. Convert 100. g of C_5H_{12} to moles:

$$\text{grams} \longrightarrow \text{moles}$$

$$(100. \text{ g } C_5H_{12}) \left(\frac{1 \text{ mol } C_5H_{12}}{72.15 \text{ g } C_5H_{12}} \right) = 1.39 \text{ mol } C_5H_{12}$$

Example 9.7

Solution

Step 2. Calculate the moles of CO_2 by the mole-ratio method:

$$\text{moles } C_5H_{12} \longrightarrow \text{moles } CO_2$$

$$(1.39 \text{ mol } C_5H_{12}) \left(\frac{5 \text{ mol } CO_2}{1 \text{ mol } C_5H_{12}} \right) = 6.95 \text{ mol } CO_2$$

Step 3. Convert moles of CO_2 to grams:

$$\text{moles } CO_2 \longrightarrow \text{grams } CO_2$$

$$(6.95 \text{ mol } CO_2) \left(\frac{44.01 \text{ g } CO_2}{1 \text{ mol } CO_2} \right) = 306 \text{ g } CO_2$$

Method 2. Continuous calculation

Remember: Round off as appropriate for your particular course.

$$\text{grams } C_5H_{12} \longrightarrow \text{moles } C_5H_{12} \longrightarrow \text{moles } CO_2 \longrightarrow \text{grams } CO_2$$

$$(100. \text{ g } C_5H_{12}) \left(\frac{1 \text{ mol } C_5H_{12}}{72.15 \text{ g } C_5H_{12}} \right) \left(\frac{5 \text{ mol } CO_2}{1 \text{ mol } C_5H_{12}} \right) \left(\frac{44.01 \text{ g } CO_2}{1 \text{ mol } CO_2} \right) = 305 \text{ g } CO_2$$

Example 9.8   How many grams of nitric acid, HNO_3, are required to produce 8.75 g of dinitrogen monoxide (N_2O) according to the following equation?

$$4 \text{ Zn}(s) + 10 \text{ HNO}_3(aq) \longrightarrow 4 \text{ Zn(NO}_3)_2(aq) + \text{N}_2\text{O}(g) + 5 \text{ H}_2\text{O}(l)$$
$$\qquad\qquad\quad\; 10 \text{ mol} \qquad\qquad\qquad\qquad\qquad\qquad 1 \text{ mol}$$

Solution

Method 1. Step-by-step

Step 1. The starting substance is 8.75 g of N_2O.

$$\text{grams } \dot{N}_2O \longrightarrow \text{moles } N_2O$$

$$(8.75 \text{ g } N_2O) \left(\frac{1 \text{ mol } N_2O}{44.02 \text{ g } N_2O} \right) = 0.199 \text{ mol } N_2O$$

Step 2. Calculate the moles of HNO_3 by the mole-ratio method:

$$\text{moles } N_2O \longrightarrow \text{moles } HNO_3$$

$$(0.199 \text{ mol } N_2O) \left(\frac{10 \text{ mol } HNO_3}{1 \text{ mol } N_2O} \right) = 1.99 \text{ mol } HNO_3$$

Step 3. Convert moles of HNO_3 to grams:

$$\text{moles } HNO_3 \longrightarrow \text{grams } HNO_3$$

$$(1.99 \text{ mol } HNO_3) \left(\frac{63.02 \text{ g } HNO_3}{1 \text{ mol } HNO_3} \right) = 125 \text{ g } HNO_3$$

CHEMISTRY IN ACTION   A Shrinking Technology

The microchip has revolutionized the field of electronics. Engineers at Bell Laboratories, Massachusetts Institute of Technology, University of California, and Stanford University are racing to produce parts for tiny machines and robots. New techniques now produce gears smaller than a grain of sand, and motors lighter than a speck of dust.

To produce ever smaller computers, calculators, and even microbots (micro-sized robots) precise quantities of chemicals in exact proportions are required. The secret to producing minute circuits is to print the entire circuit or blueprint at one time. Computers are used "to draw a

"Cleo" is a tiny surgical ▶ microrobot designed to fit inside the colon of an adult where it can be controlled by a remote joystick. "Cleo" can detect obstacles, grasp and carry objects, and remain upright.

chip." This image is then transferred into a pattern, or mask, with details finer than a human hair. In a process similar to photography, light is then shone through the mask onto a silicon-coated surface. The areas created on the silicon exhibit high or low resistance to chemical etching. Chemicals then etch away the silicon.

Micromachinery is produced in the same way. First a thin layer of silicon dioxide is applied (sacrificial material), then a layer of polysilicon is carefully

applied (structural material). A mask is then applied and the whole structure is covered with plasma (excited gas). The plasma acts as a tiny sandblaster, removing everything the mask doesn't protect. This process is repeated as the entire machine is constructed. When the entire assembly is complete, the whole machine is placed in hydrofluoric acid, which dissolves all the sacrificial material and permits the various parts of the machine to move.

To turn these micromachines into true microbots, current research is focusing on locomotion and sensing imaging systems. Possible uses for these micro-bots include "smart" pills, which could contain sensors or drug reservoirs (currently done for birth control). Tiny pumps, once inside the body, will dispense the proper amount of medication at precisely the correct site. These micro-bots are currently in production for the treatment of diabetes (to release insulin).

Method 2. Continuous calculation

grams $N_2O \longrightarrow$ moles $N_2O \longrightarrow$ moles $HNO_3 \longrightarrow$ grams HNO_3

$$(8.75 \text{ g } N_2O) \left(\frac{1 \text{ mol } N_2O}{44.02 \text{ g } N_2O} \right) \left(\frac{10 \text{ mol } HNO_3}{1 \text{ mol } N_2O} \right) \left(\frac{63.02 \text{ g } HNO_3}{1 \text{ mol } HNO_3} \right) = 125 \text{ g } HNO_3$$

Practice 9.5

How many grams of chromium(III) chloride are required to produce 75.0 g of silver chloride?

$$CrCl_3 + 3 AgNO_3 \longrightarrow Cr(NO_3)_3 + 3 AgCl$$

Practice 9.6

What mass of water is produced by the complete combustion of 225.0 g of butane (C_4H_{10})?

$$2 C_4H_{10} + 13 O_2 \longrightarrow 8 CO_2 + 10 H_2O$$

9.6   Limiting-Reactant and Yield Calculations

In many chemical processes the quantities of the reactants used are such that one reactant is in excess. The amount of the product(s) formed in such a case depends on the reactant that is not in excess. This is the **limiting reactant**—it limits the amount of product that can be formed.

Consider the case illustrated in Figure 9.2. How many bicycles can be assembled from the parts shown? The limiting part in this case is the number of pedal assembles; only three bicycles can be built because there are only three pedal assemblies. The wheels and frames are parts in excess.

Let's consider a chemical example at the molecular level in which seven molecules of H_2 are combined with four molecules of Cl_2 (Figure 9.3a). How many molecules of HCl can be produced according to this reaction?

$$H_2 + Cl_2 \longrightarrow 2\ HCl$$

If the molecules of H_2 and Cl_2 are taken apart and recombined as HCl (Figure 9.3b), we see that eight molecules of HCl can be formed before we run out of Cl_2. Therefore the Cl_2 is the limiting reactant and H_2 is in excess—three molecules of H_2 remain unreacted.

FIGURE 9.2 ▶
The number of bicycles that can be built from these parts is determined by the "limiting reactant" (the pedal assemblies).

8 wheels

4 frames

3 pedal assemblies

3 bicycles

(a) Before reaction

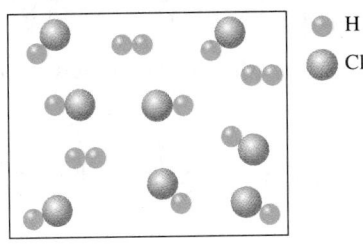

(b) After reaction

◀ **FIGURE 9.3**
Reaction of H_2 with Cl_2.

● H
● Cl

When problem statements give the amounts of two reactants, one of them is usually a limiting reactant. We can identify the limiting reactant using the following method:

1. Calculate the amount of product (moles or grams, as needed) formed from each reactant.
2. Determine which reactant is limiting. (The reactant that gives the least amount of product is the limiting reactant; the other reactant is in excess.)

Sometimes we must find the amount of excess reactant needed in the reaction, so we add a third step:

3. Calculate the amount of excess reactant required to react with the limiting reactant, then subtract this amount from the starting quantity of the excess reactant. This gives the amount of that substance that remains unreacted.

How many moles of Fe_3O_4 can be obtained by reacting 16.8 g Fe with 10.0 g H_2O? Which substance is the limiting reactant? Which substance is in excess?

Example 9.9

$$3\ Fe(s) + 4\ H_2O(g) \xrightarrow{\Delta} Fe_3O_4(s) + 4\ H_2(g)$$

Solution

Step 1. Calculate the moles of Fe_3O_4 that can be formed from each reactant:

g reactant ⟶ mol reactant ⟶ mol Fe_3O_4

The continuous calculation method is shown here. You can also use the step-by-step method to determine the requested substance.

$$(16.8\ \text{g Fe})\left(\frac{1\ \text{mol Fe}}{55.85\ \text{g Fe}}\right)\left(\frac{1\ \text{mol Fe}_3O_4}{3\ \text{mol Fe}}\right) = 0.100\ \text{mol Fe}_3O_4$$

$$(10.0\ \text{g H}_2O)\left(\frac{1\ \text{mol H}_2O}{18.02\ \text{g H}_2O}\right)\left(\frac{1\ \text{mol Fe}_3O_4}{4\ \text{mol H}_2O}\right) = 0.139\ \text{mol Fe}_3O_4$$

Step 2. Determine the limiting reactant. The limiting reactant is Fe because it produces less Fe_3O_4; the H_2O is in excess. The yield of product is 0.100 mol of Fe_3O_4.

How many grams of silver bromide (AgBr) can be formed when solutions containing 50.0 g of $MgBr_2$ and 100.0 g of $AgNO_3$ are mixed together? How many grams of the excess reactant remain unreacted?

Example 9.10

Solution $MgBr_2(aq) + 2\,AgNO_3(aq) \longrightarrow 2\,AgBr(s) + Mg(NO_3)_2(aq)$

Step 1. Calculate the grams of AgBr that can be formed from each reactant.

$$\text{g reactant} \longrightarrow \text{mol reactant} \longrightarrow \text{mol AgBr} \longrightarrow \text{g AgBr}$$

$$(50.0\text{ g MgBr}_2)\left(\frac{1\text{ mol MgBr}_2}{184.1\text{ g MgBr}_2}\right)\left(\frac{2\text{ mol AgBr}}{1\text{ mol MgBr}_2}\right)\left(\frac{187.8\text{ g AgBr}}{1\text{ mol AgBr}}\right) = 102\text{ g AgBr}$$

$$(100.0\text{ g AgNO}_3)\left(\frac{1\text{ mol AgNO}_3}{169.9\text{ g AgNO}_3}\right)\left(\frac{2\text{ mol AgBr}}{2\text{ mol AgNO}_3}\right)\left(\frac{187.8\text{ g AgBr}}{1\text{ mol AgBr}}\right) = 110.5\text{ g AgBr}$$

Step 2. Determine the limiting reactant. The limiting reactant is $MgBr_2$ because it gives less AgBr; $AgNO_3$ is in excess. The yield is 102 g AgBr.

Step 3. Calculate the grams of unreacted $AgNO_3$. Calculate the grams of $AgNO_3$ that will react with 50.0 g of $MgBr_2$:

$$\text{g MgBr}_2 \longrightarrow \text{mol MgBr}_2 \longrightarrow \text{mol AgNO}_3 \longrightarrow \text{g AgNO}_3$$

$$(50.0\text{ g MgBr}_2)\left(\frac{1\text{ mol MgBr}_2}{184.1\text{ g MgBr}_2}\right)\left(\frac{2\text{ mol AgNO}_3}{1\text{ mol MgBr}_2}\right)\left(\frac{169.9\text{ g AgNO}_3}{1\text{ mol AgNO}_3}\right) = 92.3\text{ g AgNO}_3$$

Thus 92.3 g of $AgNO_3$ reacts with 50.0 g of $MgBr_2$. The amount of $AgNO_3$ that remains unreacted is

$$100.0\text{ g AgNO}_3 - 92.3\text{ g AgNO}_3 = 7.7\text{ g AgNO}_3 \text{ unreacted}$$

The final mixture will contain 102 g AgBr(s), 7.7 g $AgNO_3$, and an undetermined amount of $Mg(NO_3)_2$ in solution.

Practice 9.7

How many grams of hydrogen chloride can be produced from 0.490 g of hydrogen and 50.0 g of chlorine?

$$H_2(g) + Cl_2(g) \longrightarrow 2\,HCl(g)$$

Practice 9.8

How many grams of barium sulfate will be formed from 200.0 g of barium nitrate and 100.0 grams of sodium sulfate?

$$Ba(NO_3)_2(aq) + Na_2SO_4(aq) \longrightarrow BaSO_4(s) + 2\,NaNO_3(aq)$$

The quantities of the products we have been calculating from equations represent the maximum yield (100%) of product according to the reaction represented by the equation. Many reactions, especially those involving organic substances, fail to give a 100% yield of product. The main reasons for this failure are the side reactions that give products other than the main product and the fact that many reactions are reversible. In addition, some product may be lost in handling and transferring from one vessel to another. The **theoretical yield** of a reaction is the calculated amount of product that can be obtained from a given amount or reactant, according to the

theoretical yield

chemical equation. The **actual yield** is the amount of product that we finally obtain.

The **percent yield** is the ratio of the actual yield to the theoretical yield multiplied by 100. Both the theoretical and the actual yields must have the same units to obtain a percent:

$$\frac{\text{actual yield}}{\text{theoretical yield}} \times 100 = \text{percent yield}$$

For example, if the theoretical yield calculated for a reaction is 14.8 g and the amount of product obtained is 9.25 g, the percent yield is

$$\text{percent yield} = \left(\frac{9.25\ \text{g}}{14.8\ \text{g}}\right)(100) = 62.5\%$$

Round off as appropriate for your particular course.

Carbon tetrachloride (CCl_4) was prepared by reacting 100. g of carbon disulfide and 100. g of chlorine. Calculate the percent yield if 65.0 g of CCl_4 was obtained from the reaction:

Example 9.11

$$CS_2 + 3\ Cl_2 \longrightarrow CCl_4 + S_2Cl_2$$

We need to determine the limiting reactant to calculate the quantity of CCl_4 (theoretical yield) that can be formed. Then we can compare that amount with the 65.0 g CCl_4 (actual yield) to calculate the percent yield.

Solution

Step 1. Determine the theoretical yield. Calculate the grams of CCl_4 that can be formed from each reactant:

g reactant $\longrightarrow$ mol reactant $\longrightarrow$ mol CCl_4 $\longrightarrow$ g CCl_4

$$(100.\ \text{g}\ CS_2)\left(\frac{1\ \text{mol}\ CS_2}{76.15\ \text{g}\ CS_2}\right)\left(\frac{1\ \text{mol}\ CCl_4}{1\ \text{mol}\ CS_2}\right)\left(\frac{153.8\ \text{g}\ CCl_4}{1\ \text{mol}\ CCl_4}\right) = 202\ \text{g}\ CCl_4$$

$$(100.\ \text{g}\ Cl_2)\left(\frac{1\ \text{mol}\ Cl_2}{70.90\ \text{g}\ Cl_2}\right)\left(\frac{1\ \text{mol}\ CCl_4}{3\ \text{mol}\ Cl_2}\right)\left(\frac{153.8\ \text{g}\ CCl_4}{1\ \text{mol}\ CCl_4}\right) = 72.3\ \text{g}\ CCl_4$$

Step 2. Determine the limiting reactant. The limiting reactant is Cl_2 because it gives less CCl_4. The CS_2 is in excess. The theoretical yield is 72.3 g CCl_4.

Step 3. Calculate the percent yield. According to the equation, 72.3 g of CCl_4 is the maximum amount or theoretical yield of CCl_4 possible from 100. g of Cl_2. Actual yield is 65.0 g of CCl_4:

$$\text{percent yield} = \left(\frac{65.0\ \text{g}}{72.3\ \text{g}}\right)\left(100\right) = 89.9\%$$

Silver bromide was prepared by reacting 200.0 g of magnesium bromide and an adequate amount of silver nitrate. Calculate the percent yield if 375.0 g of silver bromide was obtained from the reaction:

Example 9.12

$$MgBr_2 + 2\ AgNO_3 \longrightarrow Mg(NO_3)_3 + 2\ AgBr$$

Solution

Step 1. Determine the theoretical yield. Calculate the grams of AgBr that can be formed:

$$\text{g MgBr}_2 \longrightarrow \text{mol MgBr}_2 \longrightarrow \text{mol AgBr} \longrightarrow \text{g AgBr}$$

$$(200.0 \text{ g MgBr}_2)\left(\frac{1 \text{ mol MgBr}_2}{184.1 \text{ g MgBr}_2}\right)\left(\frac{2 \text{ mol AgBr}}{1 \text{ mol MgBr}_2}\right)\left(\frac{187.8 \text{ g AgBr}}{1 \text{ mol AgBr}}\right) = 408.0 \text{ g AgBr}$$

The theoretical yield is 408.0 g AgBr.

Step 2. Calculate the percent yield. According to the equation, 408.0 g AgBr is the maximum amount of AgBr possible from 200.0 g MgBr$_2$. Actual yield is 375.0 g AgBr:

$$\text{percent yield} = \frac{375.0 \text{ g AgBr}}{408.0 \text{ g AgBr}} \times 100 = 91.91\%$$

Practice 9.9

Aluminum oxide was prepared by heating 225 g of chromium(II) oxide with 125 g of aluminum. Calculate the percent yield if 100.0 g of aluminum oxide was obtained:

$$2 \text{ Al} + 3 \text{ CrO} \longrightarrow \text{Al}_2\text{O}_3 + 3 \text{ Cr}$$

When solving problems, you will achieve better results if you work in an organized manner.

1. Write data and numbers in a logical, orderly manner.
2. Make certain that the equations are balanced and that the computations are accurate and expressed to the correct number of significant figures.
3. Check the units. Units are very important; a number without units has little or no meaning.

Concepts in Review

1. Write mole ratios for any two substances involved in a chemical reaction.

2. Outline the mole-ratio method for making stoichiometric calculations.

3. Calculate the number of moles of a desired substance obtainable from a given number of moles of a starting substance in a chemical reaction (mole–mole calculations).

4. Calculate the mass of a desired substance obtainable from a given number of moles of a starting substance in a chemical reaction, and vice versa (mole–mass and mass–mole calculations).

5. Calculate the mass of a desired substance involved in a chemical reaction from a given mass of a starting substance (mass–mass calculation).

6. Deduce the limiting reactant when given the amounts of starting substances, and then calculate the moles or mass of desired substance obtainable from a given chemical reaction (limiting reactant calculation).

7. Apply theoretical yield or actual yield to any of the foregoing types of problems, or calculate theoretical and actual yields of a chemical reaction.

Key Terms

actual yield (9.6)
limiting reactant (9.6)
mole ratio (9.2)

percent yield (9.6)
stoichiometry (9.2)
theoretical yield (9.6)

Questions

1. Phosphine (PH$_3$) can be prepared by the hydrolysis of calcium phosphide, Ca$_3$P$_2$:

$$Ca_3P_2 + 6\,H_2O \longrightarrow 3\,Ca(OH)_2 + 2\,PH_3$$

Based on this equation, which of the following statements are correct? Show evidence to support your answer.
(a) One mole of Ca$_3$P$_2$ produces 2 mol of PH$_3$.
(b) One gram of Ca$_3$P$_2$ produces 2 g of PH$_3$.
(c) Three moles of Ca(OH)$_2$ are produced for each 2 mol of PH$_3$ produced.
(d) The mole ratio between phosphine and calcium phosphide is
$$\frac{2\ \text{mol PH}_3}{1\ \text{mol Ca}_3\text{P}_2}.$$
(e) When 2.0 mol of Ca$_3$P$_2$ and 3.0 mol of H$_2$O react, 4.0 mol of PH$_3$ can be formed.
(f) When 2.0 mol of Ca$_3$P$_2$ and 15.0 mol of H$_2$O react, 6.0 mol of Ca(OH)$_2$ can be formed.
(g) When 200. g of Ca$_3$P$_2$ and 100. g of H$_2$O react, Ca$_3$P$_2$ is the limiting reactant.
(h) When 200. g of Ca$_3$P$_2$ and 100. g of H$_2$O react, the theoretical yield of PH$_3$ is 57.4 g.

2. The equation representing the reaction used for the commercial preparation of hydrogen cyanide is

$$2\,CH_4 + 3\,O_2 + 2\,NH_3 \longrightarrow 2\,HCN + 6\,H_2O$$

Based on this equation, which of the following statements are correct? Rewrite incorrect statements to make them correct.
(a) Three moles of O$_2$ are required for 2 mol of NH$_3$.
(b) Twelve moles of HCN are produced for every 16 mol of O$_2$ that react.
(c) The mole ratio between H$_2$O and CH$_4$ is
$$\frac{6\ \text{mol H}_2\text{O}}{2\ \text{mol CH}_4}.$$
(d) When 12 mol of HCN are produced, 4 mol of H$_2$O will be formed.
(e) When 10 mol of CH$_4$, 10 mol of O$_2$, and 10 mol of NH$_3$ are mixed and reacted, O$_2$ is the limiting reactant.
(f) When 3 mol each of CH$_4$, O$_2$, and NH$_3$ are mixed and reacted, 3 mol of HCN will be produced.

Paired Exercises

Mole Review Exercises

3. Calculate the number of moles in these quantities:
(a) 25.0 g KNO$_3$
(b) 56 millimol NaOH
(c) 5.4 × 10^2 g (NH$_4$)$_2$C$_2$O$_4$
(d) 16.8 mL H$_2$SO$_4$ solution (d = 1.727 g/mL, 80.0% H$_2$SO$_4$ by mass)

5. Calculate the number of grams in these quantities:
(a) 2.55 mol Fe(OH)$_3$
(b) 125 kg CaCO$_3$
(c) 10.5 mol NH$_3$
(d) 72 millimol HCl
(e) 500.0 mL of liquid Br$_2$ (d = 3.119 g/mL)

4. Calculate the number of moles in these quantities:
(a) 2.10 kg NaHCO$_3$
(b) 525 mg ZnCl$_2$
(c) 9.8 × 10^{24} molecules CO$_2$
(d) 250 mL ethyl alcohol, C$_2$H$_5$OH (d = 0.789 g/mL)

6. Calculate the number of grams in these quantities:
(a) 0.00844 mol NiSO$_4$
(b) 0.0600 mol HC$_2$H$_3$O$_2$
(c) 0.725 mol Bi$_2$S$_3$
(d) 4.50 × 10^{21} molecules glucose, C$_6$H$_{12}$O$_6$
(e) 75 mL K$_2$CrO$_4$ solution (d = 1.175 g/mL, 20.0% K$_2$CrO$_4$ by mass)

7. Which contains the larger number of molecules, 10.0 g H_2O or 10.0 g H_2O_2? Show evidence for your answer.

8. Which contains the larger numbers of molecules, 25.0 g HCl or 85.0 g $C_6H_{12}O_6$? Show evidence for your answer.

Mole-Ratio Exercises

9. Given the equation for the combustion of isopropyl alcohol

$$2\,C_3H_7OH + 9\,O_2 \longrightarrow 6\,CO_2 + 8\,H_2O$$

set up the mole ratio of
(a) CO_2 to C_3H_7OH (d) H_2O to C_3H_7OH
(b) C_3H_7OH to O_2 (e) CO_2 to H_2O
(c) O_2 to CO_2 (f) H_2O to O_2

11. How many moles of CO_2 can be produced from 7.75 mol C_2H_5OH? (Balance the equation first.)

$$C_2H_5OH + O_2 \longrightarrow CO_2 + H_2O$$

13. An early method of producing chlorine was by reacting pyrolusite (MnO_2) with hydrochloric acid. How many moles of HCl will react with 1.05 mol of MnO_2? (Balance the equation first.)

$$MnO_2\,(s) + HCl\,(aq) \longrightarrow Cl_2\,(g) + \\ MnCl_2\,(aq) + H_2O\,(l)$$

15. How many grams of sodium hydroxide can be produced from 500 g of calcium hydroxide according to this equation?

$$Ca(OH)_2 + Na_2CO_3 \longrightarrow 2\,NaOH + CaCO_3$$

17. In a blast furnace, iron(III) oxide reacts with coke (carbon) to produce molten iron and carbon monoxide:

$$Fe_2O_3 + 3\,C \xrightarrow{\Delta} 2\,Fe + 3\,CO$$

How many kilograms of iron would be formed from 125 kg of Fe_2O_3?

19. Ethane gas, C_2H_6, burns in air (i.e., reacts with the oxygen in air) to form carbon dioxide and water:

$$2\,C_2H_6 + 7\,O_2 \longrightarrow 4\,CO_2 + 6\,H_2O$$

(a) How many moles of O_2 are needed for the complete combustion of 15.0 mol of ethane?
(b) How many grams of CO_2 are produced for each 8.00 g of H_2O produced?
(c) How many grams of CO_2 will be produced by the combustion of 75.0 g of C_2H_6?

10. For the reaction

$$3\,CaCl_2 + 2\,H_3PO_4 \longrightarrow Ca_3(PO_4)_2 + 6\,HCl$$

set up the mole ratio of
(a) $CaCl_2$ to $Ca_3(PO_4)_2$ (d) $Ca_3(PO_4)_2$ to H_3PO_4
(b) HCl to H_3PO_4 (e) HCl to $Ca_3(PO_4)_2$
(c) $CaCl_2$ to H_3PO_4 (f) H_3PO_4 to HCL

12. How many moles of Cl_2 can be produced from 5.60 mol HCl?

$$4\,HCl + O_2 \longrightarrow 2\,Cl_2 + 2\,H_2O$$

14. Given the equation

$$Al_4C_3 + 12\,H_2O \longrightarrow 4\,Al(OH)_3 + 3\,CH_4$$

(a) How many moles of water are needed to react with 100. g of Al_4C_3?
(b) How many moles of $Al(OH)_3$ will be produced when 0.600 mol of CH_4 is formed?

16. How many grams of zinc phosphate, $Zn_3(PO_4)_2$, are formed when 10.0 g of Zn are reacted with phosphoric acid?

$$3\,Zn + 2\,H_3PO_4 \longrightarrow Zn_3(PO_4)_2 + 3\,H_2$$

18. How many grams of steam and iron must react to produce 375 g of magnetic iron oxide, Fe_3O_4?

$$3\,Fe(s) + 4\,H_2O(g) \longrightarrow Fe_3O_4(s) + 4\,H_2(g)$$

20. Given the equation

$$4\,FeS_2 + 11\,O_2 \longrightarrow 2\,Fe_2O_3 + 8\,SO_2$$

(a) How many moles of Fe_2O_3 can be made from 1.00 mol of FeS_2?
(b) How many moles of O_2 are required to react with 4.50 mol of FeS_2?
(c) If the reaction produces 1.55 mol of Fe_2O_3, how many moles of SO_2 are produced?
(d) How many grams of SO_2 can be formed from 0.512 mol of FeS_2?
(e) If the reaction produces 40.6 g of SO_2, how many moles of O_2 were reacted?
(f) How many grams of FeS_2 are needed to produce 221 g of Fe_2O_3?

Limiting-Reactant and Percent-Yield Exercises

21. In the following equations, determine which reactant is the limiting reactant and which reactant is in excess. The amounts used are given below each reactant. Show evidence for your answers.

(a) $KOH + HNO_3 \longrightarrow KNO_3 + H_2O$
16.0 g   12.0 g

(b) $2\,NaOH + H_2SO_4 \longrightarrow Na_2SO_4 + 2\,H_2O$
10.0 g   10.0 g

23. The reaction for the combustion of propane is

$C_3H_8 + 5\,O_2 \longrightarrow 3\,CO_2 + 4\,H_2O$

(a) If 20.0 g of C_3H_8 and 20.0 g of O_2 are reacted, how many moles of CO_2 can be produced?

(b) If 20.0 g of C_3H_8 and 80.0 g of O_2 are reacted, how many moles of CO_2 are produced?

(c) If 2.0 mol of C_3H_8 and 14.0 mol of O_2 are placed in a closed container and they react to completion (until one reactant is completely used up), what compounds will be present in the container after the reaction, and how many moles of each compound are present?

*25. When a certain nonmetal whose formula is X_8 burns in air, XO_3 forms. Write a balanced equation for this reaction. If 120.0 g of oxygen gas is consumed completely, along with 80.0 g of X_8, identify element X.

.27. Aluminum reacts with bromine to form aluminum bromide:

$2\,Al + 3\,Br_2 \longrightarrow 2\,AlBr_3$

If 25.0 g of Al and 100. g of Br_2 are reacted, and 64.2 g of $AlBr_3$ product is recovered, what is the percent yield for the reaction?

*29. Carbon disulfide, CS_2, can be made from coke, C, and sulfur dioxide, SO_2:

$3\,C + 2\,SO_2 \longrightarrow CS_2 + 2\,CO_2$

If the actual yield of CS_2 is 86.0% of the theoretical yield, what mass of coke is needed to produce 950 g of CS_2?

22. In the following equations, determine which reactant is the limiting reactant and which reactant is in excess. The amounts used are given below each reactant. Show evidence for your answers.

(a) $2\,Bi(NO_3)_3 + 3\,H_2S \longrightarrow Bi_2S_3 + 6\,HNO_3$
50.0 g   6.00 g

(b) $3\,Fe + 4\,H_2O \longrightarrow Fe_3O_4 + 4\,H_2$
40.0 g   16.0 g

24. The reaction for the combustion of propane is

$C_3H_8 + 5\,O_2 \longrightarrow 3\,CO_2 + 4\,H_2O$

(a) If 5.0 mol of C_3H_8 and 5.0 mol of O_2 are reacted, how many moles of CO_2 can be produced?

(b) If 3.0 mol of C_3H_8 and 20.0 mol of O_2 are reacted, how many moles of CO_2 are produced?

(c) If 20.0 mol of C_3H_8 and 3.0 mol of O_2 are reacted, how many moles of CO_2 can be produced?

*26. When a particular metal X reacts with HCl, the resulting products are XCl_2 and H_2. Write and balance the equation. When 78.5 g of the metal reacts completely, 2.42 g of hydrogen gas results. Identify the element X.

28. Iron was reacted with a solution containing 400. g of copper(II) sulfate. The reaction was stopped after 1 hour, and 151 g of copper was obtained. Calculate the percent yield of copper obtained.

$Fe(s) + CuSO_4(aq) \longrightarrow Cu(s) + FeSO_4(aq)$

*30. Acetylene (C_2H_2) can be manufactured by the reaction of water and calcium carbide, CaC_2:

$CaC_2(s) + 2\,H_2O(l) \longrightarrow C_2H_2(g) + Ca(OH)_2(s)$

When 44.5 g of commercial grade (impure) calcium carbide is reacted, 0.540 mol of C_2H_2 is produced. Assuming that all of the CaC_2 was reacted to C_2H_2, what is the percent of CaC_2 in the commercial grade material?

Additional Exercises

31. A tool set contains 6 wrenches, 4 screwdrivers, and 2 pliers. The manufacturer has 1000 pliers, 2000 screwdrivers, and 3000 wrenches in stock. Can an order for 600 tool sets be filled? Explain briefly.

32. What is the difference between using a number as a subscript and using a number as a coefficient in a chemical equation?

33. Oxygen masks for producing O_2 in emergency situations contain potassium superoxide (KO_2). It reacts according to this equation:

$$4\,KO_2 + 2\,H_2O + 4\,CO_2 \longrightarrow 4\,KHCO_3 + 3\,O_2$$

   (a) If a person wearing such a mask exhales 0.85 g of CO_2 every minute, how many moles of KO_2 are consumed in 10.0 minutes?
   (b) How many grams of oxygen are produced in 1.0 hour?

34. Ethyl alcohol is the result of fermentation of sugar, $C_6H_{12}O_6$:

$$C_6H_{12}O_6 \longrightarrow 2\,C_2H_5OH + 2\,CO_2$$

   (a) How many grams of ethyl alcohol and how many grams of carbon dioxide can be produced from 750 g of sugar?
   (b) How many milliliters of alcohol ($d = 0.79$ g/mL) can be produced from 750 g of sugar?

35. Phosphoric acid, H_3PO_4, can be synthesized from phosphorus, oxygen, and water according to these two reactions:

$$4\,P + 5\,O_2 \longrightarrow P_4O_{10}$$

$$P_4O_{10} + 6\,H_2O \longrightarrow 4\,H_3PO_4$$

   Starting with 20.0 g P, 30.0 g O_2, and 15.0 g H_2O, what is the mass of phosphoric acid that can be formed?

36. The methyl alcohol (CH_3OH) used in alcohol burners combines with oxygen gas to form carbon dioxide and water. How many grams of oxygen are required to burn 60.0 mL of methyl alcohol ($d = 0.72$ g/mL)?

37. Hydrazine (N_2H_4) and hydrogen peroxide (H_2O_2) have been used as rocket propellants. They react according to the following equation:

$$7\,H_2O_2 + N_2H_4 \longrightarrow 2\,HNO_3 + 8\,H_2O$$

   (a) How many moles of HNO_3 are formed from 0.33 mol of hydrazine?
   (b) How many moles of hydrogen peroxide are required if 2.75 mol of water are to be produced?
   (c) How many moles of water are produced if 8.72 mol of HNO_3 are also produced?
   (d) How many grams of hydrogen peroxide are needed to react completely with 120 g of hydrazine?

38. Silver tarnishes in the presence of hydrogen sulfide (which smells like rotten eggs) and oxygen because of the reaction

$$4\,Ag + 2\,H_2S + O_2 \longrightarrow 2\,Ag_2S + 2\,H_2O$$

   How many grams of silver sulfide can be formed from a mixture of 1.1 g Ag, 0.14 g H_2S, and 0.080 g O_2?

39. After 180.0 g of zinc was dropped into a beaker of hydrochloric acid and the reaction ceased, 35 g of unreacted zinc remained in the beaker:

$$Zn + HCl \longrightarrow ZnCl_2 + H_2$$

   Balance the equation first.
   (a) How many moles of hydrogen gas were produced?
   (b) How many grams of HCl were reacted?

40. Use this equation to answer (a) and (b):

$$Fe(s) + CuSO_4(aq) \longrightarrow Cu(s) + FeSO_4(aq)$$

   (a) When 2.0 mol of Fe and 3.0 mol of $CuSO_4$ are reacted, what substances will be present when the reaction is over? How many moles of each substance are present?
   (b) When 20.0 g of Fe and 40.0 g of $CuSO_4$ are reacted, what substances will be present when the reaction is over? How many grams of each substance are present?

41. Methyl alcohol (CH_3OH) is made by reacting carbon monoxide and hydrogen in the presence of certain metal oxide catalysts. How much alcohol can be obtained by reacting 40.0 g of CO and 10.0 g of H_2? How many grams of excess reactant remain unreacted?

$$CO(g) + 2\,H_2(g) \longrightarrow CH_3OH(l)$$

*42. Ethyl alcohol (C_2H_5OH), also called grain alcohol, can be made by the fermentation of sugar:

$$C_6H_{12}O_6 \longrightarrow 2\,C_2H_5OH + 2\,CO_2$$
$$\text{glucose} \qquad \text{ethyl alcohol}$$

   If an 84.6% yield of ethyl alcohol is obtained,
   (a) what mass of ethyl alcohol will be produced from 750 g of glucose?
   (b) what mass of glucose should be used to produce 475 g of C_2H_5OH?

*43. Both $CaCl_2$ and $MgCl_2$ react with $AgNO_3$ to precipitate AgCl. When solutions containing equal masses of $CaCl_2$ and $MgCl_2$ are reacted with $AgNO_3$, which salt solution will produce the larger amount of AgCl? Show proof.

*44. An astronaut excretes about 2500 g of water a day. If lithium oxide (Li_2O) is used in the spaceship to absorb this water, how many kilograms of Li_2O must be carried for a 30-day space trip for three astronauts?

$$Li_2O + H_2O \longrightarrow 2\,LiOH$$

*45. Much commercial hydrochloric acid is prepared by the reaction of concentrated sulfuric acid with sodium chloride:

$$H_2SO_4 + 2\,NaCl \longrightarrow Na_2SO_4 + 2\,HCl$$

   How many kilograms of concentrated H_2SO_4, 96% H_2SO_4 by mass, are required to produce 20.0 L of concentrated hydrochloric acid ($d = 1.20$ g/mL, 42.0% HCl by mass)?

*46. Gastric juice contains about 3.0 g HCl per liter. If a person produces about 2.5 L of gastric juice per day, how many antacid tablets, each containing 400. mg of $Al(OH)_3$, are needed to neutralize all the HCl produced in 1 day?

$$Al(OH)_3(s) + 3\ HCl(aq) \longrightarrow AlCl_3(aq) + 3\ H_2O(l)$$

*47. When 12.82 g of a mixture of $KClO_3$ and $NaCl$ is heated strongly, the $KClO_3$ reacts according to this equation:

$$2\ KClO_3(s) \longrightarrow 2\ KCl(s) + 3\ O_2(g)$$

The NaCl does not undergo any reaction. After the heating, the mass of the residue (KCl and NaCl) is 9.45 g. Assuming that all the loss of mass represents loss of oxygen gas, calculate the percent of $KClO_3$ in the original mixture.

Answers to Practice Exercises

9.1 0.33 mol Al_2O_3

9.2 7.33 mol $Al(OH)_3$

9.3 0.8157 mol KCl

9.4 85 g $AgNO_3$

9.5 27.6 g $CrCl_3$

9.6 348.8 g H_2O

9.7 17.7 g HCl

9.8 164.3 g $BaSO_4$

9.9 88.5% yield

PUTTING IT TOGETHER
Review for Chapters 7–9

Multiple Choice: *Choose the correct answer to each of the following.*

1. 4.0 g of oxygen contains
   (a) 1.5×10^{23} atoms of oxygen
   (b) 4.0 molar masses of oxygen
   (c) 0.50 mol of oxygen
   (d) 6.022×10^{23} atoms of oxygen

2. One mole of hydrogen atoms contains
   (a) 2.0 g of hydrogen
   (b) 6.022×10^{23} atoms of hydrogen
   (c) 1 atom of hydrogen
   (d) 12 g of carbon-12

3. The mass of one atom of magnesium is
   (a) 24.31 g   (c) 12.00 g
   (b) 54.94 g   (d) 4.037×10^{-23} g

4. Avogardro's number of magnesium atoms
   (a) has a mass of 1.0 g
   (b) has the same mass as Avogadro's number of sulfur atoms
   (c) has a mass of 12.0 g
   (d) is 1 mol of magnesium atoms

5. Which of the following contains the largest number of moles?
   (a) 1.0 g Li   (c) 1.0 g Al
   (b) 1.0 g Na   (d) 1.0 g Ag

6. The number of moles in 112 g of acetylsalicylic acid (aspirin), $C_9H_8O_4$, is
   (a) 1.61   (b) 0.619   (c) 112   (d) 0.161

7. How many moles of aluminum hydroxide are in one antacid tablet containing 400 mg of $Al(OH)_3$?
   (a) 5.13×10^{-3} (c) 5.13
   (b) 0.400   (d) 9.09×10^{-3}

8. How many grams of Au_2S can be obtained from 1.17 mol of Au?
   (a) 182 g   (b) 249 g   (c) 364 g   (d) 499 g

9. The molar mass of $Ba(NO_3)_2$ is
   (a) 199.3   (b) 261.3   (c) 247.3   (d) 167.3

10. A 16-g sample of O_2
    (a) is 1 mol of O_2
    (b) contains 6.022×10^{23} molecules of O_2
    (c) is 0.50 molecule of O_2
    (d) is 0.50 molar mass of O_2

11. What is the percent composition for a compound formed from 8.15 g of zinc and 2.00 g of oxygen?
    (a) 80.3% Zn, 19.7% O   (c) 70.3% Zn, 29.7% O
    (b) 80.3% O, 19.7% Zn   (d) 65.3% Zn, 34.7% O

12. Which of these compounds contains the largest percentage of oxygen?
    (a) SO_2 (c) N_2O_3
    (b) SO_3 (d) N_2O_5

13. 2.00 mol of CO_2
    (a) has a mass of 56.0 g
    (b) contains 1.20×10^{24} molecules
    (c) has a mass of 44.0 g
    (d) contains 6.00 molar masses of CO_2

14. In Ag_2CO_3, the percent by mass of
    (a) carbon is 43.5%   (c) oxygen is 17.4%
    (b) silver is 64.2%   (d) oxygen is 21.9%

15. The empirical formula of the compound containing 31.0% Ti and 69.0% Cl is
    (a) TiCl   (c) $TiCl_3$
    (b) $TiCl_2$ (d) $TiCl_4$

16. A compound contains 54.3% C, 5.6% H, and 40.1% Cl. The empirical formula is
    (a) CH_3Cl (c) $C_2H_4Cl_2$
    (b) C_2H_5Cl (d) C_4H_5Cl

17. A compound contains 40.0% C, 6.7% H, and 53.3% O. The molar mass is 60.0 g/mol. The molecular formula is
    (a) $C_2H_3O_2$ (c) C_2HO
    (b) C_3H_8O (d) $C_2H_4O_2$

18. How many chlorine atoms are in 4.0 mol of PCl_3?
    (a) 3   (c) 12
    (b) 7.2×10^{24} (d) 2.4×10^{24}

19. What is the mass of 4.53 mol of Na_2SO_4?
    (a) 142.1 g   (c) 31.4 g
    (b) 644 g   (d) 3.19×10^{-2} g

20. The percent composition of Mg_3N_2 is
    (a) 72.2% Mg, 27.8% N
    (b) 63.4% Mg, 36.6% N
    (c) 83.9% Mg, 16.1% N
    (d) no correct answer given

21. How many grams of oxygen are contained in 0.500 mol of Na_2SO_4?
    (a) 16.0 g   (c) 64.0 g
    (b) 32.0 g   (d) no correct answer given

22. The empirical formula of a compound is CH. If the molar mass of this compound is 78.11, then the molecular formula is
    (a) C_2H_2 (c) C_6H_6
    (b) C_5H_{18} (d) no correct answer given

23. The reaction

$$BaCl_2 + (NH_4)_2CO_3 \longrightarrow BaCO_3 + 2\,NH_4Cl$$

is an example of
(a) combination
(b) decomposition
(c) single displacement
(d) double displacement

24. When the equation

$$Al + O_2 \longrightarrow Al_2O_3$$

is properly balanced, which of the following terms appears?
(a) $2\,Al$ (b) $2\,Al_2O_3$ (c) $3\,Al$ (d) $2\,O_2$

25. Which equation is *incorrectly* balanced?
(a) $2\,KNO_3 \xrightarrow{\Delta} 2\,KNO_2 + O_2$
(b) $H_2O_2 \longrightarrow H_2O + O_2$
(c) $2\,Na_2O_2 + 2\,H_2O \longrightarrow 4\,NaOH + O_2$
(d) $2\,H_2O \xrightarrow[H_2SO_4]{electrical\ energy} 2\,H_2 + O_2$

26. The reaction

$$2\,Al + 3\,Br_2 \longrightarrow 2\,AlBr_3$$

is an example of
(a) combination
(b) single displacement
(c) decomposition
(d) double displacement

27. When the equation

$$PbO_2 \xrightarrow{\Delta} PbO + O_2$$

is balanced, one term in the balanced equation is
(a) PbO_2 (b) $3\,O_2$ (c) $3\,PbO$ (d) O_2

28. When the equation

$$Cr_2S_3 + HCl \longrightarrow CrCl_3 + H_2S$$

is balanced, one term in the balanced equation is
(a) $3\,HCl$ (b) $CrCl_3$ (c) $3\,H_2S$ (d) $2\,Cr_2S_3$

29. When the equation

$$F_2 + H_2O \longrightarrow HF + O_2$$

is balanced, one term in the balanced equation is
(a) $2\,HF$ (b) $3\,O_2$ (c) $4\,HF$ (d) $4\,H_2O$

30. When the equation

$$NH_4OH + H_2SO_4 \longrightarrow$$

is completed and balanced, one term in the balanced equation is
(a) NH_4SO_4 (c) H_2OH
(b) $2\,H_2O$ (d) $2\,(NH_4)_2SO_4$

31. When the equation

$$H_2 + V_2O_5 \longrightarrow V +$$

is completed and balanced, one term in the balanced equation is
(a) $2\,V_2O_5$ (b) $3\,H_2O$ (c) $2\,V$ (d) $8\,H_2$

32. When the equation

$$Al(OH)_3 + H_2SO_4 \longrightarrow Al_2(SO_4)_3 + H_2O$$

is balanced, the sum of the coefficients will be
(a) 9 (c) 13
(b) 11 (d) 15

33. When the equation

$$H_3PO_4 + Ca(OH)_2 \longrightarrow H_2O + Ca_3(PO_4)_2$$

is balanced, the proper sequence of coefficients is
(a) 3, 2, 1, 6 (c) 2, 3, 1, 6
(b) 2, 3, 6, 1 (d) 2, 3, 3, 1

34. When the equation

$$Fe_2(SO_4)_3 + Ba(OH)_2 \longrightarrow$$

is completed and balanced, one term in the balanced equation is
(a) $Ba_2(SO_4)_3$ (c) $2\,Fe_2(SO_4)_3$
(b) $2\,Fe(OH)_2$ (d) $2\,Fe(OH)_3$

35. For the reaction

$$2\,H_2 + O_2 \longrightarrow 2\,H_2O + 572.4\,kJ$$

which of the following is not true?
(a) The reaction is exothermic.
(b) 572.4 kJ of heat are liberated for each mole of water formed.
(c) 2 mol of hydrogen react with 1 mol of oxygen.
(d) 572.4 kJ of heat are liberated for each 2 mol of hydrogen reacted.

36. How many moles is 20.0 g of Na_2CO_3?
(a) 1.89 mol (c) 212 mol
(b) 2.12×10^3 mol (d) 0.189 mol

37. What is the mass in grams of 0.30 mol of $BaSO_4$?
(a) 7.0×10^3 g (c) 70. g
(b) 0.13 g (d) 700.20 g

38. How many molecules are in 5.8 g of acetone, C_3H_6O?
(a) 0.10 molecules
(b) 6.0×10^{22} molecules
(c) 3.5×10^{24} molecules
(d) 6.0×10^{23} molecules

Problems 39–45 refer to the reaction

$$2\,C_2H_4 + 6\,O_2 \longrightarrow 4\,CO_2 + 4\,H_2O$$

39. If 6.0 mol of CO_2 are produced, how many moles of O_2 were reacted?
(a) 4.0 mol   (c) 9.0 mol
(b) 7.5 mol   (d) 15.0 mol

40. How many moles of O_2 are required for the complete reaction of 45 g of C_2H_4?
(a) 1.3×10^2 mol   (c) 112.5 mol
(b) 0.64 mol   (d) 4.8 mol

41. If 18.0 g of CO_2 are produced, how many grams of H_2O are produced?
(a) 7.37 g   (b) 3.68 g   (c) 9.00 g   (d) 14.7 g

42. How many moles of CO_2 can be produced by the reaction of 5.0 mol of C_2H_4 and 12.0 mol of O_2?
(a) 4.0 mol   (c) 8.0 mol
(b) 5.0 mol   (d) 10. mol

43. How many moles of CO_2 can be produced by the reaction of 0.480 mol of C_2H_4 and 1.08 mol of O_2?
(a) 0.240 mol   (c) 0.720 mol
(b) 0.960 mol   (d) 0.864 mol

44. How many grams of CO_2 can be produced from 2.0 g of C_2H_4 and 5.0 g of O_2?
(a) 5.5 g   (b) 4.6 g   (c) 7.6 g   (d) 6.3 g

45. If 14.0 g of C_2H_4 is reacted and the actual yield of H_2O is 7.84 g, the percent yield in the reaction is
(a) 0.56%   (b) 43.6%   (c) 87.1%   (d) 56.0%

Problems 46–48 refer to the equation

$$H_3PO_4 + MgCO_3 \longrightarrow Mg_3(PO_4)_2 + CO_2 + H_2O$$

46. The sequence of coefficients for the balanced equation is
(a) 2, 3, 1, 3, 3   (c) 2, 2, 1, 2, 3
(b) 3, 1, 3, 2, 3   (d) 2, 3, 1, 3, 2

47. If 20.0 g of carbon dioxide is produced, the number of moles of magnesium carbonate used is
(a) 0.228 mol   (c) 0.910 mol
(b) 1.37 mol   (d) 0.454 mol

48. If 50.0 g of magnesium carbonate reacts completely with H_3PO_4, the number of grams of carbon dioxide produced is
(a) 52.2 g   (c) 13.1 g
(b) 26.1 g   (d) 50.0 g

49. When 10.0 g of $MgCl_2$ and 10.0 g of Na_2CO_3 are reacted in

$$MgCl_2 + Na_2CO_3 \longrightarrow MgCO_3 + 2\,NaCl$$

the limiting reactant is
(a) $MgCl_2$ (c) $MgCO_3$
(b) Na_2CO_3 (d) NaCl

50. When 50.0 g of copper is reacted with silver nitrate solution in

$$Cu + 2\,AgNO_3 \longrightarrow Cu(NO_3)_2 + 2\,Ag$$

148 g of silver is obtained. What is the percent yield of silver obtained?
(a) 87.1%   (c) 55.2%
(b) 84.9%   (d) no correct answer given

Free Response Questions

1. Compound X requires 104 g of O_2 to produce 2 moles of CO_2 and 2.5 moles of H_2O.
(a) What is the empirical formula for X?
(b) What additional information would you need to determine the molecular formula for X?

2. Consider the reaction of sulfur dioxide with oxygen to form sulfur trioxide taking place in a closed container.

(a) Draw what you would expect to see in the box at the completion of the reaction.
(b) If you begin with 25 g of SO_2 and 5 g of oxygen gas, which is the limiting reagent?
(c) Is the following statement true or false? "When SO_2 is converted into SO_3, the percent composition of S in the compounds changes from 33% to 25%." Explain.

3. The percent composition of compound Z is 63.16% C and 8.77% H. When compound Z burns in air, the only products are carbon dioxide and water. The molar mass for Z is 114.
(a) What is the molecular formula for compound Z?
(b) What is the balanced reaction for Z burning in air?

4. Compound A decomposes at room temperature in an exothermic reaction while compound B requires heating before it will decompose in an endothermic reaction.
(a) Draw reaction profiles (potential energy vs. reaction progress) for both reactions.
(b) (i) The decomposition of 0.500 moles of $NaHCO_3$ to form sodium carbonate, water, and carbon dioxide requires 85.5 kJ of heat. How many grams of water could be collected and how many kJ of heat were absorbed when 24.0 g of CO_2 is produced?
    (ii) Could $NaHCO_3$ be compound A or compound B? Explain your reasoning.

5. Aqueous ammonium hydroxide reacts with aqueous cobalt (II) sulfate to produce aqueous ammonium sulfate and solid cobalt (II) hydroxide. When 38.0 g of one of the reactants was fully reacted with enough of the other reactant, 8.09 g of ammonium sulfate was obtained, which corresponded to a 25.0% yield.
    (a) What type of reaction took place?
    (b) Write the balanced chemical equation.
    (c) What is the theoretical yield of ammonium sulfate?
    (d) Which reactant was the limiting reagent?

6. $C_6H_{12}O_6 \longrightarrow 2\,C_2H_5OH + 2\,CO_2(g)$
    (a) If 25.0 g of $C_6H_{12}O_6$ were used, and only 11.2 g of C_2H_5OH were produced, how much reactant was left over and what volume of gas was produced? (*Note:* One mole of gas occupies 24.0 L of space at room temperature.) What assumption are you making?
    (b) What was the yield of the reaction?
    (c) What type of reaction took place?

7. When solutions containing 25 g each of lead(II) nitrate and potassium iodide are mixed, a yellow precipitate results.
    (a) What type of reaction occurred?
    (b) What is the name and formula for the solid product?
    (c) If after filtration and drying, the solid product weighed 7.66 g, what was the percent yield for the reaction?

8. Consider the following unbalanced reaction:
    $$XNO_3 + CaCl_2 \longrightarrow XCl + Ca(NO_3)_2$$
    (a) If 30.8 g of $CaCl_2$ produced 79.6 g of XCl, what is X?
    (b) Would X be able to displace hydrogen from an acid?

9. Consider the following reaction: $H_2O_2 \longrightarrow H_2O + O_2$
    (a) If at the end of the reaction there are eight water molecules and eight oxygen molecules, what was in the flask at the start of the reaction?
    (b) Does the following reaction profile indicate the reaction is exothermic or endothermic?

    (c) What type of reaction is given above?
    (d) What is the empirical formula for hydrogen peroxide?

CHAPTER

10

Modern Atomic Theory and the Periodic Table

The brilliance of these neon signs in Las Vegas is the result of electrons being transferred between energy levels.

CHAPTER 10 / OUTLINE

How do we go about studying an object that is too small to see? Think back to that birthday present you could look at but not yet open. Judging from the wrapping and size of the box was not very useful, but shaking, turning, and lifting the package all gave indirect clues to its contents. After all your experiments were done you could make a fairly good guess about the contents. But was your guess correct? The only way to know for sure would be to open the package.

Chemists have the same dilemma when they study the atom. Atoms are so very small that it isn't possible to use the normal senses to describe them. We are essentially working in the dark with this package we call the atom. However, our improvements in instruments (X-ray machines and scanning-tunneling microscopes) and measuring devices (spectrophotometers and magnetic resonance imaging, MRI) as well as in our mathematical skills are bringing us closer to revealing the secrets of the atom.

10.1 A Brief History

In the last 200 years, vast amounts of data have been accumulated to support atomic theory. When atoms were originally suggested by the early Greeks, no physical evidence existed to support their ideas. Early chemists did a variety of experiments, which culminated in Dalton's model of the atom. Because of the limitations of Dalton's model, modifications were proposed first by Thomson and then by Rutherford, which eventually led to our modern concept of the nuclear atom. These early models of the atom work reasonably well—in fact, we continue to use them to visualize a variety of chemical concepts. There remain questions that these models cannot answer, including an explanation of how atomic structure relates to the periodic table. In this chapter we will present our modern model of the atom; we will see how it varies from and improves upon the earlier atomic models.

10.2 Electromagnetic Radiation

Scientists have studied energy and light for centuries, and several models have been proposed to explain how energy is transferred from place to place. One way energy travels through space is by *electromagnetic radiation.* Examples of electromagnetic radiation include light from the sun, X-rays in your dentist's office, microwaves from your microwave oven, radio and television waves, and radiant heat from your fireplace. While these examples seem quite different, they are all similar in some important ways. Each shows wavelike behavior, and all travel at the same speed in a vacuum (3.00×10^8 m/s).

The study of wave behavior is a topic for another course, but we need some basic terminology to understand atoms. Waves have three basic characteristics: wavelength, frequency, and speed. **Wavelength** (lambda, λ) is the distance between consecutive peaks (or troughs) in a wave, as shown in Figure 10.1. **Frequency** (nu, μ)

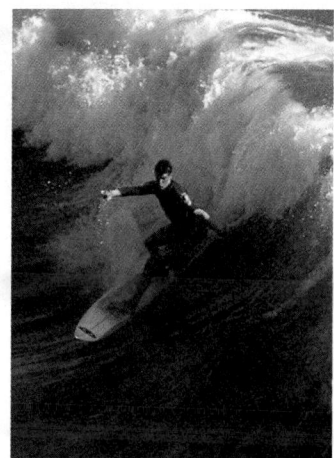

▲
Surfers judge the wavelength, frequency, and speed of waves to get the best ride.

wavelength
frequency

FIGURE 10.1 ▶
The wavelength of this wave is shown by λ. It can be measured from peak to peak or trough to trough.

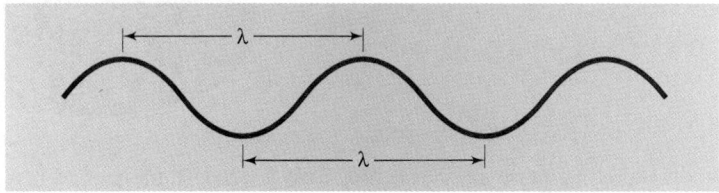

FIGURE 10.2 ▶
The electromagnetic spectrum.

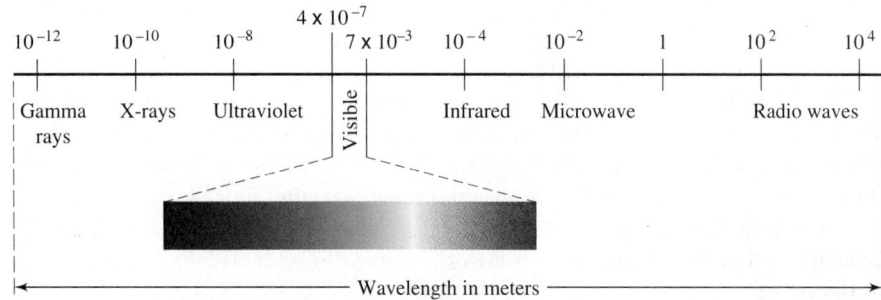

speed   tells how many waves pass a particular point per second. **Speed** (ν) tells how fast a wave moves through space.

Light is one form of electromagnetic radiation and is usually classified by its wavelength, as shown in Figure 10.2. Visible light, as you can see, is only a tiny part of the electromagnetic spectrum. Some examples of electromagnetic radiation involved in energy transfer outside the visible region are hot coals in your backyard grill, which transfer infrared radiation to cook your food; and microwaves, which transfer energy to water molecules in the food, causing them to move more quickly and thus raise the temperature of your food.

photons   We have evidence for the wavelike nature of light. We also know that a beam of light behaves like a stream of tiny packets of energy called **photons.** So what is light exactly? Is it a particle? Is it a wave? Scientists have agreed to explain the properties of electromagnetic radiation by using both wave and particle properties. Neither explanation is ideal, but currently these are our best models.

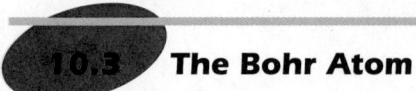

10.3 The Bohr Atom

As scientists struggled to understand the properties of electromagnetic radiation, evidence began to accumulate that atoms could radiate light. At high temperatures, or when subjected to high voltages, elements in the gaseous state give off colored light. Brightly colored neon signs illustrate this property of matter very well. When the light emitted by a gas is passed through a prism or diffraction grating, a set of

line spectrum   brightly colored lines called a **line spectrum** results (Figure 10.3). These colored lines indicate that the light is being emitted only at certain wavelengths, or frequencies, that correspond to specific colors. Each element possesses a unique set of these spectral lines that is different from the sets of all the other elements.

In 1912–1913, while studying the line spectrum of hydrogen, Niels Bohr (1885–1962), a Danish physicist, made a significant contribution to the rapidly growing knowledge of atomic structure. His research led him to believe that elec-

▲
FIGURE 10.3
Line spectrum of hydrogen. Each line corresponds to the wavelength of the energy emitted when the electron of a hydrogen atom, which has absorbed energy, falls back to a lower principal energy level.

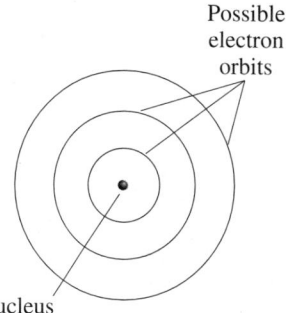

▲
FIGURE 10.4
The Bohr model of the hydrogen atom described the electron revolving in certain allowed circular orbits around the nucleus.

trons exist in specific regions at various distances from the nucleus. He also visualized the electrons as revolving in orbits around the nucleus, like planets rotating around the Sun as shown in Figure 10.4.

Bohr's first paper in this field dealt with the hydrogen atom, which he described as a single electron revolving in an orbit about a relatively heavy nucleus. He applied the concept of energy quanta, proposed in 1900 by the German physicist Max Planck (1858–1947), to the observed line spectrum of hydrogen. Planck stated that energy is never emitted in a continuous stream but only in small discrete packets called **quanta** (Latin, *quantus,* how much). From this Bohr theorized that electrons have several possible energies corresponding to several possible orbits at different distances from the nucleus. Therefore an electron has to be in one specific energy level; it cannot exist between energy levels. In other words, the energy of the electron is said to be quantized. Bohr also stated that when a hydrogen atom absorbed one or more quanta of energy, its electron would "jump" to a higher energy level.

Bohr was able to account for spectral lines of hydrogen this way. A number of energy levels are available, the lowest of which is called the **ground state.** When an electron falls from a high-energy level to a lower one (say, from the fourth to the second), a quantum of energy is emitted as light at a specific frequency, or wavelength (Figure 10.5). This light corresponds to one of the lines visible in the hydrogen spectrum (Figure 10.3). Several lines are visible in this spectrum, each one corresponding to a specific electron energy-level shift within the hydrogen atom.

The chemical properties of an element and its position in the periodic table depend on electron behavior within the atoms. In turn much of our knowledge of the behavior of electrons within atoms is based on spectroscopy. Niels Bohr contributed a great deal to our knowledge of atomic structure by (1) suggesting quantized energy levels for electrons, and (2) showing that spectral lines result from the radiation of small increments of energy (Planck's quanta) when electrons shift from one energy level to another. Bohr's calculations succeeded very well in correlating the experimentally observed spectral lines with electron energy levels for the hydrogen atom. However, Bohr's methods of calculation did not succeed for heavier atoms. More theoretical work on atomic structure was needed.

In 1924, the French physicist Louis de Broglie suggested a surprising hypothesis: All objects have wave properties. De Broglie used sophisticated mathematics to show that the wave properties for an object of ordinary size, such as a baseball, are too small to be observed. But for small objects, such as an electron, the wave prop-

quanta

ground state

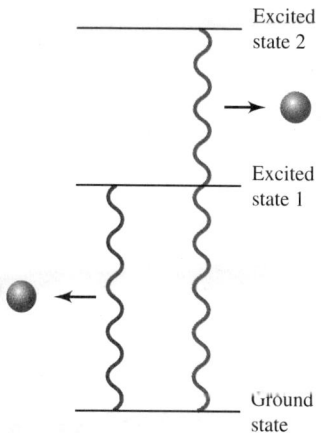

▲
FIGURE 10.5
When an excited electron returns to the ground state, energy is emitted as a photon is released. The color (wavelength) of the light is determined by the difference in energy between the two states (excited and ground).

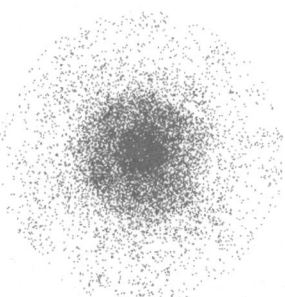

FIGURE 10.6
An orbital for a hydrogen atom. The intensity of the dots shows that the electron spends more time closer to the nucleus.

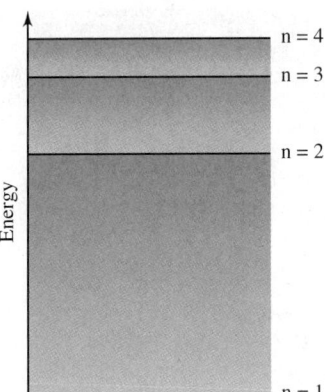

FIGURE 10.7
The first four principal energy levels in the hydrogen atom. Each level is assigned a principal quantum number n.

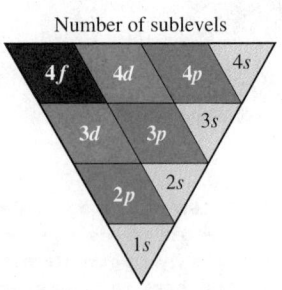

FIGURE 10.8
The types of orbitals on each of the first four principal energy levels.

erties become significant. Other scientists confirmed de Broglie's hypothesis, showing that electrons do exhibit wave properties. In 1926, Erwin Schrödinger, an Austrian physicist, created a mathematical model that described electrons as waves. Using Schrödinger's wave mechanics, we can determine the *probability* of finding an electron in a certain region around the atom.

This treatment of the atom led to a new branch of physics called *wave mechanics* or *quantum mechanics,* which forms the basis for our modern understanding of atomic structure. Although the wave-mechanical description of the atom is mathematical, it can be translated, at least in part, into a visual model. It's important to recognize that we cannot locate an electron precisely within an atom; however, it is clear that electrons are not revolving around the nucleus in orbits as Bohr postulated. The electrons

orbital   are instead found in *orbitals*. An **orbital** is pictured in Figure 10.6 as a region in space around the nucleus where there is a high probability of finding a given electron. We will have more to say about the meaning of orbitals in the next section.

    Energy Levels of Electrons

One of the ideas Bohr contributed to the modern concept of the atom was that the energy of the electron is quantized—that is, the electron is restricted to only certain allowed energies. The wave-mechanical model of the atom also predicts discrete

principal energy levels   **principal energy levels** within the atom. These energy levels are designated by the letter *n*, where *n* is a positive integer (Figure 10.7). The lowest principal energy level corresponds to $n = 1$, the next to $n = 2$, and so on. As *n* increases, the energy of the electron increases, and the electron is found on average farther from the nucleus.

sublevel   Each principal energy level is divided into **sublevels,** which are illustrated in Figure 10.8. The first principal energy level has one sublevel. The second principal energy level has two sublevels, the third energy level has three sublevels, and so on. Each of these sublevels contains spaces for electrons called orbitals.

In each sublevel the electrons are found within specified orbitals. Let's consider each principal energy level in turn. The first principal energy level ($n = 1$) has one sublevel or type of orbital. It is spherical in shape and is designated as $1s$. It is important to understand what the spherical shape of the $1s$ orbital means. The electron does *not* move around on the surface of the sphere, but rather the surface encloses a space where there is a 90% probability that the electron may be found. It might help to consider these orbital shapes in the same way we consider the atmosphere. There is no distinct dividing line between the atmosphere and "space." The boundary is quite fuzzy. The same is true for atomic orbitals. Each has a region of highest density roughly corresponding to its shape. The probability of finding the electron outside this region drops rapidly but never quite reaches zero. Scientists often speak of orbitals as electron "clouds" to emphasize the fuzzy nature of their boundaries.

How many electrons can fit into a $1s$ orbital? To answer this question we need to consider one more property of electrons. This property is called **spin.** Each electron appears to be spinning on an axis, like a globe. It can only spin in two directions. We represent this spin with an arrow, ↑ or ↓. In order to occupy the same orbital, electrons must have *opposite* spins. That is, two electrons with the same spin cannot occupy the same orbital. This gives us the answer to our question: An atomic orbital can hold a maximum of two electrons, which must have opposite spins. This rule is called the **Pauli exclusion principle.** The first principal energy level contains one type of orbital ($1s$) that holds a maximum of two electrons.

What happens with the second principal energy level ($n = 2$)? Here we find two sublevels, $2s$ and $2p$. Like $1s$ in the first principal energy level the $2s$ orbital is spherical in shape, but is larger in size and higher in energy. It also holds a maximum of two electrons. The second type of orbital is designated by $2p$. The $2p$ sublevel consists of three orbitals $2p_x$, $2p_y$, and $2p_z$. The shape of p orbitals is quite different from the s orbitals as shown in Figure 10.9.

Each p orbital has two "lobes." Remember, the space enclosed by these surfaces represents the regions of probability for finding the electrons 90% of the time. There are three separate p orbitals, each oriented in a different direction, and each p orbital can hold a maximum of two electrons. Thus the total number of electrons that can reside in all three p orbitals is six. To summarize our model, the first principal energy level of an atom has a $1s$ orbital. The second principal energy level has a $2s$ and three $2p$ orbitals labeled $2p_x$, $2p_y$, and $2p_z$ as shown in Figure 10.10.

The third principal energy level has three sublevels labeled $3s$, $3p$, and $3d$. The $3s$ orbital is spherical and larger than the $1s$ and $2s$ orbitals. The $3p_x$, $3p_y$, $3p_z$ orbitals are shaped like those of the second level, only larger. The five $3d$ orbitals have the

▲
The light in this neon sign is the result of electrons falling from one principal energy level to another.

spin

Pauli exclusion principle

Notice that there is a correspondence between the energy level and number of sublevels.

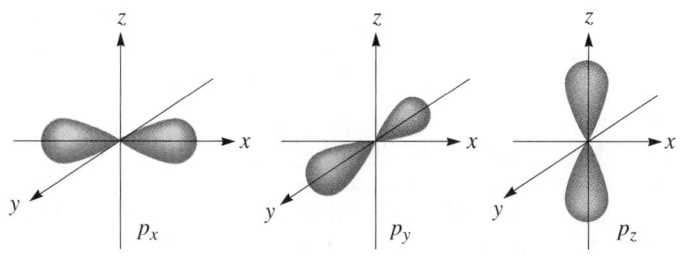

◄**FIGURE 10.9
Perspective representation of the p$_x$, p$_y$, and p$_z$ atomic orbitals.**

FIGURE 10.10 ▶
Orbitals on the second principal energy level are one 2s and three 2p orbitals.

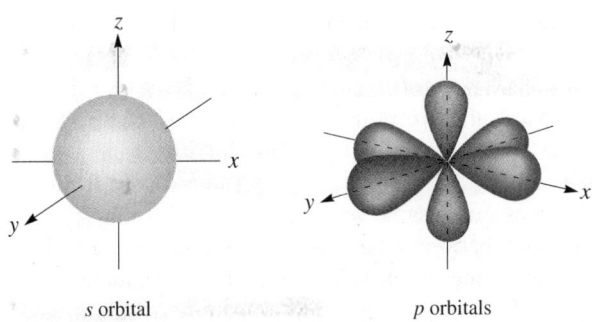

s orbital   p orbitals

FIGURE 10.11
The five d orbitals are found in the third principal energy level along with one 3s orbital and three 3p orbitals.
▼

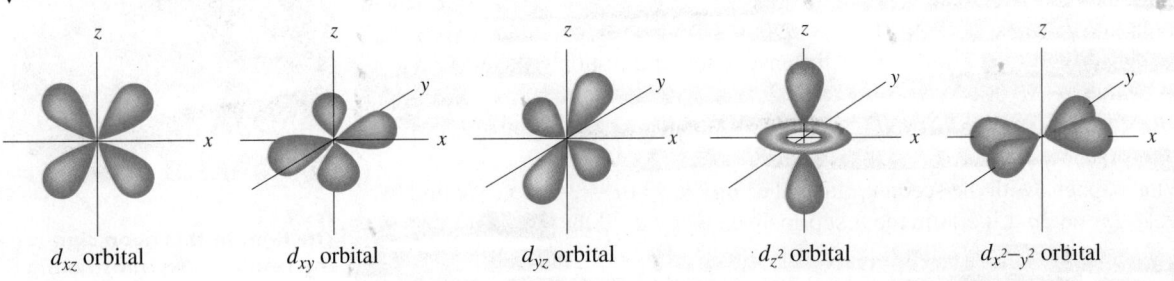

d_{xz} orbital d_{xy} orbital d_{yz} orbital d_{z^2} orbital $d_{x^2-y^2}$ orbital

shapes shown in Figure 10.11. You don't need to memorize these shapes, but notice that they look different from the *s* or *p* orbitals.

Each time a new principal energy level is added, we also add a new sublevel. This makes sense because each energy level corresponds to a larger average distance from the nucleus, which provides more room on each level for new sublevels containing more orbitals.

The pattern continues with the fourth principal energy level. It has 4*s*, 4*p*, 4*d*, and 4*f* orbitals. There are one 4*s*, three 4*p*, five 4*d*, and seven 4*f* orbitals. The shapes of the *s*, *p*, and *d* orbitals are the same as those for lower levels, only larger. We will not consider the shapes of the *f* orbitals. We summarize each principal energy level:

$n = 1$	1s			
$n = 2$	2s	2p 2p 2p		
$n = 3$	3s	3p 3p 3p	3d 3d 3d 3d 3d	
$n = 4$	4s	4p 4p 4p	4d 4d 4d 4d 4d	4f 4f 4f 4f 4f 4f 4f

The hydrogen atom consists of a nucleus (containing one proton) and an electron moving around the outside of the nucleus. In its ground state the electron occupies a 1*s* orbital, but by absorbing energy the electron can become *excited* and move to a higher energy level.

The hydrogen atom can be represented as shown in Figure 10.12. The diameter of the nucleus is about 10^{-13} cm, and the diameter of the electron orbital is about 10^{-8} cm. The diameter of the electron cloud of a hydrogen atom is about 100,000 times greater than the diameter of the nucleus.

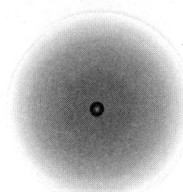

▲
FIGURE 10.12
The modern concept of a hydrogen atom consists of a proton and an electron in an s orbital. The shaded area represents a region where the electron may be found with 90% probability.

CHEMISTRY IN ACTION Atomic Clocks

Imagine a clock that keeps time to within 1 second over a million years. The National Institute of Standards and Technology in Boulder, Colorado, has an atomic clock that does just that—a little better than your average alarm, grandfather, or cuckoo clock! This atomic clock serves as the international standard for time and frequency. How does it work?

Within the glistening case are several layers of magnetic shielding. In the heart of the clock is a small oven that heats cesium metal to release cesium atoms, which are collected into a narrow beam (1 mm wide). The beam of atoms passes down a long evacuated tube while being excited by a laser until all the cesium atoms are in the same electron state.

The atoms then enter another chamber filled with reflecting microwaves. The frequency of the microwaves (9,192,631,770 cycles per second) is exactly the same frequency required to excite a cesium atom from its ground state to the next higher energy level. These excited cesium atoms then release electromagnetic radiation in a process known as fluorescence. Electronic circuits maintain the microwave frequency at precisely the right level to keep the cesium atoms moving from one level to the next. One second is equal to 9,192,631,770 of these

This clock automatically updates ▶ itself by comparing time with an atomic clock by radio signal.

vibrations. The clock is set to this frequency and can keep accurate time for over a million years.

10.5 Atomic Structures of the First 18 Elements

We have seen that hydrogen has one electron that can occupy a variety of orbitals in different principal energy levels. Now let's consider the structure of atoms with more than one electron. Because all atoms contain orbitals similar to those found in hydrogen, we can describe the structures of atoms beyond hydrogen by systematically placing electrons in these hydrogenlike orbitals. We use the following guidelines:

1. No more than two electrons can occupy one orbital.
2. Electrons occupy the lowest energy orbitals available. They enter a higher energy orbital only when the lower orbitals are filled. For the atoms beyond hydrogen, orbital energies vary as $s < p < d < f$ for a given value of n.
3. Each orbital on a sublevel is occupied by a single electron before a second electron enters. For example, all three p orbitals must contain one electron before a second electron enters a p orbital.

We can use several methods to represent the atomic structures of atoms, depending on what we are trying to illustrate. When we want to show both the nuclear makeup and the electron structure of each principal energy level (without orbital detail), we can use a diagram such as Figure 10.13.

 $n = 1 \quad 2$
$\quad 2e^- \quad 7e^-$

Fluorine atom

 $n = 1 \quad 2 \quad 3$
$\quad 2e^- \quad 8e^- \quad 1e^-$

Sodium atom

 $n = 1 \quad 2 \quad 3$
$\quad 2e^- \quad 8e^- \quad 2e^-$

Magnesium atom

FIGURE 10.13
Atomic structure diagrams of fluorine, sodium, and magnesium atoms. The number of protons and neutrons is shown in the nucleus. The number of electrons is shown in each principal energy level ◀ outside the nucleus.

CHEMISTRY IN ACTION Ripples on the Surface

Modern physicists have astounded us with the idea that elementary particles have wave properties. Protons, electrons, and all other elementary particles sometimes behave like waves and sometimes like particles. Now we have pictures to actually show this wave behavior of particles. Donald Eigler at the IBM Almaden Research Center in San Jose, California, used a scanning-tunneling microscope at 4 kelvins to produce the picture shown here. The surface of this copper crystal surprised even the physicists. The waves in the photo are produced by electrons moving around on the surface of the crystal and bouncing off impurities (the two pits in the photo) in the copper. Since each electron behaves like a wave, it interferes with itself after

reflecting off an impurity in the copper. The interference pattern is called a stand-

▲ **This scanning-tunneling microscope photo of the surface of a copper crystal beautifully illustrates the wave nature of matter. Magnification is 215,000,000×.**

ing wave, that is, a wave that vibrates up and down without visible transverse movement, similar to a plucked violin string. The crests of the waves represent regions where the electron is most probably in its particle form.

Metal atoms easily lose one or more electrons, which move about freely within the metal crystal, forming what is often called an "electron sea." At the surface of the metal these loose electrons are confined to a single layer, which is free to move in only two dimensions. Within these constraints the particles behave like waves. The electron layer responsible for this beautiful pattern is only 0.02 Å thick. Similar images of standing electron waves have also been produced with gold at room temperature.

Often we are interested in showing the arrangement of the electrons in an atom

electron configuration

in their orbitals. There are two ways to do this. The first method is called the **electron configuration.** In this method we list each type of orbital, showing the number of electrons in it as an exponent. An electron configuration is read like this:

$$2p^6$$

Number of electrons in sublevel orbitals

Principal energy level

Type of orbital

orbital diagram

We can also represent this configuration with an **orbital diagram** in which boxes represent the orbitals (containing small arrows indicating the electrons). When the orbital contains one electron, an arrow, pointing upward (↑), is placed in the box. A second arrow, pointing downward (↓), indicates the second electron in that orbital.

Let's consider each of the first 18 elements on the periodic table in turn. The order of filling for the orbitals in these elements is $1s$, $2s$, $2p$, $3s$, $3p$, and $4s$. Hydrogen, the first element, has only one electron. The electron will be in the $1s$ orbital because this is the most favorable position (where it will have the greatest attraction for the nucleus). Both representations are shown here:

H   ↑ $1s^1$

      Orbital   Electron
      diagram   configuration

Helium, with two electrons, can be shown as

He ⊞ $1s^2$
Orbital Electron
diagram configuration

The first energy level is now full. An atom with three electrons will have its third electron in the second energy level. Thus in lithium (atomic number 3), the first two electrons are in the $1s$ orbital, and the third electron is in the $2s$ orbital of the second energy level. Lithium has the following structure:

Li ⊞ ⊞ $1s^2 2s^1$

All four electrons of beryllium are s electrons:

Be ⊞ ⊞ $1s^2 2s^2$

The next six elements illustrate the filling of the p orbitals. Boron has the first p electron. Because p orbitals all have the same energy, it doesn't matter which of these orbitals fills first:

B ⊞ ⊞ ⊞⊡⊡ $1s^2 2s^2 2p^1$

Carbon is the sixth element. It has two electrons in the $1s$ orbital, two electrons in the $2s$ orbital, and two electrons to place in the $2p$ orbitals. Because it is more difficult for the p electrons to pair up than to occupy a second p orbital, the second p electron is located in a different p orbital. We could show this by writing $2p_x^1 2p_y^1$, but we usually write it as $2p^2$; it is *understood* that the electrons are in different p orbitals. The spins on these electrons are alike for reasons we will not explain here.

C ⊞ ⊞ ⊞⊞⊡ $1s^2 2s^2 2p^2$

Nitrogen has seven electrons. They occupy the $1s$, $2s$, and $2p$ orbitals. The third p electron in nitrogen is still unpaired and is found in the $2p_z$ orbital:

N ⊞ ⊞ ⊞⊞⊞ $1s^2 2s^2 2p^3$

Oxygen is the eighth element. It has two electrons in both the $1s$ and $2s$ orbitals, and four electrons in the $2p$ orbitals. One of the $2p$ orbitals is now occupied by a second electron, which has a spin opposite the electron already in that orbital:

O ⊞ ⊞ ⊞⊞⊞ $1s^2 2s^2 2p^4$

The next two elements are fluorine with nine electrons and neon with ten electrons:

F ⊞ ⊞ ⊞⊞⊞ $1s^2 2s^2 2p^5$
Ne ⊞ ⊞ ⊞⊞⊞ $1s^2 2s^2 2p^6$

With neon, the first and second energy levels are filled as shown in Table 10.1.

Sodium, element 11, has two electrons in the first energy level and eight electrons in the second energy level, with the remaining electron occupying the $3s$ orbital in the third energy level:

Na ⊞ ⊞ ⊞⊞⊞ ⊞ $1s^2 2s^2 2p^6 3s^1$
$1s$ $2s$ $2p$ $3s$

Yes, We Can See Atoms!

▲
Chromosomes are visible in an onion cell under an optical microscope.

▲
Electron microscopes reveal the DNA of an E. coli bacterium (yellow strands).

▲
DNA molecule, scanning-tunneling microscope.

For centuries, scientists have argued and theorized over the nature and existence of atoms. Today, physicists and chemists can produce pictures of atoms and even move them individually from place to place. This new-found ability to see atoms, molecules, and even watch chemical reactions occur is the direct result of the evolution of the microscope.

An optical microscope is capable of viewing objects as small as the size of a cell. To see smaller objects, an electron microscope is necessary. Since the eye cannot respond to a beam of electrons, the image is produced on a fluorescent screen or on film. These microscopes have been used for some time to photograph large molecules. To see tiny objects, however, the objects must be placed in a vacuum and the electrons must be in a high energy state. If the sample is fragile, as are most molecules, it can be destroyed before the image is formed.

In 1981, Gerd Binnig and Heinrich Rohrer, two scientists from IBM,

invented the first scanning-probe microscope. These instruments are fundamentally different from previous microscopes. In a scanning-probe microscope a probe is placed near the surface of a sample and a parameter of some sort (voltage, magnetic field, etc.) is measured. As the probe is moved across the surface, an image is produced—in the same manner a child would determine the identity of an object sealed in an opaque bag. The first of the scanning-probe instruments was called a scanning-tunneling microscope. It produced the first clear pictures of silicon atoms in January 1983. The greatest limitation for the scanning-tunneling microscope is that, in order to be viewed, organic molecules must be given a thin coating of metal so that electrons are free to jump from the surface to the probe.

In 1985, a team of physicists from Stanford University and IBM found a solution to this problem. In a new instrument known as an atomic-force

microscope, the probe measures tiny electric forces between electrons instead of the actual movement of electrons from surface to probe. The great advantage of this approach is that the probe is so gentle that even very fragile molecules remain intact. The probe is a tiny shard of a diamond attached to a tiny piece of silicon, and it works like a phonograph needle. At the University of California, Santa Barbara, a group of scientists has even succeeded in making a movie of the formation of a blood clot on the molecular level.

In industry, another type of scanning-probe instrument has been developed to check the quality of microelectronic equipment. Researchers at IBM have developed a laser-force microscope in which a tiny wire probe measures small attractive forces (surface tension of water on the sample) to show imperfections as small as 25 atoms across.

Magnesium (12), aluminum (13), silicon (14), phosphorus (15), sulfur (16), chlorine (17), and argon (18) follow in order. Table 10.2 summarizes the filling of the orbitals for elements 11–18.

The electrons in the outermost (highest) energy level of an atom are called the **valence electrons.** For example, oxygen, which has the electron configuration of $1s^2 2s^2 2p^4$, has electrons in the first and second energy levels. Therefore the second level is the valence level for oxygen. The $2s$ and $2p$ electrons are the valence electrons. In the case of magnesium ($1s^2 2s^2 2p^6 3s^2$), the valence electrons are in the

valence electrons

TABLE 10.1	Orbital Filling for the First Ten Elements*		
Number	**Element**	**Orbitals**	**Electron configuration**
		$1s$ $2s$ $2p$	
1	H	↑	$1s^1$
2	He	↑↓	$1s^2$
3	Li	↑↓ ↑	$1s^2 2s^1$
4	Be	↑↓ ↑↓	$1s^2 2s^2$
5	B	↑↓ ↑↓ ↑	$1s^2 2s^2 2p^1$
6	C	↑↓ ↑↓ ↑ ↑	$1s^2 2s^2 2p^2$
7	N	↑↓ ↑↓ ↑ ↑ ↑	$1s^2 2s^2 2p^3$
8	O	↑↓ ↑↓ ↑↓ ↑ ↑	$1s^2 2s^2 2p^4$
9	F	↑↓ ↑↓ ↑↓ ↑↓ ↑	$1s^2 2s^2 2p^5$
10	Ne	↑↓ ↑↓ ↑↓ ↑↓ ↑↓	$1s^2 2s^2 2p^6$

*Boxes represent the orbitals grouped by sublevel. Electrons are shown by arrows.

TABLE 10.2	Orbital Diagrams and Electron Configurations for Elements 11–18				
Number	**Element**	**Orbital**			**Electron configuration**
		$1s$ $2s$ $2p$ $3s$ $3p$			
11	Na	↑↓ ↑↓ ↑↓ ↑↓ ↑↓ ↑			$1s^2 2s^2 2p^6 3s^1$
12	Mg	↑↓ ↑↓ ↑↓ ↑↓ ↑↓ ↑↓			$1s^2 2s^2 2p^6 3s^2$
13	Al	↑↓ ↑↓ ↑↓ ↑↓ ↑↓ ↑↓ ↑			$1s^2 2s^2 2p^6 3s^2 3p^1$
14	Si	↑↓ ↑↓ ↑↓ ↑↓ ↑↓ ↑↓ ↑ ↑			$1s^2 2s^2 2p^6 3s^2 3p^2$
15	P	↑↓ ↑↓ ↑↓ ↑↓ ↑↓ ↑↓ ↑ ↑ ↑			$1s^2 2s^2 2p^6 3s^2 3p^3$
16	S	↑↓ ↑↓ ↑↓ ↑↓ ↑↓ ↑↓ ↑↓ ↑ ↑			$1s^2 2s^2 2p^6 3s^2 3p^4$
17	Cl	↑↓ ↑↓ ↑↓ ↑↓ ↑↓ ↑↓ ↑↓ ↑↓ ↑			$1s^2 2s^2 2p^6 3s^2 3p^5$
18	Ar	↑↓ ↑↓ ↑↓ ↑↓ ↑↓ ↑↓ ↑↓ ↑↓ ↑↓			$1s^2 2s^2 2p^6 3s^2 3p^6$

3s orbital since these are outermost electrons. Valence electrons are involved in bonding atoms to form compounds and are of particular interest to chemists, as we will see in Chapter 11.

Practice 10.1

State the valence electrons in these elements.
(a) B
(b) N
(c) Na
(d) Cl

10.6 Electron Structures and the Periodic Table

We have seen how the electrons are assigned for the atoms of elements 1–18. How do the electron structures of these atoms relate to their position on the periodic table? To answer this question, we need to look at the periodic table more closely.

The periodic table represents the efforts of chemists to organize the elements logically. Chemists of the early 19th century had sufficient knowledge of the properties of elements to recognize similarities among groups of elements. In 1869, Dimitri Mendeleev (1834–1907) of Russia and Lothar Meyer (1830–1895) of Germany independently published periodic arrangements of the elements based on increasing atomic masses. Mendeleev's arrangement is the precursor to the modern periodic table and his name is associated with it. The modern periodic table is shown in Figure 10.14.

FIGURE 10.14
The periodic table of the elements.
▼

Each horizontal row in the periodic table is called a **period,** as shown in Figure 10.14. The number of each period corresponds to the outermost energy level that contains electrons for elements in that period. Those in Period 1 contain electrons only in energy level 1 while those in Period 2 contain electrons in levels 1 and 2. In Period 3, electrons are found in levels 1, 2, and 3, and so on.

Elements that behave in a similar manner are found in **groups** or **families.** These form the vertical columns on the periodic table. Several systems exist for numbering the groups. In one system the columns are numbered from left to right using the numbers 1–18. However, we use a system that numbers the columns with Roman numerals and the letters A and B, as shown in Figure 10.14. The A groups are known as the **representative elements.** The B groups and Group VIII are called the **transition elements.** In this book we will focus on the representative elements. The groups (columns) of the periodic table often have family names. For example, the group on the far right side of the periodic table (He, Ne, Ar, Kr, Xe, and Rn) is called the *noble gases.* Group IA is called the *alkali metals,* Group IIA the *alkaline earth metals,* and Group VIIA the *halogens.*

How is the structure of the periodic table related to the atomic structures of the elements? We've just seen that the periods of the periodic table are associated with the energy level of the outermost electrons of the atoms in that period. Look at the valence electron configurations of the elements we have just examined (Figure 10.15). Do you see a pattern? The valence configuration for the elements in columns is the same. The chemical behavior and properties of elements in a particular family must therefore be associated with the electron configuration of the elements. The number for the energy level is different. This is expected since each new period is associated with a different energy level for the valence electrons.

The electron configurations for elements beyond these first 18 become long and tedious to write. We often abbreviate the electron configuration using the following notation:

Na $[Ne]3s^1$

Look carefully at Figure 10.15 and you will see that the p orbitals are full at the noble gases. By placing the symbol for the noble gas in square brackets we can abbreviate the complete electron configuration and focus our attention on the valence electrons (the electrons we will be interested in when we discuss bonding in Chapter 11). To write the abbreviated electron configuration for any element, go back to the previous

period

groups, families

representative elements
transition elements

◄ **FIGURE 10.15**
Valence electron configurations for the first 18 elements.

IA							Noble gases
1 **H** $1s^1$	IIA	IIIA	IVA	VA	VIA	VIIA	2 **He** $1s^2$
3 **Li** $2s^1$	4 **Be** $2s^2$	5 **B** $2s^22p^1$	6 **C** $2s^22p^2$	7 **N** $2s^22p^3$	8 **O** $2s^22p^4$	9 **F** $2s^22p^5$	10 **Ne** $2s^22p^6$
11 **Na** $3s^1$	12 **Mg** $3s^2$	13 **Al** $3s^23p^1$	14 **Si** $3s^23p^2$	15 **P** $3s^23p^3$	16 **S** $3s^23p^4$	17 **Cl** $3s^23p^5$	18 **Ar** $3s^23p^6$

noble gas and place its symbol in square brackets. Then list the valence electrons. Here are some examples:

B $1s^2\,2s^2\,2p^1$ $[He]2s^22p^1$

Cl $1s^22s^22p^63s^23p^5$ $[Ne]3s^23p^5$

Na $1s^22s^22p^63s^1$ $[Ne]3s^1$

The sequence for filling the orbitals is exactly as we would expect up through the $3p$ orbitals. The third energy level might be expected to fill with $3d$ electrons before electrons enter the $4s$ orbital, but this is not the case. The behavior and properties of the next two elements, potassium (19) and calcium (20), are very similar to the elements in Groups IA and IIA. They clearly belong in these groups. The other elements in Group IA and Group IIA have electron configurations that indicate valence electrons in the s orbitals. For example, since the electron configuration is connected to the element's properties, we should place the last electrons for potassium and calcium in the $4s$ orbital. Their electron configurations are

K $1s^22s^22p^63s^23p^64s^1$ or $[Ar]4s^1$

Ca $1s^22s^22p^63s^23p^64s^2$ or $[Ar]4s^2$

Elements 21–30 belong to the elements known as *transition elements*. Electrons are placed in the $3d$ orbitals for each of these elements. When the $3d$ orbitals are full, the electrons fill the $4p$ orbitals to complete the fourth period. Let's consider the overall relationship between orbital filling and the periodic table. Figure 10.16 illustrates the type of orbital filling and its location on the periodic table. The tall columns on the table (labeled IA–VIIA, and noble gases) are often called the *representative ele-*

FIGURE 10.16
Arrangement of elements according to the sublevel being filled.
▼

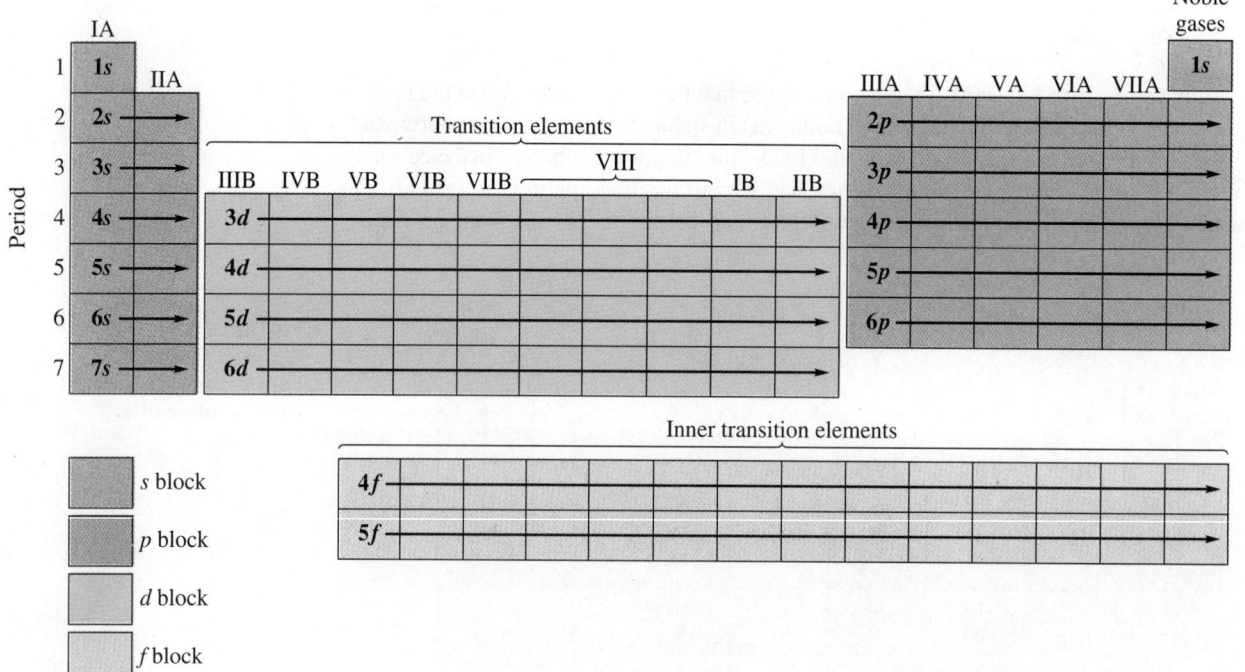

ments. Valence electrons in these elements occupy *s* and *p* orbitals. The period number corresponds to the energy level of the valence electrons. The elements in the center of the periodic table (shown in �) are the transition elements where the *d* orbitals are being filled. Notice that the number for the *d* orbitals is one behind the period number. The two rows shown at the bottom of the table in Figure 10.16 are called the *inner transition elements.* The last electrons in these elements are placed in the *f* orbitals. The number for the *f* orbitals is always two behind that of the *s* and *p* orbitals. A periodic table is almost always available to you, so if you understand the relationship between the orbitals and the periodic table, you can write the electron configuration for any element. There are several minor variations to these rules, but we won't concern ourselves with them in this course.

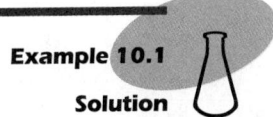

Use the periodic table to write the electron configuration for phosphorus and tin.   **Example 10.1**

Phosphorus is element 15 and is located in Period 3, Group VA. The electron configuration must have a full first and second energy level: **Solution**

$$P \qquad 1s^2 2s^2 2p^6 3s^2 3p^3 \qquad \text{or} \qquad [\text{Ne}]3s^2 3p^3$$

You can determine the electron configuration by looking across the period and counting the element blocks.

Tin is element 50 in Period 5, Group IVA, two places after the transition metals. It must have two electrons in the 5*p* series. Its electron configuration is

$$\text{Sn} \qquad 1s^2 2s^2 2p^6 3s^2 3p^6 4s^2 3d^{10} 4p^6 5s^2 4d^{10} 5p^2 \qquad \text{or} \qquad [\text{Kr}]5s^2 4d^{10} 5p^2$$

Notice that the *d* series of electrons is always one energy level behind the period number.

Practice 10.2

Use the periodic table to write the electron configuration for (a) O, (b) Ca, and (c) Ti.

The early chemists classified the elements based only on their observed properties, but modern atomic theory gives us an explanation for why the properties of elements vary periodically. For example, as we "build" atoms by filling orbitals with electrons, the same orbitals occur on each energy level. This means that the same electron configuration reappears regularly for each level. Groups of elements show similar chemical properties because of the similarity of these outermost electron configurations.

In Figure 10.17, only the electron configuration of the outermost electrons is given. This periodic table illustrates these important points:

1. The number of the period corresponds with the highest energy level occupied by electrons in that period.
2. The group numbers for the representative elements are equal to the total number of outermost electrons in the atoms of the group. For example, ele-

Group number

	IA	IIA	IIIB	IVB	VB	VIB	VIIB	VIII			IB	IIB	IIIA	IVA	VA	VIA	VIIA	Noble gases
1	1 H $1s^1$																	2 He $1s^2$
2	3 Li $2s^1$	4 Be $2s^2$											5 B $2s^22p^1$	6 C $2s^22p^2$	7 N $2s^22p^3$	8 O $2s^22p^4$	9 F $2s^22p^5$	10 Ne $2s^22p^6$
3	11 Na $3s^1$	12 Mg $3s^2$											13 Al $3s^23p^1$	14 Si $3s^23p^2$	15 P $3s^23p^3$	16 S $3s^23p^4$	17 Cl $3s^23p^5$	18 Ar $3s^23p^6$
4	19 K $4s^1$	20 Ca $4s^2$	21 Sc $4s^23d^1$	22 Ti $4s^23d^2$	23 V $4s^23d^3$	24 Cr $4s^13d^5$	25 Mn $4s^23d^5$	26 Fe $4s^23d^6$	27 Co $4s^23d^7$	28 Ni $4s^23d^8$	29 Cu $4s^13d^{10}$	30 Zn $4s^23d^{10}$	31 Ga $4s^24p^1$	32 Ge $4s^24p^2$	33 As $4s^24p^3$	34 Se $4s^24p^4$	35 Br $4s^24p^5$	36 Kr $4s^24p^6$
5	37 Rb $5s^1$	38 Sr $5s^2$	39 Y $5s^24d^1$	40 Zr $5s^24d^2$	41 Nb $5s^14d^4$	42 Mo $5s^14d^5$	43 Tc $5s^14d^6$	44 Ru $5s^14d^7$	45 Rh $5s^14d^8$	46 Pd $5s^04d^{10}$	47 Ag $5s^14d^{10}$	48 Cd $5s^24d^{10}$	49 In $5s^25p^1$	50 Sn $5s^25p^2$	51 Sb $5s^25p^3$	52 Te $5s^25p^4$	53 I $5s^25p^5$	54 Xe $5s^25p^6$
6	55 Cs $6s^1$	56 Ba $6s^2$	57 La $6s^25d^1$	72 Hf $6s^25d^2$	73 Ta $6s^25d^3$	74 W $6s^25d^4$	75 Re $6s^25d^5$	76 Os $6s^25d^6$	77 Ir $6s^25d^7$	78 Pt $6s^15d^9$	79 Au $6s^15d^{10}$	80 Hg $6s^25d^{10}$	81 Tl $6s^26p^1$	82 Pb $6s^26p^2$	83 Bi $6s^26p^3$	84 Po $6s^26p^4$	85 At $6s^26p^5$	86 Rn $6s^26p^6$
7	87 Fr $7s^1$	88 Ra $7s^2$	89 Ac $7s^26d^1$	104 Rf $7s^26d^2$	105 Db $7s^26d^3$	106 Sg $7s^26d^4$	107 Bh $7s^26d^5$	108 Hs $7s^26d^6$	109 Mt $7s^26d^7$	110 Uun $7s^16d^9$	111 Uuu $7s^16d^{10}$	112 Uub $7s^26d^{10}$						

Period

▲
FIGURE 10.17
Outermost electron configurations.

ments in Group VIIA always have the electron configuration $ns^2\,np^5$. The d and f electrons are always in a lower energy level than the highest energy level and so are not considered as outermost electrons.

3. The elements of a family have the same outermost electron configuration except that the electrons are in different energy levels.

4. The elements within each of the s, p, d, f blocks are filling the s, p, d, f orbitals, as shown in Figure 10.16.

5. Within the transition elements some discrepancies in the order of filling occur. (Explanation of these discrepancies and similar ones in the inner transition elements are beyond the scope of this book.)

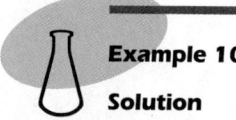

Example 10.2

Solution

Write the electron configuration of a zinc atom and a rubidium atom.

The atomic number of zinc is 30; it therefore has 30 protons and 30 electrons in a neutral atom. Using Figure 10.14, the electron configuration of a zinc atom is $1s^22s^22p^63s^23p^64s^23d^{10}$. Check by adding the superscripts, which should equal 30.

The atomic number of rubidium is 37; therefore it has 37 protons and 37 electrons in a neutral atom. With a little practice using a periodic table, the electron configuration may be written directly. The electron configuration of a rubidium atom is $1s^22s^22p^63s^23p^64s^23d^{10}4p^65s^1$. Check by adding the superscripts, which should equal 37.

Rb   Zn

Concepts in Review

1. Describe the atom as conceived by Niels Bohr.

2. Discuss the contributions to atomic theory made by Dalton, Thomson, Rutherford, Bohr, and Schrödinger. (See Chapter 5 as well.)

3. Explain what is meant by an electron orbital.

4. Explain how an electron configuration can be determined from the periodic table.

5. Write the electron configuration for any of the first 56 elements.

6. Indicate the locations of the metals, nonmetals, metalloids, and noble gases in the periodic table. (See Chapter 3 for help.)

7. Indicate the areas in the periodic table where the s, p, d, and f orbitals are being filled.

8. Determine the number of valence electrons in any atom in the Group A elements.

9. Distinguish between representative elements and transition elements.

10. Identify groups of elements by their special family names.

11. Describe the changes in valence electron structure (a) when moving from left to right in a period and (b) when going from top to bottom in a group.

12. Explain the relationship between group number and the number of valence electrons for the representative elements.

Key Terms

electron configuration (10.5)
frequency (10.2)
ground state (10.3)
groups, families (10.6)
line spectrum (10.3)

orbital (10.3, 10.4)
orbital diagram (10.5)
Pauli exclusion principle (10.4)
period (10.6)
photons (10.2)

principal energy levels (10.4)
quanta (10.3)
representative elements (10.6)
speed (10.2)

spin (10.4)
transition elements (10.6)
valence electrons (10.5)
wavelength (10.2)

Questions

1. What is an orbital?

2. Under what conditions can a second electron enter an orbital already containing one electron?

3. What is meant when we say the electron structure of an atom is in its ground state?

4. How do $1s$ and $2s$ orbitals differ? How are they alike?

5. What letters are used to designate the types of orbitals?

6. List the following orbitals in order of increasing energy: $2s, 2p, 4s, 1s, 3d, 3p, 4p, 3s$.

7. How many s electrons, p electrons, and d electrons are possible in any energy level?

8. What is the major difference between an orbital and a Bohr orbit?

9. Explain how and why Bohr's model of the atom was modified to include the cloud model of the atom.

10. Sketch the s, p_x, p_y, and p_z orbitals.

11. In the designation $3d^7$, give the significance of 3, d, and 7.

12. Describe the difference between transition and representative elements.

13. From the standpoint of electron structure, what do the elements in the s block have in common?

14. Write symbols for elements with atomic numbers 8, 16, 34, 52, and 84. What do these elements have in common?

15. Write the symbols of the family of elements that have seven electrons in their outermost energy level.

16. What is the greatest number of elements to be found in any period? Which periods have this number?

17. From the standpoint of energy level, how does the placement of the last electron in the Group A elements differ from that of the Group B elements?

18. Find the places in the periodic table where elements are not in proper sequence according to atomic mass. (See inside of front cover for periodic table.)

Paired Exercises

19. How many protons are in the nucleus of an atom of these elements?
  (a) H
  (b) B
  (c) Sc
  (d) U

20. How many protons are in the nucleus of an atom of these elements?
  (a) F
  (b) Ag
  (c) Br
  (d) Sb

21. Give the electron configuration for
  (a) B
  (b) Ti
  (c) Zn
  (d) Br
  (e) Sr

22. Give the electron configuration for
  (a) chlorine
  (b) silver
  (c) lithium
  (d) iron
  (e) iodine

23. Explain how the spectral lines of hydrogen occur.

24. Explain how Bohr used the data from the hydrogen spectrum to support his model of the atom.

25. How many orbitals exist in the third principal energy level? What are they?

26. How many electrons can be present in the fourth principal energy level?

27. Write orbital diagrams for
  (a) N
  (b) Cl
  (c) Zn
  (d) Zr
  (e) I

28. Write orbital diagrams for
  (a) Si
  (b) S
  (c) Ar
  (d) V
  (e) P

29. Which elements have these electron configurations?
  (a) $1s^2 2s^2 2p^6 3s^2$
  (b) $1s^2 2s^2 2p^6 3s^2 3p^1$
  (c) $1s^2 2s^2 2p^6 3s^2 3p^6 4s^2 3d^8$
  (d) $1s^2 2s^2 2p^6 3s^2 3p^6 4s^2 3d^5$

30. Which elements have these electron configurations?
  (a) $[Ar]4s^2 3d^1$
  (b) $[Ar]4s^2 3d^{10}$
  (c) $[Kr]5s^2 4d^{10} 5p^2$
  (d) $[Xe]6s^1$

31. Show the electron configurations for elements with these atomic numbers:
  (a) 8
  (b) 11
  (c) 17
  (d) 23
  (e) 28
  (f) 34

32. Show the electron configurations for elements with these atomic numbers:
  (a) 9
  (b) 26
  (c) 31
  (d) 39
  (e) 52
  (f) 10

33. Identify these elements from their atomic structure diagrams:
  (a) (16p 16n) $2e^-$ $8e^-$ $6e^-$
  (b) (28p 32n) $2e^-$ $8e^-$ $16e^-$ $2e^-$

34. Diagram the atomic structures (as you see in Exercise 33) for these elements:
  (a) $^{27}_{13}Al$
  (b) $^{48}_{22}Ti$

35. Why is the 11th electron of the sodium atom located in the third energy level rather than in the second energy level?

36. Why is the last electron in potassium located in the fourth energy level rather than in the third energy level?

37. What electron structure do the noble gases have in common?

38. What is unique about the noble gases, from an electron point of view?

39. How are elements in a period related to one another?

40. How are elements in a group related to one another?

41. What do the electron structures of the alkali metals have in common?

42. Why would you expect the elements zinc, cadmium, and mercury to be in the same chemical family?

43. Pick the electron structures that represent elements in the same chemical family:
(a) $1s^2 2s^1$
(b) $1s^2 2s^2 2p^4$
(c) $1s^2 2s^2 2p^2$
(d) $1s^2 2s^2 2p^6 3s^2 3p^4$
(e) $1s^2 2s^2 2p^6 3s^2 3p^6$
(f) $1s^2 2s^2 2p^6 3s^2 3p^6 4s^2$
(g) $1s^2 2s^2 2p^6 3s^2 3p^6 4s^1$
(h) $1s^2 2s^2 2p^6 3s^2 3p^6 4s^2 3d^1$

44. Pick the electron structures that represent elements in the same chemical family:
(a) $[He]2s^2 2p^6$
(b) $[Ne]3s^1$
(c) $[Ne]3s^2$
(d) $[Ne]3s^2 3p^3$
(e) $[Ar]4s^2 3d^{10}$
(f) $[Ar]4s^2 3d^{10} 4p^6$
(g) $[Ar]4s^2 3d^5$
(h) $[Kr]5s^2 4d^{10}$

45. In the periodic table, calcium, element 20, is surrounded by elements 12, 19, 21, and 38. Which of these have physical and chemical properties most resembling calcium?

46. In the periodic table, phosphorus, element 15, is surrounded by elements 14, 7, 16, and 33. Which of these have physical and chemical properties most resembling phosphorus?

47. Classify the following elements as metals, nonmetals, or metalloids (review Chapter 3 if you need help):
(a) potassium
(b) plutonium
(c) sulfur
(d) antimony

48. Classify the following elements as metals, nonmetals, or metalloids (review Chapter 3 if you need help):
(a) iodine
(b) tungsten
(c) molybdenum
(d) germanium

49. In which period and group does an electron first appear in an f orbital?

50. In which period and group does an electron first appear in a d orbital?

51. How many electrons occur in the valence level of Group VIIA and VIIB elements? Why are they different?

52. How many electrons occur in the valence level of Group IIIA and IIIB elements? Why are they different?

Additional Exercises

53. If the orbitals in an atom could each hold three electrons rather than two, what would be the atomic numbers of the first three noble gases?

54. Why does the emission spectrum for nitrogen reveal many more spectral lines than that for hydrogen?

55. Among the first 100 elements on the periodic table, how many have at least
(a) one s electron?
(b) one p electron?
(c) one d electron?
(d) one f electron?

56. For the following elements, what percent of their electrons occupy s orbitals?
(a) He   (d) Se
(b) Be   (e) Cs
(c) Xe

57. How many pairs of valence electrons do these elements have?
(a) O   (d) Xe
(b) P   (e) Rb
(c) I

58. Suppose we use a foam ball to represent a typical atom. If the radius of the ball is 1.5 cm and the radius of a typical atom is 1.0×10^{-8} cm, how much of an enlargement is this? Use a ratio to express your answer.

59. List the first element on the periodic table that satisfies each of these conditions:
(a) a completed set of p orbitals
(b) two $4p$ electrons
(c) seven valence electrons
(d) three unpaired electrons

60. Oxygen is a gas. Sulfur is a solid. What is it about their electron structures that causes them to be grouped in the same chemical family?

61. In which groups are the transition elements located?

62. How do the electron structures of the transition elements differ from the representative elements?

63. The atomic numbers of the noble gases are 2, 10, 18, 36, 54, and 86. What are the atomic numbers for the elements with six electrons in their outermost electron configuration?

64. Element number 87 is in Group IA, Period 7. Describe its outermost energy level. How many energy levels of electrons does it have?

65. If element 36 is a noble gas, in which groups would you expect elements 35 and 37 to occur?

66. Write a paragraph describing the general features of the periodic table.

67. Some scientists have proposed the existence of element 117. If it were to exist,
   (a) what would its electron configuration be?
   (b) how many valence electrons would it have?
   (c) what element would it likely resemble?
   (d) to what family and period would it belong?

68. What is the relationship between two elements if
   (a) one of them has 10 electrons, 10 protons, and 10 neutrons and the other has 10 electrons, 10 protons, and 12 neutrons?
   (b) one of them has 23 electrons, 23 protons, and 27 neutrons and the other has 24 electrons, 24 protons, and 26 neutrons?

69. Is there any pattern for the location of gases on the periodic table? for the location of liquids? for the location of solids?

Answers to Practice Exercises

10.1 (a) $2s^2 2p^1$
 (b) $2s^2 2p^3$
 (c) $3s^1$
 (d) $3s^2 3p^5$

10.2 (a) O $1s^2 2s^2 2p^4$
 (b) Ca $1s^2 2s^2 2p^6 3s^2 3p^6 4s^2$
 (c) Ti $1s^2 2s^2 2p^6 3s^2 3p^6 4s^2 3d^2$

CHAPTER

11

Chemical Bonds: The Formation of Compounds from Atoms

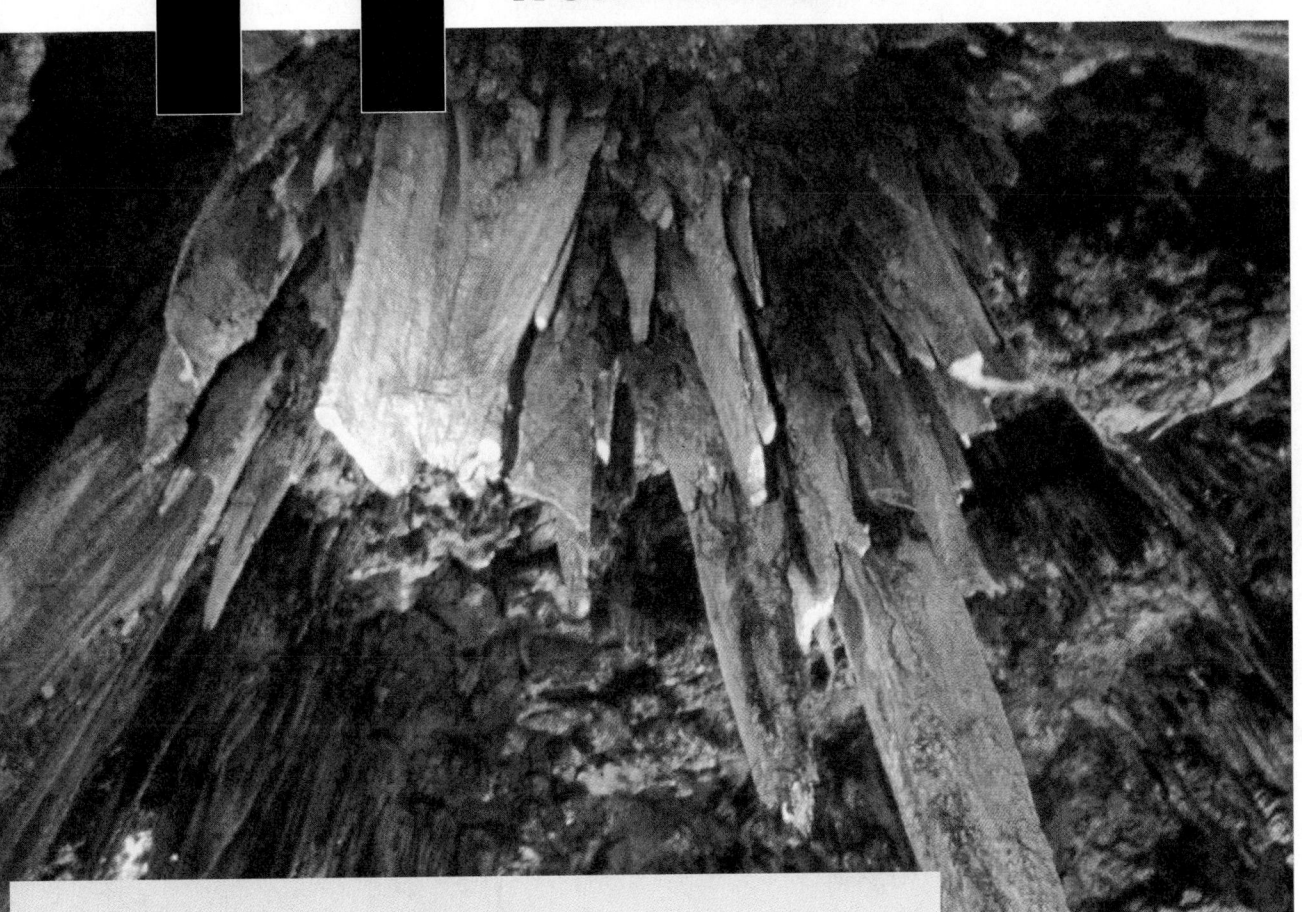

CHAPTER 11 / OUTLINE

This colorfully lighted limestone cave reveals dazzling stalactites and stalagmites, formed from calcium carbonate.

For centuries we've been aware that certain metals cling to a magnet. We've seen balloons sticking to walls. Why? Superconductors floating in air are seen in television commercials. High-speed levitation trains are heralded to be the wave of the future. How do they function? In each case, forces of attraction and repulsion are at work.

Human interactions also suggest that "opposites attract" and "likes repel." Attractions draw us into friendships and significant relationships, whereas repulsive forces may produce debate and antagonism. We form and break apart interpersonal bonds throughout our lives.

In chemistry, we also see this phenomenon. Substances form chemical bonds as a result of electrical attractions. These bonds provide the tremendous diversity of compounds found in nature.

11.1 Periodic Trends in Atomic Properties

FIGURE 11.1
The elements are classified as metals, nonmetals, and metalloids.
▼

Although atomic theory and electron configuration help us understand the arrangement and behavior of the elements, it's important to remember that the design of the periodic table is based on observing properties of the elements. Before we use the concept of atomic structure to explain how and why atoms combine to form compounds, we need to understand the characteristic properties of the elements and the trends that occur in these properties on the periodic table. These trends allow us to use the periodic table to accurately predict properties and reactions of a wide variety of substances.

1 H																	2 He
3 Li	4 Be											5 B	6 C	7 N	8 O	9 F	10 Ne
11 Na	12 Mg											13 Al	14 Si	15 P	16 S	17 Cl	18 Ar
19 K	20 Ca	21 Sc	22 Ti	23 V	24 Cr	25 Mn	26 Fe	27 Co	28 Ni	29 Cu	30 Zn	31 Ga	32 Ge	33 As	34 Se	35 Br	36 Kr
37 Rb	38 Sr	39 Y	40 Zr	41 Nb	42 Mo	43 Tc	44 Ru	45 Rh	46 Pd	47 Ag	48 Cd	49 In	50 Sn	51 Sb	52 Te	53 I	54 Xe
55 Cs	56 Ba	57 La*	72 Hf	73 Ta	74 W	75 Re	76 Os	77 Ir	78 Pt	79 Au	80 Hg	81 Tl	82 Pb	83 Bi	84 Po	85 At	86 Rn
87 Fr	88 Ra	89 Ac†	104 Rf	105 Db	106 Sg	107 Bh	108 Hs	109 Mt	110 Uun	111 Uuu	112 Uub						

Metals
Metalloids
Nonmetals

*	58 Ce	59 Pr	60 Nd	61 Pm	62 Sm	63 Eu	64 Gd	65 Tb	66 Dy	67 Ho	68 Er	69 Tm	70 Yb	71 Lu
†	90 Th	91 Pa	92 U	93 Np	94 Pu	95 Am	96 Cm	97 Bk	98 Cf	99 Es	100 Fm	101 Md	102 No	103 Lr

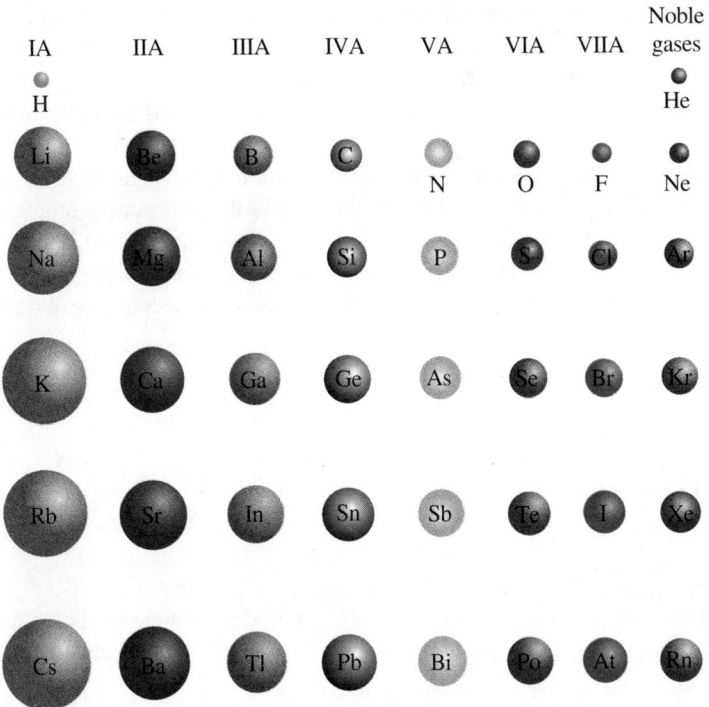

IA IIA IIIA IVA VA VIA VIIA Noble gases

H   He

Li Be B C N O F Ne

Na Mg Al Si P S Cl Ar

K Ca Ga Ge As Se Br Kr

Rb Sr In Sn Sb Te I Xe

Cs Ba Tl Pb Bi Po At Rn

◄ FIGURE 11.2
Relative atomic radii for the representative elements. Atomic radius decreases across a period and increases down a group in the periodic table.

Metals and Nonmetals

In Section 3.8, we classified elements as metals, nonmetals, or metalloids. The heavy stair-step line beginning at boron and running diagonally down the periodic table separates the elements into metals and nonmetals. Metals are usually lustrous, malleable, and good conductors of heat and electricity. Nonmetals are just the opposite—nonlustrous, brittle, and poor conductors. Metalloids are found bordering the heavy diagonal line and may have properties of both metals and nonmetals.

Most elements are classified as metals (see Figure 11.1). Metals are found on the left side of the stair-step line, while the nonmetals are located toward the upper right of the table. Note that hydrogen does not fit into the division of metals and nonmetals. It displays nonmetallic properties under normal conditions, even though it has only one outermost electron like the alkali metals. Hydrogen is considered to be a unique element.

It is the chemical properties of metals and nonmetals that interest us most. Metals tend to lose electrons and form positive ions, while nonmetals tend to gain electrons and form negative ions. When a metal reacts with a nonmetal, electrons are often transferred from the metal to the nonmetal.

Atomic Radius

The relative radii of the representative elements are shown in Figure 11.2. Notice that the radii of the atoms tend to increase down each group and that they tend to decrease from left to right across a period.

The increase in radius down a group can be understood if we consider the electron structure of the atoms. For each step down a group, an additional energy level is added to the atom. The average distance from the nucleus to the outside edge of the atom must increase as each new energy level is added. The atoms get bigger as electrons are placed in these new higher energy levels.

Understanding the decrease in atomic radius across a period requires more thought, however. As we move from left to right across a period, electrons within the same block are being added to the same energy level. Within a given energy level we expect the orbitals to have about the same size. We would then expect the atoms to be about the same size across the period. But each time an electron is added, a proton is added to the nucleus as well. The increase in positive charge (in the nucleus) pulls the electrons closer to the nucleus, which results in a gradual decrease in atomic radius across a period.

Ionization Energy

ionization energy

The **ionization energy** of an atom is the energy required to remove an electron from the atom. For example,

$$Na + \text{ionization energy} \longrightarrow Na^+ + e^-$$

The first ionization energy is the amount of energy required to remove the first electron from an atom, the second is the amount required to remove the second electron from that atom, and so on.

Table 11.1 gives the ionization energies for the removal of one to five electrons from several elements. The table shows that even higher amounts of energy are needed to remove the second, third, fourth, and fifth electrons. This makes sense because removing electrons leaves fewer electrons attracted to the positive charge in the nucleus. The data in Table 11.1 also show that an extra large ionization energy (red) is needed when an electron is removed from a noble gas structure, clearly showing the stability of this configuration.

TABLE 11.1	Ionization Energies for Selected Elements*				
	Required amounts of energy (kJ/mol)				
Element	**1st** e^-	**2nd** e^-	**3rd** e^-	**4th** e^-	**5th** e^-
H	1,314				
He	2,372	5,247			
Li	520	7,297	11,810		
Be	900	1,757	14,845	21,000	
B	800	2,430	3,659	25,020	32,810
C	1,088	2,352	4,619	6,222	37,800
Ne	2,080	3,962	6,276	9,376	12,190
Na	496	4,565	6,912	9,540	13,355

*Values are expressed in kilojoules per mole, showing energies required to remove 1 to 5 electrons per atom. Red type indicates the energy needed to remove an electron from a noble gas electron structure.

◀FIGURE 11.3
Periodic relationship of the first ionization energy for representative elements in the first four periods.

First ionization energies have been experimentally determined for most elements. Figure 11.3 plots these energies for representative elements in the first four periods. Note these important points:

1. Ionization energy in Group A elements decreases from top to bottom in a group. For example, in Group IA the ionization energy changes from 520 kJ/mol for Li to 419 kJ/mol for K.
2. Ionization energy gradually increases from left to right across a period. Noble gases have a relatively high value, confirming the nonreactive nature of these elements.

Metals don't behave in exactly the same manner. Some metals give up electrons much more easily than others. In the alkali metal family, cesium gives up its 6s electron much more easily than the metal lithium gives up its 2s electron. This makes sense when we consider that the size of the atoms increases down the group. The distance between the nucleus and the outer electrons increases and the ionization energy decreases. The most chemically active metals are located at the lower left of the periodic table.

Nonmetals have relatively large ionization energies compared to metals. Nonmetals tend to gain electrons and form anions. Since the nonmetals are located at the right side of the periodic table, it is not surprising that ionization energies tend to increase from left to right across a period. The most active nonmetals are found in the *upper* right corner of the periodic table.

11.2 Lewis Structures of Atoms

Metals tend to form cations (positively charged ions) and nonmetals form anions (negatively charged ions) in order to attain a stable valence electron structure. For many elements this stable valence level contains eight electrons (two *s* and six *p*),

FIGURE 11.4 ▶
Lewis structures of the first 20 elements. Dots represent electrons in the outermost energy level only.

IA	IIA	IIIA	IVA	VA	VIA	VIIA	Noble Gases
H·							He:
Li·	Be:	:Ḃ	:Ċ·	:N̈·	·Ö:	:F̈:	:N̈e:
Na·	Mg:	:Äl	:Ṡi·	:P̈·	·S̈:	:C̈l:	:Är:
K·	Ca:						

identical to the valence electron configuration of the noble gases. Atoms undergo re-arrangements of electron structure to lower their chemical potential energy (or to become more stable). These rearrangements are accomplished by losing, gaining, or sharing electrons with other atoms. For example, a hydrogen atom could accept a second electron and attain an electron structure the same as the noble gas helium. A fluorine atom could gain an electron and attain an electron structure like neon. A sodium atom could lose one electron to attain an electron structure like neon.

Lewis structure   The valence electrons in the outermost energy level of an atom are responsible for the electron activity that occurs to form chemical bonds. The **Lewis structure** of an atom is a representation that shows the valence electrons for that atom. American chemist Gilbert N. Lewis (1875–1946) proposed using the symbol for the element and dots for electrons. The number of dots placed around the symbol equals the number of s and p electrons in the outermost energy level of the atom. Paired dots represent paired electrons; unpaired dots represent unpaired electrons. For example, **H·** is the Lewis symbol for a hydrogen atom, $1s^1$; **:B̈** is the Lewis symbol for a boron atom, with valence electrons $2s^2 2p^1$. In the case of boron, the symbol B represents the boron nucleus and the $1s^2$ electrons; the dots represent only the $2s^2 2p^1$ electrons.

▲ A sample of boron.

Paired electrons ⟶ :**B** ⟵ Unpaired electron
Symbol of the element

The Lewis method is used not only because of its simplicity of expression but also because much of the chemistry of the atom is directly associated with the electrons in the outermost energy level. Figure 11.4 shows Lewis structures for the elements hydrogen through calcium.

Example 11.1   Write the Lewis structure for a phosphorus atom.

Solution   First establish the electron structure for a phosphorus atom, which is $1s^2 2s^2 2p^6 3s^2 3p^3$. Note that there are five electrons in the outermost energy level; they are $3s^2 3p^3$. Write the symbol for phosphorus and place the five electrons as dots around it.

$$:\overset{\displaystyle\cdot}{\underset{\displaystyle\cdot}{P}}\cdot$$

The $3s^2$ electrons are paired and are represented by the paired dots. The $3p^3$ electrons, which are unpaired, are represented by the single dots.

A quick way to determine the correct number of dots (electrons) for a Lewis structure is to use the group number. For the A groups on the periodic table, the Roman numeral is the same as the number of electrons in the Lewis structure.

Practice 11.1

Write the Lewis structure for the following elements:
(a) N (b) Al (c) Sr (d) Br

11.3 The Ionic Bond: Transfer of Electrons from One Atom to Another

The chemistry of many elements, especially the representative ones, is to attain an outer electron structure like that of the chemically stable noble gases. With the exception of helium, this stable structure consists of eight electrons in the outermost energy level (see Table 11.2).

Let's look at the electron structures of sodium and chlorine to see how each element can attain a structure of eight electrons in its outermost energy level. A sodium atom has 11 electrons: 2 in the first energy level, 8 in the second energy level, and 1 in the third energy level. A chlorine atom has 17 electrons: 2 in the first energy level, 8 in the second energy level, and 7 in the third energy level. If a sodium atom transfers or loses its $3s$ electron, its third energy level becomes vacant, and it becomes a sodium ion with an electron configuration identical to that of the noble gas neon. This process requires energy:

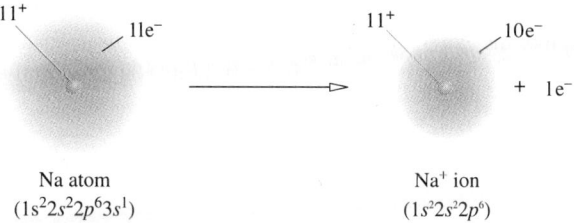

An atom that has lost or gained electrons will have a positive or negative charge, depending on which particles (protons or electrons) are in excess. Remember that a charged particle or group of particles is called an *ion*.

By losing a negatively charged electron, the sodium atom becomes a positively charged particle known as a sodium ion. The charge, $+1$, results because the nucleus still contains 11 positively charged protons, and the electron orbitals contain only 10

TABLE 11.2	Arrangement of Electrons in the Noble Gases*						
		Electron structure					
Noble gas	**Symbol**	$n = 1$	**2**	**3**	**4**	**5**	**6**
Helium	He	$1s^2$					
Neon	Ne	$1s^2$	$2s^2 2p^6$				
Argon	Ar	$1s^2$	$2s^2 2p^6$	$3s^2 3p^6$			
Krypton	Kr	$1s^2$	$2s^2 2p^6$	$3s^2 3p^6 3d^{10}$	$4s^2 4p^6$		
Xenon	Xe	$1s^2$	$2s^2 2p^6$	$3s^2 3p^6 3d^{10}$	$4s^2 4p^6 4d^{10}$	$5s^2 5p^6$	
Radon	Rn	$1s^2$	$2s^2 2p^6$	$3s^2 3p^6 3d^{10}$	$4s^2 4p^6 4d^{10} 4f^{14}$	$5s^2 5p^6 5d^{10}$	$6s^2 6p^6$

*Each gas except helium has eight electrons in its outermost energy level.

negatively charged electrons. The charge is indicated by a plus sign (+) and is written as a superscript after the symbol of the element (Na^+).

A chlorine atom with seven electrons in the third energy level needs one electron to pair up with its one unpaired $3p$ electron to attain the stable outer electron structure of argon. By gaining one electron, the chlorine atom becomes a chloride ion (Cl^-), a negatively charged particle containing 17 protons and 18 electrons. This process releases energy:

$$17+ \qquad 17e^- \qquad + \; 1e^- \longrightarrow \qquad 17+ \qquad 18e^-$$

Cl atom
($1s^2 2s^2 2p^6 3s^2 3p^5$)

Cl^- ion
($1s^2 2s^2 2p^6 3s^2 3p^6$)

Consider sodium and chlorine atoms reacting with each other. The $3s$ electron from the sodium atom transfers to the half-filled $3p$ orbital in the chlorine atom to form a positive sodium ion and a negative chloride ion. The compound sodium chloride results because the Na^+ and Cl^- ions are strongly attracted to each other by their opposite electrostatic charges. The force holding the oppositely charged ions together is an ionic bond:

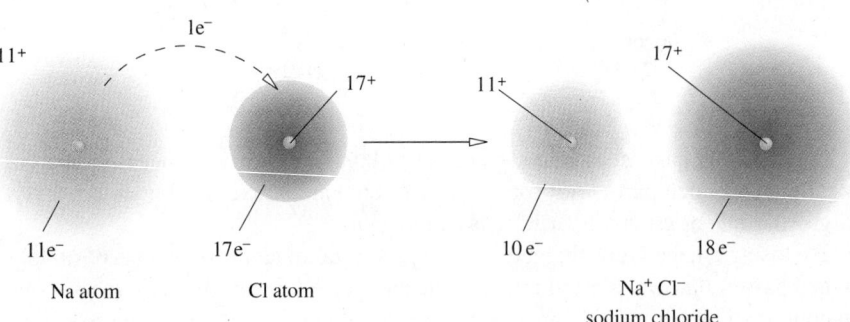

Na atom	Cl atom		Na^+ Cl^- sodium chloride

Na⁺ Cl⁻   Na⁺ Cl⁻

▲
These tiny NaCl crystals (on a penny) show the cubic structure illustrated in Figure 11.5.

▲**FIGURE 11.5**
Sodium chloride crystal. Diagram represents a small fragment of sodium chloride, which forms cubic crystals. Each sodium ion is surrounded by six chloride ions, and each chloride ion is surrounded by six sodium ions. The tiny NaCl crystals show the cubic crystal structure of salt.

The Lewis representation of sodium chloride formation is:

$$Na \cdot + \cdot \ddot{\underset{\cdot\cdot}{Cl}} \colon \longrightarrow [Na]^+ \left[\colon \ddot{\underset{\cdot\cdot}{Cl}} \colon \right]^-$$

The chemical reaction between sodium and chlorine is a very vigorous one, producing considerable heat in addition to the salt formed. When energy is released in a chemical reaction, the products are more stable than the reactants. Note that in NaCl both atoms attain a noble gas electron structure.

Sodium chloride is made up of cubic crystals in which each sodium ion is surrounded by six chloride ions and each chloride ion by six sodium ions, except at the crystal surface. A visible crystal is a regularly arranged aggregate of millions of these ions, but the ratio of sodium to chloride ions is 1:1, hence the formula NaCl. The cubic crystalline lattice arrangement of sodium chloride is shown in Figure 11.5.

Figure 11.6 contrasts the relative sizes of sodium and chlorine atoms with those of their ions. The sodium ion is smaller than the atom due primarily to two factors: (1) The sodium atom has lost its outermost electron, thereby reducing its size; and (2) the 10 remaining electrons are now attracted by 11 protons and are thus drawn closer to the nucleus. Conversely, the chloride ion is larger than the atom because (1) it has 18 electrons but only 17 protons; and (2) the nuclear attraction on each electron is thereby decreased, allowing the chlorine atom to expand as it forms an ion.

We've seen that when sodium reacts with chlorine, each atom becomes an ion. Sodium chloride, like all ionic substances, is held together by the attraction existing between positive and negative charges. An **ionic bond** is the attraction between oppositely charged ions.

Remember: A cation is always smaller than its parent atom whereas an anion is always larger than its parent atom.

ionic bond

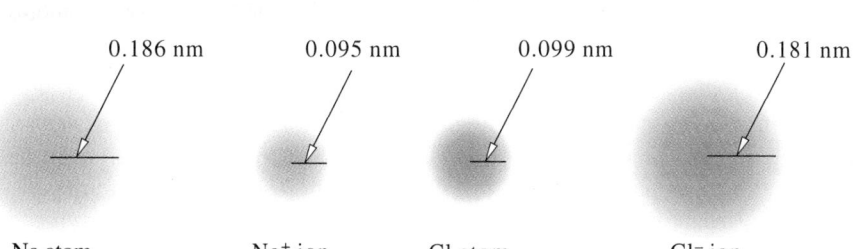

0.186 nm   0.095 nm   0.099 nm   0.181 nm

Na atom   Na⁺ ion   Cl atom   Cl⁻ ion

▲**FIGURE 11.6**
Relative radii of sodium and chlorine atoms and their ions.

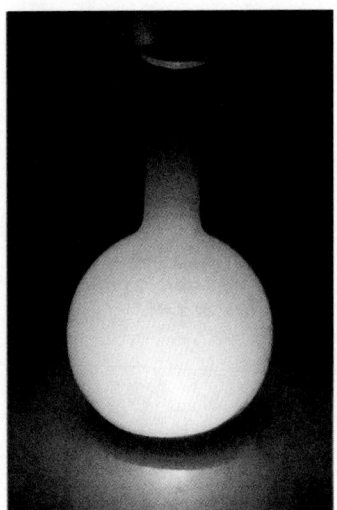

The reaction between sodium metal and chlorine gas releases both heat and light in the production of salt.

TABLE 11.3	Change in Atomic Radii (nm) of Selected Metals and Nonmetals*						
Atomic radius		**Ionic radius**		**Atomic radius**		**Ionic radius**	
Li	0.152	Li$^+$	0.060	F	0.071	F$^-$	0.136
Na	0.186	Na$^+$	0.095	Cl	0.099	Cl$^-$	0.181
K	0.227	K$^+$	0.133	Br	0.114	Br$^-$	0.195
Mg	0.160	Mg^{2+}	0.065	O	0.074	O^{2-}	0.140
Al	0.143	Al^{3+}	0.050	S	0.103	S^{2-}	0.184

*The metals shown lose electrons to become positive ions. The nonmetals gain electrons to become negative ions.

Ionic bonds are formed whenever one or more electrons are transferred from one atom to another. Metals, which have relatively little attraction for their valence electrons, tend to form ionic bonds when they combine with nonmetals.

It's important to recognize that substances with ionic bonds do not exist as molecules. In sodium chloride, for example, the bond does not exist solely between a single sodium ion and a single chloride ion. Each sodium ion in the crystal attracts six near-neighbor negative chloride ions; in turn, each negative chloride ion attracts six near-neighbor positive sodium ions (see Figure 11.5).

A metal will usually have one, two, or three electrons in its outer energy level. In reacting, metal atoms characteristically lose these electrons, attain the electron structure of a noble gas, and become positive ions. A nonmetal, on the other hand, is only a few electrons short of having a noble gas electron structure in its outer energy level and thus has a tendency to gain electrons. In reacting with metals, nonmetal atoms characteristically gain one, two, or three electrons; attain the electron structure of a noble gas; and become negative ions. The ions formed by loss of electrons are much smaller than the corresponding metal atoms; the ions formed by gaining electrons are larger than the corresponding nonmetal atoms. The dimensions of the atomic and ionic radii of several metals and nonmetals are given in Table 11.3.

Study the following examples. Note the loss and gain of electrons between atoms; also note that the ions in each compound have a noble gas electron structure.

Example 11.2   Explain how magnesium and chlorine combine to form magnesium chloride, $MgCl_2$.

Solution   A magnesium atom of electron structure $1s^2 2s^2 2p^6 3s^2$ must lose two electrons or gain six to reach a stable electron structure. If magnesium reacts with chlorine and each chlorine atom can accept only one electron, two chlorine atoms will be needed for the two electrons from each magnesium atom. The compound formed will contain one magnesium ion and two chloride ions. The magnesium atom, having lost two electrons, becomes a magnesium ion with a +2 charge. Each chloride ion will have a -1 charge. The transfer of electrons from a magnesium atom to two chlorine atoms is shown in the following illustration:

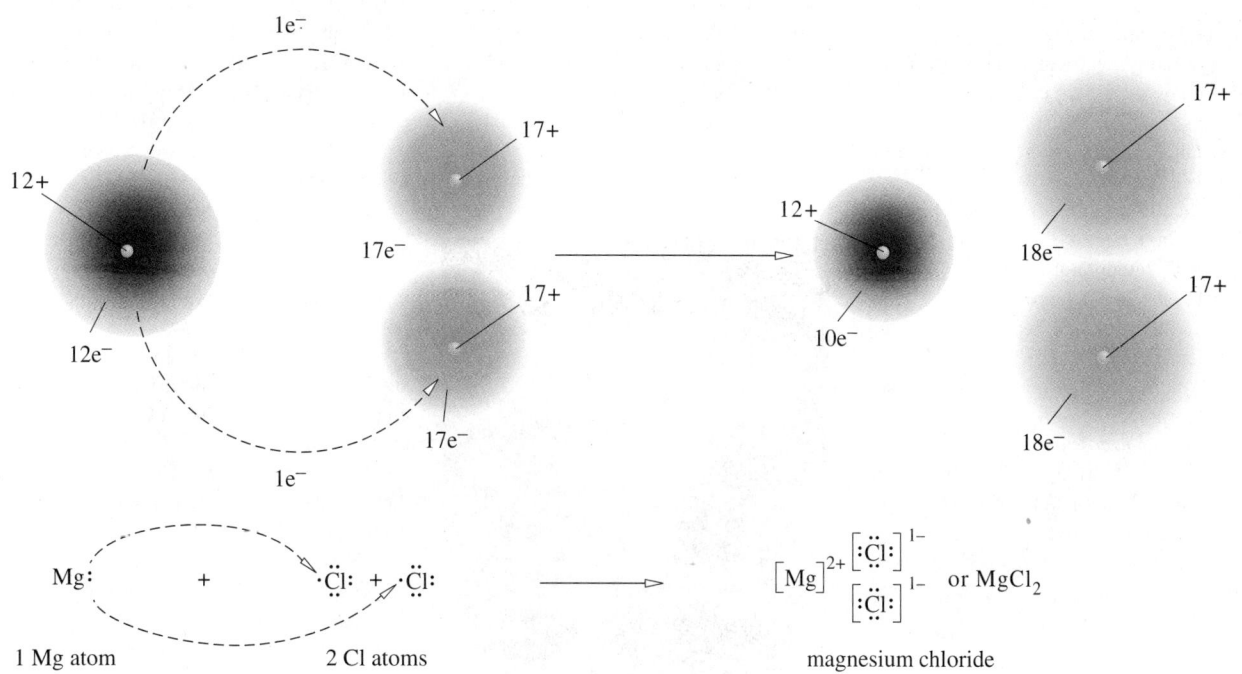

Explain the formation of sodium fluoride, NaF, from its elements.   **Example 11.3**

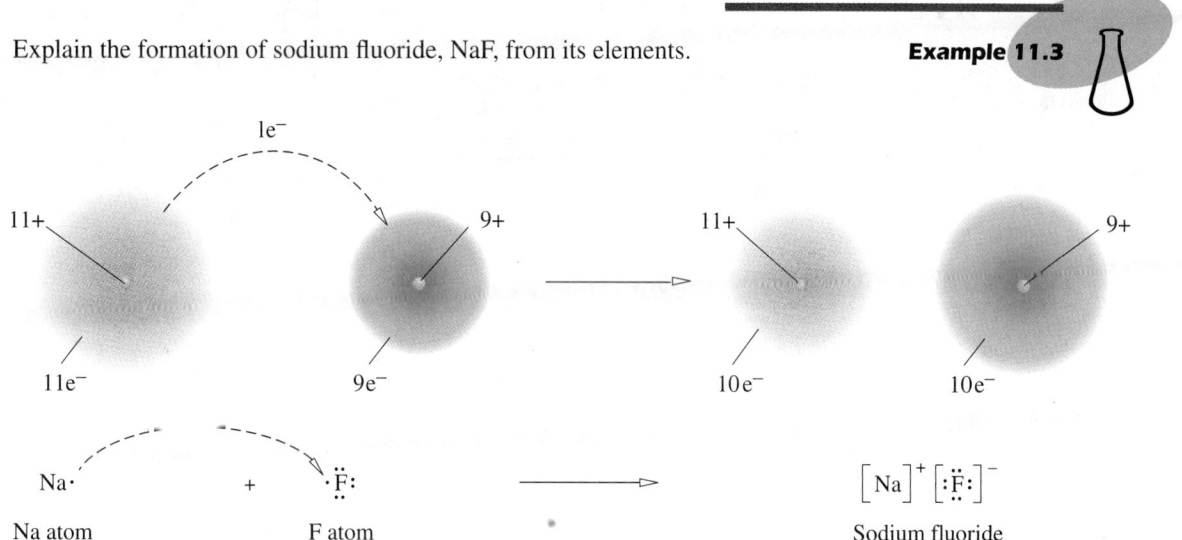

The fluorine atom, with seven electrons in its outer energy level, behaves similarly to the chlorine atom.   **Solution**

Superconductors—A New Frontier

When electric current flows through a wire, resistance slows the current and heats the wire. To keep the current flowing, this electrical "friction" must be overcome by adding energy to the system. In fact, the existence of electrical resistance limits the efficiency of all electrical devices.

In 1911, a Dutch scientist, Heike Kamerlingh Onnes, discovered that at very cold temperatures (near 0 K), electrical resistance disappears. Onnes named this phenomenon *superconductivity*. Scientists have been fascinated by it ever since. Unfortunately, because such low temperatures were required, liquid helium was necessary to cool the wires. With helium costing $3.50 per liter, commercial applications were far too expensive to be considered.

For many years scientists were convinced that superconductivity wasn't possible at higher temperatures (not even as high as 77 K, the boiling point of liquid nitrogen, a bargain at $0.39 per liter). The first high-temperature superconductor, developed in 1986, was superconducting at 30 K. The material was a complex metal oxide that has a sandwichlike crystal structure with copper and oxygen atoms on the inside and barium and lanthanum atoms on the outside.

Scientists immediately tried to develop materials that would be super-conducting at even higher temperatures. To do this they relied on their knowledge of the periodic table and the properties of chemical families. Paul Chu at the University of Houston, Texas, found that critical temperature could be raised by compressing the superconducting oxide. The pressure was too intense to be useful

▲ **When chilled in liquid nitrogen, the superconductor acts as a perfect mirror to the magnet, causing it to levitate as it "sees" its reflection in the superconductor.**

commercially, so Chu looked for another way to bring the atoms closer together. He accomplished this by replacing the barium with strontium, an element in the same family with similar chemical properties and a smaller ionic radius. The idea was successful—the critical temperature changed from 30 K to 40 K.

Chu then tried replacing the strontium with calcium (same family, smaller still) but to no avail. The new material had a lower critical temperature! But Chu persisted, and in 1987, by substituting yttrium for lanthanum (same family, smaller radius) he produced a new superconductor having a critical temperature of 95 K, well above the 77 K boiling point of liquid nitrogen. This material has the formula $YBa_2Cu_3O_7$ and is a good candidate for commercial applications.

Several barriers must be crossed before superconductors are in wide use. The materials developed so far are brittle and easily broken, nonmalleable, and do not carry as high a current per unit cross section as conventional conductors. Many researchers are currently working to surmount these problems and develop potential uses for superconductors, including high-speed levitation trains, tiny efficient electric motors, and smaller, faster computers.

Example 11.4   Explain the formation of aluminum fluoride, AlF_3, from its elements.

Solution

$$\ddot{Al}\cdot \;+\; \begin{array}{c} \cdot \ddot{F}\colon \\[4pt] \cdot \ddot{F}\colon \\[4pt] \cdot \ddot{F}\colon \end{array} \;\longrightarrow\; [Al]^{3+}\begin{array}{c}[\colon\ddot{F}\colon]^{-} \\[4pt] [\colon\ddot{F}\colon]^{-} \\[4pt] [\colon\ddot{F}\colon]^{-} \end{array} \quad \text{or} \quad AlF_3$$

1 Al atom   3 F atoms   aluminum flouride

Each fluorine atom can accept only one electron. Therefore three fluorine atoms are needed to combine with the three outer electrons of one aluminum atom. The aluminum atom has lost three electrons to become an aluminum ion, Al^{3+}, with a +3 charge.

Explain the formation of magnesium oxide, MgO, from its elements.

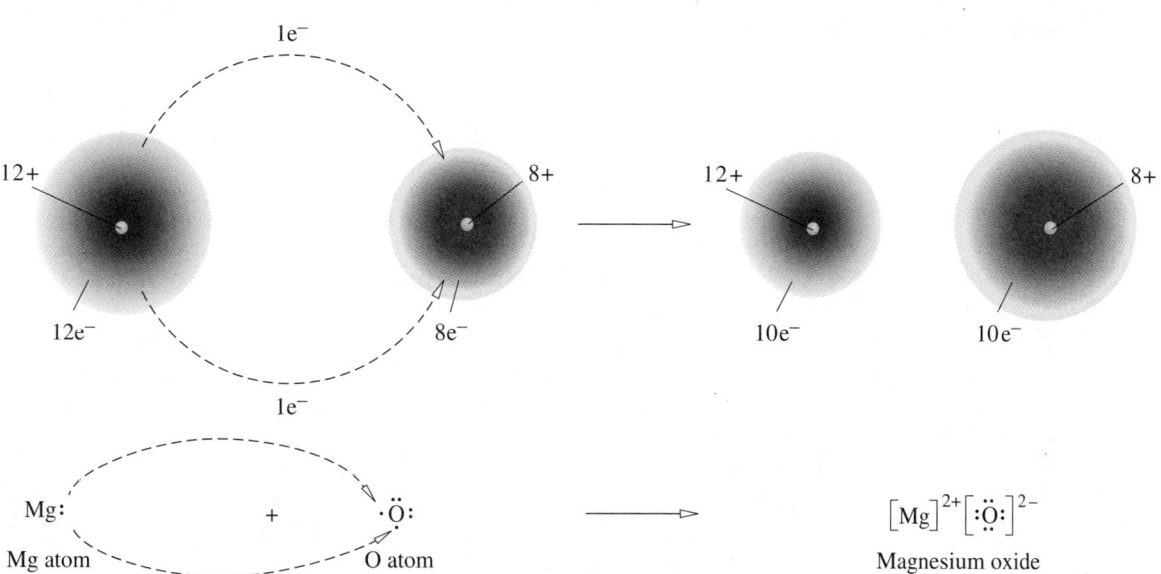

The magnesium atom, with two electrons in the outer energy level, exactly fills the need of one oxygen atom for two electrons. The resulting compound has a ratio of one magnesium atom to one oxygen atom. The oxygen (oxide) ion has a -2 charge, having gained two electrons. In combining with oxygen, magnesium behaves the same way as when it combines with chlorine—it loses two electrons.

Solution

Explain the formation of sodium sulfide, Na_2S, from its elements.

Example 11.6

Solution

$$\begin{matrix} Na\cdot \\ \\ Na\cdot \end{matrix} \quad + \quad \cdot \ddot{S}: \quad \longrightarrow \quad \begin{matrix} [Na]^+ \\ \\ [Na]^+ \end{matrix} \Big[:\ddot{S}:\Big]^{2-} \quad or \quad Na_2S$$

2 Na atoms   1 S atom   sodium sulfide

Two sodium atoms supply the electrons that one sulfur atom needs to make eight in its outer energy level.

Explain the formation of aluminum oxide, Al_2O_3, from its elements.

Example 11.7

Solution

$$[Al]^{3+} \begin{matrix}\Big[:\ddot{O}:\Big]^{2-} \\ \Big[:\ddot{O}:\Big]^{2-} \\ \Big[:\ddot{O}:\Big]^{2-}\end{matrix} \quad or \quad Al_2O_3$$

2 Al atoms   3 O atoms   aluminum oxide

One oxygen atom, needing two electrons, cannot accommodate the three electrons from one aluminum atom. One aluminum atom falls one electron short of the four electrons needed by two oxygen atoms. A ratio of two atoms of aluminum to three atoms of oxygen, involving the transfer of six electrons (two to each oxygen atom), gives each atom a stable electron configuration.

Note that in each of these examples, outer energy levels containing eight electrons were formed in all the negative ions. This formation resulted from the pairing of all the s and p electrons in these outer levels.

11.4   Predicting Formulas of Ionic Compounds

In the previous examples we've seen that when a metal and a nonmetal react to form an ionic compound, the metal loses one or more electrons to the nonmetal. In Chapter 6, where we learned to name compounds and write formulas, we saw that Group IA metals always formed +1 cations, whereas Group IIA formed +2 cations. Group VIIA elements formed −1 anions and Group VIA formed −2 anions.

It stands to reason then that this pattern is directly related to the stability of the noble gas configuration. Metals lose electrons to attain the electron configuration of a noble gas (the previous one on the periodic table). A nonmetal forms an ion by gaining enough electrons to achieve the electron configuration of the noble gas following it on the periodic table. These observations lead us to an important chemical principle:

> **In almost all stable chemical compounds of representative elements, each atom attains a noble gas electron configuration. This concept forms the basis for our understanding of chemical bonding.**

We can apply this principle in predicting the formulas of ionic compounds. To predict the formula of an ionic compound, we must recognize that chemical compounds are always electrically neutral. In addition, the metal will lose electrons to achieve noble gas configuration and the nonmetal will gain electrons to achieve noble gas configuration. Consider the compound formed between barium and sulfur. Barium has two valence electrons, whereas sulfur has six valence electrons:

Ba $[Xe]6s^2$ S $[Ne]3s^2 3p^4$

If barium loses two electrons, it will achieve the configuration of xenon. By gaining two electrons, sulfur achieves the configuration of argon. Consequently a pair of electrons is transferred between atoms. Now we have Ba^{2+} and S^{2-}. Since compounds are electrically neutral, there must be a ratio of one Ba to one S, giving the formula BaS.

The same principle works for many other cases. Since the key lies in the electron configuration, the periodic table can be used to extend the prediction even further. Because of similar electron structures, the elements in a family generally form compounds with the same atomic ratios. In general if we know the atomic ratio

TABLE 11.4	Formulas of Compounds Formed by Alkali Metals			
Lewis structure	Oxides	Chlorides	Bromides	Sulfates
Li·	Li_2O	LiCl	LiBr	Li_2SO_4
Na·	Na_2O	NaCl	NaBr	Na_2SO_4
K·	K_2O	KCl	KBr	K_2SO_4
Rb·	Rb_2O	RbCl	RbBr	Rb_2SO_4
Cs·	Cs_2O	CsCl	CsBr	Cs_2SO_4

of a particular compound, say NaCl, we can predict the atomic ratios and formulas of the other alkali metal chlorides. These formulas are LiCl, KCl, RbCl, CsCl, and FrCl (see Table 11.4).

Similarly if we know that the formula of the oxide of hydrogen is H_2O, we can predict that the formula of the sulfide will be H_2S, because sulfur has the same valence electron structure as oxygen. Recognize, however, that these are only predictions; it doesn't necessarily follow that every element in a group will behave like the others or even that a predicted compound will actually exist. For example, knowing the formulas for potassium chlorate, bromate, and iodate to be $KClO_3$, $KBrO_3$, and KIO_3, we can correctly predict the corresponding sodium compounds to have the formulas $NaClO_3$, $NaBrO_3$, and $NaIO_3$. Fluorine belongs to the same family of elements (Group VIIA) as chlorine, bromine, and iodine, so it would appear that fluorine should combine with potassium and sodium to give fluorates with the formulas KFO_3 and $NaFO_3$. However, potassium and sodium fluorates are not known to exist.

In the discussion in this section we refer only to representative metals (Groups IA, IIA, IIIA). The transition metals (Group B) show more complicated behavior (they form multiple ions), and their formulas are not as easily predicted.

The formula for calcium sulfide is CaS and that for lithium phosphide is Li_3P. Predict formulas for (a) magnesium sulfide, (b) potassium phosphide, and (c) magnesium selenide.

Example 11.8

(a) Look up calcium and magnesium in the periodic table; they are both in Group IIA. The formula for calcium sulfide is CaS, so it's reasonable to predict that the formula for magnesium sulfide is MgS.

(b) Find lithium and potassium in the periodic table; they are in Group IA. Since the formula for lithium phosphide is Li_3P, it's reasonable to predict that K_3P is the formula for potassium phosphide.

(c) Find selenium in the periodic table; it is in Group VIA just below sulfur. Therefore it's reasonable to assume that selenium forms selenide in the same way that sulfur forms sulfide. Since MgS was the predicted formula for magnesium sulfide in part (a), we can reasonably assume that the formula for magnesium selenide is MgSe.

Solution

O   C   Dry ice

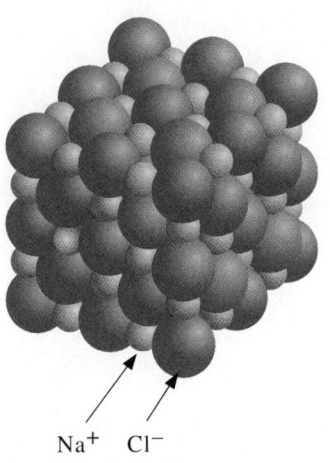

Na⁺   Cl⁻

Sodium chloride

▲
FIGURE 11.7
Solid carbon dioxide (dry ice) is composed of individual covalently bonded molecules of CO_2 closely packed together. Table salt is a large aggregate of Na⁺ and Cl⁻ ions instead of molecules.

covalent bond

FIGURE 11.8 ▶
The formation of a hydrogen molecule from two hydrogen atoms. The two 1s orbitals overlap, forming the H_2 molecule. In this molecule the two electrons are shared between the atoms, forming a covalent bond.

11.5   The Covalent Bond: Sharing Electrons

Some atoms do not transfer electrons from one atom to another to form ions. Instead they form a chemical bond by sharing pairs of electrons between them.

A **covalent bond** consists of a pair of electrons shared between two atoms. This bonding concept was introduced in 1916 by G. N. Lewis. In the millions of known compounds, the covalent bond is the predominant chemical bond.

True molecules exist in substances in which the atoms are covalently bonded. It is proper to refer to molecules of such substances as hydrogen, chlorine, hydrogen chloride, carbon dioxide, water, or sugar (Figure 11.7). These substances contain only covalent bonds and exist as aggregates of molecules. We don't use the term *molecule* when talking about ionically bonded compounds such as sodium chloride, because such substances exist as large aggregates of positive and negative ions, not as molecules (Figure 11.7).

A study of the hydrogen molecule gives us an insight into the nature of the covalent bond and its formation. The formation of a hydrogen molecule, H_2, involves the overlapping and pairing of 1s electron orbitals from two hydrogen atoms, shown in Figure 11.8. Each atom contributes one electron of the pair that is shared jointly by two hydrogen nuclei. The orbital of the electrons now includes both hydrogen nuclei, but probability factors show that the most likely place to find the electrons (the point of highest electron density) is between the two nuclei. The two nuclei are shielded from each other by the pair of electrons, allowing the two nuclei to be drawn very close to each other.

The formula for chlorine gas is Cl_2. When the two atoms of chlorine combine to form this molecule, the electrons must interact in a manner similar to that shown in the hydrogen example. Each chlorine atom would be more stable with eight electrons in its outer energy level. But chlorine atoms are identical, and neither is able to

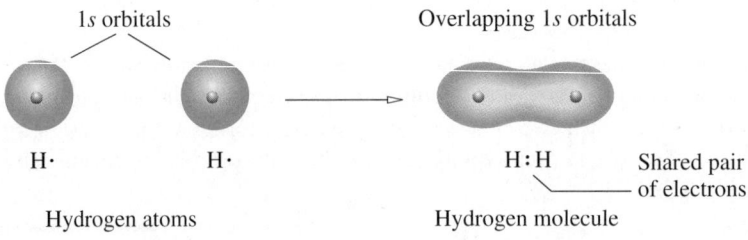

1s orbitals   Overlapping 1s orbitals

H·   H·   H:H   Shared pair of electrons

Hydrogen atoms   Hydrogen molecule

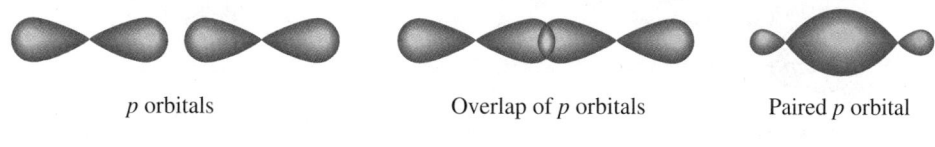

p orbitals   Overlap of p orbitals   Paired p orbital

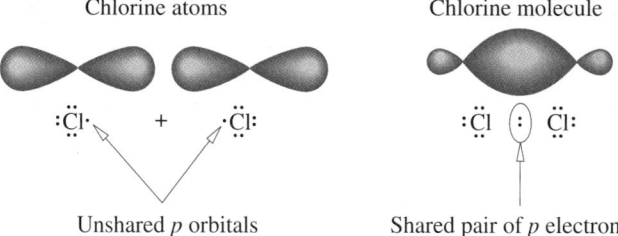

Chlorine atoms   Chlorine molecule

Unshared p orbitals   Shared pair of p electrons

◄ FIGURE 11.9
Pairing of p electrons in the formation of a chlorine molecule.

pull an electron away from the other. What happens is this: The unpaired $3p$ electron orbital of one chlorine atom overlaps the unpaired $3p$ electron orbital of the other atom, resulting in a pair of electrons that are mutually shared between the two atoms. Each atom furnishes one of the pair of shared electrons. Thus each atom attains a stable structure of eight electrons by sharing an electron pair with the other atom. The pairing of the p electrons and the formation of a chlorine molecule are illustrated in Figure 11.9. Neither chlorine atom has a positive or negative charge, because both contain the same number of protons and have equal attraction for the pair of electrons being shared. Other examples of molecules in which electrons are equally shared between two atoms are hydrogen, H_2; oxygen, O_2; nitrogen, N_2; fluorine, F_2; bromine, Br_2; and iodine, I_2. Note that more than one pair of electrons may be shared between atoms:

H:H   :F̈:F̈:   :B̈r:B̈r: :Ï:Ï: :Ö::Ö: :N⋮⋮N:

hydrogen   fluorine   bromine   iodine   oxygen   nitrogen

Remember: A dash represents a shared pair of electrons.

The Lewis structure given for oxygen does not adequately account for all the properties of the oxygen molecule. Other theories explaining the bonding in oxygen molecules have been advanced, but they are complex and beyond the scope of this book.

In writing structures we commonly replace the pair of dots used to represent a shared pair of electrons with a dash (—). One dash represents a single bond; two dashes, a double bond; and three dashes, a triple bond. The six structures just shown may be written thus:

H—H   :F̈—F̈·   :B̈r   B̈r∣   :Ï   Ï: ·Ö—Ö: :N≡N:

▲
Molecular models for F_2 (green, single bond), O_2, (black, double bond), and N_2 (blue, triple bond).

The ionic bond and the covalent bond represent two extremes. In ionic bonding the atoms are so different that electrons are transferred between them, forming a charged pair of ions. In covalent bonding two identical atoms share electrons equally. The bond is the mutual attraction of the two nuclei for the shared electrons. Between these extremes lie many cases in which the atoms are not different enough for a transfer of electrons, but are different enough that the electron pair cannot be shared equally. This unequal sharing of electrons results in the formation of a **polar covalent bond.**

polar covalent bond

11.6   Electronegativity

When two *different* kinds of atoms share a pair of electrons, a bond forms in which electrons are shared unequally. One atom assumes a partial positive charge and the other a partial negative charge with respect to each other. This difference in charge occurs because the two atoms exert unequal attraction for the pair of shared electrons. The attractive force that an atom of an element has for shared electrons in a molecule or polyatomic ion is known as its **electronegativity.** Elements differ in their electronegativities. For example, both hydrogen and chlorine need one electron to form stable electron configurations. They share a pair of electrons in hydrogen chloride, HCl. Chlorine is more electronegative and therefore has a greater attraction for the shared electrons than does hydrogen. As a result the pair of electrons is displaced toward the chlorine atom, giving it a partial negative charge and leaving the hydrogen atom with a partial positive charge. Note that the electron is not transferred entirely to the chlorine atom (as in the case of sodium chloride) and that no ions are formed. The entire molecule, HCl, is electrically neutral. A partial charge is usually indicated by the Greek letter delta, δ. Thus a partial positive charge is represented by $\delta+$ and a partial negative charge by $\delta-$.

electronegativity

hydrogen chloride

The pair of shared electrons in HCl is closer to the more electronegative chlorine atom than to the hydrogen atom, giving chlorine a partial negative charge with respect to the hydrogen atom.

A scale of relative electronegativities, in which the most electronegative element, fluorine, is assigned a value of 4.0, was developed by the Nobel laureate (1954 and 1962) Linus Pauling (1901–1994). Table 11.5 shows that the relative electronegativity of the nonmetals is high and that of the metals is low. These electronegativities indicate that atoms of metals have a greater tendency to lose electrons than do atoms of nonmetals and that nonmetals have a greater tendency to gain electrons than do metals. The higher the electronegativity value, the greater the attraction for electrons. Note that electronegativity generally increases from left to right across a period and decreases down a group for the representative elements. The highest electronegativity is 4.0 for fluorine, and the lowest is 0.7 for francium and cesium. It's important to remember that the higher the electronegativity, the stronger an atom attracts electrons.

The polarity of a bond is determined by the difference in electronegativity values of the atoms forming the bond (see Figure 11.10). If the electronegativities are the same, the bond is **nonpolar covalent** and the electrons are shared equally. If the atoms have greatly different electronegativities, the bond is very *polar.* At the extreme, one or more electrons are actually transferred and an ionic bond results.

nonpolar covalent bond

A **dipole** is a molecule that is electrically asymmetrical, causing it to be oppositely charged at two points. A dipole is often written as ⊕⊖ A hydrogen chloride molecule is polar and behaves as a small dipole. The HCl dipole may be written as

dipole

TABLE 11.5 **Three-Dimensional Representation of Electronegativity**

H+———Cl. The arrow points toward the negative end of the dipole. Molecules of H$_2$O, HBr, and ICl are polar:

H+———Cl   H+———Br   I+———Cl   H◜O◞H

How do we know whether a bond between two atoms is ionic or covalent? The difference in electronegativity between the two atoms determines the character of the bond formed between them. As the difference in electronegativity increases, the polarity of the bond (or percent ionic character) increases.

> If the electronegativity difference between two bonded atoms is greater than 1.7–1.9, the bond will be more ionic than covalent.

H$_2$ Cl$_2$

Nonpolar
molecules

HCl

Polar covalent
molecule

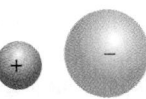

NaCl

Ionic
compound

◀ FIGURE 11.10
Nonpolar, polar covalent,
and ionic compounds.

Goal! A Spherical Molecule

One of the most diverse elements in the periodic table is carbon. Graphite and diamond, two well-known forms of elemental carbon, both contain extended arrays of carbon atoms. In graphite the carbon atoms are arranged in sheets, and the bonding between the sheets is very weak. This property makes graphite useful as a lubricant and as a writing material. Diamond consists of transparent octahedral crystals in which each carbon atom is bonded to four other carbon atoms. This three-dimensional network of bonds gives diamond the property of hardness for which it is noted. In the 1980s, a new form of carbon was discovered in which the atoms are arranged in relatively small clusters.

When Harold Kroto of the University of Sussex, England, and Richard Smalley of Rice University, Texas, discovered a strange carbon molecule of formula C_{60}, they deduced that the most stable arrangement for the atoms would be in the shape of a soccer ball. In thinking about possible arrangements, the scientists considered the geodesic domes designed by R. Buckminster Fuller in the 1960s. This cluster form of carbon was thus named *buckminsterfullerene* and is commonly called buckyballs.

Buckyballs have captured the imagination of a variety of chemists. Research on buckyballs has led to a host of possible applications for these molecules. If metals are bound to the carbon atoms, the fullerenes become superconducting; that is, they conduct electrical current without resistance, at very low temperature. Scientists are now able to make buckyball compounds that superconduct at temperatures of 45 K. Other fullerenes are being used in lubricants and optical materials.

Chemists at Yale University have managed to trap helium and neon inside buckyballs. This is the first time chemists have ever observed helium or neon in a compound of any kind. They found that at temperatures from 1000°F to 1500°F one of the covalent bonds linking neighboring carbon atoms in the buckyball breaks. This opens a window in the fullerene molecule through which a helium or neon atom can enter the buckyball. When the fullerene is allowed to cool, the broken bond between carbon atoms re-forms, shutting the window and trapping the helium or neon atom inside the buckyball. Since the trapped helium or neon cannot react or share electrons with its host, the resulting compound has forced scientists to invent a new kind of chemical formula to describe the compound. The relationship between the "prisoner" helium or neon and the host buckyball is shown with an @ sign. A helium fullerene containing 60 carbon atoms would thus be $He@C_{60}$.

Buckyballs can be tailored to fit a particular size requirement. Raymond Schinazi of the Emory University School of Medicine, Georgia, made a buckyball to fit the active site of a key HIV enzyme that paralyzes the virus, making it noninfectious in human cells. The key to making this compound was preparing a water-soluble buckyball that would fit in the active site of the enzyme. Eventually, scientists created a water-soluble fullerene molecule that has two charged arms to grasp the binding site of the enzyme. It is toxic to the virus but doesn't appear to harm the host cells.

▲ **Scandium atoms trapped in a buckyball.**

▲ **"Raspberries" of fullerene lubricant.**

If the electronegativity difference is greater than 2.0, the bond is strongly ionic. If the electronegativity difference is less than 1.5, the bond is strongly covalent.

Care must be taken to distinguish between polar bonds and polar molecules. A covalent bond between different kinds of atoms is always polar. But a molecule containing different kinds of atoms may or may not be polar, depending on its shape or geometry. Molecules of HF, HCl, HBr, HI, and ICl are all polar because each contains a single polar bond. However, CO_2, CH_4, and CCl_4 are nonpolar molecules despite the fact that all three contain polar bonds. The carbon dioxide molecule

Bond type

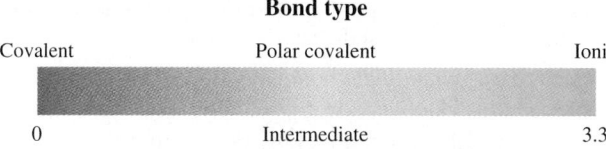

$O=C=O$ is nonpolar because the carbon–oxygen dipoles cancel each other by acting in opposite directions.

$$\overset{\longleftarrow +\quad +\longrightarrow}{O=C=O}$$

dipoles in opposite directions

Carbon tetrachloride (CCl_4) is nonpolar because the four $C-Cl$ polar bonds are identical, and since these bonds emanate from the center to the corners of a tetrahedron in the molecule, their polarities cancel one another. Methane has the same molecular structure and is also nonpolar. We will discuss the shapes of molecules later in this chapter.

We have said that water is a polar molecule. If the atoms in water were linear like those in carbon dioxide, the two $O-H$ dipoles would cancel each other, and the molecule would be nonpolar. However, water is definitely polar and has a nonlinear (bent) structure with an angle of 105° between the two $O-H$ bonds.

The relationships among types of bonds are summarized in Figure 11.11. It is important to realize that bonding is a continuum; that is, the difference between ionic and covalent is a gradual change.

11.7   Lewis Structures of Compounds

As we have seen, Lewis structures are a convenient way of showing the covalent bonds in many molecules or ions of the representative elements. In writing Lewis structures, the most important consideration for forming a stable compound is that the atoms attain a noble gas configuration.

The most difficult part of writing Lewis structures is determining the arrangement of the atoms in a molecule or an ion. In simple molecules with more than two atoms, one atom will be the central atom surrounded by the other atoms. Thus Cl_2O has two possible arrangements, $C-Cl-O$ or $Cl-O-Cl$. Usually, but not always, the single atom in the formula (except H) will be the central atom.

Although Lewis structures for many molecules and ions can be written by inspection, the following procedure is helpful for learning to write them:

Step 1. Obtain the total number of valence electrons to be used in the structure by adding the number of valence electrons in all the atoms in the molecule or ion. If you are writing the structure of an ion, add one electron for each negative charge or subtract one electron for each positive charge on the ion.

Step 2. Write the skeletal arrangement of the atoms and connect them with a single covalent bond (two dots or one dash). Hydrogen, which contains only one bonding electron, can form only one covalent bond. Oxygen atoms are not normally bonded to each other, except in compounds

Remember: The number of valence electrons of Group A elements is the same as their group number in the periodic table.

known to be peroxides. Oxygen atoms normally have a maximum of two covalent bonds, (two single bonds, or one double bond).

Step 3. Subtract two electrons for each single bond you used in Step 2 from the total number of electrons calculated in Step 1. This gives you the net number of electrons available for completing the structure.

Step 4. Distribute pairs of electrons (pairs of dots) around each atom (except hydrogen) to give each atom a noble gas structure.

Step 5. If there are not enough electrons to give these atoms eight electrons, change single bonds between atoms to double or triple bonds by shifting unbonded pairs of electrons as needed. Check to see that each atom has a noble gas electron structure (two electrons for hydrogen and eight for the others). A double bond counts as four electrons for each atom to which it is bonded.

Example 11.9 How many valence electrons are in each of these atoms: Cl, H, C, O, N, S, P, I?

Solution You can look at the periodic table to determine the electron structure, or, if the element is in Group A of the periodic table, the number of valence electrons is equal to the group number:

Atom	Group	Valence electrons
Cl	VIIA	7
H	IA	1
C	IVA	4
O	VIA	6
N	VA	5
S	VIA	6
P	VA	5
I	VIIA	7

Example 11.10 Write the Lewis structure for water, H_2O.

Solution **Step 1.** The total number of valence electrons is eight, two from the two hydrogen atoms and six from the oxygen atom.

Step 2. The two hydrogen atoms are connected to the oxygen atom. Write the skeletal structure:

H O   or   H O H
 H

Place two dots between the hydrogen and oxygen atoms to form the covalent bonds:

H:O   or   H:O:H
 ..
 H

Step 3. Subtract the four electrons used in Step 2 from eight to obtain four electrons yet to be used.

Step 4. Distribute the four electrons around the oxygen atom. Hydrogen atoms cannot accommodate any more electrons:

H—Ö: or   H—Ö—H
$\quad$ | $\quad$ ..
$\quad$ H

These arrangements are Lewis structures because each atom has a noble gas electron structure. Note that the shape of the molecule is not shown by the Lewis structure.

Write Lewis structures for a molecule of methane, CH_4.

Example 11.11

Solution

Step 1. The total number of valence electrons is eight, one from each hydrogen atom and four from the carbon atom.

Step 2. The skeletal structure contains four H atoms around a central C atom. Place two electrons between the C and each H.

$\quad$ H $\quad$ H
H C H   H:Ċ:H
$\quad$ H $\quad$ H

Step 3. Subtract the eight electrons used in Step 2 from eight to obtain zero electrons yet to be placed. Therefore the Lewis structure must be as written in Step 2:

$\qquad\qquad\qquad\qquad$ H
$\quad$ H $\quad$ |
H:Ċ:H   or   H—C—H
$\quad$ Ḧ $\quad$ |
$\qquad\qquad\qquad\qquad$ H

Write the Lewis structure for a molecule of carbon tetrachloride, CCl_4.

Example 11.12

Solution

Step 1. The total number of valence electrons to be used is 32, four from the carbon atom and seven from each of the four chlorine atoms.

Step 2. The skeletal structure contains the four Cl atoms around a central C atom. Place two electrons between the C and each Cl:

$\quad$ Cl $\quad$ Cl
Cl C Cl   Cl:Ċ:Cl
$\quad$ Cl $\quad$ Cl

Step 3. Subtract the eight electrons used in Step 2 from 32 to obtain 24 electrons yet to be placed.

Step 4. Distribute the 24 electrons (12 pairs) around the Cl atoms so that each Cl atom has eight electrons around it:

$$\text{:}\overset{\cdot\cdot}{\underset{\cdot\cdot}{Cl}}\text{:}$$
$$\text{:}\overset{\cdot\cdot}{\underset{\cdot\cdot}{Cl}}\text{:}\overset{\cdot\cdot}{\underset{\cdot\cdot}{Cl}}\text{:} \qquad \text{or} \qquad :\overset{\cdot\cdot}{\underset{\cdot\cdot}{Cl}}-C-\overset{\cdot\cdot}{\underset{\cdot\cdot}{Cl}}:$$
$$\text{:}\overset{\cdot\cdot}{\underset{\cdot\cdot}{Cl}}\text{:}$$

This arrangement is the Lewis structure; CCl_4 contains four covalent bonds.

Example 11.13 Write a Lewis structure for CO_2.

Solution

Step 1. The total number of valence electrons is 16, four from the C atom and six from each O atom.

Step 2. The two O atoms are bonded to a central C atom. Write the skeletal structure and place two electrons between the C and each O atom.

O:C:O

Step 3. Subtract the four electrons used in Step 2 from 16 to obtain 12 electrons yet to be placed.

Step 4. Distribute the 12 electrons around the C and O atoms. Several possibilities exist:

$$:\overset{\cdot\cdot}{\underset{\cdot\cdot}{O}}:C:\overset{\cdot\cdot}{\underset{\cdot\cdot}{O}}: \qquad :\overset{\cdot\cdot}{\underset{\cdot\cdot}{O}}:C:\overset{\cdot\cdot}{\underset{}{O}}: \qquad :\overset{}{\underset{\cdot\cdot}{O}}:\overset{\cdot\cdot}{\underset{}{C}}:\overset{}{\underset{\cdot\cdot}{O}}:$$
$$\text{I} \qquad\qquad \text{II} \qquad\qquad \text{III}$$

Step 5. Not all the atoms have eight electrons around them (noble gas structure). Remove one pair of unbonded electrons from each O atom in structure I and place one pair between each O and the C atom forming two double bonds:

$$:\overset{\cdot\cdot}{\underset{\cdot\cdot}{O}}::C::\overset{\cdot\cdot}{\underset{\cdot\cdot}{O}}: \qquad \text{or} \qquad :\overset{\cdot\cdot}{O}=C=\overset{\cdot\cdot}{O}:$$

Each atom now has eight electrons around it. The carbon is sharing four pairs of electrons, and each oxygen is sharing two pairs. These bonds are known as double bonds because each involves sharing two pairs of electrons.

Practice 11.4

Write the Lewis structures for the following:
(a) PBr_3 (b) $CHCl_3$ (c) HF (d) H_2CO

Although many compounds attain a noble gas structure in covalent bonding, there are numerous exceptions. Sometimes it's impossible to write a structure in which each atom has eight electrons around it. For example, in BF_3 the boron atom has only six electrons around it, and in SF_6 the sulfur atom has 12 electrons around it.

Cleaner Showers through Chemistry

Keeping the shower area sparkling clean and free of mildew is a job none of us enjoy. Now thanks to chemist Bob Black of Jacksonville, Florida, there is a new product that cleans the shower without any scrubbing! Black struggled with his home shower until he finally decided he really needed a new product that would solve the mildew and scrubbing problem.

His search was based on the following needs:

(1) a molecule to lift deposits off the walls of the shower
(2) a way to prevent hard water deposits from forming
(3) a wetting agent to wet the walls and rinse off deposits.

In all cases he limited his search to substances nontoxic and environmentally safe.

Black used a molecule called a glycol ether to lift deposits off the shower wall. This molecule is a long chain with a polar end and a nonpolar end. Substances that are nonpolar (such as grease, oils, and organic material) are attracted to the nonpolar end of the molecules. The molecules cluster together to form micelles (with the polar end pointed out). The polar sphere dissolves in the polar water from the shower washing off the organic deposits.

A micelle

Preventing the hard water deposits from forming on the shower walls required use of a molecule called EDTA. This molecule bonds to ions (like Ca^{2+}, Mg^{2+}, or Fe^{3+}) and prevents the formation of soap scum or hard water deposits on the walls.

Lastly Black added isopropyl alcohol to wet the shower wall and also to disturb the mildew fungus. He mixed all the ingredients in just the right proportions and found he no longer needed to work so hard to clean the shower.

Fortunately for all of us Black shared his solution with friends who also liked it. Black patented his new product and began mass production. You can now find Clean Shower in your local grocery store!

▲
Simply spraying your shower regularly with Clean Shower will keep it free of deposits and mildew.

Although there are exceptions, many molecules can be described using Lewis structures where each atom has a noble gas electron configuration. This is a useful model for understanding chemistry.

11.8 Complex Lewis Structures

Most Lewis structures give bonding pictures that are consistent with experimental information on bond strength and length. There are some molecules and polyatomic ions for which no single Lewis structure consistent with all characteristics and bonding information can be written. For example, consider the nitrate ion, NO_3^-. To write a Lewis structure for this polyatomic ion, we use the following steps.

Step 1. The total number of valence electrons is 24, five from the nitrogen atom, and six from each oxygen atom, plus one electron from the -1 charge.

Step 2. The three O atoms are bonded to a central N atom. Write the skeletal structure and place two electrons between each pair of atoms:

$$\left[\begin{array}{c} O \\ \ddot{O}:\ddot{N}:O \end{array} \right]^-$$

Step 3. Subtract the six electrons used in Step 2 from 24 to obtain 18 electrons yet to be placed.

Step 4. Distribute the 18 electrons around the N and O atoms:

$$:\ddot{O} \longleftarrow \text{electron deficient}$$
$$:\ddot{O}:\ddot{N}:\ddot{O}:$$

Step 5. One pair of electrons is still needed to give all the N and O atoms a noble gas structure. Move the unbonded pair of electrons from the N atom and place it between the N and the electron-deficient O atom, making a double bond.

$$\left[\begin{array}{c} :\ddot{O} \\ \| \\ :\ddot{O}-N-\ddot{O}: \end{array} \right]^{-} \quad \text{or} \quad \left[\begin{array}{c} :\ddot{O}: \\ | \\ :\ddot{O}-N=\ddot{O}: \end{array} \right]^{-} \quad \text{or} \quad \left[\begin{array}{c} :\ddot{O}: \\ | \\ \ddot{O}=N-\ddot{O}: \end{array} \right]^{-}$$

Are these all valid Lewis structures? Yes, so there really are three possible Lewis structures for NO_3^-.

A molecule or ion that has multiple correct Lewis structures shows *resonance*.
resonance structure   Each of these Lewis structures is called a **resonance structure.** In this book, however, we will not be concerned with how to choose the correct resonance structure for a molecule or ion. Therefore any of the possible resonance structures may be used to represent the ion or molecule.

Example 11.14 Write the Lewis structure for a carbonate ion, CO_3^{2-}.

Solution

Step 1. These four atoms have 22 valence electrons plus two electrons from the -2 charge, which makes 24 electrons to be placed.

Step 2. In the carbonate ion the carbon is the central atom surrounded by the three oxygen atoms. Write the skeletal structure and place two electrons between each pair of atoms:

$$\begin{array}{c} O \\ | \\ C-O \\ | \\ O \end{array}$$

Step 3. Subtract the six electrons used in Step 2 from 24 to give 18 electrons yet to be placed.

Step 4. Distribute the 18 electrons around the three oxygen atoms and indicate that the carbonate ion has a -2 charge:

$$\left[\begin{array}{c} :\ddot{O}: \\ | \\ C \\ :\ddot{O}. \quad .\ddot{O}: \end{array} \right]^{2-}$$

The difficulty with this structure is that the carbon atom has only six electrons around it instead of a noble gas octet.

Step 5. Move one of the nonbonding pairs of electrons from one of the oxygens and place them between the carbon and the oxygen. Three Lewis structures are possible:

$$\left[\begin{array}{c} :\ddot{O}: \\ | \\ \overset{\cdot\cdot}{\underset{O}{}}=C \overset{}{\underset{\ddot{O}:}{}} \end{array}\right]^{2-} \quad \text{or} \quad \left[\begin{array}{c} :\ddot{O}: \\ | \\ :\ddot{O} \quad C \quad \ddot{O}: \end{array}\right]^{2-} \quad \text{or} \quad \left[\begin{array}{c} :O: \\ || \\ :\ddot{O} \quad C \quad \ddot{O}: \end{array}\right]^{2-}$$

Practice 11.5

Write the Lewis structure for each of the following: (a) NH_3 (b) H_3O^+ (c) NH_4^+ (d) HCO_3^-.

11.9 Compounds Containing Polyatomic Ions

A polyatomic ion is a stable group of atoms that has either a positive or a negative charge and behaves as a single unit in many chemical reactions. Sodium carbonate, Na_2CO_3, contains two sodium ions and a carbonate ion. The carbonate ion, CO_3^{2-}, is a polyatomic ion composed of one carbon atom and three oxygen atoms and has a charge of -2. One carbon and three oxygen atoms have a total of 22 electrons in their outer energy levels. The carbonate ion contains 24 outer electrons and therefore has a charge of -2. In this case the two additional electrons come from the two sodium atoms, which are now sodium ions:

$$[Na]^+ \quad \left[\begin{array}{c} :\ddot{O}: \\ | \\ :\ddot{O} \quad C \overset{}{\underset{\ddot{O}\cdot}{}} \end{array}\right]^{2-} \quad [Na]^+ \qquad \left[\begin{array}{c} :\ddot{O}: \\ | \\ :\ddot{O} \quad C \overset{}{\underset{\ddot{O}}{}} \end{array}\right]^{2-}$$

sodium carbonate   carbonate ion

Sodium carbonate has both ionic and covalent bonds. Ionic bonds exist between each of the sodium ions and the carbonate ion. Covalent bonds are present between the carbon and oxygen atoms within the carbonate ion. One important difference between the ionic and covalent bonds in this compound can be demonstrated by dissolving sodium carbonate in water. It dissolves in water forming three charged particles, two sodium ions and one carbonate ion, per formula unit of Na_2CO_3:

$$Na_2CO_3 \xrightarrow{\text{water}} 2\,Na^+ + CO_3^{2-}$$

sodium carbonate   sodium ions   carbonate ion

The CO_3^{2-} ion remains as a unit, held together by covalent bonds; but where the bonds are ionic, dissociation of the ions takes place. Do not think, however, that polyatomic ions are so stable that they cannot be altered. Chemical reactions by which polyatomic ions can be changed to other substances do exist.

FIGURE 11.12▶
Geometric shapes of common molecules.

Water	Carbon dioxide	Boron trifluoride	Methane
H_2O	CO_2	BF_3	CH_4
(V-shaped)	(linear shape)	(trigonal planar)	(tetrahedral)

11.10 Molecular Shape

So far in our discussion of bonding, we have used Lewis structures to represent valence electrons in molecules and ions, but they don't indicate anything regarding the molecular or geometric shape of a molecule. The three-dimensional arrangement of the atoms within a molecule is a significant feature in understanding molecular interactions. Let's consider several examples illustrated in Figure 11.12.

Water is known to have the geometric shape known as "bent" or "V-shaped." Carbon dioxide exhibits a linear shape. BF_3 forms a third molecular shape called *trigonal planar* since all the atoms lie in one plane in a triangular arrangement. One of the more common molecular shapes is the tetrahedron illustrated by the molecule methane, CH_4.

How do we predict the geometric shape of a molecule? We will now study a model developed to assist in making predictions from the Lewis structure.

11.11 The Valence Shell Electron Pair Repulsion (VSEPR) Model

The chemical properties of a substance are closely related to the structure of its molecules. A change in a single site on a large biomolecule can make a difference in whether or not a particular reaction occurs.

Instrumental analysis can be used to determine exact spatial arrangements of atoms. Quite often, though, we only need to be able to predict the approximate structure of a molecule. A relatively simple model has been developed to allow us to make predictions of shape from Lewis structures.

Nonbonding pairs of electrons are not shown here so you can focus your attention on the shapes, not electron arrangement.

The VSEPR model is based on the idea that electron pairs will repel each other electrically and will seek to minimize this repulsion. To accomplish this minimization, the electron pairs will be arranged around a central atom as far apart as possible. Consider $BeCl_2$, a molecule with only two pairs of electrons surrounding the central atom. These electrons are arranged 180° apart for maximum separation:

$$Cl \overset{180°}{\underset{}{\frown}} Be \frown Cl$$

linear structure   This molecular structure can now be labeled as a **linear structure.** When only two pairs of electrons surround a central atom, they should be placed 180° apart to give a linear structure.

What occurs when there are three pairs of electrons on the central atom? Consider the BF_3 molecule. The greatest separation of electron pairs occurs when the angles between atoms are 120°:

This arrangement of atoms is flat (planar) and, as noted earlier, is called **trigonal planar.** When three pairs of electrons surround an atom, they should be placed 120° apart to show the trigonal planar structure.

trigonal planar

Now consider the most common situation, CH_4, with four pairs of electrons on the central carbon atom. In this case the central atom exhibits a noble gas electron structure. What arrangement best minimizes the electron pair repulsions? At first it seems that an obvious choice is a 90° angle with all the atoms in a single plane:

However, we must consider that molecules are three-dimensional. This concept results in a structure in which the electron pairs are actually 109.5° apart:

In this diagram the wedged line seems to protrude from the page whereas the dashed line recedes. Two representations of this arrangement, known as **tetrahedral structure,** are illustrated in Figure 11.13. When four pairs of electrons surround an atom, they should be placed 109.5° apart to give them a tetrahedral structure.

tetrahedral structure

The VSEPR model is based on the premise that we are counting electron pairs. It's quite possible that one or more of these electron pairs may be nonbonding or lone pairs. What happens to the molecular structure in these cases? Consider the ammonia

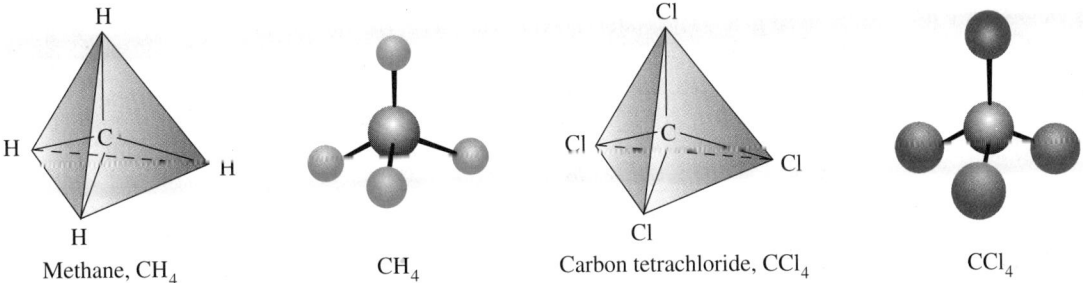

Methane, CH_4 CH_4 Carbon tetrachloride, CCl_4 CCl_4

▲
FIGURE 11.13
Ball-and-stick models of methane and carbon tetrachloride. Methane and carbon tetrachloride are nonpolar molecules because their polar bonds cancel each other in the tetrahedral arrangement of their atoms. The carbon atoms are located in the centers of the tetrahedrons.

FIGURE 11.14▶
(a) The tetrahedral arrangement of electron pairs around the N atom in the NH₃ molecule. (b) Three pairs are shared and one is unshared. (c) The NH₃ molecule is pyramidal.

(a)   (b)   (c)

FIGURE 11.15▶
(a) The tetrahedral arrangement of the four electron pairs around oxygen in the H₂O molecule. (b) Two of the pairs are shared and two are unshared. (c) The H₂O molecule is bent.

(a)   (b)   (c)

molecule. First we draw the Lewis structure to determine the number of electron pairs around the central atom:

$$H : \overset{..}{\underset{..}{N}} : H$$
$$H$$

Since there are four pairs of electrons, the arrangement of electrons around the central atom will be tetrahedral (Figure 11.14a). However, only three of the pairs are bonded to another atom, so the molecule itself is pyramidal. It is important to understand that the placement of the electron pairs determines the shape but the name for the molecule is determined by the position of the atoms themselves. Therefore, ammonia is pyramidal. See Figure 11.14c.

Now consider the effect of two unbonded pairs of electrons in the water molecule. The Lewis structure for water is

$$H - \overset{..}{O} :$$
$$|$$
$$H$$

The arrangement of electron pairs around an atom determines its shape but we name the shape of molecules by the position of the atoms.

The four electron pairs indicate a tetrahedral arrangement is necessary (see Figure 11.15a). The molecule is not called tetrahedral because two of the electron pairs are unbonded pairs. The water molecule is "bent" as shown in Figure 11.15c. Using the VSEPR model helps to explain some of the unique properties of the water molecule. Because it is bent and not linear we can see that the molecule is polar.

The properties of water that cause it to be involved in so many interesting and important roles are largely a function of its shape and polarity. We will consider water in greater detail in Chapter 13.

Example 11.15 Predict the molecular structure for these molecules: H_2S, CCl_4, AlF_3.

Solution
1. Draw the Lewis structure.
2. Count the electron pairs and determine the arrangement that will minimize repulsions.
3. Determine the positions of the atoms and name the structure.

Strong Enough to Stop a Bullet?

What do color-changing pens, bullet-resistant vests, and calculators have in common? The chemicals that make each of them work are liquid crystals. These chemicals find numerous applications; you are probably most familiar with liquid crystal displays (LCDs) and color-changing products, but these chemicals are also used to make superstrong synthetic fibers.

Molecules in a normal crystal remain in an orderly arrangement, but in a liquid crystal the molecules can flow *and* maintain an orderly arrangement at the same time. Liquid crystal molecules are linear and polar. Since the atoms tend to lie in a relatively straight line, the molecules are generally much longer than they are wide. These polar molecules are attracted to each other and are able to line up in an orderly fashion, without solidifying.

Liquid crystals with twisted arrangements of molecules give us novelty color-changing products. In these liquid crystals the molecules lie side by side in a nearly flat layer. The next layer is similar but at an angle to the one below. The closely packed flat layers have a special effect on light. As the light strikes the surface, some of it is reflected from the top layer and some from lower layers. When the same wavelength is reflected from many layers, we see a color. (This is similar to the rainbow of colors formed by oil in a puddle on the street or the film of a soap bubble.) As the temperature is increased, the molecules move faster, causing a change in the angle and the space between the layers. This results in a color change in the reflected light. Different compounds change color within different temperature ranges, allowing a variety of practical and amusing applications.

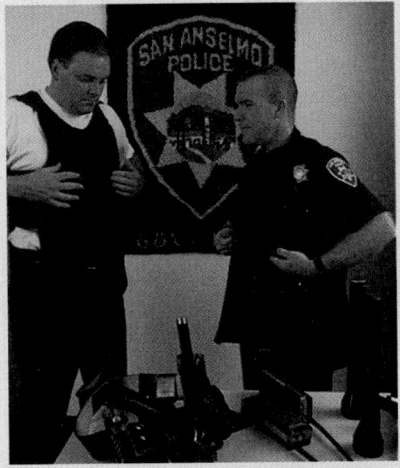

▲ **Kevlar is used to make protective vests for police.**

Liquid crystal (nematic) molecules that lie parallel to one another are used to manufacture very strong synthetic fibers. Perhaps the best example of these liquid crystals is Kevlar, a synthetic fiber used in bullet-resistant vests, canoes, and parts of the space shuttle. Kevlar is a synthetic polymer, like nylon or polyester, that gains strength by passing through a liquid crystal state during its manufacture.

In a typical polymer the long molecular chains are jumbled together, somewhat like spaghetti. The strength of the material is limited by the disorderly arrangement. The trick is to get the molecules to line up parallel to each other. Once the giant molecules have been synthesized, they are dissolved in sulfuric acid. At the proper concentration the molecules align, and the solution is forced through tiny holes in a nozzle and further aligned. The sulfuric acid is removed in a water bath, thereby forming solid fibers in near perfect alignment. One strand of Kevlar is stronger than an equal-sized strand of steel. It has a much lower density as well, making it a material of choice in bullet-resistant vests.

Molecule	Lewis structure	Number of electron pairs	Electron pair arrangement	Molecular shape
H_2S	H:S̈:H	4	tetrahedral	bent
CCl_4	:C̈l:C̈:C̈l: with :C̈l: above and :C̈l: below	4	tetrahedral	tetrahedral
AlF_3	:F̈:Al:F̈: with :F̈: above	3	trigonal planar	trigonal planar

Practice 11.6

Predict the shape for CF_4, NF_3, and BeF_2.

Concepts in Review

1. Describe how atomic radii vary (a) from left to right in a period and (b) from top to bottom in a group.

2. Describe how the ionization energies of elements vary with respect to (a) the position in the periodic table and (b) the removal of successive electrons.

3. Write Lewis structures for the representative elements from their position in the periodic table.

4. Describe (a) the formation of ions by electron transfer and (b) the nature of the chemical bond formed by electron transfer.

5. Use Lewis structures to show how ionic compounds are formed from atoms.

6. Describe a crystal of sodium chloride.

7. Predict the relative sizes of an atom and a monatomic ion for a given element.

8. Describe the covalent bond and predict whether a given covalent bond will be polar or nonpolar.

9. Draw Lewis structures for the diatomic elements.

10. Identify single, double, and triple covalent bonds.

11. Describe the changes in electronegativity in (a) moving across a period and (b) moving down a group in the periodic table.

12. Predict formulas of simple compounds formed between the representative (Group A) elements using the periodic table.

13. Describe the effect of electronegativity on the type of chemical bonds in a compound.

14. Draw Lewis structures for (a) the molecules of covalent compounds and (b) polyatomic ions.

15. Describe the difference between polar and nonpolar bonds.

16. Distinguish clearly between ionic and molecular substances.

17. Predict whether the bonding in a compound will be primarily ionic or covalent.

18. Describe the VSEPR model for molecular shape.

19. Use the VSEPR model to determine molecular structure from Lewis structure for given compounds.

Key Terms

covalent bond (11.5)
dipole (11.6)
electronegativity (11.6)
ionic bond (11.3)
ionization energy (11.1)
Lewis structure (11.2)
linear structure (11.11)
nonpolar covalent bond (11.6)
polar covalent bond (11.5)
resonance structure (11.8)
tetrahedral structure (11.11)
trigonal planar (11.11)

Questions

1. Rank these elements according to the radii of their atoms, from smallest to largest: Na, Mg, Cl, K, and Rb. (Figure 11.2)

2. Explain why much more ionization energy is required to remove the first electron from neon than from sodium. (Table 11.1)

3. Explain the large increase in ionization energy needed to remove the third electron from beryllium compared with that needed for the second electron. (Table 11.1)

4. Does the first ionization energy increase or decrease from top to bottom in the periodic table for the alkali metal family? Explain. (Figure 11.3)

5. Does the first ionization energy increase or decrease from top to bottom in the periodic table for the noble gas family? Explain. (Figure 11.3)

6. Why does barium (Ba) have a lower ionization energy than beryllium (Be)? (Figure 11.3)

7. Why is there such a large increase in the ionization energy required to remove the second electron from a sodium atom as opposed to the first? (Table 11.1)

8. Which element in the pair has the larger atomic radius? (Figure 11.2)
(a) Na or K   (c) O or F   (e) Ti or Zr
(b) Na or Mg   (d) Br or I

9. In Groups IA–VIIA, which element in each group has the smallest atomic radius? (Figure 11.2)

10. Why does the atomic size increase in going down any family of the periodic table?

11. All the atoms within each Group A family of elements can be represented by the same Lewis structure. Complete the following table, expressing the Lewis structure for each group. (Use E to represent the elements.) (Figure 11.4)

Group	IA	IIA	IIIA	IVA	VA	VIA	VIIA
	$E \cdot$						

12. Draw the Lewis structure for Cs, Ba, Tl, Pb, Po, At, and Rn. How do these structures correlate with the group in which each element occurs?

Paired Exercises

17. Which is larger, a magnesium atom or a magnesium ion? Explain.

19. Using the table of electronegativity values (Table 11.5), indicate which element is more positive and which is more negative in these compounds:
    (a) H_2O (c) NH_3 (e) NO
    (b) NaF   (d) PbS   (f) CH_4

21. Classify the bond between these pairs of elements as principally ionic or principally covalent (use Table 11.5):
    (a) sodium and chlorine
    (b) carbon and hydrogen
    (c) chlorine and carbon
    (d) calcium and oxygen

23. Explain what happens to the electron structures of Mg and Cl atoms when they react to form $MgCl_2$.

25. Use Lewis structures to show the electron transfer that enables these ionic compounds to be formed:
    (a) MgF_2 (b) K_2O

27. How many valence electrons are in each of these atoms? H, K, Mg, He, Al

29. How many electrons must be gained or lost for the following to achieve a noble gas electron structure?
    (a) a calcium atom
    (b) a sulfur atom
    (c) a helium atom

31. Which is larger? Explain.
    (a) a magnesium ion or an aluminum ion
    (b) Fe^{2+} or Fe^{3+}

13. In which general areas of the periodic table are found the elements with (a) the highest and (b) the lowest electronegativities?

14. What are valence electrons?

15. Explain why potassium usually forms a K^+ ion but not a K^{2+} ion.

16. Why does an aluminum ion have a +3 charge?

18. Which is smaller, a bromine atom or a bromide ion? Explain.

20. Using the table of electronegativity values (Table 11.5), indicate which element is more positive and which is more negative in these compounds:
    (a) HCl   (c) CCl_4 (e) MgH_2
    (b) LiH   (d) IBr   (f) OF_2

22. Classify the bond between these pairs of elements as principally ionic or principally covalent (use Table 11.5):
    (a) hydrogen and sulfur
    (b) barium and oxygen
    (c) fluorine and fluorine
    (d) potassium and fluorine

24. Write an equation representing
    (a) the change of a fluorine atom to a fluoride ion and
    (b) the change of a calcium atom to a calcium ion

26. Use Lewis structures to show the electron transfer that enables these ionic compounds to be formed:
    (a) CaO   (b) NaBr

28. How many valence electrons are in each of these atoms? Si, N, P, O, Cl

30. How many electrons must be gained or lost for the following to achieve a noble gas electron structure?
    (a) a chloride ion
    (b) a nitrogen atom
    (c) a potassium atom

32. Which is larger? Explain.
    (a) a potassium atom or a potassium ion
    (b) a bromine atom or a bromide ion

33. Let E be any representative element. Following the pattern in the table, write formulas for the hydrogen and oxygen compounds of
    (a) Na (c) Al
    (b) Ca (d) Sn

Group						
IA	**IIA**	**IIIA**	**IVA**	**VA**	**VIA**	**VIIA**
EH	EH_2	EH_3	EH_4	EH_3	H_2E	HE
E_2O	EO	E_2O_3	EO_2	E_2O_5	EO_3	E_2O_7

35. The formula for sodium sulfate is Na_2SO_4. Write the names and formulas for the other alkali metal sulfates.

37. Write Lewis structures for
    (a) Na (b) Br^- (c) O^{2-}

39. Classify the bonding in each compound as ionic or covalent:
    (a) H_2O (c) MgO
    (b) NaCl   (d) Br_2

41. Predict the type of bond that would be formed between the following pairs of atoms:
    (a) Na and N
    (b) N and S
    (c) Br and I

43. Draw Lewis structures for
    (a) H_2 (b) N_2 (c) Cl_2

45. Draw Lewis structures for
    (a) NCl_3 (c) C_2H_6
    (b) H_2CO_3 (d) $NaNO_3$

47. Draw Lewis structures for
    (a) Ba^{2+} (d) CN^-
    (b) Al^{3+} (e) HCO_3^-
    (c) SO_3^{2-}

49. Classify these molecules as polar or nonpolar:
    (a) H_2O
    (b) HBr
    (c) CF_4

51. Give the number and arrangement of the electron pairs around the central atom:
    (a) C in CCl_4
    (b) S in H_2S
    (c) Al in AlH_3

53. Use VSEPR theory to predict the structure of these polyatomic ions:
    (a) sulfate ion
    (b) chlorate ion
    (c) periodate ion

34. Let E be any representative element. Following the pattern in the table, write formulas for the hydrogen and oxygen compounds of
    (a) Sb (c) Cl
    (b) Se (d) C

Group						
IA	**IIA**	**IIIA**	**IVA**	**VA**	**VIA**	**VIIA**
EH	EH_2	EH_3	EH_4	EH_3	H_2E	HE
E_2O	EO	E_2O_3	EO_2	E_2O_5	EO_3	E_2O_7

36. The formula for calcium bromide is $CaBr_2$. Write the names and formulas for the other alkaline earth metal bromides.

38. Write Lewis structures for
    (a) Ga (b) Ga^{3+} (c) Ca^{2+}

40. Classify the bonding in each compound as ionic or covalent:
    (a) HCl   (c) NH_3
    (b) $BaCl_2$ (d) SO_2

42. Predict the type of bond that would be formed between the following pairs of atoms:
    (a) H and Si
    (b) O and F
    (c) Ca and I

44. Draw Lewis structures for
    (a) O_2 (b) Br_2 (c) I_2

46. Draw Lewis structures for
    (a) H_2S (c) NH_3
    (b) CS_2 (d) NH_4Cl

48. Draw Lewis structures for
    (a) I^- (d) ClO_3^-
    (b) S^{2-} (e) NO_3^-
    (c) CO_3^{2-}

50. Classify these molecules as polar or nonpolar:
    (a) F_2
    (b) CO_2
    (c) NH_3

52. Give the number and arrangement of the electron pairs around the central atom:
    (a) Be in BeF_2
    (b) N in NF_3
    (c) Cl in HCl

54. Use VSEPR theory to predict the structure of these polyatomic ions:
    (a) ammonium ion
    (b) sulfite ion
    (c) phosphate ion

55. Use VSEPR theory to predict the shape of these molecules:
    (a) SiH_4
    (b) PH_3
    (c) SeF_2

57. Identify this element: Element X reacts with sodium to form the compound Na_2X and is in the second period on the periodic table.

56. Use VSEPR theory to predict the shape of these molecules:
    (a) SiF_4
    (b) OF_2
    (c) Cl_2O

58. Identify this element: Element Y reacts with oxygen to form the compound Y_2O and has the lowest ionization energy of any fourth-period element on the periodic table.

Additional Exercises

59. Identify the element on the periodic table that satisfies each description:
    (a) transition metal with the largest atomic radius
    (b) alkaline earth metal with the greatest ionization energy
    (c) least dense member of the nitrogen family
    (d) alkali metal with the greatest ratio of neutrons to protons
    (e) most electronegative transition metal

60. Choose the element that fits each description:
    (a) the lower electronegativity   As or Zn
    (b) the lower chemical reactivity   Ba or Be
    (c) the fewer valence electrons   N or Ne

61. Identify two reasons why fluorine has a much higher electronegativity than neon.

62. When one electron is removed from an atom of Li, it has two left. Helium atoms also have two electrons. Why is more energy required to remove the second electron from Li than to remove the first from He?

63. Group IB elements (see the periodic table on the inside cover of your book) have one electron in their outer energy level, as do Group IA elements. Would you expect them to form compounds such as CuCl, AgCl, and AuCl? Explain.

64. The formula for lead(II) bromide is $PbBr_2$: predict formulas for tin(II) and germanium(II) bromides.

65. Why is it not proper to speak of sodium chloride molecules?

66. What is a covalent bond? How does it differ from an ionic bond?

67. Briefly comment on the structure Na:Ö:Na for the compound Na_2O.

68. What are the four most electronegative elements?

69. Rank these elements from highest electronegativity to lowest: Mg, S, F, H, O, Cs.

70. Is it possible for a molecule to be nonpolar even though it contains polar covalent bonds? Explain.

71. Why is CO_2 a nonpolar molecule, whereas CO is a polar molecule?

72. Estimate the bond angle between atoms in these molecules.
    (a) H_2S
    (b) NH_3
    (c) NH_4^+
    (d) $SiCl_4$

73. Consider the two molecules BF_3 and NF_3. Compare and contrast them in terms of
    (a) valence level orbitals on the central atom that are used for bonding
    (b) shape of the molecule
    (c) number of lone electron pairs on the central atom
    (d) type and number of bonds found in the molecule

74. With respect to electronegativity, why is fluorine such an important atom? What combination of atoms on the periodic table results in the most ionic bond?

75. Why does the Lewis structure of each element in a given group of representative elements on the periodic table have the same number of dots?

76. A sample of an air pollutant composed of sulfur and oxygen was found to contain 1.40 g sulfur and 2.10 g oxygen. What is the empirical formula for this compound? Draw a Lewis structure to represent it.

77. A dry-cleaning fluid composed of carbon and chlorine was found to have the composition 14.5% carbon and 85.5% chlorine. Its known molar mass is 166 g/mol. Draw a Lewis structure to represent it.

Answers to Practice Exercises

11.1 (a) :N·

(b) :Al

(c) Sr:

(d) :Br·

11.2 (a) Na_2S

(b) Rb_2O

11.3 (a) Mg_3N_2

(b) Ba_3As_2

11.4 (a) :Br: (b)   H
   :P:Br: :Cl:C:Cl:
   :Br: :Cl:

(c) H:F: (d)   O:
             H:C:H

11.5 (a) H:N:H   (b) [H:O:H]⁺
     H   H

(c) [H]⁺ (d) [:O:]⁻
   H:N:H   H:O:C::O:
     H

11.6 CF_4, tetrahedral; NF_3, pyramidal; BeF_2, linear

PUTTING IT TOGETHER
Review for Chapters 10–11

Multiple Choice: *Choose the correct answer to each of the following.*

1. The concept of electrons existing in specific orbits around the nucleus was the contribution of
   (a) Thomson   (c) Bohr
   (b) Rutherford   (d) Schrödinger

2. The correct electron structure for a fluorine atom, F, is
   (a) $1s^2 2s^2 2p^5$ (c) $1s^2 2s^2 2p^4 3s^1$
   (b) $1s^2 2s^2 2p^3 3s^2 3p^1$ (d) $1s^2 2s^2 2p^3$

3. The correct electron structure for $_{48}$Cd is
   (a) $1s^2 2s^2 2p^6 3s^2 3p^6 4s^2 3d^{10}$
   (b) $1s^2 2s^2 2p^6 3s^2 3p^6 4s^2 3d^{10} 4p^6 5s^2 4d^{10}$
   (c) $1s^2 2s^2 2p^6 3s^2 3p^6 4s^2 3d^{10} 4p^6 4d^4$
   (d) $1s^2 2s^2 2p^6 3s^2 3p^6 4s^2 4p^6 4d^{10} 5s^2 5d^{10}$

4. The correct electron structure of $_{23}$V is
   (a) $[Ar]4s^2 3d^3$ (c) $[Ar]4s^2 4d^3$
   (b) $[Ar]4s^2 4p^3$ (d) $[Kr]4s^2 3d^3$

5. Which of the following is the correct atomic structure for $_{22}^{48}$Ti?

   (a)
   22+
   26n
   22e−

   (b)
   22+
   48n
   22e−

   (c)
   26+
   22n
   22e−

   (d)
   22+
   26n
   48e−

6. The number of orbitals in a d sublevel is
   (a) 3   (c) 7
   (b) 5   (d) no correct answer given

7. The number of electrons in the third principal energy level in an atom having the electron structure $1s^2 2s^2 2p^6 3s^2 3p^2$ is
   (a) 2   (b) 4   (c) 6   (d) 8

8. The total number of orbitals that contain at least one electron in an atom having the structure $1s^2 2s^2 2p^6 3s^2 3p^2$ is
   (a) 5   (c) 14
   (b) 8   (d) no correct answer given

9. Which of these elements has two s and six p electrons in its outer energy level?
   (a) He   (c) Ar
   (b) O   (d) no correct answer given

10. Which element is not a noble gas?
    (a) Ra   (b) Xe   (c) He   (d) Ar

11. Which element has the largest number of unpaired electrons?
    (a) F   (b) S   (c) Cu   (d) N

12. How many unpaired electrons are in the electron structure of $_{24}$Cr, $[Ar]4s^1 3d^5$?
    (a) 2   (b) 4   (c) 5   (d) 6

13. Groups IIIA–VIIA plus the noble gases form the area of the periodic table where the electron sublevels being filled are
    (a) p sublevels   (c) d sublevels
    (b) s and p sublevels   (d) f sublevels

14. In moving down an A group on the periodic table, the number of electrons in the outermost energy level
    (a) increases regularly
    (b) remains constant
    (c) decreases regularly
    (d) changes in an unpredictable manner

15. Which of the following is an incorrect formula?
    (a) NaCl   (b) K_2O (c) AlO   (d) BaO

16. Elements of the noble gas family
    (a) form no compounds at all
    (b) have no valence electrons
    (c) have an outer electron structure of $ns^2 np^6$ (helium excepted), where n is the period number
    (d) no correct answer given

17. The lanthanide and actinide series of elements are
    (a) representative elements
    (b) transition elements
    (c) filling in d level electrons
    (d) no correct answer given

18. The element having the structure $1s^2 2s^2 2p^6 3s^2 3p^2$ is in Group
    (a) IIA   (b) IIB   (c) IVA   (d) IVB

19. In Group VA, the element having the smallest atomic radius is
    (a) Bi   (b) P   (c) As   (d) N

20. In Group IVA, the most metallic element is
    (a) C   (b) Si   (c) Ge   (d) Sn

21. Which group in the periodic table contains the least reactive elements?
    (a) IA   (b) IIA   (c) IIIA   (d) noble gases

22. Which group in the periodic table contains the alkali metals?
    (a) IA   (b) IIA   (c) IIIA   (d) IVA

23. An atom of fluorine is smaller than an atom of oxygen. One possible explanation is that, compared to oxygen, fluorine has
    (a) a larger mass number
    (b) a smaller atomic number
    (c) a greater nuclear charge
    (d) more unpaired electrons

24. If the size of the fluorine atom is compared to the size of the fluoride ion,
    (a) they would both be the same size
    (b) the atom is larger than the ion
    (c) the ion is larger than the atom
    (d) the size difference depends on the reaction

25. Sodium is a very active metal because
    (a) it has a low ionization energy
    (b) it has only one outermost electron
    (c) it has a relatively small atomic mass
    (d) all of the above

26. Which of the following formulas is not correct?
    (a) Na^+ (b) S^- (c) Al^{3+} (d) F^-

27. Which of the following molecules does not have a polar covalent bond?
    (a) CH_4 (b) H_2O (c) CH_3OH (d) Cl_2

28. Which of the following molecules is a dipole?
    (a) HBr   (b) CH_4 (c) H_2 (d) CO_2

29. Which of the following has bonding that is ionic?
    (a) H_2 (b) MgF_2 (c) H_2O (d) CH_4

30. Which of the following is a correct Lewis structure?

    (a) $:\ddot{O}:C:\ddot{O}:$ (b) $:\ddot{C}l:\overset{:\ddot{C}l:}{\underset{:\ddot{C}l:}{C}}:\ddot{C}l:$

    (c) $\ddot{C}l::\ddot{C}l$ (d) $:\dot{N}:\dot{N}:$

31. Which of the following is an incorrect Lewis structure?

    (a) $H:\ddot{N}:H$ (b) $:\ddot{O}:H \atop \ddot{}\, H$

    (c) $H:\overset{H}{\underset{H}{C}}:H$ (d) $:N:::N:$

32. The correct Lewis structure for SO_2 is
    (a) $:\ddot{O}:\ddot{S}:\ddot{O}:$ (b) $:\ddot{O}:\ddot{S}::\ddot{O}:$
    (c) $:\ddot{O}::S::\ddot{O}:$ (d) $:\ddot{O}:\ddot{S}:\ddot{O}:$

33. Carbon dioxide, CO_2, is a nonpolar molecule because
    (a) oxygen is more electronegative than carbon
    (b) the two oxygen atoms are bonded to the carbon atom
    (c) the molecule has a linear structure with the carbon atom in the middle
    (d) the carbon–oxygen bonds are polar covalent

34. When a magnesium atom participates in a chemical reaction, it is most likely to
    (a) lose 1 electron   (c) lose 2 electrons
    (b) gain 1 electron   (d) gain 2 electrons

35. If X represents an element of Group IIIA, what is the general formula for its oxide?
    (a) X_3O_4 (b) X_3O_2 (c) XO (d) X_2O_3

36. Which of the following has the same electron structure as an argon atom?
    (a) Ca^{2+} (b) Cl^0 (c) Na^+ (d) K^0

37. As the difference in electronegativity between two elements decreases, the tendency for the elements to form a covalent bond
    (a) increases
    (b) decreases
    (c) remains the same
    (d) sometimes increases and sometimes decreases

38. Which compound forms a tetrahedral molecule?
    (a) NaCl   (b) CO_2 (c) CH_4 (d) $MgCl_2$

39. Which compound has a bent (V-shaped) molecular structure?
    (a) NaCl   (b) CO_2 (c) CH_4 (d) H_2O

40. Which compound has double bonds within its molecular structure?
    (a) NaCl   (b) CO_2 (c) CH_4 (d) H_2O

41. The total number of valence electrons in a nitrate ion, NO_3^-, is
    (a) 12   (b) 18   (c) 23   (d) 24

42. The number of electrons in a triple bond is
    (a) 3   (b) 4   (c) 6   (d) 8

43. The number of unbonded pairs of electrons in H_2O is
    (a) 0   (b) 1   (c) 2   (d) 4

44. Which of the following does not have a noble gas electron structure?
    (a) Na   (b) Sc^{3+} (c) Ar   (d) O^{2-}

Free Response Questions

1. An alkaline earth metal, M, combines with a halide, X. Will the resulting compound be ionic or covalent? Why? What is the Lewis structure for the compound?

2. Is the following statement true or false? "All electrons in atoms with even atomic numbers are paired." Explain your answer using an example.

3. Discuss whether the following statement is true or false. "All nonmetals have two valence electrons in an s sublevel with the exception of the noble gases, which have at least one unpaired electron in a p sublevel."

4. The first ionization energy, IE, of potassium is lower than the first IE for calcium but the second IE of calcium is lower than the second IE of potassium. Use an electron configuration or size argument to explain this trend in IEs.

5. Chlorine has a very large first ionization energy yet it forms a chloride ion relatively easily. Explain.

6. Three particles have the same electron configuration. One is a cation of an alkali metal, one is an anion of the halide in the third period, and the third particle is an atom of a noble gas. What are the identities of the three particles (including charges)? Which particle should have the smallest atomic/ionic radius, which should have the largest, and why?

7. Why is the Lewis structure of $AlCl_3$ not written as

$$
\begin{array}{c}
\ddot{:}\ddot{C}l\ddot{:} \\
/ \\
:\ddot{C}l-Al \\
\ddot{} \qquad \backslash \\
:\ddot{C}l: \\
\ddot{}
\end{array}
$$

What is the correct Lewis structure and which electrons are shown in a Lewis structure?

8. Why does carbon have a maximum of four covalent bonds?

9. Both NCl_3 and BF_3 have a central atom bonded to three other atoms yet one is pyramidal and the other is trigonal planar. Explain.

10. Draw the Lewis structure of the atom whose electron configuration is $1s^2 2s^2 2p^6 3s^2 3p^6 4s^2 3d^{10} 4p^5$. Would you expect this atom to form an ionic, nonpolar covalent, or polar covalent bond with sulfur?

CHAPTER

12

The Gaseous State of Matter

▲
Properties of gases are
transformed to art and sport
at the hot air balloon festival
in Albuquerque, New Mexico.

CHAPTER 12 / OUTLINE

CHAPTER 12

Our atmosphere is composed of a mixture of gases, including nitrogen, oxygen, carbon dioxide, ozone, and trace amounts of others. These gases are essential to life yet they can also create hazards to us. For example, carbon dioxide is valuable when it is taken in by plants and converted to carbohydrates, but it also is associated with the potentially hazardous greenhouse effect. Ozone surrounds the earth at high altitudes and protects us from harmful ultraviolet rays, but it also destroys rubber and plastics. We require air to live, yet scuba divers must be concerned about oxygen poisoning, and the "bends."

In chemistry the study of the behavior of gases allows us to understand our atmosphere and the effects that gases have on our lives.

12.1 General Properties

In Chapter 3, solids, liquids, and gases were described briefly. In this chapter we consider the behavior of gases in greater detail.

Gases are the least dense and most mobile of the three states of matter. A solid has a rigid structure, and its particles remain in essentially fixed positions. When a solid absorbs sufficient heat, it melts and changes into a liquid. Melting occurs because the molecules (or ions) have absorbed enough energy to break out of the rigid crystal lattice structure of the solid. The molecules or ions in the liquid are more energetic than they were in the solid, as indicated by their increased mobility. Molecules in the liquid state cling to one another. When the liquid absorbs additional heat, the more energetic molecules break away from the liquid surface and go into the gaseous state—the most mobile state of matter. Gas molecules move at very high velocities and have high kinetic energy. The average velocity of hydrogen molecules at 0°C is over 1600 meters (1 mile) per second. Mixtures of gases are uniformly distributed within the container in which they are confined.

The same quantity of a substance occupies a much greater volume as a gas than it does as a liquid or a solid. For example, 1 mol of water (18.02 g) has a volume of 18 mL at 4°C. This same amount of water would occupy about 22,400 mL in the gaseous state—more than a 1200-fold increase in volume. We may assume from this difference in volume that (1) gas molecules are relatively far apart, (2) gases can be greatly compressed, and (3) the volume occupied by a gas is mostly empty space.

▲ A mole of water occupies 18 mL as a liquid but would fill this box (22.4 L) as a gas at the same temperature (25°C).

12.2 The Kinetic-Molecular Theory

Careful scientific studies of the behavior and properties of gases were begun in the 17th century by Robert Boyle (1627–1691). His work was carried forward by many investigators, and the accumulated data were used in the second half of the 19th century to formulate a general theory to explain the behavior and properties of gases. This theory is called the **kinetic-molecular theory (KMT).** The KMT has since

kinetic-molecular theory
(KMT)

been extended to cover, in part, the behavior of liquids and solids. It ranks today with the atomic theory as one of the greatest generalizations of modern science.

ideal gas

The KMT is based on the motion of particles, particularly gas molecules. A gas that behaves exactly as outlined by the theory is known as an **ideal gas.** No ideal gases exist, but under certain conditions of temperature and pressure, real gases approach ideal behavior, or at least show only small deviations from it. Under extreme conditions, such as very high pressure and low temperature, real gases deviate greatly from ideal behavior. For example, at low temperature and high pressure many gases become liquids.

The principal assumptions of the kinetic-molecular theory are:

1. Gases consist of tiny (submicroscopic) particles.
2. The distance between particles is large compared with the size of the particles themselves. The volume occupied by a gas consists mostly of empty space.
3. Gas particles have no attraction for one another.
4. Gas particles move in straight lines in all directions, colliding frequently with one another and with the walls of the container.
5. No energy is lost by the collision of a gas particle with another gas particle or with the walls of the container. All collisions are perfectly elastic.
6. The average kinetic energy for particles is the same for all gases at the same temperature, and its value is directly proportional to the Kelvin temperature.

The kinetic energy (KE) of a particle is one-half its mass times its velocity squared. It is expressed by the equation

$$KE = \frac{1}{2}mv^2$$

where m is the mass and v is the velocity of the particle.

All gases have the same kinetic energy at the same temperature. Therefore, from the kinetic energy equation we can see that, if we compare the velocities of the molecules of two gases, the lighter molecules will have a greater velocity than the heavier ones. For example, calculations show that the velocity of a hydrogen molecule is four times the velocity of an oxygen molecule.

diffusion

Due to their molecular motion, gases have the property of **diffusion,** the ability of two or more gases to mix spontaneously until they form a uniform mixture. The diffusion of gases may be illustrated by the use of the apparatus shown in Figure 12.1. Two large flasks, one containing reddish brown bromine vapors and the other dry air, are connected by a side tube. When the stopcock between the flasks is

FIGURE 12.1 ▶
Diffusion of gases. When the stopcock between the two flasks is opened, colored bromine molecules can be seen diffusing into the flask containing air.

Bromine　　　　　　　　Air

Bromine and air　　　　Bromine and air

opened, the bromine and air will diffuse into each other. After standing awhile, both flasks will contain bromine and air.

If we put a pinhole in a balloon, the gas inside will effuse or flow out of the balloon. **Effusion** is a process by which gas molecules pass through a very small orifice (opening) from a container at higher pressure to one at lower pressure.

effusion

Thomas Graham (1805–1869), a Scottish chemist, observed that the rate of effusion was dependent on the density of a gas. This observation led to **Graham's law of effusion.**

Graham's law of effusion

> The rates of effusion of two gases at the same temperature and pressure are inversely proportional to the square roots of their densities or molar masses:
>
> $$\frac{\text{rate of effusion of gas } A}{\text{rate of effusion of gas } B} = \sqrt{\frac{d\text{B}}{d\text{A}}} = \sqrt{\frac{\text{molar mass } B}{\text{molar mass } A}}$$

A major application of Graham's law occurred during World War II with the separation of the isotopes of uranium-235 (U-235) and uranium-238 (U-238). Naturally occurring uranium consists of 0.7% U-235, 99.3% U-238, and a trace of U-234. However, only U-235 is useful as fuel for nuclear reactors and atomic bombs, so the concentration of U-235 in the mixture of isotopes had to be increased.

Uranium was first changed to uranium hexafluoride, UF_6, a white solid that readily goes into the gaseous state. The gaseous mixture of $^{235}UF_6$ and $^{238}UF_6$ was then allowed to effuse through porous walls. Although the effusion rate of the lighter gas is only slightly faster than that of the heavier one,

$$\frac{\text{effusion rate } ^{235}UF_6}{\text{effusion rate } ^{238}UF_6} = \sqrt{\frac{\text{molar mass } ^{238}UF_6}{\text{molar mass } ^{235}UF_6}} = \sqrt{\frac{352}{349}} = 1.0043$$

the separation and enrichment of U-235 was accomplished by subjecting the gaseous mixture to several thousand stages of effusion.

Measurement of Pressure of Gases

  pressure

Pressure is defined as force per unit area. When a rubber balloon is inflated with air, it stretches and maintains its larger size because the pressure on the inside is greater than that on the outside. Pressure results from the collisions of gas molecules with the walls of the balloon (see Figure 12.2). When the gas is released, the force or pressure of the air escaping from the small neck propels the balloon in a rapid, irregular flight. If the balloon is inflated until it bursts, the gas escaping all at once causes an explosive noise.

The effects of pressure are also observed in the mixture of gases surrounding Earth—our atmosphere, which is composed of about 78% nitrogen, 21% oxygen, 1% argon, and other minor constituents by volume (see Table 12.1). The outer boundary of the atmosphere is not known precisely, but more than 99% of the atmosphere is below an altitude of 20 miles (32 km). Thus, the concentration of gas molecules in the atmosphere decreases with altitude, and at about 4 miles the amount of oxygen is insufficient to sustain human life. The gases in the atmosphere exert a

▲ **FIGURE 12.2**
The pressure resulting from the collisions of gas molecules with the walls of the balloon keeps the balloon inflated.

TABLE 12.1	Average Composition of Dry Air		
Gas	Percent by volume	Gas	Percent by volume
N_2	78.08	He	0.0005
O_2	20.95	CH_4	0.0002
Ar	0.93	Kr	0.0001
CO_2	0.033	Xe, H_2, and N_2O	Trace
Ne	0.0018		

atmospheric pressure

pressure known as **atmospheric pressure.** The pressure exerted by a gas depends on the number of molecules of gas present, the temperature, and the volume in which the gas is confined. Gravitational forces hold the atmosphere relatively close to Earth and prevent air molecules from flying off into outer space. Thus the atmospheric pressure at any point is due to the mass of the atmosphere pressing downward at that point.

barometer

The pressure of the gases in the atmosphere can be measured with a **barometer.** A mercury barometer may be prepared by completely filling a long tube with pure, dry mercury and inverting the open end into an open dish of mercury. If the tube is longer than 760 mm, the mercury level will drop to a point at which the column of mercury in the tube is just supported by the pressure of the atmosphere. If the tube is properly prepared, a vacuum will exist above the mercury column. The weight of mercury, per unit area, is equal to the pressure of the atmosphere. The column of mercury is supported by the pressure of the atmosphere, and the height of the column is a measure of this pressure (see Figure 12.3). The mercury barometer was invented in 1643 by the Italian physicist E. Torricelli (1608–1647), for whom the unit of pressure *torr* was named.

1 atmosphere

Air pressure is measured and expressed in many units. The standard atmospheric pressure, or simply **1 atmosphere** (atm), is the pressure exerted by a column

FIGURE 12.3 ▶
Preparation of a mercury barometer. The full tube of mercury at the left is inverted and placed in a dish of mercury.

of mercury 760 mm high at a temperature of 0°C. The normal pressure of the atmosphere at sea level is 1 atm or 760 torr or 760 mm Hg. The SI unit for pressure is the pascal (Pa), where 1 atm = 101,325 Pa or 101.3 kPa. Other units for expressing pressure are inches of mercury, centimeters of mercury, the millibar (mbar), and pounds per square inch (lb/in.² or psi). The values of these units equivalent to 1 atm are summarized in Table 12.2.

Atmospheric pressure varies with altitude. The average pressure at Denver, Colorado, 1.61 km (1 mile) above sea level, is 630 torr (0.83 atm). Atmospheric pressure is 0.5 atm at about 5.5 km (3.4 miles) altitude.

Pressure is often measured by reading the heights of mercury columns in millimeters on a barometer. Thus pressure may be recorded as mm Hg. But in many applications the torr is superceding mm Hg as a unit of pressure. In problems dealing with gases it is necessary to make interconversions among the various pressure units. Since atm, torr, and mm Hg are common pressure units, we give examples involving all three of these units.

1 atm = 760 torr = 760 mm Hg

TABLE 12.2	Pressure Units Equivalent to 1 Atmosphere
1 atm
760 torr
760 mm Hg
76 cm Hg
101.325 kPa
1013 mbar
29.9 in. Hg
14.7 lb/in.²

Example 12.1

The average atmospheric pressure at Walnut, California, is 740. mm Hg. Calculate this pressure in (a) torr and (b) atmospheres.

Solution

Let's use conversion factors that relate one unit of pressure to another.
(a) To convert mm Hg to torr, use the conversion factor 760 torr/760 mm Hg (1 torr/1 mm Hg):

$$(740. \text{ mm Hg})\left(\frac{1 \text{ torr}}{1 \text{ mm Hg}}\right) = 740. \text{ torr}$$

(b) To convert mm Hg to atm, use the conversion factor 1 atm/760. mm Hg:

$$(740. \text{ mm Hg})\left(\frac{1 \text{ atm}}{760. \text{ mm Hg}}\right) = 0.974 \text{ atm}$$

Practice 12.1

A barometer reads 1.12 atm. Calculate the corresponding pressure in (a) torr and (b) mm Hg.

12.4 Dependence of Pressure on Number of Molecules and Temperature

Pressure is produced by gas molecules colliding with the walls of a container. At a specific temperature and volume the number of collisions depends on the number of gas molecules present. The number of collisions can be increased by increasing the number of gas molecules present. If we double the number of molecules, the frequency of collisions and the pressure should double. We find, for an ideal gas, that

FIGURE 12.4 ▶
The pressure exerted by a gas is directly proportional to the number of molecules present. In each case shown, the volume is 22.4 L and the temperature is 0°C.

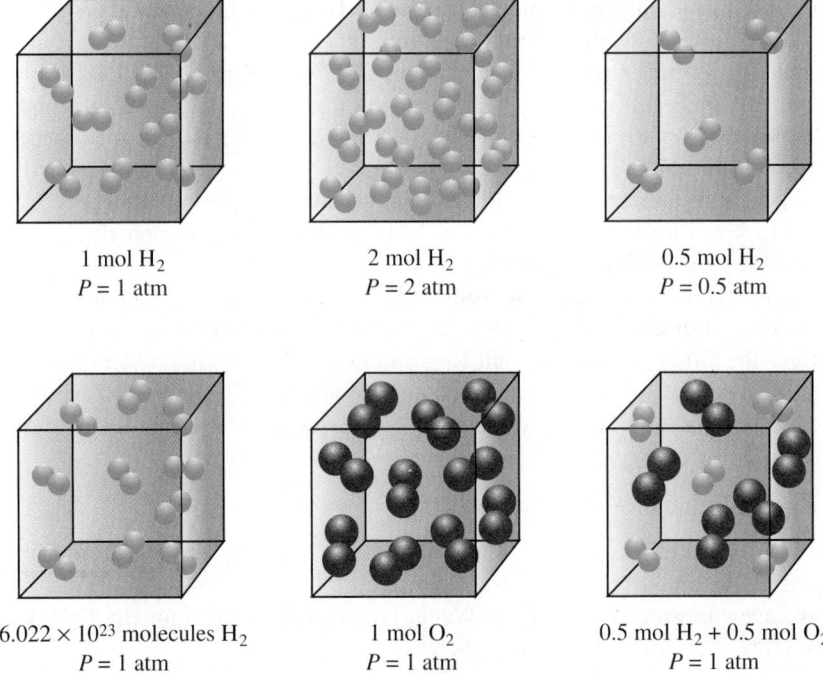

1 mol H_2
$P = 1$ atm

2 mol H_2
$P = 2$ atm

0.5 mol H_2
$P = 0.5$ atm

6.022×10^{23} molecules H_2
$P = 1$ atm

1 mol O_2
$P = 1$ atm

0.5 mol H_2 + 0.5 mol O_2
$P = 1$ atm

this doubling is actually what happens. When the temperature and mass are kept constant, the pressure is directly proportional to the number of moles or molecules of gas present. Figure 12.4 illustrates this concept.

A good example of this molecule–pressure relationship may be observed in an ordinary cylinder of compressed gas equipped with a pressure gauge. When the valve is opened, gas escapes from the cylinder. The volume of the cylinder is con-

FIGURE 12.5 ▶
The pressure of a gas in a fixed volume increases with increasing temperature. The increased pressure is due to more frequent and more energetic collisions of the gas molecules with the walls of the container at the higher temperature.

0°C
Volume = 1 liter
0.1 mole gas
$P = 2.24$ atm

100°C
Volume = 1 liter
0.1 mole gas
$P = 3.06$ atm

stant, and the decrease in quantity (moles) of gas is registered by a drop in pressure indicated on the gauge.

The pressure of a gas in a fixed volume also varies with temperature. When the temperature is increased, the kinetic energy of the molecules increases, causing more frequent and more energetic collisions of the molecules with the walls of the container. This increase in collision frequency and energy results in a pressure increase (see Figure 12.5).

12.5 Boyle's Law

Through a series of experiments, Robert Boyle (1627–1691) determined the relationship between the pressure (P) and volume (V) of a particular quantity of a gas. This relationship of P and V is known as **Boyle's law.**

Boyle's law

At constant temperature (T), the volume (V) of a fixed mass of a gas is inversely proportional to the pressure (P), which may be expressed as

$$V \propto \frac{1}{P} \qquad \text{or} \qquad P_1V_1 = P_2V_2$$

This equation says that the volume varies ($\propto$) inversely with the pressure, at constant mass and temperature. When the pressure on a gas is increased, its volume will decrease, and vice versa. The inverse relationship of pressure and volume is graphed in Figure 12.6.

When Boyle doubled the pressure on a specific quantity of a gas, keeping the temperature constant, the volume was reduced to one-half the original volume; when he tripled the pressure on the system, the new volume was one-third the original volume; and so on. His work showed that the product of volume and pressure is constant if the temperature is not changed:

$$PV = \text{constant} \qquad \text{or} \qquad PV = k \qquad \text{(mass and temperature are constant)}$$

Let's demonstrate this law using a cylinder with a movable piston so that the volume of gas inside the cylinder may be varied by changing the external pressure (see Figure 12.7). Assume that the temperature and the number of gas molecules do not change. We start with a volume of 1000 mL and a pressure of 1 atm. When we change the pressure to 2 atm, the gas molecules are crowded closer together, and the volume is reduced to 500 mL. When we increase the pressure to 4 atm, the volume becomes 250 mL.

Note that the product of the pressure times the volume is the same number in each case, substantiating Boyle's law. We may then say that

$$P_1V_1 = P_2V_2$$

where P_1V_1 is the pressure–volume product at one set of conditions, and P_2V_2 is the product at another set of conditions. In each case the new volume may be calculated by multiplying the starting volume by a ratio of the two pressures involved. Of course, the ratio of pressures used must reflect the direction in which the volume

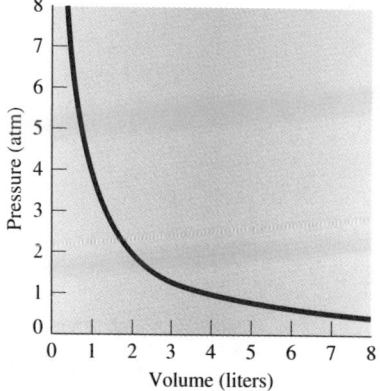

FIGURE 12.6
Graph of pressure versus volume showing the inverse PV relationship of an ideal gas.

FIGURE 12.7▶
The effect of pressure on the volume of a gas.

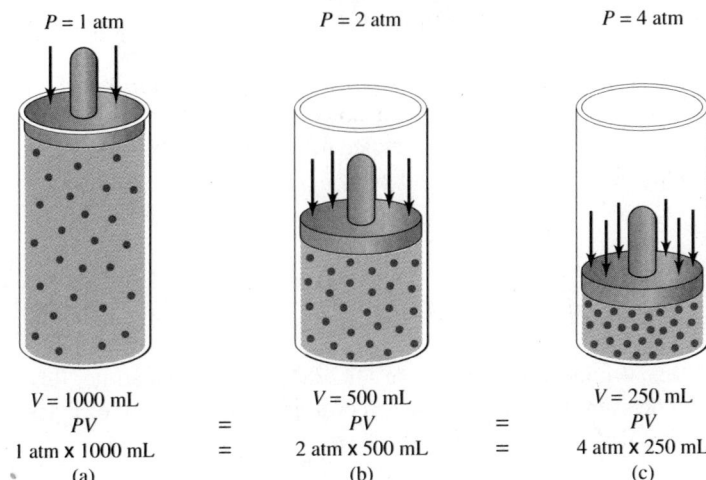

$P = 1$ atm $P = 2$ atm $P = 4$ atm

$V = 1000$ mL $V = 500$ mL $V = 250$ mL

PV = PV = PV

1 atm x 1000 mL   =   2 atm x 500 mL   =   4 atm x 250 mL

(a)   (b)   (c)

should change. When the pressure is changed from 1 atm to 2 atm, the ratio to be used is 1 atm/2 atm. Now we can verify the results given in Figure 12.7:

(a) Starting volume, 1000 mL; pressure change, 1 atm ⟶ 2 atm

$$(1000 \text{ mL})\left(\frac{1 \text{ atm}}{2 \text{ atm}}\right) = 500 \text{ mL}$$

(b) Starting volume, 1000 mL; pressure change, 1 atm ⟶ 4 atm

$$(1000 \text{ mL})\left(\frac{1 \text{ atm}}{4 \text{ atm}}\right) = 250 \text{ mL}$$

(c) Starting volume, 500 mL; pressure change, 2 atm ⟶ 4 atm

$$(500 \text{ mL})\left(\frac{2 \text{ atm}}{4 \text{ atm}}\right) = 250 \text{ mL}$$

In summary, a change in the volume of a gas due to a change in pressure can be calculated by multiplying the original volume by a ratio of the two pressures. If the pressure is increased, the ratio should have the smaller pressure in the numerator and the larger pressure in the denominator. If the pressure is decreased, the larger pressure should be in the numerator and the smaller pressure in the denominator.

new volume = original volume × ratio of pressures

We use Boyle's law in the following examples. If no mention is made of temperature, assume that it remains constant.

Example 12.2   What volume will 2.50 L of a gas occupy if the pressure is changed from 760. mm Hg to 630. mm Hg?

Solution   **Method A. Conversion Factors**

Step 1. Determine whether pressure is being increased or decreased:

pressure decreases ⟶ volume increases

Step 2. Multiply the original volume by a ratio of pressures that will result in an increase in volume:

$$V = (2.50 \text{ L})\left(\frac{760. \text{ mm Hg}}{630. \text{ mm Hg}}\right) = 3.02 \text{ L (new volume)}$$

Method B. Algebraic Equation

Decide which method is the best for you and stick with it.

Step 1. Organize the given information:

$P_1 = 760. \text{ mm Hg}$ $V_1 = 2.50 \text{ L}$

$P_2 = 630. \text{ mm Hg}$ $V_2 = ?$

Step 2. Write and solve this equation for the unknown:

$$P_1V_1 = P_2V_2 \qquad V_2 = \frac{P_1V_1}{P_2}$$

Step 3. Put the given information into this equation and calculate:

$$V_2 = \frac{(760. \text{ mm Hg})(2.50 \text{ L})}{630. \text{ mm Hg}} = 3.02 \text{ L}$$

A given mass of hydrogen occupies 40.0 L at 700. torr. What volume will it occupy at 5.00 atm pressure?

Example 12.3

Method A. Conversion Factors

Solution

Step 1. Determine whether the pressure is being increased or decreased. Note that in order to compare the values the units must be the same. We'll convert 700. torr to atm:

$$(700. \text{ torr})\left(\frac{1 \text{ atm}}{760 \text{ torr}}\right) = 0.921 \text{ atm}$$

The pressure is going from 0.921 atm to 5.00 atm.

pressure increases $\longrightarrow$ volume decreases

Step 2. Multiply the original volume by a ratio of pressures that will result in a decrease in volume:

$$V = (40.0 \text{ L})\left(\frac{0.921 \text{ atm}}{5.00 \text{ atm}}\right) = 7.37 \text{ L}$$

Method B. Algebraic Equation

Step 1. Organize the given information. Remember to make the pressure units the same.

$P_1 = 700. \text{ torr} = 0.921 \text{ atm}$ $V_1 = 40.0 \text{ L}$

$P_2 = 5.00 \text{ atm}$ $V_2 = ?$

Step 2. Write and solve this equation for the unknown:

$$P_1V_1 = P_2V_2 \qquad V_2 = \frac{P_1V_1}{P_2}$$

Step 3. Put the given information into this equation and calculate:

$$V_2 = \frac{(0.921 \text{ atm})(40.0 \text{ L})}{5.00 \text{ atm}} = 7.37 \text{ L}$$

Example 12.4   A gas occupies a volume of 200. mL at 400. torr pressure. To what pressure must the gas be subjected in order to change the volume to 75.0 mL?

Solution   **Method A. Conversion Factors**

Step 1. Determine whether volume is being increased or decreased:

volume decreases ⟶ pressure increases

Step 2. Multiply the original pressure by a ratio of volumes that will result in an increase in pressure:

new pressure = original pressure × ratio of volumes

$$P = (400. \text{ torr})\left(\frac{200. \text{ mL}}{75.0 \text{ mL}}\right) = 1067 \text{ torr} \quad \text{or} \quad 1.07 \times 10^3 \text{ torr (new pressure)}$$

Method B. Algebraic Equation

Step 1. Organize the given information. Remember to make units the same.

$$P_1 = 400. \text{ torr} \qquad V_1 = 200. \text{ mL}$$

$$P_2 = ? \qquad\qquad V_2 = 75.0 \text{ mL}$$

Step 2. Write and solve this equation for the unknown:

$$P_1V_1 = P_2V_2 \qquad P_2 = \frac{P_1V_1}{V_2}$$

Step 3. Put the given information into the equation and calculate:

$$P_2 = \frac{P_1V_1}{V_2} = \frac{(400. \text{ torr})(200. \text{ mL})}{75.0 \text{ mL}} = 1.07 \times 10^3 \text{ torr}$$

Practice 12.2

A gas occupies a volume of 3.86 L at 0.750 atm. At what pressure will the volume be 4.86 L?

 12.6   Charles' Law

The effect of temperature on the volume of a gas was observed in about 1787 by the French physicist J. A. C. Charles (1746–1823). Charles found that various gases expanded by the same fractional amount when they underwent the same change in temperature. Later it was found that if a given volume of any gas initially at 0°C was cooled by 1°C, the volume decreased by $\frac{1}{273}$; if cooled by 2°C, it decreased by $\frac{2}{273}$; if cooled by 20°C, by $\frac{20}{273}$; and so on. Since each degree of cooling reduced the volume by $\frac{1}{273}$, it was apparent that any quantity of any gas would have zero volume if it could be cooled to −273°C. Of course, no real gas can be cooled to −273°C for the simple reason that it would liquefy before that temperature is reached. However, −273°C (more precisely −273.15°C) is referred to as **absolute zero;** this temperature is the zero point on the Kelvin (absolute) temperature scale—the temperature at which the volume of an ideal, or perfect, gas would become zero.

absolute zero

The volume–temperature relationship for methane is shown graphically in Figure 12.8. Experimental data show the graph to be a straight line that, when extrapolated, crosses the temperature axis at −273.15°C, or absolute zero. This is characteristic for all gases.

In modern form, **Charles' law** is as follows:

Charles' law

> At *constant pressure* the volume of a fixed mass of any gas is directly proportional to the absolute temperature, which may be expressed as:
>
> $$V \propto T \qquad \text{or} \qquad \frac{V_1}{T_1} = \frac{V_2}{T_2}$$

Mathematically this states that the volume of a gas varies directly with the absolute temperature when the pressure remains constant. In equation form Charles' law may be written as

A capital T is usually used for absolute temperature, K, and a small t for °C.

$$V = kT \qquad \text{or} \qquad \frac{V}{T} = k \quad \text{(at constant pressure)}$$

where k is a constant for a fixed mass of the gas. If the absolute temperature of a gas is doubled, the volume will double.

To illustrate, let's return to the gas cylinder with the movable or free-floating piston (see Figure 12.9). Assume that the cylinder labeled (a) contains a quantity of gas and the pressure on it is 1 atm. When the gas is heated, the molecules move faster, and their kinetic energy increases. This action should increase the number of

◀ **FIGURE 12.8**
Volume–temperature relationship of methane (CH_4). Extrapolated portion of the graph is shown by the broken line.

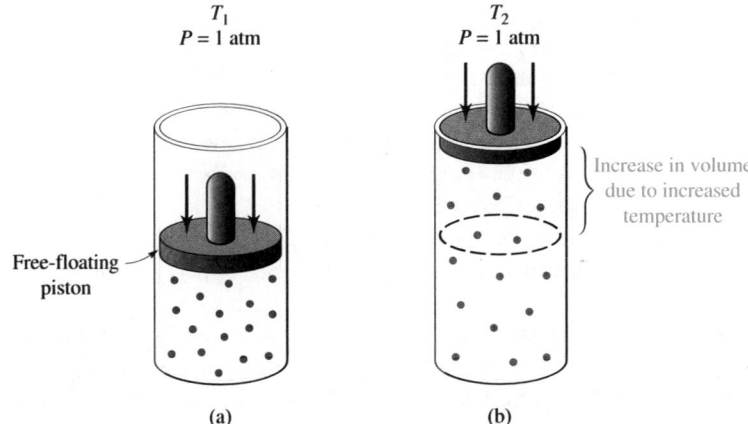

collisions per unit of time and therefore increase the pressure. However, the in-creased internal pressure will cause the piston to rise to a level at which the internal and external pressures again equal 1 atm, as we see in cylinder (b). The net result is an increase in volume due to an increase in temperature.

Another equation relating the volume of a gas at two different temperatures is

$$\frac{V_1}{T_1} = \frac{V_2}{T_2} \quad (\text{constant } P)$$

where V_1 and T_1 are one set of conditions and V_2 and T_2 are another set of conditions.

A simple experiment showing the variation of the volume of a gas with temperature is illustrated in Figure 12.10. A balloon is placed in a beaker and liquid N_2 is poured over it. The volume is reduced, as shown by the collapse of the balloon; when the balloon is removed from the liquid N_2, the gas expands and the balloon increases in size.

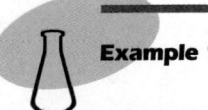

Example 12.5   Three liters of hydrogen at $-20.°C$ are allowed to warm to a room temperature of $27°C$. What is the volume at room temperature if the pressure remains constant?

Solution   **Method A. Conversion Factors**

Step 1. Determine whether temperature is being increased or decreased.

Remember temperature must
be changed to Kelvin in gas
law problems. Note that we
use 273 to convert instead of
273.15 since our original
measurements are to the
nearest degree.

$$-20.°C + 273 = 253 \text{ K}$$
$$27°C + 273 = 300. \text{ K}$$

temperature increases ⟶ volume increases

Step 2. Multiply the original volume by a ratio of temperatures that will result in an increase in volume.

$$V = (3.00 \text{ L})\left(\frac{300. \text{ K}}{253 \text{ K}}\right) = 3.56 \text{ L} \quad (\text{new volume})$$

(a)

(b)

(c)

▲
FIGURE 12.10
**The air-filled balloons in (a)
are placed in liquid nitrogen
(b). The volume of the air
decreases tremendously at
this temperature. In (c) the
balloons are removed from
the beaker and are beginning
to return to their original
volume as they warm back to
room temperature.**

Method B. Algebraic Equation

Step 1. Organize the given information. Remember to make units the same.

$$V_1 = 3.00 \text{ L} \qquad T_1 = 20.°C = 253 \text{ K}$$
$$V_2 = ? \qquad T_2 = 27°C = 300. \text{ K}$$

Step 2. Write and solve the equation for the unknown:

$$\frac{V_1}{T_1} = \frac{V_2}{T_2} \qquad V_2 = \frac{V_1 T_2}{T_1}$$

Step 3. Put the given information into the equation and calculate:

$$V_2 = \frac{V_1 T_2}{T_1} = \frac{(3.00 \text{ L})(300. \text{ K})}{253 \text{ K}} = 3.56 \text{ L}$$

If 20.0 L of oxygen are cooled from 100.°C to 0.°C, what is the new volume?

Example 12.6

Since no mention is made of pressure, assume that pressure does not change.

Solution

Method A. Conversion Factors

Step 1. Change °C to K:

$$100.°C + 273 = 373. \text{ K}$$
$$0.°C + 273 = 273. \text{ K}$$

Step 2. The ratio of temperature to be used is 273 K/373 K, because the final
volume should be smaller than the original volume. The calculation is

$$V = (20.0 \text{ L})\left(\frac{273 \text{ K}}{373 \text{ K}}\right) = 14.6 \text{ L} \quad \text{(new volume)}$$

Method B. Algebraic Equation

Step 1. Organize the given information. Remember to make units coincide.

$$V_1 = 20.0 \text{ L} \qquad T_1 = 100.°C = 373. \text{ K}$$
$$V_2 = ? \qquad T_2 = 0.°C = 273. \text{ K}$$

Step 2. Write and solve the equation for the unknown:

$$\frac{V_1}{T_1} = \frac{V_2}{T_2} \qquad V_2 = \frac{V_1 T_2}{T_1}$$

Step 3. Put the given information into the equation and calculate:

$$V_2 = \frac{V_1 T_2}{T_1} = \frac{(20.0 \text{ L})(273 \text{ K})}{373 \text{ K}} = 14.6 \text{ L}$$

Practice 12.3

A 4.50-L container of nitrogen gas at 28.0°C is heated to 56.0°C. Assuming the volume of the container can vary, what is the new volume of the gas?

12.7   Gay-Lussac's Law

J. L. Gay-Lussac (1778–1850) was a French chemist involved in the study of volume relationships of gases. The three variables (pressure, P; volume, V; and temperature, T) are needed to describe a fixed amount of a gas. Boyle's law, $PV = k$, relates pressure and volume at constant temperature; Charles' law, $V = kT$, relates volume and temperature at constant pressure. A third relationship involving pressure and temperature at constant volume is a modification of Charles' law and is some-
Gay-Lussac's law   times called **Gay-Lussac's law:**

The pressure of a fixed mass of a gas, at constant volume, is directly proportional to the Kelvin temperature:

$$P = kT \qquad \text{or} \qquad \frac{P_1}{T_1} = \frac{P_2}{T_2}$$

Example 12.7 The pressure of a container of helium is 650. torr at 25°C. If the sealed container is cooled to 0°C, what will the pressure be?

Solution   **Method A. Conversion Factors**

Step 1. Determine whether temperature is being increased or decreased.

temperature decreases ⟶ pressure decreases

Step 2. Multiply the original pressure by a ratio of Kelvin temperatures that will result in a decrease in pressure:

$$(650. \text{ torr})\left(\frac{273 \text{ K}}{298 \text{ K}}\right) = 595 \text{ torr}$$

Method B. Algebraic Equation

Step 1. Organize the given information. Remember to make units the same.

$P_1 = 650.$ torr $T_1 = 25°C = 298$ K

$P_2 = ?$ $T_2 = 0.°C = 273$ K

Step 2. Write and solve equation for the unknown:

$$\frac{P_1}{T_1} = \frac{P_2}{T_2} \qquad P_2 = \frac{P_1 T_2}{T_1}$$

Step 3. Put given information into equation and calculate:

$$P_2 = \frac{(650.\text{ torr})(273\ K)}{298\ K} = 595 \text{ torr}$$

Practice 12.4

A gas cylinder contains 40.0 L of gas at 45.0°C and has a pressure of 650. torr. What will the pressure be if the temperature is changed to 100.°C?

We may summarize the effects of changes in pressure, temperature, and quantity of a gas as follows:

1. In the case of a constant volume,
   (a) when the temperature is increased, the pressure increases.
   (b) when the quantity of a gas is increased, the pressure increases (T remaining constant).
2. In the case of a variable volume,
   (a) when the external pressure is increased, the volume decreases (T remaining constant).
   (b) when the temperature of a gas is increased, the volume increases (P remaining constant).
   (c) when the quantity of a gas is increased, the volume increases (P and T remaining constant).

12.8   Standard Temperature and Pressure

In order to compare volumes of gases, common reference points of temperature and pressure were selected and called **standard conditions** or **standard temperature and pressure** (abbreviated **STP**). Standard temperature is 273.15 K (0°C), and standard pressure is 1 atm or 760 torr or 760 mm Hg or 101.325 kPa. For purposes of comparison, volumes of gases are usually changed to STP conditions.

standard conditions
standard temperature
and pressure (STP)

---

standard temperature = 273.15 K or 0.00°C
standard pressure = 1 atm or 760 torr or 760 mm Hg or 101.325 kPa

In this text we'll use 273 K for temperature conversions and calculations. Check with your instructor for rules in your class.

12.9   Combined Gas Laws

When temperature and pressure change at the same time, the new volume may be calculated by multiplying the initial volume by the correct ratios of both pressure and temperature, as follows:

$$\text{final volume} = (\text{initial volume})\left(\begin{array}{c}\text{ratio of}\\\text{pressures}\end{array}\right)\left(\begin{array}{c}\text{ratio of}\\\text{temperatures}\end{array}\right)$$

This equation combines Boyle's and Charles' laws, and the same considerations for the pressure and temperature ratios should be used in the calculation. The four possible variations are as follows:

1. Both T and P cause an increase in volume.
2. Both T and P cause a decrease in volume.
3. T causes an increase and P causes a decrease in volume.
4. T causes a decrease and P causes an increase in volume.

The P, V, and T relationships for a given mass of any gas, in fact, may be expressed as a single equation, $PV/T = k$. For problem solving, this equation is usually written

Note, in the examples below, the use of 273 K does not change the number of significant figures in the temperature. The converted temperature is expressed to the same precision as the original measurement.

$$\frac{P_1V_1}{T_1} = \frac{P_2V_2}{T_2}$$

where P_1, V_1, and T_1 are the initial conditions and P_2, V_2, and T_2 are the final conditions.

This equation can be solved for any one of the six variables and is useful in dealing with the pressure–volume–temperature relationships of gases. Note that when T is constant ($T_1 = T_2$), Boyle's law is represented; when P is constant ($P_1 = P_2$), Charles' law is represented; and when V is constant ($V_1 = V_2$), Gay-Lussac's law is represented.

Example 12.8   Given 20.0 L of ammonia gas at 5°C and 730. torr, calculate the volume at 50.°C and 800. torr.

Solution   **Step 1.** Organize the given information, putting temperatures in Kelvin:

$$P_1 = 730.\ \text{torr} \qquad\qquad P_2 = 800.\ \text{torr}$$

$$V_1 = 20.0\ \text{L} \qquad\qquad V_2 = ?$$

$$T_1 = 5°C = 278\ \text{K} \qquad T_2 = 50.°C = 323\ \text{K}$$

Method A. Conversion Factors

Step 2. Set up ratios of T and P:

$$T\ \text{ratio} = \frac{323\ \text{K}}{278\ \text{K}} \text{ (increase in } T \text{ should increase } V\text{)}$$

$$P\ \text{ratio} = \frac{730.\ \text{torr}}{800.\ \text{torr}} \text{ (increase in } P \text{ should decrease } V\text{)}$$

Step 3. Multiply the original pressure by the ratios:

$$V_2 = (20.0 \text{ L})\left(\frac{730. \text{ torr}}{800. \text{ torr}}\right)\left(\frac{323 \text{ K}}{278 \text{ K}}\right) = 21.2 \text{ L}$$

Method B. Algebraic Equation

Step 2. Write and solve the equation for the unknown. Solve

$$\frac{P_1 V_1}{T_1} = \frac{P_2 V_2}{T_2}$$

for V_2 by multiplying both sides of the equation by T_2/P_2 and rearranging to obtain

$$V_2 = \frac{V_1 P_1 T_2}{P_2 T_1}$$

Step 3. Put the given information into the equation and calculate:

$$V_2 = \frac{(20.0 \text{ L})(730. \text{ torr})(323 \text{ K})}{(800. \text{ torr}) (278 \text{ K})} = 21.2 \text{ L}$$

To what temperature (°C) must 10.0 L of nitrogen at 25°C and 700. torr be heated in order to have a volume of 15.0 L and a pressure of 760. torr?   **Example 12.9**

Solution

Step 1. Organize the given information, putting temperatures in Kelvin:

$P_1 = 700. \text{ torr}$ $P_2 = 760. \text{ torr}$

$V_1 = 10.0 \text{ L}$ $V_2 = 15.0 \text{ L}$

$T_1 = 25°C = 298 \text{ K}$ $T_2 = ?$

Method A. Conversion Factors

Step 2. Set up ratios of V and P.

$$P \text{ ratio} = \frac{760. \text{ torr}}{700. \text{ torr}} \quad (\text{increase in } P \text{ should increase } T)$$

$$V \text{ ratio} = \frac{15.0 \text{ L}}{10.0 \text{ L}} \quad (\text{increase in } V \text{ should increase } T)$$

Step 3. Multiply the original temperature by the ratios:

$$T_2 = (298 \text{ K})\left(\frac{760. \text{ torr}}{700. \text{ torr}}\right)\left(\frac{15.0 \text{ L}}{10.0 \text{ L}}\right) = 485 \text{ K}$$

Method B. Algebraic Equation

Step 2. Write and solve the equation for the unknown:

$$\frac{P_1 V_1}{T_1} = \frac{P_2 V_2}{T_2} \qquad T_2 = \frac{T_1 P_2 V_2}{P_1 V_1}$$

Step 3. Put the given information into the equation and calculate:

$$T_2 = \frac{(298 \text{ K})(760. \text{ torr})(15.0 \text{ L})}{(700. \text{ torr})(10.0 \text{ L})} = 485 \text{ K}$$

In either method, since the problem asks for °C, we must subtract 273 from the Kelvin answer:

$$485 \text{ K} - 273 = 212°\text{C}$$

Example 12.10 The volume of a gas-filled balloon is 50.0 L at 20.°C and 742 torr. What volume will it occupy at standard temperature and pressure (STP)?

Solution

Step 1. Organize the given information, putting temperatures in Kelvin.

$P_1 = 742$ torr	$P_2 = 760.$ torr (standard pressure)
$V_1 = 50.0$ L	$V_2 = ?$
$T_1 = 20.°\text{C} = 293$ K	$T_2 = 273$ K (standard temperature)

Method A. Conversion Factors

Step 2. Set up ratios of T and P:

$$T \text{ ratio} = \frac{273 \text{ K}}{293 \text{ K}} \text{ (decrease in } T \text{ should decrease } V)$$

$$P \text{ ratio} = \frac{742 \text{ torr}}{760. \text{ torr}} \text{ (increase in } P \text{ should decrease } V)$$

Step 3. Multiply the original volume by the ratios:

$$V_2 = (50.0 \text{ L})\left(\frac{273 \text{ K}}{293 \text{ K}}\right)\left(\frac{742 \text{ torr}}{760. \text{ torr}}\right) = 45.5 \text{ L}$$

Method B. Algebraic Equation

Step 2. Write and solve the equation for the unknown:

$$\frac{P_1 V_1}{T_1} = \frac{P_2 V_2}{T_2} \qquad V_2 = \frac{P_1 V_1 T_2}{P_2 T_1}$$

Step 3. Put the given information into the equation and calculate:

$$V_2 = \frac{(742 \text{ torr})(50.0 \text{ L})(273 \text{ K})}{(760. \text{ torr})(293 \text{ K})} = 45.5 \text{ L}$$

Practice 12.5

15.00 L of gas at 45.0°C and 800. torr is heated to 400.°C, and the pressure changed to 300. torr. What is the new volume?

Practice 12.6

To what temperature must 5.00 L of oxygen at 50.°C and 600. torr be heated in order to have a volume of 10.0 L and a pressure of 800. torr?

Traditional "messenger" molecules are amino acids (also known as the building blocks for proteins). You've probably heard of endorphins, which are the messenger molecules associated with "runner's high." Until recently these molecules (called *neurotransmitters*) were thought to be specific; that is, each neurotransmitter fits into the target cell like a key in a lock. It was also thought that neurotransmitters were stored in tiny pouches where they are manufactured and then released when needed. But then two gases, nitrogen monoxide (NO) and carbon monoxide (CO), were found to act as neurotransmitters. These gases break all the "rules" for neurotransmission because as gases they are nonspecific. Gases freely diffuse into nearby cells, so NO and CO neurotransmitters must be made "on demand" since they cannot be stored. Gas neurotransmitters cannot use the lock-and-key model to act on cells, so they must use their chemical properties.

Nitrogen monoxide has been used for nearly a century to dilate blood vessels and increase blood flow, lowering blood pressure. In the late 1980s, scientists discovered that biologically produced NO is an important signaling

Viagra, the newest medication for male impotence, is the latest application of our understanding of the role of NO in the body.

molecule for nerve cells. Nitrogen monoxide mediates certain neurons that do not respond to traditional neurotransmitters. These NO-sensitive neurons are found in the cardiovascular, respiratory, digestive, and urogenital systems. Nitrogen monoxide also appears to play a role in regulating blood pressure, blood clotting, and neurotransmission. The 1998 Nobel prize in medicine was awarded to three Americans for their work on nitrogen monoxide's role in the body.

Nitrogen monoxide is a key molecule in producing erections in men.

When a man is sexually stimulated, NO is released into the penis where it activates the release of an enzyme that increases the level of a molecule called cGMP and ultimately produces relaxation of smooth muscles, allowing blood to flow into the penis. Viagra, the new impotence pill, enhances the effect of NO, inhibiting the enzyme that breaks down the cGMP molecule. This results in more cGMP molecules, and smooth muscle relaxation begins producing an erection.

After NO transmitters were discovered, researchers at Johns Hopkins Medical School reasoned that if one gas acted as a neurotransmitter so might others. They proposed CO as a possible transmitter because the enzyme used to make it is localized in specific parts of the brain (those responsible for smell and long-term memory). The enzyme used to make messengers was found in exactly the same locations! Researchers then showed that nerve cells made the messenger molecule when stimulated with CO. An inhibitor for CO blocked the messenger molecule production. Researchers are exploring the possibility of still other roles for these gaseous messenger molecules.

12.10 Dalton's Law of Partial Pressures

If gases behave according to the kinetic-molecular theory, there should be no difference in the pressure–volume–temperature relationships whether the gas molecules are all the same or different. This similarity in the behavior of gases is the basis for an understanding of **Dalton's law of partial pressures:**

Dalton's law of partial pressures

> **The total pressure of a mixture of gases is the sum of the partial pressures exerted by each of the gases in the mixture.**

Each gas in the mixture exerts a pressure that is independent of the other gases present. These pressures are called **partial pressures.** Thus if we have a mixture of

partial pressure

FIGURE 12.11 ▶
Oxygen collected over water.

Oxygen plus
water vapor

Oxygen from
generator

three gases, *A, B,* and *C,* exerting partial pressures of 50. torr, 150. torr, and 400. torr, respectively, the total pressure will be 600. torr:

$$P_{Total} = P_A + P_B + P_C$$

$$P_{Total} = 50.\ \text{torr} + 150.\ \text{torr} + 400.\ \text{torr} = 600.\ \text{torr}$$

We can see an application of Dalton's law in the collection of insoluble gases over water. When prepared in the laboratory, oxygen is commonly collected by the downward displacement of water. Thus the oxygen is not pure but is mixed with water vapor (see Figure 12.11). When the water levels are adjusted to the same height inside and outside the bottle, the pressure of the oxygen plus water vapor inside the bottle is equal to the atmospheric pressure:

$$P_{atm} = P_{O_2} + P_{H_2O}$$

To determine the amount of O_2 or any other gas collected over water, we subtract the pressure of the water vapor from the total pressure of the gas. The vapor pressure of water at various temperatures is tabulated in Appendix III.

$$P_{O_2} = P_{atm} - P_{H_2O}$$

Example 12.11 A 500.-mL sample of oxygen was collected over water at 23°C and 760. torr. What volume will the dry O_2 occupy at 23°C and 760. torr? The vapor pressure of water at 23°C is 21.2 torr.

Solution   To solve this problem, we must first determine the pressure of the oxygen alone, by subtracting the pressure of the water vapor present.

Step 1. Determine the pressure of dry O_2:

$$P_{Total} = 760.\ \text{torr} = P_{O_2} + P_{H_2O}$$

$$P_{O_2} = 760.\ \text{torr} - 21.2\ \text{torr} = 739\ \text{torr} \qquad (\text{dry } O_2)$$

Step 2. Organize the given information:

$$P_1 = 739\ \text{torr} \qquad P_2 = 760.\ \text{torr}$$

$$V_1 = 500.\ \text{mL} \qquad V_2 = ?$$

$$T \text{ is constant}$$

Step 3. Solve as a Boyle's law problem:

$$V = \frac{(500.\ \text{mL})(739\ \text{torr})}{760.\ \text{torr}} = 486\ \text{mL dry } O_2$$

Practice 12.7

Hydrogen gas was collected by downward displacement of water. A volume of
600.0 mL of gas was collected at 25.0°C and 740.0 torr. What volume will the
dry hydrogen occupy at STP?

 12.11 Avogadro's Law

Early in the 19th century, Gay-Lussac studied the volume relationships of reacting
gases. His results, published in 1809, were summarized in a statement known as
Gay-Lussac's law of combining volumes:

Gay-Lussac's law of combining
volumes

> When measured at the same temperature and pressure, the ratios of the
> volumes of reacting gases are small whole numbers.

Thus H_2 and O_2 combine to form water vapor in a volume ratio of $2:1$ (Figure
12.12); H_2 and Cl_2 react to form HCl in a volume ratio of $1:1$; and H_2 and N_2 react
to form NH_3 in a volume ratio of $3:1$.

Two years later, in 1811, Amedeo Avogadro (1776–1856) used the law of combining volumes of gases to make a simple but significant and far-reaching generalization concerning gases. **Avogadro's law** states:

> Equal volumes of different gases at the same temperature and pressure
> contain the same number of molecules.

Avogadro's law

This law was a real breakthrough in understanding the nature of gases.

1. It offered a rational explanation of Gay-Lussac's law of combining volumes
   of gases and indicated the diatomic nature of such elemental gases as hydrogen, chlorine, and oxygen.

2 volumes   1 volume   2 volumes

FIGURE 12.12
**Gay-Lussac's law of
combining volumes of
gases applied to the
reaction of hydrogen and
oxygen. When measured at
the same temperature and
pressure, hydrogen and
oxygen react in a volume
◀ ratio of 2 : 1.**

2. It provided a method for determining the molar masses of gases and for comparing the densities of gases of known molar mass (see Sections 12.12 and 12.13).

3. It afforded a firm foundation for the development of the kinetic-molecular theory.

By Avogadro's law, equal volumes of hydrogen and chlorine at the same temperature and pressure contain the same number of molecules. On a volume basis, hydrogen and chlorine react thus:

$$\text{hydrogen} + \text{chlorine} \longrightarrow \text{hydrogen chloride}$$
$$\quad\text{1 volume}\qquad\text{1 volume}\qquad\qquad\text{2 volumes}$$

Therefore, hydrogen molecules react with chlorine molecules in a 1:1 ratio. Since two volumes of hydrogen chloride are produced, one molecule of hydrogen and one molecule of chlorine must produce two molecules of hydrogen chloride. Therefore, each hydrogen molecule and each chlorine molecule must be made up of two atoms. The coefficients of the balanced equation for the reaction give the correct ratios for volumes, molecules, and moles of reactants and products:

$$H_2 \quad + \quad Cl_2 \quad \longrightarrow \quad 2\,HCl$$

1 volume	1 volume	2 volumes
1 molecule	1 molecule	2 molecules
1 mol	1 mol	2 mol

By like reasoning, oxygen molecules also must contain at least two atoms because one volume of oxygen reacts with two volumes of hydrogen to produce two volumes of water vapor.

The volume of a gas depends on the temperature, the pressure, and the number of gas molecules. Different gases at the same temperature have the same average kinetic energy. Hence, if two different gases are at the same temperature, occupy equal volumes, and exhibit equal pressures, each gas must contain the same number of molecules. This statement is true because systems with identical *PVT* properties can be produced only by equal numbers of molecules having the same average kinetic energy.

Mole–Mass–Volume Relationships of Gases

molar volume

As with many constants, the molar volume is known more exactly to be 22.414 L. We use 22.4 L in our calculations since the extra figures don't often affect the result, given the other measurements in the calculation.

Because a mole contains 6.022×10^{23} molecules (Avogadro's number), a mole of any gas will have the same volume as a mole of any other gas at the same temperature and pressure. It has been experimentally determined that the volume occupied by a mole of any gas is 22.4 L at STP. This volume, 22.4 L, is known as the **molar volume** of a gas. The molar volume is a cube about 28.2 cm (11.1 in.) on a side. The molar masses of several gases, each occupying 22.4 L at STP, are shown in Figure 12.13.

One mole of a gas occupies 22.4 L at STP.

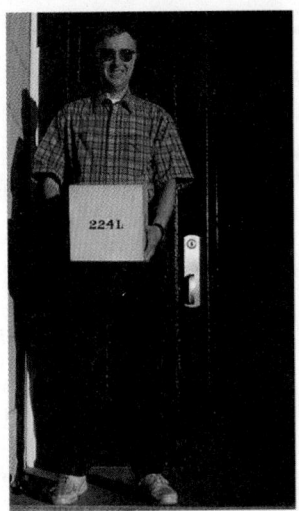

The molar volume is useful for determining the molar mass of a gas or of substances that can be easily vaporized. If the mass and the volume of a gas at STP are known, we can calculate its molar mass. For example, 1 L of pure oxygen at STP has a mass of 1.429 g. The molar mass of oxygen may be calculated by multiplying the mass of 1 L by 22.4 L/mol:

$$\left(\frac{1.429 \text{ g}}{1 \text{ L}}\right)\left(\frac{22.4 \text{ L}}{1 \text{ mol}}\right) = 32.0 \text{ g/mol} \qquad \text{(molar mass)}$$

If the mass and volume are at other than standard conditions, we change the volume to STP and then calculate the molar mass.

The molar volume, 22.4 L/mol, is used as a conversion factor to convert grams per liter to grams per mole (molar mass) and also to convert liters to moles. The two conversion factors are

**Standard conditions apply
only to pressure, temperature,
and volume. Mass is not
affected.**

$$\frac{22.4 \text{ L}}{1 \text{ mol}} \qquad \text{and} \qquad \frac{1 \text{ mol}}{22.4 \text{ L}}$$

These conversions must be done at STP except under certain special circumstances. Examples follow.

If 2.00 L of a gas measured at STP has a mass of 3.23 g, what is the molar mass of the gas?

Example 12.12

The unit of molar mass is g/mol; the conversion is from

Solution

$$\frac{\text{g}}{\text{L}} \longrightarrow \frac{\text{g}}{\text{mol}}$$

The starting amount is $\dfrac{3.23 \text{ g}}{2.00 \text{ L}}$. The conversion factor is $\dfrac{22.4 \text{ L}}{1 \text{ mol}}$.

The calculation is $\left(\dfrac{3.23 \text{ g}}{2.00 \text{ L}}\right)\left(\dfrac{22.4 \text{ L}}{1 \text{ mol}}\right) = 36.2 \text{ g/mol} \qquad \text{(molar mass)}$

Example 12.13 Measured at 40°C and 630. torr, the mass of 691 mL of ethyl ether is 1.65 g. Calculate the molar mass of ethyl ether.

Solution

Step 1. Organize the given information, converting temperatures to Kelvin. Note that we must change to STP in order to determine molar mass.

$$P_1 = 630. \text{ torr} \qquad\qquad P_2 = 760. \text{ torr}$$

$$V_1 = 691 \text{ mL} \qquad\qquad V_2 = ?$$

$$T_1 = 313 \text{ K } (40.°C) \qquad T_2 = 273 \text{ K}$$

Step 2. Use either the conversion factor method or the algebraic method and the combined gas law to correct the volume (V_2) to STP:

$$V_2 = \frac{(691 \text{ mL})(273 \text{ K})(630. \text{ torr})}{(313 \text{ K})(760. \text{ torr})} = 500. \text{ mL} = 0.500 \text{ L} \quad \text{(at STP)}$$

Step 3. In the example, V_2 is the volume for 1.65 g of the gas, so we can now find the molar mass by converting g/L to g/mol:

$$\left(\frac{1.65 \text{ g}}{0.500 \text{ L}}\right)\left(\frac{22.4 \text{ L}}{\text{mol}}\right) = 73.9 \text{ g/mol}$$

Practice 12.8

A gas with a mass of 86 g occupies 5.00 L at 25°C and 3.00 atm pressure. What is the molar mass of the gas?

12.13 Density of Gases

The density, d, of a gas is its mass per unit volume, which is generally expressed in grams per liter as follows:

$$d = \frac{\text{mass}}{\text{volume}} = \frac{\text{g}}{\text{L}}$$

Because the volume of a gas depends on temperature and pressure, both should be given when stating the density of a gas. The volume of a solid or liquid is hardly affected by changes in pressure and is changed only slightly when the temperature is varied. Increasing the temperature from 0°C to 50°C will reduce the density of a gas by about 18% if the gas is allowed to expand, whereas a 50°C rise in the temperature of water (0°C ⟶ 50°C) will change its density by less than 0.2%.

The density of a gas at any temperature and pressure can be determined by calculating the mass of gas present in 1 L. At STP, in particular,

	Molar mass (g/mol)	Density (g/L at STP)	Gas	Molar mass (g/mol)	Density (g/L at STP)
TABLE 12.3		**Density of Common Gases at STP**			

Gas	Molar mass (g/mol)	Density (g/L at STP)	Gas	Molar mass (g/mol)	Density (g/L at STP)
H_2	2.016	0.0900	H_2S	34.09	1.52
CH_4	16.04	0.716	HCl	36.46	1.63
NH_3	17.03	0.760	F_2	38.00	1.70
C_2H_2	26.04	1.16	CO_2	44.01	1.96
HCN	27.03	1.21	C_3H_8	44.09	1.97
CO	28.01	1.25	O_3	48.00	2.14
N_2	28.02	1.25	SO_2	64.07	2.86
air	(28.9)	(1.29)	Cl_2	70.90	3.17
O_2	32.00	1.43			

the density can be calculated by multiplying the molar mass of the gas by 1 mol/22.4 L:

$$d_{STP} = \text{molar mass} \left(\frac{1 \text{ mol}}{22.4 \text{ L}} \right)$$

$$\text{molar mass} = d_{STP} \left(\frac{22.4 \text{ L}}{1 \text{ mol}} \right)$$

Table 12.3 lists the densities of some common gases.

Example 12.14

Calculate the density of Cl_2 at STP.

Solution

First calculate the molar mass of Cl_2. It is 70.90 g/mol. Since d = g/L, the conversion is

$$\frac{g}{mol} \longrightarrow \frac{g}{L}$$

The conversion factor is $\frac{1 \text{ mol}}{22.4 \text{ L}}$:

$$d = \left(\frac{70.90 \text{ g}}{1 \text{ mol}} \right) \left(\frac{1 \text{ mol}}{22.4 \text{ L}} \right) = 3.17 \text{ g/L}$$

Practice 12.9

The molar mass of a gas is 20. g/mol. Calculate the density of the gas at STP.

CHEMISTRY IN ACTION Physiological Effects of Pressure Changes

The human body has a variety of methods for coping with the changes in atmospheric pressure. As we travel to the mountains, fly in an airplane, or take a high-speed elevator to the top of a sky-scraper, the pressure around us decreases. Our ears are sensitive to this because the eardrum (tympanic membrane) has air on both sides of it. The difference in pressure is relieved by yawning or moving the jaw to open the (Eustachian) tubes that connect the middle ear and throat and allow the pressure inside the eardrum to equalize with the outside.

Divers must also contend with the effects of pressure, most notably in body cavities containing air, such as the lungs, ears, and sinuses. Scuba divers don't experience a crushing effect of pressure at increased depths because the tank regulators deliver air at the same pressure as that of the surroundings. The diver must always breathe out regularly while ascending to the surface. Failure to do so can cause the lungs to expand, thus rupturing some of the alveoli, and resulting in loss of consciousness, brain damage, or heart attack. This is a clear application of Boyle's law.

Divers are also affected by consequences of **Henry's law,** which states that the amount of gas that will dissolve

▲
If a diver returns too quickly to the surface, the pressure reduction may produce bubbles in the blood— a condition known as "the bends."

in a liquid varies directly with the pressure above the liquid. This means that during a dive the gases entering the lungs are absorbed into the blood to a greater extent than at the water's surface. If the diver returns too rapidly to the surface, the swift pressure reduction can cause dissolved gases to produce bubbles in the blood, resulting in a condition known as *decompression sickness* or "the bends." The only successful method of treatment for this involves the use of a decompression chamber to increase the pressure once again and slowly decompress the diver back to normal pressure.

In the medical field, *hyperbaric units* are used to treat patients who have cells starved for oxygen. In these units the whole room may be placed at high pressure (2 or 3 atm), and the entire staff as well as the patient undergo gradual compression and, following treatment, decompression. These units are widely used to treat carbon monoxide poisoning. Oxygen is dissolved directly into the plasma giving the tissues temporary relief from oxygen deprivation. Hyperbaric units are also effective in treating other problems such as skin grafts, severe thermal burns, and radiation tissue damage.

12.14 Ideal Gas Equation

We've used four variables in calculations involving gases: the volume, V; the pressure, P; the absolute temperature, T; and the number of molecules or moles, (abbreviated n). Combining these variables into a single expression, we obtain

$$V \propto \frac{nT}{P} \quad \text{or} \quad V = \frac{nRT}{P}$$

where R is a proportionality constant known as the *ideal gas constant.* The equation is commonly written as

$$PV = nRT$$

and is known as the **ideal gas equation.** This equation states in a single expression   *ideal gas equation*
what we have considered in our earlier discussions: The volume of a gas varies di-
rectly with the number of gas molecules and the absolute temperature, and varies inversely with the pressure. The value and units of R depend on the units of P, V, and T.
We can calculate one value of R by taking 1 mol of a gas at STP conditions. Solve the
equation for R:

$$R = \frac{PV}{nT} = \frac{(1 \text{ atm})(22.4 \text{ L})}{(1 \text{ mol})(273 \text{ K})} = 0.0821 \frac{\text{L-atm}}{\text{mol-K}}$$

The units of R in this case are liter-atmospheres (L-atm) per mole Kelvin (mol-K).
When the value of $R = 0.0821$ L-atm/mol-K, P is in atmospheres, n is in moles, V is
in liters, and T is in Kelvin.

    The ideal gas equation can be used to calculate any one of the four variables
when the other three are known.

What pressure will be exerted by 0.400 mol of a gas in a 5.00-L container   **Example 12.15**
at 17°C?

Step 1. Organize the given information, converting temperatures to Kelvin: **Solution**

$P = ?$

$V = 5.00$ L

$T = 290.$ K

$n = 0.400$ mol

Step 2. Write and solve the ideal gas equation for the unknown:

$$PV = nRT \qquad \text{or} \qquad P = \frac{nRT}{V}$$

Step 3. Substitute the given information into the equation and calculate:

$$P = \frac{(0.400 \text{ mol})(0.0821 \text{ L·atm/mol·K})(290. \text{ K})}{5.00 \text{ L}} = 1.90 \text{ atm}$$

How many moles of oxygen gas are in a 50.0-L tank at 22.0°C if the pressure   **Example 12.16**
gauge reads 2000. lb/in.^{2}?

Step 1. Organize the given information, converting temperature to kelvins and   **Solution**
pressure to atmospheres:

$$P = \left(\frac{2000. \text{ lb}}{\text{in.}^2}\right)\left(\frac{1 \text{ atm}}{14.7 \text{ lb/in.}^2}\right) = 136.1 \text{ atm}$$

$V = 50.0$ L

$T = 295$ K

$n = ?$

Step 2. Write and solve the ideal gas equation for the unknown:

$$PV = nRT \quad \text{or} \quad n = \frac{PV}{RT}$$

Step 3. Substitute the given information into the equation and calculate:

$$n = \frac{(136.1 \text{ atm})(50.0 \text{ L})}{(0.0821 \text{ L·atm/mol·K})(295 \text{ K})} = 281 \text{ mol } O_2$$

Practice 12.10

A 23.8-L cylinder contains oxygen gas at 20.0°C and 732 torr. How many moles of oxygen are in the cylinder?

The molar mass of a gaseous substance can be determined using the ideal gas equation. Since molar mass = g/mol, then mol = g/molar mass. Using M for molar mass and g for grams, we can substitute g/M for n (moles) in the ideal gas equation to get

This form of the ideal gas equation is most useful in problems containing mass instead of moles.

$$PV = \frac{g}{M}RT \quad \text{or} \quad M = \frac{gRT}{PV} \quad \text{(modified ideal gas equation)}$$

which allows us to calculate the molar mass, M, for any substance in the gaseous state.

Example 12.17 Calculate the molar mass of butane gas, if 3.69 g occupies 1.53 L at 20.0°C and 1.00 atm.

Solution   Change 20°C to 293 K and substitute the data into the modified ideal gas equation:

$$M = \frac{gRT}{PV} = \frac{(3.69 \text{ g})(0.0821 \text{ L·atm/mol·K})(293 \text{ K})}{(1.00 \text{ atm})(1.53 \text{ L})} = 58.0 \text{ g/mol}$$

Practice 12.11

A sample of 0.286 g of a certain gas occupies 50.0 mL at standard temperature and 76.0 cm Hg. Determine the molar mass of the gas.

 Gas Stoichiometry

Mole–Volume and Mass–Volume Calculations

Stoichiometric problems involving gas volumes can be solved by the general mole-ratio method outlined in Chapter 9. The factors 1 mol/22.4 L and 22.4 L/1 mol are

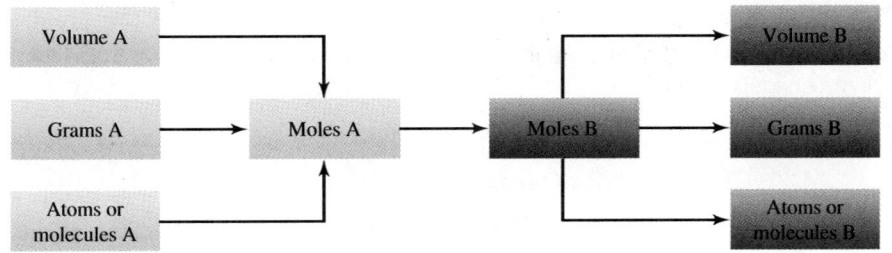

◀ FIGURE 12.14
Summary of the primary conversions involved in stoichiometry. The conversion for volumes of gases is included.

used for converting volume to moles and moles to volume, respectively. (See Figure 12.14.) These conversion factors are used under the assumption that the gases are at STP and that they behave as ideal gases. In actual practice, gases are measured at other than STP conditions, and the volumes are converted to STP for stoichiometric calculations.

In a balanced equation, the number preceding the formula of a gaseous substance represents the number of moles or molar volumes (22.4 L at STP) of that substance.

The following are examples of typical problems involving gases and chemical equations.

What volume of oxygen (at STP) can be formed from 0.500 mol of potassium **Example 12.18** chlorate?

Solution

Step 1. Write the balanced equation:

$$2\ KClO_3 \longrightarrow 2\ KCl + 3\ O_2(g)$$

Step 2. The starting amount is 0.500 mol $KClO_3$. The conversion is from

$$\text{moles } KClO_3 \longrightarrow \text{moles } O_2 \longrightarrow \text{liters } O_2$$

Step 3. Calculate the moles of O_2, using the mole-ratio method:

$$(0.500\ \text{mol } KClO_3)\left(\frac{3\ \text{mol } O_2}{2\ \text{mol } KClO_3}\right) = 0.750\ \text{mol } O_2$$

Step 4. Convert moles of O_2 to liters of O_2. The moles of a gas at STP are converted to liters by multiplying by the molar volume, 22.4 L/mol:

$$(0.750\ \text{mol } O_2)\left(\frac{22.4\ \text{L}}{1\ \text{mol}}\right) = 16.8\ \text{L } O_2$$

Setting up a continuous calculation, we obtain

$$(0.500\ \text{mol } KClO_3)\left(\frac{3\ \text{mol } O_2}{2\ \text{mol } KClO_3}\right)\left(\frac{22.4\ \text{L}}{1\ \text{mol}}\right) = 16.8\ \text{L } O_2$$

Example 12.19 How many grams of aluminum must react with sulfuric acid to produce 1.25 L of hydrogen gas at STP?

Solution   **Step 1.** The balanced equation is

$$2 \text{ Al}(s) + 3 \text{ H}_2\text{SO}_4(aq) \longrightarrow \text{Al}_2(\text{SO}_4)_3(aq) + 3 \text{ H}_2(g)$$

Step 2. We first convert liters of H_2 to moles of H_2. Then the familiar stoichiometric calculation from the equation is used. The conversion is

$$\text{L H}_2 \longrightarrow \text{mol H}_2 \longrightarrow \text{mol Al} \longrightarrow \text{g Al}$$

$$1.25 \text{ L H}_2 \left(\frac{1 \text{ mol}}{22.4 \text{ L}}\right)\left(\frac{2 \text{ mol Al}}{3 \text{ mol H}_2}\right)\left(\frac{26.98 \text{ g Al}}{1 \text{ mol Al}}\right) = 1.00 \text{ g Al}$$

Example 12.20 What volume of hydrogen, collected at 30.°C and 700. torr, will be formed by reacting 50.0 g of aluminum with hydrochloric acid?

$$2 \text{ Al}(s) + 6 \text{ HCl}(aq) \longrightarrow 2 \text{ AlCl}_3(aq) + 3 \text{ H}_2(g)$$

Solution   In this problem the conditions are not at STP, so we cannot use the method shown in Example 12.18. Either we need to calculate the volume at STP from the equation and then convert this volume to the conditions given in the problem, or we can use the ideal gas equation. Let's use the ideal gas equation.

First calculate the moles of H_2 obtained from 50.0 g of Al. Then, using the ideal gas equation, calculate the volume of H_2 at the conditions given in the problem.

Step 1. Moles of H_2: The conversion is

$$\text{grams Al} \longrightarrow \text{moles Al} \longrightarrow \text{moles H}_2$$

$$50.0 \text{ g Al}\left(\frac{1 \text{ mol Al}}{26.98 \text{ g Al}}\right)\left(\frac{3 \text{ mol H}_2}{2 \text{ mol Al}}\right) = 2.78 \text{ mol H}_2$$

Step 2. Liters of H_2: Solve $PV = nRT$ for V and substitute the data into the equation.

Convert °C to K: 30.°C + 273 = 303 K.
Convert torr to atm: (700. torr)(1 atm/760. torr) = 0.921 atm.

$$V = \frac{nRT}{P} = \frac{(2.78 \text{ mol H}_2)(0.0821 \text{ L-atm})(303 \text{ K})}{(0.921 \text{ atm})(\text{mol-K})} = 75.1 \text{ L H}_2$$

Note: The volume at STP is 62.3 L H_2.

Practice 12.12

If 10.0 g of sodium peroxide, Na_2O_2, react with water to produce sodium hydroxide and oxygen, how many liters of oxygen will be produced at 20°C and 750. torr?

$$2 \text{ Na}_2\text{O}_2(s) + 2 \text{ H}_2\text{O}(l) \longrightarrow 4 \text{ NaOH}(aq) + \text{O}_2(g)$$

Volume–Volume Calculations

When all substances in a reaction are in the gaseous state, simplifications in the calculation can be made. These are based on Avogadro's law, which states that gases under identical conditions of temperature and pressure contain the same number of molecules and occupy the same volume. Using this same law, we can also state that, under the standard conditions of temperature and pressure, the volumes of gases reacting are proportional to the numbers of moles of the gases in the balanced equation. Consider the reaction:

$$H_2(g) \; + \; Cl_2(g) \longrightarrow 2 \, HCl(g)$$

1 mol	1 mol	2 mol
22.4 L	22.4 L	2 × 22.4 L
1 volume	1 volume	2 volumes
Y volume	Y volume	2 Y volumes

In this reaction 22.4 L of hydrogen will react with 22.4 L of chlorine to give 2 (22.4) = 44.8 L of hydrogen chloride gas. This statement is true because these volumes are equivalent to the number of reacting moles in the equation. Therefore, Y volume of H_2 will combine with Y volume of Cl_2 to give 2 Y volumes of HCl. For example, 100 L of H_2 react with 100 L of Cl_2 to give 200 L of HCl; if the 100 L of H_2 and of Cl_2 are at 50°C, they will give 200 L of HCl at 50°C. When the temperature and pressure before and after a reaction are the same, volumes can be calculated without changing the volumes to STP.

For reacting gases at constant temperature and pressure:
Volume–volume relationships are the same as mole–mole relationships.

What volume of oxygen will react with 150. L of hydrogen to form water vapor? **Example 12.21** What volume of water vapor will be formed?

Assume that both reactants and products are measured at standard conditions. Calculate by using reacting volumes: **Solution**

$$2 \, H_2(g) \; + \; O_2(g) \longrightarrow 2 \, H_2O(g)$$

2 mol	1 mol	2 mol
2 × 22.4 L	22.4 L	2 × 22.4 L
2 volumes	1 volume	2 volumes
150. L	75 L	150. L

For every two volumes of H_2 that react, one volume of O_2 reacts and two volumes of $H_2O(g)$ are produced:

$$(150. \text{ L } H_2) \left(\frac{1 \text{ volume } O_2}{2 \text{ volumes } H_2} \right) = 75 \text{ L } O_2$$

$$(150. \text{ L } H_2) \left(\frac{2 \text{ volumes } H_2O}{2 \text{ volumes } H_2} \right) = 150. \text{ L } H_2O$$

Example 12.22 The equation for the preparation of ammonia is

$$3H_2(g) + N_2(g) \xrightarrow{400°C} 2\,NH_3(g)$$

Assuming that the reaction goes to completion,

(a) what volume of H_2 will react with 50.0 L of N_2?
(b) what volume of NH_3 will be formed from 50.0 L of N_2?
(c) what volume of N_2 will react with 100. mL of H_2?
(d) what volume of NH_3 will be produced from 100. mL of H_2?
(e) if 600. mL of H_2 and 400. mL of N_2 are sealed in a flask and allowed to react, what amounts of H_2, N_2, and NH_3 are in the flask at the end of the reaction?

Solution   The answers to parts (a)–(d) are shown in the boxes and can be determined from the equation by inspection, using the principle of reacting volumes:

$$3\,H_2(g) + N_2(g) \longrightarrow 2\,NH_3(g)$$
3 volumes   1 volume   2 volumes

(a) $\boxed{150.\,L}$ 50.0 L
(b)   50.0 L $\boxed{100.\,L}$
(c) 100. mL $\boxed{33.3\,mL}$
(d) 100. mL $\boxed{66.7\,mL}$

(e) Volume ratio from the equation $= \dfrac{3 \text{ volumes } H_2}{1 \text{ volume } N_2}$

$$\text{Volume ratio used} = \frac{600.\text{ mL } H_2}{400.\text{ mL } N_2} = \frac{3 \text{ volumes } H_2}{2 \text{ volumes } N_2}$$

Comparing these two ratios, we see that an excess of N_2 is present in the gas mixture. Therefore, the reactant limiting the amount of NH_3 that can be formed is H_2:

$$3\,H_2(g) + N_2(g) \longrightarrow 2\,NH_3(g)$$
600 mL   200 mL   400 mL

To have a 3 : 1 ratio of volumes reacting, 600. mL of H_2 will react with 200 mL of N_2 to produce 400. mL of NH_3, leaving 200. mL of N_2 unreacted. At the end of the reaction the flask will contain 400. mL of NH_3 and 200. mL of N_2.

Practice 12.13

What volume of oxygen will react with 15.0 L of propane (C_3H_8) to form carbon dioxide and water? What volume of carbon dioxide will be formed? What volume of water vapor will be formed?

$$C_3H_8(g) + 5\,O_2(g) \longrightarrow 3\,CO_2(g) + 4\,H_2O(g)$$

12.16   Real Gases

All the gas laws are based on the behavior of an ideal gas—that is, a gas with a behavior that is described exactly by the gas laws for all possible values of P, V, and T. Most real gases actually do behave very nearly as predicted by the gas laws over a fairly wide range of temperatures and pressures. However, when conditions are such that the gas molecules are crowded closely together (high pressure and/or low temperature), they show marked deviations from ideal behavior. Deviations occur because molecules have finite volumes and also have intermolecular attractions, which result in less compressibility at high pressures and greater compressibility at low temperatures than predicted by the gas laws. Many gases become liquids at high pressure and low temperature.

12.17   Air Pollution

Chemical reactions occur among the gases that are emitted into our atmosphere. In recent years, there has been growing concern over the effects these reactions have on our environment and our lives.

The outer portion (stratosphere) of the atmosphere plays a significant role in determining the conditions for life at the surface of the Earth. This stratosphere protects the surface from the intense radiation and particles bombarding our planet. Some of the high energy radiation from the sun acts upon oxygen molecules in the stratosphere, converting them into ozone, O_3. Different molecular forms of an element are called **allotropes** of that element. Thus oxygen and ozone are allotropic forms of oxygen:

allotrope

$$O_2 \xrightarrow{\text{sunlight}} O + O$$
$$\text{oxygen atoms}$$

$$O_2 + O \longrightarrow O_3$$
$$\text{ozone}$$

Ultraviolet radiation from the sun is highly damaging to living tissues of plants and animals. The ozone layer, however, shields the Earth by absorbing ultraviolet radiation and thus prevents most of this lethal radiation from reaching the Earth's surface. The reaction that occurs is the reverse of the preceding one:

$$O_3 \xrightarrow[\text{radiation}]{\text{ultraviolet}} O_2 + O + \text{heat}$$

Scientists have become concerned about a growing hazard to the ozone layer. Chlorofluorocarbon propellants, such as the Freons, CCl_3F and CCl_2F_2, which were used in aerosol spray cans and are used in refrigeration and air-conditioning units, are stable compounds and remain unchanged in the lower atmosphere. But when these chlorofluorocarbons are carried by convection currents to the stratosphere, they absorb ultraviolet radiation and produce chlorine atoms (chlorine free radicals), which in turn react with ozone. The following reaction sequence involving free radicals has been proposed to explain the partial destruction of the ozone layer by chlorofluorocarbons.

A free radical is a species containing an odd number of electrons. Free radicals are highly reactive.

$$CCl_3F \xrightarrow[\text{radiation}]{\text{ultraviolet}} \cdot CCl_2F + Cl \cdot \tag{1}$$

fluorocarbon   fluorocarbon   chlorine free
molecule   free radical   radical (atom)

$$Cl \cdot + O_3 \longrightarrow ClO \cdot + O_2 \tag{2}$$

$$ClO \cdot + O \longrightarrow O_2 + Cl \cdot \tag{3}$$

Because a chlorine atom is generated for each ozone molecule that is destroyed (reactions 2 and 3 can proceed repeatedly), a single chlorofluorocarbon molecule can be responsible for the destruction of many ozone molecules. During the past decade, scientists have discovered an annual thinning in the ozone layer over Antarctica. This is what we call the "hole" in the ozone layer. If this hole were to occur over populated regions of the world, severe effects would result, including a rise in the cancer rate, increased climatic temperatures, and vision problems. See Figure 12.15.

Ozone can be prepared by passing air or oxygen through an electrical discharge:

$$3\,O_2(g) + 286\text{ kJ} \xrightarrow[\text{discharge}]{\text{electrical}} 2\,O_3(g)$$

The characteristic pungent odor of ozone is noticeable in the vicinity of electrical machines and power transmission lines. Ozone is formed in the atmosphere during electrical storms and by the photochemical action of ultraviolet radiation on a mixture of nitrogen dioxide and oxygen. Areas with high air pollution are subject to high atmospheric ozone concentrations.

Ozone is not a desirable low-altitude constituent of the atmosphere because it is known to cause extensive plant damage, cracking of rubber, and the formation of eye-irritating substances. Concentrations of ozone greater than 0.1 part per million (ppm) of air cause coughing, choking, headache, fatigue, and reduced resistance to respiratory infection. Concentrations between 10 and 20 ppm are fatal to humans.

FIGURE 12.15 ▶
Satellite map showing a severe depletion or "hole" in the ozone layer over Antarctica from October 1979 to 1994. The hole is believed to be due to pollution of the atmosphere by chlorofluorocarbons used in aerosols and refrigerants.

◄ **Smog is becoming a problem in national parks like Yosemite in California as more and more people camp and drive in the parks.**

In addition to ozone, the air in urban areas contains nitrogen oxides, which are components of smog. The term *smog* refers to air pollution in urban environments. Often the chemical reactions occur as part of a *photochemical process*. Nitrogen monoxide (NO) is oxidized in the air or in automobile engines to produce nitrogen dioxide (NO_2). In the presence of light,

$$NO_2 \xrightarrow{\text{light}} NO + O$$

In addition to nitrogen oxides, combustion of fossil fuels releases CO_2, CO, and sulfur oxides. Incomplete combustion releases unburned and partially burned hydrocarbons.

Society is continually attempting to discover, understand, and control emissions that contribute to this sort of atmospheric chemistry. It is a problem that each one of us faces as we look to the future if we want to continue to support life as we know it on our planet.

Concepts in Review

1. State the principal assumptions of the kinetic-molecular theory.

2. Estimate the relative rates of effusion of two gases of known molar mass.

3. Sketch and explain the operation of a mercury barometer.

4. List two factors that determine gas pressure in a vessel of fixed volume.

5. State Boyle's, Charles' and Gay-Lussac's laws. Use all of them in problems.

6. State the combined gas law. Indicate when it is used.

7. Use Dalton's law of partial pressures and the combined gas law to determine the dry STP volume of a gas collected over water.

8. State Avogadro's law.

9. Understand the mole–mass–volume relationship of gases.

10. Determine the density of any gas at STP

11. Determine the molar mass of a gas from its density at a known temperature and pressure.

12. Solve problems involving the ideal gas equation.

13. Make mole–volume, mass–volume, and volume–volume stoichiometric calculations from balanced chemical equations.

14. State two reasons why real gases may deviate from the behavior predicted for an ideal gas.

Key Terms

absolute zero (12.6)
allotrope (12.17)
1 (one) atmosphere (12.3)
atmospheric pressure (12.3)
Avogadro's law (12.11)
barometer (12.3)
Boyle's law (12.5)
Charles' law (12.6)

Dalton's law of partial pressures (12.10)
diffusion (12.2)
effusion (12.2)
Gay Lussac's law (12.7)
Gay Lussac's law of combining volumes (12.11)
Graham's law of effusion (12.2)
Henry's law (CIA)
ideal gas (12.2)

ideal gas equation (12.14)
kinetic-molecular theory (KMT) (12.2)
molar volume (12.12)
partial pressure (12.10)
pressure (12.3)
standard conditions (12.8)
standard temperature and pressure (STP) (12.8)

Questions

1. What evidence is used to show diffusion in Figure 12.1? If H_2 and O_2 were in the two flasks, how could you prove that diffusion had taken place?

2. How does the air pressure inside the balloon shown in Figure 12.2 compare with the air pressure outside the balloon? Explain.

3. According to Table 12.1, what two gases are the major constituents of dry air?

4. How does the pressure represented by 1 torr compare in magnitude to the pressure represented by 1 mm Hg? See Table 12.2.

5. In which container illustrated in Figure 12.5 are the molecules of gas moving faster? Assume both gases to be hydrogen.

6. In Figure 12.6, what gas pressure corresponds to a volume of 4 L?

7. How do the data illustrated in Figure 12.6 substantiate Boyle's law?

8. What effect would you observe in Figure 12.9 if T_2 were lower than T_1?

9. In the diagram shown in Figure 12.11, is the pressure of the oxygen plus water vapor inside the bottle equal to, greater than, or less than the atmospheric pressure outside the bottle? Explain.

10. List five gases in Table 12.3 that are more dense than air. Explain the basis for your selection.

11. What are the basic assumptions of the kinetic-molecular theory?

12. Arrange the following gases, all at standard temperature, in order of increasing relative molecular velocities: H_2, CH_4, Rn, N_2, F_2, He. What is your basis for determining the order?

13. List, in descending order, the average kinetic energies of the molecules in Question 12.

14. What are the four parameters used to describe the behavior of a gas?

15. What are the characteristics of an ideal gas?

16. Under what condition of temperature, high or low, is a gas least likely to exhibit ideal behavior? Explain.

17. Under what condition of pressure, high or low, is a gas least likely to exhibit ideal behavior? Explain.

18. Compare, at the same temperature and pressure, equal volumes of H_2 and O_2 as to
    (a) number of molecules
    (b) mass
    (c) number of moles
    (d) average kinetic energy of the molecules
    (e) rate of effusion
    (f) density

19. How does the kinetic-molecular theory account for the behavior of gases as described by
    (a) Boyle's law?
    (b) Charles' law?
    (c) Dalton's law of partial pressures?

20. Explain how the reaction

    $$N_2(g) + O_2(g) \xrightarrow{\Delta} 2\,NO(g)$$

    proves that nitrogen and oxygen are diatomic molecules.

21. What is the reason for comparing gases to STP?

22. Is the conversion of oxygen to ozone an exothermic or endothermic reaction? How do you know?

23. When constant pressure is maintained, what effect does heating a mole of N_2 gas have on
    (a) its density?
    (b) its mass?
    (c) the average kinetic energy of its molecules?
    (d) the average velocity of its molecules?
    (e) the number of N_2 molecules in the sample?

24. Write formulas for an oxygen atom, an oxygen molecule, and an ozone molecule. How many electrons are in an oxygen molecule?

Paired Exercises

Pressure Units

25. The barometer reads 715 mm Hg. Calculate the corresponding pressure in
    (a) atmospheres
    (b) inches of Hg
    (c) $lb/in.^2$

26. The barometer reads 715 mm Hg. Calculate the corresponding pressure in
    (a) torrs
    (b) millibars
    (c) kilopascals

27. Express the following pressures in atmospheres:
    (a) 28 mm Hg
    (b) 6000. cm Hg
    (c) 795 torr
    (d) 5.00 kPa

28. Express the following pressures in atmospheres:
    (a) 62 mm Hg
    (b) 4250. cm Hg
    (c) 225 torr
    (d) 0.67 kPa

Boyle's and Charles' Laws

29. A gas occupies a volume of 400. mL at 500. mm Hg pressure. What will be its volume, at constant temperature, if the pressure is changed to (a) 760 mm Hg? (b) 250 torr?

30. A gas occupies a volume of 400. mL at 500. mm Hg pressure. What will be its volume, at constant temperature, if the pressure is changed to (a) 2.00 atm? (b) 325 torr?

31. A 500.-mL sample of a gas is at a pressure of 640. mm Hg. What must be the pressure, at constant temperature, if the volume is changed to 855 mL?

32. A 500.-mL sample of a gas is at a pressure of 640. mm Hg. What must be the pressure, at constant temperature, if the volume is changed to 450. mL?

33. Given 6.00 L of N_2 gas at $-25°C$, what volume will the nitrogen occupy at (a) 0.0°C? (b) 100. K? (Assume constant pressure.)

34. Given 6.00 L of N_2 gas at $-25°C$, what volume will the nitrogen occupy at (a) 0.0°F? (b) 345. K? (Assume constant pressure.)

Combined Gas Laws

35. A gas occupies a volume of 410 mL at 27°C and 740 mm Hg pressure. Calculate the volume the gas would occupy at STP.

36. A gas occupies a volume of 410 mL at 27°C and 740 mm Hg pressure. Calculate the volume the gas would occupy at 250.°C and 680 mm Hg pressure.

37. An expandable balloon contains 1400. L of He at 0.950 atm pressure and 18°C. At an altitude of 22 miles (temperature 2.0°C and pressure 4.0 torr), what will be the volume of the balloon?

38. A gas occupies 22.4 L at 2.50 atm and 27°C. What will be its volume at 1.50 atm and $-5.00°C$?

Dalton's Law of Partial Pressures

39. What would be the partial pressure of N_2 gas collected over water at 20°C and 720. torr pressure? (Check Appendix III for the vapor pressure of water.)

40. What would be the partial pressure of N_2 gas collected over water at 25°C and 705. torr pressure? (Check Appendix III for the vapor pressure of water.)

41. A mixture contains H_2 at 600. torr pressure, N_2 at 200. torr pressure, and O_2 at 300. torr pressure. What is the total pressure of the gases in the system?

42. A mixture contains H_2 at 325. torr pressure, N_2 at 475. torr pressure, and O_2 at 650. torr pressure. What is the total pressure of the gases in the system?

43. A sample of methane gas, CH_4, was collected over water at 25.0°C and 720. torr. The volume of the wet gas is 2.50 L. What will be the volume of the dry methane at standard pressure?

44. A sample of propane gas, C_3H_8, was collected over water at 22.5°C and 745 torr. The volume of the wet gas is 1.25 L. What will be the volume of the dry propane at standard pressure?

Mole–Mass–Volume Relationships

45. What volume will 2.5 mol of Cl_2 occupy at STP?

46. What volume will 1.25 mol of N_2 occupy at STP?

47. How many grams of CO_2 are present in 2500 mL of CO_2 at STP?

48. How many grams of NH_3 are present in 1.75 L of NH_3 at STP?

49. What volume will each of the following occupy at STP?
    (a) 1.0 mol of NO_2
    (b) 17.05 g of NO_2
    (c) 1.20×10^{24} molecules of NO_2

50. What volume will each of the following occupy at STP?
    (a) 0.50 mol of H_2S
    (b) 22.41 g of H_2S
    (c) 8.55×10^{23} molecules of H_2S

51. How many molecules of NH_3 gas are present in a 1.00-L flask of NH_3 gas at STP?

52. How many molecules of CH_4 gas are present in a 1.00-L flask of CH_4 gas at STP?

Density of Gases

53. Calculate the density of the following gases at STP:
    (a) Kr
    (b) SO_3

54. Calculate the density of the following gases at STP:
    (a) He
    (b) C_4H_8

55. Calculate the density of
    (a) F_2 gas at STP
    (b) F_2 gas at 27°C and 1.00 atm pressure

56. Calculate the density of
    (a) Cl_2 gas at STP
    (b) Cl_2 gas at 22°C and 0.500 atm pressure

Ideal Gas Equation and Stoichiometry

57. At 27°C and 750 torr pressure, what will be the volume of 2.3 mol of Ne?

58. At 25°C and 725 torr pressure, what will be the volume of 0.75 mol of Kr?

59. What volume will a mixture of 5.00 mol of H_2 and 0.500 mol of CO_2 occupy at STP?

60. What volume will a mixture of 2.50 mol of N_2 and 0.750 mol of HCl occupy at STP?

61. Given the equation:

    $$4\,NH_3(g) + 5\,O_2(g) \longrightarrow 4\,NO(g) + 6\,H_2O(g)$$

    (a) How many moles of NH_3 are required to produce 5.5 mol of NO?
    (b) How many liters of NO can be made from 12 L of O_2 and 10. L of NH_3 at STP?
    (c) At constant temperature and pressure, what is the maximum volume, in liters, of NO that can be made from 3.0 L of NH_3 and 3.0 L of O_2?

62. Given the equation:

    $$4\,NH_3(g) + 5\,O_2(g) \longrightarrow 4\,NO(g) + 6\,H_2O(g)$$

    (a) How many moles of NH_3 will react with 7.0 mol of O_2?
    (b) At constant temperature and pressure, how many liters of NO can be made by the reaction of 800. mL of O_2?
    (c) How many grams of O_2 must react to produce 60. L of NO measured at STP?

63. Given the equation:

    $$4\,FeS(s) + 7\,O_2(g) \xrightarrow{\Delta} 2\,Fe_2O_3(s) + 4\,SO_2(g)$$

    how many liters of O_2, measured at STP, will react with 0.600 kg of FeS?

64. Given the equation:

    $$4\,FeS(s) + 7\,O_2(g) \xrightarrow{\Delta} 2\,Fe_2O_3(s) + 4\,SO_2(g)$$

    how many liters of SO_2, measured at STP, will be produced from 0.600 kg of FeS?

Additional Exercises

65. Sketch a graph to show each of the following relationships:
   (a) P vs. V at constant temperature and number of moles
   (b) T vs. V at constant pressure and number of moles
   (c) T vs. P at constant volume and number of moles
   (d) n vs. V at constant temperature and pressure

66. Why is it dangerous to incinerate an aerosol can?

67. What volume does 1 mol of an ideal gas occupy at standard conditions?

68. Which of these occupies the greatest volume?
   (a) 0.2 mol of chlorine gas at 48°C and 80 cm Hg
   (b) 4.2 g of ammonia at 0.65 atm and −112°C
   (c) 21 g of sulfur trioxide at room temperature and 110 kPa

69. Which of these has the greatest density?
   (a) SF_6 at STP
   (b) C_2H_6 at room conditions
   (c) He at −80°C and 2.15 atm

70. A chemist carried out a chemical reaction that produced a gas. It was found that the gas contained 80.0% carbon and 20.0% hydrogen. It was also noticed that 1500 mL of the gas at STP had a mass of 2.01 g.
   (a) What is the empirical formula of the compound?
   (b) What is the molecular formula of the compound?
   (c) What Lewis structure fits this compound?

*71. Three gases were added to the same 2.0-L container. The total pressure of the gases was 790 torr at room temperature (25.0°C). If the mixture contained 0.65 g of oxygen gas, 0.58 g of carbon dioxide, and an unknown amount of nitrogen gas, determine:
   (a) the total number of moles of gas in the container
   (b) the number of grams of nitrogen in the container
   (c) the partial pressure of each gas in the mixture

*72. When carbon monoxide and oxygen gas react, carbon dioxide results. If 500. mL of O_2 at 1.8 atm and 15°C are mixed with 500. mL of CO at 800 mm Hg and 60°C, how many milliliters of CO_2 at STP could possibly result?

73. One of the methods for estimating the temperature at the center of the Sun is based on the ideal gas equation. If the center is assumed to be a mixture of gases whose average molar mass is 2.0 g/mol, and if the density and pressure are 1.4 g/cm^3 and 1.3×10^9 atm, respectively, calculate the temperature.

74. A soccer ball of constant volume 2.24 L is pumped up with air to a gauge pressure of 13 lb/in.^{2} at 20.0°C. The molar mass of air is about 29 g/mol.
   (a) How many moles of air are in the ball?
   (b) What mass of air is in the ball?
   (c) During the game, the temperature rises to 30.0°C. What mass of air must be allowed to escape to bring the gauge pressure back to its original value?

75. A balloon will burst at a volume of 2.00 L. If it is partially filled at 20.0°C and 65 cm Hg to occupy 1.75 L, at what temperature will it burst if the pressure is exactly 1 atm at the time that it breaks?

76. At constant temperature, what pressure would be required to compress 2500 L of hydrogen gas at 1.0 atm pressure into a 25-L tank?

77. Given a sample of a gas at 27°C, at what temperature would the volume of the gas sample be doubled, the pressure remaining constant?

78. A gas sample at 22°C and 740 torr pressure is heated until its volume is doubled. What pressure would restore the sample to its original volume?

79. A gas occupies 250 mL at 700. torr and 22°C. When the pressure is changed to 500. torr, what temperature (°C) is needed to maintain the same volume?

80. Hydrogen stored in a metal cylinder has a pressure of 252 atm at 25°C. What will be the pressure in the cylinder when the cylinder is lowered into liquid nitrogen at −196°C?

81. The tires on an automobile were filled with air to 30. psi at 71.0°F. When driving at high speeds, the tires become hot. If the tires have a bursting pressure of 44 psi, at what temperature (°F) will the tires "blow out"?

82. What volume would 5.30 L of H_2 gas at STP occupy at 70°C and 830 torr pressure?

83. What pressure will 800. mL of a gas at STP exert when its volume is 250. mL at 30°C?

84. How many gas molecules are present in 600. mL of N_2O at 40°C and 400. torr pressure? How many atoms are present? What would be the volume of the sample at STP?

85. 5.00 L of CO_2 at 500. torr and 3.00 L of CH_4 at 400. torr are put into a 10.0-L container. What is the pressure exerted by the gases in the container?

86. A steel cylinder contains 60.0 mol of H_2 at a pressure of 1500 lb/in.^{2}.
   (a) How many moles of H_2 are in the cylinder when the pressure reads 850 lb/in.^{2}?
   (b) How many grams of H_2 were initially in the cylinder?

*87. How many moles of Cl_2 are in one cubic meter (1.00 m^3) of Cl_2 gas at STP?

88. At STP, 560. mL of a gas have a mass of 1.08 g. What is the molar mass of the gas?

*89. At what temperature (°C) will the density of methane, CH_4, be 1.0 g/L at 1.0 atm pressure?

90. A gas has a density at STP of 1.78 g/L. What is its molar mass?

91. Using the ideal gas equation, $PV = nRT$, calculate:
   (a) the volume of 0.510 mol of H_2 at 47°C and 1.6 atm pressure
   (b) the number of grams in 16.0 L of CH_4 at 27°C and 600. torr pressure
   *(c) the density of CO_2 at 4.00 atm pressure and −20.0°C
   *(d) the molar mass of a gas having a density of 2.58 g/L at 27°C and 1.00 atm pressure.

92. What is the molar mass of a gas if 1.15 g occupy 0.215 L at 0.813 atm and 30.0°C?

93. What is the Kelvin temperature of a system in which 4.50 mol of a gas occupy 0.250 L at 4.15 atm?

94. How many moles of N_2 gas occupy 5.20 L at 250 K and 0.500 atm?

*95. Acetylene, C_2H_2, and hydrogen fluoride, HF, react to give difluoroethane.

$$C_2H_2(g) + 2\ HF(g) \longrightarrow C_2H_4F_2(g)$$

When 1.0 mol of C_2H_2 and 5.0 mol of HF are reacted in a 10.0-L flask, what will be the pressure in the flask at 0°C when the reaction is complete?

96. What volume of hydrogen at STP can be produced by reacting 8.30 mol of Al with sulfuric acid? The equation is

$$2\ Al(s) + 3\ H_2SO_4(aq) \longrightarrow Al_2(SO_4)_3(aq) + 3\ H_2(g)$$

97. What are the relative rates of effusion of N_2 and He?

*98. (a) What are the relative rates of effusion of CH_4 and He?
   (b) If these two gases are simultaneously introduced into opposite ends of a 100.-cm tube and allowed to diffuse toward each other, at what distance from the helium end will molecules of the two gases meet?

*99. A gas has a percent composition by mass of 85.7% carbon and 14.3% hydrogen. At STP the density of the gas is 2.50 g/L. What is the molecular formula of the gas?

*100. Assume that the reaction

$$2\ CO(g) + O_2(g) \longrightarrow 2\ CO_2(g)$$

goes to completion. When 10. mol of CO and 8.0 mol of O_2 react in a closed 10.-L vessel,
   (a) how many moles of CO, O_2, and CO_2 are present at the end of the reaction?
   (b) what will be the total pressure in the flask at 0°C?

*101. If 250 mL of O_2, measured at STP, is obtained by the decomposition of the $KClO_3$ in a 1.20-g mixture of KCl and $KClO_3$,

$$2\ KClO_3(s) \longrightarrow 2\ KCl(s) + 3\ O_2(g)$$

what is the percent by mass of $KClO_3$ in the mixture?

*102. Examine the apparatus shown. When a small amount of water is squirted into the flask containing ammonia gas (by squeezing the bulb of the medicine dropper), water from the beaker fills the flask through the long glass tubing. Explain this phenomenon. (Remember that ammonia dissolves in water.)

*103. Determine the pressure of the gas in each of the figures below:

*104. Air has a density of 1.29 g/L at STP. Calculate the density of air on Pikes Peak, where the pressure is 450 torr and the temperature is 17°C.

*105. Consider the arrangement of gases shown below. If the valve between the gases is opened and the temperature is held constant:
(a) determine the pressure of each gas
(b) determine the total pressure in the system

*106. A steel cylinder contained 50.0 L of oxygen gas under a pressure of 40.0 atm and a temperature of 25°C. What was the pressure in the cylinder during a storeroom fire that caused the temperature to rise 152°C? (Be careful!)

Answers to Practice Exercises

12.1 (a) 851 torr, (b) 851 mm Hg
12.2 0.596 atm
12.3 4.92 L
12.4 762 torr
12.5 84.7 L
12.6 861 K (588°C)
12.7 518 mL

12.8 1.4×10^2 g/mol
12.9 0.89 g/L
12.10 0.953 mol
12.11 128 g/mol
12.12 1.56 L O_2
12.13 75.0 L O_2, 45.0 L CO_2, 60.0 L H_2O

Planet Earth, that magnificent blue sphere we enjoy viewing from space, is spectacular. Over 75% of Earth is covered with water. We are born from it, drink it, bathe in it, cook with it, enjoy its beauty in waterfalls and rainbows, and stand in awe of the majesty of icebergs. Water supports and enhances life.

In chemistry, water provides the medium for numerous reactions. The shape of the water molecule is the basis for hydrogen bonds. These bonds determine the unique properties and reactions of water. The tiny water molecule holds the answers to many of the mysteries of chemical reactions.

13.1 What Is a Liquid?

In the last chapter we found that gases contain particles that are far apart, in rapid random motion, and independent of each other. The kinetic molecular theory, along with the ideal gas equation, summarizes the behavior of most gases at relatively high temperatures and low pressures.

Solids are obviously very different from gases. Solids contain particles that are very close together; solids have a high density, compress negligibly, and maintain their shape regardless of container. These characteristics indicate large attractive forces between particles. The model for solids is very different from the one for gases.

Liquids, on the other hand, lie somewhere between the extremes of gases and solids. Liquids contain particles that are close together; liquids are essentially incompressible, and have definite volume. These properties are very similar to solids. But liquids also take the shape of their containers; this is closer to the model of a gas.

Although liquids and solids show similar properties, they differ tremendously from gases. No simple mathematical relationship, like the ideal gas equation, works well for liquids or solids. Instead these models are directly related to the forces of attraction between molecules. With these general statements in mind, let's consider some specific properties of liquids.

13.2 Evaporation

When beakers of water, ethyl ether, and ethyl alcohol are allowed to stand uncovered, their volumes gradually decrease. The process by which this change takes place is called *evaporation.*

Attractive forces exist between molecules in the liquid state. Not all of these molecules, however, have the same kinetic energy. Molecules that have greater-than-average kinetic energy can overcome the attractive forces and break away from the surface of the liquid to become a gas. **Evaporation** or **vaporization** is the escape of molecules from the liquid state to the gas or vapor state.

evaporation
vaporization

In evaporation, molecules of higher-than-average kinetic energy escape from a liquid, leaving it cooler than it was before they escaped. For this reason, evaporation

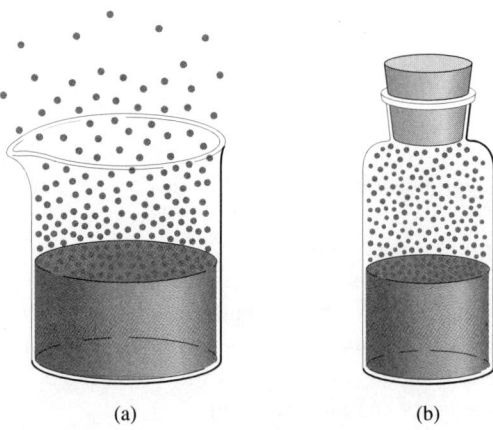

(a)　　　(b)

of perspiration is one way the human body cools itself and keeps its temperature constant. When volatile liquids such as ethyl chloride, C_2H_5Cl, are sprayed on the skin, they evaporate rapidly, cooling the area by removing heat. The numbing effect of the low temperature produced by evaporation of ethyl chloride allows it to be used as a local anesthetic for minor surgery.

Solids such as iodine, camphor, naphthalene (moth balls), and, to a small extent, even ice will go directly from the solid to the gaseous state, bypassing the liquid state. This change is a form of evaporation and is called **sublimation**:

sublimation

$$\text{liquid} \xrightarrow{\text{evaporation}} \text{vapor}$$

$$\text{solid} \xrightarrow{\text{sublimation}} \text{vapor}$$

13.3　Vapor Pressure

When a liquid vaporizes in a closed system like that shown in Figure 13.1b, some of the molecules in the vapor or gaseous state strike the surface and return to the liquid state by the process of **condensation.** The rate of condensation increases until it's equal to the rate of vaporization. At this point, the space above the liquid is said to be saturated with vapor, and an equilibrium, or steady state, exists between the liquid and the vapor. The equilibrium equation is

condensation

$$\text{liquid} \underset{\text{condensation}}{\overset{\text{vaporization}}{\rightleftarrows}} \text{vapor}$$

This equilibrium is dynamic; both processes—vaporization and condensation—are taking place, even though we cannot see or measure a change. The number of molecules leaving the liquid in a given time interval is equal to the number of molecules returning to the liquid.

At equilibrium the molecules in the vapor exert a pressure like any other gas. The pressure exerted by a vapor in equilibrium with its liquid is known as the **vapor pressure** of the liquid. The vapor pressure may be thought of as a measure of the "escaping" tendency of molecules to go from the liquid to the vapor state. The vapor pressure of a liquid is independent of the amount of liquid and vapor present, but it

vapor pressure

(a) Evacuated flask   (b) Water added at 20°C   (c) Water-vapor equilibrium at 20° C   (d) Water-vapor equilibrium at 30° C

Labels in figure: To vacuum; Manometer; Hg; 17.5 torr; 31.8 torr

FIGURE 13.2
Measurement of the vapor pressure of water at 20°C and 30°C. (a) The system is evacuated. The mercury manometer attached to the flask shows equal pressure in both legs. (b) Water has been added to the flask and begins to evaporate, exerting pressure as indicated by the manometer. (c) When equilibrium is established, the pressure inside the flask remains constant at 17.5 torr. (d) The temperature is changed to 30°C, and equilibrium is reestablished with the vapor pressure at 31.8 torr.

increases as the temperature rises. Figure 13.2 illustrates a liquid–vapor equilibrium and the measurement of vapor pressure.

When equal volumes of water, ethyl ether, and ethyl alcohol are placed in separate beakers and allowed to evaporate at the same temperature, we observe that the ether evaporates faster than the alcohol, which evaporates faster than the water. This order of evaporation is consistent with the fact that ether has a higher vapor pressure at any particular temperature than ethyl alcohol or water. One reason for this higher vapor pressure is that the attraction is less between ether molecules than between alcohol or water molecules. The vapor pressures of these three compounds at various temperatures are compared in Table 13.1.

Substances that evaporate readily are said to be **volatile.** A volatile liquid has a relatively high vapor pressure at room temperature. Ethyl ether is a very volatile liquid, water is not too volatile, and mercury, which has a vapor pressure of 0.0012 torr at 20°C, is essentially a nonvolatile liquid. Most substances that are normally in a solid state are nonvolatile (solids that sublime are exceptions).

volatile

13.4   Surface Tension

Have you ever observed water and mercury in the form of small drops? These liquids form drops because liquids have *surface tension*. A droplet of liquid that is not falling or under the influence of gravity (as on the space shuttle) will form a sphere. Spheres minimize the ratio of surface area to volume. The molecules within the liquid are attracted to the surrounding liquid molecules, but at the liquid's surface, the attraction is nearly all inward. This pulls the surface into a spherical shape. The resistance of a liquid to an increase in its surface area is called the **surface tension** of the liquid. Substances with large attractive forces between molecules have high surface tensions. The effect of surface tension in water is illustrated by floating a

surface tension

FIGURE 13.3
The meniscus of mercury (left) and water (right). The meniscus is the characteristic curve of the surface of a liquid in a narrow tube.

TABLE 13.1   Vapor Pressure of Water, Ethyl Alcohol, and Ethyl Ether at Various Temperatures

Temperature (°C)	Vapor pressure (torr)		
	Water	Ethyl alcohol	Ethyl ether*
0	4.6	12.2	185.3
10	9.2	23.6	291.7
20	17.5	43.9	442.2
30	31.8	78.8	647.3
40	55.3	135.3	921.3
50	92.5	222.2	1276.8
60	152.9	352.7	1729.0
70	233.7	542.5	2296.0
80	355.1	812.6	2993.6
90	525.8	1187.1	3841.0
100	760.0	1693.3	4859.4
110	1074.6	2361.3	6070.1

*Note that the vapor pressure of ethyl ether at temperatures of 40°C and higher exceeds standard pressure, 760 torr, which indicates that the substance has a low boiling point and should therefore be stored in a cool place in a tightly sealed container.

needle on the surface of still water. Other examples include a water strider walking across a calm pond and water beading on a freshly waxed car.

capillary action   Liquids also exhibit a phenomenon called **capillary action,** the spontaneous rising of a liquid in a narrow tube. This action results from the *cohesive forces* within the liquid, and the *adhesive forces* between the liquid and the walls of the container. If the forces between the liquid and the container are greater than those within the liquid itself, the liquid will climb the walls of the container. For example, consider the California sequoia, a tree that reaches over 200 feet in height. Although water rises only 33 feet in a glass tube (under atmospheric pressure), capillary action causes water to rise from the sequoia's roots to all its parts.

The meniscus in liquids is further evidence of cohesive and adhesive forces. When a liquid is placed in a glass cylinder, the surface of the liquid shows a curve

A water strider skims the ▶ surface of the water as a result of surface tension. At the molecular level the surface tension results from the net attraction of the water molecules toward the liquid below. In the interior of the water, the forces are balanced in all directions.

TABLE 13.2	Physical Properties of Ethyl Chloride, Ethyl Ether, Ethyl Alcohol, and Water			
Substance	**Boiling point (°C)**	**Melting point (°C)**	**Heat of vaporization J/g (cal/g)**	**Heat of fusion J/g (cal/g)**
Ethyl chloride	12.3	−139	387 (92.5)	—
Ethyl ether	34.6	−116	351 (83.9)	—
Ethyl alcohol	78.4	−112	855 (204.3)	104 (24.9)
Water	100.0	0	2259 (540)	335 (80)

called the **meniscus** (see Figure 13.3). The concave shape of water's meniscus shows that the adhesive forces between the glass and water are stronger than the cohesive forces within the water. In a nonpolar substance such as mercury, the meniscus is convex, indicating that the cohesive forces within mercury are greater than the adhesive forces between the glass wall and the mercury.

meniscus

13.5 Boiling Point

The boiling temperature of a liquid is related to its vapor pressure. We've seen that vapor pressure increases as temperature increases. When the internal or vapor pressure of a liquid becomes equal to the external pressure, the liquid boils. (By external pressure we mean the pressure of the atmosphere above the liquid.) The boiling temperature of a pure liquid remains constant as long as the external pressure does not vary.

The boiling point (bp) of water is 100°C at 1 atm pressure. Table 13.1 shows that the vapor pressure of water at 100°C is 760 torr, a figure we have seen many times before. The significant fact here is that the boiling point is the temperature at which the vapor pressure of the water or other liquid is equal to standard, or atmospheric, pressure at sea level. These relationships lead to the following definition: **Boiling point** is the temperature at which the vapor pressure of a liquid is equal to the external pressure above the liquid.

boiling point

We can readily see that a liquid has an infinite number of boiling points. When we give the boiling point of a liquid, we should also state the pressure. When we express the boiling point without stating the pressure, we mean it to be the **normal boiling point** at standard pressure (760 torr). Using Table 13.1 again, we see that the normal boiling point of ethyl ether is between 30°C and 40°C, and for ethyl alcohol it is between 70°C and 80°C because for each compound 760 torr lies within these stated temperature ranges. At the normal boiling point, 1 g of a liquid changing to a vapor (gas) absorbs an amount of energy equal to its heat of vaporization (see Table 13.2).

normal boiling point

The boiling point at various pressures can be evaluated by plotting the data of Table 13.2 on the graph in Figure 13.4, where temperature is plotted horizontally along the x-axis and vapor pressure is plotted vertically along the y-axis. The resulting curves are known as **vapor-pressure curves.** Any point on these curves represents a vapor–liquid equilibrium at a particular temperature and pressure. We can

vapor-pressure curves

FIGURE 13.4 ▶
Vapor-pressure–temperature curves for ethyl ether, ethyl alcohol, and water.

FIGURE 13.4 ▶
Vapor-pressure–temperature curves for ethyl ether, ethyl alcohol, and water.

find the boiling point at any pressure by tracing a horizontal line from the designated pressure to a point on the vapor-pressure curve. From this point we draw a vertical line to obtain the boiling point on the temperature axis. Three such points are shown in Figure 13.4; they represent the normal boiling points of the three compounds at 760 torr. By reversing this process, you can ascertain at what pressure a substance will boil at a specific temperature. The boiling point is one of the most commonly used physical properties for characterizing and identifying substances.

Practice 13.1
_____

Use the graph in Figure 13.4 to determine the boiling points of ethyl ether, ethyl alcohol, and water at 600 torr.

Practice 13.2
_____

The average atmospheric pressure in Denver is 0.83 atm. What is the boiling point of water in Denver?

13.6   Freezing Point or Melting Point

As heat is removed from a liquid, the liquid becomes colder and colder, until a temperature is reached at which it begins to solidify. A liquid changing into a solid is said to be *freezing,* or *solidifying.* When a solid is heated continuously, a temperature is reached at which the solid begins to liquefy. A solid that is changing into a liquid is said to be *melting.* The temperature at which the solid phase of a substance is in equilibrium with its liquid phase is known as the **freezing point** or **melting point** of that substance. The equilibrium equation is

freezing or melting point

$$\text{solid} \underset{\text{freezing}}{\overset{\text{melting}}{\rightleftharpoons}} \text{liquid}$$

When a solid is slowly and carefully heated so that a solid–liquid equilibrium is achieved and then maintained, the temperature will remain constant as long as both phases are present. The energy is used solely to change the solid to the liquid. The melting point is another physical property that is commonly used for characterizing substances.

The most common example of a solid–liquid equilibrium is ice and water. In a well-stirred system of ice and water, the temperature remains at 0°C as long as both phases are present. The melting point changes only slightly with pressure unless the pressure change is very large.

Ice crystals forming at the edge of a pond.

13.7 Changes of State

The majority of solids undergo two changes of state upon heating. A solid changes to a liquid at its melting point, and a liquid changes to a gas at its boiling point. This warming process can be represented by a graph called a *heating curve* (Figure 13.5). This figure shows ice being heated at a constant rate. As energy flows into the ice, the vibrations within the crystal increase and the temperature rises ($A \longrightarrow B$). Eventually, the molecules begin to break free from the crystal and melting occurs ($B \longrightarrow C$). During the melting process all energy goes into breaking down the crystal structure; the temperature remains constant.

The energy required to change exactly one gram of a solid at its melting point into a liquid is called the **heat of fusion.** When the solid has completely melted, the temperature once again rises ($C \longrightarrow D$); the energy input is increasing the molecular

heat of fusion

◄ **FIGURE 13.5**
Heating curve for a pure substance—the absorption of heat by a substance from the solid state to the vapor state. Using water as an example, the AB interval represents the ice phase; BC interval, the melting of ice to water; CD interval, the elevation of the temperature of water from 0°C to 100°C; DE interval, the boiling of water to steam; and EF interval, the heating of steam.

Comet Power

Why do comets have tails? How do comets produce such celestial wonders? A comet near the Sun sublimes some of the ice on its surface into the vacuum of space. This rapidly moving water vapor drags comet dust along with it, creating two characteristic features—the comet's tail and dusty shroud (usually called a coma). Scientists understand that the production of a tail on a comet traveling near the Sun is the result of the rapid conversion of ice to water vapor from the Sun's heat. But why do comets far from the Sun have tails when there isn't enough heat to transform ice to water vapor?

New observations from the University of Hawaii, Honolulu, suggest that the

release of carbon monoxide creates tails and shrouds in comets far from the Sun. Astronomers Matthew C. Senay and David Jewitt used two short-wavelength radio telescopes to detect the emissions

of gases from Comet Schwassmann-Wachmann 1. This particular comet never gets closer to the Sun than Jupiter. Their telescopes revealed that this comet releases a startling amount of carbon monoxide. Carbon monoxide sublimes at a temperature of 25 K. The carbon monoxide coming from the comet has about the same speed as the comet but moves in general toward the Sun. Senay and Jewitt suggest that the gas comes from the surface of the comet most illuminated by the Sun. The Sun sublimes the carbon monoxide at a rate of nearly 2000 kg/sec, producing a tail that is always oriented toward the Sun.

motion within the water. At 100°C, the water reaches its boiling point; the temperature remains constant while the added energy is used to vaporize the water to steam ($D \longrightarrow E$). The **heat of vaporization** is the energy required to change exactly one gram of liquid to vapor at its normal boiling point. The attractive forces between the liquid molecules are overcome during vaporization. Beyond this temperature all the water exists as steam and is being heated further ($E \longrightarrow F$).

heat of vaporization

Example 13.1   How many joules of energy are needed to change 10.0 g of ice at 0.00°C to water at 20.0°C?

Solution   Ice will absorb 335 J/g (heat of fusion) in going from a solid at 0°C to a liquid at 0°C. An additional 4.184 J/g°C (specific heat of water) is needed to raise the temperature of the water by 1°C.

Joules needed to melt the ice:

$$(10.0 \text{ g})\left(\frac{335 \text{ J}}{1 \text{ g}}\right) = 3.35 \times 10^3 \text{ J}$$

Joules needed to heat the water from 0.00°C to 20.0°C:

$$(10.0 \text{ g})\left(\frac{4.184 \text{ J}}{1 \text{ g°C}}\right)(20.0\text{°C}) = 837 \text{ J}$$

Thus 3350 J + 837 J = 4.19×10^3 J is needed.

Example 13.2   How many kilojoules of energy are needed to change 20.0 g of water at 20.°C to steam at 100.°C?

The specific heat of water is 4.184 J/g°C so the kilojoules needed to heat the water **Solution**
from 20.°C to 100.°C are:

$$(20.0 \text{ g})\left(\frac{4.184 \text{ J}}{1 \text{ g°C}}\right)\left(\frac{1 \text{ kJ}}{1000 \text{ J}}\right)(100. - 20.0°C) = 6.7 \text{ kJ}$$

Heat of vaporization of water is 2.26 kJ/g, so kilojoules needed to change water at
100.°C to steam at 100.°C are

$$(20.0 \text{ g})\left(\frac{2.26 \text{ kJ}}{1 \text{ g}}\right) = 45.2 \text{ kJ}$$

Thus 6.7 kJ + 45.2 kJ = 51.9 kJ is needed.

Practice 13.3

How many kilojoules of energy are required to change 50.0 g of ethyl alcohol
from 60.0°C to vapor at 78.4°C? The specific heat of ethyl alcohol is 2.138
J/g°C.

13.8 Occurrence of Water

Water is our most common natural resource. It covers about 75% of Earth's surface.
Not only is it found in the oceans and seas, in lakes, rivers, streams, and in glacial ice
deposits, it is always present in the atmosphere and in cloud formations.

About 97% of Earth's water is in the oceans. This *saline* water contains vast
amounts of dissolved minerals. More than 70 elements have been detected in the
mineral content of seawater. Only four of these—chlorine, sodium, magnesium, and
bromine—are now commercially obtained from the sea. The world's *fresh* water
comprises the other 3%, of which about two-thirds is locked up in polar ice caps and
glaciers. The remaining fresh water is found in groundwater, lakes, and the atmosphere.

Water is an essential constituent of all living matter. It is the most abundant
compound in the human body, making up about 70% of total body mass. About 92%
of blood plasma is water; about 80% of muscle tissue is water; and about 60% of a
red blood cell is water. Water is more important than food in the sense that we can
survive much longer without food than without water.

13.9 Physical Properties of Water

Water is a colorless, odorless, tasteless liquid with a melting point of 0°C and a
boiling point of 100°C at 1 atm. The heat of fusion of water is 335 J/g (80 cal/g). The
heat of vaporization of water is 2.26 kJ/g (540 cal/g). The values for water for both
the heat of fusion and the heat of vaporization are high compared with those for other
substances; this indicates strong attractive forces between the molecules.

FIGURE 13.6 ▶
Water equilibrium systems. In the beaker on the left, ice and water are in equilibrium at 0°C; in the beaker on the right, boiling water and steam are in equilibrium at 100°C.

Ice and water exist together in equilibrium at 0°C, as shown in Figure 13.6. When ice at 0°C melts, it absorbs 335 J/g in changing into a liquid; the temperature remains at 0°C. To refreeze the water, 335 J/g must be removed from the liquid at 0°C.

In Figure 13.6, both boiling water and steam are shown to have a temperature of 100°C. It takes 418 J to heat 1 g of water from 0°C to 100°C, but water at its boiling point absorbs 2.26 kJ/g in changing to steam. Although boiling water and steam are both at the same temperature, steam contains considerably more heat per gram and can cause more severe burns than hot water. Table 13.3 compares the physical properties of water with those of other hydrogen compounds of Group VIA elements.

The maximum density of water is 1.000 g/mL at 4°C. Water has the unusual property of contracting in volume as it is cooled to 4°C and then expanding when cooled from 4°C to 0°C. Therefore 1 g of water occupies a volume greater than 1 mL at all temperatures except 4°C. Although most liquids contract in volume all the way

TABLE 13.3	Physical Properties of Water and Other Hydrogen Compounds of Group VIA Elements				
Formula	**Color**	**Molar mass (g/mol)**	**Melting point (°C)**	**Boiling point, 1 atm (°C)**	**Heat of fusion J/g (cal/g)**	**Heat of vaporization J/g (cal/g)**
H_2O	Colorless	18.02	0.00	100.0	335 (80.0)	2.26×10^3 (540)
H_2S	Colorless	34.09	−85.5	−60.3	69.9 (16.7)	548 (131)
H_2Se	Colorless	80.98	−65.7	−41.3	31 (7.4)	238 (57.0)
H_2Te	Colorless	129.6	−49	−2	—	179 (42.8)

◄**FIGURE 13.7**
**Diagrams of a water
molecule: (a) electron
distribution, (b) bond
angle and O—H bond
length, (c) molecular
orbital structure, and
(d) dipole representation.**

(a)   (b)   (c)   (d)

down to the point at which they solidify, a large increase (about 9%) in volume occurs when water changes from a liquid at 0°C to a solid (ice) at 0°C. The density of ice at 0°C is 0.917 g/mL, which means that ice, being less dense than water, will float in water.

13.10   Structure of the Water Molecule

A single water molecule consists of two hydrogen atoms and one oxygen atom. Each hydrogen atom is attached to the oxygen atom by a single covalent bond. This bond is formed by the overlap of the $1s$ orbital of hydrogen with an unpaired $2p$ orbital of oxygen. The average distance between the two nuclei is known as the *bond length*. The O—H bond length in water is 0.096 nm. The water molecule is nonlinear and has a bent structure with angle of about 105 degrees between the two bonds (see Figure 13.7).

Oxygen is the second most electronegative element. As a result, the two covalent OH bonds in water are polar. If the three atoms in a water molecule were aligned in a linear structure, such as H +⟶ O ⟵+ H, the two polar bonds would be acting in equal and opposite directions and the molecule would be nonpolar. However, water is a highly polar molecule. It therefore does not have a linear structure. When atoms are bonded in a nonlinear fashion, the angle formed by the bonds is called the *bond angle*. In water the HOH bond angle is 105°. The two polar covalent bonds and the bent structure result in a partial negative charge on the oxygen atom and a partial positive charge on each hydrogen atom. The polar nature of water is responsible for many of its properties, including its behavior as a solvent.

13.11   The Hydrogen Bond

Table 13.3 compares the physical properties of H_2O, H_2S, H_2Se, and H_2Te. From this comparison it is apparent that four physical properties of water—melting point, boiling point, heat of fusion, and heat of vaporization—are extremely high and do not fit the trend relative to the molar masses of the four compounds. If the properties of water followed the progression shown by the other three compounds, we would expect the melting point of water to be below −85°C and the boiling point to be below −60°C.

Why does water exhibit these anomalies? Because liquid water molecules are held together more strongly than other molecules in the same family. The intermolecular force acting between water molecules is called a **hydrogen bond,** which acts like a very weak bond between two polar molecules. A hydrogen bond is formed between polar molecules that contain hydrogen covalently bonded to a small, highly electronegative atom such as fluorine, oxygen, or nitrogen (F—H, O—H, N—H). A

hydrogen bond

How Sweet It Is!

Did you think artificial sweeteners were a product of the post–World War II chemical industry? Not so—many of them have been around a long time, and several of the important ones were discovered quite by accident. In 1878, Ira Remsen was working late in his laboratory and realized he was about to miss a dinner with friends. In his haste to leave the lab, he forgot to wash his hands. Later at dinner he broke a piece of bread and tasted it only to discover that it was very sweet. The sweet taste had to be the chemical he had been working with in the lab. Back at the lab, he isolated saccharin—the first of the artificial sweeteners.

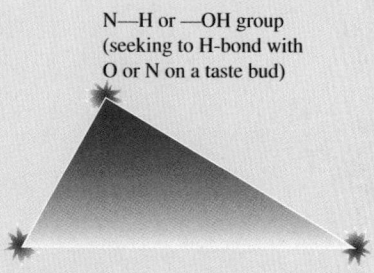

N—H or —OH group
(seeking to H-bond with
O or N on a taste bud)

O or N atom   Any hydrophobic group
(seeking to H-bond   (e.g., CH_3, C_6H_5)
with polar H on a
taste bud)

▲
Triangle of sweetness

In 1937, Michael Sveda was smoking a cigarette in his laboratory (a very dangerous practice to say the least!). He touched the cigarette to his lips and was surprised by the exceedingly sweet taste. The chemical on his hands turned out to be cyclamate, which soon became a staple of the artificial sweetener industry.

In 1965, James Schlatter was researching anti-ulcer drugs for the pharmaceutical firm G. D. Searle. In the course of his work he accidentally ingested a small amount of a preparation and found to his surprise that it had an extremely sweet taste. He had discovered aspartame, a molecule consisting of two amino acids joined together. Since only very small quantities of aspartame are necessary to produce sweetness, it proved to be an excellent low-calorie artificial sweetener. Today under the trade names of "Equal" and "Nutrasweet," aspartame is a cornerstone of the artificial sweetener industry.

More than 50 different molecules have a sweet taste, and all of them have similar molecular shapes. The triangle of sweetness theory, developed by Lamont Kier (Massachusetts College of Pharmacy) indicates that three sites on these molecules produce a structure that at-

taches to the taste bud and triggers the response that registers "sweet" in our brains.

Our taste buds are composed of proteins that can form hydrogen bonds with other molecules. The proteins contain —N—H and —OH groups (with hydrogen available to bond) as well as C=O groups (providing oxygen for hydrogen bonding). "Sweet molecules" also contain H-bonding groups including —OH, —NH_2, and O or N. These molecules not only must have the proper atoms to form hydrogen bonds, they must also contain a hydrophobic region (repels H_2O). The triangle of sweetness in the diagram shows the three necessary sites that must be located at just the proper distances.

hydrogen bond is actually the dipole–dipole attraction between polar molecules containing these three types of polar bonds.

> **Compounds that have significant hydrogen-bonding ability are those that contain H covalently bonded to F, O, or N.**

Because a hydrogen atom has only one electron, it forms only one covalent bond. When it is attached to a strong electronegative atom such as oxygen, a hydrogen atom will also be attracted to an oxygen atom of another molecule, forming a dipole–dipole attraction (H-bond) between the two molecules. Water has two types of bonds: covalent bonds that exist between hydrogen and oxygen atoms within a molecule and hydrogen bonds that exist between hydrogen and oxygen atoms in *different* water molecules.

(a)   (b)

◀ **FIGURE 13.8**
Hydrogen bonding. Water in the liquid and solid states exists as aggregates in which the water molecules are linked together by hydrogen bonds.

Hydrogen bonds are *intermolecular* bonds; that is, they are formed between atoms in different molecules. They are somewhat ionic in character because they are formed by electrostatic attraction. Hydrogen bonds are much weaker than the ionic or covalent bonds that unite atoms to form compounds. Despite their weakness, they are of great chemical importance.

The oxygen atom in water can form two hydrogen bonds—one through each of the unbonded pairs of electrons. Figure 13.8 shows (a) two water molecules linked by a hydrogen bond and (b) six water molecules linked by hydrogen bonds. A dash (—) is used for the covalent bond and a dotted line (••••) for the hydrogen bond. In water each molecule is linked to others through hydrogen bonds to form a three-dimensional aggregate of water molecules. This intermolecular hydrogen bonding effectively gives water the properties of a much larger, heavier molecule, explaining in part its relatively high melting point, boiling point, heat of fusion, and heat of vaporization. As water is heated and energy absorbed, hydrogen bonds are continually being broken until at 100°C, with the absorption of an additional 2.26 kJ/g, water separates into individual molecules, going into the gaseous state. Sulfur, selenium, and tellurium are not sufficiently electronegative for their hydrogen compounds to behave like water. The lack of hydrogen bonding is one reason why H_2S is a gas and not a liquid at room temperature.

Fluorine, the most electronegative element, forms the strongest hydrogen bonds. This bonding is strong enough to link hydrogen fluoride molecules together as *dimers*, H_2F_2, or as larger $(HF)_n$ molecular units. The dimer structure may be represented in this way:

Hydrogen bonding can occur between two different atoms that are capable of forming H-bonds. Thus we may have an O••••H—N or O—H••••N linkage in which the hydrogen atom forming the H-bond is between an oxygen and a nitrogen atom. This form of H-bond exists in certain types of protein molecules and many biologically active substances.

Example 13.3   Would you expect hydrogen bonding to occur between molecules of these substances?

(a)
$$H-\overset{\displaystyle \overset{H}{|}}{\underset{\displaystyle \underset{H}{|}}{C}}-\overset{\displaystyle \overset{H}{|}}{\underset{\displaystyle \underset{H}{|}}{C}}-\overset{..}{\underset{..}{O}}-H$$
ethyl alcohol

(b)
$$H-\overset{\displaystyle \overset{H}{|}}{\underset{\displaystyle \underset{H}{|}}{C}}-\overset{..}{\underset{..}{O}}-\overset{\displaystyle \overset{H}{|}}{\underset{\displaystyle \underset{H}{|}}{C}}-H$$
dimethyl ether

Solution   (a) Hydrogen bonding should occur in ethyl alcohol because one hydrogen atom is bonded to an oxygen atom:

$$H-\overset{\overset{H}{|}}{\underset{\underset{H}{|}}{C}}-\overset{\overset{H}{|}}{\underset{\underset{H}{|}}{C}}-\overset{..}{\underset{..}{O}}-H\cdots\overset{..}{\underset{..}{O}}-\overset{\overset{H}{|}}{\underset{\underset{H}{|}}{C}}-\overset{\overset{H}{|}}{\underset{\underset{H}{|}}{C}}-H$$

H-bond

(b) There is no hydrogen bonding in dimethyl ether because all the hydrogen atoms are bonded only to carbon atoms.

Both ethyl alcohol and dimethyl ether have the same molar mass (46.07). Although both compounds have the same molecular formula, C_2H_6O, ethyl alcohol has a much higher boiling point (78.4°C) than dimethyl ether (−23.7°C) because of hydrogen bonding between the alcohol molecules.

Practice 13.4

Would you expect hydrogen bonding to occur between molecules of these substances?

(a)
$$H-\overset{\displaystyle \overset{H}{|}}{\underset{\displaystyle \underset{H}{|}}{C}}-\overset{\displaystyle \overset{H}{|}}{\underset{\displaystyle \underset{H}{|}}{C}}-\overset{\displaystyle \overset{H}{|}}{\underset{..}{N}}-H$$

(b)
$$H-\overset{\displaystyle \overset{H}{|}}{\underset{\displaystyle \underset{H}{|}}{C}}-\overset{\displaystyle \overset{H}{|}}{\underset{..}{N}}-\overset{\displaystyle \overset{H}{|}}{\underset{\displaystyle \underset{H}{|}}{C}}-H$$

(c)
$$H-\overset{\overset{H}{|}}{\underset{\underset{H}{|}}{C}}-\overset{\overset{H}{|}}{\underset{\underset{H}{|}}{C}}-\overset{\overset{H-\overset{H}{\underset{|}{C}}-H}{|}}{\underset{\underset{H-\overset{H}{\underset{|}{C}}-H}{|}}{N:}}$$

13.12 Formation and Chemical Properties of Water

Water is very stable to heat; it decomposes to the extent of only about 1% at temperatures up to 2000°C. Pure water is a nonconductor of electricity. But when a small amount of sulfuric acid or sodium hydroxide is added, the solution is readily decomposed into hydrogen and oxygen by an electric current. Two volumes of hydrogen are produced for each volume of oxygen:

$$2\ H_2O(l) \xrightarrow[\text{H}_2\text{SO}_4\text{ or NaOH}]{\text{electrical energy}} 2\ H_2(g) + O_2(g)$$

Formation

Water is formed when hydrogen burns in air. Pure hydrogen burns very smoothly in air, but mixtures of hydrogen and air or oxygen explode when ignited. The reaction is strongly exothermic:

$$2\ H_2(g) + O_2(g) \longrightarrow 2\ H_2O(g) + 484\ \text{kJ}$$

Water is produced by a variety of other reactions, especially by (1) acid–base neutralizations, (2) combustion of hydrogen-containing materials, and (3) metabolic oxidation in living cells:

1. $HCl(aq) + NaOH(aq) \longrightarrow NaCl(aq) + H_2O(l)$
2. $2\ C_2H_2(g) + 5\ O_2(g) \longrightarrow 4\ CO_2(g) + 2\ H_2O(g) + 1212\ \text{kJ}$
   acetylene

   $CH_4(g) + 2\ O_2(g) \longrightarrow CO_2(g) + 2\ H_2O(g) + 803\ \text{kJ}$
   methane
3. $C_6H_{12}O_6(aq) + 6\ O_2(g) \xrightarrow{\text{enzymes}} 6\ CO_2(g) + 6\ H_2O(l) + 2519\ \text{kJ}$
   glucose

The combustion of acetylene shown in (2) is strongly exothermic and is capable of producing very high temperatures. It is used in oxygen–acetylene torches to cut and weld steel and other metals. Methane is known as natural gas and is commonly used as fuel for heating and cooking. The reaction of glucose with oxygen shown in (3) is the reverse of photosynthesis. It is the overall reaction by which living cells obtain needed energy by metabolizing glucose to carbon dioxide and water.

Reactions of Water with Metals and Nonmetals

The reactions of metals with water at different temperatures show that these elements vary greatly in their reactivity. Metals such as sodium, potassium, and calcium react with cold water to produce hydrogen and a metal hydroxide. A small piece of sodium added to water melts from the heat produced by the reaction, forming a silvery metal ball, which rapidly flits back and forth on the surface of the water. Caution must be used when experimenting with this reaction, because the hydrogen produced is frequently ignited by the sparking of the sodium, and it will explode, spattering sodium. Potassium reacts even more vigorously than sodium. Calcium sinks in water and liberates a gentle stream of hydrogen. The equations for these reactions are

$$2\ Na(s) + 2\ H_2O(l) \longrightarrow H_2(g) + 2\ NaOH(aq)$$

$$2\ K(s) + 2\ H_2O(l) \longrightarrow H_2(g) + 2\ KOH(aq)$$

$$Ca(s) + 2\ H_2O(l) \longrightarrow H_2(g) + Ca(OH)_2(aq)$$

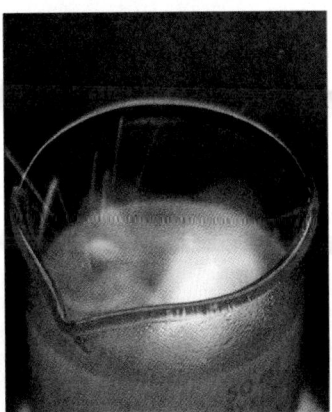

▲ **Sodium reacts vigorously with water to produce hydrogen gas and sodium hydroxide.**

Zinc, aluminum, and iron do not react with cold water but will react with steam at high temperatures, forming hydrogen and a metallic oxide. The equations are

$$Zn(s) + H_2O(g) \longrightarrow H_2(g) + ZnO(s)$$

$$2\ Al(s) + 3\ H_2O(g) \longrightarrow 3\ H_2(g) + Al_2O_3(s)$$

$$3\ Fe(s) + 4\ H_2O(g) \longrightarrow 4\ H_2(g) + Fe_3O_4(s)$$

Copper, silver, and mercury are examples of metals that do not react with cold water or steam to produce hydrogen. We conclude that sodium, potassium, and calcium are chemically more reactive than zinc, aluminum, and iron, which are more reactive than copper, silver, and mercury.

Certain nonmetals react with water under various conditions. For example, fluorine reacts violently with cold water, producing hydrogen fluoride and free oxygen. The reactions of chlorine and bromine are much milder, producing what is commonly known as "chlorine water" and "bromine water," respectively. Chlorine water contains HCl, HOCl, and dissolved Cl_2; the free chlorine gives it a yellow-green color. Bromine water contains HBr, HOBr, and dissolved Br_2; the free bromine gives it a reddish-brown color. Steam passed over hot coke (carbon) produces a mixture of carbon monoxide and hydrogen that is known as "water gas." Since water gas is combustible, it is useful as a fuel. It's also the starting material for the commercial production of several alcohols. The equations for these reactions are

$$2\ F_2(g) + 2\ H_2O(l) \longrightarrow 4\ HF(aq) + O_2(g)$$

$$Cl_2(g) + H_2O(l) \longrightarrow HCl(aq) + HOCl(aq)$$

$$Br_2(l) + H_2O(l) \longrightarrow HBr(aq) + HOBr(aq)$$

$$C(s) + H_2O(g) \xrightarrow{1000°C} CO(g) + H_2(g)$$

Reactions of Water with Metal and Nonmetal Oxides

basic anhydride

Metal oxides that react with water to form hydroxides are known as **basic anhydrides.** Examples are

$$CaO(s) + H_2O(l) \longrightarrow Ca(OH)_2(aq)$$
calcium hydroxide

$$Na_2O(s) + H_2O(l) \longrightarrow 2\ NaOH(aq)$$
sodium hydroxide

Certain metal oxides, such as CuO and Al_2O_3, do not form solutions containing OH^- ions because the oxides are insoluble in water.

acid anhydride

Nonmetal oxides that react with water to form acids are known as **acid anhydrides.** Examples are

$$CO_2(g) + H_2O(l) \rightleftharpoons H_2CO_3(aq)$$
carbonic acid

$$SO_2(g) + H_2O(l) \rightleftharpoons H_2SO_3(aq)$$
sulfurous acid

$$N_2O_5(s) + H_2O(l) \longrightarrow 2\ HNO_3(aq)$$
nitric acid

The word *anhydrous* means "without water." An anhydride is a metal oxide or a nonmetal oxide derived from a base or an oxy-acid by the removal of water. To determine the formula of an anhydride, the elements of water are removed from an acid or base formula until all the hydrogen is removed. Sometimes more than one formula unit is needed to remove all the hydrogen as water. The formula of the anhydride then consists of the remaining metal or nonmetal and the remaining oxygen atoms. In calcium hydroxide, removal of water as indicated leaves CaO as the anhydride:

$$Ca \underset{\overline{}}{\overset{O\boxed{H}}{\diagdown}} \xrightarrow{\Delta} CaO + H_2O$$

In sodium hydroxide, H_2O cannot be removed from one formula unit, so two formula units of NaOH must be used, leaving Na_2O as the formula of the anhydride:

$$\begin{matrix} NaO\boxed{H} \\ Na\boxed{OH} \end{matrix} \xrightarrow{\Delta} Na_2O + H_2O$$

The removal of H_2O from H_2SO_4 gives the acid anhydride SO_3:

$$H_2SO_4 \xrightarrow{\Delta} SO_3 + H_2O$$

The foregoing are examples of typical reactions of water but are by no means a complete list of the known reactions of water.

13.13 Hydrates

When certain solutions containing ionic compounds are allowed to evaporate, some water molecules remain as part of the crystalline compound that is left after evaporation is complete. Solids that contain water molecules as part of their crystalline structure are known as **hydrates.** Water in a hydrate is known as **water of hydration, or water of crystallization.**

> hydrate
> water of hydration
> water of crystallization

Formulas for hydrates are expressed by first writing the usual anhydrous (without water) formula for the compound and then adding a dot followed by the number of water molecules present. An example is $BaCl_2 \cdot 2\ H_2O$. This formula tells us that each formula unit of this compound contains one barium ion, two chloride ions, and two water molecules. A crystal of the compound contains many of these units in its crystalline lattice.

TABLE 13.4	Selected Hydrates		
Hydrate	**Name**	**Hydrate**	**Name**
$CaCl_2 \cdot 2\ H_2O$	calcium chloride dihydrate	$Na_2CO_3 \cdot 10\ H_2O$	sodium carbonate decahydrate
$Ba(OH)_2 \cdot 8\ H_2O$	barium hydroxide octahydrate	$(NH_4)_2C_2O_4 \cdot H_2O$	ammonium oxalate monohydrate
$MgSO_4 \cdot 7\ H_2O$	magnesium sulfate heptahydrate	$NaC_2H_3O_2 \cdot 3\ H_2O$	sodium acetate trihydrate
$SnCl_2 \cdot 2\ H_2O$	tin(II) chloride dihydrate	$Na_2B_4O_7 \cdot 10\ H_2O$	sodium tetraborate decahydrate
$CoCl_2 \cdot 6\ H_2O$	cobalt(II) chloride hexahydrate	$Na_2S_2O_3 \cdot 5\ H_2O$	sodium thiosulfate pentahydrate

Hot Ice—It's a Gas!

Ice could well become the most abundant source of energy in the 21st century. This particular type of ice has unusual properties. Even its existence has been the subject of debate since 1810 when Humphrey Day synthesized the first hydrate of chlorine. Chemists have long debated whether water could crystallize around a gas to form an unusual state of matter known as a gas hydrate. The material looks just like a chunk of ice left from a winter storm, gray and ugly. But unlike ice this gas hydrate ice pops and sizzles and a lit match causes it to burst into flames. Left alone it melts quickly into a puddle of water.

Just what is gas hydrate ice and how does it form? It turns out that gas hydrates form when water and some gases (like propane, ethane, or methane) are present at high pressure and low temperature. These conditions occur in nature at the bottom of the ocean and under permafrost in the Arctic north. Oil companies struggle with hydrate ice since it

▲ **Nuggets of gas hydrate burn as they revert to water (or ice) and gas.**

plugs pipelines and damages oil rigs in oil fields in the Arctic. A gas hydrate is a gas molecule trapped inside the crystal lattice of water molecules. The most common gas is methane produced from the decomposition of organic matter. Sonar mapping has revealed large gas hydrate deposits in polar regions such as Alaska and Siberia and off our southeastern coast.

Scientists think that gas hydrates trap huge amounts of natural gas. If this gas could be brought to the surface economically it could supply 1000 years of fuel for the United States. In Russia production of natural gas from gas hydrates has already begun. As prices increase for energy sources the costs for liberating gas hydrates become more feasible. At this point most American research is still focused on prevention of gas hydrate formation. But one day in the not too distant future we could be heating our homes and running our cars by mining gas hydrates from ice!

In naming hydrates, we first name the compound exclusive of the water and then add the term *hydrate*, with the proper prefix representing the number of water molecules in the formula. For example, $BaCl_2 \cdot 2\ H_2O$ is called *barium chloride dihydrate*. Hydrates are true compounds and follow the law of definite composition. The molar mass of $BaCl_2 \cdot 2\ H_2O$ is 244.2 g/mol; it contains 56.22% barium, 29.03% chlorine, and 14.76% water.

Water molecules in hydrates are bonded by electrostatic forces between polar water molecules and the positive or negative ions of the compound. These forces are not as strong as covalent or ionic chemical bonds. As a result water of crystallization can be removed by moderate heating of the compound. A partially dehydrated or completely anhydrous compound may result. When $BaCl_2 \cdot 2\ H_2O$ is heated, it loses its water at about 100°C:

$$BaCl_2 \cdot 2\ H_2O(s) \xrightarrow{100°C} BaCl_2(s) + 2\ H_2O(g)$$

When a solution of copper(II) sulfate ($CuSO_4$) is allowed to evaporate, beautiful blue crystals containing 5 moles water per 1 mole $CuSO_4$ are formed (Figure 13.9a). The formula for this hydrate is $CuSO_4 \cdot 5\ H_2O$; it is called copper(II) sulfate pentahydrate. When $CuSO_4 \cdot 5\ H_2O$ is heated, water is lost, and a pale green-white powder, anhydrous $CuSO_4$, is formed:

$$CuSO_4 \cdot 5\ H_2O(s) \xrightarrow{250°C} CuSO_4(s) + 5\ H_2O(g)$$

(b)

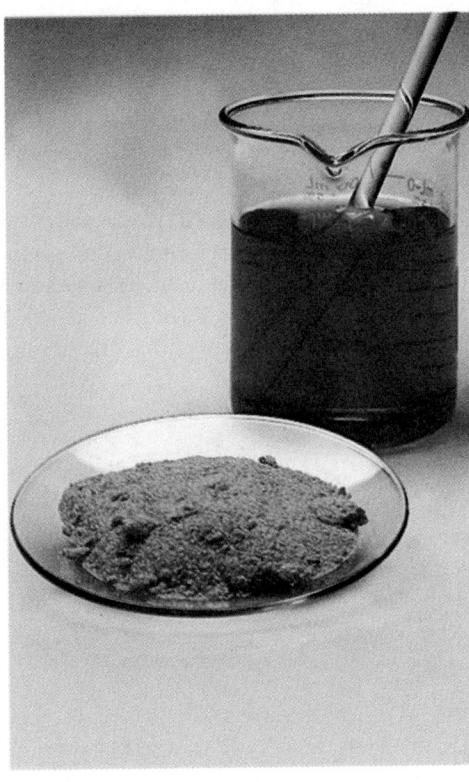

(a)

FIGURE 13.9
(a) When these blue crystals of $CuSO_4 \cdot 5\ H_2O$ are dissolved in water a blue solution forms. (b) The anhydrous crystals of $CuSO_4$ are pale green. When water is added they immediately change color to blue $CuSO_4 \cdot 5\ H_2O$ crystals

When water is added to anhydrous copper(II) sulfate, the foregoing reaction is reversed, and the compound turns blue again (Figure 13.9b). Because of this outstanding color change, anhydrous copper(II) sulfate has been used as an indicator to detect small amounts of water. The formation of the hydrate is noticeably exothermic.

The formula for plaster of paris is $(CaSO_4)_2 \cdot H_2O$. When mixed with the proper quantity of water, plaster of paris forms a dihydrate and sets to a hard mass. It is therefore useful for making patterns for the production of art objects, molds, and surgical casts. The chemical reaction is

$$(CaSO_4)_2 \cdot H_2O(s) + 3\ H_2O(l) \longrightarrow 2\ CaSO_4 \cdot 2\ H_2O(s)$$

Table 13.4 lists a number of common hydrates.

13.14 Hygroscopic Substances

Many anhydrous compounds and other substances readily absorb water from the atmosphere. Such substances are said to be **hygroscopic.** This property can be observed in this simple experiment: Spread a 10–20 g sample of anhydrous copper(II) sulfate on a watch glass and set it aside so that the compound is exposed to the air. Then determine the mass of the sample periodically for 24 hours, noting the increase in mass and the change in color. Over time water is absorbed from the atmosphere, forming the blue pentahydrate $CuSO_4 \cdot 5\ H_2O$.

Some compounds continue to absorb water beyond the hydrate stage to form solutions. A substance that absorbs water from the air until it forms a solution is said to

hygroscopic substance

deliquescence

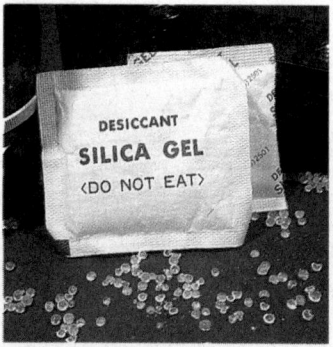

▲
Silica gel

be **deliquescent.** A few granules of anhydrous calcium chloride or pellets of sodium hydroxide exposed to the air will appear moist in a few minutes, and within an hour will absorb enough water to form a puddle of solution. Diphosphorus pentoxide (P_2O_5) picks up water so rapidly that its mass cannot be determined accurately except in an anhydrous atmosphere.

Compounds that absorb water are useful as drying agents (desiccants). Refrigeration systems must be kept dry with such agents or the moisture will freeze and clog the tiny orifices in the mechanism. Bags of drying agents are often enclosed in packages containing iron or steel parts to absorb moisture and prevent rusting. Anhydrous calcium chloride, magnesium sulfate, sodium sulfate, calcium sulfate, silica gel, and diphosphorus pentoxide are some of the compounds commonly used for drying liquids and gases that contain small amounts of moisture.

13.15   Natural Waters

Natural fresh waters are not pure, but contain dissolved minerals, suspended matter, and sometimes harmful bacteria. The water supplies of large cities are usually drawn from rivers or lakes. Such water is generally unsafe to drink without treatment. To make such water safe to drink, it is treated by some or all of the following processes (see Figure 13.10):

1. *Screening.* Removal of relatively large objects, such as trash, fish, and so on.
2. *Flocculation and sedimentation.* Chemicals, usually lime, CaO, and alum (aluminum sulfate), $Al_2(SO_4)_3$, are added to form a flocculent jellylike precipitate of aluminum hydroxide. This precipitate traps most of the fine suspended matter in the water and carries it to the bottom of the sedimentation basin.
3. *Sand filtration.* Water is drawn from the top of the sedimentation basin and passed downward through fine sand filters. Nearly all the remaining suspended matter and bacteria are removed by the sand filters.
4. *Aeration.* Water is drawn from the bottom of the sand filters and is aerated by spraying. The purpose of this process is to remove objectionable odors and tastes.
5. *Disinfection.* In the final stage chlorine gas is injected into the water to kill harmful bacteria before the water is distributed to the public. Ozone is also used in some countries to disinfect water. In emergencies water may be disinfected by simply boiling it for a few minutes.

If the drinking water of children contains an optimum amount of fluoride ion, their teeth will be more resistant to decay.

FIGURE 13.10▶
Typical municipal water treatment plant.

CHEMISTRY IN ACTION Moisturizers

Moisturizers have long been used to protect and rehydrate the skin. These products contain compounds called humectants and emollients that increase the water content of the skin in different ways. Emollients cover the skin with a layer of material that is immiscible with water. This prevents water from within the skin from evaporating. In contrast, humectants add water to the skin by attracting water vapor from the air.

The most common humectants are sorbitol, glycerin, and polypropylene glycol. Each molecule is polar, containing multiple —OH groups (see formulas).

The oxygen atom in each —OH group is considerably more electronegative than the hydrogen atom. This elec-tronegativity difference results in a partial negative charge on the oxygen, whereas the hydrogen carries a partial positive charge. This polarity and the polarity of water molecules are the basis for the attraction between the humectant and water molecules.

Emollients are composed of hydrophobic (water-insoluble) molecules. These products are made of nonpolar molecules. A great diversity of compounds fall into this category, including animal oils, vegetable oils, exotic oils (such as jojoba and aloe vera), and synthetic oils. In each case the molecules form a water-insoluble layer on the skin, which traps the skin's own moisture and feels smooth to the touch.

Many consumers believe that alcohols can dry the skin and shouldn't be used in skin care products. Is this concern justified? What is the purpose of alcohols in skin care products? The problem is not so simple as we might expect. There are two different categories of alcohol with distinctly different properties. Fatty acid alcohols are large, essentially nonpolar molecules that behave like emollients. They form a water-insoluble layer on the skin that traps the moisture. A second type of alcohol, the simple alcohol, includes ethyl and isopropyl alcohols. These substances act as solvents in the skin care product and can be drying. They absorb excess oil, dissolve one ingredient into another, and make products evaporate. Some also act to keep products from spoiling and separating. The problem with alcohol in skin care products results from overuse or use by people who don't need them,

Glycerin

Propylene glycol

Sorbitol

One of the best tests for whether a skin care product will tend to dry the skin is to examine the texture. Liquids are typically the most drying formulations and are best used on oily skin. Gels containing lightweight emollients are best for skin with varying degrees of oiliness. Creams tend to contain heavier moisturizers and are best for normal to dry skin. Ointments are very heavy, creamy products that form a barrier to the skin, acting as a humectant. Ointments are for use on severely dry or damaged skin. The key to selecting the proper moisturizer lies in understanding the way in which the product functions.

Water that contains dissolved calcium and magnesium salts is called *hard water.* Ordinary soap does not lather well in hard water; the soap reacts with the calcium and magnesium ions to form an insoluble greasy scum. However, synthetic soaps, known as detergents, have excellent cleaning qualities and do not form precipitates with hard water. Hard water is also undesirable because it causes "scale" to form on the walls of water heaters, teakettles, coffee pots, and steam irons, which greatly reduces their efficiency.

FIGURE 13.11 ▶
Simple laboratory setup for distilling liquids.

Four techniques are used to "soften" hard water:

1. **Distillation** The water is boiled, and the steam formed is condensed into a liquid again, leaving the minerals behind in the distilling vessel. Figure 13.11 illustrates a simple laboratory distillation apparatus. Commercial stills are capable of producing hundreds of liters of distilled water per hour.

2. **Calcium and magnesium precipitation** Calcium and magnesium ions are precipitated from hard water by adding sodium carbonate and lime. Insoluble calcium carbonate and magnesium hydroxide are precipitated and are removed by filtration or sedimentation.

3. **Ion exchange** Hard water is effectively softened as it is passed through a bed or tank of zeolite—a complex sodium aluminum silicate. In this process sodium ions replace objectionable calcium and magnesium ions, and the water is thereby softened:

$$Na_2(zeolite)(s) + Ca^{2+}(aq) \longrightarrow Ca(zeolite)(s) + 2\,Na^+ (aq)$$

The zeolite is regenerated by back-flushing with concentrated sodium chloride solution, reversing the foregoing reaction.

4. **Demineralization** Both cations and anions are removed by a two-stage ion-exchange system. Special synthetic organic resins are used in the ion-exchange beds. In the first stage metal cations are replaced by hydrogen ions. In the second stage anions are replaced by hydroxide ions. The hydrogen and hydroxide ions react, and essentially pure, mineral-free water leaves the second stage.

Our oceans are an enormous source of water, but seawater contains about 3.5 lb of salts per 100 lb of water. This 35,000 ppm of dissolved salts makes seawater unfit for agricultural and domestic uses. Water that contains less than 1000 ppm of salts is considered reasonably good for drinking, and safe drinking water is already being obtained from the sea in many parts of the world. Continuous research is being done in an effort to make usable water from our oceans more abundant and economical. See Figure 13.12.

◀FIGURE 13.12
Catalina Island, California,
gets most of its water from
a desalinization plant.

13.16 Water Pollution

Polluted water was formerly thought of as water that was unclear, had a bad odor or taste, and contained disease-causing bacteria. However, such factors as increased population, industrial requirements for water, atmospheric pollution, toxic waste dumps, and use of pesticides have greatly expanded the problem of water pollution.

Many of the newer pollutants are not removed or destroyed by the usual water-treatment processes. For example, among the 66 organic compounds found in the drinking water of a major city on the Mississippi River, 3 are labeled slightly toxic, 17 moderately toxic, 15 very toxic, 1 extremely toxic, and 1 supertoxic. Two are known carcinogens (cancer-producing agents), 11 are suspect, and 3 are metabolized to carcinogens. The U.S. Public Health Service classifies water pollutants under eight broad categories. These categories are shown in Table 13.5.

Many outbreaks of disease or poisoning, such as typhoid, dysentery, and cholera have been attributed directly to drinking water. Rivers and streams are an easy means for municipalities to dispose of their domestic and industrial waste products. Much of this water is used again by people downstream, and then discharged back into the water source. Then another community still farther downstream draws the same water and discharges its own wastes. Thus along waterways such as the Mississippi and Delaware rivers, water is withdrawn and discharged many times. If this water is not properly treated, harmful pollutants can build up, causing epidemics of various diseases.

The disposal of hazardous waste products adds to the water pollution problem. These products are unavoidable in the manufacture of many products we consider indispensable today. One common way to dispose of these wastes is to place them in toxic waste dumps. What has been found after many years of disposing of wastes in

TABLE 13.5	Classification of Water Pollutants	

Type of pollutant	Examples
Oxygen-demanding wastes	Decomposable organic wastes from domestic sewage and industrial wastes of plant and animal origin
Infectious agents	Bacteria, viruses, and other organisms from domestic sewage, animal wastes, and animal process wastes
Plant nutrients	Principally compounds of nitrogen and phosphorus
Organic chemicals	Large numbers of chemicals synthesized by industry, pesticides, chlorinated organic compounds
Other minerals and chemicals	Inorganic chemicals from industrial operations, mining, oil field operations, and agriculture
Radioactive substances	Waste products from mining and processing radioactive materials, airborne radioactive fallout, increased use of radioactive materials in hospitals and research
Heat from industry	Large quantities of heated water returned to water bodies from power plants and manufacturing facilities after use for cooling
Sediment from land erosion	Solid matter washed into streams and oceans by erosion, rain, and water runoff

this manner is that toxic substances have seeped into the groundwater deposits. As a result many people have become ill, and water wells have been closed until satisfactory methods of detoxifying this water are found. This problem is serious, because one-half the United States population gets its drinking water from groundwater. Cleaning up the thousands of industrial dumps and finding and implementing new and safe methods of disposing of wastes is ongoing and costly.

Many major water pollutants have been recognized and steps have been taken to eliminate them. Three that pose serious problems are lead, detergents, and chlorine-containing organic compounds. Lead poisoning, for example, has been responsible for many deaths in past years. One major toxic action of lead in the body is the inhibition of the enzyme necessary for the production of hemoglobin in the blood. The usual intake of lead into the body is through food. However, extraordinary amounts of lead can be ingested from water running through lead pipes and by using lead-containing ceramic containers for storage of food and beverages.

It has been clearly demonstrated that waterways rendered so polluted that the water is neither fit for human use nor able to sustain marine life can be successfully restored. However, keeping our lakes and rivers free from pollution is a very costly and complicated process.

Concepts in Review

1. List the common properties of liquids and solids. Explain how they are different from gases.

2. Explain the process of evaporation from the standpoint of kinetic energy.

3. Relate vapor-pressure data or vapor-pressure curves of different substances to their relative rates of evaporation and to their relative boiling points.

4. Explain the forces involved in surface tension of a liquid. Give two examples.

5. Explain why a meniscus forms on the surface of liquids in a container.

6. Explain what is occurring throughout the heating curve for water.

7. Describe a water molecule with respect to the Lewis structure, bond angle, and polarity.

8. Sketch hydrogen bonding (a) between water molecules, (b) between hydrogen fluoride molecules, and (c) between ammonia molecules.

9. Explain the effect of hydrogen bonding on the physical properties of water.

10. Determine whether a compound will or will not form hydrogen bonds.

11. Identify metal oxides as basic anhydrides and write balanced equations for their reactions with water.

12. Identify nonmetal oxides as acid anhydrides and write balanced equations for their reactions with water.

13. Deduce the formula of the acid anhydride or basic anhydride when given the formula of the corresponding acid or base.

14. Identify the product, name each reactant and product, and write equations for the complete dehydration of hydrates.

15. Outline the processes necessary to prepare safe drinking water from a contaminated river source.

16. Describe how water may be softened by distillation, chemical precipitation, ion exchange, and demineralization.

17. Complete and balance equations for (a) the reaction of water with Na, K, and Ca; (b) the reaction of steam with Zn, Al, Fe, and C; and (c) the reaction of water with halogens.

Key Terms

acid anhydride (13.12)
basic anhydride (13.12)
boiling point (13.5)
capillary action (13.4)
condensation (13.3)
deliquescence (13.14)
evaporation (13.2)
freezing or melting point (13.6)

heat of fusion (13.7)
heat of vaporization (13.7)
hydrate (13.13)
hydrogen bond (13.11)
hygroscopic substance (13.14)
meniscus (13.4)
normal boiling point (13.5)
sublimation (13.2)

surface tension (13.4)
vapor pressure (13.3)
vapor-pressure curves (13.5)
vaporization (13.2)
volatile (13.3)
water of crystallization (13.13)
water of hydration (13.13)

Questions

1. Compare the potential energy of the three states of water shown in Figure 13.6.

2. In what state (solid, liquid, or gas) would H_2S, H_2Se, and H_2Te be at 0°C? (Table 13.3)

3. The temperature of the warer in the beaker on the hotplate (Figure 13.6) reads 100°C. What is the pressure of the atmosphere?

4. Diagram a water molecule and point out the negative and positive ends of the dipole.

5. If the water molecule were linear, with all three atoms in a straight line rather than in the shape of a V, as shown in Figure 13.7, what effect would this have on the physical properties of water?

6. How do we specify 1, 2, 3, 4, 5, 6, 7, and 8 molecules of water in the formulas of hydrates? (Table 13.4)

7. Would the distillation setup in Figure 13.11 be satisfactory for separating salt and water? for separating ethyl alcohol and water? Explain.

8. If the liquid in the flask in Figure 13.11 is ethyl alcohol and the atmospheric pressure is 543 torr, what temperature will show on the thermometer? (Use Figure 13.4.)

9. If water were placed in both containers in Figure 13.1, would both have the same vapor pressure at the same temperature? Explain.

10. In Figure 13.1, in which case, (a) or (b), will the atmosphere above the liquid reach a point of saturation?

11. Suppose a solution of ethyl ether and ethyl alcohol is placed in the closed bottle in Figure 13.1. (Use Figure 13.4 for information on the substances.)
 (a) Are both substances present in the vapor?
 (b) If the answer to part (a) is yes, which has more molecules in the vapor?

12. In Figure 13.2, if 50% more water is added in part (b), what equilibrium vapor pressure will be observed in (c)?

13. At approximately what temperature would each of the substances shown in Figure 13.4 boil when the pressure is 30 torr?

14. Use the graph in Figure 13.4 to find the
 (a) boiling point of water at 500 torr
 (b) normal boiling point of ethyl alcohol
 (c) boiling point of ethyl ether at 0.50 atm

15. Consider Figure 13.5.
 (a) Why is line BC horizontal? What is happening in this interval?
 (b) What phases are present in the interval BC?
 (c) When heating is continued after point C, another horizontal line, DE, is reached at a higher temperature. What does this line represent?

16. List six physical properties of water.

17. What condition is necessary for water to have its maximum density? What is its maximum density?

18. Account for the fact that an ice–water mixture remains at 0°C until all the ice is melted, even though heat is applied to it.

19. Which contains less heat, ice at 0°C or water at 0°C? Explain.

20. Why does ice float in water? Would ice float in ethyl alcohol ($d = 0.789$ g/mL)? Explain.

21. If water molecules were linear instead of bent, would the heat of vaporization be higher or lower? Explain.

22. The heat of vaporization for ethyl ether is 351 J/g and that for ethyl alcohol is 855 J/g. Which of these compounds has hydrogen bonding? Explain.

23. Would there be more or less H-bonding if water molecules were linear instead of bent? Explain.

24. Which would you expect to show hydrogen bonding: ammonia, NH_3, or methane, CH_4? Explain.

25. In which condition are there fewer hydrogen bonds between molecules: water at 40°C or water at 80°C?

26. Which compound,
$$H_2NCH_2CH_2NH_2 \quad \text{or} \quad CH_3CH_2CH_2NH_2,$$
would you expect to have the higher boiling point? Explain. (Both compounds have similar molar masses.)

27. Explain why rubbing alcohol warmed to body temperature still feels cold when applied to your skin.

28. The vapor pressure at 20°C for the following compounds is

methyl alcohol	96 torr
acetic acid	11.7 torr
benzene	74.7 torr
bromine	173 torr
water	17.5 torr
carbon tetrachloride	91 torr
mercury	0.0012 torr
toluene	23 torr

 (a) Arrange these compounds in order of increasing rate of evaporation.
 (b) Which substance listed has the highest boiling point? the lowest?

29. Suggest a method whereby water could be made to boil at 50°C.

30. Explain why a higher temperature is obtained in a pressure cooker than in an ordinary cooking pot.

31. What is the relationship between vapor pressure and boiling point?

32. On the basis of the kinetic molecular theory, explain why vapor pressure increases with temperature.

33. Why does water have such a relatively high boiling point?

34. The boiling point of ammonia, NH_3, is −33.4°C and that of sulfur dioxide, SO_2, is −10.0°C. Which has the higher vapor pressure at −40°C?

35. Explain what is occurring physically when a substance is boiling.

36. Explain why HF (bp = 19.4°C) has a higher boiling point than HCl (bp = −85°C), whereas F_2 (bp = −188°C) has a lower boiling point than Cl_2 (bp = −34°C).

37. Why does a boiling liquid maintain a constant temperature when heat is continuously being added?

38. At what specific temperature will ethyl ether have a vapor pressure of 760 torr?

39. Why does a lake freeze from the top down?

40. What water temperature would you theoretically expect to find at the bottom of a very deep lake? Explain.

41. Is the formation of hydrogen and oxygen from water an exothermic or an endothermic reaction? How do you know?

42. (a) What is an anhydride?
  (b) What type of compound will be an acid anhydride?
  (c) What type of compound will be a basic anhydride?

Paired Exercises

43. Write the formulas for the anhydrides of these acids: $HClO_4$, H_2CO_3, H_3PO_4.

44. Write the formulas for the anhydrides of these acids: H_2SO_3, H_2SO_4, HNO_3.

45. Write the formulas for the anhydrides of these bases: $LiOH$, $NaOH$, $Mg(OH)_2$.

46. Write the formulas for the anhydrides of these bases: KOH, $Ba(OH)_2$, $Ca(OH)_2$.

47. Complete and balance these equations:
  (a) $Ba(OH)_2 \xrightarrow{\Delta}$
  (b) $CH_3OH + O_2 \longrightarrow$
      methyl alcohol
  (c) $Rb + H_2O \longrightarrow$
  (d) $SnCl_2 \cdot 2\,H_2O \xrightarrow{\Delta}$
  (e) $HNO_3 + NaOH \longrightarrow$
  (f) $CO_2 + H_2O \longrightarrow$

48. Complete and balance these equations:
  (a) $Li_2O + H_2O \longrightarrow$
  (b) $KOH \xrightarrow{\Delta}$
  (c) $Ba + H_2O \longrightarrow$
  (d) $Cl_2 + H_2O \longrightarrow$
  (e) $SO_3 + H_2O \longrightarrow$
  (f) $H_2SO_3 + KOH \longrightarrow$

49. Name these hydrates:
  (a) $BaBr_2 \cdot 2\,H_2O$
  (b) $AlCl_3 \cdot 6\,H_2O$
  (c) $FePO_4 \cdot 4\,H_2O$

50. Name these hydrates:
  (a) $MgNH_4PO_4 \cdot 6\,H_2O$
  (b) $FeSO_4 \cdot 7\,H_2O$
  (c) $SnCl_4 \cdot 5\,H_2O$

51. Distinguish between deionized water and
  (a) hard water
  (b) soft water

52. Distinguish between deionized water and
  (a) distilled water
  (b) natural water

53. How many moles of compound are in 100. g of $CoCl_2 \cdot 6\,H_2O$?

54. How many moles of compound are in 100. g of $FeI_2 \cdot 4\,H_2O$?

55. How many moles of water can be obtained from 100. g of $CoCl_2 \cdot 6\,H_2O$?

56. How many moles of water can be obtained from 100. g of $FeI_2 \cdot 4\,H_2O$?

57. When a person purchases epsom salts, $MgSO_4 \cdot 7\,H_2O$, what percent of the compound is water?

58. Calculate the mass percent of water in the hydrate $Al_2(SO_4)_3 \cdot 18\,H_2O$.

59. Sugar of lead, a hydrate of lead(II) acetate, $Pb(C_2H_3O_2)_2$, contains 14.2% H_2O. What is the empirical formula for the hydrate?

60. A 25.0-g sample of a hydrate of $FePO_4$ was heated until no more water was driven off. The mass of anhydrous sample is 16.9 g. What is the empirical formula of the hydrate?

61. How many joules are needed to change 120. g of water at 20.°C to steam at 100.°C?

62. How many joules of energy must be removed from 126 g of water at 24°C to form ice at 0°C?

*63. Suppose 100. g of ice at 0°C is added to 300. g of water at 25°C. Is this sufficient ice to lower the temperature of the system to 0°C and still have ice remaining? Show evidence for your answer.

*64. Suppose 35.0 g of steam at 100.°C is added to 300. g of water at 25°C. Is this sufficient steam to heat all the water to 100.°C and still have steam remaining? Show evidence for your answer.

*65. If 75 g of ice at 0.0°C were added to 1.5 L of water at 75°C, what would be the final temperature of the mixture?

*66. If 9560 J of energy were absorbed by 500. g of ice at 0.0°C, what would be the final temperature?

67. How many grams of water will react with the following?
  (a) 1.00 g Na
  (b) 1.00 g MgO
  (c) 1.00 g N_2O_5

68. How many grams of water will react with the following?
  (a) 1.00 mol K
  (b) 1.00 mol Ca
  (c) 1.00 mol SO_3

Additional Exercises

69. Which causes a more severe burn, liquid water at 100°C or steam at 100°C? Why?

70. You have a shallow dish of alcohol set into a tray of water. If you blow across the tray, the alcohol evaporates, while the water cools significantly and eventually freezes. Explain why.

71. Regardless of how warm the outside temperature may be, we always feels cool when stepping out of a swimming pool, the ocean, or a shower. Why is this so?

72. Sketch a heating curve for a substance X whose melting point is 40°C and whose boiling point is 65°C.
    (a) Describe what you will observe as a 60.-g sample of X is warmed from 0°C to 100°C.
    (b) If the heat of fusion of X is 80. J/g, the heat of vaporization is 190. J/g, and if 3.5 J is required to warm 1 g of X each degree, how much energy will be needed to accomplish the change in (a)?

73. Why does the vapor pressure of a liquid increase as the temperature of it is increased?

74. At the top of Mount Everest, which is just about 29,000 feet above sea level, the atmospheric pressure is about 270 torr. Use Figure 13.4 to determine the approximate boiling temperature of water on Mount Everest.

75. Explain how anhydrous copper(II) sulfate, $CuSO_4$, can act as an indicator for moisture.

76. Write formulas for magnesium sulfate heptahydrate and disodium hydrogen phosphate dodecahydrate.

77. How can soap make soft water from hard water? What objections are there to using soap for this purpose?

78. What substance is commonly used to destroy bacteria in water?

79. What chemical, other than chlorine or chlorine compounds, can be used to disinfect water for domestic use?

80. Some organic pollutants in water can be oxidized by dissolved molecular oxygen. What harmful effect can result from this depletion of oxygen in the water?

81. Why should you not drink liquids that are stored in ceramic containers, especially unglazed ones?

82. Write the chemical equation showing how magnesium ions are removed by a zeolite water softener.

83. Write an equation to show how hard water containing calcium chloride, $CaCl_2$, is softened by using sodium carbonate, Na_2CO_3.

84. How many calories are required to change 225 g of ice at 0°C to steam at 100.°C?

85. The molar heat of vaporization is the number of joules required to change 1 mol of a substance from liquid to vapor at its boiling point. What is the molar heat of vaporization of water?

*86. The specific heat of zinc is 0.096 cal/g°C. Determine the energy required to raise the temperature of 250. g of zinc from room temperature (20.0°C) to 150.°C.

*87. Suppose 150. g of ice at 0.0°C is added to 0.120 L of water at 45°C. If the mixture is stirred and allowed to cool to 0.0°C, how many grams of ice remain?

88. How many joules of energy would be liberated by condensing 50.0 mol of steam at 100.0°C and allowing the liquid to cool to 30.0°C?

89. How many kilojoules of energy are needed to convert 100. g of ice at −10.0°C to water at 20.0°C? (The specific heat of ice at −10.0°C is 2.01 J/g°C.)

90. What mass of water must be decomposed to produce 25.0 L of oxygen at STP?

*91. Suppose 1.00 mol of water evaporates in 1.00 day. How many water molecules, on the average, leave the liquid each second?

92. Compare the volume occupied by 1.00 mol of liquid water at 0°C and 1.00 mol of water vapor at STP.

*93. A quantity of sulfuric acid is added to 100. mL of water. The final volume of the solution is 122 mL and has a density of 1.26 g/mL. What mass of acid was added? Assume the density of the water is 1.00 g/mL.

94. A mixture of 80.0 mL of hydrogen and 60.0 mL of oxygen is ignited by a spark to form water.
    (a) Does any gas remain unreacted? Which one, H_2 or O_2?
    (b) What volume of which gas (if any) remains unreacted? (Assume the same conditions before and after the reaction.)

95. A student (with slow reflexes) puts his hand in a stream of steam at 100.°C until 1.5 g of water has condensed. If the water then cools to room temperature (20.0°C), how many joules have been absorbed by the student's hand?

Answers to Practice Exercises

13.1 8.5°C, 28°C, 73°C, 93°C

13.2 approximately 95°C

13.3 44.8 kJ

13.4 (a) yes, (b) yes, (c) no

CHAPTER 14

Solutions

Most substances we encounter in our daily lives are mixtures. Often they are homogeneous mixtures, which are called *solutions*. Some solutions we commonly encounter are shampoo, soft drinks, or wine. Blood plasma is a complex mixture composed of compounds and ions dissolved in water and proteins suspended in the solution. These solutions all have water as a main component, but many common items, such as air, gasoline, and steel, are also solutions that do not contain water. What are the necessary components of a solution? Why do some substances dissolve while others do not? What effect does a dissolved substance have on the properties of the solution? Answering these questions is the first step in understanding the solutions we encounter in our daily lives.

14.1 General Properties of Solutions

The term **solution** is used in chemistry to describe a system in which one or more substances are homogeneously mixed or dissolved in another substance. A simple solution has two components, a solute and a solvent. The **solute** is the component that is dissolved or is the least abundant component in the solution. The **solvent** is the dissolving agent or the most abundant component in the solution. For example, when salt is dissolved in water to form a solution, salt is the solute and water is the solvent. Complex solutions containing more than one solute and/or more than one solvent are common.

solution

solute
solvent

The three states of matter—solid, liquid, and gas—give us nine different types of solutions: solid dissolved in solid, solid dissolved in liquid, solid dissolved in gas, liquid dissolved in liquid, and so on. Of these, the most common solutions are solid dissolved in liquid, liquid dissolved in liquid, gas dissolved in liquid, and gas dissolved in gas. Some common types of solutions are listed in Table 14.1.

TABLE 14.1 Common Types of Solutions

Phase of solution	Solute	Solvent	Example
Gas	gas	gas	air
Liquid	gas	liquid	soft drinks
Liquid	liquid	liquid	antifreeze
Liquid	solid	liquid	salt water
Solid	gas	solid	H_2 in Pt
Solid	solid	solid	brass

▲
Note the beautiful purple trails of KMnO₄ as the crystals dissolve.

A true solution is one in which the particles of dissolved solute are molecular or ionic in size, generally in the range of 0.1 to 1 nm (10^{-8} to 10^{-7} cm). The properties of a true solution are as follows:

1. It is the mixture of two or more components—solute and solvent—is homogeneous, and has a variable composition; that is, the ratio of solute to solvent can be varied.
2. The dissolved solute is molecular or ionic in size.
3. It is either colored or colorless, and is usually transparent.
4. The solute remains uniformly distributed throughout the solution and will not settle out with time.
5. The solute can generally be separated from the solvent by purely physical means (e.g., by evaporation).

Let's illustrate these properties using water solutions of sugar and of potassium permanganate. We prepare two sugar solutions, the first containing 10 g of sugar added to 100 mL of water and the second containing 20 g of sugar added to 100 mL of water. Each solution is stirred until all the solute dissolves, demonstrating that we can vary the composition of a solution. Every portion of the solution has the same sweet taste because the sugar molecules are uniformly distributed throughout. If confined so that no solvent is lost, the solution will taste and appear the same a week or a month later. A solution cannot be separated into its components by filtering it. But by carefully evaporating the water, we can recover the sugar from the solution.

To observe the dissolving of potassium permanganate, $KMnO_4$, we affix a few crystals of it to paraffin wax or rubber cement at the end of a glass rod and submerge the entire rod, with the wax-permanganate end up, in a cylinder of water. Almost at once the beautiful purple color of dissolved permanganate ions, MnO_4^-, appears at the top of the rod and streams to the bottom of the cylinder as the crystals dissolve. The purple color is at first mostly at the bottom of the cylinder because $KMnO_4$ is denser than water. But after a while the purple color disperses until it's evenly distributed throughout the solution. This dispersal demonstrates that molecules and ions move about freely and spontaneously (diffuse) in a liquid or solution.

Solution permanency is explained in terms of the kinetic-molecular theory (see Section 12.2). According to the KMT both the solute and solvent particles (molecules and/or ions) are in constant random motion. This motion is energetic enough to prevent the solute particles from settling out under the influence of gravity.

14.2   Solubility

solubility

The term **solubility** describes the amount of one substance (solute) that will dissolve in a specified amount of another substance (solvent) under stated conditions. For example, 36.0 g of sodium chloride will dissolve in 100 g of water at 20°C. We say then that the solubility of NaCl in water is 36.0 g/100 g H_2O at 20°C.

Solubility is often used in a relative way. For instance, we say that a substance is very soluble, moderately soluble, slightly soluble, or insoluble. Although these terms do not accurately indicate how much solute will dissolve, they are frequently used to describe the solubility of a substance qualitatively.

Two other terms often used to describe solubility are *miscible* and *immiscible*.

miscible
immiscible

Liquids that are capable of mixing and forming a solution are **miscible**; those that do not form solutions or are generally insoluble in each other are **immiscible**.

Methyl alcohol and water are miscible in each other in all proportions. Oil and water are immiscible, forming two separate layers when they are mixed, as shown in Figure 14.1.

The general guidelines for the solubility of common ionic compounds (salts) are given in Figure 14.2. These guidelines have some exceptions, but they provide a solid foundation for the compounds considered in this course. The solubilities of over 200 compounds are given in the Solubility Table in Appendix V. Solubility data for thousands of compounds can be found by consulting standard reference sources.*

The term salt is used interchangeably with ionic compound by many chemists.

The quantitative expression of the amount of dissolved solute in a particular quantity of solvent is known as the **concentration of a solution.** Several methods of expressing concentration are described in Section 14.6.

concentration of a solution

14.3 Factors Related to Solubility

Predicting solubilities is complex and difficult. Many variables, such as size of ions, charge on ions, interaction between ions, interaction between solute and solvent, and temperature, complicate the problem. Because of the factors involved, the general rules of solubility given in Figure 14.2 have many exceptions. However, these rules are useful because they do apply to many of the more common compounds that we encounter in the study of chemistry. Keep in mind that these are rules, not laws, and are therefore subject to exceptions. Fortunately the solubility of a solute

▲
FIGURE 14.1
An immiscible mixture of oil and water.

Soluble		Insoluble
Na^+, K^+, NH_4^+		
Nitrates, NO_3^- Acetates, $C_2H_3O_2^-$		
Chlorides, Cl^- Bromides Br^- Iodides, I^-	except →	Ag^+, Hg_2^{2+}, Pb^{2+}
Sulfates, SO_4^{2-} Ag^+, Ca^{2+} are slightly soluble	except →	Ba^{2+}, Sr^{2+}, Pb^{2+}
NH_4^+ alkali metal cations	← except	Carbonates, CO_3^{2-} Phosphates, PO_4^{3-} Hydroxides, OH^- Sulfides, S^{2-}

◀**FIGURE 14.2**
The solubility of various common ions. Substances containing the ions on the left are generally soluble in cold water, while those substances containing the ions on the right are insoluble in cold water. The arrows point to the exceptions.

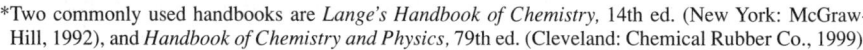

*Two commonly used handbooks are *Lange's Handbook of Chemistry,* 14th ed. (New York: McGraw-Hill, 1992), and *Handbook of Chemistry and Physics,* 79th ed. (Cleveland: Chemical Rubber Co., 1999).

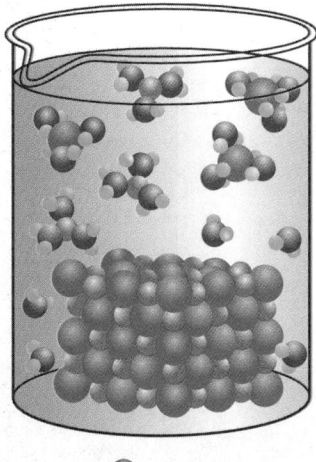

= Water

= Na⁺

= Cl⁻

▲
FIGURE 14.3
Dissolution of sodium chloride in water. Polar water molecules are attracted to Na⁺ and Cl⁻ ions in the salt crystal, weakening the attraction between the ions. As the attraction between the ions weakens, the ions move apart and become surrounded by water dipoles. The hydrated ions slowly diffuse away from the crystal to become dissolved in solution.

is relatively easy to determine experimentally. Now let's examine the factors related to solubility.

The Nature of the Solute and Solvent

The old adage "like dissolves like" has merit, in a general way. Polar or ionic substances tend to be more miscible with other polar substances. Nonpolar substances tend to be miscible with other nonpolar substances and less miscible with polar substances. Thus ionic compounds, which are polar, tend to be much more soluble in water, which is polar, than in solvents such as ether, hexane, or benzene, which are essentially nonpolar. Sodium chloride, an ionic substance, is soluble in water, slightly soluble in ethyl alcohol (less polar than water), and insoluble in ether and benzene. Pentane, C_5H_{12}, a nonpolar substance, is only slightly soluble in water but is very soluble in benzene and ether.

At the molecular level the formation of a solution from two nonpolar substances, such as hexane and benzene, can be visualized as a process of simple mixing. The nonpolar molecules, having little tendency to either attract or repel one another, easily intermingle to form a homogeneous mixture.

Solution formation between polar substances is much more complex. See, for example, the process by which sodium chloride dissolves in water (Figure 14.3). Water molecules are very polar and are attracted to other polar molecules or ions. When salt crystals are put into water, polar water molecules become attracted to the sodium and chloride ions on the crystal surfaces and weaken the attraction between Na^+ and Cl^- ions. The positive end of the water dipole is attracted to the Cl^- ions, and the negative end of the water dipole to the Na^+ ions. The weakened attraction permits the ions to move apart, making room for more water dipoles. Thus the surface ions are surrounded by water molecules, becoming hydrated ions, $Na^+(aq)$ and $Cl^-(aq)$, and slowly diffuse away from the crystals and dissolve in solution:

$$NaCl(crystal) \xrightarrow{H_2O} Na^+(aq) + Cl^-(aq)$$

Examination of the data in Table 14.2 reveals some of the complex questions relating to solubility.

The Effect of Temperature on Solubility

Temperature affects the solubility of most substances as shown by the data in Table 14.2. Most solutes have a limited solubility in a specific solvent at a fixed temperature. For most solids dissolved in a liquid, an increase in temperature results in increased solubility (see Figure 14.4). However, no single rule governs the solubility of solids in liquids with change in temperature. Some solids increase in solubility only slightly with increasing temperature (see NaCl in Figure 14.4); other solids decrease in solubility with increasing temperature (see Li_2SO_4 in Figure 14.4).

On the other hand, the solubility of a gas in water usually decreases with increasing temperature (see HCl and SO_2 in Figure 14.4). The tiny bubbles that form when water is heated are due to the decreased solubility of air at higher temperatures. The decreased solubility of gases at higher temperatures is explained in terms of the KMT by assuming that, in order to dissolve, the gas molecules must form bonds of some sort with the molecules of the liquid. An increase in temperature decreases the solubility of the gas because it increases the kinetic energy (speed) of the gas molecules and thereby decreases their ability to form "bonds" with the liquid molecules.

TABLE 14.2	Solubility of Alkali Metal Halides in Water	
	Solubility (g salt/100 g H_2O)	
Salt	0°C	100°C
LiF	0.12	0.14 (at 35°C)
LiCl	67	127.5
LiBr	143	266
LiI	151	481
NaF	4	5
NaCl	35.7	39.8
NaBr	79.5	121
NaI	158.7	302
KF	92.3 (at 18°C)	Very soluble
KCl	27.6	57.6
KBr	53.5	104
KI	127.5	208

◀ **FIGURE 14.4**
Solubility of various compounds in water. Solids are shown in red and gases are shown in blue.

The Effect of Pressure on Solubility

Small changes in pressure have little effect on the solubility of solids in liquids or liquids in liquids but have a marked effect on the solubility of gases in liquids. The solubility of a gas in a liquid is directly proportional to the pressure of that gas above the solution. Thus the amount of a gas dissolved in solution will double if the pressure of that gas over the solution is doubled. For example, carbonated beverages contain dissolved carbon dioxide under pressures greater than atmospheric pressure. When a can of carbonated soda is opened, the pressure is immediately reduced to the atmospheric pressure, and the excess dissolved carbon dioxide bubbles out of the solution.

Saturated, Unsaturated, and Supersaturated Solutions

At a specific temperature there is a limit to the amount of solute that will dissolve in a given amount of solvent. When this limit is reached, the resulting solution is said to be *saturated*. For example, when we put 40.0 g of KCl into 100 g of H_2O at 20°C, we find that 34.0 g of KCl dissolves and 6.0 g of KCl remains undissolved. The solution formed is a saturated solution of KCl.

Two processes are occurring simultaneously in a saturated solution. The solid is dissolving into solution and, at the same time, the dissolved solute is crystallizing out of solution. This may be expressed as

$$\text{solute (undissolved)} \rightleftharpoons \text{solute (dissolved)}$$

When these two opposing processes are occurring at the same rate, the amount of solute in solution is constant, and a condition of equilibrium is established between dissolved and undissolved solute. Therefore, a **saturated solution** contains dissolved solute in equilibrium with undissolved solute.

It's important to state the temperature of a saturated solution, because a solution that is saturated at one temperature may not be saturated at another. If the temperature of a saturated solution is changed, the equilibrium is disturbed, and the amount of dissolved solute will change to reestablish equilibrium.

A saturated solution may be either dilute or concentrated, depending on the solubility of the solute. A saturated solution can be conveniently prepared by dissolving a little more than the saturated amount of solute at a temperature somewhat higher than room temperature. Then the amount of solute in solution will be in excess of its solubility at room temperature, and, when the solution cools, the excess solute will crystallize, leaving the solution saturated. (In this case, the solute must be more soluble at higher temperatures and must not form a supersaturated solution.) Examples expressing the solubility of saturated solutions at two different temperatures are given in Table 14.3.

An **unsaturated solution** contains less solute per unit of volume than does its corresponding saturated solution. In other words, additional solute can be dissolved in an unsaturated solution without altering any other conditions. Consider a solution made by adding 40 g of KCl to 100 g of H_2O at 20°C (see Table 14.3). The solution formed will be saturated and will contain about 6 g of undissolved salt, because the maximum amount of KCl that can dissolve in 100 g of H_2O at 20°C is 34 g. If the solution is now heated and maintained at 50°C,

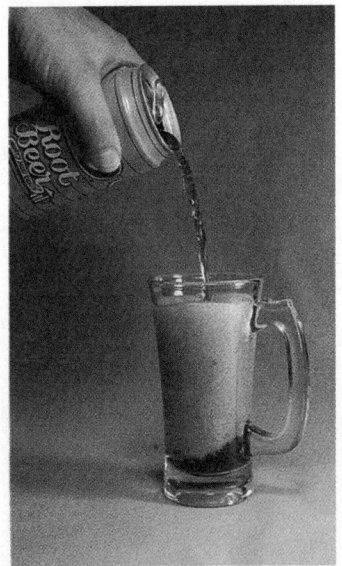

Pouring root beer into a glass illustrates the effect of pressure on solubility. The escaping CO_2 produces the foam.

saturated solution

unsaturated solution

TABLE 14.3	Saturated Solutions at 20°C and 50°C	
	Solubility (g solute/100 g H_2O)	
Solute	**20°C**	**50°C**
NaCl	36.0	37.0
KCl	34.0	42.6
$NaNO_3$	88.0	114.0
$KClO_3$	7.4	19.3
$AgNO_3$	222.0	455.0
$C_{12}H_{22}O_{11}$	203.9	260.4

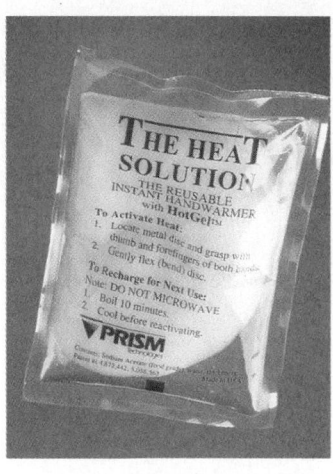

▲ **The heat released in this hot pack results from the crystallization of a supersaturated solution of sodium acetate.**

all the salt will dissolve and even more can be dissolved. The solution at 50°C is unsaturated.

In some circumstances, solutions can be prepared that contain more solute than that needed for a saturated solution at a particular temperature. These solutions are said to be **supersaturated.** However, we must qualify this definition by noting that a supersaturated solution is unstable. Disturbances, such as jarring, stirring, scratching the walls of the container, or dropping in a "seed" crystal, cause the supersaturation to return to saturation. When a supersaturated solution is disturbed, the excess solute crystallizes out rapidly, returning the solution to a saturated state.

supersaturated solution

Supersaturated solutions are not easy to prepare but may be made from certain substances by dissolving, in warm solvent, an amount of solute greater than that needed for a saturated solution at room temperature. The warm solution is then allowed to cool very slowly. With the proper solute and careful work, a supersaturated solution will result.

Will a solution made by adding 2.5 g of $CuSO_4$ to 10 g of H_2O be saturated or unsaturated at 20°C?

Example 14.1

We first need to know the solubility of $CuSO_4$ at 20°C. From Figure 14.4, we see that the solubility of $CuSO_4$ at 20°C is about 21 g per 100 g of H_2O. This amount is equivalent to 2.1 g of $CuSO_4$ per 10 g of H_2O.

Solution

Since 2.5 g per 10 g of H_2O is greater than 2.1 g per 10 g of H_2O, the solution will be saturated and 0.4 g of $CuSO_4$ will be undissolved.

Practice 14.1

Will a solution made by adding 9.0 g NH_4Cl to 20 g of H_2O be saturated or unsaturated at 50°C?

Rate of Dissolving Solids

The rate at which a solid dissolves is governed by (1) the size of the solute particles, (2) the temperature, (3) the concentration of the solution, and (4) agitation or stirring. Let's look at each of these conditions.

1. *Particle Size.* A solid can dissolve only at the surface that is in contact with the solvent. Because the surface-to-volume ratio increases as size decreases, smaller crystals dissolve faster than large ones. For example, if a salt crystal 1 cm on a side (6-cm^2 surface area) is divided into 1000 cubes, each 0.1 cm on a side, the total surface of the smaller cubes is 60 cm^2—a tenfold increase in surface area (see Figure 14.5).

2. *Temperature.* In most cases the rate of dissolving of a solid increases with temperature. This increase is due to kinetic effects. The solvent molecules move more rapidly at higher temperatures and strike the solid surfaces more often and harder, causing the rate of dissolving to increase.

3. *Concentration of the Solution.* When the solute and solvent are first mixed, the rate of dissolving is at its maximum. As the concentration of the solution increases and the solution becomes more nearly saturated with the solute, the rate of dissolving decreases greatly. The rate of dissolving is graphed in Figure 14.6. Note that about 17 g dissolves in the first 5-minute interval, but only about 1 g dissolves in the fourth 5-minute interval. Although different solutes show different rates, the rate of dissolving always becomes very slow as the concentration approaches the saturation point.

4. *Agitation or Stirring.* The effect of agitation or stirring is kinetic. When a solid is first put into water, it comes in contact only with solvent in its immediate vicinity. As the solid dissolves, the amount of dissolved solute around the solid becomes more and more concentrated, and the rate of dissolving slows down. If the mixture is not stirred, the dissolved solute diffuses very slowly through the solution; weeks may pass before the solid is entirely dis-

FIGURE 14.5 ▶
Surface area of crystals. A crystal 1 cm on a side has a surface area of 6 cm². Subdivided into 1000 smaller crystals, each 0.1 cm on a side, the total surface area is increased to 60 cm².

◄ FIGURE 14.6
Rate of dissolution of a
solid solute in a solvent.
The rate is maximum at
the beginning and decreases
as the concentration
approaches saturation.

solved. Stirring distributes the dissolved solute rapidly through the solution, and more solvent is brought into contact with the solid, causing it to dissolve more rapidly.

14.5   Solutions: A Reaction Medium

Many solids must be put into solution to undergo appreciable chemical reaction. We can write the equation for the double displacement reaction between sodium chloride and silver nitrate:

$$NaCl + AgNO_3 \longrightarrow AgCl + NaNO_3$$

But suppose we mix solid NaCl and solid $AgNO_3$ and look for a chemical change. If any reaction occurs, it is slow and virtually undetectable. In fact, the crystalline structures of NaCl and $AgNO_3$ are so different that we could separate them by tediously picking out each kind of crystal from the mixture. But if we dissolve the NaCl and $AgNO_3$ separately in water and mix the two solutions, we observe the immediate formation of a white, curdlike precipitate of silver chloride.

Molecules or ions must collide with one another in order to react. In the foregoing example, the two solids did not react because the ions were securely locked within their crystal structures. But when the NaCl and $AgNO_3$ are dissolved, their crystal lattices are broken down and the ions become mobile. When the two solutions are mixed, the mobile Ag^+ and Cl^- ions come into contact and react to form insoluble AgCl, which precipitates out of solution. The soluble Na^+ and NO_3^- ions remain mobile in solution but form the crystalline salt $NaNO_3$ when the water is evaporated:

$$NaCl(aq) + AgNO_3(aq) \longrightarrow AgCl(s) + NaNO_3(aq)$$

$$Na^+(aq) + Cl^-(aq) + Ag^+(aq) + NO_3^-(aq) \longrightarrow AgCl(s) + Na^+(aq) + NO_3^-(aq)$$

| sodium chloride | silver nitrate | silver | sodium nitrate |
| solution | solution | chloride | in solution |

The mixture of the two solutions provides a medium or space in which the Ag^+ and Cl^- ions can react. (See Chapter 15 for further discussion of ionic reactions.)

Solutions also function as diluting agents in reactions in which the undiluted reactants would combine with each other too violently. Moreover, a solution of known concentration provides a convenient method for delivering specific amounts of reactants.

14.6 Concentration of Solutions

The concentration of a solution expresses the amount of solute dissolved in a given quantity of solvent or solution. Because reactions are often conducted in solution, it's important to understand the methods of expressing concentration and to know how to prepare solutions of particular concentrations. The concentration of a solution may be expressed qualitatively or quantitatively. Let's begin with a look at the qualitative methods of expressing concentration.

Dilute and Concentrated Solutions

When we say a solution is *dilute* or *concentrated,* we are expressing, in a relative way, the amount of solute present. One gram of a compound and 2 g of a compound

Getting Clothes CO₂ Clean!

Dry cleaning has been used for many years to clean clothes that can't be placed in water (such as silk, rayon, wool) without severe damage or shrinking. The dry-cleaning process involves no water. Instead clothes are treated for stains and then washed in perchloroethylene (perc), an organic liquid. Unfortunately perc may be a health hazard and is classified as both an air pollutant and environmental contaminant. This means big trouble for small dry cleaners who now must dispose of it as hazardous waste and also worry about health risks to employees. Consequently researchers are looking for a solvent that can replace perc for dry cleaning.

One interesting alternative is to use carbon dioxide for dry cleaning. You may think of CO_2 as a greenhouse gas, the gas you breathe in and out, or the solid called dry ice. One thing's for sure—we don't often consider CO_2 as a liquid. Scientists have already started to consider liquid CO_2 as a dry cleaning alternative. Why? There is no shortage of CO_2 in the world. It can be collected from the waste generated in industry. No special disposal procedures are needed since it can be recycled and leaks would be nontoxic to humans and the environment in small quantities.

The largest obstacle for the dry-cleaning process is that CO_2 doesn't dissolve most polymers, oils, waxes, and substances that get caught in clothes. Of course, neither does water; it requires the help of detergents. So scientists have now developed polymers that can be used as detergents in liquid CO_2. Joseph DeSimone (University of North Carolina at Chapel Hill) made detergent polymers called copolymers, which are really just two polymers joined together. One end is soluble in CO_2; the other attracts oils and waxes. The new detergent polymers form micelles that trap the oils and waxes within the micelle and dissolve in CO_2.

Now machines that look like bank vaults can clean 50–70 pounds of clothes per load with CO_2 and about 5 ounces of polymer detergent. One extra bonus— since the machine operates at room temperature, stains don't require pretreating! Leather and suede can also be cleaned using CO_2. Someday soon you may walk into your local dry-cleaning shop and get your clothes CO_2 cleaned!

in solution are both dilute solutions when compared with the same volume of a solution containing 20 g of a compound. Ordinary concentrated hydrochloric acid contains 12 mol of HCl per liter of solution. In some laboratories the dilute acid is made by mixing equal volumes of water and the concentrated acid. In other laboratories the concentrated acid is diluted with two or three volumes of water, depending on its use. The term **dilute solution,** then, describes a solution that contains a relatively small amount of dissolved solute. Conversely, a **concentrated solution** contains a relatively large amount of dissolved solute.

dilute solution
concentrated solution

Mass Percent Solution

The mass percent method expresses the concentration of the solution as the percent of solute in a given mass of solution. It says that for a given mass of solution a certain percent of that mass is solute. Suppose we take a bottle from the reagent shelf that reads "sodium hydroxide, NaOH, 10%." This statement means that for every 100 g of this solution, 10 g will be NaOH and 90 g will be water. (Note that this amount of solution is 100 g and not 100 mL.) We could also make this same concentration of solution by dissolving 2.0 g of NaOH in 18 g of water. Mass percent concentrations are most generally used for solids dissolved in liquids:

$$\text{mass percent} = \frac{\text{g solute}}{\text{g solute} + \text{g solvent}} \times 100 = \frac{\text{g solute}}{\text{g solution}} \times 100$$

Note that mass percent is independent of the formula for the solute.

As instrumentation advances are made in chemistry, our ability to measure the concentration of dilute solutions is increasing as well. In addition to mass percent, chemists now commonly use **parts per million (ppm)**:

$$\text{parts per million} = \frac{\text{g solute}}{\text{g solute} + \text{g solvent}} \times 1{,}000{,}000$$

Currently, air and water contaminants, drugs in the human body, and pesticide residues are measured in parts per million.

Example 14.2   What is the mass percent of sodium hydroxide in a solution that is made by dissolving 8.00 g NaOH in 50.0 g H_2O?

Solution

grams of solute (NaOH) =   8.00 g

grams of solvent (H_2O)   = 50.0 g

$$\left(\frac{8.00 \text{ g NaOH}}{8.00 \text{ g NaOH} + 50.0 \text{ g } H_2O}\right) 100 = 13.8\% \text{ NaOH solution}$$

Example 14.3   What masses of potassium chloride and water are needed to make 250. g of 5.00% solution?

Solution   The percent expresses the mass of the solute:

250. g = total mass of solution

5.00% of 250. g = (0.0500)(250. g) = 12.5 g KCl (solute)

250. g − 12.5 g = 238 H_2O

Dissolving 12.5 g KCl in 238 g H_2O gives a 5.00% KCl solution.

Example 14.4   A 34.0% sulfuric acid solution has a density of 1.25 g/mL. How many grams of H_2SO_4 are contained in 1.00 L of this solution?

Solution   Since H_2SO_4 is the solute, we first solve the mass percent equation for grams of solute:

$$\text{mass percent} = \frac{\text{g solute}}{\text{g solution}} \times 100$$

$$\text{g solute} = \frac{\text{mass percent} \times \text{g solution}}{100}$$

The mass percent is given so we need to determine the grams of solution. The mass of the solution can be calculated from the density data. Convert density (g/mL) to grams:

1.00 L = 1.00 × 10³ mL

$$\left(\frac{1.25 \text{ g}}{\text{mL}}\right)(1.00 \times 10^3 \text{ mL}) = 1250 \text{ g}   \text{(mass of solution)}$$

Now we have all the figures to calculate the grams of solute:

$$g \text{ solute} = \frac{(34.0 \text{ g } H_2SO_4)(1250 \text{ g})}{100 \text{ g}} = 425 \text{ g } H_2SO_4$$

Thus 1.00 L of 34.0% H_2SO_4 solution contains 425 g H_2SO_4.

Practice 14.2

What is the mass percent of Na_2SO_4 in a solution made by dissolving 25.0 g Na_2SO_4 in 225.0 g H_2O?

Mass/Volume Percent (m/v)

This method expresses concentration as grams of solute per 100 mL of solution. With this system, a 10.0% (m/v) glucose solution is made by dissolving 10.0 g of glucose in water, diluting to 100 mL, and mixing. The 10.0% (m/v) solution could also be made by diluting 20.0 g to 200 mL, 50.0 g to 500 mL, and so on. Of course, any other appropriate dilution ratio may be used:

$$\text{mass/volume percent} = \frac{\text{g solute}}{\text{mL solution}} \times 100$$

Volume Percent

Solutions that are formulated from two liquids are often expressed as *volume percent* with respect to the solute. The volume percent is the volume of a liquid in 100 mL of solution. The label on a bottle of ordinary rubbing alcohol reads "isopropyl alcohol, 70% by volume." Such a solution could be made by mixing 70 mL of alcohol with water to make a total volume of 100 mL, but we cannot use 30 mL of water because the two volumes are not necessarily additive:

$$\text{volume percent} = \frac{\text{volume of liquid in question}}{\text{total volume of solution}} \times 100$$

Volume percent is used to express the concentration of alcohol in beverages. Wines generally contain 12% alcohol by volume. This translates into 12 mL of alcohol in each 100 mL of wine. The beverage industry also uses the concentration unit of *proof* (twice the volume percent). Pure alcohol is 100%, therefore 200 proof. Scotch whiskey is 86 proof or 43% alcohol.

Molarity

Mass percent solutions do not equate or express the molar masses of the solute in solution. For example, 1000. g of 10.0% NaOH solution contains 100. g NaOH;

FIGURE 14.7 ▶
Preparation of a 1 M solution.

(a)
Add 1 mole of
solute to a 1 liter
volumetric flask

(b)
Dissolve in
solvent

(c)
Add solvent to the
1 liter mark and
mix thoroughly

1000. g of 10.0% KOH solution contains 100. g KOH. In terms of moles of NaOH and KOH, these solutions contain

$$\text{mol NaOH} = (100.\ \text{g NaOH})\left(\frac{1\ \text{mol NaOH}}{40.00\ \text{g NaOH}}\right) = 2.50\ \text{mol NaOH}$$

$$\text{mol KOH} = (100.\ \text{g KOH})\left(\frac{1\ \text{mol KOH}}{56.11\ \text{g KOH}}\right) = 1.78\ \text{mol KOH}$$

From these figures we see that the two 10.0% solutions do not contain the same number of moles of NaOH and KOH. As a result we find that a 10.0% NaOH solution contains more reactive base than a 10.0% KOH solution.

We need a method of expressing concentration that will easily indicate how many moles of solute are present per unit volume of solution. For this purpose the concentration known as molarity is used.

A 1-molar solution contains 1 mol of solute per liter of solution. For example, to make a 1-molar solution of sodium hydroxide, NaOH, we dissolve 40 g NaOH (1 mol) in water and dilute the solution with more water to a volume of 1 L. The solution contains 1 mol of the solute in 1 L of solution and is said to be 1 molar in concentration. Figure 14.7 illustrates the preparation of a 1-molar solution. Note that the volume of the solute and the solvent together is 1 L.

molarity (M)   The concentration of a solution can, of course, be varied by using more or less solute or solvent; but in any case the **molarity** of a solution is the number of moles of solute per liter of solution. The abbreviation for molarity is *M*. The units of molarity are moles per liter. The expression "2.0 *M* NaOH" means a 2.0 molar solution of NaOH (2.0 mol, or 80 g, of NaOH dissolved to make 1.0 L of solution).

$$\text{molarity} = M = \frac{\text{number of moles of solute}}{\text{liter of solution}} = \frac{\text{moles}}{\text{liter}}$$

Flasks that are calibrated to contain specific volumes at a particular temperature are used to prepare solutions of a desired concentration. These *volumetric flasks* have a calibration mark on the neck that accurately indicate the measured volume. Molarity is based on a specific volume of solution and therefore will vary slightly with temperature because volume varies with temperature:

(1000 mL H_2O at 20°C = 1001 mL at 25°C)

Suppose we want to make 500 mL of 1 M solution. This solution can be prepared by determining the mass of 0.5 mol of the solute and diluting with water in a 500-mL (0.5-L) volumetric flask. The molarity will be

$$M = \frac{0.5 \text{ mol solute}}{0.5 \text{ L solution}} = 1 \text{ molar}$$

You can see that it isn't necessary to have a liter of solution to express molarity. All we need to know is the number of moles of dissolved solute and the volume of solution. Thus 0.001 mol NaOH in 10 mL of solution is 0.1 M:

$$\left(\frac{0.001 \text{ mol}}{10 \text{ mL}}\right)\left(\frac{1000 \text{ mL}}{1 \text{ L}}\right) = 0.1 \text{ } M$$

When we stop to think that a balance is not calibrated in moles but in grams, we can incorporate grams into the molarity formula. We do so by using the relationship

$$\text{moles} = \frac{\text{grams of solute}}{\text{molar mass}}$$

Substituting this relationship into our expression for molarity, we get

$$M = \frac{\text{mol}}{\text{L}} = \frac{\text{g solute}}{\text{molar mass solute} \times \text{L solution}}$$

We can now determine the mass of any amount of a solute that has a known formula, dilute it to any volume, and calculate the molarity of the solution using this formula.

Molarities of concentrated acids commonly used in the laboratory:

HCl	12 M
$HC_2H_3O_2$	17 M
HNO_3	16 M
H_2SO_4	18 M

What is the molarity of a solution containing 1.4 mol of acetic acid, $HC_2H_3O_2$, in 250. mL of solution? **Example 14.5**

By the unit conversion method we note that the concentration given in the problem statement is 1.4 mol per 250. mL (mol/mL). Since molarity = mol/L, the needed conversion is **Solution**

$$\frac{\text{mol}}{\text{mL}} \longrightarrow \frac{\text{mol}}{\text{L}} = M$$

$$\left(\frac{1.4 \text{ mol}}{250. \text{ mL}}\right)\left(\frac{1000 \text{ mL}}{\text{L}}\right) = \frac{5.6 \text{ mol}}{\text{L}} = 5.6 \text{ } M$$

Example 14.6   What is the molarity of a solution made by dissolving 2.00 g of potassium chlorate in enough water to make 150. mL of solution?

Solution   We use the unit conversion method. The steps in the conversions must lead to units of moles/liter:

$$\frac{g\ KClO_3}{mL} \longrightarrow \frac{g\ KClO_3}{L} \longrightarrow \frac{mol\ KClO_3}{L} = M$$

The data are

mass $KClO_3$ = 2.00 g   molar mass $KClO_3$ = 122.6 g/mol   volume = 150. mL

$$\left(\frac{2.00\ g\ KClO_3}{150.\ mL}\right)\left(\frac{1000\ mL}{L}\right)\left(\frac{1\ mol\ KClO_3}{122.6\ g\ KClO_3}\right) = \frac{0.109\ mol}{L} = 0.109\ M\ KClO_3$$

Example 14.7   How many grams of potassium hydroxide are required to prepare 600. mL of 0.450 M KOH solution?

Solution   The conversion is

milliliters $\longrightarrow$ liters $\longrightarrow$ moles $\longrightarrow$ grams

The data are

volume = 600. mL $M = \dfrac{0.450\ mol}{L}$ molar mass KOH = $\dfrac{56.11\ g}{mol}$

The calculation is

$$(600.\ mL)\left(\frac{1\ L}{1000\ mL}\right)\left(\frac{0.450\ mol}{L}\right)\left(\frac{56.11\ g\ KOH}{mol}\right) = 15.1\ g\ KOH$$

Practice 14.3

What is the molarity of a solution made by dissolving 7.50 g of magnesium nitrate, $Mg(NO_3)_2$, in enough water to make 25.0 mL of solution?

Practice 14.4

How many grams of sodium chloride are needed to prepare 125 mL of a 0.037 M NaCl solution?

Example 14.8   Calculate the number of moles of nitric acid in 325 mL of 16 M HNO_3 solution.

Solution   Use the equation

moles = liters $\times$ M

Substitute the data given in the problem and solve:

$$\text{moles} = (0.325 \, L)\left(\frac{16 \text{ mol HNO}_3}{1 \, L}\right) = 5.2 \text{ mol HNO}_3$$

What volume of 0.250 *M* solution can be prepared from 16.0 g of potassium carbonate?   **Example 14.9**

We start with 16.0 g K_2CO_3; we need to find the volume of 0.250 *M* solution that can be prepared from this amount of K_2CO_3. The conversion therefore is   **Solution**

$$\text{g } K_2CO_3 \longrightarrow \text{mol } K_2CO_3 \longrightarrow \text{L solution}$$

The data are

$$\text{mass } K_2CO_3 = 16.0 \text{ g} \quad M = \frac{0.250 \text{ mol}}{1 \text{ L}} \quad \text{molar mass } K_2CO_3 = \frac{138.2 \text{ g}}{1 \text{ mol}}$$

$$(16.0 \text{ g } K_2CO_3)\left(\frac{1 \text{ mol } K_2CO_3}{138.2 \text{ g } K_2CO_3}\right)\left(\frac{1 \text{ L}}{0.250 \text{ mol } K_2CO_3}\right) = 0.463 \text{ L } (463 \text{ mL})$$

Thus 463 mL of 0.250 *M* solution can be made from 16.0 g K_2CO_3.

How many milliliters of 2.00 *M* HCl will react with 28.0 g NaOH?   **Example 14.10**

Step 1. Write and balance the equation for the reaction: **Solution**

$$\text{HCl}(aq) + \text{NaOH}(aq) \longrightarrow \text{NaCl}(aq) + \text{H}_2\text{O}(aq)$$

The equation states that 1 mol of HCl reacts with 1 mol of NaOH.

Step 2. Find the number of moles NaOH in 28.0 g NaOH:

$$\text{g NaOH} \longrightarrow \text{mol NaOH}$$

$$(28.0 \text{ g NaOH})\left(\frac{1 \text{ mol}}{40.00 \text{ g}}\right) = 0.700 \text{ mol NaOH}$$

$$28.0 \text{ g NaOH} = 0.700 \text{ mol NaOH}$$

Step 3. Solve for moles and volume of HCl needed. From Steps 1 and 2 we see that 0.700 mol HCl will react with 0.700 mol NaOH, because the ratio of moles reacting is 1:1. We know that 2.00 *M* HCl contains 2.00 mol HCl per liter, and so the volume that contains 0.700 mol HCl will be less than 1 L:

$$\text{mol NaOH} \longrightarrow \text{mol HCl} \longrightarrow \text{L HCl} \longrightarrow \text{mL HCl}$$

$$(0.700 \text{ mol NaOH})\left(\frac{1 \text{ mol HCl}}{1 \text{ mol NaOH}}\right)\left(\frac{1 \text{ L HCl}}{2.00 \text{ mol HCl}}\right) = 0.350 \text{ L HCl}$$

$$(0.350 \text{ L HCl})\left(\frac{1000 \text{ mL}}{1 \, L}\right) = 350. \text{ mL HCl}$$

Therefore 350. mL of 2.00 *M* HCl contains 0.700 mol HCl and will react with 0.700 mol, or 28.0 g, of NaOH.

TABLE 14.4 Concentration Units for Solutions		
Units	**Symbol**	**Definition**
Mass percent	% m/m	$\dfrac{\text{Mass solute}}{\text{Mass solution}} \times 100$
Parts per million	ppm	$\dfrac{\text{Mass solute}}{\text{Mass solution}} \times 1{,}000{,}000$
Mass/volume percent	% m/v	$\dfrac{\text{Mass solute}}{\text{mL solution}} \times 100$
Volume percent	% v/v	$\dfrac{\text{mL solute}}{\text{mL solution}} \times 100$
Molarity	M	$\dfrac{\text{Moles solute}}{\text{L solution}}$
Molality	m	$\dfrac{\text{Moles solute}}{\text{kg solvent}}$

Practice 14.5

What volume of 0.035 M AgNO$_3$ can be made from 5.0 g of AgNO$_3$?

Practice 14.6

How many milliliters of 0.50 M NaOH are required to react completely with 25.00 mL of 1.5 M HCl?

We've now examined several ways to measure concentration of solutions quantitatively. A summary of these concentration units is found in Table 14.4.

Dilution Problems

Chemists often find it necessary to dilute solutions from one concentration to another by adding more solvent to the solution. If a solution is diluted by adding pure solvent, the volume of the solution increases, but the number of moles of solute in the solution remains the same. Thus the moles/liter (molarity) of the solution decreases. Always read a problem carefully to distinguish between (1) how much solvent must be added to dilute a solution to a particular concentration, and (2) to what volume a solution must be diluted to prepare a solution of a particular concentration.

Example 14.11 Calculate the molarity of a sodium hydroxide solution that is prepared by mixing 100. mL of 0.20 M NaOH with 150. mL of water. Assume the volumes are additive.

◀ **A serial dilution. The concentration of food coloring in well 1 (far left) is 1 part per 10 (by weight), well 2 is 1 part per 100, well 3 is 1 part per 1000, and so on. The concentration in well 6 is 1 part per million (ppm).**

Solution

This problem is a dilution problem. If we double the volume of a solution by adding water, we cut the concentration in half. Therefore, the concentration of the above solution should be less than 0.10 M. In the dilution, the moles of NaOH remain constant; the molarity and volume change. The final volume is (100. mL + 150. mL) or 250. mL.

To solve this problem, (1) calculate the moles of NaOH in the original solution, and (2) divide the moles of NaOH by the final volume of the solution to obtain the new molarity.

Step 1. Calculate the moles of NaOH in the original solution:

$$M = \frac{mol}{L} \qquad mol = L \times M$$

$$(0.100 \; \cancel{L})\left(\frac{0.20 \; mol \; NaOH}{1 \; \cancel{L}}\right) = 0.020 \; mol \; NaOH$$

Step 2. Solve for the new molarity, taking into account that the total volume of the solution after dilution is 250. mL (0.250 L):

$$M = \frac{0.020 \; mol \; NaOH}{0.250 \; L} = 0.080 \; M \; NaOH$$

Alternative Solution

When the moles of solute in a solution before and after dilution are the same, then the moles before and after dilution may be set equal to each other:

$$mol_1 = mol_2$$

where mol_1 = moles before dilution and mol_2 = moles after dilution. Then

$$mol_1 = L_1 \times M_1 \qquad mol_2 = L_2 \times M_2$$

$$L_1 \times M_1 = L_2 \times M_2$$

When both volumes are in the same units, a more general statement can be made:

$$V_1 \times M_1 = V_2 \times M_2$$

For this problem

$V_1 = 100.$ mL $M_1 = 0.20\ M$

$V_2 = 150.$ mL $+\ 100.$ mL $M_2 =$ (unknown)

Then

$$(100.\ \text{mL})(0.20\ M) = (250.\ \text{mL})M_2$$

Solving for M_2, we get

$$M_2 = \frac{(100.\ \cancel{\text{mL}})(0.20\ M)}{250.\ \cancel{\text{mL}}} = 0.080\ M\ \text{NaOH}$$

Practice 14.7

Calculate the molarity of a solution prepared by diluting 125 mL of 0.400 M $K_2Cr_2O_7$ with 875 mL of water.

Example 14.12 How many grams of silver chloride will be precipitated by adding sufficient silver nitrate to react with 1500. mL of 0.400 M barium chloride solution?

$$2\ \text{AgNO}_3(aq) + \text{BaCl}_2(aq) \longrightarrow 2\ \text{AgCl}(s) + \text{Ba(NO}_3)_2(aq)$$

Solution   This problem is a stoichiometry problem. The fact that $BaCl_2$ is in solution means that we need to consider the volume and concentration of the solution in order to determine the number of moles of $BaCl_2$ reacting.

Step 1. Determine the number of moles of $BaCl_2$ in 1500. mL of 0.400 M solution:

$$M = \frac{\text{mol}}{\text{L}} \qquad \text{mol} = \text{L} \times M \qquad 1500.\ \text{mL} = 1.500\ \text{L}$$

$$(1.500\ \cancel{\text{L}})\left(\frac{0.400\ \text{mol BaCl}_2}{\cancel{\text{L}}}\right) = 0.600\ \text{mol BaCl}_2$$

Step 2. Use the mole-ratio method to calculate the moles and grams of AgCl:

mol $BaCl_2 \longrightarrow$ mol AgCl $\longrightarrow$ g AgCl

$$(0.600\ \cancel{\text{mol BaCl}_2})\left(\frac{2\ \cancel{\text{mol AgCl}}}{1\ \cancel{\text{mol BaCl}_2}}\right)\left(\frac{143.4\ \text{g AgCl}}{\cancel{\text{mol AgCl}}}\right) = 172\ \text{g AgCl}$$

Practice 14.8

How many grams of lead(II) iodide will be precipitated by adding sufficient $Pb(NO_3)_2$ to react with 750 mL of 0.250 M KI solution?

$$2\ KI(aq) + Pb(NO_3)_2(aq) \longrightarrow PbI_2(s) + 2\ KNO_3(aq)$$

 ## 14.7 Colligative Properties of Solutions

Two solutions—one containing 1 mol (60.06 g) of urea, NH_2CONH_2, and the other containing 1 mol (342.3 g) of sucrose, $C_{12}H_{22}O_{11}$, in 1 kg of water—both have a freezing point of $-1.86°C$, not $0°C$ as for pure water. Urea and sucrose are distinctly different substances, yet they lower the freezing point of the water by the same amount. The only thing apparently common to these two solutions is that each contains 1 mol (6.022×10^{23} molecules) of solute and 1 kg of solvent. In fact, when we dissolve 1 mol of any nonionizable solute in 1 kg of water, the freezing point of the resulting solution is $-1.86°C$.

These results lead us to conclude that the freezing-point depression for a solution containing 6.022×10^{23} solute molecules (particles) and 1 kg of water is a constant, namely, $1.86°C$. Freezing-point depression is a general property of solutions. Furthermore, the amount by which the freezing point is depressed is the same for all solutions made with a given solvent; that is, each solvent shows a characteristic *freezing-point depression constant*. Freezing-point depression constants for several solvents are given in Table 14.5.

The solution formed by the addition of a nonvolatile solute to a solvent has a lower freezing point, a higher boiling point, and a lower vapor pressure than that of the pure solvent. These effects are related and are known as colligative properties. The **colligative properties** are properties that depend only on the number of solute particles in a solution and not on the nature of those particles. Freezing-point depression, boiling-point elevation, and vapor-pressure lowering are colligative properties of solutions.

colligative properties

		Freezing-point depression constant, K_f $\left(\dfrac{°C\ kg\ solvent}{mol\ solute}\right)$		Boiling-point elevation constant, K_b $\left(\dfrac{°C\ kg\ solvent}{mol\ solute}\right)$
Solvent	**Freezing point of pure solvent (°C)**		**Boiling point of pure solvent (°C)**	
Water	0.00	1.86	100.0	0.512
Acetic acid	16.6	3.90	118.5	3.07
Benzene	5.5	5.1	80.1	2.53
Camphor	178	40	208.2	5.95

TABLE 14.5 Freezing-Point Depression and Boiling-Point Elevation Constants of Selected Solvents

Microencapsulation

Producing chemical reactions so that they occur at precisely the correct moment is a common task for chemists. For this to happen, one or more of the reactants must be stored separately and released under controlled conditions precisely when the reaction is desired. One technique that accomplishes this is microencapsulation, in which reactive chemicals—solids, liquids, or gases—are sealed in tiny capsules. The material forming the wall of the capsule is carefully chosen so that the encapsulated chemicals can be released at the appropriate time. This release can be accomplished in a variety of ways—dissolving the capsules; diffusion through the capsule walls; and by mechanical, thermal, electrical, or chemical disruption of the capsules.

In one type of microencapsulation, water diffuses into the capsule and forms a solution that then diffuses into the surroundings at a constant rate. Some types of capsules contain materials that dissolve at a certain level of acidity and form pores in the capsule through which the encapsulated materials escape. Still other types of capsules dissolve completely over a given period of time.

Applications of microencapsulation are found everywhere. Carbonless paper, often used in receipts, makes use of pressure-sensitive microcapsules containing colorless dye precursors. Another reactive substance is present and converts the precursor to the colored form when pressure is applied by a pen or printer.

Microencapsulated products are found in our kitchens. Flavorings are en-

capsulated to make them easier to store in a powdered state, cut evaporation, and reduce reactions with the air. These advantages increase shelf life. Flavoring microcapsules may also be heat sensitive and release their contents during cooking or pressure sensitive (as in chewing gum) and release their contents upon chewing.

Still other encapsulated products are found in our bathrooms. Time-release microencapsulation is used in deodorants, moisturizers, colognes, and perfumes. The encapsulation process prevents evaporation, decomposition, and unwanted reactions with the air and other ingredients. Drugs and medications are frequently encapsulated to dissolve slowly over a long period of time in the body. These medications generally work in the intestinal tract, but some may also be given by injection to work within other tissues.

Fragrances have undergone microencapsulation in such products as cosmetics, health care products, detergents, and even foods. Encapsulated fragrances are responsible for the ever-present scratch-and-sniff labels found in children's books and fashion magazines. When the paper is scratched or pulled open, the fragrance is released into the air.

Other applications of microencapsulation include time-release pesticides and neutralizer for contact lenses, as well as special additives in detergents, cleaners, and paints.

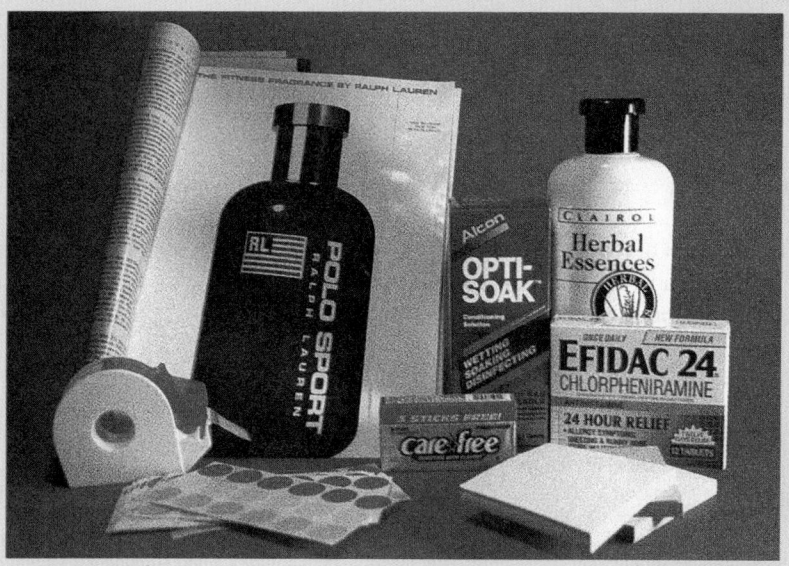

◀ **Many items we come across in our daily lives are microencapsulated. These are just a few examples.**

The colligative properties of a solution can be considered in terms of vapor pressure. The vapor pressure of a pure liquid depends on the tendency of molecules to escape from its surface. If 10% of the molecules in a solution are nonvolatile solute molecules, the vapor pressure of the solution is 10% lower than that of the pure solvent. The vapor pressure is lower because the surface of the solution contains 10% nonvolatile molecules and 90% of the volatile solvent molecules. A liquid

◀ FIGURE 14.8
Vapor-pressure curves of pure water and water solutions, showing (a) boiling-point elevation and (b) freezing-point depression effects (concentration: 1 mol solute/1 kg water).

boils when its vapor pressure equals the pressure of the atmosphere. We can thus see that the solution just described as having a lower vapor pressure will have a higher boiling point than the pure solvent. The solution with a lowered vapor pressure doesn't boil until it has been heated above the boiling point of the solvent (see Figure 14.8a). Each solvent has its own characteristic boiling-point elevation constant (Table 14.5). The boiling-point elevation constant is based on a solution that contains 1 mol of solute particles per kilogram of solvent. For example, the boiling-point elevation constant for a solution containing 1 mol of solute particles per kilogram of water is 0.512°C, which means that this water solution will boil at 100.512°C.

The freezing behavior of a solution can also be considered in terms of lowered vapor pressure. Figure 14.8b shows the vapor-pressure relationships of ice, water, and a solution containing 1 mol of solute per kilogram of water. The freezing point of water is at the intersection of the liquid and solid vapor-pressure curves (i.e., at the point where water and ice have the same vapor pressure). Because the vapor pressure of the liquid is lowered by the solute, the vapor-pressure curve of the solution does not intersect the vapor-pressure curve of the solid until the solution has been cooled below the freezing point of pure water. So the solution must be cooled below 0°C in order for it to freeze.

The foregoing discussion dealing with freezing-point depressions is restricted to *un-ionized* substances. The discussion of boiling-point elevations is restricted to *nonvolatile* and un-ionized substances. The colligative properties of ionized substances are not under consideration at this point; we will discuss them in Chapter 15.

Some practical applications involving colligative properties are (1) use of salt–ice mixtures to provide low freezing temperatures for homemade ice cream, (2) use of sodium chloride or calcium chloride to melt ice from streets, and (3) use of ethylene glycol–water mixtures as antifreeze in automobile radiators (ethylene glycol also raises the boiling point of radiator fluid, thus allowing the engine to operate at a higher temperature).

Both the freezing-point depression and the boiling-point elevation are directly proportional to the number of moles of solute per kilogram of solvent. When we deal with the colligative properties of solutions, another concentration expression,

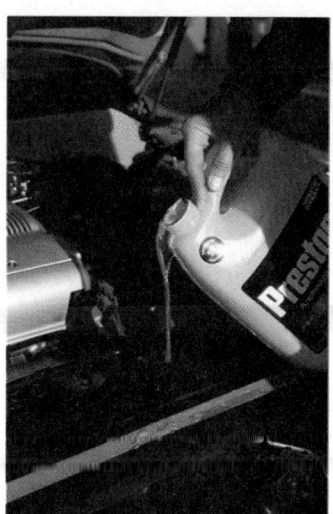

▲
Engine coolant is one application of colligative properties. The addition of coolant to the water in a radiator raises its boiling point and lowers its freezing point.

molality (m) *molality,* is used. The **molality (*m*)** of a solute is the number of moles of solute per kilogram of solvent:

$$m = \frac{\text{mol solute}}{\text{kg solvent}}$$

Note that a lowercase *m* is used for molality concentrations and a capital *M* for molarity. The difference between molality and molarity is that molality refers to moles of solute *per kilogram of solvent,* whereas molarity refers to moles of solute *per liter of solution.* For un-ionized substances, the colligative properties of a solution are directly proportional to its molality.

Molality is independent of volume. It is a mass-to-mass relationship of solute to solvent and allows for experiments, such as freezing-point depression and boiling-point elevation, to be conducted at variable temperatures.

The following equations are used in calculations involving colligative properties and molality:

$$\Delta t_f = mK_f \qquad \Delta t_b = mK_b \qquad m = \frac{\text{mol solute}}{\text{kg solvent}}$$

m = molality; mol solute/kg solvent

Δt_f = freezing-point depression; °C

Δt_b = boiling-point elevation; °C

K_f = freezing-point depression constant; °C kg solvent/mol solute

K_b = boiling-point elevation constant; °C kg solvent/mol solute

Sodium chloride or calcium ▶ chloride is used to melt ice on snowy streets and highways.

What is the molality (*m*) of a solution prepared by dissolving 2.70 g CH_3OH in 25.0 g H_2O?

Example 14.13

Solution

Since $m = \dfrac{\text{mol solute}}{\text{kg solvent}}$, the conversion is

$$\frac{2.70 \text{ g } CH_3OH}{25.0 \text{ g } H_2O} \longrightarrow \frac{\text{mol } CH_3OH}{25.0 \text{ g } H_2O} \longrightarrow \frac{\text{mol } CH_3OH}{1 \text{ kg } H_2O}$$

The molar mass of CH_3OH is (12.01 + 4.032 + 16.00) or 32.04 g/mol:

$$\left(\frac{2.70 \text{ g } CH_3OH}{25.0 \text{ g } H_2O}\right)\left(\frac{1 \text{ mol } CH_3OH}{32.04 \text{ g } CH_3OH}\right)\left(\frac{1000 \text{ g } H_2O}{1 \text{ kg } H_2O}\right) = \frac{3.37 \text{ mol } CH_3OH}{1 \text{ kg } H_2O}$$

The molality is 3.37 *m*.

Practice 14.9

What is the molality of a solution prepared by dissolving 150.0 g $C_6H_{12}O_6$ in 600.0 g H_2O?

A solution is made by dissolving 100. g of ethylene glycol, $C_2H_6O_2$, in 200. g of water. What is the freezing point of this solution?

Example 14.14

To calculate the freezing point of the solution, we first need to calculate Δt_f, the change in freezing point. Use the equation

Solution

$$\Delta t_f = mK_f = \frac{\text{mol solute}}{\text{kg solvent}} \times K_f$$

K_f (for water): $\dfrac{1.86°C \text{ kg solvent}}{\text{mol solute}}$ (from Table 14.5)

mol solute: $(100. \text{ g } C_2H_6O_2)\left(\dfrac{1 \text{ mol } C_2H_6O_2}{62.07 \text{ g } C_2H_6O_2}\right) = 1.61 \text{ mol } C_2H_6O_2$

kg solvent: $(200. \text{ g } H_2O)\left(\dfrac{1 \text{ kg}}{1000 \text{ g}}\right) = 0.200 \text{ kg } H_2O$

$$\Delta t_f = \left(\frac{1.61 \text{ mol } C_2H_6O_2}{0.200 \text{ kg } H_2O}\right)\left(\frac{1.86°C \text{ kg } H_2O}{1 \text{ mol } C_2H_6O_2}\right) = 15.0°C$$

The freezing-point depression, 15.0°C, must be subtracted from 0°C, the freezing point of the pure solvent (water):

freezing point of solution = freezing point of solvent − Δt_f

$$= 0.0°C - 15.0°C = -15.0°C$$

Therefore the freezing point of the solution is −15.0°C.

Example 14.15 A solution made by dissolving 4.71 g of a compound of unknown molar mass in 100.0 g of water has a freezing point of −1.46°C. What is the molar mass of the compound?

Solution   First substitute the data in $\Delta t_f = mK_f$ and solve for m:

$\Delta t_f = +1.46$ (since the solvent, water, freezes at 0°C)

$$K_f = \frac{1.86°C \text{ kg } H_2O}{\text{mol solute}}$$

$$1.46°C = mK_f = m \times \frac{1.86°C \text{ kg } H_2O}{\text{mol solute}}$$

$$m = \frac{1.46°C \times \text{mol solute}}{1.86°C \times \text{kg } H_2O} = \frac{0.785 \text{ mol solute}}{\text{kg } H_2O}$$

Now convert the data, 4.71 g solute/100.0 g H_2O, to g/mol:

$$\left(\frac{4.71 \text{ g solute}}{100.0 \text{ g } H_2O}\right)\left(\frac{1000 \text{ g } H_2O}{1 \text{ kg } H_2O}\right)\left(\frac{1 \text{ kg } H_2O}{0.785 \text{ mol solute}}\right) = 60.0 \text{ g/mol}$$

The molar mass of the compound is 60.0 g/mol.

Practice 14.10

What is the freezing point of the solution in Practice Exercise 14.9? What is the boiling point?

 14.8   **Osmosis and Osmotic Pressure**

When red blood cells are put into distilled water, they gradually swell and in time may burst. If red blood cells are put in a 5% urea (or a 5% salt) solution, they gradually shrink and take on a wrinkled appearance. The cells behave in this fashion

semipermeable membrane   because they are enclosed in semipermeable membranes. A **semipermeable membrane** allows the passage of water (solvent) molecules through it in either direction but prevents the passage of larger solute molecules or ions. When two solutions of different concentrations (or water and a water solution) are separated by a semipermeable membrane, water diffuses through the membrane from the solution of lower concentration into the solution of higher concentration. The diffusion of water, either from a dilute solution or from pure water, through a semipermeable membrane into a

osmosis   solution of higher concentration is called **osmosis.**

A 0.90% (0.15 M) sodium chloride solution is known as a *physiological saline solution* because it is *isotonic* with blood plasma; that is, it has the same osmotic pressure as blood plasma. Because each mole of NaCl yields about 2 mol of ions when in solution, the solute particle concentration in physiological saline solution is nearly 0.30 M. Five percent glucose solution (0.28 M) is also approximately isotonic with blood plasma. Blood cells neither swell nor shrink in an isotonic solution.

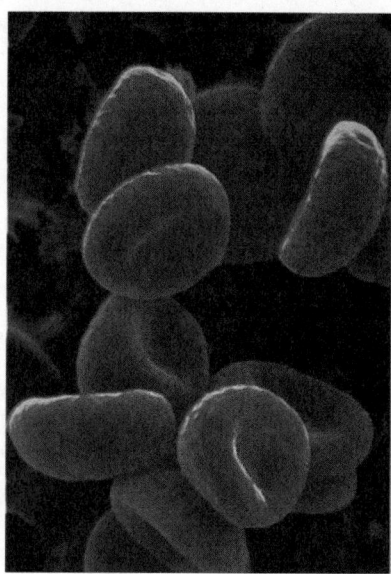

▲
Human red blood cells. Left: In an isotonic solution the concentration is the same inside and outside the cell (0.9% saline). Center: In a hypertonic solution (1.6% saline) water leaves the cells causing them to crenate (shrink). Right: In a hypotonic solution (0.2% saline) the cells swell as water moves into the cell center. Cells do not change in size. Magnification is 260,000×.

The cells described in the preceding paragraph swell in water because water is *hypotonic* to cell plasma. The cells shrink in 5% urea solution because the urea solution is *hypertonic* to the cell plasma. To prevent possible injury to blood cells by osmosis, fluids for intravenous use are usually made up at approximately isotonic concentration.

All solutions exhibit *osmotic pressure,* which is another colligative property. Osmotic pressure is dependent only on the concentration of the solute particles and is independent of their nature. The osmotic pressure of a solution can be measured by determining the amount of counterpressure needed to prevent osmosis; this pressure can be very large. The osmotic pressure of a solution containing 1 mol of solute particles in 1 kg of water is about 22.4 atm, which is about the same as the pressure exerted by 1 mol of a gas confined in a volume of 1 L at 0°C.

Osmosis has a role in many biological processes, and semipermeable membranes occur commonly in living organisms. An example is the roots of plants, which are covered with tiny structures called root hairs; soil water enters the plant by osmosis, passing through the semipermeable membranes covering the root hairs. Artificial or synthetic membranes can also be made.

Osmosis can be demonstrated with the simple laboratory setup shown in Figure 14.9. As a result of osmotic pressure, water passes through the cellophane membrane into the thistle tube, causing the solution level to rise. In osmosis the net transfer of water is always from a less concentrated to a more concentrated solution; that is, the effect is toward equalization of the concentration on both sides of the membrane. Note that the effective movement of water in osmosis is always from the region of *higher water concentration* to the region of *lower water concentration.*

Osmosis can be explained by assuming that a semipermeable membrane has passages that permit water molecules and other small molecules to pass in either direction. Both sides of the membrane are constantly being struck by water molecules in random motion. The number of water molecules crossing the membrane is proportional to the number of water molecule-to-membrane impacts per unit of time. Because the solute molecules or ions reduce the concentration of water, there are

FIGURE 14.9 ▶
Laboratory demonstration of osmosis: As a result of osmosis, water passes through the membrane causing the solution to rise in the thistle tube.

Cross section on molecular level

Semipermeable membrane

○ Sugar molecule
· Water molecule

Rising solution level

Thistle tube

Sugar solution

Water

Semipermeable membrane (cellophane)

more water molecules and thus more water molecule impacts on the side with the lower solute concentration (more dilute solution). The greater number of water molecule-to-membrane impacts on the dilute side thus causes a net transfer of water to the more concentrated solution. Again, note that the overall process involves the net transfer, by diffusion through the membrane, of water molecules from a region of higher water concentration (dilute solution) to one of lower water concentration (more concentrated solution).

This is a simplified picture of osmosis. No one has ever seen the hypothetical passages that allow water molecules and other small molecules or ions to pass through them. Alternative explanations have been proposed, but our discussion has been confined to water solutions. Osmotic pressure is a general colligative property, however, and is known to occur in nonaqueous systems.

Concepts in Review

1. Describe the types of solutions.

2. List the general properties of solutions.

3. Describe and illustrate the process by which an ionic substance dissolves in water.

4. Indicate the effects of temperature and pressure on the solubility of solids and gases in liquids.

5. Identify and explain the factors affecting the rate at which a solid dissolves in a liquid.

6. Use a solubility table or graph to determine whether a solution is saturated, unsaturated, or supersaturated at a given temperature.

7. Calculate the mass percent or volume percent for a solution.

8. Calculate the amount of solute in a given quantity of a solution when given the mass percent or volume percent of a solution.

9. Calculate the molarity of a solution from the volume and the mass, or moles, of solute.

10. Calculate the mass of a substance necessary to prepare a solution of specified volume and molarity.

11. Determine the resulting molarity in a typical dilution problem.

12. Apply stoichiometry to chemical reactions involving solutions.

13. Explain the effect of a solute on the vapor pressure of a solvent.

14. Explain the effect of a solute on boiling point and freezing point of a solution.

15. Calculate the boiling and freezing points of a solution from concentration data.

16. Calculate molality and molar mass of a solute from boiling/freezing point data.

17. Explain the process of osmosis.

Key Terms

colligative properties (14.7)
concentrated solution (14.3)
concentration of a solution (14.2)
dilute solution (14.6)
immiscible (14.2)
miscible (14.2)

molality (m) (14.7)
molarity (M) (14.6)
osmosis (14.8)
parts per million (ppm) (14.6)
saturated solution (14.3)
semipermeable membrane (14.8)

solubility (14.2)
solute (14.1)
solution (14.1)
solvent (14.1)
supersaturated solution (14.3)
unsaturated solution (14.6)

Questions

1. Sketch the orientation of water molecules (a) about a single sodium ion and (b) about a single chloride ion in solution.

2. Estimate the number of grams of sodium fluoride that would dissolve in 100 g of water at 50°C. (Table 14.2)

3. What is the solubility at 25°C of these substances? (Figure 14.4)
(a) potassium chloride   **(c)** potassium nitrate
(b) potassium chlorate

4. What is different in the solubility trend of the potassium halides compared with that of the lithium halides and the sodium halides? (Table 14.2)

5. What is the solubility, in grams of solute per 100 g of H_2O, of (a) $KClO_3$ at 60°C; (b) HCl at 20°C; (c) Li_2SO_4 at 80°C; and (d) KNO_3 at 0°C? (Figure 14.4)

6. Which substance, KNO_3 or NH_4Cl, shows the greater increase in solubility with increased temperature? (Figure 14.4)

7. Does a 2-molal solution in benzene or a 1-molal solution in camphor show the greater freezing-point depression? (Table 14.5)

8. What would be the total surface area if the 1-cm cube in Figure 14.5 were cut into cubes 0.01 cm on a side?

9. At which temperatures—10°C, 20°C, 30°C, 40°C, or 50°C—would you expect a solution made from 63 g of ammonium chloride and 150 g of water to be unsaturated? (Figure 14.4)

10. Explain why the rate of dissolving decreases. (Figure 14.6)

11. Explain how a supersaturated solution of $NaC_2H_3O_2$ can be prepared and proven to be supersaturated.

12. Assume that the thistle tube in Figure 14.9 contains 1.0 M sugar solution and that the water in the beaker has just been replaced by a 2.0 M solution of urea. Would the solution level in the thistle tube continue to rise, remain constant, or fall? Explain.

13. Name and distinguish between the two components of a solution.

14. Is it always apparent in a solution which component is the solute, for example, in a solution of a liquid in a liquid?

15. Explain why the solute does not settle out of a solution.

16. Is it possible to have one solid dissolved in another? Explain.

17. An aqueous solution of KCl is colorless, $KMnO_4$ is purple, and $K_2Cr_2O_7$ is orange. What color would you expect of an aqueous solution of $Na_2Cr_2O_7$? Explain.

18. Explain why hexane will dissolve benzene but will not dissolve sodium chloride.

19. Some drinks like tea are consumed either hot or cold, whereas others like Coca Cola are drunk only cold. Why?

20. Why is air considered to be a solution?

21. In which will a teaspoonful of sugar dissolve more rapidly, 200 mL of iced tea or 200 mL of hot coffee? Explain in terms of the KMT.

22. What is the effect of pressure on the solubility of gases in liquids? Of solids in liquids?

23. Why do smaller particles dissolve faster than large ones?

24. In a saturated solution containing undissolved solute, solute is continuously dissolving, but the concentration of the solution remains unchanged. Explain.

25. Explain why there is no apparent reaction when crystals of $AgNO_3$ and NaCl are mixed, but a reaction is apparent immediately when solutions of $AgNO_3$ and NaCl are mixed.

26. What do we mean when we say that concentrated nitric acid, HNO_3, is 16 molar?

27. Will 1 L of 1 M NaCl contain more chloride ions than 0.5 L of 1 M $MgCl_2$? Explain.

28. Champagne is usually cooled in a refrigerator prior to opening. It's also opened very carefully. What would happen if a warm bottle of champagne is shaken and opened quickly and forcefully?

29. Describe how you would prepare 750 mL of 5 M NaCl solution.

30. Explain in terms of the KMT how a semipermeable membrane functions when placed between pure water and a 10% sugar solution.

31. Which has the higher osmotic pressure, a solution containing 100 g of urea, NH_2CONH_2, in 1 kg H_2O or a solution containing 150 g of glucose, $C_6H_{12}O_6$, in 1 kg H_2O?

32. Explain why a lettuce leaf in contact with salad dressing containing salt and vinegar soon becomes wilted and limp whereas another lettuce leaf in contact with plain water remains crisp.

33. A group of shipwreck survivors floated for several days on a life raft before being rescued. Those who had drunk some seawater were found to be suffering the most from dehydration. Explain.

34. Arrange the following bases (in descending order) according to the volume of each that will react with 1 L of 1 M HCl:
    (a) 1 M NaOH
    (b) 1.5 M $Ca(OH)_2$
    (c) 2 M KOH
    (d) 0.6 M $Ba(OH)_2$

*35. Explain in terms of vapor pressure why the boiling point of a solution containing a nonvolatile solute is higher than that of the pure solvent.

36. Explain why the freezing point of a solution is lower than the freezing point of the pure solvent.

37. Which would be colder, a glass of water and crushed ice or a glass of Seven-Up and crushed ice? Explain.

38. When water and ice are mixed, the temperature of the mixture is 0°C. But, if methyl alcohol and ice are mixed, a temperature of -10°C is readily attained. Explain why the two mixtures show such different temperature behavior.

39. Which would be more effective in lowering the freezing point of 500. g of water?
    (a) 100. g of sucrose, $C_{12}H_{22}O_{11}$, or 100. g of ethyl alcohol, C_2H_5OH
    (b) 100. g of sucrose or 20.0 g of ethyl alcohol
    (c) 20.0 g of ethyl alcohol or 20.0 g of methyl alcohol, CH_3OH

40. Is the molarity of a 5 m aqueous solution of NaCl greater or less than 5 M? Explain.

Paired Exercises

41. Which of the substances listed below are reasonably soluble and which are insoluble in water? (See Figure 14.2 or Appendix V.)
    (a) KOH
    (b) $NiCl_2$
    (c) ZnS
    (d) $AgC_2H_3O_2$
    (e) Na_2CrO_4

42. Which of the substances listed below are reasonably soluble and which are insoluble in water? (See Figure 14.2 or Appendix V.)
    (a) PbI_2
    (b) $MgCO_3$
    (c) $CaCl_2$
    (d) $Fe(NO_3)_3$
    (e) $BaSO_4$

Percent Solutions

43. Calculate the mass percent of the following solutions:
    (a) 25.0 g NaBr + 100.0 g H_2O
    (b) 1.20 g K_2SO_4 + 10.0 g H_2O

44. Calculate the mass percent of the following solutions:
    (a) 40.0 g $Mg(NO_3)_2$ + 500.0 g H_2O
    (b) 17.5 g $NaNO_3$ + 250.0 g H_2O

45. How many grams of a solution that is 12.5% by mass $AgNO_3$ would contain 30.0 g of $AgNO_3$?

46. How many grams of a solution that is 12.5% by mass $AgNO_3$ would contain 0.400 mol of $AgNO_3$?

47. Calculate the mass percent of the following solutions:
    (a) 60.0 g NaCl + 200.0 g H_2O
    (b) 0.25 mol $HC_2H_3O_2$ + 3.0 mol H_2O

49. How much solute is present in 65 g of 5.0% KCl solution?

51. Calculate the mass/volume percent of a solution made by dissolving 22.0 g of CH_3OH (methanol) in C_2H_5OH (ethanol) to make 100. mL of solution.

53. What is the volume percent of 10.0 mL of CH_3OH (methanol) dissolved in water to a volume of 40.0 mL?

Molarity Problems

55. Calculate the molarity of the following solutions:
    (a) 0.10 mol of solute in 250 mL of solution
    (b) 2.5 mol of NaCl in 0.650 L of solution
    (c) 53.0 g of Na_2CrO_4 in 1.00 L of solution
    (d) 260 g of $C_6H_{12}O_6$ in 800. mL of solution

57. Calculate the number of moles of solute in each of the following solutions:
    (a) 40.0 L of 1.0 M LiCl
    (b) 25.0 mL of 3.00 M H_2SO_4

59. Calculate the grams of solute in each of the following solutions:
    (a) 150 L of 1.0 M NaCl
    (b) 260 mL of 18 M H_2SO_4

61. How many milliliters of 0.256 M KCl solution will contain the following?
    (a) 0.430 mol of KCl
    (b) 20.0 g of KCl

Dilution Problems

63. What will be the molarity of the resulting solutions made by mixing the following? Assume volumes are additive.
    (a) 100. mL 1.0 M HCl + 150 mL 2.0 M HCl
    (b) 25.0 mL 12.5 M NaCl + 75.0 mL 2.00 M NaCl

65. Calculate the volume of concentrated reagent required to prepare the diluted solutions indicated:
    (a) 15 M NH_3 to prepare 50. mL of 6.0 M NH_3
    (b) 18 M H_2SO_4 to prepare 250 mL of 10.0 M H_2SO_4

67. Calculate the molarity of the solutions made by mixing 250 mL of 0.75 M H_2SO_4 with
    (a) 150 mL of H_2O
    (b) 250 mL of 0.70 M H_2SO_4

48. Calculate the mass percent of the following solutions:
    (a) 145.0 g NaOH in 1.5 kg H_2O
    (b) 1.0 m solution of $C_6H_{12}O_6$ in water

50. How much solute is present in 250. g of 15.0% K_2CrO_4 solution?

52. Calculate the mass/volume percent of a solution made by dissolving 4.20 g of NaCl in H_2O to make 12.5 mL of solution.

54. What is the volume percent of 2.0 mL of hexane, C_6H_{14}, dissolved in benzene, C_6H_6, to a volume of 9.0 mL?

56. Calculate the molarity of the following solutions:
    (a) 0.025 mol of HCl in 10. mL of solution
    (b) 0.35 mol $BaCl_2 \cdot 2 H_2O$ in 593 mL of solution
    (c) 1.50 g of $Al_2(SO_4)_3$ in 2.00 L of solution
    (d) 0.0282 g of $Ca(NO_3)_2$ in 1.00 mL of solution

58. Calculate the number of moles of solute in each of the following solutions:
    (a) 349 mL of 0.0010 M NaOH
    (b) 5000. mL of 3.1 M $CoCl_2$

60. Calculate the grams of solute in each of the following solutions:
    (a) 0.035 L of 10.0 M HCl
    (b) 8.00 mL of 8.00 M $Na_2C_2O_4$

62. How many milliliters of 0.256 M KCl solution will contain the following?
    (a) 10.0 mol of KCl
    *(b) 71.0 g of chloride ion, Cl^-

64. What will be the molarity of the resulting solutions made by mixing the following? Assume volumes are additive.
    (a) 200. mL of 12 M HCl + 200.0 mL H_2O
    (b) 60.0 mL of 0.60 M $ZnSO_4$ + 500. mL H_2O

66. Calculate the volume of concentrated reagent required to prepare the diluted solutions indicated:
    (a) 12 M HCl to prepare 400. mL of 6.0 M HCl
    (b) 16 M HNO_3 to prepare 100. mL of 2.5 M HNO_3

68. Calculate the molarity of the solutions made by mixing 250 mL of 0.75 M H_2SO_4 with
    (a) 400. mL of 2.50 M H_2SO_4
    (b) 375 mL of H_2O

Stoichiometry Problems

69. Given the equation
$$BaCl_2(aq) + K_2CrO_4(aq) \longrightarrow BaCrO_4(s) + 2\ KCl(aq),$$
calculate
  (a) The grams of $BaCrO_4$ that can be obtained from 100.0 mL of 0.300 M $BaCl_2$
  (b) The volume of 1.0 M $BaCl_2$ solution needed to react with 50.0 mL of 0.300 M K_2CrO_4 solution

71. Given the equation
$$6\ FeCl_2(aq) + K_2Cr_2O_7(aq) + 14\ HCl(aq) \longrightarrow$$
$$6\ FeCl_3(aq) + 2\ CrCl_3(aq) + 2\ KCl(aq) + 7\ H_2O(l),$$
calculate
  (a) moles KCl produced from 2.0 mol $FeCl_2$
  (b) moles $CrCl_3$ produced from 1.0 mol $FeCl_2$
  (c) moles $FeCl_2$ required to react with 0.050 mol $K_2Cr_2O_7$
  (d) milliliters of 0.060 M $K_2Cr_2O_7$ required to react with 0.025 mol $FeCl_2$
  (e) milliliters of 6.0 M HCl required to react with 15.0 mL 6.0 M $FeCl_2$

70. Given the equation
$$3\ MgCl_2(aq) + 2\ Na_3PO_4(aq) \longrightarrow$$
$$Mg_3(PO_4)_2(s) + 6\ NaCl(aq),$$
calculate
  (a) milliliters of 0.250 M Na_3PO_4 that will react with 50.0 mL of 0.250 M $MgCl_2$
  (b) grams of $Mg_3(PO_4)_2$ that will be formed from 50.0 mL of 0.250 M $MgCl_2$

72. Given the equation
$$2\ KMnO_4(aq) + 16\ HCl(aq) \longrightarrow$$
$$2\ MnCl_2(aq) + 5\ Cl_2(g) + 8\ H_2O(l) + 2\ KCl(aq),$$
calculate
  (a) moles Cl_2 produced from 0.050 mol $KMnO_4$
  (b) moles of HCl required to react with 1.0 L of 2.0 M $KMnO_4$
  (c) milliliters of 6.0 M HCl required to react with 200. mL of 0.50 M $KMnO_4$
  (d) liters of Cl_2 gas at STP produced by the reaction of 75.0 mL of 6.0 M HCl

Molality and Colligative Properties Problems

73. Calculate the molality of these solutions:
  (a) 14.0 g CH_3OH in 100.0 g of H_2O
  (b) 2.50 mol of benzene, C_6H_6, in 250 g of hexane, C_6H_{14}

75. What is the (a) molality, (b) freezing point, and (c) boiling point of a solution containing 2.68 g of naphthalene, $C_{10}H_8$, in 38.4 g of benzene, C_6H_6?

*77. The freezing point of a solution of 8.00 g of an unknown compound dissolved in 60.0 g of acetic acid is 13.2°C. Calculate the molar mass of the compound.

74. Calculate the molality of these solutions:
  (a) 1.0 g $C_6H_{12}O_6$ in 1.0 g H_2O
  (b) 0.250 mol iodine in 1.0 kg H_2O

76. What is the (a) molality, (b) freezing point, and (c) boiling point of a solution containing 100.0 g of ethylene glycol, $C_2H_6O_2$, in 150.0 g of water?

*78. What is the molar mass of a compound if 4.80 g of the compound dissolved in 22.0 g of H_2O gives a solution that freezes at -2.50°C?

Additional Exercises

79. How many grams of solution, 10.0% NaOH by mass, are required to neutralize 150 mL of a 1.0 M HCl solution?

*80. How many grams of solution, 10.0% NaOH by mass, are required to neutralize 250.0 g of a 1.0 m solution of HCl?

*81. A sugar syrup solution contains 15.0% sugar, $C_{12}H_{22}O_{11}$, by mass and has a density of 1.06 g/mL.
  (a) How many grams of sugar are in 1.0 L of this syrup?
  (b) What is the molarity of this solution?
  (c) What is the molality of this solution?

*82. A solution of 3.84 g C_4H_2N (empirical formula) in 250.0 g of benzene depresses the freezing point of benzene 0.614°C. What is the molecular formula for the compound?

*83. Hydrochloric acid, HCl, is sold as a concentrated aqueous solution (12.0 mol/L). If the density of the solution is 1.18 g/mL, determine the molality of the solution.

*84. How many grams of KNO_3 are needed to make 450 mL of a solution that is to contain 5.5 mg/mL of potassium ion? Calculate the molarity of the solution.

85. What mass of 5.50% solution can be prepared from 25.0 g KCl?

86. Physiological saline, NaCl, solutions used in intravenous injections have a concentration of 0.90% NaCl (mass/volume).
   (a) How many grams of NaCl are needed to prepare 500.0 mL of this solution?
 *(b) How much water must evaporate from this solution to give a solution that is 9.0% NaCl (mass/volume)?

*87. A solution is made from 50.0 g KNO_3 and 175 g H_2O. How many grams of water must evaporate to give a saturated solution of KNO_3 in water at 20°C? (See Figure 14.4.)

88. What volume of 70.0% rubbing alcohol can you prepare if you have only 150 mL of pure isopropyl alcohol on hand?

89. At 20°C, an aqueous solution of HNO_3 that is 35.0% HNO_3 by mass has a density of 1.21 g/mL.
   (a) How many grams of HNO_3 are present in 1.00 L of this solution?
   (b) What volume of this solution will contain 500. g HNO_3?

*90. What is the molarity of a nitric acid solution, if the solution is 35.0% HNO_3 by mass and has a density of 1.21 g/mL?

91. To what volume must a solution of 80.0 g H_2SO_4 in 500.0 mL of solution be diluted to give a 0.10 M solution?

92. Calculate the milliliters of water that must be added to 300.0 mL of 1.40 M HCl to make a solution that is 0.500 M HCl?

93. A 10.0-mL sample of 16 M HNO_3 is diluted to 500.0 mL. What is the molarity of the final solution?

94. Given a 5.00 M KOH solution, how would you prepare 250.0 mL of 0.625 M KOH?

95. **(a)** How many moles of hydrogen will be liberated from 200.0 mL of 3.00 M HCl reacting with an excess of magnesium? The equation is

$$Mg(s) + 2\,HCl(aq) \longrightarrow MgCl_2(aq) + H_2(g)$$

   (b) How many liters of hydrogen gas, H_2, measured at 27°C and 720 torr, will be obtained? (*Hint*: Use the ideal gas equation.)

*96. What is the molarity of an HCl solution, 150.0 mL of which, when treated with excess magnesium, liberates 3.50 L of H_2 gas measured at STP?

97. Suppose you start with 100. mL of distilled water, then you add one drop (20 drops/mL) of concentrated acetic acid, $HC_2H_3O_2$ (17.8 M). What is the molarity of the resulting solution?

98. Which will be more effective in neutralizing stomach acid, HCl: a tablet containing 12.0 g $Mg(OH)_2$ or a tablet containing 10.0 g $Al(OH)_3$? Show evidence for your answer.

99. Which would be more effective as an antifreeze in an automobile radiator? A solution containing
   (a) 10 kg of methyl alcohol, CH_3OH, or 10 kg of ethyl alcohol, C_2H_5OH?
   (b) 10 m solution of methyl alcohol or 10 m solution of ethyl alcohol?

100. Automobile battery acid is 38% H_2SO_4 and has a density of 1.29 g/mL. Calculate the molality and the molarity of this solution.

101. A sugar solution made to feed hummingbirds contains 1.00 lb of sugar, $C_{12}H_{22}O_{11}$, to 4.00 lb of water. Can this solution be put outside without freezing where the temperature falls to 20.0°F at night? Show evidence for your answer.

*102. What is the (a) molality and (b) boiling point of an aqueous sugar, $C_{12}H_{22}O_{11}$, solution that freezes at −5.4°C?

103. A solution of 6.20 g $C_2H_6O_2$ in water has a freezing point of −0.372°C. How many grams of H_2O are in the solution?

104. What (a) mass and (b) volume of ethylene glycol ($C_2H_6O_2$, density = 1.11 g/mL) should be added to 12.0 L of water in an automobile radiator to protect it from freezing at −20°C? (c) To what temperature Fahrenheit will the radiator be protected?

105. Can a saturated solution ever be a dilute solution? Explain.

106. What volume of 0.65 M HCl is needed to completely neutralize 12 g NaOH?

*107. If 150 mL of 0.055 M HNO_3 are needed to completely neutralize 1.48 g of an *impure* sample of sodium hydrogen carbonate (baking soda), what percent of the sample is baking soda?

108. **(a)** How much water must be added to concentrated sulfuric acid, H_2SO_4 (17.8 M), to prepare 8.4 L of 1.5 M sulfuric acid solution?
   (b) How many moles of H_2SO_4 are in each milliliter of the original concentrate?
   (c) How many moles are in each milliliter of the diluted solution?

109. An aqueous solution freezes at −3.6°C. What is its boiling temperature?

*110. How would you prepare a 6.00 M HNO_3 solution if only 3.00 M and 12.0 M solutions of the acid are available for mixing?

*111. A 20.0-mL portion of an HBr solution of unknown strength is diluted to exactly 240 mL. If 100.0 mL of this diluted solution requires 88.4 mL of 0.37 M NaOH to achieve complete neutralization, what was the strength of the original HBr solution?

112. When 80.5 mL of 0.642 M $Ba(NO_3)_2$ is mixed with 44.5 mL of 0.743 M KOH, a precipitate of $Ba(OH)_2$ forms. How many grams of $Ba(OH)_2$ do you expect?

113. Exactly 300. g of a 5.0% sucrose solution is to be prepared. How many grams of a 2.0% solution of sucrose would contain the same number of grams of sugar?

114. A 0.25 M solution of lithium carbonate, Li_2CO_3, a drug used to treat manic depression, is prepared.
    (a) How many moles of Li_2CO_3 are present in 45.8 mL of the solution?
    (b) How many grams of Li_2CO_3 are in 750 mL of the same solution?
    (c) How many milliliters of the solution would be needed to supply 6.0 g of the solute?
    (d) If the solution has a density of 1.22 g/mL, what is its mass percent?

115. If a student accidentally mixed 400.0 mL of 0.35 M HCl with 1100 mL of 0.65 M HCl, what would be the final molarity of the hydrochloric acid solution?

Answers to Practice Exercises

14.1	unsaturated	**14.5**	0.84 L (840 mL)	**14.9**	1.387 m	
14.2	10.0% Na_2SO_4 solution	**14.6**	75 mL NaOH	**14.10**	freezing point = $-2.58°C$, boiling point = $100.71°C$	
14.3	2.02 M	**14.7**	5.00×10^{-2} M			
14.4	0.27 g NaCl	**14.8**	43 g			

PUTTING IT TOGETHER
Review for Chapters 12–14

Multiple Choice: *Choose the correct answer to each of the following.*

1. Which of these statements is *not* one of the principal assumptions of the kinetic-molecular theory for an ideal gas?
   (a) All collisions of gaseous molecules are perfectly elastic.
   (b) A mole of any gas occupies 22.4 L at STP.
   (c) Gas molecules have no attraction for one another.
   (d) The average kinetic energy for molecules is the same for all gases at the same temperature.

2. Which of the following is not equal to 1.00 atm?
   (a) 760. cm Hg (c) 760. mm Hg
   (b) 29.9 in. Hg (d) 760. torr

3. If the pressure on 45 mL of gas is changed from 600. torr to 800. torr, the new volume will be
   (a) 60 mL (c) 0.045 L
   (b) 34 mL (d) 22.4 L

4. The volume of a gas is 300. mL at 740. torr and 25°C. If the pressure remains constant and the temperature is raised to 100.°C, the new volume will be
   (a) 240. mL (c) 376 mL
   (b) 1.20 L (d) 75.0 mL

5. The volume of a dry gas is 4.00 L at 15.0°C and 745 torr. What volume will the gas occupy at 40.0°C and 700. torr?
   (a) 4.63 L (b) 3.46 L (c) 3.92 L (d) 4.08 L

6. A sample of Cl_2 occupies 8.50 L at 80.0°C and 740. mm Hg. What volume will the Cl_2 occupy at STP?
   (a) 10.7 L (b) 6.75 L (c) 11.3 L (d) 6.40 L

7. What volume will 8.00 g O_2 occupy at 45°C and 2.00 atm?
   (a) 0.462 L (b) 104 L (c) 9.62 L (d) 3.26 L

8. The density of NH_3 gas at STP is
   (a) 0.760 g/mL (c) 1.32 g/mL
   (b) 0.760 g/L (d) 1.32 g/L

9. The ratio of the relative rate of effusion of methane, CH_4, to sulfur dioxide, SO_2, is
   (a) $^{64}/_{16}$ (b) $^{16}/_{64}$ (c) ¼ (d) ²/₁

10. Measured at 65°C and 500. torr, the mass of 3.21 L of a gas is 3.5 g. The molar mass of this gas is
    (a) 21 g/mole (c) 24 g/mole
    (b) 46 g/mole (d) 130 g/mole

11. Box A contains O_2 (molar mass = 32.0) at a pressure of 200 torr. Box B, which is identical to box A in volume, contains twice as many molecules of CH_4 (molar mass = 16.0) as the molecules of O_2 in box A. The temperatures of the gases are identical. The pressure in box B is
    (a) 100 torr (c) 400 torr
    (b) 200 torr (d) 800 torr

12. A 300.-mL sample of oxygen, O_2, is collected over water at 23°C and 725 torr. If the vapor pressure of water at 23°C is 21.0 torr, the volume of dry O_2 at STP is
    (a) 256 mL (c) 341 mL
    (b) 351 mL (d) 264 mL

13. A tank containing 0.01 mol of neon and 0.04 mol of helium shows a pressure of 1 atm. What is the partial pressure of neon in the tank?
    (a) 0.8 atm (c) 0.2 atm
    (b) 0.01 atm (d) 0.5 atm

14. How many liters of NO_2 (at STP) can be produced from 25.0 g Cu reacting with concentrated nitric acid?

    $Cu(s) + 4\,HNO_3(aq) \longrightarrow$
    $$Cu(NO_3)_2(aq) + 2\,H_2O(l) + 2\,NO_2(g)$$

    (a) 4.41 L (b) 8.82 L (c) 17.6 L (d) 44.8 L

15. How many liters of butane vapor are required to produce 2.0 L CO_2 at STP?

    $2\,C_4H_{10}(g) + 13\,O_2(g) \longrightarrow 8\,CO_2(g) + 10\,H_2O(g)$
    butane

    (a) 2.0 L (b) 4.0 L (c) 0.80 L (d) 0.50 L

16. What volume of CO_2 (at STP) can be produced when 15.0 g C_2H_6 and 50.0 g O_2 are reacted?

    $2\,C_2H_6(g) + 7\,O_2(g) \longrightarrow 4\,CO_2(g) + 6\,H_2O(g)$

    (a) 20.0 L (b) 22.4 L (c) 35.0 L (d) 5.6 L

17. Which of these gases has the highest density at STP?
    (a) N_2O (b) NO_2 (c) Cl_2 (d) SO_2

18. What is the density of CO_2 at 25°C and 0.954 atm?
    (a) 1.72 g/L (c) 0.985 g/L
    (b) 2.04 g/L (d) 1.52 g/L

19. How many molecules are present in 0.025 mol of H_2 gas?
    (a) 1.5×10^{22} molecules
    (b) 3.37×10^{23} molecules
    (c) 2.40×10^{25} molecules
    (d) 1.50×10^{22} molecules

20. 5.60 L of a gas at STP has a mass of 13.0 g. What is the molar mass of the gas?
    (a) 33.2 g/mol (c) 66.4 g/mol
    (b) 26.0 g/mol (d) 52.0 g/mol

21. The heat of fusion of water is
    (a) 4.184 J/g (c) 2.26 kJ/g
    (b) 335 J/g (d) 2.26 kJ/mol

22. The heat of vaporization of water is
    (a) 4.184 J/g (c) 2.26 kJ/g
    (b) 335 J/g (d) 2.26 kJ/mol

23. The specific heat of water is
    (a) 4.184 J/g°C   (c) 2.26 kJ/g°C
    (b) 335 J/g°C   (d) 18 J/g°C

24. The density of water at 4°C is
    (a) 1.0 g/mL   (c) 18.0 g/mL
    (b) 80 g/mL   (d) 14.7 lb/in.^{3}

25. SO_2 can be properly classified as a(n)
    (a) basic anhydride   (c) anhydrous salt
    (b) hydrate   (d) acid anhydride

26. When compared to H_2S, H_2Se, and H_2Te, water is found to have the highest boiling point because it
    (a) has the lowest molar mass
    (b) is the smallest molecule
    (c) has the highest bonding
    (d) forms hydrogen bonds better than the others

27. In which of the following molecules will hydrogen bonding be important?
    (a) H—F   (c) H—Br

    (b) S—H
        |
        H

    (d)

28. Which of the following is an incorrect equation?
    (a) $H_2SO_4 + 2\,NaOH \longrightarrow Na_2SO_4 + 2\,H_2O$
    (b) $C_2H_6 + O_2 \longrightarrow 2\,CO_2 + 3\,H_2$
    (c) $2\,H_2O \xrightarrow[H_2SO_4]{electrolysis} 2\,H_2 + O_2$
    (d) $Ca + 2\,H_2O \longrightarrow H_2 + Ca(OH)_2$

29. Which of the following is an incorrect equation?
    (a) $C + H_2O(g) \xrightarrow{1000°C} CO(g) + H_2(g)$
    (b) $CaO + H_2O \longrightarrow Ca(OH)_2$
    (c) $2\,NO_2 + H_2O \longrightarrow 2\,HNO_3$
    (d) $Cl_2 + H_2O \longrightarrow HCl + HOCl$

30. How many kilojoules are required to change 85 g of water at 25°C to steam at 100.°C?
    (a) 219 kJ   (b) 27 kJ   (c) 590 kJ   (d) 192 kJ

31. A chunk of 0°C ice, mass 145 g, is dropped into 75 g of water at 62°C. The heat of fusion of water is 335 J/g. The result, after thermal equilibrium is attained, will be
    (a) 87 g ice and 133 g liquid water, all at 0°C
    (b) 58 g ice and 162 g liquid water, all at 0°C
    (c) 220 g water at 7°C
    (d) 220 g water at 17°C

32. The formula for iron(II) sulfate heptahydrate is
    (a) $Fe_2SO_4 \cdot 7\,H_2O$ (c) $FeSO_4 \cdot 7\,H_2O$
    (b) $Fe(SO_4)_2 \cdot 6\,H_2O$ (d) $Fe_2(SO_4)_3 \cdot 7\,H_2O$

33. The process by which a solid changes directly to a vapor is called
    (a) vaporization   (c) sublimation
    (b) evaporation   (d) condensation

34. Hydrogen bonding
    (a) occurs only between water molecules
    (b) is stronger than covalent bonding
    (c) can occur between NH_3 and H_2O
    (d) results from strong attractive forces in ionic compounds

35. A liquid boils when
    (a) the vapor pressure of the liquid equals the external pressure above the liquid
    (b) the heat of vaporization exceeds the vapor pressure
    (c) the vapor pressure equals 1 atm
    (d) the normal boiling temperature is reached

36. Consider two beakers, one containing 50 mL of liquid A and the other 50 mL of liquid B. The boiling point of A is 90°C and that of B is 72°C. Which of these statements is correct?
    (a) A will evaporate faster than B.
    (b) B will evaporate faster than A.
    (c) Both A and B evaporate at the same rate.
    (d) Insufficient data to answer the question.

37. 95.0 g of 0.0°C ice is added to exactly 100. g of water at 60.0°C. When the temperature of the mixture first reaches 0.0°C, the mass of ice still present is
    (a) 0.0 g   (c) 10.0 g
    (b) 20.0 g   (d) 75.0 g

38. Which of the following is not a general property of solutions?
    (a) a homogeneous mixture of two or more substances
    (b) variable composition
    (c) dissolved solute breaks down to individual molecules
    (d) the same chemical composition, the same chemical properties, and the same physical properties in every part

39. If NaCl is soluble in water to the extent of 36.0 g NaCl/100 g H_2O at 20°C, then a solution at 20°C containing 45 g NaCl/150 g H_2O would be
    (a) dilute   (c) supersaturated
    (b) saturated   (d) unsaturated

40. If 5.00 g NaCl is dissolved in 25.0 g of water, the percent of NaCl by mass is
    (a) 16.7   (c) 0.20
    (b) 20.0   (d) no correct answer given

41. How many grams of 9.0% $AgNO_3$ solution will contain 5.3 g $AgNO_3$?
    (a) 47.7   (c) 59
    (b) 0.58   (d) no correct answer given

42. The molarity of a solution containing 2.5 mol of acetic acid, $HC_2H_3O_2$, in 400. mL of solution is
    (a) 0.063 M (c) 0.103 M
    (b) 1.0 M (d) 6.3 M

43. What volume of 0.300 M KCl will contain 15.3 g KCl?
    (a) 1.46 L   (c) 61.5 mL
    (b) 683 mL   (d) 4.60 L

44. What mass of $BaCl_2$ will be required to prepare 200. mL of 0.150 M solution?
    (a) 0.750 g   (b) 156 g   (c) 6.25 g   (d) 31.2 g

Problems 45–47 relate to the reaction

$$CaCO_3 + 2\,HCl \longrightarrow CaCl_2 + H_2O + CO_2$$

45. What volume of 6.0 M HCl will be needed to react with 0.350 mol of $CaCO_3$?
    (a) 42.0 mL   (c) 117 mL
    (b) 1.17 L   (d) 583 mL

46. If 400. mL of 2.0 M HCl reacts with excess $CaCO_3$, the volume of CO_2 produced, measured at STP, is
    (a) 18 L   (b) 5.6 L   (c) 9.0 L   (d) 56 L

47. If 5.3 g $CaCl_2$ is produced in the reaction, what is the molarity of the HCl used if 25 mL of it reacted with excess $CaCO_3$?
    (a) 3.8 M (b) 0.19 M (c) 0.38 M (d) 0.42 M

48. If 20.0 g of the nonelectrolyte urea, $CO(NH_2)_2$, is dissolved in 25.0 g of water, the freezing point of the solution will be
    (a) $-2.47°C$ (c) $-24.7°C$
    (b) $-1.40°C$ (d) $-3.72°C$

49. When 256 g of a nonvolatile, nonelectrolyte unknown were dissolved in 500. g H_2O, the freezing point was found to be $-2.79°C$. The molar mass of the unknown solute is
    (a) 357   (b) 62.0   (c) 768   (d) 341

50. How many milliliters of 6.0 M H_2SO_4 must you use to prepare 500. mL of 0.20 M sulfuric acid solution?
    (a) 30   (b) 17   (c) 12   (d) 100

51. How many milliliters of water must be added to 200. mL of 1.40 M HCl to make a solution that is 0.500 M HCl?
    (a) 360. mL   (c) 140. mL
    (b) 560. mL   (d) 280. mL

52. Which procedure is most likely to increase the solubility of most solids in liquids?
    (a) stirring
    (b) pulverizing the solid
    (c) heating the solution
    (d) increasing the pressure

53. The addition of a crystal of $NaClO_3$ to a solution of $NaClO_3$ causes additional crystals to precipitate. The original solution was
    (a) unsaturated   (c) saturated
    (b) dilute   (d) supersaturated

54. Which of these anions will not form a precipitate with silver ions, Ag^+?
    (a) Cl^- (b) NO_3^- (c) Br^- (d) CO_3^{2-}

55. Which of these salts are considered to be soluble in water?
    (a) $BaSO_4$ (b) NH_4Cl (c) AgI (d) PbS

56. A solution of ethyl alcohol and benzene is 40% alcohol by volume. Which statement is correct?
    (a) The solution contains 40 mL of alcohol in 100 mL of solution.
    (b) The solution contains 60 mL of benzene in 100 mL of solution.
    (c) The solution contains 40 mL of alcohol in 100 g of solution.
    (d) The solution is made by dissolving 40 mL of alcohol in 60 mL of benzene.

Free Response Questions

1. Which solution should have a higher boiling point: 215 mL of a 10.0% (m/v) aqueous KCl solution or 224 mL of a 1.10 M aqueous NaCl solution?

2. A glass containing 345 mL of a soft drink (a carbonated beverage) was left sitting out on a kitchen counter. If the CO_2 released at room temperature (25°C) and pressure (1 atm) occupies 1.40 L, at a minimum, what is the concentration (in ppm) of the CO_2 in the original soft drink (assume the density of the original soft drink is 0.965 g/mL).

3. Dina and Murphy were trying to react 100. mL of a 0.10 M HCl solution with KOH. The procedure called for a 10% KOH solution. Dina made a 10% mass/v solution while Murphy made a 10% by mass solution. (Assume there is no volume change upon dissolving KOH.) Which solution required less volume to fully react with 100. mL of the HCl solution?

4. A flask containing 825 mL of a solution containing 0.355 moles of $CuSO_4$ was left open overnight. The next morning the flask only contained 755 mL of solution.
    (a) Which of the pathways shown below best represents evaporation and why are the others wrong?
    (b) What is the concentration (molarity) of the $CuSO_4$ solution remaining in the flask?

5. Three students at Jamston High, Zack, Gaye, and Lamont, each had the opportunity to travel over spring break. As part of a project, each of them measured the boiling point of pure water at their vacation spot. Zack found a boiling point of 93.9°C, Gaye measured 101.1°C, and Lamont read 100.°C. Which student most likely went snow skiing near Ely, Nevada, and which student most likely went water skiing in Honolulu? From the boiling point information, what can you surmise about the Dead Sea region, the location of the third student's vacation? Explain.

6. Why does a change in pressure of a gas significantly affect its volume whereas a change in pressure on a solid or liquid has negligible effect on their respective volumes? If the picture below represents a liquid at the molecular level, draw what you might expect a solid and a gas to look like.

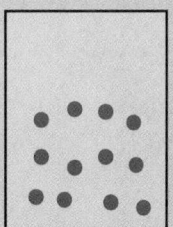

7. (a) If you filled up three balloons with equal volumes of hydrogen, argon, and carbon dioxide gas, all at the same temperature and pressure, which balloon would weigh the most? the least? Explain.

   (b) If you filled up three balloons with equal masses of nitrogen, oxygen, and neon, all to the same volume at the same temperature, which would have the lowest pressure?

8. Ray ran a double-displacement reaction using 0.050 mol $CuCl_2$ and 0.10 mol $AgNO_3$. The resulting white precipitate was removed by filtration. The filtrate was accidentally left open on the lab bench for over a week and when Ray returned the flask contained solid, blue crystals. Ray weighed the crystals and found they had a mass of 14.775 g. Was Ray expecting this number? If not, what did he expect?

9. Why is it often advantageous or even necessary to run reactions in solution rather than mixing two solids? Would you expect reactions run in the gas phase to be more similar to solutions or to solids? Why?

CHAPTER

15

Acids, Bases, and Salts

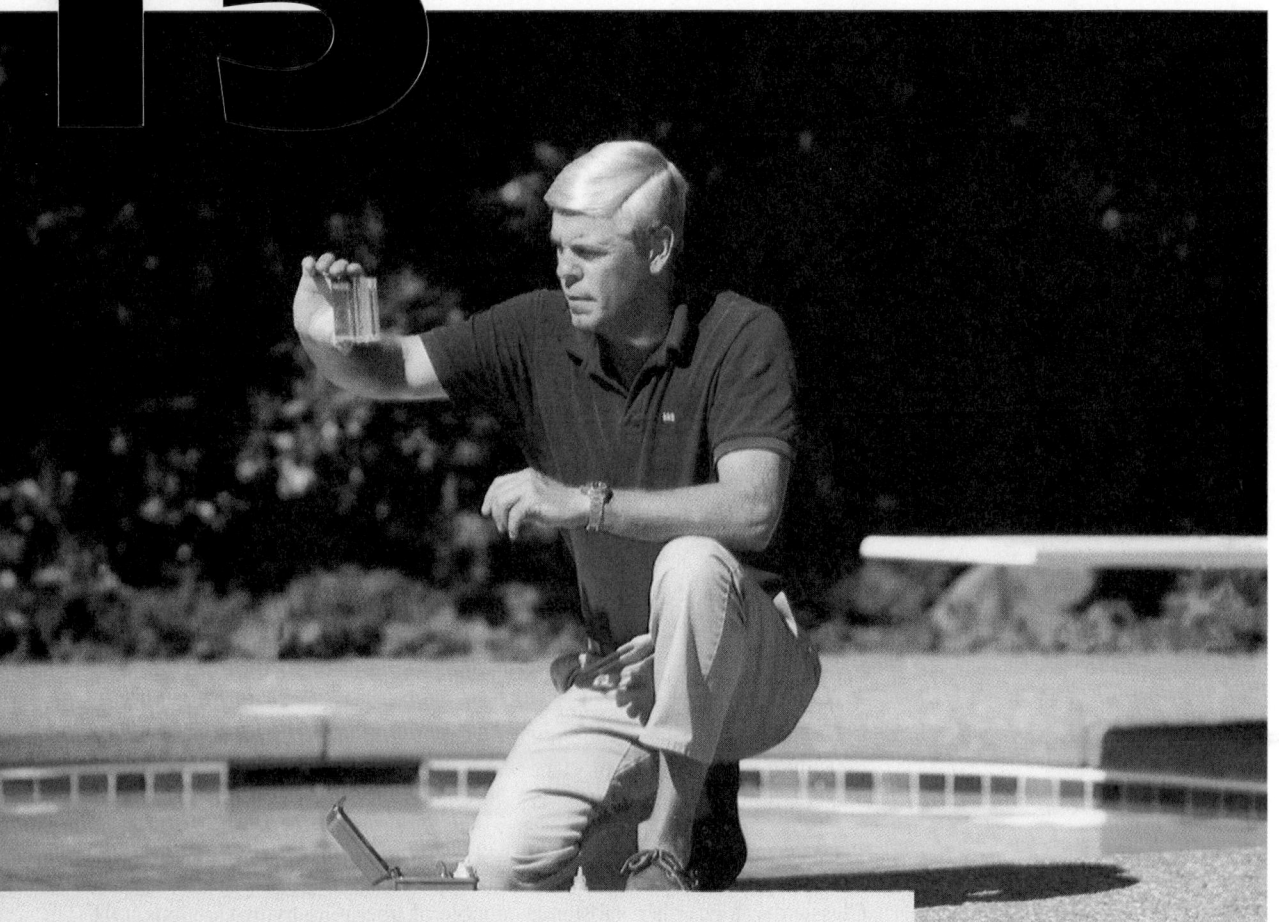

▲

Maintaining the correct water acidity allows us to safely enjoy the fun of a swimming pool.

CHAPTER
15

Acids are important chemicals. They are used in cooking to produce the surprise of tartness (from lemons) and to release CO_2 bubbles from leavening agents in baking. Vitamin C is an acid that is an essential nutrient in our diet. Our stomachs release acid to aid in digestion. Excess stomach acid can produce heartburn and indigestion. Bacteria in our mouths produce acids that can dissolve tooth enamel to form cavities. In our recreational activities we are concerned about acidity levels in swimming pools and spas. Acids are essential in the manufacture of detergents, plastics, and storage batteries. The acid base properties of substances are found in all areas of our lives. In this chapter we consider the properties of acids, bases, and salts.

15.1 Acids and Bases

The word *acid* is derived from the Latin *acidus,* meaning "sour" or "tart," and is also related to the Latin word *acetum,* meaning "vinegar." Vinegar has been around since antiquity as a product of the fermentation of wine and apple cider. The sour constituent of vinegar is acetic acid, $HC_2H_3O_2$. Characteristic properties commonly associated with acids include the following:

1. sour taste
2. change the color of litmus, a vegetable dye, from blue to red
3. react with
   • metals such as zinc and magnesium to produce hydrogen gas
   • hydroxide bases to produce water and an ionic compound (salt)
   • carbonates to produce carbon dioxide

These properties are due to the hydrogen ions, H^+, released by acids in a water solution.

Classically, a *base* is a substance capable of liberating hydroxide ions, OH^-, in water solution. Hydroxides of the alkali metals (Group IA) and alkaline earth metals (Group IIA), such as LiOH, NaOH, KOH, $Ca(OH)_2$, and $Ba(OH)_2$, are the most common inorganic bases. Water solutions of bases are called *alkaline solutions* or *base solutions.* Some of the characteristic properties commonly associated with bases include the following:

1. bitter or caustic taste
2. a slippery, soapy feeling
3. the ability to change litmus from red to blue
4. the ability to interact with acids

Several theories have been proposed to answer the question "What is an acid and a base?" One of the earliest, most significant of these theories was advanced in 1884 by Svante Arrhenius (1859–1927), a Swedish scientist, who stated that "an acid is a hydrogen-containing substance that dissociates to produce hydrogen ions, and a base is a hydroxide-containing substance that dissociates to produce hydroxide ions in aqueous solutions." Arrhenius postulated that the hydrogen ions are produced by the dissociation of acids in water, and that the hydroxide ions are produced by the dissociation of bases in water:

$$HA \longrightarrow H^+(aq) + A^-(aq)$$
acid

$$MOH \longrightarrow M^+(aq) + OH^-(aq)$$
base

An Arrhenius acid solution contains an excess of H$^+$ ions.
An Arrhenius base solution contains an excess of OH$^-$ ions.

In 1923, the Brønsted–Lowry proton transfer theory was introduced by J. N. Brønsted (1897–1947), a Danish chemist, and T. M. Lowry (1847–1936), an English chemist. This theory states that an acid is a proton donor and a base is a proton acceptor.

A Brønsted–Lowry acid is a proton (H$^+$) donor.
A Brønsted–Lowry base is a proton (H$^+$) acceptor.

Consider the reaction of hydrogen chloride gas with water to form hydrochloric acid:

$$HCl(g) + H_2O(l) \longrightarrow H_3O^+(aq) + Cl^-(aq) \tag{1}$$

In the course of the reaction, HCl donates, or gives up, a proton to form a Cl$^-$ ion, and H$_2$O accepts a proton to form the H$_3$O$^+$ ion. Thus, HCl is an acid and H$_2$O is a base, according to the Brønsted–Lowry theory.

A hydrogen ion, H$^+$, is nothing more than a bare proton and does not exist by itself in an aqueous solution. In water H$^+$ combines with a polar water molecule to form a hydrated hydrogen ion, H$_3$O$^+$, commonly called a **hydronium ion.** The H$^+$ is attracted to a polar water molecule, forming a bond with one of the two pairs of unshared electrons:

hydronium ion

$$H^+ + H \cdot \ddot{\underset{H}{O}} \cdot \longrightarrow \left[H : \ddot{\underset{H}{O}} : H \right]^+$$
hydronium ion

Note the electron structure of the hydronium ion. For simplicity we often use H$^+$ instead of H$_3$O$^+$ in equations, with the explicit understanding that H$^+$ is always hydrated in solution.

When a Brønsted–Lowry acid donates a proton, as illustrated in equation (1), it forms the conjugate base of that acid. When a base accepts a proton, it forms the conjugate acid of that base. A conjugate acid and base are produced as products. The formulas of a conjugate acid–base pair differ by one proton (H^+). Consider what happens when $HCl(g)$ is bubbled through water, as shown by this equation:

conjugate acid–base pair

$$HCl(g) + H_2O(l) \longrightarrow Cl^-(aq) + H_3O^+(aq)$$

conjugate acid–base pair

$$\qquad\text{acid}\qquad\qquad\text{base}\qquad\qquad\quad\text{base}\qquad\qquad\text{acid}$$

The conjugate acid–base pairs are HCl—Cl^- and H_3O^+—H_2O. The conjugate base of HCl is Cl^-, and the conjugate acid of Cl^- is HCl. The conjugate base of H_3O^+ is H_2O, and the conjugate acid of H_2O is H_3O^+.

Another example of conjugate acid–base pairs can be seen in this equation:

$$NH_4^+ + H_2O \longrightarrow H_3O^+ + NH_3$$

$$\quad\text{acid}\qquad\text{base}\qquad\quad\text{acid}\qquad\text{base}$$

Here the conjugate acid–base pairs are NH_4^+—NH_3 and H_3O^+—H_2O

Example 15.1   Write the formula for (a) the conjugate base of H_2O and of HNO_3, and (b) the conjugate acid of SO_4^{2-} and of $C_2H_3O_2^-$.

Solution

Remember: The difference between an acid or a base and its conjugate is one proton, H^+.

(a) To write the conjugate base of an acid, remove one proton from the acid formula:

$$H_2O \xrightarrow{-H^+} OH^- \quad \text{(conjugate base)}$$

$$HNO_3 \xrightarrow{-H^+} NO_3^- \quad \text{(conjugate base)}$$

Note that, by removing an H^+, the conjugate base becomes more negative than the acid by one minus charge.

(b) To write the conjugate acid of a base, add one proton to the formula of the base:

$$SO_4^{2-} \xrightarrow{+H^+} HSO_4^- \quad \text{(conjugate acid)}$$

$$C_2H_3O_2^- \xrightarrow{+H^+} HC_2H_3O_2 \quad \text{(conjugate acid)}$$

In each case the conjugate acid becomes more positive than the base by a $+1$ charge due to the addition of H^+.

Practice 15.1

Indicate the conjugate base for these acids:
(a) H_2CO_3 (b) HNO_2 (c) $HC_2H_3O_2$

Practice 15.2

Indicate the conjugate acid for these bases:
(a) HSO_4^- (b) NH_3 (c) OH^-

A more general concept of acids and bases was introduced by Gilbert N. Lewis. The Lewis theory deals with the way in which a substance with an unshared pair of electrons reacts in an acid–base type of reaction. According to this theory a base is any substance that has an unshared pair of electrons (electron-pair donor), and an acid is any substance that will attach itself to or accept a pair of electrons.

> **A Lewis acid is an electron-pair acceptor.**
> **A Lewis base is an electron-pair donor.**

In the reaction

$$H^+ \;+\; :\!\overset{\textstyle H}{\underset{\textstyle H}{N}}\!:\!H \;\longrightarrow\; \left[\, H\!:\!\overset{\textstyle H}{\underset{\textstyle H}{N}}\!:\!H \,\right]^+$$

acid   base

The H^+ is a Lewis acid and $:NH_3$ is a Lewis base. According to the Lewis theory, substances other than proton donors (e.g., BF_3) behave as acids:

$$\overset{\textstyle F}{\underset{\textstyle F}{F\!:\!B}} \;+\; :\!\overset{\textstyle H}{\underset{\textstyle H}{N}}\!:\!H \;\longrightarrow\; \overset{\textstyle F\; H}{\underset{\textstyle F\; H}{F\!:\!B\!:\!N\!:\!H}}$$

acid   base

These three theories, which explain how acid–base reactions occur, are summarized in Table 15.1. We will generally use the theory that best explains the reaction under consideration. Most of our examples will refer to aqueous solutions. Note that in an aqueous acidic solution the H^+ ion concentration is always greater than OH^- ion concentration. And vice versa—in an aqueous basic solution the OH^- ion concentration is always greater than the H^+ ion concentration. When the H^+ and OH^- ion concentrations in a solution are equal, the solution is neutral; that is, it is neither acidic nor basic.

TABLE 15.1	Summary of Acid–Base Definitions	
Theory	**Acid**	**Base**
Arrhenius	A hydrogen-containing substance that produces hydrogen ions in aqueous solution	A hydroxide-containing substance that produces hydroxide ions in aqueous solution
Brønsted–Lowry	A proton (H^+) donor	A proton (H^+) acceptor
Lewis	Any species that will bond to an unshared pair of electrons (electron-pair acceptor)	Any species that has an unshared pair of electrons (electron-pair donor)

Reactions of Acids

In aqueous solutions the H^+ or H_3O^+ ions are responsible for the characteristic reactions of acids. The following reactions are in an aqueous medium.

Reaction with Metals Acids react with metals that lie above hydrogen in the activity series of elements to produce hydrogen and an ionic compound (salt) (see Section 17.5):

acid + metal ⟶ hydrogen + ionic compound

$$2\ HCl(aq) + Ca(s) \longrightarrow H_2(g) + CaCl_2(aq)$$

$$H_2SO_4(aq) + Mg(s) \longrightarrow H_2(g) + MgSO_4(aq)$$

$$6\ HC_2H_3O_2(aq) + 2\ Al(s) \longrightarrow 3\ H_2(g) + 2\ Al(C_2H_3O_2)_3(aq)$$

Acids such as nitric acid (HNO_3) are oxidizing substances (see Chapter 17) and react with metals to produce water instead of hydrogen. For example,

$$3\ Zn(s) + 8\ HNO_3(dilute) \longrightarrow 3\ Zn(NO_3)_2(aq) + 2\ NO(g) + 4\ H_2O(l)$$

Reaction with Bases The interaction of an acid and a base is called a *neutralization reaction*. In aqueous solutions the products of this reaction are a salt and water:

acid + base ⟶ salt + water

$$HBr(aq) + KOH(aq) \longrightarrow KBr(aq) + H_2O(l)$$

$$2\ HNO_3(aq) + Ca(OH)_2(aq) \longrightarrow Ca(NO_3)_2(aq) + 2\ H_2O(l)$$

$$2\ H_3PO_4(aq) + 3\ Ba(OH)_2(aq) \longrightarrow Ba_3(PO_4)_2(s) + 6\ H_2O(l)$$

Reaction with Metal Oxides This reaction is closely related to that of an acid with a base. With an aqueous acid solution, the products are a salt and water:

acid + metal oxide ⟶ salt + water

$$2\ HCl(aq) + Na_2O(s) \longrightarrow 2\ NaCl(aq) + H_2O(l)$$

$$H_2SO_4(aq) + MgO(s) \longrightarrow MgSO_4(aq) + H_2O(l)$$

$$6\ HCl(aq) + Fe_2O_3(s) \longrightarrow 2\ FeCl_3(aq) + 3\ H_2O(l)$$

Reaction with Carbonates Many acids react with carbonates to produce carbon dioxide, water, and an ionic compound:

Carbonic acid (H_2CO_3) is not the product because it is unstable and spontaneously decomposes into water and carbon dioxide.

$$H_2CO_3(aq) \longrightarrow CO_2(g) + H_2O(l)$$

acid + carbonate ⟶ salt + water + carbon dioxide

$$2\ HCl(aq) + Na_2CO_3(aq) \longrightarrow 2\ NaCl(aq) + H_2O(l) + CO_2(g)$$

$$H_2SO_4(aq) + MgCO_3(s) \longrightarrow MgSO_4(aq) + H_2O(l) + CO_2(g)$$

Pucker Power

The candy counter is filled with treats that cause your lips to pucker and stimulate your tongue to send signals to your brain saying "SOUR." The human tongue has four types of taste receptors (known as "taste buds"). Sweet, bitter, salty, and sour taste buds are each concentrated in a different area of the tongue. These receptors are molecules that fit together with molecules in the candy (or other food), sending a signal to the brain, which is interpreted as a taste.

Substances that taste sour are acids. The substances that produce this sour taste in candy and other confections are often malic acid and/or citric acid.

H O
H—C—C—O—H
O
HO—C—C—O—H
O
H—C—C—O—H
H
citric acid

H O
H—C—C—O—H
O
HO—C—C—O—H
H
malic acid

Citric acid tastes more sour than malic acid. When these substances are mixed with other ingredients such as sugar, corn syrup, flavorings, and preservatives, the result is a candy both sweet and tart. The same acids can be mixed with synthetic rubber or chicle (dried latex from the sapodilla tree) to produce bubble gum with real pucker power. The

Cross section of a Mad Dawg™ gum ball.

most sour of this type of gum is called Face Slammers™. Try some on your favorite 12-year-old.

Gum manufacturers have gone a step further in incorporating acid–base chemistry into their products for an even greater surprise. In Mad Dawg™ gum, the initial taste is sour. But after a couple of minutes of chewing, brightly colored foam begins to accumulate in your mouth and ooze out over your lips. What's going on here?

The foam is a mixture of sugar and saliva mixed with carbon dioxide bubbles released when several of the gum's ingredients are mixed in the watery environment of your mouth. The citric and malic acids dissociate to form hydrogen ions while sodium hydrogen carbonate (baking soda) dissolves into sodium and hydrogen carbonate ions:

$$NaHCO_3(s) \longrightarrow Na^+(aq) + HCO_3^-(aq)$$

The hydrogen ions from the acids mix with the hydrogen carbonate ions from the baking soda to produce water and carbon dioxide gas:

$$H^+(aq) + HCO_3^-(aq) \longrightarrow H_2O(l) + CO_2(g)$$

The acids stimulate production of saliva, and the food coloring adds color to the foamy mess. The major problem for chemists in creating this treat was keeping the reactants apart until the consumer pops the gum into his or her mouth. As solids, sodium hydrogen carbonate and citric acid do not react. But introduce the slightest amount of water and the process begins. When a tablet such as Alka Seltzer™ is manufactured, the ingredients (solid citric acid, sodium hydrogen carbonate, aspirin, and flavoring) are compressed into a tablet that is sealed in a dry foil packet. When opened and dropped into water, the reaction (and the relief) begins and the bubbles are released.

In Mad Dawg™ gum balls, the center core is moist gum and the coating is applied in solution. Some of the early versions of these gum balls actually exploded as they were removed from the candy machine. To eliminate this problem, multiple coatings are now used to keep the acid in one layer (on the outside to give the first sour taste) and the sodium hydrogen carbonate in an inner layer (see diagram). When you crunch on the gum ball, the layers begin to mix in your saliva and the fun begins!

O
CH₂—C—O—O—H
O
HO—C—C—O—H   ⟶
O
CH₂—C—O—O—H
citric acid

O
CH₂—C—O
O
HO—C—C—OH   +   H⁺(aq)
O
CH₂—C—O—H
citrate ion

15.3   **Reactions of Bases**

The OH^- ions are responsible for the characteristic reactions of bases. The following reactions are in an aqueous medium.

Reaction with Acids Bases react with acids to produce a salt and water. See reaction of acids with bases in Section 15.2.

Amphoteric Hydroxides Hydroxides of certain metals, such as zinc, aluminum, and chromium, are **amphoteric**; that is, they are capable of reacting as either an acid or a base. When treated with a strong acid, they behave like bases; when reacted with a strong base, they behave like acids:

amphoteric

$$Zn(OH)_2(s) + 2\ HCl(aq) \longrightarrow ZnCl_2(aq) + 2\ H_2O(l)$$

$$Zn(OH)_2(s) + 2\ NaOH(aq) \longrightarrow Na_2Zn(OH)_4(aq)$$

Reaction of NaOH and KOH with Certain Metals Some amphoteric metals react directly with the strong bases sodium hydroxide and potassium hydroxide to produce hydrogen:

$$base + metal + water \longrightarrow salt + hydrogen$$

$$2\ NaOH(aq) + Zn(s) + 2\ H_2O(l) \longrightarrow Na_2Zn(OH)_4(aq) + H_2(g)$$

$$2\ KOH(aq) + 2\ Al(s) + 6\ H_2O(l) \longrightarrow 2\ KAl(OH)_4(aq) + 3\ H_2(g)$$

 15.4   **Salts**

Chemists use the terms ionic compound and salt interchangeably.

Salts are very abundant in nature. Most of the rocks and minerals of Earth's mantle are salts of one kind or another. Huge quantities of dissolved salts also exist in the oceans. Salts can be considered compounds derived from acids and bases. They consist of positive metal or ammonium ions combined with negative nonmetal ions (OH^- and O^{2-} excluded). The positive ion is the base counterpart and the nonmetal ion is the acid counterpart:

Salts are usually crystalline and have high melting and boiling points.

You may want to review Chapter 6 for nomenclature of acids, bases, and salts.

From a single acid such as hydrochloric acid (HCl), we can produce many chloride compounds by replacing the hydrogen with metal ions (e.g., NaCl, KCl, RbCl, $CaCl_2$, $NiCl_2$). Hence the number of known salts greatly exceeds the number of known acids and bases. If the hydrogen atoms of a binary acid are replaced by a nonmetal, the resulting compound has covalent bonding and is therefore not considered to be ionic (e.g., PCl_3, S_2Cl_2, Cl_2O, NCl_3, ICl).

These strange mineral formations called "tufa" exist at Mono Lake, California. Tufa is formed by water bubbling through sand saturated with NaCl, Na_2CO_3, and Na_2SO_4.

15.5 Electrolytes and Nonelectrolytes

We can show that solutions of certain substances are conductors of electricity with a simple conductivity apparatus, which consists of a pair of electrodes connected to a voltage source through a light bulb and switch (see Figure 15.1). If the medium between the electrodes is a conductor of electricity, the light bulb will glow when the switch is closed. When chemically pure water is placed in the beaker and the switch is closed, the light does not glow, indicating that water is a virtual nonconductor. When we dissolve a small amount of sugar in the water and test the solution, the light still does not glow, showing that a sugar solution is also a nonconductor. But, when a small amount of salt, NaCl, is dissolved in water and this solution is tested, the light glows brightly. Thus the salt solution conducts electricity. A fundamental difference exists between the chemical bonding in sugar and that in salt. Sugar is a covalently bonded (molecular) substance; common salt is a substance with ionic bonds.

FIGURE 15.1
A conductivity apparatus shows the difference in conductivity of solutions. (a) Distilled water does not conduct electricity. (b) Sugar water is a nonelectrolyte. (c) Salt water is a strong electrolyte and conducts electricity.
▼

(a)

(b)

(c)

TABLE 15.2	Representative Electrolytes and Nonelectrolytes		
Electrolytes		**Nonelectrolytes**	
H_2SO_4	$HC_2H_3O_2$	$C_{12}H_{22}O_{11}$ (sugar)	CH_3OH (methyl alcohol)
HCl	NH_3	C_2H_5OH (ethyl alcohol)	$CO(NH_2)_2$ (urea)
HNO_3	K_2SO_4	$C_2H_4(OH)_2$ (ethylene glycol)	O_2
NaOH	$NaNO_3$	$C_3H_5(OH)_3$ (glycerol)	H_2O

electrolyte
nonelectrolyte

Substances whose aqueous solutions are conductors of electricity are called **electrolytes.** Substances whose solutions are nonconductors are known as **nonelectrolytes.** The classes of compounds that are electrolytes are acids, bases, and other ionic compounds (salts). Solutions of certain oxides also are conductors because the oxides form an acid or a base when dissolved in water. One major difference between electrolytes and nonelectrolytes is that electrolytes are capable of producing ions in solution, whereas nonelectrolytes do not have this property. Solutions that contain a sufficient number of ions will conduct an electric current. Although pure water is essentially a nonconductor, many city water supplies contain enough dissolved ionic matter to cause the light to glow dimly when the water is tested in a conductivity apparatus. Table 15.2 lists some common electrolytes and nonelectrolytes.

Acids, bases, and salts are electrolytes.

 ## 15.6   Dissociation and Ionization of Electrolytes

Arrhenius received the 1903 Nobel Prize in chemistry for his work on electrolytes. He found that a solution conducts electricity because the solute dissociates immediately upon dissolving into electrically charged particles (ions). The movement of these ions toward oppositely charged electrodes causes the solution to be a conductor. According to his theory, solutions that are relatively poor conductors contain electrolytes that are only partly dissociated. Arrhenius also believed that ions exist in solution whether or not an electric current is present. In other words the electric current does not cause the formation of ions. Remember that positive ions are cations; negative ions are anions.

dissociation

We have seen that sodium chloride crystals consist of sodium and chloride ions held together by ionic bonds. **Dissociation** is the process by which the ions of a salt separate as the salt dissolves. When placed in water, the sodium and chloride ions are attracted by the polar water molecules, which surround each ion as it dissolves. In water, the salt dissociates, forming hydrated sodium and chloride ions (see Figure 15.2). The sodium and chloride ions in solution are surrounded by a specific number

◄ FIGURE 15.2
Hydrated sodium and chloride ions. When sodium chloride dissolves in water, each Na^+ and Cl^- ion becomes surrounded by water molecules. The negative end of the water dipole is attracted to the Na^+ ion, and the positive end is attracted to the Cl^- ion.

of water dipoles and have less attraction for each other than they had in the crystalline state. The equation representing this dissociation is

$$NaCl(s) + (x + y)\,H_2O \longrightarrow Na^+(H_2O)_x + Cl^-(H_2O)_y$$

A simplified dissociation equation in which the water is omitted but understood to be present is

$$NaCl(s) \longrightarrow Na^+(aq) + Cl^-(aq)$$

Remember that sodium chloride exists in an aqueous solution as hydrated ions and not as NaCl units, even though the formula NaCl (or $Na^+ + Cl^-$) is often used in equations.

The chemical reactions of salts in solution are the reactions of their ions. For example, when sodium chloride and silver nitrate react and form a precipitate of silver chloride, only the Ag^+ and Cl^- ions participate in the reaction. The Na^+ and NO_3^- remain as ions in solution:

$$Ag^+(aq) + Cl^-(aq) \longrightarrow AgCl(s)$$

Ionization is the formation of ions; it occurs as a result of a chemical reaction of certain substances with water. Glacial acetic acid (100% $HC_2H_3O_2$) is a liquid that behaves as a nonelectrolyte when tested by the method described in Section 15.5. But a water solution of acetic acid conducts an electric current (as indicated by the dull-glowing light of the conductivity apparatus). The equation for the reaction with water, which forms hydronium and acetate ions, is

ionization

$$\underset{\text{acid}}{HC_2H_3O_2} + \underset{\text{base}}{H_2O} \rightleftarrows \underset{\text{acid}}{H_3O^+} + \underset{\text{base}}{C_2H_3O_2^-}$$

or, in the simplified equation,

$$HC_2H_3O_2 \rightleftarrows H^+ + C_2H_3O_2^-$$

In this ionization reaction, water serves not only as a solvent but also as a base according to the Brønsted–Lowry theory.

Hydrogen chloride is predominantly covalently bonded, but when dissolved in water, it reacts to form hydronium and chloride ions:

$$HCl(g) + H_2O(l) \longrightarrow H_3O^+(aq) + Cl^-(aq)$$

FIGURE 15.3 ▶
**HCl solution (left) is 100%
ionized, while in HC₂H₃O₂ solution (right) almost all of the
solute is in molecular form.
HCl is a strong acid, while
HC₂H₃O₂ is a weak acid.
Note: The water molecules in
the solution are not shown in
this figure.**

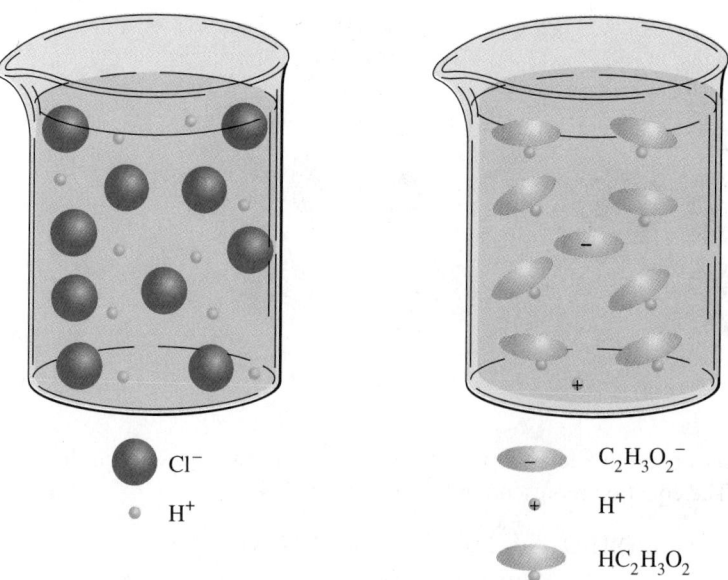

Cl^-

H^+

$C_2H_3O_2^-$

H^+

$HC_2H_3O_2$

When a hydrogen chloride solution is tested for conductivity, the light glows brilliantly, indicating many ions in the solution.

Ionization occurs in each of the preceding two reactions with water, producing ions in solution. The necessity for water in the ionization process can be demonstrated by dissolving hydrogen chloride in a nonpolar solvent such as hexane, and testing the solution for conductivity. The solution fails to conduct electricity, indicating that no ions are produced.

The terms *dissociation* and *ionization* are often used interchangeably to describe processes taking place in water. But, strictly speaking, the two are different. In the dissociation of a salt, the salt already exists as ions; when it dissolves in water, the ions separate, or dissociate, and increase in mobility. In the ionization process, ions are produced by the reaction of a compound with water.

15.7   Strong and Weak Electrolytes

strong electrolyte
weak electrolyte

Electrolytes are classified as strong or weak depending on the degree, or extent, of dissociation or ionization. **Strong electrolytes** are essentially 100% ionized in solution; **weak electrolytes** are much less ionized (based on comparing 0.1 *M* solutions). Most electrolytes are either strong or weak, with a few classified as moderately strong or weak. Most salts are strong electrolytes. Acids and bases that are strong electrolytes (highly ionized) are called *strong acids* and *strong bases*. Acids and bases that are weak electrolytes (slightly ionized) are called *weak acids* and *weak bases*.

For equivalent concentrations, solutions of strong electrolytes contain many more ions than do solutions of weak electrolytes. As a result, solutions of strong electrolytes are better conductors of electricity. Consider the two solutions, 1 *M* HCl and 1 *M* HC₂H₃O₂. Hydrochloric acid is almost 100% ionized; acetic acid is about 1% ionized. (See Figure 15.3.) Thus HCl is a strong acid and HC₂H₃O₂ is a weak acid. Hydrochloric acid has about 100 times as many hydronium ions in solution as acetic acid, making the HCl solution much more acidic.

We can distinguish between strong and weak electrolytes experimentally using the apparatus described in Section 15.5. A 1 M HCl solution causes the light to glow brilliantly, but a 1 M $HC_2H_3O_2$ solution causes only a dim glow. The strong base sodium hydroxide, NaOH, can be distinguished in a similar fashion from the weak base ammonia, NH_3. The ionization of a weak electrolyte in water is represented by an equilibrium equation showing that both the un-ionized and ionized forms are present in solution. In the equilibrium equation of $HC_2H_3O_2$ and its ions, we say that the equilibrium lies "far to the left" because relatively few hydrogen and acetate ions are present in solution:

$$HC_2H_3O_2(aq) \rightleftharpoons H^+(aq) + C_2H_3O_2^-(aq)$$

We have previously used a double arrow in an equation to represent reversible processes in the equilibrium between dissolved and undissolved solute in a saturated solution. A double arrow ($\rightleftharpoons$) is also used in the ionization equation of soluble weak electrolytes to indicate that the solution contains a considerable amount of the un-ionized compound in equilibrium with its ions in solution. (See Section 16.1 for a discussion of reversible reactions.) A single arrow is used to indicate that the electrolyte is essentially all in the ionic form in the solution. For example, nitric acid is a strong acid; nitrous acid is a weak acid. Their ionization equations in water may be indicated as

$$HNO_3(aq) \xrightarrow{H_2O} H^+(aq) + NO_3^-(aq)$$
$$HNO_2(aq) \xrightleftharpoons{H_2O} H^+(aq) + NO_2^-(aq)$$

Practically all soluble salts, acids (such as sulfuric, nitric, and hydrochloric acids), and bases (such as sodium, potassium, calcium, and barium hydroxides) are strong electrolytes. Weak electrolytes include numerous other acids and bases such as acetic acid, nitrous acid, carbonic acid, and ammonia. The terms *strong acid, strong base, weak acid,* and *weak base* refer to whether an acid or base is a strong or weak electrolyte. A brief list of strong and weak electrolytes is given in Table 15.3.

Electrolytes yield two or more ions per formula unit upon dissociation—the actual number being dependent on the compound. Dissociation is complete or nearly complete for nearly all soluble ionic compounds and for certain other strong elec-

TABLE 15.3	Strong and Weak Electrolytes		
Strong electrolytes		**Weak electrolytes**	
Most soluble salts	$HClO_4$	$HC_2H_3O_2$	$H_2C_2O_4$
H_2SO_4	NaOH	H_2CO_3	H_3BO_3
HNO_3	KOH	HNO_2	HClO
HCl	$Ca(OH)_2$	H_2SO_3	NH_3
HBr	$Ba(OH)_2$	H_2S	HF

trolytes, such as those given in Table 15.3. The following are dissociation equations for several strong electrolytes. In all cases the ions are actually hydrated:

$$NaOH \xrightarrow{H_2O} Na^+(aq) + OH^-(aq) \qquad \text{2 ions in solution per formula unit}$$

$$Na_2SO_4 \xrightarrow{H_2O} 2\,Na^+(aq) + SO_4^{2-}(aq) \qquad \text{3 ions in solution per formula unit}$$

$$Fe_2(SO_4)_3 \xrightarrow{H_2O} 2\,Fe^{3+}(aq) + 3\,SO_4^{2-}(aq)$$

$$\text{5 ions in solution per formula unit}$$

One mole of NaCl will give 1 mol of Na^+ ions and 1 mol of Cl^- ions in solution, assuming complete dissociation of the salt. One mole of $CaCl_2$ will give 1 mol of Ca^{2+} ions and 2 mol of Cl^- ions in solution:

$$NaCl \xrightarrow{H_2O} Na^+(aq) + Cl^-(aq)$$

   1 mol   1 mol   1 mol

$$CaCl_2 \xrightarrow{H_2O} Ca^{2+}(aq) + 2\,Cl^-(aq)$$

   1 mol   1 mol   2 mol

Example 15.2   What is the molarity of each ion in a solution of (a) 2.0 M NaCl, and (b) 0.40 M K_2SO_4? Assume complete dissociation.

Solution   (a) According to the dissociation equation,

$$NaCl \xrightarrow{H_2O} Na^+(aq) + Cl^-(aq)$$

   1 mol   1 mol   1 mol

the concentration of Na^+ is equal to that of NaCl (1 mol NaCl ⟶ 1 mol Na^+), and the concentration of Cl^- is also equal to that of NaCl. Therefore the concentrations of the ions in 2.0 M NaCl are 2.0 M Na^+ and 2.0 M Cl^-.

(b) According to the dissociation equation,

$$K_2SO_4 \xrightarrow{H_2O} 2\,K^+(aq) + SO_4^{2-}(aq)$$

   1 mol   2 mol   1 mol

the concentration of K^+ is twice that of K_2SO_4 and the concentration of SO_4^{2-} is equal to that of K_2SO_4. Therefore the concentrations of the ions in 0.40 M K_2SO_4 are 0.80 M K^+ and 0.40 M SO_4^{2-}.

Practice 15.3

What is the molarity of each ion in a solution of (a) 0.050 M $MgCl_2$, and (b) 0.070 M $AlCl_3$?

Colligative Properties of Electrolyte Solutions

We have learned that when 1 mol of sucrose, a nonelectrolyte, is dissolved in 1000 g of water, the solution freezes at $-1.86°C$. When 1 mol NaCl is dissolved in 1000 g of water, the freezing point of the solution is not $-1.86°C$, as might be expected, but is closer to $-3.72°C$ (-1.86×2). The reason for the lower freezing point is that 1 mol NaCl in solution produces 2 mol of particles ($2 \times 6.022 \times 10^{23}$ ions) in solution. Thus the freezing-point depression produced by 1 mol NaCl is essentially equivalent to that produced by 2 mol of a nonelectrolyte. An electrolyte such as $CaCl_2$, which yields three ions in water, gives a freezing-point depression of about three times that of a nonelectrolyte. These freezing-point data provide additional evidence that electrolytes dissociate when dissolved in water. The other colligative properties are similarly affected by substances that yield ions in aqueous solutions.

15.8 Ionization of Water

The more we study chemistry, the more intriguing the water molecule becomes. Two equations commonly used to show how water ionizes are

$$H_2O + H_2O \rightleftharpoons H_3O^+ + OH^-$$
$$\text{acid} \qquad \text{base} \qquad \text{acid} \qquad \text{base}$$

and

$$H_2O \rightleftharpoons H^+ + OH^-$$

The first equation represents the Brønsted–Lowry concept, with water reacting as both an acid and a base, forming a hydronium ion and a hydroxide ion. The second equation is a simplified version, indicating that water ionizes to give a hydrogen and a hydroxide ion. Actually, the proton, H^+, is hydrated and exists as a hydronium ion. In either case equal molar amounts of acid and base are produced so that water is neutral, having neither H^+ nor OH^- ions in excess. The ionization of water at 25°C produces an H^+ ion concentration of 1.0×10^{-7} mol/L and an OH^- ion concentration of 1.0×10^{-7} mol/L. Square brackets, [ ], are used to indicate that the concentration is in moles per liter. Thus $[H^+]$ means the concentration of H^+ is in moles per liter. These concentrations are usually expressed as

$$[H^+] \text{ or } [H_3O^+] = 1.0 \times 10^{-7} \text{ mol/L}$$

$$[OH^-] = 1.0 \times 10^{-7} \text{ mol/L}$$

These figures mean that about two out of every billion water molecules are ionized. This amount of ionization, small as it is, is a significant factor in the behavior of water in many chemical reactions.

15.9 Introduction to pH

The acidity of an aqueous solution depends on the concentration of hydrogen or hydronium ions. The pH scale of acidity gives us a simple, convenient numerical way to state the acidity of a solution. Values on the pH scale are obtained by mathematical conversion of H^+ ion concentrations to pH by the expression:

$$pH = -\log[H^+]$$

where $[H^+] = H^+$ or H_3O^+ ion concentration in moles per liter. The **pH** is defined as the *negative* logarithm of the H^+ or H_3O^+ concentration in moles per liter:

pH

$$pH = -\log[H^+] = -\log(1 \times 10^{-7}) = -(-7) = 7$$

▲
FIGURE 15.4
The pH scale of acidity and basicity.

For example, the pH of pure water at 25°C is 7 and is said to be neutral; that is, it is neither acidic nor basic, because the concentrations of H^+ and OH^- are equal. Solutions that contain more H^+ ions than OH^- ions have pH values less than 7, and solutions that contain less H^+ ions than OH^- ions have values greater than 7.

> **pH < 7.00 is an acidic solution**
> **pH = 7.00 is a neutral solution**
> **pH > 7.00 is a basic solution**

When $[H^+] = 1 \times 10^{-5}$ mol/L, pH = 5 (acidic)

When $[H^+] = 1 \times 10^{-9}$ mol/L, pH = 9 (basic)

Instead of saying that the hydrogen ion concentration in the solution is 1×10^{-5} mol/L, it's customary to say that the pH of the solution is 5. The smaller the pH value, the more acidic the solution (see Figure 15.4).

The pH scale, along with its interpretation, is given in Table 15.4, and Table 15.5 lists the pH of some common solutions. Note that a change of only one pH unit means a tenfold increase or decrease in H^+ ion concentration. For example, a solution with a pH of 3.0 is ten times more acidic than a solution with a pH of 4.0. A simplified method of determining pH from $[H^+]$ follows:

$[H^+] = 1 \times 10^{-5}$ ⟵ pH = this number (5)
pH = 5

when this number
is exactly 1

$[H^+] = 2 \times 10^{-5}$ ⟵ pH is between this number and
next lower number (4 and 5)
pH = 4.7

when this number
is between 1 and 10

Calculating the pH value for H^+ ion concentrations requires the use of logarithms, which are exponents. The **logarithm** (log) of a number is simply the power to which 10 must be raised to give that number. Thus the log of 100 is 2 ($100 = 10^2$), and the log of 1000 is 3 ($1000 = 10^3$). The log of 500 is 2.70, but you can't determine this value easily without a scientific calculator.

logarithm

Help on using calculators is found in Appendix II.

Let's determine the pH of a solution with $[H^+] = 2 \times 10^{-5}$ using a calculator. Enter 2×10^{-5} into your calculator and press the log key. The number $-4.69 . . .$ will be displayed. The pH is then

Remember: Change the sign on your calculator since
pH = −log[H⁺].

$$pH = -\log[H^+] = -(-4.69 . . .) = 4.7$$

TABLE 15.4 pH Scale for Expressing Acidity		
$[H^+]$ (mol/L)	pH	
1×10^{-14}	14	
1×10^{-13}	13	
1×10^{-12}	12	Increasing
1×10^{-11}	11	basicity
1×10^{-10}	10	
1×10^{-9}	9	
1×10^{-8}	8	
1×10^{-7}	7	Neutral
1×10^{-6}	6	
1×10^{-5}	5	
1×10^{-4}	4	
1×10^{-3}	3	Increasing
1×10^{-2}	2	acidity
1×10^{-1}	1	
1×10^{0}	0	

TABLE 15.5 The pH of Common Solutions	
Solution	pH
Gastric juice	1.0
0.1 M HCl	1.0
Lemon juice	2.3
Vinegar	2.8
0.1 M $HC_2H_3O_2$	2.9
Orange juice	3.7
Tomato juice	4.1
Coffee, black	5.0
Urine	6.0
Milk	6.6
Pure water (25°C)	7.0
Blood	7.4
Household ammonia	11.0
1 M NaOH	14.0

Next we must determine the correct number of significant figures in the logarithm. The rules for logs are different from those we use in other math operations. The number of decimal places for a log must equal the number of significant figures in the original number. Since 2×10^{-5} has one significant figure, we should round the log to one decimal place $(4.69 . . .) = 4.7$.

What is the pH of a solution with an $[H^+]$ of (a) 1.0×10^{-11}, (b) 6.0×10^{-4}, and (c) 5.47×10^{-8}?

Example 15.3

Solution

(a) $[H^+] = 1.0 \times 10^{-11}$
   (2 significant figures)
   pH $= -\log(1.0 \times 10^{-11})$
   pH $= 11.00$
   (2 decimal places)

(b) $[H^+] = 6.0 \times 10^{-4}$
   (2 significant figures)
   $\log 6.0 \times 10^{-4} = -3.22$
   pH $= -\log[H^+]$
   pH $= -(-3.22) = 3.22$
   (2 decimal places)

(c) $[H^+] = 5.47 \times 10^{-8}$
   (3 significant figures)
   $\log 5.47 \times 10^{-8} = -7.262$
   pH $= -\log[H^+]$
   pH $= -(-7.262) = 7.262$
   (3 decimal places)

Practice 15.4

What is the pH of a solution with $[H^+]$ of (a) 3.9×10^{-12} M, (b) 1.3×10^{-3} M, and (c) 3.72×10^{-6} M?

In the Pink—A Sign of Corrosion

As our fleet of commercial airplanes ages, scientists are looking for new ways to detect corrosion and stop it *before* it becomes a safety issue. The problem is that planes are subjected to heat, rain, and wind—all factors that can contribute to metal corrosion. How can a maintenance crew detect corrosion on a huge aircraft since many imaging techniques work on only small areas at a time? Imagine a plane that could tell the maintenance crew it needed repair.

Gerald Frankel and Jim Zhang from Ohio State University have developed a paint that detects changes in pH. Corrosion of metals (in contact with air and water) results in a chemical reaction that produces hydroxide ions, which increase

▲ **Phenolphthalein paint after 8 days.**

pH. Frankel and Zhang made a clear acrylic coating that they mixed with phenolphthalein. This acid–base indica-

tor turns bright fuschia when the pH increases above 8.0. The scientists can show the "paint" turns pink at corrosion sites as small as 15 μm deep.

Technicians at Wright Air Force Base in Ohio think this could lead to a whole new way to detect corrosion. William Mullins says, "You could walk down the vehicle and see there's a pink spot." This method would be especially good at detecting corrosion concealed around rivets and where metals overlap. The only limitation appears to be that the coating must be clear.

The next time your airline maintenance crew sees pink spots, you may be in for a delay or a change of aircraft!

▲ **FIGURE 15.5**
pH test paper for determining the approximate acidity of solutions.

The measurement and control of pH is extremely important in many fields. Proper soil pH is necessary to grow certain types of plants successfully. The pH of certain foods is too acidic for some diets. Many biological processes are delicately controlled pH systems. The pH of human blood is regulated to very close tolerances through the uptake or release of H^+ by mineral ions, such as HCO_3^-, HPO_4^{2-}, and $H_2PO_4^-$. Changes in the pH of the blood by as little as 0.4 pH unit result in death.

Compounds with colors that change at particular pH values are used as indicators in acid–base reactions. For example, phenolphthalein, an organic compound, is colorless in acid solution and changes to pink at a pH of 8.3. When a solution of sodium hydroxide is added to a hydrochloric acid solution containing phenolphthalein, the change in color (from colorless to pink) indicates that all the acid is neutralized. Commercially available pH test paper, such as that shown in Figure 15.5, contains chemical indicators. The indicator in the paper takes on different colors when wetted with solutions of different pH. Thus the pH of a solution can be estimated by placing a drop on the test paper and comparing the color of the test paper with a color chart calibrated at different pH values. Common applications of pH test indicators are the kits used to measure and adjust the pH of swimming pools, hot tubs, and saltwater aquariums. Electronic pH meters are used for making rapid and precise pH determinations.

15.10 Neutralization

neutralization

The reaction of an acid and a base to form a salt and water is known as **neutralization.** We've seen this reaction before, but now with our knowledge about ions and ionization, let's reexamine the process of neutralization.

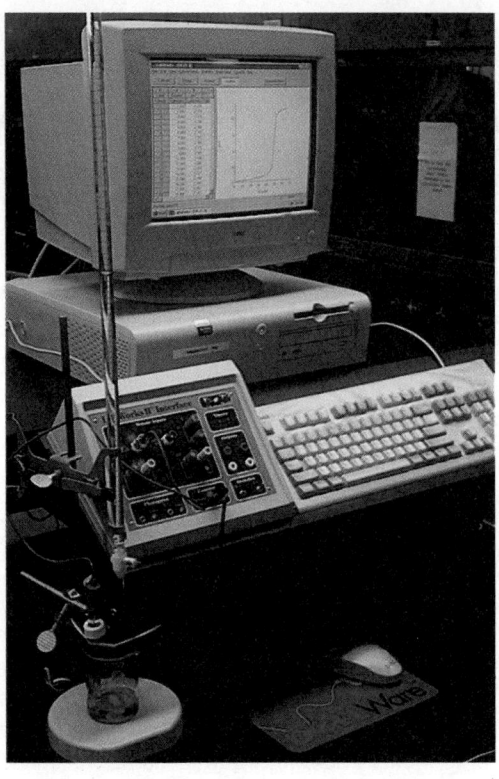

◀ The progress of a titration can be monitored by computer graphing.

Consider the reaction that occurs when solutions of sodium hydroxide and hydrochloric acid are mixed. The ions present initially are Na^+ and OH^- from the base and H^+ and Cl^- from the acid. The products, sodium chloride and water, exist as Na^+ and Cl^- ions and H_2O molecules. A chemical equation representing this reaction is

$$HCl(aq) + NaOH(aq) \longrightarrow NaCl(aq) + H_2O(l)$$

This equation, however, does not show that HCl, NaOH, and NaCl exist as ions in solution. The following total ionic equation gives a better representation of the reaction:

$$(H^+ + Cl^-) + (Na^+ + OH^-) \longrightarrow Na^+ + Cl^- + H_2O(l)$$

This equation shows that the Na^+ and Cl^- ions did not react. These ions are called **spectator ions** because they were present but did not take part in the reaction. The only reaction that occurred was that between the H^+ and OH^- ions. Therefore the equation for the neutralization can be written as this net ionic equation:

$$\underset{\text{acid}}{H^+(aq)} + \underset{\text{base}}{OH^-(aq)} \longrightarrow \underset{\text{water}}{H_2O(l)}$$

spectator ion

This simple net ionic equation represents not only the reaction of sodium hydroxide and hydrochloric acid, but also the reaction of any strong acid with any water-soluble hydroxide base in an aqueous solution. The driving force of a neutralization reaction is the ability of an H^+ ion and an OH^- ion to react and form a molecule of un-ionized water.

This meter measures the ▶ pH of garden soil.

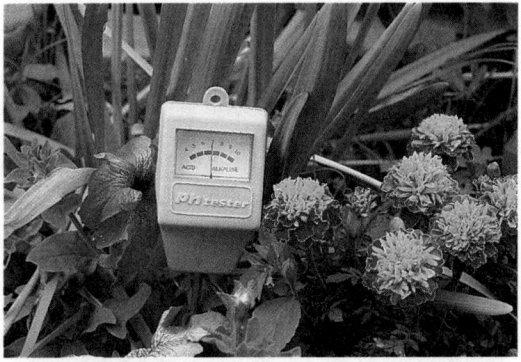

The amount of acid, base, or other species in a sample can be determined by *titration*, which measures the volume of one reagent required to react with a measured mass or volume of another reagent.

titration

Consider the titration of an acid with a base. A measured volume of acid of unknown concentration is placed in a flask, and a few drops of an indicator solution are added. Base solution of known concentration is slowly added from a buret to the acid until the indicator changes color. The indicator selected is one that changes color when the stoichiometric quantity (according to the equation) of base has been added to the acid. At this point, known as the *end point of the titration,* the titration is complete, and the volume of base used to neutralize the acid is read from the buret. The concentration or amount of acid in solution can be calculated from the titration data and the chemical equation for the reaction. Let's look at some examples.

Example 15.4   Suppose that 42.00 mL of 0.150 M NaOH solution is required to neutralize 50.00 mL of hydrochloric acid solution. What is the molarity of the acid solution?

Solution   The equation for the reaction is

$$NaOH(aq) + HCl(aq) \longrightarrow NaCl(aq) + H_2O(l)$$

In this neutralization NaOH and HCl react in a 1:1 mole ratio. Therefore the moles of HCl in solution are equal to the moles of NaOH required to react with it. First we calculate the moles of NaOH used, and from this value we determine the moles of HCl:

Data: 42.00 mL of 0.150 M NaOH   50.00 mL HCl
Molarity of acid = M (unknown)

Determine the moles of NaOH:

M = mol/L   42.00 mL = 0.04200 L

$$(0.04200 \, L)\left(\frac{0.150 \text{ mol NaOH}}{1 \, L}\right) = 0.00630 \text{ mol NaOH}$$

Since NaOH and HCl react in a 1:1 ratio, 0.00630 mol HCl was present in the 50.00 mL of HCl solution. Therefore the molarity of the HCl is

$$M = \frac{\text{mol}}{\text{L}} = \frac{0.00630 \text{ mol HCl}}{0.05000 \text{ L}} = 0.126 \, M \text{ HCl}$$

Suppose that 42.00 mL of 0.150 M NaOH solution is required to neutralize 50.00 mL of H_2SO_4 solution. What is the molarity of the acid solution?

Example 15.5

The equation for the reaction is

$$2\ NaOH(aq) + H_2SO_4(aq) \longrightarrow Na_2SO_4(aq) + 2\ H_2O(l)$$

The same amount of base (0.00630 mol NaOH) is used in this titration as in Example 15.4, but the mole ratio of acid to base in the reaction is 1:2. The moles of H_2SO_4 reacted can be calculated using the mole-ratio method:

Data: 42.00 mL of 0.150 M NaOH = 0.00630 mol NaOH

$$(0.00630\ \text{mol NaOH})\left(\frac{1\ \text{mol}\ H_2SO_4}{2\ \text{mol NaOH}}\right) = 0.00315\ \text{mol}\ H_2SO_4$$

Therefore 0.00315 mol H_2SO_4 was present in 50.00 mL of H_2SO_4 solution. The molarity of the H_2SO_4 is

$$M = \frac{\text{mol}}{\text{L}} = \frac{0.00315\ \text{mol}\ H_2SO_4}{0.05000\ \text{L}} = 0.0630\ M\ H_2SO_4$$

A 25.00-mL sample of H_2SO_4 solution required 14.26 mL 0.2240 M NaOH for complete neutralization. What is the molarity of the sulfuric acid?

Example 15.6

The equation for the reaction is

$$2\ NaOH(aq) + H_2SO_4(aq) \longrightarrow Na_2SO_4(aq) + 2\ H_2O(l)$$

The moles of NaOH needed are

$$\text{moles NaOH} = (V_{NaOH})(M_{NaOH})$$

$$= (0.01426\ \text{L})\left(0.2240\ \frac{\text{mol}}{\text{L}}\right)$$

$$= 0.003194\ \text{mol NaOH}$$

Since the mole ratio of acid to base is $\dfrac{1\ H_2SO_4}{2\ NaOH}$ the moles of acid in the sample are

$$(0.003914\ \text{mol NaOH})\left(\frac{1\ H_2SO_4}{2\ NaOH}\right) = 0.001597\ \text{mol}\ H_2SO_4$$

Now to find the molarity of the sample we divide the moles of acid by its original volume:

$$\left(\frac{0.001597\ \text{mol}\ H_2SO_4}{0.02500\ \text{L}}\right) = 0.06388\ M\ H_2SO_4$$

Practice 15.5

A 50.0-mL sample of HCl required 24.81 mL of 0.1250 M NaOH for neutralization. What is the molarity of the acid?

15.11 Writing Net Ionic Equations

In Section 15.10, we wrote the reaction of hydrochloric acid and sodium hydroxide in three different equations:

(1) $HCl(aq) + NaOH(aq) \longrightarrow NaCl(aq) + H_2O(l)$

(2) $(H^+ + Cl^-) + (Na^+ + OH^-) \longrightarrow Na^+ + Cl^- + H_2O(l)$

(3) $H^+ + OH^- \longrightarrow H_2O$

un-ionized equation

total ionic equation

net ionic equation

In the **un-ionized equation** (1), compounds are written in their molecular, or formula expressions. In the **total ionic equation** (2), compounds are written to show the form in which they are predominantly present: strong electrolytes as ions in solution; and nonelectrolytes, weak electrolytes, precipitates, and gases in their molecular (or un-ionized) forms. In the **net ionic equation** (3), only those molecules or ions that have changed are included in the equation; ions or molecules that do not change (the spectators) are omitted.

When balancing equations thus far, we've been concerned only with the atoms of the individual elements. Because ions are electrically charged, ionic equations often end up with a net electrical charge. A balanced equation must have the same net charge on each side, whether that charge is positive, negative, or zero. Therefore when balancing ionic equations, we must make sure that both the same number of each kind of atom and the same net electrical charge are present on each side.

Here is a list of rules for writing ionic equations:

1. Strong electrolytes in solution are written in their ionic form.
2. Weak electrolytes are written in their molecular (un-ionized) form.
3. Nonelectrolytes are written in their molecular form.
4. Insoluble substances, precipitates, and gases are written in their molecular forms.
5. The net ionic equation should include only substances that have undergone a chemical change. Spectator ions are omitted from the net ionic equation.
6. Equations must be balanced, both in atoms and in electrical charge.

Study the following examples. In each one the un-ionized equation is given. Write the total ionic equation and the net ionic equation for each.

Example 15.7

$HNO_3(aq) + KOH(aq) \longrightarrow KNO_3(aq) + H_2O(l)$

          un-ionized equation

Solution

$(H^+ + NO_3^-) + (K^+ + OH^-) \longrightarrow (K^+ + NO_3^-) + H_2O$

            total ionic equation

$H^+ + OH^- \longrightarrow H_2O$

    net ionic equation

The HNO_3, KOH, and KNO_3 are soluble, strong electrolytes. The K^+ and NO_3^- ions are spectator ions, have not changed, and are not included in the net ionic equation. Water is a nonelectrolyte and is written in the molecular form.

Example 15.8

$$2\,AgNO_3(aq) + BaCl_2(aq) \longrightarrow 2\,AgCl(s) + Ba(NO_3)_2(aq)$$
un-ionized equation

$$(2\,Ag^+ + 2\,NO_3^-) + (Ba^{2+} + 2\,Cl^-) \longrightarrow 2\,AgCl(s) + (Ba^{2+} + 2\,NO_3^-)$$
total ionic equation

$$Ag^+ + Cl^- \longrightarrow AgCl(s)$$
net ionic equation

Solution

Although AgCl is an ionic compound, it is written in the un-ionized form on the right side of the ionic equations because most of the Ag^+ and Cl^- ions are no longer in solution but have formed a precipitate of AgCl. The Ba^{2+} and NO_3^- ions are spectator ions.

Example 15.9

$$Na_2CO_3(aq) + H_2SO_4(aq) \longrightarrow Na_2SO_4(aq) + H_2O(l) + CO_2(g)$$
un-ionized equation

$$(2\,Na^+ + CO_3^{2-}) + (2\,H^+ + SO_4^{2-}) \longrightarrow (2\,Na^+ + SO_4^{2-}) + H_2O(l) + CO_2(g)$$
total ionic equation

$$CO_3^{2-} + 2\,H^+ \longrightarrow H_2O(l) + CO_2(g)$$
net ionic equation

Solution

Carbon dioxide, CO_2, is a gas and evolves from the solution; Na^+ and SO_4^{2-} are spectator ions.

Example 15.10

$$HC_2H_3O_2(aq) + NaOH(aq) \longrightarrow NaC_2H_3O_2(aq) + H_2O(l)$$
un-ionized equation

$$HC_2H_3O_2 + (Na^+ + OH^-) \longrightarrow (Na^+ + C_2H_3O_2^-) + H_2O$$
total ionic equation

$$HC_2H_3O_2 + OH^- \longrightarrow C_2H_3O_2^- + H_2O$$
net ionic equation

Solution

Acetic acid, $HC_2H_3O_2$, a weak acid, is written in the molecular form, but sodium acetate, $NaC_2H_3O_2$, a soluble salt, is written in the ionic form. The Na^+ ion is the only spectator ion in this reaction. Both sides of the net ionic equation have a -1 electrical charge.

Example 15.11

$$Mg(s) + 2\,HCl(aq) \longrightarrow MgCl_2(aq) + H_2(g)$$
un-ionized equation

$$Mg + (2\,H^+ + 2\,Cl^-) \longrightarrow (Mg^{2+} + 2\,Cl^-) + H_2(g)$$
total ionic equation

$$Mg + 2 H^+ \longrightarrow Mg^{2+} + H_2(g)$$

net ionic equation

Solution   The net electrical charge on both sides of the equation is +2.

Example 15.12 $H_2SO_4(aq) + Ba(OH)_2(aq) \longrightarrow BaSO_4(s) + 2 H_2O(l)$

un-ionized equation

$$(2 H^+ + SO_4^{2-}) + (Ba^{2+} + 2 OH^-) \longrightarrow BaSO_4(s) + 2 H_2O(l)$$

total ionic equation

$$2 H^+ + SO_4^{2-} + Ba^{2+} + 2 OH^- \longrightarrow BaSO_4(s) + 2 H_2O(l)$$

net ionic equation

Solution   Barium sulfate, $BaSO_4$, is a highly insoluble salt. If we conduct this reaction using the conductivity apparatus described in Section 15.5, the light glows brightly at first but goes out when the reaction is complete because almost no ions are left in solution. The $BaSO_4$ precipitates out of solution, and water is a nonconductor of electricity.

Practice 15.6

Write the net ionic equation for

$$3 H_2S(aq) + 2 Bi(NO_3)_3(aq) \longrightarrow Bi_2S_3(s) + 6 HNO_3(aq)$$

15.12   Acid Rain

Acid rain is defined as any atmospheric precipitation that is more acidic than usual. The increase in acidity might be from natural or industrial sources. Rain acidity varies throughout the world and across the United States. The pH of rain is generally lower in the eastern United States and higher in the west. Unpolluted rain has a pH of 5.6, and so is slightly acidic. This acidity results from the dissolution of carbon dioxide in the water producing carbonic acid:

$$CO_2(g) + H_2O(l) \longrightarrow H_2CO_3(aq) \rightleftharpoons H^+(aq) + HCO_3^-(aq)$$

Although the details of acid rain formation are not yet fully understood, chemists know the general process involves the following steps:

1. emission of nitrogen and sulfur oxides into the air
2. transportation of these oxides throughout the atmosphere
3. chemical reactions between the oxides and water forming sulfuric acid, H_2SO_4, and nitric acid, HNO_3
4. rain or snow, which carries the acids to the ground

The oxides may also be deposited directly on a dry surface and become acidic when normal rain falls on them.

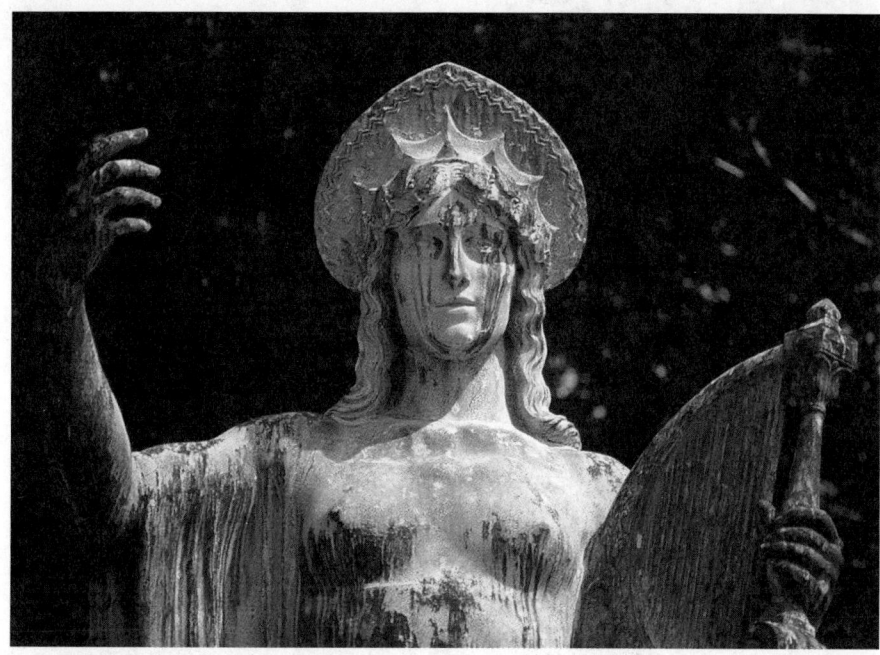

◀ **Marble masterpieces, sculpted to last forever, are slowly disappearing as acid rain dissolves the calcium carbonate (CaCO₃).**

Acid rain is not a new phenomenon. Rain was probably acidic in the early days of our planet as volcanic eruptions, fires, and decomposition of organic matter released large volumes of nitrogen and sulfur oxides into the atmosphere. Use of fossil fuels, especially since the industrial revolution about 250 years ago, has made significant changes in the amounts of pollutants being released into the atmosphere. As increasing amounts of fossil fuels have been burned, more and more sulfur and nitrogen oxides have poured into the atmosphere, thus increasing the acidity of rain.

Acid rain affects a variety of factors in our environment. For example, freshwater plants and animals decline significantly when rain is acidic; large numbers of fish and plants die when acidic water from spring thaws enters the lakes. Aluminum is leached from the soil into lakes by acidic rainwater where the aluminum compounds adversely affect the gills of fish. In addition to leaching aluminum from the soil, acid rain also causes other valuable minerals, such as magnesium and calcium, to dissolve and run into lakes and streams. It can also dissolve the waxy protective coat on plant leaves making them vulnerable to attack by bacteria and fungi.

In our cities, acid rain is responsible for extensive and continuing damage to buildings, monuments, and statues. It reduces the durability of paint and promotes the deterioration of paper, leather, and cloth. In short, we are just beginning to explore the effects of acid rain on human beings and on our food chain.

15.13 Colloids: An Introduction

When we add sugar to a flask of water and shake it, the sugar dissolves and forms a clear homogeneous *solution.* When we do the same experiment with very fine sand and water, the sand particles form a *suspension,* which settles when the shaking stops. When we repeat the experiment again using ordinary cornstarch, we find that

CHEMISTRY IN ACTION   Foam Cars—Wave of the Future?

Metal foams are a new class of materials that may revolutionize the car industry. Automotive makers know that one key to making cars that are fuel efficient is to decrease weight. Up till now that weight reduction meant higher cost (in materials like titanium and aluminum) and problems in crash testing (since light vehicles don't absorb energy well).

German automotive supplier Willhelm Karmann (whose company manufactured the Volkswagen Karmann-Ghia) developed an aluminum foam composite material with some amazing properties. Parts made of this foam composite weigh 30–50% less than an equivalent steel part and are ten times stiffer. The material is so light it floats in water and although it costs 20–25% more than steel it could be used for as much as 20% of a compact car. Since the surface of the new material is not smooth it would likely be used in structural areas of the car (firewalls, roof panels, luggage compartment walls, etc.)

How is aluminum foam made? Two layers of aluminum sheet and a middle powder layer (made of titanium metal hydride and aluminum powder) are rolled together under very high pressure

to make a single flat sheet. This sheet metal is then processed in traditional ways to make a variety of 3-D shapes. Then the sheet is placed in a 1148°F oven for two minutes. This quick bake allows the Al metal to melt and mix with H_2 gas (released from the titanium hydride) making foam. The sheet rises just like a cake (increasing five or seven

times in thickness). When the foam cools it is a rigidly formed 3-D structure between two aluminum skins.

Not only is the new aluminum foam part much lighter than its steel counterpart; it also performs well in crash tests. One of these days you may climb into a car that is really a foam colloid—with great fuel economy and crash resistance.

---

the starch does not dissolve in cold water. But if the mixture is heated and stirred, the starch forms a cloudy, opalescent *dispersion.* This dispersion does not appear to be clear and homogeneous like the sugar solution, yet it is not obviously heterogeneous and does not settle like the sand suspension. In short, its properties are intermediate between those of the sugar solution and those of the sand suspension. The starch dispersion is actually a *colloid,* a name derived from the Greek *kolla,* meaning "glue," and was coined by the English scientist Thomas Graham in 1861.

colloid

As it is now used, the word **colloid** means a dispersion in which the dispersed particles are larger than the solute ions or molecules of a true solution and smaller than the particles of a mechanical suspension. The term does not imply a gluelike quality, although most glues are colloidal materials. The size of colloidal particles ranges from a lower limit of about 1 nm (10^{-7} cm) to an upper limit of about 1000 nm (10^{-4} cm). There are eight types of colloids, which are summarized in Table 15.6.

TABLE 15.6	Types of Colloidal Dispersions	
Type	**Name**	**Examples**
Gas in liquid	foam	whipped cream, soapsuds
Gas in solid	solid foam	Styrofoam, foam rubber, pumice
Liquid in gas	liquid aerosol	fog, clouds
Liquid in liquid	emulsion	milk, vinegar in oil salad dressing, mayonnaise
Liquid in solid	solid emulsion	cheese, opals, jellies
Solid in gas	solid aerosol	smoke, dust in air
Solid in liquid	sol	india ink, gold sol
Solid in solid	solid sol	tire rubber, certain gems (e.g., rubies)

The fundamental difference between a colloidal dispersion and a true solution is the size, not the nature, of the particles. The solute particles in a solution are usually single ions or molecules that may be hydrated to varying degrees. Colloidal particles are usually aggregations of ions or molecules. However, the molecules of some polymers, such as proteins, are large enough to be classified as colloidal particles when in solution. To fully appreciate the differences in relative sizes, the volumes (not just the linear dimensions) of colloidal particles and solute particles must be compared. The difference in volumes can be approximated by assuming that the particles are spheres. A large colloidal particle has a diameter of about 500 nm, whereas a fair-sized ion or molecule has a diameter of about 0.5 nm. Thus the diameter of the colloidal particle is about 1000 times that of the solute particle. Because the volumes of spheres are proportional to the cubes of their diameters, we can calculate that the volume of a colloidal particle can be up to a billion ($10^3 \times 10^3 \times 10^3 = 10^9$) times greater than that of a solution particle.

15.14 Properties of Colloids

In 1827, while observing a strongly illuminated aqueous suspension of pollen under a high-powered microscope, Robert Brown (1773–1858) noted that the pollen grains appeared to have a trembling, erratic motion. He later determined that this erratic motion is not confined to pollen but is characteristic of colloidal particles in general. This random motion of colloidal particles is called **Brownian movement.** We can readily observe such movement by confining cigarette smoke in a small transparent chamber and illuminating it with a strong beam of light at right angles to the optical axis of the microscope. The smoke particles appear as tiny randomly moving lights because the light is reflected from their surfaces. This motion is due to the continual bombardment of the smoke particles by air molecules. Since Brownian movement can be seen when colloidal particles are dispersed in either a gaseous or a liquid medium, it affords nearly direct visual proof that matter at the molecular level is moving randomly, as postulated by the kinetic-molecular theory.

Brownian movement

FIGURE 15.6 ▶
Tyndall effect. A beam of
light is visible in a colloidal
suspension (left) but not in a
true solution (right).

FIGURE 15.6 ▶
Tyndall effect. A beam of
light is visible in a colloidal
suspension (left) but not in a
true solution (right).

When an intense beam of light is passed through an ordinary solution and viewed at an angle, the beam passing through the solution is hardly visible. A beam of light, however, is clearly visible and sharply outlined when it is passed through a *Tyndall effect* colloidal dispersion (see Figure 15.6). This phenomenon is known as the **Tyndall effect.** The Tyndall effect, like the Brownian movement, can be observed in nearly all colloidal dispersions. It occurs because the colloidal particles are large enough to scatter the rays of visible light. The ions or molecules of true solutions are too small to scatter light and therefore do not exhibit a noticeable Tyndall effect.

Another important characteristic of colloids is that the particles have relatively huge surface areas. We saw in Section 14.6 that the surface area is increased tenfold when a 1-cm cube is divided into 1000 cubes with sides of 0.1 cm. When a 1-cm cube is divided into colloidal-size cubes measuring 10^{-6} cm, the combined surface area of all the particles becomes a million times greater than that of the original cube.

Colloidal particles become electrically charged when they adsorb ions on their surfaces. *Adsorption* should not be confused with *absorption.* Adsorption refers to the adhesion of molecules or ions to a surface, whereas absorption refers to the taking in of one material by another material. Adsorption occurs because the atoms or ions at the surface of a particle are not completely surrounded by other atoms or ions as are those in the interior. Consequently these surface atoms or ions attract and adsorb ions or polar molecules from the dispersion medium onto the surfaces of the colloidal particles. This property is directly related to the large surface area presented by the many tiny particles.

15.15 Applications of Colloidal Properties

Activated charcoal has an enormous surface area, approximately 1 million square centimeters per gram in some samples. Hence, charcoal is very effective in selectively adsorbing the polar molecules of some poisonous gases and is therefore used in gas masks. Charcoal can be used to adsorb impurities from liquids as well as from gases, and large amounts are used to remove substances that have objectionable tastes and odors from water supplies. In sugar refineries activated charcoal is used to adsorb colored impurities from the raw sugar solutions.

A process widely used for dust and smoke control in many urban and industrial areas was devised by an American, Frederick Cottrell (1877–1948). The Cottrell process takes advantage of the fact that the particulate matter in dust and smoke is electrically charged. Air to be cleaned of dust or smoke is passed between electrode plates charged with a high voltage. Positively charged particles are attracted to, neutralized, and thereby precipitated at the negative electrodes. Negatively charged particles are removed in the same fashion at the positive electrodes. Large Cottrell units are fitted with devices for automatic removal of precipitated material. In some installations, particularly at cement mills and smelters, the value of the dust collected may be sufficient to pay for the precipitation equipment. Small units, designed for removing dust and pollen from air in the home, are now on the market. Unfortunately, Cottrell units remove only particulate matter; they cannot remove gaseous pollutants such as carbon monoxide, sulfur dioxide, and nitrogen oxides.

Colloidal particles become electrically charged when they adsorb ions on their surface.

Thomas Graham found that a parchment membrane would allow the passage of true solutions but would prevent the passage of colloidal dispersions. Dissolved solutes can be removed from colloidal dispersions through the use of such a membrane by a process called **dialysis.** The membrane itself is called a *dialyzing membrane.* Artificial membranes are made from such materials as parchment paper, collodion, or certain kinds of cellophane. Dialysis can be demonstrated by putting a colloidal starch dispersion and some copper(II) sulfate solution in a parchment paper bag and suspending it in running water. In a few hours the blue color of the copper(II) sulfate has disappeared, and only the starch dispersion remains in the bag.

dialysis

A life-saving application of dialysis has been the development of artificial kidneys. The blood of a patient suffering from partial kidney failure is passed through the artificial kidney machine for several hours, during which time the soluble waste products are removed by dialysis.

Concepts in Review

1. State the general characteristics of acids and bases.

2. Define an acid and base in terms of the Arrhenius, Brønsted–Lowry, and Lewis theories.

3. Identify acid–base conjugate pairs in a reaction.

4. When given the reactants, complete and balance equations for the reactions of acids with bases, metals, metal oxides, and carbonates.

5. When given the reactants, complete and balance equations of the reaction of an amphoteric hydroxide with either a strong acid or a strong base.

6. Write balanced equations for the reaction of sodium hydroxide or potassium hydroxide with zinc and with aluminum.

7. Classify common compounds as electrolytes or nonelectrolytes.

8. Distinguish between strong and weak electrolytes.

9. Explain the process of dissociation and ionization. Indicate how they differ.

10. Write equations for the dissociation and/or ionization of acids, bases, and salts in water.

11. Describe and write equations for the ionization of water.

12. Explain how pH expresses hydrogen ion concentration or hydronium ion concentration.

13. Given pH as an integer, calculate the H^+ molarity, and vice versa.

14. Use a calculator to calculate pH values from corresponding H^+ molarities.

15. Explain the process of acid–base neutralization.

16. Calculate the molarity or volume of an acid or base solution from appropriate titration data.

17. Write un-ionized, total ionic, and net ionic equations for chemical equations.

Key Terms

amphoteric (15.3)
Brownian movement (15.14)
colloid (15.13)
dialysis (15.15)
dissociation (15.6)
electrolyte (15.5)
hydronium ion (15.1)

ionization (15.6)
logarithm (15.9)
net ionic equation (15.11)
neutralization (15.10)
nonelectrolyte (15.5)
pH (15.9)
spectator ion (15.10)

strong electrolyte (15.7)
titration (15.10)
total ionic equation (15.11)
Tyndall effect (15.14)
un-ionized equation (15.11)
weak electrolyte (15.7)

Questions

1. Since a hydrogen ion and a proton are identical, what differences exist between the Arrhenius and Brønsted–Lowry definitions of an acid? (Table 15.1)

2. According to Figure 15.1, what type of substance must be in solution for the bulb to light?

3. Which of the following classes of compounds are electrolytes: acids, alcohols, bases, salts? (Table 15.2)

4. What two differences are apparent in the arrangement of water molecules about the hydrated ions as depicted in Figure 15.2?

5. The pH of a solution with a hydrogen ion concentration of 0.003 M is between what two whole numbers? (Table 15.4)

6. Which is more acidic, tomato juice or blood? (Table 15.5)

7. Use the three acid–base theories (Arrhenius, Brønsted–Lowry, and Lewis) to define an acid and a base.

8. For each acid–base theory referred to in Question 7, write an equation illustrating the neutralization of an acid with a base.

9. Write the Lewis structure for the (a) bromide ion, (b) hydroxide ion, and (c) cyanide ion. Why are these ions considered to be bases according to the Brønsted–Lowry and Lewis acid–base theories?

10. Into what three classes of compounds do electrolytes generally fall?

11. Name each compound listed in Table 15.3.

12. A solution of HCl in water conducts an electric current, but a solution of HCl in hexane does not. Explain this behavior in terms of ionization and chemical bonding.

13. How do ionic compounds exist in their crystalline structure? What occurs when they are dissolved in water?

14. An aqueous methyl alcohol, CH_3OH, solution does not conduct an electric current, but a solution of sodium hydroxide, NaOH, does. What does this information tell us about the OH group in the alcohol?

15. Why does molten NaCl conduct electricity?

16. Explain the difference between dissociation of ionic compounds and ionization of molecular compounds.

17. Distinguish between strong and weak electrolytes.

18. Explain why ions are hydrated in aqueous solutions.

19. What is the main distinction between water solutions of strong and weak electrolytes?

20. What are the relative concentrations of $H^+(aq)$ and $OH^-(aq)$ in (a) a neutral solution, (b) an acid solution, and (c) a basic solution?

21. Write the net ionic equation for the reaction of a strong acid with a water-soluble hydroxide base in an aqueous solution.

22. The solubility of HCl gas in water, a polar solvent, is much greater than its solubility in hexane, a nonpolar solvent. How can you account for this difference?

23. Pure water, containing equal concentrations of both acid and base ions, is neutral. Why?

24. Indicate the fundamental difference between a colloidal dispersion and a true solution.

25. Explain the Tyndall effect and how it may be used to distinguish between a colloidal dispersion and a true solution.

26. Explain the process of dialysis, giving a practical application in society.

Paired Exercises

27. Identify the conjugate acid–base pairs in the following equations:
  (a) $HC_2H_3O_2 + H_2SO_4 \rightleftharpoons H_2C_2H_3O_2^+ + HSO_4^-$
  (b) The two-step ionization of sulfuric acid,
$$H_2SO_4 + H_2O \longrightarrow H_3O^+ + HSO_4^-$$
$$HSO_4^- + H_2O \rightleftharpoons H_3O^+ + SO_4^{2-}$$
  (c) $HClO_4 + H_2O \longrightarrow H_3O^+ + ClO_4^-$
  (d) $CH_3O^- + H_3O^+ \longrightarrow CH_3OH + H_2O$

29. Complete and balance these equations:
  (a) $Mg(s) + HCl(aq) \longrightarrow$
  (b) $BaO(s) + HBr(aq) \longrightarrow$
  (c) $Al(s) + H_2SO_4(aq) \longrightarrow$
  (d) $Na_2CO_3(aq) + HCl(aq) \longrightarrow$
  (e) $Fe_2O_3(s) + HBr(aq) \longrightarrow$
  (f) $Ca(OH)_2(aq) + H_2CO_3(aq) \longrightarrow$

31. Which of these compounds are electrolytes? Consider each substance to be mixed with water.
  (a) HCl
  (b) CO_2
  (c) $CaCl_2$
  (d) $C_{12}H_{22}O_{11}$ (sugar)
  (e) C_3H_7OH (rubbing alcohol)
  (f) CCl_4 (insoluble)

33. Calculate the molarity of the ions present in these salt solutions. Assume each salt to be 100% dissociated:
  (a) 0.015 M NaCl
  (b) 4.25 M NaKSO$_4$
  (c) 0.20 M CaCl$_2$
  (d) 22.0 g KI in 500. mL of solution

35. In Exercise 33, how many grams of each ion would be present in 100. mL of each solution?

37. What is the molar concentration of all ions present in a solution prepared by mixing the following? Neglect the concentration of H^+ and OH^- from water. Also, assume volumes of solutions are additive.
  (a) 30.0 mL of 1.0 M NaCl and 40.0 mL of 1.0 M NaCl
  (b) 30.0 mL of 1.0 M HCl and 30.0 mL of 1.0 M NaOH
  ***(c)** 100.0 mL of 0.40 M KOH and 100.0 mL of 0.80 M HCl

39. Given the data for the following separate titrations, calculate the molarity of the HCl:

	mL HCl	Molarity HCl	mL NaOH	Molarity NaOH
(a)	40.13	M	37.70	0.728
(b)	19.00	M	33.66	0.306
(c)	27.25	M	18.00	0.555

28. Identify the conjugate acid–base pairs in the following equations:
  (a) $HCl + NH_3 \longrightarrow NH_4^+ + Cl^-$
  (b) $HCO_3^- + OH^- \rightleftharpoons CO_3^{2-} + H_2O$
  (c) $HCO_3^- + H_3O^+ \rightleftharpoons H_2CO_3 + H_2O$
  (d) $HC_2H_3O_2 + H_2O \rightleftharpoons H_3O^+ + C_2H_3O_2^-$

30. Complete and balance these equations:
  (a) $NaOH(aq) + HBr(aq) \longrightarrow$
  (b) $KOH(aq) + HCl(aq) \longrightarrow$
  (c) $Ca(OH)_2(aq) + HI(aq) \longrightarrow$
  (d) $Al(OH)_3(s) + HBr(aq) \longrightarrow$
  (e) $Na_2O(s) + HClO_4(aq) \longrightarrow$
  (f) $LiOH(aq) + FeCl_3(aq) \longrightarrow$

32. Which of these compounds are electrolytes? Consider each substance to be mixed with water.
  (a) $NaHCO_3$ (baking soda)
  (b) N_2 (insoluble gas)
  (c) $AgNO_3$
  (d) $HCOOH$ (formic acid)
  (e) $RbOH$
  (f) K_2CrO_4

34. Calculate the molarity of the ions present in these salt solutions. Assume each salt to be 100% dissociated:
  (a) 0.75 M ZnBr$_2$
  (b) 1.65 M Al$_2$(SO$_4$)$_3$
  (c) 900. g (NH$_4$)$_2$SO$_4$ in 20.0 L of solution
  (d) 0.0120 g Mg(ClO$_3$)$_2$ in 1.00 mL of solution

36. In Exercise 34, how many grams of each ion would be present in 100. mL of each solution?

38. What is the molar concentration of all ions present in a solution prepared by mixing the following? Neglect the concentration of H^+ and OH^- from water. Also, assume volumes of solutions are additive.
  (a) 100.0 mL of 2.0 M KCl and 100.0 mL of 1.0 M CaCl$_2$
  (b) 35.0 mL of 0.20 M Ba(OH)$_2$ and 35.0 mL of 0.20 M H$_2$SO$_4$
  (c) 1.00 L of 1.0 M AgNO$_3$ and 500. mL of 2.0 M NaCl

40. Given the data for the following separate titrations, calculate the molarity of the NaOH:

	mL HCl	Molarity HCl	mL NaOH	Molarity NaOH
(a)	37.19	0.126	31.91	M
(b)	48.04	0.482	24.02	M
(c)	13.13	1.425	39.39	M

41. Rewrite the following unbalanced equations, changing them into balanced net ionic equations. All reactions are in water solution.
    (a) $K_2SO_4(aq) + Ba(NO_3)_2(aq) \longrightarrow$
    $$KNO_3(aq) + BaSO_4(s)$$
    (b) $CaCO_3(s) + HCl(aq) \longrightarrow$
    $$CaCl_2(aq) + CO_2(g) + H_2O(l)$$
    (c) $Mg(s) + HC_2H_3O_2(aq) \longrightarrow$
    $$Mg(C_2H_3O_2)_2(aq) + H_2(g)$$

42. Rewrite the following unbalanced equations, changing them into balanced net ionic equations. All reactions are in water solution.
    (a) $H_2S(g) + CdCl_2(aq) \longrightarrow CdS(s) + HCl(aq)$
    (b) $Zn(s) + H_2SO_4(aq) \longrightarrow ZnSO_4(aq) + H_2(g)$
    (c) $AlCl_3(aq) + Na_3PO_4(aq) \longrightarrow$
    $$AlPO_4(s) + NaCl(aq)$$

43. In the following pairs which solution is more acidic? All are water solutions. Explain your answer.
    (a) 1 M HCl or 1 M H_2SO_4?
    (b) 1 M HCl or 1 M $HC_2H_3O_2$?

44. In the following pairs which solution is more acidic? All are water solutions. Explain your answer.
    (a) 1 M HCl or 2 M HCl?
    (b) 1 M HNO_3 or 1 M H_2SO_4?

45. What volume (in milliliters) of 0.245 M HCl will neutralize 10.0 g $Al(OH)_3$? The equation is
    $$3\,HCl(aq) + Al(OH)_3(s) \longrightarrow AlCl_3(aq) + 3\,H_2O(l)$$

46. What volume (in milliliters) of 0.245 M HCl will neutralize 50.0 mL of 0.100 M $Ca(OH)_2$? The equation is
    $$2\,HCl(aq) + Ca(OH)_2(aq) \longrightarrow CaCl_2(aq) + 2\,H_2O(l)$$

*47. A 0.200-g sample of impure NaOH requires 18.25 mL of 0.2406 M HCl for neutralization. What is the percent of NaOH in the sample?

*48. A batch of sodium hydroxide was found to contain sodium chloride as an impurity. To determine the amount of impurity, a 1.00-g sample was analyzed and found to require 49.90 mL of 0.466 M HCl for neutralization. What is the percent of NaCl in the sample?

*49. What volume of H_2 gas, measured at 27°C and 700. torr, can be obtained by reacting 5.00 g of zinc metal with 100. mL of 0.350 M HCl? The equation is
    $$Zn(s) + 2\,HCl(aq) \longrightarrow ZnCl_2(aq) + H_2(g)$$

*50. What volume of H_2 gas, measured at 27°C and 700. torr, can be obtained by reacting 5.00 g of zinc metal with 200. mL of 0.350 M HCl? The equation is
    $$Zn(s) + 2\,HCl(aq) \longrightarrow ZnCl_2(aq) + H_2(g)$$

51. Calculate the pH of solutions having these H^+ ion concentrations:
    (a) 0.01 M
    (b) 1.0 M
    (c) 6.5×10^{-9} M

52. Calculate the pH of solutions having these H^+ ion concentrations:
    (a) 1×10^{-7} M
    (b) 0.50 M
    (c) 0.00010 M

53. Calculate the pH of
    (a) orange juice, 3.7×10^{-4} $M\,H^+$
    (b) vinegar, 2.8×10^{-3} $M\,H^+$

54. Calculate the pH of
    (a) black coffee, 5.0×10^{-5} $M\,H^+$
    (b) limewater, 3.4×10^{-11} $M\,H^+$

Additional Exercises

55. What is the concentration of Ca^{2+} ions in a solution of CaI_2 having an I^- ion concentration of 0.520 M?

56. How many milliliters of 0.40 M HCl can be made by diluting 100. mL of 12 M HCl with water?

57. If 29.26 mL of 0.430 M HCl neutralizes 20.40 mL of $Ba(OH)_2$ solution, what is the molarity of the $Ba(OH)_2$ solution? The reaction is
    $$Ba(OH)_2(aq) + 2\,HCl(aq) \longrightarrow BaCl_2(aq) + 2\,H_2O(l)$$

58. A 1 m solution of acetic acid, $HC_2H_3O_2$, in water freezes at a lower temperature than a 1 m solution of ethyl alcohol, C_2H_5OH, in water. Explain.

59. At the same cost per pound, which alcohol, CH_3OH or C_2H_5OH, would be more economical to purchase as an antifreeze for your car? Why?

60. How does a hydronium ion differ from a hydrogen ion?

61. Arrange, in decreasing order of freezing points, 1 m aqueous solutions of HCl, $HC_2H_3O_2$, $C_{12}H_{22}O_{11}$ (sucrose), and $CaCl_2$. (List the one with the highest freezing point first.)

62. At 100°C the H^+ concentration in water is about 1×10^{-6} mol/L, about ten times that of water at 25°C. At which of these temperatures is
    (a) the pH of water the greater?
    (b) the hydrogen ion (hydronium ion) concentration the higher?
    (c) the water neutral?

63. What is the relative difference in H^+ concentration in solutions that differ by one pH unit?

64. A sample of pure sodium carbonate with a mass of 0.452 g was dissolved in water and neutralized with 42.4 mL of hydrochloric acid. Calculate the molarity of the acid:
$$Na_2CO_3(aq) + 2\,HCl(aq) \longrightarrow$$
$$2\,NaCl(aq) + CO_2(g) + H_2O(l)$$

65. What volume (mL) of 0.1234 M HCl is needed to neutralize 2.00 g $Ca(OH)_2$?

66. How many grams of KOH are required to neutralize 50.00 mL of 0.240 M HNO_3?

67. Two drops (0.1 mL) of 1.0 M HCl are added to water to make 1.0 L of solution. What is the pH of this solution if the HCl is 100% ionized?

68. What volume of concentrated (18.0 M) sulfuric acid must be used to prepare 50.0 L of 5.00 M solution?

69. If 3.0 g NaOH is added to 500. mL of 0.10 M HCl, will the resulting solution be acidic or basic? Show evidence for your answer.

70. If 380 mL of 0.35 M $Ba(OH)_2$ is added to 500.0 mL of 0.65 M HCl, will the mixture be acidic or basic? Find the pH of the resulting solution.

71. If 50.00 mL of 0.2000 M HCl is titrated with 0.2000 M NaOH, find the pH of the solution after the following amounts of base have been added:
(a) 0.000 mL   **(e)** 49.90 mL
(b) 10.00 mL   **(f)** 49.99 mL
(c) 25.00 mL   **(g)** 50.00 mL
(d) 49.00 mL
Plot your answers on a graph with pH on the y-axis and mL NaOH on the x-axis.

72. Sulfuric acid reacts with NaOH:
(a) Write a balanced equation for the reaction producing Na_2SO_4.
(b) How many milliliters of 0.10 M NaOH are needed to react with 0.0050 mol H_2SO_4?
(c) How many grams of Na_2SO_4 will also form?

*73. Lactic acid (found in sour milk) has an empirical formula of $HC_3H_5O_3$. A 1.0-g sample of lactic acid required 17.0 mL of 0.65 M NaOH to reach the end point of a titration. What is the molecular formula for lactic acid?

74. A 10.0-mL sample of HNO_3 was diluted to a volume of 100.00 mL. Then, 25 mL of that diluted solution was needed to neutralize 50.0 mL of 0.60 M KOH. What was the concentration of the original nitric acid?

75. The pH of a solution of a strong acid was determined to be 3. If water is then added to dilute this solution, would the pH change? Why or why not? Could enough water ever be added to raise the pH of an acid solution above 7?

76. Solution X has a pH of 2. Solution Y has a pH of 4. On the basis of this information, which of the following is true?
(a) the $[H^+]$ of X is one-half that of Y
(b) the $[H^+]$ of X is twice that of Y
(c) the $[H^+]$ of X is 100 times that of Y
(d) the $[H^+]$ of X is $^{1}/_{100}$ that of Y

77. A student is given three solutions: an acid, a base, and one that is neither acidic nor basic. The student performs tests on these solutions and records their properties. For each result, tell whether it is the property of an acid, a base, or whether you cannot decide.
(a) the solution has $[H^+] = 1 \times 10^{-7}\,M$
(b) the solution has $[OH^-] = 1 \times 10^{-2}\,M$
(c) the solution turns litmus red
(d) the solution is a good conductor of electricity

Answers to Practice Exercises

15.1 (a) HCO_3^-, (b) NO_2^-, (c) $C_2H_3O_2^-$
15.2 (a) H_2SO_4, (b) NH_4^+, (c) H_2O
15.3 (a) 0.050 M Mg^{2+}, 0.10 M Cl^-,
(b) 0.070 M Al^{3+}, 0.21 M Cl^-

15.4 (a) 11.41, (b) 2.89, (c) 5.429
15.5 (c) 0.0620 M HCl
15.6 $3\,H_2S(aq) + 2\,Bi^{3+}(aq) \longrightarrow Bi_2S_3(s) + 6\,H^+(aq)$

CHAPTER

16

Chemical Equilibrium

A coral reef is a system in ecological equilibrium with the ocean surrounding it.

CHAPTER 16 / OUTLINE

Thus far we've considered chemical change as proceeding from reactants to products. Does that mean that the change then stops? No, but often it appears to be the case at the macroscopic level. A solute dissolves until the solution becomes saturated. Once a solid remains undissolved in a container, the system appears to be at rest. The human body is a marvelous chemical factory, yet from day to day it appears to be quite the same. For example, the blood remains at a constant pH, even though all sorts of chemical reactions are taking place. Another example is a terrarium, which can be watered and sealed for long periods of time with no ill effects. Or an antacid, which absorbs excess stomach acid and does *not* change the pH of the stomach. In all of these cases reactions are proceeding, even though visible signs of chemical change are absent. Similarly when a system is at equilibrium, chemical reactions are dynamic at the molecular level. In this chapter we will consider chemical systems as they approach equilibrium conditions.

16.1   Reversible Reactions

In the preceding chapters we treated chemical reactions mainly as reactants changing to products. However, many reactions do not go to completion. Some reactions do not go to completion because they are reversible; that is, when the products are formed, they react to produce the starting reactants.

We've encountered reversible systems before. One is the vaporization of a liquid by heating and its subsequent condensation by cooling:

$$\text{liquid} + \text{heat} \longrightarrow \text{vapor}$$

$$\text{vapor} + \text{cooling} \longrightarrow \text{liquid}$$

The conversion between nitrogen dioxide, NO_2, and dinitrogen tetroxide, N_2O_4, shows us visible evidence of the reversibility of a reaction. The NO_2 is a reddish-brown gas that changes with cooling to N_2O_4, a colorless gas. The reaction is reversible by heating N_2O_4:

$$2\ NO_2(g) \xrightarrow{\text{cooling}} N_2O_4(g)$$

$$N_2O_4(g) \xrightarrow{\text{heating}} 2\ NO_2(g)$$

These two reactions may be represented by a single equation with a double arrow, $\rightleftharpoons$, to indicate that the reactions are taking place in both directions at the same time:

$$2\ NO_2(g) \rightleftharpoons N_2O_4(g)$$

This reversible reaction can be demonstrated by sealing samples of NO_2 in two tubes and placing one tube in warm water and the other in ice water (see Figure 16.1).

A **reversible chemical reaction** is one in which the products formed react to produce the original reactants. Both the forward and reverse reactions occur

reversible chemical reaction

FIGURE 16.1 ▶
Reversible reaction of NO₂
and N₂O₄. More of the
reddish-brown NO₂ molecules
are visible in the tube that is
heated (right, 80°C) than in
the tube that is cooled (left,
0°C).

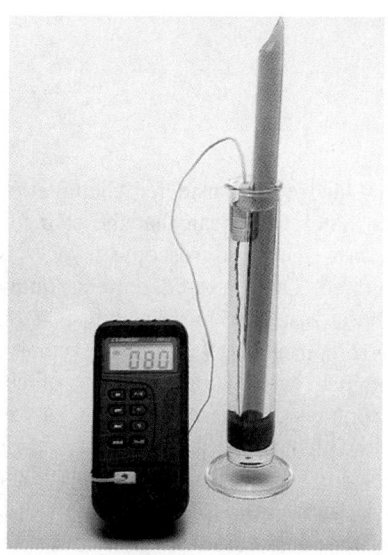

simultaneously. The forward reaction is called *the reaction to the right,* and the reverse reaction is called *the reaction to the left.* A double arrow is used in the equation to indicate that the reaction is reversible.

16.2   Rates of Reaction

chemical kinetics

Every reaction has a rate, or speed, at which it proceeds. Some are fast and some are extremely slow. The study of reaction rates and reaction mechanisms is known as **chemical kinetics.**

The rate of a reaction is variable and depends on the concentration of the reacting species, the temperature, the presence of catalysts, and the nature of the reactants. Consider the hypothetical reaction

$$A + B \longrightarrow C + D \quad \text{(forward reaction)}$$

$$C + D \longrightarrow A + B \quad \text{(reverse reaction)}$$

in which a collision between A and B is necessary for a reaction to occur. The rate at which A and B react depends on the concentration or the number of A and B molecules present; it will be fastest, for a fixed set of conditions, when they are first mixed (as shown by the height of the red line in Figure 16.2). As the reaction proceeds, the number of A and B molecules available for reaction decreases, and the rate of reaction slows down (seen as the red line flattens in Figure 16.2). If the reaction is reversible, the speed of the reverse reaction is zero at first (blue line in Figure 16.2) and gradually increases as the concentrations of C and D increase. As the number of A and B molecules decreases, the forward rate slows down because A and B cannot find one another as often in order to accomplish a reaction. To counteract this diminishing rate of reaction, an excess of one reagent is often used to keep the reaction from becoming impractically slow. Collisions between molecules may be compared to video games. When many objects are on the screen, collisions occur frequently; but if only a few objects are present, collisions can usually be avoided.

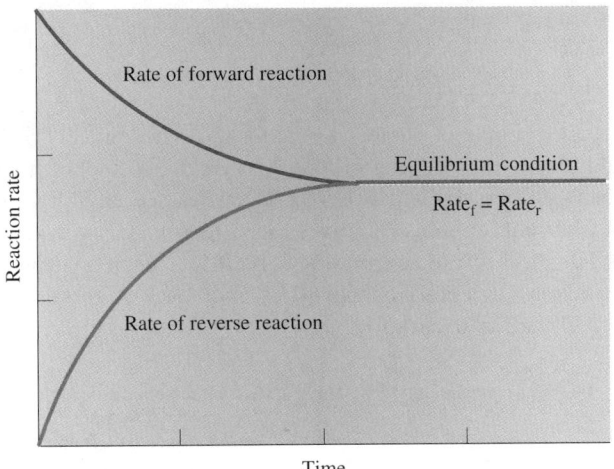

◀FIGURE 16.2
The rates of the forward
and reverse reactions
become equal at some
point in time. The forward
reaction rate (red)
decreases as a result of
decreasing amounts of
reactants. The reverse
reaction rate (blue) starts at
zero and increases as the
amount of product
increases. When the two
rates become equal
(purple), a state of chemical
equilibrium has been
reached.

16.3 Chemical Equilibrium

Any system at **equilibrium** represents a dynamic state in which two or more
opposing processes are taking place at the same time and at the same rate. A chemical equilibrium is a dynamic system in which two or more opposing chemical
reactions are going on at the same time and at the same rate. When the rate of
the forward reaction is exactly equal to the rate of the reverse reaction, a condition
of **chemical equilibrium** exists (see purple line in Figure 16.2). The concentrations
of the products and the reactants are not changing, and the system appears to be at
a standstill because the products are reacting at the same rate at which they are being
formed.

equilibrium

chemical equilibrium

> **Chemical equilibrium:**
> **rate of forward reaction = rate of reverse reaction**

A saturated salt solution is in a condition of equilibrium:

$$NaCl(s) \rightleftharpoons Na^+(aq) + Cl^-(aq)$$

At equilibrium, salt crystals are continuously dissolving, and Na^+ and Cl^- ions are
continuously crystallizing. Both processes are occurring at the same rate.

The ionization of weak electrolytes is another chemical equilibrium system:

$$HC_2H_3O_2(aq) + H_2O(l) \rightleftharpoons H_3O^+(aq) + C_2H_3O_2^-(aq)$$

In this reaction the equilibrium is established in a 1 M solution when the forward reaction has gone about 1%—that is, when only 1% of the acetic acid molecules in solution have ionized. Therefore only a relatively few ions are present, and the acid
behaves as a weak electrolyte.

The reaction represented by

$$H_2(g) + I_2(g) \underset{}{\overset{700\ K}{\rightleftharpoons}} 2\ HI(g)$$

provides another example of chemical equilibrium. Theoretically, 1.00 mol of hydrogen should react with 1.00 mol of iodine to yield 2.00 mol of hydrogen iodide. Actually, when 1.00 mol H_2 and 1.00 mol I_2 are reacted at 700 K, only 1.58 mol HI is present when equilibrium is attained. Since 1.58 is 79% of the theoretical yield of 2.00 mol HI, the forward reaction is only 79% complete at equilibrium. The equilibrium mixture will also contain 0.21 mol each of unreacted H_2 and I_2 (1.00 mol − 0.79 mol = 0.21 mol):

$$H_2(g) + I_2(g) \xrightarrow{700\ K} 2\ HI(g)$$

This equation represents the condition if the reaction were 100% complete; 2.00 mol HI would be formed and no H_2 and I_2 would be left unreacted.

$$\underset{\substack{0.21 \\ mol}}{H_2(g)} + \underset{\substack{0.21 \\ mol}}{I_2(g)} \underset{}{\overset{700\ K}{\rightleftharpoons}} \underset{\substack{1.58 \\ mol}}{2\ HI(g)}$$

This equation represents the actual equilibrium attained starting with 1.00 mol each of H_2 and I_2. It shows that the forward reaction is only 79% complete.

16.4   Le Chatelier's Principle

In 1888, the French chemist Henri Le Chatelier (1850–1936) set forth a simple, far-reaching generalization on the behavior of equilibrium systems. This generalization, known as **Le Chatelier's principle,** states

Le Chatelier's principle

> **If a stress is applied to a system in equilibrium, the system will respond in such a way as to relieve that stress and restore equilibrium under a new set of conditions.**

The application of Le Chatelier's principle helps us predict the effect of changing conditions in chemical reactions. We will examine the effect of changes in concentration, temperature, and volume.

16.5   Effect of Concentration on Equilibrium

The manner in which the rate of a chemical reaction depends on the concentration of the reactants must be determined experimentally. Many simple, one-step reactions result from a collision between two molecules or ions. The rate of such one-step reactions can be altered by changing the concentration of the reactants or products. An increase in concentration of the reactants provides more individual reacting species for collisions and results in an increase in the rate of reaction.

An equilibrium is disturbed when the concentration of one or more of its components is changed. As a result the concentration of all species will change, and a new equilibrium mixture will be established. Consider the hypothetical equilibrium represented by the equation

A + B $\rightleftharpoons$ C + D

where A and B react in one step to form C and D. When the concentration of B is increased, the following occurs:

1. The rate of the reaction to the right (forward) increases. This rate is proportional to the concentration of A times the concentration of B.
2. The rate to the right becomes greater than the rate to the left.
3. Reactants A and B are used faster than they are produced; C and D are produced faster than they are used.
4. After a period of time, rates to the right and left become equal, and the system is again in equilibrium.
5. In the new equilibrium the concentration of A is less, and the concentrations of B, C, and D are greater than in the original equilibrium.

   Conclusion: The equilibrium has shifted to the right.

Concentration	Change
$[H_2]$	?
$[I_2]$	increase
$[HI]$	?

Applying this change in concentration to the equilibrium mixture of 1.00 mol of hydrogen and 1.00 mol of iodine from Section 16.3, we find that, when an additional 0.20 mol I_2 is added, the yield of HI (based on H_2) is 85% (1.70 mol) instead of 79%. Here is how the two systems compare after the new equilibrium mixture is reached:

Original equilibrium	New equilibrium
1.00 mol H_2 + 1.00 mol I_2	1.00 mol H_2 + 1.20 mol I_2
Yield: 79% HI	Yield: 85% HI (based on H_2)
Equilibrium mixture contains:	Equilibrium mixture contains:
1.58 mol HI	1.70 mol HI
0.21 mol H_2	0.15 mol H_2
0.21 mol I_2	0.35 mol I_2

Concentration	Change
$[H_2]$	decrease
$[I_2]$	increase
$[HI]$	increase

Analyzing this new system, we see that, when 0.20 mol I_2 is added, the equilibrium shifts to the right to counteract the increase in I_2 concentration. Some of the H_2 reacts with added I_2 and produces more HI, until an equilibrium mixture is established again. When I_2 is added, the concentration of I_2 increases, the concentration of H_2 decreases, and the concentration of HI increases.

Practice 16.1

Use a chart like those in the margin to show what would happen to the concentrations of each substance in the system

   $H_2(g) + I_2(g) \rightleftharpoons 2\ HI(g)$

upon adding (a) more H_2 and (b) more HI.

The equation

$$Fe^{3+}(aq) + SCN^-(aq) \rightleftharpoons Fe(SCN)^{2+}(aq)$$

pale yellow   colorless   red

represents an equilibrium that is used in certain analytical procedures as an indicator because of the readily visible, intense red color of the complex $Fe(SCN)^{2+}$ ion. A very dilute solution of iron(III), Fe^{3+}, and thiocyanate, SCN^-, is light red. When the concentration of either Fe^{3+} or SCN^- is increased, the equilibrium shift to the right is observed by an increase in the intensity of the color, resulting from the formation of additional $Fe(SCN)^{2+}$.

If either Fe^{3+} or SCN^- is removed from solution, the equilibrium will shift to the left, and the solution will become lighter in color. When Ag^+ is added to the solution, a white precipitate of silver thiocyanate (AgSCN) is formed, thus removing SCN^- ion from the equilibrium:

$$Ag^+(aq) + SCN^-(aq) \rightleftharpoons AgSCN(s)$$

The system accordingly responds to counteract the change in SCN^- concentration by shifting the equilibrium to the left. This shift is evident by a decrease in the intensity of the red color due to a decreased concentration of $Fe(SCN)^{2+}$.

Now consider the effect of changing the concentrations in the equilibrium mixture of chlorine water. The equilibrium equation is

$$Cl_2(aq) + 2 H_2O(l) \rightleftharpoons HOCl(aq) + H_3O^+(aq) + Cl^-(aq)$$

The variation in concentrations and the equilibrium shifts are tabulated in the following table. An X in the second or third column indicates that the reagent is increased or decreased. The fourth column indicates the direction of the equilibrium shift.

| | Concentration | | |
Reagent	Increase	Decrease	Equilibrium shift
Cl_2	—	X	Left
H_2O	X	—	Right
HOCl	X	—	Left
H_3O^+	—	X	Right
Cl^-	X	—	Left

Consider the equilibrium in a 0.100 M acetic acid solution:

$$HC_2H_3O_2(aq) + H_2O(l) \rightleftharpoons H_3O^+(aq) + C_2H_3O_2^-(aq)$$

In this solution the concentration of the hydronium ion, H_3O^+, which is a measure of the acidity, is 1.34×10^{-3} mol/L, corresponding to a pH of 2.87. What will happen to the acidity when 0.100 mol of sodium acetate, $NaC_2H_3O_2$, is added to 1 L of 0.100 M acetic acid, $HC_2H_3O_2$? When $NaC_2H_3O_2$ dissolves, it dissociates into sodium ions, Na^+, and acetate ions, $C_2H_3O_2^-$. The acetate ion from the salt is a common ion to the acetic acid equilibrium system and increases the total acetate ion concentration in the solution. As a result the equilibrium shifts to the left, decreasing the hydronium ion concentration and lowering the acidity of the solution. Evidence of this decrease in acidity is shown by the fact that the pH of a solution that is 0.100 M in $HC_2H_3O_2$ and 0.100 M in $NaC_2H_3O_2$ is 4.74. The pH of several different solutions of $HC_2H_3O_2$ and $NaC_2H_3O_2$ is shown in the table that follows. Each time the acetate ion is increased, the pH increases, indicating a further shift in the equilibrium toward un-ionized acetic acid.

Concentration	Change
$[HC_2H_3O_2]$	increase
$[H_3O^+]$	decrease
$[C_2H_3O_2^-]$	increase

Solution	pH
1 L 0.100 M $HC_2H_3O_2$	2.87
1 L 0.100 M $HC_2H_3O_2$ + 0.100 mol $NaC_2H_3O_2$	4.74
1 L 0.100 M $HC_2H_3O_2$ + 0.200 mol $NaC_2H_3O_2$	5.05
1 L 0.100 M $HC_2H_3O_2$ + 0.300 mol $NaC_2H_3O_2$	5.23

In summary we can say that when the concentration of a reagent on the left side of an equation is increased, the equilibrium shifts to the right. When the concentration of a reagent on the right side of an equation is increased, the equilibrium shifts to the left. In accordance with Le Chatelier's principle the equilibrium always shifts in the direction that tends to reduce the concentration of the added reactant.

Practice 16.2

Aqueous chromate ion, CrO_4^{2-}, exists in equilibrium with aqueous dichromate ion, $Cr_2O_7^{2-}$ in an acidic solution. What effect will (a) increasing the dichromate ion, and (b) adding HCl have on the equilibrium?

$$2\,CrO_4^{2-}(aq) + 2\,H^+(aq) \rightleftharpoons Cr_2O_7^{2-}(aq) + H_2O(l)$$

16.6   Effect of Volume on Equilibrium

Changes in volume significantly affect the reaction rate only when one or more of the reactants or products is a gas, and the reaction is run in a closed container. In these cases the effect of decreasing the volume of the reacting gases is equivalent to increasing their concentrations. In the reaction

$$CaCO_3(s) \overset{\Delta}{\rightleftharpoons} CaO(s) + CO_2(g)$$

calcium carbonate decomposes into calcium oxide and carbon dioxide when heated about 825°C. Decreasing the volume of the container speeds up the reverse reaction and causes the equilibrium to shift to the left. Decreasing the volume increases the concentration of CO_2, the only gaseous substance in the reaction.

If the volume of the container is decreased, the pressure of the gas will increase. In a system composed entirely of gases, this decrease in the volume of the container will cause the reaction and the equilibrium to shift to the side that contains the smaller number of molecules. To clarify your thinking: When the container volume is decreased, the pressure in the container is increased. The system tries to lower this

▲
Gaseous ammonia is often used to add nitrogen to the fields before planting and during early growth.

Haber received the Nobel prize in chemistry for this process in 1918.

pressure by reducing the number of molecules. Let's consider an example that shows these effects.

Prior to World War I, Fritz Haber (1868–1934) invented the first major process for the fixation of nitrogen. In this process nitrogen and hydrogen are reacted together in the presence of a catalyst at moderately high temperature and pressure to produce ammonia:

$$N_2(g) \ + \ 3\,H_2(g) \rightleftharpoons 2\,NH_3(g) + 92.5\text{ kJ}$$

| 1 mol | 3 mol | 2 mol |
| 1 volume | 3 volumes | 2 volumes |

The left side of the equation in the Haber process represents 4 mol of gas combining to give 2 mol of gas on the right side of the equation. A decrease in the volume of the container shifts the equilibrium to the right. This decrease in volume results in a higher concentration of both reactants and products. The equilibrium shifts to the right toward fewer molecules.

When the total number of gaseous molecules on both sides of an equation is the same, a change in volume does not cause an equilibrium shift. The following reaction is an example:

$$N_2(g) \qquad + \qquad O_2(g) \qquad \rightleftharpoons \qquad 2\,NO(g)$$

1 mol	1 mol	2 mol
1 volume	1 volume	2 volumes
6.022×10^{23} molecules	6.022×10^{23} molecules	$2(6.022 \times 10^{23})$ molecules

When the volume of the container is decreased, the rate of both the forward and the reverse reactions will increase because of the higher concentrations of N_2, O_2, and NO. But the equilibrium will not shift because the number of molecules is the same on both sides of the equation and the effects on concentration are the same on both forward and reverse rates.

Example 16.1   What effect would a decrease in volume of the container have on the position of equilibrium in these reactions?

(a) $2\,SO_2(g) + O_2(g) \rightleftharpoons 2\,SO_3(g)$

(b) $H_2(g) + Cl_2(g) \rightleftharpoons 2\,HCl(g)$

(c) $N_2O_4(g) \rightleftharpoons 2\,NO_2(g)$

(a) The equilibrium will shift to the right because the substance on the right has a smaller number of moles than those on the left.

(b) The equilibrium position will be unaffected because the moles of gases on both sides of the equation are the same.

(c) The equilibrium will shift to the left because $N_2O_4(g)$ represents the smaller number of moles.

Solution

Practice 16.3

What effect would a decrease in the container's volume have on the position of the equilibrium in these reactions?

(a) $2 NO(g) + Cl_2(g) \rightleftharpoons 2 NOCl(g)$

(b) $COBr_2(g) \rightleftharpoons CO(g) + Br_2(g)$

16.7 Effect of Temperature on Equilibrium

When the temperature of a system is raised, the rate of reaction increases because of increased kinetic energy and more frequent collisions of the reacting species. In a reversible reaction the rate of both the forward and the reverse reactions is increased by an increase in temperature; however, the reaction that absorbs heat increases to a greater extent, and the equilibrium shifts to favor that reaction.

An increase in temperature generally increases the rate of reaction. Molecules at elevated temperatures have more kinetic energy; their collisions are thus more likely to result in a reaction.

When heat is applied to a system in equilibrium, the reaction that absorbs heat is favored. When the process, as written, is endothermic, the forward reaction is increased. When the reaction is exothermic, the reverse reaction is favored. In this sense heat may be treated as a reactant in endothermic reactions or as a product in exothermic reactions. Therefore temperature is analogous to concentration when applying Le Chatelier's principle to heat effects on a chemical reaction.

Hot coke (C) is a very reactive element. In the reaction

$$C(s) + CO_2(g) + \text{heat} \rightleftharpoons 2 CO(g)$$

very little if any CO is formed at room temperature. At 1000°C, the equilibrium mixture contains about an equal number of moles of CO and CO_2. Since the reaction is endothermic, the equilibrium is shifted to the right at higher temperatures.

When phosphorus trichloride reacts with dry chlorine gas to form phosphorus pentachloride, the reaction is exothermic:

$$PCl_3(l) + Cl_2(g) \rightleftharpoons PCl_5(s) + 88 \text{ kJ}$$

Heat must continuously be removed during the reaction to obtain a good yield of the product. According to Le Chatelier's principle, heat will cause the product, PCl_5, to decompose, re-forming PCl_3 and Cl_2. The equilibrium mixture at 200°C contains 52% PCl_5, and at 300°C it contains 3% PCl_5, verifying that heat causes the equilibrium to shift to the left.

High temperatures can cause the destruction or decomposition of the reactants or products.

▲ Light sticks. The chemical reaction that produces light in these light sticks is endothermic. Placing the light stick in hot water (right) favors this reaction, producing a brighter light than when the light stick is in ice water (left).

Example 16.2   What effect would an increase in temperature have on the position of the equilibrium in these reactions?

$$4\ HCl(g) + O_2(g) \rightleftharpoons 2\ H_2O(g) + 2\ Cl_2(g) + 95.4\ kJ \qquad (1)$$

$$H_2(g) + Cl_2(g) \rightleftharpoons 2\ HCl(g) + 185\ kJ \qquad (2)$$

$$CH_4(g) + 2\ O_2(g) \rightleftharpoons CO_2(g) + 2\ H_2O(g) + 890\ kJ \qquad (3)$$

$$N_2O_4(g) + 58.6\ kJ \rightleftharpoons 2\ NO_2(g) \qquad (4)$$

$$2\ CO_2(g) + 566\ kJ \rightleftharpoons 2\ CO(g) + O_2(g) \qquad (5)$$

$$H_2(g) + I_2(g) + 51.9\ kJ \rightleftharpoons 2\ HI(g) \qquad (6)$$

Solution   Reactions (1), (2), and (3) are exothermic; an increase in temperature will cause the equilibrium to shift to the left. Reactions (4), (5), and (6) are endothermic; an increase in temperature will cause the equilibrium to shift to the right.

Practice 16.4

What effect would an increase in temperature have on the position of the equilibrium in these reactions?

(a) $2\ SO_2(g) + O_2(g) \rightleftharpoons 2\ SO_3(g) + 198\ kJ$

(b) $H_2(g) + CO_2(g) + 41\ kJ \rightleftharpoons H_2O(g) + CO(g)$

16.8   Effect of Catalysts on Equilibrium

catalyst   A **catalyst** is a substance that influences the rate of a reaction and can be recovered essentially unchanged at the end of the reaction. A catalyst does not shift the equilibrium of a reaction; it affects only the speed at which the equilibrium is reached. It does this by lowering the activation energy for the reaction (see Figure 16.3). **Activation energy** is the minimum energy required for the reaction to occur. A catalyst

activation energy   speeds up a reaction by lowering the activation energy while not changing the energies of reactants or products. If a catalyst does not affect the equilibrium, then it follows that it must affect the rate of both the forward and the reverse reactions equally.

The reaction between phosphorus trichloride and sulfur is highly exothermic, but it's so slow that very little product, thiophosphoryl chloride, is obtained, even after prolonged heating. When a catalyst, such as aluminum chloride is added, the reaction is complete in a few seconds:

$$PCl_3(l) + S(s) \xrightarrow{\text{AlCl}_3} PSCl_3(l)$$

The lab preparation of oxygen uses manganese dioxide as a catalyst to increase the rates of decomposition of both potassium chlorate and hydrogen peroxide:

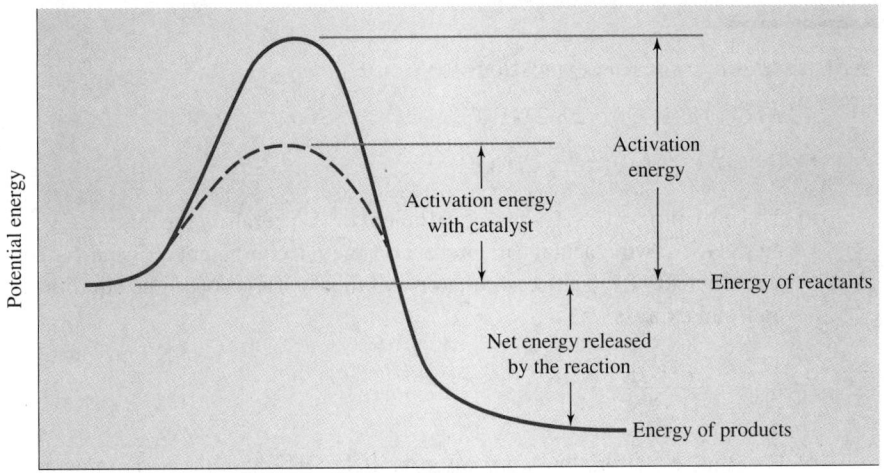

◀ **FIGURE 16.3**
Energy diagram for an exothermic reaction. Energy is put into the reaction (activation energy) to initiate the process. In the reaction shown, all of the activation energy and the net energy are released as the reaction proceeds to products. Note that the presence of a catalyst lowers the activation energy but does not change the energies of the reactants or the products.

$$2\ KClO_3(s) \xrightarrow[\Delta]{MnO_2} 2\ KCl(s) + 3\ O_2(g)$$

$$2\ H_2O_2(aq) \xrightarrow{MnO_2} 2\ H_2O(l) + O_2(g)$$

Catalysts are extremely important to industrial chemistry. Hundreds of chemical reactions that are otherwise too slow to be of practical value have been put to commercial use once a suitable catalyst was found. And in the area of biochemistry, catalysts are of supreme importance because nearly all chemical reactions in all forms of life are completely dependent on biochemical catalysts known as *enzymes.*

16.9 Equilibrium Constants

In a reversible chemical reaction at equilibrium, the concentrations of the reactants and products are constant. At equilibrium the rates of the forward and reverse reactions are equal, and an equilibrium constant expression can be written relating the products to the reactants. For the general reaction

$$aA + bB \rightleftharpoons cC + dD$$

at a given temperature, the following equilibrium constant expression can be written:

$$K_{eq} = \frac{[C]^c[D]^d}{[A]^a[B]^b}$$

where K_{eq}, the **equilibrium constant,** is constant at a particular temperature. The quantities in brackets are the concentrations of each substance in moles per liter. The superscript letters *a, b, c,* and *d* are the coefficients of the substances in the balanced equation. According to convention, we place the concentrations of the products (the substances on the right side of the equation as written) in the numerator and the concentrations of the reactants in the denominator.

equilibrium constant, K_{eq}

Note: The exponents are the same as the coefficients in the balanced equation.

Example 16.3   Write equilibrium constant expressions for

(a) $3 H_2(g) + N_2(g) \rightleftharpoons 2 NH_3(g)$

(b) $CO(g) + 2 H_2(g) \rightleftharpoons CH_3OH(g)$

Solution

(a) The only product, NH_3, has a coefficient of 2. Therefore the numerator will be $[NH_3]^2$. Two reactants are present, H_2 with a coefficient of 3 and N_2 with a coefficient of 1. The denominator will thus be $[H_2]^3[N_2]$. The equilibrium constant expression is

$$K_{eq} = \frac{[NH_3]^2}{[H_2]^3[N_2]}$$

(b) For this equation the numerator is $[CH_3OH]$ and the denominator is $[CO][H_2]^2$. The equilibrium constant expression is

$$K_{eq} = \frac{[CH_3OH]}{[CO][H_2]^2}$$

Practice 16.5

Write equilibrium constant expressions for

(a) $2 N_2O_5(g) \rightleftharpoons 4 NO_2(g) + O_2(g)$

(b) $4 NH_3(g) + 3 O_2(g) \rightleftharpoons 2 N_2(g) + 6 H_2O(g)$

The magnitude of an equilibrium constant indicates the extent to which the forward and reverse reactions take place. When K_{eq} is greater than 1, the amount of products at equilibrium is greater than the amount of reactants. When K_{eq} is less than 1, the amount of reactants at equilibrium is greater than the amount of products. A very large value for K_{eq} indicates that the forward reaction goes essentially to completion. A very small K_{eq} means that the reverse reaction goes nearly to completion and that the equilibrium is far to the left (toward the reactants). Two examples follow:

$$H_2(g) + I_2(g) \rightleftharpoons 2 HI(g) \qquad K_{eq} = 54.8 \text{ at } 425°C$$

This K_{eq} indicates that more product than reactant exists at equilibrium.

$$COCl_2(g) \rightleftharpoons CO(g) + Cl_2(g) \qquad K_{eq} = 7.6 \times 10^{-4} \text{ at } 400°C$$

This K_{eq} indicates that $COCl_2$ is stable and that very little decomposition to CO and Cl_2 occurs at 400°C. The equilibrium is far to the left.

Units are generally not included in values of K_{eq} for reasons beyond the scope of this book.

When the molar concentrations of all species in an equilibrium reaction are known, the K_{eq} can be calculated by substituting the concentrations into the equilibrium constant expression.

Calculate the K_{eq} for the following reaction based on concentrations of
$PCl_5 = 0.030$ mol/L, $PCl_3 = 0.97$ mol/L, and $Cl_2 = 0.97$ mol/L at 300°C.

Example 16.4

$$PCl_5(g) \rightleftharpoons PCl_3(g) + Cl_2(g)$$

First write the K_{eq} expression; then substitute the respective concentrations into this
equation and solve:

Solution

$$K_{eq} = \frac{[PCl_3][Cl_2]}{[PCl_5]} = \frac{(0.97)(0.97)}{(0.030)} = 31$$

**Remember: Units are not
included for K_{eq}.**

This K_{eq} is considered to be a fairly large value, indicating that at 300°C the decomposition of PCl_5 proceeds far to the right.

Practice 16.6

Calculate the K_{eq} for this reaction. Is the forward or the reverse reaction
favored?

$$2\,NO(g) + O_2(g) \rightleftharpoons 2\,NO_2(g)$$

when $[NO] = 0.050\ M$, $[O_2] = 0.75\ M$, and $[NO_2] = 0.25\ M$.

16.10 Ion Product Constant for Water

We've seen that water ionizes to a slight degree. This ionization is represented by
these equilibrium equations:

$$H_2O + H_2O \rightleftharpoons H_3O^+ + OH^- \tag{1}$$

$$H_2O \rightleftharpoons H^+ + OH^- \tag{2}$$

Equation (1) is the more accurate representation of the equilibrium because free
protons (H^+) do not exist in water. Equation (2) is a simplified and often-used representation of the water equilibrium. The actual concentration of H^+ produced in pure
water is minute and amounts to only 1.00×10^{-7} mol/L at 25°C. In pure water,

$$[H^+] = [OH^-] = 1.00 \times 10^{-7}\ \text{mol/L}$$

since both ions are produced in equal molar amounts, as shown in equation (2).

The $H_2O \rightleftharpoons H^+ + OH^-$ equilibrium exists in water and in all water solutions.
A special equilibrium constant called the **ion product constant for water, K_w,**
applies to this equilibrium. The constant K_w is defined as the product of the H^+ ion
concentration and the OH^- ion concentration, each in moles per liter:

**ion product constant for
water, K_w**

$$K_w = [H^+][OH^-]$$

The numerical value of K_w is 1.00×10^{-14}, since for pure water at 25°C,

$$K_w = [H^+][OH^-] = (1.00 \times 10^{-7})(1.00 \times 10^{-7}) = 1.00 \times 10^{-14}$$

TABLE 16.1	Relationship of H⁺ and OH⁻ Concentrations in Water Solutions				

$[H^+]$	$[OH^-]$	K_w	pH	pOH
1.00×10^{-2}	1.00×10^{-12}	1.00×10^{-14}	2.00	12.00
1.00×10^{-4}	1.00×10^{-10}	1.00×10^{-14}	4.00	10.00
2.00×10^{-6}	5.00×10^{-9}	1.00×10^{-14}	5.70	8.30
1.00×10^{-7}	1.00×10^{-7}	1.00×10^{-14}	7.00	7.00
1.00×10^{-9}	1.00×10^{-5}	1.00×10^{-14}	9.00	5.00

The value of K_w for all water solutions at 25°C is the constant 1.00×10^{-14}. It is important to realize that as the concentration of one of these ions, H^+ or OH^-, increases, the other decreases. However, the product of $[H^+]$ and $[OH^-]$ always equals 1.00×10^{-14}. This relationship can be seen in the examples shown in Table 16.1. If the concentration of one ion is known, the concentration of the other can be calculated from the K_w expression.

$$K_w = [H^+][OH^-] \qquad [H^+] = \frac{K_w}{[OH^-]} \qquad [OH^-] = \frac{K_w}{[H^+]}$$

Example 16.5 What is the concentration of (a) H^+, and (b) OH^- in a 0.001 M HCl solution? Remember that HCl is 100% ionized.

Solution (a) Since all the HCl is ionized, $H^+ = 0.001$ mol/L or 1×10^{-3} mol/L:

$$HCl \longrightarrow H^+ + Cl^-$$
$$\qquad\quad 0.001\ M \quad 0.001\ M$$

$$[H^+] = 1 \times 10^{-3} \text{ mol/L}$$

(b) To calculate the $[OH^-]$ in this solution, use the following equation and substitute the values for K_w and $[H^+]$:

$$[OH^-] = \frac{K_w}{[H^+]}$$

$$[OH^-] = \frac{1.00 \times 10^{-14}}{1 \times 10^{-3}} = 1 \times 10^{-11} \text{ mol/L}$$

Practice 16.7

Determine the $[H^+]$ and $[OH^-]$ in
(a) $5.0 \times 10^{-5}\ M$ HNO₃ (b) $2.0 \times 10^{-6}\ M$ KOH

What is the pH of a 0.010 M NaOH solution? Assume that NaOH is 100% ionized. **Example 16.6**

Since all the NaOH is ionized, $[OH^-] = 0.010$ mol/L or 1.0×10^{-2} mol/L. **Solution**

$$NaOH \longrightarrow Na^+ + OH^-$$
$$0.010\ M \quad 0.010\ M$$

To find the pH of the solution, we first calculate the H^+ ion concentration. Use the following equation and substitute the values for K_w and $[OH^-]$:

$$[H^+] = \frac{K_w}{[OH^-]} = \frac{1.00 \times 10^{-14}}{1.0 \times 10^{-2}} = 1.0 \times 10^{-12}\ \text{mol/L}$$

$$pH = -\log[H^+] = -\log(1.0 \times 10^{-12}) = 12.00$$

Practice 16.8

Determine the pH for the following solutions:
(a) $5.0 \times 10^{-5}\ M\ HNO_3$ (b) $2.0 \times 10^{-6}\ M\ KOH$

Just as pH is used to express the acidity of a solution, pOH is used to express the basicity of an aqueous solution. The pOH is related to the OH^- ion concentration in the same way that the pH is related to the H^+ ion concentration.

$$pOH = -\log[OH^-]$$

Thus a solution in which $[OH^-] = 1.0 \times 10^{-2}$, as in Example 16.6, will have pOH = 2.00.

In pure water, where $[H^+] = 1.00 \times 10^{-7}$ and $[OH^-] = 1.00 \times 10^{-7}$, the pH is 7.0, and the pOH is 7.0. The sum of the pH and pOH is always 14.0:

$$pH + pOH = 14.00$$

In Example 16.6, the pH can also be found by first calculating the pOH from the OH^- ion concentration and then subtracting from 14.00.

$$pH = 14.00 - pOH = 14.00 - 2.00 = 12.00$$

Table 16.1 summarizes the relationship between $[H^+]$ and $[OH^-]$ in water solutions.

16.11 Ionization Constants

In addition to K_w, several other equilibrium constants are commonly used. Let's consider the equilibrium constant for acetic acid in solution. Because it is a weak acid, an equilibrium is established between molecular $HC_2H_3O_2$ and its ions in solution:

$$HC_2H_3O_2(aq) \rightleftharpoons H^+(aq) + C_2H_3O_2^-(aq)$$

The ionization constant expression is the concentration of the products divided by the concentration of the reactants:

$$K_a = \frac{[H^+][C_2H_3O_2^-]}{[HC_2H_3O_2]}$$

acid ionization constant, K_a

The constant is called the **acid ionization constant, K_a,** a special type of equilibrium constant.

The concentration of water in the solution is large compared to other concentrations and does not change appreciably.

At 25°C, a 0.100 M $HC_2H_3O_2$ solution is 1.34% ionized and has an $[H^+]$ of 1.34×10^{-3} mol/L. From this information we can calculate the ionization constant for acetic acid.

A 0.100 M solution initially contains 0.100 mol of acetic acid per liter. Of this, 0.100 mol, only 1.34%, or 1.34×10^{-3} mol, is ionized, which gives an $[H^+] = 1.34 \times 10^{-3}$ mol/L. Because each molecule of acid that ionizes yields one H^+ and one $C_2H_3O_2^-$, the concentration of $C_2H_3O_2^-$ ions is also 1.34×10^{-3} mol/L. This ionization leaves $0.100 - 0.00134 = 0.099$ mol/L of un-ionized acetic acid.

Acid	Initial concentration (mol/L)	Equilibrium concentration (mol/L)
$[HC_2H_3O_2]$	0.100	0.099
$[H^+]$	0	0.00134
$[C_2H_3O_2^-]$	0	0.00134

Substituting these concentrations in the equilibrium expression, we obtain the value for K_a:

$$K_a = \frac{[H^+][C_2H_3O_2^-]}{[HC_2H_3O_2]} = \frac{(1.34 \times 10^{-3})(1.34 \times 10^{-3})}{(0.099)} = 1.8 \times 10^{-5}$$

The K_a for acetic acid, 1.8×10^{-5}, is small and indicates that the position of the equilibrium is far toward the un-ionized acetic acid. In fact, a 0.100 M acetic acid solution is 99% un-ionized.

Once the K_a for acetic acid is established, it can be used to describe other systems containing H^+, $C_2H_3O_2^-$, and $HC_2H_3O_2$ in equilibrium at 25°C. The ionization constants for several other weak acids are listed in Table 16.2.

Example 16.7   What is the $[H^+]$ in a 0.50 M $HC_2H_3O_2$ solution? The ionization constant, K_a, for $HC_2H_3O_2$ is 1.8×10^{-5}.

Solution   To solve this problem, first write the equilibrium equation and the K_a expression:

$$HC_2H_3O_2 \rightleftharpoons H^+ + C_2H_3O_2^- \qquad K_a = \frac{[H^+][C_2H_3O_2^-]}{[HC_2H_3O_2]} = 1.8 \times 10^{-5}$$

We know that the initial concentration of $HC_2H_3O_2$ is 0.50 M. We also know from the ionization equation that one $C_2H_3O_2^-$ is produced for every H^+ produced; that is,

TABLE 16.2   Ionization Constants (K_a) of Weak Acids at 25°C

Acid	Formula	K_a	Acid	Formula	K_a
Acetic	$HC_2H_3O_2$	1.8×10^{-5}	Hydrocyanic	HCN	4.0×10^{-10}
Benzoic	$HC_7H_5O_2$	6.3×10^{-5}	Hypochlorous	HClO	3.5×10^{-8}
Carbolic (phenol)	HC_6H_5O	1.3×10^{-10}	Nitrous	HNO_2	4.5×10^{-4}
Cyanic	HCNO	2.0×10^{-4}	Hydrofluoric	HF	6.5×10^{-4}
Formic	$HCHO_2$	1.8×10^{-4}			

the $[H^+]$ and the $[C_2H_3O_2^-]$ are equal. To solve, let $Y = [H^+]$, which also equals the $[C_2H_3O_2^-]$. The un-ionized $[HC_2H_3O_2]$ remaining will then be $0.50 - Y$, the starting concentration minus the amount that ionized:

$$[H^+] = [C_2H_3O_2^-] = Y \qquad [HC_2H_3O_2] = 0.50 - Y$$

	Initial	Equilibrium
$[H^+]$	0	Y
$[C_2H_3O_2^-]$	0	Y
$[HC_2H_3O_2]$	0.5	0.5 − Y

Substituting these values into the K_a expression, we obtain

$$K_a = \frac{(Y)(Y)}{0.50 - Y} = \frac{Y^2}{0.50 - Y} = 1.8 \times 10^{-5}$$

An exact solution of this equation for Y requires the use of a mathematical equation known as the *quadratic equation*. However, an approximate solution is obtained if we assume that Y is small and can be neglected compared with 0.50. Then $0.50 - Y$ will be equal to approximately 0.50. The equation now becomes

The quadratic equation is
$$y = \frac{-b \pm \sqrt{b^2 - 4ac}}{2a}$$
for the equation
$$ay^2 + by + c = 0$$

$$\frac{Y^2}{0.50} = 1.8 \times 10^{-5}$$

$$Y^2 = 0.50 \times 1.8 \times 10^{-5} = 0.90 \times 10^{-5} = 9.0 \times 10^{-6}$$

Taking the square root of both sides of the equation, we obtain

$$Y = \sqrt{9.0 \times 10^{-6}} = 3.0 \times 10^{-3} \text{ mol/L}$$

Thus the $[H^+]$ is approximately 3.0×10^{-3} mol/L in a 0.50 M $HC_2H_3O_2$ solution. The exact solution to this problem, using the quadratic equation, gives a value of 2.99×10^{-3} mol/L for $[H^+]$, showing that we were justified in neglecting Y compared with 0.50.

Practice 16.9

Calculate the hydrogen ion concentration in (a) 0.100 M hydrocyanic acid (HCN) solution, and (b) 0.0250 M carbolic acid (HC_6H_5O) solution.

Example 16.8

Solution

Calculate the percent ionization in a 0.50 M $HC_2H_3O_2$ solution.

The percent ionization of a weak acid, $HA(aq) \rightleftharpoons H^+(aq) + A^-(aq)$, is found by dividing the concentration of the H^+ or A^- ions at equilibrium by the initial concentration of HA. For acetic acid,

$$\frac{\text{concentration of } [H^+] \text{ or } [C_2H_3O_2^-]}{\text{initial concentration of } [HC_2H_3O_2]} \times 100 = \text{percent ionized}$$

Percent ionization of an acid, HA, is determined by dividing the concentration of H^+ or A^- ions at equilibrium by the initial concentration of HA and multiplying by 100.

To solve this problem, we first need to calculate the $[H^+]$. This calculation has already been done for a 0.50 M solution:

$$[H^+] = 3.0 \times 10^{-3} \text{ mol/L in a } 0.50 \ M \text{ solution} \qquad \text{(from Example 16.7)}$$

This $[H^+]$ represents a fractional amount of the initial 0.50 M $HC_2H_3O_2$. Therefore

$$\frac{3.0 \times 10^{-3} \text{ mol/L}}{0.50 \text{ mol/L}} \times 100 = 0.60\% \text{ ionized}$$

A 0.50 M $HC_2H_3O_2$ solution is 0.60% ionized.

Practice 16.10

Calculate the percent ionization for
(a) 0.100 M hydrocyanic acid (HCN)
(b) 0.0250 M carbolic acid (HC_6H_5O)

16.12   Solubility Product Constant

solubility product constant, K_{sp}

The **solubility product constant, K_{sp},** is the equilibrium constant of a slightly soluble salt. To evaluate K_{sp}, consider this example. The solubility of AgCl in water is 1.3×10^{-5} mol/L at 25°C. The equation for the equilibrium between AgCl and its ions in solution is

$$AgCl(s) \rightleftharpoons Ag^+(aq) + Cl^-(aq)$$

The equilibrium constant expression is

$$K_{eq} = \frac{[Ag^+][Cl^-]}{[AgCl(s)]}$$

The amount of solid AgCl does not affect the equilibrium system provided that some is present. In other words the concentration of solid AgCl is constant whether 1 mg or 10 g of the salt is present. Therefore the product obtained by multiplying the two constants K_{eq} and $[AgCl(s)]$ is also a constant. This is the solubility product constant, K_{sp}:

TABLE 16.3	Solubility Product Constants (K_{sp}) at 25°C		
Compound	K_{sp}	**Compound**	K_{sp}
AgCl	1.7×10^{-10}	CaF_2	3.9×10^{-11}
AgBr	5×10^{-13}	CuS	9×10^{-45}
AgI	8.5×10^{-17}	$Fe(OH)_3$	6×10^{-38}
$AgC_2H_3O_2$	2×10^{-3}	PbS	7×10^{-29}
Ag_2CrO_4	1.9×10^{-12}	$PbSO_4$	1.3×10^{-8}
$BaCrO_4$	8.5×10^{-11}	$Mn(OH)_2$	2.0×10^{-13}
$BaSO_4$	1.5×10^{-9}		

$$K_{eq} \times [AgCl(s)] = [Ag^+][Cl^-] = K_{sp}$$

$$K_{sp} = [Ag^+][Cl^-]$$

The K_{sp} is equal to the product of the $[Ag^+]$ and the $[Cl^-]$, each in moles per liter. When 1.3×10^{-5} mol/L of AgCl dissolves, it produces 1.3×10^{-5} mol/L each of Ag^+ and Cl^-. From these concentrations the K_{sp} can be calculated:

$$[Ag^+] = 1.3 \times 10^{-5} \text{ mol/L} \qquad [Cl^-] = 1.3 \times 10^{-5} \text{ mol/L}$$

$$K_{sp} = [Ag^+][Cl^-] = (1.3 \times 10^{-5})(1.3 \times 10^{-5}) = 1.7 \times 10^{-10}$$

Once the K_{sp} value for AgCl is established, it can be used to describe other systems containing Ag^+ and Cl^-.

The K_{sp} expression does not have a denominator. It consists only of the concentrations (mol/L) of the ions in solution. As in other equilibrium expressions, each of these concentrations is raised to a power that is the same number as its coefficient in the balanced equation. Here are equilibrium equations and the K_{sp} expressions for several other substances:

$$AgBr(s) \rightleftharpoons Ag^+(aq) + Br^-(aq) \qquad\qquad K_{sp} = [Ag^+][Br^-]$$

$$BaSO_4(s) \rightleftharpoons Ba^{2+}(aq) + SO_4^{2-}(aq) \qquad\qquad K_{sp} = [Ba^{2+}][SO_4^{2-}]$$

$$Ag_2CrO_4(s) \rightleftharpoons 2Ag^+(aq) + CrO_4^{2-}(aq) \qquad\qquad K_{sp} = [Ag^+]^2[CrO_4^{2-}]$$

$$CuS(s) \rightleftharpoons Cu^{2+}(aq) + S^{2-}(aq) \qquad\qquad K_{sp} = [Cu^{2+}][S^{2-}]$$

$$Mn(OH)_2(s) \rightleftharpoons Mn^{2+}(aq) + 2\,OH^-(aq) \qquad\qquad K_{sp} = [Mn^{2+}][OH^-]^2$$

$$Fe(OH)_3(s) \rightleftharpoons Fe^{3+}(aq) + 3\,OH^-(aq) \qquad\qquad K_{sp} = [Fe^{3+}][OH^-]^3$$

Table 16.3 lists K_{sp} values for these and several other substances.

When the product of the molar concentration of the ions in solution (each raised to its proper power) is greater than the K_{sp} for that substance, precipitation should occur. If the ion product is less than the K_{sp} value, no precipitation will occur.

Example 16.9   Write K_{sp} expressions for AgI and PbI_2, both of which are slightly soluble salts.

Solution   First write the equilibrium equations:

$$AgI(s) \rightleftharpoons Ag^+(aq) + I^-(aq)$$

$$PbI_2(s) \rightleftharpoons Pb^{2+}(aq) + 2\,I^-(aq)$$

Since the concentration of the solid crystals is constant, the K_{sp} equals the product of the molar concentrations of the ions in solution. In the case of PbI_2, the $[I^-]$ must be squared:

$$K_{sp} = [Ag^+][I^-] \qquad K_{sp} = [Pb^{2+}][I^-]^2$$

Example 16.10   The K_{sp} value for lead sulfate is 1.3×10^{-8}. Calculate the solubility of $PbSO_4$ in grams per liter.

Solution   First write the equilibrium equation and the K_{sp} expression:

$$PbSO_4 \rightleftharpoons Pb^{2+}(aq) + SO_4^{2-}(aq)$$

$$K_{sp} = [Pb^{2+}][SO_4^{2-}] = 1.3 \times 10^{-8}$$

Since the lead sulfate that is in solution is completely dissociated, the $[Pb^{2+}]$ or $[SO_4^{2-}]$ is equal to the solubility of $PbSO_4$ in moles per liter. Let

$$Y = [Pb^{2+}] = [SO_4^{2-}]$$

Substitute Y into the K_{sp} equation and solve:

$$[Pb^{2+}][SO_4^{2-}] = (Y)(Y) = 1.3 \times 10^{-8}$$

$$Y^2 = 1.3 \times 10^{-8}$$

$$Y = 1.1 \times 10^{-4} \text{ mol/L}$$

The solubility of $PbSO_4$ therefore is 1.1×10^{-4} mol/L. Now convert mol/L to g/L:

1 mol of $PbSO_4$ has a mass of (207.2 g + 32.07 g + 64.00 g) or 303.3 g

$$\left(\frac{1.1 \times 10^{-4} \text{ mol}}{L}\right)\left(\frac{303.3 \text{ g}}{\text{mol}}\right) = 3.3 \times 10^{-2} \text{ g/L}$$

The solubility of $PbSO_4$ is 3.3×10^{-2} g/L.

Practice 16.11

Write the K_{sp} expression for
(a) $Cr(OH)_3$ (b) $Cu_3(PO_4)_2$

Practice 16.12

The K_{sp} value for CuS is 9.0×10^{-45}. Calculate the solubility of CuS in grams per liter.

An ion added to a solution already containing that ion is called a *common ion*. When a common ion is added to an equilibrium solution of a weak electrolyte or a slightly soluble salt, the equilibrium shifts according to Le Chatelier's principle. For example, when silver nitrate, $AgNO_3$, is added to a saturated solution of AgCl, ($AgCl(s) \rightleftharpoons Ag^+ + Cl^-$), the equilibrium shifts to the left due to the increase in the $[Ag^+]$. As a result, the $[Cl^-]$ and the solubility of AgCl decreases. The AgCl and $AgNO_3$ have the common ion Ag^+. A shift in the equilibrium position upon addition of an ion already contained in the solution is known as the **common ion effect**.

common ion effect

Example 16.11

Silver nitrate is added to a saturated AgCl solution until the $[Ag^+]$ is 0.10 *M*. What will be the $[Cl^-]$ remaining in solution?

Solution

This is an example of the common ion effect. The addition of $AgNO_3$ puts more Ag^+ in solution; the Ag^+ combines with Cl^- and causes the equilibrium to shift to the left, reducing the $[Cl^-]$ in solution. After the addition of Ag^+ to the mixture, the $[Ag^+]$ and $[Cl^-]$ in solution are no longer equal.

We use the K_{sp} to calculate the $[Cl^-]$ remaining in solution. The K_{sp} is constant at a particular temperature and remains the same no matter how we change the concentration of the species involved:

$$K_{sp} = [Ag^+][Cl^-] = 1.7 \times 10^{-10} \qquad [Ag^+] = 0.10 \text{ mol/L}$$

We then substitute the $[Ag^+]$ into the K_{sp} expression and calculate the $[Cl^-]$:

$$[0.10][Cl^-] = 1.7 \times 10^{-10}$$

$$[Cl^-] = \frac{1.7 \times 10^{-10}}{0.10} = 1.7 \times 10^{-9} \text{ mol/L}$$

This calculation shows a 10,000-fold reduction of Cl^- ions in solution. It illustrates that Cl^- ions may be quantitatively removed from solution with an excess of Ag^+ ions.

Practice 16.13

Sodium sulfate, Na_2SO_4, is added to a saturated solution of $BaSO_4$ until the concentration of the sulfate ion is 2.0×10^{-2} *M*. What will be the concentration of the Ba^{2+} ions remaining in solution?

16.13 Hydrolysis

Hydrolysis is the term used for the general reaction in which a water molecule is split. For example, the net ionic hydrolysis reaction for a sodium acetate solution is

hydrolysis

$$C_2H_3O_2^-(aq) + H_2O(l) \rightleftharpoons HC_2H_3O_2(aq) + OH^-(aq)$$

TABLE 16.4	Ionic Composition of Salts and the Nature of the Aqueous Solutions They Form	
Type of salt	**Nature of aqueous solution**	**Examples**
Weak base–strong acid	Acidic	NH_4Cl, NH_4NO_3
Strong base–weak acid	Basic	$NaC_2H_3O_2$, K_2CO_3
Weak base–weak acid	Depends on the salt	$NH_4C_2H_3O_2$, NH_4NO_2
Strong base–strong acid	Neutral	NaCl, KBr

In this reaction the water molecule is split, with the H^+ combining with $C_2H_3O_2^-$ to give the weak acid $HC_2H_3O_2$ and the OH^- going into solution, making the solution more basic.

Salts that contain an ion of a weak acid undergo hydrolysis. For example, a $0.10\,M$ NaCN solution has a pH of 11.1. The hydrolysis reaction that causes this solution to be basic is

$$CN^-(aq) + H_2O(l) \rightleftharpoons HCN(aq) + OH^-(aq)$$

If a salt contains the ion of a weak base, the ion hydrolyzes to produce an acidic solution. An example is ammonium chloride, which produces the NH_4^+ and Cl^- in solution. The NH_4^+ hydrolyzes to produce an acidic solution:

$$NH_4^+(aq) + H_2O(l) \longrightarrow NH_3(aq) + H_3O^+(aq)$$

The ions of a salt derived from a strong acid and a strong base, such as NaCl, do not undergo hydrolysis and thus form neutral solutions. Table 16.4 lists the ionic composition of various salts and the nature of the aqueous solutions that they form.

Practice 16.14

Indicate whether these salts would produce an acidic, basic, or neutral aqueous solution:
(a) KCN (b) $NaNO_3$ (c) NH_4Br

16.14   Buffer Solutions: The Control of pH

The control of pH within narrow limits is critically important in many chemical applications and vitally important in many biological systems. For example, human blood must be maintained between pH 7.35 and 7.45 for the efficient transport of oxygen from the lungs to the cells. This narrow pH range is maintained by buffer systems in the blood.

buffer solution   A **buffer solution** resists changes in pH when diluted or when small amounts of acid or base are added. Two common types of buffer solutions are (1) a weak acid mixed with its conjugate base, and (2) a weak base mixed with its conjugate acid.

CHEMISTRY IN ACTION

Exchange of Oxygen and Carbon Dioxide in the Blood

The transport of oxygen and carbon dioxide between the lungs and tissues is a complex process that involves several reversible reactions, each of which behaves in accordance with Le Chatelier's principle.

The binding of oxygen to hemoglobin is a reversible reaction. The oxygen molecule must attach to the hemoglobin (Hb) and then later detach. The equilibrium equation for this reaction can be written:

$$Hb + O_2 \rightleftharpoons HbO_2$$

In the lungs the concentration of oxygen is high and favors the forward reaction. Oxygen quickly binds to the hemoglobin until it is saturated with oxygen.

In the tissues the concentration of oxygen is lower and in accordance with Le Chatelier's principle: The equilibrium position shifts to the left and the hemoglobin releases oxygen to the tissues. Approximately 45% of the oxygen diffuses out of the capillaries into the tissues, where it may be picked up by *myoglobin,* another carrier molecule.

Myoglobin functions as an oxygen-storage molecule, holding the oxygen until it is required in the energy-producing portions of the cell. The reaction between myoglobin (Mb) and oxygen can be written as an equilibrium reaction:

$$Mb + O_2 \rightleftharpoons MbO_2$$

The hemoglobin and myoglobin equations are very similar, so what accounts for the transfer of the oxygen from the hemoglobin to the myoglobin? Although both equilibria involve similar interactions, the affinity between oxygen and

▲ **Oxygen and carbon dioxide are exchanged in the red blood cells when they are in capillaries.**

hemoglobin is different from the affinity between myoglobin and oxygen. In the tissues the position of the hemoglobin equilibrium is such that it is 55% saturated with oxygen, whereas the myoglobin is at 90% oxygen saturation. Under these conditions hemoglobin will release oxygen, while myoglobin will bind oxygen. Thus oxygen is loaded onto hemoglobin in the lungs and unloaded in the tissue's cells.

Carbon dioxide produced in the cells must be removed from the tissues. Oxygen-depleted hemoglobin molecules

accomplish this by becoming carriers of carbon dioxide. The carbon dioxide does not bind at the heme site as the oxygen does, but rather at one end of the protein chain. When carbon dioxide dissolves in water, some of the CO_2 reacts to release hydrogen ions:

$$CO_2 + H_2O \rightleftharpoons HCO_3^- + H^+$$

To facilitate the removal of CO_2 from the tissues, this equilibrium needs to be moved toward the right. This shift is accomplished by the removal of H^+ from the tissues by the hemoglobin molecule. The deoxygenated hemoglobin molecule can bind H^+ ions as well as CO_2. In the lungs this whole process is reversed so the CO_2 is removed from the hemoglobin and exhaled.

Molecules that are similar in structure to the oxygen molecule can become involved in competing equilibria. Hemoglobin is capable of binding with carbon monoxide, CO, nitrogen monoxide, NO, and cyanide, CN^-. The extent of the competition depends on the affinity. Since these molecules have a greater affinity for hemoglobin than oxygen, they will effectively displace oxygen from hemoglobin. For example,

$$HbO_2 + CO \rightleftharpoons HbCO + O_2$$

Since the affinity of hemoglobin for CO is 150 times stronger than its affinity for oxygen, the equilibrium position lies far to the right. This explains why CO is a poisonous substance and why oxygen is administered to victims of CO poisoning. The hemoglobin molecules can only transport oxygen if the CO is released and the oxygen shifts the equilibrium toward the left.

425

A saltwater aquarium is a ▶ buffer system.

The action of a buffer system can be understood by considering a solution of acetic acid and sodium acetate. The weak acid, $HC_2H_3O_2$, is mostly un-ionized and is in equilibrium with its ions in solution. The sodium acetate is completely ionized:

$$HC_2H_3O_2(aq) \rightleftharpoons H^+(aq) + C_2H_3O_2^-(aq)$$

$$NaC_2H_3O_2(aq) \longrightarrow Na^+(aq) + C_2H_3O_2^-(aq)$$

Because the sodium acetate is completely ionized, the solution contains a much higher concentration of acetate ions than would be present if only acetic acid were in solution. The acetate ion represses the ionization of acetic acid and also reacts with water, causing the solution to have a higher pH (be more basic) than an acetic acid solution (see Section 16.5). Thus a 0.1 M acetic acid solution has a pH of 2.87, but a solution that is 0.1 M in acetic acid and 0.1 M in sodium acetate has a pH of 4.74. This difference in pH is the result of the common ion effect.

A buffer solution has a built-in mechanism that counteracts the effect of adding acid or base. Consider the effect of adding HCl or NaOH to an acetic acid–sodium acetate buffer. When a small amount of HCl is added, the acetate ions of the buffer combine with the H^+ ions from HCl to form un-ionized acetic acid, thus neutralizing the added acid and maintaining the approximate pH of the solution. When NaOH is added, the OH^- ions react with acetic acid to neutralize the added base and thus maintain the approximate pH. The equations for these reactions are

$$H^+(aq) + C_2H_3O_2^-(aq) \rightleftharpoons HC_2H_3O_2(aq)$$

$$OH^-(aq) + HC_2H_3O_2(aq) \rightleftharpoons H_2O(l) + C_2H_3O_2^-(aq)$$

Data comparing the changes in pH caused by adding HCl and NaOH to pure water and to an acetic acid–sodium acetate buffer solution are shown in Table 16.5.

The human body has a number of buffer systems. One of these, the hydrogen carbonate–carbonic acid buffer, $HCO_3^- - H_2CO_3$, maintains the blood plasma at a pH of 7.4. The phosphate system, $HPO_4^{2-} - H_2PO_4^-$, is an important buffer in the red blood cells as well as in other places in the body.

TABLE 16.5	Changes in pH Caused by the Addition of HCl and NaOH		
Solution		pH	Change in pH
H_2O (1000 mL)		7	—
H_2O + 0.010 mol HCl		2	5
H_2O + 0.010 mol NaOH		12	5
Buffer solution (1000 mL)			
0.10 M $HC_2H_3O_2$ + 0.10 M $NaC_2H_3O_2$		4.74	—
Buffer + 0.010 mol HCl		4.66	0.08
Buffer + 0.010 mol NaOH		4.83	0.09

Concepts in Review

1. Describe a reversible reaction.

2. Explain why the rate of the forward reaction decreases and the rate of the reverse reaction increases as a chemical reaction approaches equilibrium.

3. Describe the qualitative effect of Le Chatelier's principle.

4. Predict how the rate of a chemical reaction is affected by (a) changes in the concentration of reactants, (b) changes in volume of gaseous reactants, (c) changes in temperature, and (d) the presence of a catalyst.

5. Write the equilibrium constant expression for a chemical reaction from a balanced chemical equation.

6. Explain the meaning of the numerical constant, K_{eq}, when given the concentration of the reactants and products in equilibrium.

7. Calculate the concentration of one substance in equilibrium when given the equilibrium constant and the concentrations of all the other substances.

8. Calculate the equilibrium constant, K_{eq}, when given the concentration of reactants and products in equilibrium.

9. Calculate the ionization constant for a weak acid from appropriate data.

10. Calculate the concentrations of all the chemical species in a solution of a weak acid when given the percent ionization or the ionization constant.

11. Compare the relative strengths of acids by using their ionization constants.

12. Use the ion product constant for water, K_w, to calculate $[H^+]$, $[OH^-]$, pH, and pOH when given any one of these quantities.

13. Calculate the solubility product constant, K_{sp}, of a slightly soluble salt when given its solubility, or vice versa.

14. Compare relative solubilities of salts if solubility products are known.

15. Discuss the common ion effect on a system at equilibrium.

16. Explain hydrolysis and why some salts form acidic or basic aqueous solutions.

17. Explain how a buffer solution is able to counteract the addition of small amounts of either H^+ or OH^- ions.

Key Terms

acid ionization constant, K_a (16.11)
activation energy (16.8)
buffer solution (16.14)
catalyst (16.8)
chemical equilibrium (16.3)

chemical kinetics (16.2)
common ion effect (16.12)
equilibrium (16.3)
equilibrium constant, K_{eq} (16.9)
hydrolysis (16.13)

ion product constant for water, K_w (16.10)
Le Chatelier's principle (16.4)
reversible chemical reaction (16.1)
solubility product constant, K_{sp} (16.12)

Questions

1. How would you expect the two tubes in Figure 16.1 to appear if both are at 25°C?

2. Is the reaction $N_2O_4 \rightleftharpoons 2\ NO_2$ exothermic or endothermic? (Figure 16.1)

3. At equilibrium how do the forward and reverse reaction rates compare? (Figure 16.2)

4. For each solution in Table 16.1, what is the sum of the pH plus the pOH? What would be the pOH of a solution whose pH was -1?

5. Of the acids listed in Table 16.2, which ones are stronger than acetic acid and which are weaker?

6. Tabulate the relative order of molar solubilities of AgCl, AgBr, AgI, $AgC_2H_3O_2$, $PbSO_4$, $BaSO_4$, $BaCrO_4$, and PbS. List the most soluble first. (Table 16.3)

7. Which compound in the following pairs has the greater molar solubility? (Table 16.3)
   (a) $Mn(OH)_2$ or Ag_2CrO_4
   (b) $BaCrO_4$ or Ag_2CrO_4

8. Explain how the acetic acid–sodium acetate buffer system maintains its pH when 0.010 mol of HCl is added to 1 L of the buffer solution. (Table 16.5)

9. Explain why a precipitate of NaCl forms when HCl gas is passed into a saturated aqueous solution of NaCl.

10. Why does the rate of a reaction usually increase when the concentration of one of the reactants is increased?

11. If pure hydrogen iodide, HI, is placed in a vessel at 700 K, will it decompose? Explain.

12. Why does an increase in temperature cause the rate of reaction to increase?

13. Describe how equilibrium is reached when the substances A and B are first mixed and react as

    $$A + B \rightleftharpoons C + D$$

14. With dilution, aqueous solutions of acetic acid, $HC_2H_3O_2$, show increased ionization. For example, a 1.0 M solution of acetic acid is 0.42% ionized, whereas a 0.10 M solution is 1.34% ionized. Explain the behavior using the ionization equation and equilibrium principles.

15. A 1.0 M solution of acetic acid ionizes less and has a higher concentration of H^+ ions than a 0.10 M acetic acid solution. Explain this behavior. (See Question 14 for data.)

16. What would cause two separate samples of pure water to have slightly different pH values?

17. Why are the pH and pOH equal in pure water?

18. Explain why silver acetate is more soluble in nitric acid than in water. (*Hint*: Write the equilibrium equation first and then consider the effect of the acid on the acetate ion.) What would happen if hydrochloric acid were used in place of nitric acid?

19. Dissolution of sodium acetate, $NaC_2H_3O_2$, in pure water gives a basic solution. Why? (*Hint*: A small amount of $HC_2H_3O_2$ is formed.)

20. Describe why the pH of a buffer solution remains almost constant when a small amount of acid or base is added to it.

Paired Exercises

21. Express these reversible systems in equation form:
    (a) a mixture of ice and liquid water at 0°C
    (b) crystals of Na_2SO_4 in a saturated aqueous solution of Na_2SO_4

22. Express these reversible systems in equation form:
    (a) liquid water and vapor at 100°C in a pressure cooker
    (b) a closed system containing boiling sulfur dioxide, SO_2

23. Consider this system at equilibrium:

$$4 NH_3(g) + 3 O_2(g) \rightleftharpoons$$
$$2 N_2(g) + 6 H_2O(g) + 1531 kJ$$

(a) Is the reaction exothermic or endothermic?

(b) If the system's state of equilibrium is disturbed by the addition of O_2, in which direction, left or right, must the reaction occur to reestablish equilibrium? After the new equilibrium has been established, how will the final molar concentrations of NH_3, O_2, N_2, and H_2O compare (increase or decrease) with their concentrations before the addition of the O_2?

24. Consider this system at equilibrium:

$$4 NH_3(g) + 3 O_2(g) \rightleftharpoons$$
$$2 N_2(g) + 6 H_2O(g) + 1531 kJ$$

(a) If the system's state of equilibrium is disturbed by the addition of N_2, in which direction, left or right, must the reaction occur to reestablish equilibrium? After the new equilibrium has been established, how will the final molar concentrations of NH_3, O_2, N_2, and H_2O compare (increase or decrease) with their concentrations before the addition of the N_2?

(b) If the system's state of equilibrium is disturbed by the addition of heat, in which direction will the reaction occur, left or right, to reestablish equilibrium?

25. Consider this system at equilibrium:

$$N_2(g) + 3 H_2(g) \rightleftharpoons 2 NH_3(g) + 92.5 kJ$$

Complete the following table. Indicate changes in moles by entering I, D, N, or ? in the table. (I = increase, D = decrease, N = no change, ? = insufficient information to determine.)

Change of stress imposed on the system at equilibrium	Direction of reaction, left or right, to reestablish equilibrium	Change in number of moles		
		N_2	H_2	NH_3
(a) Add N_2				
(b) Remove H_2				
(c) Decrease volume of reaction vessel				
(d) Increase temperature				

26. Consider this system at equilibrium:

$$N_2(g) + 3 H_2(g) \rightleftharpoons 2 NH_3(g) + 92.5 kJ$$

Complete the following table. Indicate changes in moles by entering I, D, N, or ? in the table. (I = increase, D = decrease, N = no change, ? = insufficient information to determine.)

Change of stress imposed on the system at equilibrium	Direction of reaction, left or right, to reestablish equilibrium	Change in number of moles		
		N_2	H_2	NH_3
(a) Add NH_3				
(b) Increase volume of reaction vessel				
(c) Add catalyst				
(d) Add both H_2 and NH_3				

27. For the following equations, tell in which direction, left or right, the equilibrium will shift when these changes are made: The temperature is increased, the pressure is increased by decreasing the volume of the reaction vessel, and a catalyst is added.

(a) $3 O_2(g) + 271 kJ \rightleftharpoons 2 O_3(g)$
(b) $CH_4(g) + Cl_2(g) \rightleftharpoons CH_3Cl(g) + HCl(g) + 110 kJ$
(c) $2 NO(g) + 2 H_2(g) \rightleftharpoons N_2(g) + 2 H_2O(g) + 665 kJ$

28. For the following equations, tell in which direction, left or right, the equilibrium will shift when these changes are made: The temperature is increased, the pressure is increased by decreasing the volume of the reaction vessel, and a catalyst is added.

(a) $2 SO_3(g) + 197 kJ \rightleftharpoons 2 SO_2(g) + O_2(g)$
(b) $4 NH_3(g) + 3 O_2(g) \rightleftharpoons$
$2 N_2(g) + 6 H_2O(g) + 1531 kJ$
(c) $OF_2(g) + H_2O(g) \rightleftharpoons O_2(g) + 2 HF(g) + 318 kJ$

29. Utilizing Le Chatelier's principle, indicate the shift (if any) that would occur to

$$C_2H_6(g) + heat \rightleftharpoons C_2H_4(g) + H_2(g)$$

(a) if the concentration of hydrogen gas is decreased
(b) if the temperature is lowered
(c) if a catalyst is added

30. Utilizing Le Chatelier's principle, indicate the shift (if any) that would occur to

$$C_2H_6(g) + heat \rightleftharpoons C_2H_4(g) + H_2(g)$$

(a) if C_2H_6 is removed from the system
(b) if the volume of the container is increased
(c) if the temperature is raised

31. Write the equilibrium constant expression for these reactions:
  (a) $4\,HCl(g) + O_2(g) \rightleftharpoons 2\,Cl_2(g) + 2\,H_2O(g)$
  (b) $N_2(g) + 3\,H_2(g) \rightleftharpoons 2\,NH_3(g)$
  (c) $PCl_5(g) \rightleftharpoons PCl_3(g) + Cl_2(g)$

33. Write the solubility product expression, K_{sp}, for these substances:
  (a) CuS   **(c)** $PbBr_2$
  (b) $BaSO_4$ **(d)** Ag_3AsO_4

35. What effect will decreasing the $[H^+]$ of a solution have on (a) pH, (b) pOH, (c) $[OH^-]$, and (d) K_w?

37. Decide whether these salts form an acidic, basic, or neutral solution when dissolved in water:
  (a) KCl   **(c)** K_2SO_4
  (b) Na_2CO_3 **(d)** $(NH_4)_2SO_4$

39. Write hydrolysis equations for aqueous solutions of these salts:
  (a) KNO_2
  (b) $Mg(C_2H_3O_2)_2$

41. Write hydrolysis equations for these ions:
  (a) HCO_3^-
  (b) NH_4^+

43. One of the important pH-regulating systems in the blood consists of a carbonic acid–sodium hydrogen carbonate buffer:

$$H_2CO_3(aq) \rightleftharpoons H^+(aq) + HCO_3^-(aq)$$
$$NaHCO_3(aq) \longrightarrow Na^+(aq) + HCO_3^-(aq)$$

Explain how this buffer resists changes in pH when excess acid, H^+, gets into the bloodstream.

45. Calculate (a) the $[H^+]$, (b) the pH, and (c) the percent ionization of a 0.25 M solution of $HC_2H_3O_2$. ($K_a = 1.8 \times 10^{-5}$)

47. A 1.000 M solution of a weak acid, HA, is 0.52% ionized. Calculate the ionization constant, K_a, for the acid.

49. Calculate the percent ionization and pH of solutions of $HC_2H_3O_2$ ($K_a = 1.8 \times 10^{-5}$) having the following molarities: (a) 1.0 M, (b) 0.10 M, and (c) 0.010 M.

*51. A 0.37 M solution of a weak acid, HA, has a pH of 3.7. What is the K_a for this acid?

53. A common laboratory reagent is 6.0 M HCl. Calculate the $[H^+]$, $[OH^-]$, pH, and pOH of this solution.

55. Calculate the pH and the pOH of these solutions:
  (a) 0.00010 M HCl
  (b) 0.010 M NaOH
  *(c) saturated $Fe(OH)_3$ solution ($K_{sp} = 6.0 \times 10^{-38}$)

57. Calculate the $[OH^-]$ in these solutions:
  (a) $[H^+] = 1.0 \times 10^{-4}$
  (b) $[H^+] = 2.8 \times 10^{-6}$

32. Write the equilibrium constant expression for these reactions:
  (a) $HClO_2(aq) \rightleftharpoons H^+(aq) + ClO_2^-(aq)$
  (b) $HC_2H_3O_2(aq) \rightleftharpoons H^+(aq) + C_2H_3O_2^-(aq)$
  (c) $4\,NH_3(g) + 5\,O_2(g) \rightleftharpoons 4\,NO(g) + 6\,H_2O(g)$

34. Write the solubility product expression, K_{sp}, for these substances:
  (a) $Fe(OH)_3$ **(c)** CaF_2
  (b) Sb_2S_5 **(d)** $Ba_3(PO_4)_2$

36. What effect will increasing the $[H^+]$ of a solution have on (a) pH, (b) pOH, (c) $[OH^-]$, and (d) K_w?

38. Decide whether these salts form an acidic, basic, or neutral solution when dissolved in water:
  (a) $Ca(CN)_2$ **(c)** $NaNO_2$
  (b) $BaBr_2$ **(d)** NaF

40. Write hydrolysis equations for aqueous solutions of these salts:
  (a) NH_4NO_3
  (b) Na_2SO_3

42. Write hydrolysis equations for these ions:
  (a) OCl^-
  (b) ClO_2^-

44. One of the important pH-regulating systems in the blood consists of a carbonic acid–sodium hydrogen carbonate buffer:

$$H_2CO_3(aq) \rightleftharpoons H^+(aq) + HCO_3^-(aq)$$
$$NaHCO_3(aq) \longrightarrow Na^+(aq) + HCO_3^-(aq)$$

Explain how this buffer resists changes in pH when excess base, OH^-, gets into the bloodstream.

46. Calculate (a) the $[H^+]$, (b) the pH, and (c) the percent ionization of a 0.25 M solution of phenol, HC_6H_5O. ($K_a = 1.3 \times 10^{-10}$)

48. A 0.15 M solution of a weak acid, HA, has a pH of 5. Calculate the ionization constant, K_a, for the acid.

50. Calculate the percent ionization and pH of solutions of HClO ($K_a = 3.5 \times 10^{-8}$) having the following molarities: (a) 1.0 M, (b) 0.10 M, and (c) 0.010 M.

*52. A 0.23 M solution of a weak acid, HA, has a pH of 2.89. What is the K_a for this acid?

54. A common laboratory reagent is 1.0 M NaOH. Calculate the $[H^+]$, $[OH^-]$, pH, and pOH of this solution.

56. Calculate the pH and the pOH of these solutions:
  (a) 0.0025 M NaOH
  (b) 0.10 M HClO ($K_a = 3.5 \times 10^{-8}$)
  *(c) saturated $Fe(OH)_2$ solution ($K_{sp} = 8.0 \times 10^{-16}$)

58. Calculate the $[OH^-]$ in these solutions:
  (a) $[H^+] = 4.0 \times 10^{-9}$
  (b) $[H^+] = 8.9 \times 10^{-2}$

59. Calculate the $[H^+]$ in these solutions:
  (a) $[OH^-] = 6.0 \times 10^{-7}$
  (b) $[OH^-] = 1 \times 10^{-8}$

61. Given the following solubility data, calculate the solubility product constant for each substance:
  (a) $BaSO_4$, 3.9×10^{-5} mol/L
  (b) Ag_2CrO_4, 7.8×10^{-5} mol/L
  (c) $CaSO_4$, 0.67 g/L
  (d) $AgCl$, 0.0019 g/L

63. Calculate the molar solubility for these substances:
  (a) Ag_2SO_4, $K_{sp} = 1.5 \times 10^{-5}$
  (b) $Mg(OH)_2$, $K_{sp} = 7.1 \times 10^{-12}$

65. For each substance in Question 63 calculate the solubility in grams per 100. mL of solution.

*67. Solutions containing 100. mL of 0.010 M Na_2SO_4 and 100. mL of 0.001 M $Pb(NO_3)_2$ are mixed. Show by calculation whether or not a precipitate will form. Assume the volumes are additive. (K_{sp} $PbSO_4 = 1.3 \times 10^{-8}$)

69. How many moles of AgBr will dissolve in 1.0 L of 0.10 M NaBr? ($K_{sp} = 5.0 \times 10^{-13}$ for AgBr)

71. Calculate the $[H^+]$ and the pH of a buffer solution that is 0.20 M in $HC_2H_3O_2$ and contains sufficient sodium acetate to make the $[C_2H_3O_2^-]$ equal to 0.10 M. (K_a for $HC_2H_3O_2 = 1.8 \times 10^{-5}$)

73. When 1.0 mL of 1.0 M HCl is added to 50. mL of 1.0 M NaCl, the $[H^+]$ changes from 1×10^{-7} M to 2.0×10^{-2} M. Calculate the initial pH and the pH change in the solution.

60. Calculate the $[H^+]$ in these solutions:
  (a) $[OH^-] = 4.5 \times 10^{-6}$
  (b) $[OH^-] = 7.3 \times 10^{-4}$

62. Given the following solubility data, calculate the solubility product constant for each substance:
  (a) ZnS, 3.5×10^{-12} mol/L
  (b) $Pb(IO_3)_2$, 4.0×10^{-5} mol/L
  (c) Ag_3PO_4, 6.73×10^{-3} g/L
  (d) $Zn(OH)_2$, 2.33×10^{-4} g/L

64. Calculate the molar solubility for these substances:
  (a) $BaCO_3$, $K_{sp} = 2.0 \times 10^{-9}$
  (b) $AlPO_4$, $K_{sp} = 5.8 \times 10^{-19}$

66. For each substance in Question 64 calculate the solubility in grams per 100. mL of solution.

*68. Solutions containing 50.0 mL of 1.0×10^{-4} M $AgNO_3$ and 100. mL of 1.0×10^{-4} M NaCl are mixed. Show by calculation whether or not a precipitate will form. Assume the volumes are additive. (K_{sp} AgCl = 1.7×10^{-10})

70. How many moles of AgBr will dissolve in 1.0 L of 0.10 M $MgBr_2$? ($K_{sp} = 5.0 \times 10^{-13}$ for AgBr)

72. Calculate the $[H^+]$ and the pH of a buffer solution that is 0.20 M in $HC_2H_3O_2$ and contains sufficient sodium acetate to make the $[C_2H_3O_2^-]$ equal to 0.20 M. (K_a for $HC_2H_3O_2 = 1.8 \times 10^{-5}$)

74. When 1.0 mL of 1.0 M HCl is added to 50. mL of a buffer solution that is 1.0 M in $HC_2H_3O_2$ and 1.0 M in $NaC_2H_3O_2$, the $[H^+]$ changes from 1.8×10^{-5} M to 1.9×10^{-5} M. Calculate the initial pH and the pH change in the solution.

Additional Exercises

75. What is the maximum number of moles of HI that can be obtained from a reaction mixture containing 2.30 mol I_2 and 2.10 mol H_2?

76. (a) How many moles of HI are produced when 2.00 mol H_2 and 2.00 mol I_2 are reacted at 700 K? (Reaction is 79% complete.)
  (b) Addition of 0.27 mol I_2 to the system increases the yield of HI to 85%. How many moles of H_2, I_2, and HI are now present?
  (c) From the data in part (a), calculate K_{eq} for the reaction at 700 K.

*77. After equilibrium is reached in the reaction of 6.00 g H_2 with 200. g I_2 at 500. K, analysis shows that the flask contains 64.0 g of HI. How many moles of H_2, I_2, and HI are present in this equilibrium mixture?

78. What is the equilibrium constant of the reaction

$$PCl_3(g) + Cl_2(g) \rightleftharpoons PCl_5(g)$$

if a 20. L flask contains 0.10 mol PCl_3, 1.50 mol Cl_2, and 0.22 mol PCl_5?

79. If the rate of a reaction doubles for every 10°C rise in temperature, how much faster will the reaction go at 100°C than at 30°C?

80. Calculate the ionization constant for the following acids. Each acid ionizes as follows: $HA \rightleftharpoons H^+ + A^-$.

Acid	Acid concentration	$[H^+]$
Hypochlorous, HOCl	0.10 M	5.9×10^{-5} mol/L
Propanoic, $HC_3H_5O_2$	0.15 M	1.4×10^{-3} mol/L
Hydrocyanic, HCN	0.20 M	8.9×10^{-6} mol/L

81. The K_{sp} of CaF_2 is 3.9×10^{-11}. Calculate (a) the molar concentrations of Ca^{2+} and F^- in a saturated solution, and (b) the grams of CaF_2 that will dissolve in 500. mL of water.

82. The following solutions are mixed. Calculate whether or not a precipitate will form:
(a) 100 mL of 0.010 M Na_2SO_4 and 100 mL of 0.001 M $Pb(NO_3)_2$
(b) 50.0 mL of 1.0×10^{-4} M $AgNO_3$ and 100. mL of 1.0×10^{-4} M NaCl
(c) 1.0 g $Ca(NO_3)_2$ in 150 mL H_2O and 250 mL of 0.01 M NaOH
K_{sp} $PbSO_4 = 1.3 \times 10^{-8}$
K_{sp} $AgCl = 1.7 \times 10^{-10}$
K_{sp} $Ca(OH)_2 = 1.3 \times 10^{-6}$

83. If $BaCl_2$ is added to a saturated $BaSO_4$ solution until the $[Ba^{2+}]$ is 0.050 M,
(a) what concentration of SO_4^{2-} remains in solution?
(b) how much $BaSO_4$ remains dissolved in 100. mL of the solution? ($K_{sp} = 1.5 \times 10^{-9}$ for $BaSO_4$)

84. The K_{sp} for $PbCl_2$ is 2.0×10^{-5}. Will a precipitate form when 0.050 mol $Pb(NO_3)_2$ and 0.010 mol NaCl are dissolved in 1.0 L H_2O? Show evidence for your answer.

85. Suppose the concentration of a solution is 0.10 M Ba^{2+} and 0.10 M Sr^{2+}. Which sulfate, $BaSO_4$ or $SrSO_4$, will precipitate first when a dilute solution of H_2SO_4 is added dropwise to the solution? Show evidence for your answer. ($K_{sp} = 1.5 \times 10^{-9}$ for $BaSO_4$ and $K_{sp} = 3.5 \times 10^{-7}$ for $SrSO_4$)

86. Calculate the K_{eq} for the reaction

$$SO_2(g) + O_2(g) \rightleftharpoons SO_3(g)$$

when the equilibrium concentrations of the gases at 530°C are $[SO_3] = 11.0$ M, $[SO_2] = 4.20$ M, and $[O_2] = 0.60 \times 10^{-3}$ M.

87. If it takes 0.048 g BaF_2 to saturate 15.0 mL of water, what is the K_{sp} of BaF_2?

88. The K_{eq} for the formation of ammonia gas from its elements is 4.0. If the equilibrium concentrations of nitrogen gas and hydrogen gas are both 2.0 M, what is the equilibrium concentration of the ammonia gas?

89. The K_{sp} of $SrSO_4$ is 7.6×10^{-7}. Should precipitation occur when 25.0 mL of 1.0×10^{-3} M $SrCl_2$ solution is mixed with 15.0 mL of 2.0×10^{-3} M Na_2SO_4? Show proof.

*90. The solubility of Hg_2I_2 in H_2O is 3.04×10^{-7} g/L. The reaction $Hg_2I_2 \rightleftharpoons Hg_2^{2+} + 2 I^-$ represents the equilibrium. Calculate the K_{sp}.

91. Under certain circumstances, when oxygen gas is heated, it can be converted into ozone according to the following reaction equation:

$$3 O_2(g) + heat \rightleftharpoons 2 O_3(g)$$

Name three different ways that you could increase the production of the ozone.

92. One day in a laboratory, some water spilled on a table. In just a few minutes the water had evaporated. Some days later, a similar amount of water spilled again. This time, the water remained on the table after 7 or 8 hours. Name three conditions that could have changed in the lab to cause this difference.

*93. All a snowbound skier had to eat were walnuts! He was carrying a bag holding 12 dozen nuts. With his mittened hands, he cracked open the shells. Each nut that was opened resulted in one kernel and two shell halves. When he tired of the cracking and got ready to do some eating, he discovered he had 194 total pieces (whole nuts, shell halves, and kernels). What is the K_{eq} for this reaction?

94. For the reaction $CO(g) + H_2O(g) \rightleftharpoons CO_2(g) + H_2(g)$ at a certain temperature, K_{eq} is 1. At equilibrium would you expect to find
(a) only CO and H_2
(b) mostly CO_2 and H_2
(c) about equal concentrations of CO and H_2O, compared to CO_2 and H_2
(d) mostly CO and H_2O
(e) only CO and H_2O
Explain your answer briefly.

95. Write the equilibrium constant expressions for these reactions:
(a) $3 O_2(g) \rightleftharpoons 2 O_3(g)$
(b) $H_2O(g) \rightleftharpoons H_2O(l)$
(c) $MgCO_3(s) \rightleftharpoons MgO(s) + CO_2(g)$
(d) $2 Bi^{3+}(aq) + 3 H_2S(aq) \rightleftharpoons Bi_2S_3(s) + 6 H^+(aq)$

96. Reactants A and B are mixed, each initially at a concentration of 1.0 M. They react to produce C according to this equation:

$$2 A + B \rightleftharpoons C$$

When equilibrium is established, the concentration of C is found to be 0.30 M. Calculate the value of K_{eq}.

97. At a certain temperature, K_{eq} is 2.2×10^{-3} for the reaction

$$2\, ICl(g) \rightleftarrows I_2(g) + Cl_2(g)$$

Now calculate the K_{eq} value for the reaction

$$I_2(g) + Cl_2(g) \rightleftarrows 2\, ICl(g)$$

98. One drop of $1\,M$ OH^- ion is added to a $1\,M$ solution of HNO_2. What will be the effect of this addition on the equilibrium concentration of the following?

$$HNO_2(aq) \rightleftarrows H^+(aq) + NO_2^-(aq)$$

   (a) $[OH^-]$
   (b) $[H^+]$
   (c) $[NO_2^-]$
   (d) $[HNO_2]$

*99. At 500°C, the reaction

$$SO_2(g) + NO_2(g) \rightleftarrows NO(g) + SO_3(g)$$

has $K_{eq} = 90$. What will be the equilibrium concentrations of the four gases if the two reactants begin with equal concentrations of $0.50\,M$?

100. How many grams of $CaSO_4$ will dissolve in 600. mL of water? ($K_{sp} = 2.0 \times 10^{-4}$ for $CaSO_4$)

101. A student found that 0.098 g of PbF_2 was dissolved in 400. mL of saturated PbF_2. What is the K_{sp} for the lead(II) fluoride?

Answers to Practice Exercises

16.1 (a)

Concentration	Change
$[H_2]$	increase
$[I_2]$	decrease
$[HI]$	increase

(b)

Concentration	Change
$[H_2]$	increase
$[I_2]$	increase
$[HI]$	increase

16.2 (a) Equilibrium shifts left; (b) equilibrium shifts right
16.3 (a) Equilibrium shifts right; (b) equilibrium shifts left

16.4 (a) Equilibrium shifts left; (b) equilibrium shifts right
16.5 (a) $K_{eq} = \dfrac{[NO_2]^4[O_2]}{[N_2O_5]^2}$; (b) $K_{eq} = \dfrac{[N_2]^2[H_2O]^6}{[NH_3]^4[O_2]^3}$
16.6 $K_{eq} = 33$; the forward reaction is favored
16.7 (a) $[H^+] = 5.0 \times 10^{-5}$ $[OH^-] = 2.0 \times 10^{-10}$
         (b) $[H^+] = 5.0 \times 10^{-9}$ $[OH^-] = 2.0 \times 10^{-6}$
16.8 (a) 4.30; (b) 8.30
16.9 (a) 6.3×10^{-6}; (b) 1.8×10^{-6}
16.10 (a) $6.3 \times 10^{-3}\%$ ionized; (b) $7.2 \times 10^{-3}\%$ ionized
16.11 (a) $K_{sp} = [Cr^{3+}][OH^-]^3$;
          (b) $K_{sp} = [Cu^{2+}]^3[PO_4^{3-}]^2$
16.12 9.1×10^{-21} g/L
16.13 7.5×10^{-8} mol/L
16.14 (a) basic; (b) neutral; (c) acidic

CHAPTER

17

Oxidation–Reduction

▲
Rust is being removed from
this iron sculpture.

CHAPTER 17 / OUTLINE

CHAPTER 17

The variety of oxidation–reduction reactions that affect us every day is amazing. Our society runs on batteries—in our calculators, laptop computers, cars, toys, radios, televisions, and more. We paint iron railings and galvanize nails to combat corrosion. We electroplate jewelry and computer chips with very thin coatings of gold or silver. We bleach our clothes and develop our photographs in solutions using chemical reactions that involve electron transfer. We test for glucose in urine or alcohol in the breath with reactions that show vivid color changes. Plants turn energy into chemical compounds through a series of reactions called photosynthesis. These reactions all involve the transfer of electrons between substances in a chemical process called *oxidation–reaction*.

 Oxidation Number

The oxidation number of an atom (sometimes called its *oxidation state*) represents the number of electrons lost, gained, or unequally shared by an atom. Oxidation numbers can be zero, positive, or negative. An oxidation number of zero means the atom has the same number of electrons assigned to it as there are in the free neutral atom. A positive oxidation number means the atom has fewer electrons assigned to it than in the neutral atom, and a negative oxidation number means the atom has more electrons assigned to it than in the neutral atom.

The oxidation number of an atom that has lost or gained electrons to form an ion is the same as the positive or negative charge of the ion. (See Table 17.1.) In the ionic compound NaCl, the oxidation numbers are clearly $+1$ for the Na^+ ion and -1 for the Cl^- ion. The Na^+ ion has one less electron than the neutral Na atom, and the Cl^- ion has one more electron than the neutral Cl atom. In $MgCl_2$, two electrons have transferred from the Mg atom to the two Cl atoms; the oxidation number of Mg is $+2$.

In covalently bonded substances where electrons are shared between two atoms, oxidation numbers are assigned by an arbitrary system based on relative electronegativities. For symmetrical covalent molecules such as H_2 and Cl_2, each atom is assigned an oxidation number of zero because the bonding pair of electrons is shared equally between two like atoms, neither of which is more electronegative than the other:

H:H   :C̈l:C̈l:

When the covalent bond is between two unlike atoms, the bonding electrons are shared unequally because the more electronegative element has a greater attraction for them. In this case the oxidation numbers are determined by assigning both electrons to the more electronegative element.

Thus in compounds with covalent bonds such as NH_3 and H_2O,

shared pairs of electrons

Ion	Oxidation number
H^+	$+1$
Na^+	$+1$
K^+	$+1$
Li^+	$+1$
Ag^+	$+1$
Cu^{2+}	$+2$
Ca^{2+}	$+2$
Ba^{2+}	$+2$
Fe^{2+}	$+2$
Mg^{2+}	$+2$
Zn^{2+}	$+2$
Al^{3+}	$+3$
Fe^{3+}	$+3$
Cl^-	-1
Br^-	-1
F^-	-1
I^-	-1
S^{2-}	-2
O^{2-}	-2

TABLE 17.1 Oxidation Numbers for Common Ions

oxidation number
oxidation state

The oxidation number for Cu metal is 0 while the oxidation number for Cu^{2+} ions in the crystal is 2+.

the pairs of electrons are unequally shared between the atoms and are attracted toward the more electronegative elements, N and O. This causes the N and O atoms to be relatively negative with respect to the H atoms. At the same time it causes the H atoms to be relatively positive with respect to the N and O atoms. In H_2O, both pairs of shared electrons are assigned to the O atom, giving it two electrons more than the neutral O atom, and each H atom is assigned one electron less than the neutral H atom. Therefore the oxidation number of the O atom is −2, and the oxidation number of each H atom is +1. In NH_3, the three pairs of shared electrons are assigned to the N atom, giving it three electrons more than the neutral N atom, and each H atom has one electron less than the neutral atom. Therefore the oxidation number of the N atom is −3, and the oxidation number of each H atom is +1.

Assigning correct oxidation numbers to elements is essential for balancing oxidation–reduction equations.

The **oxidation number** or **oxidation state** of an element is an integer value assigned to each element in a compound or ion that allows us to keep track of electrons associated with each atom. Oxidation numbers have a variety of uses in chemistry—from writing formulas to predicting properties of compounds and assisting in the balancing of oxidation–reduction reactions in which electrons are transferred.

As a starting point, the oxidation number of an uncombined element, regardless of whether it is monatomic or diatomic, is zero. Rules for assigning oxidation numbers are summarized in Table 17.2.

Use the following steps to find the oxidation number for an element within a compound.

Step 1. Write the oxidation number of each known atom below the atom in the formula.

Step 2. Multiply each oxidation number by the number of atoms of that element in the compound.

Step 3. Write an expression indicating the sum of all the oxidation numbers in the compound. Remember: The sum of the oxidation numbers in a compound must equal zero.

TABLE 17.2	Rules for Assigning Oxidation Number

1. All elements in their free state (uncombined with other elements) have an oxidation number of zero (e.g., Na, Cu, Mg, H_2, O_2, Cl_2, N_2).
2. H is +1, except in metal hydrides, where it is −1 (e.g., NaH, CaH_2).
3. O is −2, except in peroxides, where it is −1, and in OF_2, where it is +2.
4. The metallic element in an ionic compound has a positive oxidation number.
5. In covalent compounds the negative oxidation number is assigned to the most electronegative atom.
6. The algebraic sum of the oxidation numbers of the elements in a compound is zero.
7. The algebraic sum of the oxidation numbers of the elements in a polyatomic ion is equal to the charge of the ion.

Determine the oxidation number for carbon in carbon dioxide: **Example 17.1**

$$CO_2$$

Step 1 -2
Step 2 $(-2)2$
Step 3 $C + (-4) = 0$
Step 3 $C = +4$ (oxidation number for carbon)

Determine the oxidation number for sulfur in sulfuric acid: **Example 17.2**

$$H_2SO_4$$

Step 1 $+1$ -2
Step 2 $2(+1) = +2$ $4(-2) = -8$
Step 3 $+2 + S + (-8) = 0$
                      $S = +6$ (oxidation number for sulfur)

Practice 17.1

Determine the oxidation number of (a) S in Na_2SO_4, (b) As in K_3AsO_4, and (c) C in $CaCO_3$.

Oxidation numbers in a polyatomic ion (ions containing more than one atom) are determined in a similar fashion, except that in a polyatomic ion the sum of the oxidation numbers must equal the charge on the ion instead of zero.

Determine the oxidation number for manganese in the permanganate ion MnO_4^-:   **Example 17.3**

$$MnO_4^-$$

Step 1 -2
Step 2 $(-2)4$
Step 3 $Mn + (-8) = -1$ (the charge on the ion)
             $Mn = +7$ (oxidation number for manganese)

Example 17.4   Determine the oxidation number for carbon in the oxalate ion $C_2O_4^{2-}$:

$$C_2O_4^{2-}$$
$$-2$$

Step 1 $(-2)4$
Step 2 $2C + (-8) = -2$ (the charge on the ion)
Step 3 $2C = +6$
          $C = +3$ (oxidation number for C)

Practice 17.2

Determine the oxidation numbers of (a) N in NH_4^+, (b) Cr in $Cr_2O_7^{2-}$, and (c) P in PO_4^{3-}.

Example 17.5   Determine the oxidation number of each element in (a) KNO_3, and (b) SO_4^{2-}.

Solution

(a) Potassium is a Group IA metal; therefore it has an oxidation number of $+1$. The oxidation number of each O atom is -2 (Table 17.2, Rule 3). Using these values and the fact that the sum of the oxidation numbers of all the atoms in a compound is zero, we can determine the oxidation number of N:

$$KNO_3$$
$$+1 + N + 3(-2) = 0$$
$$N = +6 - 1 = +5$$

The oxidation numbers are K, $+1$; N, $+5$; O, -2.

(b) Because SO_4^{2-} is an ion, the sum of oxidation numbers of the S and the O atoms must be -2, the charge of the ion. The oxidation number of each O atom is -2 (Table 17.2, Rule 3). Then

$$SO_4^{2-}$$
$$S + 4(-2) = -2, \quad S - 8 = -2$$
$$S = -2 + 8 = +6$$

The oxidation numbers are S, $+6$; O, -2.

Practice 17.3

Determine the oxidation number of each element in these species:
(a) $BeCl_2$ (b) $HClO$ (c) H_2O_2 (d) NH_4^+ (e) BrO_3^-

Oxidation number

$-7 \quad -6 \quad -5 \quad -4 \quad -3 \quad -2 \quad -1 \quad 0 \quad +1 \quad +2 \quad +3 \quad +4 \quad +5 \quad +6 \quad +7$

Oxidation (loss of electrons)

Reduction (gain of electrons)

◀ **FIGURE 17.1**
**Oxidation and reduction.
Oxidation results in an
increase in the oxidation
number, and reduction
results in a decrease in the
oxidation number.**

17.2 Oxidation–Reduction

Oxidation–reduction, also known as **redox,** is a chemical process in which the oxidation number of an element is changed. The process may involve the complete transfer of electrons to form ionic bonds or only a partial transfer or shift of electrons to form covalent bonds.

oxidation–reduction

redox

    Oxidation occurs whenever the oxidation number of an element increases as a result of losing electrons. Conversely, **reduction** occurs whenever the oxidation number of an element decreases as a result of gaining electrons. For example, a change in oxidation number from $+2$ to $+3$ or from -1 to 0 is oxidation; a change from $+5$ to $+2$ or from -2 to -4 is reduction (see Figure 17.1). Oxidation and reduction occur simultaneously in a chemical reaction; one cannot take place without the other.

oxidation

reduction

    Many combination, decomposition, and single-displacement reactions involve oxidation–reduction. Let's examine the combustion of hydrogen and oxygen from this point of view:

$$2\,H_2 + O_2 \longrightarrow 2\,H_2O$$

Both reactants, hydrogen and oxygen, are elements in the free state and have an oxidation number of zero. In the product (water), hydrogen has been oxidized to $+1$ and oxygen reduced to -2. The substance that causes an increase in the oxidation state of another substance is called an **oxidizing agent.** The substance that causes a decrease in the oxidation state of another substance is called a **reducing agent.** In this reaction the oxidizing agent is free oxygen, and the reducing agent is free hydrogen. In the reaction

oxidizing agent

reducing agent

$$Zn(s) + H_2SO_4(aq) \longrightarrow ZnSO_4(aq) + H_2(g)$$

metallic zinc is oxidized, and hydrogen ions are reduced. Zinc is the reducing agent, and hydrogen ions, the oxidizing agent. Electrons are transferred from the zinc metal to the hydrogen ions. The reaction is better expressed as

$$Zn^0 + 2\,H^+ + SO_4^{2-} \longrightarrow Zn^{2+} + SO_4^{2-} + H_2^0$$

Oxidation:	**Increase in oxidation number**
	Loss of electrons
Reduction:	**Decrease in oxidation number**
	Gain of electrons

▲
**In the reaction between Zn
and H_2SO_4 the Zn is oxidized
while hydrogen is reduced.**

The oxidizing agent is reduced and gains electrons. The reducing agent is oxidized and loses electrons. The transfer of electrons is characteristic of all redox reactions.

17.3   Balancing Oxidation–Reduction Equations

Many simple redox equations can be balanced readily by inspection, or by trial and error:

$$Na + Cl_2 \longrightarrow NaCl \qquad \text{(unbalanced)}$$

$$2\,Na + Cl_2 \longrightarrow 2\,NaCl \quad \text{(balanced)}$$

Balancing this equation is certainly not complicated. But as we study more complex reactions and equations such as

$$P + HNO_3 + H_2O \longrightarrow NO + H_3PO_4 \qquad \text{(unbalanced)}$$

$$3\,P + 5\,HNO_3 + 2\,H_2O \longrightarrow 5\,NO + 3\,H_3PO_4 \quad \text{(balanced)}$$

the trial-and-error method of balancing equations takes an unnecessarily long time.

One systematic method for balancing oxidation–reduction equations is based on the transfer of electrons between the oxidizing and reducing agents. Consider the first equation again:

$$Na^0 + Cl_2^0 \longrightarrow Na^+Cl^- \quad \text{(unbalanced)}$$

The superscript 0 shows the oxidation number is 0 for elements in their uncombined state

In this reaction sodium metal loses one electron per atom when it changes to a sodium ion. At the same time chlorine gains one electron per atom. Because chlorine is diatomic, two electrons per molecule are needed to form a chloride ion from each atom. These electrons are furnished by two sodium atoms. Stepwise, the reaction may be written as two half-reactions, the oxidation half-reaction and the reduction half-reaction:

$$
\begin{array}{ll}
2\,Na^0 \longrightarrow 2\,Na^+ + 2\,e^- & \text{oxidation half-reaction} \\
\underline{Cl_2^0 + 2\,e^- \longrightarrow \qquad\qquad 2\,Cl^-} & \text{reduction half-reaction} \\
Cl_2^0 + 2\,Na^0 \longrightarrow 2\,Na^+Cl^- &
\end{array}
$$

When the two half-reactions, each containing the same number of electrons, are added together algebraically, the electrons cancel out. In this reaction there are no excess electrons; the two electrons lost by the two sodium atoms are utilized by chlorine. In all redox reactions the loss of electrons by the reducing agent must equal the gain of electrons by the oxidizing agent. Here, sodium is oxidized and chlorine is reduced. Chlorine is the oxidizing agent; sodium is the reducing agent.

In the following examples we use the change-in-oxidation-number method, a system for balancing more complicated redox equations.

Example 17.6   Balance the equation

$$Sn + HNO_3 \longrightarrow SnO_2 + NO_2 + H_2O \quad \text{(unbalanced)}$$

Solution   **Step 1.** Assign oxidation numbers to each element to identify the elements being oxidized and those being reduced. Write the oxidation numbers below each element to avoid confusing them with ionic charge:

$$Sn + HNO_3 \longrightarrow SnO_2 + NO_2 + H_2O$$
$$\begin{array}{ccccc} 0 & +1\ +5\ -2 & +4\ -2 & +4\ -2 & +1\ -2 \end{array}$$

Note that the oxidation numbers of Sn and N have changed.

Step 2. Now write two new equations, using only the elements that change in oxidation number. Then add electrons to bring the equations into electrical balance. One equation represents the oxidation step; the other represents the reduction step. Remember: Oxidation produces electrons; reduction uses electrons.

$$Sn^0 \longrightarrow Sn^{4+} + 4\ e^- \qquad \text{oxidation}$$
Sn^0 loses 4 electrons

$$N^{5+} + 1\ e^- \longrightarrow N^{4+} \qquad \text{reduction}$$
N^{5+} gains 1 electron

Step 3. Multiply the two equations by the smallest whole numbers that will make the electrons lost by oxidation equal to the number of electrons gained by reduction. In this reaction the oxidation step is multiplied by 1 and the reduction step by 4. The equations become

$$Sn^0 \longrightarrow Sn^{4+} + 4\ e^- \qquad \text{oxidation}$$
Sn^0 loses 4 electrons

$$4\ N^{5+} + 4\ e^- \longrightarrow 4\ N^{4+} \qquad \text{reduction}$$
$4\ N^{5+}$ gains 4 electrons

We have now established the ratio of the oxidizing to the reducing agent as being four atoms of N to one atom of Sn.

Step 4. Transfer the coefficient in front of each substance in the balanced oxidation–reduction equations to the corresponding substance in the original equation. We need to use 1 Sn, 1 SnO_2, 4 HNO_3, and 4 NO_2:

$$Sn + 4\ HNO_3 \longrightarrow SnO_2 + 4\ NO_2 + H_2O \quad \text{(unbalanced)}$$

Step 5. In the usual manner, balance the remaining elements that are not oxidized or reduced to give the final balanced equation:

$$Sn + 4\ HNO_3 \longrightarrow SnO_2 + 4\ NO_2 + 2\ H_2O \quad \text{(balanced)}$$

In balancing the final elements, we must not change the ratio of the elements that were oxidized and reduced.

Finally, check to ensure that both sides of the equation have the same number of atoms of each element. The final balanced equation contains 1 atom of Sn, 4 atoms of N, 4 atoms of H, and 12 atoms of O on each side.

Because each new equation presents a slightly different problem and because proficiency in balancing equations requires practice, let's work through two more examples.

Balance the equation

Example 17.7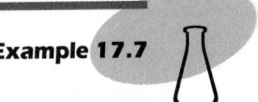

$$I_2 + Cl_2 + H_2O \longrightarrow HIO_3 + HCl \quad \text{(unbalanced)}$$

Solution

Step 1. Assign oxidation numbers:

$$I_2 + Cl_2 + H_2O \longrightarrow HIO_3 + HCl$$

$$0 \quad\quad 0 \quad\quad +1 \; -2 \quad\quad +1 \; +5 \; -2 \quad\quad +1 \; -1$$

The oxidation numbers of I_2 and Cl_2 have changed, I_2 from 0 to $+5$, and Cl_2 from 0 to -1.

Step 2. Write the oxidation and reduction steps. Balance the number of atoms and then balance the electrical charge using electrons:

$$I_2 \longrightarrow 2\, I^{5+} + 10\, e^-$$ oxidation   (10 e^- are needed to balance
I_2 loses 10 electrons   the $+10$ charge)

$$Cl_2 + 2\, e^- \longrightarrow 2\, Cl^-$$ reduction   (2 e^- are needed to balance
Cl_2 gains 2 electrons   the -2 charge)

Step 3. Adjust loss and gain of electrons so that they are equal. Multiply the oxidation step by 1 and the reduction step by 5:

$$I_2 \longrightarrow 2\, I^{5+} + 10\, e^-$$ oxidation
I_2 loses 10 electrons

$$5\, Cl_2 + 10\, e^- \longrightarrow 10\, Cl^-$$ reduction
$5\, Cl_2$ gain 10 electrons

Step 4. Transfer the coefficients from the balanced redox equations into the original equation. We need to use 1 I_2, 2 HIO_3, 5 Cl_2, and 10 HCl:

$$I_2 + 5\, Cl_2 + H_2O \longrightarrow 2\, HIO_3 + 10\, HCl \quad \text{(unbalanced)}$$

Step 5. Balance the remaining elements, H and O:

$$I_2 + 5\, Cl_2 + 6\, H_2O \longrightarrow 2\, HIO_3 + 10\, HCl \quad \text{(balanced)}$$

Check: The final balanced equation contains 2 atoms of I, 10 atoms of Cl, 12 atoms of H, and 6 atoms of O on each side.

Example 17.8 Balance the equation

$$K_2Cr_2O_7 + FeCl_2 + HCl \longrightarrow CrCl_3 + KCl + FeCl_3 + H_2O \quad \text{(unbalanced)}$$

Solution

Step 1. Assign oxidation numbers (Cr and Fe have changed):

$$K_2Cr_2O_7 + FeCl_2 + HCl \longrightarrow CrCl_3 + KCl + FeCl_3 + H_2O$$

$$+1 \; +6 \; -2 \quad +2 \; -1 \quad +1 \; -1 \quad +3 \; -1 \quad +1 \; -1 \quad +3 \; -1 \quad +1 \; -2$$

Step 2. Write the oxidation and reduction steps. Balance the number of atoms and then balance the electrical charge using electrons:

$$Fe^{2+} \longrightarrow Fe^{3+} + 1\, e^-$$ oxidation
Fe^{2+} loses 1 electron

$$2\, Cr^{6+} + 6\, e^- \longrightarrow 2\, Cr^{3+}$$ reduction
$2\, Cr^{6+}$ gain 6 electrons

Step 3. Balance the loss and gain of electrons. Multiply the oxidation step by 6 and the reduction step by 1 to equalize the transfer of electrons.

$$6\ Fe^{2+} \longrightarrow 6\ Fe^{3+} + 6\ e^{-} \qquad \text{oxidation}$$
6 Fe^{2+} lose 6 electrons

$$2\ Cr^{6+} + 6\ e^{-} \longrightarrow 2\ Cr^{3+} \qquad \text{reduction}$$
2 Cr^{6+} gain 6 electrons

Step 4. Transfer the coefficients from the balanced redox equations into the original equation. (Note that one formula unit of $K_2Cr_2O_7$ contains two Cr atoms.) We need to use 1 $K_2Cr_2O_7$, 2 $CrCl_3$, 6 $FeCl_2$, and 6 $FeCl_3$:

$$K_2Cr_2O_7 + 6\ FeCl_2 + HCl \longrightarrow$$
$$2\ CrCl_3 + KCl + 6\ FeCl_3 + H_2O \quad \text{(unbalanced)}$$

Step 5. Balance the remaining elements in this order: K, Cl, H, O.

$$K_2Cr_2O_7 + 6\ FeCl_2 + 14\ HCl \longrightarrow$$
$$2\ CrCl_3 + 2\ KCl + 6\ FeCl_3 + 7\ H_2O \quad \text{(balanced)}$$

Check: The final balanced equation contains 2 K atoms, 2 Cr atoms, 7 O atoms, 6 Fe atoms, 26 Cl atoms, and 14 H atoms on each side.

Practice 17.4

Balance these equations using the change-in-oxidation-number method:
(a) $HNO_3 + S \longrightarrow NO_2 + H_2SO_4 + H_2O$
(b) $CrCl_3 + MnO_2 + H_2O \longrightarrow MnCl_2 + H_2CrO_4$
(c) $KMnO_4 + HCl + H_2S \longrightarrow KCl + MnCl_2 + S + H_2O$

 ## Balancing Ionic Redox Equations

The main difference between balancing ionic redox equations and molecular redox equations is in how we handle the ions. In the ionic redox equations, besides having the same number of atoms of each element on both sides of the final equation, we must also have equal net charges. In assigning oxidation numbers, we must therefore remember to consider the ionic charge.

Several methods are used to balance ionic redox equations, including, with slight modification, the oxidation-number method just shown for molecular equations. But the most popular method is probably the ion–electron method.

The ion–electron method uses ionic charges and electrons to balance ionic redox equations. Oxidation numbers are not formally used, but it is necessary to determine what is being oxidized and what is being reduced. The method is as follows:

1. Write the two half-reactions that contain the elements being oxidized and reduced using the entire formula of the ion or molecule.
2. Balance the elements other than oxygen and hydrogen.

Sensitive Sunglasses

Oxidation–reduction reactions are the basis for many interesting applications. Consider photochromic glass, which is used for lenses in light-sensitive glasses. These lenses, manufactured by the Corning Glass Company, can change from transmitting 85% of light to only

◀ **An oxidation–reduction reaction causes these photochromic glasses to change from light to dark in bright sunlight.**

transmitting 22% of light when exposed to bright sunlight.

Photochromic glass is composed of linked tetrahedrons of silicon and oxygen atoms jumbled in a disorderly array, with crystals of silver chloride caught between the silica tetrahedrons. When the glass is clear, the visible light passes right through the molecules. The glass absorbs ultraviolet light, however, and this energy triggers an oxidation–reduction reaction between Ag^+ and Cl^-:

$$Ag^+ + Cl^- \xrightarrow{\text{UV light}} Ag^0 + Cl^0$$

To prevent the reaction from reversing itself immediately, a few ions of Cu^+ are incorporated into the silver chloride crystal. These Cu^+ ions react with the newly formed chlorine atoms:

$$Cu^+ + Cl^0 \longrightarrow Cu^{2+} + Cl^-$$

The silver atoms move to the surface of the crystal and form small colloidal clusters of silver metal. This metallic silver absorbs visible light, making the lens appear dark (colored).

As the glass is removed from the light, the Cu^{2+} ions slowly move to the surface of the crystal where they interact with the silver metal:

$$Cu^{2+} + Ag^0 \longrightarrow Cu^+ + Ag^+$$

The glass clears as the silver ions rejoin chloride ions in the crystals.

3. Balance oxygen and hydrogen.
   Acidic solution: For reactions in acidic solution, use H^+ and H_2O to balance oxygen and hydrogen. For each oxygen, use one H_2O. Then add H^+ as needed to balance the hydrogen atoms.
   Basic solution: For reactions in alkaline solutions, first balance as though the reaction were in an acidic solution, using Steps 1–3. Then add as many OH^- ions to each side of the equation as there are H^+ ions in the equation. Now combine the H^+ and OH^- ions into water (for example, $4\ H^+$ and $4\ OH^-$ give $4\ H_2O$). Rewrite the equation, canceling equal numbers of water molecules that appear on opposite sides of the equation.
4. Add electrons (e^-) to each half-reaction to bring them into electrical balance.
5. Since the loss and gain of electrons must be equal, multiply each half-reaction by the appropriate number to make the number of electrons the same in each half-reaction.
6. Add the two half-reactions together, canceling electrons and any other identical substances that appear on opposite sides of the equation.

Example 17.9   Balance this equation using the ion–electron method:

$$MnO_4^- + S^{2-} \longrightarrow Mn^{2+} + S^0 \quad \text{(acidic solution)}$$

Solution   **Step 1.** Write two half-reactions, one containing the element being oxidized and the other the element being reduced (use the entire molecule or ion):

$$S^{2-} \longrightarrow S^0 \qquad \text{oxidation}$$

$$MnO_4^- \longrightarrow Mn^{2+} \quad \text{reduction}$$

Step 2. Balance elements other than oxygen and hydrogen (accomplished in Step 1: 1 S and 1 Mn on each side).

Step 3. Balance O and H. Remember the solution is acidic. The oxidation requires neither O nor H, but the reduction equation needs 4 H_2O on the right and 8 H^+ on the left.

$$S^{2-} \longrightarrow S^0$$

$$8\,H^+ + MnO_4^- \longrightarrow Mn^{2+} + 4\,H_2O$$

Step 4. Balance each half-reaction electrically with electrons:

$$S^{2-} \longrightarrow S^0 + 2\,e^-$$

net charge = −2 on each side

$$5\,e^- + 8\,H^+ + MnO_4^- \longrightarrow Mn^{2+} + 4\,H_2O$$

net charge = +2 on each side

Step 5. Equalize loss and gain of electrons. In this case multiply the oxidation equation by 5 and the reduction equation by 2:

$$5\,S^{2-} \longrightarrow 5\,S^0 + 10\,e^-$$

$$10\,e^- + 16\,H^+ + 2\,MnO_4^- \longrightarrow 2\,Mn^{2+} + 8\,H_2O$$

Step 6. Add the two half-reactions together, canceling the 10 e^- from each side, to obtain the balanced equation:

$$5\,S^{2-} \longrightarrow 5\,S^0 + \cancel{10\,e^-}$$

$$\underline{\cancel{10\,e^-} + 16\,H^+ + 2\,MnO_4^- \longrightarrow 2\,Mn^{2+} + 8\,H_2O}$$

$$16\,H^+ + 2\,MnO_4^- + 5\,S^{2-} \longrightarrow 2\,Mn^{2+} + 5\,S^0 + 8\,H_2O \quad \text{(balanced)}$$

Check: Both sides of the equation have a charge of +4 and contain the same number of atoms of each element.

Balance this equation:

Example 17.10

$$CrO_4^{2-} + Fe(OH)_2 \longrightarrow Cr(OH)_3 + Fe(OH)_3 \quad \text{(basic solution)}$$

Step 1. Write the two half-reactions:

Solution

$$Fe(OH)_2 \longrightarrow Fe(OH)_3 \quad \text{oxidation}$$

$$CrO_4^{2-} \longrightarrow Cr(OH)_3 \quad \text{reduction}$$

Step 2. Balance elements other than H and O (accomplished in Step 1).

Step 3. Remember the solution is basic. Balance O and H as though the solution were acidic. Use H_2O and H^+. To balance O and H in the oxidation equation, add 1 H_2O on the left and 1 H^+ on the right side:

$$Fe(OH)_2 + H_2O \longrightarrow Fe(OH)_3 + H^+$$

Add 1 OH^- to each side:

$$Fe(OH)_2 + H_2O + OH^- \longrightarrow Fe(OH)_3 + H^+ + OH^-$$

Combine H^+ and OH^- as H_2O and rewrite, canceling H_2O on each side:

$$Fe(OH)_2 + \cancel{H_2O} + OH^- \longrightarrow Fe(OH)_3 + \cancel{H_2O}$$

$$\boxed{Fe(OH)_2 + OH^- \qquad Fe(OH)_3} \qquad \text{(oxidation)}$$

To balance O and H in the reduction equation, add 1 H_2O on the right and 5 H^+ on the left:

$$CrO_4^{2-} + 5\,H^+ \longrightarrow Cr(OH)_3 + H_2O$$

Add 5 OH^- to each side:

$$CrO_4^{2-} + 5\,H^+ + 5\,OH^- \longrightarrow Cr(OH)_3 + H_2O + 5\,OH^-$$

Combine 5 H^+ + 5 $OH^- \longrightarrow 5\,H_2O$:

$$CrO_4^{2-} + 5\,H_2O \longrightarrow Cr(OH)_3 + H_2O + 5\,OH^-$$

Rewrite, canceling 1 H_2O from each side:

$$\boxed{CrO_4^{2-} + 4\,H_2O \qquad Cr(OH)_3 + 5\,OH^-} \qquad \text{(reduction)}$$

Step 4. Balance each half-reaction electrically with electrons:

$$Fe(OH)_2 + OH^- \longrightarrow Fe(OH)_3 + e^-$$
$$\text{(balanced oxidation equation)}$$

$$CrO_4^{2-} + 4\,H_2O + 3\,e^- \longrightarrow Cr(OH)_3 + 5\,OH^-$$
$$\text{(balanced reduction equation)}$$

Step 5. Equalize the loss and gain of electrons. Multiply the oxidation reaction by 3:

$$3\,Fe(OH)_2 + 3\,OH^- \longrightarrow 3\,Fe(OH)_3 + 3\,e^-$$
$$CrO_4^{2-} + 4\,H_2O + 3\,e^- \longrightarrow Cr(OH)_3 + 5\,OH^-$$

Step 6. Add the two half-reactions together, canceling the 3 e^- and 3 OH^- from each side of the equation:

$$3\,Fe(OH)_2 + 3\,OH^- \longrightarrow 3\,Fe(OH)_3 + 3\cancel{e^-}$$
$$\underline{CrO_4^{2-} + 4\,H_2O + 3\cancel{e^-} \longrightarrow Cr(OH)_3 + 5\,OH^-}$$
$$CrO_4^{2-} + 3\,Fe(OH)_2 + 4\,H_2O \longrightarrow Cr(OH)_3 + 3\,Fe(OH)_3 + 2\,OH^- \quad \text{(balanced)}$$

Check: Each side of the equation has a charge of -2 and contains the same number of atoms of each element.

Ionic equations can also be balanced using the change-in-oxidation-number method shown in Example 17.6. To illustrate this method, let's use the equation from Example 17.10.

Balance this equation using the change-in-oxidation-number method: **Example 17.11**

$$CrO_4^{2-} + Fe(OH)_2 \longrightarrow Cr(OH)_3 + Fe(OH)_3 \text{ (basic solution)}$$

Steps 1 and 2. Assign oxidation numbers and balance the charges with electrons: **Solution**

$$Cr^{6+} + 3\, e^- \longrightarrow Cr^{3+} \quad \text{reduction}$$
$$\underset{Cr^{6+} \text{ gains } 3\, e^-}{}$$

$$Fe^{2+} \longrightarrow Fe^{3+} + e^- \quad \text{oxidation}$$
$$\underset{Fe^{2+} \text{ loses } 1\, e^-}{} \longrightarrow$$

Step 3. Equalize the loss and gain of electrons, and then multiply the oxidation step by 3:

$$Cr^{6+} + 3\, e^- \longrightarrow Cr^{3+}$$
$$\underset{Cr^{6+} \text{ gains } 3\, e^-}{}$$

$$3\, Fe^{2+} \longrightarrow 3\, Fe^{3+} + 3\, e^-$$
$$\underset{3\, Fe^{2+} \text{ lose } 3\, e^-}{}$$

Step 4. Transfer coefficients back to the original equation:

$$CrO_4^{2-} + 3\, Fe(OH)_2 \longrightarrow Cr(OH)_3 + 3\, Fe(OH)_3$$

Step 5. Balance electrically. Because the solution is basic, use OH^- to balance charges. The charge on the left side is -2, and on the right side is 0. Add 2 OH^- ions to the right side of the equation:

$$CrO_4^{2-} + 3\, Fe(OH)_2 \longrightarrow Cr(OH)_3 + 3\, Fe(OH)_3 + 2\, OH^-$$

Adding 4 H_2O to the left side balances the equation:

$$CrO_4^{2-} + 3\, Fe(OH)_2 + 4\, H_2O \longrightarrow$$
$$Cr(OH)_3 + 3\, Fe(OH)_3 + 2\, OH^- \quad \text{(balanced)}$$

Check: Each side of the equation has a charge of -2 and contains the same number of atoms of each element.

FIGURE 17.2
A coil of copper placed in a
silver nitrate solution forms
silver crystals on the wire.
The pale blue of the solution
indicates the presence of
copper ions.

TABLE 17.3   Activity Series of Metals

Ease of oxidation		
K	$\longrightarrow$ K$^+$	+ e$^-$
Ba	$\longrightarrow$ Ba^{2+}	+ 2 e$^-$
Ca	$\longrightarrow$ Ca^{2+}	+ 2 e$^-$
Na	$\longrightarrow$ Na$^+$	+ e$^-$
Mg	$\longrightarrow$ Mg^{2+}	+ 2 e$^-$
Al	$\longrightarrow$ Al^{3+}	+ 3 e$^-$
Zn	$\longrightarrow$ Zn^{2+}	+ 2 e$^-$
Cr	$\longrightarrow$ Cr^{3+}	+ 3 e$^-$
Fe	$\longrightarrow$ Fe^{2+}	+ 2 e$^-$
Ni	$\longrightarrow$ Ni^{2+}	+ 2 e$^-$
Sn	$\longrightarrow$ Sn^{2+}	+ 2 e$^-$
Pb	$\longrightarrow$ Pb^{2+}	+ 2 e$^-$
H$_2$	$\longrightarrow$ **2 H$^+$**	**+ 2 e$^-$**
Cu	$\longrightarrow$ Cu^{2+}	+ 2 e$^-$
As	$\longrightarrow$ As^{3+}	+ 3 e$^-$
Ag	$\longrightarrow$ Ag$^+$	+ e$^-$
Hg	$\longrightarrow$ Hg^{2+}	+ 2 e$^-$
Au	$\longrightarrow$ Au^{3+}	+ 3 e$^-$

activity series of metals

Practice 17.6

Balance these equations using the change-in-oxidation-number method:
(a) $Zn \longrightarrow Zn(OH)_4^{2-} + H_2$ (basic solution)
(b) $H_2O_2 + Sn^{2+} \longrightarrow Sn^{4+}$ (acidic solution)
(c) $Cu + Cu^{2+} \longrightarrow Cu_2O$ (basic solution)

17.5   Activity Series of Metals

Knowledge of the relative chemical reactivities of the elements helps us predict the
course of many chemical reactions. For example, calcium reacts with cold water to
produce hydrogen, and magnesium reacts with steam to produce hydrogen. Therefore calcium is considered a more reactive metal than magnesium:

$$Ca(s) + 2 H_2O(l) \longrightarrow Ca(OH)_2(aq) + H_2(g)$$
$$Mg(s) + \underset{\text{steam}}{H_2O(g)} \longrightarrow MgO(s) + H_2(g)$$

The difference in their activity is attributed to the fact that calcium loses its two
valence electrons more easily than magnesium and is therefore more reactive and/or
more readily oxidized than magnesium.

When a coil of copper is placed in a solution of silver nitrate ($AgNO_3$), free
silver begins to plate out on the copper. (See Figure 17.2.) After the reaction has continued for some time, we can observe a blue color in the solution, indicating the presence of copper(II) ions. The equations are

$$Cu^0(s) + 2 AgNO_3(aq) \longrightarrow 2 Ag^0(s) + Cu(NO_3)_2(aq)$$
$$Cu^0(s) + 2 Ag^+(aq) \longrightarrow 2 Ag^0(s) + Cu^{2+}(aq) \qquad \text{net ionic equation}$$
$$Cu^0(s) \longrightarrow Cu^{2+}(aq) + 2 e^- \qquad \text{oxidation of } Cu^0$$
$$Ag^+(aq) + e^- \longrightarrow Ag^0(s) \qquad \text{reduction of } Ag^+$$

If a coil of silver is placed in a solution of copper(II) nitrate, $Cu(NO_3)_2$, no reaction
is visible.

$$Ag^0(s) + Cu(NO_3)_2(aq) \longrightarrow \text{no reaction}$$

In the reaction between Cu and $AgNO_3$, electrons are transferred from Cu^0 atoms to
Ag^+ ions in solution. Copper has a greater tendency than silver to lose electrons, so an
electrochemical force is exerted upon silver ions to accept electrons from copper
atoms. When an Ag^+ ion accepts an electron, it is reduced to an Ag^0 atom and is no
longer soluble in solution. At the same time, Cu^0 is oxidized and goes into solution as
Cu^{2+} ions. From this reaction we can conclude that copper is more reactive than silver.

Metals such as sodium, magnesium, zinc, and iron that react with solutions of
acids to liberate hydrogen are more reactive than hydrogen. Metals such as copper,
silver, and mercury that do not react with solutions of acids to liberate hydrogen are
less reactive than hydrogen. By studying a series of reactions such as these, we can
list metals according to their chemical activity, placing the most active at the top and
the least active at the bottom. This list is called the **activity series of metals.** Table
17.3 lists some of the common metals in the series. The arrangement corresponds to

the ease with which the elements are oxidized or lose electrons, with the most easily oxidizable element listed first. More extensive tables are available in chemistry reference books.

The general principles governing the arrangement and use of the activity series are as follows:

1. The reactivity of the metals listed decreases from top to bottom.
2. A free metal can displace the ion of a second metal from solution, provided that the free metal is above the second metal in the activity series.
3. Free metals above hydrogen react with nonoxidizing acids in solution to liberate hydrogen gas.
4. Free metals below hydrogen do not liberate hydrogen from acids.
5. Conditions such as temperature and concentration may affect the relative position of some of these elements.

Here are two examples using the activity series of metals.

Will zinc metal react with dilute sulfuric acid?

Example 17.12

Solution

From Table 17.3, we see that zinc is above hydrogen; therefore zinc atoms will lose electrons more readily than hydrogen atoms. Hence zinc atoms will reduce hydrogen ions from the acid to form hydrogen gas and zinc ions. In fact, these reagents are commonly used for the laboratory preparation of hydrogen. The equation is

$$Zn(s) + H_2SO_4(aq) \longrightarrow ZnSO_4(aq) + H_2(g)$$

$$Zn(s) + 2\,H^+(aq) \longrightarrow Zn^{2+}(aq) + H_2(g) \qquad \text{(net ionic equation)}$$

Will a reaction occur when copper metal is placed in an iron(II) sulfate solution?

Example 17.13

Solution

No, copper lies below iron in the series, loses electrons less easily than iron, and therefore will not displace iron(II) ions from solution. In fact, the reverse is true. When an iron nail is dipped into a copper(II) sulfate solution, it becomes coated with free copper. The equations are

$$Cu(s) + FeSO_4(aq) \longrightarrow \text{no reaction}$$

$$Fe(s) + CuSO_4(aq) \longrightarrow FeSO_4(aq) + Cu(s)$$

From Table 17.3, we may abstract the following pair in their relative position to each other:

$$Fe \longrightarrow Fe^{2+} + 2\,e^-$$

$$Cu \longrightarrow Cu^{2+} + 2\,e^-$$

According to Principle 2 on the use of the activity series, we can predict that free iron will react with copper(II) ions in solution to form free copper metal and iron(II) ions in solution:

$$Fe(s) + Cu^{2+}(aq) \longrightarrow Fe^{2+}(aq) + Cu(s) \quad \text{(net ionic equation)}$$

Electrolytic and Voltaic Cells

electrolysis
electrolytic cell

The process in which electrical energy is used to bring about chemical change is known as **electrolysis.** An **electrolytic cell** uses electrical energy to produce a chemical reaction. The use of electrical energy has many applications in industry—for example, in the production of sodium, sodium hydroxide, chlorine, fluorine, magnesium, aluminum, and pure hydrogen and oxygen, and in the purification and electroplating of metals.

What happens when an electric current is passed through a solution? Let's consider a hydrochloric acid solution in a simple electrolytic cell, as shown in Figure 17.3. The cell consists of a source of direct current (a battery) connected to two electrodes that are immersed in a solution of hydrochloric acid. The negative electrode is called the **cathode** because cations are attracted to it. The positive electrode is called the **anode** because anions are attracted to it. The cathode is attached to the negative pole and the anode to the positive pole of the battery. The battery supplies electrons to the cathode.

cathode
anode

When the electric circuit is completed, positive hydronium ions (H_3O^+) migrate to the cathode where they pick up electrons and evolve as hydrogen gas. At the same time the negative chloride ions (Cl^-) migrate to the anode, where they lose electrons and evolve as chlorine gas.

Reaction at the cathode:

$$H_3O^+ + 1\,e^- \longrightarrow H^0 + H_2O \qquad \text{(reduction)}$$
$$H^0 + H^0 \longrightarrow H_2$$

FIGURE 17.3 ▶
During the electrolysis of a hydrochloric acid solution, positive hydronium ions are attracted to the cathode, where they gain electrons and form hydrogen gas. Chloride ions migrate to the anode, where they lose electrons and form chlorine gas. The equation for this process is $2\,HCl(aq) \longrightarrow H_2(g) + Cl_2(g)$.

Reaction at the anode:

$$Cl^- \longrightarrow Cl^0 + 1\ e^- \qquad \text{(oxidation)}$$

$$Cl^0 + Cl^0 \longrightarrow Cl_2$$

$$2\ HCl(aq) \xrightarrow{\text{electrolysis}} H_2(g) + Cl_2(g) \qquad \text{net reaction}$$

Note that oxidation–reduction has taken place. Chloride ions lost electrons (were oxidized) at the anode, and hydronium ions gained electrons (were reduced) at the cathode.

Oxidation always occurs at the anode and reduction at the cathode.

When concentrated sodium chloride solutions (brines) are electrolyzed, the products are sodium hydroxide, hydrogen, and chlorine. The overall reaction is

$$2\ Na^+(aq) + 2\ Cl^-(aq) + 2\ H_2O(l) \xrightarrow{\text{electrolysis}}$$
$$2\ Na^+(aq) + 2\ OH^-(aq) + H_2(g) + Cl_2(g)$$

The net ionic equation is

$$2\ Cl^-(aq) + 2\ H_2O(l) \longrightarrow 2\ OH^-(aq) + H_2(g) + Cl_2(g)$$

During electrolysis, Na^+ ions move toward the cathode and Cl^- ions move toward the anode. The anode reaction is similar to that of hydrochloric acid; the chlorine is liberated:

$$2\ Cl^-(aq) \longrightarrow Cl_2(g) + 2\ e^-$$

Even though Na^+ ions are attracted by the cathode, the facts show that hydrogen is liberated there. No evidence of metallic sodium is found, but the area around the cathode tests alkaline from the accumulated OH^- ions. The reaction at the cathode is

$$2\ H_2O(l) + 2\ e^- \longrightarrow H_2(g) + 2\ OH^-(aq)$$

If electrolysis is allowed to continue until all the chloride is reacted, the solution remaining will contain only sodium hydroxide, which on evaporation yields solid NaOH. Large amounts of sodium hydroxide and chlorine are made by this process.

When molten sodium chloride (without water) is subjected to electrolysis, metallic sodium and chlorine gas are formed:

$$2\ Na^+(l) + 2\ Cl^-(l) \xrightarrow{\text{electrolysis}} 2\ Na(l) + Cl_2(g)$$

An important electrochemical application is the electroplating of metals. Electroplating is the art of covering a surface or an object with a thin adherent electrodeposited metal coating. Electroplating is done for protection of the surface of the base metal or for a purely decorative effect. The layer deposited is surprisingly thin, varying from as little as 5×10^{-5} cm to 2×10^{-3} cm, depending on the metal and the intended use. The object to be plated is set up as the cathode and is immersed in a solution containing ions of the plating metal. When an electric current passes through the solution, metal ions that migrate to the cathode are reduced, depositing on the object as the free metal. In most cases the metal deposited on the object is replaced

Electroplating is a process ▶
involving oxidation–reduction
reactions. It can be used to
protect an object or for decorative effect.

in the solution by using an anode of the same metal. The following equations show
the chemical changes in the electroplating of nickel:

Reaction at the cathode: $Ni^{2+}(aq) + 2\,e^- \longrightarrow Ni(s)$ Ni plated out
on an object

Reaction at the anode: $Ni(s) \longrightarrow Ni^{2+}(aq) + 2\,e^-$ Ni replenished
in solution

Metals commonly used in commercial electroplating are copper, nickel, zinc, lead,
cadmium, chromium, tin, gold, and silver.

In the electrolytic cell shown in Figure 17.3, electrical energy from the voltage
source is used to bring about nonspontaneous redox reactions. The hydrogen and
chlorine produced have more potential energy than was present in the hydrochloric
acid before electrolysis.

Conversely, some spontaneous redox reactions can be made to supply useful
amounts of electrical energy. When a piece of zinc is put in a copper(II) sulfate solution, the zinc quickly becomes coated with metallic copper. We expect this coating to
happen because zinc is above copper in the activity series; copper(II) ions are therefore reduced by zinc atoms:

$$Zn^0(s) + Cu^{2+}(aq) \longrightarrow Zn^{2+}(aq) + Cu^0(s)$$

This reaction is clearly a spontaneous redox reaction, but simply dipping a zinc rod
into a copper(II) sulfate solution will not produce useful electric current. However,
when we carry out this reaction in the cell shown in Figure 17.4, an electric current
is produced. The cell consists of a piece of zinc immersed in a zinc sulfate solution
and connected by a wire through a voltmeter to a piece of copper immersed in
copper(II) sulfate solution. The two solutions are connected by a salt bridge. Such
a cell produces an electric current and a potential of about 1.1 volts when both

◀ **FIGURE 17.4**
Zinc–copper voltaic cell.
The cell has a potential of
1.1 volts when ZnSO₄ and
CuSO₄ solutions are 1.0 M.
The salt bridge provides
electrical contact between
the two half-cells.

solutions are 1.0 M in concentration. A cell that produces electric current from a spontaneous chemical reaction is called a **voltaic cell.** A voltaic cell is also known as a *galvanic cell.*

voltaic cell

The driving force responsible for the electric current in the zinc–copper cell originates in the great tendency of zinc atoms to lose electrons relative to the tendency of copper(II) ions to gain electrons. In the cell shown in Figure 17.4, zinc atoms lose electrons and are converted to zinc ions at the zinc electrode surface; the electrons flow through the wire (external circuit) to the copper electrode. Here copper(II) ions pick up electrons and are reduced to copper atoms, which plate out on the copper electrode. Sulfate ions flow from the $CuSO_4$ solution via the salt bridge into the $ZnSO_4$ solution (internal circuit) to complete the circuit. The equations for the reactions of this cell are

anode	$Zn^0(s) \longrightarrow Zn^{2+}(aq) + 2\,e^-$	(oxidation)
cathode	$Cu^{2+}(aq) + 2\,e^- \longrightarrow Cu^0(s)$	(reduction)
net ionic	$Zn^0(s) + Cu^{2+}(aq) \longrightarrow Zn^{2+}(aq) + Cu^0(s)$	
overall	$Zn(s) + CuSO_4(aq) \longrightarrow ZnSO_4(aq) + Cu(s)$	

The redox reaction, the movement of electrons in the metallic or external part of the circuit, and the movement of ions in the solution or internal part of the circuit of the copper–zinc cell are very similar to the actions that occur in the electrolytic cell of Figure 17.3. The only important difference is that the reactions of the zinc–copper cell are spontaneous. This spontaneity is the crucial difference between all voltaic and electrolytic cells.

Voltaic cells use chemical reactions to produce electrical energy, and electrolytic cells use electrical energy to produce chemical reactions.

Calculators, watches, radios, walkmen, and portable CD players are powered by small efficient voltaic cells called *dry cell batteries*. Called dry cells because they do not contain a liquid electrolyte (like the voltaic cells discussed earlier), dry cell batteries are found in several different versions.

The *acid*-type dry cell battery contains a zinc inner case that functions as the anode. A carbon (graphite) rod runs through the center and is in contact with the zinc case at one end and a moist paste of solid MnO_2, NH_4Cl, and carbon that functions as the cathode. (See Figure 17.5a.) The cell produces about 1.5 volts. The *alkaline*-type dry cell battery is the same as the acid type except the NH_4Cl is replaced by either KOH or NaOH. These dry cells typically last longer because the zinc anode corrodes more slowly in basic conditions. A third type of dry cell is the *zinc–mercury* cell shown in Figure 17.5b. The reactions occurring in this cell are

anode $Zn^0 + 2\ OH^- \longrightarrow ZnO + H_2O + 2\ e^-$ (oxidation)

cathode $HgO + H_2O + 2\ e^- \longrightarrow Hg^0 + 2\ OH^-$ (reduction)

net ionic $Zn^0 + Hg^{2+} \longrightarrow Zn^{2+} + Hg^0$

overall $Zn^0 + HgO \longrightarrow ZnO + Hg^0$

To offset the relatively high initial cost, this cell (a) provides current at a very steady potential of about 1.5 volts; (b) has an exceptionally long service life—that is, high energy output to weight ratio; (c) is completely self-contained; and (d) can be stored for relatively long periods of time when not in use.

An automobile storage battery is an energy reservoir. The charged battery acts as a voltaic cell and through chemical reactions furnishes electrical energy to operate the starter, lights, radio, and so on. When the engine is running, a generator or alternator produces and forces an electric current through the battery and, by electrolytic chemical action, restores it to the charged condition.

The cell unit consists of a lead plate filled with spongy lead and a lead dioxide plate, both immersed in dilute sulfuric acid solution, which serves as the electrolyte (see Figure 17.6). When the cell is discharging, or acting as a voltaic cell, these reactions occur:

Pb plate (anode): $Pb^0 \longrightarrow Pb^{2+} + 2\ e^-$ (oxidation)

PbO_2 plate (cathode): $PbO_2 + 4\ H^+ + 2\ e^- \longrightarrow Pb^{2+} + 2\ H_2O$ (reduction)

◀ **Figure 17.6**
**Cross-sectional diagram of a
lead storage battery cell.**

Net ionic redox reaction: $Pb^0 + PbO_2 + 4\,H^+ \longrightarrow 2\,Pb^{2+} + 2\,H_2O$

Precipitation reaction on
  plates: $Pb^{2+}(aq) + SO_4^{2-}(aq) \longrightarrow PbSO_4(s)$

Because lead(II) sulfate is insoluble, the Pb^{2+} ions combine with SO_4^{2-} ions to form
a coating of $PbSO_4$ on each plate. The overall chemical reaction of the cell is

$$Pb(s) + PbO_2(s) + 2\,H_2SO_4(aq) \xrightarrow[\text{cycle}]{\text{discharge}} 2\,PbSO_4(s) + 2\,H_2O(l)$$

The cell can be recharged by reversing the chemical reaction. This reversal is
accomplished by forcing an electric current through the cell in the opposite direction.
Lead sulfate and water are reconverted to lead, lead (IV) oxide, and sulfuric acid:

$$2\,PbSO_4(s) + 2\,H_2O(l) \xrightarrow[\text{cycle}]{\text{charge}} Pb(s) + PbO_2(s) + 2\,H_2SO_4(aq)$$

The electrolyte in a lead storage battery is a 38% by mass sulfuric acid solution
having a density of 1.29 g/mL. As the battery is discharged, sulfuric acid is removed,
thereby decreasing the density of the electrolyte solution. The state of charge or discharge of the battery can be estimated by measuring the density (or specific gravity)
of the electrolyte solution with a hydrometer. When the density has dropped to about
1.05 g/mL, the battery needs recharging.

In a commercial battery, each cell consists of a series of cell units of alternating
lead–lead(IV) oxide plates separated and supported by wood, glass wool, or fiber-
glass. The energy storage capacity of a single cell is limited, and its electrical potential is only about 2 volts. Therefore a bank of six cells is connected in series to
provide the 12-volt output of the usual automobile battery.

CHEMISTRY IN ACTION   A New Look for a Great Lady

Corrosion has been a problem for the world since the earliest use of metals. The complexities of corrosion are exemplified by our famous Statue of Liberty. The beautiful lady was given to the United States by France in 1886. She is constructed of a skeleton of iron bars (ribs) that were bent to precisely conform to her shape and covered with a thin skin of copper. Her skin is connected to the skeleton with special copper "saddles" or bands that overlap the iron bars and are riveted in place.

Iron will corrode (oxidize) in moist air and, if in contact with copper and an electrolyte (seawater), the corrosion rate can increase 100-fold. The harbor environment of New York fosters accelerated corrosion, and within less than a century the iron ribs had corroded to a point where the lady was in serious danger of collapse.

To repair and reinforce the statue, each of the iron bars (over 1300 in all) were replaced one at a time with stainless

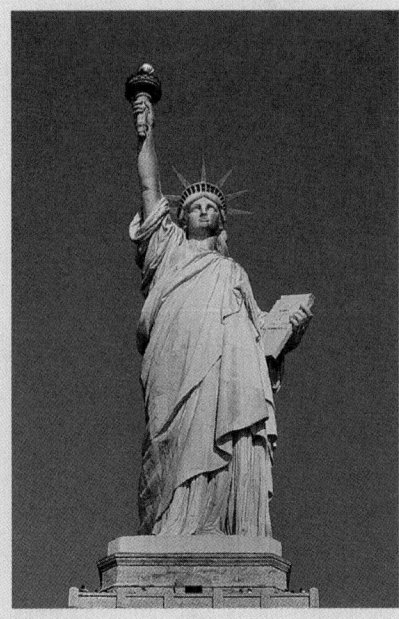

▲ **The Statue of Liberty was in serious danger of collapse before renovation.**

steel, which corrodes much less easily. At the same time, the copper saddles were coated with Teflon to insulate them from contact with the iron and further reduce future corrosion.

The external appearance of the statue was left essentially unchanged. The bluish green patina is a natural product formed as copper undergoes atmospheric corrosion. Patina may take several forms and exists as a protective film that adheres tightly to the surface of the copper and reduces the rate of corrosion underneath.

The patina of the Statue of Liberty is primarily $CuSO_4 \cdot 3\,Cu(OH)_2$. It has protected the lady well—during her hundred years, the copper skin has thinned only about 4%. Unfortunately, acid rain appears to be changing the patina to $CuSO_4 \cdot 2\,Cu(OH)_2$, which doesn't bond as closely to the copper surface. Chemists are concerned that this change will result in the loss of patina and increased corrosion of the copper skin.

Concepts in Review

1. Assign oxidation numbers to all the elements in a compound or ion.
2. Determine which element is being oxidized and which element is being reduced in an oxidation–reduction reaction.
3. Identify the oxidizing agent and the reducing agent in an oxidation–reduction reaction.
4. Balance oxidation–reduction equations in molecular and ionic forms.
5. Outline the general principles concerning the activity series of the metals.
6. Use the activity series to determine whether a proposed single-displacement reaction will occur.
7. Distinguish between an electrolytic and a voltaic cell.
8. Draw a voltaic cell that will produce electric current from an oxidation–reduction reaction involving two metals and their salts.
9. Identify the anode reaction and the cathode reaction in an electrolytic or voltaic cell.

Key Terms

activity series of metals (17.5)
anode (17.6)
cathode (17.6)
electrolysis (17.6)

electrolytic cell (17.6)
oxidation (17.2)
oxidation number (17.1)
oxidation–reduction (17.2)

oxidation state (17.1)
oxidizing agent (17.2)
redox (17.2)

reducing agent (17.2)
reduction (17.2)
voltaic cell (17.6)

Questions

1. In the equation

$$I_2 + 5 Cl_2 + 6 H_2O \longrightarrow 2 HIO_3 + 10 HCl$$

   (a) has iodine been oxidized or has it been reduced?
   (b) has chlorine been oxidized or has it been reduced? (Figure 17.1)

2. Which element of each pair is more active? (Table 17.3)
   (a) Ag or Al
   (b) Na or Ba
   (c) Ni or Cu

3. Will the following combinations react in aqueous solution? (Table 17.3)
   (a) $Zn + Cu^{2+}$ (e) $Ba + FeCl_2$
   (b) $Ag + H^+$ (f) $Pb + NaCl$
   (c) $Sn + Ag^+$ (g) $Ni + Hg(NO_3)_2$
   (d) $As + Mg^{2+}$ (h) $Al + CuSO_4$

4. The reaction between powdered aluminum and iron(III) oxide (in the thermite process) producing molten iron is very exothermic.
   (a) Write the equation for the chemical reaction that occurs.
   (b) Explain in terms of Table 17.3 why a reaction occurs.
   (c) Would you expect a reaction between powdered iron and aluminum oxide?
   (d) Would you expect a reaction between powdered aluminum and chromium(III) oxide?

5. Write equations for the chemical reaction of aluminum, chromium, gold, iron, copper, magnesium, mercury, and zinc with dilute solutions of (a) hydrochloric acid and (b) sulfuric acid. If a reaction will not occur, write "no reaction" as the product. (Table 17.3)

6. An $NiCl_2$ solution is placed in the apparatus shown in Figure 17.3, instead of the HCl solution shown. Write equations for
   (a) the anode reaction
   (b) the cathode reaction
   (c) the net electrochemical reaction

7. What is the major distinction between the reactions occurring in Figures 17.3 and 17.4?

8. In the cell shown in Figure 17.4,
   (a) what would be the effect of removing the voltmeter and connecting the wires shown coming to the voltmeter?
   (b) what would be the effect of removing the salt bridge?

9. Why are oxidation and reduction said to be complementary processes?

10. When molten $CaBr_2$ is electrolyzed, calcium metal and bromine are produced. Write equations for the two half-reactions that occur at the electrodes. Label the anode half-reaction and the cathode half-reaction.

11. Why is direct current used instead of alternating current in the electroplating of metals?

12. What property of lead(IV) oxide and lead(II) sulfate makes it unnecessary to have salt bridges in the cells of a lead storage battery?

13. Explain why the density of the electrolyte in a lead storage battery decreases during the discharge cycle.

14. In one type of alkaline cell used to power devices such as portable radios, Hg^{2+} ions are reduced to metallic mercury when the cell is being discharged. Does this reduction occur at the anode or the cathode? Explain.

15. Differentiate between an electrolytic cell and a voltaic cell.

16. Why is a porous barrier or a salt bridge necessary in some voltaic cells?

Paired Exercises

17. What is the oxidation number of the underlined element in each compound?
    (a) NaCl
    (b) FeCl$_3$
    (c) PbO$_2$
    (d) NaNO$_3$
    (e) H$_2$SO$_3$
    (f) NH$_4$Cl

18. What is the oxidation number of the underlined element in each compound?
    (a) KMnO$_4$
    (b) I$_2$
    (c) NH$_3$
    (d) KClO$_3$
    (e) K$_2$CrO$_4$
    (f) K$_2$Cr$_2$O$_7$

19. What is the oxidation number of the underlined elements?
    (a) $\underline{S}^{2-}$
    (b) $\underline{N}O_2^-$
    (c) $Na_2\underline{O}_2$
    (d) $\underline{Bi}^{3+}$

20. What is the oxidation number of the underlined elements?
    (a) $\underline{O}_2$
    (b) $\underline{As}O_4^{3-}$
    (c) $Fe(\underline{OH})_3$
    (d) $\underline{I}O_3^-$

21. In the following half-reactions, which element is changing oxidation state? Is the half-reaction an oxidation or a reduction? Supply the proper number of electrons to each side to balance each equation.
    (a) $Zn^{2+} \longrightarrow Zn$
    (b) $2\,Br^- \longrightarrow Br_2$
    (c) $MnO_4^- + 8\,H^+ \longrightarrow Mn^{2+} + 4\,H_2O$
    (d) $Ni \longrightarrow Ni^{2+}$

22. In the following half-reactions, which element is changing oxidation state? Is the half-reaction an oxidation or a reduction? Supply the proper number of electrons to each side to balance each equation:
    (a) $SO_3^{2-} + H_2O \longrightarrow SO_4^{2-} + 2\,H^+$
    (b) $NO_3^- + 4\,H^+ \longrightarrow NO + 2\,H_2O$
    (c) $S_2O_4^{2-} + 2\,H_2O \longrightarrow 2\,SO_3^{2-} + 4\,H^+$
    (d) $Fe^{2+} \longrightarrow Fe^{3+}$

23. In the following unbalanced equations, identify
    (a) the oxidized element and the reduced element
    (b) the oxidizing agent and the reducing agent
        (1) $Cr + HCl \longrightarrow CrCl_3 + H_2$
        (2) $SO_4^{2-} + I^- + H^+ \longrightarrow H_2S + I_2 + H_2O$

24. In the following unbalanced equations, identify
    (a) the oxidized element and the reduced element
    (b) the oxidizing agent and the reducing agent
        (1) $AsH_3 + Ag^+ + H_2O \longrightarrow H_3AsO_4 + Ag + H^+$
        (2) $Cl_2 + NaBr \longrightarrow NaCl + Br_2$

25. Balance these equations using the change-in-oxidation-number method:
    (a) $Zn + S \longrightarrow ZnS$
    (b) $AgNO_3 + Pb \longrightarrow Pb(NO_3)_2 + Ag$
    (c) $Fe_2O_3 + CO \longrightarrow Fe + CO_2$
    (d) $H_2S + HNO_3 \longrightarrow S + NO + H_2O$
    (e) $MnO_2 + HBr \longrightarrow MnBr_2 + Br_2 + H_2O$

26. Balance these equations using the change-in-oxidation-number method:
    (a) $Cl_2 + KOH \longrightarrow KCl + KClO_3 + H_2O$
    (b) $Ag + HNO_3 \longrightarrow AgNO_3 + NO + H_2O$
    (c) $CuO + NH_3 \longrightarrow N_2 + Cu + H_2O$
    (d) $PbO_2 + Sb + NaOH \longrightarrow PbO + NaSbO_2 + H_2O$
    (e) $H_2O_2 + KMnO_4 + H_2SO_4 \longrightarrow$
    $$O_2 + MnSO_4 + K_2SO_4 + H_2O$$

27. Balance these ionic redox equations using the ion–electron method. These reactions occur in acidic solution.
    (a) $Zn + NO_3^- \longrightarrow Zn^{2+} + NH_4^+$
    (b) $NO_3^- + S \longrightarrow NO_2 + SO_4^{2-}$
    (c) $PH_3 + I_2 \longrightarrow H_3PO_2 + I^-$
    (d) $Cu + NO_3^- \longrightarrow Cu^{2+} + NO$
    *(e) $ClO_3^- + Cl^- \longrightarrow Cl_2$

28. Balance these ionic redox equations using the ion–electron method. These reactions occur in acidic solution.
    (a) $ClO_3^- + I^- \longrightarrow I_2 + Cl^-$
    (b) $Cr_2O_7^{2-} + Fe^{2+} \longrightarrow Cr^{3+} + Fe^{3+}$
    (c) $MnO_4^- + SO_2 \longrightarrow Mn^{2+} + SO_4^{2-}$
    (d) $H_3AsO_3 + MnO_4^- \longrightarrow H_3AsO_4 + Mn^{2+}$
    *(e) $Cr_2O_7^{2-} + H_3AsO_3 \longrightarrow Cr^{3+} + H_3AsO_4$

29. Balance these ionic redox equations using the ion–electron method. These reactions occur in basic solutions.
    (a) $Cl_2 + IO_3^- \longrightarrow Cl^- + IO_4^-$
    (b) $MnO_4^- + ClO_2^- \longrightarrow MnO_2 + ClO_4^-$
    (c) $Se \longrightarrow Se^{2-} + SeO_3^{2-}$
    *(d) $Fe_3O_4 + MnO_4^- \longrightarrow Fe_2O_3 + MnO_2$
    *(e) $BrO^- + Cr(OH)_4^- \longrightarrow Br^- + CrO_4^{2-}$

30. Balance these ionic redox equations using the ion–electron method. These reactions occur in basic solutions.
    (a) $MnO_4^- + SO_3^{2-} \longrightarrow MnO_2 + SO_4^{2-}$
    (b) $ClO_2 + SbO_2^- \longrightarrow ClO_2^- + Sb(OH)_6^-$
    (c) $Al + NO_3^- \longrightarrow NH_3 + Al(OH)_4^-$
    *(d) $P_4 \longrightarrow HPO_3^{2-} + PH_3$
    *(e) $Al + OH^- \longrightarrow Al(OH)_4^- + H_2$

Additional Exercises

31. The chemical reactions taking place during discharge in a lead storage battery are

    $$Pb + SO_4^{2-} \longrightarrow PbSO_4$$
    $$PbO_2 + SO_4^{2-} + 4\,H^+ \longrightarrow PbSO_4 + 2\,H_2O$$

    (a) Complete each half-reaction by supplying electrons.
    (b) Which reaction is oxidation and which is reduction?
    (c) Which reaction occurs at the anode of the battery?

32. Use this unbalanced redox equation

    $$KMnO_4 + HCl \longrightarrow KCl + MnCl_2 + H_2O + Cl_2$$

    to indicate
    (a) the oxidizing agent
    (b) the reducing agent
    (c) the number of electrons that are transferred per mole of oxidizing agent

33. How many moles of NO gas will be formed by the reaction of 25.0 g of silver with nitric acid?

$$Ag + HNO_3 \longrightarrow AgNO_3 + NO + H_2O \text{ (acid solution)}$$

34. What volume of chlorine gas, measured at STP, is required to react with excess KOH to form 0.300 mol $KClO_3$?

$$Cl_2 + KOH \longrightarrow KCl + KClO_3 + H_2O \text{ (acid solution)}$$

35. What mass of $KMnO_4$ is needed to react with 100. mL H_2O_2 solution? ($d = 1.031$ g/mL, 9.0% H_2O_2 by mass)

$$H_2O_2 + KMnO_4 + H_2SO_4 \longrightarrow$$
$$O_2 + MnSO_4 + K_2SO_4 + H_2O \qquad \text{(acid solution)}$$

*36. What volume of 0.200 M $K_2Cr_2O_7$ will be required to oxidize 5.00 g H_3AsO_3?

$$Cr_2O_7^{2-} + H_3AsO_3 \longrightarrow Cr^{3+} + H_3AsO_4 \text{ (acid solution)}$$

*37. What volume of 0.200 M $K_2Cr_2O_7$ will be required to oxidize the Fe^{2+} ion in 60.0 mL of 0.200 M $FeSO_4$ solution?

$$Cr_2O_7^{2-} + Fe^{2+} \longrightarrow Cr^{3+} + Fe^{3+} \qquad \text{(acid solution)}$$

*38. A sample of crude potassium iodide was analyzed using this reaction (not balanced):

$$I^- + SO_4^{2-} \longrightarrow I_2 + H_2S \text{ (acid solution)}$$

If a 4.00-g sample of crude KI produced 2.79 g of iodine, what is the percent purity of the KI?

*39. What volume of NO gas, measured at 28°C and 744 torr, will be formed by the reaction of 0.500 mol Ag reacting with excess nitric acid?

$$Ag + HNO_3 \longrightarrow AgNO_3 + NO + H_2O \text{ (acid solution)}$$

40. How many moles of H_2 can be produced from 100.0 g Al according to this reaction?

$$Al + OH^- \longrightarrow Al(OH)_4^- + H_2 \text{ (basic solution)}$$

41. There is something incorrect about these half-reactions:
(a) $Cu^+ + e^- \longrightarrow Cu^{2+}$
(b) $Pb^{2+} + e^{2-} \longrightarrow Pb$
Identify what is wrong, and correct it.

42. Why can oxidation *never* occur without reduction?

43. The following observations were made concerning metals A, B, C, and D.
(a) When a strip of metal A is placed in a solution of B^{2+} ions, no reaction is observed.
(b) Similarly, A in a solution containing C^+ ions produces no reaction.
(c) When a strip of metal D is placed in a solution of C^+ ions, black metallic C deposits on the surface of D, and the solution tests positively for D^{2+} ions.

(d) When a piece of metallic B is placed in a solution of D^{2+} ions, metallic D appears on the surface of B and B^{2+} ions are found in the solution.
Arrange the ions, A^+, B^{2+}, C^+, and D^{2+}, in order of their ability to attract electrons. List them in order of increasing ability.

44. Tin normally has oxidation numbers of 0, +2, and +4. Which of these species can be an oxidizing agent, which can be a reducing agent, and which can be both? In each case what product would you expect as the tin reacts?

45. Manganese is an element that can exist in numerous oxidation states. In each of these compounds identify the oxidation number of the manganese. Which compound would you expect to be the best oxidizing agent and why?
(a) $Mn(OH)_2$
(b) MnF_3
(c) MnO_2
(d) K_2MnO_4
(e) $KMnO_4$

46. Which equations represent oxidations?
(a) $Mg \longrightarrow Mg^{2+}$
(b) $SO_2 \longrightarrow SO_3$
(c) $KMnO_4 \longrightarrow MnO_2$
(d) $Cl_2O_3 \longrightarrow Cl^-$

*47. In the following equation, note the reaction between manganese(IV) oxide and bromide ions:

$$MnO_2 + Br^- \longrightarrow Br_2 + Mn^{2+}$$

(a) Balance this redox reaction in acidic solution.
(b) How many grams of MnO_2 would be needed to produce 100.0 mL of 0.05 M Mn^{2+}?
(c) How many liters of bromine vapor at 50°C and 1.4 atm would also result?

48. Use the table shown to complete the following reactions. If no reaction occurs, write NR:

(a) $F_2 + Cl^- \longrightarrow$
(b) $Br_2 + Cl^- \longrightarrow$
(c) $I_2 + Cl^- \longrightarrow$
(d) $Br_2 + I^- \longrightarrow$

Activity
ease of reduction ↑ F_2
Cl_2
Br_2
I_2

49. Manganese metal reacts with HCl to give hydrogen gas and the Mn^{2+} ion in solution. Write a balanced equation for the reaction.

50. If zinc is allowed to react with dilute nitric acid, zinc is oxidized to the +2 ion, while the nitrate ion can be reduced to ammonium, NH_4^+. Write a balanced equation for the reaction in acidic solution.

51. In the following equations, identify the
   (a) atom or ion oxidized
   (b) atom or ion reduced
   (c) oxidizing agent
   (d) reducing agent
   (e) change in oxidation number associated with each oxidizing process
   (f) change in oxidation number associated with each reducing process

   (1) $C_3H_8 + O_2 \longrightarrow CO_2 + H_2O$
   (2) $HNO_3 + H_2S \longrightarrow NO + S + H_2O$
   (3) $CuO + NH_3 \longrightarrow N_2 + H_2O + Cu$
   (4) $H_2O_2 + Na_2SO_3 \longrightarrow Na_2SO_4 + H_2O$
   (5) $H_2O_2 \longrightarrow H_2O + O_2$

52. In the galvanic cell shown in the diagram, a strip of silver is placed in a solution of silver nitrate, and a strip of lead is placed in a solution of lead(II) nitrate. The two beakers are connected with a salt bridge. Determine
   (a) the anode
   (b) the cathode
   (c) where oxidation occurs
   (d) where reduction occurs
   (e) which direction electrons flow through the wire
   (f) which direction ions flow through the solution

Answers to Practice Exercises

17.1 (a) $S = +6$, (b) $As = +5$, (c) $C = +4$
17.2 (*Note*: $H = +1$ even though it comes second in the formula; N is a nonmetal.)
   (a) $N = -3$, (b) $Cr = +6$, (c) $P = +5$
17.3 (a) $Be = +2; Cl = -1$, (b) $H = +1; Cl = +1; O = -2$,
   (c) $H = +1; O = -1$, (d) $N = -3; H = +1$,
   (e) $Br = +5; O = -2$
17.4 (a) $6\,HNO_3 + S \longrightarrow 6\,NO_2 + H_2SO_4 + 2\,H_2O$
   (b) $2\,CrCl_3 + 3\,MnO_2 + 2\,H_2O \longrightarrow 3\,MnCl_2 + 2\,H_2CrO_4$,
   (c) $2\,KMnO_4 + 6\,HCl + 5\,H_2S \longrightarrow$
   $\qquad\qquad\qquad 2\,KCl + 2\,MnCl_2 + 5\,S + 8\,H_2O$

17.5 (a) $4\,H^+ + 2\,I^- + 2\,NO_2^- \longrightarrow I_2 + 2\,NO + 2\,H_2O$
   (b) $2\,OH^- + Cl_2 + IO_3^- \longrightarrow IO_4^- + H_2O + 2\,Cl^-$
   (c) $AuCl_4^- + Sn^{2+} \longrightarrow Sn^{4+} + AuCl + 3\,Cl^-$
17.6 (a) $Zn + 2\,H_2O + 2\,OH^- \longrightarrow Zn(OH)_4^{2-} + H_2$
   (b) $H_2O_2 + Sn^{2+} + 2\,H^+ \longrightarrow Sn^{4+} + 2\,H_2O$
   (c) $Cu + Cu^{2+} + 2\,OH^- \longrightarrow Cu_2O + H_2O$
17.7 (a) yes, (b) no, (c) no

Multiple Choice: *Choose the correct answer to each of the following.*

1. When the reaction

   $$Al + HCl \longrightarrow$$

   is completed and balanced, this term appears in the balanced equation:
   (a) $3\,HCl$ (b) $AlCl_2$ (c) $3\,H_2$ (d) $4\,Al$

2. When the reaction

   $$CaO + HNO_3 \longrightarrow$$

   is completed and balanced, this term appears in the balanced equation:
   (a) H_2 (b) $2\,H_2$ (c) $2\,CaNO_3$ (d) H_2O

3. When the reaction

   $$H_3PO_4 + KOH \longrightarrow$$

   is completed and balanced, this term appears in the balanced equation:
   (a) H_3PO_4 (c) KPO_4
   (b) $6\,H_2O$ (d) $3\,KOH$

4. When the reaction

   $$HCl + Cr_2(CO_3)_3 \longrightarrow$$

   is completed and balanced, this term appears in the balanced equation:
   (a) Cr_2Cl (b) $3\,HCl$ (c) $3\,CO_2$ (d) H_2O

5. Which of these is not a salt?
   (a) $K_2Cr_2O_7$ (c) $Ca(OH)_2$
   (b) $NaHCO_3$ (d) $Na_2C_2O_4$

6. Which of these is not an acid?
   (a) H_3PO_4 (b) H_2S (c) H_2SO_4 (d) NH_3

7. Which of these is a weak electrolyte?
   (a) NH_4OH (c) K_3PO_4
   (b) $Ni(NO_3)_2$ (d) $NaBr$

8. Which of these is a nonelectrolyte?
   (a) $HC_2H_3O_2$ (c) $KMnO_4$
   (b) $MgSO_4$ (d) CCl_4

9. Which of these is a strong electrolyte?
   (a) H_2CO_3 (c) NH_4OH
   (b) HNO_3 (d) H_3BO_3

10. Which of these is a weak electrolyte?
    (a) $NaOH$ (c) $HC_2H_3O_2$
    (b) $NaCl$ (d) H_2SO_4

11. A solution has an H^+ concentration of $3.4 \times 10^{-5}\,M$. The pH is
    (a) 4.47 (b) 5.53 (c) 3.53 (d) 5.47

12. A solution with a pH of 5.85 has an H^+ concentration of
    (a) $7.1 \times 10^{-5}\,M$ (c) $3.8 \times 10^{-4}\,M$
    (b) $7.1 \times 10^{-6}\,M$ (d) $1.4 \times 10^{-6}\,M$

13. If 16.55 mL of 0.844 M NaOH is required to titrate 10.00 mL of a hydrochloric acid solution, the molarity of the acid solution is
    (a) 0.700 M (c) 1.40 M
    (b) 0.510 M (d) 0.255 M

14. What volume of 0.462 M NaOH is required to neutralize 20.00 mL of 0.391 M HNO_3?
    (a) 23.6 mL (c) 9.03 mL
    (b) 16.9 mL (d) 11.8 mL

15. 25.00 mL of H_2SO_4 solution requires 18.92 mL of 0.1024 M NaOH for complete neutralization. The molarity of the acid is
    (a) 0.1550 M (c) 0.07750 M
    (b) 0.03875 M (d) 0.06765 M

16. Dilute hydrochloric acid is a typical acid, as shown by its
    (a) color (b) odor (c) solubility (d) taste

17. What is the pH of a 0.00015 M HCl solution?
    (a) 4.0 (c) between 3 and 4
    (b) 2.82 (d) no correct answer given

18. The chloride ion concentration in 300. mL of 0.10 M $AlCl_3$ is
    (a) 0.30 M (c) 0.030 M
    (b) 0.10 M (d) 0.90 M

19. The amount of $BaSO_4$ that will precipitate when 100. mL of 0.10 M $BaCl_2$ and 100. mL of 0.10 M Na_2SO_4 are mixed is
    (a) 0.010 mol (c) 23 g
    (b) 0.10 mol (d) no correct answer given

20. The freezing point of a 0.50 m NaCl aqueous solution will be about
    (a) $-1.86°C$ (c) $-2.79°C$
    (b) $-0.93°C$ (d) no correct answer given

21. The equation

    $$HC_2H_3O_2 + H_2O \rightleftharpoons H_3O^+ + C_2H_3O_2^-$$ implies that

    (a) If you start with 1.0 mol $HC_2H_3O_2$, 1.0 mol H_3O^+ and 1.0 mol $C_2H_3O_2^-$ will be produced.
    (b) An equilibrium exists between the forward reaction and the reverse reaction.
    (c) At equilibrium, equal molar amounts of all four substances will exist.
    (d) The reaction proceeds all the way to the products, then reverses, going all the way back to the reactants.

22. If the reaction A + B $\rightleftharpoons$ C + D is initially at equilibrium, and then more A is added, which of the following is not true?
(a) More collisions of A and B will occur; the rate of the forward reaction will thus be increased.
(b) The equilibrium will shift toward the right.
(c) The moles of B will be increased.
(d) The moles of D will be increased.

23. What will be the H^+ concentration in a 1.0 M HCN solution? ($K_a = 4.0 \times 10^{-10}$)
(a) $2.0 \times 10^{-5} M$ (c) $4.0 \times 10^{-10} M$
(b) 1.0 M (d) $2.0 \times 10^{-10} M$

24. What is the percent ionization of HCN in Exercise 23?
(a) 100%   (c) $2.0 \times 10^{-3}\%$
(b) $2.0 \times 10^{-8}\%$ (d) $4.0 \times 10^{-8}\%$

25. If $[H^+] = 1 \times 10^{-5} M$, which of the following is not true?
(a) pH = 5   (c) $[OH^-] = 1 \times 10^{-5} M$
(b) pOH = 9   (d) The solution is acidic.

26. If $[H^+] = 2.0 \times 10^{-4} M$, then $[OH^-]$ will be
(a) $5.0 \times 10^{-9} M$ (c) $2.0 \times 10^{-4} M$
(b) 3.70   (d) $5.0 \times 10^{-11} M$

27. The solubility product of $PbCrO_4$ is 2.8×10^{-13}. The solubility of $PbCrO_4$ is
(a) $5.3 \times 10^{-7} M$ (c) $7.8 \times 10^{-14} M$
(b) $2.8 \times 10^{-13} M$ (d) 1.0 M

28. The solubility of AgBr is $6.3 \times 10^{-7} M$. The value of the solubility product is
(a) 6.3×10^{-7} (c) 4.0×10^{-48}
(b) 4.0×10^{-13} (d) 4.0×10^{-15}

29. Which of these solutions would be the best buffer solution?
(a) $0.10 M HC_2H_3O_2 + 0.10 M NaC_2H_3O_2$
(b) $0.10 M$ HCl
(c) $0.10 M$ HCl + $0.10 M$ NaCl
(d) pure water

30. For the reaction $H_2(g) + I_2(g) \rightleftharpoons 2 HI(g)$, at 700 K, $K_{eq} = 56.6$. If an equilibrium mixture at 700 K was found to contain 0.55 M HI and 0.21 M H_2, the I_2 concentration must be
(a) 0.046 M (c) 22 M
(b) 0.025 M (d) 0.21 M

31. The equilibrium constant for the reaction
$2 A + B \rightleftharpoons 3 C + D$ is
(a) $\dfrac{[C]^3[D]}{[A]^2[B]}$ (c) $\dfrac{[3C][D]}{[2A][B]}$
(b) $\dfrac{[2A][B]}{[3C][D]}$ (d) $\dfrac{[A]^2[B]}{[C]^3[D]}$

32. In the equilibrium represented by
$N_2(g) + O_2(g) \rightleftharpoons 2 NO_2(g)$
as the pressure is increased, the amount of NO_2 formed
(a) increases   (b) decreases
(c) remains the same
(d) increases and decreases irregularly

33. Which factor will not increase the concentration of ammonia as represented by this equation?
$3 H_2(g) + N_2(g) \rightleftharpoons 2 NH_3(g) + 92.5$ kJ
(a) increasing the temperature
(b) increasing the concentration of N_2
(c) increasing the concentration of H_2
(d) increasing the pressure

34. If $HCl(g)$ is added to a saturated solution of AgCl, the concentration of Ag^+ in solution
(a) increases
(b) decreases
(c) remains the same
(d) increases and decreases irregularly

35. The solubility of $CaCO_3$ at 20°C is 0.013 g/L. What is the K_{sp} for $CaCO_3$?
(a) 1.3×10^{-8} (c) 1.7×10^{-8}
(b) 1.3×10^{-4} (d) 1.7×10^{-4}

36. The K_{sp} for $BaCrO_4$ is 8.5×10^{-11}. What is the solubility of $BaCrO_4$ in grams per liter?
(a) 9.2×10^{-6} (c) 2.3×10^{-3}
(b) 0.073   (d) 8.5×10^{-11}

37. What will be the $[Ba^{2+}]$ when 0.010 mol Na_2CrO_4 is added to 1.0 L of saturated $BaCrO_4$ solution? See Exercise 36 for K_{sp}.
(a) $8.5 \times 10^{-11} M$ (c) $9.2 \times 10^{-6} M$
(b) $8.5 \times 10^{-9} M$ (d) $9.2 \times 10^{-4} M$

38. Which would occur if a small amount of sodium acetate crystals, $NaC_2H_3O_2$, were added to 100 mL of 0.1 M $HC_2H_3O_2$ at constant temperature?
(a) The number of acetate ions in the solution would decrease.
(b) The number of acetic acid molecules would decrease.
(c) The number of sodium ions in solution would decrease.
(d) The H^+ concentration in the solution would decrease.

39. If the temperature is decreased for the endothermic reaction
$A + B \rightleftharpoons C + D$
which of the following is true?
(a) The concentration of A will increase.
(b) No change will occur.
(c) The concentration of B will decrease.
(d) The concentration of D will increase.

40. In K_2SO_4, the oxidation number of sulfur is
(a) +2 (b) +4 (c) +6 (d) −2

41. In $Ba(NO_3)_2$, the oxidation number of N is
(a) +5 (b) −3 (c) +4 (d) −1

42. In the reaction

$$H_2S + 4\,Br_2 + 4\,H_2O \longrightarrow H_2SO_4 + 8\,HBr$$

the oxidizing agent is
(a) H_2S (b) Br_2 (c) H_2O (d) H_2SO_4

43. In the reaction

$$VO_3^- + Fe^{2+} + 4\,H^+ \longrightarrow VO^{2+} + Fe^{3+} + 2\,H_2O$$

the element reduced is
(a) V (b) Fe (c) O (d) H

Questions 44–46 pertain to the activity series.

K Ca Mg Al Zn Fe H Cu Ag

44. Which of these pairs will not react in water solution?
(a) Zn, $CuSO_4$ (c) Fe, $AgNO_3$
(b) Cu, $Al_2(SO_4)_3$ (d) Mg, $Al_2(SO_4)_3$

45. Which element is the most easily oxidized?
(a) K (b) Mg (c) Zn (d) Cu

46. Which element will reduce Cu^{2+} to Cu but will not reduce Zn^{2+} to Zn?
(a) Fe (b) Ca (c) Ag (d) Mg

47. In the electrolysis of fused (molten) $CaCl_2$, the product at the negative electrode is
(a) Ca^{2+} (b) Cl^- (c) Cl_2 (d) Ca

48. In its reactions, a free element from Group IIA in the periodic table is most likely to
(a) be oxidized   (c) be unreactive
(b) be reduced   (d) gain electrons

49. In the partially balanced redox equation

$$3\,Cu + HNO_3 \longrightarrow 3\,Cu(NO_3)_2 + 2\,NO + H_2O$$

the coefficient needed to balance H_2O is
(a) 8 (b) 6 (c) 4 (d) 3

50. Which reaction does not involve oxidation–reduction?
(a) burning sodium in chlorine
(b) chemical union of Fe and S
(c) decomposition of $KClO_3$
(d) neutralization of NaOH with H_2SO_4

51. How many moles of Fe^{2+} can be oxidized to Fe^{3+} by 2.50 mol Cl_2 according to this equation?

$$Fe^{2+} + Cl_2 \longrightarrow Fe^{3+} + Cl^-$$

(a) 2.50 mol   (c) 1.00 mol
(b) 5.00 mol   (d) 22.4 mol

52. How many grams of sulfur can be produced in this reaction from 100 mL of 6.00 *M* HNO_3?

$$HNO_3 + H_2S \longrightarrow S + NO + H_2O$$

(a) 28.9 g (b) 19.3 g (c) 32.1 g (d) 289 g

53. Which of these ions can be reduced by H_2?
(a) Hg^{2+} (b) Sn^{2+} (c) Zn^{2+} (d) K^+

54. Which of the following is *not* true of a zinc–mercury cell?
(a) It provides current at a steady potential.
(b) It has a short service life.
(c) It is self-contained.
(d) It can be stored for long periods of time.

Balancing Oxidation–Reduction Equations
Balance each equation.

55. $P + HNO_3 \longrightarrow HPO_3 + NO + H_2O$

56. $MnSO_4 + PbO_2 + H_2SO_4 \longrightarrow$
$$HMnO_4 + PbSO_4 + H_2O$$

57. $Cr_2O_7^{2-} + Cl^- \longrightarrow Cr^{3+} + Cl_2$ (acidic solution)

58. $MnO_4^- + AsO_3^{3-} \longrightarrow Mn^{2+} + AsO_4^{3-}$ (acidic solution)

59. $S^{2-} + Cl_2 \longrightarrow SO_4^{2-} + Cl^-$ (basic solution)

60. $Zn + NO_3^- \longrightarrow Zn(OH)_4^{2-} + NH_3$ (basic solution)

61. $KOH + Cl_2 \longrightarrow KCl + KClO + H_2O$ (basic solution)

62. $As + ClO_3^- \longrightarrow H_3AsO_3 + HClO$ (acidic solution)

63. $MnO_4^- + Cl^- \longrightarrow Mn^{2+} + Cl_2$ (acidic solution)

64. $H_2O_2 + Cl_2O_7 \longrightarrow ClO_2^- + O_2$ (basic solution)

Free Response Questions

1. You are investigating the properties of two new metallic elements found on Pluto, Bz and Yz. Bz reacts with aqueous HCl to produce hydrogen gas and $BzCl_3$. Yz has no reaction with aqueous HCl. However, the formula for the compound it forms with chlorine is $YzCl_2$. Write the balanced reaction that should occur if a galvanic cell is set up with Bz and Yz electrodes in solutions containing the metallic ions.

2. Suppose that 25 mL of an iron(II) nitrate solution is added to a beaker containing aluminum metal. Assume the reaction went to completion with no excess reagents. The solid iron produced was removed by filtration.
(a) Write a balanced redox equation for the reaction.
(b) For which solution, the initial iron(II) nitrate or the solution after the solid iron was filtered out, would the freezing point be lower? Explain your answer.

3. 50. mL of a 0.10 M HCl solution is poured equally into two flasks.
   (a) What is the pH of the HCl solution?
   (b) Next, 0.050 mol Zn is added to Flask A and 0.050 mol Cu is added to Flask B. Determine the pH of each solution after approximately 20 minutes.

4. (a) Write a balanced acid–base reaction that produces Na_2S.
   (b) If Na_2S is added to an aqueous solution of H_2S ($K_a = 9.1 \times 10^{-8}$), will the pH of the solution rise or fall? Explain.

5. (a) Would you expect a reaction to take place between HCN (aq) and $AgNO_3$ (aq) (K_{sp} for AgCN = 5.97×10^{-17})? Explain, and if a reaction occurs, write the net ionic equation.
   (b) If NaCN is added to distilled water, would you expect the solution to be acidic, basic, or neutral? Explain using any chemical equations that may be appropriate.

6. For each set of beakers below, draw a picture you might expect to see when the contents of the beakers are mixed together and allowed to react.

Na_2SO_4 (aq) $Pb(NO_3)_2$ (aq)

(a)

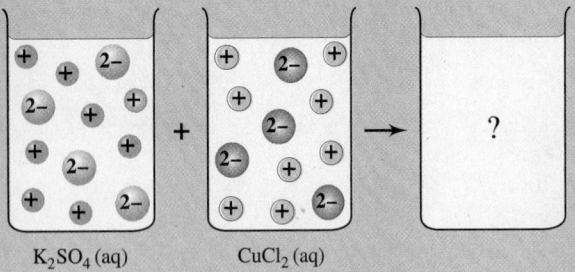

K_2SO_4 (aq) $CuCl_2$ (aq)

7. The picture below represents the equilibrium condition of
$$2A_3X \rightleftharpoons 2A_2X + A_2$$

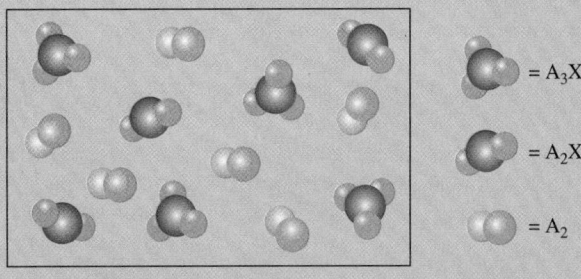

= A_3X

= A_2X

= A_2

   (a) What is the equilibrium constant?
   (b) Does the equilibrium lie to the left or to the right?
   (c) Do you think the reaction is a redox reaction? Explain your answer.

8. The picture below represents the equilibrium condition of the reaction $X_2 + 2G \rightleftharpoons X_2G_2$.

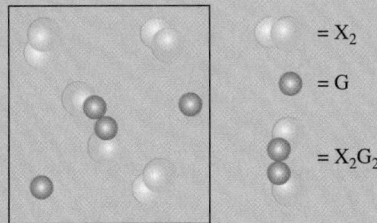

= X_2

= G

= X_2G_2

   (a) What is the equilibrium constant?
   (b) If the ratio of reactants to products increased when the temperature was raised, was the reaction exothermic or endothermic?
   (c) Provide a logical explanation for why the equilibrium shifts to the right when the pressure is increased at constant temperature.

9. The hydroxide ion concentration of a solution is 3.4×10^{-10} M. Balance the following equation:
$$Fe^{2+}(aq) + MnO_4^-(aq) \longrightarrow Fe^{3+}(aq) + Mn^{2+}(aq)$$

▲
Carbon dating is used to determine the age of civilizations who lived in Cliff Palace, at Mesa Verde, Colorado.

The nucleus of the atom is a source of tremendous energy. Harnessing this energy has enabled us to fuel power stations, treat cancer, and preserve food. We use isotopes in medicine to diagnose illness and to detect minute quantities of drugs or hormones. Researchers use radioactive tracers to sequence the human genome. We also use nuclear processes to detect explosives in luggage and to establish the age of objects such as human artifacts and rocks. In this chapter we'll consider properties of nuclei and their applications in our lives.

18.1   Discovery of Radioactivity

▲
X-ray technology opened the door for a new world of diagnosis and treatment.

In 1895, Wilhelm Konrad Roentgen (1845–1923) made an important breakthrough that eventually led to the discovery of radioactivity. Roentgen discovered X-rays when he observed that a vacuum discharge tube enclosed in a thin, black cardboard box caused a nearby piece of paper coated with the salt barium platinocyanide to glow with a brilliant phosphorescence. From this and other experiments he concluded that certain rays, which he called X-rays, were emitted from the discharge tube, penetrated the box, and caused the salt to glow. Roentgen's observations that X-rays could penetrate other bodies and affect photographic plates led to the development of X-ray photography.

Shortly after this discovery, Antoine Henri Becquerel (1852–1908) attempted to show a relationship between X-rays and the phosphorescence of uranium salts. In one experiment he wrapped a photographic plate in black paper, placed a sample of uranium salt on it, and exposed it to sunlight. The developed photographic plate showed that rays emitted from the salt had penetrated the paper. When Becquerel attempted to repeat the experiment, the sunlight was intermittent, so he placed the entire setup in a drawer. Several days later he developed the photographic plate, expecting to find it only slightly affected. To his amazement he found an intense image on the plate. He repeated the experiment in total darkness and obtained the same results, proving that the uranium salt emitted rays that affected the photographic plate without its being exposed to sunlight. Thus did the discovery of radioactivity come about, a combination of numerous experiments by the finest minds of the day—and serendipity. Becquerel later showed that the rays coming from uranium are able to ionize air and are also capable of penetrating thin sheets of metal.

The name *radioactivity* was coined two years later (in 1898) by Marie Curie. **Radioactivity** is the spontaneous emission of particles and/or rays from the nucleus of an atom. Elements having this property are said to be radioactive.

radioactivity

In 1898, Marie Sklodowska Curie (1867–1934) and her husband Pierre Curie (1859–1906) turned their research interests to radioactivity. In a short time the Curies discovered two new elements, polonium and radium, both of which are radioactive. To confirm their work on radium, they processed 1 ton of pitchblende residue ore to obtain 0.1 g of pure radium chloride, which they used to make further studies on the properties of radium and to determine its atomic mass.

In 1899, Ernest Rutherford began to investigate the nature of the rays emitted from uranium. He found two particles, which he called *alpha* and *beta particles*.

TABLE 18.1	Isotopic Notation for Several Particles (and Small Isotopes) Associated with Nuclear Chemistry		
Particle	**Symbol**	**Atomic number** Z	**Mass number** A
Neutron	$_{0}^{1}n$	0	1
Proton	$_{1}^{1}H$	1	1
Beta particle (electron)	$_{-1}^{0}e$	−1	0
Positron (positive electron)	$_{+1}^{0}e$	1	0
Alpha particle (helium nucleus)	$_{2}^{4}He$	2	4
Deuteron (heavy hydrogen nucleus)	$_{1}^{2}H$	1	2

Soon he realized that uranium, while emitting these particles, was changing into another element. By 1912, over 30 radioactive isotopes were known, and many more are known today. The *gamma ray,* a third type of emission from radioactive materials similar to an X-ray, was discovered by Paul Villard (1860–1934) in 1900. Rutherford's description of the nuclear atom led scientists to attribute the phenomenon of radioactivity to reactions taking place in the nuclei of atoms.

The symbolism and notation we described for isotopes in Chapter 5 is also very useful in nuclear chemistry:

For example, $_{92}^{238}U$ represents a uranium isotope with an atomic number of 92 and a mass number of 238. This isotope is also designated as U-238 or uranium-238 and contains 92 protons and 146 neutrons. The protons and neutrons collectively are known as **nucleons.** The mass number is the total number of nucleons in the nucleus. Table 18.1 shows the isotopic notations for several particles associated with nuclear chemistry.

nucleon

When we speak of isotopes, we mean atoms of the same element with different masses, such as $_{8}^{16}O$, $_{8}^{17}O$, $_{8}^{18}O$. In nuclear chemistry we use the term **nuclide** to mean any isotope of any atom. Thus $_{8}^{16}O$ and $_{92}^{235}U$ are referred to as nuclides. Nuclides that spontaneously emit radiation are referred to as *radionuclides.*

nuclide

18.2 Natural Radioactivity

Radioactive elements continuously undergo **radioactive decay,** or disintegration, to form different elements. The chemical properties of an element are associated with its electronic structure, but radioactivity is a property of the nucleus. Therefore

radioactive decay

neither ordinary changes of temperature and pressure nor the chemical or physical state of an element has any effect on its radioactivity.

The principal emissions from the nuclei of radionuclides are known as alpha rays (or particles), beta rays (or particles), and gamma rays. Upon losing an alpha or beta particle, the radioactive element changes into a different element. We will explain this process in detail later.

half-life

Each radioactive nuclide disintegrates at a specific and constant rate, which is expressed in units of half-life. The **half-life** ($t_{1/2}$) is the time required for one-half of a specific amount of a radioactive nuclide to disintegrate. The half-lives of the elements range from a fraction of a second to billions of years. To illustrate, suppose we start with 1.0 g of $^{226}_{88}$Ra, ($t_{1/2} = 1620$ years):

Element	$t_{1/2}$
$^{238}_{92}$U	4.5×10^9 years
$^{226}_{88}$Ra	1620 years
$^{15}_{6}$C	2.4 seconds

$$1.0 \text{ g } ^{226}_{88}\text{Ra} \xrightarrow[1620 \text{ years}]{t_{1/2}} 0.50 \text{ g } ^{226}_{88}\text{Ra} \xrightarrow[1620 \text{ years}]{t_{1/2}} 0.25 \text{ g } ^{226}_{88}\text{Ra}$$

The half-lives of the various radioisotopes of the same element differ dramatically. Half-lives for certain isotopes of radium, carbon, and uranium are listed in Table 18.2.

Example 18.1 The half-life of $^{131}_{53}$I is 8 days. How much $^{131}_{53}$I from a 32-g sample remains after five half-lives?

Solution Using the following graph, we can find the number of grams of ^{131}I remaining after one half-life:

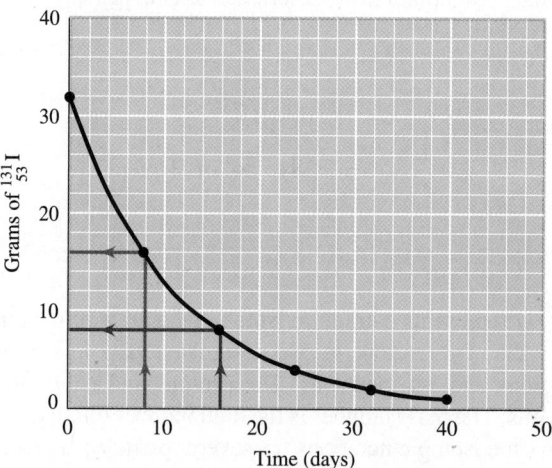

Trace a perpendicular line from 8 days on the *x*-axis to the line on the graph. Now trace a horizontal line from this point on the plotted line to the *y*-axis and read the corresponding grams of ^{131}I. Continue this process for each half-life, adding 8 days to the previous value on the *x*-axis:

Half-lives	0	1	2	3	4	5
Number of days		8	16	24	32	40
Amount remaining	32 g	16 g	8 g	4 g	2 g	1 g

Starting with 32 g, 1 g ^{131}I remains after five half-lives (40 days).

TABLE 18.2	Half-Lives for Radium, Carbon, and Uranium Isotopes		
Isotope	**Half-life**	**Isotope**	**Half-life**
Ra-223	11.7 days	C-14	5668 years
Ra-224	3.64 days	C-15	2.4 seconds
Ra-225	14.8 days	U-235	7.1×10^8 years
Ra-226	1620 years	U-238	4.5×10^9 years
Ra-228	6.7 years		

In how many half-lives will 10.0 g of a radioactive nuclide decay to less than 10% of its original value?   **Example 18.2**

We know that 10% of the original amount is 1.0 g. After the first half-life, half the original material remains and half has decayed (5.00 g). After the second half-life, one-fourth of the original material remains (i.e., one-half of 5.00 g). This progression continues, reducing the quantity remaining by half for each half-life that passes.   **Solution**

Half-lives	0	1	2	3	4
Percent remaining	100%	50%	25%	12.5%	6.25%
Amount remaining	10.0 g	5.00 g	2.50 g	1.25 g	0.625 g

Therefore the amount remaining will be less than 10% sometime between the third and the fourth half-lives.

Practice 18.1

The half-life of $^{14}_{6}C$ is 5668 years. How much $^{14}_{6}C$ will remain after six half-lives in a sample that initially contains 25.0 g?

Nuclides are said to be either *stable* (nonradioactive) or *unstable* (radioactive). Elements that have atomic numbers greater than 83 (bismuth) are naturally radioactive, although some of the nuclides have extremely long half-lives. Some of the naturally occurring nuclides of elements 81, 82, and 83 are radioactive, and some are stable. Only a few naturally occurring elements that have atomic numbers less than 81 are radioactive. However, no stable isotopes of element 43 (technetium) or of element 61 (promethium) are known.

Radioactivity is believed to be a result of an unstable ratio of neutrons to protons in the nucleus. Stable nuclides of elements up to about atomic number 20 generally have about a 1:1 neutron-to-proton ratio. In elements above number 20, the neutron-to-proton ratio in the stable nuclides gradually increases to about 1.5:1 in element number 83 (bismuth). When the neutron-to-proton ratio is too high or too low, alpha, beta, or other particles are emitted to achieve a more stable nucleus.

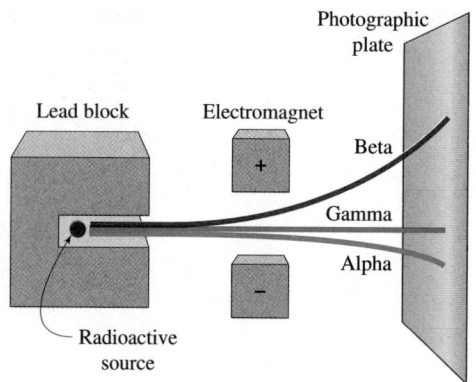

18.3　Alpha Particles, Beta Particles, and Gamma Rays

The classical experiment proving that alpha and beta particles are oppositely charged was performed by Marie Curie (see Figure 18.1). She placed a radioactive source in a hole in a lead block and positioned two poles of a strong electromagnet so that the radiations that were given off passed between them. The paths of three different kinds of radiation were detected by means of a photographic plate placed some distance beyond the electromagnet. The lighter beta particles were strongly deflected toward the positive pole of the electromagnet; the heavier alpha particles were less strongly deflected and in the opposite direction. The uncharged gamma rays were not affected by the electromagnet and struck the photographic plates after traveling along a path straight out of the lead block.

Alpha Particles

alpha particle

An **alpha particle** (α) consists of two protons and two neutrons, has a mass of about 4 amu, and a charge of $+2$. It is a helium nucleus which is usually given one of the following symbols: α or ^4_2He. When an alpha particle is emitted from the nucleus, a different element is formed. The atomic number of the new element is 2 less, and the mass is 4 amu less, than that of the starting element.

> **Loss of an alpha particle from the nucleus results in**
> **loss of 4 in the mass number (A)**
> **loss of 2 in the atomic number (Z)**

For example, when $^{238}_{92}\text{U}$ loses an alpha particle, $^{234}_{90}\text{Th}$ is formed, because two neutrons and two protons are lost from the uranium nucleus. This disintegration may be written as a nuclear equation:

$$^{238}_{92}\text{U} \longrightarrow {}^{234}_{90}\text{Th} + \alpha \qquad \text{or} \qquad {}^{238}_{92}\text{U} \longrightarrow {}^{234}_{90}\text{Th} + {}^4_2\text{He}$$

For the loss of an alpha particle from $^{226}_{88}\text{Ra}$, the equation is

$$^{226}_{88}\text{Ra} \longrightarrow {}^{222}_{86}\text{Rn} + {}^{4}_{2}\text{He} \quad \text{or} \quad {}^{226}_{88}\text{Ra} \longrightarrow {}^{222}_{86}\text{Rn} + \alpha$$

A nuclear equation, like a chemical equation, consists of reactants and products and must be balanced. To have a balanced nuclear equation, the sum of the mass numbers (superscripts) on both sides of the equation must be equal, and the sum of the atomic numbers (subscripts) on both sides of the equation must be equal:

What new nuclide will be formed when $^{230}_{90}\text{Th}$ loses an alpha particle? The new nuclide will have a mass of 226 amu and will contain 88 protons, so its atomic number is 88. Locate the corresponding element on the periodic chart—in this case, $^{226}_{88}\text{Ra}$ or radium-226.

Beta Particles

The **beta particle** (β) is identical in mass and charge to an electron; its charge is -1. Both a beta particle and a proton are produced by the decomposition of a neutron:

$$^{1}_{0}\text{n} \longrightarrow {}^{1}_{1}\text{p} + {}^{0}_{-1}\text{e}$$

beta particle

The beta particle leaves, and the proton remains in the nucleus. When an atom loses a beta particle from its nucleus, a different element is formed that has essentially the same mass but an atomic number that is 1 greater than that of the starting element. The beta particle is written as β or $^{0}_{-1}\text{e}$.

> Loss of a beta particle from the nucleus results in
> no change in the mass number (A)
> increase of 1 in the atomic number (Z)

Examples of equations in which a beta particle is lost are

$$^{234}_{90}\text{Th} \longrightarrow {}^{234}_{91}\text{Pa} + \beta$$

$$^{234}_{91}\text{Pa} \longrightarrow {}^{234}_{92}\text{U} + {}^{0}_{-1}\text{e}$$

$$^{210}_{82}\text{Pb} \longrightarrow {}^{210}_{83}\text{Bi} + \beta$$

Gamma Rays

Gamma rays (γ) are photons of energy. A gamma ray is similar to an X-ray but is more energetic. They have no electrical charge and no measurable mass. Gamma rays are released from the nucleus in many radioactive changes along with either

gamma ray

alpha or beta particles. Gamma radiation does not result in a change of atomic number or the mass of an element.

> **Loss of a gamma ray from the nucleus results in no change in mass number (A) or atomic number (Z)**

Example 18.3　(a) Write an equation for the loss of an alpha particle from the nuclide $^{194}_{78}\text{Pt}$.

(b) What nuclide is formed when $^{228}_{88}\text{Ra}$ loses a beta particle from its nucleus?

Solution　(a) Loss of an alpha particle, $^{4}_{2}\text{He}$, results in a decrease of 4 in the mass number and a decrease of 2 in the atomic number:

Mass of new nuclide:　　　　　　$A - 4$　or　$194 - 4 = 190$

Atomic number of new nuclide:　$Z - 2$　or　$78 - 2 = 76$

Looking up element number 76 on the periodic table, we find it to be osmium, Os. The equation then is

$$^{194}_{78}\text{Pt} \longrightarrow {}^{190}_{76}\text{Os} + {}^{4}_{2}\text{He}$$

(b) The loss of a beta particle from a $^{228}_{88}\text{Ra}$ nucleus means a gain of 1 in the atomic number with no essential change in mass. The new nuclide will have an atomic number of ($Z + 1$) or 89, which is actinium, Ac:

$$^{228}_{88}\text{Ra} \longrightarrow {}^{228}_{89}\text{Ac} + {}^{0}_{-1}\text{e}$$

Example 18.4　What nuclide will be formed when $^{214}_{82}\text{Pb}$ successively emits two beta particles, then one alpha particle from its nucleus? Write successive equations showing these changes.

Solution　The changes brought about in the three steps outlined are as follows:

β loss:　Increase of 1 in the atomic number; no change in mass

β loss:　Increase of 1 in the atomic number; no change in mass

α loss:　Decrease of 2 in the atomic number; decrease of 4 in the mass

The equations are

$$^{214}_{82}\text{Pb} \longrightarrow {}^{214}_{83}\text{X} + \beta \longrightarrow {}^{214}_{84}\text{X} + \beta \longrightarrow {}^{210}_{82}\text{X} + \alpha$$

where X stands for the new nuclide formed. Looking up each of these elements by their atomic numbers, we rewrite the equations

$$^{214}_{82}\text{Pb} \xrightarrow{\;\beta\;} {}^{214}_{83}\text{Bi} \xrightarrow{\;\beta\;} {}^{214}_{84}\text{Po} \xrightarrow{\;\alpha\;} {}^{210}_{82}\text{Pb}$$

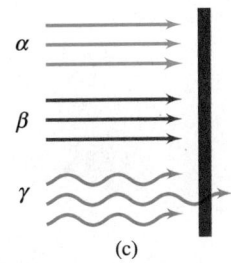

(a)   (b)   (c)

◀ **FIGURE 18.2**
Relative penetrating ability of alpha, beta, and gamma radiation. (a) Thin sheet of paper; (b) thin sheet of aluminum; (c) 5-cm lead block.

TABLE 18.3   Characteristics of Nuclear Radiation

Radiation	Symbol	Mass (amu)	Electrical charge	Velocity	Composition	Ionizing power
Alpha	α, $^4_2$He	4	+2	Variable, less than 10% the speed of light	He nucleus	High
Beta	β, $^0_{-1}$e	$\dfrac{1}{1837}$	−1	Variable, up to 90% the speed of light	Identical to an electron	Moderate
Gamma	γ	0	0	Speed of light	Photons or electromagnetic waves of energy	Almost none

Practice 18.2

What nuclide will be formed when $^{222}_{86}$Rn emits an alpha particle?

The ability of radioactive rays to pass through various objects is in proportion to the speed at which they leave the nucleus. Gamma rays travel at the velocity of light (186,000 miles per second) and are capable of penetrating several inches of lead. The velocities of beta particles are variable, the fastest being about nine-tenths the velocity of light. Alpha particles have velocities less than one-tenth the velocity of light. Figure 18.2 illustrates the relative penetrating power of these rays. A few sheets of paper will stop alpha particles; a thin sheet of aluminum will stop both alpha and beta particles; and a 5-cm block of lead will reduce, but not completely stop, gamma radiation. In fact, it is difficult to stop all gamma radiation. Table 18.3 summarizes the properties of alpha, beta, and gamma radiation.

18.4   Radioactive Disintegration Series

The naturally occurring radioactive elements with a higher atomic number than lead (Pb) fall into three orderly disintegration series. Each series proceeds from one

These three series begin with the elements uranium, thorium, and actinium.

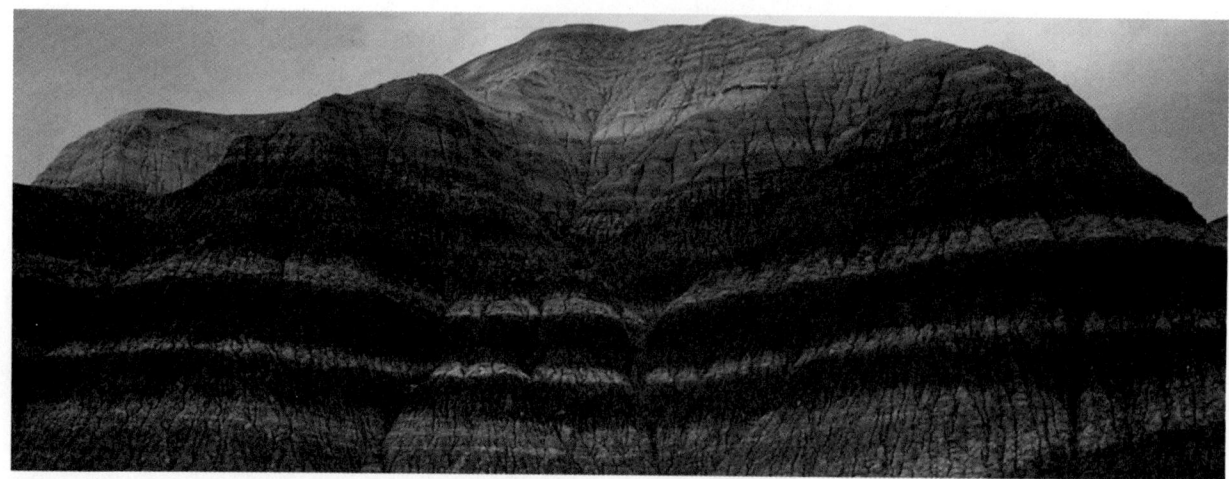

The age of geologic
formations like these found in
Petrified Forest National Park
can be determined using disintegration series.

FIGURE 18.3 ▶
**The uranium disintegration
series. $^{238}_{92}$U decays by a
series of alpha (α) and beta
(β) emissions to the stable
nuclide $^{206}_{82}$Pb.**

element to the next by the loss of either an alpha or a beta particle, finally ending in
a nonradioactive nuclide. The uranium series starts with $^{238}_{92}$U and ends with $^{206}_{82}$Pb.
The thorium series starts with $^{232}_{90}$Th and ends with $^{208}_{82}$Pb. The actinium series starts
with $^{235}_{92}$U and ends with $^{207}_{82}$Pb. A fourth series, the neptunium series, starts with the
synthetic element $^{241}_{94}$Pu and ends with the stable bismuth nuclide $^{209}_{83}$Bi. The uranium
series is shown in Figure 18.3. Gamma radiation, which accompanies alpha and beta
radiation, is not shown.

By using these a series and the half-lives of its members, scientists have been
able to approximate the age of certain geologic deposits. This approximation is done
by comparing the amount of $^{238}_{92}$U with the amount of $^{206}_{82}$Pb and other nuclides in the
series that are present in a particular geologic formation. Rocks found in Canada and
Finland have been calculated to be about 3.0×10^9 (3 billion) years old. Some me-
teorites have been determined to be 4.5×10^9 years old.

Practice 18.3

What nuclides are formed when $^{238}_{92}U$ undergoes the following decays?
(a) an alpha particle and a beta particle
(b) three alpha particles and two beta particles

18.5 Transmutation of Elements

Transmutation is the conversion of one element into another by either natural or artificial means. Transmutation occurs spontaneously in natural radioactive disintegrations. Alchemists tried for centuries to convert lead and mercury into gold by artificial means, but transmutation by artificial means was not achieved until 1919, when Ernest Rutherford succeeded in bombarding the nuclei of nitrogen atoms with alpha particles and produced oxygen nuclides and protons. The nuclear equation for this transmutation can be written as

transmutation

$$^{14}_{7}N + \alpha \longrightarrow {}^{17}_{8}O + {}^{1}_{1}H \quad \text{or} \quad {}^{14}_{7}N + {}^{4}_{2}He \longrightarrow {}^{17}_{8}O + {}^{1}_{1}H$$

It is believed that the alpha particle enters the nitrogen nucleus, forming $^{18}_{9}F$ as an intermediate, which then decomposes into the products.

Rutherford's experiments opened the door to nuclear transmutations of all kinds. Atoms were bombarded by alpha particles, neutrons, protons, deuterons ($^{2}_{1}H$), electrons, and so forth. Massive instruments were developed for accelerating these particles to very high speeds and energies to aid their penetration of the nucleus. The famous cyclotron was developed by E. O. Lawrence (1901–1958) at the University of California; later instruments include the Van de Graaf electrostatic generator, the betatron, and the electron and proton synchrotrons. With these instruments many nuclear transmutations became possible. Equations for a few of these are as follows:

$$^{7}_{3}Li + {}^{1}_{1}H \longrightarrow 2\,{}^{4}_{2}He$$

$$^{40}_{18}Ar + {}^{1}_{1}H \longrightarrow {}^{40}_{19}K + {}^{1}_{0}n$$

$$^{23}_{11}Na + {}^{1}_{1}H \longrightarrow {}^{23}_{12}Mg + {}^{1}_{0}n$$

$$^{114}_{48}Cd + {}^{2}_{1}H \longrightarrow {}^{115}_{48}Cd + {}^{1}_{1}H$$

$$^{2}_{1}H + {}^{2}_{1}H \longrightarrow {}^{3}_{1}H + {}^{1}_{1}H$$

$$^{209}_{83}Bi + {}^{2}_{1}H \longrightarrow {}^{210}_{84}Po + {}^{1}_{0}n$$

$$^{16}_{8}O + {}^{1}_{0}n \longrightarrow {}^{13}_{6}C + {}^{4}_{2}He$$

$$^{238}_{92}U + {}^{12}_{6}C \longrightarrow {}^{244}_{98}Cf + 6\,{}^{1}_{0}n$$

18.6 Artificial Radioactivity

Irene Joliot-Curie (daughter of Pierre and Marie Curie) and her husband Frederic Joliot-Curie observed that when aluminum-27 is bombarded with alpha particles, neutrons and positrons (positive electrons) are emitted as part of the products. When

▲
Aerial view of Stanford University's Linear Accelerator.

CHEMISTRY IN ACTION How Old Is This Object?

When archaeologists discover a fossil or artifact, how do they determine its age? They couldn't count rings in fossils (as we do with trees), so an American chemist developed a technique called radiocarbon dating. W. F. Libby, who received the Nobel Prize in chemistry in 1960 for his work, based his method on the decay rate of C-14.

Radiocarbon dating works this way: Carbon dioxide in the atmosphere contains a fixed ratio of radioactive C-14 to ordinary C-12.

Plants that consume carbon dioxide during photosynthesis and animals that eat the plants contain the same proportion of C-14 to C-12 as long as they are alive. When an organism dies, the amount of C-12 remains fixed, but the C-14 content diminishes according to its half-life (5668 years). By comparing the ratio of C-14 to C-12 in an object to the same ratio in living plants, we can estimate the age of the object being evaluated. In 5668 years, one-half of the radiocarbon initially present will have undergone decomposition. In 11,336 years, one-fourth of the original C-14 will be left. The age of fossil material, archaeological specimens, and old wood is then determined by measuring the remaining C-14.

The age of specimens from ancient Egyptian tombs calculated by radiocarbon dating correlates closely with the chronological age established by Egyptologists. Charcoal samples obtained at Darrington Walls, a woodhenge in Great Britain, were determined to be about 4000 years old. Radiocarbon dating instruments currently in use enable researchers to date specimens back as far as 70,000 years.

Radioactive decay has also been used to date samples other than those containing carbon. For example, the age of rock formations containing uranium has been approximated by determining the ratio of U-238 to Pb-206. Lead-206 is the last isotope formed in the U-238 disintegration series. Thus a geologic deposit containing a 1 : 1 ratio of U-238 to Pb-206 corresponds to a time lapse of one half-life of U-238, which is 4.5×10^9 years, assuming that all the lead came from the decay of U-238. The age of Moon rocks returned to Earth by the *Apollo* missions were calculated by similar techniques.

◀ **Radiocarbon dating can be used to verify the age of artifacts from Aztec civilizations.**

the source of alpha particles is removed, neutrons cease to be produced, but positrons continue to be emitted. This observation suggested that the neutrons and positrons come from two separate reactions. It also indicated that a product of the first reaction is radioactive. After further investigation they discovered that, when aluminum-27 is bombarded with alpha particles, phosphorus-30 and neutrons are produced. Phosphorus-30 is radioactive, has a half-life of 2.5 minutes, and decays to silicon-30 with the emission of a positron. The equations for these reactions are

$$^{27}_{13}\text{Al} + {}^{4}_{2}\text{He} \longrightarrow {}^{30}_{15}\text{P} + {}^{1}_{0}\text{n}$$

$$^{30}_{15}\text{P} \longrightarrow {}^{30}_{14}\text{Si} + {}^{0}_{+1}\text{e}$$

artificial radioactivity
induced radioactivity

The radioactivity of nuclides produced in this manner is known as **artificial radioactivity** or **induced radioactivity.** Artificial radionuclides behave like natural radioactive elements in two ways: They disintegrate in a definite fashion and they have a specific half-life. The Joliot-Curies received the Nobel Prize in chemistry in 1935 for the discovery of artificial, or induced, radioactivity.

18.7 Measurement of Radioactivity

Radiation from radioactive sources is so energetic that it is called *ionizing radiation*. When it strikes an atom or a molecule, one or more electrons are knocked off, and an ion is created. The Geiger counter, an instrument commonly used to detect and measure radioactivity, depends on this fact. The instrument consists of a Geiger–Müller detecting tube and a counting device. The detector tube is a pair of oppositely charged electrodes in an argon gas-filled chamber fitted with a thin window. When radiation, such as a beta particle, passes through the window into the tube, some argon is ionized, and a momentary pulse of current (discharge) flows between the electrodes. These current pulses are electronically amplified in the counter and appear as signals in the form of audible clicks, flashing lights, meter deflections, or numerical readouts (Figure 18.4).

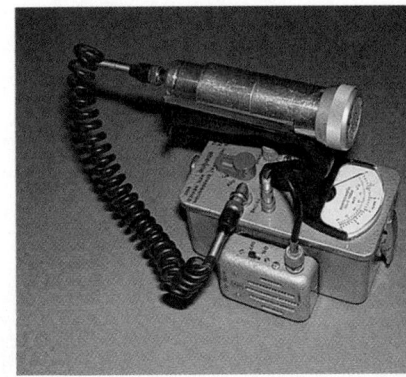

▲ **FIGURE 18.4**
Geiger–Müller survey meter.

The amount of radiation that an individual encounters can be measured by a film badge. This badge contains a piece of photographic film in a light-proof holder and is worn in areas where radiation might be encountered. The silver grains in the film will darken when exposed to radiation. The badges are processed after a predetermined time interval to determine the amount of radiation the wearer has been exposed to.

A scintillation counter is used to measure radioactivity for biomedical applications. A scintillator is composed of molecules that emit light when exposed to ionizing radiation. A light-sensitive detector counts the flashes and converts them to a numerical readout.

The *curie* is the unit used to express the amount of radioactivity produced by an element. One **curie** (Ci) is defined as the quantity of radioactive material giving 3.7×10^{10} disintegrations per second. The basis for this figure is pure radium, which has an activity of 1 Ci/g. Because the curie is such a large quantity, the millicurie and microcurie, representing one-thousandth and one-millionth of a curie, respectively, are more practical and more commonly used.

curie (Ci)

The curie only measures radioactivity emitted by a radionuclide. Different units are required to measure exposure to radiation. The **roentgen (R)** quantifies exposure to gamma or X-rays; 1 roentgen is defined as the amount of radiation required to produce 2.1×10^9 ions/cm^3 of dry air. The **rad (radiation *absorbed dose*)** is defined as the amount of radiation that provides 0.01 J of energy per kilogram of matter. The amount of radiation absorbed will change depending on the type of matter. The roentgen and the rad are numerically similar; 1 roentgen of gamma radiation provides 0.92 rad in bone tissue.

roentgen (R)

rad (radiation absorbed dose)

Neither the rad nor the roentgen indicates the biological damage caused by radiation. One rad of alpha particles has the ability to cause ten times more damage than 1 rad of gamma rays or beta particles. Another unit, **rem (roentgen *equivalent to man*)** takes into account the degree of biological effect caused by the type of radiation exposure; 1 rem is equal to the dose in rads multiplied by a factor specific to the form of radiation. The factor is 10 for alpha particles and 1 for both beta particles and gamma rays. Units of radiation are summarized in Table 18.4.

rem (roentgen equivalent to man)

TABLE 18.4	Radiation Units	
Unit	**Measure**	**Equivalent**
curie (Ci)	rate of decay of a radioactive substance	1 Ci = 3.7×10^{10} disintegrations/sec
roentgen (R)	exposure based on the quantity of ionization produced in air	1 R = 2.1×10^9 ions/cm^3
rad	absorbed dose of radiation	1 rad = 0.01 J/kg matter
rem	radiation dose equivalent	1 rem = 1 rad $\times$ factor
gray (Gy) (SI unit)	energy absorbed by tissue	1 Gy = 1 J/kg tissue (1 Gy = 100 rad)

18.8   Nuclear Fission

nuclear fission

In **nuclear fission** a heavy nuclide splits into two or more intermediate-sized fragments when struck in a particular way by a neutron. The fragments are called *fission products*. As the atom splits, it releases energy and two or three neutrons, each of which can cause another nuclear fission. The first instance of nuclear fission was reported in January 1939 by the German scientists Otto Hahn (1879–1968) and Fritz Strassmann (1902–1980). Detecting isotopes of barium, krypton, cerium, and lanthanum after bombarding uranium with neutrons led scientists to believe that the uranium nucleus had been split.

Characteristics of nuclear fission are as follows:

1. Upon absorption of a neutron, a heavy nuclide splits into two or more smaller nuclides (fission products).
2. The mass of the nuclides formed ranges from about 70–160 amu.
3. Two or more neutrons are produced from the fission of each atom.
4. Large quantities of energy are produced as a result of the conversion of a small amount of mass into energy.
5. Most nuclides produced are radioactive and continue to decay until they reach a stable nucleus.

One process by which this fission takes place is illustrated in Figure 18.5. When a heavy nucleus captures a neutron, the energy increase may be sufficient to cause deformation of the nucleus until the mass finally splits into two fragments, releasing energy and usually two or more neutrons.

In a typical fission reaction, a $^{235}_{92}$U nucleus captures a neutron and forms unstable $^{236}_{92}$U. This $^{236}_{92}$U nucleus undergoes fission, quickly disintegrating into two fragments, such as $^{139}_{56}$Ba and $^{94}_{36}$Kr, and three neutrons. The three neutrons in turn may be captured by three other $^{235}_{92}$U atoms, each of which undergoes fission, producing nine neutrons, and so on. A reaction of this kind, in which the products

chain reaction

cause the reaction to continue or magnify, is known as a **chain reaction.** For a chain reaction to continue, enough fissionable material must be present so that each atomic fission causes, on average, at least one additional fission. The minimum quantity of an element needed to support a self-sustaining chain reaction is called

CHEMISTRY IN ACTION   Does Your Food Glow in the Dark?

Over the past 25 years the Food and Drug Administration (FDA) has approved irradiation to delay ripening or kill microbes and insects in wheat, potatoes, fresh fruits, and poultry as well as spices. But the only regular use of irradiation on food in the United States has been on spices.

Now the FDA has approved irradiation of red meat and new legislation allows the labels about irradiated food to be in much smaller type. The general public has not yet accepted the use of radiation to reduce microbes and help preserve food. Why? Many people believe that irradiating food makes it radioactive. This is not true; in fact, the most that irradiation of food does is to produce compounds similar to those created by cooking and also to reduce the vitamin content of some food.

How does irradiation of food work? The radioactive source currently used is cobalt-60. It is contained in a concrete cell (see diagram) with 6-foot-thick walls. Inside the cell is a pool of water

with racks of thin Co-60 rods suspended above. When the rods are not being used they are submerged in the water, which absorbs the gamma radiation. When food to be irradiated moves into the cells the Co-60 rods are lifted from the water, and the boxes of food move among them on a conveyor being irradiated from all sides.

Scientists who have investigated the process say that food irradiation is safe. The Centers for Disease Control and Pre-

vention in Atlanta estimates that food-borne illness causes as many as 9,000 deaths per year. Irradiation provides one way to reduce microbial contamination in food. It does not solve the problem of careless handling of food by processors, and long-term studies on humans have not yet been concluded regarding irradiated food supplies. Once again we face the issue of balancing benefit and risk.

the **critical mass.** Since energy is released in each atomic fission, chain reactions provide a steady supply of energy. (A chain reaction is illustrated in Figure 18.6.) Two of the many possible ways in which $^{235}_{92}U$ may fission are shown by:

critical mass

$$^{235}_{92}U + ^{1}_{0}n \longrightarrow ^{139}_{56}Ba + ^{94}_{36}Kr + 3\,^{1}_{0}n$$

$$^{235}_{92}U + ^{1}_{0}n \longrightarrow ^{144}_{54}Xe + ^{90}_{38}Sr + 2\,^{1}_{0}n$$

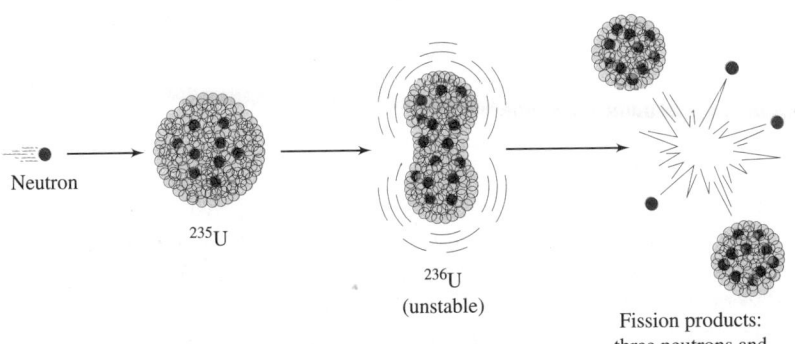

◀ **FIGURE 18.5**
The fission process. When a neutron is captured by a heavy nucleus, the nucleus becomes more unstable. The more energetic nucleus begins to deform, resulting in fission. Two nuclear fragments and three neutrons are produced by this fission process.

FIGURE 18.6 ▶
Fission and chain reaction of $^{235}_{92}$U. Each fission produces two major fission fragments and three neutrons, which may be captured by other $^{235}_{92}$U nuclei, continuing the chain reaction.

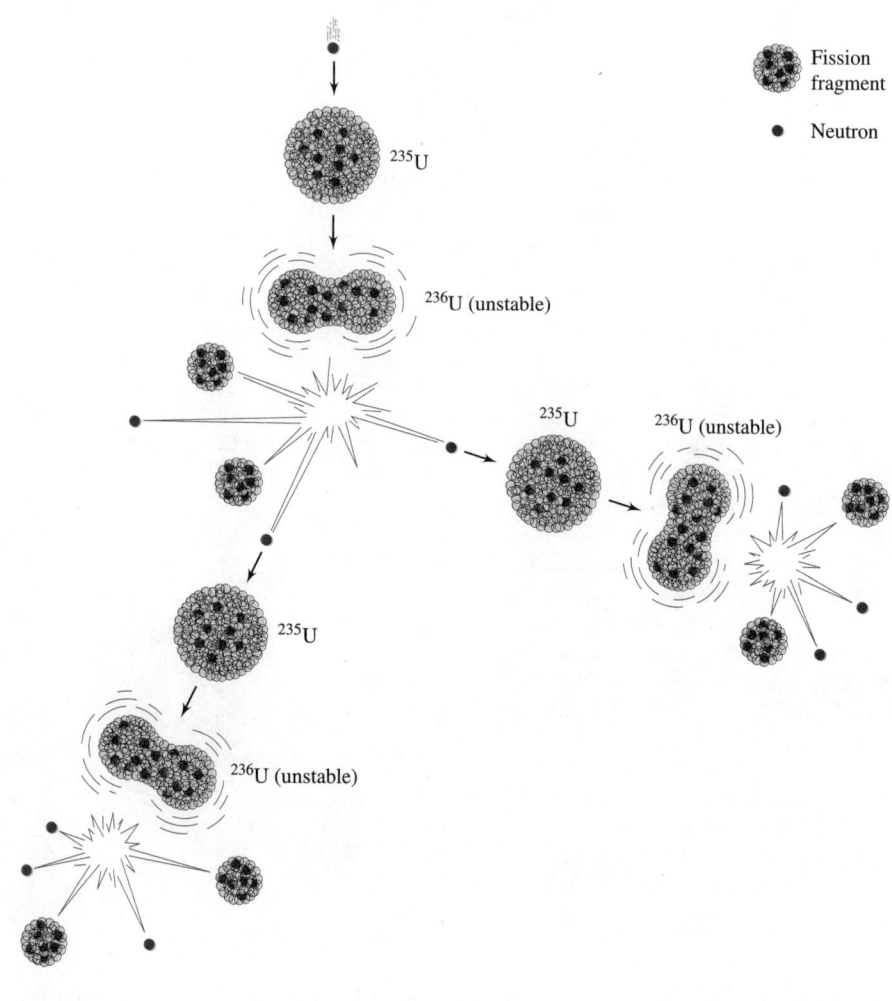

18.9   Nuclear Power

Nearly all electricity for commercial use is produced by machines consisting of a turbine linked by a drive shaft to an electrical generator. The energy required to run the turbine can be supplied by falling water, as in hydroelectric power plants, or by steam as in thermal power plants.

The world's demands for energy, largely from fossil fuels, is heavy. At the present rates of consumption, the estimated world supply of fossil fuels is sufficient for only a few centuries. Although the United States has large coal and oil shale deposits, it currently imports over 40% of its oil supply. We clearly need to develop alternative energy sources. At present, uranium is the most productive alternative energy source, and about 17% of the electrical energy used in the United States is generated from power plants using uranium fuel.

A nuclear power plant is a thermal power plant in which heat is produced by a nuclear reactor instead of by combustion of fossil fuel. The major components of a nuclear reactor are

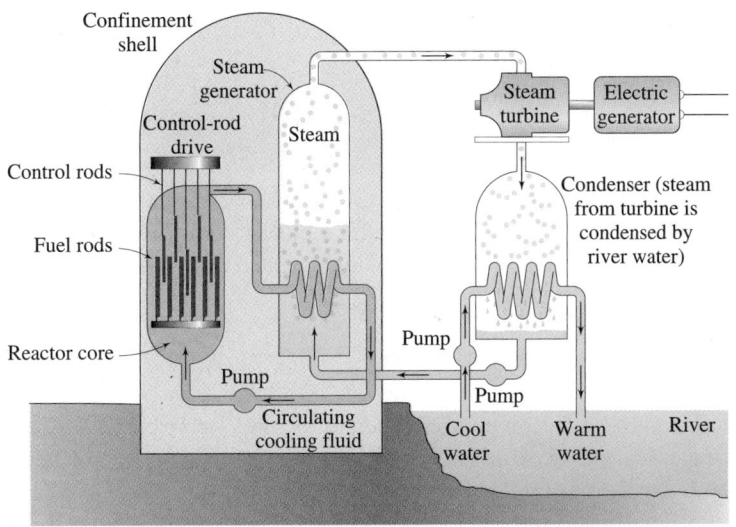

Diagram of a nuclear power plant. Heat produced by the reactor core is carried by a cooling fluid to a steam generator, which turns a turbine that produces electricity.

1. an arrangement of nuclear fuel, called the reactor core
2. a control system, which regulates the rate of fission and thereby the rate of heat generation
3. a cooling system, which removes the heat from the reactor and also keeps the core at the proper temperature

One type of reactor uses metal slugs containing uranium enriched from the normal 0.7% U-235 to about 3% U-235. The self-sustaining fission reaction is moderated, or controlled, by adjustable control rods containing substances that slow down and capture some of the neutrons produced. Ordinary water, heavy water, and molten sodium are typical coolants used. Energy obtained from nuclear reactions in the form of heat is used in the production of steam to drive turbines for generating electricity. (See Figure 18.7.)

The potential dangers of nuclear power were tragically demonstrated by the accidents at Three Mile Island, Pennsylvania (1979), and Chernobyl in the former U.S.S.R. (1986). Both accidents resulted from the loss of coolant to the reactor core. The reactors at Three Mile Island were covered by concrete containment buildings and therefore released a relatively small amount of radioactive material into the atmosphere. But because the Soviet Union did not require containment structures on nuclear power plants, the Chernobyl accident resulted in 31 deaths and the resettlement of 135,000 people. The release of large quantities of I-131, Cs-134, and Cs-137 appears to be causing long-term health problems in that exposed population.

Another major disadvantage of nuclear power is its highly radioactive waste products, some of which have half-lives of thousands of years. As yet, no technology has been developed that disposes of these dangerous wastes in complete safety.

In the United States, reactors designed for commercial power production use uranium oxide, U_3O_8, that is enriched with the relatively scarce fissionable U-235 isotope. Because the supply of U-235 is limited, a new type of reactor known as the *breeder reactor* has been developed. Breeder reactors produce additional fissionable material at the same time that the fission reaction is occurring. In a breeder reactor, excess neutrons convert nonfissionable isotopes, such as U-238 or Th-232, to fissionable isotopes, Pu-239 or U-233 shown on the next page.

▲
Interior view of a reactor hall at a nuclear power plant in France. The core of the reactor is situated in the water pool (blue area).

$$\underset{92}{^{238}}U + \underset{0}{^{1}}n \longrightarrow \underset{92}{^{239}}U \overset{}{\underset{\beta}{\searrow}} \underset{93}{^{239}}Np \overset{}{\underset{\beta}{\searrow}} \underset{94}{^{239}}Pu$$

$$\underset{90}{^{232}}Th + \underset{0}{^{1}}n \longrightarrow \underset{90}{^{233}}Th \overset{}{\underset{\beta}{\searrow}} \underset{91}{^{233}}Pa \overset{}{\underset{\beta}{\searrow}} \underset{92}{^{233}}U$$

These transmutations make it possible to greatly extend the supply of fuel for nuclear reactors. No breeder reactors are presently in commercial operation in the United States, but a number of them are being operated in Europe and Great Britain.

18.10   The Atomic Bomb

The atomic bomb is a fission bomb; it operates on the principle of a very fast chain reaction that releases a tremendous amount of energy. An atomic bomb and a nuclear reactor both depend on self-sustaining nuclear fission chain reactions. The essential difference is that in a bomb the fission is "wild," or uncontrolled, whereas in a nuclear reactor the fission is moderated and carefully controlled. A minimum critical mass of fissionable material is needed for a bomb, or a major explosion will not occur. When a quantity smaller than the critical mass is used, too many neutrons formed in the fission step escape without combining with another nucleus, and a chain reaction does not occur. Therefore the fissionable material of an atomic bomb must be stored as two or more subcritical masses and brought together to form the critical mass at the desired time of explosion. The temperature developed in an atomic bomb is believed to be about 10 million degrees Celsius.

The nuclides used in atomic bombs are U-235 and Pu-239. Uranium deposits contain about 0.7% of the U-235 isotope, the remainder being U-238. Uranium-238 does not undergo fission except with very high energy neutrons. It was discovered, however, that U-238 captures a low energy neutron without undergoing fission and that the product, U-239, changes to Pu-239 (plutonium) by a beta-decay process. Plutonium-239 readily undergoes fission upon capture of a neutron and is therefore useful for nuclear weapons. The equations for the nuclear transformations are

$$\underset{92}{^{238}}U + \underset{0}{^{1}}n \longrightarrow \underset{92}{^{239}}U \overset{}{\underset{\beta}{\searrow}} \underset{93}{^{239}}Np \overset{}{\underset{\beta}{\searrow}} \underset{94}{^{239}}Pu$$

The hazards of an atomic bomb explosion include not only shock waves from the explosive pressure and tremendous heat, but also intense radiation in the form of alpha particles, beta particles, gamma rays, and ultraviolet rays. Gamma rays and X-rays can penetrate deeply into the body, causing burns, sterilization, and gene mutation, which can adversely affect future generations. Both radioactive fission products and unfissioned material are present after the explosion. If the bomb explodes near the ground, many tons of dust are lifted into the air. Radioactive material adhering to this dust, known as *fallout,* is spread by air currents over wide areas of the land and constitutes a lingering source of radiation hazard.

Today nuclear war is probably the most awesome threat facing civilization. Only two rather primitive fission-type atom bombs were used to destroy the Japanese cities of Hiroshima and Nagasaki and bring World War II to an early end. The threat of nuclear war is increased by the fact that the number of nations possessing nuclear weapons is steadily increasing.

◀ **The mushroom cloud is a signature of uncontrolled fission in an atomic bomb.**

18.11 Nuclear Fusion

The process of uniting the nuclei of two light elements to form one heavier nucleus is known as **nuclear fusion.** Such reactions can be used for producing energy, because the masses of the two nuclei that fuse into a single nucleus are greater than the mass of the nucleus formed by their fusion. The mass differential is liberated in the form of energy. Fusion reactions are responsible for the tremendous energy output of the sun. Thus aside from relatively small amounts from nuclear fission and radioactivity, fusion reactions are the ultimate source of our energy, even the energy from fossil fuels. They are also responsible for the devastating power of the thermonuclear, or hydrogen, bomb.

nuclear fusion

Fusion reactions require temperatures on the order of tens of millions of degrees for initiation. Such temperatures are present in the Sun but have been produced only momentarily on Earth. For example, the hydrogen, or fusion, bomb is triggered by the temperature of an exploding fission bomb. Two typical fusion reactions are

$$\underset{\text{tritium}}{^3_1\text{H}} + \underset{\text{deuterium}}{^2_1\text{H}} \longrightarrow {}^4_2\text{He} + {}^1_0\text{n} + \text{energy}$$

$$\underset{\substack{3.0150 \\ \text{amu}}}{^3_1\text{H}} + \underset{\substack{1.0079 \\ \text{amu}}}{^1_1\text{H}} \longrightarrow \underset{\substack{4.0026 \\ \text{amu}}}{^4_2\text{He}} + \text{energy}$$

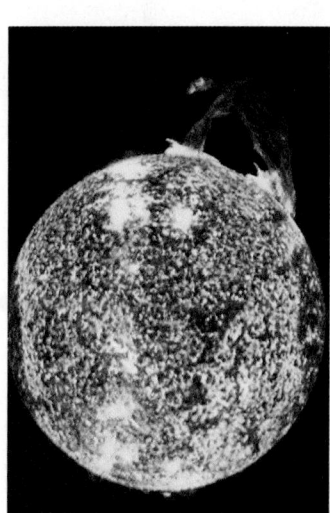

▲ **Solar flares such as these are indications of fusion reactions occurring at temperatures of millions of degrees.**

The total mass of the reactants in the second equation is 4.0229 amu, which is 0.0203 amu greater than the mass of the product. This difference in mass is manifested in the great amount of energy liberated.

During the past 50–55 years, a great deal of research in the United States and in other countries, especially the former Soviet Union, has focused on controlled nuclear fusion reactions. The goal of controlled nuclear fusion has not yet been attained, although the required ignition temperature has been reached in several devices. Evidence to date leads us to believe that we can develop a practical fusion power reactor. Fusion power, if we can develop it, will be far superior to fission power for the following reasons:

1. Virtually infinite amounts of energy are possible from fusion. Uranium supplies for fission power are limited, but heavy hydrogen, or deuterium (the most likely fusion fuel), is abundant. It is estimated that the deuterium in a cubic mile of seawater used as fusion fuel can provide more energy than the petroleum reserves of the entire world.

2. From an environmental viewpoint, fusion power is much "cleaner" than fission power because fusion reactions (in contrast to uranium and plutonium fission reactions) do not produce large amounts of long-lived and dangerously radioactive isotopes.

18.12   Mass–Energy Relationship in Nuclear Reactions

Large amounts of energy are released in nuclear reactions; thus significant amounts of mass are converted to energy. We stated earlier that the amount of mass converted to energy in chemical changes is insignificant compared to the amount of energy released in a nuclear reaction. In fission reactions about 0.1% of the mass is converted into energy. In fusion reactions as much as 0.5% of the mass may be changed into energy. The Einstein equation, $E = mc^2$, can be used to calculate the energy liberated, or available, when the mass loss is known. For example, in the reaction

$$\underset{\text{7.016 g}}{^{7}_{3}\text{Li}} \ + \ \underset{\text{1.008 g}}{^{1}_{1}\text{H}} \ \longrightarrow \ \underset{\text{4.003 g}}{^{4}_{2}\text{He}} \ + \ \underset{\text{4.003 g}}{^{4}_{2}\text{He}} \ + \text{energy}$$

the mass difference between the reactants and products (8.024 g − 8.006 g) is 0.018 g. The energy equivalent to this amount of mass is 1.62×10^{12} J. By comparison, this is more than 4 million times greater than the 3.9×10^5 J of energy obtained from the complete combustion of 12.01 g (1 mol) of carbon.

mass defect

nuclear binding energy

The mass of a nucleus is actually less than the sum of the masses of the protons and neutrons that make up that nucleus. The difference between the mass of the protons and the neutrons in a nucleus and the mass of the nucleus is known as the **mass defect.** The energy equivalent to this difference in mass is known as the **nuclear binding energy.** This energy is the amount that would be required to break a nucleus into its individual protons and neutrons. The higher the binding energy, the more stable the nucleus. Elements of intermediate atomic masses have high binding energies. For example, iron (element number 26) has a very high binding energy and therefore a very stable nucleus. Just as electrons attain less energetic and more stable arrangements through ordinary chemical reactions, neutrons and protons attain less energetic and more stable arrangements through nuclear fission or fusion reactions. Thus when uranium undergoes fission, the products have less mass (and greater binding energy) than the original uranium. In like manner, when hydrogen and lithium fuse to form helium, the helium has less mass (and greater binding energy) than the hydrogen and lithium. It is this conversion of mass to energy that accounts for the very large amounts of energy associated with both fission and fusion reactions.

18.13   Transuranium Elements

transuranium element

The elements that follow uranium on the periodic table and that have atomic numbers greater than 92 are known as the **transuranium elements.** They are synthetic radioactive elements; none of them occur naturally.

The first transuranium element, number 93, was discovered in 1939 by Edwin M. McMillan (1907–1991) at the University of California while he was investigating the fission of uranium. He named it neptunium for the planet Neptune. In 1941, element 94, plutonium, was identified as a beta-decay product of neptunium:

$$^{238}_{93}Np \longrightarrow {}^{238}_{94}Pu + {}^{0}_{-1}e$$

$$^{239}_{93}Np \longrightarrow {}^{239}_{94}Pu + {}^{0}_{-1}e$$

Plutonium is one of the most important fissionable elements known today.

Since 1964, the discoveries of eight new transuranium elements, numbers 104–112, have been announced. These elements have been produced in minute quantities by high energy particle accelerators.

18.14   Biological Effects of Radiation

Radiation with energy to dislocate bonding electrons and create ions when passing through matter is classified as **ionizing radiation.** Alpha particles, beta particles, gamma rays, and X-rays fall into this classification. Ionizing radiation can damage or kill living cells and can be particularly devastating when it strikes the cell nuclei and affects molecules involved in cell reproduction. The effects of radiation on living organisms fall into these general categories: (1) acute or short-term effects, (2) long-term effects, and (3) genetic effects.

ionizing radiation

Acute Radiation Damage

High levels of radiation, especially from gamma rays or X-rays, produce nausea, vomiting, and diarrhea. The effect has been likened to a sunburn throughout the body. If the dosage is high enough, death will occur in a few days. The damaging effects of radiation appear to be centered in the nuclei of the cells, and cells that are undergoing rapid cell division are most susceptible to damage. It is for this reason that cancers are often treated with gamma radiation from a Co-60 source. Cancerous cells multiply rapidly and are destroyed by a level of radiation that does not seriously damage normal cells.

Long-Term Radiation Damage

Protracted exposure to low levels of any form of ionizing radiation can weaken an organism and lead to the onset of malignant tumors, even after fairly long time delays. The largest exposure to synthetic sources of radiation is from X-rays. Evidence suggests that the lives of early workers in radioactivity and X-ray technology may have been shortened by long-term radiation damage.

Strontium-90 isotopes are present in the fallout from atmospheric testing of nuclear weapons. Strontium is in the same periodic-table group as calcium, and its chemical behavior is similar to that of calcium. Hence when foods contaminated with Sr-90 are eaten, Sr-90 ions are laid down in the bone tissue along with ordinary calcium ions. Strontium-90 is a beta emitter with a half-life of 28 years. Blood cells manufactured in bone marrow are affected by the radiation from Sr-90. Hence there is concern that Sr-90 accumulation in the environment may cause an increase in the incidence of leukemia and bone cancers.

CHEMISTRY IN ACTION

A Window into Living Organisms

Imagine viewing a living process as it is occurring. This was the dream of many scientists in the past as they tried to extract this knowledge from dead tissue. Today, because of innovations in nuclear chemistry, this dream is a common, everyday occurrence.

Compounds containing a radionuclide are described as being *labeled* or *tagged*. These compounds undergo their normal chemical reactions, but their location can be detected because of their radioactivity. When such compounds are given to a plant or an animal, the movement of the nuclide can be traced through the organism by the use of a Geiger counter or other detecting device.

In an early use of the tracer technique, the pathway by which CO_2 becomes fixed into carbohydrate ($C_6H_{12}O_6$) during photosynthesis was determined. The net equation for photosynthesis is

$$6\ CO_2 + 6\ H_2O \longrightarrow C_6H_{12}O_6 + 6\ O_2$$

Radioactive $^{14}CO_2$ was injected into a colony of green algae, and the algae was then placed in the dark and killed at selected time intervals. When the radioactive compounds were separated by paper chromatography and analyzed, the results elucidated a series of light-independent photosynthetic reactions.

Biological research using tracer techniques have determined

1. the rate of phosphate uptake by plants, using radiophosphorus
2. the flow of nutrients in the digestive tract using radioactive barium compounds
3. the accumulation of iodine in the thyroid gland, using radioactive iodine
4. the absorption of iron by the hemoglobin of the blood, using radioactive iron

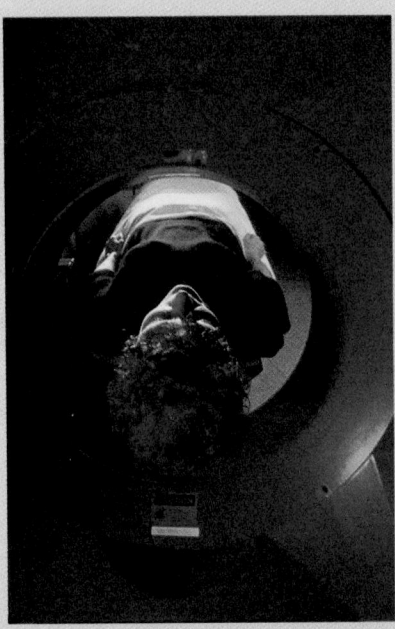

▲ **A radioactive tracer is injected into this patient and absorbed by the brain. The PET scanner detects photons emitted by the tracer and produces an image used in medical diagnosis and research.**

In chemistry, uses for tracers are unlimited. The study of reaction mechanisms, the measurement of the rates of chemical reactions, and the determination of physical constants are just a few of the areas of application.

Radioactive tracers are commonly used in medical diagnosis. The radionuclide must be effective at a low concentration and have a short half-life to reduce the possibility of damage to the patient.

Radioactive iodine (I-131) is used to determine thyroid function, where the body concentrates iodine. In this process a small amount of radioactive potassium or sodium iodide is ingested. A detector is focused on the thyroid gland and measures the amount of iodine in the gland. This picture is then compared to that of a normal thyroid to detect any differences.

Doctors examine the heart's pumping performance and check for evidence of obstruction in coronary arteries by *nuclear scanning*. The radionuclide Tl-201, when injected into the bloodstream, lodges in healthy heart muscle. Thallium-201 emits gamma radiation, which is detected by a special imaging device called a *scintillation camera*. The data obtained are simultaneously translated into pictures by a computer. With this technique doctors can observe whether heart tissue has died after a heart attack and whether blood is flowing freely through the coronary passages.

One of the most recent applications of nuclear chemistry is the use of positron emission tomography (PET) in the measurement of dynamic processes in the body, such as oxygen use or blood flow. In this application a compound is made that contains a positron-emitting nuclide such as C-11, O-15, or N-13. The compound is injected into the body, and the patient is placed in an instrument that detects the positron emission. A computer produces a three-dimensional image of the area.

PET scans have been used to locate the areas of the brain involved with epileptic seizures. Glucose tagged with C-11 is injected, and an image of the brain is produced. Since the brain uses glucose almost exclusively for energy, diseased areas that use glucose at a rate different than normal tissue can then be identified.

Genetic Effects

The information needed to create an individual of a particular species, be it a bacterial cell or a human being, is contained within the nucleus of a cell. This genetic information is encoded in the structure of DNA (deoxyribonucleic acid) molecules, which make up genes. The DNA molecules form precise duplicates of themselves when cells divide, thus passing genetic information from one generation to the next. Radiation can damage DNA molecules. If the damage is not severe enough to prevent the individual from reproducing, a mutation may result. Most mutation-induced traits are undesirable. Unfortunately, if the bearer of the altered genes survives to reproduce, these traits are passed along to succeeding generations. In other words the genetic effects of increased radiation exposure are found in future generations, not in the present generation.

Because radioactive rays are hazardous to health and living tissue, special precautions must be taken in designing laboratories and nuclear reactors, in disposing of waste materials, and in monitoring the radiation exposure of people working in this field.

Concepts in Review

1. Outline the historical development of nuclear chemistry, including the major contributions of Henri Becquerel, Marie Curie, Ernest Rutherford, Irene Joliet-Curie, Otto Hahn, Fritz Strassmann, and Edwin McMillen.

2. Write balanced nuclear chemical equations using isotopic notation.

3. Determine the amount of radionuclide remaining after a given period of time when the starting amount and half-life are given.

4. List the characteristics that distinguish alpha particles, beta particles, and gamma rays from the standpoint of mass, charge, relative velocities, and penetrating power.

5. Describe the effect of a magnetic field on alpha particles, beta particles, and gamma rays.

6. Describe a radioactive disintegration series, and predict which isotope would be formed by the loss of specified numbers of alpha and beta particles from a given radionuclide.

7. Discuss the transmutation of elements.

8. Indicate the methods used for the detection of radiation.

9. Distinguish between radioactive disintegration and nuclear fission reactions.

10. Explain how the fission of U-235 can lead to a chain reaction and why a critical mass is necessary.

11. Explain how the energy from nuclear fission is converted to electrical energy.

12. Explain the difference between fission reactions in a nuclear reactor and those of an atomic bomb.

13. Explain what is meant by the term *nuclear fusion* and why a massive effort to develop controlled nuclear fusion is in progress.

14. Indicate the significance of mass defect and nuclear binding energy.

15. Indicate the major effects of radiation on living organisms.

16. Explain how the age of objects can be determined using radioactivity.

17. Indicate several current uses for radioactive tracers.

Key Terms

alpha particle (18.3)
artificial radioactivity (18.6)
beta particle (18.3)
chain reaction (18.8)
critical mass (18.8)
curie (Ci) (18.7)
gamma ray (18.3)
half-life (18.2)

induced radioactivity (18.6)
ionizing radiation (18.14)
mass defect (18.12)
nuclear binding energy (18.12)
nuclear fission (18.8)
nuclear fusion (18.11)
nucleon (18.1)
nuclide (18.1)

rad (radiation absorbed dose) (18.7)
radioactive decay (18.2)
radioactivity (18.1)
rem (roentgen equivalent to man) (18.7)
roentgen (R) (18.7)
transmutation (18.5)
transuranium element (18.13)

Questions

Questions refer to tables, figures, and key words and concepts defined in the chapter. A particularly challenging question or exercise is indicated with an asterisk.

1. To afford protection from radiation injury, which kind of radiation requires (a) the most shielding? (b) the least shielding?

2. Why is an alpha particle deflected less than a beta particle in passing through an electromagnetic field?

3. Name three pairs of nuclides that might be obtained by fissioning U-235 atoms.

4. Identify these people and their associations with the early history of radioactivity:
   (a) Antoine Henri Becquerel
   (b) Marie and Pierre Curie
   (c) Wilhelm Roentgen
   (d) Ernest Rutherford
   (e) Otto Hahn and Fritz Strassmann

5. Why is the radioactivity of an element unaffected by the usual factors that affect the rate of chemical reactions, such as ordinary changes of temperature and concentration?

6. Distinguish between the terms *isotope* and *nuclide*.

7. The half-life of Pu-244 is 76 million years. If Earth's age is about 5 billion years, discuss the feasibility of finding this nuclide as a naturally occurring nuclide.

8. Tell how alpha, beta, and gamma radiation are distinguished from the standpoint of
   (a) charge
   (b) relative mass
   (c) nature of particle or ray
   (d) relative penetrating power

9. Distinguish between natural and artificial radioactivity.

10. What is a radioactive disintegration series?

11. Briefly discuss the transmutation of elements.

12. Stable Pb-208 is formed from Th-232 in the thorium disintegration series by successive, α, β, β, α, α, α, α, β, β, α particle emissions. Write the symbol (including mass and atomic number) for each nuclide formed in this series.

13. The nuclide Np-237 loses a total of seven alpha particles and four beta particles. What nuclide remains after these losses?

14. Bismuth-211 decays by alpha emission to give a nuclide that in turn decays by beta emission to yield a stable nuclide. Show these two steps with nuclear equations.

15. What was Otto Hahn and Fritz Strassmann's contribution to nuclear physics?

16. What is a breeder reactor? Explain how it accomplishes the "breeding."

17. What is the essential difference between the nuclear reactions in a nuclear reactor and those in an atomic bomb?

18. Why must a certain minimum amount of fissionable material be present before a self-supporting chain reaction can occur?

19. What is mass defect and nuclear binding energy?

20. Explain why radioactive rays are classified as ionizing radiation.

21. Give a brief description of the biological hazards associated with radioactivity.

22. Strontium-90 has been found to occur in radioactive fallout. Why is there so much concern about this radionuclide being found in cow's milk? (Half-life of Sr-90 is 28 years.)

23. What is a radioactive tracer? How is it used?

24. Describe the radiocarbon method for dating archaeological artifacts.

25. How might radioactivity be used to locate a leak in an underground pipe?

26. Anthropologists have found bones whose age suggests that the human line may have emerged in Africa as much as 4 million years ago. If wood or charcoal were found with such bones, would C-14 dating be useful in dating the bones? Explain.

Paired Exercises

27. Indicate the number of protons, neutrons, and nucleons in these nuclei:
    (a) $^{35}_{17}Cl$ (b) $^{226}_{88}Ra$

28. Indicate the number of protons, neutrons, and nucleons in these nuclei:
    (a) $^{235}_{92}U$ (b) $^{82}_{35}Br$

29. How are the mass and the atomic number of a nucleus affected by the loss of an alpha particle?

30. How are the mass and the atomic number of a nucleus affected by the loss of a beta particle?

31. Write nuclear equations for the alpha decay of
(a) $^{218}_{85}At$ (b) $^{221}_{87}Fr$

32. Write nuclear equations for the alpha decay of
(a) $^{192}_{78}Pt$ (b) $^{210}_{84}Po$

33. Write nuclear equations for the beta decay of
(a) $^{14}_{6}C$ (b) $^{137}_{55}Cs$

34. Write nuclear equations for the beta decay of
(a) $^{239}_{93}Np$ (b) $^{90}_{38}Sr$

35. Write nuclear equations for the conversion of $^{13}_{6}C$ to $^{14}_{6}C$.

36. Write nuclear equations for the conversion of $^{30}_{15}P$ to $^{30}_{14}Si$.

37. Complete and balance these nuclear equations by supplying the missing particles:
(a) $^{27}_{13}Al + ^{4}_{2}He \longrightarrow ^{30}_{15}P +$ ____
(b) $^{27}_{14}Si \longrightarrow ^{0}_{+1}e +$ ____
(c) ____ $+ ^{2}_{1}H \longrightarrow ^{13}_{7}N + ^{1}_{0}n$
(d) ____ $\longrightarrow ^{82}_{36}Kr + ^{0}_{-1}e$

38. Complete and balance these nuclear equations by supplying the missing particles:
(a) $^{66}_{29}Cu \longrightarrow ^{66}_{30}Zn +$ ____
(b) $^{0}_{-1}e +$ ____ $\longrightarrow ^{7}_{3}Li$
(c) $^{27}_{13}Al + ^{4}_{2}He \longrightarrow ^{30}_{14}Si +$ ____
(d) $^{85}_{37}Rb +$ ____ $\longrightarrow ^{82}_{35}Br + ^{4}_{2}He$

39. Strontium-90 has a half-life of 28 years. If a 1.00-mg sample were stored for 112 years, what mass of Sr-90 would remain?

40. Strontium-90 has a half-life of 28 years. If a sample was tested in 1980 and found to be emitting 240 counts/min, in what year would the same sample be found to be emitting 30 counts/min? How much of the original Sr-90 would be left?

*41. Consider the fission reaction

$$^{235}_{92}U + ^{1}_{0}n \longrightarrow ^{94}_{38}Sr + ^{139}_{54}Xe + 3\,^{1}_{0}n + energy$$

Calculate the following using this mass data (1.0 g is equivalent to 9.0×10^{13} J):

U-235 = 235.0439 amu   Sr-94 = 93.9154 amu
Xe-139 = 138.9179 amu   n = 1.0087 amu
(a) the energy released in joules for a single event (one uranium atom splitting)
(b) the energy released in joules per mole of uranium splitting
(c) the percentage of mass lost in the reaction

42. Consider the fusion reaction

$$^{1}_{1}H + ^{2}_{1}H \longrightarrow ^{3}_{2}He + energy$$

Calculate the following using this mass data (1.0 g is equivalent to 9.0×10^{13} J):
$^{1}_{1}H$ = 1.00794 amu
$^{2}_{1}H$ = 2.01410 amu
$^{3}_{2}He$ = 3.01603 amu
(a) the energy released in joules per mole of He-3 formed
(b) the percentage of mass lost in the reaction

Additional Exercises

43. If radium costs $50,000 a gram, how much will 0.0100 g of $^{226}RaCl_2$ cost if the price is based only on the radium content?

44. An archaeological specimen was analyzed and found to be emitting only 25% as much C-14 radiation per gram of carbon as newly cut wood. How old is this specimen?

45. Barium-141 is a beta emitter. What is the half-life if a 16.0-g sample of the nuclide decays to 0.500 g in 90 minutes?

*46. Calculate (a) the mass defect and (b) the binding energy of $^{7}_{3}Li$ using the mass data:
$^{7}_{3}Li$ = 7.0160 g   n = 1.0087 g
p = 1.0073 g e^{-} = 0.00055 g
1.0 g ≡ 9.0×10^{13} J (from $E = mc^2$)

*47. In the disintegration series $^{235}_{92}U \longrightarrow ^{207}_{82}Pb$, how many alpha and beta particles are emitted?

48. List three devices used for radiation detection and explain their operation.

49. The half-life of I-123 is 13 hours. If 10 mg of I-123 is administered to a patient, how much I-123 remains after 3 days and 6 hours?

50. Clearly distinguish between fission and fusion. Give an example of each.

51. Starting with 1 g of a radioactive isotope whose half-life is 10 days, sketch a graph showing the pattern of decay for that material. On the x-axis, plot time (you may want to simply show multiples of the half-life), and on the y-axis, plot mass of material remaining. Then after completing the graph, explain why a sample never really gets to the point where all of its radioactivity is considered to be gone.

52. Identify each missing product (name the element and give its atomic number and mass number) by balancing the following nuclear equations:
  (a) $^{235}U + \, ^1_0n \longrightarrow \, ^{143}Xe + 3 \, ^1_0n + \underline{\quad}$
  (b) $^{235}U + \, ^1_0n \longrightarrow \, ^{102}Y + 3 \, ^1_0n + \underline{\quad}$
  (c) $^{14}N + \, ^1_0n \longrightarrow \, ^1H + \underline{\quad}$

53. Consider these reactions:
  (a) $H_2O(l) \longrightarrow H_2O(g)$
  (b) $2 \, H_2(g) + O_2(g) \longrightarrow 2 \, H_2O(g)$
  (c) $^2_1H + \, ^2_1H \longrightarrow \, ^3_1H + \, ^1_1H$
  The following energy values belong to one of these equations:

  energy$_1$ 115.6 kcal released
  energy$_2$ 10.5 kcal absorbed
  energy$_3$ 7.5×10^7 kcal released

  Match the equation to the energy value and briefly explain your choices.

54. When $^{235}_{92}U$ is struck by a neutron, the unstable isotope $^{236}_{92}U$ results. When that daughter isotope undergoes fission, there are numerous possible products. If strontium-90 and three neutrons are the results of one such fission, what is the other product?

55. Write balanced nuclear equations for
  (a) beta emission by $^{29}_{12}Mg$
  (b) alpha emission by $^{150}_{60}Nd$
  (c) positron emission by $^{72}_{33}As$

56. Rubidium-87, a beta emitter, is the product of positron emission. Identify
  (a) the product of rubidium-87 decay
  (b) the precursor of rubidium-87

57. Potassium-42 is used to locate brain tumors. Its half-life is 12.5 hours. Starting with 15.4 mg, what fraction will remain after 100 hours? If it was necessary to have at least 1 μg for a particular procedure, could you hold the original sample for 200 hours before using it?

58. How much of a sample of cesium-137 ($t_{1/2} = 30$ years) must have been present originally if, after 270 years, 15.0 g remains?

59. Suppose that the existence of element 114 were confirmed and reported. What element would this new substance fall beneath in the periodic table? Would it be a metal? What typical ion might you expect it to form in solution?

60. Cobalt-60 has a half-life of 5.26 years. If 1.00 g of ^{60}Co was allowed to decay, how many grams would be left after
  (a) one half-life?
  (b) two half-lives?
  (c) four half-lives?
  (d) ten half-lives?

61. Write balanced equations to show these changes:
  (a) alpha emission by boron-11
  (b) beta emission by strontium-88
  (c) neutron absorption by silver-107
  (d) proton emission by potassium-41
  (e) electron absorption by antimony-116

62. The $^{14}_6C$ content of an ancient piece of wood was found to be one-sixteenth of that in living trees. How many years old is this piece of wood, if the half-life of carbon-14 is 5668 years?

63. The curie is equal to 3.7×10^{10} disintegrations/sec, and the becquerel is equivalent to just 1 disintegration/sec. Suppose a hospital has a 150-g radioactive source with an activity of 1.24 Ci. What is its activity in becquerels?

Answers to Practice Exercises

18.1 0.391 g
18.2 $^{218}_{84}Po$
18.3 (a) $^{234}_{91}Pa$
  (b) $^{226}_{88}Ra$

PUTTING IT TOGETHER
Review for Chapter 18

Multiple Choice: *Choose the correct answer to each of the following.*

1. If $^{238}_{92}U$ loses an alpha particle, the resulting nuclide is
   (a) $^{237}_{92}U$ (b) $^{234}_{90}Th$ (c) $^{238}_{93}Np$ (d) $^{236}_{90}Th$

2. If $^{210}_{82}Pb$ loses a beta particle, the resulting nuclide is
   (a) $^{209}_{83}Bi$ (b) $^{210}_{81}Ti$ (c) $^{206}_{80}Hg$ (d) $^{210}_{83}Bi$

3. In the equation $^{209}_{83}Bi + ? \longrightarrow ^{210}_{84}Po + ^{1}_{0}n$, the missing bombarding particle would be
   (a) $^{2}_{1}H$ (b) $^{1}_{0}n$ (c) $^{4}_{2}He$ (d) $^{0}_{-1}e$

4. Which of the following is not a characteristic of nuclear fission?
   (a) Upon absorption of a proton, a heavy nucleus splits into two or more smaller nuclei.
   (b) Two or more neutrons are produced from the fission of each atom.
   (c) Large quantities of energy are produced.
   (d) Most nuclei formed are radioactive.

5. The half-life of Sn-121 is 10 days. If you started with 40 g of this isotope, how much would you have left 30 days later?
   (a) 10 g (b) none (c) 15 g (d) 5 g

6. $^{241}_{94}Pu$ successively emits β, α, α, β, α, α. At that point, the nuclide has become
   (a) $^{225}_{94}Pu$ (b) $^{225}_{88}Ra$ (c) $^{207}_{84}Po$ (d) $^{219}_{84}Po$

7. The roentgen is the unit of radiation that measures
   (a) an absorbed dose of radiation
   (b) exposure to X-rays
   (c) the dose from a different type of radiation
   (d) the rate of decay of a radioactive substance

8. The radioactivity ray with the greatest penetrating ability is
   (a) alpha (c) gamma
   (b) beta (d) proton

9. In a nuclear reaction,
   (a) mass is lost
   (b) mass is gained
   (c) mass is converted into energy
   (d) energy is converted into mass

10. As the temperature of a radionuclide increases, its half-life
    (a) increases (c) remains the same
    (b) decreases (d) fluctuates

11. The nuclide that has the longest half-life is
    (a) $^{238}_{92}U$ (b) $^{210}_{82}Pb$ (c) $^{234}_{90}Th$ (d) $^{222}_{88}Ra$

12. Which of the following is not a unit of radiation?
    (a) curie (b) roentgen (c) rod (d) rem

13. When $^{235}_{92}U$ is bombarded by a neutron, the atom can fission into
    (a) $^{124}_{53}I + ^{109}_{47}Ag + 2\ ^{1}_{0}n$
    (b) $^{123}_{50}Sn + ^{110}_{42}Mo + 2\ ^{1}_{0}n$
    (c) $^{134}_{56}Ba + ^{128}_{36}Xe + 2\ ^{1}_{0}n$
    (d) $^{90}_{38}Sr + ^{143}_{58}Ce + 2\ ^{1}_{0}n$

14. In the nuclear equation
    $$^{45}_{21}Sc + ^{1}_{0}n \longrightarrow X + ^{1}_{1}H$$
    the nuclide X that is formed is
    (a) $^{45}_{22}Ti$ (b) $^{45}_{20}Ca$ (c) $^{46}_{22}Ti$ (d) $^{45}_{20}K$

15. What type of radiation is a very energetic form of photon?
    (a) alpha (c) gamma
    (b) beta (d) positron

16. When $^{239}_{92}U$ decays to $^{239}_{93}Np$, what particle is emitted?
    (a) positron (c) alpha particle
    (b) neutron (d) beta particle

Free Response Questions

1. The nuclide $^{223}_{87}Fr$ emits three radioactive particles losing 8 mass units and 3 atomic number units. Propose a radioactive decay series and write the symbol for the resulting nuclide.

2. How many half-lives are required to change 96 g of a sample of a radioactive isotope to 1.5 g over approximately 24 days? What is the approximate half-life of this isotope?

3. (a) What type of process is $^{6}_{3}Li + ^{1}_{0}n \longrightarrow ^{3}_{1}H + ^{4}_{2}He$?
   (b) How is this process different from radioactive decay?
   (c) Could nuclear fission be classified as the same type process?

4. Calculate the nuclear binding energy of $^{56}_{26}Fe$.
   Mass data:
   $^{56}_{26}Fe = 55.9349$ g/mol
   $n = 1.0087$ g/mol $e^{-} = 0.00055$ g/mol
   $p = 1.0073$ g/mol 1.0 g $= 9.0 \times 10^{13}$ J

CHAPTER

19

Organic Chemistry: Saturated Hydrocarbons

▲ Organic molecules are studied with advanced technology. The instrumentation pictured here separates one chemical from another and is commonly used in the pharmaceutical industry.

CHAPTER 19 / OUTLINE

Many substances throughout nature contain silicon or carbon within their molecular structures. Silicon is the staple of the geologist—it combines with oxygen in a variety of ways to produce silica and a family of compounds known as the silicates. These compounds form the chemical foundation of most types of sand, rocks, and soil, essential materials of the construction industry.

In the living world, carbon, in combination with hydrogen, oxygen, nitrogen, and sulfur, forms the basis for millions of organic compounds. Carbon compounds provide us with energy sources in the form of hydrocarbons and their derivatives that allow us to heat and light our homes, drive our automobiles to work, and fly off to Hawaii for vacation. Small substitutions in these carbon molecules can produce chlorofluorocarbons, compounds used in plastics and refrigerants. An understanding of these molecules and their effect upon our global environment is vital in the continuing search to find ways to maintain our lifestyles while preserving the planet.

19.1 Organic Chemistry: History and Scope

During the late 18th and the early 19th centuries, chemists were baffled by the fact that compounds obtained from animal and vegetable sources defied the established rules for inorganic compounds—namely, that compound formation is due to a simple attraction between positively and negatively charged elements. In their experience with inorganic chemistry, groups of two or three elements formed only one, or at most a few, different compounds. However, they observed that one group—carbon, hydrogen, oxygen, and nitrogen—gave rise to a large number of different compounds that often were remarkably stable.

Because no organic compounds had been synthesized from inorganic substances and because there was no other explanation for the complexities of organic compounds, chemists believed that organic compounds were formed by some "vital force." The **vital force theory** held that organic substances could originate only from living material. In 1828, a German chemist Friedrich Wöhler (1800–1882) did a simple experiment that eventually proved to be the death blow to this theory. In attempting to prepare ammonium cyanate (NH_4CNO) by heating cyanic acid (HCNO) and ammonia, Wöhler obtained a white crystalline substance that he identified as urea. Wöhler knew that urea must be an authentic organic substance because it is a product of metabolism that can be isolated from urine. Although Wöhler's discovery was not immediately and generally recognized, the vital force theory was overthrown by this simple observation that an organic compound had been made from nonliving materials.

After the work of Wöhler, it was apparent that no vital force other than skill and knowledge was needed to make organic chemicals in the laboratory and that inorganic as well as organic substances could be used as raw materials. Today, **organic chemistry** designates the branch of chemistry that deals with carbon compounds but does not imply that these compounds must originate from some form of life. A few special kinds of carbon compounds (e.g., carbon oxides, metal carbides, and metal

vital force theory

The formula for urea is

$$H_2N-\overset{\overset{\displaystyle O}{\|}}{C}-NH_2$$

organic chemistry

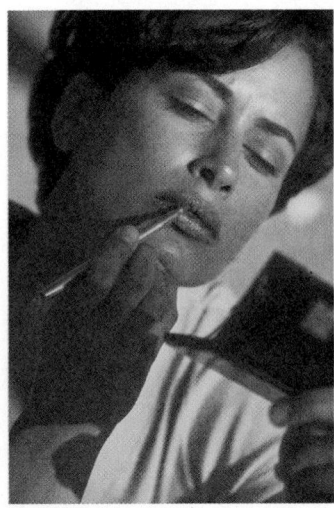

▲
Organic molecules form the products used in cosmetics and perfumes.

carbonates) are often excluded from the organic classification because their chemistry is more conveniently related to that of inorganic substances.

The field of organic chemistry is vast, for it includes not only the composition of all living organisms, but also that of a great many other materials that we use daily. Examples of organic materials are nutrients (fats, proteins, carbohydrates); fuels; fabrics (cotton, wool, rayon, nylon); wood and paper products; paints and varnishes; plastics; dyes; soaps and detergents; cosmetics; medicinals; rubber products; and explosives.

What makes carbon special and different from the other elements in the periodic table? Carbon has the unique ability to bond to itself in long chains and rings of varying size. The greater the number of carbon atoms in a molecule, the more ways there are to link these atoms in different arrangements. This flexibility in the arrangement of atoms produces compounds with the same chemical composition and different structures. There is no theoretical limit on the number of organic compounds that can exist. In addition to this unique bonding property, carbon forms strong covalent bonds with a variety of elements, especially hydrogen, nitrogen, oxygen, sulfur, phosphorus, and the halogens. These are the elements most commonly found in organic compounds.

 The Carbon Atom: Bonding and Shape

The carbon atom is central to all organic compounds. The atomic number of carbon is 6, and its electron structure is $1s^2 2s^2 2p^2$. Two stable isotopes of carbon exist, ^{12}C and ^{13}C. In addition, there are several radioactive isotopes; ^{14}C is the most widely known of these because of its use in radiocarbon dating. With four electrons in its outer shell, carbon has oxidation numbers ranging from $+4$ to -4, and it forms predominantly covalent bonds. Carbon occurs as the free element in diamond, graphite, coal, coke, carbon black, charcoal, lampblack, and buckminsterfullerene.

A carbon atom usually forms four covalent bonds; each bond results from two atoms sharing a pair of electrons (see Section 11.5). The number of electron pairs that two atoms share determines whether the bond is single or multiple. In a single bond, only one pair of electrons is shared by the atoms. If both atoms have the same electronegativity (see Section 11.6) the bond is classified as nonpolar. In this type of bond there is no separation of positive and negative charge between the atoms.

Carbon also forms multiple bonds by sharing two or three pairs of electrons between two atoms. The double bond formed by sharing two electron pairs is stronger than a single bond but not twice as strong. It is also shorter than a single bond. Similarly, the triple bond formed by sharing three electron pairs is stronger and shorter than a double bond. An organic compound is classified as **saturated** if it contains only single bonds and as **unsaturated** if the molecules possess one or more multiple carbon–carbon bonds.

saturated compound
unsaturated compound

Lewis structures are useful in representing the bonding between atoms in a molecule, but these representations tell us little about the geometry of the molecules. A number of bonding theories predict the shape of molecules; one common theory is the valence shell electron pair repulsion (VSEPR) theory (see Section 11.11). This is a fairly simple, yet accurate method for determining the shape of a molecule.

VSEPR theory states that electron pairs repel each other since they have like charges. The electron pairs will therefore try to spread out as far as possible around an atom. In addition, unshared pairs of electrons occupy more space than shared electron pairs.

◀ FIGURE 19.1
Tetrahedral structure of carbon: (a) a regular tetrahedron; (b) a carbon atom with tetrahedral bonds; (c) a carbon atom within a regular tetrahedron; (d) a methane molecule, CH_4.

What does VSEPR theory tell us about the shapes of carbon-containing compounds? Consider the simplest organic molecule, methane, CH_4. It contains one carbon atom with four single bonds to hydrogen atoms. The four shared electron pairs must be placed as far apart as possible in three dimensions. This results in the hydrogen atoms forming the corners of a tetrahedron with the carbon atom in the center (see Figure 19.1). The angle between these tetrahedral bonds is 109.5°, which is the least strained of the possible angles.

Double or triple bonds have a significant effect on the shape of the molecule. Because of increased repulsion between the pairs, additional electron pairs that are in close proximity take up more space than those in a single bond. Consider the Lewis structure for C_2H_4 (see Figure 19.2a). Each carbon atom has three separate regions for shared electrons. To place them as far apart as possible requires placing each atom at the corner of a triangle (see Figure 19.2b). The bond angles around the carbon atoms are 120°.

In a triple bond the carbon has only two regions for shared electrons. To be placed as far apart as possible, a linear arrangement is required. The bond angle is 180°.

(a) H:C::C:H
         ̈H ̈H

(b) H   120° H
      \   ⟳   /
        C ⚌ C
      /   \
    H   H

▲
FIGURE 19.2
(a) Lewis structure for C_2H_4.
(b) Shape of a molecule with a carbon–carbon double bond. The hydrogens and carbon form the vertices of a triangle. The bond angles around the carbon are 120°.

$$\overset{180°}{H:C:::C:H \qquad H-C\equiv C-H}$$

We will study more about the carbon–carbon double and triple bonds in Chapter 20.

Practice 19.1

Estimate the bond angles around the starred carbon (C*) for the following:

(a)   H
                    |
       N≡C—C*—H
                    |
                   H

(b) Cl—C=C*—H
            | |
           H   H

19.3 Molecular Models

Models are often used in organic chemistry to illustrate molecules. There are many ways to represent molecules, and each has its advantages and disadvantages. Two common methods are ball-and-stick models and space-filling models. Both types of models are illustrated in Figure 19.3 for the methane molecule.

FIGURE 19.3 ▶
*Molecular models of methane:
(a) ball-and-stick model; (b)
space-filling model.*

(a)   (b)

H : C : C : H (a)   H—C—C—H (b)   (c)   (d)

▲
FIGURE 19.4
*Models of ethane: (a) Lewis
structure; (b) bond-line
drawing; (c) ball-and-stick
model; (d) space-filling model.*

In the ball-and-stick model, different atoms are represented by different colored balls, and bonds are represented by sticks. The space-filling model gives a more accurate picture of the actual molecule but is not as clear in showing the chemical bonds between the atoms. Since most organic molecules contain many atoms, it is often difficult or inconvenient to draw either of these types of models on paper every time we wish to represent a molecule. Chemists often translate a three-dimensional model into a two-dimensional representation (a Lewis structure or bond-line drawing). Examples of the four models are shown in Figure 19.4. In both Lewis and bond-line drawings, it is important to remember that the bond angle is not 90° the way it appears in the drawing. Rather it is 109.5° because the molecule is actually three-dimensional.

Formulas of organic molecules are also represented differently than those of inorganic compounds. A formula gives information about the composition of a compound. Inorganic compounds contain relatively few groups of atoms and are frequently represented by *empirical formulas,* which give only the simplest ratio of the atoms in a molecule. Larger molecules are often represented by a *molecular formula,* which gives more information. The molecular formula gives the actual number of atoms in a molecule. For inorganic compounds the empirical and molecular formulas are often the same. The models we've just considered represent yet another even more informative type of formula. In a *structural formula,* the arrangement of the atoms within the molecule is clearly shown. Organic chemists often shorten these structural formulas into a final type called *condensed structural formulas.* In these formulas each carbon is grouped with the hydrogens bonded to it and then written as a formula—for example, CH_3CH_3 or $CH_3CH_2CH_2CH_3$. The diversity of methods for writing organic formulas is illustrated in Table 19.1.

TABLE 19.1 **Classes of Organic Compounds**

Class of compound	General formula^a	IUPAC name^{b,c}	Molecular formula	Condensed formula	Structural formula
Alkane	RH	Eth*ane* (Ethane)	C_2H_6	CH_3CH_3	
Alkene	$R-CH=CH_2$	Eth*ene* (Ethylene)	C_2H_4	$H_2C=CH_2$	
Alkyne	$R-C\equiv C-H$	Eth*yne* (Acetylene)	C_2H_2	$HC\equiv CH$	$H-C\equiv C-H$
Alkyl halide	RX	Chloroethane (Ethyl chloride)	C_2H_5Cl	CH_3CH_2Cl	
Alcohol	ROH	Ethan*ol* (Ethyl alcohol)	C_2H_6O	CH_3CH_2OH	
Ether	$R-O-R$	Methoxymethane (Dimethyl ether)	C_2H_6O	CH_3OCH_3	
Aldehyde	$R-\overset{\mid}{\underset{H}{C}}=O$	Ethan*al* (Acetaldehyde)	C_2H_4O	CH_3CHO	
Ketone	$R-\underset{O}{\overset{\|}{C}}-R$	Propan*one* (Dimethyl ketone)	C_3H_6O	CH_3COCH_3	
Carboxylic acid	$R-\underset{O}{\overset{\|}{C}}-OH$	Ethan*oic acid* (Acetic acid)	$C_2H_4O_2$	CH_3COOH	
Ester	$R-\underset{O}{\overset{\|}{C}}-OR$	Methyl ethan*oate* (Methyl acetate)	$C_3H_6O_2$	CH_3COOCH_3	
Amide	$R-\underset{O}{\overset{\|}{C}}-NH_2$	Ethan*amide* (Acetamide)	C_2H_5ON	CH_3CONH_2	
Amine	$R-CH_2-NH_2$	Aminoethane (Ethylamine)	C_2H_7N	$CH_3CH_2NH_2$	

^{a}The letter R is used to indicate any of the many possible alkyl groups.
^{b}Class name ending in italic.
^{c}Common name in parentheses.

▲
Fossil fuels are mined from many sources: An oil exploration rig in Texas, a coal miner in Pennsylvania, and a natural gas operation in Washington (left to right).

 Classifying Organic Compounds

It is impossible for anyone to study the properties of each of the millions of organic compounds. Organic compounds with similar structures are grouped into classes as shown in Table 19.1. The members of each class of compounds contain a characteristic *functional group* atom or group of atoms called a **functional group** shown as the red portion of the structural formula in Table 19.1. Organic compounds from different classes may have the same molecular formulas but completely different chemical and physical properties.

19.5   Hydrocarbons

hydrocarbon **Hydrocarbons** are compounds that are composed entirely of carbon and hydrogen atoms bonded to each other by covalent bonds. Hydrocarbons are classified into two major categories, aliphatic and aromatic. The term *aromatic* refers to compounds that contain benzene rings. All hydrocarbons that are not aromatic are often described as *aliphatic* **aliphatic** (from the Greek word *aleiphar,* meaning "fat"). The aliphatic hydrocarbons include the alkanes, alkenes, alkynes, and cycloalkanes (see Figure 19.5).

Fossil fuels—natural gas, petroleum, and coal—are the principal sources of hydrocarbons. Natural gas is primarily methane with small amounts of ethane, propane, and butane. Petroleum is a mixture of hydrocarbons from which gasoline, kerosene, fuel oil, lubricating oil, paraffin wax, and petrolatum (themselves mixtures of hydrocarbons) are separated. Coal tar, a volatile by-product of making coke from coal for use in the steel industry, is the source of many valuable chemicals including the aromatic hydrocarbons benzene, toluene, and naphthalene.

The fossil fuels provide a rich resource of hydrocarbons for human society. In the past these resources have been used primarily as a source of heat (via combustion). We now see that extensive combustion can have severe environmental consequences (e.g., air pollution and global warming). Fossil fuels also serve as the raw materials for much of today's chemical industry. One theme that runs through the study of organic chemistry is the synthetic relationship between compounds; for example, acids are often formed from alcohols. Fossil fuels are the starting materials for many of these synthetic sequences. And, in the long run, the fossil fuels may well prove to be more valuable to us as a source of organic chemicals than as a source of heat.

 Saturated Hydrocarbons: Alkanes

The **alkanes,** also known as *paraffins* or *saturated hydrocarbons,* are straight- or branched-chain hydrocarbons with only single covalent bonds between the carbon atoms. We shall study the alkanes in some detail because many other classes of organic compounds can be considered as derivatives of these substances. For example, it is necessary to learn the names of the first ten members of the alkane series because these names are used as a basis for naming other classes of compounds.

Methane, CH_4, is the first member of the alkane series. Alkanes with two-, three-, and four-carbon atoms are ethane, propane, and butane, respectively. The names of the first four alkanes are of common or trivial origin and must be memorized; but the names beginning with the fifth member, pentane, are derived from Greek numbers and are relatively easy to recall. The names and formulas of the first ten members of the series are given in Table 19.2.

Successive compounds in the alkane series differ from each other in composition by one carbon and two hydrogen atoms. When each member of a series differs from the next member by a CH_2 group, the series is called a **homologous series.** The members of a homologous series are similar in structure but have a regular difference in formula. All common classes of organic compounds exist in homologous series. Each homologous series can be represented by a general formula. For all open-chain alkanes, the general formula is C_nH_{2n+2}, where n corresponds to the number of carbon atoms in the molecule. The molecular formula of any specific alkane is easily determined from this general formula. Thus, for pentane, $n = 5$ and $2n + 2 = 12$, so the formula is C_5H_{12}. For hexadecane, the 16-carbon alkane, the formula is $C_{16}H_{34}$.

alkane

$CH_3CH_2CH_2CH_2CH_3$
straight carbon chain

CH_3
|
$CH_3CHCH_2CH_3$
branched chain of carbon atoms

homologous series

C_nH_{2n+2}
**general formula for
open-chain alkanes**

TABLE 19.2	Names, Formulas, and Physical Properties of Straight-Chain Alkanes			
Name	Molecular formula C_nH_{2n+2}	Condensed structural formula	Boiling point (°C)	Melting point (°C)
Methane	CH_4	CH_4	−161	−183
Ethane	C_2H_6	CH_3CH_3	−88	−172
Propane	C_3H_8	$CH_3CH_2CH_3$	−45	−187
Butane	C_4H_{10}	$CH_3CH_2CH_2CH_3$	−0.5	−138
Pentane	C_5H_{12}	$CH_3CH_2CH_2CH_2CH_3$	36	−130
Hexane	C_6H_{14}	$CH_3CH_2CH_2CH_2CH_2CH_3$	69	−95
Heptane	C_7H_{16}	$CH_3CH_2CH_2CH_2CH_2CH_2CH_3$	98	−90
Octane	C_8H_{18}	$CH_3CH_2CH_2CH_2CH_2CH_2CH_2CH_3$	125	−57
Nonane	C_9H_{20}	$CH_3CH_2CH_2CH_2CH_2CH_2CH_2CH_2CH_3$	151	−54
Decane	$C_{10}H_{22}$	$CH_3CH_2CH_2CH_2CH_2CH_2CH_2CH_2CH_2CH_3$	174	−30

Practice 19.2

(a) Write molecular formulas for the alkanes that contain 12, 14, and 20 carbons.

(b) Write the condensed structural formulas for the straight-chain alkanes that contain 12 and 14 carbons.

19.7   Carbon Bonding in Alkanes

A carbon atom is capable of forming single covalent bonds with one, two, three, or four other atoms. To understand this remarkable bonding ability, we must look at the electron structure of carbon. The valence electrons of carbon in their ground state are $2s^2 2p_x^1 2p_y^1$. All of carbon's valence electrons can be shared to make a total of four bonds. When a carbon atom is bonded to other atoms by single bonds (e.g., to four hydrogen atoms in CH_4), it would appear at first that there should be two different types of bonds—bonds involving the $2s$ electrons and bonds involving the $2p$ electrons of the carbon atom. However, this is not the case. All four carbon–hydrogen bonds are identical.

If the carbon atom is to form four equivalent bonds, its electrons in the $2s$ and $2p$ orbitals must rearrange to four equivalent orbitals. To form the four equivalent orbitals, imagine that a $2s$ electron is promoted to a $2p$ orbital, giving carbon an outer shell electron structure of $2s^1 2p_x^1 2p_y^1 2p_z^1$. The $2s$ orbital and the three $2p$ orbitals then hybridize to form four equivalent hybrid orbitals, which are designated sp^3 orbitals. The orbitals formed (sp^3) are neither s orbitals nor p orbitals, but are instead a hybrid of those orbitals, having one-fourth s orbital character and three-fourths p orbital character. This process is illustrated in Figure 19.6. It is these sp^3 orbitals that are directed toward the corners of a regular tetrahedron (see Figure 19.7).

(a)   (b)   (c)

A single bond is formed when one of the sp^3 orbitals overlaps an orbital of another atom. Thus each C—H bond in methane is the result of the overlapping of a carbon sp^3 orbital and a hydrogen s orbital, Figure 19.7c. Once the bond is formed, the pair of bonding electrons constituting it are said to be in a molecular orbital. In a similar way, a C—C single bond results from the overlap of sp^3 orbitals between two carbon atoms. This type of bond (C—H and C—C) is called a sigma (σ) bond. A **sigma bond** exists if the electron cloud formed by the pair of bonding electrons lies on a straight line drawn between the nuclei of the bonded atoms.

sigma bond

19.8 Isomerism

The properties of an organic substance are dependent on its molecular structure. The majority of organic compounds are made from relatively few elements—carbon, hydrogen, oxygen, nitrogen, and the halogens. The valence bonds or points of attachment may be represented in structural formulas by a corresponding number of dashes attached to each atom:

$$-\overset{|}{\underset{|}{C}}- \qquad H- \qquad -O- \qquad -\overset{}{\underset{|}{N}}- \qquad Cl- \qquad Br- \qquad I- \qquad F-$$

Thus carbon has four bonds to each atom, nitrogen three bonds, oxygen two bonds, and hydrogen and the halogens one bond to each atom.

In an alkane, each carbon atom is joined to four other atoms by four single covalent bonds. These bonds are separated by angles of 109.5° (the angles correspond to those formed by lines drawn from the center of a regular tetrahedron to its corners). Alkane molecules contain only carbon–carbon and carbon–hydrogen bonds and are essentially nonpolar. Alkane molecules are nonpolar because (1) carbon–carbon bonds are nonpolar since they are between like atoms; (2) carbon–hydrogen bonds are only slightly polar because the difference in electronegativity between carbon and hydrogen atoms is small; and (3) the bonds in an alkane are symmetrically directed toward the corners of a tetrahedron. Because of their low polarity, alkane molecules have very little intermolecular attraction and therefore have relatively low boiling points compared with other organic compounds of similar molar mass.

Without the use of models or perspective drawings, the three-dimensional character of atoms and molecules is difficult to portray accurately. However, concepts of structure can be conveyed to some extent by structural formulas.

To write the correct structural formula for propane, C_3H_8, we must determine how to place each atom in the molecule. An alkane contains only single bonds. Therefore, each carbon atom must be bonded to four other atoms by either C—C or C—H bonds. Hydrogen must be bonded to only one carbon atom by a C—H bond, since C—H—C bonds do not occur and an H—H bond would simply represent a hydrogen molecule. Applying this information, we find that the only possible structure for propane is

$$
\begin{array}{ccccccc}
 & H & & H & & H & \\
 & | & & | & & | & \\
H- & C & - & C & - & C & -H \\
 & | & & | & & | & \\
 & H & & H & & H &
\end{array}
$$

propane

However, it is possible to write two structural formulas corresponding to the molecular formula C_4H_{10}:

"Isobutane" is a common name; a more systematic nomenclature will be introduced in Section 19.9.

$$
\begin{array}{ccccccccc}
 & H & & H & & H & & H & \\
 & | & & | & & | & & | & \\
H- & C & - & C & - & C & - & C & -H \\
 & | & & | & & | & & | & \\
 & H & & H & & H & & H &
\end{array}
$$

butane

and

$$
\begin{array}{c}
H \\
| \\
H-C-H \\
\\
\text{H—C—C—C—H}
\end{array}
$$

isobutane

Two C_4H_{10} compounds with the structural formulas shown here actually exist. The butane with the unbranched carbon chain is called *butane* or *normal butane* (abbreviated *n*-butane); it boils at −0.5°C and melts at −138.3°C. The common name for the branched-chain butane is *isobutane;* it boils at −11.7°C and melts at −159.5°C. These differences in physical properties are sufficient to establish that the two compounds are different substances even though they have the same molecular formula. Models illustrating the structural arrangement of the atoms in methane, ethane, propane, butane, and isobutane are shown in Figure 19.8.

This phenomenon of two or more compounds having the same molecular formula but different structural arrangements of their atoms is called **isomerism.** The various individual compounds are called **isomers.** For example, there are two isomers of butane, C_4H_{10}. Isomerism is very common among organic compounds and is another reason for the large number of known compounds. There are 3 isomers of pentane, 5 isomers of hexane, 9 isomers of heptane, 18 isomers of octane, 35 isomers of nonane, and 75 isomers of decane. The phenomenon of isomerism is a compelling reason for the use of structural formulas.

isomerism

isomer

▲
FIGURE 19.8
Ball-and-stick models illustrating structural formulas of methane, ethane, propane, butane, and isobutane.

Isomers are compounds that have the same molecular formula but different structural formulas.

To save time and space in writing, condensed structural formulas are often used. In the condensed structural formulas, atoms and groups that are attached to a carbon atom are generally written to the right of that carbon atom.

Let's interpret the condensed structural formula for propane:

$$\overset{1}{CH_3} - \overset{2}{CH_2} - \overset{3}{CH_3} \quad \text{or} \quad CH_3CH_2CH_3$$

Carbon 1 has three hydrogen atoms attached to it and is bonded to carbon 2, which has two hydrogen atoms on it and is bonded to carbon 3, which has three hydrogen atoms bonded to it.

When structures are written as in butane, $CH_3CH_2CH_2CH_3$, the bonds (long dashes) are often not shown. But when a group is shown above or below the main carbon chain, as in isobutane (2-methylpropane), a vertical dash must be used to indicate the point of attachment of the group to the main carbon chain.

Example 19.1 There are three isomers of pentane, C_5H_{12}. Write structural formulas and condensed structural formulas for these isomers.

Solution In a problem of this kind it is best to start by writing the carbon skeleton of the compound containing the longest continuous carbon chain. In this case, it is five carbon atoms (from pentane):

$$C-C-C-C-C$$

Now complete the structure by attaching hydrogen atoms around each carbon atom so that each carbon atom has four bonds attached to it. The carbon atoms at each end of the chain need three hydrogen atoms. The three inner carbon atoms each need two hydrogen atoms to give them four bonds.

$$
\begin{array}{ccccc}
H & H & H & H & H \\
| & | & | & | & | \\
H-C- & C- & C- & C- & C-H \\
| & | & | & | & | \\
H & H & H & H & H
\end{array}
\qquad CH_3CH_2CH_2CH_2CH_3
$$

For the next isomer, start by writing a four-carbon chain and attach the fifth carbon atom to either of the middle carbon atoms—do not use the end ones.

$$
\begin{array}{cc}
\quad C & \quad\quad C \\
\quad | & \quad\quad | \\
C-C-C-C & \quad C-C-C-C
\end{array}
\qquad
\begin{array}{l}
\text{Both of these structures} \\
\text{represent the same compound.}
\end{array}
$$

Now add the 12 hydrogen atoms to complete the structure:

$$
\begin{array}{ccccc}
 & & H & & \\
 & H & | & H & \\
 & \diagdown & C & \diagup & \\
H & H & | & H & \\
| & | & | & | & \\
H-C- & C- & C- & C-H & \\
| & | & | & | & \\
H & H & H & H &
\end{array}
\qquad
\begin{array}{c}
CH_3 \\
| \\
CH_3CH_2CHCH_3
\end{array}
\quad \text{or} \quad CH_3CH_2CH(CH_3)_2
$$

This is the only four-chain compound that can be written.

For the third isomer, start with a three-carbon chain, attach the other two carbon atoms to the central carbon atom, and complete the structure by adding the 12 hydrogen atoms:

$$
\begin{array}{cc}
C & \\
| & \\
C-C-C & \\
| & \\
C &
\end{array}
\qquad
\begin{array}{ccc}
 & H & \\
 H & | & H \\
 & C & \\
 H & | & H \\
 | & | & | \\
H-C- & C- & C-H \\
 | & | & | \\
 H & | & H \\
 & C & \\
 H & | & H \\
 & H &
\end{array}
\qquad
\begin{array}{c}
CH_3 \\
| \\
CH_3CCH_3 \\
| \\
CH_3
\end{array}
\quad \text{or} \quad C(CH_3)_4
$$

No more isomers of C_5H_{12} can be written.

Practice 19.3

Write structural and condensed formulas for the isomers of hexane, C_6H_{14}.

19.9 Naming Organic Compounds

In the early development of organic chemistry, each new compound was given a name, usually by the person who had isolated or synthesized it. Names were not systematic but often conveyed some information—usually about the origin of the substance. Wood alcohol (methanol), for example, was so named because it was obtained by destructive distillation or pyrolysis of wood. Methane, formed during underwater decomposition of vegetable matter in marshes, was originally called marsh gas. A single compound was often known by several names. For example, the active ingredient in alcoholic beverages has been called alcohol, ethyl alcohol, methyl carbinol, grain alcohol, spirit, and ethanol.

A meeting in Geneva in 1892 initiated the development of an international system for naming compounds. In its present form the method recommended by the International Union of Pure and Applied Chemistry is systematic, generally unambiguous, and internationally accepted. It is called the **IUPAC System.** Despite the existence of the official IUPAC System, a great many well-established common, or trivial, names and abbreviations (e.g., TNT and DDT) are used because of their brevity and/or convenience. It is thus necessary to have a knowledge of both the IUPAC System and many common names.

IUPAC System

In order to name organic compounds systematically, you must be able to recognize certain common alkyl groups. **Alkyl groups** have the general formula C_nH_{2n+1} (one less hydrogen atom than the corresponding alkane). The missing H atom may have been detached from any carbon in the alkane. The name of the group is formed from the name of the corresponding alkane by simply dropping *-ane* and substituting a *-yl* ending. The names and formulas of selected alkyl groups up to and including four carbon atoms are given in Table 19.3 on page 507. The letter R is often used in formulas to mean any of the many possible alkyl groups.

alkyl group

$$R = C_nH_{2n+1} \qquad \text{Any alkyl group}$$

Three prefixes commonly indicate structural information. They are *iso-, sec-* (for secondary) and *tert-* or *t-* (for tertiary) (see Table 19.3). *Iso-* indicates that a compound or alkyl group has the following structure at one end of the carbon chain:

$$
\begin{array}{c}
\quad\ \ \ CH_3 \\
\quad\ \ \ | \\
CH_3\!-\!CH\!-
\end{array}
$$

A carbon atom bonded to only one other carbon is a primary carbon atom (1°). A carbon atom bonded to two other carbons is a secondary (*sec-*) carbon atom (2°). Thus, the name *sec-*butyl indicates that this group contains four carbons (butyl) and the carbon that connects this group to other atoms is bonded to two other carbon atoms. A carbon atom bonded to three other carbons is a tertiary carbon atom (3°). So the name *tert-*butyl indicates that this group contains four carbons (butyl) and the carbon that connects this group to other atoms is bonded to three other carbons (see Table 19.3).

$$1° \rightarrow CH_3$$
$$\underset{1°\quad\quad 3°\quad\quad 2°\quad\quad 2°\quad\quad 1°}{CH_3-CH-CH_2-CH_2-CH_3}$$

Practice 19.4

Locate the primary, secondary, and tertiary carbon atoms in this compound:

$$CH_3-\underset{\overset{|}{CH_3}}{CH}-CH_2-CH_2-\underset{\overset{|}{CH_3}}{CH}-CH_2-\underset{\underset{CH_3}{|}}{\overset{\overset{CH_3}{|}}{C}}-CH_2-\underset{\overset{|}{CH_2CH_3}}{CH}-CH_3$$

A few relatively simple rules are all that are needed to name a great many alkanes according to the IUPAC System. In later sections these rules will be extended to cover other classes of compounds, but advanced texts or references must be consulted for the complete system.

IUPAC Rules for Naming Alkanes

1. Select the longest continuous chain of carbon atoms as the parent compound, and consider all alkyl groups attached to it as branch chains that have replaced hydrogen atoms of the parent hydrocarbon. The name of the alkane consists of the name of the parent compound prefixed by the names of the branch-chain alkyl groups attached to it.
2. Number the carbon atoms in the parent carbon chain from one end to the other, starting from the end closest to the first carbon atom that has a branch chain.

$$\underset{1\quad 2\quad 3\quad 4\quad 5\quad 6}{CH_3CH_2\overset{\overset{CH_3}{|}}{CH}CH_2CH_2CH_3} \quad (not \underset{6\quad 5\quad 4\quad 3\quad 2\quad 1}{CH_3CH_2\overset{\overset{CH_3}{|}}{CH}CH_2CH_2CH_3})$$

3. Name each branch-chain alkyl group and designate its position on the parent carbon chain by a number (e.g., 3-methyl means a methyl group attached to carbon 3).
4. When the same alkyl-group branch chain occurs more than once, indicate this by a prefix (*di-*, *tri-*, *tetra-*, etc.) written in front of the alkyl-group name (e.g., *dimethyl* indicates two methyl groups). The numbers indicating the positions of these alkyl groups are separated by a comma, followed by a hyphen, and placed in front of the name (e.g., 2,3-dimethyl).
5. When several different alkyl groups are attached to the parent compound, list them in alphabetical order (e.g., *ethyl* before *methyl* as in 3-ethyl-4-methyloctane).

Adding numerical prefixes (di-, tri-, tetra-, etc.) to the alkyl name does not change the alphabetical order. For example, ethyl is still named before dimethyl.

TABLE 19.3	Names and Formulas of Selected Alkyl Groups		
Formula	**Name**	**Formula**	**Name**
CH_3-	methyl	CH_3 $\vert$ CH_3CH-	isopropyl
CH_3CH_2-	ethyl		
$CH_3CH_2CH_2-$	propyl	CH_3 $\vert$ CH_3CHCH_2-	isobutyl
$CH_3CH_2CH_2CH_2-$	butyl		
$CH_3(CH_2)_3CH_2-$	pentyl		
$CH_3(CH_2)_4CH_2-$	hexyl	CH_3 $\vert$ CH_3CH_2CH-	sec-butyl (secondary butyl)
$CH_3(CH_2)_5CH_2-$	heptyl		
$CH_3(CH_2)_6CH_2-$	octyl	CH_3 $\vert$ CH_3C- $\vert$ CH_3	tert-butyl or t-butyl (tertiary butyl)
$CH_3(CH_2)_7CH_2-$	nonyl		
$CH_3(CH_2)_8CH_2-$	decyl		

The compound shown here is commonly called isopentane. Let's name it using the IUPAC System:

$$\overset{4}{CH_3}-\overset{3}{CH_2}-\overset{2}{CH}-\overset{1}{CH_3}$$
$$\vert$$
$$CH_3$$

2-methylbutane
(isopentane)

Explanation The longest continuous chain contains four carbon atoms. Therefore, the parent compound is butane and the compound is named as a butane (Rule 1). The methyl group (CH_3-) attached to carbon 2 is named as a prefix to butane, the "2-" indicating the point of attachment of the methyl group on the butane chain (Rule 3).

How would we write the structural formula for 2-methylpentane? An analysis of its name gives us this information.

1. The parent compound, pentane, contains five carbons. Write and number the five-carbon skeleton of pentane. (At this point, numbering can be either from right to left or from left to right.)

$$\overset{5}{C}-\overset{4}{C}-\overset{3}{C}-\overset{2}{C}-\overset{1}{C}$$

2. Put a methyl group on carbon 2 ("2-methyl" in the name gives this information):

$$\overset{5}{C}-\overset{4}{C}-\overset{3}{C}-\overset{2}{C}-\overset{1}{C}$$
$$\vert$$
$$CH_3$$

3. Add hydrogens to give each carbon four bonds. The structural formula is

$$CH_3-CH_2-CH_2-CH-CH_3$$
$$| $$
$$CH_3$$

2-methylpentane

Should this compound be called 4-methylpentane? No, the IUPAC System specifically states that the parent carbon chain is numbered starting from the end nearest to the side or branch chain.

It is important to understand that it is the longest continuous sequence of carbon atoms and alkyl groups that determines the name of a compound, and not the way the sequence is written. Each of the following formulas represents 2-methylpentane:

$$\overset{1}{C}H_3-\overset{2}{C}H-\overset{3}{C}H_2-\overset{4}{C}H_2-\overset{5}{C}H_3$$
$$|$$
$$CH_3$$

$$\overset{5}{C}H_3-\overset{4}{C}H_2-\overset{3}{C}H_2-\overset{2}{C}H-\overset{1}{C}H_3$$
$$\overset{CH_3}{\overset{|2}{}}$$

$$\overset{5}{C}H_3 \quad\quad \overset{1}{C}H_3$$
$$| \quad\quad \overset{2|}{}$$
$$\underset{4}{C}H_2-\underset{3}{C}H_2-CH$$
$$|$$
$$CH_3$$

$$\overset{CH_3}{\overset{2|}{}}$$
$$\overset{1}{C}H_3-CH-CH_2$$
$$|$$
$$\underset{4}{C}H_2-\underset{5}{C}H_3$$

$$\overset{CH_3}{\overset{2|}{}}$$
$$\overset{1}{\underset{3}{}}CH \quad \overset{4}{C}H_2$$
$$\underset{1}{C}H_3 \quad \underset{3}{C}H_2 \quad \underset{5}{C}H_3$$

$$\overset{5}{} \quad\quad\quad \overset{5}{}$$
$$CH_3 \quad\quad\quad CH_3$$
$$| \quad\quad\quad |$$
$$\underset{1}{C}H_3-\underset{2}{C}H-\underset{3}{C}H_2-\underset{4}{C}H_2$$

The following formulas and names demonstrate other aspects of the official nomenclature system:

$$\overset{4}{C}H_3-\overset{3}{C}H-\overset{2}{C}H-\overset{1}{C}H_3$$
$$| \quad |$$
$$CH_3 \quad CH_3$$

2,3-dimethylbutane

Explanation The name of this compound is 2,3-dimethylbutane. The longest carbon chain is four, indicating butane; "dimethyl" indicates two methyl groups; "2,3-" means that one CH_3 is on carbon 2 and one is on carbon 3 (Rule 4).

$$\overset{CH_3}{\overset{|2}{}}$$
$$\overset{4}{C}H_3-\overset{3}{C}H_2-\overset{}{C}-\overset{1}{C}H_3$$
$$|$$
$$C_3$$

, -dimethylbutane

Explanation In 2,2-dimethylbutane, both methyl groups are on the same carbon atom; both numbers are required (Rule 4).

$$\overset{CH_3}{\overset{|4}{}}$$
$$\overset{1}{C}H_3-\overset{2}{C}H-\overset{3}{C}H_2-\overset{}{C}H-\overset{5}{C}H_2-\overset{6}{C}H_3$$
$$|$$
$$CH_3$$

2,4-dimethylhexane

Explanation In 2,4-dimethylhexane, molecule numbered from left to right (Rule 2).

$$CH_3-\overset{3}{C}H-\overset{4}{C}H_2-\overset{5}{C}H_2-\overset{6}{C}H_3$$
$$\overset{2|}{C}H_2$$
$$\overset{1|}{C}H_3$$

3-methylhexane

Explanation In 3-methylhexane, there are six carbons in the longest continuous chain (Rule 1).

$$CH_2-CH_3$$
$$\overset{8}{C}H_3-\overset{7}{C}H_2-\overset{6}{C}H_2-\overset{5}{C}H_2-C-CH-CH-\overset{1}{C}H_3$$
$$CH_3 \quad Cl \quad CH_3$$

3-chloro-4-ethyl-2,4-dimethyloctane

Explanation In this compound the longest carbon chain is eight. The groups that are attached or substituted for H on the octane chain are named in alphabetical order (Rule 5).

Example 19.2

Write the formulas for
(a) 3-ethylpentane   (b) 2,2,4-trimethylpentane

Solution

(a) The name pentane indicates a five-carbon chain. Write five connected carbon atoms and number them:

$$\overset{1}{C}-\overset{2}{C}-\overset{3}{C}-\overset{4}{C}-\overset{5}{C}$$

An ethyl group is written as CH_3CH_2-. Attach this group to carbon 3:

$$\overset{1}{C}-\overset{2}{C}-\overset{3}{C}-\overset{4}{C}-\overset{5}{C}$$
$$CH_2CH_3$$

Now add hydrogen atoms to give each carbon atom four bonds. Carbons 1 and 5 each need three H atoms; carbons 2 and 4 each need two H atoms; and carbon 3 needs one H atom. The formula is complete:

$$CH_3CH_2CHCH_2CH_3$$
$$CH_2CH_3$$

(b) Pentane indicates a five-carbon chain. Write five connected carbon atoms and number them:

$$\overset{1}{C}-\overset{2}{C}-\overset{3}{C}-\overset{4}{C}-\overset{5}{C}$$

There are three methyl groups (CH_3-) in the compound (*trimethyl*), two attached to carbon 2 and one attached to carbon 4. Attach these three methyl groups to their respective carbon atoms:

$$
\overset{\displaystyle CH_3}{\underset{\displaystyle CH_3}{\overset{1}{C}-\overset{2}{C}-\overset{3}{C}-\overset{4}{C}-\overset{5}{C}}}\quad CH_3
$$

Now add H atoms to give each carbon atom four bonds. Carbons 1 and 5 each need three H atoms; carbon 2 does not need any H atoms; carbon 3 needs two H atoms; and carbon 4 needs one H atom. The formula is complete:

$$
CH_3\underset{\displaystyle CH_3}{\overset{\displaystyle CH_3\ \ CH_3}{C}CH_2CHCH_3}
$$

Example 19.3   Name the following compounds:

(a)
$$
CH_3CH_2CH_2CH_2\overset{\displaystyle CH_3}{C}HCH_3
$$

(b)
$$
CH_3CH_2CH_2\overset{\displaystyle CH_2CH_3}{C}HCH_2\underset{\displaystyle CH_2CH_3}{C}HCH_3
$$

Solution

(a) The longest continuous carbon chain contains six carbon atoms (Rule 1). Thus the parent name of the compound is hexane. Number the carbon chain from right to left so that the methyl group attached to carbon 2 is given the lowest possible number (Rule 2). With a methyl group on carbon 2, the name of the compound is 2-methylhexane (Rule 3).

(b) The longest continuous carbon chain contains eight carbon atoms (Rule 1). Number the chain from right to left (Rule 2).

$$
\overset{8}{C}-\overset{7}{C}-\overset{6}{C}-\overset{5}{C}-\overset{4}{C}-\overset{3}{C}-C
$$
with $C-C$ above carbon 5 and $\overset{2}{C}-\overset{1}{C}$ below carbon 3.

Thus the parent name is octane. As the chain is numbered, there is a methyl group on carbon 3 and an ethyl group on carbon 5, so the name of the compound is 5-ethyl-3-methyloctane. Note that ethyl is named before methyl (alphabetical order, Rule 5).

Practice 19.5

Write the formula for (a) 2-methylhexane, (b) 3,4-dimethylheptane, (c) 2-chloro-3-ethylpentane.

Practice 19.6

Name the following compounds:

$$
\text{(a)} \quad
\begin{array}{c}
 \text{CH}_3{-}\text{CH}{-}\text{CH}_3 \\
 | \\
\text{CH}_3\text{CHCH}_2\text{CCH}_2\text{CH}_2\text{CH}_3 \\
| | \\
\text{CH}_3 \text{CH}_3
\end{array}
$$

$$
\text{(b)} \quad
\begin{array}{c}
\text{CH}_2\text{CH}_3 \\
| \\
\text{CH}_3\text{CH}_2\text{CHCH}_2\text{CHCH}_3 \\
| \\
\text{CH}_3
\end{array}
$$

19.10 Introduction to the Reactions of Carbon

One reason that organic chemistry is so important in our modern world is the versatility of carbon. Many diverse reactions are possible with this element. One of the most important carbon reactions is oxidation–reduction. *When carbon atoms are oxidized, they often form additional bonds to oxygen.* In the following example, carbon in methane is oxidized to carbon dioxide:

$$
\begin{array}{c}
\text{H} \\
| \\
\text{H}{-}\text{C}{-}\text{H} \;+\; 2\,\text{O}_2 \;\longrightarrow\; \underset{\text{carbon dioxide}}{\text{O}{=}\text{C}{=}\text{O}} \;+\; 2\,\text{H}_2\text{O} \\
| \\
\underset{\text{methane}}{\text{H}}
\end{array}
$$

When carbon atoms are reduced, they often form additional bonds to hydrogen. During this reaction, the carbon may decrease the number of bonds to oxygen and increase the number of bonds to hydrogen. The following reaction shows the reduction of a carbon compound:

$$
\underset{\text{carbon dioxide}}{\text{O}{=}\text{C}{=}\text{O}} \;+\; 3\,\text{H}_2 \;\longrightarrow\;
\begin{array}{c}
\text{H} \\
| \\
\text{H}{-}\text{C}{-}\text{OH} \;+\; \text{H}_2\text{O} \\
| \\
\underset{\text{methanol}}{\text{H}}
\end{array}
$$

Each class of organic compounds can undergo important oxidation–reduction reactions.

Although countless different carbon-containing molecules exist, there are relatively few types of organic reactions, and only a few carbon atoms, at most, are involved in any common reaction. These changes can be usefully categorized by counting the number of atoms bonded to a reactive carbon. Carbon can be bonded to a maximum of four atoms.

▲ **Space filling model of amoxicillin.**

Plants use carbon reduction to build cell walls.

substitution reaction

1. If in a reaction one atom in a molecule is exchanged by another atom or group of atoms, a **substitution reaction** has taken place. For example,

$$H{-}\underset{\underset{\displaystyle H}{|}}{\overset{\overset{\displaystyle H}{|}}{C}}{-}H \ + \ Br_2 \longrightarrow H{-}\underset{\underset{\displaystyle H}{|}}{\overset{\overset{\displaystyle H}{|}}{C}}{-}Br \ + \ HBr$$

methane,　　　　　　bromomethane,
colorless gas　　　　colorless liquid

elimination reaction

2. An **elimination reaction** is a reaction in which a single reactant is split into two products, and one of the products is eliminated. For example,

$$H{-}\underset{\underset{\displaystyle H}{|}}{\overset{\overset{\displaystyle H}{|}}{C}}{-}\underset{\underset{\displaystyle H}{|}}{\overset{\overset{\displaystyle H}{|}}{C}}{-}Br \longrightarrow \underset{\underset{\displaystyle H}{|}}{\overset{\overset{\displaystyle H}{|}}{C}}{=}\underset{\underset{\displaystyle H}{|}}{\overset{\overset{\displaystyle H}{|}}{C}} \ + \ HBr$$

bromoethane,　　　　ethene,
colorless liquid　　　colorless gas

Elimination reactions form multiple bonds.

addition reaction

3. Two reactants adding together to form a single product is called an **addition reaction.** An addition reaction can be thought of as the reverse of an elimination reaction.

$$\underset{\underset{\displaystyle H}{|}}{\overset{\overset{\displaystyle H}{|}}{C}}{=}\underset{\underset{\displaystyle H}{|}}{\overset{\overset{\displaystyle H}{|}}{C}} \ + \ HBr \longrightarrow H{-}\underset{\underset{\displaystyle H}{|}}{\overset{\overset{\displaystyle H}{|}}{C}}{-}\underset{\underset{\displaystyle H}{|}}{\overset{\overset{\displaystyle H}{|}}{C}}{-}Br$$

ethene,　　　　　　　bromoethane,
colorless gas　　　　colorless liquid

Practice 19.7

The following reactions form ethanol. Categorize each reaction as substitution, elimination, or addition.

(a) $CH_2 = CH_2 + H_2O \xrightarrow{\ H^+\ } CH_3CH_2OH$
(b) $CH_3CH_2I + H_2O \longrightarrow CH_3CH_2OH + HI$

 Reactions of Alkanes

One type of alkane reaction has inspired people to explore equatorial jungles, endure the heat and sandstorms of the deserts of Africa and the Middle East, mush across the frozen Arctic, and drill holes—some more than 30,000 feet deep—on land and on the ocean floor! These strenuous and expensive activities have been undertaken because alkanes, as well as other hydrocarbons, undergo combustion with oxygen with the evolution of large amounts of heat energy. Methane, for example, reacts with oxygen:

$$CH_4(g) + 2\ O_2(g) \longrightarrow CO_2(g) + 2\ H_2O(g) + 802.5\ \text{kJ}\ (191.8\ \text{kcal})$$

When carbon dioxide is formed, the alkane has undergone *complete* oxidation. The resulting thermal energy can be converted to mechanical and electrical energy. Combustion reactions overshadow all other reactions of alkanes in economic importance. But combustion reactions are not usually of great interest to organic chemists because carbon dioxide and water are the only chemical products of complete combustion.

Aside from their combustibility, alkanes are relatively sluggish and limited in reactivity. But with proper activation, such as high temperature and/or catalysts, alkanes can be made to react in a variety of ways. Some important noncombustion reactions of alkanes include the following:

▲ Space filling model of ethane.

1. **Halogenation** (a *substitution* reaction). A halogen is substituted for a hydrogen atom in halogenation. When a specific halogen such as chlorine is used, the reaction is called chlorination; RH is an alkane (alkyl group + H atom) that reacts with halogens in this manner.

    Example: $\quad RH + X_2 \longrightarrow RX + HX \qquad (X = Cl\ or\ Br)$
    $\qquad\qquad\quad CH_3CH_3 + Cl_2 \longrightarrow CH_3CH_2Cl + HCl$
    $\qquad\qquad\qquad\qquad\qquad\qquad\quad$ chloroethane

    This reaction yields alkyl halides, RX, which are useful as intermediates for the manufacture of other substances.
2. **Dehydrogenation** (an *elimination* reaction). Hydrogen is lost from an organic compound during dehydrogenation.

▲ Space filling model of chlorethane.

Example: $C_nH_{2n+2} \xrightarrow{700-900°C} C_nH_{2n} + H_2$
  alkane   alkene

$CH_3CH_2CH_3 \xrightarrow{\Delta} CH_3CH{=}CH_2 + H_2$
  propane   propene

This reaction yields alkenes, which, like alkyl halides, are useful chemical intermediates. Hydrogen is a valuable by-product.

3. **Cracking** (breaking up large molecules to form smaller ones).

Example: $C_{16}H_{34} \xrightarrow{\Delta} C_8H_{18} + C_8H_{16}$ (one set of many
  alkane   alkane   alkene   possible products)

4. **Isomerization** (rearrangement of molecular structures).

Example: $CH_3CH_2CH_2CH_2CH_3 \xrightarrow[\Delta, \text{pressure}]{\text{Catalyst}} CH_3CH_2CHCH_3$
                                                                                                    |
                                                                                                  CH_3

Halogenation is used extensively in the manufacture of petrochemicals (chemicals derived from petroleum and used for purposes other than fuels). The other three reactions—dehydrogenation, cracking, and isomerization—singly or in combination, are of great importance in the production of motor fuels and petrochemicals.

A well-known reaction of methane and chlorine is shown by the equation

$CH_4 + Cl_2 \longrightarrow CH_3Cl + HCl$
                          chloromethane
                        (methyl chloride)

The reaction of methane and chlorine gives a mixture of mono-, di-, tri-, and tetra-substituted chloromethanes.

$CH_4 \xrightarrow{Cl_2} CH_3Cl \xrightarrow{Cl_2} CH_2Cl_2 \xrightarrow{Cl_2} CHCl_3 \xrightarrow{Cl_2} CCl_4 + 4\,HCl$

However, if an excess of chlorine is used, the reaction can be controlled to give all tetrachloromethane (carbon tetrachloride). On the other hand, if a large ratio of methane to chlorine is used, the product will be predominantly chloromethane (methyl chloride). Table 19.4 lists the formulas and names for all the chloromethanes. The names for the other halogen-substituted methanes follow the same pattern as for the chloromethanes; for example, CH_3Br is bromomethane, or methyl bromide, and CHI_3 is triiodomethane, or iodoform.

TABLE 19.4 Chlorination Products of Methane

Formula	IUPAC name	Common name
CH_3Cl	Chloromethane	Methyl chloride
CH_2Cl_2	Dichloromethane	Methylene chloride
$CHCl_3$	Trichloromethane	Chloroform
CCl_4	Tetrachloromethane	Carbon tetrachloride

Chloromethane is a monosubstitution product of methane. The term **monosubstitution** refers to the fact that one hydrogen atom in an organic molecule is substituted by another atom or by a group of atoms. In hydrocarbons, for example, when we substitute one chlorine atom for a hydrogen atom, the new compound is a monosubstitution (monochlorosubstitution) product. In a like manner we can have di-, tri-, tetra-, and so on, substitution products.

monosubstitution

This kind of chlorination (or bromination) is general with alkanes. There are nine different chlorination products of ethane. See if you can write the structural formulas for all of them.

When propane is chlorinated, two isomeric monosubstitution products are obtained because a hydrogen atom can be replaced on either the first or second carbon as shown here:

$$CH_3CH_2CH_3 + Cl_2 \xrightarrow[25°C]{light} CH_3CH_2CH_2Cl + CH_3CHClCH_3 + HCl$$

1-chloropropane   2-chloropropane

The letter X is commonly used to indicate a halogen atom. The formula RX indicates a halogen atom attached to an alkyl group and represents the class of compounds known as the **alkyl halides.** When R is CH_3, then CH_3X can be CH_3F, CH_3Cl, CH_3Br, or CH_3I.

alkyl halide

Alkyl halides are named systematically in the same general way as alkanes. Halogen atoms are identified as *fluoro-, chloro-, bromo-,* or *iodo-* and are named as substituents like branch-chain alkyl groups. Study these examples:

Names for halogen atoms in organic compounds.
F fluoro
Cl chloro
Br bromo
I iodo

$CH_3-CHCl-CH_2-CH_3$ $CH_2Cl-CHBr-CH_3$

2-chlorobutane   2-bromo-1-chloropropane

$CH_3-CH_2-CH-CHCl-CH_3$ $CH_3-CH_2-CHCl_2$
|
CH_3

1,1-dichloropropane

2-chloro-3-methylpentane

How many monochlorosubstitution products can be obtained from pentane?

Example 19.4

Solution

First write the formula for pentane:

$$\overset{5}{C}H_3\overset{4}{C}H_2\overset{3}{C}H_2\overset{2}{C}H_2\overset{1}{C}H_3$$

Now rewrite the formula five times substituting a Cl atom for an H atom on each C atom:

I $CH_3CH_2CH_2CH_2\overset{1}{C}H_2Cl$ Cl on carbon 1
II $CH_3CH_2CH_2\overset{2}{C}HClCH_3$ Cl on carbon 2
III $CH_3CH_2\overset{3}{C}HClCH_2CH_3$ Cl on carbon 3
IV $CH_3\overset{4}{C}HClCH_2CH_2CH_3$ Cl on carbon 4
V $\overset{5}{C}H_2ClCH_2CH_2CH_2CH_3$ Cl on carbon 5

Compounds I and V are identical. By numbering compound V from left to right, we find that both compounds (I and V) are 1-chloropentane. Compounds II and IV are identical. By numbering compound IV from left to right, we find that both compounds (II and IV) are 2-chloropentane. Thus there are three monochlorosubstitution products of pentane: 1-chloropentane, 2-chloropentane, and 3-chloropentane.

Practice 19.8

How many dichlorosubstitution isomers can be written for hexane?

You may be familiar with Teflon® nonstick surfaces. This material is fluorocarbon polymer. The fluorine atoms cause the surface to be both unreactive and slick. Alkyl halides also have many uses in medicine. Inhalational general anesthetics are mostly carbon compounds that contain chlorine, fluorine, and/or bromine. Adding halogens improves the anesthetic by stabilizing and limiting metabolism of the compound. Several examples of common inhalational anesthetics follow (isoflurane and desflurane also contain an oxygen and are classified as ethers):

$CF_3CHClBr$ $CF_3CHCl—O—CHF_2$ $CF_3CHF—O—CHF_2$

   halothane   isoflurane   desflurane

Medical implants such as pacemakers and infusion ports are often coated with alkyl halide polymers. The coating minimizes both bacterial attachment (and the possibility of infection) and blood clotting.

 Sources of Alkanes

The two main sources of alkanes are natural gas and petroleum. Natural gas is formed by the anaerobic decay of plants and animals. The composition of natural gas varies in different locations. Its main component is methane (80–95%), the balance being varying amounts of other hydrocarbons, hydrogen, nitrogen, carbon monoxide, carbon dioxide, and in some locations, hydrogen sulfide. Economically significant amounts of methane are now obtained by the decomposition of sewage, garbage, and other organic waste products.

Petroleum, also called *crude oil,* is a viscous black liquid consisting of a mixture of hydrocarbons with smaller amounts of nitrogen and sulfur-containing organic compounds. Petroleum is formed by the decomposition of plants and animals over millions of years. The composition of petroleum varies widely from one locality to another. Crude oil is refined into such useful products as gasoline, kerosene, diesel fuel, jet fuel, lubricating oil, heating oil, paraffin wax, petroleum jelly (petrolatum), tars, and asphalt.

At the rate that natural gas and petroleum are being used, these sources of hydrocarbons are destined to be in short supply and virtually exhausted in the not-too-distant future. Alternative sources of fuels must be developed.

◀ Petroleum is converted to
the starting materials for
many chemicals at a refinery.

19.13 Gasoline: A Major Petroleum Product

A large fraction of all petroleum is burned as fuel (see Figure 19.9). Historically, fuels were used for heating and cooking; today most fuels provide power for transportation. Because different engines require different fuels, there are many different fuels. However, befitting the importance of the automobile in transportation, gasoline is the fuel produced in the largest quantity.

Gasoline, aside from the additives put into it, consists primarily of hydrocarbons. Gasoline, as it is distilled from crude oil, causes "knocking" when burned in high-compression automobile engines. Knocking, caused by a too-rapid combustion or detonation of the air–gasoline mixture, is a severe problem in high-compression engines. The knock-resistance of gasolines, a quality that varies widely, is usually expressed in terms of *octane number,* or *octane rating.*

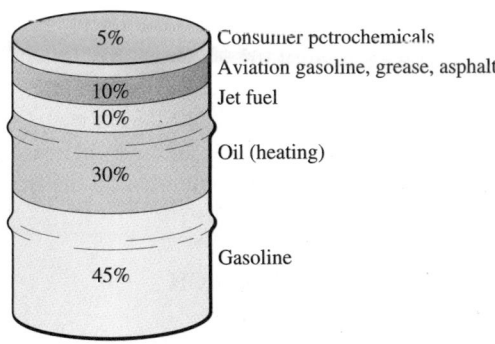

5% Consumer petrochemicals
Aviation gasoline, grease, asphalt
10% Jet fuel
10%
Oil (heating)
30%
Gasoline
45%

1 barrel crude oil = 42 gallons = 159 liters

◀ **FIGURE 19.9**
Uses of petroleum.

Unleaded gasolines areformulated to different octane ratings so they can be used in different engines. ▶

Minimum Octane Rating R+M/2 Method Shown on Buttons
All Grades Unleaded

Because of its highly branched chain structure, isooctane (2,2,4-trimethylpentane), is a motor fuel that is resistant to knocking. Mixtures of isooctane and heptane, a straight-chain alkane that knocks badly, have been used as standards to establish octane ratings of gasolines. Isooctane is arbitrarily assigned an octane number of 100 and heptane an octane number of 0. To determine the octane rating, a gasoline is compared with mixtures of isooctane and heptane in a test engine. The octane number of the gasoline corresponds to the percentage of isooctane present in the isooctane–heptane mixture that matches the knocking characteristics of the gasoline being tested. Thus a 90-octane gasoline has knocking characteristics matching those of a mixture of 90% isooctane and 10% heptane.

When first used to establish octane numbers, isooctane was the most knock-resistant substance available. However, because technological advances have resulted in engines with greater power and compression ratios, higher-quality fuels were subsequently necessary. Fuels containing more highly branched hydrocarbons, unsaturated hydrocarbons, or aromatic hydrocarbons burn more smoothly than isooctane and have a higher octane rating than 100.

Practice 19.9

━━━━━━━━━

Which of the following molecules would be predicted to yield a higher octane rating? (Remember that more highly branched alkanes commonly yield higher octane numbers.) Draw each structure before reaching your conclusion.
(a) 3-isopropyl-2,4-dimethylpentane
(b) 3,4-diethylhexane

In recent years manufacturers have "reformulated" gasoline to solve environmental problems as well as to boost the octane rating. Tetraethyl lead, $(C_2H_5)_4Pb$, a gasoline additive that very effectively increased octane ratings, has been eliminated from gasoline because of environmental dangers from lead. Compounds that contain oxygen have replaced tetraethyl lead. These additives increase the octane ratings and decrease the hydrocarbon pollution from automobile exhaust. Two examples are

$$CH_3-\overset{\displaystyle CH_3}{\underset{\displaystyle CH_3}{\overset{|}{\underset{|}{C}}}}-O-CH_3 \qquad CH_3CH_2OH$$

methyl-*tert*-butyl ether (MTBE)   ethanol

$$\begin{array}{cccc}
\text{CH}_2 & \text{CH}_2\!-\!\text{CH}_2 & \text{CH}_2 & \text{CH}_2 \\
\end{array}$$

$$
\begin{array}{cccc}
& & \text{CH}_2 & \\
\text{CH}_2 & \text{CH}_2\!-\!\text{CH}_2 & \text{CH}_2\quad\text{CH}_2 & \text{CH}_2\quad\text{CH}_2 \\
\big/ \ \big\backslash & |\qquad\; | & \big\backslash\qquad\big/ & |\qquad\quad| \\
\text{CH}_2\!-\!\text{CH}_2 & \text{CH}_2\!-\!\text{CH}_2 & \text{CH}_2\!-\!\text{CH}_2 & \text{CH}_2\quad\text{CH}_2 \\
& & & \text{CH}_2
\end{array}
$$

cyclopropane	cyclobutane	cyclopentane	cyclohexane
C_3H_6	C_4H_8	C_5H_{10}	C_6H_{12}

Unfortunately, environmental issues are often very complex. Use of MTBE is a case in point. While this additive has improved air quality, MTBE has leaked from underground gasoline storage tanks into, and polluted, groundwater. Thus, scientists are now looking for new, environmentally benign additives to improve gasoline combustion.

Although scientists have worked long and hard to improve gasoline to make a better fuel and to decrease air pollution, it is an unavoidable fact that even the cleanest burning gasoline increases atmospheric carbon dioxide, which can contribute to the "greenhouse effect." Furthermore, combustion uses hydrocarbons in a nonrecyclable way. Burning gasoline pollutes the air and consumes a valuable resource. As petroleum chemists look to the future, they recommend (1) development of nonpolluting energy sources, and (2) shifting the major use of petroleum from fuel to the manufacture of other organic compounds.

Cycloalkanes

Cyclic, or closed-chain, alkanes also exist. These substances, called **cycloalkanes,** *cycloparaffins,* or *naphthenes,* have the general formula C_nH_{2n}. Their names are formed by adding the prefix *cyclo* to the name of the alkane with the same number of carbon atoms. For example, the six-carbon cycloalkane is called cyclohexane. Note that this series of compounds has two fewer hydrogen atoms than the open-chain alkanes. The bonds for the two missing hydrogen atoms are accounted for by an additional carbon–carbon bond in the cyclic ring of carbon atoms. Structures for the four smallest cycloalkanes are shown in Figure 19.10.

cycloalkane

With the following exception noted for cyclopropane and cyclobutane, cycloalkanes are generally similar to open-chain alkanes in both physical properties and chemical reactivity. Cycloalkanes are saturated hydrocarbons; they contain only single bonds between carbon atoms.

The reactivity of cyclopropane, and to a lesser degree that of cyclobutane, is greater than that of other alkanes. This greater reactivity exists because the carbon–carbon bond angles in these substances deviate substantially from the normal tetrahedral angle. The carbon atoms form a triangle in cyclopropane, and in cyclobutane they approximate a square. Cyclopropane molecules therefore have carbon–carbon bond angles of 60°, and in cyclobutane the bond angles are about

cyclopropane

hexane $CH_3CH_2CH_2CH_2CH_2CH_3$

FIGURE 19.11
Ball-and-stick models illustrating cyclopropane, hexane, and cyclohexane. In cyclopropane all the carbon atoms are in one plane. The angle between the carbon atoms is 60°, not the usual 109.5°; therefore, the cyclopropane ring is strained. In cyclohexane the carbon–carbon bonds are not strained. This is because the molecule is puckered (as shown in the chair conformation) with carbon–carbon bond angles about 109.5°, as found in hexane.

cyclohexane

90°. In the open-chain alkanes and in larger cycloalkanes, the carbon atoms are in a three-dimensional zigzag pattern in space and have normal (tetrahedral) bond angles of about 109.5°.

Bromine adds to cyclopropane readily and to cyclobutane to some extent. In this reaction the ring breaks and an open-chain dibromopropane is formed:

$$CH_2—CH_2 \; (CH_2) + Br_2 \longrightarrow BrCH_2CH_2CH_2Br$$

cyclopropane   1,3-dibromopropane

Cyclopropane and cyclobutane react in this way because their carbon–carbon bonds are strained and therefore weakened. Cycloalkanes whose rings have more than four carbon atoms do not react in this way because their molecules take the shape of non-planar puckered rings. These rings can be considered to be formed by simply joining the end carbon atoms of the corresponding normal alkanes. The resulting cyclic molecules are nearly strain-free, with carbon atoms arranged in space so that the bond angles are close to 109.5° (Figure 19.11).

conformation   The **conformation** of a molecule is its three-dimensional shape in space. Molecular models show that cyclohexane can assume two distinct nonplanar conformations. One form is shaped like a chair, while the other is shaped like a boat (see Figure 19.12). In the chair form the hydrogen atoms are separated as effectively as

◀FIGURE 19.12
Conformations of cyclo-
hexane: (a) chair conformation; (b) boat conformation.
Axial hydrogens are shown
in blue in the chair conformation.

(a)   (b)

possible, so this is the more stable conformation. Six of the hydrogens in "chair" cy-clohexane lie approximately in the same plane as the carbon ring. These are called **equatorial hydrogens.** The other six hydrogen atoms are approximately at right angles above or below the plane of the ring. These are called **axial hydrogens.** Sub-stituent groups are usually found in equatorial positions where they are farthest from hydrogens and other groups. Five- and six-membered rings are a common occur-rence in organic chemistry and biochemistry. These conformations and isomers are evident in carbohydrates as well as in nucleic acids. For simplicity, we draw cyclic pentanes and hexanes as flat pentagon and hexagon rings, respectively (see Figure 19.10).

equatorial hydrogen
axial hydrogen

Cyclopropane is a useful general anesthetic and, along with certain other cy-cloalkanes, is used as an intermediate in some chemical syntheses. The high reactiv-ity of cyclopropane requires great care in its use as an anesthetic because it is an extreme fire and explosion hazard. The cyclopentane and cyclohexane ring struc-tures are present in many naturally occurring molecules such as prostaglandins, steroids (e.g., cholesterol and sex hormones), and some vitamins.

Practice 19.10

Draw the structures for the two cycloalkanes that have the molecular formula, C_4H_8.

Naming Cycloalkanes

Substituted cycloalkanes are named by naming and numbering the location of the substituent(s) on the carbon ring, followed by the name of the parent cycloalkane. For example, for a monosubstituted cycloalkane it does not matter on which carbon atom the substituted group is placed:

$C_6H_{11}CH_3$ $C_6H_{11}CH_3$

are both named methylcyclohexane. However, when a cycloalkane has two or more substituted groups, the ring is numbered (clockwise or counterclockwise) to give the location of the substituted groups the smallest numbers. All the other rules for

CHEMISTRY IN ACTION — Molecules to Communicate, Refrigerate, and Save Lives

By far the most important use of alkanes in our world is as a source of energy for industrial and consumer use. Hydrocarbons do not generally function as an energy source in the physiological world of living organisms. The compounds that supply the fuel necessary for most life contain oxygen as well as carbon and hydrogen.

Alkanes do find other significant uses in our lives. High-molar-mass alkanes soften or curtail evaporation from the skin (see Section 13.15). Petroleum jelly and mineral oil are both mixtures of hydrocarbons used as skin protectants and as lubricants.

Certain insects use hydrocarbons (as well as other organic compounds) as chemical communication devices. These compounds, called pheromones, are secreted by an insect and recognized by other members of the species as a message. The function of the pheromone varies with its composition. It can be a sex attractant, an alarm, or an indication of the path to a source of food. When ants are disturbed, they release alarm pheromones that have been identified as undecane, $CH_3(CH_2)_9CH_3$, and tridecane, $CH_3(CH_2)_{11}CH_3$. Our growing understanding of these molecules is beginning to lead to their use in insect abatement. Commercial traps are baited with sex pheromone, and the insects are captured without the use of pesticides.

Freons and other chlorofluorocarbons (CFCs) are useful because they are

▲ Ants communicate to each other through the secretion of pheromones.

nontoxic, nonflammable, and noncorrosive. Many of these compounds have low boiling points and make excellent refrigerants. They have also been used as propellants in aerosol sprays and in the production of some fast-food containers. Unfortunately, the use of CFCs is a major factor in the destruction of the ozone layer (see Section 12.17). Use of these compounds is currently being discontinued by major industries in an effort to protect our atmosphere.

Some fluorinated hydrocarbons are used as blood substitutes. All the hydrogens in the hydrocarbon are replaced with fluorine, producing a compound known as a fluorocarbon. Dispersions of these compounds in water can absorb nearly three times the oxygen per unit as whole blood. Organisms receiving blood substitutes continue to produce whole blood. The substitute permits the tissues to absorb oxygen while it remains chemically inert and is excreted over a period of time.

naming alkanes also apply. Study these examples for dimethylcycloalkane, $C_6H_{10}(CH_3)_2$:

    A   B   C   D   E   F   G

A is 1,1-dimethylcyclohexane
B is 1,2-dimethylcyclohexane
C is 1,3-dimethylcyclohexane
D is 1,4-dimethylcyclohexane
E is 1,3-dimethylcyclohexane (numbered counterclockwise)
F is 1,2-dimethylcyclohexane (numbered counterclockwise)
G is 1,2-dimethylcyclohexane (same as B and F)

Note that B, F, and G are the same compound and C and E are the same compound. Therefore, there are four isomers of dimethylcyclohexane.

Practice 19.11

Write structural formulas for
(a) 1,4-diethylcyclohexane
(b) 1-ethyl-3-methylcyclohexane
(c) 1,2,4-trimethylcyclopentane

Concepts in Review

1. Describe the tetrahedral nature of the carbon atom.
2. Explain why the concept of hybridization is used to describe the bonding of carbon in simple compounds such as methane.
3. List and describe three common classes of organic reactions.
4. Explain the bonding in alkanes.
5. Write the Lewis structures for alkanes and halogenated alkanes.
6. Write the names and formulas for the first ten normal alkanes.
7. Understand the concept of isomerization.

8. Write structural formulas and IUPAC names for the isomers of an alkane or a halogenated alkane.
9. Give the IUPAC name of a hydrocarbon or a halogenated hydrocarbon when given the structural formula, and vice versa.
10. Write equations for the halogenation of an alkane, giving all possible monohalosubstitution products.
11. Write structural formulas and names for simple cycloalkanes.
12. Draw the two major conformations for cyclohexane.
13. Discuss the advantages and disadvantages of gasoline as a fuel, including octane rating.

Key Terms

The terms listed here have been defined within this chapter.
Section numbers are referenced in parentheses for each term.

addition reaction (19.10)
aliphatic (19.5)
alkane (19.6)
alkyl group (19.9)
alkyl halide (19.11)
axial hydrogen (19.14)
conformation (19.14)
cycloalkane (19.14)

elimination reaction (19.10)
equatorial hydrogen (19.14)
functional group (19.4)
homologous series (19.6)
hydrocarbon (19.5)
isomerism (19.8)
isomer (19.8)
IUPAC System (19.9)

monosubstitution (19.11)
organic chemistry (19.1)
sigma bond (19.7)
saturated compound (19.2)
substitution reaction (19.10)
unsaturated compound (19.2)
vital force theory (19.1)

Questions

Questions refer to tables, figures, and key words and concepts defined within the chapter. A particularly challenging question is indicated with an asterisk.

1. What are the main reasons for the large number of organic compounds?

2. Why is it believed that a carbon atom must form hybrid electron orbitals when it bonds to hydrogen atoms to form methane?

3. Write the names and formulas for the first ten normal alkanes.

4. State the number of sigma bonds in a molecule of
   (a) ethane   (b) butane   (c) 2-methylpropane

5. What is a major advantage and a major disadvantage to the industrial use of the Freons?

6. What is the difference between the meaning of the prefixes, *iso-*, *sec-*, and *tert-*?

Paired Exercises

These exercises are paired. Each odd-numbered exercise is followed by a similar even-numbered exercise. Answers to the even-numbered exercises are given in Appendix VI.

7. Write the Lewis structures for
   (a) CCl_4
   (b) C_2Cl_6
   (c) $CH_3CH_2CH_3$

8. Write the Lewis structures for
   (a) methane
   (b) propane
   (c) pentane

9. Which of these formulas represent isomers?
   (a) $CH_3CH_2CH_2CH_3$

   (b) $CH_3CH_2CH_2CH_2CH_3$

   (c) CH_3CHCH_3
       CH_3CH_2

   (d) $CH_3CH_2CH_2CH_2CH_2CH_3$

   (e) CH_3 CH_3
       $CH-CH$
       CH_3 CH_3

   (f) CH_2-CH_2 (g) $CH_3CHCH_2CH_2CH_3$
       CH_2 CH_2 CH_3
          CH_2

   (h) CH_2 (i) CH_3CH_2
       $CHCH_2CH_3$ CH_2CH_3
       CH_2

11. How many methyl groups are in each formula of Exercise 9?

13. Write the condensed structural formulas for the nine isomers of heptane.

15. Draw structural formulas for all the isomers of
    (a) CH_2Cl_2 (b) C_3H_7Br
    (c) $C_3H_6Cl_2$ (d) $C_4H_8Cl_2$

17. Give IUPAC names for the following:
    (a) $CH_3CH_2CH_2Cl$

    (b) $CH_3CHClCH_3$

    (c) $(CH_3)_3CCl$

    (d) $CH_3CH_2CHCH_3$
                CH_3

    (e) CH_3CHCH_3
        $CH_3CH_2CH_2CHCH_3$

10. Which of these formulas represent the same compound?
    (a) $CH_3CHCH_2CHCH_3$
         CH_3 CH_3

    (b) CH_3
        $CH_2CHCH_2CHCH_3$
         CH_3 CH_3

    (c) CH_3
        $CH_3CHCH_2CH_2CHCH_3$
                 CH_3

    (d) CH_3
        $CH_3CHCHCH_2CH_2$
         CH_3 CH_3

    (e) CH_3
        CH_3CHCH_2
        $CH_3CH_2CHCH_3$

    (f) CH_3
        CH_3CH
        CH_2CHCH_3
        CH_2CH_3

12. How many methyl groups are in each formula of Exercise 10?

14. Write the condensed structural formulas for the five isomers of hexane.

16. Draw structural formulas for all the isomers of
    (a) CH_3Br (b) C_2H_5Cl
    (c) C_4H_9I (d) C_3H_6BrCl

18. Give the IUPAC name for each of the following compounds:
    (a) CH_3CH_2Cl

    (b) $(CH_3)_2CHCH_2Cl$

    (c) $CH_3CHClCH_2CH_3$

    (d) CH_2
        $CHCH_3$
        CH_2

    (e) $(CH_3)_2CHCH_2CH(CH_3)_2$

19. Draw structural formulas for the following compounds:
    (a) 2,4-dimethylpentane
    (b) 2,2-dimethylpentane
    (c) 3-isopropyloctane
    (d) 5,6-diethyl-2,7-dimethyl-5-propylnonane
    (e) 2-ethyl-1,3-dimethylcyclohexane

21. The following names are incorrect. Explain why the name is wrong and give the correct name.
    (a) 3-methylbutane
    (b) 2-ethylbutane
    (c) 2-dimethylpentane
    (d) 1,4-dimethylcyclopentane

23. Draw the structures for the ten dichlorosubstituted isomers, $C_5H_{10}Cl_2$, of 2-methylbutane.

25. Complete the equations for (a) the monochlorination, and (b) the complete combustion of butane.
    (a) $CH_3CH_2CH_2CH_3 + Cl_2 \xrightarrow{h\nu}$
    (b) $CH_3CH_2CH_2CH_3 + O_2 \xrightarrow{\Delta}$

27. Write names for

    (a)

    (b)

    (c)

20. Draw structural formulas for the following compounds:
    (a) 4-ethyl-2-methylhexane
    (b) 4-tert-butylheptane
    (c) 4-ethyl-7-isopropyl-2,4,8-trimethyldecane
    (d) 3-ethyl-2,2-dimethyloctane
    (e) 1,3-diethylcyclohexane

22. The following names are incorrect. Explain why the name is wrong and give the correct name.
    (a) 3-methyl-5-ethyloctane
    (b) 3,5,5-triethylhexane
    (c) 4,4-dimethyl-3-ethylheptane
    (d) 1,6-dimethylcyclohexane

24. The structure for hexane is $CH_3CH_2CH_2CH_2CH_2CH_3$

Draw the structural formulas for all the monochlorohexanes, $C_6H_{13}Cl$, that have the same linear carbon structure as hexane.

26. Complete the equations for (a) the monobromination, and (b) the complete combustion of propane.
    (a) $CH_3CH_2CH_3 + Br_2 \xrightarrow{h\nu}$
    (b) $CH_3CH_2CH_3 + O_2 \xrightarrow{\Delta}$

28. Write structural formulas for
    (a) 1,4-di(*tert*-butyl)cyclohexane
    (b) 1-methyl-3-propylcyclohexane
    (c) 1,3-dichloro-2,5 diethylcyclohexane

Additional Exercises

These exercises are not paired and provide additional practice on concepts covered in this chapter.

29. A hydrocarbon sample of formula C_4H_{10} is brominated, and four different monobromo compounds of formula

C_4H_9Br are isolated. Is the sample a pure compound or a mixture of compounds? Explain your answer.

30. The name of the compound of formula $C_{11}H_{24}$ is undecane. What is the formula for dodecane, the next higher homologue in the alkane series?

31. The newer, more environmentally safe substitutes for chlorofluorocarbons (CFCs) are termed hydrochlorofluorocarbons (HCFCs). Write an equation for the monochlorination of 1,2-difluoroethane to produce an HCFC.

*32. High-efficiency cars use gasoline at about the rate of 60 miles per gallon. One gallon of gasoline contains about 19 mol of octane. How many moles of carbon dioxide will be exhausted to travel 60 miles (assuming complete combustion of the octane)? How many liters of carbon dioxide will be exhausted (at 20.°C and 1.0 atm pressure)?

33. Write the structures and names for all the cycloalkanes that have the molecular formula C_5H_{10}.

34. The following reactions are important in biochemistry. Classify them as either substitution, elimination, or addition reactions.

(a)
$$HO-\overset{\overset{O}{\|}}{C}-\underset{\underset{CH_2OH}{|}}{CH(H_2PO_4)} \longrightarrow$$

$$HO-\overset{\overset{O}{\|}}{C}-\underset{\underset{CH_2}{\|}}{C(H_2PO_4)} + H_2O$$

(b) $CH_3-\overset{\overset{O}{\|}}{C}-S-Coenzyme\ A\ +\ H_2O \longrightarrow$

$$CH_3-\overset{\overset{O}{\|}}{C}-OH\ +\ HS-Coenzyme\ A$$

(c) $HO-\overset{\overset{O}{\|}}{C}-CH=\underset{\underset{\underset{O}{\|}}{C-OH}}{CCH_2}-\overset{\overset{O}{\|}}{C}-OH\ +\ H_2O\ \xrightarrow{H^+}$

$$HO-\overset{\overset{O}{\|}}{C}-\underset{\underset{-\ OH}{|}}{CH}-\underset{\underset{\underset{O}{\|}}{C-OH}}{CHCH_2}-\overset{\overset{O}{\|}}{C}-OH$$

35. Assume that you have two test tubes, one containing hexane and the other containing 3-methylheptane. Can

you tell which one is in which tube by testing their solubility in water? Why or why not?

36. Gasoline is a mixture of hydrocarbons. The components must be blended so that the gasoline has the correct volatility, and gasoline must have enough highly volatile compounds so that sufficient vapors are available to burn when the spark plug fires. In general, the lower the boiling point of a substance, the higher the volatility. Place the following compounds in their order of increasing volatility: butane, propane, and hexane.

37. In each of the following, state whether the compounds given are (i) isomers of one another, (ii) the same compound, or (iii) neither isomers nor the same compound.

(a) $CH_3CH_2CHBrCH_2Br$ and $CH_2BrCH_2CH_2CH_2Br$

(b)
$$\underset{\underset{CH_3}{|}}{CH_2}-\overset{\overset{CH_3}{|}}{CH}-\underset{\underset{CH_3}{|}}{CH_2}\ \ and\ \ 2\text{-methylhexane}$$

(c)
$$\underset{\underset{\underset{CH_3}{|}}{CH_2}}{CH_2}-\overset{\overset{CH_2-CH_3}{|}}{CH}-\underset{\underset{CH_3}{|}}{CH_2}\ \ and\ \ CH_3(CH_2)_5CH_3$$

(d) Pentane and cylcopentane

38. Freon-12 is nontoxic, nonflammable, and noncorrosive. It has a boiling point of $-30°C$ and is commonly used as a refrigerant. Its formula is CCl_2F_2.
   (a) What types of hybrid orbitals would you expect to be present in this compound?
   (b) Would you expect Freon-12 to exhibit structural isomerism? Why or why not?
   (c) Give an appropriate IUPAC name for Freon-12.
   (d) Based on Table 19.1, to which class of compounds does Freon-12 belong?

39. Pheromones are chemical communicators. When an ant is disturbed, it can release undecane, $C_{11}H_{24}$, and tridecane, $C_{13}H_{28}$, to send an alarm signal to other ants.
   (a) Write the structural formulas for undecane and tridecane.
   (b) Are either of these compounds alkanes? How can you tell?

40. Write the structure for, and comment on, the name 1,3,5-trimethylcyclopentane.

Answers to Practice Exercises

19.1 (a) 109.5°
  (b) 120°

19.2 (a) $C_{12}H_{26}$, $C_{14}H_{30}$, $C_{20}H_{42}$
  (b) $CH_3CH_2CH_2CH_2CH_2CH_2CH_2CH_2CH_2CH_2CH_2CH_3$
      $CH_3CH_2CH_2CH_2CH_2CH_2$ $CH_2CH_2CH_2CH_2CH_2$ $CH_2CH_2CH_3$

19.3

$CH_3CH_2CH_2CH_2CH_2CH_3$

$CH_3CH_2CH_2CHCH_3$ (with CH₃ branch)

$CH_3CH_2CHCH_2CH_3$ (with CH₃ branch)

$CH_3CHCHCH_3$ (with CH₃, CH₃ branches)

$CH_3CH_2C-CH_3$ (with CH₃, CH₃ branches)

19.4

$\underset{1}{CH_3}\underset{2}{CH}CH_2CH_2CHCH_2CCH_2CHCH_3$

Primary carbons: 1 10 11 13 14 15 16
Secondary carbons: 3 4 6 8 12
Tertiary carbons: 2 5 9

19.5

  (a) $CH_3CHCH_2CH_2CH_2CH_3$ (with CH₃)
  (c) $CH_3CHCHCH_2CH_3$ (with Cl, CH₂CH₃)

  (b) $CH_3CH_2CHCHCH_2CH_2CH_3$ (with CH₃, CH₃)

19.6 (a) 4-isopropyl-2,4-dimethylheptane
  (b) 3,5-dimethylheptane

19.7 (a) addition
  (b) substitution

19.8 There are 12 dichloro products of hexane

19.9
  (a) $CH_3CHCHCHCH_3$ more branching, predict higher octane rating (with CH₃, CH₃, CH₃ branches)

  (b) $CH_3CH_2CHCHCH_2CH_3$ (with CH₂CH₃, CH₂CH₃)

19.10

C_4H_8 C_4H_8

19.11

  (a)

  (b)

  (c)

CHAPTER
20

Unsaturated Hydrocarbons

CHAPTER 20 / OUTLINE

▲ Perfumes and colognes interact with chemicals in our skin to produce unique fragrances.

CHAPTER
20

Think of the many images a particular fragrance or smell can evoke. A favorite perfume triggers memories of a romantic evening, while the aroma of a light-bodied red wine reminds us of a favorite restaurant. The compounds responsible for such evocative odors are the essential oils in plants. Many of these molecules are classified as unsaturated hydrocarbons.

Unsaturated hydrocarbons enhance our lives in many ways. The polymers, manufactured from unsaturated molecules, would be difficult to do without. There are large numbers of these; you are probably familiar with polyethylene plastic bags and bottles and polystyrene Styrofoam cups, and plastic wrap. The molecules associated with the essential oils in plants often contain multiple bonds beween carbon atoms. These oils are widely used in cosmetics, medicines, flavorings, and perfumes.

Hydrocarbons also form rings of carbon atoms. These ring molecules are the basis for many consumer products such as detergents, insecticides, and dyes. Such molecules are known as aromatic carbon compounds and are also found in living organisms.

20.1 Bonding in Unsaturated Hydrocarbons

alkene
alkyne
aromatic compound

The unsaturated hydrocarbons consist of three families of homologous compounds that contain multiple bonds between carbon atoms. In each family every compound contains fewer hydrogens than the alkane with the corresponding number of carbons. Compounds in the first family, known as the **alkenes,** contain carbon–carbon double bonds. Those in the second family, known as the **alkynes,** contain carbon–carbon triple bonds, and those in the third family, known as **aromatic compounds,** contain benzene rings.

The double bonds in alkenes are different from the sp^3 hybrid bonds found in the alkanes. The hybridization of the carbons that form the double bond in alkenes may be visualized in the following way. One of the $2s$ electrons of carbon is promoted to a $2p$ orbital to form the four half-filled orbitals, $2s^1 2p_x^1 2p_y^1 2p_z^1$. Three of these orbitals $(2s^1 2p_x^1 2p_y^1)$ hybridize, thereby forming three equivalent orbitals designated as sp^2. Thus the four orbitals available for bonding are three sp^2 orbitals and one p orbital. This process is illustrated in Figure 20.1.

The three sp^2 hybrid orbitals form angles of 120° with each other and lie in a single plane. The remaining $2p$ orbital is oriented perpendicular to this plane, with

FIGURE 20.1 ▶
Schematic hybridization of $2s^2 2p_x^1 2p_y^1$ orbitals of carbon to form three sp^2 electron orbitals and one p electron orbital.

Four carbon electrons in their ground-state orbitals

Three sp^2 orbitals and one p orbital availible for bonding

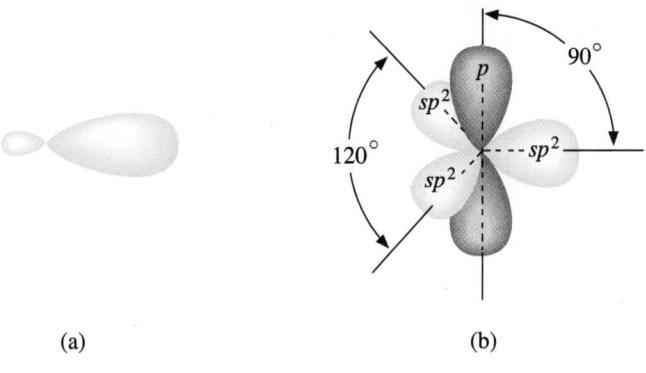

◀ **FIGURE 20.2**
(a) A single sp² electron orbital and (b) a side view of three sp² orbitals all lying in the same plane with a p orbital perpendicular to the three sp² orbitals.

(a)   (b)

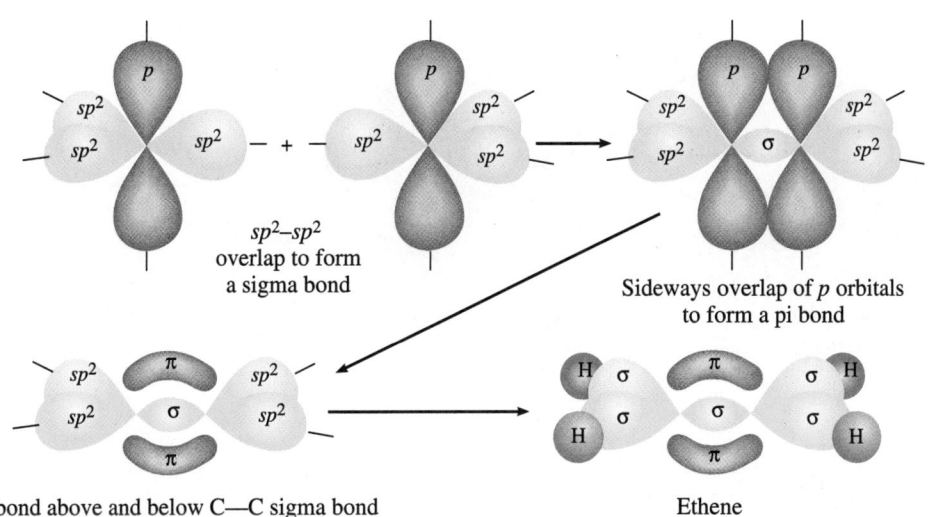

sp^2–sp^2 overlap to form a sigma bond

Sideways overlap of p orbitals to form a pi bond

Pi bond above and below C—C sigma bond

Ethene

▲
FIGURE 20.3
Pi (π) and sigma (σ) bonding in ethene.

one lobe above and one lobe below the plane (see Figure 20.2). In the formation of a double bond, an sp^2 orbital of one carbon atom overlaps an identical sp^2 orbital of another carbon to form a sigma bond. At the same time the two perpendicular p orbitals (one on each carbon atom) overlap to form a **pi (π) bond** between the two carbon atoms. This pi bond consists of two electron clouds, one above and one below the sigma bond (see Figure 20.3). The CH_2=CH_2 molecule is completed as the remaining sp^2 orbitals (two on each carbon atom) overlap hydrogen s orbitals to form sigma bonds between the carbon and hydrogen atoms. Thus there are five sigma bonds and one pi bond in an ethene molecule.

    In the formula commonly used to represent ethene, CH_2=CH_2, no distinction is made between the sigma bond and the pi bond in the carbon–carbon double bond. However, these bonds are actually very different from each other. The sigma bond is formed by the overlap of sp^2 orbitals; the pi bond is formed by the overlap of p orbitals. The sigma bond electron cloud is distributed about a line joining the carbon nuclei, but the pi bond electron cloud is distributed above and below the sigma bond region (see Figure 20.3). The carbon–carbon pi bond is much weaker and, as a consequence, much more reactive than the carbon–carbon sigma bond.

pi bond

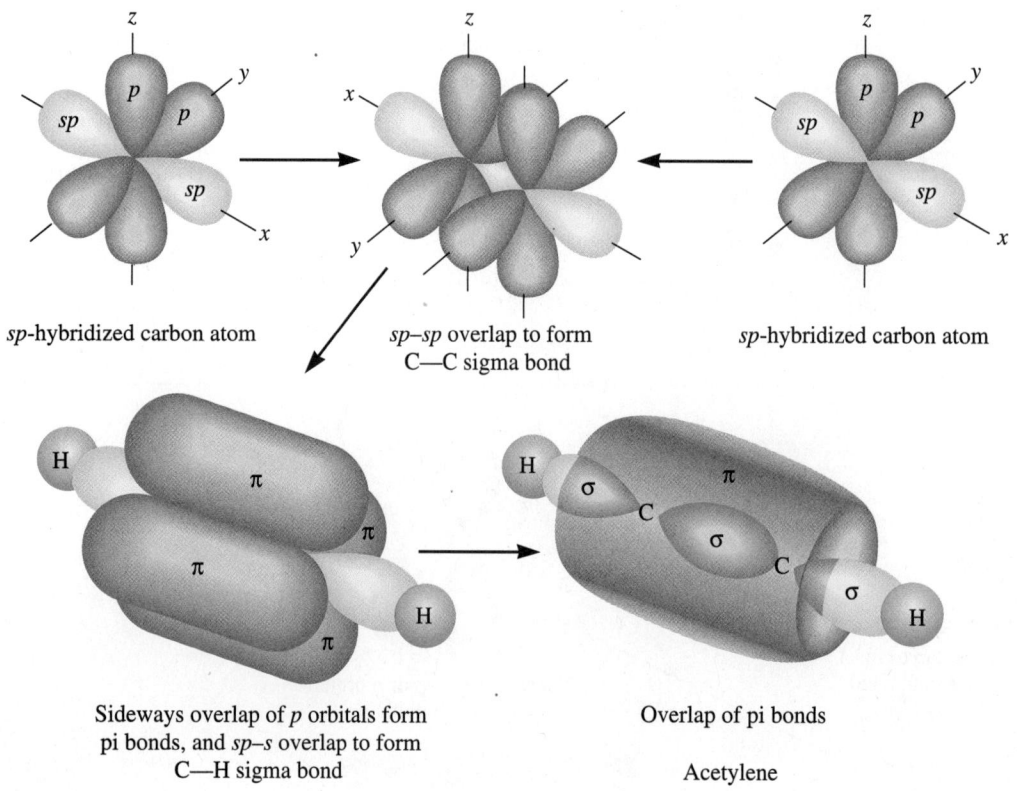

sp-hybridized carbon atom   *sp–sp* overlap to form   *sp*-hybridized carbon atom
                                          C—C sigma bond

Sideways overlap of *p* orbitals form   Overlap of pi bonds
pi bonds, and *sp–s* overlap to form
          C—H sigma bond   Acetylene

▲
FIGURE 20.4
Pi (π) and sigma (σ) bonding
in acetylene.

The formation of a triple bond between carbon atoms, as in acetylene, CH≡CH, may be visualized as follows:

1. A carbon atom 2s electron is promoted to a 2p orbital ($2s^1 2p_x^1 2p_y^1 2p_z^1$)
2. The 2s orbital hybridizes with one of the 2p orbitals to form two equivalent orbitals known as *sp* orbitals. These two hybrid orbitals lie on a straight line that passes through the center of the carbon atom. The remaining two unhybridized 2p orbitals are oriented at right angles to these *sp* orbitals and to each other.

2p ↿⇂ ↿ __   A 2s electron   2p ↿ ↿ ↿   Hybridization   2p ↿ ↿ __
2s ↿⇂   is promoted to   2s ↿   sp ↿ ↿
                a 2p electron

Four carbon   Two *sp* orbitals
atoms in their   and 2p orbitals
ground state   available for bonding
orbitals

3. In forming carbon–carbon bonds, one carbon *sp* orbital overlaps an identical *sp* orbital on another carbon atom to establish a sigma bond between the two carbon atoms.

4. The remaining *sp* orbitals (one on each carbon atom) overlap *s* orbitals on hydrogens to form sigma bonds and establish the H—C—C—H bond sequence. Because all the atoms forming this sequence lie in a straight line, the acetylene molecule is linear.

5. The two 2*p* orbitals on each carbon overlap simultaneously to form two pi bonds. These two pi bond orbitals occupy sufficient space to overlap each other and form a continuous tubelike electron cloud surrounding the sigma bond between the carbon atoms (Figure 20.4). These pi bond electrons (as in ethene) are not as tightly held by the carbon nuclei as the sigma bond electrons. Acetylene, consequently, is a very reactive substance.

20.2 Nomenclature of Alkenes

The names of alkenes are derived from the names of corresponding alkanes.

IUPAC Rules for Naming Alkenes

1. Select the longest continuous carbon–carbon chain that contains the double bond.
2. Name this parent compound as you would an alkane but change the *-ane* ending to *-ene;* for example, propane is changed to propene.

$$CH_3CH_2CH_3 \qquad CH_3CH=CH_2$$
propane   propene

3. Number the carbon chain of the parent compound starting with the end nearer to the double bond. Use the smaller of the two numbers on the double-bonded carbon atoms to indicate the position of the double bond. Place this number in front of the alkene name; for example, 2-butene means that the carbon–carbon double bond is between carbons 2 and 3.
4. Branch chains and other groups are treated as in naming alkanes, by numbering and assigning them to the carbon atom to which they are bonded.

Study the following examples of named alkenes:

$$\overset{4}{C}H_3\overset{3}{C}H_2\overset{2}{C}H=\overset{1}{C}H_2 \qquad \overset{1}{C}H_3\overset{2}{C}H=\overset{3}{C}HC\overset{4}{H}_3$$
1-butene   2-butene

$$\overset{4}{C}H_3\overset{3}{\underset{\underset{CH_3}{|}}{C}}H\overset{2}{C}H=\overset{1}{C}H_2 \qquad \overset{6}{C}H_3\overset{5}{C}H_2\overset{4}{C}H_2\overset{3}{\underset{\underset{CH_2CH_2CH_3}{|}}{C}}H\overset{2}{C}H=\overset{1}{C}H_2$$
3-methyl-1-butene   3-propyl-1-hexene

General formula for alkenes is C_nH_{2n}

To write a structural formula from a systematic name, the naming process is reversed. For example, how would we write the structural formula for 4-methyl-2-pentene? The name indicates

1. Five carbons in the longest chain
2. A double bond between carbons 2 and 3
3. A methyl group on carbon 4

Write five carbon atoms in a row. Place a double bond between carbons 2 and 3, and place a methyl group on carbon 4:

$$\overset{1}{C}-\overset{2}{C}=\overset{3}{C}-\overset{4}{C}-\overset{5}{C}$$
$$| $$
$$CH_3$$

carbon skeleton

Now add hydrogen atoms to give each carbon atom four bonds. Carbons 1 and 5 each need three H atoms; carbons 2, 3, and 4 each need one H atom. The complete formula is

$$CH_3CH=CHCHCH_3$$
$$|$$
$$CH_3$$

4-methyl-2-pentene

Example 20.1 Write structural formulas for (a) 2-pentene and (b) 7-methyl-2-octene.

Solution

(a) The stem *pent-* indicates a five-carbon chain; the suffix *-ene* indicates a carbon–carbon double bond; the number 2 locates the double bond between carbons 2 and 3. Write five carbon atoms in a row and place a double bond between carbons 2 and 3:

$$\overset{1}{C}-\overset{2}{C}=\overset{3}{C}-\overset{4}{C}-\overset{5}{C}$$

Add hydrogen atoms to give each carbon atom four bonds. Carbons 1 and 5 each need three H atoms; carbons 2 and 3 each need one H atom; carbon 4 needs two H atoms. The complete formula is

$$CH_3CH=CHCH_2CH_3$$

2-pentene

(b) Octene, like octane, indicates an eight-carbon chain. The chain contains a double bond between carbons 2 and 3 and a methyl group on carbon 7. Write eight carbon atoms in a row, place a double bond between carbons 2 and 3, and place a methyl group on carbon 7:

$$\overset{1}{C}-\overset{2}{C}=\overset{3}{C}-\overset{4}{C}-\overset{5}{C}-\overset{6}{C}-\overset{7}{C}-\overset{8}{C}$$
$$|$$
$$CH_3$$

Now add hydrogen atoms to give each carbon atom four bonds. The complete formula is

$$CH_3CH{=}CHCH_2CH_2CH_2CHCH_3$$
$$\overset{|}{CH_3}$$

7-methyl-2-octene

Name this compound.

$$CH_3CH_2CCH_2CH_2CH_3$$
$$\overset{\|}{CH_2}$$

Example 20.2

The longest carbon chain contains six carbons. However, since the compound is an alkene, we must include the double bond in the chain. The longest carbon chain containing the double bond has five carbons. Therefore, the compound is named as a pentene (Rules 1 and 2).

Solution

$$\overset{2\ 3\quad\ 4\ \ \ 5}{CH_3CH_2CCH_2CH_2CH_3}$$
$$\overset{\underset{1}{\|}}{CH_2}$$

Attached to carbon 2 is an ethyl group. The name is 2-ethyl-1-pentene.

Practice 20.1

Write structural formulas for (a) 3-hexene, (b) 4-ethyl-2-heptene, (c) 3,4-dimethyl-2-pentene, and (d) 2-cyclohexyl-3-hexene.

Practice 20.2

Name these compounds.

(a) $CH_3C{=}CHCH_3$
$\qquad\ \ \overset{|}{CH_2CH_3}$

(b) $CH_3C{=}CHCHCH_3$
$\qquad\ \ \ \overset{|}{CH_3}\quad\overset{|}{CH_3}$

Practice 20.3 Application to Biochemistry

(a) Synthesis of cholesterol starts with a close relative of the following compound:

$$\overset{CH_2}{\overset{\|}{CH_3CCH_2CH_3}}$$

Name this compound.

(b) Cholesterol biosynthesis then forms an isomer similar to the following:

$$CH_3$$
$$\mid$$
$$CH_3C{=}CH_2CH_3$$

Name this compound.

The common name for a compound is often more widely used than the IUPAC name. For example, "ethylene" is the common name for ethene. Ethylene is used to make many important products such as the plastic polyethylene and the major component of automobile antifreeze, ethylene glycol.

20.3   Geometric Isomerism in Alkenes

Only two dichloroethanes are known: 1,1-dichloroethane, $CHCl_2CH_3$, and 1,2-dichloroethane, CH_2ClCH_2Cl. But surprisingly there are three dichloroethenes—namely, 1,1-dichloroethene, $CCl_2{=}CH_2$, and *two* isomers of 1,2-dichloroethene, $CHCl{=}CHCl$. There is only one 1,2-dichloroethane because carbon atoms rotate freely about a single bond. Thus the structural formulas I and II that follow represent the same compound. The chlorine atoms are simply shown in different conformations in the two formulas due to rotation of the CH_2Cl group about the carbon–carbon single bond.

$$
\begin{array}{cc}
\underset{\underset{H}{|}}{\overset{\overset{Cl}{|}}{H-C}}\ominus\underset{\underset{H}{|}}{\overset{\overset{Cl}{|}}{C-H}} &
\underset{\underset{H}{|}}{\overset{\overset{Cl}{|}}{H-C}}-\underset{\underset{Cl}{|}}{\overset{\overset{H}{|}}{C-H}} \\
\text{I} & \text{II}
\end{array}
$$

1,2-dichloroethane

Compounds containing a carbon–carbon double bond (pi bond) have restricted rotation about that double bond. Rotation cannot occur unless the pi bond breaks—a difficult process. Restricted rotation in a molecule gives rise to a type of isomerism known as *geometric isomerism*. Isomers that differ from each other only in the geometry of their molecules and not in the order of their atoms are known as **geometric isomers.** They are also called cis–trans isomers. Two isomers of 1,2-dichloroethene exist because of geometric isomerism.

geometric isomer

For a further explanation, let's look at the geometry of an ethene molecule. This molecule is planar, or flat, with all six atoms lying in a single plane as in a rectangle:

$$
\begin{array}{c}
H \qquad\qquad H \\
\diagdown \qquad\quad \diagup \\
C{=}C \\
\diagup \qquad\quad \diagdown \\
H \qquad\qquad H
\end{array}
$$

Because the hydrogen atoms are identical, only one structural arrangement is possible for ethene. But if one hydrogen atom on each carbon atom is replaced by chlorine, for example, two different geometric isomers are possible:

cis-1,2-dichloroethene   trans-1,2-dichloroethene

$$
\begin{array}{c}
\text{Cl} \qquad \text{Cl} \\
\diagdown \qquad \diagup \\
\text{C}=\text{C} \\
\diagup \qquad \diagdown \\
\text{H} \qquad \text{H}
\end{array}
\quad \text{and} \quad
\begin{array}{c}
\text{H} \qquad \text{Cl} \\
\diagdown \qquad \diagup \\
\text{C}=\text{C} \\
\diagup \qquad \diagdown \\
\text{Cl} \qquad \text{H}
\end{array}
$$

cis-1,2-dichloroethene   trans-1,2-dichloroethene
(bp = 60.1°C)   (bp = 48.4°C)

Both of these isomers are known and have been isolated. The fact that they have different boiling points as well as other different physical properties is proof that they are not the same compound. Note that, in naming these geometric isomers, the prefix *cis-* is used to designate the isomer with the substituent groups (chlorine atoms) on the same side of the double bond, and the prefix *trans-* designates the isomer with the substituent groups on opposite sides of the double bond.

Molecules of *cis-* and *trans-*1,2-dichloroethene are not superimposable. That is, we cannot pick up one molecule and place it over the other in such a way that all the atoms in each molecule occupy the same relative positions in space. Nonsuperimposability is a general test for isomerism that all kinds of isomers must meet.

An alkene shows cis–trans isomerism when each carbon atom of the double bond has two different kinds of groups attached to it.

$$
\begin{array}{c}
\text{a} \qquad \text{a} \\
\diagdown \qquad \diagup \\
\text{C}=\text{C} \\
\diagup \qquad \diagdown \\
\text{b} \qquad \text{b}
\end{array}
\qquad\qquad
\begin{array}{c}
\text{a} \qquad \text{b} \\
\diagdown \qquad \diagup \\
\text{C}=\text{C} \\
\diagup \qquad \diagdown \\
\text{b} \qquad \text{a}
\end{array}
$$

cis isomer   trans isomer

An alkene does not show cis–trans isomerism if one carbon atom of the double bond has two identical groups attached to it. Thus there are no geometric isomers of ethene or propene.

$$
\left.
\begin{array}{c}
\text{H} \qquad \text{H} \\
\diagdown \qquad \diagup \\
\text{C}=\text{C} \\
\diagup \qquad \diagdown \\
\text{H} \qquad \text{H}
\end{array}
\right\}
\xleftarrow[\text{the same}]{\text{Two groups}}
\left\{
\begin{array}{c}
\text{H} \qquad \text{CH}_3 \\
\diagdown \qquad \diagup \\
\text{C}=\text{C} \\
\diagup \qquad \diagdown \\
\text{H} \qquad \text{H}
\end{array}
\right.
$$

Four structural isomers of butene, C_4H_8, are known. Two of these, 1-butene and 2-methylpropene, do not show geometric isomerism.

$$CH_3CH_2 \quad H \atop C=C \atop H \quad H \qquad \xleftarrow[\text{the same}]{\text{Two groups}} \qquad CH_3 \quad H \atop C=C \atop CH_3 \quad H$$

1-butene   2-methylpropene

The other two butenes are the cis–trans isomers shown here.

$$CH_3 \quad CH_3 \atop C=C \atop H \quad H \qquad \text{and} \qquad CH_3 \quad H \atop C=C \atop H \quad CH_3$$

cis-2-butene   *trans*-2-butene

Example 20.3   Draw structures for (a) *trans*-3-heptene and (b) *cis*-5-chloro-2-hexene.

Solution   (a) The compound contains seven carbon atoms with a double bond between carbons 3 and 4. First draw a C=C double bond in a planar arrangement:

$$\diagdown \atop C=C \atop \diagup \qquad$$

In the trans positions attach a two-carbon chain to one carbon atom and a three-carbon chain to the other carbon atom to give a continuous chain of seven carbon atoms:

$$\overset{1}{C}-\overset{2}{C} \atop \overset{3}{C}=\overset{4}{C} \atop \overset{5}{C}-\overset{6}{C}-\overset{7}{C}$$

Now attach hydrogen atoms to give each carbon atom four bonds. Carbons 3 and 4 need only one hydrogen atom apiece; these two hydrogen atoms are also trans to each other. The structure is

$$CH_3CH_2 \quad H \atop C=C \atop H \quad CH_2CH_2CH_3$$

(b) The compound contains six carbons, a double bond between carbons 2 and 3, and a Cl atom on carbon 5. First, draw the straight-chain compound.

$$\overset{1}{C}H_3\overset{2}{C}H=\overset{3}{C}H\overset{4}{C}H_2\overset{5}{C}HCl\overset{6}{C}H_3$$

Each carbon of the double bond has two different groups attached to it, a requirement of cis–trans isomerism. Now draw the C=C double bond in a planar arrangement.

$$\overset{2}{\underset{}{C}}=\overset{3}{\underset{}{C}}$$

In the cis position place the CH_3 group on carbon 2 and the $CH_2CHClCH_3$ group on carbon 3.

$$\underset{}{\overset{H_3C}{\diagdown}}\underset{}{\overset{CH_2CHClCH_3}{\diagup}} \\ C=C$$

Add H atoms to give each carbon atom four bonds. The structure is

$$\overset{H_3C}{\diagdown}\quad\overset{CH_2CHClCH_3}{\diagup} \\ C=C \\ \overset{}{\diagup}\quad\overset{}{\diagdown} \\ H\qquad H$$

Draw structural formulas and names for all the isomers of pentene, C_5H_{10}. Identify all geometric isomers.

Example 20.4

Start by drawing the isomers of the five-carbon chain, placing the C=C in all possible positions (there are two possible isomers, 1-pentene and 2-pentene). Then proceed to the four-carbon chains with a methyl group branch chain. Locate the C=C and the CH_3 group in all possible positions. No three-carbon chain pentenes are possible. Check for duplications from the names of the compounds.

Solution

$$CH_2{=}CHCH_2CH_2CH_3 \qquad\qquad CH_3CH{=}CHCH_2CH_3$$
1-pentene $\qquad\qquad\qquad\qquad$ 2-pentene

$$\overset{CH_3}{\underset{|}{}} \\ CH_2{=}CCH_2CH_3$$
2-methyl-1-butene

$$\overset{CH_3}{\underset{|}{}} \\ CH_3C{=}CHCH_3$$
2-methyl-2-butene

$$\overset{CH_3}{\underset{|}{}} \\ CH_2{=}CHCHCH_3$$
3-methyl-1-butene

Of these five compounds, only 2-pentene can have cis–trans isomers. The others have two identical groups on one carbon atom of the double bond. Draw the cis–trans isomers.

$$\overset{H}{\diagdown}\quad\overset{H}{\diagup} \\ C=C \\ \overset{CH_3}{\diagup}\quad\overset{CH_2CH_3}{\diagdown}$$
cis-2-pentene

$$\overset{H}{\diagdown}\quad\overset{CH_2CH_3}{\diagup} \\ C=C \\ \overset{CH_3}{\diagup}\quad\overset{H}{\diagdown}$$
trans-2-pentene

Example 20.5   Is the compound below the cis or the trans isomer?

$$
\begin{array}{ccc}
\overset{1}{C}H_3 & & CH_3 \\
\diagdown 2 & & 3\diagup \\
& C=C & \\
\diagup & & \diagdown 4\ 5 \\
H & & CH_2CH_3
\end{array}
$$

Solution

In branched-chain alkenes, the cis–trans designation is ordinarily given to the structure containing the longest continuous carbon chain that includes the carbon–carbon double bond. In this case the longest chain is the 2-pentene, in which the methyl group on carbon 2 and the ethyl group on carbon 3 are trans to each other. Thus the name of the compound shown is *trans*-3-methyl-2-pentene. The cis isomer is

$$
\begin{array}{ccc}
CH_3 & & CH_2CH_3 \\
\diagdown & & \diagup \\
& C=C & \\
\diagup & & \diagdown \\
H & & CH_3
\end{array}
$$

Practice 20.4

Draw structures (a) *cis*-3-methyl-2-pentene and (b) *trans*-2-bromo-2-pentene.

Practice 20.5

Determine whether geometric isomers exist for the following compounds. Draw structures for the cis and trans isomers.
(a) 1-chloro-3-methyl-2-pentene   (b) 3-hexene   (c) 2,3-dimethyl-2-pentene

Practice 20.6

Identify each of the following as the cis or trans isomer.

(a)
$$
\begin{array}{ccc}
H & & CH_3 \\
\diagdown & & \diagup \\
& C=C & \\
\diagup & & \diagdown \\
CH_3CH_2 & & Cl
\end{array}
$$

(b)
$$
\begin{array}{ccc}
CH_3CH_2 & & CH_2CH_3 \\
\diagdown & & \diagup \\
& C=C & \\
\diagup & & \diagdown \\
CH_3 & & Br
\end{array}
$$

Practice 20.7 Application to Biochemistry

Biochemical processes are very specific. For example, only the trans isomer of the following two compounds can be used in the synthesis of fatty acids:

$$
\begin{array}{ccc}
H & & H \\
\diagdown & & \diagup \\
& C=C & \\
\diagup & & \diagdown \\
CH_3CH_2CH_2 & & \underset{\parallel}{\overset{}{C}}-SACP \\
& & O
\end{array}
\qquad
\begin{array}{ccc}
CH_3CH_2CH_2 & & H \\
\diagdown & & \diagup \\
& C=C & \\
\diagup & & \diagdown \\
H & & \underset{\parallel}{\overset{}{C}}-SACP \\
& & O
\end{array}
$$

(ACP means acyl carrier protein)
Which isomer, A or B, will be used in fatty acid synthesis?

Living cells are sensitive to, and respond to, specific isomers. For example, the eye only responds to light, when a specific *cis* isomer is converted to its trans form. The molecule involved, retinal, is derived from vitamin A and β-carotene (found in vegetables such as carrots). One of the double bonds in retinal is converted to the *cis* form.

11-*cis*-retinal
— *cis* double bond

▲
A three-dimensional representation of 11-cis-retinal.

The 11-*cis*-retinal is combined with a protein (opsin) to form the visual pigment, rhodopsin. When light is absorbed, the cis double bond changes to a *trans* double bond.

all-*trans*-retinal
— *trans* double bond

The isomerization of 11-*cis*-retinal to all-*trans*-retinal starts a cascade of events that our brains interpret as "seeing."

Many compounds have more than one carbon–carbon double bond. Compounds with two (*di*) double bonds are called *dienes;* those with three (*tri*) double bonds are *trienes,* and so forth. The location of the double bonds are numbered like in alkenes. One notable example is 2-methyl-1,3-butadiene, the component of natural rubber.

Study the names and structures of the following compounds.

$$\overset{1}{C}H_2 = \overset{2}{C}H - \overset{3}{C}H = \overset{4}{C}H_2$$

1,3-butadiene

Position of   Four   Inserted   Two
the double   carbon   to ease   double
bonds   atoms   pronunciation   bonds

$$CH_2 = CH - CH - CH = CH_2$$
1,4-pentadiene

$$CH_2 = \overset{\overset{\displaystyle CH_3}{|}}{C} - CH = CH_2$$
2-methyl-1,3-butadiene

20.4 Cycloalkenes

cycloalkene

The general formula for cyclohexenes is C_nH_{2n-2}

As the name implies, **cycloalkenes** are cyclic compounds that contain a carbon–carbon double bond in the ring. The two most common cycloalkenes are cyclopentene and cyclohexene. The double bond may be placed between any two carbon atoms:

cyclopentene (C_5H_8)   cyclohexene (C_6H_{10})

In cycloalkenes the carbons of the double bond are assigned numbers 1 and 2. Thus the positions of the double bond need not be indicated in the name of the compound. Other substituents on the ring are named in the usual manner, and their positions on the ring are indicated with the smallest possible numbers.

In the following examples, note that a Cl or a CH_3 has replaced one H atom in the molecule. The ring is numbered either clockwise or counterclockwise starting with the carbon–carbon double bond so that the substituted group(s) have the smallest possible numbers.

3-chlorocyclohexene   1-methylcyclopentene   1, 3-dimethylcyclohexene
(C_6H_9Cl)   ($C_5H_7CH_3$)   [$C_6H_8(CH_3)_2$]

No cis–trans designation is necessary for cycloalkenes that contain up to seven carbon atoms in the ring. Cyclooctene has been shown to exist in both cis and trans-forms.

20.5 Preparation and Physical Properties of Alkenes

Common preparation methods for alkenes start with saturated organic molecules; each carbon in these molecules is bonded to four other atoms. Then, to form double bonds, atoms must be removed. The product contains double-bonded carbons that are connected to only three atoms each. Alkene synthesis commonly means "getting rid" of some atoms—an elimination reaction. Watch for evidence of elimination in the following two examples of alkene preparation.

Cracking

Ethene can be produced by the cracking of petroleum. **Cracking,** or pyrolysis, is the process in which saturated hydrocarbons are heated to very high temperatures in the presence of a catalyst (usually silica–alumina). This results in large molecules breaking into smaller ones, with the elimination of hydrogen, to form alkenes and small hydrocarbons like methane and ethane. Unfortunately, cracking always results in mixtures of products and is therefore not used often in the laboratory.

cracking

$$\text{Alkane } (C_nH_{2n+2}) \xrightarrow[\text{catalyst}]{\Delta} \text{Mixture of alkenes + Alkanes + Hydrogen gas}$$

$$2 \text{ CH}_3\text{CH}_2\text{CH}_3 \xrightarrow{\sim 500^{\circ}\text{C}} \text{CH}_3\text{CH}{=}\text{CH}_2 + \text{CH}_2{=}\text{CH}_2 + \text{CH}_4 + \text{H}_2$$

Dehydration of Alcohols

Dehydration involves the elimination of a molecule of water from a reactant molecule. Dehydration reactions are very common in organic chemistry as well as in biochemistry.

dehydration

To produce an alkene by dehydration, an alcohol is heated in the presence of concentrated sulfuric acid.

$$\underset{\overset{|}{\underset{H}{\overset{|}{\text{CH}_3\text{C}}}}}{\overset{H}{\overset{|}{}}}{-}\underset{\overset{|}{\underset{\text{OH}}{}}}{\overset{H}{\overset{|}{\text{CCH}_3}}} \xrightarrow[\Delta]{\text{conc. H}_2\text{SO}_4} \underset{}{\overset{H \quad H}{\overset{| \quad |}{\text{CH}_3\text{C}{=}\text{CCH}_3}}} + \text{ H}_2\text{O}$$

The alkene is formed as a result of elimination of H and OH (shown in blue) from adjacent carbon atoms.

Practice 20.8 Application to Biochemistry

Living cells use common organic reactions. An important beginning to the citric acid cycle dehydrates citric acid to form an alkene-containing compound aconitic acid. Draw the *cis* and *trans* structures of this acid.

$$\underset{}{\overset{\text{CH}_2{-}\text{COOH}}{\overset{|}{}}}$$
$$\text{HO}{-}\text{C}{-}\text{COOH}$$
$$\underset{}{\overset{|}{\text{CH}_2{-}\text{COOH}}}$$

citric acid

Physical Properties

Alkenes have physical properties very similar to the corresponding alkanes. This is not surprising since the difference between an alkane and an alkene is simply two hydrogen atoms.

General formula for alkanes	C_nH_{2n+2}
General formula for alkenes	C_nH_{2n}

Since alkenes have slightly smaller molar masses, their boiling points are slightly lower than the corresponding alkanes. The smaller alkenes (to 5 carbons) are gases at room temperature. As the chain lengthens (5–17 carbons), the alkenes are liquid, and above 17 carbons they are solid. The alkenes are nonpolar, like the other hydrocarbons, and so are insoluble in water but soluble in organic solvents. The densities of most alkenes are much less than water. Table 20.1 shows the properties of some alkenes. Notice that isomers (C_4H_8) have similar boiling points although they differ significantly in melting points. Melting-point differences occur because the isomers have different shapes and therefore fit into their crystalline structures in significantly different ways.

 20.6   Chemical Properties of Alkenes

Alkene chemistry is especially important in the biochemistry of fats.

What type of reaction might be expected for an alkene (or an alkyne)? Note that ethene ($CH_2{=}CH_2$) has only three atoms bonded to each carbon; acetylene ($CH{\equiv}CH$) has only two atoms bonded to each carbon. Both alkenes and alkynes have fewer than the maximum of four atoms bonded per carbon. These molecules are more reactive than the corresponding alkanes and readily undergo addition reactions.

Addition

Addition at the carbon–carbon double bond is the most common reaction of alkenes. Hydrogen, halogens (Cl_2 or Br_2), hydrogen halides, sulfuric acid, and water are some

TABLE 20.1   Physical Properties of Alkenes

Molecular formula	Structural formula	IUPAC name	Density, $20\frac{°C}{4°C}$ (g/mL)	Melting point (°C)	Boiling point (°C)
C_2H_4	$CH_2{=}CH_2$	Ethene	—	−169	−104
C_3H_6	$CH_3CH{=}CH_2$	Propene	—	−185	−48
C_4H_8	$CH_3CH_2CH{=}CH_2$	1-Butene	0.595	−185	−6
C_4H_8	$(CH_3)_2C{=}CH_2$	2-Methylpropene	0.594	−14	−7
C_5H_{10}	$CH_3(CH_2)_2CH{=}CH_2$	1-Pentene	0.641	−138	30

of the reagents that can be added to unsaturated hydrocarbons. Ethene, for example, reacts in the presence of a platinum catalyst in this fashion:

$$CH_2{=}CH_2 \ + \ H_2 \ \xrightarrow[\text{1 atm}]{\text{Pt, 25°C}} \ CH_3{-}CH_3$$

ethene   ethane

$$CH_2{=}CH_2 \ + \ Br{-}Br \ \longrightarrow \ CH_2Br{-}CH_2Br \quad (Cl_2 \text{ reacts similarly})$$

colorless   reddish   1,2-dibromoethane
                   brown   colorless

The disappearance of the reddish-brown bromine color provides visible evidence of reaction. Other reactions of ethene include the following.

$$CH_2{=}CH_2 \ + \ HCl \ \longrightarrow \ CH_3CH_2Cl$$

chloroethane
(ethyl chloride)

$$CH_2{=}CH_2 \ + \ HOSO_3H \, (conc.) \ \longrightarrow \ CH_3CH_2OSO_3H$$

sulfuric acid   ethyl hydrogen sulfate

$$CH_2{=}CH_2 \ + \ HOH \ \xrightarrow{\ H^+\ } \ CH_3CH_2OH$$

ethanol
(ethyl alcohol)

Unsaturated alkenes are readily converted to saturated molecules by addition. Also, note that addition is the reverse of elimination. For example, by the addition of water, an alkene can be converted to an alcohol, and by the elimination of water, an alcohol can be converted to an alkene.

The preceding examples dealt with ethene, but reactions of this kind can be made to occur on almost any molecule that contains a carbon–carbon double bond. If a symmetrical molecule such as Cl_2 is added to propene, only one product, 1,2-dichloropropane, is formed:

$$CH_2{=}CH{-}CH_3 + Cl_2 \ \longrightarrow \ CH_2Cl{-}CHCl{-}CH_3$$

1,2-dichloropropane

But if an unsymmetrical molecule such as HCl is added to propene, two products are theoretically possible, depending upon which carbon atom adds the hydrogen. The two possible products are 1-chloropropane and 2-chloropropane. Experimentally we find that 2-chloropropane is formed almost exclusively:

$$CH_3{-}CH{=}CH_2 \ + \ HCl$$

$$\nearrow CH_3CHClCH_3$$
(about 100%)

$$\searrow CH_3CH_2CH_2Cl$$
(trace)

The double bond is broken, and unsaturated alkene molecules become saturated by an addition reaction.

▲
Bromine has the characteristic orange color (left), but when added to an alkene the color disappears (right).

The H^+ indicates that the reaction is carried out under acidic conditions.

A single product is obtained because the reaction follows specific steps. The sum of these steps is termed a *reaction mechanism*. The reaction between propene and HCl proceeds by the following addition mechanism:

1. A proton (H^+) from HCl bonds to carbon 1 of propene utilizing the pi bond electrons. The intermediate formed is a positively charged alkyl group, or carbocation. The positive charge is localized on carbon 2 of this carbocation.

$$\overset{1}{C}H_2=\overset{2}{C}H-\overset{3}{C}H_3 \ + \ HCl \ \longrightarrow \ CH_3-\overset{+}{C}H-CH_3 \ + \ Cl^-$$

   isopropyl carbocation

2. The chloride ion then adds to the positively charged carbon atom to form a molecule of 2-chloropropane:

$$CH_3-\overset{+}{C}H-CH_3 \ + \ Cl^- \ \longrightarrow \ CH_3-CHCl-CH_3$$

   2-chloropropane

carbocation   An ion in which a carbon atom has a positive charge is known as a **carbocation.** There are four types of carbocations: methyl, primary (1°), secondary (2°), and tertiary (3°). Examples of these four types are as follows:

$$H-\overset{\overset{\displaystyle H}{|}}{\underset{\underset{\displaystyle H}{|}}{C^+}} \qquad CH_3-\overset{\overset{\displaystyle H}{|}}{\underset{\underset{\displaystyle H}{|}}{C^+}} \qquad CH_3CH_2-\overset{\overset{\displaystyle H}{|}}{\underset{\underset{\displaystyle H}{|}}{C^+}} \qquad CH_3-\overset{+}{\underset{\underset{\displaystyle H}{|}}{C}}-CH_3 \qquad CH_3-\overset{\overset{\displaystyle CH_3}{|}}{\underset{\underset{\displaystyle CH_3}{|}}{C^+}}$$

| methyl carbocation | ethyl carbocation (primary) | propyl carbocation (primary) | isopropyl carbocation (secondary) | *t*-butyl carbocation (tertiary) |

A carbon atom is designated as primary if it is bonded to one carbon atom, secondary if it is bonded to two carbon atoms, and tertiary if it is bonded to three carbon atoms. Thus in a primary carbocation the positive carbon atom is bonded to only one carbon atom. In a secondary carbocation the positive carbon atom is bonded to two carbon atoms. In a tertiary carbocation the positive carbon atom is bonded to three carbon atoms. There are no quaternary carbocations.

The order of stability of carbocations and hence the ease with which they are formed is tertiary > secondary > primary. Thus, in the reaction of propene and HCl, the more stable isopropyl carbocation (secondary) is formed as an intermediate in preference to n-propyl carbocation (primary).

Stability of carbocations: $3° > 2° > 1° > \overset{+}{C}H_3$

In the middle of the 19th century, a Russian chemist, V. Markovnikov, observed reactions of this kind, and in 1869 he formulated a useful generalization now known **Markovnikov's rule**   as **Markovnikov's rule.** This rule in essence states:

> **When an unsymmetrical molecule such as HX(HCl) adds to a carbon–carbon double bond, the hydrogen from HX goes to the carbon atom that has the greater number of hydrogen atoms.**

As you can see, the addition of HCl to propene, discussed previously, follows Markovnikov's rule. The addition of HI to 2-methylpropene is another example illustrating this rule:

$$CH_3-\underset{\underset{}{\overset{\overset{CH_3}{|}}{C}}}=CH_2 \ + \ HI \ \longrightarrow \ CH_3-\underset{\underset{I}{|}}{\overset{\overset{CH_3}{|}}{C}}-CH_3$$

2-iodo-2-methylpropane
(*tert*-butyl iodide)

In this reaction, the $=CH_2$ is the alkene carbon with the most H atoms (two); it thus adds the H^+ followed by the addition of I^- to the other alkene carbon atom.

General rules of this kind help us predict the products of reactions; however, exceptions are known for most such rules.

Write formulas for the organic products formed when 2-methyl-1-butene reacts with (a) H_2, Pt/25°C; (b) Cl_2; (c) HCl; and (d) H_2O, H^+.

Example 20.6

First write the formula for 2-methyl-1-butene:

Solution

$$\overset{1}{CH_2}=\overset{2}{\underset{\overset{|}{CH_3}}{C}}-\overset{3}{CH_2}-\overset{4}{CH_3}$$

(a) The double bond is broken when a hydrogen molecule adds. One H atom adds to each carbon atom of the double bond. Platinum, Pt, is a necessary catalyst in this reaction. The product is

$$CH_3\underset{\overset{|}{CH_3}}{CH}CH_2CH_3$$

2-methylbutane

(b) The Cl_2 molecule adds to the carbons of the double bond. One Cl atom adds to each carbon atom of the double bond. The product is

$$\underset{\overset{|}{Cl}}{CH_2}\underset{\overset{|}{Cl}}{\overset{\overset{CH_3}{|}}{C}}CH_2CH_3$$

1,2-dichloro-2-methylbutane

(c) HCl adds to the double bond according to Markovnikov's rule. The H^+ goes to carbon 1 (the more stable 3° carbocation is formed as an intermediate product), and the Cl^- goes to carbon 2. The product is

$$CH_3\underset{\overset{|}{Cl}}{\overset{\overset{CH_3}{|}}{C}}CH_2CH_3$$

2-chloro-2-methylbutane

(d) The net result of this reaction is a molecule of water added across the double bond. The H adds to carbon 1 (the carbon with the greater number of hydrogen atoms), and the OH adds to carbon 2 (the carbon of the double bond with the lesser number of hydrogen atoms). The product is

$$
\begin{array}{c}
\text{CH}_3 \\
| \\
\text{CH}_3\text{CCH}_2\text{CH}_3 \\
| \\
\text{OH}
\end{array}
$$

2-methyl-2-butanol
(an alcohol)

Example 20.7   Write equations for the addition of HCl to (a) 1-pentene and (b) 2-pentene.

Solution   (a) In the case of 1-pentene, $CH_3CH_2CH_2CH{=}CH_2$, the proton from HCl adds to carbon 1 to give the more stable secondary carbocation, followed by the addition of Cl^- to give the product 2-chloropentane. The addition is directly in accordance with Markovnikov's rule.

$$CH_3CH_2CH_2CH{=}CH_2 \xrightarrow{\text{HCl}} CH_3CH_2CH_2\overset{+}{C}HCH_3 \xrightarrow{Cl^-} CH_3CH_2CH_2CHClCH_3$$

2° carbocation   2-chloropentane

(b) In 2-pentene, $CH_3CH_2CH{=}CHCH_3$, each carbon of the double bond has one H atom, and the addition of a proton to either one forms a secondary carbocation. After the addition of Cl^-, the result is two isomeric products that are formed in almost equal quantities:

$$CH_3CH_2CH{=}CHCH_3 \begin{cases} \xrightarrow{\text{HCl}} CH_3CH_2CH_2\overset{+}{C}HCH_3 \xrightarrow{Cl^-} CH_3CH_2CH_2CHClCH_3 \\ \\ \xrightarrow{\text{HCl}} CH_3CH_2\overset{+}{C}HCH_2CH_3 \xrightarrow{Cl^-} CH_3CH_2CHClCH_2CH_3 \end{cases}$$

2-chloropentane

3-chloropentane

Practice 20.9

Write formulas for the organic products formed when 3-methyl-2-pentene reacts with (a) H_2/Pt; (b) Br_2; (c) HCl; and (d) H_2O, H^+.

Practice 20.10

Write equations for (a) addition of water to 1-methylcyclopentene and (b) addition of HI to 2-methyl-2-butene.

The nutritionally essential amino acid, threonine, can be made by hydration of the following alkene:

$$
\begin{array}{c}
CH_2 \\
\parallel \\
CH \\
| \\
NH_2-CH-COOH
\end{array}
$$

The hydration follows Markovnikov's rule. Show the structure of the alcohol formed.

Oxidation

Another typical reaction of alkenes is oxidation at the double bond. For example, when shaken with a cold, dilute solution of potassium permanganate, $KMnO_4$, an alkene is converted to a glycol (glycols are dihydroxy alcohols). Ethene reacts in this manner:

$$CH_2{=}CH_2 \ + \ KMnO_4(aq) \ + \ H_2O \ \longrightarrow \ \underset{\underset{OH \quad OH}{|\qquad|}}{CH_2{-}CH_2} \ + \ MnO_2 \ + \ KOH$$

ethene (ethylene) (purple)   1,2 ethanediol (ethylene glycol)   (brown)

The *Baeyer test* makes use of this reaction to detect or confirm the presence of double (or triple) bonds in hydrocarbons. Evidence of reaction (positive Baeyer test) is the disappearance of the purple color of permanganate ions. The Baeyer test is not specific for detecting unsaturation in hydrocarbons because other classes of compounds may also give a positive Baeyer test.

Carbon–carbon double bonds are found in many different kinds of molecules. Most of these substances react with potassium permanganate and undergo somewhat similar reactions with other oxidizing agents including oxygen in the air and, especially, with ozone. Such reactions are frequently troublesome. For example, premature aging and cracking of automobile tires in smoggy atmospheres occur because ozone attacks the double bonds in rubber molecules. Cooking oils and fats sometimes develop disagreeable odors and flavors because the oxygen of the air reacts with the double bonds present in these materials. Potato chips, because of their large surface area, are especially subject to flavor damage caused by oxidation of the unsaturated cooking oils that they contain.

20.7   Alkynes: Nomenclature and Preparation

Nomenclature

IUPAC Rules for Naming Alkynes

The rules for naming alkynes are the same as those for alkenes, but the ending *-yne* is used to indicate the presence of a triple bond. Table 20.2 lists names and formulas for some common alkynes.

Practice 20.12

Write formulas for
(a) 3-methyl-1-butyne
(b) 4-ethyl-4-methyl-2-hexyne
(c) cyclohexylethyne

Preparation

Although triple bonds are very reactive, it is relatively easy to synthesize alkynes. Acetylene can be prepared inexpensively from calcium carbide and water.

$$CaC_2 + 2\,H_2O \longrightarrow HC \equiv CH + Ca(OH)_2$$

Acetylene is also prepared by the cracking of methane in an electric arc.

$$2\,CH_4 \xrightarrow{1500°C} HC \equiv CH + 3\,H_2$$

The general formula for alkynes is C_nH_{2n-2}

TABLE 20.2	Nomenclature for Some Common Alkynes	
Molecular formula	Structural formula	IUPAC name*
C_2H_2	$H-C \equiv C-H$	Ethyne*
C_3H_4	$CH_3-C \equiv C-H$	Propyne
C_4H_6	$CH_3CH_2-C \equiv C-H$	1-Butyne
C_4H_6	$CH_3-C \equiv C-CH_3$	2-Butyne

*Ethyne is commonly known as acetylene.

20.8 Physical and Chemical Properties of Alkynes

Physical Properties

Acetylene is a colorless gas, with little odor when pure. The disagreeable odor we associate with it is the result of impurities (usually PH_3). Acetylene is insoluble in water and is a gas at normal temperature and pressure (bp = $-84°C$). As a liquid, acetylene is very sensitive and may decompose violently (explode), spontaneously, or from a slight shock.

Alkynes are only rarely found in living organisms.

$$HC \equiv CH \longrightarrow H_2 + 2\ C + 227\ kJ\ (54.3\ kcal)$$

To eliminate the danger of explosions, acetylene is dissolved under pressure in acetone and is packed in cylinders that contain a porous inert material.

Chemical Properties

Acetylene is used mainly (1) as fuel for oxyacetylene cutting and welding torches and (2) as an intermediate in the manufacture of other substances. Both uses are dependent upon the great reactivity of acetylene. Acetylene and oxygen mixtures produce flame temperatures of about 2800°C. Acetylene readily undergoes addition reactions rather similar to those of ethene. It reacts with chlorine and bromine and decolorizes a permanganate solution (Baeyer's test). Either one or two molecules of bromine or chlorine can be added:

$$HC \equiv CH + Br_2 \longrightarrow CHBr = CHBr$$
$$\text{1,2-dibromoethene}$$

or

$$HC \equiv CH + 2\ Br_2 \longrightarrow CHBr_2 - CHBr_2$$
$$\text{1,1,2,2-tetrabromoethane}$$

It is apparent that either unsaturated or saturated compounds can be obtained as addition products of acetylene. Often, unsaturated compounds capable of undergoing further reactions are made from acetylene. For example, vinyl chloride, which is used to make the plastic polyvinyl chloride (PVC), can be made by simple addition of HCl to acetylene:

$$CH_2 \equiv CH + HCl \longrightarrow CH_2 = CHCl$$
$$\text{chloroethene}$$
$$\text{(vinyl chloride)}$$

(*Note:* The common name for the $CH_2 = CH-$ group is *vinyl.*) If the reaction is not properly controlled, another HCl adds to the chloroethene. This reaction follows Markovinkov's rule.

$$CH_2 = CHCl + HCl \longrightarrow CH_3CHCl_2$$
$$\text{1,1-dichloroethane}$$

Addition reactions are common for all alkynes.

Superabsorbants used in disposable diapers increase the amount of liquid the diaper will absorb without leaking. ▶

Hydrogen chloride reacts with other alkynes in a similar fashion to form substituted alkenes. The addition follows Markovnikov's rule. Alkynes can react with 1 or 2 mol of HCl. Consider the reaction of propyne with HCl:

$$CH_3C{\equiv}CH \ + \ HCl \longrightarrow CH_3CCl{=}CH_2$$
$$2\text{-chloropropene}$$

$$CH_3CCl{=}CH_2 \ + \ HCl \longrightarrow CH_3CCl_2CH_3$$
$$2,2\text{-dichloropropane}$$

Certain reactions are unique to alkynes. Alkynes are capable of reacting at times when alkenes will not. Acetylene, with certain catalysts, reacts with HCN to form $CH_2{=}CHCN$ (acrylonitrile). This chemical is used industrially to manufacture Orlon, a polymer commonly found in clothing. It is also used to form the superabsorbants, which are capable of retaining up to 2000 times their mass of water. These superabsorbants are used in disposable diapers as well as in soil additives to retain water.

The reactions of other alkynes resemble those of acetylene. Although many other alkynes are known, acetylene is by far the most important industrially.

Practice 20.13

Write formulas for the products formed by the reaction of HCl with
(a) 1-butyne and (b) 2-pentyne to form saturated compounds.

 20.9   Aromatic Hydrocarbons: Structure

Benzene and all substances with structures and chemical properties that resemble benzene are classified as *aromatic compounds*. The word *aromatic* originally referred to the rather pleasant odor possessed by many of these substances, but this meaning has been dropped. Benzene, the parent substance of the aromatic hydrocarbons, was first isolated by Michael Faraday in 1825; its correct molecular formula, C_6H_6, was established a few years later. The establishment of a reasonable structural formula that would account for the properties of benzene was a difficult problem for chemists in the mid-19th century.

Finally, in 1865, August Kekulé proposed that the carbon atoms in a benzene molecule are arranged in a six-membered ring with one hydrogen atom bonded to each carbon atom and with three carbon–carbon double bonds, as shown in the following structure:

Kekulé soon realized that there should be two dibromobenzenes, based on double- and single-bond positions relative to the two bromine atoms:

Since only one dibromobenzene (with bromine atoms on adjacent carbons) could be produced, Kekulé suggested that the double bonds are in rapid oscillation within the molecule. He therefore proposed representing the structure of benzene in this fashion:

Kekulé's concepts are a landmark in the history of chemistry. They are the basis of the best representation of the benzene molecule devised in the 19th century, and they mark the beginning of our understanding of structure in aromatic compounds.

Kekulé's formulas have one serious shortcoming: They represent benzene and related substances as highly unsaturated compounds. Yet benzene does not react like a typical alkene; it does not decolorize bromine solutions rapidly, nor does it destroy the purple color of permanganate ions (Baeyer's test). Instead, benzene behaves

FIGURE 20.5 ▶
**Bonding in a benzene mole-
cule: (a) sp^2–sp^2 orbital
overlap to form the carbon
ring structure; (b) carbon–
hydrogen bonds formed by
sp^2–s orbital overlap and
overlapping of p orbitals;
(c) pi electron clouds above
and below the plane of the
carbon ring.**

FIGURE 20.5 ▶
Bonding in a benzene molecule: (a) sp^2–sp^2 orbital overlap to form the carbon ring structure; (b) carbon–hydrogen bonds formed by sp^2–s orbital overlap and overlapping of p orbitals; (c) pi electron clouds above and below the plane of the carbon ring.

chemically like a typical alkane. Its reactions are usually the substitution type, wherein a hydrogen atom is replaced by some other group; for example,

$$C_6H_6 + Cl_2 \xrightarrow{Fe} C_6H_5Cl + HCl$$

This problem was not fully resolved until the technique of X-ray diffraction, developed in the years following 1912, permitted chemists to determine the actual distances between the nuclei of carbon atoms in molecules. The center-to-center distances between carbon atoms in different kinds of hydrocarbon molecules are

Ethane (single bond)	0.154 nm
Ethene (double bond)	0.134 nm
Benzene	0.139 nm

Because only one carbon–carbon distance (bond length) is found in benzene, it is apparent that alternating single and double bonds do not exist in the benzene molecule.

Modern theory accounts for the structure of the benzene molecule in this way: The orbital hybridization of the carbon atoms is sp^2 (see the structure of ethene in Section 20.1). A planar hexagonal ring is formed by the overlapping of two sp^2 orbitals on each of six carbon atoms. The other sp^2 orbital on each carbon atom overlaps an s orbital of a hydrogen atom, bonding the carbon to the hydrogen by a sigma bond. The remaining six p orbitals, one on each carbon atom, overlap each other and form doughnut-shaped pi electron clouds above and below the plane of the ring (see Figure 20.5).* The electrons composing these clouds are not attached to particular

---

*The detailed pi molecular orbital theory of the benzene structure is beyond the scope of this book.

carbon atoms but are delocalized and associated with the entire molecule. This electronic structure imparts unusual stability to benzene and is responsible for many of the characteristic properties of aromatic compounds. Because of this stable electronic structure, benzene does not readily undergo addition or elimination reactions, but it does undergo substitution reactions.

For convenience, present-day chemists usually write the structure of benzene as one or the other of these abbreviated forms:

In all of these representations, it is understood that there is a carbon atom and a hydrogen atom at each corner of the hexagon. The classical Kekulé structures are represented by formulas A and B. However, neither of these Kekulé structures actually exists, although in some circumstances they are still used. The real benzene molecule is a hybrid of these structures and is commonly represented by either formula C or D. The circle or the dashed circle indicates the special nature of the benzene pi bonds, as shown in Figure 20.5. We will use the hexagon with the solid circle to represent a benzene ring.

Hexagons are used to represent the structural formulas of benzene derivatives—that is, substances in which one or more hydrogen atoms in the ring have been replaced by other atoms or groups. Chlorobenzene, for example, is written in this fashion:

chlorobenzene, C_6H_5Cl

This notation indicates that the chlorine atom has replaced a hydrogen atom and is bonded directly to a carbon atom in the ring. Thus the correct formula for chlorobenzene is C_6H_5Cl, not C_6H_6Cl.

20.10 Naming Aromatic Compounds

Substituted benzenes are the most common benzene derivatives because substitution is the most common reaction type for benzene. A substituted benzene is derived by replacing one or more hydrogen atoms of benzene by another atom or group of atoms. Thus a monosubstituted benzene has the formula C_6H_5G, where G is the group replacing a hydrogen atom.

Monosubstituted Benzenes

Some monosubstituted benzenes are named by adding the name of the substituent group as a prefix to the word *benzene*. The name is written as one word. Note that the

Substituted benzenes are found in biochemical pigments, hormones, and nerve transmitters.

position of the substituent is not important here as all the positions in the hexagon are equivalent. Examples include the following:

| nitrobenzene | ethylbenzene | chlorobenzene | bromobenzene |

Certain monosubstituted benzenes have special names. These are used as parent names for further substituted compounds and should be memorized.

toluene
(methylbenzene)

phenol
(hydroxybenzene)

styrene
(vinylbenzene)

benzoic acid
(benzene carboxylic acid)

benzaldehyde
(benzene carboxaldehyde)

aniline
(aminobenzene)

The C_6H_5- group is known as the phenyl group (pronounced *fen-il*), and the name *phenyl* is used to name compounds that cannot easily be named as benzene derivatives. For example, the following compounds are named as derivatives of alkanes:

3-chloro-2-phenylpentane

diphenylmethane

Disubstituted Benzenes

When two substituent groups replace two hydrogen atoms in a benzene molecule, three different isomeric compounds are possible. The prefixes *ortho-*, *meta-* and *para-* (abbreviated *o-*, *m-*, and *p-*) are used to name disubstituted benzenes. In ortho disubstituted compounds, the two substituents are located on adjacent carbon atoms of the benzene ring (the 1,2-positions). In meta disubstituted compounds, the two substituents are one carbon apart (the 1,3-position). In para disubstituted compounds, the two substituents are located on opposite points of the hexagon (the 1,4-position). Thus, with respect to a group G on the benzene ring, there are two ortho positions, two meta positions, and one para position.

Consider the dichlorobenzenes, $C_6H_4Cl_2$. Note that the three isomers have different physical properties, indicating that they are truly different substances:

ortho-dichlorobenzene
(mp –17.2°C, bp 180.4°C)

meta-dichlorobenzene
(mp –24.8°C, bp 172°C)

para-dichlorobenzene
(mp 53.1°C, bp 174.4°C)

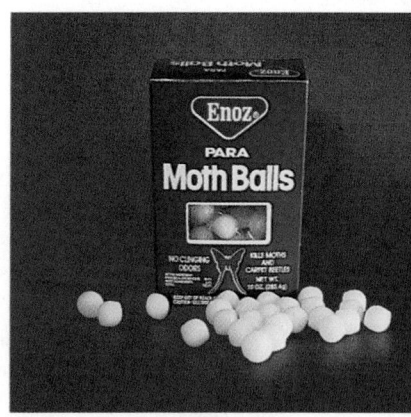

▲
Some mothballs are made from para-dichlorobenzene.

When the two substituents are different and neither is part of a compound with a special name, the names of the two substituents are given in alphabetical order, followed by the word *benzene*. For example,

o-bromochlorobenzene

m-ethylnitrobenzene

The dimethylbenzenes have the special name *xylene*.

ortho-xylene

meta-xylene

para-xylene

When one of the substituents corresponds to a monosubstituted benzene that has a special name, the disubstituted compound is named as a derivative of that parent compound. In the following examples the parent compounds are phenol, aniline, and toluene:

o-nitrophenol

p-bromoaniline

m-nitrotoluene

The numbering system as described under polysubstituted benzenes is also used to name disubstituted benzenes.

Polysubstituted Benzenes

When there are more than two substituents on a benzene ring, the carbon atoms in the ring are numbered starting at one of the substituted groups. Numbering may be either clockwise or counterclockwise but must be done in the direction that gives the lowest possible numbers to the substituent groups. When the compound is named as a derivative of one of the special parent compounds, the substituent of the parent compound is considered to be on carbon 1 of the ring (e.g., the CH_3 group is on carbon 1 in 2,4,6-trinitrotoluene). The following examples illustrate this system:

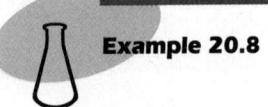

1,3,5-trinitrobenzene   1,2,4-tribromobenzene   2,4,6-trinitrotoluene   5-bromo-2-chlorophenol
                                      (not 1,4,6-)   (TNT)

Example 20.8 Write formulas and names for all the possible isomers of (a) chloronitrobenzene, $C_6H_4Cl(NO_2)$, and (b) tribromobenzene, $C_6H_3Br_3$.

Solution   (a) The name and formula indicate a chloro group (Cl) and a nitro group (NO_2) attached to a benzene ring. There are six positions in which to place these two groups. They can be ortho, meta, or para to each other.

   o-chloronitrobenzene   *m*-chloronitrobenzene   *p*-chloronitrobenzene

(b) For tribromobenzene, start by placing the three bromo groups in the 1-, 2-, and 3-positions; then the 1-, 2-, and 4-positions; and so on until all possible isomers are formed. The name of each isomer will allow you to check that the formulas have not been duplicated.

   1,2,3-tribromobenzene   1,2,4-tribromobenzene   1,3,5-tribromobenzene

There are only three isomers of tribromobenzene. If one erroneously writes the 1,2,5- compound, a further check will show that, by numbering the rings as indicated, it is in reality the 1,2,4- isomer.

Br
Br
6 1
5 2
4 3
Br

1,2,5-tribromobenzene
(incorrect name)

Br
2
3 Br
4 1
5 6
Br

1,2,4-tribromobenzene
(correct name)

Practice 20.14

Write formulas and names for all possible isomers of chlorophenol.

Practice 20.15

Name the following:

(a) COOH

Br

(b) NH$_2$

F

(c) O || C—H

Cl

NO$_2$

Practice 20.16 Application to Biochemistry

Substituted benzenes are important parts of some amino acids and hormones. We focus only on the biologically important substituted benzenes. Name the following:

(a) OH

OH

found in the hormone
adrenaline

(b) OH

found in the amino
acid tyrosine

(c) OH

I   I

found in the thyroid
hormones

20.11 Polycyclic Aromatic Compounds

There are many other aromatic ring systems besides benzene. Their structures consist of two or more rings in which two carbon atoms are common to two rings. These compounds are known as **polycyclic or fused aromatic ring systems.** Three of the most common hydrocarbons in this category are naphthalene, anthracene, and phenanthrene. One hydrogen is attached to each carbon atom except at the carbons common to two rings.

*polycyclic or fused
aromatic ring system*

naphthalene, $C_{10}H_8$

anthracene, $C_{14}H_{10}$

phenanthrene, $C_{14}H_{10}$

All three of these substances can be obtained from coal tar. Naphthalene is known as moth balls and has been used as a moth repellant for many years.

A number of the polycyclic aromatic hydrocarbons (and benzene) have been shown to be carcinogenic (cancer-producing). Formulas for some of the more notable ones, found in coal tar, tar from cigarette smoke, and soot in urban environments are as follows:

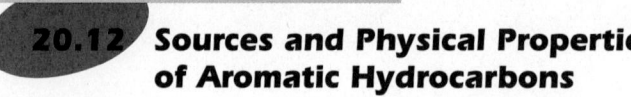

1,2-benzanthracene   1,2,5,6-dibenzanthracene   3,4-benzpyrene

These compounds are processed in living cells and become potent carcinogens. For example, the lung cells absorb polycyclic aromatic hydrocarbons from inhaled cigarette smoke. The aromatic rings are chemically modified, and the compounds become active carcinogens.

20.12   Sources and Physical Properties of Aromatic Hydrocarbons

When coal is heated to high temperatures (450–1200°C) in the absence of air to produce coke (C), coal gas and a complex mixture of condensable substances called *coal tar* are driven off:

$$\text{Coal} \xrightarrow{\Delta} \text{Coke} + \text{Coal gas} + \text{Coal tar}$$

The aromatic hydrocarbons, such as benzene, toluene, xylene, naphthalene, and anthracene, were first obtained in significant quantities from coal tar. Since coal tar itself is a by-product of the manufacture of coke, the total amount of aromatics that can be obtained from this source is limited. The demand for aromatic hydrocarbons, which are used in the production of a vast number of materials such as drugs, dyes, detergents, explosives, insecticides, plastics, and synthetic rubber, became too great to be obtained from coal tar alone. Processes were devised to make aromatic hydrocarbons from the relatively inexpensive alkanes found in petroleum. Currently, about one-third of our benzene supply and the greater portion of our toluene and xylene supplies are obtained from petroleum.

Aromatic hydrocarbons are essentially nonpolar substances, insoluble in water but soluble in many organic solvents. They are liquids or solids and usually have densities less than that of water. Aromatic hydrocarbons burn readily, usually with smoky yellow flames as a result of incomplete carbon combustion. Some are blended with other hydrocarbons to make good motor fuels with excellent antiknock properties.

20.13   Chemical Properties of Aromatic Hydrocarbons

▲
Yellow flames and smoke result from incomplete carbon combustion of burning gasoline.

Substitution

The most characteristic reactions of aromatic hydrocarbons involve the substitution of some group for a hydrogen on one of the ring carbons. Recall that a substitution reaction does not change the number of atoms bonded to each carbon. For example, each carbon in benzene will start and end a substitution reaction bonded to three other atoms.

The following are examples of typical aromatic substitution reactions. In each of these reactions, a functional group is substituted for a hydrogen atom.

1. Halogenation (chlorination or bromination)   When benzene reacts with chlorine or bromine in the presence of a catalyst such as iron(III) chloride or iron(III) bromide, a Cl or a Br atom replaces an H atom to form the products.

$$\bigcirc + X_2 \xrightarrow{FeX_3} \bigcirc\!\!-X + HX$$

benzene   chlorine   chlorobenzene
          or bromine   or bromobenzene

2. Nitration When benzene reacts with a mixture of concentrated nitric acid and concentrated sulfuric acid at about 50°C, nitrobenzene is formed. In this reaction a nitro group, $-NO_2$, is substituted for an H atom of benzene.

$$\bigcirc + HO-NO_2 \xrightarrow{H_2SO_4} \bigcirc\!\!-NO_2 + H_2O$$

benzene   nitric acid   nitrobenzene

3. Alkylation (Friedel–Crafts reaction)   There are many variations of the Friedel–Crafts reaction. In this type the alkyl group from an alkyl halide (RX), in the presence of $AlCl_3$ catalyst, substitutes for an H atom on the benzene ring.

Plants Reduce "PAH-lution"

Polycyclic aromatic hydrocarbons (PAHs) are by-products of combustion released into the air by traffic, residential heating, industrial processes, and other human activities. The bad news is that a number of these PAHs are known or suspected carcinogens and our homes and offices appear to have significant accumulations of these chemicals. The good news is that houseplants remove significant amounts of these pollutants from the air.

A recent study (Staci Simonich and Ronald Hites, Indiana University) of the concentrations of ten representative PAHs in the vegetation and soil concludes that houseplants absorb PAHs by accumulating the chemicals in their waxy outer tissues. Grasses, shrubs, and trees provide the same sort of cleanup outdoors. Apparently, the waxiness of a plant controls its PAH uptake; the plant's ability to absorb PAHs varies with season and temperature.

Simonich and Hites modeled the removal of PAHs by plants in the region of the United States encompassing Kansas, Missouri, Kentucky, and Virginia north to the Canadian border where about 52% of the U.S. population lives. They concluded that plants absorb a large fraction of this "PAH-lution," about 43.5% in this region of the country. Of the remainder, 10% falls directly on the soil, 5% into the water, and 41% is chemically transformed or drifts into Canada or the ocean (see pie graph).

Although plants are useful in controlling PAHs indoors and out, they don't completely solve the PAH-lution problem. John Roberts, a Seattle-based consulting engineer, found that carpets are a significant reservoir for PAHs. His team found that the levels of several PAH compounds in carpets equaled or exceeded the PAH levels in yard soils by as much as 20 times. Such levels are well above the PAH levels tolerated by hazardous waste sites in the state of Washington. Carpets shield the PAHs from sunlight, moisture, and microbes that would normally break down or transform them.

Thus far, there is no obvious connection between cigarette smoking or

▲
Indoor plants absorb polycyclic aromatic hydrocarbons in their waxy outer tissue.

fireplace use and PAH levels in indoor air. Since PAH concentrations in the dust in outside doormats correlate with those found in carpet dust, one theory is that we track the pollutants in on our shoes. While adding plants to our houses can remove PAHs from the air, we may also need to consider removing carpets and/or removing our shoes at the door to lower the levels of these chemicals in our homes.

43.5%   10%   5%   41%

☐ Absorbed by plants
☐ Chemically transformed
■ Directly to soil
■ Directly into water

benzene   chloroethane   ethylbenzene

From about 1860 onward, especially in Germany, a great variety of useful substances such as dyes, explosives, and drugs were synthesized from aromatic hydrocarbons by reactions of the types just described. These early syntheses were developed by trial-and-error methods. A good picture of the reaction mechanism, or step-by-step sequence of intermediate stages in the overall reaction, was not obtained until about 1940.

We now recognize that aromatic substitution reactions usually proceed by a mechanism called *electrophilic substitution*. Three steps are involved:

Step 1. An electrophile (electron-seeking group) is formed.

Step 2. The electrophile is attached to the benzene ring forming a positively charged carbocation intermediate.

Step 3. A hydrogen ion is lost from the carbocation to form the product. This reaction mechanism is illustrated in the chlorination of benzene catalyzed by iron(III) chloride.

$$FeCl_3 + Cl_2 \longrightarrow FeCl_4^- + Cl^+$$

Step 1. iron(III) chloride   chloronium ion
(electrophile)

Step 2.

a carbocation

Step 3.

chlorobenzene

In Step 1 the electrophile (chloronium ion) is formed. In Step 2 the chloronium ion adds to benzene to form an intermediate carbocation, which loses a proton (H^+) in Step 3 to form the products, C_6H_5Cl and HCl. The catalyst, $FeCl_3$, is regenerated in Step 3. Nitration and alkylation also proceed by an electrophilic mechanism.

This same mechanism is used by living organisms when aromatic rings gain or lose substituents. For example, the thyroxines (thyroid gland hormones) contain aromatic rings that are iodinated by following an electrophilic substitution mechanism. Iodination is a key step in producing these potent hormones. In general, scientists find that most of life's reactions follow mechanisms that have been elucidated in the organic chemist's laboratory.

Practice 20.17 Application to Biochemistry

In the thyroid gland, a part of the amino acid tyrosine is iodinated to begin thyroid hormone formation. (*Note*: R represents the remainder of each biochemical.)

a part of tyrosine

Show Step 2 of the electrophilic substitution mechanism for this reaction (assume that iodine reacts in the same way as chlorine).

Side-Chain Oxidation

Carbon chains attached to an aromatic ring are fairly easy to oxidize. Reagents most commonly used to accomplish this in the laboratory are $KMnO_4$ or $K_2Cr_2O_7$ + H_2SO_4. No matter how long the side chain is, the carbon atom attached to the aromatic ring is oxidized to a carboxylic acid group, $-COOH$. For example, toluene, ethylbenzene, and propylbenzene are all oxidized to benzoic acid. For this oxidation to occur, there must be at least one H atom on the carbon bonded to the aromatic ring.

toluene   benzoic acid

ethylbenzene

propylbenzene

Concepts in Review

1. Explain the sp^2 and sp hybridization of carbon atoms.
2. Explain the formation of a pi bond.
3. Explain the formation of double and triple bonds.
4. Distinguish, by formulas, the difference between saturated and unsaturated hydrocarbons.
5. Name and write structural formulas of alkenes, alkynes, cycloalkenes, and aromatic compounds.
6. Determine from structural formulas whether a compound can exist as geometric isomers.
7. Name geometric isomers by the cis–trans method.
8. Write equations for addition reactions of the alkenes and alkynes.
9. Explain the formation of carbocations and the role they play in chemical reactions.
10. Apply Markovnikov's rule to the addition of HCl, HBr, HI, and H_2O/H^+ to alkenes and alkynes.
11. Explain the Baeyer test for unsaturation.
12. Distinguish, using simple chemical tests, among alkanes, alkenes, and alkynes.
13. Describe the nature of benzene and how its properties differ from open-chain unsaturated compounds.
14. Name monosubstituted, disubstituted, and polysubstituted benzene compounds.
15. Draw structural formulas of substituted benzene compounds.
16. Recognize the more common fused aromatic ring compounds.
17. Write equations for the following reactions of benzene and substituted benzenes: halogenation (chlorination or bromination), nitration, alkylation (Friedel–Crafts reaction), and side-chain oxidation.
18. Describe and write equations for the mechanism by which benzene compounds are brominated in the presence of $FeBr_3$ or chlorinated in the presence of $FeCl_3$.

Key Terms

The terms listed here have been defined within this chapter. Section numbers are referenced in parentheses for each term.

alkene (20.1)	carbocation (20.6)	dehydration (20.5)	pi bond (20.1)
alkyne (20.1)	cracking (20.5)	geometric isomer (20.3)	polycyclic or fused aromatic
aromatic compound (20.1)	cycloalkene (20.4)	Markovnikov's rule (20.6)	ring system (20.11)

Questions

Questions refer to tables, figures, and key words and concepts defined within the chapter. A particularly challenging question or exercise is indicated with an asterisk.

1. The double bond in ethene, C_2H_4, is made up of a sigma bond and a pi bond. Explain how the pi bond differs from the sigma bond.
2. Why is it possible to obtain cis and trans isomers of 1,2-dichloroethene but not of 1,2-dichloroethane?
3. Why do many rubber products deteriorate rapidly in smog-ridden areas?
4. Explain the two different types of explosion hazards present when acetylene is being handled.
5. Why is an elimination reaction often considered to be the reverse of an addition reaction?
6. In terms of historical events, why did the major source of aromatic hydrocarbons shift from coal tar to petroleum during the 10-year period 1935–1945?
7. Explain how the reactions of benzene provide evidence that its structure does not include double bonds like those found in alkenes.
8. How is the retinal structure changed when light is absorbed by the visual pigment rhodopsin?

Paired Exercises

These exercises are paired. Each odd-numbered exercise is
followed by a similar even-numbered exercise. Answers to the
even-numbered exercises are given in Appendix VI.

9. Draw Lewis structures to represent the following
   molecules:
   (a) ethane (b) ethene (c) ethyne

10. Draw Lewis structures to represent the following
    molecules:
    (a) propane (b) propene (c) propyne

11. There are eleven possible isomeric iodobutenes, C_4H_7I,
    including cis–trans isomers.
    (a) Write the structural formula for each isomer.
    (b) Name each isomer and include the prefix *cis-* or *trans-*
        where apppropriate.

12. Four isomeric chloropropenes, C_3H_5Cl, are possible.
    (a) Draw and name the structural formula for each
        isomer.
    (b) There is another compound with this same molecular
        formula. What is its structure and name?

13. Draw structural formulas for the following:
    (a) 2,5-dimethyl-3-hexene
    (b) *cis*-4-methyl-2-pentene
    (c) 3-pentene-1-yne
    (d) *trans*-3-hexene
    (e) 3-methyl-1-pentyne
    (f) 3-methyl-2-phenylhexane

14. Draw structural formulas for the following:
    (a) 3-ethyl-3-methyl-1-pentene
    (b) *cis*-1,2-diphenylethene
    (c) 3-phenyl-1-butyne
    (d) cyclopentene
    (e) 1-methylcyclohexene
    (f) 3-isopropylcyclopentene

15. The following names are incorrect. State why each name
    is wrong and give the correct name.
    (a) 3-methyl-3-butene
    (b) *cis*-3-pentene
    (c) *cis*-2-methyl-2-pentene

16. The following names are incorrect. State why each name
    is wrong and give the correct name.
    (a) 3-ethyl-1-butene
    (b) 2-chlorocyclohexene
    (c) *trans*-4-hexene

17. Name the following compounds:

    (a)
    $$CH_3CH_2 \quad CH_3$$
    $$\diagdown \quad \diagup$$
    $$C=C$$
    $$\diagup \quad \diagdown$$
    $$H \quad CH_2CH_3$$

    (b)
    (benzene ring)—$CH_2C{\equiv}CH$

    (c) $CH_3CHBrCHBrC{\equiv}CCH_3$

18. Name the following compounds:

    (a)
    $$H \quad H$$
    $$\diagdown \quad \diagup$$
    $$C=C$$
    $$\diagup \quad \diagdown$$
    $$CH_3 \quad CHCH_2CH_3$$
    $$\qquad\qquad |$$
    $$\qquad\qquad CH_3$$

    (b)
    $$CH_3 \quad CH_3$$
    $$\diagdown \quad \diagup$$
    $$C=C$$
    $$\diagup \quad \diagdown$$
    $$CH_3 \quad CH_3$$

    (c)
    $$CH_3CH_2CHCH=CH_2$$
    $$\qquad\quad |$$
    $$\qquad\quad CH$$
    $$\qquad H_3C \diagdown \; \diagup CH_3$$

19. Write the structural formulas and IUPAC names for all the
    hexynes.

20. Write the structural formulas and IUPAC names for all the
    pentynes.

21. Which of the following molecules have structural
    formulas that permit cis–trans isomers to exist?
    (a) $(CH_3)_2C{=}CHCH_3$
    (b) $CH_3CH{=}CHCl$
    (c) $CCl_2{=}CBr_2$

22. Which of the following molecules have structural
    formulas that permit cis–trans isomers to exist?
    (a) $CH_2{=}CHCl$
    (b) $CH_3CH_2C{\equiv}CCH_3$
    (c) $CH_2ClCH{=}CHCH_2Cl$

23. Complete the following equations:
(a) $CH_3CH_2CH_2CH=CH_2 + Br_2 \longrightarrow$

(b) $CH_3CH_2\underset{\underset{CH_3}{|}}{C}=CHCH_3 + HI \longrightarrow$

(c) $CH_3CH_2CH=CH_2 + H_2O \xrightarrow{H^+}$

(d)
$\text{C}_6\text{H}_5\!-\!CH=CH_2 + H_2 \xrightarrow[\text{1 atm}]{\text{Pt, 25°C}}$

(e) $CH_3CH=CHCH_3 + KMnO_4 \xrightarrow[\text{cold}]{H_2O}$

25. Complete the following equations:
(a) $CH_3C\equiv CCH_3 + Br_2 \text{ (1 mol)} \longrightarrow$
(b) The two step reaction

$CH\equiv CH + HCl \longrightarrow \underset{\text{product}}{\text{First}} \xrightarrow{HCl}$

(c) $CH_3CH_2CH\equiv CH + H_2 \text{ (1 mol)} \xrightarrow[\text{25°C, 1 atm}]{\text{Pt}}$

27. Write the formula and name for the product when cyclohexene reacts with
(a) Br_2 (c) H_2O, H^+
(b) HI (d) $KMnO_4(aq)$ (cold)

29. Write equations to show how 2-butyne can be converted into
(a) 2,3-dibromobutane
(b) 2,2-dibromobutane
(c) 2,2,3,3-tetrabromobutane

31. Write structural formulas for
(a) benzene (c) benzoic acid
(b) aniline (d) naphthalene

33. Write structural formulas for
(a) 1,3,5-tribromobenzene
(b) o-bromochlorobenzene
(c) tert-butylbenzene
(d) p-xylene

35. Write structural formulas and names for all the isomers of
(a) dichlorobromobenzene, $C_6H_3Cl_2Br$
(b) the toluene derivatives of formula C_9H_{12}

37. Write the structures and names for all the isomers that can be written by substituting an additional chlorine atom in o-chlorobromobenzene.

24. Complete the following equations:
(a) $CH_3CH_2CH_2CH=CH_2 + H_2O \xrightarrow{H^+}$
(b) $CH_3CH_2CH=CHCH_3 + HBr \longrightarrow$
(c) $CH_2=CHCl + Br_2 \longrightarrow$

(d)
$\text{C}_6\text{H}_5\!-\!CH=CH_2 + HCl \longrightarrow$

(e) $CH_2=CHCH_2CH_3 + KMnO_4 \xrightarrow[\text{cold}]{H_2O}$

26. Complete the following equations:
(a) $CH_3C\equiv CH + H_2 \text{ (1 mol)} \xrightarrow[\text{25°C, 1 atm}]{\text{Pt}}$

(b) $CH_3C\equiv CCH_3 + Br_2 \text{ (2 mol)} \longrightarrow$
(c) The two step reaction

$CH_3C\equiv CH + HCl \longrightarrow \underset{\text{product}}{\text{First}} \xrightarrow{HCl}$

28. Write the formula and name for the product when cyclopentene reacts with
(a) Cl_2 (c) H_2, Pt
(b) HBr (d) H_2O, H^+

30. Write equations to show how 2-pentyne can be converted into
(a) 2,3-dichloropentane
(b) 2,2-dichloropentane
(c) 2,2,3,3-tetrachloropentane

32. Write structural formulas for
(a) toluene (c) phenol
(b) ethylbenzene (d) anthracene

34. Write structural formulas for
(a) 1,3-dichloro-5-nitrobenzene
(b) m-dinitrobenzene
(c) 1,1-diphenylethane
(d) styrene

36. Write structural formulas and names for all the isomers of
(a) trichlorobenzene, $C_6H_3Cl_3$
(b) the benzene derivatives of formula C_8H_{10}

38. Write the structures and names for all the isomers that can be written by substituting a third chlorine atom in o-dichlorobenzene.

39. Name the following compounds:

(a) CH$_2$CH$_3$ ... Cl

(b) CH$_2$CH$_2$CH$_3$

(c) NH$_2$ ... NO$_2$

(d) OH ... Br

(e) CH

40. Name the following compounds:

(a) CH=CH$_2$

(b) CH$_3$ ... NO$_2$

(c) COOH ... Br, Br

(d) CH—CH$_3$, CH$_3$

(e) Br ... Br, Br ... OH

41. Complete the following equations and name the organic products:

(a) + Br$_2$ $\xrightarrow{FeBr_3}$

(b) CH$_3$ + HNO$_3$ $\xrightarrow{H_2SO_4}$ CH$_3$

42. Complete the following equations and name the organic products:

(a) + CH$_3$CHCH$_3$ (Cl) $\xrightarrow{AlCl_3}$

(b) CH$_3$ + KMnO$_4$ $\xrightarrow{H_2O}$

Additional Exercises

These exercises are not paired or labeled by topic and provide additional practice on concepts covered in this chapter.

*43. Two alkyl bromides are possible when 2-methyl-1-pentene is reacted with HBr. Which one will predominate? Why?

*44. Cyclohexane and 2-hexene both have the formula C$_6$H$_{12}$. How could you distinguish one from the other by chemical tests?

45. Write the structures and names of the methyl pentenes that show geometric isomerism.

46. Write structural formulas for (a) methyl carbocation, (b) propyl carbocation, (c) *tert*-butyl carbocation, and (d) pentyl carbocation.

47. Complete the following equations and name the products:

(a)

+ Br_2 $\longrightarrow$

(b)

+ HCl $\longrightarrow$

(c)

+ HCl $\longrightarrow$

(d)

+ HCl $\longrightarrow$

48. Describe the reaction mechanism by which benzene is brominated in the presence of $FeBr_3$. Show the equations.

*49. You are given three colorless liquid samples that are known to be benzene, 1-hexene, and 1-hexyne. Using simple chemical tests, devise a scheme to identify these samples.

50. Another method for making alkenes is by dehydrohalogenation (removal of HX) from an alkyl halide. What alkenes will be formed when HCl is removed from (a) 2-chlorobutane, (b) 1-chloropentane, and (c) chlorocyclohexane?

51. An aliphatic hydrocarbon, C_4H_8, reacts with Br_2, but no HBr is formed. Will there be a color change? Explain. Give possible structures for the original molecule.

52. Heptane and 1-heptene are both colorless liquids that boil in the range of 90–100°C. Describe a simple chemical test that would distinguish between the two liquids.

53. Describe the bonding in ethane, ethene, and ethyne. Are they different or the same? Explain.

54. An unknown compound has the molecular formula C_5H_{10}. This compound does not show addition reactions with hydrogen or water. Write one possible structure for this compound. Draw four other structures that have the formula C_5H_{10}.

55. Draw structures for all benzene compounds with the formula C_8H_9Cl. (There are 14 isomers.)

Answers to Practice Exercises

20.1 (a) $CH_3CH_2CH{=}CHCH_2CH_3$

(b) $CH_3CH{=}CHCHCH_2CH_2CH_3$ with CH_2CH_3 substituent

(c) $CH_3CH{=}CCHCH_3$ with CH_3 and CH_3 substituents

(d) $CH_3CHCH{=}CHCH_2CH_3$ with cyclohexyl substituent

20.2 (a) 3-methyl-2-pentene
(b) 2,4-dimethyl-2-pentene

20.3 2-methyl-1-butene
2-methyl-2-butene

20.4 (a) (b)

20.5 (a) Yes

trans

cis

(b) Yes

cis

trans

(c) No geometric isomers

20.6 (a) trans (b) cis

20.7 Isomer B will be used in fatty acid synthesis.

20.8

20.9 (a) $CH_3CH_2CHCH_2CH_3$ with CH_3 substituent

(b) $CH_3CHCCH_2CH_3$ with Br Br substituents

(c) $CH_3CH_2CCH_2CH_3$ with CH_3 and Cl substituents

(d) $CH_3CH_2CCH_2CH_3$ with CH_3 and OH substituents

20.10 (a)

(b)

20.11

20.12 (a) $CH_3CHC \equiv CH$
                $\quad\quad\quad |$
                $\quad\quad\quad CH_3$

(b) $CH_3C \equiv C - \underset{\underset{CH_2CH_3}{|}}{\overset{\overset{CH_3}{|}}{C}} - CH_2CH_3$

(c)

$-C \equiv CH$

20.13 (a) $CH_3CH_2CCl_2CH_3$
(b) $CH_3CH_2CCl_2CH_2CH_3 + CH_3CCl_2CH_2CH_2CH_3$

20.14

o-chlorophenol   *m*-chlorophenol   *p*-chlorophenol

20.15
(a) COOH ... (b) NH$_2$ ... (c) H—C=O

m-bromobenzoic   *o*-fluoroaniline   2-chloro-3-nitrobenzaldehyde
acid

20.16 (a) *o*-hydroxyphenol or *o*-dihydroxybenzene
(b) phenol
(c) 2,6-diiodophenol

20.17

CHAPTER

21

Polymers: Macromolecules

CHAPTER 21 / OUTLINE

▲
The helmets protecting the football players' heads are made of plastic, as are many other parts of their uniforms.

What is a "mer"? The terms *polymer* and *monomer* are part of our everyday speech, but we are usually familiar only with the prefixes, *poly* meaning "many" and *mono* meaning "one." Although "mer" sounds like something from a child's cartoon show, scientists use this small three-letter root to convey an important concept. It is derived from the Greek *meros,* meaning "part." So, a monomer is a "one part" and a polymer is a "many part."

Look around you and observe the objects that could be described in these terms. A brick building could be described as a polymer, with each brick representing a monomer. A string of pearls is an elegant polymer, with each pearl a monomer. Starting with simple chemical monomers, chemists have learned to construct both utilitarian and elegant polymers. As you will see in this chapter, many of the polymers are of great commercial importance.

21.1   Macromolecules

Thus far we have dealt mainly with rather small organic molecules that contain up to 50 atoms and some (fats) that contain up to about 150 atoms. But there exist in nature some very large molecules (macromolecules) that contain tens of thousands of atoms. Some of these, such as starch, glycogen, cellulose, proteins, silk, and DNA, have molar masses in the millions and are central to many of our life processes. Synthetic macromolecules touch every phase of our lives. Today it is hard to imagine a world without polymers. Textiles for clothing, carpeting, draperies, shoes, toys, automobile parts, construction materials, synthetic rubber, chemical equipment, medical supplies, cooking utensils, synthetic leather, recreational equipment—the list could go on and on. All these and a host of others that we consider to be essential in our daily lives are wholly or partly synthetic polymers. Until about 1930 most of these polymers were unknown. The vast majority of these polymeric materials are based on petroleum. Because petroleum is a nonreplaceable resource, our dependence on polymers is another good reason for not squandering the limited world supply of petroleum.

Polyethylene is an example of a synthetic polymer. Ethylene, derived from petroleum, is made to react with itself to form polyethylene (or polythene). Polyethylene is a long-chain hydrocarbon made from many ethylene units:

$$\text{Many units } CH_2{=}CH_2 \longrightarrow -\overbrace{CH_2CH_2}-\overbrace{CH_2CH_2}-\overbrace{CH_2CH_2}-\overbrace{CH_2CH_2}-$$

$$\underset{\text{ethylene}}{n\ CH_2{=}CH_2} \xrightarrow{\text{catalyst}} \underset{\text{polyethylene}}{(CH_2CH_2)_n}$$

A typical polyethylene molecule contains anywhere from 2500 to 25,000 ethylene units joined in a continuous chain.

The process of forming very large, high-molar-mass molecules from smaller units is called **polymerization.** The large molecule, or unit, is called the **polymer**

polymerization
polymer

and the small unit, the **monomer.** The term *polymer* is derived from the Greek word *polumerēs,* meaning "having many parts." Ethylene is a monomer, and polyethylene is a polymer. Because of their large size, polymers are often called *macromolecules.* Another commonly used term is *plastics.* The word *plastic* means "to be capable of being molded, or pliable." Although not all polymers are pliable and capable of being remolded, *plastics* has gained general acceptance and has come to mean any of a variety of polymeric substances.

monomer

21.2 Synthetic Polymers

Some of the early commercial polymers were merely modifications of naturally occurring substances. One chemically modified natural polymer, nitrated cellulose, was made and sold as Celluloid late in the 19th century. But the first commercially successful, fully synthetic polymer, Bakelite, was made from phenol and formaldehyde by Leo Baekeland in 1909. This was the beginning of the modern plastics industry. Chemists began to create many synthetic polymers in the late 1920s. Since then, ever-increasing numbers of synthetic macromolecular materials have transformed the modern world. Even greater numbers of polymers have been made and discarded for technical or economic reasons. Polymers are used extensively in nearly every industry. For example, the electronics industry uses "plastics" in applications ranging from microchip production to fabrication of heat and impact-resistant cases for the assembled products. In the auto industry, huge amounts of polymers are used as body, engine, transmission, and electrical system components, as well as for tires. Vast quantities of polymers with varied and sometimes highly specialized characteristics are used for packaging; for example, packaging for frozen foods must withstand subfreezing temperatures as well as those met in either conventional or microwave ovens.

▲
Synthetic polymers are used in the production of microchips.

Although there is a great variety of synthetic polymers, they can be classified into the following general groups based on their properties and uses:

1. Rubberlike materials or elastomers
2. Flexible films
3. Synthetic textiles and fibers
4. Resins (or plastics) for casting, molding, and extruding
5. Coating resins for dip-, spray-, or solvent-dispersed applications
6. Miscellaneous (e.g., hydraulic fluids, foamed insulation, ion-exchange resins)

21.3 Polymer Types

Organic chemistry has had perhaps the greatest impact on our lives via the polymers (plastics) that we use. Although these polymers often have complex structures, the underlying chemistry of polymerization is relatively simple. Two classes of organic reactions are routinely used, the addition reaction and the substitution reaction. An **addition polymer** is a polymer that is produced by successive *addition reactions.* Polyethylene is an example of an addition polymer. A **condensation polymer** is a polymer that is formed when monomers combine and split out water or some other simple substance—a *substitution reaction.* Nylon is a condensation polymer.

addition polymer
condensation polymer

FIGURE 21.1 ▶
Diagrams of thermoplastic and thermosetting polymer structures.

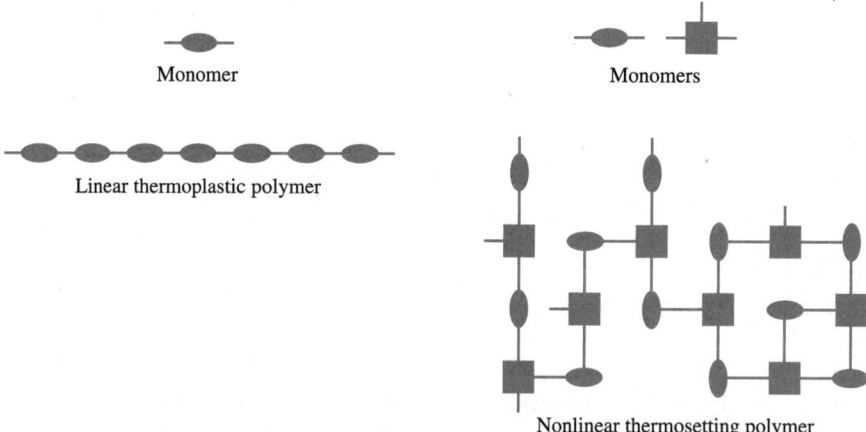

Monomer

Monomers

Linear thermoplastic polymer

Nonlinear thermosetting polymer

Polymers are either thermoplastic or thermosetting. Those that soften on reheating are **thermoplastic polymers;** those that set to an infusible solid and do not soften on reheating are **thermosetting polymers.** Thermoplastic polymers are formed when monomer molecules join end to end in a linear chain with little or no cross-linking between the chains. Thermosetting polymers are macromolecules in which the polymeric chains are cross-linked to form a network structure. The structures of thermoplastic and thermosetting polymers are illustrated in Figure 21.1

thermoplastic polymers
thermosetting polymers

21.4   Addition Polymerization

Addition polymerization starts with monomers that contain C=C double bonds. When these double bonds react, each alkene carbon bonds to another monomer.

$$\underset{|}{\overset{|}{C}}=\underset{|}{\overset{|}{C}} \ + \ \underset{|}{\overset{|}{C}}=\underset{|}{\overset{|}{C}} \ \longrightarrow \ -\underset{|}{\overset{|}{C}}-\underset{|}{\overset{|}{C}}-\underset{|}{\overset{|}{C}}-\underset{|}{\overset{|}{C}}-$$

This reaction is typical of many alkene addition reactions (see Chapter 20).

Ethylene polymerizes to form polyethylene according to the reaction shown in Section 21.1. Polyethylene is one of the most important and also the most widely used polymer on the market today. It is a tough, inert, but flexible thermoplastic material. Over 1.1×10^{10} kg of polyethylene is produced annually in the United States alone. Polyethylene is made into hundreds of different articles, many of which we use every day. Low-density polyethylene is stretchable and is used in objects such as food wrap and plastic squeeze bottles; high-density polyethylene is used in items that require more strength, such as plastic grocery bags and cases for electronic components.

The double bond is the key structural feature involved in the polymerization of ethylene. Ethylene derivatives, in which one or more hydrogen atoms have been replaced by other atoms or groups, can also be polymerized. This is often called *vinyl polymerization.* Many of our commercial synthetic polymers are made from such modified ethylene monomers. The names, structures, and uses of some of these polymers are given in Table 21.1.

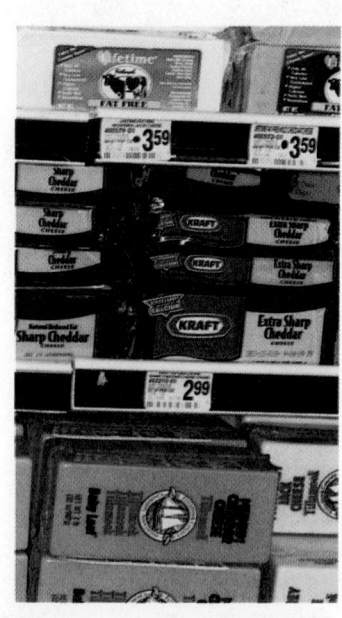

▲
Polyethylene is used to wrap foods both in stores and at home.

TABLE 21.1 Polymers Derived from Modified Ethylene Monomers

Monomer	Polymer	Identification Code for Most Commonly Recycled Polymers	Uses
$CH_2=CH_2$ ethylene	$+CH_2-CH_2)_n$ polyethylene	4 LDPE (Low Density Polyethylene) 2 HDPE (High Density Polyethylene)	Packing material, molded articles, containers, toys
$CH_2=CH$ $\quad\mid$ $\quad CH_3$ propylene	$\left(CH_2-CH\right)_n$ $\qquad\mid$ $\qquad CH_3$ polypropylene	5 PP	Textile fibers, molded articles, lightweight ropes, autoclavable biological equipment
$CH_2=C$ with CH_3, CH_3 isobutylene	$+CH_2-C$ with CH_3, CH_3 $)_n$ polyisobutylene		Pressure-sensitive adhesives, butyl rubber (contains some isoprene as copolymer)
$CH_2=CH$ $\quad\mid$ $\quad Cl$ vinyl chloride	$\left(CH_2-CH\right)_n$ $\qquad\mid$ $\qquad Cl$ polyvinyl chloride (PVC)	3 V	Garden hoses, pipes, molded articles, floor tile, electrical insulation, vinyl leather
$CH_2=CCl_2$ vinylidene chloride	$+CH_2-CCl_2)_n$ Saran		Food packaging, textile fibers, pipes, tubing (contains some vinyl chloride as copolymer)
$CH_2=CH$ $\quad\mid$ $\quad CN$ acrylonitrile	$\left(CH_2-CH\right)_n$ $\qquad\mid$ $\qquad CN$ Orlon, Acrilan		Textile fibers
$CF_2=CF_2$ tetrafluoroethylene	$+CF_2-CF_2)_n$ Teflon		Gaskets, valves, insulation, heat-resistant and chemical-resistant coatings, linings for pots and pans
$CH_2=CH$ (phenyl) styrene	$\left(CH_2-CH\right)_n$ (phenyl) polystyrene	6 PS	Molded articles, Styrofoam, insulation, toys, disposable food containers
$CH_2=CH$ $\quad\mid$ $\quad OC-CH_3$ $\qquad\parallel$ $\qquad O$ vinyl acetate	$\left(CH_2-CH\right)_n$ $\qquad\mid$ $\qquad OC\ CH_3$ $\qquad\parallel$ $\qquad O$ polyvinyl acetate		Adhesives, paint, and varnish
$CH_2=C-CH_3$ $\qquad\mid$ $\qquad C-O-CH_3$ $\qquad\parallel$ $\qquad O$ methylmethacrylate	$\left(CH_2-C\right)_n$ with CH_3 $\qquad\mid$ $\qquad C-O-CH_3$ $\qquad\parallel$ $\qquad O$ Lucite, Plexiglas (acrylic resins)		Contact lenses, clear sheets for windows and optical uses, molded articles, automobile finishes

Free radicals catalyze or initiate many addition polymerizations. Organic peroxides, ROOR, are frequently used for this purpose. The reaction proceeds as shown in the following three steps:

Step 1. *Free-radical formation.*   The peroxide splits into free radicals:

$$RO{:}OR \longrightarrow 2\,RO\cdot$$

Step 2. *Propagation of polymeric chain.*   The initial free radical adds to ethylene to form a new free radical. The chain continues to elongate (polymerize) as long as free radicals continue to add to ethylene.

$$RO\cdot + CH_2{=}CH_2 \longrightarrow ROCH_2CH_2\cdot$$

$$ROCH_2CH_2\cdot + CH_2{=}CH_2 \longrightarrow ROCH_2CH_2CH_2CH_2\cdot$$

$$ROCH_2CH_2CH_2CH_2\cdot + CH_2{=}CH_2 \longrightarrow RO(CH_2CH_2)_n\cdot$$

Step 3. *Termination.*   Polymerization stops when the free radicals are used up. This occurs when free radicals combine to form a stable compound.

$$RO(CH_2CH_2)_n\cdot + \cdot OR \longrightarrow RO(CH_2CH_2)_nOR$$

$$RO(CH_2CH_2)_n\cdot + \cdot(CH_2CH_2)_nOR \longrightarrow RO(CH_2CH_2)_n(CH_2CH_2)_nOR$$

Addition (or vinyl) polymerization of ethylene and its substituted derivatives yields saturated polymers—that is, polymer chains without carbon–carbon double bonds. The pi bond is eliminated when a free radical adds to an ethylene molecule. One electron of the pi bond pairs with the unpaired electron of the free radical, thus bonding the radical to the ethylene unit. The other pi bond electron remains unpaired, generating a new and larger free radical. This new free radical then adds another ethylene molecule, continuing the building of the polymeric chain. This process is illustrated by the following electron-dot diagram:

Practice 21.1

Complete this free-radical addition equation:

$$ROCH_2CH_2\cdot + CH_2{=}CHCH_3 \longrightarrow$$

21.5   Addition Polymers: Use and Reuse (Recycling)

Over 50% of all polymers produced and used in the United States are made from ethylene and related compounds. These polymers are inexpensive to produce and can be shaped and molded into almost any form. Their very utility makes disposal a severe problem.

▲
Plastic is ground up during the recycling process before it is used to make new articles such as these play structures.

For millions of years, microorganisms have broken down large molecules to smaller ones, starting the biosynthetic process over again. Unfortunately, the C—C single bonds of common addition polymers, like polyethylene, are different from those of most natural polymers and cannot be metabolized by many microorganisms. Consequently, while paper and cardboard in our landfills are very slowly disintegrated by microorganisms, most of the plastic waste is not.

Recycling seems to be the best solution to the problem of disposing of these long-lived addition polymers. Reuse of polyethylene (or any other of these plastics in relatively pure form) is fairly easy—simply melt and re-form into the desired product. The big problems of recycling lie with separating the different types of plastic; some plastics are not compatible with others and must be separated before reuse. The symbols on plastic goods (see Table 21.1) are one step toward plastic identification and separation. Each plastic container used in commerce carries a number that identifies its constituent polymer. It is possible (although difficult) to manually separate containers based on this code. Current research has focused on machine-separation techniques that depend on properties of plastics such as density. We recycle a relatively small amount of all plastics (about 20% of all soft-drink bottles) compared with about 30% of aluminum; we clearly could recycle much more.

21.6 Butadiene Polymers

A diene is a compound that contains two carbon–carbon double bonds. Another type of addition polymer is based on the compound 1,3-butadiene or its derivatives:

$$\overset{1}{C}H_2=\overset{2}{C}H-\overset{3}{C}H=\overset{4}{C}H_2$$

1,3-butadiene

Natural rubber is a polymer of isoprene (2-methyl-1,3-butadiene). Many synthetic elastomers or rubberlike materials are polymers of isoprene or of butadiene.

The tires of these cars are made of rubber that has been vulcanized. This process improves the temperature range for use and resistance to abrasion. ▶

Unlike the saturated ethylene polymers, these polymers are unsaturated; that is, they have double bonds in their polymeric structures.

$$n \; CH_2{=}\overset{\overset{\displaystyle CH_3}{|}}{C}{-}CH{=}CH_2 \longrightarrow (CH_2{-}\overset{\overset{\displaystyle CH_3}{|}}{C}{=}CH{-}CH_2)_n$$

isoprene   rubber polymer chain
(2-methyl-1,3-butadiene)   (polyisoprene)

$$n \; CH_2{=}CH{-}CH{=}CH_2 \longrightarrow (CH_2{-}CH{=}CH{-}CH_2)_n$$

1,3-butadiene   butadiene polymer chain

$$n \; CH_2{=}\overset{\overset{\displaystyle Cl}{|}}{C}{-}CH{=}CH_2 \longrightarrow (CH_2{-}\overset{\overset{\displaystyle Cl}{|}}{C}{=}CH{-}CH_2)_n$$

2-chloro-1,3-butadiene   neoprene polymer chain
(chloroprene)   (polychloroprene)

In this kind of polymerization, the free radical adds to the butadiene monomer at carbon 1 of the carbon–carbon double bond. At the same time, a double bond is formed between carbons 2 and 3, and a new free radical is formed at carbon 4. This process is illustrated in the following diagram:

$$H{-}\overset{\overset{\displaystyle H}{|1}}{C}\!:\!\overset{\overset{\displaystyle H}{|2}}{C}\!:\!\overset{\overset{\displaystyle H}{|3}}{C}\!:\!\overset{\overset{\displaystyle H}{|4}}{C}{-}H \longrightarrow RO\!:\!\overset{\overset{\displaystyle H}{|}}{C}\!:\!\overset{\overset{\displaystyle H}{|}}{C}\!:\!:\!\overset{\overset{\displaystyle H}{|}}{C}\!:\!\overset{\overset{\displaystyle H}{|}}{C}\!\cdot$$

RO·   1,3-butadiene

free radical   free-radical chain lengthened
                                 by four carbon atoms

The process continues adding butadiene molecules until the polymeric chain is terminated.

One of the outstanding synthetic rubbers (styrene–butadiene rubber, SBR) is made from two monomers, styrene and 1,3-butadiene. These substances form a

copolymer—that is, a polymer containing two different kinds of monomer units. **copolymer**
Styrene and butadiene do not necessarily have to combine in a 1:1 ratio. In the
actual manufacture of SBR polymers, about 3 mol of butadiene is used per mole
of styrene. Thus the butadiene and styrene units are intermixed, but in a ratio of
about 3:1.

$$-CH_2CH=CHCH_2 + CH_2CH + CH_2CH=CHCH_2 + CH_2CH=CHCH_2-$$

styrene   butadiene
unit   unit

segment of styrene–butadiene rubber (SBR)

The presence of double bonds at intervals along the chains of rubber and rubberlike synthetic polymers designed for use in tires is almost a necessity and, at the
same time, a disadvantage. On the positive side, double bonds make vulcanization
possible. On the negative side, double bonds afford sites where ozone, present especially in smoggy atmospheres, can attack the rubber, causing "age hardening" and
cracking. Vulcanization extends the useful temperature range of rubber products and
imparts greater abrasion resistance to them. The vulcanization process is usually accomplished by heating raw rubber with sulfur and other auxiliary agents. It consists
of introducing sulfur atoms that connect or cross-link the long strands of polymeric
chains. Vulcanization was devised through trial-and-error experimentation by the
American inventor Charles Goodyear in 1839, long before the chemistry of the
process was understood. Goodyear's patent on "Improvement in India Rubber" was
issued on June 15, 1844. In the segment of vulcanized rubber shown here, the chains
of polymerized isoprene are cross-linked by sulfur–sulfur bonds, giving the polymer
more strength and elasticity.

$$
\begin{array}{ccc}
CH_3 & CH_3 & CH_3 \\
| & | & | \\
-CH_2C=CHCHCH_2C=CHCH_2CHC=CHCH_2- \\
| & | \\
S & S \\
| & | \\
S & S \\
| & | \\
-CH_2C=CHCHCH_2C=CHCH_2CHC=CHCH_2- \\
| & | & | \\
CH_3 & CH_3 & CH_3
\end{array}
$$

segment of vulcanized rubber

21.7 Geometric Isomerism in Polymers

The recurring double bonds in isoprene and butadiene polymers make it possible to
have polymers with specific spatial orientation as a result of cis–trans isomerism.
Recall from Section 20.3 that two carbon atoms joined by a double bond are not free
to rotate and thus give rise to cis–trans isomerism. An isoprene polymer can have all-
cis, all-trans, or a random distribution of cis and trans configurations about the
double bonds.

Natural rubber is *cis*-polyisoprene with an all-cis configuration about the carbon–carbon double bonds. Gutta-percha, also obtained from plants, is a *trans*-polyisoprene with an all-trans configuration. Although these two polymers have the same composition, their properties are radically different. The cis natural rubber is a soft, elastic material, whereas the trans gutta-percha is a tough, nonelastic, hornlike substance. Natural rubber has many varied uses. Some uses of gutta-percha include electrical insulation, dentistry, and golf balls.

all-cis configuration of natural rubber

all-trans configuration of natural rubber

Chicle is another natural substance containing polyisoprenes. It is obtained by concentrating the latex from the sapodilla tree, which contains about 5% *cis*-polyiso-prene and 12% *trans*-polyisoprene. The chief use of chicle is in chewing gum.

Only random or nonstereospecific polymers are obtained by free-radical poly-merization. Synthetic polyisoprenes made by free-radical polymerization are much inferior to natural rubber, since they contain both the cis and the trans isomers. But in the 1950s, Karl Ziegler (1898–1973) of Germany and Giulio Natta (1903–1979) of Italy developed catalysts [e.g., $(C_2H_5)_3Al/TiCl_4$] that allowed polymerization to proceed by an ionic mechanism, producing stereochemically controlled polymers. Ziegler–Natta catalysts made possible the synthesis of polyisoprene with an all-cis configuration and with properties fully comparable to those of natural rubber. This material is known by the odd but logical name *synthetic natural rubber*. In 1963, Natta and Ziegler were jointly awarded the Nobel prize for their work on stereo-chemically controlled polymerization reactions.

Other types of polymers will be discussed in later chapters.

Practice 21.2

Random free-radical polymerization of isoprene produces a polymer with a mixture of cis and trans double bonds. Draw four units of a polymer that con-tains both the cis and trans forms.

Concepts in Review

1. Write formulas for addition polymers when given the monomer(s).

2. Describe the properties of a thermoplastic polymer and a thermosetting polymer.

3. Explain the free-radical mechanism for polymer formation.

4. Identify polymers from their trade names (e.g., Saran and Teflon).

5. Explain the effect of cross-linking in polymers.

6. Draw a segment of the structural formula of natural rubber to illustrate the all-cis and all-trans configurations.

7. Explain how butadiene-type polymers are formed.

8. Explain vulcanization and its effect on rubber.

CHEMISTRY IN ACTION Slowing Soil Erosion

One of the least expensive ways to irrigate farm fields is by using furrows—ditches that run along each planted row. The major problem with furrow irrigation is that, as water runs along a ditch, it erodes valuable topsoil. Losses can run between 2.8 and 28 tons per acre. This erosion also robs fields of fertilizers and pesticides, dumping them as pollutants into nearby lakes and streams. In addition, tiny clay particles eventually clog soil pores near the end of the furrow, thus reducing the amount of water that can filter through to the roots of the plants.

Soil scientists in Idaho have tested special crystals that can be added to the irrigation water to stop erosion. Just what are these magic crystals and how do they work? The crystals are polyacrylamide (PAM), a long-chain polymer commonly used to clean wastewater. There are thousands of different PAM molecules, varying by electrical charge, number of units, and cross-linking.

▲
This field has been treated with polyacrylamide (PAM) crystals, which reduce erosion caused by irrigation.

$$\begin{array}{ccc} O & O & O \\ \| & \| & \| \\ C-NH_2 & C-NH_2 & C-NH_2 \\ | & | & | \\ -CH_2-CH-CH_2-CH-CH_2-CH- \end{array}$$

short section of PAM polymer

The PAM polymer that interested Robert Sojka, of the Agricultural Research Service in Kimberly, Idaho, is one in which the molecules average 30,000 units in length and are negatively charged. It has very low toxicity and is primarily used in wastewater-treatment facilities to aggregate fine solids into particles large enough to settle out or be captured by filters. Scientists think the PAM seeks out and binds to the broken edges of crystalline clay particles, increasing their cohesiveness. This makes the dirt more resistant to the shearing forces exerted by water flowing down the furrow.

It takes very little PAM to dramatically reduce erosion. Just 10 ppm added to water during the first hour of irrigation can reduce erosion 70–99%. If the soil is not disturbed, the next irrigation can be done without additional PAM, giving about 50% erosion reduction. Or farmers can refresh their irrigation furrows by using 1 ppm PAM in the water. On steeply sloping fields, the use of PAM can improve water infiltration into soils so much that the crop yields are dramatically boosted.

The new erosion treatment still has some drawbacks. The polymer works well in clean water, but when added to water clouded with sediment, it causes the particles to flocculate and settle out, filling the head of the furrow. It also requires extra steps to measure and mix proper concentrations of PAM. The cost of treating a field is about $2–3 per acre, which over a growing season can seriously affect profits.

Farmers view PAM treatment of soil as a long-term investment. By reducing topsoil loss, PAM preserves the value of their farm land. Scientists think PAM can help farmers all over the United States. For example, PAM could improve the absorption of water into soils in the Central Valley of California, increasing harvests by as much as 20%. This polymer gives farmers an opportunity to hold onto the earth and improve their crop yields at the same time.

Key Terms

The terms listed here have been defined within this chapter. Section numbers are referenced in parentheses for each term.

addition polymer (21.3)
condensation polymer (21.3)
copolymer (21.6)

monomer (21.1)
polymer (21.1)
polymerization (21.1)
thermoplastic polymers (21.3)
thermosetting polymers (21.3)

Questions

Questions refer to tables, figures, and key words and concepts defined within the chapter. A particularly challenging question or exercise is indicated with an asterisk.

1. How does condensation polymerization differ from addition polymerization?

2. What property distinguishes a thermoplastic polymer from a thermosetting polymer?

3. Why must at least some monomer molecules be trifunctional to form a thermosetting polymer?

4. How are rubber molecules modified by vulcanization?

Paired Exercises

These exercises are paired. Each odd-numbered exercise is followed by a similar even-numbered exercise. Answers to the even-numbered exercises are given in Appendix VI.

5. How many ethylene units are in a polyethylene molecule that has a molar mass of approximately 25,000?

6. Calculate the molar mass of a polystyrene molecule consisting of 3000 monomer units.

7. Write formulas showing the structure of
   (a) polyvinyl chloride   (c) polypropylene
   (b) poly-2-phenyl-2-butene (d) Lucite

8. Write structural formulas for the following polymers:
   (a) Saran   (c) Teflon
   (b) Orlon   (d) polystyrene

9. Draw the structure of the free radical that is formed when the first two propylene units combine during propagation of the polypropylene chain.

10. Draw the structure of the free radical that is formed when the first two ethylene units combine during propagation of the polyethylene chain.

11. Write a structural formula showing the polymer that can be formed from
    (a) propylene
    (b) 2-methylpropene
    (c) 2-butene

12. Write a structural formula showing the polymer that can be formed from
    (a) ethylene
    (b) chloroethene
    (c) 1-butene

13. Write structures showing two possible ways in which vinyl chloride can polymerize to form polyvinyl chloride. Show four units in each structure.

14. Write structures showing two possible ways in which acrylonitrile can polymerize to form an "acrylic" fiber (Orlon, Acrilan). Show four units in each structure.

15. Natural rubber is the all-cis polymer of isoprene. Write the structure of the polymer showing at least three isoprene units.

16. Gutta-percha is the all-trans polymer of isoprene. Write the structure of the polymer showing at least three isoprene units.

17. Is the useful life of natural rubber shortened in smoggy atmospheres? Briefly explain your answer.

18. Is the useful life of styrene–butadiene rubber (SBR) shortened in smoggy atmospheres? Briefly explain your answer.

19. Write the chemical structures for the monomers of
    (a) natural rubber
    (b) synthetic (SBR) rubber

20. Write the chemical structures for the monomers of
    (a) synthetic natural rubber
    (b) neoprene rubber

21. Nitrile rubber (Buna N) is a copolymer of two parts of 1,3-butadiene and one part of acrylonitrile (CH_2=CHCN). Write a structure for this synthetic rubber, showing at least two units of the polymer.

22. Ziegler–Natta catalysts can orient the polymerization of propylene to form isotactic polypropylene—that is, polypropylene with all the methyl groups on the same side of the long carbon chain. Write the structure for (a) isotactic polypropylene and (b) another possible geometric form of polypropylene.

Additional Exercises

These exercises are not paired or labeled by topic and provide additional practice on concepts covered in this chapter.

23. Of polyethylene, polypropylene, and polystyrene, which polymer contains the highest mass percent of carbon? Calculate the percentage.

24. Draw three units of the polymer that can be made from

$$CH_2 = CHC \overset{\overset{\displaystyle O}{\|}}{{}} - OCH_2CH_3$$

25. Superglue contains the following monomer:

$$CH_2 = \underset{\underset{\displaystyle CN}{|}}{C} - \overset{\overset{\displaystyle O}{\|}}{C} - OCH_3$$

cyanoacrylate ester

When this monomer is spread between two surfaces, it rapidly polymerizes to give a very strong bond. Write the structure of the polymer formed.

26. (a) Write the structure for a polymer that can be formed from 2,3-dimethyl-1,3-butadiene. (b) Can 2,3-dimethyl-1,3-butadiene form cis and trans isomers? Explain.

27. Would you expect it to be easier to recycle a thermosetting polymer or a thermoplastic polymer? Why?

28. A copolymer formed from ethylene and tetrafluoroethylene can be used to make wire insulation. Write the structural formulas for the monomers. Write a portion of the copolymer that you would expect from an addition polymerization reaction.

29. How many isoprene units are in a natural rubber that has a molar mass of 250,000 g/mol? Show your calculation.

30. The polymer used in hard contact lenses is poly methyl methacrylate. The structures of the monomer methyl methacrylate and a section of the polymer are shown. Would you expect this polymer to result from an addition reaction or a condensation reaction? Why?

methyl methacrylate

polymer

Answers to Practice Exercises

21.1 $ROCH_2CH_2CH_2CH\cdot$
            |
            CH_3

21.2

cis   trans   trans   cis

PUTTING IT TOGETHER
Review for Chapters 19–21

Multiple Choice: *Choose the correct answer to each of the following.*

1. The element present in every organic compound is
   (a) hydrogen (b) oxygen (c) carbon (d) nitrogen

2. Compounds containing only carbon and hydrogen are known as
   (a) methane   (c) isomers
   (b) hydrocarbons (d) carbohydrates

3. The simplest hydrocarbon compound is
   (a) CH_2 (b) CH_3 (c) CH_4 (d) C_2H_4

4. The carbon–hydrogen sigma bond in methane is made by the overlap of
   (a) an s orbital and an sp^3 orbital
   (b) two sp^3 orbitals
   (c) two s orbitals
   (d) electron orbitals from two carbon atoms

5. In the alkane homologous series, the formula of each member differs from the preceding member by
   (a) CH_3 (c) CH_2
   (b) CH (d) no correct answer given

6. The general formula for a saturated open chain hydrocarbon is
   (a) C_nH_{2n} (b) C_nH_n (c) C_nH_{2n+1} (d) C_nH_{2n+2}

7. The general formula for a saturated cycloalkane is
   (a) C_nH_{2n} (b) C_nH_n (c) C_nH_{2n+1} (d) C_nH_{2n+2}

8. The general formula for an alkene is
   (a) C_nH_{2n} (b) C_nH_n (c) C_nH_{2n+1} (d) C_nH_{2n+2}

9. The name for the alkane C_5H_{12} is
   (a) propane (b) butane (c) nonane (d) pentane

10. Compounds made up of the same number and kinds of atoms but differing in their molecular structure are known as
    (a) isotopes (c) isomers
    (b) homologs (d) hydrocarbons

11. The products of the complete combustion of hydrocarbons are
    (a) carbon and hydrogen
    (b) carbon monoxide and water
    (c) carbon dioxide and water
    (d) no correct answer given

12. When pentane is chlorinated, the number of possible monochlorosubstitution products is
    (a) one (b) three (c) five (d) six

13. When hexane is chlorinated, the number of possible monochlorosubstitution products is
    (a) one (b) three (c) five (d) six

14. The IUPAC name for $CH_3CH_2CH_2CHClCH_3$ is
    (a) 4-chloropentane (c) 2-chloropentane
    (b) 4-chlorohexane   (d) 2-chloropropane

15. How many methyl and ethyl groups are in this formula?
$$CH_3$$
$$|$$
$$CH_3CH_2CHCHCH_2CH_3$$
$$|$$
$$CH_2CH_3$$
    (a) four (b) five (c) six (d) seven

16. The molecular formula for chlorocyclohexane is
    (a) $C_6H_{11}Cl$ (c) $C_6H_{13}Cl$
    (b) $C_6H_{12}Cl$ (d) no correct formula given

17. Which formula is iodoform?
    (a) CH_3I (b) CH_2I_2 (c) CHI_3 (d) CI_4

18. Which compound can be both a cycloalkane and a cycloalkene?
    (a) C_3H_6 (c) C_5H_8
    (b) C_4H_8 (d) no correct answer given

19. What is the correct name for:
$$CH_2CH_3 \qquad CH_2CH_3$$
$$| \qquad\qquad |$$
$$CH_3CH - CH_2 - CHCH_3$$
    (a) 4-ethyl-2-methylhexane
    (b) 3,5-dimethylheptane
    (c) 2-ethyl-4-methylhexane
    (d) 2,4-diethylpentane

20. A pi bond is formed from
    (a) two sp^3 electron orbitals
    (b) two sp^2 electron orbitals
    (c) an sp^3 and an sp^2 electron orbital
    (d) two p orbitals

21. A carbon–carbon triple bond consists of
    (a) a sigma bond and two pi bonds
    (b) three pi bonds
    (c) three sigma bonds
    (d) two sigma bonds and one pi bond

22. The compound of formula C_6H_{12} cannot be
    (a) an alkene   (c) an alkyne
    (b) a cycloalkane (d) no correct answer given

23. How many hydrogen atoms does the open chain compound C_8H_{10} need to become a saturated hydrocarbon?
    (a) 2 (b) 4 (c) 6 (d) 8

24. Which formula represents an alkyne hydrocarbon?
    (a) C_nH_{2n} (b) C_nH_n (c) C_nH_{2n-2} (d) C_nH_{2n-4}

25. Which formula represents a cycloalkene hydrocarbon?
    (a) C_nH_n (b) C_nH_{2n} (c) C_nH_{2n-2} (d) C_nH_{2n-4}

26. Which statement is <u>not</u> true? The compound C_6H_{10} can have in its structure
   (a) two carbon–carbon double bonds
   (b) one carbon–carbon triple bond
   (c) one cyclic ring and one carbon–carbon double bond
   (d) one cyclic ring and two carbon–carbon double bonds

27. The name of this compound is

$$CH_3CH=CHCHCH_3$$
$$\qquad\qquad |$$
$$\qquad\quad CH_3$$

   (a) 4-methyl-2-pentene   **(c)** isohexene
   (b) 2-methyl-3-pentene   **(d)** no correct name given

28. The name of this compound is

$$CH_3C\equiv C-CHCH_3$$
$$\qquad\qquad\quad |$$
$$\qquad\qquad CH_2CH_3$$

   (a) 3-methyl-4-hexyne   **(c)** 4-methyl-2-hexyne
   (b) 4-ethyl-2-pentyne   **(d)** 4-methyl-2-hexene

29. Which is an incorrect name for the compounds shown?
   (a) $CH_3CH_2CH_2Cl$
       1-chloropropane

   (b) $CH_3CH_2CH=CHCH_3$
       2-pentene

   (c)

       1,3-dimethylcyclopropane   **(d)** 3-methylcyclopentene

30. The formula $CH_3CH_2CH_2^+$ is a
   (a) primary carbocation   **(c)** tertiary carbocation
   (b) secondary carbocation   **(d)** quaternary carbocation

31. Which of the following compounds has a structure that can have cis–trans isomers?
   (a) $CH_3CH=CHCl$
   (b) $CH_3CH_2C\equiv CCH_3$
   (c) $(CH_3)_2C=CHCl$
   (d) $CH_3CH_2C=C-CH_2Cl$
$$\qquad\qquad\qquad |\quad\ |$$
$$\qquad\qquad\quad CH_3\ CH_2Cl$$

32. Which compound below is *cis*-2,3–dichloro-2-butene?

33. Which compound below is aniline?

34. Which compound below is 2-chloro-4-nitrotoluene?

35. Which compound below is *m*-xylene?

36. Which compound below is benzaldehyde?

37. Which compound below is *p*-dihydroxybenzene?

38. Which compound below is naphthalene?

(a)

(b)

(c)

(d) CH=CH₂

39. How many benzene isomers can be written for the formula C_8H_9Br?
(a) 6   (b) 10   (c) 12   (d) 14

Write the products for the reactions given in Questions 40–46.

40. $(CH_3)_2C=CHCH_3 + H_2O \xrightarrow{H^+}$
(a) $(CH_3)_2CH_2CHCH_3$
   $\quad\quad\quad\quad\quad |$
   $\quad\quad\quad\quad\quad OH$
(c) $(CH_3)_2C-CHCH_3$
   $\quad\quad\quad\quad |\;\;|$
   $\quad\quad\quad OH\;OH$
(b) $(CH_3)_2CCH_2CH_3$
   $\quad\quad\quad\quad |$
   $\quad\quad\quad\quad OH$
(d) $(CH_3)_2CH-O-CH_2CH_3$

41.
   [benzene ring]–$CH_2CH_2CH_3 + KMnO_4$ (hot) $\xrightarrow{H_2O}$
(a) [benzene ring]–$CH-CHCH_3$
   $\quad\quad\quad\quad\quad |\;\;\;\;\;|$
   $\quad\quad\quad\quad\quad OH\;\;OH$
(b) [benzene ring]–$CHCH_2CH_3$
   $\quad\quad\quad\quad\quad |$
   $\quad\quad\quad\quad\quad OH$
(c) [benzene ring]–CH_2CH_2COOH
(d) [benzene ring]–$COOH$

42. $CH_3CH_2CH=CHCH_2CH_3 + HCl \longrightarrow$
(a) $CH_3CH_2CH_2CH_2CH_2CH_3$
(b) $CH_3CH_2CHClCHClCH_2CH_3$
(c) $CH_3CH_2CH_2CHClCH_2CH_3$
(d) no correct answer given

43. $CH_3C\equiv CH + 2 HBr \longrightarrow$
(a) $CH_3CBr_2CH_3$ (c) $CH_3CHBrCHBr_2$
(b) $CH_3CH_2CHBr_2$ (d) $CH_2BrCH_2CH_2Br$

44.
$\quad\quad\quad CH_3$
$\quad\quad\quad |$
$CH_3C=CHCH_3 + Br_2 \longrightarrow$
(a) $\quad CH_3$
   $\quad\quad |$
   $CH_3CBrCHBrCH_3$
(c) $\quad CH_3$
   $\quad\quad |$
   $CH_3C=CHCH_2Br$
(b) $\quad CH_3$
   $\quad\quad |$
   $CH_3CHCBr_2CH_3$
(d) $\quad CH_3$
   $\quad\quad |$
   $CH_3CBrCH_2CH_3$

45.
   Cl–[benzene ring]–Cl $+ HNO_3 \xrightarrow{H_2SO_4}$
(a) [benzene ring with Cl, NO₂, NO₂, Cl]
(c) [benzene ring with Cl, Cl, NO₂]
(b) [benzene ring with NO₂]
(d) [benzene ring with Cl, NO₂, O₂N, Cl]

46.
   [benzene ring] + [benzene ring]–$CH_2Cl \xrightarrow{AlCl_3}$
(a) [benzene ring]–CH_2CH_2–[benzene ring]
(b) [benzene ring]–CH_2–[benzene ring]
(c) [benzene ring]–Cl
(d) no correct answer given

47. Macromolecules are made from small units called
(a) telemers   (c) monomers
(b) polymers   (d) dimers

48. Which of the following is not a natural macromolecule?
(a) cellulose (b) protein (c) orlon (d) glycogen

49. Polymers such as polyethylene, polystyrene, and polyvinyl chloride are known as
(a) condensation polymers
(b) substitution polymers
(c) addition polymers
(d) transition polymers

50. Which of the following is a free radical?
(a) CH_3^+ (c) $CH_2{=}CH^-$

(b)   (d) $CH_3CH_2^-$

51. Which of these polymers contains the highest mass percent of carbon?
(a) polyethylene   (c) polypropylene
(b) polystyrene   (d) polyvinyl chloride

52. The monomer of natural rubber is
(a) 2-chloro-1,3-butadiene
(b) 1,3-butadiene
(c) styrene
(d) 2-methyl-1,3-butadiene

53. Who invented the process of vulcanization?
(a) Charles Goodyear   (c) Charles Goodman
(b) Charles Goodrich   (d) Charles Granville

54. How many vinyl chloride units ($CH_2{=}CHCl$) are in 1.0 mol of a polyvinyl chloride polymer that has a molar mass of 32,000 g/mol?
(a) 4.2×10^2 (c) 6.2×10^2
(b) 5.1×10^2 (d) 2.0×10^6

55. Which of the following formulas is the polymer Teflon?
(a) $+CF_2{-}CF_2{\rightarrow}_n$

(b) $+CH_2{-}CCl_2{\rightarrow}_n$

(c) $+CH_2{-}CH{\rightarrow}_n$
           |
          CN

(d) $+CH_2{-}CH{\rightarrow}_n$
           |
          CH_3

56. Which of the following formulas is the polymer Saran?
(a) $+CF_2CF_2{\rightarrow}_n$ (b) $+CH_2CCl_2{\rightarrow}_n$

(c) $+CH_2{-}CH{\rightarrow}_n$ (d) $+CH_2CH{\rightarrow}_n$
           | |
          CN CH_3

57. Which of the following formulas is the polymer polypropylene?
(a) $+CH_2CH_2CH_2{\rightarrow}_n$ (c) $+CH_2CH{\rightarrow}_n$

(b) $+CH_2CCl_2{\rightarrow}_n$ (d) $+CH_2CH{\rightarrow}_n$
                                              |
                                             CH_3

58. Which of the following formulas is a diene?
(a) $CH_3CH{=}CHCH_3$ (c)

(b) $CH_2{=}C{=}CH_2$ (d) no correct answer given

59. The formula for the polymer that can be formed from 2-butene is
(a) $+CH_2CH_2CH_2CH_2{\rightarrow}_n$ (c)
                                              CH_3 CH_3
                                               | |
                                              $+CH{-}CH{\rightarrow}_n$

(b) $+CH_2CH{=}CHCH_2{\rightarrow}_n$ (d)
                                              CH_2CH_3
                                               |
                                              $+CH_2{-}CH{\rightarrow}_n$

60. A polymer that is made from two monomer units is known as a
(a) double polymer   (c) dipolymer
(b) copolymer   (d) bipolymer

Free Response Questions

1. Why are alkanes unable to undergo addition reactions?

2. What general type of polymer would be most useful for making the handles for pots that can be used on a stove?

3. What is the difference in name, functional group, and reactivity (common type of reaction) between the following three molecules?

4. A bottle on the shelf in your lab is missing part of its label. What is left of the label indicates the formula for the compound is a six-carbon hydrocarbon. You are pretty sure it is one of the compounds below. Among the other lab chemicals readily available is bromine. Suggest what test(s) you could perform that would give quick visual results to determine the identity of the compound.

        $CH_3CH_2CH_2CH_2CH_2CH_3$

I   **II**   **III**

5. Can alkynes have geometric isomers? Why or why not? Are there any groups of alkanes that could have geometric isomers?

6. Each of the following pairs of compounds have similar molar masses. Would you prefer to use physical properties (e.g., boiling point or solubility) or chemical properties to distinguish between the compounds in each pair? Explain and give examples where appropriate.
   (a) benzene and cyclohexane
   (b) hexane and 1-hexyne
   (c) cyclohexene and cyclohexane

7. Plexiglass is the trademarked name for a polymer often used as a glass substitute partly because of its transparent qualities. From the structure of Plexiglass shown below, do you think Plexiglass is an addition or a condensation polymer and what is(are) the monomer(s)?

$$\left(\begin{array}{c} \underset{|}{CH_3} \quad\quad \underset{|}{CH_3} \quad\quad \underset{|}{CH_3} \quad\quad \underset{|}{CH_3} \quad\quad \underset{|}{CH_3} \\ -C-CH_2-C-CH_2-C-CH_2-C-CH_2-C-CH_2- \\ \underset{|}{CO_2CH_3} \quad \underset{}{CO_2CH_3} \quad \underset{}{CO_2CH_3} \quad \underset{}{CO_2CH_3} \quad \underset{}{CO_2CH_3} \end{array}\right)_n$$

8. Draw and name the structures for compounds I–III obtained in the following reaction sequence.

$$CH_3CH_3 \xrightarrow[\text{light}]{Cl_2} \textbf{I} \xrightarrow[\text{AlCl}_3]{\text{benzene}} \textbf{II} \xrightarrow[\text{H}_2\text{SO}_4]{K_2Cr_2O_7} \textbf{III}$$

Alcohols, Ethers, Phenols, and Thiols

▲
These Indy cars have just refueled with methanol during the annual race at Elkhart Lake, Wisconsin.

CHAPTER 22 / OUTLINE

CHAPTER
22

We tend to play as strenuously as we work. Our "vacations" often involve driving to areas where we can indulge in vigorous physical activities like hiking, cycling, skiing, or swimming. After a hard day "torturing" our bodies, we relax and commiserate over our sore muscles. During these activities, organic molecules that contain an —OH group play a significant role. Ethylene glycol acts as a coolant in the radiator of our car, sugar and other carbohydrates provide the biochemical energy for our physical activities, ethyl alcohol is a component in any alcoholic beverage we might consume, and phenolic compounds are an active ingredient in the muscle rubs and analgesics that relieve our sore muscles.

Just what changes does the addition of an —OH group produce in the physical and chemical properties of an organic molecule? In this chapter we begin to examine the effect of various functional groups on organic molecules.

22.1 Functional Groups

Organic compounds were originally obtained from plants and animals, which, even today, are still the direct sources of many important chemicals. As a case in point, millions of tons of sucrose (table sugar) are obtained from sugar cane and sugar beet juices each year. However, as our knowledge of chemistry has increased, we have been able to synthesize many naturally occurring compounds, often at far less cost than the natural products. Of even greater significance than the cheaper manufacture of natural substances has been the synthesis of new substances totally unlike any natural product. These syntheses have been greatly aided by the realization that organic chemicals can be divided into a relatively small number of classes and studied on the basis of similar chemical properties (see Table 19.1). The various classes of compounds are identified by the presence of certain characteristic groups called functional groups. For example, if a hydroxyl group (—OH) is substituted for a hydrogen atom in an alkane molecule, the resulting compound is an **alcohol.** Thus alcohols are a class of compounds in which the functional group is the hydroxyl group.

alcohol

Through the chemical reactions of functional groups, it is possible to create or synthesize new substances. The synthesis of new and possibly useful compounds or the more economical synthesis of known compounds is a main concern of modern organic chemistry. Most chemicals used today do not occur in nature but are synthesized from naturally occurring materials. The chemical and physical properties of an organic compound depend on (1) the kinds and number of functional groups present and (2) the shape and size of the molecule.

The structures of alcohols, ethers, and phenols may be derived from the structure of water by replacing the hydrogen atoms of water with alkyl groups (R) or aromatic rings:

$$H\!-\!\overset{\displaystyle O}{}\!-\!H \qquad R\!-\!\overset{\displaystyle O}{}\!-\!H \qquad R\!-\!\overset{\displaystyle O}{}\!-\!R \qquad \bigcirc\!-\!\overset{\displaystyle O}{}\!-\!H$$

water   alcohol   ether   phenol

The R— groups in ethers can be the same or different and can be alkyl groups or aromatic rings.

 ## Classification of Alcohols

primary alcohol
secondary alcohol
tertiary alcohol

Structurally, an alcohol is derived from an aliphatic hydrocarbon by the replacement of at least one hydrogen atom with a hydroxyl group (—OH). Alcohols are represented by the general formula ROH, with methanol (CH_3OH) being the first member of the homologous series (R represents an alkyl or substituted alkyl group). Models illustrating the structural arrangements of the atoms in methanol and ethanol are shown in Figure 22.1.

Alcohols are classified as **primary** (1°), **secondary** (2°), or **tertiary** (3°), depending on whether the carbon atom to which the —OH group is attached is directly bonded to one, two, or three other carbon atoms, respectively. (The terms secondary and tertiary are used as they were for alkyl groups in Chapter 19.) Generalized formulas for 1°, 2°, and 3° alcohols are as follows:

$$
\begin{array}{ccc}
\text{H} & \text{R} & \text{R} \\
| & | & | \\
\text{R—C—OH} & \text{R—C—OH} & \text{R—C—OH} \\
| & | & | \\
\text{H} & \text{H} & \text{R}
\end{array}
$$

primary alcohol   secondary alcohol   tertiary alcohol

Formulas of specific examples of these classes of alcohols are shown in Table 22.1. Methanol (CH_3OH) is grouped with the primary alcohols.

Molecular structures with more than one —OH group attached to a single carbon atom are generally not stable. But an alcohol molecule can contain two or more —OH groups if each —OH is attached to a different carbon atom. Accordingly, alcohols are also classified as monohydroxy, dihydroxy, trihydroxy, and so on, on the basis of the number of hydroxyl groups per molecule. **Polyhydroxy alcohols** and *polyols* are general terms for alcohols that have more than one —OH group per molecule. Polyhydroxy compounds are very important molecules in living cells as they include the carbohydrate class of biochemicals.

polyhydroxy alcohol

An alcohol such as 2-butanol can be written in a single-line formula by enclosing the —OH group in parentheses and placing it after the carbon to which it is bonded. For example, the following two formulas represent the same compound:

$$
\begin{array}{ll}
CH_3CH_2CHCH_3 & CH_3CH_2CH(OH)CH_3 \\
\quad\quad\quad | & \\
\quad\quad\quad OH &
\end{array}
$$

Practice 22.1 Application to Biochemistry

Blood sugar (glucose) contains five alcohol groups. Using the structure of glucose shown here, label each alcohol group as 1°, 2° or 3°.

$O{=}CHCH(OH)CH(OH)CH(OH)CH(OH)CH_2OH$

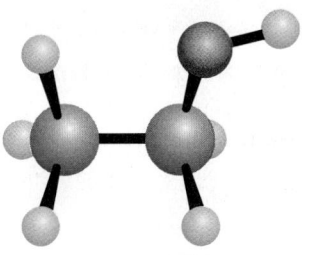

▶ **FIGURE 22.1**
Ball-and-stick models illustrating structural formulas of methanol and ethanol.

CH₃OH
Methanol

CH_3OH Methanol

CH_3CH_2OH Ethanol

TABLE 22.1   Names and Classification of Alcohols

Class	Formula	IUPAC name	Common name*	Boiling point (°C)
Primary	CH_3OH	Methanol	Methyl alcohol	65.0
Primary	CH_3CH_2OH	Ethanol	Ethyl alcohol	78.5
Primary	$CH_3CH_2CH_2OH$	1-Propanol	n-Propyl alcohol	97.4
Primary	$CH_3CH_2CH_2CH_2OH$	1-Butanol	n-Butyl alcohol	118
Primary	$CH_3(CH_2)_3CH_2OH$	1-Pentanol	n-Pentyl alcohol (n-amyl)	138

Primary	CH_3CHCH_2OH
	     |
	    CH_3
		2-Methyl-1-propanol	Isobutyl alcohol	108

Primary	⬡— CH_2OH	Phenylmethanol	Benzyl alcohol	205.0

Secondary	CH_3CHCH_3
	     |
	    OH
		2-Propanol	Isopropyl alcohol	82.5

Secondary	$CH_3CH_2CHCH_3$
	       |
	      OH
		2-Butanol	sec-Butyl alcohol	91.5

Tertiary	CH_3
	      |
	 CH_3—C—OH
	      |
	     CH_3
		2-Methyl-2-propanol	t-Butyl alcohol	82.9

Dihydroxy	$HOCH_2CH_2OH$	1,2-Ethanediol	Ethylene glycol	197

Trihydroxy	$HOCH_2CHCH_2OH$
	       |
	      OH
		1,2,3-Propanetriol	Glycerol or glycerine	290

*The abbreviations n, sec, and t stand for normal, secondary, and tertiary, respectively.

22.3   Naming Alcohols

If you know how to name alkanes, it is easy to name alcohols by the IUPAC System:

IUPAC Rules for Naming Alcohols

1. Select the longest continuous chain of carbon atoms containing the hydroxyl group.
2. Number the carbon atoms in this chain so that the one bearing the —OH group has the lowest possible number.
3. Form the parent alcohol name by replacing the final -*e* of the corresponding alkane name by -*ol*. When isomers are possible, locate the position of the —OH by placing the number (hyphenated) of the carbon atom to which the —OH is bonded immediately before the parent alcohol name.
4. Name each alkyl side chain (or other group) and designate its position by number.

Now we will use the IUPAC rules given above to name the alcohol $CH_3CH_2CH_2CH_2OH$.

Rule 1. The longest carbon chain containing the —OH group has four carbons.

Rule 2. Number the carbon atoms, giving the carbon bonded to the —OH the number 1.

$$\overset{4}{C}-\overset{3}{C}-\overset{2}{C}-\overset{1}{C}-OH$$

Rule 3. The name of the four-carbon alkane is butane. Replace the final *e* in butane with *ol*, forming the name *butanol*. Since the —OH is on carbon 1, place a *1* before butanol to give the complete alcohol name 1-butanol.

Rule 4. No groups of atoms other than hydrogen are attached to the butanol chain, so the name of this alcohol is 1-butanol.

Study how these examples and those shown in Table 22.1 are named using the IUPAC System.

$$\overset{3}{CH_3}-\overset{2}{CH_2}-\overset{1}{CH_2OH}$$
1-propanol

$$\overset{1}{CH_3}-\overset{2}{CH}-\overset{3}{CH_3}$$
$$|$$
$$OH$$
2-propanol

cyclohexanol

1,3-cyclohexanediol

$$\overset{4}{CH_3}-\overset{3}{CH}-\overset{2}{CH_2}-\overset{1}{CH_2OH}$$
$$|$$
$$CH_3$$
3-methyl-1-butanol

$$\overset{2}{HOCH_2}-\overset{1}{CH_2OH}$$
1,2-ethanediol

Name this alcohol by the IUPAC method.   **Example 22.1**

CH₃CH₂CHCH₂CHCH₃
        | |
       CH₃   OH

Rule 1. The longest continuous carbon chain containing the —OH group has   **Solution**
six carbon atoms.

Rule 2. This carbon chain is numbered from right to left so that the —OH
group has the lowest possible number:

6   5   4   3   2   1
C — C — C — C — C — C
          | |
         CH₃   OH

In this case, the —OH is on carbon 2.

Rule 3. The name of the six-carbon alkane is hexane. Replace the final *e*
in hexane by *ol,* to form the name *hexanol.* Since the —OH is on
carbon 2, place a 2- before hexanol to give the parent alcohol name
2-hexanol.

Rule 4. A methyl group (—CH₃) is located on carbon 4. Therefore, the full
name of the compound is 4-methyl-2-hexanol.

Write the structural formula of 3,3-dimethyl-2-hexanol.   **Example 22.2**

Rule 1. The "2-hexanol" refers to a six-carbon chain with an —OH group on   **Solution**
carbon 2. Write six carbons in a row and place an —OH on carbon 2.

1   2   3   4   5   6
C — C — C — C — C — C
     |
    OH

Rule 2. Place the two methyl groups ("3,3-dimethyl") on carbon 3:

                    CH₃
                     |
1   2   3   4   5   6
C — C — C — C — C — C
     | |
    OH CH₃

Rule 3. Finally, add H atoms to give each carbon atom four bonds.

                    CH₃
                     |
CH₃CH — C — CH₂CH₂CH₃
     | |
    OH   CH₃

3,3-dimethyl-2-hexanol

Practice 22.2

Write the IUPAC name for each of the following:

(a) $CH_3CH_2CHCH_2CH_2CH_3$
                 |
                 CH_2OH

(b) $CH_3CH_2CHCH_2CCH_3$
                 | |
                 OH CH_3 (with Br above the last carbon)

Practice 22.3

Write the structural formula for each of the following:
(a) 3-methylcyclohexanol   (b) 4-ethyl-2-methyl-3-heptanol

Several of the alcohols are generally known by their common names, so it may be necessary to know more than one name for a given alcohol. The common name is usually formed from the name of the alkyl group that is attached to the —OH group, followed by the word *alcohol*. (See the examples in Table 22.1.)

22.4   Physical Properties of Alcohols

Sugars are easily dissolved and transported through the bloodstream because they contain many hydroxyl groups.

The physical properties of alcohols are related to those of both water and alkane hydrocarbons. This is easily understandable if we recall certain facts about water and the alkanes. Water molecules are quite polar. The properties of water, such as its high boiling point and its ability to dissolve many polar substances, are largely due to the polarity of its molecules. Alkane molecules possess almost no polarity. The properties of the alkanes reflect this lack of polarity—for example, their relatively low boiling points and inability to dissolve water and other polar substances. An alcohol molecule is made up of a water-like hydroxyl group joined to a hydrocarbon-like alkyl group.

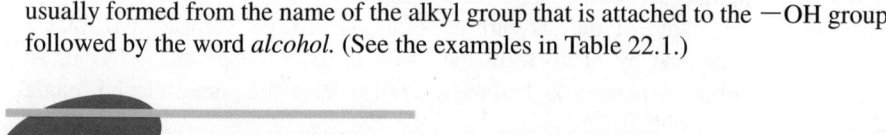

water   alcohol

One striking property of alcohols is their relatively high boiling points. The simplest alcohol, methanol, boils at 65°C. But methane, the simplest hydrocarbon, boils at −162°C. The boiling points of the normal alcohols increase in a regular fashion with increasing number of carbon atoms. Branched-chain alcohols have lower boiling points than the corresponding straight-chain alcohols (see Table 22.1).

Alcohols containing up to three carbon atoms are infinitely soluble in water. With one exception (2-methyl-2-propanol), alcohols with four or more carbon atoms have limited solubility in water. In contrast, all hydrocarbons are essentially insoluble in water.

The hydroxyl group on the alcohol molecule is responsible for both the water solubility and the relatively high boiling points of the low-molar-mass alcohols. Hydrogen bonding (see Section 13.11) between water and alcohol molecules accounts for the solubility, and hydrogen bonding between alcohol molecules accounts for their high boiling points.

O—H····H—O—H····
····H—H····O—H   R—O—H····O—R
           H   H
····H   R
   O····H—O—R   O····H—O—R
····H   ····H

water–alcohol   alcohol–alcohol

As the length of the hydrocarbon chain increases, the polar effect of the hydroxyl group becomes relatively less important. Alcohols with 5–11 carbons are oily liquids of slight water solubility, and in physical behavior they resemble the corresponding alkane hydrocarbons. However, the effect of the —OH group is still noticeable in that their boiling points are higher than those of alkanes with similar molar masses (see Table 22.2). Alcohols containing 12 or more carbons are waxlike solids that resemble solid alkanes in physical appearance.

In alcohols with two or more hydroxyl groups, the effect of the hydroxyl groups on intermolecular attractive forces is, as we might suspect, even more striking than in monohydroxy alcohols. Ethanol, CH_3CH_2OH, boils at 78°C, but the boiling point of 1,2-ethanediol, $CH_2(OH)CH_2OH$, is 197°C. Comparison of the boiling points of ethanol, 1-propanol, and 1,2-ethanediol shows that increased molar mass does not account for the high boiling point of 1,2-ethanediol (see Table 22.3). The higher boiling point is primarily a result of additional hydrogen bonding due to the two —OH groups in the 1,2-ethanediol molecule.

TABLE 22.2	Comparison of Boiling Points of Alkanes and Monohydroxy Alcohols		
Name	Boiling point (°C)	Name	Boiling point (°C)
Pentane	36	1-Pentanol	138
Hexane	69	1-Hexanol	157
Octane	126	1-Octanol	195
Decane	174	1-Decanol	233

1-propanol   1-octanol

◄ Space filling models of 1-propanol (left) and 1-octanol (right).

TABLE 22.3	Comparison of the Boiling Points of Ethanol, 1,2-Ethanediol, and 1-Propanol		
Name	**Formula**	**Molar mass**	**Boiling point (°C)**
Ethanol	CH_3CH_2OH	46	78
1,2-Ethanediol	$CH_2(OH)CH_2OH$	62	197
1-Propanol	$CH_3CH_2CH_2OH$	60	97

Example 22.3   Glucose is one of the most important carbohydrates in biochemistry. It has six carbons and five alcohol groups (molar mass = 180.2 g). How would you predict the water solubility of glucose will differ from that of 1-hexanol?

Solution   Each polar alcohol group attracts water molecules and increases the solubility of organic compounds in water. Because glucose contains five —OH groups whereas hexanol contains only one, glucose should dissolve to a much greater extent than 1-hexanol. In fact, only about 0.6 g of hexanol will dissolve in 100 g of water (at 20°C). In contrast, about 95 g of glucose will dissolve in 100 g of water. The high water solubility of glucose is important because this molecule is transported in a water solution to the body's cells.

Practice 22.4

List the following alcohols in order of increasing solubility in water:
(a) 1-hexanol   (b) 1-pentanol   (c) 1,3-propanediol   (d) 1-butanol

 Chemical Properties of Alcohols

An alcohol contains the hydroxyl functional group (—OH). Chemists have chosen the term *functional group* to indicate that this group brings a "function" to an organic molecule. Just as adding new electronic chips to a calculator adds new functions, attaching a hydroxyl group to an alkyl chain allows a molecule to function in new ways. As we have seen, the —OH group interacts with water so that when a molecule adds a hydroxyl group, its solubility in water increases. In general, the "functions" of a hydroxyl group help determine the properties of an alcohol.

Acidic and Basic Properties

Aliphatic alcohols are similar to water in their acidic/basic properties. If an alcohol is mixed with a strong acid, it will accept a proton (act as a Brønsted–Lowry base) to
oxonium ion   form a protonated alcohol or **oxonium ion.**

$$CH_3-\ddot{O}H + H_2SO_4 \longrightarrow CH_3-\overset{+}{\ddot{O}}\diagup{}^{H}_{H} + HSO_4^-$$

Methanol and ethanol have approximately the same acid strength as water, while the larger alcohols are weaker acids than water reflecting the properties of the longer alkenelike carbon chains. Both water and alcohols react with alkali metals to release hydrogen gas and an anion.

$$2\ H_2O + 2\ Na \longrightarrow 2\ Na^+\ {}^-OH + H_2(g)$$
$$\text{sodium hydroxide}$$

$$2\ CH_3CH_2OH + 2\ Na \longrightarrow 2\ Na^+\ {}^-OCH_2CH_3 + H_2(g)$$
$$\text{sodium ethoxide}$$

The resulting anion in the alcohol reaction is known as an **alkoxide ion** (RO^-). *alkoxide ion* Alkoxides are strong bases (stronger than hydroxide) and so are used in organic chemistry when a strong base is required in a nonaqueous solution.

The order of reactivity of alcohols with sodium or potassium is primary > secondary > tertiary. Alcohols do not react with sodium as vigorously as water. Reactivity decreases with increasing molar mass since the —OH group becomes a relatively smaller, less significant part of the molecule.

As you will see in later chapters, the —OH group is also part of the carboxylic acid functional group $(-\overset{O}{\overset{\|}{C}}-OH)$. The —OH behaves much differently in the carboxylic acid group than it does in an alcohol.

Oxidation

We will consider only a few of the many reactions that alcohols are known to undergo. One important reaction is oxidation. We saw in Chapter 17 that the oxidation number of an element increases as a result of oxidation. Carbon can exist in several oxidation states, ranging from -4 to $+4$. In the -4 oxidation state, such as in methane, the carbon atom is considered to be completely reduced. In carbon dioxide the carbon atom is completely oxidized; that is, it is in its highest oxidation state $(+4)$. In many cases, oxidation reactions in organic chemistry and biochemistry can be considered in a simple manner without the use of oxidation numbers. Oxidation is the loss of hydrogen or the gain of bonds to oxygen by the organic reactant. Table 22.4 illustrates the progression of oxidation states for various compounds containing one carbon atom.

Carbon atoms exist in progressively higher stages of oxidation in different functional-group compounds:

$$\text{Alkanes} \longrightarrow \text{Alcohols} \longrightarrow \left\{\begin{array}{c}\text{Aldehydes}\\\text{Ketones}\end{array}\right\} \longrightarrow \begin{array}{c}\text{Carboxylic}\\\text{acids}\end{array} \longrightarrow \begin{array}{c}\text{Carbon}\\\text{dioxide}\end{array}$$

Increasing oxidation state $\longrightarrow$

The synthesis of alcohols enables us to make many compounds that are more oxidized. In general, it is not practical to convert alkanes directly to alcohols, but once

	Compound	Number of C—O bonds	Oxidation state
TABLE 22.4	**Oxidation States of Carbon in One-Carbon Compounds**		
CH_4	Methane	0	−4
CH_3OH	Methanol	1	−2
$H_2C{=}O$	Methanal (formaldehyde)	2	0
$\overset{\displaystyle O}{\underset{\displaystyle HC-OH}{\|\|}}$	Methanoic acid (formic acid)	3	+2
$O{=}C{=}O$	Carbon dioxide	4	+4

the —OH functional group has been attached, further oxidations are possible. The hydroxyl group gives an organic compound the capability ("function") of forming an aldehyde, ketone, or carboxylic acid. Industrial chemists use this hydroxyl-group function often; compounds as diverse as plastics, antibiotics, and fertilizers are synthesized starting with alcohols. The following equations represent generalized oxidation reactions in which the oxidizing agent is represented by [O].

Tertiary alcohols do not have a hydrogen on the —OH carbon and so cannot react with oxidizing agents except by such drastic procedures as combustion. Both primary and secondary alcohols are oxidized as shown in the equations by the loss of the blue hydrogen atoms.

Alcohols in biochemistry react in a similar manner. For example, after a meal that includes an alcoholic beverage, our liver oxidizes the ethanol. Think about how ethanol might react given that ethanol is a primary alcohol.

First, ethanol is converted to an aldehyde:

$$CH_3-CH_2OH \xrightarrow{[O]} CH_3-\overset{\overset{\displaystyle O}{\|}}{C}-H$$

ethanol   acetaldehyde

Unfortunately, the aldehyde is toxic and often damages the liver. Excess alcohol consumption can cause liver cirrhosis. Second, the aldehyde is converted to a carboxylic acid:

$$CH_3-\overset{\overset{\displaystyle O}{\|}}{CH} \xrightarrow{[O]} CH_3-\overset{\overset{\displaystyle O}{\|}}{C}-OH$$

acetaldehyde   acetic acid

Acetic acid can be used as a source of biochemical energy.

Some common oxidizing agents used for specific reactions are potassium permanganate, $KMnO_4$, in an alkaline solution, potassium dichromate, $K_2Cr_2O_7$, in an acid solution, or oxygen of the air. A complete equation for an alcohol oxidation is fairly complex. Since our main interest is in the changes that occur in the functional groups, we can convey this information in abbreviated form:

$$CH_3CH_2OH \xrightarrow[\Delta]{K_2Cr_2O_7/H_2SO_4} CH_3\overset{\overset{\displaystyle O}{\|}}{C}-H + H_2O$$

ethanol   ethanal

Visible evidence that a reaction occurred is a change in color of $Cr_2O_7^{2-}$. When reduced in an acid medium, orange colored $Cr_2O_7^{2-}$. changes to green (Cr^{3+}).

Although the abbreviated equation lacks some of the details, it does show the overall reaction involving the organic compounds. Additional information is provided by notations above and below the arrow, which indicate that this reaction is carried out in heated potassium dichromate–sulfuric acid solution. Abbreviated equations of this kind will be used frequently in the remainder of this book.

What are the products when (a) 1-propanol, (b) 2-propanol, and (c) cyclohexanol are oxidized with $K_2Cr_2O_7/H_2SO_4$?

Example 22.4

(a) The formula for 1-propanol is $CH_3CH_2CH_2OH$. Since it is a primary alcohol, it can be oxidized to an aldehyde or a carboxylic acid. The oxidation occurs at the carbon bonded to the $-OH$ group, and this carbon atom becomes an aldehyde or a carboxylic acid. The rest of the molecule remains the same.

Solution

$$CH_3CH_2CH_2OH \begin{cases} \xrightarrow{[O]} CH_3CH_2\overset{\overset{\displaystyle H}{|}}{C}=O \\ \\ \xrightarrow{[O]} CH_3CH_2COOH \end{cases}$$

1-propanol

propanal

propanoic acid

(b) 2-Propanol is a secondary alcohol and is oxidized to a ketone. Ketones resist further oxidation. The oxidation occurs at the carbon bonded to the —OH group.

$$\underset{\underset{\text{2-propanol}}{\overset{|}{OH}}}{CH_3CHCH_3} \xrightarrow{[O]} \underset{\underset{\text{propanone (acetone)}}{\overset{||}{O}}}{CH_3CCH_3}$$

(c) Cyclohexanol is also a secondary alcohol and is oxidized to cyclohexanone, a ketone.

cyclohexanol   cyclohexanone

Practice 22.5

Write the structure for the products of the oxidation of these alcohols with $K_2Cr_2O_7$ and H_2SO_4. What evidence would you see that a reaction took place?
(a) 1-butanol   (b) 2-methyl-2-butanol   (c) cyclopentanol

Practice 22.6 Application to Biochemistry

Many cells produce energy via the citric acid cycle. One citric acid cycle reaction oxidizes the following acid. Write the structure of the oxidized product.

$$\underset{\underset{\text{malic acid}}{CH_2-COOH}}{HO-\overset{|}{C}H-COOH}$$

Dehydration

The term *dehydration* implies the elimination of water. Alcohols can be dehydrated to form alkenes (Section 20.5) or ethers. One of the more effective dehydrating agents is sulfuric acid. Whether an ether or an alkene is formed depends on the ratio of alcohol to sulfuric acid, the reaction temperature, and the type of alcohol (1°, 2°, or 3°).

Alkenes: Intramolecular Dehydration   The formation of alkenes from alcohols requires a relatively high temperature. Water is removed from within a *single* alcohol molecule and a carbon–carbon double bond forms as shown here:

$$H-\overset{\overset{H}{|}}{\underset{\underset{H}{|}}{C}}-\overset{\overset{H}{|}}{\underset{\underset{OH}{|}}{C}}-H \xrightarrow[180°C]{96\% \ H_2SO_4} CH_2{=}CH_2 + H_2O$$

(only possible alkene product)

For many alcohols there is more than one way to remove water. Therefore, the double bond can be located in different positions. The major product in such cases is the alkene in which the carbon–carbon double bond has the greatest number of alkyl substituents on it (or lesser number of hydrogens). For example, 2-butanol can lose water from either carbon 1 and carbon 2 or from carbon 2 and carbon 3.

$$CH_3CH_2CHCH_3 \xrightarrow[100°C]{60\% \ H_2SO_4} CH_3CH=CHCH_3 + CH_3CH_2CH=CH_2 + H_2O$$

OH

2-butanol

2-butene
(major product)

1-butene

To predict the major product in an intramolecular dehydration, follow these steps: (1) Remove H and OH from adjacent carbons forming a carbon–carbon double bond; (2) if there are choices (multiple hydrogen-containing neighboring carbon atoms adjacent to the —OH carbon), remove the hydrogen from the carbon with fewer hydrogens. This is known as **Saytzeff's rule.** This rule in essence states:

Saytzeff's rule

> **During intramolecular dehydration, if there is a choice of positions for the carbon–carbon double bond, the preferred location is the one that generally gives the more highly substituted alkene—that is, the alkene with the most alkyl groups attached to the double-bond carbons.**

Ethers: Intermolecular Dehydration A dehydration reaction can take place between two alcohol molecules to produce an ether. However, the dehydration to make an ether is only useful for primary alcohols since secondary and tertiary alcohols predominantly yield alkenes.

$$\begin{matrix} CH_3CH_2O\boxed{H} \\ CH_3CH_2\boxed{OH} \end{matrix} \xrightarrow[140°C]{96\% \ H_2SO_4} CH_3CH_2OCH_2CH_3 + H_2O$$

diethyl ether

A reaction in which *two* molecules are combined by removing a small molecule is known as a **condensation reaction.** There are many examples of other condensation reactions in the formation of biochemical molecules.

condensation reaction

The type of dehydration that occurs depends on the temperature and the number of reactant molecules. Lower temperatures and *two* alcohol molecules produce ethers, while higher temperatures and a *single* alcohol produce alkenes.

Dehydration reactions are often used in industry to make relatively expensive ethers from lower priced alcohols. Alkenes are less expensive than alcohols and so are not produced industrially by this method.

Esterification (Conversion of Alcohols to Esters)

An alcohol can react with a carboxylic acid to form an ester and water. The reaction is represented as follows:

$$R-\overset{\overset{\displaystyle O}{\|}}{C}-\boxed{OH} + \boxed{H}-\boxed{OR'} \underset{\longleftarrow}{\overset{H^+}{\longrightarrow}} R\overset{\overset{\displaystyle O}{\|}}{C}-OR' + HOH$$

carboxylic acid   alcohol   ester

$$CH_3\overset{\overset{\displaystyle O}{\|}}{C}-OH + HOCH_2CH_3 \underset{\longleftarrow}{\overset{H^+}{\longrightarrow}} CH_3\overset{\overset{\displaystyle O}{\|}}{C}-OCH_2CH_3 + HOH$$

acetic acid   ethanol   ethyl acetate

Esterification is an important reaction of alcohols and is discussed in greater detail in Chapter 24.

Utility of the Hydroxyl Functional Group

The hydroxyl group is a particularly important functional group; it introduces a myriad of possible reactions leading to a variety of other valuable organic compounds (see Figure 22.2). An organic chemist might term the hydroxyl group as a valuable intermediate, a "gateway" functional group.

The hydroxyl group has a special place in biochemical reactions as well. Commonly, an hydroxyl is created to "open the gateway" for other reactions. For example, fats are degraded using reactions that are fundamentally the same as those employed by industrial organic chemists. First, a portion of the fat is converted to an alkene:

$$W-CH_2-CH_2-Z \longrightarrow W-CH=CH-Z$$

part of a fat   an alkene
(W and Z abbreviate the
rest of the fat molecule)

Then, the alkene is hydrated to form an alcohol:

$$W-CH=CH-Z + H_2O \overset{H^+}{\longrightarrow} W-\overset{\overset{\displaystyle OH}{|}}{CH}-CH_2-Z$$

an alcohol

FIGURE 22.2 ▶
Hydroxyl group reactions.

Oxidation follows to convert the alcohol into a ketone:

$$
\underset{\text{a ketone}}{W-\overset{\overset{\displaystyle OH}{|}}{C}H-CH_2-Z \xrightarrow{[O]} W-\overset{\overset{\displaystyle O}{||}}{C}-CH_2-Z}
$$

Forming the ketone allows fats to be broken down into smaller pieces as degradation continues.

Serious physiological consequences result when cells cannot degrade fats. About 10% of the infants who die from SIDS (Sudden Infant Death Syndrome) are unable to degrade fats. Therefore, their bodies cannot depend on fat for energy. It is theorized that the infants' lack of energy leads to sudden death.

22.6 Common Alcohols

Three general methods for making alcohols are

1. *Hydrolysis of an ester*

$$
\underset{\text{ester}}{R\overset{\overset{\displaystyle O}{||}}{C}-OR'} + HOH \xrightarrow[\Delta]{H^+} \underset{\text{carboxylic acid}}{R\overset{\overset{\displaystyle O}{||}}{C}-OH} + \underset{\text{alcohol}}{R'OH}
$$

Hydrolysis is a reaction of water with another species in which the water molecule is split (see Section 16.13). The hydrolysis of an ester is the reverse reaction of esterification. A carboxylic acid and an alcohol are formed as products. The reaction can be conducted in an acid or an alkaline medium.

2. *Alkaline hydrolysis of an alkyl halide* (1° and 2° alcohols only)

$$
\underset{\text{alkyl halide}}{RX} + NaOH(aq) \longrightarrow \underset{\text{alcohol}}{ROH} + NaX
$$

$$
CH_3CH_2Cl + NaOH(aq) \longrightarrow CH_3CH_2OH + NaCl
$$

3. *Catalytic reduction of aldehydes and ketones* to produce primary and secondary alcohols. These reactions are discussed in Chapter 23.

In theory, these general methods provide a way to make almost any desired alcohol, but they may not be practical for a specific alcohol because the necessary starting material—ester, alkyl halide, aldehyde, or ketone—cannot be obtained at a reasonable cost. Hence, for economic reasons, most of the widely used alcohols are made on an industrial scale by special methods that have been developed for specific alcohols. The preparation and properties of several of these alcohols are described in the following paragraphs.

Methanol

When wood is heated to a high temperature in an atmosphere lacking oxygen, methanol (wood alcohol) and other products are formed and driven off. The process is called *destructive distillation,* and until about 1925, nearly all methanol was obtained in this way. In the early 1920s, the synthesis of methanol by high-pressure catalytic hydrogenation of carbon monoxide was developed in Germany. The reaction is

$$CO + 2\,H_2 \xrightarrow[\text{300–400°C, 200 atm}]{\text{ZnO–Cr}_2\text{O}_3} CH_3OH$$

The most economical nonpetroleum source of carbon monoxide for making methanol is coal. In addition to coal, burnable materials such as wood, agricultural wastes, and sewage sludge also are potential sources of methanol.

Methanol is a volatile (bp 65°C), highly flammable liquid. It is poisonous and capable of causing blindness or death if taken internally. Exposure to methanol vapors for even short periods of time is dangerous. Despite this danger, over 5.1×10^9 kg is manufactured annually. Methanol is used for

1. Conversion to formaldehyde (methanal), primarily for use in the manufacture of polymers
2. Manufacture of other chemicals, especially various kinds of esters
3. Denaturing ethyl alcohol (rendering it unfit as a beverage)
4. An industrial solvent

The use of about 10% methanol or ethanol in gasoline (*gasohol*) has shown promising results in reducing the amount of air pollutants emitted in automobile exhausts. Another benefit of using methanol in gasoline is that methanol can be made from nonpetroleum sources.

Ethanol

Ethanol is without doubt the earliest and most widely known alcohol. It is or has been known by a variety of other names such as ethyl alcohol, "alcohol," grain alcohol, and spirit. Huge quantities of this substance are prepared by fermentation. Starch and sugar are the raw materials. Starch is first converted to sugar by enzyme- or acid-catalyzed hydrolysis. (An enzyme is a biological catalyst, as discussed in Chapter 30.) Conversion of simple sugars to ethanol is accomplished by yeast:

$$\underset{\text{glucose}}{C_6H_{12}O_6} \xrightarrow[\text{H}_2\text{O}]{\text{yeast}} \underset{\text{ethanol}}{2\,CH_3CH_2OH} + 2\,CO_2$$

Notable by-products of the fermentation of sugars are glycerol, lactic acid, succinic acid, and inositol (a B vitamin). Pure ethanol (100%) is very hygroscopic. It takes up water rapidly to a stable concentration of 95.6% ethanol.

For legal use in beverages, ethanol is made by fermentation, but a large part of the alcohol for industrial uses (5.9×10^8 kg annually) is made by the acid-catalyzed addition of water to ethylene (made from petroleum).

$$CH_2{=}CH_2 + H_2O \xrightarrow{\text{H}^+} CH_3CH_2OH$$

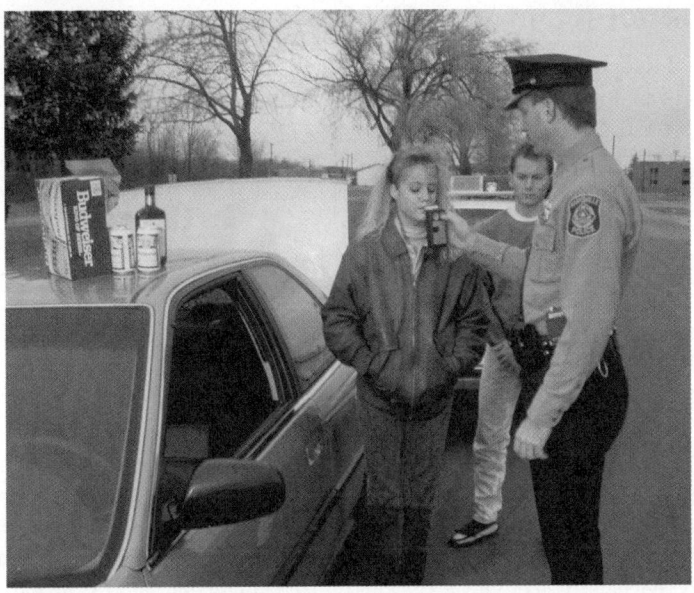

◀ **Chemical breathalyzer tests that determine alcohol level can be administered to drivers at the roadside.**

Some of the economically significant uses of ethanol include the following:

1. Intermediate in the manufacture of other chemicals such as acetaldehyde, acetic acid, ethyl acetate, and diethyl ether
2. Solvent for many organic substances
3. Compounding ingredient for pharmaceuticals, perfumes, flavorings, and so on
4. Essential ingredient of alcoholic beverages

Ethanol acts physiologically as a food, as a drug, and as a poison. It is a food in the limited sense that the body is able to metabolize it to carbon dioxide and water with the production of energy. As a drug, ethanol is often mistakenly considered to be a stimulant, but it is in fact a depressant. In moderate quantities ethanol causes drowsiness and depresses brain functions so that activities requiring skill and judgment (such as automobile driving) are impaired. In larger quantities ethanol causes nausea, vomiting, impaired perception, and incoordination. If a very large amount is consumed, unconsciousness and ultimately death may occur.

Authorities maintain that the effects of ethanol on automobile drivers are a factor in about half of all fatal traffic accidents in the United States. When you realize that traffic accidents are responsible for many thousands of deaths each year in the United States, the gravity of the problem cannot be overstated.

Heavy taxes are imposed on alcohol in beverages. A gallon of pure alcohol costs only a few dollars to produce, but in a distilled beverage it bears a U.S. federal tax of about 27 dollars.

Ethanol for industrial use is often denatured (rendered unfit for drinking) and thus is not taxed. Denaturing is done by adding small amounts of methanol and other denaturants that are extremely difficult to remove. Denaturing is required by the federal government to protect the beverage-alcohol tax source. Special tax-free use permits are issued to scientific and industrial users who require pure ethanol for non-beverage uses.

▲
Before bottling the wine, a vintner checks alcohol content, color, and sediment.

2-Propanol (Isopropyl Alcohol)

2-Propanol (isopropyl alcohol) is made from propene derived from petroleum. This synthesis is analogous to that used for making ethanol from ethylene:

$$CH_3CH\!=\!CH_2 + H_2O \xrightarrow{\ H^+\ } \underset{\underset{\displaystyle OH}{|}}{CH_3CHCH_3}$$

$$\text{propene} \qquad\qquad\qquad \text{2-propanol}$$

Note that 2-propanol, not 1-propanol, is produced. This is because, in the first step of the reaction, an H^+ adds to carbon 1 of propene according to Markovnikov's rule (Section 20.6). The —OH group then ends up on carbon 2 to give the final product.

2-Propanol is a relatively low-cost alcohol that is manufactured in large quantities, about 6.3×10^8 kg annually. It is not a potable alcohol, and merely breathing large quantities of the vapor may cause dizziness, headache, nausea, vomiting, mental depression, and coma. Isopropyl alcohol is used (1) to manufacture other chemicals (especially acetone), (2) as an industrial solvent, and (3) as the principal ingredient in rubbing alcohol formulations.

Ethylene Glycol (1,2-Ethanediol)

Ethylene glycol is the simplest alcohol containing two —OH groups. Like most other relatively cheap, low-molar-mass alcohols, it is commercially derived from petroleum. One industrial synthesis is from ethylene via ethylene oxide (oxirane).

$$2\,CH_2\!=\!CH_2 + O_2 \xrightarrow[\text{200–300°C}]{\text{Ag catalyst}} 2\,\overset{\displaystyle O}{\overset{\diagup\ \diagdown}{CH_2\!-\!CH_2}}$$

$$\text{ethylene} \qquad\qquad\qquad \text{ethylene oxide}$$
$$\text{(oxirane)}$$

$$\overset{\displaystyle O}{\overset{\diagup\ \diagdown}{CH_2\!-\!CH_2}} + H_2O \xrightarrow{\ H^+\ } HOCH_2CH_2OH$$

$$\text{1,2-ethanediol}$$
$$\text{(ethylene glycol)}$$

This alcohol is commonly referred to as ethylene glycol in commercial products. Major uses of ethylene glycol are (1) in the preparation of the synthetic polyester fibers (Dacron) and film (Mylar), (2) as a major ingredient in "permanent-type" antifreeze for cooling systems, (3) as a solvent in the paint and plastics industries, and (4) in the formulations of printing ink and ink for ballpoint pens.

The low molar mass, complete water solubility, low freezing point, and high boiling point make ethylene glycol a nearly ideal antifreeze. A 58% by mass aqueous solution of this alcohol freezes at −48°C. Its high boiling point and high heat of vaporization prevent it from being boiled away and permit higher, and therefore more efficient, engine-operating temperatures than are possible with water alone. The U.S. production of ethylene glycol amounts to about 2.4×10^9 kg annually. Ethylene glycol is extremely toxic when ingested.

Glycerol (1,2,3-Propanetriol)

Glycerol, also known as *glycerine* or 1,2,3-propanetriol, is an important trihydroxy alcohol. Glycerol is a syrupy liquid with a sweet, warm taste. It is about 0.6 times as sweet as cane sugar. It is obtained as a by-product of processing animal and vegetable fats to make soap and other products, and it is also synthesized commercially from propene. The major uses of glycerol are (1) as a raw material in the manufacture of polymers and explosives, (2) as an emollient in cosmetics, (3) as a humectant in tobacco products, and (4) as a sweetener. Each use is directly related to the three —OH groups on glycerol.

The —OH groups provide sites through which the glycerol unit bonds to other molecules to form a polymer. The explosive nitroglycerine, or glyceryltrinitrate, is made by reacting the —OH groups with nitric acid:

$$
\begin{array}{c}
\text{CH}_2\text{OH} \\
| \\
\text{CHOH} \\
| \\
\text{CH}_2\text{OH}
\end{array}
\; + \; 3\,\text{HONO}_2 \;\longrightarrow\;
\begin{array}{c}
\text{CH}_2\text{ONO}_2 \\
| \\
\text{CHONO}_2 \\
| \\
\text{CH}_2\text{ONO}_2
\end{array}
\; + \; 3\,\text{H}_2\text{O}
$$

       glycerol   nitric acid   glyceryltrinitrate
                                         (nitroglycerine)

The three polar —OH groups on the glycerol molecule are able to hold water molecules by hydrogen bonding. Consequently, glycerol is a hygroscopic substance; that is, it has the ability to take up water vapor from the air. It is therefore used as a skin moisturizer in cosmetic preparations. Glycerol is also used as an additive in tobacco products; by taking up moisture from the air, it prevents the tobacco from becoming excessively dry and crumbly.

22.7 Phenols

The term **phenol** is used for the class of compounds that have a hydroxy group attached to an aromatic ring. The parent compound is also called phenol, C_6H_5OH (pronounced *fee-nol*).

phenol

Naming Phenols

Many phenols are named as derivatives of the parent compound using the general methods for naming aromatic compounds. For example,

    phenol   *m*-bromophenol   *p*-aminophenol   2,4,6-trinitrophenol
                                                      (picric acid)

▲ **Space-filling model of phenol.**

Common Phenols

Many natural substances have phenolic groups in their structures. The formulas and brief descriptions of several examples include the following:

catechol	resorcinol	hydroquinone
(*o*-dihydroxybenzene)	(*m*-dihydroxybenzene)	(*p*-dihydroxybenzene)

The *ortho*-, *meta*-, and *para*-dihydroxybenzenes have the special names catechol, resorcinol, and hydroquinone, respectively. The catechol structure occurs in many natural substances, and hydroquinone, a manufactured product, is commonly used as a photographic reducer and developer.

vanillin	eugenol	thymol

Vanillin is the principal odorous component of the vanilla bean. It is one of the most widely used flavorings and is also used for masking undesirable odors in many products such as paints.

Eugenol is the essence of oil of cloves. It is used in the manufacture of synthetic vanillin. Thymol occurs in the oil of thyme. It has a pleasant odor and flavor and is used as an antiseptic in preparations such as mouthwashes. Thymol is the starting material for the synthesis of menthol, the main constituent of oil of peppermint. Thymol is a widely used flavoring and pharmaceutical.

2,6-di-*t*-butyl-4-methylphenol
(butylated hydroxytoluene, BHT)

urushiols

Butylated hydroxytoluene (BHT) is used in small amounts as an antioxidant preservative for food, synthetic rubber, vegetable oils, soap, and some plastics.

The active irritants in poison ivy and poison oak are called urushiols. They are catechol derivatives with an unbranched 15-carbon side chain in position 3 on the phenol ring.

o-cresol

m-cresol

p-cresol

The *ortho-*, *meta-*, and *para*-methylphenols are present in coal tar and are known as cresols. They are all useful disinfectants.

tetrahydrocannabinol
(from marijuana)

phenolphthalein

adrenalin
(epinephrine)

The principal active component of marijuana is tetrahydrocannabinol. It is obtained from the dried leaves and flowering tops of the hemp plant and has been used since antiquity for its physiological effects. The common acid–base indicator phenolphthalein is a phenol derivative. Epinephrine (adrenalin) is secreted by the adrenal gland in response to stress, fear, anger, or other heightened emotional states. It stimulates the conversion of glycogen to glucose in the body. Phenol is the starting material for the manufacture of aspirin, one of the most widely used drugs for self-medication.

22.8 Properties of Phenols

In the pure state, phenol is a colorless crystalline solid with a melting point at about 41°C and a characteristic odor. Phenol is highly poisonous. Ingestion of even small amounts of it may cause nausea, vomiting, circulatory collapse, and death from respiratory failure.

Phenol is a weak acid; it is more acidic than alcohols and water but less acidic than acetic and carbonic acids. The pH values are as follows: 0.1 *M* acetic acid, 2.87; water, 7.0; 0.1 *M* phenol, 5.5. Thus phenol reacts with sodium hydroxide solution to form a salt but does not react with sodium hydrogen carbonate. The salt formed is called sodium phenoxide or sodium phenolate. Sodium hydroxide does not remove a hydrogen atom from an alcohol because alcohols are weaker acids than water.

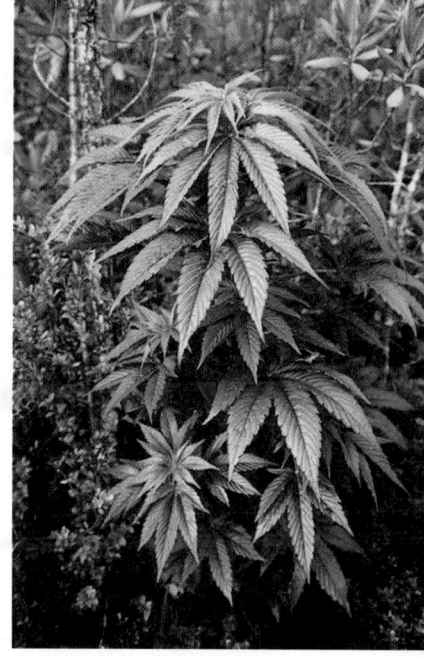

▲
Marijuana is a member of the hemp family. It contains the physiologically active molecule tetrahydrocannabinol.

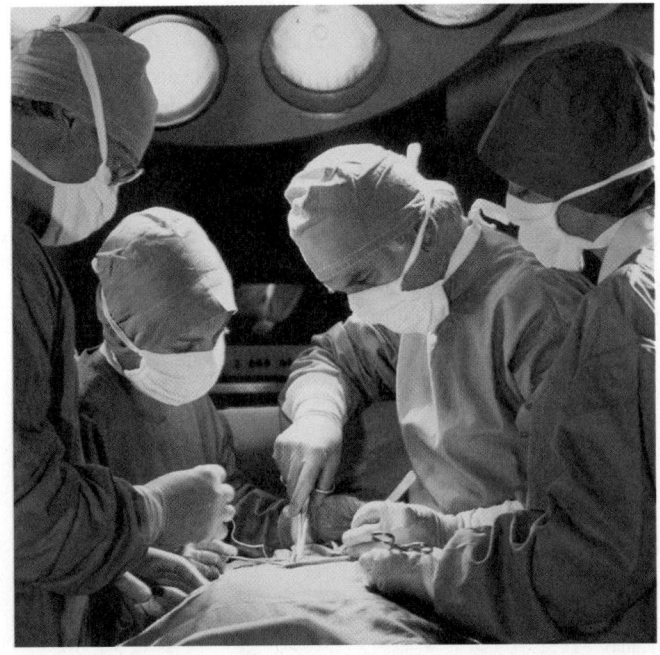

In general, the phenols are toxic to microorganisms. They are widely used as antiseptics and disinfectants. Phenol was the first compound to be used extensively as an operating room disinfectant. Joseph Lister (1827–1912) first used phenol for this purpose in 1867. The antiseptic power of phenols is increased by substituting alkyl groups (up to six carbons) in the benzene ring. For example, 4-hexylresorcinol is used as an antiseptic in numerous pharmaceuticals.

4-hexylresorcinol

22.9   Production of Phenol

Phenol is obtained from coal tar. In addition, several commercial methods are used to produce phenol synthetically. The most economical of these methods starts with benzene and propene, which react to form cumene. Cumene is then oxidized by air to cumene hydroperoxide, which is treated with dilute sulfuric acid to obtain phenol and acetone. The economic feasibility of the process is due to the fact that two important commercial products are produced. The equations for the reactions are

Over 1.9×10^9 kg of synthetic phenol is produced annually in the United States. The chief use of phenol is for the manufacture of phenol–formaldehyde resins and plastics (see Section 23.6).

22.10 Ethers

Ethers have the general formula ROR'. The groups R and R' can be derived from saturated, unsaturated, or aromatic hydrocarbons, and, for a given ether, R and R' may be alike or different. Cyclic ethers are formed by joining the ends of a single hydrocarbon chain through an oxygen atom to form a ring structure. Table 22.5 shows structural formulas and names for some of the different kinds of ethers.

Naming Ethers

Individual ethers, like alcohols, may be known by several names. The ether having the formula $CH_3CH_2-O-CH_2CH_3$ and formerly widely used as an anesthetic is called diethyl ether, ethyl ether, ethoxyethane, or simply ether. Common names of ethers are formed from the names of the groups attached to the oxygen atom, followed by the word *ether*. The IUPAC allows this method for naming simple ethers.

To name ethers by the IUPAC System, we need to learn the name of another group, the alkoxy group (RO—). An alkoxy group consists of an alkyl or aryl group

TABLE 22.5	Names and Structural Formulas of Ethers	
Name*	**Formula**	**Boiling point (°C)**
Dimethyl ether (methoxymethane)	CH_3-O-CH_3	−25
Ethyl methyl ether (methoxyethane)	$CH_3CH_2-O-CH_3$	8
Diethyl ether (ethoxyethane)	$CH_3CH_2-O-CH_2CH_3$	35
Ethyl isopropyl ether (2-ethoxypropane)	$CH_3CH_2-O-\underset{\underset{CH_3}{\|}}{CH}CH_3$	54
Divinyl ether (ethenyloxyethene)	$CH_2{=}CH-O-CH{=}CH_2$	39
Anisole, methyl phenyl ether (methoxybenzene)	⬡—OCH_3	154
Diphenyl ether (phenoxybenzene)	⬡—O—⬡	259
Ethylene oxide (oxirane)	CH_2-CH_2 over O	11
Tetrahydrofuran (THF) (1,4-epoxybutane)	CH_2-CH_2 / CH_2 CH_2 / O	66

*The IUPAC names are in parentheses.

and an oxygen atom. It is named by dropping the -*yl* of the alkyl name and adding the suffix -*oxy*. For example,

CH_3O- is called methoxy (meth + oxy)

CH_3CH_2O- is called ethoxy (eth + oxy)

⬡—$O-$ is called phenoxy (phen + oxy)

In the IUPAC System, ethers are named as alkoxy ($RO-$) derivatives of the alkane corresponding to the longest carbon–carbon chain in the molecule.

IUPAC Rules for Naming Ethers

1. Select the longest carbon continuous chain and label it with the name of the corresponding alkane.
2. Change the *-yl* ending of the other hydrocarbon group to *-oxy* to obtain the alkoxy group name. For example, CH_3O- is called *methoxy*.
3. Combine the two names from Steps 1 and 2, giving the alkoxy name and its position on the longest carbon chain first, to form the ether name.

$$\boxed{CH_3-O}\boxed{CH_2CH_3}$$

This is the longest C—C chain, so call it *ethane*.

CH_3O- is the alkoxy group.

Modify the name for the CH_3O- group from *methyl* to *methoxy* and combine with *ethane* to obtain the name of the ether, methoxyethane. Thus

$CH_3CH_2-O-CH_2CH_3$ is ethoxyethane

$CH_3CH_2CH_2-O-CH_2CH_2CH_2CH_3$ is 1-propoxybutane

$$\begin{array}{ccc} CH_3 & CH_3 & CH_3 \\ | & | & | \\ CH_3CH-O-CHCH_2CHCH_3 \end{array}$$ is 2-isopropoxy-4-methylpentane

Additional examples are found in Table 22.5.

Practice 22.7

Give IUPAC names for the following ethers:

(a) $CH_3-O-CH_2CHCH_2CH_3$
$\qquad\qquad\qquad\quad |$
$\qquad\qquad\qquad CH_3$

(c) $O-CH_2CH_3$; Cl

(b) $CH_3CH_2-O-C(CH_3)_3$

Practice 22.8

Give common names for the following ethers:

(a) $CH_3-O-CH_2CH_2CH_3$

(c) $O-CH_2CH_3$

(b) $(CH_3)_3C-O-C(CH_3)_3$

Structures and Properties of Ethers

An oxygen atom linking two carbon atoms is the key structural feature of an ether molecule. This oxygen atom causes ether molecules to have a bent shape somewhat like that of water and alcohol molecules:

$$H \overset{\ddot{\ddot{O}}}{\diagup\diagdown} H \qquad R \overset{\ddot{\ddot{O}}}{\diagup\diagdown} H \qquad R \overset{\ddot{\ddot{O}}}{\diagup\diagdown} R$$

water   alcohol   ether

Ethers are somewhat more polar than alkanes, because alkanes lack the oxygen atom with its exposed, nonbonded electrons. But ethers are much less polar than alcohols, since no hydrogen is attached to the oxygen atom in an ether. The solubility and boiling point (vapor pressure) characteristics of ethers are related to the $C-O-C$ structure. Alkanes have virtually no solubility in water or acid. But about 7.5 g of diethyl ether will dissolve in 100 g of water at 20°C. Diethyl ether also dissolves in sulfuric acid. Hydrogen bonding between ether and water molecules and between ether and acid molecules is responsible for this solubility.

ether ··· water   ether ··· acid

Because no $-OH$ group is present, hydrogen bonding does not occur between ether molecules. This lack of hydrogen bonding can be seen by comparing the boiling points of a hydrocarbon, an ether, and an alcohol of similar molar mass, as in Table 22.6. The boiling point of the ether is somewhat above that of the hydrocarbon but much lower than that of the more polar alcohol.

Ethers, especially diethyl ether, are exceptionally good solvents for organic compounds. Many polar compounds, including water, acids, alcohols, and other oxygenated organic compounds, dissolve, at least to some extent, in ethers. This solubility is a result of intermolecular attractions between the slightly polar ether mole-

TABLE 22.6	Boiling Points of Ethers, Alkanes, and Alcohols		
Name	**Formula**	**Molar mass**	**Boiling point (°C)**
Dimethyl ether	CH_3OCH_3	46	−24
Propane	$CH_3CH_2CH_3$	44	−42
Ethanol	CH_3CH_2OH	46	78
Ethyl methyl ether	$CH_3OCH_2CH_3$	60	8
Butane	$CH_3CH_2CH_2CH_3$	58	−0.6
1-Propanol	$CH_3CH_2CH_2OH$	60	97
2-Propanol	$CH_3CH(OH)CH_3$	60	83

CHEMISTRY IN ACTION Anesthetics

The most widely known use of ethyl ether has been as a general anesthetic for surgery. The introduction of ether for this purpose is one of the great landmarks of medicine. Two Americans, Crawford W. Long and William T. Morton, played important roles in this development. Long, a physician, used ether in a surgical operation as early as 1842 but did not publish his discovery until 1849. Morton, a dentist, used ether as an anesthetic for dental work in 1846. He publicly demonstrated its effectiveness by administering ether to a patient undergoing surgery at the Massachusetts General Hospital on October 16, 1846.

The word *anesthesia* is from the Greek, meaning insensibility, and was suggested to Morton by the poet and physician Oliver Wendell Holmes. A *general anesthetic* is a substance or combination of substances that produces both unconsciousness and insensitivity to pain.

Ether produces unconsciousness by depressing the activity of the central nervous system. Other ethers such as divinyl ether (Vinethene) and methoxyflurane (Penthrane) were also eventually used as general inhalation anesthetics.

$$CH_2{=}CH{-}O{-}CH{=}CH_2$$
divinyl ether

$$CHCl_2CF_2{-}O{-}CH_3$$
methoxyflurane

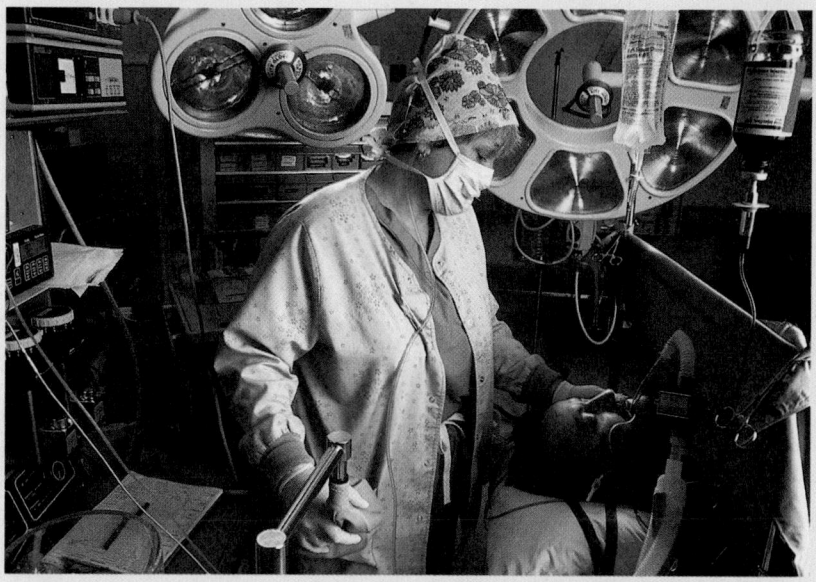

▲
An anesthetist prepares to give a general anesthetic to her patient.

The major disadvantages of ether include flammability, irritation of respiratory passages, and occurrence of nausea and vomiting after its use.

These hazards posed by the ethers have resulted in the use of other substances (such as nitrous oxide, N_2O, or halogenated compounds like halothane, $CF_3CHClBr$) as general anesthetics. These substances are superior in some respects and, in recent years, have largely replaced the ethers; however, these sub-

stitute compounds pose other hazards and must also be used with caution.

A number of theories explaining the biochemical activity of anesthetics have been proposed though none has been proved conclusively. All are based on the nonpolar nature of the compounds. This property leads to high solubility in fats and membranes. One theory suggests that anesthetics interfere with electrical activity in nerve impulses by dissolving in brain cells.

cules and the molecules of the other polar substance. Nonpolar compounds such as hydrocarbons and alkyl halides also dissolve in ethers. These substances dissolve because the ether molecules are not very polar and therefore are not strongly attracted either to one another or to the other kinds of molecules. Thus ether molecules are able to intermingle freely with the molecules of a nonpolar substance and form a solution by simple mixing.

In summary, ethers are polar enough to dissolve some polar substances, but their polarity is so slight that they act as nonpolar solvents for a great many nonpolar substances.

Ethers have little chemical reactivity, but because a great many organic substances dissolve readily in ethers, they are often used as solvents in laboratory and manufacturing operations. Their use can be dangerous, since low-molar-mass ethers are volatile, and their highly flammable vapors form explosive mixtures with air. Another hazard of ethers is that, despite their generally low chemical reactivity, oxygen of the air slowly reacts with them to form unstable peroxides that are subject to explosive decomposition:

$$CH_3CH_2-O-CH_2CH_3 + O_2 \longrightarrow CH_3CH-O-CH_2CH_3$$
$$|$$
$$O-O-H$$

diethyl ether hydroperoxide

22.12 Preparation of Ethers

Organic chemistry is replete with reactions named after scientists who discovered the reactions.

We have seen that ethers can be made by intermolecular dehydration of alcohols by heating in the presence of an acid (see Section 22.5). Ethers are also made from alkyl halides and sodium alkoxides or sodium phenoxides via a substitution reaction (the Williamson synthesis).

$$RX + R'ONa \longrightarrow ROR' + NaX$$

alkyl halide   alkoxide   ether

The alkyl halide, RX, may be a methyl or primary alkyl group, but not a secondary or tertiary alkyl group or an aryl group. The alkoxide, R'ONa, may be methyl, primary, secondary, tertiary, or an aryl group.

Alkyl halides are especially capable of substitution reactions. Here, the halide is replaced by an alkoxide or phenoxide to form an ether. This synthesis is generally useful in the preparation of mixed ethers (where R ≠ R') and aromatic ethers:

$$CH_3CH_2ONa + CH_3Br \longrightarrow CH_3CH_2-O-CH_3 + NaBr$$

sodium ethoxide   bromomethane   ethyl methyl ether

sodium phenoxide   bromomethane   methoxybenzene (anisole)

Practice 22.9

Write equations for the preparation of (a) 1-propoxybutane, (b) methyl phenyl ether, and (c) benzyl ethyl ether, starting with an alcohol, a phenol, and an alkyl halide.

22.13 Thiols

Sulfur and oxygen are found next to each other in the same family on the periodic table. This indicates some similarity in the formulas of their compounds. Organic

compounds that contain the —SH group are analogs of alcohols. The —SH-contain-
ing compounds are known as **thiols,** or **mercaptans.**

Thiols are named by the same system as alcohols except that the suffix *-thiol* is
used in place of *-ol.*

thiol

mercaptan

$$CH_3SH \qquad\qquad CH_3CH_2CHCH_3$$
$$\qquad\qquad\qquad\qquad\qquad | $$
methanethiol $\qquad\qquad\qquad\qquad SH$
(methyl mercaptan)

2-butanethiol
(*sec*-butyl mercaptan)

Thiols have a higher molar mass than corresponding alcohols but boil at lower
temperatures (ethanol, 78°C; ethanethiol, 36°C). The reason for this discrepancy lies
in the fact that alcohols form hydrogen bonds while thiols do not.

The major important properties of thiols are summarized as follows:

1. Foul odors: Some of these compounds smell so awful that companies make
   special labels to warn consumers. The odor given off by a frightened skunk
   has thiols as the active ingredient. Natural gas is odorized to be detectable by
   adding small amounts of methanethiol.
2. Oxidation to disulfides:

$$2\,RSH \xrightarrow{\text{[O]}} R\!-\!S\!-\!S\!-\!R$$

thiol $\qquad\qquad$ disulfide

This reaction can be accomplished using many oxidizing agents. The sulfur
is being oxidized just as the carbon was being oxidized in alcohols.

Thiols serve important functions in living systems. The disulfide structure often
binds proteins into biologically useful three-dimensional shapes (see Chapter 29).
The thiol is also a key part of a molecule (coenzyme A) that plays a central role in the
metabolism of carbohydrates, lipids, and proteins (see Chapter 34).

Concepts in Review

1. Name alcohols by common and IUPAC methods.
2. Write the structural formula when given the name of an
   alcohol.
3. Write the structural formulas for all the isomeric alcohols
   of a given molecular formula.
4. Recognize and identify primary, secondary, and tertiary
   alcohols.
5. Indicate a class of biochemicals that has many of the
   same properties as the polyhydroxy alcohols.
6. Summarize the acidic/basic properties of alcohols and
   alkoxides.
7. Write equations for the oxidation and dehydration of
   alcohols.
8. Understand the importance of the hydroxyl group as an
   intermediate for the synthesis of other organic molecules.

9. Differentiate between intramolecular and intermolecular dehydration, indicating proper conditions for
   each.
10. Write equations for the synthesis of alcohols from alkyl
    halides and from alkenes.
11. Explain the relative reactivities of primary, secondary,
    and tertiary alcohols.
12. Understand the common methods of preparing methyl
    alcohol, ethyl alcohol, isopropyl alcohol, ethylene glycol,
    and glycerol.
13. Name phenols and write their formulas.
14. Understand the differences in properties of the hydroxyl
    group when bonded to an aromatic ring (a phenol) and to
    an aliphatic group (an alcohol).
15. Be familiar with the general properties of phenols.

16. Name ethers and write their formulas.

17. Write equations for preparing ethers by the Williamson synthesis.

18. List major properties for thiols.

19. Discuss relative boiling points and water solubilities of comparable hydrocarbons, alcohols, dihydroxy alcohols, ethers, and thiols.

20. Discuss the hazards of using ethers in the laboratory.

Key Terms

The terms listed here have been defined within this chapter. Section numbers are referenced in parentheses for each term.

alcohol (22.1)
alkoxide ion (22.5)
condensation reaction (22.5)
ether (22.10)
mercaptan (22.13)
oxonium ion (22.5)
polyhydroxy alcohol (22.2)

phenol (22.7)
primary alcohol (22.2)
Saytzeff's rule (22.5)
secondary alcohol (22.2)
tertiary alcohol (22.2)
thiol (22.13)

Questions

Questions refer to tables, figures, and key words and concepts defined within the chapter. A particularly challenging question or exercise is indicated with an asterisk.

1. Write the structural formulas for and give an example of
   (a) an alkyl halide   (e) a ketone
   (b) a phenol   (f) a carboxylic acid
   (c) an ether   (g) an ester
   (d) an aldehyde   (h) a thiol

2. Although it is possible to make alkenes from alcohols, alkenes are seldom, if ever, made in this way on an industrial scale. Why not?

3. Isopropyl alcohol is usually used in rubbing alcohol formulations. Why is this alcohol used in preference to normal propyl alcohol?

4. What classes of compounds can be formed by the oxidation of primary alcohols? Cite examples.

5. Why is 1,2-ethanediol (ethylene glycol) superior to methanol as an antifreeze for automobile radiators?

6. Briefly outline the physiological effects of (a) methanol and (b) ethanol.

7. Write equations for the cumene hydroperoxide synthesis of phenol and acetone.

8. What two hazards may be present when working with low-molar-mass ethers?

9. Explain, in terms of molecular structure, why ethanol, CH_3CH_2OH (molar mass = 46), is a liquid at room temperature and dimethyl ether, CH_3OCH_3 (molar mass = 46), is a gas.

10. What is one possible cause of Sudden Infant Death Syndrome (SIDS)? Relate this cause to the chemistry of alcohols.

Paired Exercises

These exercises are paired. Each odd-numbered exercise is followed by a similar even-numbered exercise. Answers to the even-numbered exercises are given in Appendix VI.

11. Write structural formulas for the following:
    (a) methanol
    (b) 3-methyl-1-hexanol
    (c) 1,2-propanediol
    (d) 1-phenylethanol
    (e) 2,3-butanediol
    (f) 2-propanethiol

12. Write structural formulas for the following:
    (a) 2-butanol
    (b) 2-methyl-2-butanol
    (c) 2-propanol
    (d) cyclopentanol
    (e) 1-pentanethiol
    (f) 4-ethyl-3-hexanol

13. There are eight open-chain isomeric alcohols that have the formula $C_5H_{11}OH$. Write the structural formula and the IUPAC name for each of these alcohols.

14. Write structures for all the isomers (alcohols and ethers) with the formula
    (a) C_3H_8O
    (b) $C_4H_{10}O$

15. Which of the isomers in Exercise 13 are
    (a) primary alcohols?
    (b) secondary alcohols?
    (c) tertiary alcohols?

16. Which of the isomers in Exercise 14 are
    (a) primary alcohols?
    (b) secondary alcohols?
    (c) tertiary alcohols?

17. Name the following compounds:
    (a) $CH_3CH_2CH_2CH_2OH$

    (b) $CH_3CH(OH)CH_3$

    (c) CH_2CHCH_2OH with CH_3

    (d) H_2C—CH_2 with O bridge

    (e) $CH_3\overset{\overset{CH_2CH_3}{|}}{\underset{\underset{OH}{|}}{C}}CH_3$

    (f) cyclohexane with OH and CH_3

    (g) $HOCH_2\overset{\overset{CH_3}{|}}{\underset{\underset{CH_3}{|}}{CH}}CHCH_2OH$

18. Name the following compounds:
    (a) CH_3CH_2OH

    (b) benzene ring with CH_2CH_2OH

    (c) $CH_3\overset{\overset{CH_2CH_3}{|}}{\underset{\underset{OH}{|}}{C}}CH_2CH_3$

    (d) cyclopentane with OH and CH_3

    (e) $CH_3CH_2CH(OH)CH_2CH_3$

    (f) $CH_3\overset{}{\underset{\underset{HO}{|}}{CH}}\overset{}{\underset{\underset{OH}{|}}{CH_2}}$

    (g) $CH_3CH_2\overset{}{\underset{\underset{CH_2CH_3}{|}}{CH}}CH_2\overset{}{\underset{\underset{}{|}}{CH}}CH_3$ with OH

19. Write the formula and the name of the chief product when the following alcohols are dehydrated to alkenes:
    (a) $CH_3\overset{\overset{CH_3}{|}}{\underset{\underset{OH}{|}}{CH}}CHCH_3$

    (b) $CH_3\overset{}{\underset{\underset{OH}{|}}{CH}}CH_2CH_2CH_3$

    (c) cyclohexane with OH

20. Write the formula and the name of the chief product when the following alcohols are dehydrated to alkenes:
    (a) cyclopentane with CH_3 and OH

    (b) $CH_3CH_2\overset{}{\underset{\underset{OH}{|}}{CH}}CH_3$

    (c) $CH_3\overset{\overset{CH_3}{|}}{\underset{\underset{OH}{|}}{CH}}CHCH_2CH_3$

21. Write the formula and the name of the alcohol that, when dehydrated, gives rise to the following alkene <u>only</u>:

(a) $CH_3CH_2CH_2CH{=}CHCH_2CH_3$

(b)

(c)

23. 3-Ethyl-1-hexanol is a primary alcohol. Write the formulas of two different organic compounds that can be obtained by oxidizing this alcohol.

25. Write the equation for the preparation of alcohols by reacting each of the following alkenes with sulfuric acid and water:

(a) propene
(b) 1-butene
(c) 2-pentene

27. Alcohols can be made by reacting alkyl halides with sodium hydroxide as follows:

$$RX + NaOH \longrightarrow ROH + NaX$$

Give the names and formulas of the alkyl bromides (RBr) needed to prepare the following alcohols by this method:

(a) 2-propanol
(b) cyclohexanol
(c) 3-methyl-1-butanol

29. Complete the following equations and name the principal organic product(s) formed in each case:

(a) $2\ CH_3CH_2OH \xrightarrow[140°C]{96\%\ H_2SO_4}$

(b) $CH_3CH_2CH_2OH \xrightarrow[180°C]{96\%\ H_2SO_4}$

(c) $CH_3CH(OH)CH_2CH_3 \xrightarrow[\Delta]{K_2Cr_2O_7/H_2SO_4}$

(d) $CH_3CH_2\overset{\overset{\displaystyle O}{\|}}{C}{-}OCH_2CH_3\ +\ H_2O \xrightarrow{H^+}$

31. Write structural formulas for each of the following:

(a) *o*-methylphenol
(b) *m*-dihydroxybenzene
(c) 4-hydroxy-3-methoxybenzaldehyde (vanillin)

22. Write the formula and the name of the alcohol that, when dehydrated, gives rise to the following alkene <u>only</u>:

(a)

(b)

(c) $\underset{CH_3CHCH{=}CCH_3}{\overset{\overset{\displaystyle CH_3}{|}\qquad\overset{\displaystyle CH_3}{|}}{}}$

24. 3-Methyl-2-pentanol is a secondary alcohol. Write the formula of one organic compound that can be obtained by oxidizing this alcohol.

26. Write the equation for the preparation of alcohols by reacting each of the following alkenes with sulfuric acid and water:

(a) 2-butene
(b) 1-pentene
(c) 2-methyl-2-butene

28. Alcohols can be made by reacting alkyl halides with sodium hydroxide as follows:

$$RX + NaOH \longrightarrow ROH + NaX$$

Give the names and formulas of the alcohols produced from the following alkyl bromides by this method:

(a) 2-bromobutane
(b) 2-bromo-3-ethylpentane
(c) bromocyclopentane

30. Complete the following equations giving only the major organic products:

(a) $CH_3CH_2OH\ +\ Na \longrightarrow$

(b) $CH_3CH_2CH_2CH_2OH \xrightarrow[H_2SO_4]{K_2Cr_2O_7}$

(c) $CH{=}CH_2 \xrightarrow{H_2O/H_2SO_4}$

(d) $CH_3CH_2\overset{\overset{\displaystyle O}{\|}}{C}{-}OCH_3\ +\ NaOH \longrightarrow$

32. Write structural formulas for each of the following:

(a) *p*-nitrophenol
(b) 2,6-dimethylphenol
(c) *o*-dihydroxybenzene

33. Name the following compounds:

(a)

(b) HO ⬡ CH$_3$

(c)

NO$_2$ / OH / CH$_2$CH$_3$

(d)

OH / Cl / Br

34. Name the following compounds:

(a)

OH / OH

(b)

OH / H$_3$C / CH$_3$

(c)

OH / NO$_2$ / NO$_2$

(d) HO ⬡ CH$_2$(CH$_2$)$_4$CH$_3$

35. Arrange the following substances in order of increasing solubility in water:

(a) $CH_3CH_2OCH_2CH_2CH_3$
(b) $CH_3CH(OH)CH_2CH_2CH_3$
(c) $CH_3CH_2CH_2CH_2CH_3$
(d) $CH_3CH(OH)CH(OH)CH_2CH_3$

36. Arrange the following substances in order of decreasing solubility in water:

(a) $CH_3CH(OH)CH(OH)CH_2OH$
(b) $CH_3CH_2OCH_2CH_3$
(c) $CH_3CH_2CH_2CH_2OH$
(d) $CH_3CH(OH)CH_2CH_2OH$

37. Fourteen isomeric saturated ethers have the formula $C_6H_{14}O$. Write the structural formula for each of these ethers.

38. Six isomeric saturated ethers have the formula $C_5H_{12}O$. Write the structural formula and name for each of these ethers.

39. Write the balanced equation for the complete combustion of 2-butanol.

40. Write the balanced equation for the complete combustion of diethyl ether.

41. Write formulas for all possible combinations of RONa and RCl for making each of these ethers:

(a) $CH_3CH_2OCH_3$

(b)

CH$_2$OCH$_2$CH$_3$

(c)

O—CH$_2$CH$_3$

42. Write formulas for all possible combinations of RONa and RCl for making each of these ethers:

(a) $CH_3CH_2CH_2OCH_2CH_2CH_3$

(b)

CH$_3$ / HCOCH$_2$CH$_2$CH$_3$ / CH$_3$

(c)

O

Additional Exercises

These exercises are not paired or labeled by topic and provide additional practice on concepts covered in this chapter.

43. Phenol and its derivatives have been used as antiseptics since the middle 1800s. One common present-day antiseptic is related to phenol:

OH

OH

$CH_2CH_2CH_2CH_2CH_2CH_3$

Name this compound.

*44. Cyclic glycols show cis–trans isomerism because of restricted rotation about the carbon–carbon bonds in the ring. Draw and label the structures for *cis-* and *trans*-1,2-cyclopentanediol.

45. Arrange these three compounds in order of increasing acidity:

HO ⬡ CH_3 H_2O ⬡ CH_2OH

46. A deficiency in the production of dopamine by the brain is believed to be the cause of Parkinson's disease. Dopamine is part of a class of compounds called catecholamines. Draw the structure of the common phenol derivative that is a part of the dopamine molecule.

*47. When 1-butanol is dehydrated in an acid medium to an alkene, it yields mainly 2-butene rather than 1-butene. This indicates that the dehydration to an alkene is at least a two-step reaction. Suggest a mechanism to explain the reaction. (*Hint:* A primary carbocation is formed initially.)

48. Methyl alcohol is toxic partly because it is metabolized to formaldehyde ($O=CH_2$) after ingestion. What type of reaction converts methanol to formaldehyde?

49. Write equations to show how the following transformations can be accomplished. Some conversions may require more than one step, and some reactions studied in previous chapters may be needed.

(a) $CH_3CHCH_3 \longrightarrow CH_3CCH_3$
   $\quad\quad |\quad\quad\quad\quad\quad\quad ||$
   $\quad\quad OH\quad\quad\quad\quad\quad O$

(b) $CH_3CH_2CH_2CH=CH_2 \longrightarrow CH_3CH_2CH_2CHCH_3$
   $\quad\quad\quad\quad\quad\quad\quad\quad\quad\quad\quad\quad\quad\quad | $
   $\quad\quad\quad\quad\quad\quad\quad\quad\quad\quad\quad\quad\quad OH$

(c) $CH_3CH_2OH \longrightarrow CH_3CH_2O^-Na^+$

(d) $CH_3CH_2CH=CH_2 \longrightarrow CH_3CH_2CCH_3$
   $\quad\quad\quad\quad\quad\quad\quad\quad\quad\quad\quad\quad ||$
   $\quad\quad\quad\quad\quad\quad\quad\quad\quad\quad\quad\quad O$

(e) $CH_3CH_2CH_2CH_2OH \longrightarrow CH_3CH_2CHCH_3$
   $\quad\quad\quad\quad\quad\quad\quad\quad\quad\quad\quad\quad\quad | $
   $\quad\quad\quad\quad\quad\quad\quad\quad\quad\quad\quad\quad\quad Cl$

(f) $CH_3CH_2CH_2Cl \longrightarrow CH_3CH_2C=O$
   $\quad\quad\quad\quad\quad\quad\quad\quad\quad\quad\quad\quad\quad\quad |$
   $\quad\quad\quad\quad\quad\quad\quad\quad\quad\quad\quad\quad\quad\quad H$

50. Starting with *p*-methylphenol and ethane, show the equations for the synthesis of *p*-ethoxytoluene.

*51. Give a simple chemical test that will distinguish between the compounds in each of the following pairs:
(a) ethanol and dimethyl ether
(b) 1-pentanol and 1-pentene
(c) *p*-methylphenol and methoxybenzene

52. In what important ways do phenols differ from alcohols?

53. Draw all the isomeric compounds that contain a benzene ring and have the molecular formula $C_8H_{10}O$.

54. Which of the following compounds would you expect to react with sodium hydroxide? Write equations for those that react.

(a) ⬡ OH (b) ⬡ CH_2OH

(c) ⬡ O ⬡

55. Arrange these compounds in order of *increasing* boiling points. Give reasons for your answer.
1-pentanol   1,2-pentanediol   1-octanol

Answers to Practice Exercises

22.1 O=CHCH(OH)CH(OH)CH(OH)CH(OH)CH$_2$OH

           ↑   ↑   ↑   ↑   ↑

         2°   2°   2°   2°   1°

22.2 (a) 2-ethyl-1-pentanol (b) 5-bromo-5-methyl-3-hexanol

22.3 (a) OH

(structure: cyclohexane ring with OH and CH$_3$)

(b) CH$_3$CHCHCHCH$_2$CH$_2$CH$_3$ with CH$_3$, CH$_2$CH$_3$, and OH substituents

22.4 1-hexanol, 1-pentanol, 1-butanol, 1,3-propanediol

22.5 (a) CH$_3$CH$_2$CH$_2$CH$_2$OH $\xrightarrow{[O]}$ CH$_3$CH$_2$CH$_2$C(=O)—H

CH$_3$CH$_2$CH$_2$CH$_2$OH $\xrightarrow{[O]}$ CH$_3$CH$_2$CH$_2$C(=O)—OH

(b) No reaction (2-methyl-2-butanol is a tertiary alcohol)

(c) (cyclopentanone structure)

Evidence of reaction is that orange colored $Cr_2O_7^{2-}$ turned green (Cr^{3+}).

22.6 O=C—COOH

       |

    CH$_2$—COOH

22.7 (a) 1-methoxy-2-methylbutane

(b) 2-ethoxy-2-methylpropane

(c) o-chloroethoxybenzene

22.8 (a) methyl propyl ether

(b) di-*tert*-butyl ether

(c) ethyl phenyl ether

22.9 (a) 2 CH$_3$CH$_2$CH$_2$OH + 2 Na ⟶ 2 CH$_3$CH$_2$CH$_2$ONa + H$_2$

CH$_3$CH$_2$CH$_2$ONa + CH$_3$CH$_2$CH$_2$CH$_2$Br ⟶ CH$_3$CH$_2$CH$_2$—O—CH$_2$CH$_2$CH$_2$CH$_3$ + NaBr

OR

(a) 2 CH$_3$CH$_2$CH$_2$CH$_2$OH + 2 Na ⟶ 2 CH$_3$CH$_2$CH$_2$CH$_2$ONa + H$_2$

CH$_3$CH$_2$CH$_2$CH$_2$ONa + CH$_3$CH$_2$CH$_2$Br ⟶ CH$_3$CH$_2$CH$_2$CH$_2$—O—CH$_2$CH$_2$CH$_3$ + NaBr

(b) 2 (phenol, OH) + 2 Na ⟶ 2 (ONa benzene) + H$_2$

(ONa benzene) + CH$_3$Br ⟶ (OCH$_3$ benzene) + NaBr

(c) 2 (benzyl alcohol, CH$_2$OH) + 2 Na ⟶ 2 (CH$_2$ONa benzene) + H$_2$

(CH$_2$ONa benzene) + CH$_3$CH$_2$Br ⟶ (CH$_2$—O—CH$_2$CH$_3$ benzene) + NaBr

OR

(c) 2 CH$_3$CH$_2$OH + 2 Na ⟶ 2 CH$_3$CH$_2$ONa + H$_2$

CH$_3$CH$_2$ONa + (CH$_2$Br benzene) ⟶ (CH$_2$—O—CH$_2$CH$_3$ benzene) + NaBr

23

Aldehydes and Ketones

▲ How do you wish to pay for your purchase: cash, check, or plastic? Credit cards and many other plastic products are made with molecules known as aldehydes.

CHAPTER 23 / OUTLINE

Many organic molecules contain a carbon atom double-bonded to an oxygen atom. This grouping of atoms is particularly reactive and is present in both aldehydes and ketones. Formaldehyde is by far the most common aldehyde molecule. It is used commercially as a preservative for animal specimens and in the formation of formaldehyde polymers. Examples of these polymers are found in Melmac plastic dishes, Formica table tops, and in the adhesives used in the manufacture of plywood and fiberboard. Other aldehydes are found in nature as spices.

Acetone is the simplest and most common ketone in our lives. It is used in nail polish remover, paints, varnishes, and resins. Acetone is also produced in the body during lipid metabolism. It is generally metabolized, but in diabetic patients more is formed than can be oxidized. The presence of acetone in a urine sample or on the breath is a positive indicator of diabetes. An understanding of aldehydes and ketones forms the basis for a discussion of many important organic and biochemical reactions.

23.1 Structures of Aldehydes and Ketones

The aldehydes and ketones are closely related classes of compounds. Their structures contain the **carbonyl group,** $\diagup C{=}O$, a carbon–oxygen double bond. **Aldehydes** have at least one hydrogen atom bonded to the carbonyl group, whereas **ketones** have only alkyl or aryl (aromatic, denoted Ar) groups bonded to the carbonyl group.

carbonyl group
aldehyde
ketone

$$\underset{\text{aldehydes}}{R-\overset{\overset{\displaystyle O}{\|}}{C}-H \qquad Ar-\overset{\overset{\displaystyle O}{\|}}{C}-H}$$

$$\underset{\text{ketones}}{R-\overset{\overset{\displaystyle O}{\|}}{C}-R \qquad R-\overset{\overset{\displaystyle O}{\|}}{C}-Ar \qquad Ar-\overset{\overset{\displaystyle O}{\|}}{C}-Ar}$$

In a linear expression, the aldehyde group is often written as CHO. For example,

$$CH_3CHO \quad \text{is equivalent to} \quad CH_3\overset{\overset{\displaystyle O}{\|}}{C}-H$$

In the linear expression of a ketone, the carbonyl group is written as CO; for example,

$$CH_3COCH_3 \quad \text{is equivalent to} \quad CH_3\overset{\overset{\displaystyle O}{\|}}{C}CH_3$$

The general formula for the saturated homologous series of aldehydes and ketones is $C_nH_{2n}O$.

23.2   Naming Aldehydes and Ketones

Aldehydes

> **IUPAC Rules for Naming Aldehydes**
>
> 1. To establish the parent name, select the longest continuous chain of carbon atoms that contains the aldehyde group.
> 2. The carbons of the parent chain are numbered starting with the aldehyde group. Since the aldehyde group is at the beginning (or end) of a chain, it is understood to be number 1.
> 3. Form the parent aldehyde name by dropping the *-e* from the corresponding alkane name and adding the suffix *-al*.
> 4. Other groups attached to the parent chain are named and numbered as we have done before.

The first member of the homologous series, $H_2C{=}O$, is methanal. The name *methanal* is derived from the hydrocarbon methane, which contains one carbon atom. The second member of the series is ethanal; the third member of the series is propanal; and so on.

$$CH_4 \qquad H{-}\overset{\displaystyle O}{\overset{\|}{C}}{-}H \qquad CH_3CH_3 \qquad CH_3\overset{\displaystyle O}{\overset{\|}{C}}{-}H$$

methane   methanal   ethane   ethanal
          (from methane + *al*)   (from ethane + *al*)

The longest carbon chain containing the aldehyde group is the parent compound. Other groups attached to this chain are numbered and named as before. For example,

$$\overset{6}{C}H_3\overset{5}{C}H_2\overset{\overset{\displaystyle CH_3}{|}}{\underset{4}{C}}H\overset{3}{C}H_2\overset{2}{C}H_2\overset{\overset{\displaystyle O}{\|}}{\underset{1}{C}}{-}H$$

4-methylhexanal

The naming of aldehydes is also illustrated in Table 23.1. [The common names for some aldehydes are widely used. Common names for the aliphatic aldehydes are derived from the common names of the carboxylic acids (see Table 24.1).] The *-ic acid* or *-oic acid* ending of the acid name is dropped and is replaced with the suffix *-aldehyde*. Thus the name of the one-carbon acid, formic acid, becomes formaldehyde for the one-carbon aldehyde, and the name for the two-carbon acid, acetic acid, becomes acetaldehyde for the two-carbon aldehyde.

$$H{-}\overset{\displaystyle O}{\overset{\|}{C}}{-}OH \qquad H{-}\overset{\displaystyle O}{\overset{\|}{C}}{-}H \qquad CH_3{-}\overset{\displaystyle O}{\overset{\|}{C}}{-}OH \qquad CH_3{-}\overset{\displaystyle O}{\overset{\|}{C}}{-}H$$

formic acid   formaldehyde   acetic acid   acetaldehyde

TABLE 23.1	IUPAC Names of Selected Aldehydes
Formula	**IUPAC name**
$H-\overset{O}{\overset{\|}{C}}-H$	Methanal
$CH_3\overset{O}{\overset{\|}{C}}-H$	Ethanal
$CH_3CH_2\overset{O}{\overset{\|}{C}}-H$	Propanal
$CH_3CH_2CH_2\overset{O}{\overset{\|}{C}}-H$	Butanal
$CH_3\underset{CH_3}{\overset{O}{\overset{\|}{CHC}}}-H$	2-Methylpropanal

Aromatic aldehydes contain an aldehyde group bonded to an aromatic ring. These aldehydes are named after the corresponding carboxylic acids. Thus the name benzaldehyde is derived from benzoic acid, and the name *p*-tolualdehyde is derived from *p*-toluic acid.

benzaldehyde   benzoic acid   *p*-tolualdehyde   *p*-toluic acid

In dialdehydes the suffix *-dial* is added to the corresponding hydrocarbon name; for example,

$$H-\overset{O}{\overset{\|}{C}}CH_2CH_2\overset{O}{\overset{\|}{C}}-H$$

is named butanedial.

▲
Aldehydes are used in the manufacture of building materials.

Ketones

IUPAC Rules for Naming Ketones

1. To establish the parent name, select the longest continuous chain of carbon atoms that contain the ketone group.
2. Form the parent name by dropping the *-e* from the corresponding alkane name and adding the suffix *-one*.
3. If the chain is longer than four carbons, it is numbered so that the carbonyl group has the smallest number possible; this number is prefixed to the parent name of the ketone.
4. Other groups attached to the parent chain are named and numbered as we have done before.

For example,

$$\underset{\text{propanone}}{CH_3-\overset{\overset{\displaystyle O}{\|}}{C}-CH_3} \qquad \underset{\text{2-pentanone}}{\overset{5\quad4\quad3\quad2\quad1}{CH_3CH_2CH_2-\overset{\overset{\displaystyle O}{\|}}{C}-CH_3}} \qquad \underset{\text{4-methyl-3-hexanone}}{\overset{1\quad2\quad3\quad4\quad5\quad6}{CH_3CH_2-\overset{\overset{\displaystyle O}{\|}}{C}-\underset{\underset{\displaystyle CH_3}{|}}{CH}CH_2CH_3}}$$

Note that in 4-methyl-3-hexanone the carbon chain is numbered from left to right to give the carbonyl group the lowest possible number.

An alternative non-IUPAC method often used to name simple ketones lists the names of the alkyl or aromatic groups attached to the carbonyl carbon together with the word *ketone*. Thus butanone ($CH_3COCH_2CH_3$) is methyl ethyl ketone:

$$\underset{\underset{\displaystyle \text{methyl}}{\uparrow}}{CH_3}-\underset{\underset{\displaystyle \text{ketone}}{\uparrow}}{\overset{\overset{\displaystyle O}{\|}}{C}}-\underset{\underset{\displaystyle \text{ethyl}}{\uparrow}}{CH_2CH_3}$$

Two of the most widely used ketones have special common names: Propanone is called acetone, and butanone is known as methyl ethyl ketone, or MEK.

Aromatic ketones are named in a fashion similar to that for aliphatic ketones and often have special names as well.

1-phenylethanone (IUPAC)
(methyl phenyl ketone)
(acetophenone)

1-phenyl-1-propanone (IUPAC)
(ethyl phenyl ketone)
(propiophenone)

Write both the formulas and the names for the straight-chain five- and six-carbon aldehydes.

Example 23.1

The IUPAC names are based on the five- and six-carbon alkanes. Drop the -e of the alkane name and add the suffix -al. Pentane, C_5, becomes pentanal and hexane, C_6, becomes hexanal. (The aldehyde group does not need to be numbered; it is understood to be on carbon 1.) The common names (in parentheses) are derived from valeric acid and caproic acid, respectively.

$$CH_3CH_2CH_2CH_2\overset{\displaystyle O}{\overset{\|}{C}}-H$$

pentanal (valeraldehyde)

$$CH_3CH_2CH_2CH_2CH_2\overset{\displaystyle O}{\overset{\|}{C}}-H$$

hexanal (caproaldehyde)

Give two names for each of the following ketones:

Example 23.2

(a) $CH_3CH_2\overset{\displaystyle O}{\overset{\|}{C}}CH_2\overset{\displaystyle CH_3}{\overset{|}{C}}HCH_3$

(b) $CH_3CH_2CH_2\overset{\displaystyle O}{\overset{\|}{C}}$ —⟨⟩

(a) The parent carbon chain that contains the carbonyl group has six carbons. Number this chain from the end nearer to the carbonyl group. The ketone group is on carbon 3, and a methyl group is on carbon 5. The six-carbon alkane is hexane. Drop the -e from hexane and add -one to give the parent name, hexanone. Prefix the name hexanone with a 3- to locate the ketone group and with 5-methyl- to locate the methyl group. The name is 5-methyl-3-hexanone. The common name is ethyl isobutyl ketone since the C=O has an ethyl group and an isobutyl group bonded to it.

(b) The longest aliphatic chain has four carbons. The parent ketone name is butanone, derived by dropping the e of butane and adding one. The butanone has a phenyl group attached to carbon 1. The IUPAC name is therefore 1-phenyl-1-butanone. The common name for this compound is phenyl n-propyl ketone, since the C=O group has a phenyl and an n-propyl group bonded to it.

Practice 23.1

Write structures for the following carbonyl compounds:
(a) 4-bromo-5-hydroxyhexanal
(b) phenylethanal
(c) 3-buten-2-one
(d) diphenylmethanone (diphenyl ketone)

Practice 23.2

Name each compound using the IUPAC System:

(a) ⬡=O

(c) $CH_3CHCH_2\overset{\displaystyle O}{\overset{\displaystyle \|}{C}}-H$
     |
     CH_3

(b) $ClCH_2CH_2\overset{\displaystyle O}{\overset{\displaystyle \|}{C}}-CH_3$

(d) $CH=CH-\overset{\displaystyle O}{\overset{\displaystyle \|}{C}}-H$ (with phenyl group)

Practice 23.3 Application to Biochemistry

Honey is especially sweet because of the sugar, fructose. Give the IUPAC name for fructose.

$$CH_2(OH)\overset{\displaystyle O}{\overset{\displaystyle \|}{C}}CH(OH)CH(OH)CH(OH)CH_2OH$$
fructose

23.3   Bonding and Physical Properties

The carbon atom of the carbonyl group is sp^2-hybridized and is joined to three other atoms by sigma bonds. The fourth bond is made by overlapping p electrons of carbon and oxygen to form a pi bond between the carbon and oxygen atoms. Since the carbon has a double bond, the possibility of adding another bonded atom (an addition reaction) exists for carbonyl carbon atoms.

Because the oxygen atom is considerably more electronegative than carbon, the carbonyl group is polar, with the electrons shifted toward the oxygen atom. This makes the oxygen atom partially negative (δ^-) and leaves the carbon atom partially positive (δ^+). Many of the chemical reactions of aldehydes and ketones are due to this polarity.

The hydrogen in a hydrogen bond must also be covalently bonded to a nitrogen, oxygen, or fluorine.

H-bonds

no H-bonds

$$\overset{\delta^+ \quad \delta^-}{>C=\ddot{O}\!:}$$
polarity

Unlike alcohols, aldehydes and ketones cannot hydrogen-bond to themselves because no hydrogen atom is attached to the oxygen atom of the carbonyl group. Aldehydes and ketones, therefore, have lower boiling points than alcohols of comparable molar mass (Table 23.2).

Low-molar-mass aldehydes and ketones are soluble in water, but for five or more carbons, the solubility decreases markedly. Ketones are very efficient organic solvents.

CHEMISTRY IN ACTION   Familiar Aldehydes and Ketones

Aldehydes and ketones are all around you. For example, aldehydes are used commercially in the production of plastics. The common polymers known as Formica, used on counters and in telephones; Melmac, used in plastic tableware; and Bakelite, found in handles on kitchen utensils, are all derived from formaldehyde.

Then there are the aromatic aldehydes. They are responsible for the characteristic flavor and aroma of many common flavorings and spices. Examples include vanillin (see Chemistry in Action, Chapter 5), cinnamaldehyde, citral, and benzaldehyde (see Figure 23.1).

Higher-molar-mass ketones are also useful because of their unique aromas. Camphor (Figure 23.1) is a moth repellant. Muscone, which is extracted from the musk glands of the male musk deer, is frequently used to give the musky aroma to perfumes. Tiny amounts of these chemicals produce long-lasting aromas. Perfume manufacturers closely guard the secret of their ingredients to keep their formulations unique and selling well.

In living organisms, aldehydes and ketones are found in many different compounds. Sugars are often classified as containing an aldehyde group, as in glucose or ribose, or a ketone group, as in fructose (see Figure 23.1). The compounds known as steroids, belonging to the lipid family, also often contain aldehyde or ketone groups. These lipids may function as hormones, as in the case of cortisone.

▲ Formica, one of the most familiar formaldehyde-derived polymers, comes in an array of colors and designs.

Name	Molar mass	Boiling point (°C)
TABLE 23.2 **Boiling Points of Selected Aldehydes and Ketones and Corresponding Alcohols**		
1-Propanol	60	97
Propanal	58	49
Propanone	58	56
1-Butanol	74	118
Butanal	72	76
Butanone	72	80
1-Pentanol	88	138
Pentanal	86	103
2-Pentanone	86	102

The lower-molar-mass aldehydes have a penetrating, disagreeable odor and are partially responsible for the taste of some rancid and stale foods. As the molar mass increases, the odor of both aldehydes and ketones, especially the aromatic ones, becomes more fragrant. Some are even used in flavorings and perfumes. A few of these and other selected aldehydes and ketones are shown in Figure 23.1.

benzaldehyde
(oil of bitter almonds)

cinnamaldehyde
(oil of cinnamon)

carvone
(chief component of spearmint oil)

muscone
(gland of male musk deer, used in perfume)

civetone
(secretion of the civet cat, used in perfume)

camphor
(from the camphor tree)

cortisone
(hormone; regulation of carbohydrate
and protein metabolism; used to
reduce inflammation)

glucose
(sugar)

ribose
(sugar)

fructose
(sugar)

citral
(oil of lemon)

vitamin K$_1$
(antihemorrhagic vitamin)

▲
FIGURE 23.1
Selected naturally occurring
aldehydes and ketones

23.4 Chemical Properties of Aldehydes and Ketones

The carbonyl group ($>C=O$) is the functional group of both aldehydes and ketones. Associated with every functional group are characteristic reactions. Aldehydes undergo both oxidation and reduction reactions; ketones undergo reduction reactions.

The carbon atom in the carbonyl group also has a double bond. What kind of reactivity does this suggest for aldehydes and ketones? This question can be answered by reviewing the chemistry of the alkenes in Chapter 20. Like alkenes, the aldehydes and ketones have a carbon atom that is bonded to three other atoms—one less than the usual maximum of four. Such a carbon atom can easily bond to one more atom. Alkenes readily undergo addition reactions. This suggests that the aldehydes and ketones also undergo addition reactions. In fact, addition is the characteristic reaction of aldehydes and ketones.

An addition reaction for aldehydes and ketones can be envisioned as

$$\underset{C}{\overset{O}{\|}} + HX \longrightarrow -\underset{\underset{X}{|}}{\overset{\overset{OH}{|}}{C}}-$$

Oxidation

Aldehydes are easily oxidized to carboxylic acids by a variety of oxidizing agents, including (under some conditions) oxygen of the air. Oxidation is the reaction in which aldehydes differ most from ketones. In fact, aldehydes and ketones may be separated into classes by their relative susceptibilities to oxidation. Aldehydes are easily oxidized to carboxylic acids by $K_2Cr_2O_7 + H_2SO_4$ and by mild oxidizing agents such as Ag^+ and Cu^{2+} ions; ketones are unaffected by such reagents. Ketones can be oxidized under drastic conditions—for example, by treatment with hot potassium permanganate solution. However, under these conditions carbon–carbon bonds are broken, and a variety of products are formed. Equations for the oxidation of aldehydes by dichromate are

$$3\,\underset{}{R\overset{O}{\overset{\|}{C}}-H} + Cr_2O_7^{2-} + 8\,H^+ \longrightarrow 3\,R\overset{O}{\overset{\|}{C}}-OH + 2\,Cr^{3+} + 4\,H_2O$$
carboxylic acid

$$3\,CH_3\overset{O}{\overset{\|}{C}}-H + Cr_2O_7^{2-} + 8\,H^+ \longrightarrow 3\,CH_3\overset{O}{\overset{\|}{C}}-OH + 2\,Cr^{3+} + 4\,H_2O$$
acetic acid

Tollens test

The **Tollens test** (silver-mirror test) for aldehydes is based on the ability of silver ions to oxidize aldehydes. The Ag^+ ions are thereby reduced to metallic silver. In practice, a little of the suspected aldehyde is added to a solution of silver nitrate and ammonia in a clean test tube. The appearance of a silver mirror on the inner wall of the tube is a positive test for the aldehyde group. The abbreviated equation is

$$R\overset{O}{\overset{\|}{C}}-H + 2\,Ag^+ \xrightarrow[H_2O]{NH_3} R\overset{O}{\overset{\|}{C}}-O^-\,NH_4^+ + 2\,Ag(s) \quad \text{(general reaction)}$$

$$CH_3\overset{O}{\overset{\|}{C}}-H + 2\,Ag^+ \xrightarrow[H_2O]{NH_3} CH_3\overset{O}{\overset{\|}{C}}-O^-\,NH_4^+ + 2\,Ag(s)$$

Fehling and Benedict solutions contain Cu^{2+} ions in an alkaline medium. In the **Fehling and the Benedict tests,** the aldehyde group is oxidized to an acid by Cu^{2+} ions. The blue Cu^{2+} ions are reduced and form brick-red copper(I) oxide (Cu_2O), which precipitates during the reaction. These tests can be used for detecting carbohydrates that have an available aldehyde group. The abbreviated equation is

$$\underset{\text{(blue)}}{\overset{\displaystyle O \atop \displaystyle \|}{RC-H} + 2\,Cu^{2+}} \xrightarrow[H_2O]{NaOH} \underset{\text{(brick-red)}}{\overset{\displaystyle O \atop \displaystyle \|}{RC-O^-Na^+} + Cu_2O(s)}$$

Because most ketones do not give a positive test with Tollens, Fehling, or Benedict solutions, these tests are used to distinguish between aldehydes and ketones.

$$\overset{\displaystyle O \atop \displaystyle \|}{R-C-R} + Ag^+ \xrightarrow[H_2O]{NH_3} \text{No reaction}$$

$$\overset{\displaystyle O \atop \displaystyle \|}{R-C-R} + Cu^{2+} \xrightarrow[H_2O]{OH^-} \text{No reaction}$$

Aldehydes and ketones are often involved in industrial syntheses because oxidation is a relatively simple reaction that can start with inexpensive reactants. Petroleum, natural gas, and the oxygen in the air are inexpensive and abundant materials; many aldehydes and ketones can be prepared from these starting materials. In turn, the aldehydes and ketones are useful for making even more valuable products. For example, simple hydrocarbons are oxidized to form acetaldehyde, which can be oxidized by air to acetic acid, the most widely used carboxylic acid. Air oxidation of 2-propanol yields acetone, an important commercial solvent. Methanol is air oxidized to formaldehyde, which is the precursor for a wide variety of plastics and glues.

The oxidation of aldehydes is an important reaction in biochemistry. When our cells "burn" carbohydrates, they take advantage of the aldehyde reactivity. The aldehyde group is oxidized to a carboxylic acid and is eventually converted to carbon dioxide, which is then exhaled. This stepwise oxidation provides some of the energy necessary to sustain life.

Practice 23.4
_____

Write the structures of the products (or indicate "no reaction") for the following:
(a) pentanal in the Tollens test
(b) 4-methyl-2-hexanone in the Tollens test
(c) 3-pentanone in the Fehling test
(d) 4-methylhexanal in the Fehling test

Reduction

Aldehydes and ketones are easily reduced to alcohols, either by elemental hydrogen in the presence of a catalyst or by chemical reducing agents such as lithium aluminum hydride ($LiAlH_4$) or sodium borohydride ($NaBH_4$). Aldehydes yield primary alcohols; ketones yield secondary alcohols:

$$
\underset{\displaystyle R-\overset{\displaystyle O}{\overset{\displaystyle \|}{C}}-H}{} \xrightarrow[\Delta]{H_2/Ni} \underset{\text{1° alcohol}}{RCH_2OH} \qquad \text{(general reaction)}
$$

$$
\underset{\displaystyle R-\overset{\displaystyle O}{\overset{\displaystyle \|}{C}}-R}{} \xrightarrow[\Delta]{H_2/Ni} \underset{\text{2° alcohol}}{R-\overset{\displaystyle OH}{\overset{\displaystyle |}{C}H}-R} \qquad \text{(general reaction)}
$$

$$
CH_3\overset{\displaystyle O}{\overset{\displaystyle \|}{C}}-H \xrightarrow[\Delta]{H_2/Ni} CH_3CH_2OH
$$

$$
CH_3\overset{\displaystyle O}{\overset{\displaystyle \|}{C}}CH_3 \xrightarrow[\Delta]{H_2/Ni} CH_3\overset{\displaystyle OH}{\overset{\displaystyle |}{C}H}CH_3
$$

In biochemistry, aldehyde/ketone reduction is commonly reversible. For example, hard-working muscles generate an excess of a specific ketone, which is then reduced to an alcohol before being secreted into the bloodstream.

$$
\underset{\text{pyruvic acid}}{CH_3\overset{\displaystyle O}{\overset{\displaystyle \|}{C}}-COOH} \xrightarrow{\text{reduction}} \underset{\text{lactic acid}}{CH_3\overset{\displaystyle OH}{\overset{\displaystyle |}{C}H}-COOH}
$$

The product, lactic acid, contributes to the burning sensation of hard-working muscles and is removed from the blood by the liver. The first reaction for lactic acid in the liver is a conversion *back* to the ketone.

$$
\underset{\text{lactic acid}}{CH_3\overset{\displaystyle OH}{\overset{\displaystyle |}{C}H}-COOH} \xrightarrow{\text{oxidation}} \underset{\text{pyruvic acid}}{CH_3\overset{\displaystyle O}{\overset{\displaystyle \|}{C}}-COOH}
$$

This reversible reaction completes a cycle; the ketone is reduced to lactic acid, which is then oxidized to the ketone once again. Of what advantage is a cycle? This is a common question in biochemistry. In this case, the oxidation of lactic acid by the liver accelerates glucose production—glucose that the muscle needs to do more work.

Practice 23.5 Application to Biochemistry

Human starvation leads to a condition called "ketosis." A specific ketone is formed by the liver and released into the bloodstream.

$$
CH_3-\overset{\displaystyle O}{\overset{\displaystyle \|}{C}}-CH_2COOH
$$

At the same time the liver reduces some of this ketone. The product alcohol also is released into the blood. Show the structure of the alcohol.

Addition Reactions

Addition of Alcohols Compounds derived from aldehydes and ketones that contain an alkoxy and a hydroxy group on the same carbon atom are known as **hemiacetals** and **hemiketals.** In a like manner, compounds that have two alkoxy groups on the same carbon atom are known as **acetals** and **ketals.**

hemiacetal
hemiketal
acetal
ketal

The alkoxy group in these compounds forms an ether linkage.

$$
\underset{\text{hemiacetal}}{\overset{R}{\underset{H}{>}}C\overset{OH}{\underset{OR'}{<}}}
\qquad
\underset{\text{hemiketal}}{\overset{R}{\underset{R}{>}}C\overset{OH}{\underset{OR'}{<}}}
\qquad
\underset{\text{acetal}}{\overset{R}{\underset{H}{>}}C\overset{OR'}{\underset{OR'}{<}}}
\qquad
\underset{\text{ketal}}{\overset{R}{\underset{R}{>}}C\overset{OR'}{\underset{OR'}{<}}}
$$

Most open-chain hemiacetals and hemiketals are so unstable that they cannot be isolated. On the other hand, acetals and ketals are stable in alkaline solutions but are unstable in acid solutions, in which they are hydrolyzed back to the original aldehyde or ketone.

In the following reactions we show only aldehydes in the equations, but keep in mind that ketones behave in a similar fashion, although they are not as reactive.

Aldehydes react with alcohols in the presence of a trace of acid to form hemiacetals:

$$
\underset{\text{propanal}}{CH_3CH_2\overset{\overset{\displaystyle O}{\|}}{C}-H} \;+\; \underset{\text{methanol}}{CH_3OH} \;\overset{H^+}{\rightleftharpoons}\; CH_3CH_2\underset{\underset{\displaystyle OCH_3}{|}}{\overset{\overset{\displaystyle OH}{|}}{CH}}
$$

1-methoxy-1-propanol
(propionaldehyde methyl hemiacetal)

In the presence of excess alcohol and a strong acid such as dry HCl, aldehydes or hemiacetals react with a second molecule of the alcohol to give an acetal:

$$
CH_3CH_2\underset{\underset{\displaystyle OCH_3}{|}}{\overset{\overset{\displaystyle OH}{|}}{CH}} \;+\; CH_3OH \;\overset{\text{dry HCl}}{\rightleftharpoons}\; CH_3CH_2\underset{\underset{\displaystyle OCH_3}{|}}{\overset{\overset{\displaystyle OCH_3}{|}}{CH}} \;+\; H_2O
$$

1,1-dimethoxypropane
(propionaldehyde dimethyl acetal)

A hemiacetal has both an alcohol and an ether group attached to the aldehyde carbon. An acetal has two ether groups attached to the aldehyde carbon.

If the alcohol and carbonyl groups are within the same molecule, the result is the formation of a cyclic hemiacetal (or hemiketal). This *intramolecular* cyclization is particularly significant in carbohydrate chemistry during the study of monosaccharides (Chapter 27).

$$
\begin{array}{c}
CH_2\!-\!O^{\diagup H} \\
CH_2 \qquad C{\overset{\diagup O}{\diagdown H}} \\
CH_2\!-\!CH_2
\end{array}
\;\overset{H^+}{\rightleftharpoons}\;
\begin{array}{c}
CH_2\!-\!O \\
CH_2 \qquad C{\overset{\diagup OH}{\diagdown H}} \\
CH_2\!-\!CH_2
\end{array}
$$

5-hydroxypentanal   stable hemiacetal

◄ Riot control police use
plastic Lucite shields for
protection.

Practice 23.6

Write structures and IUPAC names for the hemiacetal and acetal and the
hemiketal and ketal formed by reacting (a) ethanal and ethanol, and
(b) butanone and methanol.

Addition of Hydrogen Cyanide The addition of hydrogen cyanide, HCN, to aldehydes and ketones forms a class of compounds known as cyanohydrins. **Cyanohydrins** have a cyano ($-CN$) group and a hydroxyl group on the same carbon atom.
This reaction is typical of carbonyl addition reactions. It is catalyzed by a small
amount of base:

cyanohydrin

$$CH_3\overset{\overset{\displaystyle O}{\|}}{C}-H \ + \ HCN \ \xrightarrow{OH^-} \ CH_3\overset{\overset{\displaystyle OH}{|}}{CH}CN$$

acetaldehyde   acetaldehyde cyanohydrin

$$CH_3\underset{\underset{\displaystyle O}{\|}}{C}CH_3 \ + \ HCN \ \xrightarrow{OH^-} \ CH_3\overset{\overset{\displaystyle CH_3}{|}}{\underset{\underset{\displaystyle OH}{|}}{C}}-CN$$

acetone   acetone cyanohydrin

In the cyanohydrin reaction, the more positive H atom of HCN adds to the
oxygen of the carbonyl group, and the $-CN$ group adds to the carbon atom of
the carbonyl group. In the aldehyde addition, the length of the carbon chain is
increased by one carbon. The ketone addition product also contains an additional
carbon atom.

Cyanohydrins are useful intermediates for the synthesis of several important compounds. For example, the hydrolyses of cyanohydrins produce α-hydroxy acids:

$$CH_3CH-CN + H_2O \xrightarrow{H^+} CH_3CHCOOH + NH_4^+$$
$$\quad\;\; |OH \qquad\qquad\qquad\qquad\quad |OH$$

lactic acid
(an α-hydroxy acid)

Acetaldehyde can also be converted into other important biochemical compounds such as the amino acid alanine:

$$CH_3-\overset{O}{\overset{\|}{C}}-H \xrightarrow{HCN} CH_3-\overset{OH}{\overset{|}{\underset{|}{C}}}-CN \xrightarrow{NH_3} CH_3-\overset{NH_2}{\overset{|}{\underset{|}{C}}}-CN \xrightarrow[H^+]{H_2O} CH_3-\overset{H_2N}{\overset{|}{\underset{|}{C}}}-\overset{O}{\overset{\|}{C}}-OH$$
$$\qquad\qquad\qquad\qquad\quad H \qquad\qquad\quad H \qquad\qquad\qquad H$$

alanine

Some commercial reactions also involve the use of cyanohydrins. Acetone cyanohydrin can be converted to methyl methacrylate when refluxed with methanol and a strong acid. The methyl methacrylate can then be polymerized to Lucite or Plexiglas, both transparent plastics. (See Table 21.1.)

Aldol Condensation (Self-addition) In a carbonyl compound the carbon atoms are labeled alpha (α), beta (β), gamma (γ), delta (δ), and so on, according to their positions with respect to the carbonyl group. The α-carbon is adjacent to the carbonyl carbon, the β-carbon is next, the γ-carbon is third, and so forth. The hydrogens attached to the α-carbon atom are therefore called α-hydrogens, and so on, as shown here:

$$\overset{\delta}{-C}-\overset{\gamma}{C}-\overset{\beta}{C}-\overset{\alpha}{C}-C=O$$

β-hydrogen atom H H α-hydrogen atom

The hydrogen atoms attached to the α-carbon atom have the unique ability to be more easily released as protons than other hydrogens within the molecule.

An aldehyde or ketone that contains α-hydrogens may add to itself or to another α-hydrogen containing aldehyde or ketone. The product of this reaction contains both a carbonyl group and an alcohol group within the same molecule. The reaction is known as an **aldol condensation** and is catalyzed by dilute base. The aldol condensation is similar to the other carbonyl addition reactions. An α-hydrogen adds to the carbonyl oxygen, and the remainder of the molecule adds to the carbonyl carbon.

aldol
(3-hydroxybutanal)

In this reaction an alpha hydrogen first transfers from one molecule to the oxygen of the other molecule. This breaks the $C{=}O$ pi bond, leaving intermediates in which a carbon atom of each molecule has three bonds.

$$CH_3\overset{\overset{\displaystyle O}{\|}}{C}{-}H \ + \ \textcircled{H}{-}CH_2\overset{\overset{\displaystyle O}{\|}}{C}{-}H \ \longrightarrow \ CH_3\overset{\overset{\displaystyle OH}{|}}{C^+}{-}H \ + \ {^-}CH_2\overset{\overset{\displaystyle O}{\|}}{C}{-}H$$

The two intermediates then bond to each other, forming the product.

$$CH_3\overset{\overset{\displaystyle OH}{|}}{C^+}{-}H \ + \ {^-}CH_2\overset{\overset{\displaystyle O}{\|}}{C}{-}H \ \longrightarrow \ CH_3\overset{\overset{\displaystyle OH}{|}}{C}HCH_2\overset{\overset{\displaystyle O}{\|}}{C}{-}H$$

> **For a molecule to undergo an aldol condensation reaction, the molecule must have at least one H atom on the carbon alpha to the carbonyl group.**

Acetone also undergoes the aldol condensation:

$$CH_3\overset{\overset{\displaystyle O}{\|}}{C}CH_3 \ + \ \textcircled{H}{-}CH_2\overset{\overset{\displaystyle O}{\|}}{C}CH_3 \ \xrightarrow[\text{NaOH}]{\text{dilute}} \ CH_3\overset{\overset{\displaystyle OH}{|}}{\underset{\underset{\displaystyle CH_3}{|}}{C}}{-}CH_2\overset{\overset{\displaystyle O}{\|}}{C}CH_3$$

acetone   acetone   diacetone alcohol
                                             (4-hydroxy-4-methyl-2-pentanone)

Write the equation for the aldol condensation of propanal.

Example 23.3

First write the structure for propanal and locate the α-hydrogen atoms:

Solution

$$CH_3\overset{\overset{\displaystyle H}{|}}{\underset{\underset{\displaystyle \textcircled{H}\leftarrow\alpha H}{|}}{C}}HC{=}O$$

Now write two propanal molecules and transfer an α-hydrogen from one molecule to the oxygen of the second molecule. After the pi bond breaks, the two carbon atoms that are bonded to only three other atoms bond to each other to form the product.

$$CH_3CH_2\overset{\overset{\displaystyle O}{\|}}{C}{-}H \atop \textcircled{H}\ CH_3\overset{\overset{\displaystyle O}{\|}}{C}HC{-}H \ \xrightarrow[\text{NaOH}]{\text{dilute}} \ CH_3CH_2\overset{\overset{\displaystyle OH}{|}}{\underset{\underset{\displaystyle H}{|}}{C^+}} \atop {^-}\overset{\overset{\displaystyle O}{\|}}{C}HC{-}H \atop \underset{\displaystyle CH_3}{|} \ \longrightarrow \ CH_3CH_2\overset{\overset{\displaystyle OH}{|}}{C}H\overset{}{\underset{\underset{\displaystyle CH_3}{|}}{C}}H\overset{\overset{\displaystyle O}{\|}}{C}{-}H$$

3-hydroxy-2-methylpentanal

Practice 23.7

(a) Write the equation for the aldol condensation of butanal.
(b) Write the equation for the aldol condensation of 3-pentanone.

Practice 23.8 Application to Biochemistry

One step in the biochemical process that makes blood sugar (glucose) depends on an aldol condensation to connect an aldehyde to a ketone.

$$
\begin{array}{ccc}
& & \overset{\displaystyle O}{\underset{\displaystyle \|}{}} \\
\overset{\displaystyle W}{\underset{\displaystyle |}{}} & & \overset{\displaystyle CH}{\underset{\displaystyle |}{}} \\
C{=}O & + & CHOH \longrightarrow \\
\underset{\displaystyle *CH_2OH}{|} & & \underset{\displaystyle W'}{|}
\end{array}
$$

(W and W′ represent the remainder of the molecules)

Write the structure of the product given that the hydrogen on the starred carbon (*C) reacts with the aldehyde.

Both the aldol condensation and the addition of hydrogen cyanide form new C—C bonds. These reactions are used in industry to build larger molecules from smaller precursors. For example, nifedipine (Procardia, Adalat) is an important heart medication (part of a class of drugs known as calcium-channel blockers) used to treat various forms of angina. Although nifedipine is a complex molecule, it can be synthesized from the simpler compound, o-nitrobenzaldehyde (which contains a reactive aldehyde group).

o-nitrobenzaldehyde   nifedipine

By using an aldol condensation, the drug is synthesized in one step. The carbonyl functional group provides the reactivity needed by pharmaceutical chemists to create a complex molecule with important biological activity.

Aldol condensations are common in biochemistry. For example, the strength of collagen depends on aldol condensation reactions. Collagen is the most abundant protein in humans (see Chapter 29). It is found (1) in the bone matrix around which the mineral calcium is deposited, (2) in the material that gives skin its tear-resistant quality, and (3) in tendons for strength. After the collagen molecule is formed, aldehyde groups are added along its length. Then, when collagen molecules are layered next to each other, aldol condensations occur naturally.

The cross-linking bonds form a strong network. Without cross-links, collagen fibers are weak. Genetic diseases that limit cross-linking commonly result in malformed bones, hyperextendable joints, and an early death from disrupted blood vessels. Of course, too much cross-linking can also pose problems. A loss of joint flexibility for older adults has been partly attributed to increased collagen cross-linking.

23.5 Common Aldehydes and Ketones

Numerous methods have been devised for making aldehydes and ketones. The oxidation of alcohols is a very general method. Special methods are often used for the commercial production of individual aldehydes and ketones.

Formaldehyde (Methanal) This aldehyde is made from methanol by reaction with oxygen (air) in the presence of a silver or copper catalyst:

$$2\ CH_3OH + O_2 \xrightarrow[400°C]{Ag\ or\ Cu} 2\ H_2C{=}O + 2\ H_2O$$

formaldehyde
(methanal)

Formaldehyde is a poisonous, irritating gas that is very soluble in water. It is marketed as a 37% aqueous solution called *formalin,* which also contains 10–15% methanol to keep formaldehyde from polymerizing. By far the largest use of formaldehyde is in the manufacture of polymers. About 1.33×10^9 kg of formaldehyde is manufactured annually in the United States.

Formaldehyde vapors are intensely irritating to the mucous membranes. Ingestion may cause severe abdominal pains, leading to coma and death.

It is of interest that formaldehyde may have had a significant role in chemical evolution. Formaldehyde is believed to have been a component of the primitive atmosphere of the earth. Scientists theorize that the reactivity of this single-carbon aldehyde enabled it to form more complex organic molecules—molecules that were precursors of the still more complicated substances that today are essential components of every living organism.

Formaldehyde readily polymerizes to form a linear solid polymer, paraformaldehyde $+CH_2O+_n$, which releases formaldehyde when heated. Formaldehyde has been used to preserve biological specimens, in fumigation, and in a number of other polymers.

Acetaldehyde (Ethanal) Acetaldehyde is a volatile liquid (bp 21°C) with a pungent, irritating odor. It has a general narcotic action and in large doses may cause respiratory paralysis. Its principal use is as an intermediate in the manufacture of other chemicals such as acetic acid and 1-butanol. Acetic acid, for example, is made by air oxidation of acetaldehyde:

$$2\ CH_3\overset{O}{\overset{\|}{C}}-H\ +\ O_2\ \xrightarrow[\Delta]{Mn^{2+}}\ 2\ CH_3\overset{O}{\overset{\|}{C}}-OH$$

Acetaldehyde undergoes reactions in which three or four molecules combine or polymerize to form the cyclic compounds paraldehyde and metaldehyde:

$$3\ CH_3-\overset{O}{\overset{\|}{C}}-H\ \xrightarrow[\Delta]{H^+}$$

paraldehyde
(bp 125°C)

Paraldehyde is a controlled substance and has been used as a sedative.

$$4\ CH_3\overset{O}{\overset{\|}{C}}-H\ \xrightarrow[-20°C]{Ca(NO_3)_2+HBr}$$

metaldehyde
(mp 246°C)

Metaldehyde is very attractive and very poisonous to slugs and snails. For this reason, it is an active ingredient in some pesticides that are sold for lawn and garden use. Metaldehyde is also used as a solid fuel.

Acetone and Methyl Ethyl Ketone Ketones are widely used organic solvents. Acetone, in particular, is used in very large quantities for this purpose. The U.S.

production of acetone is about 1.26×10^8 kg annually. It is used as a solvent in the manufacture of drugs, chemicals, and explosives; for removal of paints, varnishes, and fingernail polish; and as a solvent in the plastics industry. Methyl ethyl ketone (MEK) is also widely used as a solvent, especially for lacquers. Both acetone and MEK are made by oxidation (dehydrogenation) of secondary alcohols. Acetone is also a coproduct in the manufacture of phenol (see Section 22.9).

$$CH_3CHCH_3 \xrightarrow[250-300°C]{Cu} CH_3CCH_3 + H_2$$
$$\underset{OH}{|} \qquad\qquad \underset{O}{\|}$$

2-propanol   acetone
                              (propanone)

$$CH_3CH_2CHCH_3 \xrightarrow[250-300°C]{Cu} CH_3CH_2CCH_3 + H_2$$
$$\underset{OH}{|} \qquad\qquad \underset{O}{\|}$$

2-butanol   methyl ethyl ketone
                        (2-butanone)

Acetone is also formed in the human body as a by-product of lipid metabolism. Normal concentrations of acetone in the body are less than 1 mg/100 mL of blood. In patients with diabetes mellitus, the concentration of acetone may rise, and it is then excreted in the urine where it can be easily detected. Sometimes the odor of acetone can be detected on the breath of these patients.

23.6 Condensation Polymers

Condensation polymers are formed by reactions between functional groups on adjacent monomers. As a rule, a smaller molecule, usually water, is eliminated in the reaction. Each monomer loses a small group of atoms and gains a larger group, a substitution reaction that allows the polymer to continue growing. Condensation polymerization uses functional groups that favor substitution reactions. For the polymer to continue to grow, each monomer must undergo at least two substitution reactions and must thus be at least bifunctional. If cross-linking is to occur, there must be more than two functional groups on some monomer molecules.

Many different condensation polymers have been synthesized. Important classes include the polyesters, polyamides, phenol–formaldehyde polymers, and polyurethanes. Most biochemical polymers are the condensation type. We will only consider phenol–formaldehyde polymers at this time.

Phenol–Formaldehyde Polymers

As noted earlier, a phenol–formaldehyde condensation polymer (Bakelite) was first marketed over 90 years ago. Polymers of this type are still widely used, especially in electrical equipment, because of their insulating and fire-resistant properties. Polymers made from phenol are known as *phenolics*.

Each phenol molecule can react with formaldehyde to lose an H atom from the para position and from each ortho position (indicated by arrows):

Each formaldehyde molecule reacts with two phenol molecules to eliminate water:

Similar reactions can occur at the other two reactive sites on each phenol molecule, leading to the formation of the polymer. This polymer is thermosetting because it has an extensively cross-linked network structure. The following is a typical section of this structure:

phenol–formaldehyde polymer

Concepts in Review

1. Recognize aldehydes and ketones from their formulas.

2. Relate the general reactivity of aldehydes and ketones to the carbonyl functional group properties.

3. Give IUPAC and common names of aldehydes and ketones.

4. Write formulas of aldehydes and ketones when given their names.

5. Understand why aldehydes and ketones have lower boiling points than alcohols.

6. Write equations showing the oxidation of alcohols to aldehydes and ketones.

7. Write the structure of the alcohol formed when an aldehyde or ketone is reduced.

8. Discuss the Tollens, Benedict, and Fehling tests, including the reagents used, evidence of a positive test, and the equations of the reactions that occur in positive tests; distinguish between the reactions of aldehydes and ketones.

9. Recognize whether an aldehyde or ketone undergoes the aldol condensation.

10. Write equations showing the aldol condensation of aldehydes and ketones.

11. Write equations for the formation and hydrolysis of cyanohydrins.

12. Write equations for the formation and decomposition of hemiacetals, hemiketals, acetals, and ketals.

13. Be familiar with the polymers made from formaldehyde.

Key Terms

The terms listed here have been defined within this chapter. Section numbers are referenced in parentheses for each term.

acetal (23.4)
aldehyde (23.1)
aldol condensation (23.4)
carbonyl group (23.1)
cyanohydrin (23.4)
Fehling and Benedict tests (23.4)

hemiacetal (23.4)
hemiketal (23.4)
ketal (23.4)
ketone (23.1)
Tollens test (23.4)

Questions

Questions refer to tables, figures, and key words and concepts defined within the chapter. A particularly challenging question or exercise is indicated with an asterisk.

1. Write generalized structures for
   (a) an aldehyde   (e) a hemiketal
   (b) a ketone   (f) an acetal
   (c) a dialdehyde   (g) a ketal
   (d) a hemiacetal   (h) a cyanohydrin

2. Explain, in terms of structure, why aldehydes and ketones have lower boiling points than alcohols of similar molar masses.

3. Write structural formulas for propanal and propanone. Judging from these formulas, do you think that aldehydes and ketones are isomeric with each other? Show evidence and substantiate your answer by testing with a four-carbon aldehyde and a four-carbon ketone.

4. How are collagen fibers impacted by the aldol condensation reaction?

5. Why are high-molar-mass aldehydes and ketones not very soluble in water?

6. What type of orbital bonds make up the carbonyl group?

Paired Exercises

These exercises are paired. Each odd-numbered exercise is followed by a similar even-numbered exercise. Answers to the even-numbered exercises are given in Appendix VI.

7. Give the IUPAC name for each of these aldehydes unless otherwise indicated:

(a) $H_2C{=}O$ (give both the IUPAC name and the common name)

(b) $CH_3CHCH_2\overset{O}{\overset{\|}{C}}{-}H$ with CH_3 on the second carbon

(c) $H{-}\overset{O}{\overset{\|}{C}}CH_2CH_2\overset{O}{\overset{\|}{C}}{-}H$

(d) C₆H₅$CH{=}CH\overset{O}{\overset{\|}{C}}{-}H$

(e) CH_3 and H on one carbon $C{=}C$ with $\overset{O}{\overset{\|}{C}}{-}H$ and H

8. Give the IUPAC name for each of these aldehydes unless otherwise indicated:

(a) $CH_3\overset{O}{\overset{\|}{C}}{-}H$ (give both the IUPAC name and the common name)

(b) $CH_3CH_2CH_2\overset{O}{\overset{\|}{C}}{-}H$

(c) C₆H₅$\overset{O}{\overset{\|}{C}}{-}H$

(d) chlorobenzaldehyde with $\overset{O}{\overset{\|}{C}}{-}H$, Cl, and $CH(CH_3)_2$ ($H_3C{-}CH{-}CH_3$) substituents

(e) $CH_3CHCH_2\overset{O}{\overset{\|}{C}}{-}H$ with OH on the second carbon

9. Give the IUPAC name for each of these ketones unless otherwise indicated:
   (a) CH_3COCH_3 (three names)

(b) C₆H₅$\overset{O}{\overset{\|}{C}}{-}CH_2CH_3$ (two names)

(c) cyclopentanone (cyclopentane ring with $=O$)

(d) $CH_3\overset{CH_3}{\underset{OH}{\overset{|}{C}}}CH_2\overset{O}{\overset{\|}{C}}CH_3$

10. Give the IUPAC name for each of these ketones unless otherwise indicated:
   (a) $CH_3CH_2COCH_3$ (two names)

(b) $CH_3\overset{O}{\overset{\|}{C}}{-}\overset{CH_3}{\underset{CH_3}{\overset{|}{C}}}CH_3$ (two names)

(c) $CH_3\overset{O}{\overset{\|}{C}}CH_2CH_2\overset{O}{\overset{\|}{C}}CH_3$

(d) C₆H₅$CH_2\overset{O}{\overset{\|}{C}}CH_3$ (two names)

11. Write the structural formulas for
    (a) 1,3-dichloropropanone
    (b) 3-butenal
    (c) 4-phenyl-3-hexanone
    (d) hexanal
    (e) 3-ethyl-2-pentanone

12. Write the structural formulas for
    (a) 3-hydroxypropanal
    (b) 4-methyl-3-hexanone
    (c) cyclohexanone
    (d) 2,4,6-trichloroheptanal
    (e) 3-pentenal

13. Which compound in each of the following pairs has the higher boiling point? (Try to answer without consulting tables.)
    (a) 2-hexanone or 2,5-hexanedione
    (b) hexane or hexanal
    (c) 2-pentanone or 2-pentanol
    (d) propanone or butanone

14. Which compound in each of the following pairs has the higher boiling point? (Try to answer without consulting tables.)
    (a) pentane or pentanal
    (b) benzaldehyde or benzyl alcohol
    (c) 2-hexanone or butanone
    (d) 1-butanol or butanal

15. Which compound in each of the following pairs has the higher aqueous solubility? (Try to answer without consulting tables.)
    (a) 2-hexanone or 2,5-hexanedione
    (b) propane or propanal
    (c) heptanal or ethanal

16. Which compound in each of the following pairs has the higher aqueous solubility? (Try to answer without consulting tables.)
    (a) acetaldehyde or ethane
    (b) 2-pentanone or 2,4-pentanediol
    (c) propanal or hexanal

17. Write equations to show how the following compounds are oxidized by (1) $K_2Cr_2O_7 + H_2SO_4$ and (2) air + Cu or Ag + heat:
    (a) 3-pentanol
    (b) 3-ethyl-1-hexanol

18. Write equations to show how the following compounds are oxidized by (1) $K_2Cr_2O_7 + H_2SO_4$ and (2) air + Cu or Ag + heat:
    (a) 1-propanol
    (b) 2,3-dimethyl-2-butanol

19. (a) What functional group is present in a compound that gives a positive Tollens test?
    (b) What is the visible evidence for a positive Tollens test?
    (c) Write an equation showing the reaction involved in a positive Tollens test.

20. (a) What functional group is present in a compound that gives a positive Fehling test?
    (b) What is the visible evidence for a positive Fehling test?
    (c) Write an equation showing the reaction involved in a positive Fehling test.

21. Give the products for the reaction of the following compounds with Tollens reagent:
    (a) butanal
    (b) benzaldehyde
    (c) methyl ethyl ketone

22. Give the products for the reaction of the following compounds with Fehling reagent:
    (a) propanal
    (b) acetone
    (c) 3-methylpentanal

23. Write equations showing the aldol condensation for the following compounds:
    (a) butanal (b) phenylethanal

24. Write equations showing the aldol condensation for the following compounds:
    (a) 3-pentanone (b) propanal

25. Complete the following equations:

    (a) $CH_3\underset{\underset{O}{\|}}{C}CH_3 \ + \ CH_2CH_2 \ \underset{\underset{OH\ OH}{|\ \ |}}{} \ \overset{dry\ HCl}{\rightleftharpoons}$

    (b) $CH_3CH_2\overset{\overset{O}{\|}}{C}-H \ + \ CH_3CH_2OH \ \overset{H^+}{\rightleftharpoons}$

    (c) $CH_3CH(CH_3)CH_2CH(OCH_3)_2 \ \overset{H_2O}{\underset{H^+}{\longrightarrow}}$

26. Complete the following equations:

    (a) $CH_3CH_2\overset{\overset{O}{\|}}{C}-H \ + \ CH_3CH_2CH_2OH \ \overset{dry\ HCl}{\rightleftharpoons}$

    (b) $+ \ CH_3OH \ \overset{H^+}{\rightleftharpoons}$

    (c) $CH_3CH_2CH_2CH(OCH_3)_2 \ \overset{H_2O}{\underset{H^+}{\longrightarrow}}$

27. Write equations for the following sequence of reactions:
    (a) propanone + HCN $\longrightarrow$
    (b) product of part (a) + H_2O $\longrightarrow$
    (c) product of part (b) + acetaldehyde + dry HCl $\longrightarrow$

28. Write equations for the following sequence of reactions:
    (a) benzaldehyde + HCN $\longrightarrow$
    (b) product of part (a) + H_2O $\longrightarrow$
    (c) product of part (b) + $K_2Cr_2O_7 + H_2SO_4$ $\longrightarrow$

Additional Exercises

These exercises are not paired or labeled by topic and provide additional practice on concepts covered in this chapter.

29. 3-Hydroxypropanal can form an intramolecular cyclic hemiacetal. What is the structure of the hemiacetal?

30. The following cyanohydrin is used in a multistep synthesis to make the artificial fiber Orlon (polyacrylonitrile):

$$\underset{\displaystyle CH_3CHC\equiv N}{\overset{\displaystyle OH}{|}}$$

From what carbonyl-containing compound is this cyanohydrin synthesized? (Give both the structural formula and name.)

*31. Write the structure for each aldol condensation product that is possible when a mixture of ethanal and propanal is reacted with dilute NaOH.

32. The following molecules ("ketone bodies") are used by the human body as an emergency energy supply primarily for the muscles:

$$\underset{\displaystyle CH_3CHCH_2COOH}{\overset{\displaystyle OH}{|}} \qquad \underset{\displaystyle CH_3CCH_2COOH}{\overset{\displaystyle O}{\|}}$$

compound I   compound II

These two molecules can be interconverted in a single biochemical reaction. What chemical change takes place when compound I is converted to compound II?

33. Give a simple visible chemical test that will distinguish between the compounds in each of the following pairs:

(a) $CH_3CH_2\overset{\displaystyle O}{\overset{\displaystyle \|}{C}}-H$ and $CH_3\overset{\displaystyle O}{\overset{\displaystyle \|}{C}}CH_3$

(b) $CH_3CH_2\overset{\displaystyle O}{\overset{\displaystyle \|}{C}}-H$ and $CH_2{=}CH\overset{\displaystyle O}{\overset{\displaystyle \|}{C}}-H$

(c) a benzene ring with CH_2CH_2OH and a benzene ring with $\overset{\displaystyle CHCH_3}{\underset{\displaystyle OH}{|}}$

34. Write equations to show how you could prepare lactic acid, $CH_3CH(OH)COOH$, from acetaldehyde through a cyanohydrin intermediate.

35. The millipede carries the following cyanohydrin as a chemical defense weapon:

$$\underset{\displaystyle CHC\equiv N}{\overset{\displaystyle OH}{|}}$$

When attacked, the millipede reverses the reaction that formed this cyanohydrin, and deadly hydrogen cyanide gas is released. One millipede is said to be able to release enough poisonous gas to kill a small mouse. Write the reaction that releases hydrogen cyanide and name the aldehyde product.

36. Pyruvic acid is formed during muscle exertion. If the muscles are working strenuously, this acid is converted to lactic acid.

$$\underset{\displaystyle CH_3CCOOH}{\overset{\displaystyle O}{\|}} \qquad \underset{\displaystyle CH_3CHCOOH}{\overset{\displaystyle OH}{|}}$$

pyruvic acid   lactic acid

What chemical change takes place as pyruvic acid is converted to lactic acid?

*37. During the biological synthesis of glucose, two 3-carbon compounds are connected via a new carbon–carbon single bond to form a 6-carbon straight-chain molecule (a sugar). Given the reactants in a simplified version of this reaction, use your knowledge of the aldol condensation reaction to write the structure of the product.

$$\underset{\displaystyle OH\ OH}{\overset{\displaystyle O}{\|}}CH_2CHCH + \underset{\displaystyle OH\ OH}{\overset{\displaystyle O}{\|}}CH_2CCH_2 \longrightarrow$$

38. Ketones are prepared by oxidation of secondary alcohols. Name the alcohol that should be used to prepare
(a) 3-pentanone
(b) methyl ethyl ketone
(c) 4-phenyl-2-butanone

39. Write structures for all isomeric aldehydes and ketones with the molecular formula of $C_6H_{12}O$.

40. Write structures for all the benzaldehyde isomers with the molecular formula of $C_9H_{10}O$.

41. (a) Starting with ethanal, write equations showing the synthesis of 1-butanol.
(b) Choose your starting material from methanal, ethanal, propanal, and butanal, and write equations showing the synthesis of 1,3-propanediol.

42. Write a structure showing four units of a linear polymer made from acetaldehyde.

43. Using *p*-cresol in place of phenol to form a phenol–formaldehyde polymer results in a thermoplastic rather than a thermosetting polymer. Explain why this occurs.

OH

p-cresol

CH_3

Answers to Practice Exercises

23.1 (a) $CH_3CHCHCH_2CH_2C-H$ (OH, Br, O)

(c) $CH_3CCH=CH_2$ (O)

(b) phenyl$-CH_2C-H$ (O)

(d) diphenyl ketone (O)

23.2 (a) cyclohexanone **(c)** 3-methylbutanal
(b) 4-chloro-2-butanone **(d)** 3-phenylpropenal
23.3 1,3,4,5,6-pentahydroxy-2-hexanone

23.4 (a) $CH_3CH_2CH_2CH_2C-O^-NH_4^+ + Ag(s)$
(b) No reaction
(c) No reaction
(d) $CH_3CH_2CHCH_2CH_2C-O^-Na^+ + Cu_2O(s)$ (CH_3, O)

23.5 $CH_3-CH-CH_2-COOH$ (OH)

23.6 (a) CH_3CH (OH / OCH_2CH_3) 1-ethoxyethanol; CH_3CH (OCH_2CH_3 / OCH_2CH_3) 1,1-diethoxyethane
(b) $CH_3CH_2CCH_3$ (OH / OCH_3) 2-methoxy-2-butanol; $CH_3CH_2CCH_3$ (OCH_3 / OCH_3) 2,2-dimethoxybutane

23.7 (a) $2 CH_3CH_2CH_2C-H \xrightarrow{\text{dilute } OH^-} CH_3CH_2CH_2C-CHC-H$ (OH, O, H, CH_2CH_3)
(b) $2 CH_3CH_2CCH_2CH_3 \xrightarrow{\text{dilute } OH^-} CH_3CH_2C-CH_2CH_3$ (OH) with $CH_3CHCHCH_2CH_3$ (O)

23.8 W, C=O, CHOH, CHOH, CHOH, W'

PUTTING IT TOGETHER
Review for Chapters 22–23

Multiple Choice: *Choose the correct answer to each of the following.*

1. Which of the following is <u>not</u> a primary alcohol?
   (a) CH_3CHCH_2OH
       |
       CH_3
   (c)
   (b) ⬠—OH
   (d) $CH_3CH_2CH_2CH_2OH$

2. Which of the following formulas is a tertiary alcohol?
   (a) CH_3CHCH_3
       |
       OH
   (c) CH_2CHCH_2
       | | |
       OH OHOH
   (b) $CH_3CH_2CHCH_3$
           |
           OH
   (d) [cyclohexane with CH₃ and OH]

3. Arrange the following compounds in order of increasing solubility in water:
   (A) C_5H_{12}
   (C) CH_3CHCH_2OH
            |
            OH
   (B) $CH_3CH_2OCH_2CH_3$
   (D) $CH_3CH_2CH_2CH_2OH$
   (a) BDCA (b) DCAB (c) BDAC (d) ABDC

4. Arrange the following alcohols in order of increasing boiling points:
   (A) CH_3CH_2OH
   (C) $CH_3CH(OH)CH_3$
   (B) $CH_3CH_2CH_2OH$
   (D) $HOCH_2CH_2OH$
   (a) ABCD (b) ABDC (c) ACBD (d) DCBA

5. Which of the following formulas is isoamyl alcohol?
   (a) CH_3
          |
     CH_3CHCH_2OH
   (c) [benzene ring with CH_2OH]
   (b) $CH_3CH_2CH_2CH_2OH$
   (d) CH_3
          |
     $CH_3CHCH_2CH_2OH$

6. Give the IUPAC name for
       CH_3
        |
   $CH_3C—CHCH_3$
       | |
      CH_3 OH

   (a) 1,1,1-trimethyl-2-propanol
   (b) 1-*tert*-butylethanol
   (c) 3,3-dimethyl-2-butanol
   (d) 2,2-dimethyl-3-butanol

7. Which of the following formulas is 3-methylcyclohexanol?
   (a) [cyclohexane with OH and CH_3]
   (c) [cyclohexane with OH and CH_3]
   (b) [cyclohexane with OH and CH_3]
   (d) no correct formula given

8. Give the correct IUPAC name for
        CH_3 OH
         | |
   $CH_3CH—C—CH_2CH_2CH_3$
           |
         CH_2CH_3

   (a) 2-methyl-3-ethyl-3-hexanol
   (b) 4-ethyl-5-methyl-4-hexanol
   (c) 3-ethyl-2-methyl-3-hexanol
   (d) 4-isopropyl-4-hexanol

9. Methanol has the following properties <u>except</u>
   (a) colorless liquid
   (b) poisonous
   (c) boils below 100°C
   (d) does not hydrogen-bond

10. The alcohol present in alcoholic beverages is
    (a) methyl alcohol (c) isopropyl alcohol
    (b) ethyl alcohol   (d) ethylene glycol

11. What is the oxidation number of the tertiary carbon atom in this compound?
          CH_3
           |
    $CH_3CH_2C—OH$
           |
          CH_3

    (a) −1 (b) +1 (c) 0 (d) +2

12. What is the product of this chemical reaction?
    $CH_3CH_2OH + Na \longrightarrow$
    (a) CH_3CH_2ONa
    (b) $CH_3CH_2OCH_2CH_3$
    (c) $CH_3CH_2CH_2CH_3$
    (d) no reaction

13. Oxidation of 2-pentanol yields
    (a) pentanal   (c) pentanoic acid
    (b) 2-pentanone   (d) 3-pentanone

14. Name the major alkene formed by the dehydration of

$$CH_3CH_2CHCHCH_3$$
with CH_3 on the third carbon and OH on the fourth carbon

(a) $CH_3CH_2CHCH=CH_2$ with CH_3

(c) $CH_3CH_2C=CHCH_3$ with CH_3

(b) $CH_3CH_2CCH_2CH_3$ with CH_2 double bond

(d) $CH_3CH=CCH=CH_2$ with CH_3

15. Which of the following compounds is ethylene glycol?
    (a) $HOCH_2OCH_2OH$
    (b) $HOCH_2CH_2OH$
    (c) CH_2CH_2 with O bridge
    (d) $HOCH=CHOH$

16. What is the name for $CH_3CH_2-O-CH_2CH_2CH_3$?
    (a) propoxyethane
    (b) 1-propoxyethane
    (c) ethyl propyl ether
    (d) diethyl oxide

17. Which of the following formulas is methyl-*tert*-butyl ether (MTBE)?

(a) $CH_3C-O-CCH_3$ with CH_3 groups

(c) $CH_3OCCH_2CH_3$ with CH_3

(b) $CH_3-O-CCH_3$ with CH_3 groups

(d) $CH_3-O-CHCH_2CH_3$ with CH_3

18. Give the name for

$$CH_3CH-O-CH_2CH_3$$ with CH_3

(a) 2-methyl diethyl ether   (c) 2-pentyl ether
(b) ethyl isopropyl ether   (d) ethoxy propyl ether

19. Which of the following is the structure of the ether formed by the dehydration of 2-methyl-1-propanol?

(a) $CH_3CH-O-CHCH_3$ with CH_3 groups

(b) $CH_3CH_2CH_2-O-CH_2CH_2CH_3$

(c) $CH_3CHCH_2-O-CH_2CHCH_3$ with CH_3 groups

(d) $CH_3C=CH-O-CH=CCH_3$ with CH_3 groups

20. How many isomeric ethers are possible for $C_5H_{10}O$?
    (a) 4   (b) 6   (c) 8   (d) 9

21. Which of the following compounds is phenol?
    (a) OH — cyclohexane ring
    (b) OH — benzene ring with OH
    (c) OH — benzene ring
    (d) CH_3 with OH benzene ring

22. Which of the following compounds is catechol?
    (a) OH with CH_3
    (b) OH with OH
    (c) OCH_3
    (d) OH with OH

23. Give the name for

benzene ring with OH, Br, CH_3

(a) *m*-methyl-*p*-bromophenol
(b) 4-bromo-5-methylphenol
(c) 4-bromo-3-methylphenol
(d) 3-bromo-2-methylphenol

In Problems 24–28, which product is formed in the following chemical reactions?

24. $CH_3CH_2\underset{\underset{OH}{|}}{\overset{\overset{CH_3}{|}}{C}}HCHCH_3 \xrightarrow[180°C]{96\% H_2SO_4}$ alkene

(a) $\underset{\overset{CH_3}{|}}{CH_3CH}=CHCH_3$

(b) $CH_3CH_2CH=\underset{\overset{CH_3}{|}}{C}CH_3$

(c) $CH_3CH_2\underset{\overset{CH_3}{|}}{C}=CHCH_3$

(d) $CH_3CH_2\underset{\overset{CH_2}{||}}{C}CH_2CH_3$

25. $CH_3CH_2CH_2CH=CH_2 + H_2O \xrightarrow{H^+}$
   (a) $CH_3CH_2CH_2CH(OH)CH_3$
   (b) $CH_3CH_2CH_2CH_2CH_2OH$
   (c) $CH_3CH_2CH_2CH(OH)CH_2OH$
   (d) no correct answer given

26. ⬠—OH $\xrightarrow[H^+]{Cr_2O_7^{2-}}$

   (a) (cyclopentene)   (c) (two cyclopentane rings joined by —O—)
   (b) (cyclopentanone) =O   (d) (cyclopentane with epoxide) O

27. $CH_3CH_2Br + CH_3\underset{\overset{CH_3}{|}}{C}HCH_2ONa \longrightarrow$
   (a) $CH_3\underset{\overset{CH_3}{|}}{C}HCH_2-O-CH_2\underset{\overset{CH_3}{|}}{C}HCH_3$
   (b) $CH_3CH_2-O-CH_2CH_2CH_3$
   (c) $CH_3CH_2-O-\underset{\overset{CH_3}{|}}{C}HCH_3$
   (d) $CH_3CH_2-O-CH_2\underset{\overset{CH_3}{|}}{C}HCH_3$

28. (cyclohexane with Br) + $CH_3CH_2ONa \longrightarrow$

   (a) (cyclohexane with OCH₂CH₃)   (c) (cyclohexane with CH₂CH₃)
   (b) (cyclohexane with Br and CH₂CH₃)   (d) no reaction

29. The carbonyl group, $\diagdown C=O$, contains
   (a) two sigma bonds
   (b) two pi bonds
   (c) a sigma bond and a pi bond
   (d) an sp^3 electron orbital

30. Which statement is <u>not</u> true?
   (a) Aliphatic aldehydes and ketones are isomeric.
   (b) Aliphatic aldehydes and ethers are isomeric.
   (c) Aromatic aldehydes contain a benzene ring.
   (d) Ketones are less reactive than aldehydes.

31. Arrange the following compounds in order of increasing boiling points:

   (A) $CH_3CH_2\underset{}{\overset{\overset{H}{|}}{C}}=O$ 　 (C) $CH_3CH_2CH_2CH_2OH$
   (B) $CH_3CH_2CH_2OH$ 　 (D) $CH_3\underset{\overset{}{\underset{O}{||}}}{C}CH_2CH_3$

   (a) ADBC (b) BCDA (c) ACBD (d) DCBA

32. Which of these compounds has the lowest vapor pressure?
   (a) ethanal (b) butanal (c) pentanal (d) propanal

33. Which of these compounds has the highest boiling point?
   (a) ethanal (b) butanal (c) pentanal (d) propanal

34. Give the name for

   $CH_3CH_2CH_2CH_2\underset{}{\overset{\overset{H}{|}}{C}}=O$

   (a) 1-pentanol　(c) pentanal
   (b) propanal　(d) pentanaldehyde

35. Give the name for

   (benzene ring)—$\underset{\overset{CH_3}{|}}{C}H\overset{\overset{CH_3CH_2\ H}{\ \ \ \ |}}{\underset{}{C}}HC=O$

   (a) 3-ethyl-2-methyl-3-phenylpropanal
   (b) 4-methyl-3-phenylpentanal
   (c) 2-methyl-3-phenylhexanal
   (d) no correct name given

36. Give the name for

   $CH_3CH_2\underset{\underset{O}{||}}{\overset{\overset{CH_3}{|}}{C}}CHCH_2CH_3$

   (a) 4-methyl-3-hexanone　(c) 3-methyl-4-hexanone
   (b) ethyl isobutyl ketone　(d) butyl ethyl ketone

37. Which of the following compounds is benzaldehyde?

(a) [benzene ring with CCH₃ and =O]   (c) [benzene ring with C=O and OH]

(b) [benzene ring with H, C=O]   (d) $CH_3CH_2CCH_3$ with =O

38. A 37% aqueous solution of formaldehyde is called
(a) acetone   (b) aldol   (c) formalin   (d) methanal

39. What is the correct name for this compound?

$$O = CCH_2CH_2CH_2C = O$$
(with H on each end carbon)

(a) 1,5-pentanedial   (c) ethanedial
(b) 1,5-propanedial   (d) pentanedial

40. What is the correct name for this compound?

$$CH_3CH_2CH_2CCH_2CH_3$$
(with =O below the fourth carbon)

(a) 4-hexanone   (c) 3-hexanone
(b) butyl ethyl ketone   (d) dipropyl ketone

41. What is the correct name for this compound?

$$CH_3CHCH_2CCH_2 \text{—(phenyl)}$$
(with CH₃ and O branches)

(a) benzyl sec-butyl ketone
(b) 4-methyl-1-phenyl-2-pentanone
(c) 2-methyl-5-phenyl-4-pentanone
(d) no correct name given

42. Choose the incorrect name for

[benzene ring with C—CH₃ and =O]

(a) methyl phenyl ketone   (c) 1-phenylethanone
(b) benzophenone   (d) acetophenone

43. Which of the following compounds is a cyanohydrin?
(a) CH_3CH_2CN (c) $CH_3 — O — CH_2CN$

(b) CH_3CHCN with OH   (d) [benzene ring with CN]

44. Give the name of the following compounds that, when oxidized, will yield

$$CH_3CH_2CH_2CH_2C = O$$
(with H on terminal carbon)

(a) 1-pentanol   (c) 3-pentanol
(b) 2-pentanol   (d) 1-propanol

45. What is the evidence that a reaction occurred when an aldehyde is oxidized by $Cr_2O_7^{2-} + H^+$?
(a) orange-colored solution formed
(b) silver mirror formed
(c) brick red Cu_2O formed
(d) green solution formed

46. Reduction of an aldehyde produces a
(a) ketone   (c) secondary alcohol
(b) primary alcohol   (d) hydrocarbon

47. What is the product of this reaction?

$$CH_3CH_2C = O + HCN \longrightarrow$$
(with H on the carbonyl carbon)

(a) CH_3CHCN with OH   (c) CH_3CH_2CHCN with OH

(b) $CH_3CH_2 — O — CH_2CN$ (d) CH_3CHCH_2CN with OH

48. Which of the following compounds will give a positive Tollens test?
(a) $CH_3CHC = O$ with H and OH   (c) $CH_3C = O$ with OH

(b) [cyclopentane ring with =O]   (d) $CH_3CCH_2CH_3$ with =O

In Problems 49 and 50, use these formulas:

(A) $CH_3CH_2CH — OCH_2CH_3$ with OH

(B) $(CH_3CH_2)_2C — CH_2CH_3$ with OCH_2CH_3

(C) $CH_3C = OCH_2CH_3$ with OCH_2CH_3 and CH_3

(D) $(CH_3CH_2)_2C — OCH_2CH_3$ with OH

49. Which formula represents a hemiacetal?
(a) A   (b) B   (c) C   (d) D

50. Which formula represents a ketal?
(a) A   (b) B   (c) C   (d) D

51. Which of the following compounds will <u>not</u> undergo an aldol condensation?

(a) $H_2C=O$

(c) $CH_3CH_2CCH_3$ (with $\parallel$ O below)

(b) $CH_3C=O$ with H above and C=O

(d) (benzene ring)—$CH_2C=O$ with H above

52. Which of the following compounds will be isolated from this reaction?

CH_3CCH_3 (with $\parallel$ O) $\xrightarrow[H^+]{Cr_2O_7^{2-}}$

(a) CH_3CH_2COOH

(c) CH_3CCOOH (with $\parallel$ O)

(b) CH_3CCH_3 (with $\parallel$ O)

(d) CH_3CHCH_3 (with OH below)

53. Ketones can be prepared by oxidation of
(a) primary alcohols
(b) secondary alcohols
(c) tertiary alcohols
(d) cannot be prepared by oxidation reactions

54. What is the product of this aldol condensation reaction?

$CH_3CHC=O$ (with H above, CH_3 below) $\xrightarrow[NaOH]{dilute}$

(a) $CH_3CH—COONa$ (with CH_3 above)

(c) $CH_3CH—CH—C—C=O$ (with CH_3, CH_3 H above and OH, CH_3 below)

(b) $CH_3CHCHCHCH_3$ (with CH_3 CH_3 above and OH below)

(d) no reaction

55. What is the product of this aldol condensation reaction?

$CH_3C—C=O$ (with CH_3 H above and CH_3 below) $\xrightarrow[NaOH]{dilute}$

(a) $CH_3C—CH—C—CH_3$ (with CH_3 and CH_3 above and CH_3 OH CH_3 below)

(b) $CH_3C—COOH$ (with CH_3 above and CH_3 below)

(c) $CH_3C—CH—C—C=O$ (with CH_3 and CH_3 H above and CH_3 OH CH_3 below)

(d) no reaction

56. The polymer of phenol and formaldehyde is known as
(a) paraformaldehyde   (c) paraldehyde
(b) Bakelite   (d) meta

57. When hydrolyzed, cyanohydrins form
(a) α-keto acids   (c) glycols
(b) α-hydroxy acids   (d) α-hydroxy aldehydes

58. What is the product of this reaction?

(cyclohexanone with =O) $+ 2 CH_3OH \xrightarrow[HCl]{dry}$

(a) (cyclohexane with OH and OCH$_3$)

(c) (cyclohexane with OCH$_3$ and OCH$_3$)

(b) (cyclohexane with OCH$_3$, =O, OCH$_3$)

(d) (cyclohexane with OCH$_3$)

59. The substance that precipitates in a positive Benedict test is
(a) Ag (b) Ag_2O (c) CuO (d) Cu_2O

60. Which of the following compounds will give both a positive Tollens and Benedict test?
(a) CH_3CH_2OH

(c) $CH_3CH_2CCH_3$ (with $\parallel$ O)

(b) (cyclohexane with C—CH$_3$ and =O)

(d) $CH_3C=O$ (with H above)

61. Oxidation of an aldehyde will yield a
(a) primary alcohol   (c) ketone
(b) carboxylic acid   (d) secondary alcohol

62. How many isomeric aldehydes and ketones are possible for $C_5H_{10}O$?
(a) 5 (b) 6 (c) 7 (d) 9

Free Response Questions

1. An open-chain compound has the formula $C_6H_{12}O$. Chemical tests show that there is no alkene or aromatic group in the compound. How do you identify whether the compound is an alcohol, an ether, an aldehyde, or a ketone? You are allowed to perform only one chemical test.

2. Why are aldehydes and ketones easily able to undergo addition reactions but alcohols, ethers, and thiols are not?

3. A compound has the formula
   $CH_3OCH_2CH(OH)CH_2CH_2COCH(CH_3)CH_2CHO$
   What functional group(s) is (are) present? Draw the compound showing the relevant functional group(s).

4. Why is it possible to have straight chain triol, triether, trithiol, or trione, but it is not possible to have a straight-chain trialdehyde? Draw a structure for each type of compound.

5. Which of the following would you expect to be most soluble in water? least soluble in water? Explain.

   methoxyethane, butene, 1-propanol, propanal

6. Draw and name three structural isomers of the following compound, each containing a <u>different</u> functional group.

7. Draw the organic products I–III for the following reaction sequence.

8. Fill in the missing reagents or organic structures, I-III for the following reactions. Give the IUPAC names for compounds **I, III,** and **IV.**

9. Which of the following compounds can easily be identified from the others by taking a boiling point? Briefly explain. For those that cannot be easily identified by boiling point, indicate how you may distinguish among them.

   2-hexanone, 1-hexanol, hexanal, dipropyl ether, 1-pentanethiol

10. Suggest at least two visual chemical tests that could be used to distinguish between the members of the following pairs of compounds:
    (a) phenol and 1-hexanol
    (b) phenol and hexanal
    (c) 1-hexanol and 3-hexanone

11. Acetals and ketals are very commonly used as protecting groups in organic synthesis (converting the carbonyl into a functional group that is unreactive to the conditions used to react with another functional group). Draw the structure of the <u>acetal</u> produced when the following compound is reacted with ethylene glycol in the presence of an acid catalyst.

CHAPTER

24

Carboxylic Acids and Esters

All citrus fruits contain a variety of carboxylic acids.

Whenever we eat food with a sour or tart taste, it is very likely that at least one carboxylic acid is present in that food. For example, citric acid is present in lemons, acetic acid is present in vinegar, and lactic acid is present in sour milk. Carboxylic acids are also important compounds in biochemistry. Citric acid is found in our blood, and lactic acid is produced in our muscles during the breakdown of glucose. In living systems, these acids are most often found in the form of salts or acid derivatives. In the world around us, we use carboxylic acid salts as preservatives, especially in cheeses and breads, and to treat skin irritations like diaper rash and athlete's foot.

Carboxylic acid derivatives known as esters are also responsible for the sweet and pleasant odors and tastes of our food. We frequently use these compounds as artificial flavors in foods in place of more expensive natural extracts. In biochemistry esterlike molecules act as the energy carriers in many cells. As you will see, the properties of carboxylic acids are quite distinct from those of aldehydes and alcohols.

24.1 Carboxylic Acids

The functional group of the carboxylic acids is called a **carboxyl group** and is represented in the following ways:

carboxyl group

$$\overset{O}{\underset{}{\underset{\parallel}{-C}}}-OH \quad \text{or} \quad -COOH \quad \text{or} \quad -CO_2H$$

Carboxylic acids can be either aliphatic or aromatic:

$$\overset{O}{\underset{RC}{\parallel}}-OH \qquad \overset{O}{\underset{CH_3C}{\parallel}}-OH \qquad \overset{O}{\underset{ArC}{\parallel}}-OH$$

aliphatic

aromatic

"R" is often used to symbolize an aliphatic group; "Ar" symbolizes an aromatic group.

24.2 Nomenclature and Sources of Aliphatic Carboxylic Acids

Aliphatic carboxylic acids form a homologous series. The carboxyl group is always at the beginning of a carbon chain, and the carbon atom in this group is understood to be carbon number 1 when the compound is named.

IUPAC Rules for Naming Carboxylic Acids

1. To establish the parent name, identify the longest carbon chain that includes the carboxyl group.
2. Drop the final *e* from the corresponding hydrocarbon name.
3. Add the suffix *-oic* acid.

Thus the names corresponding to the one-, two-, and three-carbon acids are methanoic acid, ethanoic acid, and propanoic acid. These names are derived from methane, ethane, and propane.

CH_4	methane	HCOOH	methanoic acid
CH_3CH_3	ethane	CH_3COOH	ethanoic acid
$CH_3CH_2CH_3$	propane	CH_3CH_2COOH	propanoic acid

Other groups bonded to the parent chain are numbered and named as we have done previously. For example,

$$\overset{5}{C}H_3\overset{4}{C}H_2\overset{3}{C}H\overset{2}{C}H_2\overset{1}{C}OOH$$
$$|$$
$$CH_3$$

3-methylpentanoic acid

▲
The sting of an ant bite is caused by formic acid.

Common and IUPAC nomenclature systems should not be intermixed.

Unfortunately, the IUPAC method is neither the only nor the most used method for naming acids. Organic acids are usually known by common names. Methanoic, ethanoic, and propanoic acids are called formic, acetic, and propionic acids, respectively. These names often refer to a natural source of the acid and are not systematic. Formic acid gets its name from the Latin word *formica,* meaning ant. This acid contributes to the stinging sensation of ant bites. Acetic acid is found in vinegar and gets its name from the Latin word for vinegar. The name butyric acid comes from the Latin term for butter, since it is a constituent of butterfat. The 6-, 8-, and 10-carbon acids are found in goat fat; their names are derived from the Latin word for goat. These three acids—caproic, caprylic, and capric—along with butyric acid have characteristic and disagreeable odors. In a similar way the names of the 12-, 14-, and 16-carbon acids—lauric, myristic, and palmitic—come from plants from which the corresponding acid has been isolated. The name stearic acid comes from a Greek word meaning beef fat or tallow, which is a good source of this acid. Many of the carboxylic acids, principally those with even numbers of carbon atoms ranging from 4 to about 20, exist in combined form in plant and animal fats. These are called *fatty acids* (see Chapter 28). Table 24.1 lists the common and IUPAC names, together with some of the physical properties, of the more important saturated aliphatic acids.

Another nomenclature method uses letters of the Greek alphabet ($\alpha, \beta, \gamma, \delta, \ldots$) to name certain acid derivatives, especially hydroxy, amino, and halogen acids. When Greek letters are used, the carbon atoms, beginning with the one adjacent to the carboxyl group, are labeled $\alpha, \beta, \gamma, \delta, \ldots$. When numbers are used (the IUPAC System), the numbers begin with the carbon in the $-COOH$ group.

$$\underset{5}{\overset{\delta}{C}}-\underset{4}{\overset{\gamma}{C}}-\underset{3}{\overset{\beta}{C}}-\underset{2}{\overset{\alpha}{C}}-\underset{1}{\overset{O}{\overset{\|}{C}}}-OH$$

TABLE 24.1 Names, Formulas, and Physical Properties of Saturated Aliphatic Carboxylic Acids

Common name (IUPAC name)	Formula	Melting point (°C)	Boiling point (°C)	Solubility in water[a]
Formic acid (methanoic acid)	HCOOH	8.4	100.8	∞
Acetic acid (ethanoic acid)	CH₃COOH	16.6	118	∞
Propionic acid (propanoic acid)	CH₃CH₂COOH	−21.5	141.4	∞
Butyric acid (butanoic acid)	CH₃(CH₂)₂COOH	−6	164	∞
Valeric acid (pentanoic acid)	CH₃(CH₂)₃COOH	−34.5	186.4	3.3
Caproic acid (hexanoic acid)	CH₃(CH₂)₄COOH	−3.4	205	1.1
Caprylic acid (octanoic acid)	CH₃(CH₂)₆COOH	16.3	239	0.1
Capric acid (decanoic acid)	CH₃(CH₂)₈COOH	31.4	269	Insoluble
Lauric acid (dodecanoic acid)	CH₃(CH₂)₁₀COOH	44.1	225[b]	Insoluble
Myristic acid (tetradecanoic acid)	CH₃(CH₂)₁₂COOH	54.2	251[b]	Insoluble
Palmitic acid (hexadecanoic acid)	CH₃(CH₂)₁₄COOH	63	272[b]	Insoluble
Stearic acid (octadecanoic acid)	CH₃(CH₂)₁₆COOH	69.6	287[b]	Insoluble
Arachidic acid (eicosanoic acid)	CH₃(CH₂)₁₈COOH	77	298[b]	Insoluble

[a]Grams of acid per 100 g of water.
[b]Boiling point is given at 100 mm Hg pressure instead of atmospheric pressure because thermal decomposition occurs before this acid reaches its boiling point at atmospheric pressure.

$$CH_3CH_2CHCOOH \qquad CH_3CHCOOH \qquad CH_2ClCH_2COOH$$
$$\quad\quad\;\; | \qquad\qquad\qquad\; |$$
$$\quad\quad\; OH \qquad\qquad\quad\; NH_2$$

Common name: α-hydroxybutyric acid   α-aminopropionic acid   β-chloropropionic acid

IUPAC name: 2-hydroxybutanoic acid   2-aminopropanoic acid   3-chloropropanoic acid

Write formulas for the following:
   (a) 3-chloropentanoic acid   (b) γ-hydroxybutyric acid   (c) phenylacetic acid

Example 24.1

(a) This is an IUPAC name, as indicated by the use of position number. *Pentanoic* indicates a five-carbon acid. Substituted on carbon 3 is a chlorine atom. Write five carbon atoms in a row. Make carbon 1 a carboxyl group, place a Cl on carbon 3, and add hydrogens to give each carbon four bonds. The formula is

Solution

$$CH_3CH_2CHClCH_2COOH$$

(b) This is a common name as indicated by the Greek letter. *Butyric* indicates a four-carbon acid. The γ-position is three carbons removed from the carboxyl group. Therefore, the formula is

$$\overset{\gamma\text{-carbon}}{HO-CH_2CH_2CH_2COOH}$$

(c) *Acetic acid* (a common name) is the familiar two-carbon acid. There is only one place to substitute the phenyl group and still call the compound an acid—that is, at the CH_3 group. Substitute a phenyl group for one of the three H atoms to give the formula:

Practice 24.1

Write formulas for (a) 2-methylpropanoic acid, (b) β-chlorocaproic acid, and (c) cyclohexanecarboxylic acid.

Practice 24.2 Application to Biochemistry

Uncontrolled diabetics have high blood levels of the following carboxylic acid. Give the IUPAC name.

$$\overset{OH}{\underset{|}{CH_3-CH}}-CH_2-\overset{O}{\overset{||}{C}}-OH$$

24.3   Physical Properties of Carboxylic Acids

Each aliphatic carboxylic acid molecule is polar and consists of a carboxyl group and a hydrocarbon group (R). These two unlike parts have great bearing on the physical, as well as chemical, behavior of the molecule as a whole. The first four acids, formic through butyric, are completely soluble (miscible) in water (Table 24.1). Beginning with pentanoic acid (valeric acid), the water solubility falls sharply and is only about 0.1 g of acid per 100 g of water for octanoic acid (caprylic acid). Acids of this series with more than eight carbons are virtually insoluble in water. The water-solubility characteristics of the first four acids are determined by the highly soluble polar carboxyl group. Thereafter, the water insolubility of the nonpolar hydrocarbon chain is dominant.

The polarity due to the carboxyl group is evident in the boiling-point data. Formic acid (HCOOH) boils at about 101°C. Carbon dioxide, a nonpolar substance of similar molar mass, remains in the gaseous state until it is cooled to −78°C. In like manner, the boiling point of acetic acid (molar mass 60 g/mol)is 118°C, whereas nonpolar butane (molar mass 58 g/mol) boils at −0.6°C. The comparatively high

▲
Space-filling models of acetic acid (top) and octanoic acid (bottom).

boiling points for carboxylic acids are due to intermolecular attractions resulting from hydrogen bonding. In fact, molar-mass determinations on gaseous acetic acid (near its boiling point) show a value of about 120 g/mol indicating that two molecules are joined together to form a *dimer*, $(CH_3COOH)_2$.

hydrogen bonding in   acetic acid dimer
carboxylic acids

Saturated monocarboxylic acids that have fewer than ten carbon atoms are liquids at room temperature, whereas those with more than ten carbon atoms are waxlike solids.

Carboxylic acids and phenols, like mineral acids such as HCl, ionize in water to produce hydronium ions and anions (Chapter 15). Carboxylic acids are generally weak acids; that is, they are only slightly ionized in water. Phenols are, in general, even weaker acids than carboxylic acids. For example, the ionization constant for acetic acid is 1.8×10^{-5} (1.3% ionized for a 0.10 M solution) and that for phenol is 1.3×10^{-10} (0.00036% ionized for a 0.10 M solution). Equations illustrating these ionizations are given here:

HCl + H_2O $\longrightarrow$ H_3O^+ + Cl^-

hydrogen chloride   hydronium ion   chloride ion

$CH_3\overset{O}{\overset{\|}{C}}{-}OH$ + H_2O $\rightleftharpoons$ H_3O^+ + $CH_3\overset{O}{\overset{\|}{C}}{-}O^-$

acetic acid   hydronium ion   acetate ion

phenol   hydronium ion   phenoxide ion

A weak monopratic acid (HA) ionizes according to the following equilibrium:
HA $\rightleftharpoons$ H$^+$ + A$^-$
The acid equilibrium constant expression is
$$K_a = \frac{[H^+][A^-]}{[HA]}$$

Carboxylic acids are very common in biochemistry. When these molecules ionize, both the carboxylate anion and the H^+ can have significant effects. For example, if the liver releases too many carboxylic acids, the blood becomes too acidic. This happens during uncontrolled diabetes, and the resulting ketoacidosis can be deadly.

Under biological conditions, most carboxylic acids form anions. In turn, living cells respond to changes in ionic charge. A good example starts with a positively charged molecule, dopamine. When dopamine enters the brain, it eases the effects of Parkinson's disease, a chronic disease of the central nervous system. Unfortunately, dopamine is not able to cross the blood–brain barrier. So, scientists added a carboxylic acid group to dopamine, creating Dopa. Dopa contains both a negative and a positive charge when it enters the blood. The charges neutralize each other, and Dopa thus passes through the blood–brain barrier. Once into the brain, Dopa is converted to dopamine, which alleviates the symptoms of Parkinson's disease.

24.4   Classification of Carboxylic Acids

Thus far our discussion has dealt mainly with a single type of acid—that is, saturated monocarboxylic acids. But various other kinds of carboxylic acids are known. We discuss some of the more important ones here.

Unsaturated Carboxylic Acids

An unsaturated acid contains one or more carbon–carbon double bonds. The first member of the homologous series of unsaturated carboxylic acids, containing one carbon–carbon double bond, is acrylic acid, $CH_2=CHCOOH$. The IUPAC name for $CH_2=CHCOOH$ is propenoic acid. Derivatives of acrylic acid are used to manufacture a class of synthetic polymers known as the acrylates (see Chapter 21). These polymers are widely used as textiles and in paints and lacquers. Unsaturated carboxylic acids undergo the reactions of both an unsaturated hydrocarbon and a carboxylic acid.

Even one carbon–carbon double bond in the molecule exerts a major influence on the physical and chemical properties of an acid. The effect of a double bond can be seen when comparing the two 18-carbon acids, stearic and oleic. Stearic acid, $CH_3(CH_2)_{16}COOH$, a solid that melts at 70°C, shows only the reactions of a carboxylic acid. On the other hand, oleic acid,

$$CH_3(CH_2)_7CH=CH(CH_2)_7COOH \text{ (mp 16°C)}$$

with one double bond, is a liquid at room temperature and shows the reactions of an unsaturated hydrocarbon as well as those of a carboxylic acid.

Aromatic Carboxylic Acids

In an aromatic carboxylic acid, the carbon of the carboxyl group ($-COOH$) is bonded directly to a carbon in an aromatic ring. The parent compound of this series is benzoic acid. Other common examples are the three isomeric toluic acids:

benzoic acid   o-toluic acid   m-toluic acid   p-toluic acid

Dicarboxylic Acids

Acids of both the aliphatic and aromatic series that contain two or more carboxyl groups are known. Those with two $-COOH$ groups are called dicarboxylic acids. The simplest member of the aliphatic series is oxalic acid. The next member in the

TABLE **24.2**	Names and Formulas of Selected Dicarboxylic Acids	
Common name[a]	**IUPAC name**	**Formula**
Oxalic acid	Ethanedioic acid	HOOCCOOH
Malonic acid	Propanedioic acid	HOOCCH$_2$COOH
Succinic acid	Butanedioic acid	HOOC(CH$_2$)$_2$COOH
Glutaric acid	Pentanedioic acid	HOOC(CH$_2$)$_3$COOH
Adipic acid	Hexanedioic acid	HOOC(CH$_2$)$_4$COOH
Pimelic acid	Heptanedioic acid	HOOC(CH$_2$)$_5$COOH
Fumaric acid	*trans*-2-Butenedioic acid	HOOCCH=CHCOOH
Maleic acid	*cis*-2-Butenedioic acid	HOOCCH=CHCOOH

[a]A mnemonic for remembering the common names of the saturated dicarboxylic acid uses the first letter of the acid name:
Oh My Such Good Apple Pie

homologous series is malonic acid. Several dicarboxylic acids and their names are listed in Table 24.2.

The IUPAC names for dicarboxylic acids are formed by modifying the corresponding hydrocarbon names to end in *-dioic acid.* Thus the two-carbon acid is ethanedioic acid (derived from ethane). However, the common names for dicarboxylic acids are frequently used.

Oxalic acid is found in various plants including spinach, cabbage, and rhubarb. Among its many uses are bleaching straw and leather and removing rust and ink stains. Although oxalic acid is poisonous, the amounts present in the previously mentioned vegetables are usually not harmful.

Malonic acid is made synthetically but was originally prepared from malic acid, which is commonly found in apples and many fruit juices. Malonic acid is one of the major compounds used in the manufacture of the class of drugs known as barbiturates. When heated above their melting points, malonic acid and substituted malonic acids lose carbon dioxide to give monocarboxylic acids. Thus malonic acid yields acetic acid when strongly heated:

$$\underset{\text{malonic acid}}{\begin{array}{c} \text{COOH} \\ | \\ \text{CH}_2 \\ | \\ \text{COOH} \end{array}} \xrightarrow{150°C} \underset{\text{acetic acid}}{\text{CH}_3\text{COOH}} + \text{CO}_2(g)$$

Malonic acid is the biological precursor for the synthesis of fatty acids. In living cells, when malonic acid loses carbon dioxide, the acetic acid units are linked together to begin forming long-chain fatty acids.

Succinic acid has been known since the 16th century, when it was obtained as a distillation product of amber. Succinic, fumaric, and citric acids are among the important acids in the energy-producing metabolic pathway known as the citric acid cycle (see Chapter 34). Citric acid is a tricarboxylic acid that is widely distributed in

This South African vendor is selling fresh spinach, which contains oxalic acid in addition to important vitamins. ▶

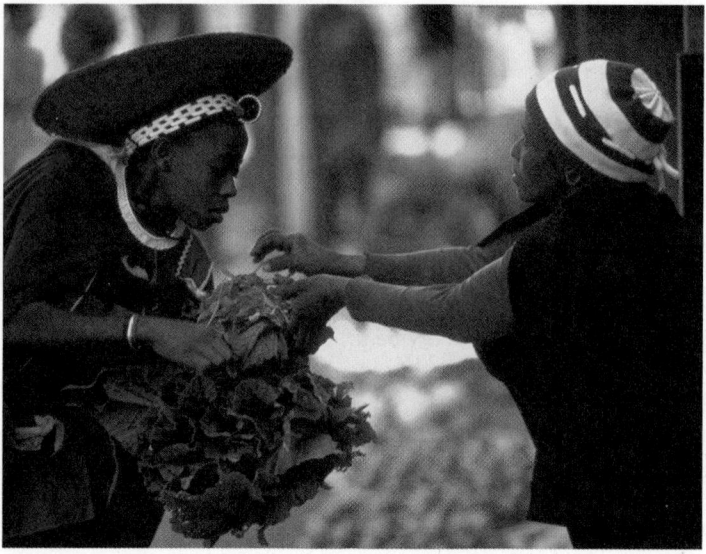

plant and animal tissue, especially in citrus fruits (lemon juice contains 5–8%). The formula for citric acid is

$$CH_2COOH$$
$$|$$
$$HO-C-COOH$$
$$|$$
$$CH_2COOH$$

citric acid

When succinic acid is heated, it loses water, forming succinic anhydride, an acid anhydride. Glutaric acid behaves similarly, forming glutaric anhydride, a six-membered ring.

succinic acid   succinic anhydride

Adipic acid is the most important commercial dicarboxylic acid. It is made from benzene by converting it first to cyclohexene and then by oxidation to adipic acid. About 8.2×10^8 kg of adipic acid is produced annually in the United States. Most of the adipic acid is used to produce nylon (Chapter 25). It is also used in polyurethane foams, plasticizers, and lubricating-oil additives.

Aromatic dicarboxylic acids contain two carboxyl groups attached directly to an aromatic ring.

Examples are the three isomeric phthalic acids, $C_6H_4(COOH)_2$:

o-phthalic acid
(phthalic acid)

m-phthalic acid
(isophthalic acid)

p-phthalic acid
(terephthalic acid)

Dicarboxylic acids are *bifunctional;* that is, they have two sites where reactions can occur. Therefore, they are often used as monomers in the preparation of synthetic polymers such as Dacron polyester (Section 24.9).

Hydroxy Acids

Lactic acid, found in sour milk, sauerkraut, and dill pickles, has the functional groups of both a carboxylic acid and an alcohol. Lactic acid is the end product when our muscles use glucose for energy in the absence of oxygen, a process called *glycolysis* (see Chapter 34). Salicylic acid is both a carboxylic acid and a phenol. It is of special interest because a family of useful drugs—the salicylates—are derivatives of this acid. The salicylates include aspirin and function as *analgesics* (pain relievers) and as *antipyretics* (fever reducers). The structural formulas of several hydroxy acids are given here:

lactic acid
(2-hydroxypropanoic acid)

malic acid
(2-hydroxybutanedioic acid)

salicylic acid
(o-hydroxybenzoic acid)

tartaric acid
(2,3-dihydroxybutanedioic acid)

Amino Acids

Naturally occuring amino acids have this general formula; the amino group is in the alpha position:

Each amino acid molecule has a carboxyl group that acts as an acid and an amino group that acts as a base. About 20 biologically important amino acids, each with a different group represented by R, are found in nature. (In amino acids, R does not always represent an alkyl group.) The immensely complicated protein molecules, found in every form of life, are built from amino acids. Some protein molecules contain more than 10,000 amino acid units. Amino acids and proteins are discussed in more detail in Chapter 29.

Practice 24.3

Write a structural formula for the smallest molecule of each of the following acids:
(a) a carboxylic acid, (b) an α-amino acid, (c) an α-hydroxy acid,
(d) a β-bromo acid, (e) a dicarboxylic acid, (f) an aromatic carboxylic acid

Practice 24.4 Application to Biochemistry

The food additive monosodium glutamate (MSG) is derived from 2-aminopentanedioic acid. Write the structural formula for this acid.

24.5   Preparation of Carboxylic Acids

Many different methods of preparing carboxylic acids are known. We consider only a few examples.

Oxidation of an Aldehyde or a Primary Alcohol This general method is used to convert an aldehyde or primary alcohol to the corresponding carboxylic acid:

$$RCH_2OH \xrightarrow{[O]} RCOOH$$

Butyric acid (butanoic acid) can be obtained by oxidizing either 1-butanol or butanal with potassium dichromate in the presence of sulfuric acid. Aromatic acids may be prepared by the same general method. For example, benzoic acid is obtained by oxidizing benzyl alcohol.

$$CH_3CH_2CH_2CH_2OH \xrightarrow[\substack{H_2SO_4 \\ \Delta}]{Cr_2O_7^{2-}} CH_3CH_2CH_2COOH$$

1-butanol   butanoic acid

$$CH_3CH_2CH_2\overset{\overset{\displaystyle O}{\displaystyle \|}}{C}-H \xrightarrow[\substack{H_2SO_4 \\ \Delta}]{Cr_2O_7^{2-}} CH_3CH_2CH_2COOH$$

butanal   butanoic acid

benzyl alcohol   benzoic acid

Carboxylic acids can also be obtained by the hydrolysis or saponification of esters (see Section 24.10).

Practice 24.5

Name the compound formed when (a) 1-hexanol and (b) 2-hexanol are oxidized with $K_2Cr_2O_7/H_2SO_4$.

Oxidation of Alkyl Groups Attached to Aromatic Rings When reacted with a strong oxidizing agent (alkaline permanganate solution or potassium dichromate and sulfuric acid), alkyl groups bonded to aromatic rings are oxidized to carboxyl groups. Regardless of the size or length of the alkyl group, the carbon atom adjacent to the ring remains bonded to the ring and is oxidized to a carboxyl group. The remainder of the alkyl group goes either to carbon dioxide or to a salt of a carboxylic acid. Thus sodium benzoate is obtained when toluene, ethylbenzene, or propylbenzene is heated with alkaline permanganate solution:

toluene   sodium benzoate

ethylbenzene   sodium benzoate

propylbenzene   sodium benzoate   sodium acetate

Since the reaction is conducted in an alkaline medium, a salt of the carboxylic acid (sodium benzoate) is formed instead of the free acid. To obtain the free carboxylic acid, the reaction mixture is acidified with a strong mineral acid (HCl or H_2SO_4) in a second step.

sodium benzoate   benzoic acid

Practice 24.6

What aromatic hydrocarbon of formula C_9H_{12}, when oxidized with $NaMnO_4/NaOH$, will give (a) benzoic acid and (b) o-phthalic acid?

Hydrolysis of Nitriles Nitriles, RCN, which can be prepared by adding HCN to aldehydes and ketones (Section 23.4) or by reacting alkyl halides with KCN, can be hydrolyzed to carboxylic acids:

$$\underset{\text{alkyl halide}}{RX} + KCN \longrightarrow \underset{\text{a nitrile}}{RCN} + KX$$

$$RCN + 2\,H_2O \xrightarrow{\ H^+\ } RCOOH + NH_4^+$$

$$CH_3CN + 2\,H_2O \xrightarrow{\ H^+\ } CH_3COOH + NH_4^+$$

Practice 24.7

Ethanal is reacted with HCN. Write equations for the nitrile addition product and the subsequent hydrolysis to the acid. Name the acid.

The biological importance of carboxylic acids has resulted in numerous biochemical syntheses. Most living cells form carboxylic acids by oxidation of aldehydes—a common reaction in organic chemistry. In contrast, the vitamin biotin provides a unique synthesis. This vitamin traps and then transfers carbon dioxide to an acceptor molecule to form a carboxylic acid. Malonic acid (see Section 24.4) is formed using biotin, and the synthesis of fatty acids (see Section 24.2) depends on biotin. In fact, biotin is essential for human good health.

24.6   Chemical Properties of Carboxylic Acids

The structural formula of the carboxyl group consists of a carbonyl group and a hydroxyl group in combination. From such a combination, carboxylic acids might well be expected to have reactions characteristic of both aldehydes and alcohols. But such is not the case; carboxylic acids have their own unique properties.

Hybrids are not uncommon in our world. The nectarines available in our markets are large, sweet, firm-fleshed fruits. The shopper may not be aware that the nectarine is a hybrid created from a peach and a plum because this hybrid does not have the flavor and texture characteristics of either parent fruit. In the same way, the carboxyl group is a new combination of atoms—a new functional group with properties different from those of both the carbonyl and hydroxyl groups.

The carboxyl group is the most acidic of any organic functional group discussed so far. The —OH group acts as a proton donor, even more effectively than phenol.

The carboxyl group commonly is involved in substitution-type reactions. The C=O retains its double-bond character while the —OH is replaced by another group or atom. Thus carboxylic acids react with many substances to produce derivatives. In these reactions the hydroxyl group is replaced by a halogen (—Cl), an acyloxy group (—OOCR), an alkoxy group (—OR), or an amino group (—NH_2). The general reactions are summarized here:

Substitution reactions require an exchange of atoms or groups of atoms:
$$A—X + Y \rightarrow A—Y + X$$

$$\underset{R—C—OH}{\overset{O}{\parallel}} \xrightarrow{-Cl} \underset{R—C—Cl}{\overset{O}{\parallel}} \quad \text{(acyl halide)}$$

$$\underset{R—C—OH}{\overset{O}{\parallel}} \xrightarrow{-OOCR'} \underset{R—C—O—C—R'}{\overset{O \qquad O}{\parallel \qquad \parallel}} \quad \text{(acid anhydride)}$$

$$\underset{R—C—OH}{\overset{O}{\parallel}} \xrightarrow{-OR'} \underset{R—C—OR'}{\overset{O}{\parallel}} \quad \text{(ester)}$$

$$\underset{R—C—OH}{\overset{O}{\parallel}} \xrightarrow{-NH_2} \underset{R—C—NH_2}{\overset{O}{\parallel}} \quad \text{(amide)}$$

Acid–Base Reactions Because of their ability to form hydrogen ions in solution, acids in general have the following properties:

1. Sour taste
2. Change blue litmus to red and affect other suitable indicators
3. Form water solutions with pH values of less than 7
4. Undergo neutralization reactions with bases to form water and a salt

All of the foregoing general properties of an acid are readily seen in low-molar-mass carboxylic acids such as acetic acid. However, these general properties are greatly influenced by the size of the hydrocarbon chain attached to the carboxyl group. In stearic acid, for example, taste, effect on indicators, and pH are not detectable because the long hydrocarbon chain makes the acid insoluble in water. But stearic acid reacts with a base to form water and a salt. With sodium hydroxide, the equation for the reaction is

$$\underset{\text{stearic acid}}{C_{17}H_{35}COOH} + NaOH \longrightarrow \underset{\text{sodium stearate}}{C_{17}H_{35}COONa} + H_2O$$

The salts formed from this neutralization reaction have different properties than the acids. Sodium and potassium salts are soluble in water and dissociate completely in solution. These properties assist in the separation of carboxylic acids from other nonpolar compounds. For example, when a base like NaOH is added to a reaction mixture containing a carboxylic acid, the carboxylic acid reacts, forming its sodium salt and water. The salt dissolves in water while the remaining nonpolar molecules stay in the organic layer. Once the layers are separated, some mineral acid like HCl can be added to the carboxylic acid salt to re-form and to recover the acid.

Carboxylic acids generally react with sodium hydrogen carbonate to release carbon dioxide. This reaction can be used to distinguish a carboxylic acid from a phenol (also a weak acid). Phenols do not react with sodium hydrogen carbonate, although they will be neutralized by a strong base.

Acid Chloride Formation Thionyl chloride ($SOCl_2$) reacts with carboxylic acids to form acid chlorides. In this substitution reaction, a chlorine atom replaces an $-OH$ group.

$$\underset{\text{acid}}{RC\!-\!OH} + \underset{\substack{\text{thionyl}\\\text{chloride}}}{SOCl_2} \longrightarrow \underset{\text{acid chloride}}{RC\!-\!Cl} + SO_2 + HCl$$

$$CH_3C\!-\!OH + SOCl_2 \longrightarrow \underset{\text{acetyl chloride}}{CH_3C\!-\!Cl} + SO_2 + HCl$$

Acid chlorides undergo substitution reactions.

Acid chlorides are extremely reactive substances. They must be kept away from moisture, or they will hydrolyze back to the carboxylic acid:

$$RC\!-\!Cl + H_2O \longrightarrow RC\!-\!OH + HCl$$

Acid chlorides are more reactive than acids and can be used in place of acids to prepare esters and amides:

$$CH_3C\!-\!Cl + CH_3OH \longrightarrow \underset{\text{methyl acetate}}{CH_3C\!-\!OCH_3} + HCl$$

$$CH_3C\!-\!Cl + 2\,NH_3 \longrightarrow \underset{\text{acetamide}}{CH_3C\!-\!NH_2} + NH_4Cl$$

Acid Anhydride Formation Inorganic anhydrides are formed by the elimination of a molecule of water from an acid or a base:

$$H_2SO_3 \longrightarrow SO_2 + H_2O$$

$$Ba(OH)_2 \xrightarrow{\Delta} BaO + H_2O$$

An organic anhydride is formed by the elimination of a molecule of water from two molecules of a carboxylic acid.

$$R\!-\!C\!-\!OH + HO\!-\!C\!-\!R' \longrightarrow R\!-\!C\!-\!O\!-\!C\!-\!R' + H_2O$$

The most commonly used organic anhydride is acetic anhydride. It can be prepared by the reaction of acetyl chloride with sodium acetate.

$$CH_3\overset{\overset{O}{\|}}{C}-Cl \;+\; Na^{+\;-}O-\overset{\overset{O}{\|}}{C}-CH_3 \longrightarrow CH_3\overset{\overset{O}{\|}}{C}-O-\overset{\overset{O}{\|}}{C}-CH_3 \;+\; NaCl$$

acetic anhydride

Acid anhydrides are very reactive and can be used to synthesize amides and esters, although not as often as the acid chlorides in organic synthesis. In living cells, acid anhydrides commonly activate carboxylic acids for further reaction.

The "energy currency" for living cells is the acid anhydride (see Chapter 33).

Ester Formation   An **ester** is an organic compound formed by the reaction of an acid and an alcohol or a phenol. Water is also a product in this reaction.

ester

$$RC-\boxed{OH} \;+\; R'O\boxed{H} \;\overset{H^+}{\rightleftharpoons}\; RC-OR' \;+\; H_2O$$

carboxylic acid   alcohol   ester
(R can be H, alkyl, or Ar,
but R' cannot be H)

$$H-\overset{\overset{O}{\|}}{C}-OH \;+\; CH_3CH_2OH \;\overset{H^+}{\rightleftharpoons}\; H-\overset{\overset{O}{\|}}{C}-OCH_2CH_3 \;+\; H_2O$$

formic acid   ethanol   ethyl formate

At first glance, this looks like the familiar acid–base neutralization reaction. But this is not the case, because the alcohol does not yield OH^- ions, and the ester, unlike a salt, is a molecular, not an ionic, substance. The forward reaction of an acid and an alcohol is called *esterification;* the reverse reaction of an ester with water is called *hydrolysis.* Chemists frequently modify reaction conditions to favor the formation of either esters or their component parts, alcohols and acids.

Esterification is one of the most important reactions of carboxylic acids. Many biologically significant substances are esters.

24.7   Nomenclature of Esters

The general formula for an ester is RCOOR', where R may be a hydrogen, alkyl group, or aryl group, and R' may be an alkyl group, or aryl group, but *not* a hydrogen. Esters are found throughout nature. The ester linkage is particularly important in the study of fats and oils, both of which are esters. Esters of phosphoric acid are of vital importance to life as well.

Esters are alcohol derivatives of carboxylic acids. They are named in much the same way as salts. The alcohol part is named first, followed by the name of the acid modified to end in *-ate.* The *-ic* ending of the organic acid name is replaced by the ending *-ate.* Thus in the IUPAC System *ethanoic acid* becomes *ethanoate.* In the common names, *acetic acid* becomes *acetate.* To name an ester, it is necessary to recognize the portion of the ester molecule that comes from the acid and the portion

The odor of pineapple is due to the ester ethyl butyrate, and the odor of ripe bananas is due to isoamyl acetate. ▶

that comes from the alcohol. In the general formula for an ester, the $RC=O$ comes from the acid, and the $R'O$ comes from the alcohol:

Ester name: methyl ethanoate or methyl acetate

The R' in $R'O$ is named first, followed by the name of the acid modified by replacing *-ic* acid with *-ate*. The ester derived from ethyl alcohol and acetic acid is called ethyl acetate or ethyl ethanoate. Consider the ester formed from CH_3CH_2COOH and CH_3OH:

$$CH_3CH_2C\overset{O}{\underset{}{\|}}\boxed{\text{—OH}} + H\text{—OCH}_3 \overset{H^+}{\rightleftharpoons} CH_3CH_2C\overset{O}{\underset{}{\|}}\text{—OCH}_3 + H_2O$$

propanoic acid   methanol   methyl propanoate
(propionic acid)   (methyl alcohol)   (methyl propionate)

Esters of aromatic acids are named in the same general way as those of aliphatic acids. For example, the ester of benzoic acid and isopropyl alcohol is

isopropyl benzoate

Formulas and names for additional esters are given in Table 24.3.

| TABLE 24.3 | Formulas, Names, and Odors of Selected Esters | | |

Formula	IUPAC name	Common name	Odor or flavor
$\underset{\text{CH}_3\overset{\displaystyle\text{O}}{\overset{\|}{\text{C}}}-\text{OCH}_2\text{CH}_2\overset{\overset{\displaystyle\text{CH}_3}{\|}}{\text{CH}}\text{CH}_3}{}$	Isopentyl ethanoate	Isoamyl acetate	Banana, pear
$\text{CH}_3\text{CH}_2\text{CH}_2\overset{\displaystyle\text{O}}{\overset{\|}{\text{C}}}-\text{OCH}_2\text{CH}_3$	Ethyl butanoate	Ethyl butyrate	Pineapple
$\text{HC}\overset{\displaystyle\text{O}}{\overset{\|}{}}-\text{OCH}_2\overset{\overset{\displaystyle}{\|}}{\text{CH}}\text{CH}_3$ CH_3	Isobutyl methanoate	Isobutyl formate	Raspberry
$\text{CH}_3\overset{\displaystyle\text{O}}{\overset{\|}{\text{C}}}-\text{OCH}_2(\text{CH}_2)_6\text{CH}_3$	Octyl ethanoate	Octyl acetate	Orange
benzene ring with $\overset{\displaystyle\text{O}}{\overset{\|}{\text{C}}}-\text{OCH}_3$ and OH	2-Hydroxymethylbenzoate	Methyl salicylate	Wintergreen

Name the following esters:

Example 24.2

Solution

(a) $\text{H}-\overset{\displaystyle\text{O}}{\overset{\|}{\text{C}}}-\text{OCH}_2\text{CH}_2\text{CH}_3$

(b) benzene ring $-\overset{\displaystyle\text{O}}{\overset{\|}{\text{C}}}-\text{OCH}_2\text{CH}_3$

(c) $\begin{array}{l}\text{O}=\text{C}-\text{OCH}_2\text{CH}_3\\ \quad|\\ \quad\text{CH}_2\\ \quad|\\ \text{O}=\text{C}-\text{OCH}_2\text{CH}_3\end{array}$

(a) First identify the acid and alcohol components. The acid contains one carbon and is formic acid. The alcohol is propyl alcohol.

$\text{H}-\overset{\displaystyle\text{O}}{\overset{\|}{\text{C}}}\overbrace{-\text{OCH}_2\text{CH}_2\text{CH}_3}$

formic acid   propyl alcohol

Change the -*ic* ending of the acid to -*ate*, making the name formate or methanoate. The name of the ester then is propyl formate or propyl methanoate.

(b) The acid is benzoic acid; the alcohol is ethyl alcohol. Using the same procedure as in part (a), the name of the ester is ethyl benzoate.

(c) The acid is the three-carbon dicarboxylic acid, malonic acid. The alcohol is ethyl alcohol. Both acid groups are in the ester form. The name, therefore, is diethyl malonate.

Practice 24.8

Name the following esters:

24.8   Occurrence and Physical Properties of Esters

Since many acids and many alcohols are known, the number of esters theoretically possible is very large. In fact, both natural and synthesized esters exist in almost endless variety. Simple esters derived from monocarboxylic acids and monohydroxy alcohols are colorless, generally nonpolar liquids or solids. The low polarity of ester molecules is substantiated by the fact that both their water solubility and boiling points are lower than those of either acids or alcohols of similar molar masses.

Low- and intermediate-molar-mass esters (from both acids and alcohols up to about ten carbons) are liquids with characteristic (usually fragrant or fruity) odors. The distinctive odor and flavor of many fruits are caused by one or more of these esters. The difference in properties between an acid and its esters is remarkable. For example, in contrast to the extremely unpleasant odor of butyric acid, ethyl butyrate has the pleasant odor of pineapple and methyl butyrate the odor of rum. Esters are used in flavoring and scenting agents (see Table 24.3). They are generally good solvents for organic substances, and those that have relatively low molar masses are volatile. Therefore, esters such as ethyl acetate, butyl acetate, and isoamyl acetate are extensively used in paints, varnishes, and lacquers.

High-molar-mass esters (formed from acids and alcohols of 16 or more carbons) are waxes and are obtained from various plants. They are used in furniture wax and automobile wax preparations; for example, carnauba wax contains esters of 24- and 28-carbon fatty acids and 32- and 34-carbon alcohols. Polyesters with very high molar masses, such as Dacron, are widely used in the textile industries.

24.9 Polyesters: Condensation Polymers

Condensation polymers are formed by substitution reactions between neighboring monomers. The polyesters are joined by ester linkages between carboxylic acid and alcohol groups; the macromolecule formed may be linear or cross-linked. A linear polyester is obtained from the bifunctional monomers terephthalic acid and ethylene glycol. Esterification occurs between the alcohol and acid groups on both ends of both monomers, forming long-chain macromolecules:

Diacid + Diol $\longrightarrow$ Polyester

$$HOOC(CH_2)_xCOOH + HO(CH_2)_yOH \longrightarrow \left[\begin{matrix} C(CH_2)_xC-O(CH_2)_yO \\ \| \quad\quad \| \\ O \quad\quad O \end{matrix} \right]_n$$

$$HOOC-\langle\bigcirc\rangle-COOH \quad HOCH_2CH_2OH \quad \text{(monomers)}$$

terephthalic acid   ethylene glycol

$$-OCH_2CH_2O\left[\begin{matrix} O \\ \| \\ C \end{matrix} -\langle\bigcirc\rangle- \begin{matrix} O \\ \| \\ C \end{matrix} -OCH_2CH_2O \right]_n \begin{matrix} O \\ \| \\ C \end{matrix} -\langle\bigcirc\rangle- \begin{matrix} O \\ \| \\ C \end{matrix} -$$

This polymer may be drawn into fibers or formed into transparent films of great strength. Dacron and Terylene synthetic textiles and Mylar films are made from this polyester, as is the familiar 2-L soft-drink bottle. The symbol on objects made of this material is shown in the margin. In actual practice the dimethyl ester of terephthalic acid is used, and the molecule split out is methyl alcohol instead of water.

PETE

When trifunctional acids or alcohols are used as monomers, cross-linked thermosetting polyesters are obtained (see Chapter 21). One common example is the reaction of glycerol and *o*-phthalic acid. The polymer formed is one of a group of polymers known as alkyd resins. Glycerol has three functional —OH groups, and phthalic acid has two functional —COOH groups:

$$\begin{matrix} OH \\ | \\ HOCH_2CHCH_2OH \end{matrix} \qquad \langle\bigcirc\rangle \begin{matrix} COOH \\ \\ COOH \end{matrix}$$

glycerol   *o*-phthalic acid

A cross-linked macromolecular structure is formed that, with modifications, has proved to be one of the most outstanding materials used in the coatings industry.

24.10 Chemical Properties of Esters

The most important reaction of esters is *hydrolysis*. Hydrolysis is the splitting of molecules through the addition of water. The majority of organic and biochemical substances react only very slowly, if at all, with water. In order to increase the rate of

CHEMISTRY IN ACTION   Aspirin

From the earliest days of medicine, people have obtained pain relief by chewing willow bark. In 1840, the active compound was isolated from the bark and identified as salicylic acid. Unfortunately, salicylic acid has several undesirable side effects, including a very sour taste and irritation of the stomach lining.

In 1883, an organic chemist reacted salicylic acid with acetic anhydride to form the ester acetylsalicylic acid.

acetylsalicylic acid

The chemist happened to work for the Bayer Company, and the drug was aspirin.

Aspirin is the most widely used drug in the world and for good reason: it acts as a fever reducer (antipyretic), a pain reliever (analgesic), and an anti-inflammatory agent. In large doses (lethal is between 30 and 40 g), it is also a poison. Tablets are manufactured by mixing 0.32 g of aspirin with an inert binder (often starch) to hold the tablet together. Approximately half of the aspirin manufactured in the United States (more than 2.0×10^7 kg) is made into aspirin tablets. The remainder is used in combination pain relievers and cold remedies.

Until relatively recently it was not understood how aspirin acted within the body. Chemists have now determined that aspirin inhibits an enzyme necessary for the synthesis of prostaglandins (Chapter 28). Functions of prostaglandins include elevation of the blood pressure, tissue inflammation, and activation of the pain receptors in tissues. A reduction in the synthesis of prostaglandins produces both analgesic and anti-inflammatory effects.

Aspirin usage can have undesirable side effects as it can irritate the stomach lining, inhibit blood clotting, and prolong labor in childbirth. It is also associated with the development of Reyes syndrome, a brain disease that occurs in children recovering from chicken pox or flu. Some people are allergic to aspirin.

Publicity over Reyes syndrome and allergic reactions resulted in the develop-

ment of alternatives to aspirin. The most common alternatives are acetaminophen and ibuprofen. Acetaminophen (Tylenol) acts an an analgesic and antipyretic but is not anti-inflammatory. Ibuprofen (Advil, Motrin), a fairly recent addition to the nonprescriptive drug category, acts as a prostaglandin inhibitor in a manner similar to aspirin.

acetaminophen

ibuprofen

Other esters of salicylic acid also show medicinal effects. Methyl salicylate is an oil with the odor of wintergreen. It is commonly found as the active ingredient in pain-relieving liniments. It has the unusual property of penetrating the skin surface where it hydrolyzes, releasing salicylic acid and relieving the pain. Phenyl salicylate is an ester that is not hydrolyzed by acids. It is used to coat pills in order to permit them to pass through the stomach to the intestine before hydrolyzing and releasing their contents.

these reactions, a catalyst is required. In the laboratory chemists often employ an acid or base as a catalyst for hydrolysis. In living systems enzymes act as catalysts.

Acid Hydrolysis

Hydrolysis of an ester involves reaction with water to form a carboxylic acid and an alcohol. The hydrolysis is catalyzed by strong acids (H_2SO_4 and HCl) or by certain enzymes.

$$\underset{\text{ester}}{RC\!-\!OR'} + H_2O \xrightarrow[\text{or enzyme}]{H^+} \underset{\text{acid}}{RC\!-\!OH} + \underset{\text{alcohol}}{R'OH}$$

(with C=O double bonds)

$$\underset{\text{methyl propanoate}}{CH_3CH_2C\!-\!OCH_3} + H_2O \xrightarrow{H^+} \underset{\text{propanoic acid}}{CH_3CH_2COOH} + \underset{\text{methanol}}{CH_3OH}$$

methyl salicylate $+ H_2O \xrightarrow{H^+}$ salicylic acid $+ CH_3OH$ methanol

Alkaline Hydrolysis (Saponification)

Saponification is the hydrolysis of an ester by a strong base (NaOH or KOH) to produce an alcohol and a salt (or soap if the salt formed is from a high-molar-mass acid): *saponification*

$$\underset{\text{ester}}{RC\!-\!OR'} + NaOH \xrightarrow[\Delta]{H_2O} \underset{\text{salt}}{RC\!-\!O^-Na^+} + \underset{\text{alcohol}}{R'OH}$$

$$\underset{\text{ethyl stearate}}{CH_3(CH_2)_{16}C\!-\!OCH_2CH_3} + NaOH \xrightarrow[\Delta]{H_2O} \underset{\text{sodium stearate}}{CH_3(CH_2)_{16}CONa} + CH_3CH_2OH$$

The carboxylic acid may be obtained by reacting the salt with a strong acid:

$$CH_3(CH_2)_{16}COONa + HCl \longrightarrow \underset{\text{stearic acid}}{CH_3(CH_2)_{16}COOH} + NaCl$$

Notice that in saponification the base is a reactant and not a catalyst.

Practice 24.9

Write equations showing (a) the acid hydrolysis and (b) the alkaline hydrolysis of isobutyl pentanoate.

24.11 Glycerol Esters

Fats and **oils** are esters of glycerol and predominantly long-chain fatty acids. Fats and oils are also called **triacylglycerols** or **triglycerides,** since each molecule is derived from one molecule of glycerol and three molecules of fatty acid. *fat, oil triacylglycerol triglyceride*

Glycerol portion →

General formula for a triacylglycerol	Typical triacylglycerol containing three different fatty acids

$$CH_2-O-\overset{\overset{\displaystyle O}{\|}}{C}-R$$

$$CH-O-\overset{\overset{\displaystyle O}{\|}}{C}-R'$$

$$CH_2-O-\overset{\overset{\displaystyle O}{\|}}{C}-R''$$

$$CH_2-O-\overset{\overset{\displaystyle O}{\|}}{C}-C_{17}H_{35} \quad \text{(Stearic)}$$

$$CH-O-\overset{\overset{\displaystyle O}{\|}}{C}-C_{15}H_{31} \quad \text{(Palmitic)}$$

$$CH_2-O-\overset{\overset{\displaystyle O}{\|}}{C}-C_{11}H_{23} \quad \text{(Lauric)}$$

The structural formulas of triacylglycerol molecules differ because

1. The length of the fatty acid chain varies from 4 to 20 carbons, but the number of carbon atoms in the chain is nearly always even.
2. Each fatty acid may be saturated or unsaturated and may contain one, two, or three carbon–carbon double bonds.
3. A triacylglycerol may, and frequently does, contain three different fatty acids.

The most abundant saturated fatty acids in fats and oils are lauric, myristic, palmitic, and stearic acids (Table 24.4). The most abundant unsaturated acids in fats and oils contain 18 carbon atoms and have one, two, or three carbon–carbon double bonds. In all of these naturally occurring unsaturated acids, the configuration about the double bond is cis. Their formulas are

$$CH_3(CH_2)_7CH\!=\!CH(CH_2)_7COOH$$

oleic acid

$$CH_3(CH_2)_4CH\!=\!CHCH_2CH\!=\!CH(CH_2)_7COOH$$

linoleic acid

$$CH_3CH_2CH\!=\!CHCH_2CH\!=\!CHCH_2CH\!=\!CH(CH_2)_7COOH$$

linolenic acid

The major physical difference between fats and oils is that fats are solid and oils are liquid at room temperature (see Section 28.2). Since the glycerol part of the structure is the same for a fat and an oil, the difference must be due to the fatty acid end of the molecule. Fats contain a larger proportion of saturated fatty acids, whereas oils contain greater amounts of unsaturated fatty acids. The term *polyunsaturated* has been popularized in recent years; it means that each molecule of fat in a particular product contains several double bonds.

Fats and oils are obtained from natural sources. In general, fats come from animal sources and oils from vegetable sources. Thus lard is obtained from hogs and tallow from cattle and sheep. Olive, cottonseed, corn, peanut, soybean, canola, linseed, and other oils are obtained from the fruit or seed of their respective vegetable sources. Table 24.4 shows the major constituents of several fats and oils.

Triacylglycerols are the principal form in which energy is stored in the body. The caloric value per unit of mass is over twice as great as that for carbohydrates and proteins. As a source of energy, triacylglycerols can be completely replaced by either

The caloric value per unit mass for triacylglycerols is about 9 kcal/g, while proteins and carbohydrates yield about 4 kcal/g.

TABLE 24.4 **Fatty Acid Composition of Selected Fats and Oils**

Fat or oil	Fatty acid (%)				
	Myristic acid	Palmitic acid	Stearic acid	Oleic acid	Linoleic acid
Animal fat					
Butter[a]	7–10	23–26	10–13	30–40	4–5
Lard	1–2	28–30	12–18	41–48	6–7
Tallow	3–6	24–32	14–32	35–48	2–4
Vegetable oil					
Olive	0–1	5–15	1–4	49–84	4–12
Peanut	—	6–9	2–6	50–70	13–26
Corn	0–2	7–11	3–4	43–49	34–42
Cottonseed	0–2	19–24	1–2	23–33	40–48
Soybean	0–2	6–10	2–4	21–29	50–59
Linseed[b]	—	4–7	2–5	9–38	3–43

[a]Butyric acid, 3–4%
[b]Linolenic acid, 25–58%

carbohydrates or proteins. However, some minimum amount of fat is needed in the diet, because fat supplies the nutritionally essential unsaturated fatty acids (linoleic, linolenic, and arachidonic) required by the body (see Section 28.2).

Hydrogenation of Glycerides Addition of hydrogen is a characteristic reaction of the carbon–carbon pi bonds. Industrially, low-cost vegetable oils are partially hydrogenated to obtain solid fats that are useful as shortening in baking or in making margarine. In this process, hydrogen gas is bubbled through hot oil that contains a finely dispersed nickel catalyst. The hydrogen adds to the carbon–carbon double bonds of the oil to saturate the double bonds and form fats:

$$H_2 + -CH=CH- \xrightarrow{Ni} -CH_2-CH_2-$$
in oil or fat

In actual practice, only some of the double bonds are allowed to become saturated. The degree of hydrogenation can be controlled to obtain a product of any desired degree of saturation. The products resulting from the partial hydrogenation of oils are marketed as solid shortening (Crisco, Spry, etc.) and are used for cooking and baking. Oils and fats are also partially hydrogenated to improve their keeping qualities. Rancidity in fats and oils results from air oxidation at points of unsaturation, producing low-molar-mass aldehydes and acids of disagreeable odor and flavor.

Hydrogenation can also change the double-bond geometric isomers (see Chapter 20). Naturally occurring unsaturated oils commonly contain cis double bonds. During hydrogenation some of these double bonds convert to the trans isomer. Humans have difficulty metabolizing these "unnatural" fats and oils. Recent evidence indicates that "trans fats" increase the risk of atherosclerosis (heart disease).

Hydrogenolysis Triacylglycerols can be split and reduced in a reaction called hydrogenolysis (splitting by hydrogen). Hydrogenolysis requires higher temperatures and pressures and a different catalyst (copper chromite) than does hydrogenation of double bonds. Each triacylglycerol molecule yields a molecule of glycerol and three primary alcohol molecules. The hydrogenolysis of glyceryl trilaurate is represented as follows:

$$
\begin{array}{c}
\text{CH}_2\text{—O—}\overset{\displaystyle \overset{\text{O}}{\|}}{\text{C}}\text{—C}_{11}\text{H}_{23} \\
\text{CH—O—}\overset{\displaystyle \overset{\text{O}}{\|}}{\text{C}}\text{—C}_{11}\text{H}_{23} \quad + \quad 6\,\text{H}_2 \quad \xrightarrow[\Delta,\ \text{pressure}]{\text{copper chromite}} \quad 3\,\text{CH}_3(\text{CH}_2)_{10}\text{CH}_2\text{OH} \\
\text{CH}_2\text{—O—}\overset{\displaystyle \overset{\text{O}}{\|}}{\text{C}}\text{—C}_{11}\text{H}_{23}
\end{array}
$$

glyceryl trilaurate

lauryl alcohol (1-dodecanol)

$$
\begin{array}{c}
\text{CH}_2\text{OH} \\
+ \quad \text{CHOH} \\
\text{CH}_2\text{OH}
\end{array}
$$

glycerol

Long-chain primary alcohols obtained by this reaction are important, since they are used to manufacture other products, especially synthetic detergents (see Section 24.12).

Hydrolysis Triacylglycerols can be hydrolyzed, yielding fatty acids and glycerol. The hydrolysis is catalyzed by digestive enzymes at room temperatures and by mineral acids at high temperatures:

$$
\begin{array}{c}
\text{CH}_2\text{—O—}\overset{\displaystyle \overset{\text{O}}{\|}}{\text{C}}\text{—R} \\
\text{CH—O—}\overset{\displaystyle \overset{\text{O}}{\|}}{\text{C}}\text{—R}' \quad + \quad 3\,\text{H}_2\text{O} \quad \xrightarrow[\text{enzymes}]{\text{H}^+ \text{ or}} \\
\text{CH}_2\text{—O—}\overset{\displaystyle \overset{\text{O}}{\|}}{\text{C}}\text{—R}''
\end{array}
$$

triacylglycerol

$$
\begin{array}{c}
\text{RCOOH} \\
\text{R}'\text{COOH} \\
\text{R}''\text{COOH}
\end{array}
\quad + \quad
\begin{array}{c}
\text{CH}_2\text{OH} \\
\text{CHOH} \\
\text{CH}_2\text{OH}
\end{array}
$$

fatty acids (3 molecules)

glycerol

The enzyme-catalyzed reaction occurs in digestive reactions and in biological degradation (or metabolic) processes. The acid-catalyzed reaction is employed in the commercial preparation of fatty acids and glycerol.

Saponification Saponification of a fat or oil involves the alkaline hydrolysis of a triester. The products formed are glycerol and the alkali metal salts of fatty acids, which are called soaps. As a specific example, glyceryl tripalmitate reacts with sodium hydroxide to produce sodium palmitate and glycerol:

$$\underset{\text{glyceryl tripalmitate}}{\begin{array}{l} CH_2-O-\overset{\displaystyle O}{\overset{\|}{C}}-C_{15}H_{31} \\[2mm] CH-O-\overset{\displaystyle O}{\overset{\|}{C}}-C_{15}H_{31} \\[2mm] CH_2-O-\overset{\displaystyle O}{\overset{\|}{C}}-C_{15}H_{31} \end{array}} + 3\,NaOH \xrightarrow{\Delta} \underset{\substack{\text{sodium palmitate}\\ \text{(a soap)}}}{3\,C_{15}H_{31}COONa} + \underset{\text{glycerol}}{\begin{array}{l} CH_2OH \\ CHOH \\ CH_2OH \end{array}}$$

Glycerol, fatty acids, and soaps are valuable articles of commerce, and the processing of fats and oils to obtain these products is a major industry.

Practice 24.10

Write the structure for a triacylglycerol that has one unit each of myristic, palmitic, and oleic acids.

24.12 Soaps and Synthetic Detergents

In the broadest sense possible, a detergent is simply a cleansing agent. Soap has been used as a cleansing agent for at least 2000 years and is thereby classified as a detergent under this definition. However, beginning about 1930, a number of new cleansing agents that were superior in many respects to ordinary soap began to appear on the market. Because they were both synthetic organic products and detergents, they were called **synthetic detergents,** or syndets. A soap is distinguished from a synthetic detergent on the basis of chemical composition and not on the basis of function or usage.

synthetic detergent

Soaps

In former times soap-making was a crude operation. Surplus fats were boiled with wood ashes or with some other alkaline material. Today, soap is made in large manufacturing plants under controlled conditions. Salts of long-chain fatty acids are called **soaps.** However, only the sodium and potassium salts of carboxylic acids containing 12–18 carbon atoms are of great value as soaps, because of their abundance in fats.

soap

Fat or oil + NaOH $\longrightarrow$ Soap + Glycerol

To understand how a soap works as a cleansing agent, let's consider sodium palmitate, $CH_3(CH_2)_{14}COONa$, as an example of a typical soap. In water this substance exists as sodium ions, Na^+, and palmitate ions, $CH_3(CH_2)_{14}COO^-$. The sodium ion is an ordinary hydrated metal ion. The cleansing property, then, must be centered in the palmitate ion. The palmitate ion contains both a **hydrophilic** (water-loving) and a **hydrophobic** (water-fearing) group. The hydrophilic end is the polar,

hydrophilic
hydrophobic

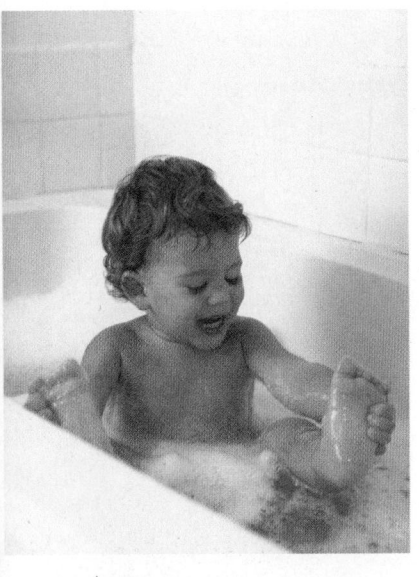

▲
The cleansing action of soap is of little interest to this baby who is enjoying the water!

negatively charged carboxylate group. The hydrophobic end is the long hydrocarbon group. The hydrocarbon group is soluble in oils and greases but not in water. The hydrophilic carboxylate group is soluble in water.

The cleansing action of a soap is explained in this fashion: When the soap comes in contact with grease on a soiled surface, the hydrocarbon end of the soap dissolves in the grease, leaving the negatively charged carboxylate end exposed on the grease surface. Because the negatively charged carboxylate groups are strongly attracted by water, small droplets are formed, and the grease is literally lifted or floated away from the soiled object (see Figure 24.1).

The cleansing property of a soap is due to its ability to act as an emulsifying agent between water and water-insoluble greases and oils. The grease–soap emulsion is stable because the oil droplets repel each other due to the negatively charged carboxyl groups on their surfaces. Some insoluble particulate matter is carried away with the grease; the remainder is wetted and mechanically washed away in the water. Synthetic detergents function in a similar way.

Ordinary soap is a good cleansing agent in soft water, but it is not satisfactory in hard water because insoluble calcium, magnesium, and iron(III) salts are formed. Palmitate ions, for example, are precipitated by calcium ions:

$$Ca^{2+}(aq) + 2\ CH_3(CH_2)_{14}COO^-\ (aq) \longrightarrow [CH_3(CH_2)_{14}COO]_2Ca(s)$$

palmitate ion   calcium palmitate

These precipitates are sticky substances and are responsible for "bathtub ring" and the sticky feel of hair after being shampooed with soap in hard water.

Soaps are ineffective in acidic solutions because water-insoluble molecular fatty acids are formed:

$$CH_3(CH_2)_{14}COO^- + H^+ \longrightarrow CH_3(CH_2)_{14}COOH$$

palmitic acid molecule

Synthetic Detergents

Once it was recognized that the insoluble hydrocarbon radical joined to a highly polar group was the key to the detergent action of soaps, chemists set out to make new substances that would have similar properties. About 1930, synthetic detergents (syndets) began to replace soaps, and now about 4 lb of syndets are sold for each pound of soap.

Although hundreds of substances with detergent properties are known, an idea of their general nature can be obtained from consideration of sodium lauryl sulfate and sodium p-dodecylbenzene sulfonate.

$$CH_3(CH_2)_{10}CH_2OSO_3^- Na^+$$

        sodium lauryl sulfate

$$CH_3(CH_2)_{10}CH_2 \text{---}\bigcirc\text{---} SO_3^- Na^+$$

        sodium p-dodecylbenzene sulfonate

Sodium lauryl sulfate and sodium p-dodecylbenzene sulfonate act in water in much the same way as sodium palmitate. Like the palmitate ion, the negative lauryl sulfate ion has a long hydrocarbon chain that is soluble in grease and a sulfate group that is attracted to water:

$$\underbrace{CH_3CH_2CH_2CH_2CH_2CH_2CH_2CH_2CH_2CH_2CH_2CH_2}_{\substack{\text{nonpolar hydrophobic end,}\\ \text{grease soluble}}}\text{---}\underbrace{OSO_3^-}_{\substack{\text{polar hydrophilic end,}\\ \text{water soluble}}}$$

The one great advantage these synthetic detergents have over soap is that their calcium, magnesium, and iron(III) salts, as well as their sodium salts, are soluble in water. Therefore, they are nearly as effective in hard water as in soft water.

The foregoing are anionic detergents, because they contain long-chain negatively charged ions. Other detergents, both cationic and nonionic, have been developed for special purposes. A cationic detergent has a long hydrocarbon chain and a positive charge.

$$\underbrace{CH_3(CH_2)_{14}CH_2}_{\substack{\text{grease soluble,}\\ \text{hydrophobic}}}\text{---}\underbrace{\overset{+}{N}(CH_3)_3}_{\substack{\text{water soluble,}\\ \text{hydrophilic}}}$$

Nonionic detergents are molecular substances. The molecule of a nonionic detergent contains a grease-soluble component and a water-soluble component. Some of these substances are especially useful in automatic washing machines because they have good detergent but low sudsing properties. The structure of a representative nonionic detergent is

$$\underbrace{CH_3(CH_2)_{10}CH_2}_{\substack{\text{grease soluble,}\\ \text{hydrophobic}}}\text{---}O\text{---}\underbrace{(CH_2CH_2O)_7\text{---}CH_2CH_2OH}_{\substack{\text{water soluble,}\\ \text{hydrophilic}}}$$

Biodegradability

biodegradable

Organic substances that are readily decomposed by microorganisms in the environment are said to be **biodegradable.** All naturally occurring organic substances are eventually converted to simple inorganic molecules and ions such as CO_2, H_2O, N_2, Cl^-, and SO_4^{2-}. Most of these conversions are catalyzed by enzymes produced by microorganisms. These enzymes attack only certain specific molecular configurations that are found in substances occurring in nature.

A number of years ago a serious environmental pollution problem arose in connection with synthetic detergents. Some of the early syndets, which contained highly branched chain hydrocarbons, had no counterparts in nature. Therefore, enzymes capable of degrading them did not exist, and the detergents were essentially nonbiodegradable and broke down very, very slowly. As a result, these syndets accumulated in water supplies, where they caused severe pollution problems due to excessive foaming and other undesirable effects.

Detergent manufacturers, acting on the recommendations of chemists and biologists, changed from a branched-chain alkyl benzene to a straight-chain alkyl benzene raw material. Detergents that contain the straight-chain alkyl groups are biodegradable.

$$CH_3CHCH_2CHCH_2CHCH_2CH\!\!-\!\!\bigcirc\!\!-SO_3^-Na^+$$
$$\quad\;\; | \qquad\; | \qquad\; | \qquad\; |$$
$$\quad\;\; CH_3 \quad CH_3 \quad CH_3 \quad CH_3$$

a nonbiodegradable detergent

$$CH_3CH_2CH_2CH_2CH_2CH_2CH_2CH_2CH_2CH_2CH_2CH_2\!\!-\!\!\bigcirc\!\!-SO_3^-Na^+$$

a biodegradable detergent

24.13   Esters and Anhydrides of Phosphoric Acid

Phosphoric acid has a Lewis structure similar to that of a carboxylic acid.

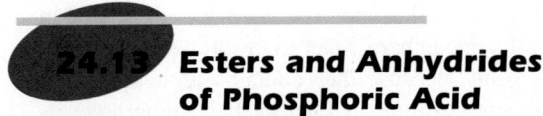

carboxylic acid   phosphoric acid

In both molecules an —OH is attached to an element that is double-bonded to an oxygen. In fact, phosphoric acid has three such —OH groups. This similarity in structure permits phosphoric acid to behave as a carboxylic acid in reaction with an alcohol. The product of the esterification reaction is called a phosphate ester.

$$
\begin{array}{ccccc}
& O & & & O \\
& \| & & & \| \\
HO\!-\!P\!-\!OH & + & HOCH_2CH_3 & \xrightarrow{\;H^+\;} & HO\!-\!P\!-\!OCH_2CH_3 \;+\; H_2O \\
& | & & & | \\
& OH & & & OH
\end{array}
$$

phosphoric acid   ethanol   monoethyl phosphate

This phosphate ester still has two —OH groups that can form additional esters. The result of one further esterification reaction is a molecule with two phosphate ester linkages. The diester then can contain two different alcohol groups in the same molecule. This structure is common in biochemistry specifically in nucleic acids (Chapter 28) and phospholipids (Chapter 31).

Pyrophosphoric acid is an acid anhydride formed by splitting out a molecule of water between two molecules of phosphoric acid.

$$
\underset{\text{phosphoric acid}}{\text{HO}-\overset{\overset{\text{O}}{\|}}{\underset{\underset{\text{OH}}{|}}{\text{P}}}-\text{OH} \;+\; \underset{\text{phosphoric acid}}{\text{HO}-\overset{\overset{\text{O}}{\|}}{\underset{\underset{\text{OH}}{|}}{\text{P}}}-\text{OH} \;\longrightarrow\; \underset{\text{pyrophosphoric acid}}{\text{HO}-\overset{\overset{\text{O}}{\|}}{\underset{\underset{\text{OH}}{|}}{\text{P}}}-\text{O}-\overset{\overset{\text{O}}{\|}}{\underset{\underset{\text{OH}}{|}}{\text{P}}}-\text{OH} \;+\; \text{H}_2\text{O}
$$

In general, both phosphate esters and phosphoric acid anhydrides have significant biological importance. The phosphate esters "tag" many biochemicals, labeling them for specific biological purposes. The phosphoric acid anhydrides serve as a temporary store of metabolic energy (see Chapter 33).

Concepts in Review

1. Give the common and IUPAC names of carboxylic acids.
2. Write structural formulas for saturated carboxylic acids, unsaturated carboxylic acids, amino acids, hydroxy acids, aromatic carboxylic acids, and dicarboxylic acids.
3. Tell how water solubility of carboxylic acids varies with increasing molar mass.
4. Relate the boiling points of carboxylic acids to their structure.
5. Write equations for the preparation of carboxylic acids by (a) oxidation of alcohols and aldehydes, (b) hydrolysis or saponification of esters and fats, (c) oxidation of aromatic hydrocarbons, and (d) hydrolysis of nitriles.
6. Write equations showing the effect of heat on malonic, succinic, and glutaric acids.
7. Write equations for the reactions of carboxylic acids that form (a) salts, (b) esters, and (c) acid chlorides.
8. Write an equation to show the formation of an ester from an acid chloride.
9. Write the common names, IUPAC names, and formulas of esters.
10. Identify the portion of an ester that is derived from a carboxylic acid and the portion derived from an alcohol.
11. Write the structure of a triacylglycerol (triglyceride) when given the fatty acid composition.
12. Describe several uses for the polyesters.
13. Write formulas for (a) polyesters from their monomers, and (b) monomers from a polyester formula.
14. Explain the differences between a fat and an oil.
15. Write equations illustrating the (a) hydrogenation, (b) hydrogenolysis, (c) hydrolysis, and (d) saponification of a fat or oil.
16. Write the structural formulas for the three principal unsaturated carboxylic acids found in fats and oils.
17. Explain how a soap or synthetic detergent acts as a cleansing agent.
18. Explain why syndets are effective and soaps are not effective as cleansing agents in hard water.
19. Differentiate among cationic, anionic, and nonionic detergents.
20. Explain the similarities between a carboxylic acid and phosphoric acid.
21. Write equations showing the formation of a phosphate ester and the formation of a phosphoric acid anhydride.
22. Indicate the significant role of phosphate esters and anhydrides in living organisms.
23. Indicate the ways in which aspirin acts within the body.
24. Explain how aspirin and ibuprofen differ from acetaminophen in medicinal use within the body.

Key Terms

The terms listed here have been defined within this chapter.
Section numbers are referenced in parenthesis for each term.

biodegradable (24.12)
carboxyl group (24.1)
ester (24.6)
fat (24.11)

hydrophilic (24.12)
hydrophobic (24.12)
oil (24.11)
saponification (24.10)

soap (24.12)
synthetic detergent (24.12)
triacylglycerol (24.11)
triglyceride (24.11)

Questions

Questions refer to tables, figures, and key words and concepts
defined within the chapter. A particularly challenging question
or exercise is indicated with an asterisk.

1. Using a specific compound in each case, write structural formulas for the following:
   (a) an aliphatic carboxylic acid
   (b) an aromatic carboxylic acid
   (c) an α-hydroxy acid
   (d) an α-amino acid
   (e) a β-chloro acid
   (f) a dicarboxylic acid
   (g) an unsaturated carboxylic acid
   (h) an ester
   (i) a nitrile
   (j) a sodium salt of a carboxylic acid
   (k) an acid halide
   (l) a triacylglycerol
   (m) a soap

2. Which of the following would have the more objectionable odor? Briefly explain.
   (a) a 1% solution of butyric acid (C_3H_7COOH) or
   (b) a 1% solution of sodium butyrate (C_3H_7COONa)

3. Which has the greater solubility in water?
   (a) methyl propanoate or propanoic acid
   (b) sodium palmitate or palmitic acid
   (c) sodium stearate or barium stearate
   (d) phenol or sodium phenoxide

4. Explain the difference between
   (a) a fat and an oil
   (b) a soap and a syndet
   (c) hydrolysis and saponification

5. Explain the cleansing action of detergents.

6. Cite the principal advantages that synthetic detergents (syndets) have over soaps.

7. List the medicinal effects and the risks associated with aspirin.

8. Besides aspirin, give two examples of other esters of salicylic acid that have medicinal uses.

9. What substances are commonly substituted for aspirin? Indicate their medical effects.

10. Why is a polyester defined as a condensation polymer?

Paired Exercises

These exercises are paired. Each odd-numbered exercise is followed by a similar even-numbered exercise. Answers to the even-numbered exercises are given in Appendix VI.

11. Name the following compounds:

   (a) $CH_3(CH_2)_4COOH$ (IUPAC name)

   (b) $HOOCCH=CHCOOH$ (common name for the trans isomer)

   (c) (common name)

   (d) $CH_3(CH_2)_{16}COOH$ (common name)

   (e) $CH_3CH_2COO^-Na^+$ (IUPAC name)

13. Write structures for the following compounds:
   (a) hexanoic acid
   (b) malonic acid
   (c) sodium benzoate
   (d) *o*-toluic acid
   (e) stearic acid
   (f) 2-chloropropanoic acid
   (g) potassium butyrate

15. Assume that you have 0.01 *M* solutions of the following substances:
   (a) NH_3 **(c)** NaCl   **(e)** CH_3COOH
   (b) HCl   **(d)** NaOH   **(f)** OH

Arrange them in order of increasing pH (list the most acidic solution first).

17. Give IUPAC and common names for the following:

   (a) $CH_2=CHC(=O)-OCH_3$

   (b) $CH_3CH_2CH_2C(=O)-OCH_2CH_3$

   (c)

12. Name the following compounds:

   (a) COOH, Cl (IUPAC name)

   (b) $CH_3(CH_2)_7CH=CH(CH_2)_7COOH$ (common name)

   (c) COOH, CH_3 (common name)

   (d) $CH_3CH_2CHCOOH$ with OH (IUPAC name)

   (e) $CH_3CH_2COO^-NH_4^+$ (IUPAC name)

14. Write structures for the following compounds:
   (a) oxalic acid
   (b) pentanoic acid
   (c) *o*-phthalic acid
   (d) linolenic acid
   (e) sodium *p*-aminobenzoate
   (f) ammonium propanoate
   (g) β-hydroxybutyric acid

16. Assume that you have 0.04 *M* solutions of the following substances:
   (a) NH_3 **(c)** HBr   **(e)** KBr
   (b) KOH   **(d)** HCOOH   **(f)** OH

Arrange them in order of decreasing pH (list the most basic solution first).

18. Give IUPAC and common names for the following:

   (a) $HC(=O)-OCH_3$

   (b) $C(=O)-OCH_2CH_2CH_3$

   (c) $CH_3CH_2C(=O)-OCH_2CH_3$

19. Write the structural formulas for
(a) methyl formate   (c) ethyl benzoate
(b) butyl butanoate

20. Write the structural formulas for
(a) propyl acetate   (c) ethyl hexanoate
(b) methyl benzoate

21. Write the structural formula and name of the principal organic product for the following reactions:

(a) $CH_3(CH_2)_7CH=CH(CH_2)_7COOH + H_2 \xrightarrow{Ni}$

(b) $\xrightarrow[H_2SO_4]{Na_2Cr_2O_7}$

(c) $CH_3CH_2COOH + NaOH \longrightarrow$

(d) $CH_3(CH_2)_3CH_2\overset{O}{\overset{\|}{C}}-OCH_2CH_2CH_3 + NaOH \xrightarrow{\Delta}$

(e) $\xrightarrow[\Delta]{NaMnO_4/NaOH}$

22. Write the structural formula and name of the principal organic product for the following reactions:

(a) $\xrightarrow[\Delta]{NaMnO_4/NaOH}$

(b) $HOOCCH=CHCOOH + H_2 \xrightarrow{Ni}$

(c) $CH_3(CH_2)_7CH=CH(CH_2)_7COOH + NaOH \longrightarrow$

(d) $CH_3CH_2CH_2\overset{O}{\overset{\|}{C}}-H \xrightarrow[H_2SO_4]{Na_2Cr_2O_7}$

(e) $+ NaOH \xrightarrow{\Delta}$

23. Write the structural formula of the ester that when hydrolyzed would yield
(a) methanol and acetic acid
(b) ethanol and formic acid
(c) 2-propanol and benzoic acid

24. Write the structural formula of the ester that when hydrolyzed would yield
(a) methanol and propanoic acid
(b) 1-octanol and acetic acid
(c) ethanol and butanoic acid

25. Write structural formulas for the reactants that will yield the following esters:
(a) methyl palmitate
(b) phenyl propionate
(c) dimethyl succinate

26. Write structural formulas for the reactants that will yield the following esters:
(a) isopropyl formate
(b) diethyl adipate
(c) benzyl benzoate

27. Write structural formulas for the organic products of the following reactions:

(a) $+ H_2O \longrightarrow$

(b) $CH_2=CHCOOH + Br_2 \longrightarrow$

(c) $+ H_2O \xrightarrow{H^+}$

(d) $CH_3CH_2OH + CH_3\overset{O}{\overset{\|}{C}}-Cl \longrightarrow$

(e) $CH_3\overset{O}{\overset{\|}{C}}-Cl + NH_3 \longrightarrow$

28. Write structural formulas for the organic products of the following reactions:

(a) $+ NH_3 \longrightarrow$

(b) $CH_3CH_2COOH + SOCl_2 \longrightarrow$

(c) $CH_3CH_2\overset{O}{\overset{\|}{C}}-Cl + CH_3OH \longrightarrow$

(d) $CH_3CH_2CH_2C\equiv N + H_2O \xrightarrow{H^+}$

(e) $+ SOCl_2 \longrightarrow$

29. What simple tests can be used to distinguish between the following pairs of compounds?
(a) benzoic acid and sodium benzoate
(b) maleic acid and malonic acid

30. What simple tests can be used to distinguish between the following pairs of compounds?
(a) benzoic acid and ethyl benzoate
(b) succinic acid and fumaric acid

31. Write structural formulas and names for the organic products of the following reactions:

(a) $CH_3\underset{\underset{COOH}{|}}{\overset{\overset{COOH}{|}}{CH}}$ $\xrightarrow{150°C}$

(b) $+ \ H_2O \ \xrightarrow[\Delta]{H^+}$

(c) $CH_3CH_2COOH \ +$ $\xrightarrow{H^+}$

32. Write structural formulas and names for the organic products of the following reactions:

(a) $\underset{\underset{\underset{\underset{COOH}{\diagdown}}{CH_2}}{|}}{\overset{\overset{\overset{COOH}{\diagup}}{CH_2}}{\underset{|}{CH_2}}}$ $\xrightarrow{\Delta}$

glutaric acid

(b) $\underset{\underset{COOH}{|}}{\overset{\overset{COOH}{|}}{CH_2}} \ + \ 2\ CH_3CH_2OH \xrightarrow{H^+}$

(c) $+ \ H_2O \ \xrightarrow[\Delta]{H^+}$

33. Trans isomers of naturally occurring fatty acids can act as metabolic inhibitors. Draw the structural formulas of *cis,cis,cis*-linolenic acid and its all-trans isomer.

34. The geometric configuration of naturally occurring unsaturated 18-carbon acids is all cis. Draw structural formulas for
(a) *cis*-oleic acid
(b) *cis,cis*-linoleic acid

35. Would $CH_3(CH_2)_{12}COOH$ or $CH_3(CH_2)_{12}COONa$ be the more useful cleansing agent in soft water? Explain.

36. Would $CH_3(CH_2)_{11}OSO_3Na$ (sodium lauryl sulfate) or $CH_3CH_2CH_2OSO_3Na$ (sodium propyl sulfate) be the more effective detergent in hard water? Explain.

37. Which one of the following substances is a good detergent in water? Is this substance a nonionic, anionic, or cationic detergent?
(a) $C_{16}H_{33}N(CH_3)_3^+Cl^-$
   hexadecyltrimethyl ammonium chloride
(b) $C_{16}H_{34}$
   hexadecane
(c) $C_{15}H_{31}COOH$
   palmitic acid (hexadecanoic acid)
(d) $C_{15}H_{31}\overset{\overset{O}{\|}}{C}-OC_{16}H_{33}$
   cetyl palmitate (hexadecyl hexadecanoate)

38. Which one of the following substances is a good detergent in water? Is this substance a nonionic, anionic, or cationic detergent?
(a) $CH_3(CH_2)_{11}O-\overset{\overset{O}{\|}}{C}(CH_2)_{14}CH_3$
(b) $HO-\overset{\overset{O}{\|}}{C}(CH_2)_{14}CH_3$
(c) $CH_3(CH_2)_{10}CH_2O(CH_2CH_2O)_7CH_2CH_2OH$
(d) $CH_3(CH_2)_{14}CH_2-$

39. Show the products of the reaction of 1 mol of phosphoric acid with
(a) 1 mol of ethanol to make a phosphate monoester
(b) 1 mol of ethanol followed by 1 mol of 1-propanol to make a phosphate diester

40. Show the products of the reaction of 1 mol of phosphoric acid with
(a) 3 mol of methanol to make a phosphate ester
(b) 1 mol of methanol followed by 1 mol of 2-butanol to make a phosphate diester

Additional Exercises

These exercises are not paired and provide additional practice on concepts covered in this chapter.

41. The FDA allows the addition of up to 0.1% by weight of sodium benzoate to baked goods to retard spoilage. How many grams of sodium benzoate can be included in 1 lb of dinner rolls (454 g = 1 lb)? How many moles?

42. Aspirin is synthesized on an industrial scale by combining the following reactants:

COOH

OH

$+$ $CH_3\overset{\overset{O}{\|}}{C}-O-\overset{\overset{O}{\|}}{C}CH_3 \longrightarrow$

Show the organic products from this reaction and name both reactants and products.

43. Upon hydrolysis, an ester of formula $C_6H_{12}O_2$ yields an acid A and an alcohol B. When B is oxidized, it yields a product identical to A. What is the structure of the ester? Explain your answer.

44. The gas, phosgene,

$Cl-\overset{\overset{O}{\|}}{C}-Cl$

was used in chemical warfare during World War I. Today this chemical is used extensively when forming amide linkages during industrial syntheses. Identify the functional groups on this molecule and name the acid from which phosgene is derived.

45. Most plant oils are high in unsaturated fatty acids; the average soybean triacylglycerol has two linoleic acid esters and one oleic acid ester. However, several plant oils are considered less healthy because they contain higher amounts of saturated fatty acids. For example, an average triacylglycerol from palm oil contains two palmitic acid esters and one oleic acid ester. Give structural formulas for both the soybean and the palm triacylglycerols.

46. Write the structural formula of a triacylglycerol that contains one unit each of lauric acid, palmitic acid, and oleic acid. How many other triacylglycerols, each containing all three of these acids, are possible?

47. Choose one of the triacylglycerols from Exercise 46. Write the names and formulas of all products expected when this triacylglycerol is
(a) reacted with water at high temperature and pressure in the presence of mineral acid

(b) reacted with hydrogen at relatively high pressure and temperature in the presence of a copper chromite catalyst
(c) boiled with potassium hydroxide
(d) reacted with hydrogen in the presence of Ni

48. Write a balanced chemical equation for the reaction of KOH with glyceryl tristearate. Discuss any changes in solubility that might occur as a result of this reaction.

*49. Starting with ethyl alcohol as the only source of organic material and using any other reagents you desire, write equations to show the synthesis of
(a) acetic acid
(b) ethyl acetate
(c) β-hydroxybutyric acid

50. Waxes can be formed from the reaction between a fatty acid and a long-chain alcohol. A major component of beeswax is a combination of a 16-carbon acid and a 30-carbon alcohol. Write the condensed formula for this component of beeswax.

51. Compound A, C_7H_8, reacted with $KMnO_4$ to form compound B, $C_7H_6O_2$. Compound B then combined with methanol to yield a sweet smelling liquid, compound C. Compound A did not react with Br_2 in CCl_4. Write structural formulas of A, B, and C.

52. A good lipstick is uniform in color and sticks to the lips once it is applied. The stickiness of a lipstick is controlled by the addition of a compound such as isopropyl myristate. Write the structural formula of isopropyl myristate.

53. An ester A of formula $C_{10}H_{12}O_2$ contains a benzene ring. When hydrolyzed, A gives an alcohol B, C_7H_8O, and an acid C, $C_3H_6O_2$. What are the structures of A, B, and C?

54. You are given two unlabeled bottles, one of which contains butanoic acid and the other ethyl butanoate. Describe a simple test that you could do to determine which compound is in which bottle.

*55. If 1.00 kg of triolein (glycerol trioleate) is converted to tristearin (glycerol tristearate) by hydrogenation:
(a) How many liters of hydrogen (at STP) are required?
(b) What is the mass of the tristearin that is produced?

56. Consider this statement: "When methyl propanoate is hydrolyzed, formic acid and propanol are formed." If this statement is true, write a balanced chemical reaction for it. If it is false, explain why.

57. Write structural formulas for all the esters that have a molecular formula of $C_5H_{10}O_2$.

58. (a) Calculate the mass percent of oxygen in Dacron.
    (b) The number of units in Dacron polyester varies from 80 to 130. Calculate the molar mass of a Dacron that consists of 105 units.

59. (a) Write the structure of a polyester that is made from the monomers *m*-phthalic acid and 1,3-propanediol.
    (b) Write the structures for the monomers of

This polyester is known as Kodel and produces a fabric that has good crease resistance.

60. Alkyd polyesters are made from glycerol and phthalic acid. Would this kind of polymer more likely be thermosetting or thermoplastic? Explain.

Answers to Practice Exercises

24.1 (a) $CH_3CHCOOH$ (with CH_3 substituent)

    (c) cyclohexane-COOH

    (b) $CH_3CH_2CH_2CHCH_2COOH$ (with Cl substituent)

24.2 3-hydroxybutanoic acid

24.3 (a) $HC-OH$ (with O double bond)
    (d) CH_2BrCH_2COOH

    (b) NH_2CH_2COOH
    (e) $HO-C-C-OH$ (with two O double bonds)

    (c) $HOCH_2COOH$
    (f) benzene-COOH

24.4 CH_2COOH
$|$
CH_2
$|$
$H_2NCHCOOH$

24.5 (a) caproic acid or hexanoic acid
    (b) 2-hexanone

24.6 (a) benzene-$CH_2CH_2CH_3$
    (b) benzene with CH_3 and CH_2CH_3

24.7 CH_3C-H (with O double bond) $+ HCN \xrightarrow{OH^-}$ CH_3CHCN (with OH)

CH_3CHCN (with OH) $+ H_2O \xrightarrow{H^+} CH_3CHCOOH$ (with OH) $+ NH_4^+$

lactic acid, 2-hydroxypropanioc acid

24.8 (a) methyl propanoate **(b)** phenyl ethanoate (phenyl acetate) **(c)** dimethyl succinate **(d)** methyl benzoate **(e)** methyl salicylate (wintergreen)

24.9
(a) $CH_3CH_2CH_2CH_2\overset{O}{\overset{\|}{C}}-OCH_2\overset{CH_3}{\overset{|}{C}}HCH_3 + H_2O \xrightarrow{H^+}$

$CH_3CH_2CH_2CH_2\overset{O}{\overset{\|}{C}}-OH + CH_3\overset{CH_3}{\overset{|}{C}}HCH_2OH$

(b) $CH_3CH_2CH_2CH_2\overset{O}{\overset{\|}{C}}-OCH_2\overset{CH_3}{\overset{|}{C}}HCH_3 + NaOH \xrightarrow[\Delta]{H_2O}$

$CH_3CH_2CH_2CH_2\overset{O}{\overset{\|}{C}}-O^-Na^+ + CH_3\overset{CH_3}{\overset{|}{C}}HCH_2OH$

24.10 $CH_2-O-\overset{O}{\overset{\|}{C}}(CH_2)_7CH=CH(CH_2)_7CH_3$
$|$
$CH-O-\overset{O}{\overset{\|}{C}}(CH_2)_{12}CH_3$
$|$
$CH_2-O-\overset{O}{\overset{\|}{C}}(CH_2)_{14}CH_3$

The sequence of the three esters in the triacylglycerol may vary.

Nylon is one of the materials used to give these colorful sails their strength and durability.

694

Many organic compounds contain nitrogen. The amines and amides are the two major classes of nitrogen-containing compounds. Amines isolated from plants form a group of compounds called alkaloids. Thousands of alkaloids have been isolated. Many of these compounds exhibit physiological activity. Examples of common alkaloid compounds include quinine, used in the treatment of malaria; strychnine, a poison; morphine, a narcotic; and caffeine, a stimulant. Many other drugs are also nitrogen-containing compounds.

Amides are nitrogen derivatives of carboxylic acids. These compounds are found as polymers, both commercially as in nylon, and biologically as in proteins. An understanding of the chemistry of organic nitrogen compounds is the cornerstone of genetics and is essential to unlocking the chemical secrets of living organisms.

25.1 Amides: Nomenclature and Physical Properties

Carboxylic acids react with ammonia to form ammonium salts:

$$\underset{\text{carboxylic acid}}{RC-OH} + \underset{\text{ammonia}}{NH_3} \longrightarrow \underset{\text{ammonium salt}}{RC-O^- NH_4^+}$$

$$\underset{\text{acetic acid}}{CH_3C-OH} + NH_3 \longrightarrow \underset{\text{ammonium acetate}}{CH_3C-O^- NH_4^+}$$

Ammonium salts of carboxylic acids are ionic substances. Ammonium acetate, for example, is ionized and exists as ammonium ions and acetate ions, both in the crystalline form and when dissolved in water.

When heated, ammonium salts of carboxylic acids lose a molecule of water and are converted to *amides:*

$$\underset{\text{ammonium salt}}{R-C-O^- NH_4^+} \overset{\Delta}{\longrightarrow} \underset{\text{amide}}{RC-NH_2} + H_2O$$

$$\underset{\text{ammonium acetate}}{CH_3C-O^- NH_4^+} \overset{\Delta}{\longrightarrow} \underset{\substack{\text{ethanamide} \\ \text{(acetamide)}}}{CH_3C-NH_2} + H_2O$$

amide   **Amides** are neutral (nonbasic) molecular substances; they exist as molecules (not ions) both in the crystalline form and when dissolved in water. An amide contains the following characteristic structure:

$$\text{R}-\overset{\displaystyle \overset{\text{O}}{\|}}{\text{C}}-\text{NH}_2$$

amide structure

In amides the carbon atom of a carbonyl group is bonded directly to a nitrogen atom of an $-\text{NH}_2$, $-\text{NHR}$, or $-\text{NR}_2$ group. The amide structure occurs in numerous substances, including proteins and some synthetic polymers such as nylon.

Two systems are used to name amides.

IUPAC Rules for Naming Amides

1. The IUPAC name is based on the longest carbon chain that includes the amide group.
2. Drop the *-oic acid* ending from the corresponding IUPAC acid name.
3. Add the suffix *-amide*.

Using the IUPAC rules,

methanoic acid	becomes	methanamide
HCOOH		$\text{HC}-\text{NH}_2$ (with O double-bonded to C)

ethanoic acid	becomes	ethanamide
CH_3COOH		$\text{CH}_3\text{C}-\text{NH}_2$ (with O double-bonded to C)

In a like manner, the common names for amides are formed from the common names of the corresponding carboxylic acids by dropping the *-ic* or *-oic acid* ending and adding the suffix *-amide*. Thus

formic acid	becomes	formamide
HCOOH		$\text{HC}-\text{NH}_2$ (with O double-bonded to C)

butyric acid	becomes	butyramide
$\text{CH}_3\text{CH}_2\text{CH}_2\text{COOH}$		$\text{CH}_3\text{CH}_2\text{CH}_2\text{C}-\text{NH}_2$ (with O double-bonded to C)

▲
Space filling model of butanamide
$\text{CH}_3\text{CH}_2\text{CH}_2\text{C}-\text{NH}_2$
with O double-bonded to C

benzoic acid   becomes   benzamide

When the nitrogen of an amide is connected to an alkyl or aryl group, the group is named as a prefix preceded by the letter *N*.

CH₃—C—N—CH₃   is   *N*-methylacetamide
(*N*-methylethanamide)

CH₃CH₂—C—N—CH₂CH₃   is   *N,N*-diethylpropionamide
(*N,N*-diethylpropanamide)

Formulas and names of selected amides are shown in Table 25.1.

Practice 25.1

Write formulas for
(a) *N*-isopropylmethanamide   (b) *N*-ethylbutanamide
(c) *N*-methyl-2-methylpropanamide   (d) *N,N*-diethylhexanamide
(e) *p*-aminobenzamide

Except for formamide, a liquid, all other unsubstituted amides are solids at room temperature. Many are odorless and colorless. Low-molar-mass amides are soluble in water, but solubility decreases quickly as molar mass increases. The amide functional group is polar, and nitrogen is capable of hydrogen bonding. The solubility of these molecules and their exceptionally high melting points and boiling points are the result of this polarity and hydrogen bonding between molecules, shown in Figure 25.1.

(a) Hydrogen bonding between amides and water molecules

(b) Intermolecular hydrogen bonding

◀ FIGURE 25.1
Hydrogen bonding in amides.

Formula	IUPAC name	Common name
$\overset{\displaystyle O}{\overset{\|}{HC}}-NH_2$	Methanamide	Formamide
$CH_3\overset{\displaystyle O}{\overset{\|}{C}}-NH_2$	Ethanamide	Acetamide
$CH_3CH_2\overset{\displaystyle O}{\overset{\|}{C}}-NH_2$	Propanamide	Propionamide
$CH_3\underset{\underset{\textstyle CH_3}{\|}}{CH}\overset{\displaystyle O}{\overset{\|}{C}}-NH_2$	2-Methylpropanamide	Isobutyramide
$CH_3\overset{\displaystyle O}{\overset{\|}{C}}-\underset{}{\overset{\overset{\textstyle H}{\|}}{N}}-\bigcirc$	N-Phenylethanamide	Acetanilide
$CH_3\overset{\displaystyle O}{\overset{\|}{C}}-\underset{\underset{\textstyle H}{\|}}{N}-\bigcirc-OH$	N-p-Hydroxyphenylethanamide	Acetaminophen
$H_2N-\overset{\displaystyle O}{\overset{\|}{C}}-\bigcirc$	Benzamide	Benzamide
$CH_3NH-\overset{\displaystyle O}{\overset{\|}{C}}-\bigcirc$	N-Methylbenzamide	N-Methylbenzamide

TABLE 25.1   Formulas and Names of Selected Amides

 ## 25.2   Chemical Properties of Amides

Hydrolysis is one of the more important reactions of amides. This type of reaction is analogous to the hydrolysis of carboxylic acid esters. The amide is cleaved into two parts, the carboxylic acid portion and the nitrogen-containing portion. As in ester hydrolysis, this reaction requires the presence of a strong acid or a strong base for it to occur in the laboratory. Amide hydrolysis is accomplished in living systems during the degradation of proteins by enzymatic reactions under much milder conditions (Chapter 30). Hydrolysis of an unsubstituted amide in an acid solution produces a carboxylic acid and an ammonium salt.

$$CH_3CH_2CH_2\overset{\displaystyle O}{\overset{\|}{C}}-NH_2 \ + \ H_2O \ + \ HCl \ \xrightarrow{\Delta} \ CH_3CH_2CH_2\overset{\displaystyle O}{\overset{\|}{C}}-OH \ + \ NH_4Cl$$

Basic hydrolysis results in the production of ammonia and the salt of a carboxylic acid.

$$CH_3-\overset{\displaystyle O}{\overset{\|}{C}}-NH_2 \ + \ NaOH \ \xrightarrow{\Delta} \ CH_3\overset{\displaystyle O}{\overset{\|}{C}}-O^-\,Na^+ \ + \ NH_3(g)$$

Show the products of (a) the acid hydrolysis and (b) the basic hydrolysis of

Example 25.1

$$\text{C}_6\text{H}_5-\overset{\displaystyle O}{\overset{\|}{C}}-NH_2$$

(a) In acid hydrolysis the C—N bond is cleaved, and the carboxylic acid is formed. The —NH₂ group is converted into an ammonium ion.

Solution

$$\text{C}_6\text{H}_5-\overset{\displaystyle O}{\overset{\|}{C}}-NH_2 \ + \ H_2O \ + \ H^+ \ \longrightarrow \ \text{C}_6\text{H}_5-\overset{\displaystyle O}{\overset{\|}{C}}-OH \ + \ NH_4^+$$

(b) In basic solution the C—N bond is also cleaved, but since the solution is basic, the salt of the carboxylic acid is formed along with ammonia.

$$\text{C}_6\text{H}_5-\overset{\displaystyle O}{\overset{\|}{C}}-NH_2 \ + \ NaOH \ \longrightarrow \ \text{C}_6\text{H}_5-\overset{\displaystyle O}{\overset{\|}{C}}-O^-\,Na^+ \ + \ NH_3$$

Practice 25.2

Give the products of the acidic and basic hydrolysis of

(a) $CH_3CH_2\overset{\displaystyle O}{\overset{\|}{C}}-NH_2$ (b) $CH_3\overset{\displaystyle O}{\overset{\|}{C}}-N(CH_3)_2$

Practice 25.3 Application to Biochemistry

Many digestive enzymes are produced by the pancreas and then activated when secreted into the small intestine. Thus, for our own safety, these enzymes become functional only when our body needs them to digest food. The activation process is hydrolysis of specific amide bonds. For example, chymotrypsin becomes active when the amide bond between amino acid segments #15 and #16 is hydrolyzed. Show the product of this reaction:

25.3   Polyamides: Condensation Polymers

Polyamides are condensation polymers. They can be formed from two monomers, a dicarboxylic acid and a diamine, or from a single monomer having a carboxylic acid and an amine in the same molecule. Proteins are biological polyamides (see Practice 25.3 and Chapter 29). Perhaps the most famous of the synthetic polyamides is Nylon-66.

Although there are several nylons, Nylon-66 is one of the best known and the first commercially successful polyamide. This polymer was so named because it is made from two 6-carbon monomers, adipic acid, $HOOC(CH_2)_4COOH$, and 1,6-hexanediamine, $H_2N(CH_2)_6NH_2$. The polymer chains of polyamides contain recurring amide linkages. The amide linkage can be made by reacting a carboxylic acid group with an amine group:

The repeating structural unit of the Nylon-66 chain consists of one adipic acid unit and one 1,6-hexanediamine unit:

$$HOOC-(CH_2)_4-COOH \qquad H_2N-(CH_2)_6-NH_2$$

adipic acid

1,6-hexanediamine
(hexamethylenediamine;
1,6-diaminohexane)

segment of Nylon-66 polyamide

$$-NH(CH_2)_5C{\overset{O}{\underset{\parallel}{}}} \left[NH(CH_2)_5C{\overset{O}{\underset{\parallel}{}}} \right]_n NH(CH_2)_5C{\overset{O}{\underset{\parallel}{}}}-$$

segment of Nylon-6 polyamide

Nylon was developed as a synthetic fiber for stockings and other wearing apparel and introduced to the public at the New York World's Fair in 1939. Today Nylon is used to make fibers for clothing and carpeting, filaments for fishing lines and ropes, bristles for brushes, tents, shower curtains, umbrellas, surgical sutures, tire cords, sports equipment, artificial human hair, synthetic animal fur, and molded objects such as gears and bearings. For the latter application, no lubrication is required because nylon surfaces are inherently slippery.

25.4 Urea

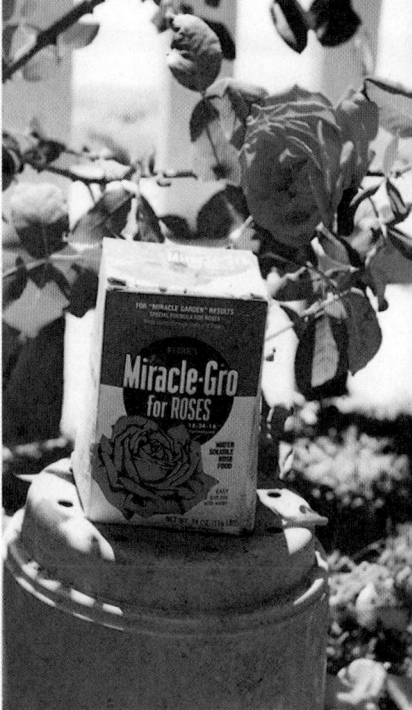

The body disposes of nitrogen by the formation of a diamide known as urea.

$$H_2N-\overset{O}{\underset{\parallel}{C}}-NH_2$$

urea

Urea is a white solid that melts at 133°C. It is soluble in water and therefore is excreted from the body in the urine. The normal adult excretes about 30 g of urea daily.

Urea is a common commercial product as well. It is widely used in fertilizers to add nitrogen to the soil, or as a starting material in the production of plastics and barbiturates.

▲
Urea is one of the major sources of nitrogen in this rose food. It stimulates growth while supplying essential nitrogen.

$$O=C{\overset{NH_2}{\underset{NH_2}{}}} + {\overset{CH_3CH_2O-C=O}{\underset{CH_3CH_2O-C=O}{CH_2}}} \longrightarrow O=C{\overset{\overset{H}{N}-C{\overset{O}{}}}{\underset{\underset{H}{N}-C{\overset{O}{}}}{CH_2}}} + 2\ CH_3CH_2OH$$

urea   diethylmalonate   barbituric acid

The various barbiturate drugs have certain organic groups substituted for the hydrogen atoms on the CH_2. When two ethyl groups replace these hydrogen atoms, the compound is known as *barbital,* a controlled substance (depressant) and a hypnotic. The unsubstituted barbituric acid has no hypnotic properties. There are many barbiturate drugs.

25.5 Amines: Nomenclature and Physical Properties

An **amine** is a substituted ammonia molecule with basic properties and has the general formula RNH_2, R_2NH, or R_3N, where R is an alkyl or an aryl group. Amines are classified as primary (1°), secondary (2°), or tertiary (3°), depending on the

amine

number of hydrocarbon groups attached to the nitrogen atom. Some examples include the following:

$$H-N\overset{\displaystyle H}{\underset{\displaystyle H}{}}$$

ammonia

$$\overset{\displaystyle H}{\underset{\displaystyle H}{}}N-CH_3$$

methylamine
(1° amine)

$$\overset{\displaystyle CH_3}{\underset{\displaystyle H}{}}N-CH_2CH_3$$

methylethylamine
(2° amine)

$$CH_3CH_2-N\overset{\displaystyle CH_2CH_3}{\underset{\displaystyle CH_2CH_3}{}}$$

triethylamine
(3° amine)

aniline
(1° amine)

cyclohexylamine
(1° amine)

Practice 25.4

Identify these compounds as 1°, 2°, or 3° amines.

(a) $CH_3CH_2CH_2NH_2$

(b) $CH_3\overset{\displaystyle CH_3}{\underset{\displaystyle CH_3}{C}}-NH_2$

(c) $\boxed{}N-H$

(d) $\boxed{}-NH_2$

(e) $CH_3CH_2\overset{}{\underset{\displaystyle CH_3}{N}}CH_2CH_3$

(f) $CH_3CH_2NHCH_2CH_3$

(g) $\bigcirc-CH_2NH_2$

The IUPAC System names simple amines using the name of the alkane (omitting the final -e) and adding the ending -amine. The NH$_2$ and side chains are located by number on the parent carbon chain.

$CH_3CH_2NH_2$

ethanamine

$CH_3CH_2CH_2NH_2$

1-propanamine

$CH_3\overset{}{\underset{\displaystyle NH_2}{CH}}CH_3$

2-propanamine

$CH_3\overset{\displaystyle CH_3}{\underset{\displaystyle CH_3}{C}}-NH_2$

2-methyl-2-propanamine

$\bigcirc-NH_2$

cyclohexanamine

Secondary and tertiary amines are named as *N*-substituted primary amines using the longest carbon chain as the parent name. A capital *N* in the name indicates the alkyl groups attached to the nitrogen atom.

CH_3—NH—$CH_2CH_2CH_3$ *N*-methyl-1-propanamine

CH_3—N—$CH_2CH_2CH_2CH_3$ *N*-ethyl-*N*-methyl-1-butanamine
     |
   CH_2CH_3

$(CH_3)_2N$—$CH_2CH_2CHCH_2CH_3$ *N,N*-dimethyl-3-ethyl-1-pentanamine
             |
          CH_2CH_3

N,N-dimethylcyclopentanamine

For naming diamines, the final -*e* of the alkane name is not omitted.

H_2N—$CH_2CH_2CH_2$—NH_2 1,3-propanediamine
H_2N—$(CH_2)_6$—NH_2 1,6-hexanediamine

Simple amines are most often referred to by their common names. The names for aliphatic amines are formed by naming the alkyl group or groups attached to the nitrogen atom followed by the ending -*amine*. Thus CH_3NH_2 is methylamine, $(CH_3)_2NH$ is dimethylamine, and $(CH_3)_3N$ is trimethylamine. Other examples are

$CH_3CH_2NH_2$ $CH_3CH_2CH_2NH_2$ CH_3CHCH_3 $CH_3CHCH_2CH_3$

ethylamine   propylamine NH_2 NH_2

                              isopropyl amine   *sec*-butylamine

         CH_3
         |
CH_3C—NH_2 CH_3
         | |
         CH_3 $CH_3NCH_2CH_3$

tert-butylamine   ethyldimethylamine

The most important aromatic amine is aniline ($C_6H_5NH_2$). Derivatives are named as substituted anilines. To identify a substituted aniline in which the substituent group is attached to the nitrogen atom, an *N*- is placed before the group name to indicate that the substituent is bonded to the nitrogen atom and not to a carbon atom in the ring. For example, the following compounds are called aniline, *N*-methylaniline, and *N,N*-dimethylaniline:

aniline   *N*-methylaniline   *N,N*-dimethylaniline

When a group is substituted for a hydrogen atom in the ring, the resulting ring-substituted aniline is named as we have previously done with aromatic compounds. The monomethyl ring-substituted anilines are known as toluidines. Study the names for the following substituted anilines:

o-toluidine
(o-methylaniline)

m-toluidine
(m-methylaniline)

p-toluidine
(p-methylaniline)

N-ethylaniline

m-ethylaniline

2,3-dimethylaniline

p-chloroaniline

Physiologically, aniline is a toxic substance. It is easily absorbed through the skin and affects both the blood and the nervous system. Aniline reduces the oxygen-carrying capacity of the blood by converting hemoglobin to methemoglobin. Methemoglobin is the oxidized form of hemoglobin in which the iron has gone from a $+2$ to a $+3$ oxidation state.

Example 25.2   Name the two compounds given.

(a) $CH_3CHCH_2-NH_2$ with CH_3 attached to middle carbon

(b) benzene ring with COOH and NH_2 (para)

Solution

(a) The alkyl group attached to NH_2 is an isobutyl group. Thus the common name is isobutylamine. The longest chain containing the NH_2 has three carbons. Therefore, the parent carbon chain is propane, and the IUPAC name is 2-methyl-1-propanamine.

(b) The parent compound on which the name is based is benzoic acid. With an amino group in the para position, the name is *p*-aminobenzoic acid. The acronym for *p*-aminobenzoic acid is PABA. Esters of PABA are some of the most effective ultraviolet screening agents and are used in suntanning lotions. PABA is a component of folic acid, an essential vitamin for humans.

Practice 25.5

Name the following compounds.

(a) $CH_3CH_2CH_2NH_2$

(b) $CH_3-N-CH(CH_3)_2$

(c) $CH_3CHCH_2CH_3$
      |
      NH_2

(d) [benzene ring]$-CH_2NH_2$ (common name)

(e) $CH_3CH_2NCH_2CH_2CH_2CH_3$
             |
             CH_3

Practice 25.6 Application to Biochemistry

Many neurotransmitters are amines. For example, 4-aminobutanoic acid is a neurotransmitter that is found in about 30% of all the brain's nerve cells. Draw the structure of this amino acid.

Ring compounds in which the atoms in the ring are not all alike are known as **heterocyclic compounds.** The most common heteroatoms are oxygen, nitrogen, and sulfur. A number of the nitrogen-containing heterocyclic compounds are present in naturally occurring biological substances such as DNA, which controls heredity. The structural formulas of several nitrogen-containing heterocyclics are as follows:

heterocyclic compound

pyrrole	pyridine	piperidine	pyrimidine	purine
(C_4H_5N)	(C_5H_5N)	$(C_5H_{11}N)$	$(C_4H_4N_2)$	$(C_5H_4N_4)$

Amines are capable of hydrogen bonding with water. As a result, the aliphatic amines with up to six carbons are quite soluble in water. Methylamine and ethylamine are flammable gases with a strong ammoniacal odor. Trimethylamine has a "fishy" odor. Higher-molar-mass amines have obnoxious odors. The foul odors arising from dead fish and decaying flesh are due to amines released by bacterial decomposition. Two of these compounds are diamines, 1,4-butanediamine and 1,5-pentanediamine. Each compound contains two amino groups:

$H_2NCH_2CH_2CH_2CH_2NH_2$ $H_2NCH_2CH_2CH_2CH_2CH_2NH_2$
1,4-butanediamine   1,5-pentanediamine
(putrescine)   (cadaverine)

Simple aromatic amines are all liquids or solids. They are colorless or almost colorless when freshly prepared but become dark brown or red when exposed to air and light.

25.6   Preparation of Amines

Alkylation of Ammonia and Amines The substitution of alkyl groups for hydrogen atoms of ammonia can be done by reacting ammonia with alkyl halides. Thus, in successive reactions, a primary, a secondary, and a tertiary amine can be formed.

$$NH_3 \xrightarrow{CH_3Br} \underset{\substack{\text{methylamine} \\ (1°)}}{CH_3NH_2} \xrightarrow{CH_3Br} \underset{\substack{\text{dimethylamine} \\ (2°)}}{(CH_3)_2NH} \xrightarrow{CH_3Br} \underset{\substack{\text{trimethylamine} \\ (3°)}}{(CH_3)_3N}$$

quaternary ammonium salt

Tertiary amines can be further alkylated so that four organic groups bond to the nitrogen atom. Compounds of this type are called **quaternary ammonium salts.** The positive charge is on the nitrogen atom as in ammonium salts. For example,

$$CH_3-\overset{\underset{|}{CH_3}}{\underset{\underset{|}{CH_3}}{\ddot{N}}}-CH_3 \ + \ CH_3Br \longrightarrow CH_3-\overset{\underset{|}{CH_3}}{\underset{\underset{|}{CH_3}}{N^+}}-CH_3 \ Br^-$$

tetramethylammonium bromide

Quaternary ammonium salts are well known in biologically active compounds and in many popular medicinals. For example, acetylcholine, an active neurotransmitter in the brain, is a quaternary ammonium salt. Choline is an important component of many biological membranes. The quaternary ammonium salt thiamine hydrochloride is marketed as vitamin B_1. Many well-known fabric softening agents used in laundering clothes are quaternary ammonium salts.

$$CH_3-\overset{\underset{|}{CH_3}}{\underset{\underset{|}{CH_3}}{N^+}}-CH_2CH_2OH \qquad CH_3-\overset{\underset{|}{CH_3}}{\underset{\underset{|}{CH_3}}{N^+}}-CH_2CH_2\overset{\underset{\|}{O}}{O}CCH_3$$

$$\text{choline} \qquad\qquad\qquad \text{acetylcholine}$$

Reduction of Amides and Nitriles Amides can be reduced with lithium aluminum hydride to give amines. For example, acetamide can be reduced to ethylamine; and when *N,N*-diethylacetamide is reduced, triethylamine is formed:

$$CH_3\overset{\underset{\|}{O}}{C}-NH_2 \xrightarrow{\text{LiAlH}_4} \underset{\text{ethylamine}}{CH_3CH_2NH_2}$$

$$CH_3\overset{\underset{\|}{O}}{C}-N\overset{CH_2CH_3}{\underset{CH_2CH_3}{<}} \xrightarrow{\text{LiAlH}_4} \underset{\text{triethylamine}}{(CH_3CH_2)_3N}$$

Nitriles, RCN, are also reducible to amines using hydrogen and a metal catalyst:

$$\underset{\text{propionitrile}}{CH_3CH_2C\equiv N} \xrightarrow{\text{H}_2/\text{Ni}} \underset{\text{propylamine}}{CH_3CH_2CH_2NH_2}$$

Reduction of Aromatic Nitro Compounds Aniline, the most widely used aromatic amine, is made by reducing nitrobenzene. The nitro group can be reduced by several reagents; Fe and HCl, or Sn and HCl, are commonly used.

nitrobenzene   phenylammonium chloride   aniline
(anilinium chloride)

Practice 25.7

Write the structure of an amide or nitrile that when reduced forms
(a) ethylmethylamine
(b) 2-methylpropanamine
(c) 3-ethyl-2-methylpentanamine

25.7 Chemical Properties of Amines

Alkaline Properties of Amines

In many respects, amines resemble ammonia in their reactions. Thus amines are bases and, like ammonia, produce OH^- ions in water:

$$\ddot{N}H_3 + (H)OH \rightleftharpoons NH_4^+ + OH^-$$

ammonia molecule   ammonium ion   hydroxide ion

A Brønsted–Lowry base is a proton (H^+) acceptor.

Methylamine and aniline react in the same manner:

$$CH_3\ddot{N}H_2 + (H)OH \rightleftharpoons CH_3\overset{+}{N}H_3 + OH^-$$

methylamine molecule   methylammonium ion   hydroxide ion

aniline molecule   anilinium ion   hydroxide ion

The ions formed are substituted ammonium ions. They are named by replacing the amine ending by ammonium and, for the aromatic amines, by replacing the aniline name by anilinium.

dimethylammonium ion   o-methylanilinium ion   N-methylanilinium ion

Like ammonia, amines are weak bases. Methylamine is a slightly stronger base than ammonia, and aniline is considerably weaker than ammonia. The pH values for 0.1 M solutions are methylamine, 11.8; ammonia, 11.1; and aniline, 8.8.

Amines are much more basic than amides, and our bodies recognize this difference. The liver converts ammonia, which is basic and toxic, into the safer diamide, urea (see Chapter 35, Section 35.9):

$$2\ NH_3 + H_2CO_3 \longrightarrow NH_2{-}\overset{\overset{\displaystyle O}{\|}}{C}{-}NH_2 + 2\ H_2O$$

(a partial summary of the urea cycle)

Urea does not react with water to form the hydroxide ion. Thus we normally excrete nitrogen without making the blood or urine too basic. In contrast, blood ammonia levels increase during liver cirrhosis, and the resulting increased blood pH can be life-threatening.

Because amine groups form substituted ammonium ions under physiological conditions, they provide the positive charge for biological molecules. For example, neurotransmitters are often positively charged. The structures of two such transmitter compounds, dopamine and serotonin, are shown here:

dopamine

serotonin

Salt Formation

An amine reacts with a strong acid to form a salt; for example, methylamine and hydrogen chloride react in this fashion:

$$\underset{\substack{\text{methylamine} \\ \text{molecule}}}{CH_3NH_2(g)} + \underset{\substack{\text{hydrogen chloride} \\ \text{molecule}}}{HCl(g)} \longrightarrow \underset{\substack{\text{methylammonium} \\ \text{chloride (salt)}}}{CH_3\overset{+}{N}H_3Cl^-}$$

Methylammonium chloride is made up of methylammonium ions, $CH_3N^+H_3$, and chloride ions, Cl^-. It is a white crystalline salt that in physical appearance resembles ammonium chloride very closely.

Aniline reacts in a similar manner, forming anilinium chloride:

anilinium chloride
(aniline hydrochloride salt)

Many amines or amino compounds are more stable in the form of the hydrochloride salt. When the free amine is wanted, the HCl is neutralized to liberate the free amine. Thus

$$\overset{+}{R}NH_3 \ Cl^- \ (\text{or } RNH_2 \cdot HCl) + NaOH \longrightarrow RNH_2 + NaCl + H_2O$$

an amine hydrochloride salt   free amine

Formation of Amides

Primary and secondary amines react with acid chlorides to form substituted amides. For example,

$$\underset{}{CH_3\overset{\overset{\displaystyle O}{\|}}{C}-Cl} + 2\,(CH_3CH_2)_2NH \longrightarrow CH_3\overset{\overset{\displaystyle O}{\|}}{C}-N\overset{\displaystyle CH_2CH_3}{\underset{\displaystyle CH_2CH_3}{}} + (CH_3CH_2)_2\overset{+}{N}H_2Cl^-$$

N,N-diethylacetamide

As seen from this example, amide formation creates larger molecules from smaller precursors. Building larger, more complex molecules is often a goal in both industrial chemistry and biochemistry. Amide formation plays an important role in this process. The barbiturates provide a simple example from pharmaceutical chemistry. Amide bonds form the central six-membered ring that is characteristic of barbiturates (the amide bonds are shown in blue).

Seconal

Living cells also use amide linkages to create proteins. In proteins, small reactants (amino acids) are connected via amide bonds. It is not uncommon for several hundred amino acids to be linked together to form a single protein molecule.

Practice 25.8 Application to Biochemistry

The pyrimidine heterocycles are a very important part of a cell's genetic information (nucleic acids). The pyrimidine ring is closed by amide formation. Show the products of the following reaction:

Sources and Uses of Selected Amines

Nitrogen compounds are found throughout the plant and animal kingdoms. Amines, substituted amines, and amides occur in every living cell. Many of these compounds have important physiological effects. Several examples of well-known nitrogen compounds follow.

Many antibacterial agents contain nitrogen. Common examples include the synthetic sulfa drugs and penicillin-related antibiotics, which are synthesized by molds:

sulfanilamide

ampicillin

Amines and amides are part of the structures of the B-complex vitamins. Examples are thiamine (vitamin B_1) and nicotinamide (niacin).

thiamine
(vitamin B_1)

nicotinamide (niacin)

A wide variety of nitrogen-containing compounds affect both the central and peripheral nervous system. Local anesthetics like the common drug, procaine, contain amines:

procaine hydrochloride (novocaine)

alkaloid   Basic compounds that are derived from plants and show physiological activity are known as **alkaloids.** These substances are usually amines. Nicotine, an alkaloid derived from tobacco leaves, acts to stimulate the nervous system:

nicotine

The opium alkaloids are often called opiates and include both compounds derived from the opium poppy and synthetic compounds that have morphinelike activity (sleep-inducing and analgesic properties). These drugs are classified as narcotics because they produce physical addiction, and they are strictly regulated by federal law.

methadone

Another common narcotic alkaloid is obtained from the leaves of the coca plant, cocaine:

cocaine

Amphetamines, as the name implies, are amine-containing compounds. These drugs act to stimulate the central nervous system by mimicking the action of compounds such as epinephrine, which is produced naturally by the body. They are used to treat depression, narcolepsy, and obesity. Use of amphetamines produces a feeling of well-being, loss of fatigue, and increasing alertness. The most widely abused amphetamine is methamphetamine, commonly called "speed."

epinephrine
(adrenalin)

methamphetamine
(Methedrine)

Barbiturates are synthetic drugs classified as sedatives. They are prepared from urea and substituted malonic acid, and they contain amide groups. Barbiturates act to depress the activity of brain cells. For this reason, they are often called "downers"; they are one of the more widely abused drugs.

pentobarbital
(Nembutal)

Coffee Talk

Coffee is one of the most widely consumed beverages in the world. Coffee drinkers rely on it to supply them a regular dose of caffeine, a stimulant sometimes called the world's most popular drug. Caffeine is an alkaloid.

caffeine

Caffeine is found in the seeds and leaves of certain plants, especially trees or shrubs of the genus *Coffea*, widely cultivated in the tropics.

A number of foods and drugs including chocolate, soft drinks, tea, and pain relievers contain caffeine, but society mainly gets its caffeine from coffee.

The primary effect of caffeine occurs within 30–60 minutes after its consumption when caffeine levels peak in the bloodstream to produce alertness and heightened concentration. Caffeine also postpones exhaustion and augments the capacity for physical work. It is precisely this last effect that has caused the U.S. Olympic committee to consider caffeine a "performance enhancer" and to routinely screen athletes for the drug. Caffeine also constricts blood vessels, relieving some headaches, which is why it is found in certain pain relievers. Too much caffeine results in "coffee nerves," which can include lightheadedness, irritability, unsteady hands, and diarrhea. Excess caffeine can also cause insomnia.

Caffeine used on a regular basis results in tolerance and dependence. Tolerance can develop very quickly, usually within days. Dependence is not a function of large amounts of caffeine. Researchers at Johns Hopkins University found signs of dependence in people who drank less than three cups of coffee per day. Caffeine-dependent people who try

▲ **The ambience of coffee houses encourages relaxation and conversation.**

to quit the coffee habit are miserable; they suffer headaches, nausea, and runny noses, and, in the worst cases, vomiting.

Studies over the last two decades have shown that coffee doesn't trigger heart attacks or raise blood cholesterol levels, at least for Americans. Americans tend to drink filtered coffee, which eliminates oils and other substances thought to trigger cholesterol problems. Caffeine does, however, raise blood pressure slightly in some people. In 1981, a study linked coffee to cancer of the pancreas, but since then several extensive studies have found no correlation between coffee and a variety of cancers.

Women who are pregnant or trying to become pregnant may want to carefully control their caffeine intake. Studies have found a link between the amount of caffeine consumed and infertility. A Yale University study (1993) showed 300 mg of caffeine per day reduced a woman's chance of conceiving by 25%.

Caffeine is not an innocuous drug. It acts powerfully on both the mind and the body, but it does not often pose a serious health risk.

How Much Caffeine?

Item	Caffeine content (mg)	
	Average	Range
Hot drinks (per 8 fluid oz)		
Drip-brewed coffee	100	70–215
Instant coffee	70	35–169
Tea, U.S. brands	50	25–110
Cocoa	5	2–25
Decaf coffee (brewed and instant)	4	2–8
Soft drinks (per 12 fluid oz)		
Mountain Dew	54	
Coca-Cola (Diet, regular)	46	
Dr. Pepper	41	
Pepsi-Cola (Diet, regular)	35	
Chocolate (per 1 oz)		
Dark chocolate	20	5–35
Milk chocolate	6	1–15
Over-the-counter drugs (per tablet or capsule)		
Vivarin	200	
No Doz	100	
Excedrin	65	
Anacin	32	

Sources: U.S. Food and Drug Administration; soft drink and instant coffee manufacturers.

Many common tranquilizers contain amines and amides. These drugs are used to modify psychotic behavior without inducing sleep, or to reduce anxiety or restlessness. Psychotic behavior is treated with strong tranquilizers such as Thorazine. Milder tranquilizers such as diazepam (Valium) are often used to relieve the pressure and anxiety of daily life.

diazepam
(Valium)

Other amine-containing drugs are being developed as scientists learn more about the central nervous system. Fluoxetine (Prozac) is a widely used and very effective antidepressant.

F_3C—⬡—O—$CHCH_2CH_2NHCH_3$

fluoxetine (Prozac)

Recently Prozac has been found useful in treating other central nervous system disorders.

Concepts in Review

1. Name and write structural formulas for amides.

2. Explain the high melting and boiling points of the amides, compared to alkanes of similar molar mass.

3. Write equations for both acidic and basic hydrolysis of amides.

4. Write formulas for polyamides from their monomers, and identify the monomers of a polyamide from its structure.

5. Name and write structural formulas for amines.

6. Distinguish among primary, secondary, and tertiary amines.

7. Show that amines are bases in their reactions with water and with acids.

8. Write equations for reactions of amines to form substituted amides.

9. Write equations for the formation of amines from (a) alkyl halides plus ammonia, (b) reduction of amides with $LiAlH_4$, (c) reduction of nitriles with H_2 and Ni, and (d) reduction of aromatic nitro compounds.

10. Identify and write equations for the formation of quaternary ammonium salts.

11. Indicate the major physiological responses to barbiturates, tranquilizers, amphetamines, and alkaloids.

12. Identify amine and amide functional groups in selected naturally occurring nitrogen compounds.

Key Terms

The terms listed here have been defined within this chapter.
Section numbers are referenced in parentheses for each term.

alkaloid (25.8)
amide (25.1)
amine (25.5)

heterocyclic compound (25.5)
quaternary ammonium salt (25.6)

Questions

Questions refer to tables, figures, and key words and concepts
defined within the chapter.

1. Write the structural formula and name of (a) a typical
amine and (b) a typical amide. Explain how the difference
in functional groups affects the chemical properties of
amines and amides.

2. Contrast the physical properties of amides with those of
amines.

3. Explain why amines have approximately the same water
solubility as alcohols of similar molar mass.

4. (a) What is a heterocyclic compound?
   (b) How many heterocyclic rings are present in (i) purine,
   (ii) ampicillin, (iii) methadone, and (iv) nicotine?

5. Indicate the functional groups present in
   (a) procaine hydrochloride   (c) nicotinamide
   (b) cocaine   (d) methamphetamine

6. Explain why it is advantageous for the body to excrete
urea rather than ammonia.

Paired Exercises

These exercises are paired. Each odd-numbered exercise is
followed by a similar even-numbered exercise.

7. Draw structural formulas for the following:
   (a) p-methylaniline
   (b) 2-butanamine
   (c) N-methyl-p-bromobenzamide

9. Name the following compounds:

   (a) $CH_3\overset{\displaystyle O}{\overset{\|}{C}}—NH_2$ (common name)

   (b) $CH_3CHCH_2CHCH_3$ (IUPAC name)
   $\quad\quad\ \ \ |\quad\quad\ |$
   $\quad\quad\ OH\quad NH_2$

   (c) (IUPAC name)
   [structure: benzene ring with $\overset{O}{\overset{\|}{C}}$—NHCH$_3$ and CH$_3$ substituents]

   (d) $CH_3—\overset{\displaystyle CH_3}{\underset{\displaystyle CH_2CH_3}{N}}—CH_2\underset{\displaystyle CH_3}{C}CH_2CH_3$ (IUPAC name)

8. Draw structural formulas for the following:
   (a) 4-chloro-2-pentanamine
   (b) N,N-diethylbenzamide
   (c) m-chloroaniline

10. Name the following compounds:

   (a) $CH_3\underset{\displaystyle |}{\overset{\displaystyle CH_3}{C}}HCH_2CH_2NH_2$ (IUPAC name)

   (b) $HC—NH_2$ (common name)
   $\quad\ \|$
   $\quad\ O$

   (c) $CH_3CH_2\overset{\displaystyle O}{\overset{\|}{C}}—NH—$ [benzene ring]   (IUPAC name)

   (d) $CH_3CH—\overset{\displaystyle CH_2CH_3}{\underset{\displaystyle CH_3}{N}}—CHCH_3$ (common and
   $\quad\quad\ |\quad\quad\quad\quad |$
   $\quad\quad CH_3\quad\quad\ CH_3$ IUPAC names)

11. Arrange this set of compounds in order of increasing solubility in water.

$$CH_3CH_2\overset{\displaystyle O}{\overset{\displaystyle \|}{C}}-NHCH_3$$

$$CH_3CH_2\overset{\displaystyle O}{\overset{\displaystyle \|}{C}}-N(CH_3)_2$$

$$CH_3CH_2\overset{\displaystyle O}{\overset{\displaystyle \|}{C}}-NH_2$$

13. Acetamide is soluble in water. Show the possible hydrogen bonding that helps explain this property.

15. Classify the following compounds as an acid, a base, or neither:
    (a) CH_3CH_2OH
    (b) $CH_3CH_2NH_2$
    (c) benzyl CH_2NH_2
    (d) phenol OH
    (e) $CH_3CH_2\overset{\displaystyle O}{\overset{\displaystyle \|}{C}}-NH_2$
    (f) $CH_3CH_2\overset{\displaystyle O}{\overset{\displaystyle \|}{C}}-OH$

17. Predict the organic products for the following reactions:
    (a) $CH_3NH-\overset{\displaystyle O}{\overset{\displaystyle \|}{C}}CH_3 + H_2O + H^+ \xrightarrow{\Delta}$

    (b) Acetic acid + isopropylamine $\xrightarrow{\Delta}$

    (c) $+ NaOH \xrightarrow{\Delta}$

19. Draw structural formulas for all the amines having the formula $C_4H_{11}N$. Classify each as either a primary amine, a secondary amine, or a tertiary amine.

12. Arrange this set of compounds in order of increasing solubility in water.

$$CH_3\overset{\displaystyle O}{\overset{\displaystyle \|}{C}}-NH_2$$

$$CH_3(CH_2)_4\overset{\displaystyle O}{\overset{\displaystyle \|}{C}}-NH_2$$

$\overset{\displaystyle O}{\overset{\displaystyle \|}{C}}-NH_2$

14. Urea is soluble in water. Show the possible hydrogen bonding that helps explain this property.

16. Classify the following compounds as an acid, a base, or neither:
    (a) $CH_3CH_2OCH_2CH_3$
    (b) $CH_3CH_2NHCH_2CH_3$
    (c) $CH_2CH_2CH_2NH_2$
    (d) $CH_2CH_2CH_2OH$
    (e) $\overset{\displaystyle O}{\overset{\displaystyle \|}{C}}-NH_2$; Br
    (f) $\overset{\displaystyle O}{\overset{\displaystyle \|}{C}}-OH$; Cl ; Cl

18. Predict the organic products for the following reactions:
    (a) Acetic acid + diethylamine $\xrightarrow{\Delta}$

    (b) $CH_2\overset{\displaystyle O}{\overset{\displaystyle \|}{C}}-N(CH_3)_2 + H_2O + H^+ \xrightarrow{\Delta}$

    (c) $CH_3CHCH_2\overset{\displaystyle O}{\overset{\displaystyle \|}{C}}-OH + CH_3NH_2 \xrightarrow{\Delta}$; CH_3

20. Draw structural formulas for all the amines having the formula C_3H_9N. Classify each as either a primary amine, a secondary amine, or a tertiary amine.

21. Classify the following amines as primary, secondary, or tertiary:
  (a) $CH_3CH_2CH_2NH_2$

  (b)

  (c)

  (d) $H_2NCH_2CH_2CH_2NH_2$

22. Classify the following amines as primary, secondary, or tertiary:
  (a) CH_3NHCH_3
  (b) $CH_3CH_2N(CH_3)_2$

  (c)

  (d) $CH_3 - N - CH_2CH_2CH_2CH_3$
         $\quad\quad\quad |$
         $\quad\quad CH_2CH_2$

23. Low-molar-mass aliphatic amines generally have odors suggestive of ammonia and/or stale fish. Which of the following solutions would have the stronger odor? Explain.
  (a) a 1% trimethylamine solution in 1.0 M sulfuric acid or
  (b) a 1% trimethylamine solution in 1.0 M NaOH

24. Low-molar-mass aliphatic amines generally have odors suggestive of ammonia and/or stale fish. Which of the following solutions would have the stronger odor? Explain.
  (a) a 1% isopropylamine solution in 1.0 M hydrochloric acid or
  (b) a 1% isopropylamine solution in 1.0 M KOH

25. Name the following compounds:
  (a) CH_3NHCH_3 (common name)
  (b)   (IUPAC name)
  (c)   (IUPAC name)
  (d) $(C_2H_5)_4N^+ I^-$ (common name)
  (e)   (IUPAC name)
  (f) $(CH_3CH_2)_2N-$⟨ ⟩   (common name)

26. Name the following compounds:
  (a)   (IUPAC name)
  (b) $CH_3CH_2CHCH_3$ (IUPAC name)
         $\quad\quad\quad |$
         $\quad\quad NH_2$
  (c)   (IUPAC name)
  (d) $CH_3CH_2NH_3^+ Br^-$ (common name)
  (e)   (common name, which is also IUPAC name)
  (f) $CH_3\overset{\displaystyle O}{\overset{\|}{C}}-NHCH_2CH_3$ (common name)

27. Draw structural formulas for the following compounds:
  (a) *N*-methylethanamine
  (b) aniline
  (c) 1,4-butanediamine
  (d) ethylisopropylmethylamine
  (e) pyridine
  (f) triethylammonium chloride

28. Draw structural formulas for the following compounds:
  (a) tributylamine
  (b) *N*-methylanilinium chloride
  (c) ethylammonium chloride
  (d) 2-amino-1-pentanol
  (e) 3,3-dimethyl-*N*-phenyl-2-aminohexane
  (f) 1,2-ethanediamine

29. Write equations to show how each conversion may be accomplished. (Some conversions may require more than one step.)

(a) $CH_3CH_2CH_2Br \longrightarrow CH_3CH_2CH_2NH_2$

(b) $CH_3CH_2CH_2Br \longrightarrow CH_3CH_2CH_2CH_2NH_2$

(c)

$\longrightarrow$

(d) $CH_3CH_2CH_2NH_2 \longrightarrow$

30. Show structural formulas for the organic products from the following reactions:

(a) $CH_3CH_2\overset{\displaystyle O}{\overset{\|}{C}}-O^-\ NH_4^+ \xrightarrow{\Delta}$

(b) $CH_3CH_2CH_2C\equiv N\ +\ H_2 \xrightarrow{Ni}$

(c)

$\xrightarrow[HCl]{Sn}$

(d)

$+\ CH_3\overset{\displaystyle O}{\overset{\|}{C}}-Cl \longrightarrow$

Additional Exercises

These exercises are not paired or labeled by topic and provide additional practice on concepts covered in this chapter.

31. Propylamine, ethylmethylamine, and trimethylamine have the same formula, C_3H_9N. Explain why the boiling point of trimethylamine is considerably lower than the boiling points of the other two compounds.

32. List three classes of drugs that are also amines and indicate the major physiological effects of each.

33. When you experience a sudden fright, your blood epinephrine (adrenalin) level rises quickly to $1.0 \times 10^{-7}\ M$. Given a blood volume of 5.0 L, how many grams of epinephrine are in circulation when you are in this condition?

34. Show the structure known as an amide linkage. Indicate some important classes of biochemicals that contain this linkage.

35. After cleaning or packing fish, workers often use lemon juice to clean their hands. What is the purpose of the lemon juice? Explain.

36. Cephalosporins are a newer generation of antibiotics that are often effective against penicillin-resistant bacteria.

(The blue ring is critical to the cephalosporin's antibiotic activity.) Identify the amide and ester groups.

37. Write a chemical reaction in which aniline acts as a base.

38. Putrescine, a product of decay of organic matter, is necessary for the growth of cells and occurs in animal tissues. Draw the structure and identify the type of nitrogen bonds that are present.

39. Why are many drugs given as an ammonium salt?

40. Aspartame is 200 times sweeter than sucrose. Circle any amine or amide in this structure.

41. (a) Starting with ethanol and using any other reagents, write reactions for the synthesis of 1,4-butanediamine.
(b) Starting with acetyl chloride, $CH_3\overset{\displaystyle O}{\overset{\|}{C}}-Cl$ and using any other reagents, write reactions for the synthesis of a primary, a secondary, and a tertiary amine.

42. The polymer Kevlar is many times stronger than steel and is often used to make bulletproof vests. This polymer is formed by linking terephthalic acid and *p*-aminoaniline via amide bonds. Show the structure of the polymer (two units).

43. (a) Write structures for the monomers used to make the polymer shown.

$$-\overset{\overset{\displaystyle O}{\|}}{C}-(CH_2)_8-\overset{\overset{\displaystyle O}{\|}}{C}-NH-(CH_2)_6-NH-\overset{\overset{\displaystyle O}{\|}}{C}-(CH_2)_8-\overset{\overset{\displaystyle O}{\|}}{C}-NH-(CH_2)_6-NH-$$

(b) The polymer Quiana is spun into a soft silky fabric. The structure of the polymer is

$$-\overset{\overset{\displaystyle O}{\|}}{C}-(CH_2)_6-\overset{\overset{\displaystyle O}{\|}}{C}-NH-\bigcirc-CH_2-\bigcirc-NH-\overset{\overset{\displaystyle O}{\|}}{C}-(CH_2)_6-\overset{\overset{\displaystyle O}{\|}}{C}-NH-\bigcirc-CH_2-\bigcirc-NH-$$

Draw the structures of the two monomers used to make Quiana.

Answers to Practice Exercises

25.1

(a) $HC\overset{\overset{\displaystyle O}{\|}}{}\!-\!NH\overset{\overset{\displaystyle CH_3}{|}}{C}HCH_3$

(b) $CH_3CH_2CH_2\overset{\overset{\displaystyle O}{\|}}{C}-NHCH_2CH_3$

(c) $CH_3\overset{\underset{\displaystyle |}{CH_3}}{C}H\overset{\overset{\displaystyle O}{\|}}{C}-NHCH_3$

(d) $CH_3CH_2CH_2CH_2CH_2\overset{\overset{\displaystyle O}{\|}}{C}\!-\!\overset{\underset{\displaystyle |}{CH_2CH_3}}{N}\!-\!CH_2CH_3$

(e) $H_2N-\bigcirc-\overset{\overset{\displaystyle O}{\|}}{C}-NH_2$

25.2

(a) acid hydrolysis $\quad CH_3CH_2\overset{\overset{\displaystyle O}{\|}}{C}-OH + NH_4{}^+$

basic hydrolysis $\quad CH_3CH_2\overset{\overset{\displaystyle O}{\|}}{C}-O^- + NH_3$

(b) acid hydrolysis $\quad CH_3\overset{\overset{\displaystyle O}{\|}}{C}-OH + \overset{+}{N}H_2(CH_3)_2$

basic hydrolysis $\quad CH_3\overset{\overset{\displaystyle O}{\|}}{C}-O^- + NH(CH_3)_2$

25.3

$$\overset{\diamondsuit\text{#14}}{\sim\!CH}-\overset{\overset{\displaystyle O}{\|}}{C}-NH-\overset{\diamondsuit\text{#15}}{CH}-\overset{\overset{\displaystyle O}{\|}}{C}-OH + NH_2-\overset{\diamondsuit\text{#16}}{CH}-\overset{\overset{\displaystyle O}{\|}}{C}-NH-\overset{\diamondsuit\text{#17}}{CH}\sim$$

25.4 (a) 1° (b) 1° (c) 2° (d) 1° (e) 3° (f) 2° (g) 1°

25.5 (a) propanamine (propylamine)
(b) N-isopropyl-N-methylaniline
(c) 2-methylpropanamine (sec-butylamine)
(d) benzylamine
(e) N-ethyl-N-methylbutanamine

25.6 $NH_2CH_2CH_2CH_2COOH$

25.7

(a) $CH_3\overset{\overset{\displaystyle O}{\|}}{C}-NHCH_3 \quad$ or $\quad HC\overset{\overset{\displaystyle O}{\|}}{}\!-\!NHCH_2CH_3$

(b) $CH_3\overset{\underset{\displaystyle |}{CH_3}}{C}HC\!\equiv\!N \quad$ or $\quad CH_3\overset{\underset{\displaystyle |}{CH_3}}{C}H\overset{\overset{\displaystyle O}{\|}}{C}-NH_2$

(c) $CH_3CH_2\overset{\underset{\displaystyle |}{CH_3}}{C}HCHC\!\equiv\!N \quad$ or $\quad CH_3CH_2\overset{\underset{\displaystyle |}{CH_3}}{C}HCH\overset{\overset{\displaystyle O}{\|}}{C}-NH$

25.8

$$+ H_2O$$

PUTTING IT TOGETHER
Review for Chapters 24–25

Multiple Choice: Choose the correct answer to each of the following.

1. The functional group for carboxylic acids is
   (a) $-COOH$ (c) $-C=O$
   (b) $>CHOH$ (d) (benzaldehyde structure: phenyl ring with $C=O$ and H)

2. The IUPAC name for $CH_3CH_2CH_2COOH$ is
   (a) butyric acid   (c) butanoic acid
   (b) butonic acid   (d) propanoic acid

3. The common name for the six-carbon dioic acid is
   (a) oxalic acid   (c) adipic acid
   (b) succinic acid   (d) malonic acid

4. How many carbon atoms are in the compound diethylmalonate?
   (a) 4   (b) 5   (c) 7   (d) 8

5. What is the name for this compound?

   (structure: benzene ring with COONa and CH_2CH_3 substituents, meta)

   (a) *m*-ethylbenzoic acid
   (b) sodium *m*-ethylbenzoate
   (c) sodium *m*-toluate
   (d) sodium *m*-methyltoluate

6. The name for $CH_3CHBrCH_2COOH$ is
   (a) α-bromobutyric acid   (c) γ-bromobutyric acid
   (b) β-bromobutyric acid   (d) 2-bromobutanoic acid

7. The formula for stearic acid is
   (a) $CH_3(CH_2)_{10}COOH$ (c) $CH_3(CH_2)_{14}COOH$
   (b) $CH_3(CH_2)_{12}COOH$ (d) $CH_3(CH_2)_{16}COOH$

8. Which compound can show cis–trans isomerism?
   (a) succinic acid   (c) adipic acid
   (b) fumaric acid   (d) *m*-toluic acid

9. Which compound is oxalic acid?
   (a) (benzene ring with COOH and COOH)
   (c) $COOH$ — $COOH$
   (b) (benzene ring with COOH and COOH)
   (d) $COOH$ — CH_2 — $COOH$

10. Which of these acids is the least soluble in water?
    (a) HCOOH   (c) $CH_3CH_2CH_2COOH$
    (b) CH_3COOH (d) $CH_3(CH_2)_6COOH$

11. Which class of compounds generally has pleasant odors?
    (a) carboxylic acids   (c) amides
    (b) acyl halides   (d) esters

12. Which of these compounds does <u>not</u> have intermolecular hydrogen bonding?
    (a) $CH_3\overset{O}{\overset{||}{C}}-OCH_3$ (c) $CH_3\overset{O}{\overset{||}{C}}-NHCH_3$
    (b) CH_3COOH (d) CH_3CH_2OH

13. Which compound, when oxidized, yields terphthalic acid?
    (a) (benzene ring with CH_3 and OH, para)
    (c) (benzene ring with CH_3 and CH_2CH_3, para)
    (b) (benzene ring with CH_3 and CH_3, ortho)
    (d) (benzene ring with OH and OH, meta)

14. What product is formed in this reaction?

    $$CH_3CH_2COOH + CH_3CH_2OH \xrightarrow{H^+}$$

    (a) $CH_3\overset{O}{\overset{||}{C}}-OCH_2CH_2CH_3$
    (b) $CH_3CH_2\overset{O}{\overset{||}{C}}-OCH_2CH_3$
    (c) $CH_3CH_2CH_3-O-CH_2CH_3$
    (d) $CH_3CH_2\overset{O}{\overset{||}{C}}-O-\overset{O}{\overset{||}{C}}CH_2CH_3$

15. What products are formed when this compound is hydrolyzed in an alkaline solution?

    $$CH_3CH_2\overset{O}{\overset{||}{C}}-OCH_2CH_3$$

    (a) $CH_3CH_2COOH + CH_3CH_2OH$
    (b) $CH_3COOH + CH_3CH_2CH_2OH$
    (c) $CH_3CH_2COONa + CH_3CH_2OH$
    (d) $CH_3CH_2COONa + CH_3CH_2ONa$

16. What is the name for this compound?

$$CH_3\overset{\overset{\displaystyle O}{||}}{C}-O-\overset{\overset{\displaystyle O}{||}}{C}CH_3$$

    (a) dimethylcarbonyl ether   (c) diacetyl ether
    (b) acetic anhydride   (d) no correct name given

17. Which compound compound will be formed in this reaction?

    (a)
    (c)
    (b)
    (d)

18. Which ester has the odor of pineapple?
    (a) octyl acetate   (c) ethyl butyrate
    (b) isoamyl acetate   (d) methyl salicylate

19. Which compound is aspirin?
    (a)
    (c)
    (b)
    (d)

20. Dacron polyester is a copolymer of
    (a) oxalic acid and ethylene glycol
    (b) terphthalic acid and ethylene glycol
    (c) malonic acid and ethylene glycol
    (d) adipic acid and 1,6-hexanediamine

21. Which formula is linoleic acid?
    (a) $CH_3(CH_2)_{16}COOH$
    (b) $CH_3(CH_2)_4CH=CH(CH_2)_7COOH$
    (c) $CH_3(CH_2)_7CH=CH(CH_2)_7COOH$
    (d) $CH_3(CH_2)_4CH=CHCH_2CH=CH(CH_2)_7COOH$

22. If glyceryl tripalmitate is saponified with NaOH, the products are
    (a) glycerol and tripalmitate
    (b) glycerol and $CH_3(CH_2)_{16}COONa$
    (c) glycerol and $CH_3(CH_2)_{14}COONa$
    (d) glycerol and $CH_3(CH_2)_{14}CH_2ONa$

23. Which of these compounds is a soap?
    (a) $CH_3CH_2CH_2COONa$
    (b) $CH_3(CH_2)_{14}CH_2ONa$
    (c) $CH_3(CH_2)_{14}COONa$
    (d) $[CH_3(CH_2)_{14}COO]_2Ca$

Consider formulas **A, B, C,** *and* **D** *for Questions 24–26.*
    (A) $CH_3(CH_2)_{12}CH_2OSO_3Na$
    (B) $CH_3(CH_2)_{14}CH_2N(CH_3)_3Cl$
    (C) $CH_3(CH_2)_{10}CH_2-O-(CH_2CH_2O)_7-CH_2CH_2OH$

    (D)

24. Which formula is a cationic detergent?
    (a) A   (b) B   (c) C   (d) D

25. Which formula is an anionic detergent?
    (a) A   (b) B   (c) C   (d) D

26. Which formula is a nonionic detergent?
    (a) A   (b) B   (c) C   (d) D

27. How many ester linkages can be made from one molecule of phosphoric acid?
    (a) 1   (b) 2   (c) 3   (d) 4

28. Which formula is an acyl halide?
    (a)
    (c) $$CH_3\overset{\overset{\displaystyle O}{||}}{C}-Cl$$
    (b) $CH_3CHClCOOH$
    (d) $$CH_3\overset{\overset{\displaystyle O}{||}}{C}-CH_2Cl$$

29. The amide structure in a compound is
    (a) neutral   (c) basic
    (b) acidic   (d) both acidic and basic

30. In an amide, the nitrogen atom is bonded to
    (a) an alkyl group   (c) a carbonyl group
    (b) a benzene ring   (d) a secondary carbon atom

31. The formula for 2-methylbutanamide is

(a) $CH_3CCH_2NH_2$ (with =O)

(c) $CH_3CH_2CHC-NH_2$ (with CH_3 on CH and =O below C)

(b) $CH_3CHC-NH_2$ (with CH_3 above, =O below)

(d) CH_3C-NH_2 (with CH_3 above, CH_3 below)

32. What is the name of this compound?

(ring)$-C(=O)-N(CH_3)-CH_2CH_3$

(a) *N*-isopropylbenzamide
(b) ethylmethylbenzamide
(c) *N*-ethyl-*N*-methylbenzoamide
(d) *N*-ethyl-*N*-methylbenzamide

33. What is the name of this compound?

(ring)$-NH-CCH_3$ (with =O)

(a) aniline ethanoate   (c) acetanilide
(b) methanilide   (d) benzanilide

34. What is the name of this compound?

$CH_3CHCH_2CH_2C-NH_2$ (with CH_2CH_3 and O)

(a) 4-ethylpentanamide   (c) 4-methylhexanamide
(b) isohexanamide   (d) 3-methylcaproamide

35. What is the IUPAC name for this compound:

$H_2NCH_2CH_2CH_2CH_2CH_2CH_2CH_2C-NH_2$ (with =O)

(a) 7-aminooctanamide   (c) 1,8-aminooctanamide
(b) 1,8-octanediamine   (d) 8-aminooctanamide

36. What is the name of this compound?

$CH_3CHCH_2C-N(CH_3)_2$ (with CH_3 and O)

(a) trimethylbutyramide
(b) *N,N*-dimethyl-3-methylbutanamide
(c) *N,N*-dimethyl-2-methylbutanamide
(d) no correct answer given

37. When $CH_3CH_2C-NH_2$ (with =O) is hydrolyzed in an acid solution, the products are

(a) $CH_3CH_2COOH + NH_4^+$
(b) $CH_3COOH + HC-NH_2$ (with =O)
(c) $CH_3CH_2CH_2NH_2$
(d) $CH_3CH_2COOH + NH_4OH$

38. When $CH_3CH_2C-NHCH_3$ (with =O) is hydrolyzed in a basic solution, the products are

(a) $CH_3CH_2CHNHCH_3$ (with OH)
(b) $CH_3CH_2COOH + HC-NHCH_3$ (with =O)
(c) $CH_3CH_2COONa + CH_3NH_2$
(d) $CH_3CH_2C-H + CH_3NH_2$ (with =O)

What are the product(s) for the reactions shown in Questions 39 and 40?

39. $CH_3CH_2C-OH + NH_3 \longrightarrow$ (with =O)

(a) $CH_3CH_2C-ONH_2$ (with OH)
(b) $CH_3CH_2C-ONH_4$ (with =O)
(c) $CH_3CH_2CHNH_2$ (with OH)
(d) $CH_3CHCOOH$ (with NH_2)

40. $CH_3CH_2C-Cl + NH_3 \longrightarrow$ (with =O)

(a) $CH_3CH_2C-NH_2 + NH_4Cl$ (with =O)
(b) $CH_3CH_2CHNH_2$ (with OCl)
(c) $CH_3CH_2CH_2NH_2 + NH_4Cl$
(d) $CH_3CH_2CClNH_2$ (with OH)

41. Which of the following compounds is the most acidic?

(a) CH_3C-H (with =O)   (c) CH_3C-NH_2 (with =O)
(b) CH_3COOH (d) $CH_3CH_2NH_2$

42. Which of the following compounds is a secondary amine?

(a) $CH_3CH_2NH_2$

(c) ⬡—CH_2NH_2

(b) $CH_3CH_2NHCH_3$

(d) $CH_3CH_2\overset{\displaystyle O}{\overset{\|}{C}}-N(CH_3)_2$

43. Which class of amine is this compound?

⬡—NH_2

(a) primary   (c) tertiary
(b) secondary   (d) quaternary

44. Which compound is <u>not</u> capable of hydrogen bonding with water?

(a) CH_3NHCH_3

(c) [pyrrolidine ring with N—H]

(b) $(CH_3CH_2)_2\overset{+}{\underset{\displaystyle CH_3}{N}}CH_3$

(d) $CH_3CH_2NH_2$

45. The formula for urea is

(a) $CH_3\overset{\displaystyle O}{\overset{\|}{C}}-NHCH_3$

(c) ⬡—$\overset{\displaystyle }{\underset{\displaystyle O}{\overset{\|}{C}}}-NH_2$

(b) $H_2NCH_2CH_2NH_2$

(d) $H_2N-\overset{\displaystyle O}{\overset{\|}{C}}-NH_2$

46. The formula for anilinium bromide is

(a) ⬡—NHBr

(c) ⬡ with Br and NH_2

(b) ⬡ with NH_2 and Br

(d) ⬡—$\overset{+}{N}H_3Br^-$

47. Which formula is a quarternary ammonium ion?
(a) $[CH_3NH_3]^+$ (c) $[(CH_3)_3NH]^+$
(b) $[CH_3NH_2CH_3]^+$ (d) $[(CH_3)_4N]^+$

What is the product of the reactions shown in Questions 48–50?

48. $CH_3CH_2CN \xrightarrow{\text{H}_2/\text{Ni}}$

(a) $CH_3CH_2NH_2$ (c) $CH_3CH_2\overset{\displaystyle O}{\overset{\|}{C}}-NH_2$
(b) $CH_3CH_2CH_2NH_2$ (d) $CH_3CH_2CH_2NHCH_2CH_2CH_3$

49. $CH_3CH_2Cl + CH_3NH_2 \longrightarrow$

(a) $CH_3CH_2NHCH_3$

(c) $CH_3\underset{\displaystyle NH_2}{\overset{\displaystyle CH_3}{\underset{|}{\overset{|}{C}}}}Cl$

(b) $CH_3CHClNHCH_3$

(d) $CH_3\underset{\displaystyle CH_3}{\overset{\displaystyle CH_3}{\underset{|}{\overset{|}{N}}}}CH_3$

50. ⬡—$\overset{\displaystyle O}{\overset{\|}{C}}-NH_2 \xrightarrow{\text{LiAlH}_4}$

(a) ⬡—CN

(b) ⬡—NH_2

(c) ⬡—CH_2NH_2

(d) ⬡—$\overset{\displaystyle O}{\overset{\|}{C}}-NH-\overset{\displaystyle O}{\overset{\|}{C}}-$⬡

51. Which set of reagents will convert nitrobenzene to aniline?
(a) $AlCl_3 + NH_3$ (c) $Sn + HCl/NaOH$
(b) $NaOH + NH_3$ (d) $LiAlH_4$

52. Which of the following compounds is nicotine?

(a) ⬡—$CH_2\underset{\displaystyle }{\overset{\displaystyle NHCH_3}{\underset{}{\overset{|}{C}}}HCH_3}$

(b) HO—[ring with HO and OH]—$CHCH_2NHCH_3$

(c) [pyridine ring fused to pyrrolidine ring with CH_3 on N]

(d) ⬡—$\overset{\displaystyle }{\underset{\displaystyle O}{\overset{\|}{C}}}-NHCH_3$

53. Which statement does <u>not</u> apply to Nylon-66?
  (a) It is a copolymer.
  (b) It is a condensation polymer.
  (c) It is a thermosetting polymer.
  (d) It contains carbon, hydrogen, oxygen, and nitrogen.

54. Which formula is the repeating structure for Nylon-66?

(a) $-[C-(CH_2)_4NH(CH_2)_6NH]_n-$ (with C=O)

(b) $-[C-(CH_2)_4-C-NH(CH_2)_6-NH]_n-$ (with two C=O)

(c) $-[C-(CH_2)_6C-NH(CH_2)_6-NH]_n-$ (with two C=O)

(d) $-[C-(CH_2)_6-NH]_n-$ (with C=O)

55. How many isomers (amines) can be written that have a molecular formula of $C_4H_{11}N$?
  (a) 2 **(b)** 4 **(c)** 6 **(d)** 8

Free Response Questions

1. Amino acids contain both an amine group and a carboxylic acid group. Explain why amino acids are often written as

$$H_3N^+ - CH - C - O^-$$
(with C=O above and R below CH)

(R changes depending on the specific amino acid)

2. Which of the following fatty acids is most likely synthetically produced? Explain.

I (trans isomer)

II (cis isomer)

3. In the Krebs cycle succinic acid is converted to fumaric acid which in turn is converted to malic acid. Draw the reaction sequence and indicate what type of reaction is occurring in each of these two processes. In the body, enzymes facilitate these reactions but we can also carry

them out synthetically in the lab. Suggest how the second step could be done in the lab.

4. What are the formulas and names of the products expected when phenyl benzoate is saponified using NaOH. Would the products be different if a weaker base were used?

5. Which of the following compounds contain(s) a carboxylic acid functional group?

$$CH_3CH_2CH_2CHC-H$$
(with C=O and OH below CH)

$$CH_3CH_2CH_2CH_2COOH$$

$$HOCH_2CCH_2CH_2CH_2C-OH$$
(with two C=O)

6. Would you expect phosphate monoesters to be more or less soluble in water than carboxylic esters? Explain. Which would you expect to have a higher melting and boiling point? (Assume R is such that the molar masses are similar.)

7. Draw the organic compounds I–III in the following reaction sequence. Also indicate what type of reaction is occurring in each step.

8. Although amides can be prepared by the reaction of a carboxylic acid with ammonia or an amine, the acid halide (RCOCl) is more commonly used in place of the carboxylic acid. Suggest two advantages of using the acid halide over the carboxylic acid.

9. Draw structures I–III for the following reaction sequence.

I (pentanoic acid with a double bond between the γ and δ carbons) $\xrightarrow{H_2O/H^+}$ **II**

$\xrightarrow[\text{heat}]{H^+}$ **III** (product of a condensation reaction)

10. Spinosyn D is one of two active molecules making up the insecticide Spinosad. This insecticide has been hailed for its high selectivity in targeting many chewing insect pests such as caterpillars, worms, flies and beetles while being harmless against most beneficial insects. In addition, Spinosad, which appears to work by causing neurotoxicity in its targets does not appear to affect mammals or birds, is easily broken down by sunlight, and does not leach through soil to groundwater. Spinosad is considered to be a reduced-risk pesticide by the EPA.

Identify as many functional groups as you can in spinosyn D (stereochemistry not specified).

Spinosyn D

▲

The mirror image of the ballet dancer appears to have her left arm raised. Molecules that are nonsuperimposable mirror images of each other are said to be chiral and are stereoisomers.

CHAPTER
26

Many of us grew up hearing such comments as, "Can't you tell your left from your right?" Have you watched a small child try to differentiate between a right and left shoe? Not surprisingly, the distinction between right and left is difficult. After all, our bodies are reasonably symmetrical. For example, both hands are made up of the same components (four fingers, a thumb, and a palm) ordered in the same way (from thumb through little finger). Yet, there is a difference if we try to put a left-handed glove on our right hand or a right shoe on our left foot.

Molecules possess similar, subtle structural differences, which can have a major impact on their chemical reactivity. For example, although there are two forms of blood sugar, related as closely as our left and right hands, only one of these structures can be used by our bodies for energy. Stereoisomerism is a subject that attempts to define these subtle differences in molecular structure.

26.1 Review of Isomerism

Isomerism is the phenomenon of two or more compounds having the same number and kinds of atoms (see Section 19.9). There are two types of isomerism. In the first type, known as structural isomerism, the difference between isomers is due to different structural arrangements of the atoms that form the molecules. For example, butane and isobutane, ethanol and dimethyl ether, and 1-chloropropane and 2-chloropropane are structural isomers:

$CH_3CH_2CH_2CH_3$
butane

CH_3CH_2OH
ethanol

$CH_3CH_2CH_2Cl$
1-chloropropane

CH_3CHCH_3
|
CH_3

2-methylpropane
(isobutane)

CH_3OCH_3
methoxymethane
(dimethyl ether)

$CH_3CHClCH_3$
2-chloropropane

In the second type of isomerism, the isomers have the same structural formulas but differ in the spatial arrangement of the atoms. This type of isomerism is known as **stereoisomerism.** Thus compounds that have the same structural formulas but differ in their spatial arrangement are called **stereoisomers.** There are two types of stereoisomers: cis–trans or geometric isomers, which we have already considered, and optical isomers, the subject of this chapter. The outstanding feature of optical isomers is that they have the ability to rotate the plane of plane-polarized light.

stereoisomerism
stereoisomer

26.2 Plane-Polarized Light

plane-polarized light

Plane-polarized light is light that is vibrating in only one plane. Ordinary (unpolarized) light consists of electromagnetic waves vibrating in all directions (planes) perpendicular to the direction in which it is traveling. When ordinary light passes

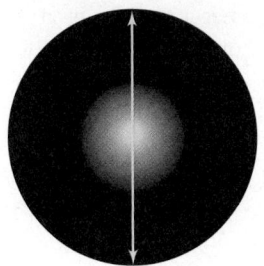

(a) Ordinary (unpolarized) light   (b) Plane-polarized light

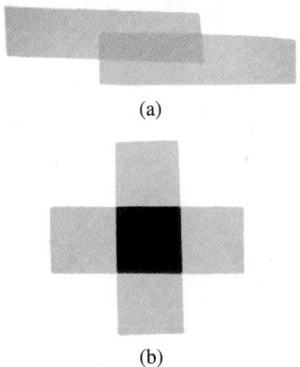

◀ FIGURE 26.1
(a) Diagram of ordinary light vibrating in all possible directions (planes) and (b) diagram of plane-polarized light vibrating in a single plane. The beam of light is coming toward the reader.

(a)

(b)

▲
FIGURE 26.2
Two Polaroid filters (top) with axes parallel and (bottom) with axes at right angles. In (a), light passes through both filters and emerges polarized. In (b), the polarized light that emerges from one filter is blocked and does not pass through the second filter, which is at right angles to the first. With no light emerging, the filters appear black.

through a polarizer, it emerges vibrating in only one plane and is called plane-polarized light (Figure 26.1).

Polarizers are made from calcite or tourmaline crystals or from a Polaroid filter, which is a transparent plastic that contains properly oriented embedded crystals. Two Polaroid filters with parallel axes allow the passage of plane-polarized light. But when one filter is placed so that its axis is at right angles to that of the other filter, the passage of light is blocked and the filters appear black (Figure 26.2).

Specific Rotation

The rotation of plane-polarized light is quantitatively measured with a polarimeter. The essential features of this instrument are (1) a light source (usually a sodium lamp), (2) a polarizer, (3) a sample tube, (4) an analyzer (which is another matched polarizer), and (5) a calibrated scale (360°) for measuring the number of degrees the plane of polarized light is rotated. The calibrated scale is attached to the analyzer (see Figure 26.3). When the sample tube contains a solution of an optically inactive material, the axes of the polarizer and the analyzer are parallel and the scale is at zero degrees; the light passing through is at maximum intensity. When a solution of an optically active substance is placed in the sample tube, the plane in which the polarized light is vibrating is rotated through an angle (α). The analyzer is then rotated to

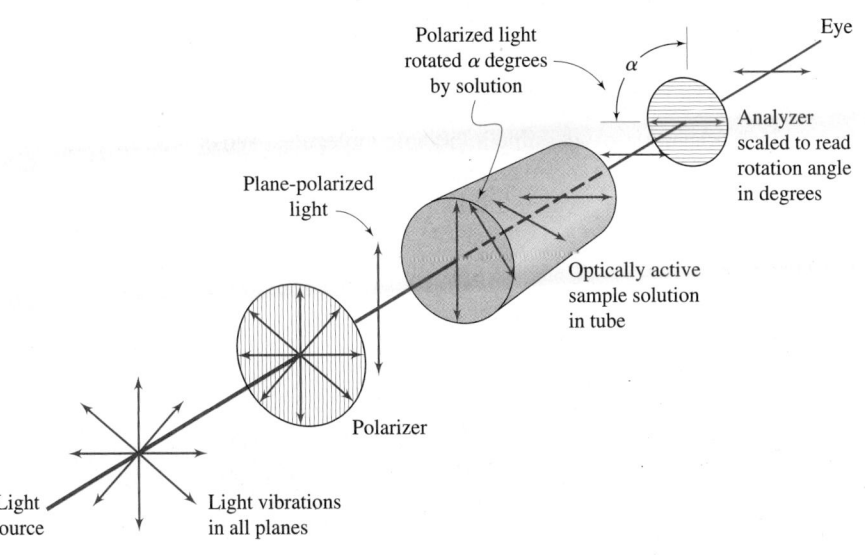

Polarized light rotated α degrees by solution

α

Eye

Analyzer scaled to read rotation angle in degrees

Plane-polarized light

Optically active sample solution in tube

Polarizer

Light source

Light vibrations in all planes

◀ FIGURE 26.3
Schematic diagram of a polarimeter. This instrument measures the angle α through which an optically active substance rotates the plane of polarized light.

▲
Polarized lenses in sunglasses reduce glare from the sun.

specific rotation

the position where the emerging light is at maximum intensity. The number of degrees and the direction of rotation by the solution are then read from the scale as the observed rotation.

The **specific rotation,** [α], of a compound is the number of degrees that polarized light would be rotated by passing through 1 decimeter (dm) of a solution of the substance at a concentration of 1 g/mL. The specific rotation of optically active substances is listed in chemical handbooks along with other physical properties. The following formula is used to calculate specific rotation from polarimeter data:

$$[\alpha] = \frac{\text{Observed rotation in degrees}}{\left(\begin{array}{c}\text{Length of}\\\text{sample tube in decimeters}\end{array}\right)\left(\begin{array}{c}\text{Sample concentration in}\\\text{grams per milliliter}\end{array}\right)}$$

26.3   Optical Activity

optical activity

dextrorotatory
levorotatory

▲
Quartz is optically active.

asymmetric carbon atom

FIGURE 26.4 ▶
Three-dimensional representation of an asymmetric carbon atom with four different groups bonded to it. The carbon atom is a sphere. Bonds to A and B project from the sphere toward the observer. Bonds to C and D project from the sphere away from the observer.

Many naturally occurring substances are able to rotate the plane of polarized light. Because of this ability to rotate polarized light, such substances are said to be **optically active.** When plane-polarized light passes through an optically active substance, the plane of the polarized light is rotated. If the rotation is to the right (clockwise), the substance is said to be **dextrorotatory;** if the rotation is to the left (counterclockwise), the substance is said to be **levorotatory.**

Some minerals, notably quartz, rotate the plane of polarized light (in fact, optical activity was discovered in minerals). However, when such mineral crystals are melted, the optical activity disappears. This means that the optical activity of these crystals must be due to an ordered arrangement within the crystals.

In 1848, Louis Pasteur (1822–1895) observed that sodium ammonium tartrate, a salt of tartaric acid, exists as a mixture of two kinds of crystals. Pasteur carefully hand-separated the two kinds of crystals. Investigating their properties, he found that solutions made from either kind of crystal would rotate the plane of polarized light, but in opposite directions. Since this optical activity was present in a solution, it could not be caused by a specific arrangement within a crystal. Instead, Pasteur concluded that this difference in optical activity occurred because of a difference in molecular symmetry.

The tetrahedral arrangement of single bonds around a carbon atom makes asymmetry (lack of symmetry) possible in organic molecules. When four different atoms or functional groups are bonded to a carbon atom, the molecule formed is asymmetric, and the carbon atom is called an **asymmetric carbon atom** (see Figure 26.4). In 1874, J. H. van't Hoff (1852–1911) and J. A. Le Bel (1847–1930) concluded that the presence of at least one asymmetric carbon atom in a molecule of an optically active

D

A   B

C

◯ = Carbon atom

A, B, C, D = Four different atoms or groups of atoms

substance is a key factor for optical activity. The first Nobel prize in chemistry was awarded to van't Hoff in 1901.

Molecules of optically active substances must have at least one center of dissymmetry. Although there are optically active compounds that do not contain asymmetric carbon atoms (their center of dissymmetry is due to some other structural feature), most optically active organic substances contain one or more asymmetric carbon atoms.

Your right and your left hands are mirror images of each other; that is, your left hand is a mirror reflection of your right hand, and vice versa. Furthermore, your hands are not superimposable on each other (see Figure 26.5). **Superimposable** means that, when we lay one object upon another, all parts of both objects coincide exactly.

A molecule that is not superimposable on its mirror image is said to be **chiral.** The word *chiral* comes from the Greek word *cheir,* meaning hand. Chiral molecules have the property of "handedness"; that is, they are related to each other in the same manner as the right and left hands. An asymmetric carbon atom is also called a **chiral carbon atom** or chiral center. Molecules or objects that are superimposable on each other are **achiral.** A molecule cannot be chiral if it has a plane of symmetry. For example, if you can pass a plane through a molecule (or an object) in such a way that one-half of the molecule is the mirror image of the other half, the molecule is not chiral. Some chiral and achiral objects are shown in Figure 26.6.

▲
FIGURE 26.5
The left hand is the same as the mirror image of the right hand. Right and left hands are not superimposable; hence they are chiral.

superimposable
chiral
chiral carbon atom
achiral

26.4 Projection Formulas

Molecules of a compound that contain one chiral carbon atom occur in two optically active isomeric forms. This is because the four different groups bonded to the chiral carbon atom can be oriented in space in two different configurations. It is important to understand how we represent such isomers on paper.

(a) chiral

(b) achiral

◀ **FIGURE 26.6**
Can you explain why these objects are chiral or achiral?

Let's consider the spatial arrangement of a lactic acid molecule, $CH_3CH(OH)COOH$, that contains one chiral carbon atom:

$$
\overset{1}{C}OOH \\
| \\
H - \overset{2}{C}* - OH \qquad C* = \text{chiral carbon atom} \\
\overset{3}{|} \\
CH_3
$$

lactic acid

Three-dimensional models are the best means of representing such a molecule, but by adopting certain conventions and using imagination, we can formulate the images on paper. The geometrical arrangement of the four groups about the chiral carbon (carbon 2) is the key to the stereoisomerism of lactic acid. The four bonds attached to carbon 2 are separated by angles of about 109.5°. Diagram I in Figure 26.7 is a three-dimensional representation of lactic acid in which the chiral carbon atom is represented as a sphere, with its center in the plane of the paper. The —H and —OH groups are projected forward from the paper (toward the observer), and the —COOH and —CH$_3$ groups are projected back from the paper (away from the observer).

FIGURE 26.7 ▶
Methods of representing three-dimensional formulas of a compound that contains one chiral carbon atom. All three structures represent the same molecule.

For convenience simpler diagrams such as II and III are used for writing structures of chiral compounds. These are much easier and faster to draw. In II, it is understood that the groups (—H and —OH) attached to the horizontal bonds are coming out of the plane of the paper toward the observer, and the groups attached to the vertical bonds are projected back from the paper. The molecule represented by formula III is made by drawing a cross and attaching the four groups in their respective positions, as in formula II. The chiral carbon atom is understood to be located where the lines cross. Formulas II and III are called **projection formulas.**

projection formula

It is important to be careful when comparing projection formulas. Two rules apply: (1) Projection formulas must not be turned 90°; (2) projection formulas must not be lifted or flipped out of the plane of the paper. Projection formulas may, however, be turned 180° in the plane of the paper without changing the spatial arrangement of the molecule. Consider the following projection formulas:

Formulas I, II, III, IV, and V represent the same molecule. Formula IV was obtained by turning formula III 180°. Formula V is formula IV drawn in a three-dimensional representation. If formula III is turned 90°, the other stereoisomer of lactic acid is represented, as shown in formulas VI and VII.

$$CH_3 \;—\!\!\!\!\!\begin{array}{c} H \\ | \\ | \\ OH \end{array}\!\!\!\!\!— COOH$$

VI

$$\begin{array}{c} H \\ \end{array}$$

CH₃ — COOH

OH

VII

(a) Redraw the three-dimensional formula A as a projection formula. (b) Draw the three-dimensional formula represented by the projection formula B.

Example 26.1

(A)

$$\begin{array}{c} CH_3 \\ \end{array}$$

H — CH₂NH₂

Br

CH₃CHBrCH₂NH₂

(B) $$CH_3 \\ H —\!\!\!\!\!|\!\!\!\!\!— CH_2CH_3 \\ OH$$

CH₃CH(OH)CH₂CH₃

(a) Draw a vertical and a horizontal line crossing each other. Place the CH₃ at the top and the Br at the bottom of the vertical line. Place the H at the left and the CH₂NH₂ at the right of the horizontal line to complete the projection formula:

Solution

$$CH_3 \\ H —\!\!\!\!\!|\!\!\!\!\!— CH_2NH_2 \qquad \text{(projection formula of A)} \\ Br$$

(b) Draw a small circle to represent the chiral carbon atom (carbon 2) that is located where the two lines cross in the projection formula. Draw a short line extending from the top and from the bottom of the circle. Place the CH₃ on the top and the OH on the bottom at the end of these lines. Now draw two short lines from within the circle coming toward your left and right arms. Place the H at the end of the left line and the CH₂CH₃ at the end of the right line. The finished formula should look like this:

$$\begin{array}{c} CH_3 \\ \end{array}$$

H — CH₂CH₃   (three-dimensional formula of B)

OH

Practice 26.1

Redraw this projection formula as a three-dimensional formula.

$$
\begin{array}{c}
\text{Br} \\
| \\
\text{Cl}-\!\!\!\!\!\!\!-\text{CH}_3 \\
| \\
\text{F}
\end{array}
$$

Practice 26.2

Redraw this three-dimensional formula as a projection formula.

$$
\text{COOH}
$$

$$
\text{H} \quad \text{NH}_2
$$

$$
\text{CH}_2\text{CH}_3
$$

26.5   Enantiomers

The formulas developed in Section 26.4 dealt primarily with a single kind of lactic acid molecule. But two stereoisomers of lactic acid are known, one that rotates the plane of polarized light to the right and one that rotates it to the left. These two forms of lactic acid are shown in Figure 26.8. If we examine these two structural formulas carefully, we can see that they are mirror images of each other. The reflection of either molecule in a mirror corresponds to the structure of the other molecule. Even though the two molecules have the same molecular formula and the same four groups attached to the central carbon atom, they are not superimposable. Therefore, the two molecules are not identical but are isomers. One molecule rotates the plane of polarized light to the left and is termed *levorotatory;* the other molecule rotates it to the right and is *dextrorotatory.* A plus ($+$) or a minus ($-$) sign written in parentheses and placed in front of a name or formula indicates the direction of rotation of polarized light and becomes part of the name of the compound. Plus ($+$) indicates rotation to the right and minus ($-$) to the left. Using projection formulas, we write the two lactic acids as shown here.

$$
\begin{array}{cc}
\text{COOH} & \text{COOH} \\
| & | \\
\text{H}-\text{C}-\text{OH} & \text{HO}-\text{C}-\text{H} \\
| & | \\
\text{CH}_3 & \text{CH}_3 \\
(-)\text{-lactic acid} & (+)\text{-lactic acid}
\end{array}
$$

Originally, it was not known which lactic acid structure was the ($+$) or the ($-$) compound. However, we now know that they are as shown. Chiral molecules that are **enantiomer** mirror images of each other are stereoisomers and are called **enantiomers.** The word *enantiomer* comes from the Greek word *enantios,* which means opposite.

Many, but not all, molecules that contain a chiral carbon are chiral. To decide whether a molecule is chiral and has an enantiomer, make models of the molecule and

FIGURE 26.8
Mirror-image isomers of lactic acid. Each isomer is the mirror reflection of the other. (−)-Lactic acid rotates plane-polarized light to the left, and (+)-lactic acid rotates plane-polarized light to the right.

Mirror   Mirror

COOH

H   OH

CH₃

(−)-lactic acid

COOH

HO   H

CH₃

(+)-lactic acid

◀ Enantiomers of lactic acid.

(−)-lactic acid   (+)-lactic acid

of its mirror image and see if they are superimposable. This is the ultimate test, but instead of making models every time, first examine the formula to see if it has a chiral carbon atom. If a chiral carbon atom is found, draw a cross and attach the four groups on the chiral carbon to the four ends of the cross. The chiral carbon is understood to be located where the lines cross. Remember that a chiral carbon atom has four different groups attached to it. Let's test the compounds 2-butanol and 2-chloropropane:

$$CH_3CH_2CHCH_3 \qquad CH_3CHCH_3$$
$$\qquad | \qquad\qquad\qquad\quad |$$
$$\qquad OH \qquad\qquad\qquad\quad Cl$$

2-butanol   2-chloropropane

In 2-butanol, carbon 2 is chiral. The four groups attached to carbon 2 are H, OH, CH₃, and CH₂CH₃. Draw the structure and its mirror image:

Mirror

CH₃

H —— OH

CH₂

CH₃

VIII

CH₃

HO —— H

CH₂

CH₃

IX

CH₃

CH₂

H —— OH

CH₃

X (IX turned 180°)

enantiomers

Turning structure IX 180° in the plane of the paper allows H and OH to coincide with their position in VIII, but CH₃ and CH₂CH₃ do not coincide. Therefore, we con-

▲
Overuse of muscle tissue results in the buildup of (+)-lactic acid.

clude that the mirror-image structures VIII and IX are enantiomers since they are not superimposable.

In 2-chloropropane, the four groups attached to carbon 2 are H, Cl, CH_3, and CH_3. Note that two groups are the same. Draw the structure and its mirror image:

Mirror

$$CH_3 \qquad\qquad CH_3 \qquad\qquad CH_3$$
$$H-\!\!\!-\!\!\!-Cl \qquad Cl-\!\!\!-\!\!\!-H \qquad H-\!\!\!-\!\!\!-Cl$$
$$CH_3 \qquad\qquad CH_3 \qquad\qquad CH_3$$

XI   XII   XIII (XII turned 180°)

When we turn structure XII 180° in the plane of the paper, the two structures XI and XIII are superimposable, proving that 2-chloropropane, which does not have a chiral carbon, does not exist in enantiomeric forms.

Example 26.2   Draw mirror-image isomers for any of the following compounds that can exist as enantiomers:

(a) $CH_3CH_2CH_2OH$ (c) $CH_3CH_2CHClCH_2CH_2CH_3$
(b) $CH_3CH_2CHClCH_2CH_3$

Solution   First check each formula for chiral carbon atoms:

(a) No chiral carbon atoms; each carbon has at least two groups that are the same.
(b) No chiral carbon atoms; carbon 3 has H, Cl, and two CH_3CH_2 groups.
(c) Carbon 3 is chiral; the four groups on carbon 3 are H, Cl, CH_3CH_2, and $CH_3CH_2CH_2$. Draw mirror images:

Mirror

$$\begin{array}{c} H \\ | \\ CH_3CH_2\!-\!C\!-\!CH_2CH_2CH_3 \\ | \\ Cl \end{array} \qquad\qquad \begin{array}{c} H \\ | \\ CH_3CH_2CH_2\!-\!C\!-\!CH_2CH_3 \\ | \\ Cl \end{array}$$

Practice 26.3

Draw the mirror-image isomers for the following compounds that can exist as enantiomers:
(a) $NH_2CH_2CHBrCH_2OH$ (c) $NH_2CH_2CH_2CH_2OH$
(b) $NH_2CH_2CHBrCH_2NH_2$

Practice 26.4

Draw structural formulas for the pentyl alcohols, $C_5H_{11}OH$, that will show optical activity.

Practice 26.5 Application to Biochemistry

Two important compounds in our body's energy metabolism are citric acid and isocitric acid.

```
        COOH                    COOH
         |                       |
    H — C — H            HO — C — H
         |                       |
  HO — C — COOH           H — C — COOH
         |                       |
    H — C — H              H — C — H
         |                       |
        COOH                    COOH

     citric acid            isocitric acid
```

Star any carbons that are chiral and draw an enantiomer for any compound that is chiral.

The relationship between enantiomers is such that if we change the position of any two groups attached to a chiral carbon atom on a compound that contains only one chiral carbon atom, we obtain the structure of its enantiomer. If we make a second change, the structure of the original isomer is obtained again. In both cases shown here, (+)-lactic acid is formed by interchanging the positions of two groups on (−)-lactic acid:

```
      CH₃                             COOH                           COOH
       |          change position of   |        change position of    |
  H —— OH    ←———————————————    H —— OH   ———————————————→   HO —— H
       |          CH₃ and COOH         |          H and OH            |
      COOH                            CH₃                            CH₃

  (+)-lactic acid                 (−)-lactic acid                (+)-lactic acid
```

To test whether two projection formulas are the same structure or are enantiomers, (1) turn one structure 180° in the plane of the paper and compare to see if they are superimposable, or (2) make successive group interchanges until the formulas are identical. If an odd number of interchanges are made, the two original formulas represent enantiomers; if an even number of interchanges are made, the two formulas represent the same compound. The following two examples illustrate this method. Are structures XIV and XV the same compound?

```
          CH₃                 H
           |                  |
   Br —— H          Br —— Cl
           |                  |
          Cl                 CH₃

         XIV                  XV
```

```
      H                            H                           CH₃
      |          interchange       |        interchange         |
 Br —— Cl   ———————————→   Br —— CH₃  ———————————→   Br —— H
      |          CH₃ and Cl        |        CH₃ and H           |
     CH₃                          Cl                           Cl

      XV                                                       XIV
```

Two interchanges were needed to make structure XV identical to structure XIV. Therefore, structures XIV and XV represent the same compound.

Do structures XVI and XVII represent the same compound?

Three interchanges were needed to make structure XVII identical to structure XVI. Therefore, structures XVI and XVII do not represent the same compound; they are enantiomers.

Enantiomers ordinarily have the same chemical properties, and other than optical rotation, they also have the same physical properties (see Table 26.1 on page 741). They rotate plane-polarized light the same number of degrees but in opposite directions.

Enantiomers usually differ in their biochemical properties. In fact, most living cells are able to use only one isomer of a mirror-image pair. As an example, (+)-glucose ("blood sugar") is used in the body for metabolic energy, whereas (−)-glucose is not. Enantiomers are truly different molecules and are treated as such by most organisms. The key factors of enantiomers and optical isomerism can be summarized as follows:

1. A carbon atom that has four different groups bonded to it is called an asymmetric or a chiral carbon atom.
2. A compound with one chiral carbon atom can exist in two isomeric forms called enantiomers.
3. Enantiomers are nonsuperimposable mirror-image isomers.
4. Enantiomers are optically active; that is, they rotate plane-polarized light.
5. One isomer of an enantiomeric pair rotates polarized light to the left (counterclockwise). The other isomer rotates polarized light to the right (clockwise). The degree of rotation is the same but in opposite directions.
6. Rotation of polarized light to the right is indicated by (+), placed in front of the name of the compound. Rotation to the left is indicated by (−), for example, (+)-lactic acid and (−)-lactic acid.

26.6 Racemic Mixtures

A mixture containing equal amounts of a pair of enantiomers is known as a **racemic mixture.** Such a mixture is optically inactive and shows no rotation of polarized light when tested in a polarimeter. Each enantiomer rotates the plane of polarized light by the same amount but in opposite directions. Thus the rotation by each isomer is canceled. The ($\pm$) symbol is often used to designate racemic mixtures. For example, a racemic mixture of lactic acid is written as ($\pm$)-lactic acid because this mixture contains equal molar amounts of ($+$)-lactic acid and ($-$)-lactic acid.

racemic mixture

Racemic mixtures are usually obtained in laboratory syntheses of compounds in which a chiral carbon atom is formed. Thus catalytic reduction of pyruvic acid (an achiral compound) to lactic acid produces a racemic mixture containing equal amounts of ($+$)- and ($-$)-lactic acid:

$$CH_3CCOOH + H_2 \xrightarrow{Ni} CH_3CHCOOH$$
$$\underset{O}{\|} \qquad\qquad\qquad \underset{OH}{|}$$

pyruvic acid   ($\pm$)-lactic acid

As a general rule, in the biological synthesis of optically active compounds, only one of the isomers is produced. For example, ($+$)-lactic acid is produced by reactions occurring in muscle tissue, and ($-$)-lactic acid is produced by lactic acid bacteria in the souring of milk. These stereospecific reactions occur because biochemical syntheses are enzyme catalyzed. The preferential production of one isomer over another is often due to the configuration (shape) of the specific enzyme involved. Returning to the hand analogy, if the "right-handed" enantiomer is produced, then the enzyme responsible for the product can be likened to a right-handed glove.

The mirror-image isomers (enantiomers) of a racemic mixture are alike in all ordinary physical properties except in their action on polarized light. It is possible to separate or resolve racemic mixtures into their optically active components. In fact, Pasteur's original work with sodium ammonium tartrate involved such a separation. But a general consideration of the methods involved in such separations is beyond the scope of our present discussion.

Many pharmaceuticals are synthesized as racemic mixtures because organic synthetic processes are often not stereospecific. Unfortunately, often only one enantiomer has biological activity because most biomolecules are almost always stereospecific. Thus the contents of a racemic mixture drug may be one-half bioinactive. Furthermore, the inactive enantiomer may cause adverse effects. The most famous example of drug-related problems with racemic mixtures involved the drug, thalidomide.

thalidomide racemic mixture
(the chiral carbon is marked with a star, *)

One isomer is effective at suppressing some immune responses while the other isomer causes birth defects. While the racemic mixture was on sale in Europe (from about 1957 to 1962), about 10,000 babies were born with serious malformations.

Today, almost every new approved drug is a single isomer (not a racemic mixture). Single isomer drugs are not only safer but often more effective. For example, the most common asthma drug, albuterol, is a racemic mixture of the following two isomers:

active isomer
(levalbuterol)

inactive isomer

(the chiral carbon is marked with a star,*)

Albuterol is a bronchodilator that causes relaxing of the bronchial airways, but it also frequently causes nervousness and irregular heartbeat. By using only the active isomer, levalbuterol, a lower dosage is effective and fewer side reactions are noted. It is estimated that about 80% of chiral drugs will be marketed as single isomers rather than racemic mixtures by the year 2000.

The formulas for several more chiral drugs are shown in Figure 26.9.

FIGURE 26.9 ▶
Some examples of common chiral drugs.

metoprolol (Lopressor®)
treats high blood pressure

3-(3,4-dihydroxyphenyl)-alanine (L-Dopa)
treats Parkinson's disease

pseudoephedrine (Sudafed®)
treats nasal congestion

methylphenidate (Ritalin®)
treats attention deficit disorder

Practice 26.6 Application to Biochemistry

The chirality of drugs is very important because the reactions of biomolecules are generally stereospecific. Star the chiral carbons in the drug formulas in Figure 26.9.

26.7   Diastereomers and Meso Compounds

The enantiomers discussed in the preceding sections are stereoisomers. That is, they differ only in the spatial arrangement of the atoms and groups within the molecule. The number of stereoisomers increases as the number of chiral carbon atoms increases. The maximum number of stereoisomers for a given compound is obtained by the formula 2^n, where n is the number of chiral carbon atoms.

2^n = **Maximum number of stereoisomers for a given chiral compound**
n = **Number of chiral carbon atoms in a molecule**

As we have seen, there are two ($2^1 = 2$) stereoisomers of lactic acid. But for a substance with two nonidentical chiral carbon atoms, such as 2-bromo-3-chlorobutane ($CH_3CHBrCHClCH_3$), four stereoisomers are possible ($2^2 = 4$). These four possible stereoisomers are written as projection formulas in this way (carbons 2 and 3 are chiral):

XVIII	XIX	XX	XXI
enantiomers		enantiomers	

Remember that, for comparison, projection formulas may be turned 180° in the plane of the paper, but they cannot be lifted (flipped) out of the plane. Formulas XVIII and XIX, and formulas XX and XXI, represent two pairs of nonsuperimposable mirror-image isomers and are therefore two pairs of enantiomers. All four compounds are optically active. But the properties of XVIII and XIX differ from the properties of XX and XXI because they are not mirror-image isomers of each other. Stereoisomers that are not enantiomers (not mirror images of each other) are called **diastereomers.** There are four different pairs of diastereomers of 2-bromo-3-chlorobutane: They are XVIII and XX, XVIII and XXI, XIX and XX, and XIX and XXI.

diastereomer

The 2^n formula indicates that four stereoisomers of tartaric acid are possible. The projection formulas of these four possible stereoisomers are written in this way (carbons 2 and 3 are chiral):

$$
\begin{array}{cccc}
 & \text{COOH} & & \text{COOH} \\
\text{HO}\!-\!\!\!\!& |\!\!\!\!&-\text{H} & \text{H}\!-\!\!\!\!& |\!\!\!\!&-\text{OH} \\
\text{H}\!-\!\!\!\!& |\!\!\!\!&-\text{OH} & \text{HO}\!-\!\!\!\!& |\!\!\!\!&-\text{H} \\
 & \text{COOH} & & \text{COOH}
\end{array}
$$

Formulas XXII and XXIII represent nonsuperimposable mirror-image isomers and are thus enantiomers. Formulas XXIV and XXV are also mirror images. But by turning XXV 180°, we see that it is exactly superimposable on XXIV. Therefore, XXIV and XXV represent the same compound, and only *three* stereoisomers of tartaric acid actually exist. Compound XXIV is achiral and does not rotate polarized light. A plane of symmetry can be passed between carbons 2 and 3 so that the top and bottom halves of the molecule are mirror images:

Thus the molecule is internally compensated. The rotation of polarized light in one direction by half of the molecule is exactly compensated by an opposite rotation by the other half. Stereoisomers that contain chiral carbon atoms and are superimposable on their own mirror images are called **meso compounds,** or **meso structures.** All meso compounds are optically inactive.

meso compound
meso structure

The term *meso* comes from the Greek word *mesos,* meaning middle. It was first used by Pasteur to name a kind of tartaric acid that was optically inactive and could not be separated into different forms by any means. Pasteur called it *meso*-tartaric acid, because it seemed intermediate between the (+)- and (−)-tartaric acid. The three stereoisomers of tartaric acid are represented and designated in this fashion:

(−)-tartaric acid   (+)-tartaric acid   *meso*-tartaric acid

TABLE 26.1 Properties of Tartaric Acid, HOOCCH(OH)CH(OH)COOH				
Name	Specific gravity	Melting point (°C)	Solubility (g/100 g H_2O)	Specific rotation [α]
(+)-Tartaric acid	1.760	170	$147^{20°C}$	+12°
(−)-Tartaric acid	1.760	170	$147^{20°C}$	−12°
(±)-Tartaric acid (racemic mixture)	1.687	206	$20.6^{20°C}$	0°
meso-Tartaric acid	1.666	140	$125^{15°C}$	0°

The physical properties of tartaric acid stereoisomers are given in Table 26.1. Note that the properties of (+)-tartaric acid and (−)-tartaric acid are identical except for opposite rotation of polarized light. However, *meso*-tartaric acid has properties that are entirely different from those of the other isomers. But most surprising is the fact that the racemic mixture, though composed of equal parts of the (+) and (−) enantiomers, differs from them in specific gravity, melting point, and solubility. Why, for example, is the melting point of the racemic mixture higher than that of any of the other forms? The melting point of any substance is largely dependent on the attractive forces holding the ions or molecules together. The melting point of the racemic mixture is higher than that of either enantiomer. Therefore, we can conclude that the attraction between molecules of the (+) and (−) enantiomers in the racemic mixture is greater than the attraction between molecules of the (+) and (+) or the (−) and (−) enantiomers.

How many stereoisomers can exist for the following compounds? Write their structures and label any pairs of enantiomers and meso compounds. Point out any diastereomers.

Example 26.3

(a) $CH_3CHBrCHBrCH_2CH_3$ (b) $CH_2BrCHClCHClCH_2Br$

(a) Carbons 2 and 3 are chiral, so there can be a maximum of four stereoisomers ($2^2 = 4$). Write structures around the chiral carbons:

Solution

There are four stereoisomers: two pairs of enantiomers (I and II, and III and IV) and no meso compounds. Structures I and III, I and IV, II and III, and II and IV are diastereomers.

(b) Carbons 2 and 3 are chiral, so there can be a maximum of four stereoisomers. Write structures around the chiral carbons:

$$
\begin{array}{cccc}
\text{CH}_2\text{Br} & \text{CH}_2\text{Br} & \text{CH}_2\text{Br} & \text{CH}_2\text{Br} \\
| & | & | & | \\
\text{H}-\text{C}-\text{Cl} & \text{Cl}-\text{C}-\text{H} & \text{Cl}-\text{C}-\text{H} & \text{H}-\text{C}-\text{Cl} \\
| & | & | & | \\
\text{H}-\text{C}-\text{Cl} & \text{Cl}-\text{C}-\text{H} & \text{H}-\text{C}-\text{Cl} & \text{Cl}-\text{C}-\text{H} \\
| & | & | & | \\
\text{CH}_2\text{Br} & \text{CH}_2\text{Br} & \text{CH}_2\text{Br} & \text{CH}_2\text{Br} \\
\text{V} & \text{VI} & \text{VII} & \text{VIII}
\end{array}
$$

meso compound   enantiomers

There are three stereoisomers: one pair of enantiomers (VII and VIII) and one meso compound (V). Structures V and VI represent the meso compound because there is a plane of symmetry between carbons 2 and 3, and turning VI 180° makes it superimposable on V. Structures V and VII and V and VIII are diastereomers.

Practice 26.7

Write all stereoisomer structures and label any pairs of enantiomers and meso compounds for the following compound. Also, point out any diastereomers.

HOOCCHClCH$_2$CH$_2$CHClCOOH

Practice 26.8 Application to Biochemistry

D-2-Deoxyribose is a sugar that is very important in the structure of nucleic acids, the cell's genetic information:

$$
\begin{array}{c}
\text{O} \\
\| \\
\text{C}-\text{H} \\
| \\
\text{H}-\text{C}-\text{H} \\
| \\
\text{H}-\text{C}-\text{OH} \\
| \\
\text{H}-\text{C}-\text{OH} \\
| \\
\text{CH}_2\text{OH}
\end{array}
$$

Write all the stereoisomers of D-2-deoxyribose. Label any pairs of enantiomers.

Concepts in Review

1. Identify all chiral (asymmetric) carbon atoms in a given formula.

2. Explain the use of a polarimeter.

3. Explain how polarized light is obtained.

4. Explain the phenomenon of optical isomerism.

5. Determine whether a compound is chiral or achiral.

6. Draw projection formulas for all possible stereoisomers of a given compound. Label enantiomers, diastereomers, and meso compounds.

7. Given the formula of a compound, calculate the maximum possible number of optical isomers.

8. Understand the meaning of $(+)$ and $(-)$ relative to the optical activity of a compound.

9. Draw the mirror image of a given structure.

10. Compare projection formulas to ascertain whether they represent identical compounds or enantiomers.

11. Compare the physical properties of enantiomers, diastereomers, and racemic mixtures.

12. Explain why meso compounds are optically inactive.

13. Determine whether optical isomers are formed in simple chemical reactions.

Key Terms

The terms listed here have been defined within this chapter. Section numbers are referenced in parentheses for each term.

achiral (26.3)
asymmetric carbon atom (26.3)
chiral (26.3)
chiral carbon atom (26.3)

dextrorotatory (26.3)
diastereomer (26.7)
enantiomer (26.5)
levorotatory (26.3)
meso compound (26.7)

meso structure (26.7)
optical activity (26.3)
plane-polarized light (26.2)
projection formula (26.4)
racemic mixture (26.6)

specific rotation (26.2)
stereoisomer (26.1)
stereoisomerism (26.1)
superimposable (26.3)

Questions

Questions refer to tables, figures, and key words and concepts defined within the chapter. A particularly challenging question or exercise is indicated with an asterisk.

1. What is a chiral carbon atom? Draw structural formulas of three different compounds that contain one chiral carbon atom and mark the chiral carbon in each with an asterisk.

2. How can you tell when the axes of two Polaroid filters are parallel? When one filter has been rotated by 90°?

3. What is a necessary and sufficient condition for a compound to show enantiomerism?

4. Differentiate between an enantiomer and a diastereomer.

5. The physical properties for $(+)$-2-methyl-1-butanol are: specific rotation, $+5.76°$; bp, 129°C; density, 0.819 g/mL. What are these same properties for $(-)$-2-methyl-1-butanol?

6. Why are single-isomer drugs commonly more effective than racemic mixtures?

Paired Exercises

These exercises are paired. Each odd-numbered exercise is followed by a similar even-numbered exercise. Answers to the even-numbered exercises are given in Appendix VI.

7. Explain why it is not possible to separate enantiomers by ordinary chemical and physical means.

9. Which of these objects is chiral?
   (a) your ear
   (b) a pair of pliers
   (c) a coiled spring
   (d) the letter b

8. Explain why it is usually possible to separate diastereomers by ordinary chemical and physical means.

10. Which of these objects is chiral?
   (a) a wood screw
   (b) the letter o
   (c) the letter g
   (d) this textbook

11. How many chiral carbon atoms are present in each of the following compounds?

(a)
$$\text{Cl}-\overset{\overset{\displaystyle H}{|}}{\underset{\underset{\displaystyle H}{|}}{C}}-\overset{\overset{\displaystyle Cl}{|}}{\underset{\underset{\displaystyle H}{|}}{C}}-\text{Br}$$

(b) $CH_3CH_2CH_2CHClCH_3$

(c)
$$\begin{array}{c} H-C=O \\ H-\underset{|}{\overset{|}{C}}-OH \\ H-\underset{|}{\overset{|}{C}}-OH \\ H-\underset{|}{\overset{|}{C}}-OH \\ H \end{array}$$

(d)
$$\text{HO}-\overset{\overset{\displaystyle H}{|}}{\underset{\underset{\displaystyle H}{|}}{C}}-\overset{\overset{\displaystyle H}{|}}{\underset{\underset{\displaystyle NH_2}{|}}{C}}-\overset{\overset{\displaystyle O}{\|}}{C}-\text{OH}$$

12. How many chiral carbon atoms are present in each of the following compounds?

(a)
$$\text{H}-\overset{\overset{\displaystyle Br}{|}}{\underset{\underset{\displaystyle H}{|}}{C}}-\overset{\overset{\displaystyle Cl}{|}}{\underset{\underset{\displaystyle H}{|}}{C}}-\text{H}$$

(b) $CH_3CH_2\overset{\overset{\displaystyle OH}{|}}{C}HCH_2CH_3$

(c)
$$\begin{array}{c} H \\ H-\underset{|}{\overset{|}{C}}-OH \\ C=O \\ HO-\underset{|}{\overset{|}{C}}-H \\ H-\underset{|}{\overset{|}{C}}-OH \\ H-\underset{|}{\overset{|}{C}}-OH \\ H-\underset{|}{\overset{|}{C}}-OH \\ H \end{array}$$

(d) $CH_3CH_2CHBrCHClCH_3$

13. Write the formulas and decide which of the following compounds will show optical activity:
(a) 1-chloropentane
(b) 3-chloropentane
(c) 2-chloro-2-methylpentane
(d) 4-chloro-2-methylpentane

14. Write the formulas and decide which of the following compounds will show optical activity:
(a) 2-chloropentane
(b) 1-chloro-2-methylpentane
(c) 3-chloro-2-methylpentane
(d) 3-chloro-3-methylpentane

15. Glucose, $C_6H_{12}O_6$, has four chiral carbon atoms. How many stereoisomers of glucose are theoretically possible?

16. Fructose, $C_6H_{12}O_6$, has three chiral carbon atoms. How many stereoisomers of fructose are theoretically possible?

17. Do structures A and B represent enantiomers or the same compound? Justify your answer.

(A) $\text{Br}-\overset{\overset{\displaystyle Cl}{|}}{\underset{\underset{\displaystyle F}{|}}{C}}-\text{H}$ (B) $\text{H}-\overset{\overset{\displaystyle Br}{|}}{\underset{\underset{\displaystyle F}{|}}{C}}-\text{Cl}$

18. Do structures A and B represent enantiomers or the same compound? Justify your answer.

(A) $\text{Br}-\overset{\overset{\displaystyle H}{|}}{\underset{\underset{\displaystyle CH_3}{|}}{C}}-\text{F}$ (B) $\text{Br}-\overset{\overset{\displaystyle CH_3}{|}}{\underset{\underset{\displaystyle F}{|}}{C}}-\text{H}$

19. Which of these projection formulas represent $(-)$-lactic acid and which represent $(+)$-lactic acid?

$$
\begin{array}{ccc}
\text{COOH} & & \text{COOH} \\
\text{H}-\!\!\!\!\!\!-\text{OH} & & \text{HO}-\!\!\!\!\!\!-\text{H} \\
\text{CH}_3 & & \text{CH}_3 \\
(-)\text{-lactic acid} & & (+)\text{-lactic acid}
\end{array}
$$

(a) $\begin{array}{c}\text{CH}_3 \\ \text{HO}-\!\!\!\!-\text{H} \\ \text{COOH}\end{array}$ **(d)** $\begin{array}{c}\text{CH}_3 \\ \text{H}-\!\!\!\!-\text{OH} \\ \text{COOH}\end{array}$

(b) $\begin{array}{c}\text{OH} \\ \text{H}-\!\!\!\!-\text{COOH} \\ \text{CH}_3\end{array}$ **(e)** $\begin{array}{c}\text{COOH} \\ \text{CH}_3-\!\!\!\!-\text{H} \\ \text{OH}\end{array}$

(c) $\begin{array}{c}\text{COOH} \\ \text{H}-\!\!\!\!-\text{CH}_3 \\ \text{OH}\end{array}$ **(f)** $\begin{array}{c}\text{H} \\ \text{CH}_3-\!\!\!\!-\text{OH} \\ \text{COOH}\end{array}$

20. Which of these projection formulas represent the naturally occurring amino acid, $(+)$-alanine and which represent $(-)$-alanine?

$$
\begin{array}{ccc}
\text{COOH} & & \text{COOH} \\
\text{H}_2\text{N}-\!\!\!\!\!\!-\text{H} & & \text{H}-\!\!\!\!\!\!-\text{NH}_2 \\
\text{CH}_3 & & \text{CH}_3 \\
(+)\text{-alanine} & & (-)\text{-alanine}
\end{array}
$$

(a) $\begin{array}{c}\text{CH}_3 \\ \text{H}_2\text{N}-\!\!\!\!-\text{H} \\ \text{COOH}\end{array}$ **(d)** $\begin{array}{c}\text{CH}_3 \\ \text{H}-\!\!\!\!-\text{NH}_2 \\ \text{COOH}\end{array}$

(b) $\begin{array}{c}\text{NH}_2 \\ \text{H}-\!\!\!\!-\text{COOH} \\ \text{CH}_3\end{array}$ **(e)** $\begin{array}{c}\text{COOH} \\ \text{CH}_3-\!\!\!\!-\text{H} \\ \text{NH}_2\end{array}$

(c) $\begin{array}{c}\text{COOH} \\ \text{H}-\!\!\!\!-\text{CH}_3 \\ \text{NH}_2\end{array}$ **(f)** $\begin{array}{c}\text{H} \\ \text{CH}_3-\!\!\!\!-\text{NH}_2 \\ \text{COOH}\end{array}$

21. Draw projection formulas for all possible stereoisomers of the following compounds. Label pairs of enantiomers and meso compounds.
 (a) 1,2-dibromopropane
 (b) 2-butanol
 (c) 3-chlorohexane

22. Draw projection formulas for all possible stereoisomers of the following compounds. Label pairs of enantiomers and meso compounds.
 (a) 2,3-dichlorobutane
 (b) 2,4-dibromopentane
 (c) 3-hexanol

23. Draw projection formulas for all the stereoisomers of 1,2,3-trihydroxybutane. Point out enantiomers, meso compounds, and diastereomers, where present.

24. Draw projection formulas for all the stereoisomers of 3,4-dichloro-2-methylpentane. Point out enantiomers, meso compounds, and diastereomers, where present.

*25. Write structures for the four stereoisomers of 2-hydroxy-3-pentene.

*26. Write structures for the four stereoisomers of 2-chloro-3-hexene.

27. **(a)** Draw the nine structural isomers of $C_4H_8Cl_2$.
 (b) Identify which structures represent chiral molecules, and draw all possible pairs of enantiomers and meso compounds, if any.

28. **(a)** Draw the four structural isomers of $C_3H_6Br_2$.
 (b) Identify which structures represent chiral molecules, and draw all possible pairs of enantiomers and meso compounds, if any.

29. $(+)$-2-Bromopentane is further brominated to give dibromopentanes, $C_5H_{10}Br_2$. Write structures for all the possible isomers formed and indicate which of these isomers will be optically active. [Remember that $(+)$-2-bromopentane is optically active.]

30. $(+)$-2-Chlorobutane is further chlorinated to give dichlorobutanes, $C_4H_8Cl_2$. Write structures for all the possible isomers formed and indicate which of these isomers will be optically active. [Remember that $(+)$-2-chlorobutane is optically active.]

31. In the chlorination of propane, 1-chloropropane and 2-chloropropane are obtained as products. After separation by distillation, neither product rotates the plane of polarized light. Explain these results.

32. In the chlorination of butane, 1-chlorobutane and 2-chlorobutane are obtained as products. After separation by distillation, neither product rotates the plane of polarized light. Explain these results.

33. Which, if any, of the following are meso compounds?

35. Some substituted cycloalkanes are chiral. Draw the structures and enantiomers for any of the following that are chiral:

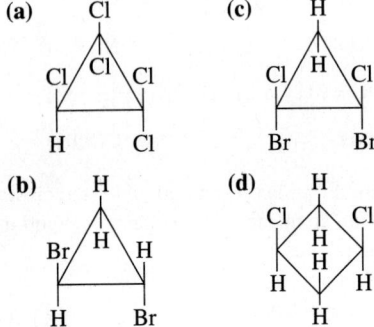

34. Which, if any, of the following are meso compounds?

36. Some substituted cycloalkanes are chiral. Draw the structures and enantiomers for any of the following that are chiral:

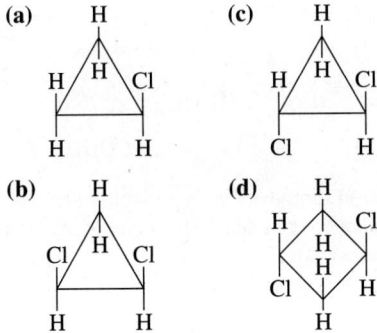

Additional Exercises

These exercises are not paired and provide additional practice on concepts covered in this chapter.

37. Suppose a carbon atom is located at the center of a square with four different groups attached to the corners in a planar arrangement. Would the compound rotate polarized light? Explain.

38. Ibuprofen is a common pain reliever found in medications such as Motrin and Advil. It is sold as a racemic mixture. The structure of ibuprofen is as follows:

$(CH_3)_2CHCH_2$⬡$CHCOOH$ (with CH₃ on the CH)

(a) Identify the chiral carbon atom(s) in ibuprofen.
(b) Using projection formulas, draw the enantiomers of this drug.

39. (a) A chiral substance was identified as a primary alcohol of the formula $C_5H_{12}O$. What is its structure?
(b) The compound $C_6H_{14}O_3$ has three primary alcohol groups and is chiral. What is its structure?

40. The following dextrorotatory isomer of carvone gives caraway seeds their distinctive smell.
(a) Identify any chiral carbon atoms in this molecule.
(b) The enantiomer of this molecule is responsible for the refreshing smell of spearmint. How does the "spearmint" molecule differ from the "caraway seed" molecule?

*41. What is the structure of a substance of formula $C_3H_8O_2$ that is (a) chiral and contains two —OH groups; (b) chiral and contains one —OH group; (c) achiral and contains two —OH groups? (Only one —OH group can be bonded to a carbon atom.)

42. Ephedrine is a potent drug originally found in a Chinese herbal medicine. It acts as a bronchial dilator and is used to treat asthma. The structure of ephedrine is

⬡—CH(OH)CH(CH₃)NHCH₃

How many stereoisomers are possible for this drug? Briefly explain your reasoning.

43. One significant property of most amino acids is their chirality. Identify the chiral carbon in the amino acid alanine shown here:

$$\underset{\underset{CH_3}{|}}{\overset{\overset{O}{\overset{||}{C}}-OH}{H_2N-\overset{|}{C}-H}}$$

Draw the mirror image of this compound.

44. When biological interaction of a drug involves optical isomers, drug design becomes complicated because often only one isomer will have the desired property. For example, levomethorphan is an addictive opiate and dextromethorphan is a nonaddictive cough suppressant. What symbols would you use to differentiate between the two? Can you tell which one rotates light to the right? Explain why or why not.

45. Write structures for all stereoisomers and label the enantiomers and meso compounds for
(a) 2-bromo-3-chlorobutane
(b) 2,3,4-trichloro-1-pentanol.

46. Draw all possible meso structures for alkanes that have the following molecular formulas:
(a) C_8H_{18}
(b) C_9H_{20}
(c) $C_{10}H_{22}$

47. Draw structures for all optically active alcohols having the formula $C_6H_{14}O$. (Draw only one enantiomer of each structure.)

Answers to Practice Exercises

26.1

Br, Cl, CH₃, F (Newman projection)

26.2

COOH
H—NH₂
CH₂CH₃

26.3 (a)

⟵—Mirror
NH₂CH₂—C(H)(Br)—CH₂OH | HOCH₂—C(H)(Br)—CH₂NH₂

(b), (c) have no asymmetric carbons

26.4

CH₃CH₂CH₂CHCH₃
         |
         OH

         CH₃
         |
CH₃CH₂CHCH₂OH

         CH₃
         |
CH₃CHCHCH₃
         |
         OH

26.5

COOH
HO—C*—H
H—C*—COOH
H—C—H
COOH
isocitric acid

COOH
H—C*—OH
HOOC—C*—H
H—C—H
COOH
enantiomer

26.6

$CH_3OCH_2CH_2$—[benzene ring]—$OCH_2\overset{*}{C}H$—CH_2—NH—CH—CH_3 (with OH on the *CH, and CH_3 on the CH)

[catechol ring with two OH groups]—$CH_2\overset{*}{C}HCOOH$ (with NH_2)

[benzene ring]—$\overset{*}{C}H$—$\overset{*}{C}H$—NH—CH_3 (with OH on first *CH, CH_3 on second *CH)

[piperidine ring with N—H]—$\overset{*}{C}H$—$\overset{\parallel}{\underset{*}{C}}$—$OCH_3$ (with phenyl on *CH, C=O)

26.7

I	II	III	IV
meso compound		enantiomers	

Structures I and III, I and IV, II and III, and II and IV are diasteromers.

26.8

| I | II | III | IV |

Enantiomers are structures I and II, and III and IV.

27

Carbohydrates

CHAPTER 27 / OUTLINE

Green plants turn H_2O, CO_2, and sunlight into carbohydrates.

749

CHAPTER

27

What is the most abundant organic chemical in the world? The answer is not petroleum products, plastics, or drugs. Rather, it is cellulose. An amazing 10 billion tons of cellulose are formed daily in the biosphere. Aggregates of cellulose allow the California redwoods to stretch hundreds of feet toward the sky and make a Brazil nut a "hard nut to crack." Products as diverse as the paper in this book and cotton in clothing are derived from cellulose. So, perhaps it is not surprising that this carbohydrate is the most widespread organic chemical in the world.

Carbohydrates are molecules of exceptional utility. These molecules provide basic diets for many of us (starch and sugar), roofs over our heads, and clothes for our bodies (cellulose). They also thicken our ice cream, stick postage stamps to our letters, and provide biodegradable plastic trash sacks. From relatively simple components (carbon, hydrogen, and oxygen), nature has created one of the premier classes of biochemicals.

27.1 Carbohydrates: A First Class of Biochemicals

Carbohydrates are among the most widespread and important biochemicals. Most of the matter in plants, except water, consists of these substances. Carbohydrates are one of the three principal classes of energy-yielding nutrients; the other two are fats and proteins. Because of their widespread distribution and their role in many vital metabolic processes such as photosynthesis, carbohydrates have been scrutinized by scientists for 150 years.

The name *carbohydrates* was given to this class of compounds many years ago by French scientists who called them *hydrates de carbone* because their empirical formulas approximated $(C \cdot H_2O)_n$. It was found later that not all substances classified as carbohydrates conform to this formula (e.g., rhamnose, $C_6H_{12}O_5$, and deoxyribose, $C_5H_{10}O_4$). It seems clear that carbohydrates are not simply hydrated carbon; they are complex substances that contain from three to many thousands of carbohydrate carbon atoms. **Carbohydrates** are generally defined as polyhydroxy aldehydes or polyhydroxy ketones or substances that yield these compounds when hydrolyzed.

The simplest carbohydrates are glyceraldehyde and dihydroxyacetone:

glyceraldehyde   dihydroxyacetone

These substances are "polyhydroxy" because each molecule has more than one hydroxyl group. Glyceraldehyde contains a carbonyl carbon in a terminal position and

◀ **When sugar (sucrose) and sulfuric acid (left) are combined, water is removed from the sugar, leaving a black column of carbon (right).**

is therefore an aldehyde. The internal carbonyl of dihydroxyacetone identifies it as a ketone. Much of the chemistry and biochemistry of carbohydrates can be understood from a basic knowledge of the chemistry of the hydroxyl and carbonyl functional groups (see Chapters 22 and 23 for review).

 ## Classification of Carbohydrates

A carbohydrate is classified as a monosaccharide, a disaccharide, an oligosaccharide, or a polysaccharide, depending on the number of monosaccharide units linked to form the molecule. A **monosaccharide** is a carbohydrate that cannot be hydrolyzed to simpler carbohydrate units. The monosaccharide is the basic carbohydrate unit of cellular metabolism. A **disaccharide** yields two monosaccharides—either alike or different—when hydrolyzed:

monosaccharide

disaccharide

$$\text{Disaccharide} + \text{Water} \xrightarrow{\text{H}^+ \text{ or enzymes}} 2 \text{ Monosaccharides}$$

Disaccharides are often used by plants or animals to transport monosaccharides from one cell to another. The monosaccharides and disaccharides generally have names ending in -*ose*—for example, glucose, sucrose, and lactose. These water-soluble carbohydrates, which have a characteristically sweet taste, are also called *sugars*.

An **oligosaccharide** has two to six monosaccharide units linked together. *Oligo* comes from the Greek word *oligos,* which means small or few. Free oligosaccharides that contain more than two monosaccharide units are rarely found in nature.

oligosaccharide

A **polysaccharide** is a macromolecular substance that can be hydrolyzed to yield many monosaccharide units:

polysaccharide

$$\text{Polysaccharide} + \text{Water} \xrightarrow{\text{H}^+ \text{ or enzymes}} \text{Many monosaccharide units}$$

Sugar cane is a major source for sucrose, or table sugar.

Remember, enantiomers are mirror-image isomers.

Polysaccharides are important as structural supports (particularly in plants) and also serve as a storage depot for monosaccharides (which cells use for energy).

Carbohydrates can also be classified in other ways. A monosaccharide might be described with respect to several of these categories:

1. As a triose, tetrose, pentose, hexose, or heptose. Theoretically, a monosaccharide can have any number of carbons greater than three, but only monosaccharides of three to seven carbons are commonly found in the biosphere.

Trioses $C_3H_6O_3$ Hexoses $C_6H_{12}O_6$
Tetroses $C_4H_8O_4$ Heptoses $C_7H_{14}O_7$
Pentoses $C_5H_{10}O_5$

2. As an aldose or ketose, depending on whether an aldehyde group ($-CHO$) or keto group ($>C=O$) is present. For ketoses, the $>C=O$ is normally located on carbon 2.

3. As a D or L isomer, depending on the spatial orientation of the $-H$ and $-OH$ groups attached to the carbon atom adjacent to the terminal primary alcohol group. When the $-OH$ is written to the right of this carbon in the projection formula (see Section 26.4), the D isomer is represented. When this $-OH$ is written to the left, the L isomer is represented. The reference compounds for this classification are the trioses D-glyceraldehyde and L-glyceraldehyde, whose formulas follow. Also shown are two aldohexoses (D- and L-glucose) and a ketohexose (D-fructose).

D-Configuration → ; Terminal 1° ROH → ; D-glyceraldehyde ; L-glyceraldehyde ; ← L-Configuration ; Terminal 1° ROH

L-Configuration → ; Terminal 1° ROH → ; L-glucose ; D-glucose ; D-fructose ; ← D-Configuration ; Terminal 1° ROH →

The letters D and L do not in any way refer to the direction of optical rotation of a carbohydrate. The D and L forms of any specific compound are enantiomers (e.g., D- and L-glucose).

4. As a (+) or (−) isomer, depending on whether the monosaccharide rotates the plane of polarized light to the right (+) or to the left (−). (See Section 26.5.)

5. As a furanose or a pyranose, depending on whether the cyclic structure of the carbohydrate is related to that of the five- or six-membered heterocyclic ring compound furan or pyran (a heterocyclic ring contains more than one kind of atom in the ring):

furan, C_4H_4O pyran, C_5H_6O
(five-membered ring with   (six-membered ring with
oxygen in the ring)   oxygen in the ring)

6. As having an alpha (α) or beta (β) configuration, based on the orientation of the —H and —OH groups about a specific chiral carbon in the cyclic form of the monosaccharide (Section 27.6).

Write projection formulas for (a) an L-aldotriose, (b) a D-ketotetrose, and (c) a D-aldopentose.

Example 27.1

(a) Triose indicates a three-carbon carbohydrate; aldo indicates that the compound is an aldehyde; L- indicates that the —OH on carbon 2 (adjacent to the terminal CH_2OH) is on the left. The aldehyde group is carbon 1.

Solution

$$H-\overset{1}{C}{=}O$$
$$HO-\overset{2}{C}-H$$
$$\overset{3}{C}H_2OH$$

an L-aldotriose

(b) Tetrose indicates a four-carbon carbohydrate; keto indicates a ketone group (on carbon 2); D- indicates that the —OH on carbon 3 (adjacent to the terminal CH_2OH) is on the right. Carbons 1 and 4 have the configuration of primary alcohols.

$$\overset{1}{C}H_2OH$$
$$\overset{2}{C}{=}O$$
$$H-\overset{3}{C}-OH$$
$$\overset{4}{C}H_2OH$$

a D-ketotetrose

(c) Pentose indicates a five-carbon carbohydrate; aldo indicates an aldehyde group (on carbon 1); D- indicates that the —OH on carbon 4 (adjacent to the terminal CH_2OH) is on the right. The orientation of the —OH groups on carbons 2 and 3 is not specified here and therefore can be written in either direction for this problem.

$$H-\overset{1}{C}{=}O$$
$$H-\overset{2}{C}-OH$$
$$H-\overset{3}{C}-OH$$
$$H-\overset{4}{C}-OH$$
$$\overset{5}{C}H_2OH$$

a D-aldopentose

Practice 27.1

Write the projection formula for an L-ketopentose.

Practice 27.2

In a projection formula for a D-aldotriose, is the —OH of the secondary alcohol carbon written on the right or the left side?

27.3   Importance of Carbohydrates

If a carbohydrate carbon has less than 4 bonds to oxygen, it can be oxidized further.

As we stated earlier, carbohydrates are the most abundant organic chemical—they must be molecules of exceptional utility. But what makes them so special? Why are they so important in our daily lives?

Carbohydrates are very effective energy-yielding nutrients. In general, a molecule can provide biological energy if it (1) contains carbon, (2) is relatively reactive (contains some polar bonds), and (3) contains carbons that have less than four bonds to oxygen. Carbohydrates fit this description very well. The average carbohydrate contains about 40% carbon by mass. Carbohydrates are relatively reactive because they contain many polar groups (e.g., hydroxyl groups). Finally, each carbohydrate may be oxidized further.

Carbohydrates can serve as very effective building materials. These materials need to be water-soluble and strong but not brittle. As you will see in Section 27.14, large, water-insoluble polymers (polysaccharides) are formed from simple carbohydrates. Cellulose polymers can form durable fibers by hydrogen bonding. These fibers provide building materials for strong, relatively rigid wood construction; for flexible, comfortable cotton clothing; and for flexible yet smooth paper writing material. Cellulose also forms at least part of the cell wall for all higher plants.

Carbohydrates are important water-soluble molecules. Many smaller ones, especially the mono- and disaccharides, are very water soluble. These carbohydrates are often mixed into foods to give them a sweet taste. Water solubility allows these molecules to pass easily from living cell to living cell.

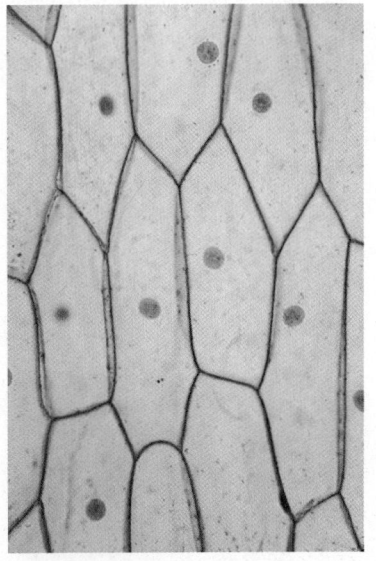

▲
Carbohydrates are important building blocks in the formation of cell walls in higher plants. Shown here are the cell walls of an onion.

27.4   Monosaccharides

Although a great many monosaccharides have been synthesized, only a very few appear to be of much biological significance. One pentose monosaccharide (ribose) and its deoxy derivative are essential components of ribonucleic acid (RNA) and of deoxyribonucleic acid (DNA) (see Chapter 31). However, the hexose monosaccharides are the most important carbohydrate sources of cellular energy. Three hexoses—glucose, galactose, and fructose—are of major significance in nutrition. All three have the same molecular formula, $C_6H_{12}O_6$, and thus deliver the same amount of cellular energy. They differ in structure but are biologically interconvertible. Glucose plays a central role in carbohydrate energy utilization. Other carbohydrates are usually converted to glucose before cellular utilization. The structure of glucose is considered in detail in Section 27.5.

Glucose

Glucose is the most important of the monosaccharides. It is an aldohexose and is found in the free state in plant and animal tissue. Glucose is also known as *dextrose* or *grape sugar.* It is a component of the disaccharides sucrose, maltose, and lactose, and is the monomer of the polysaccharides starch, cellulose, and glycogen. Among the common sugars, glucose is of intermediate sweetness (see Section 27.12).

Glucose is the key sugar of the body and is carried by the bloodstream to all body parts. The concentration of glucose in the blood is normally 80–100 mg per 100 mL of blood. Because glucose is the most abundant carbohydrate in the blood, it is commonly called *blood sugar.* Glucose requires no digestion and may therefore be given intravenously to patients who cannot take food by mouth. Glucose is found in the urine of those who have diabetes mellitus (sugar diabetes). The condition in which glucose is excreted in the urine is called glycosuria.

Galactose

Galactose is also an aldohexose and occurs, along with glucose, in lactose and in many oligo- and polysaccharides such as pectin, gums, and mucilages. Galactose is an isomer of glucose, differing only in the spatial arrangement of the —H and —OH groups around carbon 4 (see Section 27.5). Galactose is synthesized in the mammary glands to make the lactose of milk. It is also a constituent of glycolipids and glyco-proteins in many cell membranes such as those in nervous tissue. Galactose is less than half as sweet as glucose.

The inability of infants to metabolize galactose is an inherited condition called galactosemia. In galactosemia, the galactose concentration increases markedly in the blood and also appears in the urine. Galactosemia causes vomiting, diarrhea, enlarge-ment of the liver, and often mental retardation. If not recognized within a few days after birth, it can lead to death. If diagnosis is made early and lactose is excluded from the diet, the symptoms disappear and normal growth may be resumed.

Fructose

Fructose, also known as *levulose,* is a ketohexose that occurs in fruit juices, honey, and (along with glucose) as a constituent of sucrose. Fructose is the major con-stituent of the polysaccharide inulin, a starchlike substance present in many plants such as dahlia tubers, chicory roots, and Jerusalem artichokes. Fructose is the sweet-est of all the common sugars, being about twice as sweet as glucose. This accounts for the sweetness of high-fructose corn syrup and honey. The enzyme invertase, present in bees, splits sucrose into glucose and fructose. Fructose is metabolized di-rectly but is also readily converted to glucose in the liver.

27.5   Structure of Glucose and Other Aldoses

In one of the classic feats of research in organic chemistry, Emil Hermann Fischer (1852–1919), working in Germany, established the structural configuration of glucose along with that of many other sugars. For his work he received the Nobel prize in chemistry in 1902. Fischer devised projection formulas that relate the structure of a sugar to one or the other of the two enantiomeric forms of glyceraldehyde. These projection formulas represent three-dimensional stereoisomers (see Chapter 26) in a two-dimensional plane. (Remember that stereoisomers cannot be interconverted without breaking and reforming covalent bonds. Stereoisomers differ in their chemical reactivity.) Each carbohydrate isomer has a different shape and thus reacts differently in biological systems.

In Fischer projection formulas, the molecule is represented with the aldehyde (or ketone) group at the top. The —H and —OH groups attached to interior carbons are written to the right or to the left as they would appear when projected toward the observer. The two glyceraldehydes are represented thus:

A projection formula is commonly written in either of two ways, with or without the chiral carbon atoms.

H—C=O

H   OH

CH₂OH

D-glyceraldehyde
(three-dimensional
representation)

H—¹C=O
H—²C—OH
³CH₂OH

H—¹C=O
H—²——OH
³CH₂OH

D-glyceraldehyde
(Fischer projection
formulas)

H—C=O

HO   H

CH₂OH

L-glyceraldehyde
(three-dimensional
representation)

H—¹C=O
HO—²C—H
³CH₂OH

H—¹C=O
HO—²——H
³CH₂OH

L-glyceraldehyde
(Fischer projection
formulas)

In the three-dimensional molecules represented by these formulas, the carbon 2 atoms are in the plane of the paper. The —H and —OH groups project forward (toward the observer); the —CHO and —CH₂OH groups project backward (away from the observer). Any two monosaccharides that differ only in the configuration around a single carbon atom are called **epimers.** Thus D- and L-glyceraldehyde are epimers.

epimer

Fischer recognized that there are two enantiomeric forms of glucose. To these forms he assigned the following structures and names:

Enantiomers appear to have similar structures, but they differ at every chiral carbon.

H—¹C=O
H—²C—OH
HO—³C—H
H—⁴C—OH
H—⁵C—OH
⁶CH₂OH

H—¹C=O
H—²——OH
HO—³——H
H—⁴——OH
H—⁵——OH
⁶CH₂OH

D-glucose

H—¹C=O
HO—²C—H
H—³C—OH
HO—⁴C—H
HO—⁵C—H
⁶CH₂OH

H—¹C=O
HO—²——H
H—³——OH
HO—⁴——H
HO—⁵——H
⁶CH₂OH

L-glucose

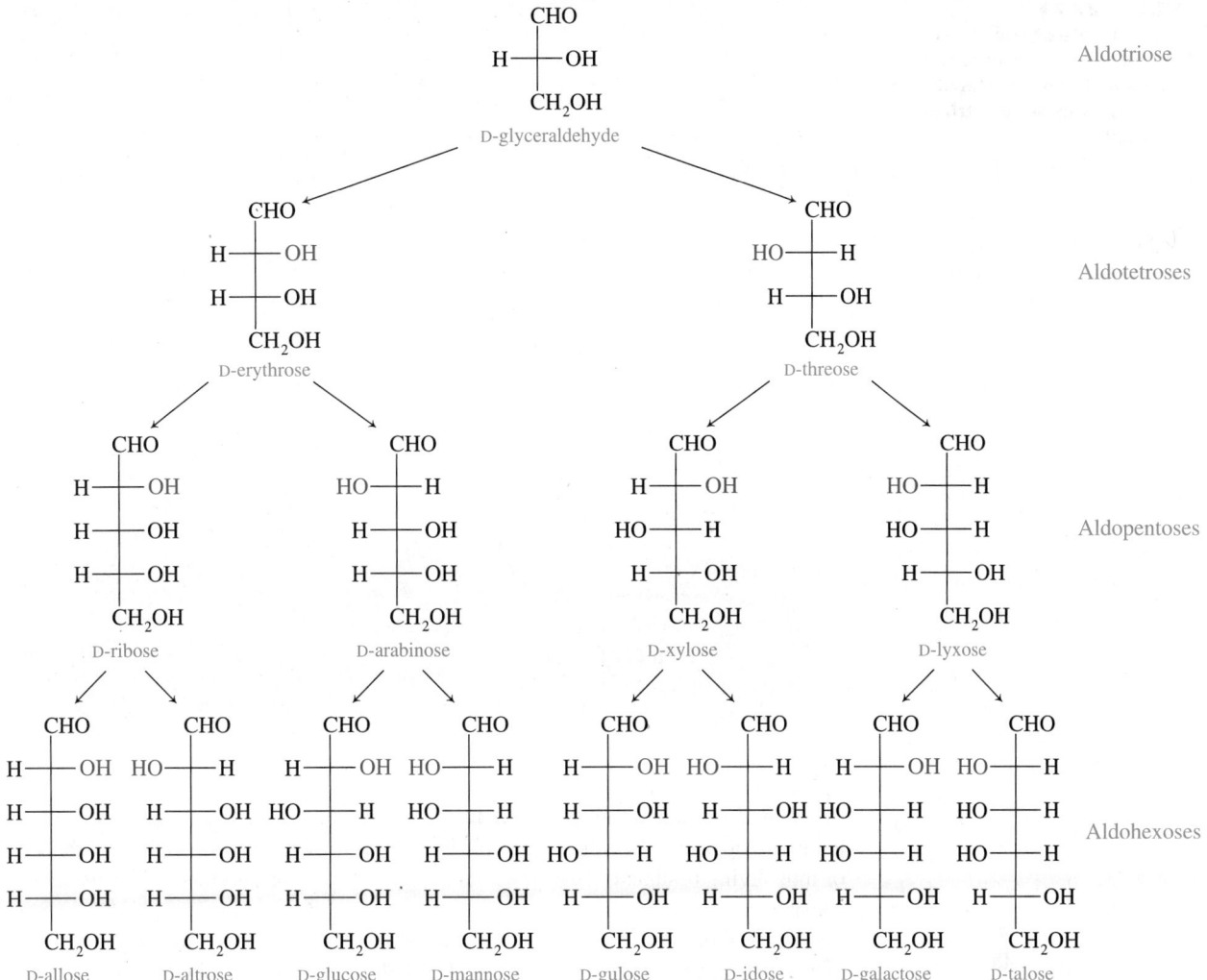

FIGURE 27.1
Configurations of the D-family of aldoses. The hydroxyl group on the new chiral carbon atom added in going from triose to tetrose to pentose to hexose is shown in red.

The structure called D-glucose is so named because the —H and —OH on carbon 5 are in the same configuration as the —H and —OH on carbon 2 in D-glyceraldehyde. The configuration of the —H and —OH on carbon 5 in L-glucose corresponds to the —H and —OH on carbon 2 in L-glyceraldehyde.

Fischer recognized that 16 different aldohexoses, 8 with the D configuration and 8 with the L configuration, were possible. This follows our formula 2^n for optical isomers (see Section 26.7). Glucose has four chiral carbon atoms and should have 16 stereoisomers (2^4). The configurations of the D-aldose family are shown in Figure 27.1. In this family, new chiral carbon atoms are formed as we go from triose to tetrose to pentose to hexose. Each time a new chiral carbon is added, a pair of epimers is formed that differ only in the structure around carbon 2. This sequence continues until eight D-aldohexoses are created. A similar series starting with L-glyceraldehyde is known, making a total of 16 aldohexoses.

FIGURE 27.2 ▶
**An example of the Kiliani–
Fischer synthesis in which
two aldotetrose molecules are
formed from an aldotriose
molecule.**

The aldehyde carbon of
D-glyceraldehyde undergoes
an addition reaction during
Kiliani–Fischer synthesis.

Practice 27.3

Refer to Figure 27.1 and draw the structure for
(a) the enantiomer of D-idose
(b) the enantiomer of D-xylose

The 16 aldohexoses have all been synthesized, but only D-glucose and D-galactose appear to be of considerable biological importance. Since the metabolism of most living organisms revolves about D-glucose, our discussion will focus on this substance.

The laboratory conversion of one aldose into another aldose containing one more carbon atom is known as the Kiliani–Fischer synthesis. This synthesis makes use of the aldehyde's ability to bond to an additional group (see Section 23.5). Heinrich Kiliani (1855–1945) discovered that the addition of HCN to an aldose formed a cyanohydrin. Emil Fischer subsequently published a method for converting the cyanohydrin nitrile group ($-CN$) into an aldehyde. The Kiliani–Fischer synthesis involves (1) the addition of HCN to form a cyanohydrin; (2) hydrolysis of the $-CN$ group to $-COOH$; and (3) reduction with sodium amalgam, Na(Hg), to form the aldehyde. As an example, the formation of D-erythrose and D-threose from D-glyceraldehyde is shown in Figure 27.2.

Practice 27.4

Write the structures of the two aldoheptoses formed when D-glucose is subjected to the Kiliani–Fischer synthesis.

α-D-(+)-glucopyranose
[α] = +112°

on same side

D-(+)-glucose
(open-chain form)

β-D-(+)-glucopyranose
[α] = +18.7°

-OH on opposite side of -o bridge

(a) Modified Fischer projection formulas

The aldehyde carbon of the open-chain D-glucose undergoes an addition reaction when the cyclic structure of D-glucose forms.

α-D-(+)-glucopyranose

-OH on opposite side

D-(+)-glucose
(open-chain form)

β-D-(+)-glucopyranose

beta (on same side)

(b) Haworth perspective formulas

CH₂OH

above D — CH₂OH above ring

below L — " " " below ring

27.6 Cyclic Structure of Glucose; Mutarotation

Straight open-chain D-glucose is so reactive that almost all molecules quickly rearrange their bonds to form two new structures. These two forms are diastereomers and differ with respect to their rotation of polarized light. One form, labeled α-D-glucopyranose, has a specific rotation, [α], of +112°; the other, labeled β-D-glucopyranose, has a specific rotation of +18.7°. An interesting phenomenon occurs when these two forms of glucose are put into separate solutions and allowed to stand for several hours. The specific rotation of each solution changes to +52.7°. This phenomenon is known as mutarotation. An explanation of mutarotation is that D-glucose exists in solution as an equilibrium mixture of two cyclic forms and the open-chain form (see Figure 27.3). The two cyclic molecules are optical isomers, differing only in the orientation of the —H and —OH groups about carbon 1. When dissolved, some α-D-glucopyranose molecules are transformed into β-D-glucopyranose, and vice versa, until an equilibrium is reached between the α and β forms. (Note that no other chiral centers in D-glucose are altered when this sugar is dissolved.) The equilibrium solution contains about 36% α molecules and 64% β molecules, with a trace of open-chain molecules. When two cyclic isomers differ only in their stereo arrangement about the carbon involved in mutarotation, they are called **anomers.** For example, α- and β-D-glucopyranose are anomers (see Figure 27.3). **Mutarotation** is the process by which anomers are interconverted.

anomer

mutarotation

The cyclic forms of D-glucose may be represented by either Fischer projection formulas or by Haworth perspective formulas. These structures are shown in Figure 27.3. In the cyclic Fischer projection formulas of the D-aldoses, the α form has the

FIGURE 27.4 ▶
Three-dimensional representa-
tions of the chair form of α-D-
glucopyranose.

(a) Ball-and-stick model

(b) Space-filling model

—OH on carbon 1 written to the right; in the β form the —OH on carbon 1 is on the left. The Haworth formula represents the molecule as a flat hexagon with the —H and —OH groups above and below the plane of the hexagon. In the α form the —OH on carbon 1 is written below the plane; in the β form the —OH on carbon 1 is above the plane. In converting the projection formula of a D-aldohexose to the Haworth formula, the —OH groups on carbons 2, 3, and 4 are written below the plane if they project to the right and above the plane if they project to the left. Carbon 6 is written above the plane.

Haworth formulas are sometimes shown in abbreviated schematic form. For example, α-D-(+)-glucopyranose is shown as

The four open bonds are understood to have hydrogen atoms on them.

α-D-(+)-glucopyranose
(abbreviated form)

Although both the Fischer projection formula and the Haworth formula provide useful representations of carbohydrate molecules, it is important to understand that these structures only approximate the true molecular shapes. We know, for example, that the pyranose ring is not flat but, rather, can assume either a chair or boat conformation like the cycloalkanes (see Section 19.14). Most naturally occurring monosaccharides are found in the chair form as shown in Figure 27.4 for α-D-glucopyranose. Even this three-dimensional structure does not truly capture how a sugar molecule must appear. For, unlike this representation, we know that atoms move as close together as possible when they form molecules. Perhaps the most accurate representation of a sugar molecule is the space-filling model. A space-filling model of α-D-glucopyranose is shown in Figure 27.4. At best, any two-dimensional representation is a compromise in portraying the three-dimensional configuration of such molecules. Structural models are much more effective, especially if constructed by the student.

The two cyclic forms of D-glucose differ only in the relative positions of the —H and —OH groups attached to carbon 1. Yet, this seemingly minor structural difference has important biochemical consequences because the physical shape of a molecule often determines its biological use. For example, the fundamental structural difference between starch and cellulose is that starch is a polymer of α-D-glucopyranose, whereas cellulose is a polymer of β-D-glucopyranose. As a consequence, starch is easily digested by humans, but we are totally unable to digest cellulose.

Example 27.2

Write the pyranose Haworth perspective formulas and names for the two anomers of D-mannose.

Solution

First, write the open-chain Fischer projection formula for D-mannose (you must memorize this structure or know where to find it in the text) and number the carbons from top (the aldehyde group) to bottom (the primary alcohol group).

```
      1
  H — C = O
      2
 HO —|— H
      3
 HO —|— H
      4
  H —|— OH
      5
  H —|— OH
      6
     CH₂OH
```

Next, draw the structure of the Haworth pyranose ring. Number the carbons from the right-hand point of the hexagon clockwise around the cyclic form, placing the CH_2OH (C6) group in the up position on the ring. Then, refer to the open-chain Fischer projection formula. All the hydroxyl groups on the right of the open chain should be written *down* in the Haworth formula, and all the hydroxyl groups on the left should be written *up*. Since this rule only applies to chiral centers, we can ignore the hydroxyl on carbon 6.

The carbon involved in mutarotation, carbon 1, can have either of two configurations, the α anomer when the hydroxyl is pointed down or the β anomer when pointed up. The last step is to add the hydroxyl group at carbon 1 and name the anomers.

α-D-mannopyranose β-D-mannopyranose

27.7   Hemiacetals and Acetals

In Chapter 23 we studied the reactions of aldehydes (and ketones) to form hemiacetals and acetals. The hemiacetal structure consists of an ether linkage and an alcohol linkage on the same carbon atom (shown in red), whereas the acetal structure has two ether linkages to the same carbon atom:

hemiacetal structures

acetal structures

Cyclic structures of monosaccharides are intramolecular hemiacetals. Five- or six-membered rings are especially stable.

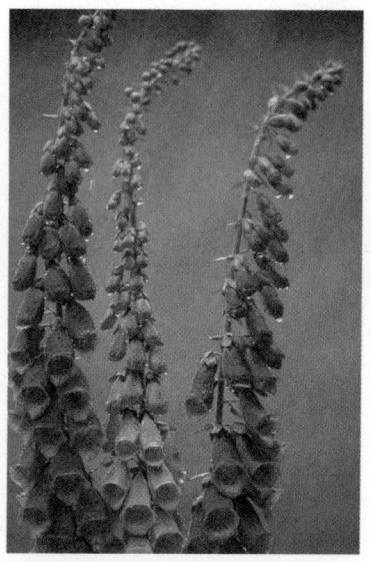

▲
Digitalis, a well-known heart stimulant, is produced in nature by the foxglove plant.

However, in an aqueous solution, the ring often opens and the hemiacetal momentarily reverts to the open-chain aldehyde. When the open chain closes, it forms either the α or the β anomer. Mutarotation results from this opening and closing of the hemiacetal ring (see Figure 27.3).

When an alcohol, ROH, reacts with another alcohol, R′OH to split out H_2O, the product formed can be an ether, ROR′. Carbohydrates are alcohols and behave in a similar manner. When a monosaccharide hemiacetal reacts with an alcohol, the product is an acetal. In carbohydrate terminology this acetal structure is called a **glycoside** (derived from the Greek word *glykys,* meaning sweet). In the case of glucose, it would be a glucoside; if galactose, a galactoside; and so on.

glycoside

CH$_2$OH

— Acetal structure

— Glycosidic linkage

an α-glycoside —sweet
(R = a variety of groups)

A glycoside differs significantly from a monosaccharide with respect to chemical reactivity.

When α-D-glucopyranose is heated with methyl alcohol and a small quantity of hydrogen chloride is added, two optically active isomers are formed—methyl α-D-glucopyranoside and methyl β-D-glucopyranoside:

CH$_2$OH

$\xrightleftharpoons[\text{HCl}]{\text{CH}_3\text{OH}}$

CH$_2$OH

+

CH$_2$OH

α-D-glucopyranose

methyl α-D-glucopyranoside
(mp 165°C, [α] = + 158°)

methyl β-D-glucopyranoside
(mp 107°C, [α] = − 33°)

Unlike D-glucose, the two glycoside products no longer undergo mutarotation. They do not form open-chain compounds in aqueous solution. Acetals tend to be more stable and less reactive than hemiacetals.

The glycosidic linkage occurs in a wide variety of natural substances. All carbohydrates other than monosaccharides are glycosides. Heart stimulants such as digitalis and ouabain are known as heart glycosides. Several antibiotics such as streptomycin and erythromycin are also glycosides.

27.8 Structures of Galactose and Fructose

Galactose, like glucose, is an aldohexose and differs structurally from glucose only in the configuration of the —H and —OH group on carbon 4:

Differs from
D-glucose here ⟶

H—^{1}C=O
H—^{2}—OH
HO—^{3}—H
HO—^{4}—H
H—^{5}—OH
^{6}CH$_2$OH

D-galactose

H—^{1}C=O
H—^{2}—OH
HO—^{3}—H
H—^{4}—OH
H—^{5}—OH
^{6}CH$_2$OH

D-glucose

To identify anomers in the Haworth formula, focus on carbon 1 for aldoses and carbon 2 for ketoses. If the —OH on this carbon is written below the plane of the ring, the molecule is an α anomer; if the —OH is written above the plane of the ring, the molecule is a β anomer.

Galactose, like glucose, also exists primarily in two cyclic pyranose forms that have hemiacetal structures and undergo mutarotation:

α-D-galactopyranose

β-D-galactopyranose

Fructose is a ketohexose. The open-chain form may be represented in a Fischer projection formula:

D-fructose

Like glucose and galactose, fructose exists in both cyclic and open-chain forms. One common cyclic structure is a five-membered furanose ring in the β configuration:

β-D-fructofuranose

27.9   Pentoses

An open-chain aldopentose has three chiral carbon atoms. Therefore, eight (2^3) isomeric aldopentoses are possible. The four possible D-pentoses are shown in Figure 27.1. Arabinose and xylose occur in some plants as polysaccharides called pentosans. D-Ribose and its derivative, D-2-deoxyribose, are the most interesting pentoses because of their relationship to nucleic acids and the genetic code (Chapter 31). Note the difference between the two names, D-ribose and D-2-deoxyribose. In the latter name, the 2-deoxy means that oxygen is missing from the D-ribose molecule at carbon 2. Check the formulas that follow to verify this difference.

◄ Glucose is often given intravenously to patients suffering from major trauma.

H—¹C=O
H—²—OH
H—³—OH
H—⁴—OH
⁵CH₂OH

D-ribose

⁵CH₂OH ... OH ... H ... H ... OH ... OH   or   CH₂OH ... O ... OH ... OH ... OH

β-D-ribofuranose

H—¹C=O
H—²—H
H—³—OH
H—⁴—OH
⁵CH₂OH

D-2-deoxyribose

⁵CH₂OH ... O ... OH ... H ... H ... OH ... H   or   CH₂OH ... O ... OH ... OH

β-D-2-deoxyribofuranose

The ketose that is closely related to D-ribose is named D-ribulose. (Ketose names are often derived from the corresponding aldose name by modifying the suffix -*ose* to -*ulose*.) This ketose is an intermediate that allows cells to make many other monosaccharides. In photosynthetic organisms, D-ribulose is used to capture carbon dioxide and thus make new carbohydrates.

CH₂OH
C=O
H——OH
H——OH
CH₂OH

D-ribulose

Practice 27.6

Write Haworth formulas in the furanose ring form for the following:
(a) α-D-fructofuranose
(b) α-D-ribofuranose

27.10 Disaccharides

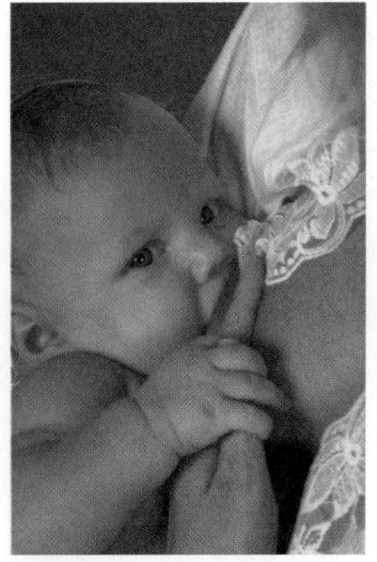

▲
Milk sugar, or lactose, is found mainly in the milk of mammals.

Disaccharides are carbohydrates composed of two monosaccharide residues united by a glycosidic linkage. The two important disaccharides found in the free state in nature are sucrose and lactose ($C_{12}H_{22}O_{11}$). Sucrose, commonly known as *table sugar,* exists throughout the plant kingdom. Sugar cane contains 15–20% sucrose, and sugar beets 10–17%. Maple syrup and sorghum are also good sources of sucrose. Lactose, also known as *milk sugar,* is found free in nature mainly in the milk of mammals. Human milk contains about 6.7% lactose and cow milk about 4.5% lactose.

Unlike sucrose and lactose, several other important disaccharides are derived directly from polysaccharides by hydrolysis (see Section 27.14). For example, maltose, isomaltose, and cellobiose are formed when specific polysaccharides are hydrolyzed. Of this group, maltose is the most common and is found as a constituent of sprouting grain.

Upon hydrolysis, disaccharides yield two monosaccharide molecules. The hydrolysis is catalyzed by hydrogen ions (acids), usually at elevated temperatures, or by certain enzymes that act effectively at room or body temperatures. A different enzyme is required for the hydrolysis of the three disaccharides:

$$\text{Sucrose} + \text{Water} \xrightarrow{\text{H}^+ \text{ or sucrase}} \text{Glucose} + \text{Fructose}$$

$$\text{Lactose} + \text{Water} \xrightarrow{\text{H}^+ \text{ or lactase}} \text{Galactose} + \text{Glucose}$$

$$\text{Maltose} + \text{Water} \xrightarrow{\text{H}^+ \text{ or maltase}} \text{Glucose} + \text{Glucose}$$

The enzyme lactase is present in the small intestine of infants and allows them to easily digest lactose from their milk diet. Unfortunately, as people mature their intestines often stop producing the lactase enzyme, and they lose the ability to digest lactose. Instead, this sugar is metabolized by common bacteria that live in the large intestine. The gas and intestinal discomfort that results is termed milk or lactose intolerance and is a condition that afflicts many adults. Carefully note the small structural differences between lactose and the other common disaccharides in the next section. Small differences in molecular shape often determine what our bodies do with specific molecules.

27.11 Structures and Properties of Disaccharides

Disaccharides contain an acetal structure (glycosidic linkage), and some also contain a hemiacetal structure. The acetal structure in maltose may be considered as being derived from two glucose molecules by the elimination of a molecule of water between the —OH group on carbon 1 of one glucose unit and the —OH group of carbon 4 on the other glucose unit. This is an α-1,4-glycosidic linkage since the glucose units have the α configuration and are joined at carbons 1 and 4. In a more systematic nomenclature, this form of maltose is known as α-D-glucopyranosyl-(1,4)-α-D-glucopyranose. If the structure of glucose is known, this name provides a complete description for drawing the maltose formula.

never find αD βD in a straight chain only cyclic

equilir. rotation of 52.7

α-D-glucopyranose unit α-D-glucopyranose unit

α-1,4-Glycosidic linkage

Acetal structure

$+ H_2O$

maltose
α-D-glucopyranosyl-(1,4)-α-D-glucopyranose

Lactose consists of a β-D-galactopyranose unit linked to an α-D-glucopyranose unit. These are joined by a β-1,4-glycosidic linkage from carbon 1 on galactose to carbon 4 on glucose. The more systematic name for lactose is β-D-galactopyranosyl-(1,4)-α-D-glucopyranose.

Two conventions are used for writing Haworth formulas connecting two or more monosaccharide units through a glycoside bond. Using the first convention for lactose, for example, the glycosidic bond is shown in a bent fashion to provide ste reochemical information.

^{1}C   O   C ^{4} (bent structure)

The bond from carbon 1 in galactose is the β configuration pointing up and is bonded through oxygen to carbon 4 of glucose whose bond to oxygen points down.

In the second convention, the two monosaccharides are shown in a "stacked" position. In lactose the glycosidic bond from carbon 1 of galactose in the β position is pointed up and is directly bonded through oxygen to carbon 4 of glucose whose bond to oxygen is pointed down. These two conventions are illustrated below:

β-1,4-Glycosidic linkage

β-1,4-Glycosidic linkage

β-D-galactopyranose unit   α-D-glucopyranose unit   β-D-galactopyranose unit   α-D-glucopyranose unit

lactose

β-D-galactopyranosyl-(1,4)-α-D-glucopyranose

Sucrose consists of an α-D-glucopyranose unit and a β-D-fructofuranose unit. These monosaccharides are joined by an oxygen bridge from carbon 1 on glucose to carbon 2 on fructose—that is, by an α-1,2-glycosidic linkage.

α-D-glucopyranose unit

O ← glycosidic linkage

β-D-fructofuranose unit

α-D-glucopyranosyl-(1,2)-β-D-fructofuranose

In this perspective formula, fructose is bonded to glucose through the β-2 position of fructose to the α-1 position of glucose.

Example 27.3   Write the Haworth formula for isomaltose, which is α-D-glucopyranosyl-(1,6)-α-D-glucopyranose.

Solution   Recognize that this disaccharide is composed of two α-D-glucopyranose units linked between carbon 6 of one unit and carbon 1 of the other. First, write the Haworth formula for the monosaccharides and number their carbons.

The two α-D-glucopyranose units must be linked in such a way that the stereochemistry at carbon 1 is preserved (carbon 6 is not a chiral center). One correct way to write the isomaltose structure is as follows:

isomaltose (α 1,6-glycosidic linkage)

Practice 27.7

Write the structure for cellobiose (a disaccharide that can be derived from plants), β-D-glucopyranosyl-(1,4)-β-D-glucopyranose.

Lactose and maltose both show mutarotation, which indicates that one of the monosaccharide units has a hemiacetal ring that opens and closes to interchange anomers. Sucrose has no hemiacetal structure and hence does not mutarotate.

The three disaccharides sucrose, lactose, and maltose have physical properties associated with large polar molecules. All three are crystalline solids and are quite soluble in water; the solubility of sucrose amounts to 200 g per 100 g of water at 0°C. Hydrogen bonding between the polar —OH groups on the sugar molecules and the water molecules is a major factor in this high solubility. These sugars are not easily melted. In fact, lactose is the only one with a clearly defined melting point (201.6°C). Sucrose and maltose begin to decompose when heated to 186°C and 103°C, respectively. When sucrose is heated to melting, it darkens and undergoes partial decomposition. The resulting mixture is known as caramel, or burnt sugar, and is used as coloring and as a flavoring agent in foods.

27.12 Sweeteners and Diet

Carbohydrates have long been valued for their ability to sweeten foods. Fructose is the sweetest of the common sugars (a scale of relative sweetness is given in Table 27.1), although sucrose (table sugar) is the most commonly used sweetener. Astonishingly large amounts of sucrose are produced from sugar beets and cane: World production is on the order of 90 million tons annually. There are no essential chemical differences between cane and beet sugar. In the United States, approximately 20–30% of the average caloric intake is sucrose (about 150 g/day per person). Low price and sweet taste are the major reasons for high sucrose consumption. Note that sucrose is only 58% as sweet as fructose (Table 27.1). However, because sucrose is inexpensive to produce and is amenable to a variety of food processing techniques, approximately 60–80% of all sweeteners are sucrose.

Sucrose has a tendency to crystallize from concentrated solutions or syrups. Therefore, in commercial food preparations (e.g., candies, jellies, and canned fruits) the sucrose is often hydrolyzed.

$$\text{Sucrose} + H_2O \xrightarrow{H^+} \text{Glucose} + \text{Fructose}$$

invert sugar   The resulting mixture of glucose and fructose, usually in solution, is called **invert sugar.** Invert sugar has less tendency to crystallize than sucrose, and it has greater sweetening power than an equivalent amount of sucrose. The nutritive value of the sucrose is not affected in any way by the conversion to invert sugar because the same hydrolysis reaction occurs in normal digestion.

High-fructose syrups, derived from cornstarch, are used as sweeteners in products such as soft drinks. Starch is a polymer of glucose, contains no fructose, and is not sweet. But through biotechnology a method was developed to convert starch to a

▲
This sugar substitute is far sweeter than sucrose and is nonnutritive.

TABLE 27.1	Relative Sweetness of Sugars and Sugar Substitutes (Based on fructose = 100)		
Sugars	**Relative sweetness**	**Sugar substitutes (common brand names)**	**Relative sweetness**
Fructose	100	Sucralose (Splenda®)	3.5×10^4
Invert sugar	75	Saccharin (Sweet 'N Low®)	1.7×10^4
Sucrose	58	Acesulfame potassium (Sunette®, Sweet One®)	1.2×10^4
Glucose	43	Aspartame (Equal, NutraSweet®)	1.2×10^4
Maltose	19		
Galactose	19		
Lactose	9.2		

very sweet syrup of high fructose content. The starch polymers are hydrolyzed to glucose; then part of the glucose is enzymatically converted to fructose yielding the syrup. High-fructose corn syrups are used because they are more economical sweetening agents than either cane or beet sugar.

Unfortunately, high sugar consumption presents health problems. For many people, sucrose is a source of too many calories. Oral bacteria also find sucrose easy to metabolize, increasing the incidence of dental caries. Finally, because the monosaccharides, fructose and glucose, are quickly absorbed from the small intestine, sugar consumption leads to a rapid increase in blood sugar. Such a sharp rise can be dangerous for people with impaired carbohydrate metabolism—for example, with diabetes mellitus.

Scientists have searched diligently for sugar substitutes and successfully have identified both natural and artificial sweeteners. Several artificial sweeteners are currently approved as additives in U.S. foods (other countries may have different guidelines):

acesulfame potassium   aspartame   saccharin   sucralose

Saccharin was the first purely artificial, noncarbohydrate sweetener to be developed (discovered in 1879). Much sweeter than the natural sugars (see Table 27.1), saccharin has a distinctive aftertaste. Sugar substitutes need to be as sweet as possible so that only small amounts have to be added to foods. Note that in all cases these chemicals are much sweeter than fructose (see Table 27.1). Each chemical may have slightly different uses; for example, aspartame intensifies fruit flavors while acesulfame potassium can be used for baking.

Notice the similarities between sucralose and sucrose.

This group of sweeteners has been shown to be safe in many animal studies. Yet, safety concerns are difficult to answer completely. The U.S. Food and Drug Administration sets an Acceptable Daily Intake (ADI) for each chemical—a level of consumption that is deemed safe over a lifetime. For example, the ADI for aspartame is 50 mg aspartame per kilogram of body weight. This corresponds to about 15 cans of aspartame-containing soft drink per day over the average adult lifespan.

The history of sodium cyclamate illustrates the difficulties of health risk assessment.

$H-N-SO_3^- Na^+$

sodium cyclamate

This sweetener was discovered in the 1930s and was approved for food use in the 1960s. However, animal studies indicated some cancer risk from cyclamate, and the sweetener was banned in 1970. Since that time many studies have rebutted the earlier findings, and the FDA may approve cyclamate once again.

Partly because of continued health concerns, natural substitutes for sucrose are also being developed. Again, the aim is to find a chemical that sweetens without the calories of sucrose or the risk of dental caries. Several products that may gain wide approval in the United States as food additives include (1) nonnutritive sweeteners derived from citrus fruits (the dihydrochalcones, 300–2000 times sweeter than sucrose); (2) a nonnutritive extract from licorice root (glycyrrhizin, 50–100 times sweeter than sucrose); (3) an extract from a South American plant leaf (stevioside, 300 times sweeter than sucrose); (4) a mixture of sweet-tasting proteins from an African fruit (thaumatin, 2000–3000 times sweeter than sucrose).

27.13   Redox Reactions of Monosaccharides

Oxidation

The aldehyde groups in monosaccharides can be oxidized to monocarboxylic acids by mild oxidizing agents such as bromine water. The carboxylic acid group is formed at carbon 1. The name of the resulting acid is formed by changing the *-ose* ending to *-onic acid*. Glucose yields gluconic acid; galactose, galactonic acid; and so on.

Carbohydrate oxidation increases the number of bonds between carbon and oxygen.

$$
\begin{array}{c}
\text{H—C=O} \\
\text{H——OH} \\
\text{HO——H} \\
\text{H——OH} \\
\text{H——OH} \\
\text{CH}_2\text{OH}
\end{array}
\;+\; \text{Br}_2 \;+\; \text{H}_2\text{O} \longrightarrow
\begin{array}{c}
\text{COOH} \\
\text{H——OH} \\
\text{HO——H} \\
\text{H——OH} \\
\text{H——OH} \\
\text{CH}_2\text{OH}
\end{array}
\;+\; 2\,\text{HBr}
$$

D-glucose   D-gluconic acid

Dilute nitric acid, a vigorous oxidizing agent, oxidizes both carbon 1 and carbon 6 of aldohexoses to form dicarboxylic acids. The resulting acid is named by changing the *-ose* sugar suffix to *-aric acid*. Glucose yields glucaric acid (saccharic acid); galactose, galactaric acid (mucic acid).

Notice that galactaric acid is a meso compound—it is superimposable on its mirror image.

$$
\begin{array}{c}
\text{H—C=O} \\
\text{H——OH} \\
\text{HO——H} \\
\text{H——OH} \\
\text{H——OH} \\
\text{CH}_2\text{OH}
\end{array}
\xrightarrow{\text{warm HNO}_3}
\begin{array}{c}
\text{COOH} \\
\text{H——OH} \\
\text{HO——H} \\
\text{H——OH} \\
\text{H——OH} \\
\text{COOH}
\end{array}
\qquad
\begin{array}{c}
\text{COOH} \\
\text{H——OH} \\
\text{HO——H} \\
\text{HO——H} \\
\text{H——OH} \\
\text{COOH}
\end{array}
$$

D-glucose   glucaric acid   galactaric acid
                     (saccharic acid)   (mucic acid)

Reduction

Monosaccharides can be reduced to their corresponding polyhydroxy alcohols by reducing agents such as H_2/Pt or sodium amalgam, Na(Hg). For example, glucose yields sorbitol (glucitol), galactose yields galactitol (dulcitol), and mannose yields mannitol; all of these are hexahydric alcohols (containing six —OH groups).

H—C=O H——OH HO——H H——OH H——OH CH₂OH D-glucose	$\xrightarrow{\text{H}_2/\text{Pt}}$

CH₂OH
H——OH
HO——H
H——OH
H——OH
CH₂OH

D-glucitol
(sorbitol)

Carbohydrate reduction decreases the number of bonds between carbon and oxygen.

Hexahydric alcohols have properties resembling those of glycerol (Section 22.7). Because of their affinity for water, they are used as moisturizing agents in food and cosmetics. Sorbitol, galactitol, and mannitol occur naturally in a variety of plants.

Redox Tests for Carbohydrates

Some sugars reduce silver ions to free silver, and copper(II) ions to copper(I) ions, under prescribed conditions. Such sugars are called **reducing sugars.** This reducing ability, which is useful in classifying sugars and in certain clinical tests, is dependent on the presence of (1) aldehydes, (2) α-hydroxyketone groups (—CH₂COCH₂OH) such as in fructose, or (3) hemiacetal structures in cyclic molecules such as maltose. These groups are easily oxidized to carboxylic acid (or carboxylate ion) groups; the metal ions are thereby reduced ($Ag^+ \rightarrow Ag^0$; $Cu^{2+} \rightarrow Cu^+$). Several different reagents, including Tollens, Fehling, Benedict, and Barfoed reagents, are used to detect reducing sugars (see Section 23.5). The Benedict, Fehling, and Barfoed tests depend on the formation of copper(I) oxide precipitate to indicate a positive reaction (see Figure 27.5).

$$\underset{\substack{\text{aldehyde} \\ \text{group}}}{\text{RC}-\text{H}} + 2\ Cu^{2+} + 5\ OH^- \longrightarrow \underset{\substack{\text{carboxylate} \\ \text{ion group}}}{\text{RC}-\text{O}^-} + \underset{\substack{\text{copper(I)} \\ \text{oxide (brick red)}}}{Cu_2O(s)} + 3\ H_2O$$

(blue)

Barfoed reagent contains Cu^{2+} ions in the presence of acetic acid. It is used to distinguish reducing monosaccharides from reducing disaccharides. Under the same reaction conditions, the reagent is reduced more rapidly by monosaccharides.

Glucose and galactose contain aldehyde groups; fructose contains an α-hydroxyketone group. Therefore, all three of these monosaccharides are reducing sugars.

reducing sugar

▲
Sorbitol occurs naturally in plants and has an affinity for water. It is an ingredient in many moisturizers and lotions.

A carbohydrate molecule need not have a free aldehyde or α-hydroxyketone group to be a reducing sugar. A hemiacetal structure (see below) also reacts. Maltose and the cyclic form of glucose are examples of molecules with the hemiacetal structure.

glucose   maltose

Under mildly alkaline conditions, the rings open at the points indicated by the arrows to form aldehyde groups:

glucose (ring structure)   glucose (open-chain structure)

▲
FIGURE 27.5
Benedict test for reducing sugars. The test tube on the left contains Benedict reagent. The test tube on the right shows the results of a positive Benedict test with glucose [formation of a brick-red precipitate of copper(I) oxide].

Among the disaccharides, lactose and maltose have hemiacetal structures and are therefore reducing sugars. Sucrose is not a reducing sugar because it does not have the hemiacetal structure.

Many clinical tests monitor glucose as a reducing sugar. For example, Benedict and Fehling reagents are used to detect the presence of glucose in urine. Initially, the reagents are deep blue in color. A positive test is indicated by a color change to greenish-yellow, yellowish-orange, or brick-red, corresponding to an increasing glucose (reducing sugar) concentration. These tests are used to estimate the amount of glucose in the urine of diabetics in order to adjust the amount of insulin needed for proper glucose utilization.

Alternatively, a clinical test (glucose oxidase test) uses an enzyme-catalyzed oxidation of glucose to test for urine sugar. The inclusion of an enzyme ensures a reaction that is specific for the glucose structure, allowing a more selective test for glucose in the urine.

Color reactions such as those mentioned in the preceding paragraphs provide the basis for many home glucose monitoring kits. These kits allow diabetics and others to measure blood glucose levels easily and on a regular basis. Perhaps the biggest disadvantage to the kits now in use is the need to draw a drop of blood from a finger. However, this disadvantage may be soon solved. Glucose tests have become so sensitive that they can detect glucose levels in the liquid between the skin cells (the interstitial fluid); these tests no longer require blood. Also, an implant is in development that may provide "real-time" continuous glucose monitoring.

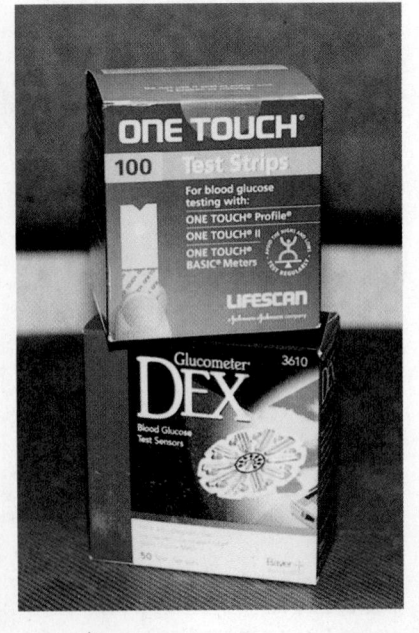

▲
Home glucose test kits

Two samples labeled A and B are known to be D-threose and D-erythrose. Water solutions of each sample are optically active. However, when each solution was warmed with nitric acid, the solution from sample A became optically inactive while that from sample B was still optically active. Determine which sample (A or B) contains D-threose and which sample contains D-erythrose.

Example 27.4

In this problem we need to examine the structures of D-threose and D-erythrose, write equations for the reaction with nitric acid, and examine the products to see why one is optically active and the other optically inactive. Start by writing the formulas for D-threose and D-erythrose:

Solution

$$
\begin{array}{ccc}
\text{H}-\text{C}=\text{O} & & \text{H}-\text{C}=\text{O} \\
\text{HO}-\!\!\!-\!\!\!-\text{H} & & \text{H}-\!\!\!-\!\!\!-\text{OH} \\
\text{H}-\!\!\!-\!\!\!-\text{OH} & & \text{H}-\!\!\!-\!\!\!-\text{OH} \\
\text{CH}_2\text{OH} & & \text{CH}_2\text{OH} \\
\text{D-threose} & & \text{D-erythrose}
\end{array}
$$

Oxidation of these tetroses with HNO_3 yields dicarboxylic acids:

$$
\begin{array}{ccccccc}
\text{H}-\text{C}=\text{O} & & \text{COOH} & \Big| & \text{H}-\text{C}=\text{O} & & \text{COOH} \\
\text{HO}-\text{H} & \xrightarrow[\text{HNO}_3]{\text{warm}} & \text{HO}-\text{H} & \Big| & \text{H}-\text{OH} & \xrightarrow[\text{HNO}_3]{\text{warm}} & \text{H}-\text{OH} \\
\text{H}-\text{OH} & & \text{H}-\text{OH} & \Big| & \text{H}-\text{OH} & & \text{H}-\text{OH} \\
\text{CH}_2\text{OH} & & \text{COOH} & \Big| & \text{CH}_2\text{OH} & & \text{COOH} \\
\text{D-threose} & & \text{I} & \Big| & \text{D-erythrose} & & \text{II}
\end{array}
$$

Product I is a chiral molecule and is optically active. Product II is a meso compound and is optically inactive. Therefore, sample A is D-erythrose, since oxidation yields the meso acid. Sample B then must be D-threose.

Practice 27.8

A disaccharide yields no copper(I) oxide when reacted with Benedict reagent. This carbohydrate is composed of two α-D-galactopyranose units. Identify the carbon from each monosaccharide involved in the acetal linkage.

Practice 27.9

Write the structure of the product formed when D-galactose is reduced with H_2/Pt. Is this compound optically active?

27.14 Polysaccharides Derived from Glucose

Although many naturally occurring polysaccharides are known, three—starch, cellulose, and glycogen—are of major importance. All three, when hydrolyzed, yield D-glucose as the only product, according to this approximate general equation:

$$(C_6H_{10}O_5)_n + n\ H_2O \rightarrow n\ C_6H_{12}O_6$$

polysaccharide molecule
(approximate formula)

D-glucose

This hydrolysis reaction establishes that all three polysaccharides are polymers made up of glucose monosaccharide units. It also means that the differences in properties among the three polysaccharides must be due to differences in the structures and/or sizes of these molecules.

Many years of research were required to determine the detailed structures for polysaccharide molecules. Consideration of all this work is beyond the scope of our discussion, so what follows is an abbreviated summary of the results.

Starch

Starch is found in plants, mainly in the seeds, roots, or tubers (see Figure 27.6). Corn, wheat, potatoes, rice, and cassava are the chief sources of dietary starch. The two main components of starch are amylose and amylopectin. Amylose molecules are unbranched chains composed of about 25–1300 α-D-glucose units joined by α-1,4-glycosidic linkages, as shown in Figure 27.7. The stereochemistry of the α anomer causes amylose to coil into a helical conformation. Partial hydrolysis of this linear polymer yields the disaccharide maltose.

Amylopectin is a branched-chain polysaccharide with much larger molecules than those of amylose. Amylopectin molecules consist, on average, of several thousand α-D-glucose units with molar masses ranging up to 1 million or more. The main chain contains glucose units connected by α-1,4-glycosidic linkages. Branch chains are linked to the main chain through α-1,6-glycosidic linkages about every 25 glucose units, as shown in Figure 27.7. This molecule has a characteristic treelike structure because of its many branch chains. Partial hydrolysis of amylopectin yields both maltose and the related disaccharide isomaltose, α-D-glucopyranosyl-(1,6)-α-D-glucopyranose.

Despite the presence of many polar —OH groups, starch molecules are insoluble in cold water, apparently because of their very large size. Starch readily forms colloidal dispersions in hot water. Such starch "solutions" form an intense blue-black color in the presence of free iodine. Hence, a starch solution can be used to detect free iodine, or a dilute iodine solution can be used to detect starch.

Starch is readily converted to glucose by heating with water and a little acid (e.g., hydrochloric or sulfuric acid). It is also readily hydrolyzed at room temperature by certain digestive enzymes. The hydrolysis of starch to maltose and glucose is shown in the following equation:

$$\text{Starch} \xrightarrow[\substack{\text{or salivary and}\\\text{pancreatic amylase}}]{\text{acid} + \Delta} \text{Dextrins} + \text{Maltose} \xrightarrow[\substack{\text{or maltase and other}\\\text{intestinal enzymes}}]{\text{acid} + \Delta} \text{D-Glucose}$$

The hydrolysis of starch can be followed qualitatively by periodically testing samples from a mixture of starch and saliva with a very dilute iodine solution. The change of color sequence is blue-black $\longrightarrow$ blue $\longrightarrow$ purple $\longrightarrow$ pink $\longrightarrow$ colorless as the starch molecules are broken down into smaller and smaller fragments.

▲
FIGURE 27.6
Polarized light micrograph of starch granules in potato tuber cells.

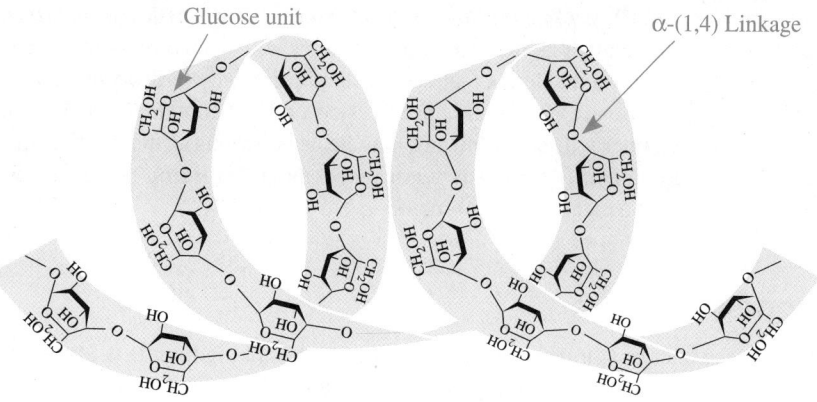

(a) Molecular structure of amylose chain

(b) Array of glucose units (dots) in amylose

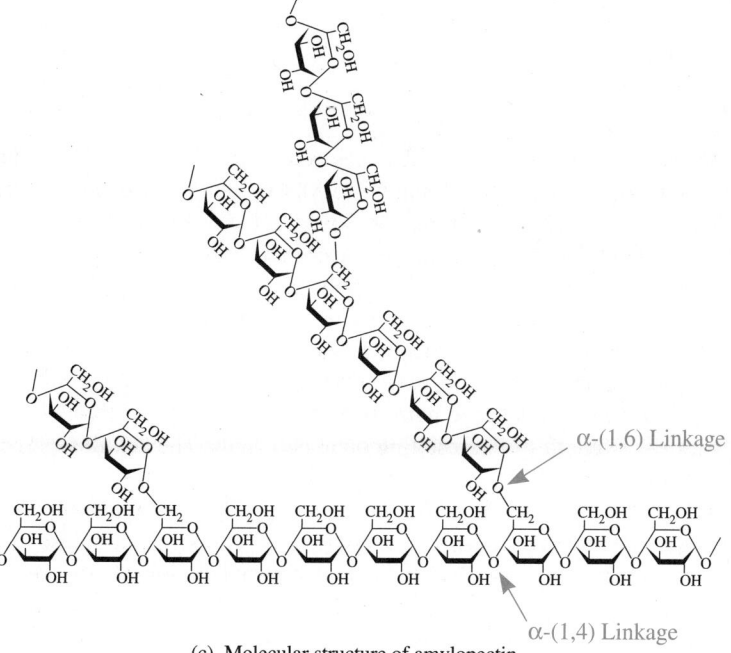

(c) Molecular structure of amylopectin

(d) Array of glucose units (dots) in amylopectin

▲
FIGURE 27.7
Representations of amylose and amylopectin.

Hydrolysis is a key chemical reaction in the digestion of starchy foods. If these foods are well chewed, salivary amylase normally decreases the starch polymer chain length from on the order of a thousand glucose units to about eight per chain. In the small intestine, pancreatic amylase continues digestion to form maltose. Enzymes in the small intestine membranes complete the conversion of starch to glucose, which is then absorbed into the bloodstream.

Starch is the most important energy-storage carbohydrate of the plant kingdom. In turn, humans and other animals consume huge quantities of starch. This polymer is an important food source because it has the appropriate structure to be readily broken down to D-glucose. The carbons from starch provide much of our daily energy needs as they are oxidized to carbon dioxide.

New biodegradable packaging materials are now being made from starch. Starch foam cushioning such as the "foam peanuts" packed around breakable items can be washed down the sink. Boxes made from starch are easily hydrolyzed in contrast to paper and cardboard. Research is progressing on food containers made from starch; the food container will be edible and may be as nutritious as the food within! Perhaps most importantly, as scientists learn more about the starch polymers, we make progress toward a cleaner environment.

Glycogen

Glycogen is the energy-storage carbohydrate of the animal kingdom. It is formed by polymerization of glucose and is stored in the liver and in muscle tissues. Structurally, it is very similar to the amylopectin fraction of starch except that it is more highly branched. The α-1,6-glycosidic linkages occur on one of every 12–18 glucose units.

Cellulose

Cellulose is the most abundant organic substance found in nature. It is the chief structural component of plants and wood (Figure 27.8). Cotton fibers are almost pure cellulose; wood, after removal of moisture, consists of about 50% cellulose. Cellulose is an important substance in the textile and paper industries.

Cellulose, like starch and glycogen, is a polymer of glucose. But cellulose differs from starch and glycogen because the glucose units are joined by β-1,4-glycosidic linkages instead of α-1,4-glycosidic linkages. The stereochemistry of the β anomer allows this polymer to form an extended chain that can hydrogen-bond to adjacent cellulose molecules. The large number of hydrogen bonds so formed partially accounts for the strength of the resulting plant cell walls. The cellulose structure is illustrated in Figure 27.9.

When partial hydrolysis of cellulose occurs it produces the disaccharide cellobiose, β-D-glucopyranosyl-(1,4)-β-D-glucopyranose. However, cellulose has greater resistance to hydrolysis than either starch or glycogen. It is not appreciably hydrolyzed when boiled in a 1% sulfuric acid solution. It does not show a color reaction with iodine. Humans cannot digest cellulose because they have no enzymes capable of catalyzing its hydrolysis. Fortunately, some microorganisms found in soil and in the digestive tracts of certain animals produce enzymes that do catalyze the breakdown of cellulose. The presence of these microorganisms explains why cows and other herbivorous animals thrive on grass—and also why termites thrive on wood.

The —OH groups of starch and cellulose can be reacted without destruction of the macromolecular structures. For example, nitric acid converts an —OH group to a nitrate group in this fashion:

$$\text{Cellulose} \underline{- \boxed{OH \ + \ H} ONO_2} \longrightarrow \text{Cellulose} - ONO_2 \ + \ H_2O$$

hydroxyl group on cellulose molecule / nitric acid / nitrate group on cellulose molecule / water

If only a portion of the —OH groups on the cellulose molecules are nitrated, a plastic nitrocellulose material known as celluloid or pyroxylin is obtained. This material has been used to make such diverse articles as billiard balls, celluloid shirt collars,

Packing material shown here is made from starch. It cushions like plastic but dissolves in water like flour. Scientists are developing more starch packaging because it is easily biodegradable.

▲ **FIGURE 27.8**
Scanning electron micrograph of cellulose fibers in a plant cell wall.

(a) Haworth formula

(b) Three-dimensional representation

and photographic film. By nitration of nearly all —OH groups, a powerful high explosive is obtained. This highly nitrated cellulose, or "guncotton," is the basic ingrediant in modern "smokeless" gunpowder.

Another modified cellulose, cellulose acetate, is made by esterification of —OH groups with acetic acid (acetic anhydride). About two-thirds of the —OH groups are esterified:

Cellulose—O H + CH$_3$—C—O—C—CH$_3$
hydroxyl group on cellulose molecule   acetic anhydride

⟶ Cellulose—O—C—CH$_3$ + CH$_3$COOH
acetate group on cellulose molecule

Cellulose acetate, unlike the dangerously flammable cellulose nitrate, can be made to burn only with difficulty. For this reason, cellulose acetate has displaced cellulose

CHEMISTRY IN ACTION Polysaccharides

Many carbohydrate polymers are composed of monosaccharides other than glucose and serve a variety of purposes for both plants and animals. Some polysaccharides are made up of only one type of monosaccharide. For example, D-mannose–containing polymers are found in many plants, including pine trees and orchid tubers. Polysaccharides composed of the pentose D-xylose are also found throughout the vegetable kingdom.

More complex polysaccharides are often found in animals. These carbohydrates are linked together by glycosidic bonds at various positions on the monosaccharides. Thus a sugar may be bonded to its neighbor from carbon 1 to carbon 3, or carbon 1 to carbon 4, or carbon 1 to carbon 6, and so on. Because of these numerous linkages, the polysaccharides often have a complex, branching structure. Also, it is not uncommon for these molecules to contain many different kinds of monosaccharides. These complex carbohydrates serve a variety of functions in animals.

Mucopolysaccharides (or glycosaminoglycans) make up part of the connective tissue and are found in the joints and the skin. The slimy, mucuslike con-

▲ **The connective tissue inside the knee contains mucopolysaccharides.**

sistency of these molecules derives from their special chemical properties. About half of the sugar units are acidic in these carbohydrates. These acid groups become negatively charged under physiological conditions, causing the polymer chains to repel each other.

Water fills the space between the polymer chains and gives the mucopolysaccharide a spongelike consistency. One gram of some mucopolysaccharides can absorb up to 20 L of water. This natural shock absorber and lubricant is necessary for animal locomotion.

Even more complicated polysaccharides are found on the surfaces of almost all cells. These carbohydrates serve as "labels" or antigens allowing organisms to distinguish their own cells from invading bacteria, for example. Antigen recognition illustrates a very important biochemical principle: *Molecular shape carries information that guides the reactions of life.*

In humans, the polysaccharides on the surface of the red blood cells give rise to many blood types, often classified by the ABO system. Cells of different blood types have surface polysaccharides with different structures. Cells carrying one carbohydrate structure are commonly not tolerated by an individual of another blood type. For example, if type AB blood is transfused into someone with type O red blood cells, the new cells will be attacked by the immune system. Red cell destruction can lead to serious injury or death.

nitrate in almost all kinds of photographic films. The textile known as acetate rayon is made from cellulose acetate. Cellulose acetate is also used as a clear, transparent packaging film. In another process, cellulose reacts with carbon disulfide in the presence of sodium hydroxide to form a soluble cellulose derivative called cellulose xanthate, from which cellulose can be regenerated. Viscose rayon textiles and cellophane packaging materials are made of regenerated cellulose prepared by this process.

Concepts in Review

1. Why is the oxidation of carbon in carbohydrates important?

2. Compare the number of carbon–oxygen bonds in carbohydrates to those in carbon dioxide.

3. Classify carbohydrates as mono-, di-, oligo-, or polysaccharides.

4. Explain the use of D, L, (+), and (−) in naming carbohydrates.

5. Identify and write pyranose and furanose ring structures of carbohydrates.

6. Identify and write Fischer projection and Haworth formulas for carbohydrates.

7. Identify the structural feature of a carbohydrate that makes it a reducing sugar.

8. Explain the phenomenon of mutarotation.

9. Distinguish between hemiacetal and acetal structures in a carbohydrate.

10. Understand what a glycoside linkage is.

11. Understand disaccharide composition and the manner in which monosaccharides are linked together in sucrose, lactose, and maltose.

12. Identify monosaccharides that are epimers.

13. List the major sources of glucose, galactose, fructose, sucrose, lactose, and maltose.

14. Describe some disadvantages of a diet high in sucrose and some alternatives to this sweetener.

15. Identify three disaccharides that are derived from polysaccharides.

16. Describe the Benedict test and state what evidence must be seen to indicate a positive test.

17. Write chemical equations for the oxidation of monosaccharides by bromine and by nitric acid.

18. Write chemical equations for the reduction of monosaccharides.

19. Identify monosaccharides that give optically inactive (meso) products when reduced to polyhydroxy alcohols or when oxidized to dicarboxylic acids.

20. Discuss the structural differences between amylose, amylopectin, and cellulose.

Key Terms

The terms listed here have been defined within this chapter. Section numbers are referenced in parentheses for each term.

anomer (27.6)	epimer (27.5)	monosaccharide (27.2)	polysaccharide (27.2)
carbohydrate (27.1)	glycoside (27.7)	mutarotation (27.6)	reducing sugar (27.13)
disaccharide (27.2)	invert sugar (27.12)	oligosaccharide (27.2)	

Questions

Questions refer to tables, figures, and key words and concepts defined within the chapter. A particularly challenging question or exercise is indicated with an asterisk.

1. Why is the oxidation state of a carbohydrate carbon important in metabolism?

2. What is the significance of the notations D and L in the name of a carbohydrate?

3. What is the significance of the notations (+) and (−) in the name of a carbohydrate?

4. What is galactosemia and what are its effects on humans?

5. Which of the D-aldohexoses in Figure 27.1 are epimers?

6. Explain how a carbohydrate with a pyranose structure differs from one with a furanose structure.

7. Explain how α-D-glucopyranose differs from β-D-glucopyranose.

8. Are the cyclic forms of monosaccharides hemiacetals or glycosides?

9. Explain the phenomenon of mutarotation.

10. What is (are) the major sources of each of these carbohydrates?
    (a) sucrose (b) lactose (c) maltose

11. Consider the eight aldohexoses given in Figure 27.1 and answer the following:
    (a) The eight aldohexoses are oxidized by nitric acid to dicarboxylic acids. Which of these give meso (optically inactive) dicarboxylic acids?
    (b) Write the structures and names for the enantiomers of D-altrose and D-idose.

12. Explain why invert sugar is sweeter than sucrose.

13. What are the structural differences between amylose and cellulose?

14. What visual difference would you expect between a dilute glucose solution and a concentrated glucose solution in the Benedict test?

15. What are the two main components of starch? How are they alike and how do they differ?

16. What change in the digestive system leads to lactose intolerance?

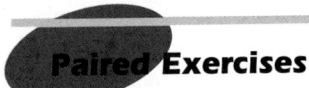

Paired Exercises

These exercises are paired. Each odd-numbered exercise is followed by a similar even-numbered exercise. Answers to the even-numbered exercises are given in Appendix VI.

17. Dihydroxyacetone is the simplest ketose.
  (a) Write a Fischer projection formula for this ketose. Are there any chiral carbon atoms in this molecule?
  (b) Write the Fischer projection formula of the product obtained by reacting dihydroxyacetone with hydrogen in the presence of platinum.

19. Give the structure of an epimer of D-mannose.

21. Write structures and names for the enantiomers of D-galactose, D-mannose, and D-ribose.

23. Write the cyclic structures for α-D-glucopyranose, β-D-galactopyranose, and α-D-mannopyranose.

25. Briefly explain why starch is easily biodegradable in human digestion.

27. Starting with the proper D-tetrose, show the steps for the synthesis of D-glucose by the Kiliani–Fischer synthesis.

29. Is D-2-deoxymannose the same as D-2-deoxyglucose? Briefly explain.

31. What is the monosaccharide composition of the following?
  (a) sucrose   **(c)** amylose
  (b) glycogen   **(d)** maltose

33. How does the structure of cellobiose differ from that of isomaltose?

35. Of the disaccharides sucrose and lactose, which show mutarotation? Explain why.

37. Draw structural formulas for cellobiose and isomaltose. Point out the portion of the structure that is responsible for making these sugars reducing.

39. Give systematic names for isomaltose and cellobiose.

41. Describe the principal structural differences and similarities between the members of the following pairs:
  (a) D-glucose and D-fructose
  (b) maltose and sucrose
  (c) cellulose and glycogen

43. Write the structural formula and the name of the dicarboxylic acid that is formed when D-galactose is oxidized with nitric acid.

45. Write the formulas for the four L-ketohexoses. Indicate which pair (or pairs) of the L-ketohexoses are epimers.

*47. Draw the Haworth formulas for
  (a) β-D-glucopyranosyl-(1,4)-α-D-galactopyranose
  (b) β-D-galactopyranosyl-(1,6)-β-D-mannopyranose

18. Glyceraldehyde is the simplest aldose.
  (a) Write Fischer projection formulas for and identify the D and L forms of this aldose.
  (b) Write the Fischer projection formula of the product obtained by reacting D-glyceraldehyde with hydrogen in the presence of platinum.

20. Give the structure of an epimer of D-galactose.

22. Write structures and names for the enantiomers of D-glucose, D-fructose, and D-2-deoxyribose.

24. Write the cyclic structures of β-D-glucopyranose, α-D-galactopyranose, and β-D-mannopyranose.

26. Briefly explain why cellulose is difficult to biodegrade in human digestion.

28. Starting with the proper D-triose, show the steps for the synthesis of D-ribose by the Kiliani–Fischer synthesis.

30. Is D-2-deoxygalactose the same as D-2-deoxyglucose? Briefly explain.

32. What is the monosaccharide composition of the following?
  (a) lactose   **(c)** cellulose
  (b) amylopectin   **(d)** sucrose

34. How does the structure of maltose differ from that of isomaltose?

36. Of the disaccharides maltose and isomaltose, which show mutarotation? Explain why.

38. Draw structural formulas for maltose and lactose. Point out the portion of the structure that is responsible for making these sugars reducing.

40. Give systematic names for maltose and lactose.

42. Describe the principal structural differences and similarities between the members of the following pairs:
  (a) D-ribose and D-2-dexoyribose
  (b) amylose and amylopectin
  (c) lactose and isomaltose

44. Write the structural formula and the name of the dicarboxylic acid that is formed when D-mannose is oxidized with nitric acid.

46. Write the formulas for the four L-aldopentoses. Indicate which pair (or pairs) of the L-aldopentoses are epimers.

*48. Draw the Haworth formulas for
  (a) β-D-mannopyranosyl-(1,4)-β-D-galactopyranose
  (b) β-D-galactopyranosyl-(1,6)-α-D-glucopyranose

Additional Exercises

These exercises are not paired and provide additional practice on concepts covered in this chapter.

49. Why is glucose sometimes called "blood sugar"?

50. Cite two advantages of aspartame, as a sweetener, over sucrose.

*51. Candy makers who want an especially sweet treat may add some lemon juice (an acid) to their boiling sucrose solution before letting the candy harden. What chemical reaction will the lemon juice aid? How will this chemical reaction make the candy sweeter? (See Section 27.12.)

52. What changes must take place to convert cornstarch into the sweetener high-fructose corn syrup?

53. A thickener commonly used in ice cream, alginic acid, is a seaweed polymer composed of the following monomer:

   (a) From what monosaccharide might this monomer be derived?

   (b) Alginic acid is made by linking these monomers together via β-(1,4)-glycosidic bonds. Draw a short section (three units) of this polymer.

54. A reddish color is obtained when compound A (a disaccharide) is reacted with Benedict solution. Is this compound more likely to be maltose or sucrose? Briefly explain.

*55. Refer to this compound to answer the questions that follow:

(a) How many chiral carbons are present in this compound as it is written? Label them.

(b) Draw the α anomer pyranose ring form of this compound.

(c) How many chiral carbons are present in the structure you drew in part (b)? Label them.

56. If the compound shown below rotates light 25° to the right, draw the structure of a closely related compound that rotates light 25° to the left. Are the two compounds epimers of one another? Why or why not?

*57. Draw the structure of a nonreducing disaccharide composed of two molecules of α-D-galactopyranose.

58. Why is cellulose considered "fiber" in your diet but starch is not? Refer specifically to the structures of cellulose, amylose, and amylopectin in your answer.

59. When lactose is metabolized, the first step separates it into glucose and galactose. Galactose must be changed into glucose before it can be further metabolized in the body.

(a) What changes have to occur in the structure of galactose to make it into glucose?

(b) Which of these terms best describes the relationship of glucose to galactose: disastereomers, enantiomers, or epimers?

60. One of your classmates has told you that the optical rotation of D-glucose and D-mannose are equal and opposite because they are epimers of one another. Should you believe your classmate? Why or why not?

Answers to Practice Exercises

27.1

```
    CH₂OH
     |
    C=O
     |
H ——— OH        The orientation of the —OH on
     |          carbon 3 is not specified and
HO ——— H        can be written in either direction.
     |
    CH₂OH
```

27.2 The right side

27.3 (a)

```
 H—C=O
     |
H ——— OH
     |
HO ——— H
     |
H ——— OH
     |
HO ——— H
     |
    CH₂OH
```

(b)

```
 H—C=O
     |
HO ——— H
     |
H ——— OH
     |
HO ——— H
     |
    CH₂OH
```

27.4

```
 H—C=O
     |
H ——— OH
     |
H ——— OH
     |
HO ——— H          and
     |
H ——— OH
     |
H ——— OH
     |
    CH₂OH
```

```
 H—C=O
     |
HO ——— H
     |
H ——— OH
     |
HO ——— H
     |
H ——— OH
     |
H ——— OH
     |
    CH₂OH
```

27.5 (a)

(b)

27.6 (a)

(b)

27.7

27.8 Carbon 1 from each α-D-galactopyranose

27.9

```
    CH₂OH
     |
H—C—OH
     |
HO—C—H       This compound is not optically active,
     |       it is a meso compound.
HO—C—H
     |
H—C—OH
     |
    CH₂OH
```

▲
Polar bears have a large
reserve of lipids.

The behavior of lipids in water—the fact that they are insoluble—is a key to their importance in nature. For example, an oil slick can spread for many square miles on the surface of the ocean partly because oil and water don't mix. This same principle enables cells to surround themselves with a thin film of lipid, the cell membrane. We protect a fine wood floor with wax, another lipid, because we can depend on this material to adhere to the floor and not dissolve in water. A lipid's stickiness and water insolubility also creates diseases such as atherosclerosis, where arteries become partially clogged by cholesterol-containing lipids. Lipids bring both benefits and problems—these molecules are truly a mixed blessing.

28.1 Lipids: Hydrophobic Molecules

lipid

Lipids are water insoluble, oily, or greasy biochemical compounds that can be extracted from cells by nonpolar solvents such as ether, chloroform, or benzene. Unlike carbohydrates, lipids share no common chemical structure.

What makes a molecule such as a lipid insoluble in water? To answer this question, we must establish two important principles about water solutions: A compound dissolves in water (1) if the water molecules bond well to the potential solute and (2) if the water molecules can still move relatively freely around the dissolved compound. For example, salt (sodium chloride) dissolves because it forms ions to which water molecules can bond. Also, water molecules are free to move around these small ions. Sugar (sucrose) dissolves because it forms hydrogen bonds with water and because it is still a relatively small molecule.

Lipid structures and solubilities differ from both salts and carbohydrates. Lipid molecules are big enough to substantially affect the free movement of water molecules. In addition, lipids cannot hydrogen-bond to the extent that carbohydrates can, nor do they form the large number of positive and negative charges found in a salt solution. What makes a lipid water insoluble? Lipids are *large* and relatively *nonpolar* molecules and thus are water insoluble.

Many carbohydrates and salts are said to be *hydrophilic* ("water loving"). In contrast, lipids are said to be *hydrophobic* ("water fearing").

Consider fatty acids, which are common components of lipids. As shown in Table 28.1, when the number of atoms in a fatty acid molecule increases, the water solubility of the fatty acid decreases dramatically. Water molecules can easily maneuver around smaller compounds like butyric acid, which is infinitely soluble in water. However, these same water molecules run into a huge barrier when they encounter the 18-carbon chain of stearic acid, and so only a little of this fatty acid dissolves in water (0.0003 g/100 g of water).

The hydrophobic nature of lipids contributes significantly to the biological functions of these molecules. Their water insolubility allows lipids to serve as barriers to aqueous solutions. This property, as we shall see later, is of great importance when lipids form cellular membranes.

28.2 Classification of Lipids

Lipids are hydrophobic molecules; their structures are relatively large and nonpolar. Yet, within this broad description lipid structures vary markedly. The following classification scheme recognizes important structural differences.

1. **Simple lipids**
   (a) *Fats and oils:* esters of fatty acids and glycerol
   (b) *Waxes:* esters of high-molar-mass fatty acids and high-molar-mass alcohols
2. **Compound lipids**
   (a) *Phospholipids:* substances that yield glycerol, phosphoric acid, fatty acids, and a nitrogen-containing base upon hydrolysis
   (b) *Sphingolipids:* substances that yield an unsaturated amino alcohol (sphingosine), a long-chain fatty acid, and either a carbohydrate or phosphate and a nitrogen base upon hydrolysis
   (c) *Glycolipids:* substances that yield sphingosine, a fatty acid, and a carbohydrate upon hydrolysis
3. **Steroids**
   Substances that possess the steroid nucleus, which is a 17-carbon structure consisting of four fused carbocyclic rings. Cholesterol and several hormones are in this class.
4. **Miscellaneous lipids**
   Substances that do not fit into the preceding classifications; these include the fat-soluble vitamins A, D, E, and K, and lipoproteins.

> An ester is formed by splitting out a molecule of water between an alcohol and an acid.

The most abundant lipids are the fats and oils. These substances constitute one of the three important classes of foods. In the following discussion, we examine fats and oils as they pertain to biochemistry. A more complete consideration of the properties and composition of various fats and oils is given in Section 24.10.

28.3 Simple Lipids

Fatty Acids

Fatty acids, which form part of most lipids, are carboxylic acids with long, hydrophobic carbon chains. The formulas for some of the most common fatty acids are shown in Table 28.1. All these fatty acids are straight-chain compounds with an even number of carbon atoms. Five of the fatty acids in this table palmitoleic, oleic, linoleic, linolenic, and arachidonic—are unsaturated, having carbon–carbon double bonds in their structures. Animal and higher plant cells produce lipids in which palmitic, oleic, linoleic, and stearic acids predominate. Over one-half of plant and animal fatty acids are unsaturated, plant lipids tending to be more unsaturated than their animal counterparts.

Double bonds impart some special characteristics to the unsaturated fatty acids. Remember that the presence of double bonds raises the possibility of geometric isomerism (Section 20.3). Unsaturated fatty acids may be either cis or trans isomers.

Fatty acid	Number of C atoms	Formula	Solubility (g/100 g water)	Melting point (°C)
Saturated acids				
Butyric acid	4	$CH_3CH_2CH_2COOH$	∞	−4.7
Caproic acid	6	$CH_3(CH_2)_4COOH$	1.08	−1.5
Caprylic acid	8	$CH_3(CH_2)_6COOH$	0.07	16
Capric acid	10	$CH_3(CH_2)_8COOH$	0.015	32
Lauric acid	12	$CH_3(CH_2)_{10}COOH$	0.006	48
Myristic acid	14	$CH_3(CH_2)_{12}COOH$	0.002	57
Palmitic acid	16	$CH_3(CH_2)_{14}COOH$	0.0007	63
Stearic acid	18	$CH_3(CH_2)_{16}COOH$	0.0003	70
Arachidic acid	20	$CH_3(CH_2)_{18}COOH$	—	77
Unsaturated acids*				
Palmitoleic acid	16	$CH_3(CH_2)_5CH{=}CH(CH_2)_7COOH$	—	0.5
Oleic acid	18	$CH_3(CH_2)_7CH{=}CH(CH_2)_7COOH$	—	13
Linoleic acid	18	$CH_3(CH_2)_4CH{=}CHCH_2CH{=}CH(CH_2)_7COOH$	—	−5
Linolenic acid	18	$CH_3CH_2CH{=}CHCH_2CH{=}CHCH_2CH{=}CH(CH_2)_7COOH$	—	−11
Arachidonic acid	20	$CH_3(CH_2)_4(CH{=}CHCH_2)_4CH_2CH_2COOH$	—	−50

TABLE 28.1 Some Naturally Occurring Fatty Acids

*Omega (ω) is the last letter in the Greek alphabet. Correspondingly, the last carbon atom in a carbon chain of a compound is often referred to as the omega carbon. In reference to unsaturated carboxylic acids, omega plus a number (e.g., ω-3) indicates the location of the first carbon–carbon double bond, counting from the omega carbon.

The omega designation of these acids are: palmitoleic (ω-7); oleic (ω-9); linoleic (ω-6); linolenic (ω-3); arachidonic (ω-6).

To illustrate the effect of these double bonds on fatty acid structure, the two fatty acids in Figure 28.1 are portrayed in a simplified manner with each of the many —CH₂— groups as an apex at the intersection between two single bonds. Note that the trans isomer is almost a linear molecule while the double bond in the cis isomer introduces a kink into the fatty acid structure. Unsaturated fatty acids found in nature are almost always cis isomers. These kinked fatty acids cannot stack closely together and hence do not solidify easily. As shown in Table 28.1, unsaturated fatty acids have lower melting points than saturated fatty acids of a similar size. Cooking oils are liquids at room temperature because a high percentage of their fatty acids are unsaturated. In like manner, biological membranes are very fluid because of the presence of fatty acid cis isomers (see Section 28.6).

Practice 28.1

Draw the cis and trans isomers for palmitoleic acid having the formula $CH_3(CH_2)_5CH{=}CH(CH_2)_7COOH$.

Scientists continue to learn more about the nutritional importance of unsaturated fatty acids. If unsaturated fatty acids are converted from cis to trans, they lose

◀ FIGURE 28.1
Cis- and trans-oleic acids.
Notice the bent structure of
the cis isomer.

cis-oleic acid

trans-oleic acid

much of their special dietary importance. For example, trans isomers are formed when margarine is produced by partially hydrogenating vegetable oils. Many health organizations suggest the use of softer and less hydrogenated fats where feasible.

Three common polyunsaturated fatty acids are linoleic, linolenic, and arachidonic acids. These fatty acids represent two nutrient classes that are metabolized differently. The difference lies with how close the double bonds are to the terminal methyl group. Linolenic acid has a double bond on the third carbon from the end of the carbon chain or the omega (ω) position. Thus linolenic acid is an ω-3 fatty acid. In contrast, linoleic and arachidonic acids have a double bond on the sixth carbon from the end of the chain; they are ω-6 fatty acids. Good health requires a diet that supplies both classes of polyunsaturated fatty acids.

The three unsaturated fatty acids—linoleic, linolenic, and arachidonic—are essential for animal nutrition and must be present in the diet. Diets lacking these fatty acids lead to impaired growth and reproduction, and skin disorders such as eczema and dermatitis. A dermatitis disorder can be attributed to an unsaturated fatty acid deficiency if the symptoms clear up when that fatty acid is supplied in the diet.

Certain fatty acids, as well as other lipids, are biochemical precursors of several classes of hormones. The well-known steroid hormones are synthesized from cholesterol and will be discussed later in this chapter. Arachidonic acid and, to a lesser extent, linolenic acid are also used by the body to make hormonelike substances. The biochemicals derived from arachidonic acid are collectively termed eicosanoids using a derivative of the Greek word for twenty *(eikosi)* to indicate that these compounds have 20 carbon atoms. Prostaglandins are perhaps the best known of the eicosanoid class, which also includes the leukotrienes, prostacyclins, and thromboxanes. Cell membranes release arachidonic acid in response to a variety of circumstances, including infection and allergic reactions. In turn, enzymes in the surrounding fluid convert this fatty acid to specific eicosanoids by adding oxygen

Polyunsaturated means the compound has more than one double bond.

▲
Prostaglandins are synthesized at the site in response to allergic reactions such as the dermatitis shown here.

FIGURE 28.2
Several examples of eicosanoids. Each of these molecules is derived from arachidonic acid.

to the arachidonic double bonds. Some examples of eicosanoids are shown in Figure 28.2.

Unlike true hormones, eicosanoids are not transported via the bloodstream to their site of action, but rather take effect where they are synthesized. Prostaglandins are a primary cause of the swelling, redness, and pain associated with tissue inflammation. Platelets in the bloodstream form the thromboxanes, which act as vasoconstrictors and stimulate platelet aggregation, as an initial step in blood clotting. Leukotrienes are formed by a variety of white blood cells as well as other tissues and cause many of the symptoms associated with an allergy attack. For example, asthma is thought to be mediated by the leukotrienes.

Nonsteroidal anti-inflammatory drugs (NSAIDs) block the oxidation of arachidonic acid to form prostaglandins and thromboxanes. These drugs include the common pain relievers aspirin; ibuprofen (e.g., Advil, Motrin, Nuprin); naproxon (e.g., Aleve); and ketoprofen (e.g., Orudis KT, Actron). Recent research indicates that low levels of aspirin may prevent heart attacks and strokes, possibly by blocking synthesis of the thromboxanes that participate in blood clotting. Cortisone acts as an anti-inflammatory drug by decreasing the release of arachidonic acid from the cell membranes.

Diets may have a significant effect on eicosanoid formation. For example, eating fish can increase the intake of long-chain, ω-3 polyunsaturated fatty acids. These fatty acids substitute for arachidonic acid (an ω-6 fatty acid) and form less potent eicosanoids. It has been suggested that cultures for which fish is a dietary staple (such as the Greenland Eskimos) have a low incidence of heart disease partly

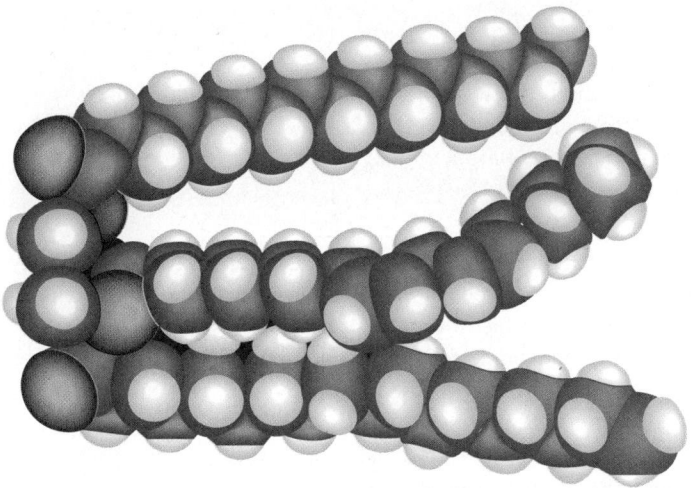

▶ **FIGURE 28.3**
Space-filling model of a tria-clyglycerol formed by react-ing glycerol with one palmitic acid, one oleic acid, and one stearic acid. Note the kink introduced into oleic acid by the cis double bond.

because of a decrease in the physiological effects of some of the eicosanoids (e.g., the thromboxanes).

Fats and Oils

Chemically, fats and oils are esters of glycerol and the higher-molar-mass fatty acids. They have the general formula

$$CH_2-O-\overset{\underset{\displaystyle O}{||}}{C}-R_1$$
$$CH-O-\overset{\underset{\displaystyle O}{||}}{C}-R_2$$
$$CH_2-O-\overset{\underset{\displaystyle O}{||}}{C}-R_3$$

where the R's can be either long-chain saturated or unsaturated hydrocarbon groups. Figure 28.3 is a space-filling model of a typical fat.

Fats may be considered to be triesters formed from the trihydroxy alcohol glycerol and three molecules of fatty acids. Most of the fatty acids in these esters have 14–18 carbons. Because there are three ester groups per glycerol, these molecules are called triacylglycerols, or triglycerides (an older name that is still commonly used). The three R groups are usually different.

glycerol   fatty acids   a triacylglycerol

▲
Cultures that use fish as a dietary staple have a low level of heart disease.

> **Practice 28.2**
>
> ──────────
>
> Write the balanced equation (showing structures of reactants and products) for the reaction between a molecule of glycerol, a molecule of stearic acid, and two palmitic acid molecules to form a triacylglycerol.

Fats and oils fit the general description of a lipid. They are large molecules, averaging more than 50 carbon atoms per molecule, with many nonpolar, uncharged groups. Triacylglycerols are insoluble in water.

Fats are an important food source for humans and normally account for about 25–50% of their caloric intake. When oxidized to carbon dioxide and water, fats supply about 40 kJ of energy per gram (9.5 kcal/g), which is more than twice the amount obtained from carbohydrates or proteins. This is partly because triacylglycerol carbons are more reduced than those of most other foods.

In addition, the average fat contains about 75% carbon by mass, whereas the average carbohydrate contains only about 40% carbon. Fats are indeed a rich source of biochemical energy.

In a calorie-conscious society, fatty foods can be too much of a good thing. Thus scientists have developed fat substitutes. In general, fat substitutes need to give a creamy texture and taste to foods without the energy content of fats. A cellulose-based substitute was introduced in the 1960s. Since that time a form of processed protein (microparticulated) with a creamy texture has also been used as a fat substitute. Both carbohydrate- and protein-based fat substitutes are naturally lower in calories than fats. Yet, neither cooks like a fat. Thus the development of olestra represents an important advance for food scientists. Olestra contains long-chain fatty acids, so it tastes and cooks like a fat. But, because the fatty acids are bonded to sucrose and not to glycerol, olestra is indigestable. Olestra has been approved for use in snack foods; medical studies continue on potential long-term impacts of this fat substitute. For more detail on olestra, see the Chemistry in Action on the next page.

Waxes

Waxes are esters of high-molar-mass fatty acids and high-molar-mass alcohols. They have the general formula

$$R'-\overset{\displaystyle O}{\overset{\displaystyle \|}{C}}-O-R$$

in which the alcohol (ROH) contributes up to about 30 carbons, and the fatty acid (R'COOH) also provides an equivalent number of carbons. Waxes are very large

Nutritionists recommend that less than 30% of the daily dietary caloric intake be derived from fats.

The name, wax, derives from the old English word, "weax," which means "material from a honeycomb."

CHEMISTRY IN ACTION   Designing Low-Calorie Fats

Fatty foods provide eating enjoyment for many people. A baked potato just isn't the same without butter or sour cream. A doughnut or french fries should be deep-fried to be tasty. Yet, high-fat diets are a primary cause of obesity and heart disease. The choice of whether or not to eat fatty foods is a dilemma. A solution may be offered by a fat substitute called olestra. Approved by the U.S. Food and Drug Administration in 1996 for snack foods, this compound cooks and tastes like fat but is not digestable. Olestra is marketed under the trade name, Olean.

Olestra is a synthetic compound that combines a sugar (sucrose) and six to eight fatty acids. The fatty acids bond to the sucrose through ester bonds. Like a triacylglycerol, olestra is hydrophobic. When mixed with water, olestra aggregates just as a natural fat does. These aggregates make for the creamy flavor and texture that we have come to associate with fats.

However, the olestra molecule has a different structure than a common triacylglycerol. A triacylglycerol is formed from a small alcohol (glycerol) and three long-chain fatty acids.

In olestra, sucrose provides the alcohol groups to which fatty acids are bonded.

The much greater number of fatty acid esters give olestra a significantly different shape than a triacylglycerol, and molecular shape is important in digestion. Digestive enzymes only break down molecules of specific shapes. The enzymes that digest fats look for a molecular shape similar to a triacylglycerol. Olestra, escapes digestion because it does not have that shape.

Sucrose

Olestra

▲ **We can greatly reduce our fat intake by using an olestra product.**

Olestra's nondigestibility has some important consequences. This fat substitute isn't absorbed by the body. So, olestra doesn't provide energy to the cells, and it has zero caloric content. Also, olestra doesn't raise the level of lipids in the bloodstream and thus helps to avoid atherosclerosis.

Sound too good to be true? As always, there is a down side. Since olestra is not digested, aggregates of this fatty substance continue to pass through the digestive tract. These aggregates may trap fat-soluble nutrients (e.g., the fat-soluble vitamins A, D, E, and, K) leading to a nutritional deficiency. Also, for some people, these fatty aggregates cause lower intestinal problems such as diarrhea.

On balance, olestra may bring healthy and pleasurable eating for many people. Science has given us a new dietary option. We can decide to avoid the excess calories and health risks of a high-fat diet by using olestra, but we must also be aware of potential side effects. As with all new food additives, we need to make informed choices when we plan our diets.

molecules with almost no polar groups. They represent one of the most hydrophobic classes of lipids.

Their extreme water insolubility allows waxes to serve a protective function. Leaves, feathers, fruit, and fur are often naturally coated with a wax. Hardwood floors, cars, and leather goods are just a few of the many products that can be protected by a wax. Waxes tend to be the hardest of the lipids because their carbon chains are long and have very few double bonds. As with fats and oils, the size of the wax molecule and the number of double bonds contained in its carbon chains determine whether the wax will be a liquid or a solid.

Practice 28.3

Write the formula for a wax formed from stearic acid and 1-octacosanol [$CH_3(CH_2)_{26}CH_2OH$].

28.4   Compound Lipids

Phospholipids

phospholipid   The **phospholipids** are a group of compounds that yield one or more fatty acid molecules, a phosphate group, and usually a nitrogenous base upon hydrolysis. The phosphate group and the nitrogenous base, which are found at one end of the phospholipid molecule, often have negative and positive charges. Consequently, in contrast to the triacylglycerols, phospholipids have a hydrophilic end that interacts with water.

$$
\begin{array}{l}
CH_2-O-\boxed{\text{Fatty acid}} \\
\ \ | \\
CH-O-\boxed{\text{Fatty acid}} \\
\ \ | \\
CH_2-O-\boxed{\text{Fatty acid}}
\end{array}
\left.\rule{0pt}{1.1cm}\right\} \text{All hydrophobic}
$$

a triacylglycerol

$$
\begin{array}{l}
CH_2-O-\boxed{\text{Fatty acid}} \\
\ \ | \\
CH-O-\boxed{\text{Fatty acid}}
\end{array}
\left.\rule{0pt}{0.7cm}\right\} \text{Hydrophobic}
$$
$$
\ CH_2-O-\boxed{\text{Phosphate } + \ \substack{\text{Nitrogen}\\ \text{base}}}\left.\rule{0pt}{0.5cm}\right\} \text{Hydrophilic}
$$

a phospholipid

As will be seen later in this chapter, a lipid with both hydrophobic and hydrophilic character is needed to make membranes. It is not surprising that phospholipids are one of the most important membrane components.

Phospholipids are also involved in the metabolism of other lipids and nonlipids. Although they are produced to some extent by almost all cells, most of the phospholipids that enter the bloodstream are formed in the liver. Descriptions of representative phospholipids follow.

◀ **FIGURE 28.4**
Space-filling model of a phosphatidic acid that is esterified to palmitic acid at the top glycerol carbon, to oleic acid at the middle carbon, and to a phosphate group at the bottom glycerol carbon. Note the kink introduced into the oleic acid by the cis double bond.

Phosphatidic Acids Phosphatidic acids are glyceryl esters of fatty acids and phosphoric acid. The phosphatidic acids are important intermediates in the synthesis of triacylglycerols and other phospholipids.

$$CH_2-O-\underset{\underset{O}{\|}}{C}-R_1 \left.\right\rangle \text{Hydrophobic}$$

$$CH-O-\underset{\underset{O}{\|}}{C}-R_2$$

$$CH_2-O-\underset{\underset{O^-}{\|}}{P}-O^- \left.\right\rangle \text{Hydrophilic}$$

a phosphatidic acid

Figure 28.4 is a space-filling model of a typical phosphatidic acid. As with all common phospholipids, the fatty acid chains are large relative to the rest of this molecule. Other phospholipids are formed from a phosphatidic acid when specific nitrogen-containing compounds are linked to the phosphate group by an ester bond. Three commonly used nitrogen compounds are choline, ethanolamine, and L-serine:

$$\underset{\text{choline}}{HOCH_2CH_2\overset{\overset{\displaystyle CH_3}{|}}{\underset{\underset{\displaystyle CH_3}{|}}{N^+}}-CH_3} \qquad \underset{\text{ethanolamine}}{HOCH_2CH_2\overset{+}{N}H_3} \qquad \underset{\text{L-serine}}{HOCH_2\underset{\underset{NH_3}{|+}}{CH}COO^-}$$

(handwritten note: 4-carbons attached)

Because other phospholipids are structurally related to phosphatidic acids, their names are also closely related.

(handwritten structures)

$HO-\overset{\overset{H}{|}}{\underset{\underset{H}{|}}{C}}-\overset{\overset{H}{|}}{\underset{\underset{H}{|}}{C}}-\overset{\overset{CH_3}{}}{N}-CH_3 \qquad H_2NCH\ \overset{\overset{OH}{|}}{C}^-H \qquad HOH_2C-\overset{\overset{H}{|}}{\underset{\underset{NH_3}{|}}{C}}COOH$

▲
Soybean oil is the main commercial source of phosphatidyl choline.

Phosphatidyl Cholines (Lecithins) Phosphatidyl cholines (lecithins) are glyceryl esters of fatty acids, phosphoric acid, and choline. The synonym *lecithin* is an older term that is still used, particularly in commercial products that contain phosphatidyl choline. Phosphatidyl cholines are synthesized in the liver and are present in considerable amounts in nerve tissue and brain substance. Most commercial phosphatidyl choline is obtained from soybean oil and contains palmitic, stearic, palmitoleic, oleic, linoleic, linolenic, and arachidonic acids. Phosphatidyl choline is an edible and digestible emulsifying agent used extensively in the food industry. For example, chocolate and margarine are generally emulsified with phosphatidyl choline. Phosphatidyl choline is also used as an emulsifier in many pharmaceutical preparations.

The single most important biological function for phosphatidyl choline is as a membrane component. This phospholipid makes up between 10 and 20% of many membranes.

$$
\begin{array}{l}
CH_2\!-\!O\!-\!\underset{\underset{O}{\|}}{C}\!-\!R_1 \\[1.2em]
CH\!-\!O\!-\!\underset{\underset{O}{\|}}{C}\!-\!R_2 \\[1.2em]
CH_2\!-\!O\!-\!\underset{\underset{O^-}{\|}}{\overset{O}{P}}\!-\!O\!-\!CH_2CH_2\overset{+}{N}{}^{+}\!-\!CH_3 \\
\end{array}
$$

choline

Hydrophobic

Hydrophilic

a phosphatidyl choline (lecithin) molecule

Phosphatidyl Ethanolamines (Cephalins) Another important constituent of biological membranes is the phosphatidyl ethanolamines (cephalins). These lipids are glyceryl esters of fatty acids, phosphoric acid, and ethanolamine ($HOCH_2CH_2NH_2$). They are found in essentially all living organisms.

$$
\begin{array}{l}
CH_2\!-\!O\!-\!\underset{\underset{O}{\|}}{C}\!-\!R_1 \\[1.2em]
CH\!-\!O\!-\!\underset{\underset{O}{\|}}{C}\!-\!R_2 \\[1.2em]
CH_2\!-\!O\!-\!\underset{\underset{O^-}{\|}}{\overset{O}{P}}\!-\!OCH_2CH_2NH_3^{+} \\
\end{array}
$$

ethanolamine

Hydrophobic

Hydrophilic

a phosphatidyl ethanolamine (cephalin) molecule

Example 28.1 Write the formula for a phosphatidyl ethanolamine that contains two palmitic acid groups.

Solution Phosphatidyl ethanolamine is a phospholipid and so contains glycerol, phosphate, fatty acids, and a nitrogen base. First, write the structure for glycerol. Then add two palmitic acid units, which are linked by ester bonds to the top two carbons of glycerol.

$$\begin{array}{ll}
\text{CH}_2\text{OH} & \text{CH}_2\text{—O—C(CH}_2)_{14}\text{CH}_3 \\
| & | \qquad \| \\
\text{CHOH} & \qquad \quad \text{O} \\
| & | \\
\text{CH}_2\text{OH} & \text{CH—O—C(CH}_2)_{14}\text{CH}_3 \\
& | \qquad \| \\
& \qquad \quad \text{O} \\
& | \\
& \text{CH}_2\text{OH}
\end{array}$$

Next, connect a phosphate group via an ester bond to the bottom glycerol carbon to form a phosphatidic acid. Finally, link the ethanolamine to the phosphate group to yield phosphatidyl ethanolamine.

$$\begin{array}{ll}
\text{CH}_2\text{—O—C(CH}_2)_{14}\text{CH}_3 & \text{CH}_2\text{—O—C(CH}_2)_{14}\text{CH}_3 \\
| \qquad \| & | \qquad \| \\
\quad \quad \text{O} & \quad \quad \text{O} \\
| & | \\
\text{CH—O—C(CH}_2)_{14}\text{CH}_3 & \text{CH—O—C(CH}_2)_{14}\text{CH}_3 \\
| \qquad \| & | \qquad \| \\
\quad \quad \text{O} & \quad \quad \text{O} \\
| \qquad \text{O} & | \qquad \text{O} \\
| \qquad \| & | \qquad \| \\
\text{CH}_2\text{—O—P—O}^- & \text{CH}_2\text{—O—P—OCH}_2\text{CH}_2\overset{+}{\text{N}}\text{H}_3 \\
\quad \quad | & \quad \quad | \\
\quad \quad \text{O}^- & \quad \quad \text{O}^-
\end{array}$$

Practice 28.4

Write the structure of a phosphatidyl choline that contains palmitic acid and stearic acid. How many structures are possible for this compound?

Sphingolipids

Sphingolipids are compounds that, when hydrolyzed, yield a hydrophilic group *sphingolipid* (either phosphate and choline or a carbohydrate), a long-chain fatty acid (18–26 carbons), and sphingosine (an unsaturated amino alcohol). When drawn as follows, the similarities between sphingosine and glycerol esterified to one fatty acid become recognizable:

$$\begin{array}{ll}
\text{OH} & \qquad \qquad \text{O} \\
| & \qquad \qquad \| \\
\text{CH—CH}=\text{CH(CH}_2)_{12}\text{CH}_3 & \text{CH}_2\text{—O—C(CH}_2)_n\text{CH}_3 \\
| \quad * & | \qquad * \\
\text{CH—NH}_2 & \text{CH—OH} \\
| \quad * & | \qquad * \\
\text{CH}_2\text{—OH} & \text{CH}_2\text{—OH} \\
\qquad \text{sphingosine} & \text{glycerol esterified with one fatty acid}
\end{array}$$

The starred atoms on sphingosine react further to make sphingolipids, just as the starred atoms on the glycerol compound react further to give triacylglycerols or phospholipids.

Sphingolipids are common membrane components because they have both hydrophobic and hydrophilic character. For example, sphingomyelins are found in the myelin sheath membranes that surround nerves:

OH
|
CH—CH=CH(CH$_2$)$_{12}$CH$_3$
CH—NH—C—R   — F.A.
               ||
               O
|
CH$_2$—O—P—O—CH$_2$CH$_2$—N$^+$—CH$_3$ Hydrophilic
        || |
        O$^-$ CH$_3$

                                    CH$_3$

} Hydrophobic   *water hating (does not dissolve)*

water loving (dissolves in H$_2$O)

phosphate   a sphingomyelin   *choline*

Notice the hydrophobic and hydrophilic parts of this molecule. Sphingomyelins can also be classified as phospholipids.

Glycolipids

glycolipid   Sphingolipids that contain a carbohydrate group are also known as **glycolipids.** The two most important classes of glycolipids are cerebrosides and gangliosides. These substances are found mainly in cell membranes of nerve and brain tissue. A cerebroside may contain either D-galactose or D-glucose. The following formula of a galactocerebroside shows the typical structure of cerebrosides:

OH
|
CH—CH=CH(CH$_2$)$_{12}$CH$_3$
CH—NH—C—R   } Hydrophobic
        ||
        O
|
CH$_2$—O

CH$_2$OH
HO   O
   OH   } Hydrophilic

      OH

a β-galactocerebroside

Gangliosides resemble cerebrosides in structure but contain complex oligosaccharides instead of simple monosaccharides.

Practice 28.5

Write the structure of a sphingomyelin that contains stearic acid.

28.5   Steroids

Steroids are compounds that have the steroid nucleus, which consists of four fused   *steroids*
carbocyclic rings. This nucleus contains 17 carbon atoms in one five-membered and
three six-membered rings. Modifications of this nucleus in the various steroid compounds include added side chains, hydroxyl groups, carbonyl groups, ring double
bonds, and so on.

steroid ring nucleus

Steroids are closely related in structure but are highly diverse in function. Examples of steroids and steroid-containing materials are (1) cholesterol, the most
abundant steroid in the body, which is widely distributed in all cells and serves as a
major membrane component; (2) bile salts, which aid in the digestion of fats; (3) ergosterol, a yeast steroid, which is converted to vitamin D by ultraviolet radiation; (4)
digitalis and related substances called cardiac glycosides, which are potent heart
drugs; (5) the adrenal cortex hormones, which are involved in metabolism; and (6)
the sex hormones, which control sexual characteristics and reproduction. The formulas for several steroids are given in Figure 28.5.

Cholesterol is the parent compound from which the steroid hormones are synthesized. As we will see, small changes in steroid structure can lead to large changes
in hormonal action. Cholesterol is first converted to progesterone, a compound that
helps control the menstrual cycle and pregnancy. This hormone is, in turn, the parent
compound from which testosterone and the adrenal corticosteroids are produced.
Notice that the long side chain on carbon 17 in cholesterol (Figure 28.5) is smaller in
progesterone and is eliminated when testosterone is formed. Interestingly, testosterone is the precursor for the female sex hormones such as estradiol. These sex hormones are produced by the gonads, either the male testes or the female ovaries. The
small structural differences between testosterone and estradiol trigger vastly different physiological responses.

Practice 28.6

Identify the functional groups in (a) progesterone, (b) cholic acid, and (c)
estradiol.

Cholesterol is also used to build cell membranes, many of which contain about
25% by mass of this steroid. In fact, often there is as much cholesterol as there is
phospholipid, sphingolipid, or glycolipid. These latter three lipid classes cause the

FIGURE 28.5

Structures of selected steroids. Arrows show the biosynthetic relationship between steroids derived from cholesterol. Ergosterol and digoxin are from plant sources.

membrane to be more oily; cholesterol solidifies the membrane. This different be-havior arises from an important structural difference. Cholesterol's four fused rings make it a rigid molecule while other membrane lipids are more flexible. When bio-logical membranes are synthesized, their cholesterol level is adjusted to achieve an appropriate balance between a solid and liquid consistency.

28.6   Hydrophobic Lipids and Biology

The hydrophobic nature of lipids has many important biological consequences. The water insolubility of lipids results in (1) lipid aggregation that causes atherosclerosis, and (2) lipid aggregation that forms biological membranes. When a lipid is sur-rounded by water, it is in a hostile environment. The lipid molecules aggregate to minimize their contact with water. This explains why olive oil (a mixture of triacyl-glycerols) separates from vinegar if it is allowed to stand. This same process of sep-aration occurs continuously in biological solutions.

Let's look at this process on a molecular level. The hydrophilic part of lipid mol-ecules is attracted to water and forms an interface with it, but the hydrophobic part distances itself from water molecules. Depending on the general shape of the lipid molecules, different-shaped aggregates form. Smaller lipids like fatty acids will come together to make micelles. If the fatty acid is shown as follows,

$$CH_3(CH_2)_n COOH$$

hydrophobic   hydrophilic

then a micelle can be visualized as in Figure 28.6. Notice that the hydrophilic car-boxyl groups coat the aggregate and protect the hydrophobic alkyl groups from water.

More complex lipids such as phospholipids and sphingolipids are shaped differ-ently than fatty acids and will aggregate differently in water solutions. These lipids have two hydrophobic alkyl groups and can be represented as

hydrophobic   hydrophilic

Having two hydrophobic chains, these lipids are not as thin as fatty acids. They don't aggregate to form micelles in aqueous solution but rather form liposomes. As shown in Figure 28.7, a simple liposome is bounded by two layers of lipid. The hydropho-bic alkyl chains are covered by hydrophilic groups on both the liposome's inside and outside. Thus, unlike micelles, liposomes have a water core.

Pharmaceutical companies are actively developing liposomes for use in drug delivery systems. These systems consist of specially prepared liposomes carrying drug molecules inside. If the liposome surface is modified appropriately, then the drug will be absorbed only by a specific organ or tissue. Current research aims at

▲
Olive oil is a hydrophobic lipid. Notice how the oil mole-cules (top) separate from the vinegar.

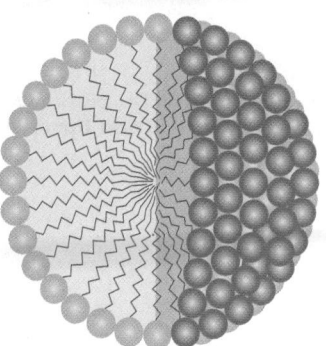

▲
**FIGURE 28.6
Cutaway of a micelle. Note that the hydrophobic chains extend to the center of this aggregate.**

▲
FIGURE 28.7
A cutaway of a single lipo-
some. Note the bilayer shell
and the water solution core.

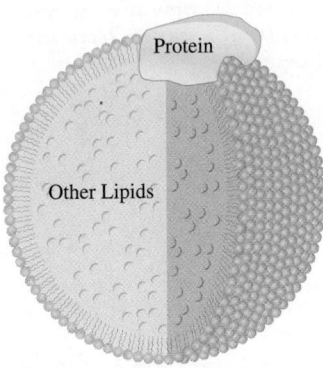

▲
FIGURE 28.8
A cutaway of VLDL. Note the
single-layer shell and the lipid
core.

finding the correct lipids (and other hydrophobic molecules) to make liposomes that target a variety of tissues.

Atherosclerosis

Atherosclerosis is a metabolic disease that leads to deposits of cholesterol and other lipids on the inner walls of the arteries. The name *athere* is Greek for "mush" and nicely describes the appearance of these fatty deposits (called *plaque*). As plaque accumulates, the arterial passages become progressively narrower. The walls of the arteries also lose their elasticity and their ability to expand to accommodate the volume of blood pumped by the heart. Blood pressure increases as the heart works harder to pump sufficient blood through the narrowed passages; this may eventually lead to a heart attack. The accumulation of plaque also causes the inner walls to have a rough rather than a normal smooth surface. This condition may lead to coronary thrombosis (heart attack due to blood clots).

Plaque formation begins because of a lipid's natural tendency to aggregate. In fact, atherosclerosis might be described as a disease partly caused by unwanted lipid aggregation. Scientists have searched for the causes of this aggregation. High serum cholesterol levels often lead to plaque deposits. This makes sense since cholesterol is a large lipid molecule and higher lipid concentrations cause more aggregation. Physicians often recommend that serum cholesterol levels be lowered by restricting intake of major dietary sources such as red meat, liver, and eggs.

Another cause of atherosclerosis is improper transportation of cholesterol through the blood. Cholesterol (and other lipids) must be packaged for transport because lipids aggregate in the aqueous bloodstream. The liver packages dietary lipid into aggregates known as *very-low-density lipoproteins* (VLDL). A cutaway of a VLDL is shown in Figure 28.8. The VLDL surface is hydrophilic and contains a single layer of lipids (and some protein) while the interior provides space for the more hydrophobic triacylglycerols and cholesterol. The VLDLs travel through the bloodstream delivering triacylglycerols to fat cells (the adipose tissue) (see Figure 28.9). The VLDL packages are converted to smaller *low-density lipoproteins* (LDL), which deliver cholesterol to peripheral tissues. Proper circulation of cholesterol also depends on a third lipoprotein, the *high-density lipoprotein* (HDL). The HDL acts as a cholesterol scavenger by collecting cholesterol and returning it to the liver, essentially opposing the action of LDL.

People with high plasma LDL concentrations are prone to atherosclerosis even though they may be on low-cholesterol diets. If LDLs are damaged by oxidation, they selectively accumulate in plaque causing the deposits to grow. In contrast, large amounts of plasma HDL appear to prevent plaque formation, and HDL levels can be increased by strenuous exercise, weight loss, and administration of estradiol.

However, the most common drug treatments for preventing atherosclerosis still involve the reduction of serum cholesterol levels by (1) decreasing triglyceride production (often associated with increasing HDL levels); (2) inhibiting metabolic synthesis of cholesterol; and/or (3) increasing excretion of the cholesterol derivitives, the bile acids. Note that each of these drug treatments assumes a low-cholesterol diet and so the treatment focuses on lowering the body's cholesterol production and storage. The fibrate drugs (e.g., Lopid® and Tricor®) directly lower triglyceride levels and may increase HDL. The statins (e.g., Mevacor,® Lipitor,® and Zocor®)

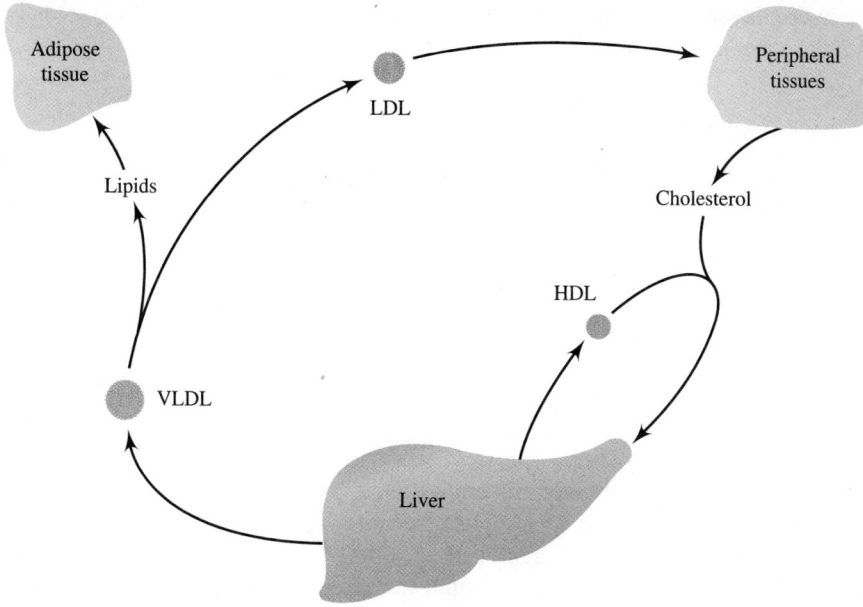

FIGURE 28.9
The lipid distribution system through the blood-stream.

block metabolic production of cholesterol. These drugs are especially effective at lowering serum cholesterol since up to 80% of the total body cholesterol is metabolically produced. Resin drugs (e.g., Colestid® and Questran®) bind the bile acids so they will be excreted. This can lead to a net decrease in body cholesterol levels because bile acids are synthesized from cholesterol. Much the same effect is caused by soluble dietary fiber. Thus diets with good sources of soluble fiber (e.g., oat and barley meal) can significantly decrease serum cholesterol.

Another route to lower serum cholesterol is to decrease intestinal absorption of cholesterol. Certain plant sterols have such an effect. Recently, scientists have modified plant sterols and included them in special margarines. Sitostanol ester (see Figure 28.5) is included in Benecol®, while sitosterol is included in Take Control®. When these margarines are eaten daily (1- to 2-g servings), they have been shown to decrease serum cholesterol levels by about 10–20%.

Biological Membranes

Biological membranes are thin, semipermeable cellular barriers. The general function of these barriers is to exclude dangerous chemicals from the cell while allowing nutrients to enter. Membranes also confine special molecules to specific sections of the cell. Because almost all the dangerous chemicals, nutrients, and special molecules are water soluble, the membranes can act as effective barriers only if they impede the movement of hydrophilic (water-soluble) molecules.

To act as such a barrier, a membrane must have some special properties. To exclude water and water solutes, the bulk of a membrane must be hydrophobic. But a membrane necessarily touches water both inside and outside the cell. Therefore, the surface of a membrane must be hydrophilic. Thus a membrane can be visualized as being layered much like a piece of laminated plywood.

This clogged artery shows plaque formation on the artery wall.

The hydrophobic interior provides the barrier while the hydrophilic exterior interacts with the aqueous environment:

Plywood model of a simple membrane

The cell uses lipids to give the membrane its hydrophobic nature. In fact, by selecting the right lipids, both the hydrophobic and hydrophilic portions of a membrane can be assembled. There are several classes of membrane lipids. The most important of these are the phospholipids and sphingolipids. Remember that these lipids have both a hydrophobic and a hydrophilic section.

lipid bilayer   Membrane lipids naturally aggregate to form lipid bilayers. A **lipid bilayer** is composed of two adjoining layers of lipid molecules aligned so that their hydrophobic portions form the bilayer interior while their hydrophilic portions form the bilayer exterior. A lipid bilayer has the necessary properties of a membrane—a hydrophobic barrier and a hydrophilic surface.

Lipids give this barrier an oily or fluid appearance. The more unsaturated fatty acids in the membrane, the more fluid it will be. Other lipids, most importantly cholesterol, cause the membrane to be less fluid.

Membrane fluidity can have significant effects on cell function. It is thought that general anesthetics (e.g., ether, halothane) are effective partly because they dissolve in membranes, altering the fluidity of the lipid bilayer. During severe cirrhosis of the liver, red cells are forced to take abnormally large amounts of cholesterol into their membranes, causing these membranes to be less fluid. The red cells become more rigid, have greater difficulty passing through narrow capillaries, and are destroyed more easily.

It is truly amazing that lipid bilayers form spontaneously, simply because of the ability of lipids to aggregate. With no complicated planning, a barrier forms with both a hydrophobic interior and a hydrophilic exterior. This barrier provides the basis of the biological membrane. In fact, scientists have proposed that once lipid-like molecules were formed on primal earth, lipid bilayers must have spontaneously aggregated. These bilayers may have provided the boundaries needed as primitive cells first developed.

All known cells in today's world need a membrane that is more complicated than a simple lipid bilayer. A membrane must function as more than just a barrier. Tasks such as passing molecules from one side of a bilayer to the other are an essential part of life. Yet many molecules are hydrophilic and have difficulty traversing a

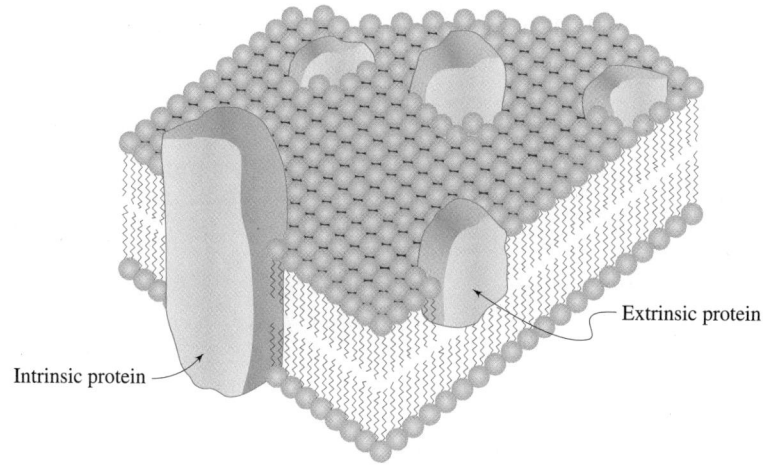

Extrinsic protein

Intrinsic protein

◀ **FIGURE 28.11**
The fluid-mosaic model of a membrane.

lipid bilayer. The cell is faced with a dilemma. How can it selectively allow some hydrophilic molecules to cross the lipid bilayer while excluding others?

Proteins in the fluid bilayer solve this dilemma. These proteins allow specific molecular transport through the hydrophobic interior. (For a general discussion of proteins, see Chapter 29.) They recognize specific molecules on the exterior of a cell membrane and shuttle these molecules into the cell. This process can be as simple as providing a tunnel through the membrane for selected nutrients. If the protein helps (facilitates) transport without using energy, the process is called **facilitated diffusion.** Some transport requires energy, such as when molecules are moved from areas of low concentration to areas of high concentration (the opposite direction from that of diffusion). This energy-requiring transport is termed **active transport** (see Figure 28.10).

Thus a complete cellular membrane must have both lipid and protein. A typical membrane includes about 60% protein, 25% phospholipid, 10% cholesterol, and 5% sphingolipid. The fluid lipid bilayer is studded with many solid proteins. The proteins form a random pattern on the outer surface of the oily lipid. This general membrane is called the **fluid-mosaic model** (see Figure 28.11). An *intrinsic membrane protein* is one in which the bulk of the protein is inside the lipid bilayer; a protein that is mainly on the surface is an *extrinsic membrane protein*.

facilitated diffusion

active transport

fluid-mosaic model

Concepts in Review

1. Describe several general features of lipid structures.

2. Describe the classes of lipids and their functions.

3. Briefly explain why long-carbon-chain fatty acids are hydrophobic.

4. State which fatty acids commonly occur in fats and oils.

5. State which fatty acids are essential to human diets.

6. Explain why unsaturated fatty acids have lower melting points than the corresponding saturated fatty acids.

7. Explain the use of the Greek letter omega (ω) in conjunction with fatty acids.

8. Contrast structure of a fat with a fat substitute like olestra.

9. Briefly discuss the biological importance of eicosanoids.

10. Define the major biological function of waxes.

11. Describe the general structural makeup of phospholipids.

12. Describe the similarities between phospholipids and sphingolipids.

13. Draw the structural feature common to all steroids.

14. Draw the structures for cholesterol and several other common steroids.

15. Describe the biochemical relationships between common steroid hormones.

16. Discuss various components of the steroid distribution system.

17. Discuss atherosclerosis and the factors that affect it.

18. Draw a schematic of a lipid bilayer.

19. Discuss some important characteristics of a membrane lipid.

20. Describe two forms of membrane transport.

Key Terms

The terms listed here have been defined within this chapter. Section numbers are referenced in parentheses for each term.

active transport (28.6)
facilitated diffusion (28.6)
fluid-mosaic model (28.6)
glycolipid (28.4)

lipid bilayer (28.6)
lipid (28.1)
phospholipid (28.4)
sphingolipid (28.4)

steroid (28.5)
wax (28.3)

Questions

Questions refer to tables, figures, and key words and concepts defined within the chapter. A particularly challenging question or exercise is indicated with an asterisk.

1. Why are the lipids, which are dissimilar substances, classified as a group?

2. Briefly, explain why caproic acid is more water soluble than stearic acid.

3. Briefly explain why arachidonic acid is of special biological importance.

4. What are the three essential fatty acids? What are the consequences of their being absent from the diet?

5. List two reasons why fats contain more biochemical energy than carbohydrates.

6. How is aspirin thought to relieve inflammation?

7. Wax serves a protective function on many types of leaves. How does it do this?

8. What two properties must a membrane lipid possess?

9. In what organ in the body are phospholipids mainly produced?

10. List the four classes of eicosanoids.

11. What two structural features allow sphingomyelin to serve as a membrane lipid?

12. What is atherosclerosis? In general, how is it produced and what are its symptoms?

13. Briefly describe the body's cholesterol distribution system.

14. How does a diet that contains a large amount of fish possibly decrease the risk of a heart attack?

15. Why is HDL a potential aid in controlling serum cholesterol levels?

16. Why can a lipid bilayer be described as a barrier?

17. What advantages does a liposome provide as a vehicle for drug delivery?

18. Distinguish active transport from facilitated diffusion.

19. What common structural feature do all steroids possess? Write the structural formulas for two steroids.

20. Why are the statin drugs so effective at reducing serum cholesterol levels?

Paired Exercises

These exercises are paired. Each odd-numbered exercise is followed by a similar even-numbered exercise. Answers to the even-numbered exercises are given in Appendix VI.

21. Write the structural formula of a triacylglycerol that contains one unit each of palmitic, stearic, and oleic acids. How many other triacylglycerols are possible using one unit of each of these acids?

22. Write the structural formula of a triacylglycerol that contains two units of palmitic acid and one unit of oleic acid. How many other triacylglycerols are possible using this same set of fatty acids?

23. Would a triacylglycerol that contains three units of lauric acid be more hydrophobic than a triacylglycerol that contains three units of stearic acid? Explain.

24. Would a triacylglycerol that contains three units of stearic acid be more hydrophobic than a triacylglycerol that contains three units of myristic acid? Explain.

25. Draw formulas for the products of a hydrolysis reaction involving a triacylglycerol that contains palmitic acid, oleic acid, and linoleic acid.

26. Draw formulas for the products of a hydrolysis reaction involving a triacylglycerol that contains oleic acid, stearic acid, and palmitic acid.

27. Give the structure of a phospholipid that is formed from glycerol, phosphoric acid, palmitic acid, and ethanolamine.

28. Give the structure of a phospholipid that is formed from glycerol, phosphoric acid, stearic acid, and choline.

29. Given the following components, draw the structure of a sphingolipid: sphingosine, oleic acid, phosphoric acid, choline.

30. Given the following components, draw the structure of a sphingolipid: sphingosine, stearic acid, phosphoric acid, ethanolamine.

31. Given the following components, draw the structure of a glycolipid: sphingosine, palmitic acid, D-glucose.

32. Given the following components, draw the structure of a glycolipid: sphingosine, oleic acid, D-galactose.

33. Sketch the portion of a micelle that forms from palmitic acid. Show the structural formula for one palmitic acid in your sketch.

34. Sketch the portion of a micelle that forms from myristic acid. Show the structural formula for one myristic acid in your sketch.

35. How does LDL differ in both structure and function from VLDL?

36. How does HDL differ in both structure and function from LDL?

37. In what ways is sphingosine similar to a glycerol molecule that has been esterified to one fatty acid?

38. In what ways is a diacylglycerol similar to a sphingosine that has been linked to a fatty acid by an amide bond?

39. If sodium ion is moved from a solution (0.1 *M*) on one side of a membrane to a solution (0.001 *M*) on the other side, is this process facilitated diffusion or active transport? Briefly explain.

40. If phosphate ion is moved from a solution (0.1 *M*) on one side of a membrane to a solution (0.5 *M*) on the other side, is this process facilitated diffusion or active transport? Briefly explain.

41. Differentiate between the function of a thromboxane and a prostaglandin.

42. Differentiate between the function of a thromboxane and a leukotriene.

Additional Exercises

These exercises are not paired and provide additional practice on concepts covered in this chapter.

43. What biochemical effect of ibuprofen decreases inflammation, redness, and swelling?

44. How does olestra differ structurally from natural fat? Explain why olestra passes through the body without being digested.

45. (a) Draw structural formulas for the three essential fatty acids.
 (b) What are the consequences of these fatty acids being absent from the diet?

*46. Suppose a simple meal is composed of 10 g fat, 17 g carbohydrate, and 19 g protein. The total energy from this meal equals approximately 234 Calories (kcal). If the amount of fat intake is increased to 15 g, what is the energy content now?

47. Beeswax is a mixture of long-chain hydrocarbons and esters. The esters consist of straight-chain alcohols with 24–36 carbon atoms esterified with straight-chain carboxylic acids having up to 36 carbon atoms. Write the structure for triacontanyl hexadecanoate, a C_{30} alcohol and a C_{16} acid.

48. Write the structure for the 18-carbon, ω-9 fatty acid. Give another name for this compound.

Answers to Practice Exercises

28.1

cis-isomer   *trans*-isomer

28.2

or

28.3 $CH_3(CH_2)_{16}\overset{\displaystyle O}{\overset{\displaystyle \|}{C}}-OCH_2(CH_2)_{26}CH_3$

28.4 Two possible structures:

and

28.5

28.6 (a) Progesterone: two ketone functional groups and one C=C double bond
 (b) Cholic acid: three alcohol functional groups, one carboxylic acid functional group
 (c) Estradiol: one alcohol functional group, one phenol functional group

PUTTING IT TOGETHER
Review of Chapters 26–28

Multiple Choice: *Choose the correct answer to each of the following.*

1. The instrument that measures the rotation of plane-polarized light is a
   (a) polariscope (c) polarimeter
   (b) polaroid   (d) polar light meter

2. cis-trans Isomers that do not have a chiral carbon atom
   (a) will rotate polarized light
   (b) are not stereoisomers
   (c) will not be enantiomers
   (d) are optically active

3. A compound that rotates plane-polarized light is said to be
   (a) optically active   (c) polarized
   (b) optically reactive (d) achiral

4. An object that cannot be superimposed on its mirror image is said to be
   (a) achiral (b) chiral (c) meso (d) dextrorotatory

5. A substance in solution that rotates plane-polarized light to the left is
   (a) levorotatory   (c) achiral
   (b) dextrorotatory (d) left-handed

6. Which of these items is chiral?
   (a) a square box (c) a glove
   (b) CH_3Cl (d) a bowling pin

7. How many chiral carbon atoms are in this structure?

   $$CH_3-CH-CH-CH-CH_3$$
   $$\quad\quad | \quad\quad | \quad\quad |$$
   $$\quad\quad OH \quad Br \quad Cl$$

   (a) 3 (b) 4 (c) 5 (d) 6

8. Compound X has an optical rotation of $-12.4°$. What is the rotation of its enantiomer?
   (a) $+24.8°$ (b) $+12.4°$ (c) $+6.2°$ (d) $-12.4°$

9. Which physical property will be different for the isomers in a pair of enantiomers?
   (a) boiling point   (c) melting point
   (b) optical rotation (d) density

10. What is the maximum number of stereoisomers for glucose, which has this structure?

    $$CH_2-CH-CH-CH-CH-C=O$$
    $$| \quad\quad | \quad\quad | \quad\quad | \quad\quad | \quad\quad |$$
    $$OH \quad OH \quad OH \quad OH \quad OH \quad H$$

    (a) 4 (b) 8 (c) 16 (d) 32

11. Which statement is <u>not</u> true about a meso compound?
    (a) A meso compound has a plane of symmetry in the molecule.
    (b) A meso compound does not rotate the plane of polarized light.
    (c) A meso compound is superimposable on its mirror image.
    (d) A meso compound is a racemic mixture.

12. How many meso forms does the compound shown below have?

    $$CH_3-CH-CH-CH-CH_3$$
    $$\quad\quad | \quad\quad | \quad\quad |$$
    $$\quad\quad OH \quad Br \quad OH$$

    (a) 1 (b) 2 (c) 3 (d) 4

13. The compound shown in Question 12 has how many pairs of enantiomers?
    (a) 0 (b) 1 (c) 2 (d) 3

14. Which of the following structures is <u>not</u> a mirror image of the compound shown below?

    $$\quad\quad H$$
    $$\quad\quad |$$
    $$CH_3C-CH_2CH_3$$
    $$\quad\quad |$$
    $$\quad\quad OH$$

15. What do these two structures represent?

    $$\begin{array}{cc} COOH & COOH \\ H-\!\!\!\mid\!\!\!-OH & CH_3-\!\!\!\mid\!\!\!-H \\ CH_3 & OH \end{array}$$
    $$H\quad OH \quad and \quad CH_3 \quad H$$

    (a) a pair of enantiomers   (c) a pair of diastereomers
    (b) a racemic mixture   (d) the same compound

16. Three of these projection structures are the same compound; the other one is their enantiomer. Which structure is the enantiomer?

    (a) $Br-\!\!\!\overset{CH_3}{\underset{H}{\mid}}\!\!\!-Cl$ (c) $H-\!\!\!\overset{Cl}{\underset{CH_3}{\mid}}\!\!\!-Br$

    (b) $Br-\!\!\!\overset{Cl}{\underset{H}{\mid}}\!\!\!-CH_3$ (d) $Cl-\!\!\!\overset{CH_3}{\underset{H}{\mid}}\!\!\!-Br$

17. A solution of a racemic mixture shows no optical rotation because
(a) the two enantiomers contain two chiral carbon atoms
(b) it contains equal amounts of the (+) and (−) forms of the two enantiomers
(c) the two enantiomers cannot be separated
(d) the two enantiomers have the same boiling and melting points

18. In the reaction shown below, a chiral carbon atom is formed, but the product shows no optical activity. Why?

$$CH_3CH_2CH=CH_2 + HCl \longrightarrow CH_3CH_2CHClCH_3$$

(a) The reaction follows Markovnikov's rule.
(b) The product contains a mixture of 50% dextro- and 50% levorotatory stereoisomers.
(c) No optical activity can result from a chemical reaction.
(d) A meso compound is formed.

19. Which of the following two sugars are related as enantiomers?
(a) D-glucose and L-mannose
(b) D-galactose and D-mannose
(c) L-glucose and L-mannose
(d) L-galactose and D-galactose

20. Which of the following two sugars are related as epimers?
(a) D-glucose and L-mannose
(b) D-galactose and D-mannose
(c) L-glucose and L-mannose
(d) L-galactose and D-galactose

21. α-D-Glucopyranose is related to β-D-glucopyranose as
(a) a pair of epimers   (c) a pair of diastereomers
(b) a pair of anomers   (d) no correct answer given

22. D-Ribose is <u>not</u> classified as a
(a) tetrose   (c) aldose
(b) monosaccharide   (d) pentose

23. Sucrose is classified as a
(a) monosaccharide   (c) reducing sugar
(b) trisaccharide   (d) nonreducing sugar

24. Which of the following is an example of a ketose?
(a) D-glyceraldehyde   (c) L-galactose
(b) α-D-glucofuranose   (d) dihydroxyacetone

25. What is the name of the following carbohydrate?

CH$_2$OH

OH   O   OH

OH

OH

(a) α-D-glucopyranose   (c) β-D-galactopyranose
(b) lactose   (d) β-D-glucopyranose

26. What is the name of the mirror image of the following sugar?

$$
\begin{array}{c}
CH_2OH \\
| \\
C=O \\
| \\
H-C-OH \\
| \\
HO-C-H \\
| \\
HO-C-H \\
| \\
CH_2OH
\end{array}
$$

(a) D-fructose   (c) sucrose
(b) D-ketose   (d) L-fructose

27. The ketone carbon in the sugar structure of Question 26 has an oxidation number of
(a) −2   (c) +3
(b) +2   (d) −1

28. What is the name of the following carbohydrate?

(a) β-D-glucopyranose   (c) α-D-galactopyranose
(b) α-D-mannopyranose   (d) no correct answer given

29. The conversion of β-D-mannopyranose to α-D-mannopyranose is called
(a) mutarotation   (c) reduction
(b) oxidation   (d) epimerization

30. Lactose is also known as
(a) levulose   (c) grape sugar
(b) invert sugar   (d) milk sugar

31. In a Benedict test, which of the following functional groups does <u>not</u> react:
(a) hemiacetal   (c) acetal
(b) hemiketal   (d) aldehyde

32. In a Benedict test, D-glucose is converted to
(a) D-glucitol   (c) glucaric acid
(b) D-gluconic acid   (d) saccharic acid

33. Amylopectin can be described as
(a) a linear oligosaccharide
(b) a branching oligosaccharide
(c) a component of glycogen
(d) a component of starch

34. When maltose is reacted in the Benedict test,
(a) Cu$_2$O is produced   (c) no change is observed
(b) the hemiacetal is reduced   (d) the acetal is cleaved

35. Cellulose can be described as a polymer of
    (a) α-D-glucopyranose   (c) β-D-glucofuranose
    (b) α-D-glucofuranose   (d) no correct answer given

36. Invert sugar is formed from
    (a) α-D-glucopyranosyl-(1,4)-α-D-glucopyranose
    (b) β-D-glucopyranosyl-(1,4)-β-D-glucopyranose
    (c) α-D-glucopyranosyl-(1,2)-β-D-fructofuranose
    (d) β-D-galactopyranosyl-(1,4)-α-D-glucopyranose

37. The FDA sets dietary levels for sugar substitutes. These
    levels are called
    (a) recommended dietary allowance
    (b) acceptable daily intake
    (c) acceptable dietary consumption
    (d) recommended daily consumption

38. Breaking the acetal bond of the following disaccharide
    produces what two compounds?

    (a) D-galactose and D-mannose
    (b) D-mannose and D-glucose
    (c) D-galactose and D-glucose
    (d) D-glucose and D-ribose

39. Dilute nitric acid converts D-galactose into
    (a) glucaric acid   (c) sorbitol
    (b) D-galactonic acid   (d) galactaric acid

40. When D-glucose reacts with H_2/Pt, the aldehyde carbon
    changes from an oxidation number of $+1$ to an oxidation
    number of
    (a) 0   (b) $+3$ (c) -1 (d) -2

41. Which of the following lipids contain a total of three ester
    functional groups?
    (a) phospholipid   (c) wax
    (b) triacylglycerol   (d) sphingolipid

42. Stearic acid should have a higher melting point than oleic
    acid because
    (a) stearic acid has no carbon–carbon double bonds while
        oleic acid has one double bond
    (b) oleic acid is only 14 carbons long while stearic acid
        contains 18 carbons
    (c) oleic acid has a trans double bond while stearic acid
        has a cis double bond
    (d) stearic acid has two double bonds while oleic has only
        one

43. Which of the following is an example of an ω-3 fatty acid?
    (a) oleic acid   (c) linolenic acid
    (b) linoleic acid   (d) no correct answer given

44. Aspirin blocks oxidation of which fatty acid?
    (a) arachidonic acid   (c) oleic acid
    (b) stearic acid   (d) arachidic acid

45. The only common lipid class that contains a carbohydrate
    is a
    (a) phospholipid   (c) triacylglycerol
    (b) wax   (d) glycolipid

46. An unsaturated fatty acid contains some carbon atoms
    with an oxidation number of
    (a) 0   (b) -1 (c) $+4$ (d) $+1$

47. Commercial products refer to which lipid class as lecithin?
    (a) phosphatidyl serine   (c) phosphatidyl choline
    (b) cholesterol   (d) phosphatidyl ethanolamine

48. Which of the following compounds is <u>not</u> an eicosanoid:
    (a) prostaglandin   (c) leukotriene
    (b) cephalin   (d) prostacyclin

49. The lipoprotein that acts as a "cholesterol scavenger" is
    abbreviated
    (a) HDL   (b) VLDL   (c) LDL   (d) MLDL

50. The fluid-mosaic model of a biological membrane has
    (a) a hydrophilic interior
    (b) a hydrophobic exterior
    (c) no cholesterol in the lipid bilayer
    (d) a hydrophobic interior

51. A transport process that requires an energy input is called
    (a) facilitated diffusion   (c) passive transport
    (b) active diffusion   (d) active transport

52. Which of the following classes of lipids contains an amide
    bond?
    (a) triacylglycerol   (c) phospholipid
    (b) sphingolipid   (d) wax

53. Which of the following classes of lipids is the most
    hydrophobic?
    (a) phospholipid   (c) wax
    (b) sphingolipid   (d) glycolipid

54. Bile salts are derived from
    (a) arachidonic acid   (c) cholesterol
    (b) lecithin   (d) prostaglandin

55. Because of their shape, fatty acids tend to form one of the
    following in aqueous solution:
    (a) liposomes   (c) HDLs
    (b) micelles   (d) LDLs

56. Which of the following classes of lipids contains four
    ester bonds?
    (a) triacylglycerol   (c) sphingolipid
    (b) phospholipid   (d) glycolipid

Free Response Questions

1. (a) How are the following molecules classified? Are they enantiomers or epimers of each other? Explain.

(b) Based on your knowledge of the anomers of carbohydrates, which molecule has an α-OH and which has a β-OH group?

pregnenolone

2. The partial hydrolysis of a glucose polymer (cellulose, glycogen, amylose, or amylopectin) yielded the following disaccharide only. Give the systematic name for the disaccharide, state which polysaccharide was hydrolyzed, and indicate if a dilute solution of sulfuric acid would have accomplished this hydrolysis.

3. L-Rhamnose is a common 6-deoxy sugar. The Haworth structure is given below.

(a) What is the maximum possible number of diastereomers of this sugar?

(b) Is D-rhamnose an enantiomer, diastereomer, or anomer of L-rhamnose?

(c) Draw the disaccharide α-D-glucopyranosyl-(1,4)-α-L-rhamnopyranose.

(d) Is the disaccharide you drew a reducing sugar?

4. Are the triacylglycerols shown below structural isomers, optical isomers, geometric isomers, or identical molecules? Explain.

5. (a) Does the lipid below contain a triacylglycerol or a sphingosine structure?

(b) Does the substituent on the phosphate group represent an aldohexose? Explain.

6. In the sphingomyelin shown, give the name of the fatty acid that is present, the functional group that resulted from its bonding to sphingosine, and the type of reaction that most likely occurred between sphingosine and the fatty acid.

$$
\begin{array}{l}
\overset{\displaystyle OH}{\underset{\displaystyle |}{HC}}-CH=CH(CH_2)_{12}CH_3 \\[2mm]
\underset{\displaystyle |}{HC}-NH-\overset{\displaystyle O}{\overset{\displaystyle ||}{C}}-CH_2(CH_2)_{17}CH_3 \\[2mm]
H_2C-O-\overset{\displaystyle |}{\underset{\displaystyle |}{\overset{\displaystyle ||}{P}}}-O-CH_2CH_2\overset{+}{N}(CH_3)_3 \\[2mm]
O^-
\end{array}
$$

7. Name the fatty acids used in the triacylglycerol shown below and indicate which one could be classified as an ω-6 fatty acid. How many optical isomers are possible for this lipid?

$$
\begin{array}{l}
H_2C-O-\overset{\displaystyle O}{\overset{\displaystyle ||}{C}}-(CH_2)_{12}CH_3 \\[2mm]
HC-O-\overset{\displaystyle O}{\overset{\displaystyle ||}{C}}-(CH_2)_7CH=CHCH_2CH=CH(CH_2)_4CH_3 \\[2mm]
H_2C-O-\overset{\displaystyle O}{\overset{\displaystyle ||}{C}}-(CH_2)_7CH=CH(CH_2)_5CH_3
\end{array}
$$

8. If the ketone in L-psicose is reduced to an alcohol, explain why two molecules could be expected to form, what their relationship to each other is, and whether either or both will have optical activity.

$$
\begin{array}{c}
CH_2OH \\
| \\
C=O \\
HO-\!\!\!-H \\
HO-\!\!\!-H \longrightarrow \\
HO-\!\!\!-H \\
| \\
CH_2OH
\end{array}
$$

L-psicose

9. The following biological compounds exhibit different solubility properties when water is the solvent. Classify each compound by type—for example, as a lipid (be specific if possible). Note whether they are soluble, partially soluble, or insoluble in water, and explain why.

I (n is very large)

II

$$
CH_3(CH_2)_{18}CH=CH(CH_2)_{10}\overset{\displaystyle O}{\overset{\displaystyle ||}{C}}-O(CH_2)_{35}CH_3
$$

III

10. In the glycolipid below, what functional group results from the connection of the carbohydrate to the sphingosine structure? Name the carbohydrate used and indicate how many chiral carbons are in the entire lipid.

$$
\begin{array}{l}
\overset{\displaystyle OH}{\underset{\displaystyle |}{HC}}-CH=CH(CH_2)_{12}CH_3 \\[2mm]
HC-NH-\overset{\displaystyle O}{\overset{\displaystyle ||}{C}}-(CH_2)_{16}CH_3 \\[2mm]
H_2C-O
\end{array}
$$

▲
The spider has three sets of spinnerets that produce protein-containing fluids that harden as they are drawn out to form silk threads.

Proteins are present in every living cell. Their very name, derived from the Greek word *proteios,* which means holding first place, signifies the importance of these substances. Think of the startling properties of these molecules. Spider-web protein is many times stronger than the toughest steel; hair, feathers, and hooves are all made from one related group of proteins; another protein provides the glass—clear lens material found in the eyes and needed for vision. Very small quantities (milligram amounts) of certain proteins missing from the blood can signify that a person's metabolic processes are out of control. Juvenile-onset diabetes mellitus results from a lack of the insulin protein. Dwarfism can arise when the growth hormone protein is lacking. A special "antifreeze" blood protein allows Antarctic fish to survive at body temperatures below freezing.

This list could go on and on, but what is perhaps most amazing is that the great variety of proteins are made from the same, relatively small, group of amino acids. By using various amounts of these amino acids in different sequences, nature creates biochemical compounds that are essential to the many functions needed to sustain life.

Another aspect of proteins is their importance in our nourishment. Proteins are one of the three major classes of foods. The other two, carbohydrates and fats, are needed for energy; proteins are needed for growth and maintenance of body tissue. Some common foods with high (over 10%) protein content are fish, beans, nuts, cheese, eggs, poultry, and meat. These foods tend to be scarce and relatively expensive. Therefore, proteins are the class of foods that is least available to the under-nourished people of the world. Hence the question of how to secure an adequate supply of high-quality protein for an ever-increasing population is one of the world's more critical problems (see Chapter 32).

29.1 The Structure–Function Connection

Proteins function as structural materials and as enzymes (catalysts) that regulate the countless chemical reactions taking place in every living organism, including the reactions involved in the decomposition and synthesis of proteins.

All proteins are polymeric substances that yield amino acids on hydrolysis. Those that yield only amino acids when hydrolyzed are classified as **simple proteins;** those that yield amino acids and one or more additional products are classified as **conjugated proteins.** There are approximately 200 different known amino acids in nature. Some are found in only one particular species of plant or animal, others in only a few life forms. But 20 of these amino acids are found in almost all proteins. Furthermore, these same 20 amino acids are used by all forms of life in the synthesis of proteins.

All proteins contain carbon, hydrogen, oxygen, and nitrogen. Some proteins contain additional elements, usually sulfur, phosphorus, iron, copper, or zinc. The significant presence of nitrogen in all proteins sets them apart from carbohydrates and lipids. The average nitrogen content of proteins is about 16%.

simple protein

conjugated protein

Proteins are highly specific in their functions. The amino acid units in a given protein molecule are arranged in a definite sequence. An amazing fact about proteins is that in some cases if just one of the hundreds or thousands of amino acid units is missing or out of place, the biological function of that protein is seriously damaged or destroyed. The sequence of amino acids in a protein establishes the function of that protein.

This relationship between structure and function contrasts sharply with that for other classes of biochemicals. For example, carbohydrates provide cellular energy because they contain one particular type of atom, reduced carbon, that is readily oxidizable. This important function does not depend directly on the sequence in which the atoms are arranged. On the other hand, an appropriate sequence of amino acids produces a protein strong enough to form a horse's hoof, a different sequence produces a protein capable of absorbing oxygen in the lungs and releasing it to needy cells; yet another sequence produces a hormone capable of directing carbohydrate metabolism for an entire organism. Full understanding of the function of a protein requires an understanding of its structure.

29.2   The Nature of Amino Acids

Each amino acid has two functional groups, an amino group ($-NH_2$) and a carboxyl group ($-COOH$). The amino acids found in proteins are called alpha (α) amino acids because the amino group is attached to the first or α-carbon atom adjacent to the carboxyl group. The beta (β) position is the next adjacent carbon; the gamma (γ) position the next carbon; and so on. The following formula represents an α-amino acid:

$$CH_3CH_2CHCOOH$$
$$|$$
$$NH_2$$

α-amino butyric acid

Amino acids are represented by this general formula:

Variable group ⟶ R—C—COOH ⟵ Carboxyl group

H (top), NH₂ (bottom) ⟵ Amino group

The portion of the molecule designated R is commonly referred to as the *amino acid side chain*. It is not restricted to alkyl groups and may contain (a) open-chain, cyclic, or aromatic hydrocarbon groups; (b) additional amino or carboxyl groups; (c) hydroxyl groups; or (d) sulfur-containing groups.

Amino acids are divided into three groups: neutral, acidic, and basic. They are classified as

1. *Neutral* when their molecules have the same number of amino and carboxyl groups
2. *Acidic* when their molecules have more carboxyl groups than amino groups
3. *Basic* when their molecules have more amino groups than carboxyl groups

The names, formulas, and abbreviations of the common amino acids are given in Table 29.1 on pages 818 and 819. Two of the these (aspartic acid and glutamic acid) are classified as acidic, three (lysine, arginine, and histidine) as basic, and the remainder are classified as neutral amino acids.

▲ **Cerebrum nerve synapse. Neurotransmitters, many of which are amino acids, cross these synapses to send messages between cells.**

Perhaps the most important role played by amino acids is as the building blocks for proteins. However, selected amino acids also have physiological importance on their own. Many neurotransmitters are amino acids or their derivatives. Glycine and glutamic acid function as chemical messengers between nerve cells in some organisms. Tyrosine is converted to the very important neurotransmitter dopamine. A deficiency of this amino acid derivative causes Parkinson's disease, which can be relieved by another compound formed from tyrosine, L-dopa. Tyrosine is also the parent compound for the "flight-or-fight" hormone, epinephrine (adrenalin), and the metabolic hormone thyroxine. Histidine, which is converted in the body to histamine, is yet another amino acid with an important physiological role. Histamine causes the stomach lining to secrete HCl but is probably best known for causing many of the symptoms associated with tissue inflammation and colds and is the reason antihistamines are such important over-the-counter medications.

▲ **Antihistamines lessen the symptoms of colds and allergies by reducing tissue inflammation.**

Practice 29.1

Write the structures for the simplest α-amino acid and β-amino acid.

 29.3 Essential Amino Acids

Protein is broken down during digestion into its constituent amino acids, which supply much of the body's need for amino acids (see Chapter 32). Eight of the amino acids are **essential amino acids** (see Table 29.2 on page 819) because they are essential to the normal functioning of the human body. Since the body is not capable of synthesizing them, they must be supplied in our diets if we are to enjoy normal health. Some animals require other amino acids in addition to those listed for humans. Rats, for example, require two additional amino acids—arginine and histidine—in their diets.

essential amino acids

On a nutritional basis, proteins are classified as *complete* or *incomplete*. A complete protein supplies all the essential amino acids; an incomplete protein is deficient in one or more essential amino acids. Many proteins, especially those from vegetable sources, are incomplete. For example, protein from corn (maize) is deficient in lysine. The nutritional quality of such vegetable proteins can be greatly improved by supplementing them with the essential amino acids that are lacking, if these can be synthesized at reasonable costs. Lysine, methionine, and tryptophan are now used to enrich human food and livestock feed as a way to extend the world's limited supply of high-quality food protein. In still another approach to the problem of obtaining more high-quality protein, plant breeders have developed maize varieties with greatly improved lysine content. Genetic engineering may hold the key to further significant improvements in plant protein quality.

Vegetarians must carefully choose combinations of foods that include all essential amino acids.

Practice 29.2

Which of the following are essential amino acids for humans?
Gly, Ala, Leu, Lys, Tyr, Trp, His

▲
Adding essential amino acids to livestock feed is a way to extend high-quality protein.

TABLE 29.1	Common Amino Acids Derived from Proteins	
Name	**Abbreviation**	**Formula**
Alanine	Ala	$CH_3CHCOOH$ with NH_2
Arginine	Arg	$NH_2-C-NH-CH_2CH_2CH_2CHCOOH$ with NH and NH_2
Asparagine	Asn	$NH_2C-CH_2CHCOOH$ with O and NH_2
Aspartic acid	Asp	$HOOCCH_2CHCOOH$ with NH_2
Cysteine	Cys	$HSCH_2CHCOOH$ with NH_2
Glutamic acid	Glu	$HOOCCH_2CH_2CHCOOH$ with NH_2
Glutamine	Gln	$NH_2CCH_2CH_2CHCOOH$ with O and NH_2
Glycine	Gly	$HCHCOOH$ with NH_2
Histidine	His	$N=CH$, HC $C-CH_2CHCOOH$, $N-H$, NH_2
Isoleucine*	Ile	$CH_3CH_2CH-CHCOOH$ with CH_3 and NH_2
Leucine*	Leu	$(CH_3)_2CHCH_2-CHCOOH$ with NH_2
Lysine*	Lys	$NH_2CH_2CH_2CH_2CH_2CHCOOH$ with NH_2
Methionine*	Met	$CH_3SCH_2CH_2CHCOOH$ with NH_2
Phenylalanine*	Phe	$C_6H_5-CH_2CHCOOH$ with NH_2

*Amino acids essential in human nutrition

TABLE 29.1	Common Amino Acids Derived from Proteins (Continued)	
Name	Abbreviation	Formula
Proline	Pro	![structure] — COOH with pyrrolidine ring N—H
Serine	Ser	$HOCH_2CHCOOH$ with NH_2
Threonine*	Thr	CH_3CH—$CHCOOH$ with OH and NH_2
Tryptophan*	Trp	indole ring—C—$CH_2CHCOOH$ with CH, N—H, NH_2
Tyrosine	Tyr	HO—⟨ring⟩—$CH_2CHCOOH$ with NH_2
Valine*	Val	$(CH_3)_2CHCHCOOH$ with NH_2

*Amino acids essential in human nutrition

TABLE 29.2	Essential Amino Acids for Humans
Isoleucine
Leucine
Lysine
Methionine
Phenylalanine
Threonine
Tryptophan
Valine

29.4 D-Amino Acids and L-Amino Acids

All amino acids, except glycine, have at least one asymmetric carbon atom. For example, two stereoisomers of alanine are possible:

D-(−)-alanine   L-(+)-alanine

Refer to Chapter 26 to review stereoisomerism.

Fischer projection formulas illustrate the D and L configurations of amino acids in the same way they illustrate the configurations of D- and L- glyceraldehyde (Section

27.2). The —COOH group is written at the top of the projection formula, and the D configuration is indicated by writing the alpha —NH$_2$ to the right of carbon 2. The L configuration is indicated by writing the alpha —NH$_2$ to the left of carbon 2. Although some D-amino acids occur in nature, only L-amino acids occur in proteins. The (+) and (−) signs in the name indicate the direction of rotation of plane-polarized light by the amino acid. Most amino acids have relatively complex structures that make use of the projection formula difficult. Thus, unless stereochemical information is explicitly considered, amino acids will be shown using a condensed, structural formula.

29.5   Amphoterism

A Brønsted–Lowry acid is a proton donor; a Brønsted–Lowry base is a proton acceptor.

Amino acids are *amphoteric* (or *amphiprotic*); that is, they can react either as an acid or as a base. For example, with a strong base such as sodium hydroxide, alanine reacts as an acid:

$$CH_3CHCOOH \ + \ NaOH \ \longrightarrow \ CH_3CHCOO^- \, Na^+ \ + \ H_2O$$

$$\underset{\text{alanine}}{\overset{|}{NH_2}} \qquad\qquad\qquad \underset{\text{sodium alanate}}{\overset{|}{NH_2}}$$

With a strong acid such as HCl, alanine reacts as a base:

$$CH_3CHCOOH \ + \ HCl \ \longrightarrow \ CH_3CHCOOH$$

$$\underset{\text{alanine}}{\overset{|}{NH_2}} \qquad\qquad\qquad \underset{\text{alanyl ammonium chloride}}{\overset{|}{NH_3^+ \, Cl^-}}$$

Even in neutral biological solutions, amino acids do not actually exist in the molecular form shown in the preceding equations. Instead, they exist mainly as dipolar ions called **zwitterions.** Again, using alanine as an example, the proton on the carboxyl group transfers to the amino group, forming a zwitterion by an acid–base reaction within the molecule:

zwitterion

$$CH_3CHCOO\,\textcircled{H} \ \longrightarrow \ CH_3CHCOO^-$$

$$\underset{\text{alanine molecule}}{\overset{|}{H_2N\!:\!\!\leftarrow}} \qquad\qquad \underset{\text{alanine zwitterion}}{\overset{|}{NH_3^+}}$$

On an ionic basis, the reaction of alanine with OH$^-$ and H$^+$ is

$$CH_3CHCOO^- \ + \ OH^- \ \longrightarrow \ CH_3CHCOO^- \ + \ H_2O \qquad (1)$$

$$\underset{\text{alanine zwitterion}}{\overset{|}{NH_3^+}} \qquad\qquad\qquad \underset{\text{alanate anion}}{\overset{|}{NH_2}}$$

$$CH_3CHCOO^- \ + \ H^+ \ \longrightarrow \ CH_3CHCOOH \qquad\qquad (2)$$

$$\underset{\text{alanine zwitterion}}{\overset{|}{NH_3^+}} \qquad\qquad\qquad \underset{\text{alanyl ammonium cation}}{\overset{|}{NH_3^+}}$$

Most other amino acids behave like alanine. Together with protein molecules that contain $-COOH$ and $-NH_2$ groups, they help to buffer or stabilize the pH of the blood at about 7.4. The pH is maintained close to 7.4 because any excess acid or base in the blood is neutralized by reactions such as shown in equations (1) and (2).

$$\underset{\substack{| \\ NH_3^+ \\ \text{cation form} \\ II}}{RCHCOOH} \underset{H^+}{\overset{OH^-}{\rightleftarrows}} \underset{\substack{| \\ NH_3^+ \\ \text{zwitterion form} \\ I}}{RCHCOO^-} \underset{H^+}{\overset{OH^-}{\rightleftarrows}} \underset{\substack{| \\ NH_2 \\ \text{anion form} \\ III}}{RCHCOO^-} \qquad (3)$$

When an amino acid in solution has equal positive and negative charges, as in formula I of equation (3), it is electrically neutral and does not migrate toward either the positive or negative electrode when placed in an electrolytic cell. The pH at which there is no migration toward either electrode is called the **isoelectric point** (see Table 29.3). If acid (H^+) is added to an amino acid at its isoelectric point, the equilibrium is shifted toward formula II, and the cation formed migrates toward the negative electrode. When base (OH^-) is added, the anion formed (formula III) migrates toward the positive electrode. Differences in isoelectric points are important in isolating and purifying amino acids and proteins, since their rates and directions of migration can be controlled in an electrolytic cell by adjusting the pH. This method of separation is called *electrophoresis*.

Amino acids are classified as basic, neutral, or acidic depending on whether the ratio of $-NH_2$ to $-COOH$ groups in the molecules is greater than $1:1$, equal to $1:1$, or less than $1:1$, respectively. Furthermore, this ratio differs from $1:1$ only if the amino acid side chain $(R-)$ contains an additional amino or carboxyl group. For example, if the side chain contains a carboxyl group, the amino acid is considered acidic. Thus the $R-$ group determines whether an amino acid is classified as basic, neutral, or acidic.

Isoelectric points are found at pH values ranging from 7.8 to 10.8 for basic amino acids, 4.8 to 6.3 for neutral amino acids, and 2.8 to 3.3 for acidic amino acids. It is logical that a molecule such as glutamic acid, with one amino group and two carboxyl groups, would be classified as acidic and that its isoelectric point would be at a pH lower than 7.0. It might also seem that the isoelectric point of an amino acid classified as neutral, such as alanine, with one amino and one carboxyl group, would have an isoelectric point of 7.0. However, the isoelectric point of alanine is 6.0, not 7.0. This is because the carboxyl group and the amino group are not equally ionized. The carboxyl group of alanine ionizes to a greater degree as an acid than the amino group ionizes as a base.

isoelectric point

TABLE 29.3	Isoelectric Points of Selected Amino Acids
Amino acid	**pH at isoelectric point**
Arginine	10.8
Lysine	9.7
Alanine	6.0
Glycine	6.0
Serine	5.7
Glutamic acid	3.2
Aspartic acid	2.9

Draw the structure of L-serine in a strongly acidic solution.

Example 29.1

Solution

First, draw the structure for L-serine as the molecule would exist in a neutral solution. The α-amino and carboxylic acid groups form a zwitterion.

$$\underset{\substack{| \\ NH_3^+}}{HOCH_2CHCOO^-}$$

When the solution is made acidic (the concentration of hydrogen ions is increased), the amine group is unaffected because it is already protonated. However, the carboxylate anion bonds to a hydrogen ion, resulting in an L-serine structure with a net positive charge.

$$HOCH_2\underset{\underset{NH_3^+}{|}}{CH}COOH$$

Practice 29.3

Draw the structure of L-valine in a strongly basic solution.

29.6   Formation of Polypeptides

Proteins are polyamides consisting of amino acid units joined through amide structures. If we react two glycine molecules, with the elimination of a molecule of water, we form a compound containing the amide structure, also called the **peptide linkage,** or peptide bond. The elimination of water occurs between the carboxyl group of one amino acid and the α-amino group of a second amino acid (see Section 25.1). The product formed from two glycine molecules is called glycylglycine (abbreviated Gly-Gly). Because it contains two amino acid units, it is called a dipeptide.

peptide linkage

An amide is formed by reaction between a carboxylic acid and an amine.

glycylglycine
(Gly-Gly)

If three amino acid units are included in a molecule, it is a tripeptide; if four, a tetrapeptide; if five, a pentapeptide; and so on. Peptides containing up to about 40–50 amino acid units in a chain are called **polypeptides.** The units making up the peptide are amino acids less the elements of water and are referred to as *amino acid residues* or, simply, residues. Still larger chains of amino acids are known as proteins.

polypeptides

When amino acids form a polypeptide chain, a carboxyl group and an α-amino group are involved in each peptide bond (amide bond). While these groups are joined in peptide bonds, they cannot ionize as acids or bases. Consequently, the properties of a polypeptide/protein are determined to a large extent by the side chains of the amino acid residues.

In linear peptides one end of the chain has a free amino group and the other end a free carboxyl group. The amino-group end is called the *N-terminal residue* and the other end the *C-terminal residue:*

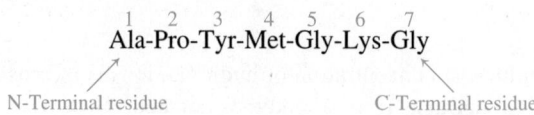

The sequence of amino acids in a chain is numbered starting with the N-terminal residue, which is written to the left with the C-terminal residue at the right. Any segment of the sequence that is not specifically known is placed in parentheses. Thus in the heptapeptide above, if the order of tyrosine and methionine were not known, the structure would be written as

Ala-Pro-(Met, Tyr)-Gly-Lys-Gly

Peptides are named as acyl derivatives of the C-terminal amino acid with the C-terminal unit keeping its complete name. The -ine ending of all but the C-terminal amino acid is changed to -yl, and these listed in the order in which they appear, starting with the N-terminal amino acid.

Thus Ala-Tyr-Gly is called alanyltyrosylglycine. The name of Arg-Glu-His-Ala is arginylglutamylhistidylalanine.

Alanine and glycine can form two different dipeptides, Gly-Ala and Ala-Gly, using each amino acid only once:

If three different amino acids react—for example, glycine, alanine, and threonine—six tripeptides in which each amino acid appears only once are possible.

Gly-Ala-Thr	Ala-Thr-Gly	Thr-Ala-Gly
Gly-Thr-Ala	Ala-Gly-Thr	Thr-Gly-Ala

The number of possible peptides rises very rapidly as the number of amino acid units increases. For example, there are 120 ($1 \times 2 \times 3 \times 4 \times 5 = 120$) different ways to combine five different amino acids to form a pentapeptide, using each amino acid only once in each molecule. If the same constraints are applied to 15 different amino acids, the number of possible combinations is greater than 1 trillion (10^{12})! Since a protein molecule can contain several hundred amino acid units, with individual amino acids occurring several times, the number of possible combinations from 20 amino acids is simply beyond imagination.

There are a number of small, naturally occurring polypeptides with significant biochemical functions. In general, these substances serve as hormones or nerve transmitters. Their functions range from controlling pain and pleasure responses in the brain to controlling smooth muscle contraction or kidney fluid excretion rates (see Table 29.4).

The amino acid sequence and chain length give a polypeptide its biological effectiveness and specificity. For example, recent research has shown that the effects of opiates (opium derivatives) on the brain are also exhibited by two naturally occurring pentapeptides, Leu-enkephalin (Tyr-Gly-Gly-Phe-Leu) and Met-enkephalin (Tyr-Gly-Gly-Phe-Met). These two pentapeptides are natural painkillers. Alterations of the amino acid sequence—which alter the side-chain characteristics—cause drastic changes in the analgesic effects of these pentapeptides. The substitution of L-alanine for either of the glycine residues in these compounds (simply changing one side-chain group from $-H$ to $-CH_3$) causes an approximately 1000-fold decrease in effectiveness as a painkiller! The substitution of L-tyrosine for L-phenylalanine causes a comparable loss of activity. Even the substitution of D-tyrosine for the L-tyrosine residue causes a considerable loss in the analgesic effectiveness of the pentapeptides.

It is clearly evident that a particular sequence of amino acid residues is essential for proper polypeptide function. This sequence aligns the side-chain characteristics (large or small; polar or nonpolar; acidic, basic, or neutral) in the proper positions for a specific polypeptide function.

Oxytocin and vasopressin are similar nonapeptides, differing only at two positions in their primary structure (see Table 29.4). Yet their biological functions differ dramatically. Oxytocin controls uterine contractions during labor in childbirth and also causes contraction of the smooth muscles of the mammary glands, resulting in

TABLE 29.4 Primary Structures and Functions of Some Biological Polypeptides*

Name	Primary structure	General biological function
Substance P	Arg-Pro-Lys-Pro-Gln-Gln-Phe-Phe-Gly-Leu-Met-NH$_2$	Is a pain-producing agent
Bradykinin	Arg-Pro-Pro-Gly-Phe-Ser-Pro-Phe-Arg	Affects tissue inflammation and blood pressure
Angiotensin II	Asp-Arg-Val-Tyr-Val-His-Pro-Phe	Maintains water balance and blood pressure
Leu-enkephalin Met-enkephalin	Tyr-Gly-Gly-Phe-Leu Tyr-Gly-Gly-Phe-Met	Relieves pain, produces sense of well-being
Vasopressin	┌──────S─S──────┐ Cys-Tyr-Phe-Gln-Asn-Cys-Pro-Arg-Gly-NH$_2$	Increases blood pressure, decreases kidney water excretion
Oxytocin	┌──────S─S──────┐ Cys-Tyr-Ile-Gln-Asn-Cys-Pro-Leu-Gly-NH$_2$	Initiates childbirth labor, causes mammary gland milk release, affects kidney excretion of water and sodium

*Where $-NH_2$ is indicated at the end of the sequence, the C-terminal amino acid has an amide structure rather than a free $-COOH$.

◄ **The nonapeptide oxytocin helps to control uterine contractions during labor.**

milk excretion. Vasopressin in high concentration raises the blood pressure and has been used to treat surgical shock. Vasopressin is also an antidiuretic, regulating the excretion of fluid by the kidneys. The absence of vasopressin leads to diabetes insipidus. This condition is characterized by excretion of up to 30 L of urine per day but can be controlled by administration of vasopressin or its derivatives.

The isolation and synthesis of oxytocin and vasopressin was accomplished by Vincent du Vigneaud (1901–1978) and co-workers at Cornell University. Du Vigneaud was awarded the Nobel prize in chemistry in 1955 for this work. Synthetic oxytocin is indistinguishable from the natural material. It is available commercially and is used for the induction of labor in the late stages of pregnancy.

Write the structure of the tripeptide Ser-Gly-Ala.

First, write the structures of the three amino acids in this tripeptide.

Example 29.2

Solution

$$HOCH_2CHCOOH \qquad CH_2COOH \qquad CH_3CHCOOH$$
$$\qquad | \qquad\qquad\qquad | \qquad\qquad\qquad | $$
$$\qquad NH_2 \qquad\qquad\quad NH_2 \qquad\qquad\quad NH_2$$

serine (Ser)   glycine (Gly)   alanine (Ala)

By convention, the amino acid residue written at the left end of the tripeptide has a free amino group while the residue at the right end has a free carboxylic acid group. Now split out a water molecule in two places: between the carboxyl group of serine and the amino group of glycine, and between the carboxyl group of glycine and the amino group of alanine. When the amino acids are connected by peptide linkages, the following structure results:

$$\qquad\quad O \qquad\quad O$$
$$\qquad\quad \| \qquad\quad \|$$
$$HOCH_2CHC-NHCH_2C-NHCHCOOH$$
$$\qquad | \qquad\qquad\qquad\qquad\quad |$$
$$\qquad NH_2 \qquad\qquad\qquad\qquad CH_3$$

Practice 29.4

Write the structure of the pentapeptide Gly-Leu-Asp-Ser-Cys.

 29.7 **Protein Structure**

By 1940, a great deal of information concerning proteins had been assembled. Their elemental composition was known, and they had been carefully classified according to solubility in various solvents. Proteins were known to be polymers of amino acids, and the different amino acids had, for the most part, been isolated and identified. Protein molecules were known to be very large in size, with molar masses ranging from several thousand to several million.

Knowledge of protein structure would answer many chemical and biological questions. But for a while the task of determining the actual structure of molecules of such colossal size appeared to be next to impossible. Then Linus Pauling, at the California Institute of Technology, attacked the problem by a new approach. Using X-ray diffraction techniques, Pauling and his collaborators painstakingly determined the bond angles and dimensions of amino acids and of dipeptides and tripeptides. After building accurate scale models of the dipeptides and tripeptides, they determined how they could fit together into likely polypeptide configurations. Based on this work, Pauling and Robert Corey (1897–1971) proposed in 1951 that two different conformations—the *α-helix* and the *β-pleated sheet*—were the most probable stable polypeptide chain configurations of protein molecules. These two macromolecular structures are illustrated in Figure 29.1. Within a short time it was established that many proteins do have structures corresponding to those predicted by Pauling and Corey. This work was a very great achievement. Pauling received the 1954 Nobel prize in chemistry for this work on protein structure. Pauling's and Corey's work inspired another great biochemical breakthrough—the concept of the double-helix structure for deoxyribonucleic acid, DNA (see Section 31.6).

Proteins are very large molecules. But just how many amino acid units must be present for a substance to be a protein? There is no universally agreed-upon answer to this question. Some authorities state that a protein must have a molar mass of at least 6000 or contain about 50 amino acid units. Smaller amino acid polymers, containing from 5 to 50 amino acid units, are classified as polypeptides and are not proteins. In reality, there is no clearly defined lower limit to the molecular size of proteins. The distinction is made to emphasize (1) that proteins usually serve structural or enzymatic functions, while polypeptides usually serve hormone-related functions, and (2) that the three-dimensional conformation of proteins is directly related to function, while the relationship is not so distinct with polypeptides.

In general, for a protein molecule to serve a specific biological function, it must have a closely defined overall shape. Chemists typically describe large proteins on several levels: (1) a primary structure, (2) a secondary structure, (3) a tertiary structure, and, for the most complex proteins, (4) a quaternary structure.

primary structure   The **primary structure** of a protein is established by the number, kind, and sequence of amino acid units composing the polypeptide chain or chains making up the molecule. The primary structure determines the alignment of side-chain characteristics, which, in turn, determines the three-dimensional shape into which the

(a) α-Helix

(b) Collagen

◀── Chain 1

◀── Chain 2

(c) β-Pleated sheet

▲
FIGURE 29.1
Three common protein
secondary structures.

protein folds. In this sense the amino acid sequence is of primary importance in establishing protein shape.

Determining the sequence of the amino acids in even one protein molecule was a formidable task. The amino acid sequence of beef insulin was announced in 1955 by the British biochemist Frederick Sanger. This accomplishment required several years of effort by a team under Sanger's direction. He was awarded the 1958 Nobel prize in chemistry for this work. Insulin is a hormone that regulates the blood-sugar level. A deficiency of insulin leads to the condition of diabetes. Beef insulin consists of 51 amino acid units in two polypeptide chains. The two chains are connected by disulfide linkages ($-S-S-$) of two cysteine residues at two different sites. The primary structure is shown in Figure 29.2. Insulins from other animals, including humans, differ slightly by one, two, or three amino acid residues in chain A.

secondary structure

The **secondary structure** of proteins can be characterized as a regular, three-dimensional structure held together by hydrogen bonding between the oxygen of the

$\diagup\!\!\diagup$C$=$O and the hydrogen of the H$-$N$\diagdown$ groups in the polypeptide chains:

$$\diagup\!\!\diagup C=O \cdots H-N\diagdown$$
$$\text{Hydrogen bond}$$

The α-helical and β-pleated-sheet structures of Pauling and Corey are two examples of secondary structure. As shown in Figure 29.1, essentially every peptide bond is involved in at least one hydrogen bond in these structures. Proteins having α-helical or β-pleated-sheet secondary structures are strongly held in particular conformations by virtue of the large number of hydrogen bonds.

tertiary structure

The **tertiary structure** of a protein refers to the distinctive and characteristic conformation, or shape, of a protein molecule. This overall three-dimensional conformation is held together by a variety of interactions between amino acid side chains. These interactions include (1) hydrogen bonding, (2) ionic bonding, and (3) disulfide bonding. Here are some examples.

1. Glutamic acid–tyrosine hydrogen bonding

$$\text{Protein}-CH_2CH_2\overset{\overset{\displaystyle OH}{|}}{C}=O \cdots HO-\langle\bigcirc\rangle-\text{Protein}$$

2. Glutamic acid–lysine ionic bonding

$$\text{Protein}-CH_2CH_2\overset{\overset{\displaystyle O}{\|}}{C}-O^-\ \overset{+}{N}H_3-CH_2CH_2CH_2CH_2-\text{Protein}$$

3. Cysteine–cysteine disulfide bonding

$$\text{Protein}-CH_2-S-S-CH_2-\text{Protein}$$

The tertiary structure depends on the number and location of these interactions, variables that are fixed when the primary structure is synthesized. Thus the tertiary structure depends on the primary structure. For example, there are three locations in the insulin molecule (Figure 29.2) where the primary sequence permits disulfide bonding. This specificity has an obvious bearing on the shape of insulin.

$$
\begin{array}{c}
\overset{15}{\text{Gln}}\text{—Leu—Glu—Asn—Tyr—}\overset{20}{\text{Cys}}\overset{21}{\text{—Asn}} \\
\end{array}
$$

Tyr
|
Leu
|
Ser
|
Cys—Val ⁱ⁰
| |
S Ser
| |
S Ala
|
Gly—Ile—Val—Glu—Gln—Cys—Cys
Chain A

S
|
S

Cys—Gly ²⁰
| |
Val Glu
| |
Leu Arg
| |
Tyr Gly
| |
¹⁵Leu Phe
| |
Ala Phe ²⁵
| |
Glu Tyr
| |
Phe—Val—Asn—Gln—His—Leu—Cys—Gly—Ser—His—Leu—Val Thr
Chain B Pro
|
Lys
|
Ala ³⁰

◀ **FIGURE 29.2**
Amino acid sequence of beef insulin.

Hair is especially rich in disulfide bonds. These can be broken by certain reducing agents and restored by an oxidizing agent. This fact is the key to "cold" permanent waving of hair. Some of the disulfide bonds are broken by applying a reducing agent to the hair. The hair is then styled with the desired curls or waves. These are then permanently set by using an oxidizing agent to reestablish the disulfide bonds at different points.

$$-CH_2-S-S-CH_2- \xrightarrow{\text{reducing agent}}$$

bonds in
normal hair

$$-CH_2-S-H \quad H-S-CH_2- \xrightarrow{\text{oxidizing agent}} -CH_2-S-S-CH_2-$$

broken bonds
of "reduced" form

reestablished
bonds of permanently
waved hair

If we examine Table 29.1 closely, we can conclude that most amino acids have side chains that form neither hydrogen, ionic, nor disulfide bonds. What then do the majority of amino acids do to hold together the tertiary structure of a protein? This is an important question and leads to an equally important answer. The uncharged relatively nonpolar amino acids form the center or core of most proteins. These amino acids have side chains that do not bond very well to water; they are like saturated hydrocarbons or lipids and are hydrophobic. When a protein is synthesized, the uncharged, nonpolar amino acids turn inward toward each other, excluding water and forming the core of the protein structure.

A fourth type of structure, called **quaternary structure,** is found in some complex proteins. These proteins are made of two or more smaller protein subunits (polypeptide chains). Nonprotein components may also be present. The quaternary

▲
In a permanent wave, disulfide bonds in the hair are broken with reducing agents, set to a curling pattern, and then restored by an oxidizing agent.

quaternary structure

structure refers to the shape of the entire complex molecule and is determined by the way in which the subunits are held together by *noncovalent* bonds—that is, by hydrogen bonding, ionic bonding, and so on.

Practice 29.5

Show a tertiary structure for the following pentapeptide *after* a disulfide bond forms: Ala-Cys-Gly-Gly-Cys (use the amino acid structures given in Table 29.1).

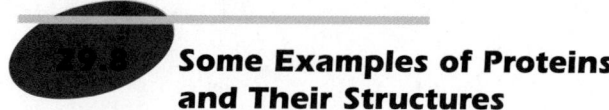

Some Examples of Proteins and Their Structures

Fibrous Proteins

fibrous protein

The **fibrous proteins** are an important class of proteins that contain highly developed secondary structures. As their name suggests, fibrous proteins have a "fiberlike" or elongated shape. Because secondary structures provide strength, these proteins tend to function in support roles. Several of these proteins are shown in Figure 29.3. The α-keratins depend on the α-helix for strength and are found in such diverse structures as hair, feathers, horns, and hooves. The silk protein, fibroin, is folded into a β-pleated-sheet secondary structure.

"Collegen" derives from the Greek word, kolla, meaning glue.

The most abundant protein in the animal kingdom, collagen, is a fibrous protein with a unique secondary structure (Figure 29.3). Collagen forms the bone matrix around which the calcium phosphate mineral crystallizes. Ligaments, tendons, and skin are composed of a large proportion of collagen. The structure of this protein allows it to provide a strong framework for each of these tissues. Collagen is a long, slender protein whose three strands are wrapped one around another as a rope would be woven, and, like a rope, the finished product is much stronger than a single strand. These triple helices are in turn stacked like cordwood to form collagen fibers. The resulting tough material can be aligned to form tendons or deposited as a network to form a bone or skin structure. Next time you take a stride, think about the force exerted on your Achilles tendon and how your being able to function depends on the strength of collagen.

Collagen takes on an added importance for artificial biological implants. In fact, collagen is an almost ideal biomaterial; it is strong, stable, and does not generally cause a dangerous allergic reaction. Artificial materials such as plastics or metals are often coated with collagen before implantation. Collagen can be used directly in artificial skin, in bone graft substitutes, and as vascular or tissue sealants.

Globular Proteins

globular protein

The **globular proteins** share a similar compact shape. Although it may not be apparent at first, this roughly spherical structure allows each globular protein its own complex tertiary structure. Each unique structure is associated with a particular function in the biological world. Some proteins function to bind specific biochemicals. This function is important for biological transport (e.g., oxygen transport by hemo-

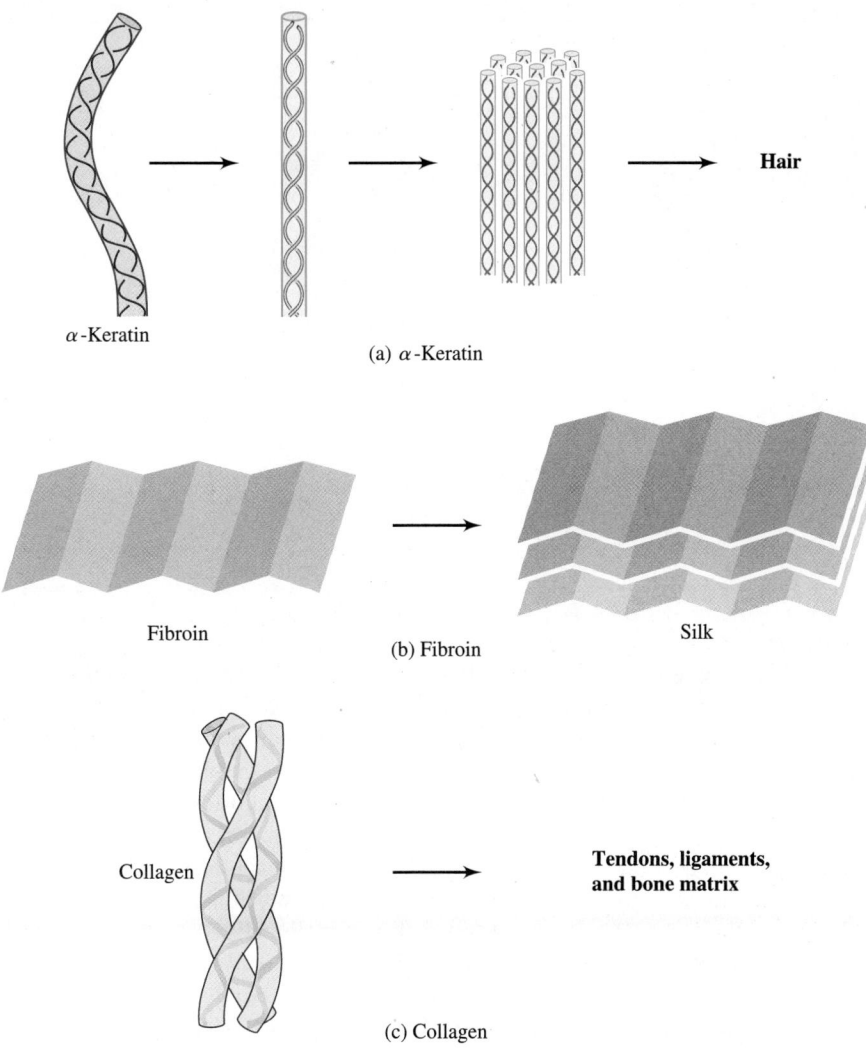

Hair

α-Keratin

(a) α-Keratin

Fibroin

(b) Fibroin

Silk

Collagen

Tendons, ligaments, and bone matrix

(c) Collagen

globin) and for processes such as the immune response (e.g., immunoglobulins). Other globular proteins bind biochemicals and then catalyze a chemical reaction (e.g., enzymes). A catalyst is required for essentially all metabolic reactions.

We will consider several protein structures in detail. These proteins contain hundreds to thousands of atoms and look very complicated at first glance. However, scientists have learned to look for much simpler, major structural features. These features are often related to a protein's function. To help visualize these molecular relationships, we will use drawing techniques that emphasize major structural features.

Myoglobin—a binding protein Myoglobin is one of the simplest of the globular proteins (153 amino acid residues with a molar mass of about 17,000 g/mol). It functions as a *binding protein,* carrying oxygen for muscle tissue. The space-filling model in Figure 29.4a shows myoglobin to be a globular protein. To better visualize the details of myoglobin's structure, we must strip away the amino acid side chains and focus only on the covalent chain linking the amino acids as seen in Figure 29.4b. This simplified view shows myoglobin with a folded-sausage structure, where each section of the sausage represents a segment of an α-helix. The α-helical secondary structure

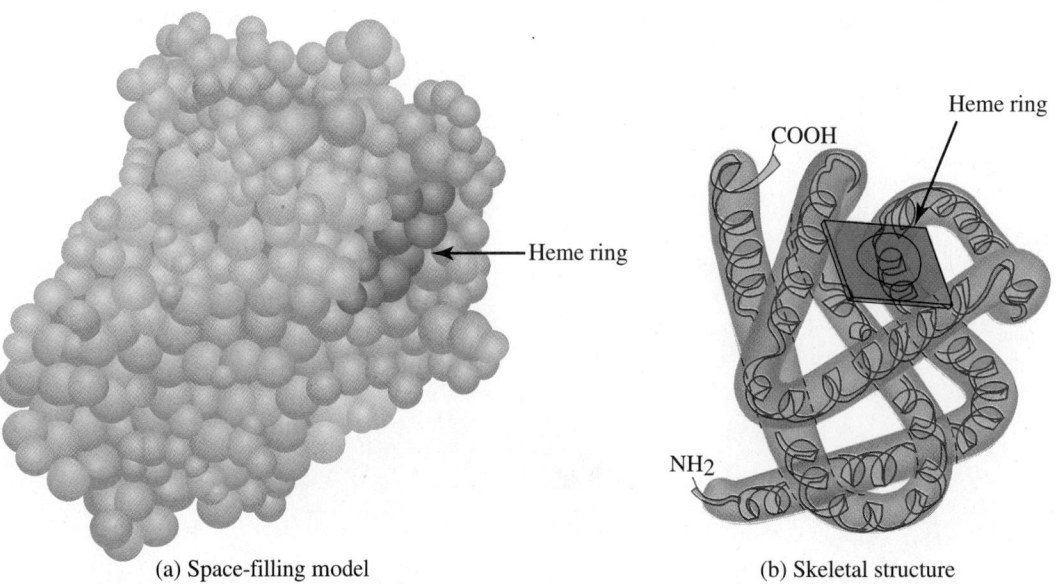

(a) Space-filling model   (b) Skeletal structure

▲
**FIGURE 29.4
Representations of
myoglobin.**

provides stability for myoglobin. The folds of the myoglobin structure are held firmly
in place by hydrogen bonds and other interactions between amino acid side chains.
Within the center of this structure is an organically bound iron atom (in the heme
group, shown in Figure 29.4b by a rectangular solid) that allows myoglobin to store
and transport oxygen. Because of the heme group, myoglobin is a *conjugated*
protein—that is, a protein that contains groups other than amino acid residues.

Jaundice is a disease caused by the improper processing of the heme group. As
the heme is degraded, bilirubin forms. If the liver does not continuously excrete this
yellow pigment, bilirubin builds up in body tissues. Jaundice is characterized by a
yellowing of the skin and may be indicative of liver malfunction.

Carboxypeptidase A—an enzyme Carboxypeptidase A is a small enzyme
(307 amino acid residues with a molar mass of 34,500 g/mol) that catalyzes protein
digestion in the small intestine. The space-filling model shows carboxypeptidase A
to be a globular protein (Figure 29.5a). The major structural features of carboxypeptidase A are shown with a "ribbon structure" (Figure 29.5b). As we saw in myoglobin, this structure shows the position of the protein chain and leaves out the location
of the amino acid side chains. An α-helix is shown with a curling line,

while each strand of β-pleated sheet is a broader line with an arrowhead at its end to
show which way the protein chain is running (from the amino end toward the carboxyl terminus):

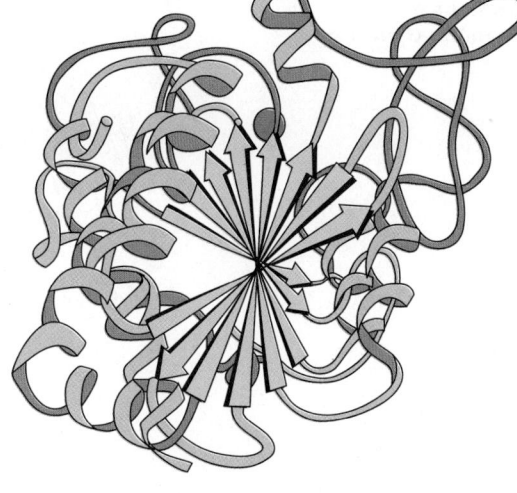

(a) Space-filling model

(b) Ribbon structure

▲
FIGURE 29.5
Representations of
carboxypeptidase A.

A β-pleated sheet provides a stable core for carboxypeptidase A. This sheet is often called a fan structure to describe the appearance of the twisted β-pleated sheet. A set of α-helices surround carboxypeptidase A's β-pleated sheet and further strengthen the protein structure.

Proteins share common skeletal structures just as various animal species have common bone structures. Some proteins use the α-helix like myoglobin (see Figure 29.4); spiral tubes are twisted around to support the bulk of these proteins. In other proteins the β-pleated sheet is configured to provide a rigid protein framework. Some proteins have a fan of β-pleated sheet (like carboxypeptidase A), while for other molecules the β-pleated sheet has wrapped around on itself to form a barrel. The β-pleated sheets often act as the core of a protein with the α-helices on the outside.

Larger proteins are commonly folded into a number of compact globular units. These units are about the size of myoglobin or carboxypeptidase A and are called **protein domains.** In this way, a larger protein is like several myoglobin-sized proteins connected together by peptide bonds. Each protein domain has its own secondary structure that usually performs a discrete task in a protein's overall function.

protein domain

Immunoglobulin G—a binding protein Immunoglobulins (antibodies) are large binding proteins (for immunoglobulin G, about 1320 amino acid residues with a molar mass of about 145,000 g/mol) that serve a critical biological function. These proteins bind molecules foreign to the body as a defense against disease. Figure 29.6 shows immunoglobulins to be complex molecules. They are composed of more than one polypeptide chain and have a quaternary structure as well as a primary, a secondary, and a tertiary structure. Each polypeptide chain has several protein domains, and β-pleated sheets form a stable core within each protein domain.

Diagrams of an immunoglobulin are often used to describe the binding function of these proteins (Figure 29.7). A successful immunological response requires that our bodies be prepared with a protein whose binding site is closely complementary to each of millions of different invading molecules (antigens). Just how our body accomplishes this feat appears to start with the shape of a typical

FIGURE 29.6 ▶
Representations of an immunoglobulin G protein.

(a) Space-filling molecule

(b) Ribbon structure

immunoglobulin G (Figure 29.7). Antigen binding regions are located at the ends of two arms. These are hinged together so that the distance between binding sites varies with the size of the invading particle. The immunoglobulin protein chains are endowed with special hypervariable regions; every different immunoglobulin has a distinct amino acid sequence in its antigen binding sites and therefore has sites of unique size and shape. The body ensures that there will be immunoglobulins for many different antigens by producing millions of proteins with different hypervariable regions. Note that two different protein chains make up the antigen binding site. Since each chain has a hypervariable region that can be changed independently, the body has the ability to produce upwards of 10^{10} (ten billion) immunoglobulins with different binding sites, more than enough to handle exposure to invading particles.

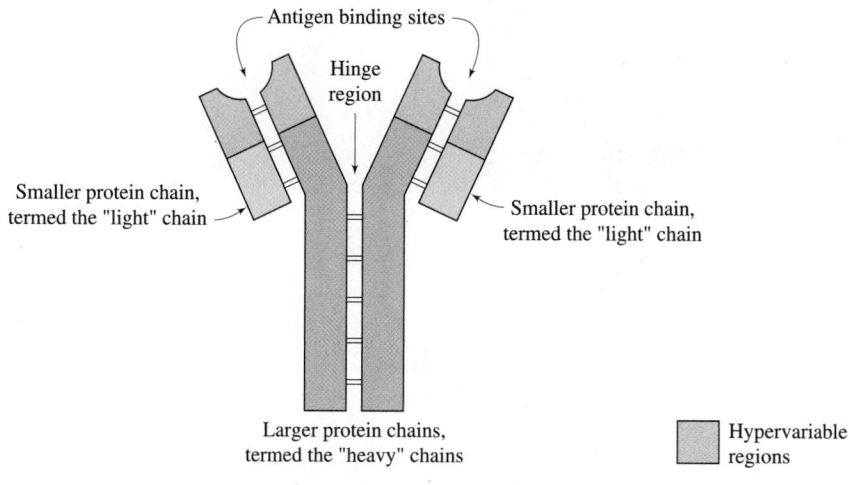

◀ FIGURE 29.7
**Diagrammatic shape of a
typical immunoglobulin G
protein.**

Hemoglobin—a binding protein Quaternary structure is important in proteins that are involved in the control of metabolic processes. For example, hemoglobin (574 amino acid residues with a molar mass of 64,500 g/mol), the oxygen transport protein of the blood, is composed of four subunits (Figure 29.8). Each subunit resembles myoglobin in that a set of helical segments surround the oxygen-binding heme group. The important structural difference is that hemoglobin has a quaternary structure. This leads to an equally important functional difference. Whereas myoglobin always binds and releases oxygen in the same way, hemoglobin changes its oxygen-binding characteristics depending on the available oxygen. When hemoglobin binds oxygen on one subunit, the protein's conformation changes to facilitate the binding of three additional oxygen molecules (see Figure 29.9). Conversely, the loss of one oxygen from hemoglobin facilitates the release of oxygen from the other sites. The oxygen binding is said to be *cooperative.* Cooperativity implies a shared action: When one O_2 binds, three more are encouraged to bind. When one O_2 leaves, three more are encouraged to leave.

Even minor alterations in primary structure may have drastic effects on the three-dimensional structural function of a protein. Sickle-cell anemia is a graphic example of this fact. This crippling genetic disease is due to red blood cells assuming a sickle shape. Sickled cells have impaired vitality and function. Sickle-cell anemia has been traced to a small change in the primary structure, or amino acid residue sequence, of hemoglobin. Normal hemoglobin molecules contain four polypeptide subunits—two identical α-polypeptide chains and two identical β-polypeptide chains. Sickle-cell and normal β-polypeptide chains of hemoglobin each contain 146 amino acid residues and differ only in the residue at the sixth position. In the sickle-cell hemoglobin chain, a glutamic acid residue has been replaced by a valine residue at the sixth position. This change is sufficient, under some circumstances, to cause hemoglobin to aggregate into long filaments—a different quaternary structure. Large amounts of hemoglobin are present in red blood cells, and the changed quaternary structure causes sickling of the cell and greatly diminishes cell vitality (see Figure 29.10). This cell affliction has led to the premature deaths of many affected individuals.

FIGURE 29.8 ▶
Representations of
hemoglobin.

(a) Space-filling model

(b) Skeletal structure

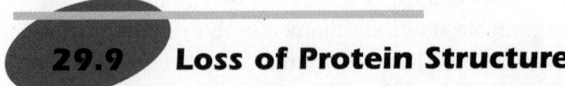

29.9   Loss of Protein Structure

Because protein structure is so important to life's functions, the loss of protein structure can be crucial. If a protein loses only its natural three-dimensional conforma-
denaturation   tion, the process is referred to as **denaturation.** In contrast, hydrolysis of peptide

◄ FIGURE 29.9
The oxygen–hemoglobin binding process. The circles and the squares represent two different conformations of the hemoglobin molecule. Oxygen binding or release causes hemoglobin to change its conformation.

bonds ultimately converts proteins into their constituent amino acids. Denaturation often precedes hydrolysis.

Denaturation involves the alteration or disruption of the secondary, tertiary, or quaternary—but not primary—structure of proteins (see Figure 29.11). Because a protein's function depends on its natural conformation, biological activity is lost with denaturation. This process may involve changes ranging from the subtle and reversible alterations caused by a slight pH shift to the extreme alterations involved in tanning a skin to form leather.

Environmental changes may easily disrupt natural protein structure, which is held together predominantly by noncovalent, relatively weak bonds. As gentle an act as pouring a protein solution can cause denaturation. Purified proteins must often be stored under ice-cold conditions because room temperature denatures them. It is not surprising that a wide variety of chemical and physical agents can also denature proteins. To name a few: strong acids and strong bases, salts (especially those of heavy metals), certain specific reagents such as tannic acid and picric acid, alcohol and other organic solvents, detergents, mechanical action such as whipping, high temperature, and ultraviolet radiation. Denatured proteins are generally less soluble than natural proteins and often coagulate or precipitate from solution. Cooks have taken advantage of this for many years. When egg white, which is a concentrated solution of egg albumin protein, is stirred vigorously (as with an egg beater), an unsweetened meringue forms; the albumin denatures and coagulates. A cooked egg solidifies partially because egg proteins including albumin are denatured by heat (see Figure 29.12).

In clinical laboratories the analysis of blood serum for small molecules such as glucose and uric acid is hampered by the presence of serum protein. This problem is resolved by first treating the serum with an acid to denature and precipitate the protein. The precipitate is removed, and the protein-free liquid is then analyzed.

Loss of protein structure also occurs with hydrolysis of the peptide bonds to produce free amino acids. This chemical reaction destroys the protein's primary structure. Proteins can be hydrolyzed by boiling in a solution containing a strong

▲
FIGURE 29.10
Scanning electron micrograph of sickled red blood cell (top) and normal red blood cell (bottom).

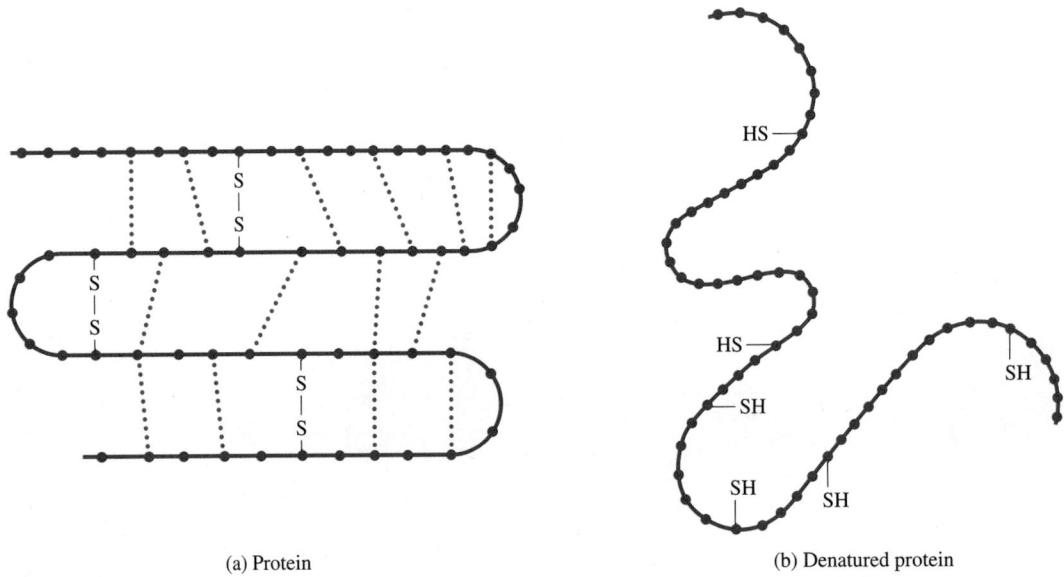

(a) Protein   (b) Denatured protein

▲
FIGURE 29.11
Structural change when a protein molecule is denatured: The relatively weak hydrogen, electrostatic, and disulfide bonds are broken, resulting in a change of structure and properties (. . . denotes the noncovalent bonds that stabilize a protein's natural conformation).

acid such as HCl or a strong base such as NaOH. At ordinary temperatures, proteins can be hydrolyzed using enzymes (see Chapter 30). These molecules, called proteolytic enzymes, are themselves proteins that function to catalyze or speed up the hydrolysis reaction. The essential reaction of hydrolysis is the breaking of a peptide linkage and the addition of the elements of water:

$$\underset{\substack{|\\H}}{\overset{\substack{H \quad R \quad O\\|\quad\;\;|\quad\;\;\|}}{\text{www}N-C-C}}-\underset{\substack{|\\H}}{\overset{\substack{H\\|}}{N}}-\underset{\substack{|\\R}}{\overset{\substack{H\\|}}{C}}-\underset{\substack{\|\\O}}{\overset{}{C}}\text{www} + H_2O \longrightarrow \underset{\substack{|\\H}}{\overset{\substack{H \quad R \quad O\\|\quad\;\;|\quad\;\;\|}}{\text{www}N-C-C}}-OH + H_2N-\underset{\substack{|\\R}}{\overset{\substack{H\\|}}{C}}-\underset{\substack{\|\\O}}{\overset{}{C}}\text{www}$$

Peptide linkage in protein chain

(www = the rest of the protein chain)

FIGURE 29.12
Left: Raw egg white is concentrated protein. Middle: In a cooked egg, the proteins are denatured by heat. Right: If egg whites are beaten, the protein denatures and coagulates.
▼

Any molecule with one or more peptide bonds, from the smallest dipeptide to the largest protein molecule, can be hydrolyzed. During hydrolysis, proteins are broken down into smaller and smaller fragments until all component amino acids are liberated.

Age and Memory: A Protein Connection

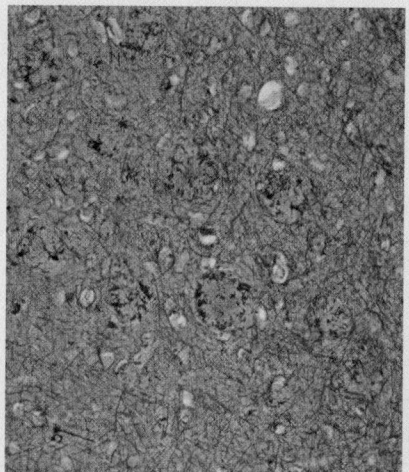

▲ *Alzheimer brain showing senile plaque.*

▲ *Normal neuron from the cortex of a human brain.*

▲ *Neuron with "tangles."*

Each of us worries about losing our ability to remember as we age. Alzheimer's disease has become a symbol of this fear for many families. We watch and worry as our elders slowly become debilitated and lose contact with daily life.

On a biochemical level, Alzheimer's disease is caused by protein malfunctions. This disease affects many older people (17–20 million worldwide), causing increasing memory loss and dementia. Alzheimer's disease has complex biochemical roots, but it is now recognized that this disease results from the death of brain cells (neurons) and is associated with a buildup of "plaques" and "tangles." These changes reflect a serious problem with specific neuronal proteins.

"Senile plaques" are found through-

out the brains of patients who suffer from Alzheimer's disease. These plaques are composed of dead and dying brain tissue and contain a large amount of a neurotoxic protein, β-amyloid. Understanding why this buildup occurs is central to understanding Alzheimer's disease.

Alzheimer's disease can be caused by an abnormal β-amyloid protein. In some types of Alzheimer's disease, genetic mutations of the β-amyloid protein are found. A minimal change of one amino acid is enough to cause a toxic buildup of β-amyloid protein. In other types of Alzheimer's disease, the β-amyloid protein may be incorrectly formed from its larger precursor protein. In turn, this can lead to the wrong three-dimensional β-amyloid shape and a buildup of this neurotoxin.

"Neurofibrillary tangles" are made up of a denatured protein called *tau*. The tau protein in Alzheimer patients has too many attached phosphate groups. The protein gains too large a negative charge, which causes it to denature and precipitate. The resulting "tangles" block normal transport of nutrients through the neurons.

It appears that changes associated with two neuronal proteins, β-amyloid and tau, are enough to disrupt the precise balance required in nerve metabolism. Cell death follows and the Alzheimer patient slowly loses mental capacity. As we discover more details of this disease, we will hopefully reach a point where it will become treatable. Treatment may be only a matter of correcting neuronal protein synthesis.

Dietary protein must have its primary structure completely destroyed before it can provide nutrition for the body. Thus digestion involves both the process of denaturation and hydrolysis. Stomach acid causes most proteins to denature. Then proteolytic enzymes in the stomach and the small intestine hydrolyze the proteins to smaller and smaller fragments until the free amino acids are formed and they can then be absorbed through the intestinal membranes into the bloodstream.

 Tests for Proteins and Amino Acids

Many tests have been devised to detect and distinguish among amino acids, peptides, and proteins. Some examples are described here.

Xanthoproteic Test Proteins containing a benzene ring—for example, the amino acids phenylalanine, tryptophan, and tyrosine—react with concentrated nitric acid to give yellow reaction products. Nitric acid on skin produces a positive xanthoproteic test, as skin proteins are modified by this reaction.

Biuret Test A violet color is produced when dilute copper(II) sulfate is added to an alkaline solution of a peptide or protein. At least two peptide bonds must be present, as the color change occurs only when peptide bonds can surround the Cu^{2+} ion. Thus amino acids and dipeptides do not give a positive biuret test.

Ninhydrin Test Triketohydrindene hydrate, generally known as *ninhydrin,* is an extremely sensitive reagent for amino acids:

ninhydrin

All amino acids, except proline and hydroxyproline, give a blue solution with ninhydrin. Proline and hydroxyproline produce a yellow solution. Less than 1 μg (10^{-6}g) of an amino acid can be detected with ninhydrin.

Chromatographic Separation

Complex mixtures of amino acids are readily separated by thin-layer, paper, or column chromatography. In chromatographic methods the components of a mixture are separated by means of differences in their distributions between two phases. Separation depends on the relative tendencies of the components to remain in one phase or the other. In *thin-layer chromatography* (TLC), for example, a liquid and a solid phase are used. The procedure is as follows: A tiny drop of a solution containing a mixture of amino acids (obtained by hydrolyzing a protein) is spotted on a strip (or sheet) coated with a thin layer of dried alumina or some other adsorbant. After the spot has dried, the bottom edge of the strip is placed in a suitable solvent. The solvent ascends the strip (by diffusion), carrying the different amino acids upward at different rates. When the solvent front nears the top, the strip is removed from the solvent and dried. The locations of the different amino acids are established by spraying the strip with ninhydrin solution and noting where colored spots appear.

The pattern of colored spots is called a chromatogram. The identities of the amino acids in an unknown mixture can be established by comparing the chromatogram of the mixture with a chromatogram produced by known amino acids. A typical chromatogram of amino acids is shown in Figure 29.13.

Since proteins tend to denature during thin-layer chromatography, they are often separated by column chromatography. In this technique a solution containing a mixture of proteins (liquid phase) is allowed to percolate through a column packed with beads of a suitable polymer (solid phase). Separation of the mixture is accomplished as the proteins partition between the solid and liquid phases. The separated proteins move at different rates and are collected as they leave the column. Unlike amino acid chromatography, mild conditions must be maintained to limit protein denaturation. In the past it was not uncommon for a protein column chromatography to take from 6 to 12 hours. Recent developments in the field of high-performance liquid chromatography (HPLC) have shortened this time to minutes.

Electrophoresis

Perhaps the most precise separation technique available to the protein chemist is electrophoresis. This technique separates proteins based on both a difference in size and a difference in charge. Electrophoresis is based on the principle that charged particles move in an electric field. Since different proteins contain differing numbers of positively and negatively charged amino acids, proteins will separate in an electric field (see Figure 29.14). Sometimes proteins with an overall positive charge are separated from proteins with an overall negative charge (Figure 29.14a). But, most often, conditions are chosen so that all proteins have either a net positive or a net negative charge. Proteins with a greater charge are attracted more strongly to an electrode and move faster during electrophoresis (see Figure 29.14b). The faster moving proteins separate from the slower moving proteins.

Protein electrophoresis commonly is carried out in a gel matrix. This gel may be a natural polymer like starch or agarose or, it may be a purely synthetic polymer like polyacrylamide. No matter which gel is used, the proteins are forced to move through a sievelike material. They experience friction; the larger the protein, the more friction and the slower the protein will tend to move. Thus the speed at which a protein moves depends on both charge *and* size during standard electrophoresis. Tens of proteins may be separated precisely using one gel electrophoresis.

This technique has proved not only valuable, but also versatile. Several modifications of the standard procedure are used routinely.

1. *SDS (sodium dodecyl sulfate) electrophoresis:* In this procedure, a negatively charged detergent (SDS) is added to the protein solution and acts to mask differences in charges between proteins. The electrophoretic separation depends primarily on size differences. SDS electrophoresis is often used to determine the approximate molar mass of proteins.

2. *Isoelectric focusing:* During this type of electrophoresis, proteins are forced to move through a pH gradient. The proteins stop moving when they reach their *isoelectric point* pH because at this pH the protein carries no net charge (see Section 29.5). The electrophoresis is completed once all the proteins have been "focused" at their isoelectric points. This technique separates pro-

Slide I

Slide IA

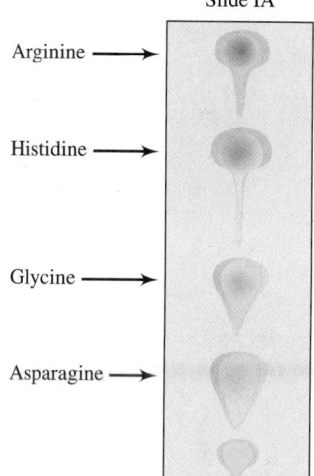

Arginine ⟶

Histidine ⟶

Glycine ⟶

Asparagine ⟶

▲
FIGURE 29.13
Chromatogram showing separation of selected amino acids. On slide I, spotted on the chromatographic strip, is an amino acid mixture containing arginine, histidine, glycine, and asparagine. On slide IA is the developed chromatogram showing the separated amino acids after treatment with ninhydrin. The solvent is a 250:60:250 volume ratio of 1-butanol–acetic acid–water.

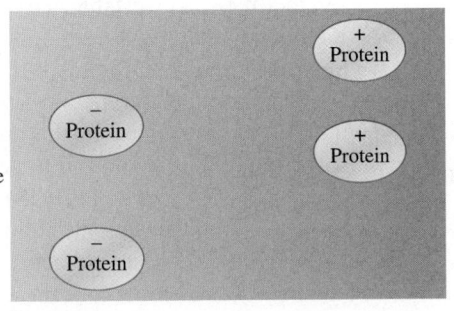

(a) Separating negatively charged proteins
from positively charged proteins

(b) Separating the more positively charged proteins
from the less positively charged ones.

▲
FIGURE 29.14
Two forms of electrophoresis.

teins based on the number and kind of titratable amino acid side chains that they contain.

Today, two-dimensional electrophoresis is not uncommon. This procedure multiplies the benefits of an electrophoretic separation by running one separation in the y-direction followed by a second electrophoresis in the x-direction using the same gel slab. Figure 29.15 shows a separation of all proteins from a bacterium (*E. coli*). The proteins were first separated by molar mass using SDS electrophoresis in the y-direction. Each band was then separated in the x-direction using isoelectric focusing, a technique that separates proteins based on charge. Over 1000 proteins have been separated on this one gel!

Electrophoresis can help a medical diagnosis that depends on protein identification. If this identification distinguishes proteins by their size and charge, electrophoresis is an ideal technique. For example, genetic diseases that affect hemoglobin can be diagnosed using isoelectric focusing. These diseases (e.g., sickle-cell anemia, the α- and β-thalassemias) may change only one amino acid in hemoglobin, yet they can be extremely serious. The thalassemias lead to an early death as a consequence of severe anemia. Often these hemoglobinopathies are characterized by changes in hemoglobin's size and charge. Standard electrophoresis can distinguish these changes; however, isoelectric focusing provides added sensitivity that is espe-

◀ FIGURE 29.15
Two-dimensional elec-
trophoresis of proteins from
the E. coli bacterium. Separation by molar mass has
been used in the y-direction
and separation by charge
has been used in the x-di-
rection.

cially important when using small amounts of blood (e.g., with newborn infant). For adults, these screening techniques have been especially useful as a genetic counseling tool.

29.11 Determination of the Primary Structure of Polypeptides

Sanger's Reagent The pioneering work of Frederick Sanger gave us the first complete primary structure of a protein, insulin, in 1955. He used specific enzymes to hydrolyze insulin into smaller peptides and amino acids, then separated and identified the hydrolytic products by various chemical reactions.

Sanger's reagent, 2-4-dinitrofluorobenzene (DNFB), reacts with the α-amino group of the N-terminal amino acid of a polypeptide chain. The carbon–nitrogen bond between the amino acid and the benzene ring of Sanger's reagent is more resistant to hydrolysis than are the remaining peptide linkages. Thus, when the substituted polypeptide is hydrolyzed, the terminal amino acid remains with the dinitrobenzene group and can be isolated and identified. The remaining peptide chain is hydrolyzed to free amino acids in the process. This method marked an important step in determining the amino acid sequence in a protein. The Sanger's reagent reaction is illustrated in the following equations.

DNFB   polypeptide chain

DNFB polypeptide derivative

DNFB-terminal
amino acid derivative

mixture of amino acids

From the DNFB hydrolysis, Sanger learned which amino acids are present in insulin. By less drastic hydrolysis, he split the insulin molecule into peptide fragments consisting of two, three, four, or more amino acid residues. After analyzing vast numbers of fragments utilizing the N-terminal method, he pieced them together in the proper sequence by combining fragments with overlapping structures at their ends, finally elucidating the entire insulin structure. As an example, consider the overlap that occurs between the hexapeptide and heptapeptide shown here:

Gly-Glu-Arg-Gly-Phe-Phe   Hexapeptide
              Gly-Phe-Phe-Tyr-Thr-Pro-Lys   Heptapeptide
Gly-Glu-Arg-Gly-Phe-Phe-Tyr-Thr-Pro-Lys   Decapeptide

The three residues Gly-Phe-Phe at the end of the hexapeptide match the three residues at the beginning of the heptatpeptide. By using these three residues in common, the structure of the decapeptide shown, which occurs in chain B of insulin (residues 20–29), is determined. (See Figure 29.2.)

Edman Degradation   More recently, the Edman degradation method has been developed to split off amino acids one at a time from the N-terminal end of a polypeptide chain. In this procedure the reagent, phenylisothiocyanate, is first added to the N-terminal amino group. The N-terminal amino acid–phenylisothiocyanate addition product is then cleaved from the polypeptide chain. The resultant substituted phenylthiohydantoin is isolated and identified. The shortened polypeptide chain is then ready to undergo another Edman degradation. By repeating this set of reactions, chemists determine the amino acid sequence of a polypeptide directly. A machine, the protein sequenator, which carries out protein sequencing by Edman degradation automatically, is now available.

phenylisothiocyanate   polypeptide chain

phenylthiohydantoin   shortened polypeptide chain

29.12   Synthesis of Peptides and Proteins

Since each amino acid has two functional groups, it is not difficult to form dipeptides or even fairly large polypeptide molecules. As mentioned in Section 29.6, glycine and alanine react to form two different dipeptides, glycylalanine and alanylglycine. If threonine, alanine, and glycine react, six tripeptides, each made up of three different amino acid units are obtained. By reacting a mixture of several amino acids, polypeptides are produced. This clarifies the fact that even a small protein like insulin cannot be produced by simply reacting a mixture of the required amino acids. When such a large polypeptide is synthesized, it is formed by joining amino acids one by one in the proper sequence. Within cells, the proper sequence is maintained by careful genetic control (see Section 31.10).

When a polypeptide chain of known primary structure is synthesized *in vitro* (synthesis in "glass" without the aid of living tissue), the process must be controlled so that only one particular amino acid is added at each stage. Remarkable progress has been made in developing the necessary techniques. Amino acids are used in which either the amino or carboxylic acid group has been inactivated or blocked with a suitable *blocking agent*. The blocked amino or carboxylic acid group is reactivated by removing the blocking agent in the next stage of the synthesis. For example, if the tripeptide Thr-Ala-Gly is to be made, amino-blocked threonine is reacted with carboxyl-blocked alanine to make the doubly blocked Thr-Ala dipeptide. The carboxyl-blocked alanine end of the dipeptide can then be reactivated and reacted with carboxyl-blocked glycine to make the blocked Thr-Ala-Gly tripeptide. The free tripeptide can then be obtained by removing the blocking agents from both ends. Using $\boxed{B}$ to designate the blocking agents, this synthesis is represented schematically as follows:

$$\boxed{B}\ H_2NCHCOOH\ +\ H_2NCHCOOH\ \boxed{B}\ \longrightarrow\ \boxed{B}\ H_2NCHC\overset{\displaystyle O}{\overset{\|}{-}}NHCHCOOH\ \boxed{B}$$

amino-blocked threonine + carboxyl-blocked alanine → double-blocked Thr-Ala dipeptide

$$\boxed{B}\ H_2NCHC\overset{\displaystyle O}{\overset{\|}{-}}NHCHCOOH\ +\ H_2NCH_2COOH\ \boxed{B}$$

amino-blocked Thr-Ala dipeptide + carboxyl-blocked glycine

$$\boxed{B}\ H_2NCHC\overset{\displaystyle O}{\overset{\|}{-}}NHCHC\overset{\displaystyle O}{\overset{\|}{-}}NHCH_2COOH\ \boxed{B}\ \longrightarrow\ \text{Thr-Ala-Gly tripeptide}$$

double-blocked Thr-Ala-Gly tripeptide

Longer polypeptide chains of specified amino acid sequence could be made by this general technique, but the procedure was very tedious. The synthesis of insulin required an effort equivalent to one person working steadily for several years.

Fortunately, it is not necessary to prepare an entire synthetic protein polypeptide chain by starting at one end and adding amino acids one at a time. Previously prepared shorter polypeptide chains of known structure can be joined to form a long polypeptide chain. This method was used to make synthetic insulin. The insulin molecule, shown in Figure 29.2, consists of two polypeptide chains bonded together by two disulfide bonds. A third disulfide bond, between two amino acids in one chain, makes a small loop in that chain. Once the two chains had been assembled from fragments, a seemingly formidable problem remained—how to form the disulfide bonds in the correct positions. As it turned out, it was necessary only to bring the two chains together, with the cysteine side chains in the reduced condition, and to treat them with an oxidizing agent. Disulfide bonds formed at the right places, and biologically active insulin molecules were obtained.

In the mid-1960s, a machine capable of automatically synthesizing polypeptide chains of known amino acid sequence was designed by R. Bruce Merrifield of Rockefeller University. The starting amino acid is bonded to a plastic surface (polystyrene bead) in the reaction chamber of the apparatus. Various reagents needed for building a chain of predetermined structures are automatically delivered to the reaction chamber in a programmed sequence. Twelve reagents and about 100 operations are needed to lengthen the chain by a single amino acid residue. But the machine is capable of adding residues to the chain at the rate of six a day. Such a machine makes it possible to synthesize complex molecules like insulin in a few days. The first large polypeptide (pancreatic ribonuclease), containing 124 amino acid residues, was synthesized by this method in 1969. More recently, a human growth hormone (HGH) containing 188 amino acid residues was synthesized using this technique.

Over the past 20 years, techniques have been developed that use biological systems to create and produce new proteins. These procedures start with genetic material that codes for the protein of interest. Then this material is properly modified and introduced into rapidly growing cells that are treated to overproduce this foreign protein. The result can be a harvest of much needed protein, such as the human insulin used by diabetics or the human growth hormone used to treat dwarfism. These procedures constitute a part of genetic engineering and will be discussed in more detail in Chapter 31. The advantages of these new techniques become evident when we compare the rate of biological protein synthesis with that achieved by the machine designed by Merrifield. If a bacterium such as *E. coli* is used to produce a human protein such as growth hormone, one amino acid residue can be added every 0.01 second; a growth hormone molecule can be produced every 10–20 seconds. In contrast, the same growth hormone would take days to produce in the chemistry laboratory.

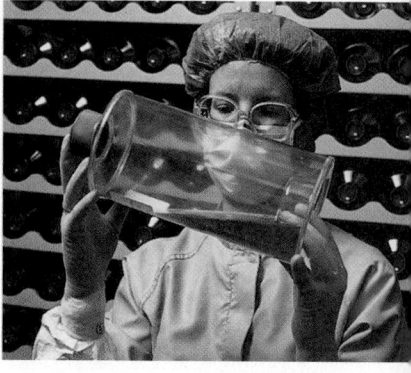

▲
Production of a cloned protein. The technician is holding a container with cultured cells in a nutrient medium.

Concepts in Review

1. List five foods that are major sources of proteins.
2. List the elements that are usually contained in proteins.
3. Write formulas for and know the abbreviations for the 20 common amino acids.
4. Write the zwitterion formula of an amino acid and know how it behaves in a dilute acidic and a dilute basic solution.
5. Understand why a protein in solution at its isoelectric pH does not migrate in an electrolytic cell.
6. Combine amino acids into polypeptide chains.
7. Understand how peptide chains are named and numbered.
8. Understand what is meant by the N-terminal and C-terminal residues of a peptide chain.
9. Briefly explain the primary, secondary, and tertiary structure of a protein.
10. Explain the meaning of the statement, "The primary structure helps establish a protein's three-dimensional structure."
11. Define quaternary structure and explain its importance in the function of hemoglobin.
12. Understand the bonding in the α-helical and β-pleated-sheet structure of a protein.
13. Give an example of a structural protein.

14. Explain the role cysteine plays in the structure of polypeptides and proteins.
15. Describe how the α-helix and β-pleated sheet are important in the structure of proteins.
16. List the most important protein functions.
17. Relate the structure of collagen to its biological function.
18. Discuss the role of the immunoglobulins in the body's defense mechanisms and identify important structural elements of these proteins.
19. Describe Sanger's work on determining the primary structure of proteins.
20. Reconstruct a peptide chain from a knowledge of its hydrolysis products.
21. Explain how Edman degradation allows the primary structure to be determined.
22. Describe chromatographic methods of separating amino acids.
23. Describe how electrophoresis is used to separate protein.
24. State the physical evidence observed in a positive reaction in the (a) xanthoproteic test, (b) biuret test, and (c) ninhydrin test.
25. Know what is meant by denaturation and the various methods by which proteins can be denatured.

Key Terms

The terms listed here have been defined within this chapter.
Section numbers are referenced in parentheses for each term.

conjugated protein (29.1)
denaturation (29.9)
essential amino acid (29.3)
fibrous protein (29.8)

globular protein (29.8)
isoelectric point (29.5)
peptide linkage (29.6)
polypeptide (29.6)

primary structure (29.7)
protein domain (29.8)
quaternary structure (29.7)
secondary structure (29.7)

simple protein (29.1)
tertiary structure (29.7)
zwitterion (29.5)

Questions

Questions refer to tables, figures, and key words and concepts defined within the chapter. A particularly challenging question or exercise is indicated with an asterisk.

1. Why are the common amino acids in proteins called α-amino acids?

2. What elements are present in amino acids and proteins?

3. Why are proteins from some foods of greater nutritional value than others?

4. Write the names of the amino acids that are essential to humans.

5. Why are amino acids amphoteric? Why are they optically active?

6. What can you say about the number of positive and negative charges on a protein molecule at its isoelectric point?

7. Explain what is meant by (a) the primary structure, (b) the secondary structure, (c) the tertiary structure, and (d) the quaternary structure of a protein.

8. What special role does the sulfur-containing amino acid cysteine have in protein structure?

9. How do myoglobin and hemoglobin differ in structure and function?

10. Differentiate between an α-helix and a β-pleated sheet structure of a protein.

11. Explain how hydrolysis of a protein differs from denaturation.

12. Which amino acids give a positive xanthoproteic test?

13. What is the visible evidence observed in a positive reaction for the following tests?
    (a) xanthoproteic test
    (b) biuret test
    (c) ninhydrin test

14. Briefly describe separation of proteins by column chromatography.

15. (a) What is thin-layer chromatography?
    (b) Describe how amino acids are separated using this technique.
    (c) What reagent is commonly used to locate the amino acids in the chromatogram?

16. How does SDS electrophoresis differ from standard electrophoresis?

Paired Exercises

These exercises are paired. Each odd-numbered exercise is followed by a similar even-numbered exercise. Answers to the even-numbered exercises are given in Appendix VI.

17. Write the structural formulas for the D and L forms of alanine

    CH$_3$CHCOOH
    |
    NH$_2$

    Which form is found in proteins?

18. Write the structural formulas for the D and L forms of serine

    HOCH$_2$CHCOOH
    |
    NH$_2$

    Which form is found in proteins?

19. A 100.0-g sample of a food product is analyzed and found to contain 6.0 g of nitrogen. If protein contains an average of 16% nitrogen, what percentage of the food is protein?

20. A 250.0-g sample of hamburger is analyzed and found to contain 5.2 g of nitrogen. If protein contains an average of 16% nitrogen, how many grams of protein are contained in the hamburger?

21. Write the structural formula representing the following amino acid at its isoelectric point:

$$CH_3CH-CHCOOH$$
$$\qquad | \qquad |$$
$$\qquad OH \quad NH_2$$

threonine

22. Write the structural formula representing the following amino acid at its isoelectric point:

$$NH_2CCH_2CHCOOH$$
$$\qquad || \qquad |$$
$$\qquad O \qquad NH_2$$

asparagine

23. For the amino acid

phenylalanine

write:
(a) the zwitterion formula
(b) the formula in 0.1 M H_2SO_4
(c) the formula in 0.1 M NaOH

24. For the amino acid

tryptophan

write:
(a) the zwitterion formula
(b) the formula in 0.1 M H_2SO_4
(c) the formula in 0.1 M NaOH

25. Write ionic equations to show how alanine

$$CH_3CHCOOH$$
$$\qquad |$$
$$\qquad NH_2$$

acts as a buffer toward
(a) H^+ **(b)** OH^-

26. Write ionic equtions to show how leucine

$$(CH_3)_2CHCH_2CHCOOH$$
$$\qquad\qquad\qquad |$$
$$\qquad\qquad\qquad NH_2$$

acts as a buffer toward
(a) H^+ **(b)** OH^-

27. Write the structural formula representing the following amino acid at its isoelectric point:

$$CH_3SCH_2CH_2CHCOOH$$
$$\qquad\qquad\qquad |$$
$$\qquad\qquad\quad NH_2$$

methionine

28. Write the structural formula representing the following amino acid at its isoelectric point:

$$(CH_3)_2CHCHCOOH$$
$$\qquad\qquad\quad |$$
$$\qquad\qquad\quad NH_2$$

valine

29. Write the full structural formula of the two dipeptides containing

$$HOCH_2CHCOOH \quad and \quad CH_3CHCOOH$$
$$\qquad\quad | \qquad\qquad\qquad\qquad\quad |$$
$$\qquad\quad NH_2 \qquad\qquad\qquad\qquad NH_2$$

serine   alanine

Indicate the location of the peptide bonds.

30. Write the full structural formula of the two dipeptides containing

$$\qquad\qquad\qquad\qquad\qquad\qquad OH$$
$$\qquad\qquad\qquad\qquad\qquad\qquad |$$
$$CH_2COOH \quad and \quad CH_3CHCHCOOH$$
$$\quad |\qquad\qquad\qquad\qquad\qquad\qquad\qquad |$$
$$\quad NH_2 \qquad\qquad\qquad\qquad\qquad\qquad NH_2$$

glycine   threonine

Indicate the location of the peptide bonds.

31. Use the following amino acids

CH_2COOH	$HOCH_2CHCOOH$	$CH_3CHCOOH$
NH_2	NH_2	NH_2
glycine	serine	alanine

to write structures for
(a) glycylglycine
(b) alanylglycylserine
(c) glycylserylglycine

33. Using amino acid abbreviations, write all the possible tripeptides that contain one unit each of glycine, phenylalanine, and leucine.

35. What part of the following structure bonds when a β-pleated sheet forms? Identify the type of bonds that form.

$$\text{—CHCNCH—}$$
with H, O, CH$_3$ above and H below

37. At pH = 7, draw the bond that might form between two serine side chains ($HOCH_2-$) to hold together a tertiary protein structure.

39. Would this compound

$$NH_2CH_2CNHCHCNHCHCOOH$$
with O, CH$_3$, CHOH, CH$_3$ groups, O

react with
(a) copper(II) sulfate in the biuret test?
(b) concentrated HNO_3 to give a positive xanthoproteic test?
(c) ninhydrin?

41. Write the products for the complete hydrolysis of

$$NH_2CHCNHCHCNHCHCOOH$$
with CH$_3$, CH$_2$, CH$_2$, COOH, phenyl, O, O

*43. Human cytochrome c contains 0.43% iron by mass. If each cytochrome c contains one iron atom, what is the molar mass of cytochrome c?

32. Use the following amino acids

CH_2COOH	$HOCH_2CHCOOH$	$CH_3CHCOOH$
NH_2	NH_2	NH_2
glycine	serine	alanine

to write structures for
(a) alanylalanine
(b) serylglycylglycine
(c) serylglycylalanine

34. Using amino acid abbreviations, write all the possible tripeptides that contain one unit each of tyrosine, aspartic acid, and alanine.

36. What part of the following structure bonds when an α-helix forms? Identify the type of bonds that form.

$$\text{—CHCNCH—}$$
with H, O, CH$_3$ above and H below

38. At pH = 7, draw the bond that might form between lysine side chain ($NH_2CH_2CH_2CH_2CH_2-$) and an aspartic acid side chain ($HOOCCH_2-$) to hold together a tertiary protein structure.

40. Would this compound

$$NH_2CH_2CNHCHCNHCHCOOH$$
with O, CH$_2$, CH$_2$, OH, COOH groups, O

react with
(a) copper(II) sulfate in the biuret test?
(b) concentrated HNO_3 to give a positive xanthoproteic test?
(c) ninhydrin?

42. Write the products for the complete hydrolysis of

$$NH_2CHCNHCHCNHCHCOOH$$
with HOOC, CH$_2$, CH$_2$, OH, phenyl, CH$_3$, CH$_2$, O, O

*44. Human hemoglobin contains 0.33% iron by mass. If each hemoglobin contains four iron atoms, what is the molar mass of hemoglobin?

45. What is the amino acid sequence of a heptapeptide, which upon hydrolysis yields the tripeptides Gly-Phe-Leu, Phe-Ala-Gly, and Leu-Ala-Tyr? The heptapeptide contains one residue each of Gly, Leu, and Tyr and two residues each of Ala and Phe.

46. What is the amino acid sequence of a heptapeptide, which upon hydrolysis yields the tripeptides Phe-Gly-Tyr, Phe-Ala-Ala, and Ala-Leu-Phe? The heptapeptide contains one residue each of Gly, Leu, and Tyr and two residues each of Ala and Phe.

Additional Exercises

These exercises are not paired and provide additional practice on concepts covered in this chapter.

47. A nonapeptide is obtained from blood plasma by treatment with the enzyme trypsin. Analysis shows that both terminal amino acids are arginine. Total hydrolysis of this peptide yields Gly, Ser, 2 Arg, 2 Phe, 3 Pro. Partial hydrolysis gives Phe-Ser, Phe-Arg, Arg-Pro, Pro-Pro, Pro-Gly-Phe, Ser-Pro-Phe. What is the amino acid sequence of the nonapeptide?

48. (a) At what pH will arginine not migrate to either electrode in an electrolytic cell?
(b) In what pH range will it migrate toward the positive electrode?

49. Is it possible for a protein to have a primary structure if it does not have a secondary structure? Is it possible for a protein to have a secondary structure if it does not have a primary structure? Explain.

50. Ribonuclease is a protein that is slightly smaller than myoglobin. How many protein domains would you predict for the structure of ribonuclease? Explain your answer.

51. Threonine has two chiral centers. Write Fischer projection formulas for its stereoisomers.

52. How are the hypervariable regions in immunoglobulin chains important to this protein's function?

53. The isoelectric point for alanine is 6.00 and for lysine is 9.74.
(a) Write the structure for alanine as it would be in a buffer with a pH of 9.0.
(b) Write the structure for lysine as it would be in a buffer with a pH of 9.0.
(c) Would the charge on lysine be positive, negative or neutral when it is in a buffer with a pH of 9.0?

54. How many dipeptides containing glycine can be written using the 20 amino acids from Table 29.1?

55. The structures of vasopressin and oxytocin are given in Table 29.3. Which one would you expect to have the higher isoelectric point? Why?

56. Thin-layer chromatography is used to separate amino acids from one another. The separation is based on the relative polarity of the amino acids. Which of these amino acids would you expect to be the most polar? Why?

leucine, alanine, glutamic acid

Answers to Practice Exercises

29.1 α-Amino acid: CH_2COOH
$\qquad\qquad\qquad\quad\ |$
$\qquad\qquad\qquad\quad NH_2$

$\quad\ \ \ \beta$-Amino acid: CH_2CH_2COOH
$\qquad\qquad\qquad\qquad\quad |$
$\qquad\qquad\qquad\qquad\ NH_2$

29.2 Leu (leucine), Lys (lysine), Trp (tryptophan)

29.3 $(CH_3)_2CHCHCOO^-$
$\qquad\qquad\quad\ |$
$\qquad\qquad\ NH_2$

29.4

$$\begin{array}{ccccc} & O & O & O & O \\ & \| & \| & \| & \| \\ CH_2C & -NHCHC & -NHCHC & -NHCHC & -NHCHCOOH \\ | & | & | & | & | \\ NH_2 & CH_2 & CH_2 & CH_2 & CH_2 \\ & | & | & | & | \\ & CH & COOH & OH & SH \\ & / \backslash \\ & CH_3\ \ CH_3 \end{array}$$

29.5

$$\begin{array}{ccc} O & O & O \\ \| & \| & \| \\ CH_3CHC-NHCHC-NHCH_2C \\ | & | & | \\ NH_2 & CH_2 & NH \\ & | & | \\ & S & CH_2 \\ & | & | \\ & & C=O \\ & | & | \\ & & NH \\ & | & | \\ S-CH_2-CH \\ & & | \\ & & COOH \end{array}$$

The tertiary structure must contain a cyclic structure joined by the disulfide bond.

30

Enzymes

Yeast provides the enzymes
necessary to make bread rise.

CHAPTER 30 / OUTLINE

CHAPTER
30

Perhaps you have seen laundry detergents advertised as "containing enzymes." Enzymes are added to detergents to clean especially hard-to-remove spots. In living organisms, enzymes also take on the "tough" metabolic tasks—in fact, life cannot continue in the absence of enzymes.

Enzymes are important because they can accelerate a chemical reaction by 1 million to 100 million times. Imagine what would happen if some of our tasks were accelerated by that amount! A 2-hour daily homework assignment would be completed in about one-thousandth of a second. A flight from Los Angeles to New York would take about one-hundredth of a second. The building of Hoover Dam, a monumental task requiring 5 years of earth moving and complex steel and concrete work, would occur in about 30 seconds.

Scientists now understand the general characteristics of enzymes. And, as we will see in this chapter, these molecules achieve almost miraculous results by following some very basic chemical principles.

30.1   Molecular Accelerators

Enzymes are the catalysts of biochemical reactions. Enzymes catalyze nearly all the myriad reactions that occur in living cells. Uncatalyzed reactions that require hours of boiling in the presence of a strong acid or a strong base can occur in a fraction of a second in the presence of the proper enzyme at room temperature and nearly neutral pH. The catalytic functions of enzymes are directly dependent on their three-dimensional structures. It was believed until quite recently that all enzymes were proteins. However, research under way since about 1980 has shown that certain ribonucleic acids (RNAs, see Chapter 31) can also function as enzymes.

enzyme

Louis Pasteur was one of the first scientists to study enzyme-catalyzed reactions. He believed that living yeasts or bacteria were required for these reactions, which he called *fermentations*—for example, the conversion of glucose to alcohol by yeast. In 1897, Eduard Büchner (1860–1917) made a cell-free filtrate that contained enzymes prepared by grinding yeast cells with very fine sand. The enzymes in this filtrate converted glucose to alcohol, thus proving that the presence of living cells was not required for enzyme activity. For this work Büchner received the Nobel prize in chemistry in 1907.

Enzymes are essential to life. The critical biochemical reactions occur too slowly in the absence of enzymes for life to be maintained. The typical biochemical reaction occurs more than a million times faster when catalyzed by an enzyme. For example, we know that the reduced carbons of carbohydrates can react with oxygen to produce carbon dioxide and energy. Yet the sucrose in the sugar bowl at home never reacts significantly with the oxygen in the air. These sucrose molecules must overcome an energy barrier (activation energy) before reaction can occur (see Figure 30.1). In the sugar bowl, the energy barrier is too large. But in the cell, enzymes lower the activation energy and enable sucrose to react rapidly enough to provide the energy needed for life processes.

FIGURE 30.1 ▶
A typical reaction-energy profile: The lower activation energy in the cell is due to the catalytic effect of enzymes.

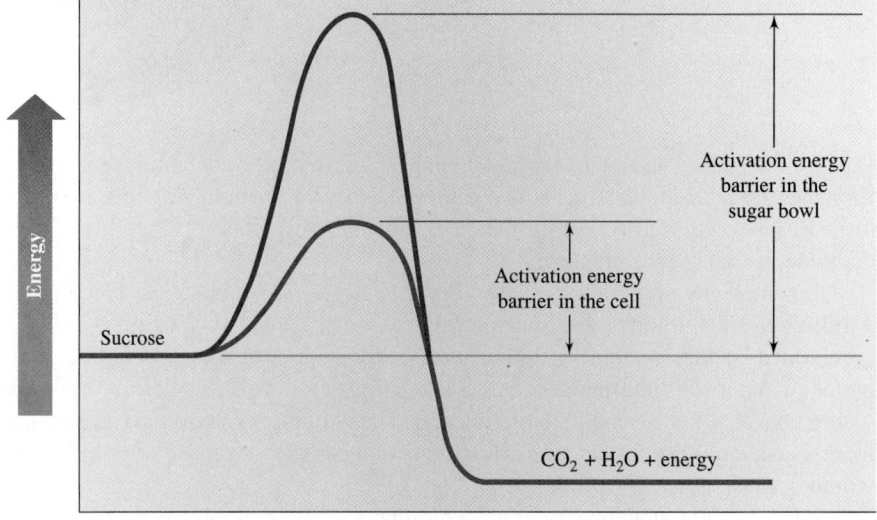

Reaction progress

Each organism contains thousands of enzymes. Some are simple proteins consisting only of amino acid units. Others are conjugated and consist of a protein part, or *apoenzyme,* and a nonprotein part, or *coenzyme.* Both parts are essential, and a functioning enzyme consisting of both the protein and nonprotein parts is called a *holoenzyme:*

Apoenzyme + Coenzyme = Holoenzyme

Often the coenzyme is derived from a vitamin, and one coenzyme may be associated with many different enzymes.

For some enzymes, an inorganic component such as a metal ion (e.g., Ca^{2+}, Mg^{2+}, or Zn^{2+}) is required. This inorganic component is an *activator.* From the standpoint of function, an activator is analogous to a coenzyme, but inorganic components are not called coenzymes.

Another remarkable property of enzymes is their specificity of reaction; that is, a certain enzyme catalyzes the reaction of a specific type of substance. For example, the enzyme maltase catalyzes the reaction of maltose and water to form glucose. Maltase has no effect on the other two common disaccharides, sucrose and lactose. Each of these sugars requires a specific enzyme—sucrase to hydrolyze sucrose, lactase to hydrolyze lactose. These reactions are indicated by the following equations:

$$\underset{\text{maltose}}{C_{12}H_{22}O_{11}} + H_2O \xrightarrow{\text{maltase}} \underset{\text{glucose}}{C_6H_{12}O_6} + \underset{\text{glucose}}{C_6H_{12}O_6}$$

$$\underset{\text{sucrose}}{C_{12}H_{22}O_{11}} + H_2O \xrightarrow{\text{sucrase}} \underset{\text{glucose}}{C_6H_{12}O_6} + \underset{\text{fructose}}{C_6H_{12}O_6}$$

$$\underset{\text{lactose}}{C_{12}H_{22}O_{11}} + H_2O \xrightarrow{\text{lactase}} \underset{\text{glucose}}{C_6H_{12}O_6} + \underset{\text{galactose}}{C_6H_{12}O_6}$$

The substance acted on by an enzyme is called the *substrate*. Sucrose is the substrate of the enzyme sucrase. Enzymes have been named by adding the suffix *-ase* to the root of the substrate name. Note the derivations of maltase, sucrase, and lactase from maltose, sucrose, and lactose. Many enzymes, especially digestive enzymes, have common names such as pepsin, rennin, trypsin, and so on. These names have no systematic significance.

In the International Union of Biochemistry (IUB) System, enzymes are assigned to one of six classes, the names of which clearly describe the nature of the reaction they catalyze. The six main classes of enzymes are

1. *Oxidoreductases:* Enzymes that catalyze the oxidation–reduction between two substrates
2. *Transferases:* Enzymes that catalyze the transfer of a functional group between two substrates
3. *Hydrolases:* Enzymes that catalyze the hydrolysis of esters, carbohydrates, and proteins (polypeptides)
4. *Lyases:* Enzymes that catalyze the removal of groups from substrates by mechanisms other than hydrolysis
5. *Isomerases:* Enzymes that catalyze the interconversion of stereoisomers and structural isomers
6. *Ligases:* Enzymes that catalyze the linking of two compounds by breaking a phosphate anhydride bond in adenosine triphosphate (ATP, see Chapter 31)

Because the systematic name is usually long and often complex, working or practical names are used for enzymes. For example, adenosine triphosphate creatine phosphotransferase is called creatine kinase, and acetylcholine acylhydrolase is called acetylcholine esterase.

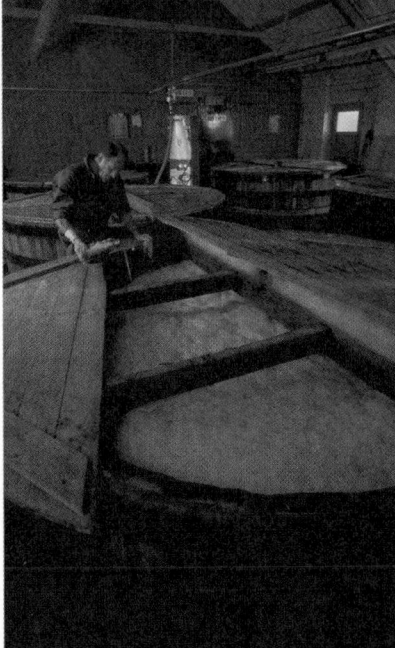

▲
A worker adds yeast to a vat in a whiskey distillery.

30.2 Rates of Chemical Reactions

Enzymes catalyze biochemical reactions and thus increase the rate of these chemical reactions—but how does this process take place? To answer this question, we must first consider some general properties of a chemical reaction.

Every chemical reaction starts with at least one reactant and finishes with a minimum of one product. As the reaction proceeds, the reactant concentration decreases and the product concentration increases. We often plot these changes as a function of time, as shown in Figure 30.2 for the hypothetical conversion of reactant A into product B.

A → B

A reaction rate is defined as a change in concentration with time—the rate at which the reactants of a chemical reaction disappear and the products form.

The reactant must pass through a high-energy **transition state** to be converted into a product. This transition state is an unstable structure with characteristics of both the reactant and the product. The energy necessary to move a reactant to the transition state is termed the activation energy. The larger this energy barrier, the slower the reaction rate.

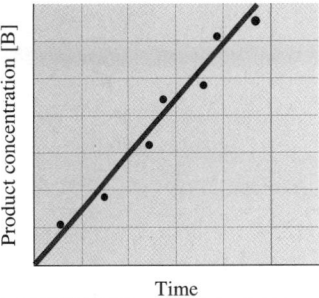

▲
**FIGURE 30.2
The change in product concentration [B] as a function of time. The reaction rate is determined by measuring the slope of this line.**

transition state

FIGURE 30.3 ▶
An energy profile for the reaction between water and
carbon dioxide: The dashed
lines in the transition-state
formula indicate bonds being
formed and broken.

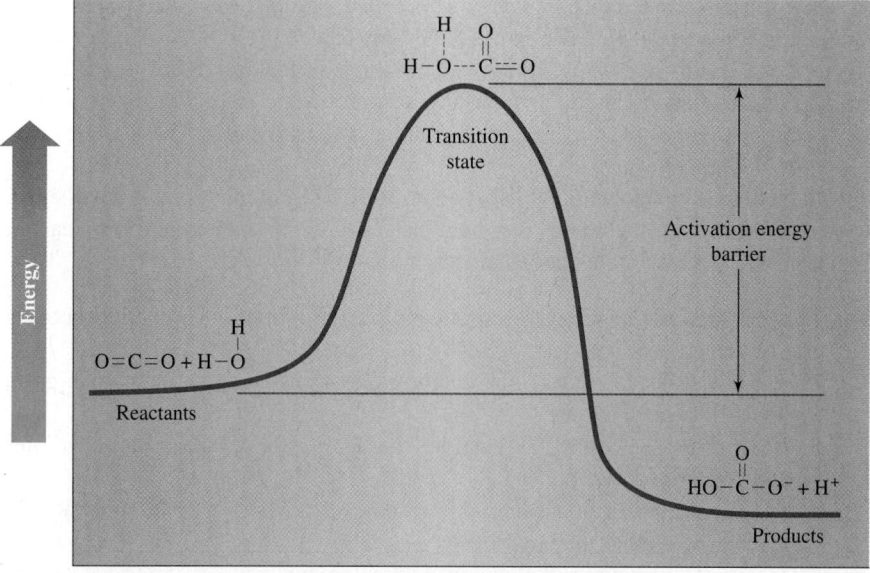

For example, carbon dioxide can be reacted with water to yield hydrogen carbonate ion.

$$CO_2 + H_2O \longrightarrow HCO_3^- + H^+$$

Yet, as shown in Figure 30.3, energy is required to form the transition state. Once the transition state is created, the remainder of the process proceeds easily. As with almost all chemical reactions, reaching the transition state is difficult and limits the rate at which reactants are converted to products.

There are three common ways to increase a reaction rate:

1. *Increasing the reactant concentration:* When the reactant concentrations are made larger, the number of reactant molecules with the necessary activation energy also increases. For simple reactions in the absence of a catalyst, the reaction rate increases with reactant concentration.
2. *Increasing the reaction temperature:* An increase in temperature generally means that each reactant molecule becomes more energetic. A larger fraction of the reactants have the activation energy necessary to be converted to products, and the reaction rate increases.
3. *Adding a catalyst:* A catalyst lowers the activation energy by allowing a new, lower-energy transition state. Since this new process has a lower activation energy, more reactants have the energy to become products. The reaction rate increases.

Biological systems can rarely change reactant concentration or temperature upon demand. Thus, to alter reaction rates, the system developed a set of superb catalysts—the enzymes.

A Michaelis–Menten plot showing the rate of an enzyme-catalyzed reaction as a function of substrate concentration. The lower left portion of the graph marks the approximate area where an enzyme responds best to concentration changes.

Practice 30.1

For the carbon dioxide + water reaction shown in Figure 30.3, predict which set of conditions will yield a faster reaction rate. Briefly explain.

(a) CO_2 pressure = 100 torr
   $T = 37°C$ *vs.*
   Activation energy = 31 kcal/mol

      CO_2 pressure = 100 torr
      $T = 37°C$
      Activation energy = 26 kcal/mol

(b) CO_2 pressure = 100 torr
   $T = 37°C$ *vs.*
   Activation energy = 31 kcal/mol

      CO_2 pressure = 130 torr
      $T = 37°C$
      Activation energy = 31 kcal/mol

30.3 Enzyme Kinetics

In 1913, two German researchers, Leonor Michaelis (1875–1949) and Maud Menten (1879–1960) measured enzyme-catalyzed reaction rates as a function of substrate (reactant) concentration. They observed that most enzyme-catalyzed reactions show an increasing rate with increasing substrate concentration *but* only to a specific maximum velocity, V_{max}. A graph like that in Figure 30.4 is often called a Michaelis–Menten plot.

Scientists have learned much about the nature of enzymes by studying Michaelis–Menten plots. First, because the rate approaches a maximum, we can conclude that enzymes have a limited catalytic ability. Once these enzymes are operating at a maximum, a further increase in substrate (reactant) concentration does not change the reaction rate. By analogy, an enzyme shuttles reactants to the transition state (and on to products) like an usher seating patrons at a theater. No matter how many people arrive at the entrance, a conscientious usher can only lead the ticket holders to their places at a set pace. Similarly, each specific enzyme has a set, maximum catalytic rate.

FIGURE 30.5 ▶
Michaelis–Menten plots for
two glucose metabolic
enzymes.

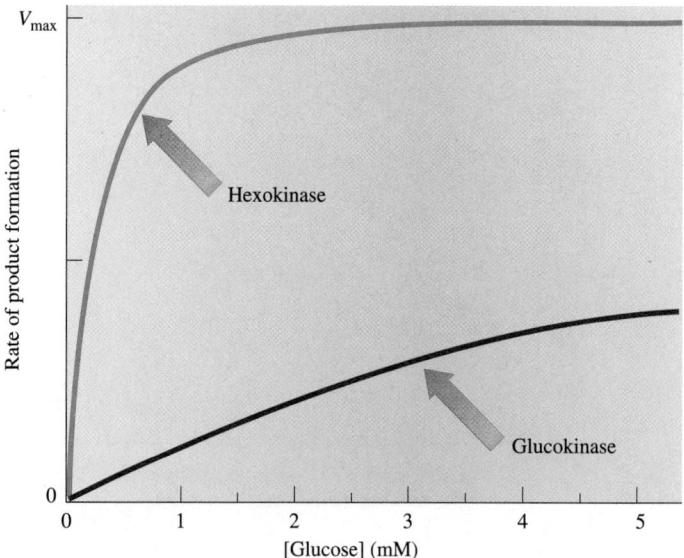

Scientists believe that specific enzymes are tailored to fit specific metabolic needs. Some enzymes work well at low substrate concentrations, whereas others require much higher concentrations before they operate efficiently. Some enzymes catalyze reactions very quickly while others fit a biochemical need by reacting much more slowly. Several examples will illustrate these concepts.

Glucose metabolism is partly controlled by two mammalian enzymes that work effectively in different concentration ranges: Hexokinase starts the conversion of glucose to energy at low glucose concentrations; glucokinase begins the storage of glucose in glycogen only at higher glucose concentrations (see Figure 30.5). Thus hexokinase has first priority to use the available glucose because the enzyme reacts at lower glucose concentrations. Only if there is excess glucose will the glucokinase enzyme start to convert glucose into glycogen.

turnover number   An enzyme's catalytic speed is also matched to an organism's metabolic needs. This catalytic speed is commonly measured as a **turnover number**—the number of molecules an enzyme can react or "turn over" in a given time span. For example, the enzyme catalase has a large turnover number, 10,000,000/s. This enzyme protects us from the toxin hydrogen peroxide and effectively destroys 10 million hydrogen peroxide molecules per second. In contrast, the enzyme chymotrypsin has a much smaller turnover number of about 0.2/s. This enzyme digests 2 protein peptide bonds per 10 seconds as a meal is passed slowly through the small intestine. Since the digestive process is slow, chymotrypsin is not required to have the catalytic speed of catalase. In general, an enzyme's turnover number matches the speed required for a biological process.

Example 30.1   Two enzymes react with the same substrate under identical conditions. Enzyme A is found to have a turnover number of 1500/s whereas enzyme B shows a turnover number of 500/s. Compare the catalytic efficiency of the two enzymes.

The turnover number is a measure of catalytic efficiency. Under identical conditions, the larger the turnover number, the more efficient an enzyme is as a catalyst. Because enzyme A has a larger turnover number than enzyme B, enzyme A is the more efficient catalyst.

Solution

Practice 30.2

Under optimal conditions, the digestive enzyme pepsin has a turnover number of about 30/min while a second digestive enzyme, trypsin, has a turnover number of 12/min. Both of these enzymes digest proteins. Which would you judge to be the more efficient? Briefly explain.

30.4 Industrial-Strength Enzymes

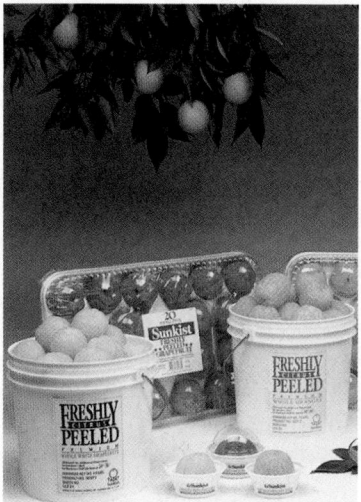

▲
The enzyme pectinase can effectively peel an orange, leaving the fruit in perfect condition.

Not only are enzymes important in biology, they are increasingly important in industry. Enzymes offer two major advantages to manufacturing processes and in commercial products: First, enzymes cause very large increases in reaction rates even at room temperature; second, enzymes are relatively specific and can be used to target selected reactants. Perhaps the biggest disadvantage to industrial enzymes is their relative short supply (and therefore higher cost as compared to traditional chemical treatments). Recent developments in biotechnology offer supplies of less expensive enzymes through genetic engineering.

The following sections focus on common industrial enzymes. Many of these enzymes have a digestive or "breakdown" function. They are hydrolases but are known by common names: **proteases** (proteolytic enzymes) break down proteins; **lipases** digest lipids; cellulases, amylases, lactases, and pectinases break down the carbohydrates, cellulose, amylose, lactose, and pectin, respectively.

proteases
lipases

Food Processing

Enzymes have long been used in food processing. The citrus industry has perfected a process to remove peel from oranges or grapefruits by using the enzyme pectinase. The pectinase penetrates the peel in a vacuum infusion process. There it dissolves the albedo (the white stringy material) that attaches the peel to the fruit. When the fruit is removed from the solution, the skin can be peeled easily by machine or hand. The industry markets prepeeled citrus to hospitals, airlines, and restaurants.

The enzyme lactase is used on an industrial scale to convert lactose to glucose and galactose. This change gives people who are lactose-intolerant the ability to consume milk products. Among the great number of different molecules in milk, lactase only reacts with lactose. Since the glucose and galactose formed are sweeter than lactose and differ in solubility, lactase treatment produces milk with different properties. Ice cream made from lactase-treated milk, for example, tends to be creamier.

▲
Customer is looking at a pair of denim jeans that have been softened by "biostoning" with a cellulase enzyme.

About 25% of all industrial enzymes are used to convert cornstarch into syrups that are equivalent in sweetness and in calories to ordinary table sugar. More than 5 billion pounds of such syrups are produced annually. The process uses three enzymes: The first, α-amylase, catalyzes the liquefaction of starch to dextrins; the second, a glucoamylase, catalyzes the breakdown of dextrins to glucose; the third, glucose isomerase, converts glucose to fructose.

$$\text{Starch} \xrightarrow{\alpha\text{-amylase}} \text{Dextrins}$$

$$\text{Dextrins} \xrightarrow{\text{glucoamylase}} \text{Glucose}$$

$$\text{Glucose} \xrightarrow[\text{isomerase}]{\text{glucose}} \text{Fructose}$$

The product is a high-fructose syrup equivalent in sweetness to sucrose. One of these syrups, sold commercially since 1968, contains by dry weight about 42% fructose, 50% glucose, and 8% other carbohydrates.

Technology

Industrial enzymes offer solutions to environmental pollution problems for some industries. Cotton has traditionally been "scoured" with the strong base sodium hydroxide to prepare it for dyeing. Unfortunately, wastewater from this process can be harmful to the environment. Recently, a pectinase has successfully replaced this toxic chemical. The enzyme operates under milder conditions and achives the same "scoured" result.

A number of industries use cellulases. Denim manufacturers use it to give a soft appearance to this cotton fabric, "stonewashing" or "biostoning" the cloth. In this process, the cellulases digest some of the cellulose on the surface of the cotton. Paper producers use cellulases to complete the breakdown of wood chips to paper pulp. This enzyme treatment forms a better pulp and paper can be produced with smaller amounts of bleach, again easing environmental pollution.

Consumer Goods

Industrial enzymes are used most often as detergent additives. These enzymes clean under relatively mild conditions and can remove spots not accessible to soaps. Proteases digest the most difficult clothing stains (e.g., grass, blood, and sweat, which contain proteins). Lipases team up with soaps to remove greasy stains. Finally, amylases are often added to digest starchy residues from foods such as oatmeal, chocolate, and mashed potatoes. As a side benefit, detergents that are effective in mild conditions also save on water heating bills.

Medical Uses

Enzymes are used in medicine primarily because of their specificity. For example, several enzymes are used to dissolve blood clots in patients with such diseases as lung embolism (clot in the lung), stroke (clot in the brain), and heart attack (clot in the heart). These proteases are specific for the blood protein plasminogen, which is

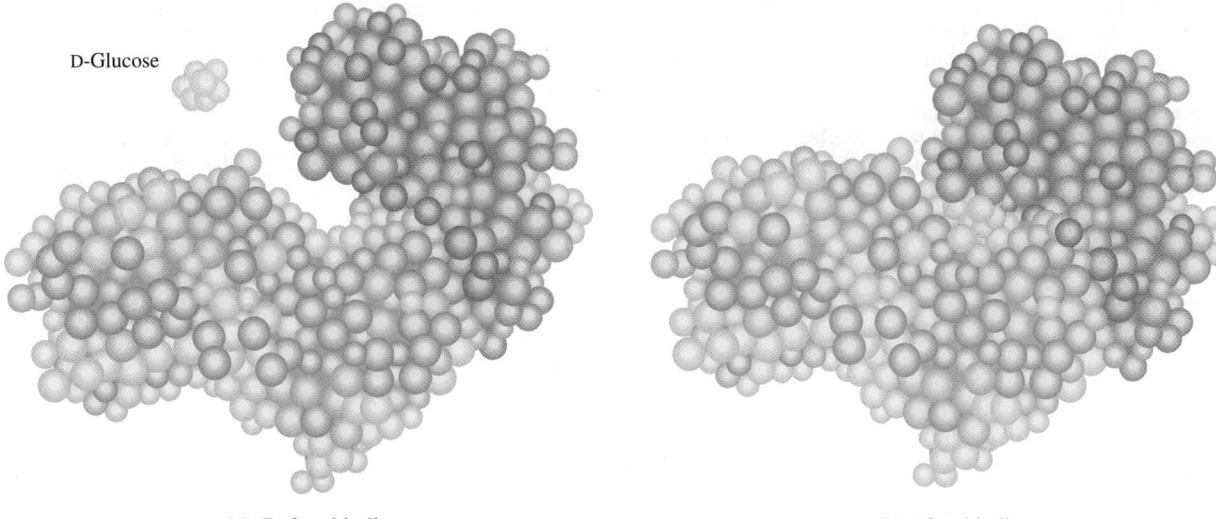

D-Glucose

(a) Before binding   (b) After binding

▲
FIGURE 30.6
A space-filling model of the enzyme hexokinase (a) before and (b) after it binds to the substrate D-glucose. Note the two protein domains for this enzyme, which are colored differently.

converted to plasmin and degrades the blood clot. The proteases are called plasminogen activators; treatment can dissolve a clot in 30–60 minutes.

Practice 30.3
_____

Briefly explain why cellulases are not used to "biostone" polyester fabrics.

30.5 Enzyme Active Site

Catalysis takes place on a small portion of the enzyme structure called the *enzyme active site*. Often this is a crevice or pocket on the enzyme that represents only 1–5% of the total surface area. Figure 30.6 shows a space-filling model of the enzyme hexokinase, which catalyzes a first step in the breakdown of glucose to provide metabolic energy. Notice that the active site is located in a crevice. Glucose enters this site and is bound. The enzyme then must change shape before the reaction takes place. Thus, although catalysis occurs at the small active site, the entire three-dimensional structure of the enzyme is important.

By examining values such as the turnover number and the effective substrate concentration range, scientists have gained a basic understanding of what takes place at an enzyme active site. To function effectively, an enzyme must attract and bind the substrate. Once the substrate is bound, a chemical reaction is catalyzed. This two-step process is described by the following general sequence. Enzyme (E) and substrate (S) combine to form an enzyme–substrate intermediate (E–S). This intermediate decomposes to give the product (P) and regenerate the enzyme:

$$E + S \xrightarrow[\text{binding}]{} E\text{--}S \xrightarrow[\text{catalysis}]{} E + P$$

FIGURE 30.7 ▶
Enzyme–substrate interaction
illustrating both the lock-and-
key hypothesis and the
induced-fit model. The correct
substrate (▢–●) fits the active
site (lock-and-key hypothesis).
This substrate also causes an
enzyme conformation change
that positions a catalytic
group (*) to cleave the appropriate bond (induced-fit
model).

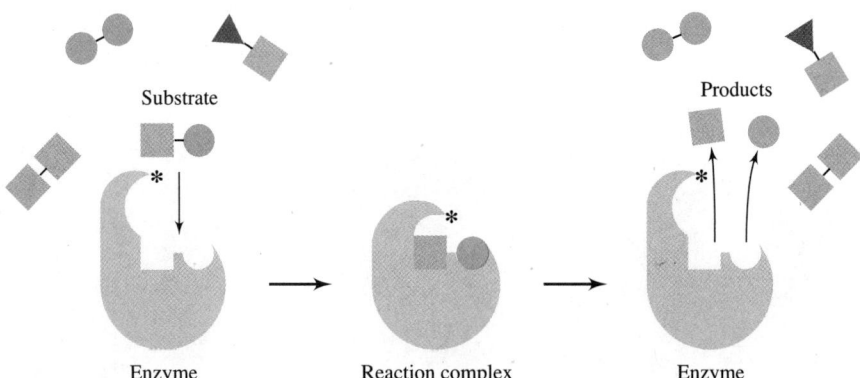

For the hydrolysis of maltose by the enzyme maltase, the sequence is

$$\underset{\text{E}}{\text{Maltase}} + \underset{\text{S}}{\text{Maltose}} \xrightarrow{\text{binding}} \underset{\text{E–S}}{\text{Maltase–Maltose}}$$

$$\underset{\text{E–S}}{\text{Maltase–Maltose}} + H_2O \xrightarrow{\text{catalysis}} \underset{\text{E}}{\text{Maltase}} + \underset{\text{P}}{\text{2 Glucose}}$$

Each different enzyme has its own unique active site whose shape determines, in part, which substrates can bind. Enzymes are said to be stereospecific; that is, each enzyme catalyzes reactions for only a limited number of different reactant structures. For example, maltase binds to maltose (two glucose units linked by an α-1,4-glycosidic bond) but not to lactose (galactose coupled to glucose). In fact, this enzyme even distinguishes maltose from cellobiose (glucose coupled to glucose by a β-1,4-linkage), in which only the glycosidic linkage is different. Enzyme stereospecificity is a very important means by which the cell controls its biochemistry.

Enzyme specificity is partially due to a complementary relationship between the active site and substrate structures. This **lock-and-key hypothesis** envisions the substrate as a key that fits into the appropriate active site, the lock. Although this hypothesis describes a fundamental property of enzyme–substrate binding, scientists have known for some time that enzyme active sites are not rigid as a lock would be. Instead, the active site bends to a certain degree when the appropriate substrate binds. The **induced-fit model** proposes that the active site adjusts its structure in order to prepare the substrate–enzyme complex for catalysis. Enzyme stereospecificity is thus explained in terms of an active site having a somewhat flexible shape. This shape is rigid enough to exclude very dissimilar substrates (lock-and-key hypothesis) but flexible enough to accommodate (induced-fit model) and allow catalysis of appropriate substrates (see Figure 30.7).

The fact that an enzyme is flexible helps explain enzyme-binding specificity and also how the enzyme converts reactants into products. *An enzyme is a dynamic catalyst.* As the enzyme attracts the substrate into the active site, the enzyme's shape and the reactant's shape both begin to change. Note the change in hexokinase structure when glucose is bound (Figure 30.6). This alteration of enzyme shape aids the transformation of the reactant into the product. Figure 30.7 depicts an enzyme where a shape change leads to catalysis. This only exemplifies the numerous alterations that enzymes bring about, for example, bonds broken or formed, charges moved, and new molecular substituents added or removed.

An actual metabolic reaction will illustrate some of the basic features of enzyme catalysis. For example, let's take the enzyme hexokinase, which catalyzes the transformation of glucose to glucose-6-phosphate.

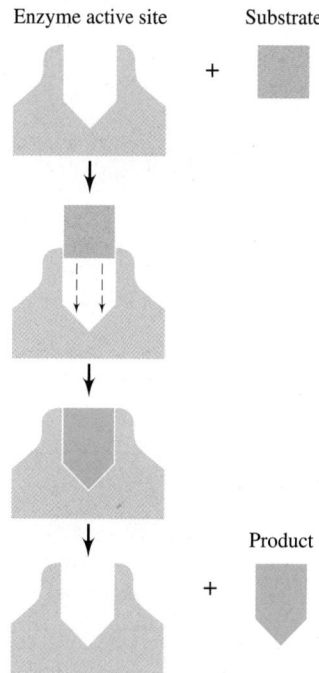

Enzyme active site   Substrate

▲
FIGURE 30.8
Strain hypothesis: The substrate being forced toward the product shape by enzyme binding.

glucose $\xrightarrow[\text{Mg}^{2+}, \text{ATP}]{\text{hexokinase}}$ glucose-6-phosphate

Like many enzymes, hexokinase requires a second reactant. In this case, adenosine-5′-triphosphate (ATP) supplies the phosphate group that is transferred to the glucose. ATP is a very important molecule within the cell because it often transfers energy as well as phosphate groups from enzyme to enzyme (see Chapter 31). Additionally, the metal ion, Mg^{2+}, is required as an activator in this reaction. With all of these components present, the enzyme can start the catalytic process.

When glucose binds, the hexokinase shape changes to bring the ATP close to the carbon 6 of the glucose, thus forcing the transfer of phosphate to this specific carbon. Hexokinase speeds this reaction in several ways. First, this enzyme acts to bring the reactants close together, a process termed **proximity catalysis.** Second, hexokinase positions the reactants so the proper bonds will break and form. (The enzyme ensures that a phosphate is added to carbon 6 of glucose and not to one of the other carbons.) This is often termed the **productive binding hypothesis** because reactants are bound/oriented in such a way that products result. Hexokinase is a successful catalyst—glucose reacts with ATP 10 billion (10^{10}) times faster than it would in the absence of this enzyme. Increases in reaction rate of this magnitude are essential to life. In fact, without enzymes, cellular reactions are too slow to keep cells alive.

proximity catalysis

productive binding hypothesis

Biotechnologists make use of our understanding of enzyme catalysis to design (with nature's help) completely new enzymes. Antibodies (immunoglobulin proteins) are produced that bind tightly to a molecule like in the transition state. When these antibodies bind reactant molecules, the strong attractive forces "strain" the reactants as illustrated in Figure 30.8. The reactant molecule is impelled to change shape to fit the binding site. Catalysis occurs, and these antibodies act as enzymes! Scientists have termed this mode of catalysis the **strain hypothesis,** and it is thought to be important in many natural enzymes as well as these "antibody enzymes." In the not-too-distant future, such artificial enzymes may serve important industrial and medical applications.

strain hypothesis

30.6   Temperature and pH Effects on Enzyme Catalysis

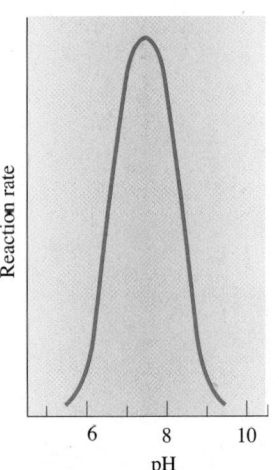

▲
FIGURE 30.9
A plot of the enzyme-catalyzed rate as a function of pH.

Essentially any change that affects protein structure also affects an enzyme's catalytic function. If an enzyme is denatured, its activity will be lost. Thus strong acids and bases, organic solvents, mechanical action, and high temperature are examples of treatments that decrease an enzyme-catalyzed rate of reaction.

Even slight changes in the pH can have profound effects on enzyme catalysis. Remember that some of the amino acids that make up enzymes have side chains whose charge depends on pH. For example, side chains with carboxylic acid functional groups may be either neutral or negatively charged; those with amino groups may be either neutral or positively charged. Thus as the pH changes, the charges on an enzyme, or more specifically at the active site of an enzyme, also change. Enzyme catalysis is affected. For each enzyme there is an optimal pH; shifts to more acidic or more basic conditions decrease the enzyme activity as shown in Figure 30.9.

Because enzymes are so sensitive to small pH changes, our bodies have developed elaborate mechanisms to control the amount of acid and base present in cellular fluids. The kidneys and lungs share the responsibility for maintaining blood acid and base levels. If the blood pH shifts outside of the narrow range 6.8–7.8, death often results. The body's metabolic processes can no longer maintain life partially because of a pH-induced change in enzyme activities.

Body temperature is also carefully controlled partially because enzyme activities are particularly temperature-sensitive. A typical plot of temperature versus enzyme activity is shown in Figure 30.10. At very low temperatures, few reactant molecules have enough energy to overcome the activation energy barrier and the reaction occurs slowly. As the temperature increases, more reactants have the necessary activation energy and the rate increases. But a point is reached where the high temperature causes enzyme denaturation, and the rate of reaction starts to decrease. Thus both high and low temperatures cause slow reaction rates, and, as with pH, an optimal temperature exists for each enzyme. Not surprisingly, the optimal temperature for many human enzymes is approximately body temperature (about 37°C).

Not all enzymes operate optimally around the physiological temperature (37°C) and pH (7). In fact, chemists have long sought enzymes that have optimal activity at higher temperatures and more acidic or basic pHs, conditions that favor industrial reactions. Such enzymes can be harvested from microorganisms. It has been estimated that fewer than 5% of all fungal and bacterial species have been discovered. Thus scientists travel the world to find new microorganisms (and new enzymes). Especially valuable species have been found in hot springs, deep ocean vents, and tropical rain forests.

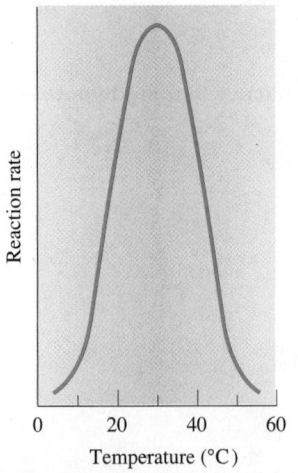

▲
FIGURE 30.10
A plot of the temperature dependence of an enzyme-catalyzed reaction.

Table 30.1 gives some characteristics for bacterial proteases commonly used with detergents. Each protease has different optimum conditions. Both Alcalase® and Savinase® are ideal for warm-water washes with temperature optima at about 55°C. However, Savinase® has a better pH optimum for detergent powders, which commonly have a pH around 9–10. In contrast, Alcalase® has a pH optimum of about 8, which is better adapted for liquid soaps. Esperase® is frequently used in industrial/institutional laundries. This enzyme remains active at temperatures up to 70°C and a pH of 12. In each case, knowing the conditions needed for washing enabled scientists to find the right enzyme.

TABLE 30.1	Some Examples of Enzymes Used with Detergents		
Enzyme name (the most common trademarked name)	Optimum pH	Optimum temperature	Common use
Alcalase®	7–9	40–60°C	Liquid soaps
Savinase®	8–11	40–60°C	Powder detergents
Esperase®	10–12	50–70°C	Industrial/institutional laundries

30.7 Enzyme Regulation

Since enzymes are vital to life, enzyme catalysis is under careful cellular control. The cell uses a variety of mechanisms to change the rate of substrate conversion to product. Sometimes a new group of atoms is covalently joined to the enzyme in a process called *covalent modification.* In other cases, another molecule is noncovalently bound to the enzyme. The protein structural change that results causes a decrease in enzyme activity, **enzyme inhibition,** or an increase in activity, **enzyme activation.**

    Hexokinase is an example of an enzyme that is under cellular control. Because this enzyme begins the breakdown of glucose to yield cellular energy, the hexokinase-catalyzed reaction is often not needed when cellular energy levels are high. Control occurs in the following way: (1) If glucose is no longer needed for cellular energy, the glucose-6-phosphate concentration increases; (2) this compound binds to the hexokinase enzyme and inhibits the enzyme activity. Because glucose-6-phosphate is a product of the reaction, this form of enzyme control is called **product inhibition.** Hexokinase responds to the overall cellular energy state and does not use more glucose than needed.

    Feedback inhibition and **feedforward activation** are two other common forms of enzyme control. To understand these mechanisms, visualize the various cellular processes as assembly lines with a different enzyme at each step from raw material to finished product. Feedback inhibition affects enzymes at the beginning of the molecular assembly line. A final product acts as "feedback" and inhibits an enzyme from using too many molecular starting materials. In contrast, feedforward activation often controls enzymes at the end of the molecular assembly line. If there is an excess of starting materials, these molecules will "feedforward" and activate enzymes, which in turn cause the whole process to move faster. As with other control mechanisms, both feedback inhibition and feedforward activation serve to coordinate enzyme processes within the cell.

    A variety of drug therapies make use of enzyme control to selectively affect target cells. Careful drug design can create a molecule that binds to only one type of enzyme. By binding, the drug blocks normal catalysis and causes enzyme inhibition.

enzyme inhibition
enzyme activation

product inhibition

feedback inhibition
feedforward activation

This pharmacist is ▶ dispensing AZT, the AIDS drug that impedes virus reproduction.

A revolution in mental health treatments has followed the introduction of drugs to inhibit specific brain processes. For example, Prozac® and related drugs specifically inhibit reuptake of serotonin, a neurotransmitter. More serotonin remains to activate neurons that in turn alleviate mental depression. Other neurotransmitters are also amenable to treatment by specific enzyme inhibitor drugs.

The statin drugs (e.g., lovastatin, pravastatin, simvastatin, fluvastatin, atorvastatin) cause a marked decrease in blood cholesterol levels. These drugs specifically inhibit a key enzyme in the cholesterol synthesis pathway. Although the human body contains thousands of different enzymes, the statins specifically impact one.

These examples of drug therapy illustrate a very important principle: Because biological processes depend on enzymes, enzyme control often has a major impact on life.

Concepts in Review

1. Briefly explain why living cells need catalysts.

2. List the six main classes of enzymes and their functions.

3. Compare a simple, uncatalyzed reaction with an enzyme-catalyzed process.

4. Define the symbol V_{max} and relate it to the Michaelis–Menten plot.

5. Discuss the relationship between the reaction rate and the enzyme concentration.

6. Describe an enzyme active site.

7. State what is meant by the specificity of an enzyme.

8. List several ways that an enzyme can facilitate the conversion of substrate to product.

9. Summarize the effects of pH and temperature on an enzyme-catalyzed reaction.

10. Describe how inhibition and activation are important in enzyme control.

11. Discuss the potential role of enzymes in medicine and industry.

Key Terms

The terms listed here have been defined within this chapter.
Section numbers are referenced in parentheses for each term.

enzyme (30.1)
enzyme activation (30.7)
enzyme inhibition (30.7)
feedback inhibition (30.7)
feedforward activation (30.7)

induced-fit model (30.5)
lipases (30.4)
lock-and-key hypothesis
  (30.5)
product inhibition (30.7)

proteases (30.4)
proximity catalysis (30.5)
productive binding
  hypothesis (30.5)
strain hypothesis (30.5)

transition state (30.2)
turnover number (30.3)

Questions

Questions refer to tables, figures, and key words and concepts
defined within the chapter. A particularly challenging question
or exercise is indicated with an asterisk.

1. What is activation energy and how is this energy affected by enzymes?

2. What is the general role of enzymes in the body?

3. Distinguish between a coenzyme and an apoenzyme.

4. Give the names of enzymes that catalyze the hydrolysis of (a) sucrose, (b) lactose, and (c) maltose.

5. What are the six general classes of enzymes?

6. A catalyst increases the rate of a chemical reaction. List two other means of increasing reaction rates.

7. Differentiate between the lock-and-key hypothesis and the induced-fit model for enzymes.

8. List three ways that substrate binding to the active site helps the reactants convert to products.

9. How does an enzyme inhibitor differ from an enzyme substrate?

10. Describe the process and function of the enzyme pectinase, which is used in food processing.

Paired Exercises

These exercises are paired. Each odd-numbered exercise is
followed by a similar even-numbered exercise. Answers to the
even-numbered exercises are given in Appendix VI.

11. An enzyme reacts 0.005 *M* substrate every 3.5 min. What is the reaction rate in units of molar per second?

13. Lysozyme, an enzyme that cleaves the cell wall of many bacteria, has a turnover number of 0.5/s. How many reactants can be converted to products by lysozyme in 1 min?

15. Two enzymes are being studied; the first enzyme uses 1-butanol, $C_4H_{10}O$, as a substrate while the second enzyme uses 2-methyl-2-propanol, $C_4H_{10}O$, as a substrate. Based on the lock-and-key hypothesis, how might the shapes for the enzyme active sites differ?

17. Why does an enzyme-catalyzed reaction rate decrease at low temperatures?

19. If the V_{max} for an enzyme is decreased, would you predict activation or inhibition? Briefly explain.

12. An enzyme reacts 0.02 *M* of substrate every 8 min. What is the reaction rate in units of molar per second?

14. Pepsin, a digestive enzyme found in the stomach, has a turnover number of 1.2/s for a specific protein substrate. How many proteins can be digested by three pepsin molecules in 5 min?

16. Two enzymes are being studied; the first enzyme uses butanoic acid, $C_4H_8O_2$, as a substrate while the second enzyme uses acetic acid, $C_2H_4O_2$, as a substrate. Based on the lock-and-key hypothesis, how might the shapes for the enzyme active sites differ?

18. Why does an enzyme-catalyzed reaction rate decrease at high temperatures?

20. If the turnover number for an enzyme is increased, would you predict activation or inhibition? Briefly explain.

21. A scientist studies two enzymes that catalyze the same reaction. Enzyme A has a turnover number of 225/s, while enzyme B has a turnover number of 120/s. The scientist concludes that enzyme B is more effective than enzyme A. Do you agree? Briefly explain.

22. A scientist studies two enzymes that catalyze the same reaction. Enzyme A has a turnover number of 0.05/s while enzyme B has a turnover number of 9.8×10^{-1}/s. The scientist concludes that enzyme B is more effective than enzyme A. Do you agree? Briefly explain.

23. The amino acid glutamine, is produced by a metabolic pathway in the liver. As the concentration of glutamine goes up, the metabolic pathway slows down. Is this control feedforward activation or feedback inhibition? Briefly explain.

24. The amino acid glutamine, is used by a liver metabolic pathway to make urea. As the concentration of glutamine goes up, the metabolic pathway speeds up. Is this control feedforward activation or feedback inhibition? Briefly explain.

25. Briefly explain why cellulase is used in paper manufacturing.

26. Briefly explain why proteases are used to treat stroke.

Additional Exercises

These exercises are not paired and provide additional practice on concepts covered in this chapter.

27. Chymotrypsin has a turnover number for a glycine-containing substrate of 0.05/s and a turnover number for an L-tyrosine–containing substrate of 200/s. For which substrate is chymotrypsin a more efficient catalyst? Briefly explain.

28. Feedback inhibition is an important form of enzyme regulation. Based on what you know about this control mechanism, how might a different process, "feedback activation," cause regulatory problems for a cell?

29. Explain the meaning of V_{max} by drawing a Michaelis–Menten plot.

30. The enzyme lactase is used in the food industry to modify milk products. What chemical change does lactase catalyze?

Answers to Practice Exercise

30.1 (a) The condition set with the lower activation energy (26 kcal/mol) yields the faster reaction rate because the energy barrier to reaction is smaller. (b) The condition set with the larger CO_2 pressure yields the faster reaction rate because there are more reactant molecules.

30.2 Under these conditions, pepsin is the more efficient because it has the larger turnover number, converting 30 reactant molecules to products per minute.

30.3 You would not use cellulase to "biostone" polyester fabric because cellulase is an enzyme that acts specifically on cellulose. Since polyesters are not made of cellulose, the enzyme would have no effect on them.

CHAPTER

31

Nucleic Acids and Heredity

▲

The science of genetics began when Gregor Mendel studied pea plants in the mid-nineteenth century.

The plight of Doctor Frankenstein's monster touches a chord in all of us. A scientist has given this monster life but cannot control his creation. The monster is "unnatural" and the story ends tragically.

The advent of genetic engineering has raised a similar specter. Genetic engineers work with the molecules that code life—the nucleic acids. By changing the code, they can produce new life forms. Already, bacteria have been altered to make needed human proteins. Recently, both cows and goats have been genetically engineered to produce a human protein in their milk.

The scientists involved in these initial programs have followed careful protocols and have produced valued medicines. However, the day may come when we can decide whether humans should be made smarter or stronger via genetic engineering. How this decision will be made and what it will be is the topic of ongoing heated discussions and much controversy; the potential ramifications are enormous—and not only in the scientific community. The power to even consider such decisions and to possibly open Pandora's box in ways that were once only the stuff of science fiction is the result of our understanding of the biochemistry of nucleic acids.

31.1 Molecules of Heredity—A Link

The question of how hereditary material duplicates itself was one of the most baffling problems of biology for many years. Generations of biologists attempted in vain to solve this problem and to answer the question "Why are the offspring of a species undeniably of that species?" Many thought the chemical basis for heredity lay in the structure of the proteins. But no one was able to provide evidence showing how protein could reproduce itself. The answer to the heredity problem was finally found in the structure of the nucleic acids.

The unit structure of all living things is the cell. Suspended in the nuclei of cells are chromosomes, which consist largely of proteins and nucleic acids. A simple

nucleoprotein
nucleic acid

protein bonded to a nucleic acid is called a **nucleoprotein.** There are two types of **nucleic acids**—those that contain the sugar deoxyribose and those that contain the sugar ribose. Accordingly, they are called deoxyribonucleic acid (DNA) and ribonucleic acid (RNA). Although many of us think of DNA as a recent discovery, it was actually discovered in 1869 by the Swiss physiologist Friedrich Miescher (1844–1895), who extracted it from the nuclei of cells.

31.2 Bases and Nucleosides

Nucleic acids are complex chemicals that combine several different classes of smaller molecules. As with many complex structures, it is easier to understand the whole by first studying its component parts. We begin our examination of nucleic

adenine
(6-aminopurine)

guanine
(2-amino-6-oxypurine)

cytosine
(2-oxy-4-aminopyrimidine)

thymine
(2,4-dioxy-5-methylpyrimidine)

uracil
(2,4-dioxypyrimidine)

▲
FIGURE 31.1
Purine and pyrimidine bases found in living matter.

acids by learning about a critical part of these molecules, two classes of heterocyclic bases called the purines and the pyrimidines:

purine, $C_5H_4N_4$ pyrimidine, $C_4H_4N_2$

These parent compounds are related in structure; the pyrimidine is a six-membered heterocyclic ring while the purine contains both a five- and six-membered ring. The nitrogen atoms cause these compounds to be known as heterocycles (the rings are made up of more than just carbon atoms) and also as bases. Like the ammonia nitrogen, these heterocycles react with hydrogen ions to make a solution more basic.

Five major bases are commonly found in nucleic acids—two purine bases (adenine and guanine) and three pyrimidine bases (cytosine, thymine, and uracil). Figure 31.1 gives one stable form for each compound. These bases must be available to the cell in order to reproduce genetic information (DNA). The importance of these bases is underscored by their use in cancer chemotherapy. Specific "modified" bases are used to kill fast-growing cancer cells. Perhaps the most common anticancer drug in this category is 5-fluorouracil.

5-fluorouracil

Note that 5-fluorouracil is similar to thymine—it acts to inhibit the enzyme that catalyzes the formation of thymine. Cancer cells require a rapid synthesis of DNA. Without sufficient thymine to form DNA, cancer cells die.

The natural bases differ one from another in their ring substituents. Each base has a lower-most nitrogen, which is bonded to a hydrogen as well as two carbons. This specific —NH shares chemical similarities with an alcohol (—OH) group.

adenosine

deoxyadenosine

cytidine

deoxycytidine

Just as two sugars molecules can be linked when an alcohol of one monosaccharide reacts with a second monosaccharide (see Chapter 27), so a purine or pyrimidine molecule can be bonded to a sugar molecule by a reaction with the —NH group.

nucleoside   A **nucleoside** is formed when either a purine or pyrimidine base is linked to a sugar molecule, usually D-ribose or D-2′-deoxyribose.

D-ribose

D-2′-deoxyribose

The base and sugar are bonded together between carbon 1′ of the sugar and either the purine nitrogen at position 9 or the pyrimidine nitrogen at position 1 by splitting out a molecule of water. Typical structures of nucleosides are shown in Figure 31.2. A prime is added to the position number to differentiate the sugar numbering system from the purine or pyrimidine numbering system.

The name of each nucleoside emphasizes the importance of the base to the chemistry of the molecule. Thus adenine and D-ribose react to yield adenosine, whereas cytosine and D-2′-deoxyribose yield deoxycytidine. The root of the nucleoside name derives from the purine or pyrimidine name. The compositions of the common ribonucleosides and the deoxyribonucleosides are given in Table 31.1.

TABLE 31.1	Composition of Ribonucleosides and Deoxyribonucleosides	
Name	**Composition**	**Abbreviation**
Adenosine	Adenine–ribose	A
Deoxyadenosine	Adenine–deoxyribose	dA
Guanosine	Guanine–ribose	G
Deoxyguanosine	Guanine–deoxyribose	dG
Cytidine	Cytosine–ribose	C
Deoxycytidine	Cytosine–deoxyribose	dC
Thymidine	Thymine–ribose	T
Deoxythymidine	Thymine–deoxyribose	dT
Uridine	Uracil–ribose	U
Deoxyuridine	Uracil–deoxyribose	dU

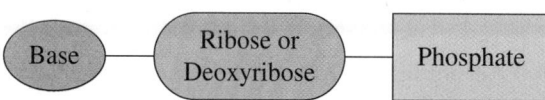

adenosine-5′-monophosphate   deoxyadenosine-5′-monophosphate
(AMP)   (dAMP)

◀ **FIGURE 31.3**
Examples of nucleotides.

31.3 Nucleotides: Phosphate Esters

A more complex set of biological molecules is formed by linking phosphate groups to nucleosides. Phosphate esters of nucleosides are termed **nucleotides.** These molecules consist of a purine or a pyrimidine base linked to a sugar, which in turn is bonded to at least one phosphate group.

nucleotide

```
( Base )—( Ribose or
           Deoxyribose )—[ Phosphate ]
```

The ester may be a monophosphate, a diphosphate, or a triphosphate. When two or more phosphates are linked together, a high-energy phosphate anhydride bond is formed (see Section 31.4). The ester linkage may be to the hydroxyl group of position 2′, 3′, or 5′ of ribose or to position 3′ or 5′ of deoxyribose. Examples of nucleotide structures are shown in Figure 31.3.

adenosine-5'-diphosphate
(ADP)

adenosine-5'-triphosphate
(ATP)

▲
FIGURE 31.4
Structures of ADP and ATP.

Nucleotide abbreviations start with the corresponding nucleoside abbreviation (see Table 31.1). The letters MP (monophosphate) can be added to any of these to designate the corresponding nucleotide. Thus GMP is guanosine monophosphate. A lowercase d is placed in front of GMP if the nucleotide contains the deoxyribose sugar (dGMP). When the letters such as AMP or GMP are given, it is generally understood that the phosphate group is attached to position 5' of the ribose unit (5'-AMP). If attachment is elsewhere, it will be designated, for example, as 3'-AMP.

Two other important adenosine phosphate esters are adenosine diphosphate (ADP) and adenosine triphosphate (ATP). Note that the letters DP are used for diphosphate and TP for triphosphate. In these molecules the phosphate groups are linked together. The structures are similar to AMP except that they contain two and three phosphate residues, respectively (see Figure 31.4). All the nucleosides form mono-, di-, and triphosphate nucleotides.

31.4   High-Energy Nucleotides

The nucleotides have a central role in the energy transfers in many metabolic processes. ATP and ADP are especially important in these processes since these two nucleotides store and release energy to the cells and tissues. The source of energy is the foods we eat, particularly carbohydrates and fats. Energy is released as the carbons from these foods are oxidized (see Chapter 33). Part of this energy is used to maintain body temperature, and part is stored in the phosphate anhydride bonds of such molecules as ADP and ATP. Because a relatively large amount of energy is stored in these bonds, they are known as high-energy phosphate anhydride bonds.

High-energy phosphate anhydride bonds
(ATP)

High-energy anhydride bond
(ADP)

Energy is released during the hydrolysis of high-energy phosphate anhydride bonds in ADP and ATP. In the hydrolysis, ATP forms ADP and inorganic phosphate (P_i) yielding about 35 kJ of energy per mole of ATP:

◀ **Muscle movements, including heartbeats, depend on energy from ATP.**

$$ATP + H_2O \underset{\substack{\text{energy} \\ \text{storage}}}{\overset{\substack{\text{energy} \\ \text{utilization}}}{\rightleftharpoons}} ADP + P_i + \sim 35 \text{ kJ}$$

The hydrolysis reaction is reversible, with ADP being converted to ATP by still higher energy molecules. In this manner, energy is supplied to the cells from ATP, and energy is stored by the synthesis of ATP. Processes such as muscle movement, nerve sensations, vision, and even the maintenance of our heartbeats are all dependent on energy from ATP.

31.5   Polynucleotides; Nucleic Acids

Starting with two nucleotides, a dinucleotide is formed by splitting out a molecule of water between the $-$OH of the phosphate group of one nucleotide and the $-$OH on carbon 3′ of the ribose or deoxyribose of the other nucleotide. Then another and another nucleotide can be added in the same manner until a polynucleotide chain is formed. Each nucleotide is linked to its neighbors by phosphate ester bonds (see Section 24.12).

Two series of polynucleotide chains are known, one containing D-ribose and the other D-2′-deoxyribose. One polymeric chain consists of the monomers AMP, GMP, CMP, and UMP, and is known as a polyribonucleotide. The other chain contains the monomers dAMP, dGMP, dCMP, and dTMP, and is known as a polydeoxyribonucleotide:

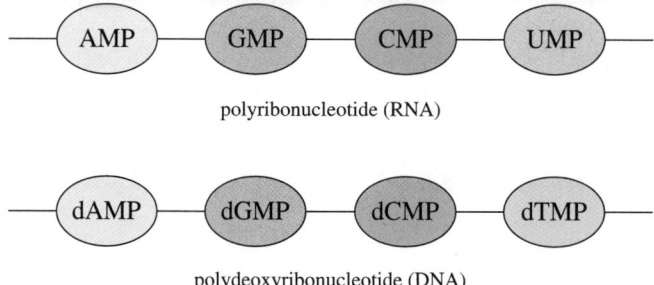

polyribonucleotide (RNA)

polydeoxyribonucleotide (DNA)

To another phosphate

(U)

(G)

To another phosphate

UMP

GMP

CMP

AMP

To another phosphate

(C)

(A)

To another phosphate

FIGURE 31.5 ▶
A segment of ribonucleic acid (RNA) consisting of the four nucleotides: adenosine monophosphate, cytidine monophosphate, guanosine monophosphate, and uridine monophosphate.

ribonucleic acid (RNA)

deoxyribonucleic acid (DNA)

    The nucleic acids DNA and RNA are polynucleotides. **Ribonucleic acid (RNA)** is a polynucleotide that upon hydrolysis yields ribose, phosphoric acid, and the four purine and pyrimidine bases adenine, guanine, cytosine, and uracil. **Deoxyribonucleic acid (DNA)** is a polynucleotide that yields D-2′-deoxyribose, phosphoric acid, and the four bases adenine, guanine, cytosine, and thymine. Note that RNA and DNA contain one different pyrimidine nucleotide: RNA contains uridine, whereas DNA contains thymidine. A segment of a ribonucleic acid chain is shown in Figure 31.5. As will be described later, RNA and DNA also commonly differ in function: DNA serves as the storehouse for genetic information; RNA aids in expressing genetic characteristics.

31.6   Structure of DNA

Deoxyribonucleic acid (DNA) is a polymeric substance made up of the four nucleotides dAMP, dGMP, dCMP, and dTMP. The size of the DNA polymer varies with

TABLE **31.2**	Relative Amounts of Purines and Pyrimidines in Samples of DNA					
Source	Adenine	Thymine	Ratio A/T	Guanine	Cytosine	Ratio G/C
Beef thymus	29.0	28.5	1.02	21.2	21.2	1.00
Beef liver	28.8	29.0	0.99	21.0	21.1	1.00
Beef sperm	28.7	27.2	1.06	22.2	22.0	1.01
Human thymus	30.9	29.4	1.05	19.9	19.8	1.00
Human liver	30.3	30.3	1.00	19.5	19.9	0.98
Human sperm	30.9	31.6	0.98	19.1	18.4	1.04
Hen red blood cells	28.8	29.2	0.99	20.5	21.5	0.96
Herring sperm	27.8	27.5	1.01	22.2	22.6	0.98
Wheat germ	26.5	27.0	0.98	23.5	23.0	1.02
Yeast	31.7	32.6	0.97	18.3	17.4	1.05
Vaccinia virus	29.5	29.9	0.99	20.6	20.0	1.03
Bacteriophage T_2	32.5	32.6	1.00	18.2	18.6	0.98

the complexity of the organism; more complex organisms tend to have larger DNAs. For example, simple bacteria like *Escherichia coli* have about 8 million nucleotides in their DNA while a human DNA contains up to 500 million nucleotides. The order in which these nucleotides occur varies in different DNA molecules, and it is within this order that the genetic information in a cell is stored.

Scientists have tried to understand DNA structure by determining the nucleotide composition of this molecule from many different sources. For a long time it was thought that the four nucleotides occurred in equal amounts in DNA. However, more refined analyses showed that the amounts of purine and pyrimidine bases vary in different DNA molecules. Surprisingly, careful consideration of these data also showed that the ratios of adenine to thymine and guanine to cytosine are always essentially 1:1. This observation served as an important key to unraveling the structure of DNA. The analyses of DNA from several species are shown in Table 31.2.

A second important clue to the special configuration and structure of DNA came from X-ray diffraction studies. Most significant was the work of Maurice Wilkins (b. 1916) of Kings College of London. Wilkins's X-ray pictures implied that the nucleotide bases were stacked one on top of another like a stack of saucers. From his work as well as that of others, the American biologist James D. Watson (b. 1928) and British physicist Francis H.C. Crick (b. 1916), working at Cambridge University, designed and built a scale model of a DNA molecule. In 1953, Watson and Crick announced their now famous double-stranded helical structure for DNA. This was a milestone in the history of biology, and in 1962 Watson, Crick, and Wilkins were awarded the Nobel prize in medicine and physiology for their studies of DNA.

The structure of DNA, according to Watson and Crick, consists of two polymeric strands of nucleotides in the form of a double helix, with both nucleotide strands coiled around the same axis (see Figure 31.6). Along each strand are alternate phosphate and deoxyribose units with one of the four bases adenine, guanine, cytosine, or thymine attached to deoxyribose as a side group. The double helix is held together by hydrogen bonds extending from the base on one strand of the double helix to a complementary base on the other strand. The structure of DNA has

FIGURE 31.6 ▶
**Right: Double-stranded helical
structure of DNA (·····
denotes a hydrogen bond
between adjoining bases).
Left: Space filling model of
DNA.**

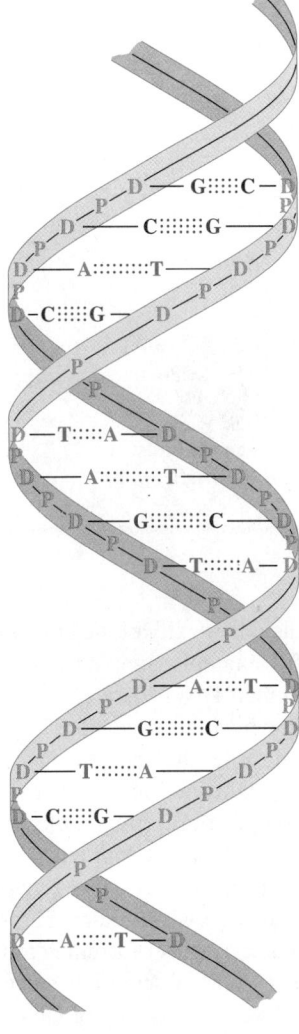

P = Phosphate
D = Deoxyribose
A = Adenine
T = Thymine
C = Cytosine
G = Guanine

▲
**James Watson and Francis
Crick with their DNA model.**

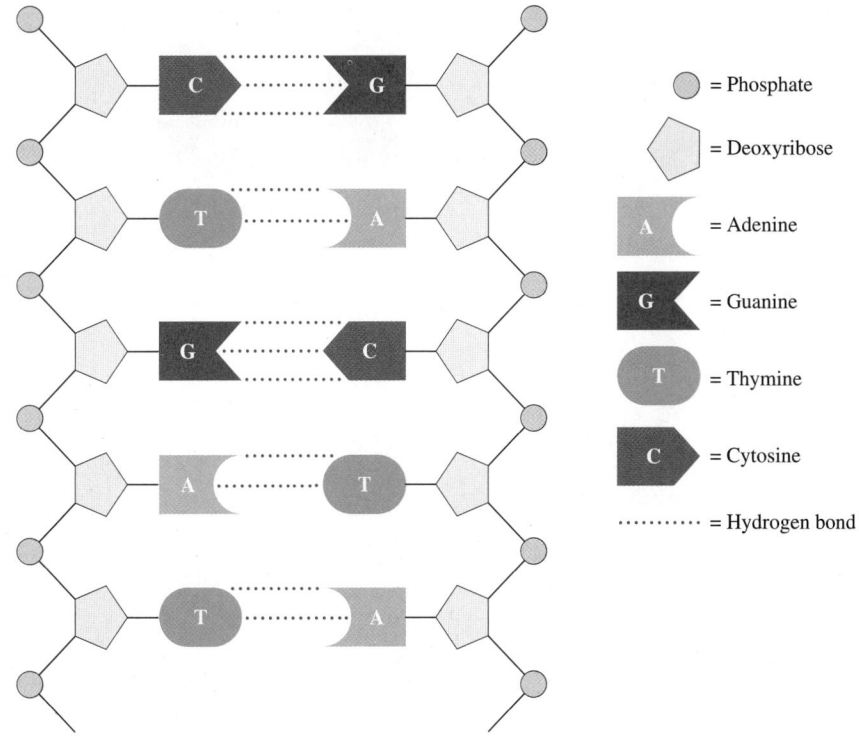

◀ **FIGURE 31.7**
**DNA segment showing
phosphate, deoxyribose,
and complementary base
pairings held together by
hydrogen bonds.**

⬤ = Phosphate

⬠ = Deoxyribose

A = Adenine

G = Guanine

T = Thymine

C = Cytosine

········· = Hydrogen bond

been likened to a ladder that is twisted into a double helix, with the rungs of the ladder kept perpendicular to the twisted railings. The phosphate and deoxyribose units alternate along the two railings of the ladder, and two nitrogen bases form each rung of the ladder.

In the Watson–Crick model of DNA, the two polynucleotide strands fit together best when a purine base is adjacent to a pyrimidine base. Although this allows for four possible base pairings, A–T, A–C, G–C, and G–T, only the adenine–thymine and guanine–cytosine base pairs can effectively hydrogen-bond together. Under normal conditions, an adenine on one polynucleotide strand is paired with a thymine on the other strand; a guanine is paired with a cytosine (see Figure 31.7). This pairing results in A:T and G:C ratios of 1:1, as substantiated by the data in Table 31.2. The hydrogen bonding of complementary base pairs is shown in Figure 31.8. Note that if the sequence of one strand is known, the sequence of the other strand can be determined. The two DNA polymers are said to be *complementary* to each other. As will be discussed in Section 31.7, the cell is able to chemically "read" one strand in order to synthesize its complementary partner.

The double helix is an important part of DNA structure. It does not explain how the very large DNA can be packed into a cell or the even smaller cell nucleus. For example, a human DNA molecule can be extended to almost 10 cm in length and yet is contained in a nucleus with a diameter about a hundred thousand times smaller. To begin the necessary packing, the DNA is looped around small aggregates of positively charged histone proteins. Hundreds of these aggregates are associated with each DNA molecule so that the DNA is foreshortened and has the appearance of a string of pearls. Further condensation is achieved by wrapping this structure into a tight coil called a solenoid, as shown in Figure 31.9. Finally, the solenoid nucleopro-

FIGURE 31.8 ▶
Hydrogen bonding between the complementary bases thymine and adenine (T · · · · A) and cytosine and guanine (C · · · · G). Note that one pair of bases has two hydrogen bonds and the other pair has three hydrogen bonds.

tein complex is wound around a protein scaffold within the nucleus. By following this complete procedure, human DNA can be reduced in length to about 10 μm in condensed chromosomes. Like a high-density computer disk, DNA takes up only a little space relative to the large amount of genetic information it contains.

31.7　DNA Replication

The foundations of our present concepts of heredity and evolution were laid within the span of a decade. Charles Robert Darwin (1809–1882), in *The Origin of Species* (1859), presented evidence supporting the concept of organic evolution and his theory of natural selection. Gregor Johann Mendel (1822–1884) discovered the basic laws of heredity in 1866, and Friedrich Miescher discovered nucleic acid in 1869. Although Darwin's views were widely discussed and generally accepted by biologists within a few years, Mendel's and Miescher's work went unnoticed for many years.

Mendel's laws were rediscovered about 1900 and led to our present understanding of heredity and the science of genetics. Interest in nucleic acids lagged until nearly the 1950s, when chemical and X-ray data led to Watson and Crick's idea that DNA exists in a double helix and that DNA has the possible copying mechanism for genetic material.

Heredity is the process by which the physical and mental characteristics of parents are transferred to their offspring. For this to occur, it is necessary for the material responsible for genetic transfer to be able to make exact copies of itself. The polymeric DNA molecule is the chemical basis for heredity. The genetic information needed for transmitting a species' characteristics is coded along the polymeric chain. Although the chain is made from only four different nucleotides, the information content of DNA resides in the sequence of these nucleotides.

genome　　　The **genome** is the sum of all hereditary material contained in a cell. Within the

Histone protein

DNA

Nucleoprotein

◀ **FIGURE 31.9**
Condensed form of DNA.
After the DNA polymer
wraps twice around the
histone protein aggregates
(shown as round balls), it
coils around a central axis
to form a solenoid struc-
ture.

eucaryotic genome are chromosomes, which are long, threadlike bodies composed of nucleic acids and proteins that contain the fundamental units of heredity, called genes. A **gene** is a segment of the DNA chain that controls formation of a molecule of RNA. In turn, many RNAs determine the amino acid sequences for specific polypeptides or proteins. One gene commonly directs the synthesis of only one polypeptide or protein molecule. The cell has the capability of producing a multitude of different proteins because each DNA molecule contains a large number of different genes.

    For life to continue relatively unchanged, genetic information must be reproduced exactly each time a cell divides. **Replication,** as the name implies, is the biological process for duplicating the DNA molecule. The DNA structure of Watson and Crick holds the key to replication; because of the complementary nature of DNA's nitrogen bases, adenine bonds only to thymine and guanine only to cytosine. Nucleotides with complementary bases can hydrogen-bond to each single strand of DNA and hence be incorporated into a new DNA double helix. Every double-stranded DNA molecule that is produced contains one template strand and one newly formed, complementary strand. This form of DNA synthesis, known as *semiconservative* replication, is illustrated in Figure 31.10.

    Replication is one of the most complicated enzyme-catalyzed processes in life. Enzymes are required to unwind the DNA before replication and to repackage the DNA after synthesis. The two template strands are copied differently. One daughter

gene

replication

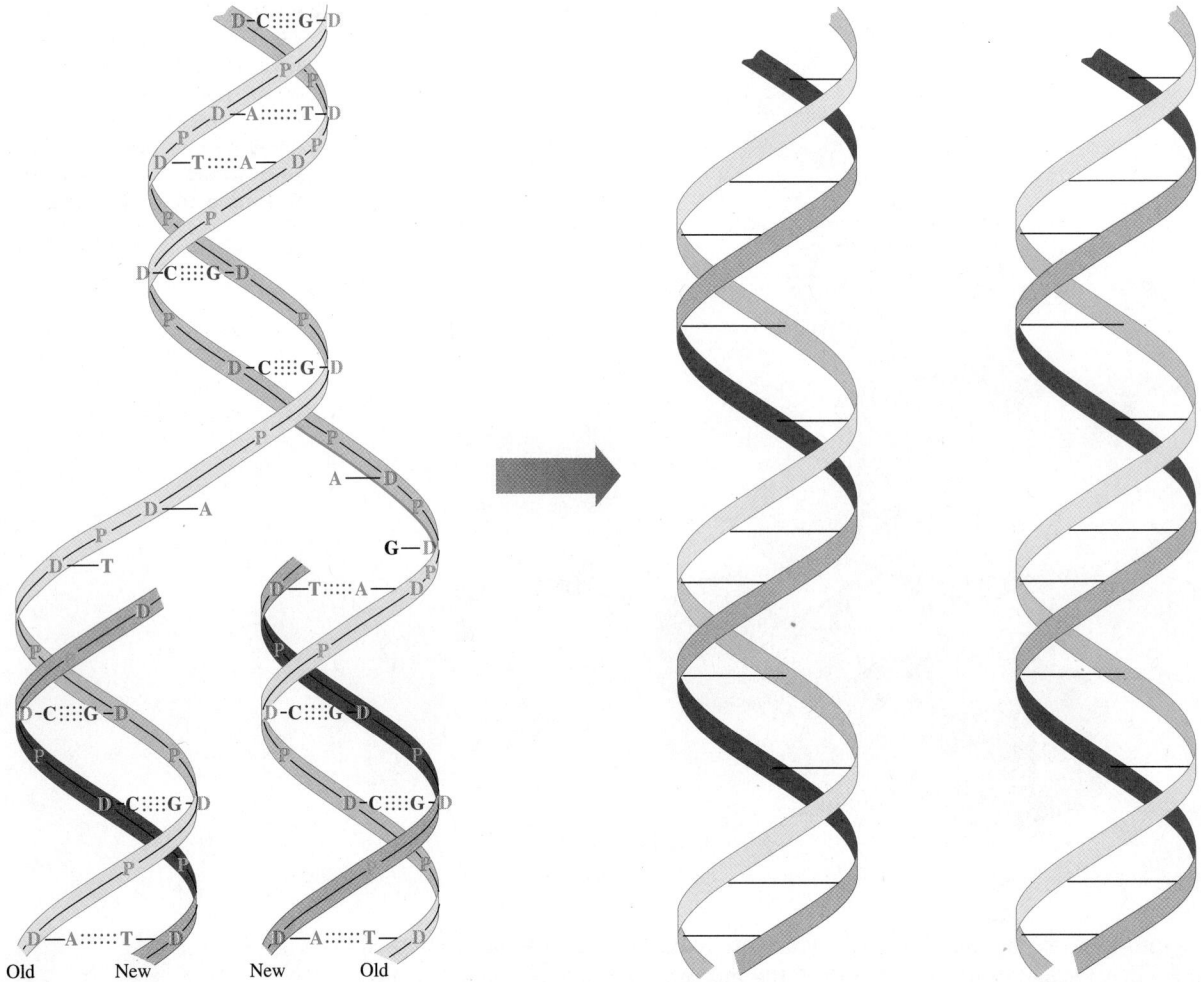

Old   New   New   Old

▲
FIGURE 31.10
Semiconservative replication
of DNA: The two helices
unwind, separating at the
hydrogen bonds. Each strand
then serves as a template,
recombining with the proper
nucleotides forming two new
double-stranded helices. The
newly synthesized DNA
strands are shown in red.

strand grows directly toward the point at which the templates are unwinding while
the other daughter strand is synthesized away from this point. In this way, the same
enzyme-catalyzed reaction can be used to synthesize both strands. However, the one
daughter strand that has been created in small fragments must be connected before
replication is complete. So, while one DNA strand is formed by continuous synthesis, the other new strand is formed by a repetition of fragment synthesis followed by
a coupling reaction (see Figure 31.11).

Other processes "proofread" the new polymers to check for errors. Amazingly,
replication is so carefully coordinated that a mistake is passed on to the new strands
only once in about a billion times. Of course, this important fact means that new cells
retain the genetic characteristics of their parents.

DNA Cell Division, and Cloning

The DNA content of cells doubles just before the cell divides, and one-half of the
DNA goes to each daughter cell. After cell division is completed, each daughter cell
contains DNA and the full genetic code that was present in the original cell. This

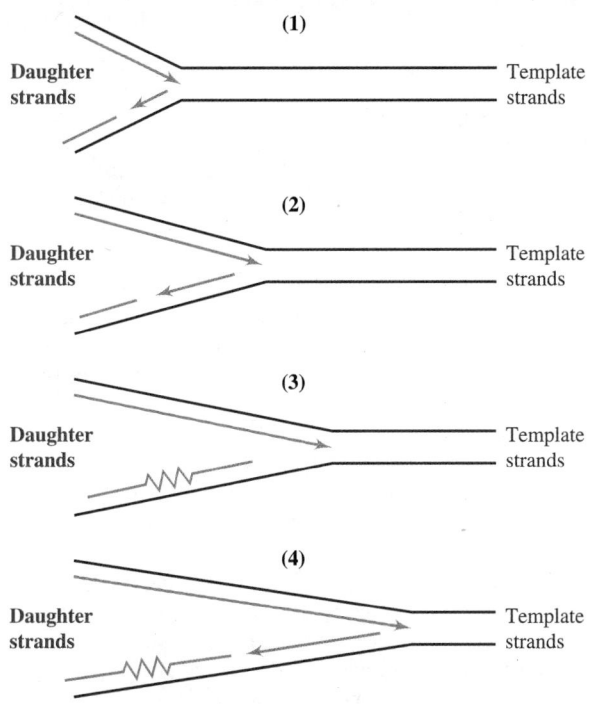

(1)

Daughter
strands   Template
                                                             strands

(2)

Daughter
strands   Template
                                                             strands

(3)

Daughter
strands   Template
                                                             strands

(4)

Daughter
strands   Template
                                                             strands

◄ FIGURE 31.11
The basic replication process. Arrows indicate the direction of DNA synthesis. Between (1) and (2), the template DNA strands unwind, and some synthesis occurs for both daughter strands. Moving to (3), the DNA fragments are connected on one daughter strand (⌁) while DNA synthesis continues on the other strand and the template strands unwind further. DNA synthesis again takes place for both daughter strands in (4).

process of ordinary cell division is known as **mitosis** and occurs in all the cells of the body except the reproductive cells.

mitosis

As we have indicated before, DNA is an integral part of the chromosomes. Each species carries a specific number of chromosomes in the nucleus of each of its cells. The number of chromosomes varies with different species. Humans have 23 pairs, or 46 chromosomes. The fruit fly has 4 pairs, or 8 chromosomes. Each chromosome contains DNA molecules. Mitosis produces cells with the same chromosomal content as the parent cell. However, in the sexual reproductive cycle, cell division occurs by a different process known as meiosis.

In sexual reproduction, two cells, the sperm cell from the male and the egg cell (or ovum) from the female, unite to form the cell of the new individual. If reproduction took place with mitotic cells, the normal chromosome content would double when two cells united. However, in **meiosis** the cell splits in such a way as to reduce the number of chromosomes to one-half of the number normally present (23 in humans). The sperm cell carries half of the chromosomes from its original cell, and the egg cell also carries half of the chromosomes from its original cell. When the sperm and the egg cells unite during fertilization, the cell once again contains the correct number of chromosomes and the hereditary characteristics of the species. Thus the offspring derives half its genetic characteristics from the father and half from the mother.

meiosis

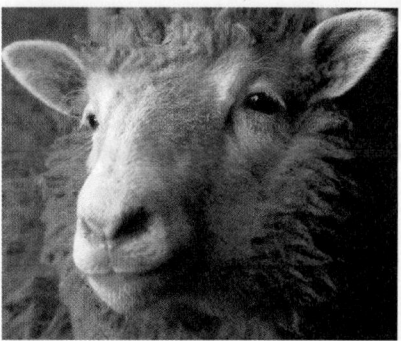

Although sexual reproduction allows for genetic diversity, there are times when reproduction of an exact copy might be useful. For example, to replicate a sheep that grows especially fine wool or a cow that gives especially good milk could be quite beneficial. Recently, such copies, called clones, have been produced from adult animals (sheep and cattle). Molecular biologists first transplant an animal cell nucleus into an egg cell. Once the egg has started to divide, it is implanted into a

▲
"Dolly," the first sheep cloned from adult cells.

Fingerprints in the Genes

Just as for a set of fingerprints, no two individuals' DNA sequences are exactly the same. More than 100 years ago, detectives first started to use fingerprints from crime scenes to identify criminals. Today, forensic chemists use biochemistry to unlock the DNA sequence to identify possible criminals.

▲ **DNA fingerprint matching. Note that Suspect S2 matches evidence blood sample E(vs).**

Because the human genome contains approximately 3 billion base pairs, it is not practical for forensic chemists to examine the entire genome sequence. Instead, they identify randomly selected pieces of DNA. If enough DNA is available (e.g., from a 1-cm-diameter blood stain), this genetic material can be broken into pieces and separated on a gel by electrophoresis. The gel is then treated with radioactive molecules that bind to specific sites and serve as tags on the DNA fragments. These tags form a pattern on the gel known as a DNA "fingerprint."

A second technique uses the enzyme DNA polymerase to replicate the DNA from a very small sample (only several nanograms of DNA are needed). The replication process multiplies the amount of DNA available until there are enough DNA pieces to separate and visualize in gel electrophoresis. Again, a pattern is

formed that can be treated like a fingerprint.

Unfortunately, DNA fingerprinting uses only a small segment of the total genome. Although each person has a unique total genome, there is an extremely small chance that two people will have the same DNA sequence for a particular segment. So, scientists must estimate the possibility of a match by chance, and courtroom decisions often hinge on complex statistical reasoning.

Still, DNA fingerprinting, a technique only available since 1990, continues to make courtroom history. Since most body fluids contain cells that carry DNA, DNA "fingerprints" can often be found at crime scenes, and both "DNA" convictions and exonerations have occurred because of this. Many experts believe that DNA fingerprinting will become more and more widely used in court proceedings in the future.

female animal and gestation proceeds. Each cloned offspring carries a DNA sequence matched to the DNA from the original adult animal donor. The first cloned animal was a sheep named Dolly, who was produced in Scotland in 1996. Cloning is not a routine procedure; there are still many technical difficulties. Yet, the scientific problems are being solved. It will be much harder to solve the societal and ethical implications of routine cloning.

DNA, Cancer, and Chemotherapy

Scientists understood a key principle about tissue growth once they realized that cells must replicate DNA to reproduce. For example, fast-growing cancer cells must make DNA much faster than normal cells. Thus many common cancer chemotherapies are designed to interfere with DNA replication. Some drugs (e.g., topotecan, originally derived from the campotheca tree in Japan) block the unwinding of DNA that is necessary before replication can proceed. Some flat molecules like cisplatin (Platinol-AQ®) or adriamycin slip in between the flat DNA bases and disrupt replication.

$$\begin{array}{c} Cl \qquad Cl \\ \diagdown \; Pt \; \diagup \\ \diagup \qquad \diagdown \\ H_2N \qquad NH_2 \end{array}$$

cisplatin

Still other drugs (e.g., gemcitabine, Gemzar®) are unnatural nucleosides that cause an abortive DNA replication. Note that the sugar ring of gemcitabine has been altered so that it no longer forms a 3′ phosphate ester to the next nucleotide in a growing DNA polymer.

gemcitabine

Of course, the "downside" to these treatments is that normal cells are impacted to a certain extent as well. Fast-growing cells such as those in bone marrow are almost always affected adversely by cancer chemotherapies. Thus scientists continue to search for more specific and better treatments for cancer.

31.8 RNA: Genetic Transcription

One of the main functions of DNA is to direct the synthesis of ribonucleic acids (RNAs). RNA differs from DNA in the following ways: (1) It consists of a single polymeric strand of nucleotides rather than a double helix; (2) it contains the pentose D-ribose instead of D-2′-deoxyribose; (3) it contains the pyrimidine base uracil instead of thymine; and (4) some types of RNA have a significant number of modified bases in addition to the common four. RNA also differs functionally from DNA. Whereas DNA serves as the storehouse of genetic information, RNA is used to process this information into proteins. Three types of RNA are needed to produce proteins: ribosomal RNA (rRNA), messenger RNA (mRNA), and transfer RNA (tRNA).

More than 80% of the cellular RNA is ribosomal RNA. It is found in the ribosomes, where it is associated with protein in proportions of about 60–65% protein to 30–35% rRNA. Ribosomes are the sites for protein synthesis.

Messenger RNA carries genetic information from DNA to the ribosomes. It is a template made from DNA and carries the code that directs the synthesis of proteins. The size of mRNA varies according to the length of the polypeptide chain it will encode.

The primary function of tRNA is to bring amino acids to the ribosomes for incorporation into protein molecules. Consequently, at least one tRNA exists for each of the 20 amino acids required for proteins. Transfer RNA molecules have a number of structural features in common. The end of the chain of all tRNA molecules terminates with the three nucleotides C, C, A to which is attached the amino acid to be transferred to a protein chain. The primary structure of tRNA allows extensive folding of the molecule such that complementary bases are hydrogen-bonded to each other to form a structure that appears like a cloverleaf. The cloverleaf model of

▲
These ribosomes are translating mRNA strands to produce proteins.

▲
FIGURE 31.12
Representations of tRNA. The anticodon triplet (UUC) located at the lower loop is complementary to GGA (which is the code for glycine) on mRNA.

tRNA has an anticodon loop consisting of seven unpaired nucleotides. Three of these nucleotides make up an anticodon (see Figure 31.12). The anticodon is complementary to, and hydrogen-bonds with, three bases on an mRNA. The other two loops in the cloverleaf structure enable the tRNA to bind to the ribosome and other specific enzymes during protein synthesis (see Section 31.11).

transcription

The making of RNA from DNA is called **transcription.** The verb *transcribe* literally means to copy, often into a different format. When the nucleotide sequence of one strand of DNA is transcribed into a single strand of RNA, genetic information is copied from DNA to RNA. This transcription occurs in a complementary fashion and depends on hydrogen-bonded pairing between appropriate bases (see Figure 31.13). For example, where a DNA segment has a guanine base, RNA has a cytosine base. Cytosine is transcribed to guanine, thymine to adenine, and adenine to uracil (the thyminelike base that is found in RNA).

Because transcription is the initial step in the expression of genetic information, this process is under stringent cellular control. Only a small fraction of the total information stored in DNA is used at any one time. In procaryotic cells (see Section 33.5), related genes are often located together so that they can be transcribed in concert. The control of eucaryotic gene expression is more complex.

oncogenes

The importance of transcription control is emphasized by the discovery of the **oncogenes.** These genes are present in cancerous or malignant cells and code for

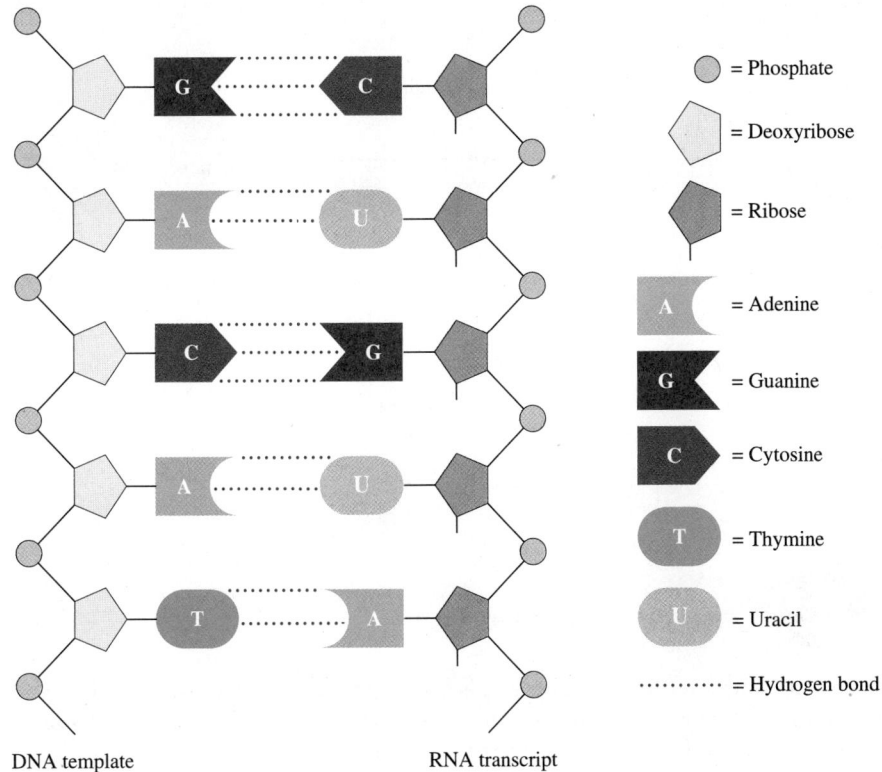

DNA template   RNA transcript

= Phosphate

= Deoxyribose

= Ribose

A = Adenine

G = Guanine

C = Cytosine

T = Thymine

U = Uracil

= Hydrogen bond

◀ FIGURE 31.13
Transcription of RNA from DNA. The sugar in RNA is ribose. The complementary base of adenine is uracil. After transcription is complete, the new RNA separates from its DNA template and travels to another location for further use.

proteins that control cell growth. To the surprise of many investigators, it was found that these same oncogenes are present in many normal mammalian cells. The question arose, "Why do some cells become cancerous and some remain normal?" The answer: Cancerous cells lose control of oncogene transcription. Control is lost when the DNA structure is altered by, for example, a viral infection or a chemical carcinogen. The oncogenes are then transcribed too often, cell growth is uncontrolled—and a cancer develops.

Following transcription, the RNA molecules produced from DNA are often modified before they are put to use. Ribosomal RNA and tRNA molecules form as larger precursors and are then trimmed to the correct size. After transcription, many of the bases in tRNA are modified by methylation, saturation of a double bond, or by isomerization of the ribose-base linkage. This posttranscriptional modification or processing changes the information content of the RNA.

Eucaryotic mRNA undergoes considerable alteration before the message that it carries is ready to guide protein synthesis. These changes may include the elimination of portions of the mRNA molecule, the splicing together of two or more mRNA molecules, or the addition of new bases to one end of the mRNA. During this processing, the message encoded in the mRNA molecule is apparently refined so that it can be correctly read by the ribosome.

Figure 31.14 summarizes the processes by which genetic information is transferred within a cell. These steps culminate in the conversion of a coded nucleotide message into a protein amino acid sequence.

31.9 The Genetic Code

For a long time after the structure of DNA was elucidated, scientists struggled with the problem of how the information stored in DNA could specify the synthesis of so many different proteins. Since the backbone of the DNA molecule contains a regular structure of repeated and identical phosphate and deoxyribose units, the key to the code had to lie with the four bases—adenine, guanine, cytosine, and thymine.

The code, using only the four nucleotides A, G, C, and T, must at least be capable of coding for the 20 amino acids that occur in proteins. If each nucleotide codes one amino acid, only four amino acids can be represented. If the code uses two nucleotides to specify an amino acid, 16 (4×4) combinations are possible—still not enough. But, using three nucleotides, we get 64 ($4 \times 4 \times 4$) possible combinations—which is more than enough to specify the 20 common amino acids in proteins. We now know that each code word requires a sequence of three nucleotides. The code is therefore a triplet

codon

code. Each triplet of three nucleotides is called a **codon,** and, in general, each codon specifies one amino acid. Thus, to describe a protein containing 200 amino acid units, a gene containing at least 200 codons, or 600 nucleotides, is required.

In the sequence of biological events, the code from a gene in DNA is first transcribed to a coded RNA, which, in turn, directs the synthesis of a protein. The 64 possible codons for mRNA are given in Table 31.3. In this table a three-letter sequence (first nucleotide–second nucleotide–third nucleotide) specifies a particular amino acid. For example, the codon CAC (cytosine–adenine–cytosine) is the code for the amino acid histidine (His). Note that three codons in the table, marked TC, do not encode any amino acids. These are called *nonsense* or *termination codons.* They act as signals to indicate where the synthesis of a protein molecule is to end. The other 61 codons identify 20 amino acids. Methionine and tryptophan have only one codon each. For the other amino acids, the code is redundant; that is, each amino acid is specified by at least two, and sometimes by as many as six, codons.

Scientists believe that the genetic code is a universal code for all living organisms (with a few exceptions); that is, the same nucleotide triplet specifies a given amino acid regardless of whether that amino acid is synthesized by a bacterial cell, a pine tree, or a human being.

TABLE 31.3	The Genetic Code for Messenger RNA				
First nucleotide	**Second nucleotide**	**Third nucleotide and amino acid coded**			
		U	**C**	**A**	**G**
U	U	Phe	Phe	Leu	Leu
	C	Ser	Ser	Ser	Ser
	A	Tyr	Tyr	TC*	TC*
	G	Cys	Cys	TC*	Trp
C	U	Leu	Leu	Leu	Leu
	C	Pro	Pro	Pro	Pro
	A	His	His	Gln	Gln
	G	Arg	Arg	Arg	Arg
A	U	Ile	Ile	Ile	Met
	C	Thr	Thr	Thr	Thr
	A	Asn	Asn	Lys	Lys
	G	Ser	Ser	Arg	Arg
G	U	Val	Val	Val	Val
	C	Ala	Ala	Ala	Ala
	A	Asp	Asp	Glu	Glu
	G	Gly	Gly	Gly	Gly

*Termination or nonsense codon

This table shows the sequence of nucleotides in the triplet codons of messenger RNA that specify a given amino acid. For example, UUU or UUC is the codon for Phe, UCU is the codon for Ser; CAU or CAC is the codon for His.

31.10 Genes and Medicine

Two facts have been known for many years: (1) Most human traits can be traced to the genetic makeup of the individual, and (2) some diseases (e.g., sickle-cell anemia and Tay-Sach's disease) occur mainly within certain ethnic groups, clearly indicating a genetic connection for these maladies. These two facts motivated scientists to learn more about the human genome. By the early 1980s, techniques were available to "read" genetic material—that is, to determine the sequence of bases in the strings of DNA-comprising genes.

A complete sequencing or "mapping" of the human genome was desirable but daunting—given that this genome contains about 3 billion base pairs. The time required to sequence or decode the entire genome in a single laboratory was estimated to be about 1000 years! However, after much discussion and correspondence among many of the world's leading scientists, the Human Genome Project was organized. This project got under way in 1988 and is being carried forward in many laboratories in several countries with the object of mapping nothing less than the entire human genome.

The first five years of the project were used to locate genetic markers—"benchmarks" on the different human chromosome pairs. And as these genetic markers have been identified, the genes associated with a variety of diseases have been located. Genes have been found that trigger muscular dystrophy, Huntington's disease, cystic fibrosis, and some breast cancers, to name a few.

Recent research has turned to sequencing the chromosomes. This involves breaking a selected segment of the genome into pieces, sequencing each piece, and then matching overlapping sequences—like fitting a giant jigsaw puzzle together. Computer programs for fitting the pieces together have facilitated the project. The first billion bases in the human genome were sequenced by November 1999. The complete mapping of the human genome (90%+) was announced in June, 2000.

Enormous benefits from the Human Genome Project will arise because the DNA sequence codes for all life processes. Genetic counseling will be more accurate and birth defects more predictable. The same Human Genome Project techniques applied to microbes, plants, and animals will aid agriculture and industry. One primary benefit is that the genetic basis of inheritable diseases will be better understood, and, this understanding will eventually be used to correct genetic errors, that is, gene therapy.

For example, one chromosome that is already sequenced (chromosome 19) has been found to contain the genetic defect for nine related diseases, including Huntington's disease. Knowing the DNA sequence error and where to locate this error gives gene therapists a target to work on. Now, scientists are working on ways to delete the error and add a corrected DNA sequence. The techniques used in gene therapy are in the development stages; yet clinical trials are already being conducted because of this therapy's potential benefits. These early trials have mostly been unsuccessful, but scientists have high hopes for gene therapy in the immediate future.

▲
Scientist studies a gene map as part of the Human Genome Project.

31.11 Biosynthesis of Proteins

The biosynthesis of proteins is extremely complex, and the following is only a cursory description of the overall process. The production of a polypeptide using an mRNA template is called **translation.** This term is used because it literally means a change from one language to another. The genetic code is *translated* into the primary structure of a polypeptide or a protein.

translation

The biosynthesis of proteins begins when messenger RNA leaves the cellular nucleus and travels to the cytoplasm. Each mRNA is then bound to five or more ribosomes, the bodies responsible for protein synthesis. Amino acids must also be transferred to the ribosomes. To accomplish this step, cellular energy in the form of ATP is used to couple amino acids to tRNAs to form aminoacyl–tRNA complexes that can bind to the ribosomes.

$$\underset{\underset{NH_2}{|}}{R-CH}-\overset{\overset{O}{\|}}{C}-OH \;+\; tRNA \;+\; ATP \;\xrightarrow[\text{synthetase}]{\text{aminoacyl–tRNA}}$$

$$\underset{\underset{NH_2}{|}}{R-CH}-\overset{\overset{O}{\|}}{C}-tRNA \;+\; AMP \;+\; HO-\overset{\overset{O}{\|}}{\underset{\underset{OH}{|}}{P}}-O-\overset{\overset{O}{\|}}{\underset{\underset{OH}{|}}{P}}-OH$$

aminoacyl–tRNA

A different specific enzyme is utilized for binding each of the 20 amino acids to a corresponding tRNA. Although only 20 amino acids are involved, there are about 60 different tRNA molecules in the cells. Thus the function of tRNA is to bring amino acids to the ribosome synthesis site.

Initiation The next step in the process is the initiation of polypeptide synthesis. Two codons signal for the start of protein synthesis, AUG and GUG (AUG being the more common). As shown in Table 31.3, these codons also code for incorporation of the amino acids methionine and valine. Most mRNAs have more than one AUG and GUG, and the ribosome must choose at which codon to begin. It appears that the ribosome uses information in addition to the AUG or GUG triplet to choose the correct starting codon for protein synthesis.

Once the correct starting point has been identified, a special initiator tRNA binds to the ribosome. This specific tRNA carries an *N*-formyl methionine in procaryotic cells (e.g., bacterial cells; see Section 33.5):

$$\begin{array}{c} \text{HC}{=}\text{O} \\ | \\ \text{NH} \quad \text{O} \\ | \quad \quad || \\ \text{CH}_3{-}\text{S}{-}\text{CH}_2\text{CH}_2\text{CH}{-}\text{C}{-}\text{tRNA} \end{array}$$

N-formyl methionine–tRNA

Because of the attached formyl group (CHO), a second amino acid will react only with the carboxyl group of methionine, the correct direction for protein synthesis. In eucaryotic cells, the initiator tRNA carries a methionine group.

Elongation The next stage involves the elongation or growth of the peptide chain, which is assembled one amino acid at a time. After the initiator tRNA attaches to the mRNA codon, the elongation of the polypeptide chain involves the following steps (see Figure 31.15):

1. The next aminoacyl–tRNA enters the ribosome and attaches to mRNA through the hydrogen bonding of the tRNA anticodon to the mRNA codon.
2. The peptide bond between the two amino acids forms by transferring the amino acid from the initial aminoacyl–tRNA to the incoming aminoacyl–tRNA. In this step, which is catalyzed by the enzyme peptidyl transferase, the carboxyl group of the first amino acid separates from its tRNA and forms a peptide bond with the free amino group of the incoming aminoacyl–tRNA.
3. The tRNA carrying the peptide chain (now known as a peptidyl–tRNA) moves over in the ribosome, the free tRNA is ejected, and the next aminoacyl–tRNA enters the ribosome.

The peptide chain is transferred to the incoming amino acid, and the sequence is repeated over and over as the ribosome moves over to the next codon on the mRNA chain. Each time the ribosome moves to the next mRNA codon, the entire peptide chain is transferred to the incoming amino acid.

The initiation and elongation of polypeptide chains requires cellular energy. The primary source of this energy is the nucleotide guanosine-5′-triphosphate (GTP). At various steps in the growth of the protein chain, a high-energy GTP phosphate anhydride bond is hydrolyzed, yielding GDP and a phosphate group. The energy that is released drives protein synthesis. This reaction is analogous to that involving ATP (see Section 31.4).

Termination The termination of the polypeptide chain occurs when a "nonsense" or termination codon appears. In normal cells no tRNAs have complementary anti-codons to the termination codons. Because there is no new aminoacyl–tRNA to bind to the ribosome, the peptidyl–tRNA is hydrolyzed, and the free polypeptide (protein) is released. All of these amazing, coordinated steps are accomplished at a high rate of speed—about 1 minute for a 146 amino acid chain of human hemoglobin and 10–20 seconds for a 300–500 amino acid chain in the bacterium *Escherichia coli (E. coli)*. This mechanism of protein synthesis is illustrated in Figure 31.15.

31.12   Changing the Genome: Mutations and Genetic Engineering

mutation

From time to time, a new trait appears in an individual that is not present in either its parents or its ancestors. These traits, which are generally the result of genetic or chromosomal changes, are called **mutations.** Some mutations are beneficial, but most are harmful. Because mutations are genetic, they can be passed on to the next and future generations.

mutagen

Mutations occur spontaneously or are caused by chemical agents or various types of radiation such as X-rays, cosmic rays, and ultraviolet rays. The agent that causes the mutation is called a **mutagen.** Exposure to mutagens can produce changes in the DNA of the sperm or ova. The likelihood of such changes is increased by the intensity and length of exposure to the mutagen. Mutations may then show up as birth defects in the next generation. Common types of genetic alterations include the substitution of one purine or pyrimidine for another during DNA replication. Such a substitution changes the genetic code and causes misinformation to be transcribed from the DNA. A mutagen can also alter genetic material by causing a chromosome or chromosome fragment to be added or removed.

In recent years, laboratory techniques for controlling genetic change have been developed. These techniques, collectively known as genetic engineering, are responsible for considerable progress in medicine and biology. Genetic engineers are now able to insert specific genes into the genome of a host cell and thus program it to produce new and different proteins.

Genetic engineering is a reality because of several basic advances in nucleic acid biochemistry. First, scientists are now able to isolate, identify, and then synthesize multiple copies of specific genetic messages. A key to this process is the *DNA polymerase chain reaction.* DNA polymerase is an enzyme that replicates DNA when supplied a starting fragment (the primer) and a DNA strand to copy (the template). By supplying the appropriate primer, the polymerase can be induced to synthesize a specific gene, even in the presence of a wide variety of other genetic material. What makes this reaction especially valuable is that the polymerase uses the newly synthesized DNA as a template for further replication. Thus, after 20 synthetic cycles, almost a million copies of the original gene can be produced. Like a radioactive chain reaction, the polymerase chain reaction leads to an explosion in the numbers of copied genes.

A second important process in genetic engineering is the insertion of genetic material into a "foreign" genome. Special enzymes, the *restriction endonucleases,*

▲
These genetically engineered tomatoes are disease-resistant and ripen slowly so that they can be picked later in the growing season.

FIGURE 31.15
Biosynthesis of proteins. Step 1, initiation: mRNA enters and complexes with the ribosomes. tRNA carrying an amino acid (aminoacyl–tRNA) enters the ribosome and attaches to mRNA at its complementary codon. Step 2, elongation: Another aminoacyl–tRNA enters the ribosome and attaches to the next codon on mRNA. The peptide chain elongates when a peptide bond is formed by the transfer of the peptide chain from the initial (Met) to the incoming amino acid. This sequence is repeated until the protein chain of amino acids is complete. Step 3, termination: The sequence ends when a termination codon appears on the mRNA.

provide a key step during gene insertion. These enzymes split double-stranded DNA at very specific locations.

original unbroken DNA chain
(from host genome)

broken DNA chain with "sticky ends"

Often the newly formed break has what are termed "sticky ends"; that is, one end of the break "sticks" to the end of another break via hydrogen bonding between complementary bases. When both a gene and a foreign genome are processed in this way, they bond together. Then, with the aid of other enzymes (ligases), the gene can be covalently bonded into place.

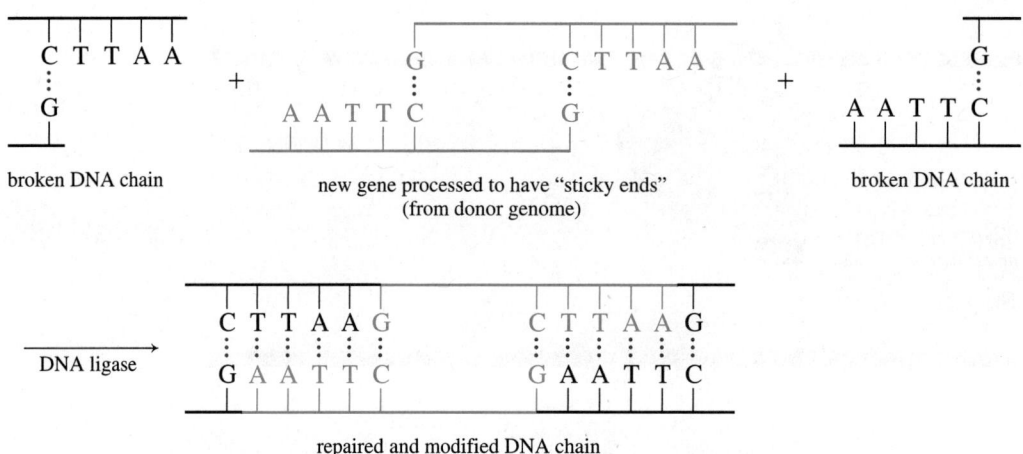

broken DNA chain

new gene processed to have "sticky ends"
(from donor genome)

broken DNA chain

repaired and modified DNA chain
containing new gene

recombinant DNA

The modified and repaired foreign genome now contains a new gene. The result of this process is a form of recombinant DNA. The term **recombinant DNA** refers to DNA whose genes have been rearranged to contain new or different hereditary information.

Agriculture is greatly impacted by genetic engineering. Planners estimate that the amount of arable land worldwide will not change much but the population will continue to increase. Genetic engineering can improve land use by, for example, increasing yield and decreasing crop losses due to insects.

Increased yields can be achieved in a variety of ways. Presently, much effort is going into aquaculture; fish are genetically engineered to produce larger amounts of the fish growth hormone. These fish grow much faster than wild types (coho salmon up to 37 times faster; catfish up to 60% faster). Since fish continue to grow throughout their lifespan, this faster growth rate should result in bigger fish and more protein for the world's population. A modification of this procedure has been used to increase the milk production in cows. Bovine growth hormone is made by genetically modified microorganisms. This hormone is then injected into the cows, causing a 10% increase in milk production.

◀ **Bovine growth hormone injected into cows causes a 10% increase in milk production.**

Efforts to make plants more resistant to plant diseases, insect pests, and other hazards comprise another promising avenue of research. One approach to this goal is to modify the plant so that it makes its own insecticide. For example, the most damaging insect to potatoes in the United States is the potato beetle. A bacterial gene that codes for a toxin that kills this pest has been successfully inserted into potato plants. A related technique is used to modify plants so that they are not affected by specific herbicides. A genetically altered soybean that tolerates the common herbicide glyphosate (Roundup) is now commercially available.

Genetic engineering is also being used to change the quality of the final products. For example, slow-ripening tomatoes are now widely available. Because they ripen slowly, the tomatoes can be picked later in the growing season and should thus be better tasting when purchased by the consumer. An important oil-producing plant, rapeseed (canola), has been genetically altered to produce up to 40% lauric acid, an important component of many detergents. In the near future, this same plant may be producing another oil that is used to make a nylon.

The techniques used in genetic engineering are becoming routine. It is clear that many benefits can be drawn from this technology. It is equally clear that there may be unknown negative consequences of these procedures. Genetic engineering remains controversial because it is difficult to foresee the future consequences of such a major innovation.

The Double Helix Revisited

By James Watson
The man who launched the Human Genome Project celebrates its success

None of us so privileged few who first saw the double helix in the spring of 1953 ever contemplated that we might in our lifetime see it completely decoded. All our dreams at the time centered on the next big objective—finding how the four letters of the DNA alphabet (A, T, G and C) spell out the linear sequences of amino acids in the synthesis of proteins, the main actors in the drama of cellular life. As it turns out, the essence of the genetic code and of the molecular machinery that reads it was solidly established by 1966, only 13 years after Francis Crick and I discovered the double helix.

Then the creative juices of science turned to how to read the messages of DNA. To our surprise, Frederick Sanger at Cambridge University and Walter Gilbert at Harvard, working independently, needed less than a decade to develop powerful methods for determining the order of DNA letters. At roughly the same time, Herbert Boyer and Stanley Cohen devised elegantly simple procedures for cutting and rejoining DNA molecules to produce "recombinant DNA."

Then voices of doom proclaimed that these procedures would create life forms as threatening to our existence as nuclear weapons. Such false alarms held us back only a few years, however. By 1980 the immense powers of recombinant DNA were let loose for the public good. Soon they were to change irreversibly the faces of biology and medicine and bring modern biotechnology into existence.

It was my desire to help speed up human genetics that drove me in 1986 to become an early partisan of the Human Genome Project, whose ultimate objective was to sequence the roughly 3 billion DNA letters that comprise our genetic code. Though many young hot-

▲
Watson, in 1957, with the world's most famous molecule.

shots argued that the time for the project had not yet arrived, those of us a generation older were seeing at too close hand our parents and spouses falling victim to diseases of genetic predisposition. And virtually all of us knew couples rearing children whose future was clouded by a bad throw of the genetic dice. So the National Academy of Sciences assembled an expert committee that reported one year later that the human instruction book could be established within 15 years—if we assembled the appropriate scientific leadership and give it $3 billion to spend wisely over those 15 years.

Our first report emphasized that we should begin by sequencing the relatively tiny genomes (1 million to 13 million letters) of bacteria and yeast and then move on to the 100 million-letter-size genomes of worms and flies. We were confident that by the time we were done, sequencing technology would cost less than 50¢ a letter, and that by then, we would be ready to tackle the human genome. We were also confident that genomics would pay scientific and medical dividends long before the final letters of the human genome were in place.

So when I went up to Capitol Hill in May 1987 with the N.A.S. report in

hand, I promised that long before the genome project was completed we would have cloned many of the key genes predisposing humans to Alzheimer's or to cancers of the breast and colon, all diseases known to run in families. Happily, time has seen our science so move.

Congress accepted this message much faster than many of my fellow molecular biologists, soon appropriating moneys that let the genome project start with a bang. In October 1988, I went to Washington to direct the National Institutes of Health's major role in the effort. From the start, I worked to ensure that the project was an international one, supported by all the major countries of the developed world. That way no one nation or private body would be perceived as controlling the human genome. We also wanted all the data placed on the Internet so that they would be available, free of cost, all over the world.

Today those involved in the international consortium take pride in posting their new DNA sequences on the Web within 24 hours of assembly. Twelve years ago, no one could have imagined that nearly 500 million base pairs of assembled DNA would be posted in just one month.

With the project essentially complete three years ahead of schedule, we must note the change of heart of its early enemies. Instead of wanting us shut down, as they did as late as 1991, they now beg us to move quickly on the genomes of the mouse, the rat and the dog. Equally important, we note that no other big science project, save possibly the Manhattan Project, has been carried out with such zeal for the common good. In sharing their sequences so freely and so quickly, members of our genome community have little time left to promote their scientific reputations.

By contrast, the large infusions of private capital over the past two years support companies that aim to find and patent key DNA sequences before they

continued ▶

The Double Helix Revisited, continued

become publicly available. Not surprisingly, the leaders of these companies have implied that those of us who started the project were no longer needed. To our vast relief, the publicly supported effort received not less but more money. Our backers want to ensure that all the essential features of the human genome are available without cost to all the people of the world. The events of the past few weeks have shown that those who work for the public good do not necessarily fall behind those driven by personal gain.

"The Double Helix Revisited," *Time*, July 3, 2000, p. 30. Copyright © 2000 Time Inc. Reprinted by permission.

Concepts in Review

1. Write the structural formulas for the two purine and three pyrimidine bases found in nucleotides.

2. Distinguish between ribonucleotides and deoxyribonucleotides.

3. List the compositions, abbreviations, and structures for the ten nucleotides in DNA and RNA.

4. Write the structural formulas for ADP and ATP.

5. Identify where energy is stored in ADP and ATP.

6. Write a structural formula of a segment of a polynucleotide that contains four nucleotides.

7. Describe the double-helix structure of DNA according to Watson and Crick.

8. Explain the concept of complementary bases.

9. Describe and illustrate the replication process of DNA.

10. Understand how heredity factors are stored in DNA molecules.

11. Distinguish between mitosis, meiosis, and cloning.

12. Describe how oncogenes are related to cancer.

13. Describe the genetic code.

14. State the functions of the three different kinds of RNA.

15. Describe the transcription of the genetic code from DNA to RNA.

16. Describe the biosynthesis of proteins.

17. Understand how mutations are caused.

18. Briefly describe genetic engineering.

19. Explain some of the advantages of gene therapy.

20. Describe DNA fingerprinting.

21. Describe the Human Genome Project and the benefits to be derived from it.

Key Terms

The terms listed here have been defined within this chapter. Section numbers are referenced in parentheses for each term.

codon (31.9)
deoxynucleic acid (DNA) (31.5)
gene (31.7)
genome (31.7)

meiosis (31.7)
mitosis (31.7)
mutagen (31.12)
mutation (31.12)
nucleic acid (31.1)

nucleoprotein (31.1)
nucleoside (31.2)
nucleotide (31.3)
oncogene (31.8)
recombinant DNA (31.12)

ribonucleic acid (RNA) (31.5)
replication (31.7)
transcription (31.8)
translation (31.11)

Questions

Questions refer to tables, figures, and key words and concepts defined within the chapter. A particularly challenging question or exercise is indicated with an asterisk.

1. Write the names and structural formulas for the five nitrogen bases found in nucleotides.

2. What is the difference between a nucleoside and a nucleotide?

3. What are the principal structural differences between DNA and RNA?

4. What is the major function of ATP in the body?

5. Briefly describe the structure of DNA as proposed by Watson and Crick.

6. What is meant by the term complementary bases?

7. What is the genetic code?

8. Why are at least three nucleotides needed for one unit of the genetic code?

9. Starting with DNA, briefly outline the biosynthesis of proteins.

10. How does a codon differ from an anticodon?

11. Explain the role of *N*-formyl methionine in procaryotic protein synthesis.

12. What is a mutation?

13. What is meant by a "DNA fingerprint"?

14. Why will gene therapy benefit from the Human Genome Project?

Paired Exercises

These exercises are paired. Each odd-numbered exercise is followed by a similar even-numbered exercise. Answers to the even-numbered exercises are given in Appendix VI.

15. Identify the compounds represented by the following letters: A, AMP, dADP, UTP.

16. Identify the compounds represented by the following letters: G, GMP, dGDP, CTP.

17. Write structural formulas for the substances represented by
(a) A   (c) CDP
(b) AMP   (d) dGMP

18. Write structural formulas for the substances represented by
(a) U   (c) CTP
(b) UMP   (d) dTMP

19. Draw the structure for a three-nucleotide segment of single-stranded DNA using the bases C, T, and A.

20. Draw the structure for a three-nucleotide segment of RNA using the bases G, U, and A.

21. Use structural formulas to show the hydrogen bonding between adenine and uracil.

22. Use structural formulas to show the hydrogen bonding between guanine and cytosine.

23. What is the role of RNA in the genetic process?

24. What is the role of DNA in the genetic process?

25. Differentiate between replication and transcription.

26. Differentiate between transcription and translation.

27. How does the function of tRNA differ from that of mRNA?

28. How does the function of mRNA differ from that of rRNA?

29. There are 146 amino acid residues in the β-polypeptide chain of hemoglobin. How many nucleotides are needed to code for the amino acids in this chain?

30. There are 573 amino acid residues in the phosphoglycerate kinase enzyme. How many nucleotides are needed to code for the amino acids in this chain?

31. A segment of a DNA strand consists of TCAATACCCGCG.
(a) What is the nucleotide order in the complementary mRNA?
(b) What is the anticodon order in the tRNA?
(c) What is the sequence of amino acids coded by the DNA?

32. A segment of a DNA strand consists of GCTTAGACCTGA.
(a) What is the nucleotide order in the complementary mRNA?
(b) What is the anticodon order in the tRNA?
(c) What is the sequence of amino acids coded by the DNA?

33. Describe the bond (e.g., the atoms involved, the functional groups that react) that forms during transcription to link nucleotides together.

34. Describe the bond (e.g., the atoms involved, the functional groups that react) that forms during translation to link amino acids together.

35. Briefly describe the events that occur during translation termination.

36. Briefly describe the events that occur during translation initiation.

37. What is the tRNA anticodon if the codon in mRNA is as follows?
    (a) GUC   (c) UUU
    (b) AGG   (d) CCA

38. What is the tRNA anticodon if the codon in mRNA is as follows?
    (a) CGC   (c) GAU
    (b) ACA   (d) UUC

Additional Exercises

These exercises are not paired and provide additional practice on concepts covered in this chapter.

39. Explain why the ratio of thymine to adenine in DNA is 1 : 1, but the ratio of thymine to guanine is not necessarily 1 : 1.

40. In RNA does the guanine content have to be equal to the cytosine content? Explain. Do they have to be equal in DNA? Explain.

41. Mutations in DNA can occur when one base is substituted for another or when an extra base is inserted into the DNA strand. Explain why an insertion mutation is apt to be more harmful than a substitution mutation.

42. The sequence of bases in a segment of mRNA is UUUCAUAAG. Answer the following questions:
    (a) What amino acids would be coded for by this segment of mRNA?
    (b) What is the sequence of DNA that would produce the given segment of mRNA?

43. Describe the use of DNA polymerase in DNA finger-printing.

44. What is the Human Genome Project?

PUTTING IT TOGETHER
Review for Chapters 29–31

Multiple Choice: *Choose the correct answer for each of the following.*

1. Which of the following describes this structure?

    $$H-\underset{\underset{NH_2}{|}}{\overset{\overset{H}{|}}{C}}-COOH$$

    (a) an α-amino acid
    (b) an amino acid with one chiral center
    (c) an α-hydroxy acid
    (d) a β-hydroxy acid

2. How is this structure classified?

    $$CH_3-\underset{\underset{NH_2}{|}}{\overset{\overset{H}{|}}{C}}-COOH$$

    (a) as a basic amino acid
    (b) as an acidic hydroxy acid
    (c) as a neutral amino acid
    (d) no correct answer given

3. How is this structure classified?

    $$HOOC-CH_2-\underset{\underset{NH_2}{|}}{\overset{\overset{H}{|}}{C}}-COOH$$

    (a) as a basic hydroxy acid
    (b) as an acidic amino acid
    (c) as a basic amino acid
    (d) as an acidic hydroxy acid

4. A complete protein provides
    (a) all essential enzymes
    (b) all essential nutrients
    (c) all essential amino acids
    (d) all neutral amino acids

5. What does this structure represent?

    $$H-\underset{\underset{CH_3}{|}}{\overset{\overset{COOH}{|}}{C}}-NH_2$$

    (a) a biologically common amino acid
    (b) a D-amino acid
    (c) an L-amino acid
    (d) a β-amino acid

6. What does this structure represent?

    $$H-\underset{\underset{H}{|}}{\overset{\overset{COOH}{|}}{C}}-NH_2$$

    (a) a D-amino acid
    (b) an L-amino acid
    (c) an α-amino acid
    (d) no correct answer given

7. Under strongly basic conditions, the following molecule will form an ion of

    $$CH_3-\underset{}{\overset{\overset{OH}{|}}{C}}H-\underset{\underset{NH_2}{|}}{\overset{\overset{H}{|}}{C}}-COOH$$

    (a) 0 charge
    (b) −2 charge
    (c) +1 charge
    (d) −1 charge

8. In an electrophoresis cell, to which electrode will the molecule shown migrate at its isoelectric point?

    (a) the positive electrode
    (b) neither electrode
    (c) the negative electrode
    (d) both electrodes

9. Under strongly acidic conditions, which electrode(s) does this molecule migrate toward?

    $$H_2NCH_2CH_2CH_2CH_2-\underset{\underset{NH_2}{|}}{\overset{\overset{H}{|}}{C}}-COOH$$

    (a) the positive electrode
    (b) neither electrode
    (c) the negative electrode
    (d) both electrodes

10. Using Table 29.1, choose the correct abbreviation for the amino acids in the structure shown.

    (a) gly-ala-ser
    (b) ala-thr-gly
    (c) gly-thr-ser-asp
    (d) ser-cys-ala

11. Using Table 29.1, choose the correct abbreviation for the amino acids in the structure shown.

    (a) gly-asp-thr-cys
    (b) gly-ala-ser-glu
    (c) asp-ala-ser-cys
    (d) ser-cys-ala-pro

12. The structure in Question 10 is an example of
    (a) a secondary structure   (c) a primary structure
    (b) a tertiary structure   (d) a quaternary structure

13. The structure in Question 11 is an example of
    (a) a tripeptide   (c) a pentapeptide
    (b) a dipeptide   (d) no correct answer given

14. When a polypeptide is formed from seven amino acids molecules, how many molecules of water are formed?
    (a) seven   (c) six
    (b) five   (d) none

15. Which of the following is an example of a pain-producing polypeptide?
    (a) bradykinin   (c) oxytocin
    (b) substance P   (d) leu-enkephalin

16. Which of the following polypeptides is used to induce labor in childbirth?
    (a) bradykinin   (c) oxytocin
    (b) substance P   (d) leu-enkephalin

17. Which of the following protein structures is held together by hydrogen bonds between the amide "N—H" and the amide "C=O"?
    (a) primary structure   (c) secondary structure
    (b) tertiary structure   (d) quaternary structure

18. A high percentage of the secondary structure, the β-pleated sheet, is found in
    (a) myoglobin   (c) hemoglobin
    (b) collagen   (d) fibroin

19. A high percentage of α-helix is found in
    (a) carboxypeptidase A   (c) α-keratin
    (c) fibroin   (d) no correct answer given

20. When two cysteine amino acid side chains form a bond, it is called
    (a) an ionic bond   (c) a hydrophobic bond
    (b) a hydrogen bond   (d) a disulfide bond

21. Bonds between amino acid side chains commonly hold together
    (a) tertiary structures   (c) primary structures
    (b) secondary structures   (d) delta structures

22. Myoglobin is an example of
    (a) a fibrous protein
    (b) a β-pleated sheet containing protein
    (c) a conjugated protein
    (d) an enzyme

23. Hemoglobin is said to be "cooperative" because
    (a) four O_2 bind, each successive binding being more difficult
    (b) four O_2 bind, each successive binding being easier
    (c) one O_2 binds and hemoglobin works with other proteins
    (d) eight O_2 bind, each successive binding being easier

24. Which of the following proteins is *not* classified as a binding protein?
    (a) myoglobin   (c) immunoglobulin G
    (b) hemoglobin   (d) carboxypeptidase A

25. In electrophoresis, when proteins are separated by moving through a pH gradient, the process is called
    (a) SDS electrophoresis
    (b) isoelectric focusing
    (c) thin-layer chromatography
    (d) high-performance liquid chromatography

26. Denaturation of proteins does *not* cause a loss of
    (a) primary structure   (c) tertiary structure
    (b) quaternary structure   (d) secondary structure

27. Enzymes increase the rate of a reaction by
    (a) decreasing the reaction temperature
    (b) decreasing the activation energy
    (c) increasing the activation energy
    (d) increasing the reactant concentration

28. An enzyme that is also a conjugated protein can be called
    (a) a holoenzyme   (c) a coenzyme
    (b) an activator enzyme (d) an apoenzyme

29. The enzymes that digest proteins are classed as
    (a) lyases   (c) ligases
    (b) hydrolases   (d) transferases

30. When an enzyme lowers the activation energy of the transition state,
    (a) the reaction rate also decreases
    (b) the substrate concentration increases
    (c) the product concentration also decreases
    (d) no correct answer given

31. An enzyme reaction that results in a large turnover number means
    (a) the reaction rate is small
    (b) the reaction temperature is low
    (c) the enzyme is an efficient catalyst
    (d) the substrate concentration is low

32. In commercial products, proteases are *not* used to
    (a) soften denim   (c) dissolve blood clots
    (b) remove clothing stains (d) no correct answer given

33. Statin drugs that lower blood cholesterol levels can be classified as
    (a) proteases   (c) enzyme inhibitors
    (b) enzyme denaturants   (d) hydrolases

34. Industrial detergents depend upon enzymes that are stable at
    (a) low pH and high temperatures
    (b) low pH and low temperatures
    (c) high pH and low temperatures
    (d) high pH and high temperatures

35. Nucleic acids
- **(a)** do not contain deoxyribose
- **(b)** do not contain uracil
- **(c)** are polymers
- **(d)** are also called nucleotides

36. What is the name for this compound?

- **(a)** adenine   **(c)** thymine
- **(b)** uracil   **(d)** cytosine

37. What is the name for this compound?

- **(a)** cytidine   **(c)** thymine
- **(b)** adenine   **(d)** no correct answer given

38. Thymidine 5′-monophosphate is an example of a
- **(a)** nucleoside   **(c)** nucleotide
- **(b)** purine base   **(d)** pyrimidine base

39. Which of the following is the structure for deoxyguanosine?

(a)

(b)

(c)

(d)

40. Which of the following is the abbreviation for this compound?

- **(a)** UMP   **(c)** CMP
- **(b)** dCMP   **(d)** dTMP

41. What is the name for this compound?

(a) adenosine-5′-diphosphate
(b) adenosine-4′,5′-diphosphate
(c) adenosine-3′,5′-diphosphate
(d) adenidine-3′,5′-diphosphate

42. ATP is a high-energy
(a) nucleoside   (c) nucleotide
(b) nucleic acid   (d) purine base

43. Which of the following is *not* a complementary base pair?
(a) A-G   (c) A-U
(b) A-T   (d) C-G

44. Semiconservative replication means
(a) RNA is synthesized from DNA
(b) each new DNA has one new strand and one template strand
(c) some DNA is lost during each replication
(d) no correct answer given

45. Which of the following scientist(s) discovered the basic laws of heredity?
(a) Darwin   (c) Watson and Crick
(b) Mendel   (d) Miescher

46. The ribonucleic acid that carries an amino acid to a ribosome is
(a) a transfer RNA   (c) a ribosomal RNA
(b) a messenger RNA   (d) a nuclear RNA

47. When transcription takes place,
(a) a U codes for a T   (c) a T codes for a U
(b) an A codes for a T   (d) an A codes for a U

48. The biosynthesis of proteins is termed
(a) replication   (c) initiation
(b) translation   (d) transcription

49. A key role in genetic engineering is played by
(a) restriction endonucleases
(b) 3′ to 5′ exonucleases
(c) retardant exonucleases
(d) 5′ to 3′ exonucleases

50. A nonsense codon is an example of a
(a) mutation   (c) termination signal
(b) mutagen   (d) a codon that nature has not used

Free Response Questions

1. Ingestion of ethylene glycol $HOCH_2CH_2OH$ (commonly found in antifreeze) can lead to death. The ethylene glycol undergoes the following enzyme-catalyzed reactions to produce oxalic acid, which is actually the lethal material. What type of enzyme catalyzes the first step of the process?

$$HOCH_2CH_2OH \xrightarrow{\text{Step 1}} HOCH_2C\overset{\displaystyle O}{\overset{\|}{-}}H \rightarrow \rightarrow \rightarrow HOOCCOOH$$

ethylene glycol   oxalic acid

2. Photosynthesis involves the conversion of CO_2 and H_2O in the presence of sunlight to carbohydrates (primarily glucose) and oxygen. The following transformations are part of the process occurring during the dark reactions of photosynthesis. What type of enzyme is needed to catalyze each of the reactions given below?

sedoheptulose-1,7-bisphosphate   sedoheptulose-7-phosphate

xylulose-5-phosphate   ribulose-5-phosphate   ribulose-1,5-bisphosphate

3. Name and draw the peptide resulting from the following codon sequence. UCAAUGCCGUAG

4. As the name implies, globular proteins tend to have a roundish, globular shape. What types of side chains might you expect to find on the amino acid residues that face the outside of a globular protein if it is primarily found in an aqueous environment in the body?

5. If a peptide were synthesized according to the codon below, how many carboxylate groups would be present?
AUCGAUCAGGCAGAGUAA

6. Carboxypeptidase A is a digestive enzyme that cleaves amino acids from a peptide chain from the C-terminal end. If the following peptide were being digested, what type of reaction would take place if carboxypeptidase A were to act upon it? Draw and name the residue that would be the first product of the reaction.

$$
\overset{+}{H_3}N-CH-\overset{\overset{O}{\|}}{C}-NH-CH-\overset{\overset{O}{\|}}{C}-NH-CH-\overset{\overset{O}{\|}}{C}-NH-CH-\overset{\overset{O}{\|}}{C}-O^-
$$

with side chains:
- first residue: $CH_2-CH_2-C(=O)-NH_2$
- second residue: $CHOH-CH_3$
- third residue: CH_2-SH
- fourth residue: CH_2—(phenol ring with OH)

7. Suggest a reason why digestive enzymes that work in the stomach would not be efficient in an environment similar to that of saliva in the mouth.

8. Is the dipeptide aspartylphenylalanine the same as phenylalanylaspartic acid? Will either (or both) give a positive xanthoproteic test? Give a possible codon sequence for aspartylphenylalanine.

9. In the biosynthesis of proteins, amino acids are coupled to tRNAs to form aminoacyl–tRNA complexes using ATP. This is actually a two-step process where the first step is the formation of an aminoacyl–adenylate as shown below. In the aminoacyl–adenylate, identify the following portions of the molecule: the ribose sugar, the nucleic acid base, the amino acid (and state which it is), and the phosphate group.

$$
H_2N-CH-\overset{\overset{O}{\|}}{C}-O-\overset{\overset{O}{\|}}{\underset{\underset{O^-}{|}}{P}}-O-CH_2
$$

(with indole-containing side chain H_2C- attached to the amino acid, a ribose sugar bearing two OH groups, and an adenine base with NH_2 substituent)

CHAPTER

32

Nutrition

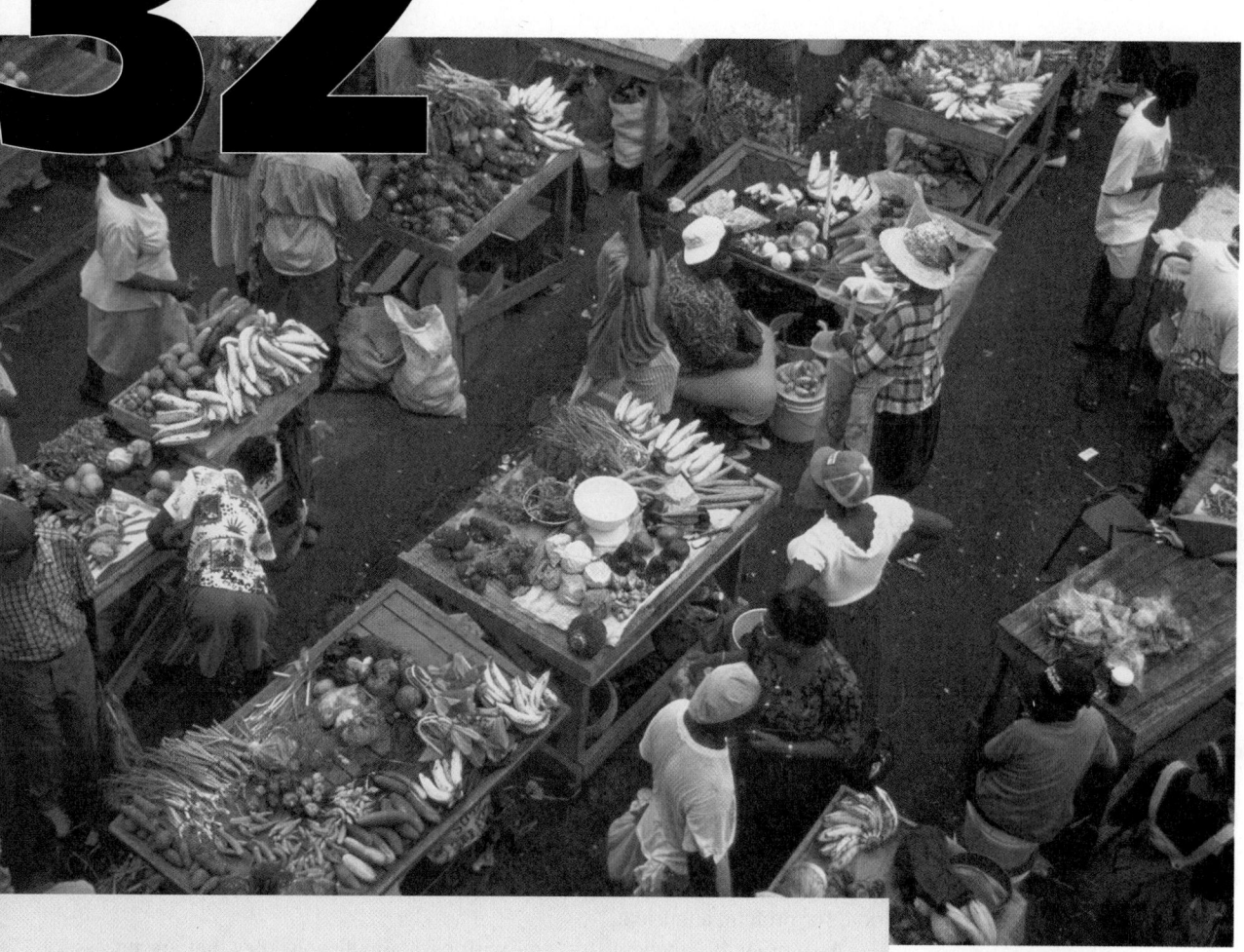

▲ A healthy assortment of fresh vegetables is necessary to provide adequate nutrition.

CHAPTER 32 / OUTLINE

Jack Sprat could eat no fat,
his wife could eat no lean,
yet twixt the two of them,
they licked the platter clean.

—MOTHER GOOSE

For hundreds of years we have known that diet is important to good health and that dietary needs may vary from one individual to another. Yet only recently have we been able to understand nutritional needs on a molecular level. In the nursery rhyme, Jack Sprat is clearly on a low-fat diet. Today, we can point to evidence that fat intake (especially saturated fat) is related to heart disease. Fat molecules accumulate on the blood vessel walls, which leads to hardening of the arteries.

Nutrition is a science of practical importance. Every person is concerned with diet. As children we often heard "Finish your milk" and "Please eat your vegetables." Do you remember being told that chocolate causes complexion problems? How many of your friends watch their weight?

An understanding of nutrition and digestion is closely coupled with an understanding of biochemistry. As we learn more about how diet impacts health, we can make better choices concerning which foods to purchase and to eat.

32.1 Nutrients

nutrient

The need to choose the right diet has given rise to the science of nutrition, the study of nutrients—how they are digested, absorbed, metabolized, and excreted. **Nutrients** are components of the food we eat that makes body growth, maintenance, and repair possible. Milk is a food, but the calcium from that milk is a nutrient. We eat meat, a food, for protein, a nutrient.

As you study biochemistry, you will learn about molecules that are necessary for life. In future chapters, you will learn how the cell uses and produces these molecules. Unfortunately, our cells are not self-sufficient. They require a constant input of energy, a source of carbon and nitrogen, as well as a variety of minerals and special molecules. Cells can synthesize many molecules, but they need starting materials. Nutrients, when digested and absorbed, provide the building blocks that enable cells to make carbohydrates, lipids, proteins, and nucleic acids, as well as provide the energy we need to live.

Nutritionists divide nutrients into six broad classes: (1) carbohydrates, (2) lipids, (3) proteins, (4) vitamins, (5) minerals, and (6) water. The first three classes—carbohydrates, lipids, and proteins—are the major sources of the building materials, replacement parts, and energy needs of the cells. These nutrients are used in rela-

macronutrient

tively large amounts and are known as **macronutrients.** The next three classes of nutrients—vitamins, minerals, and water—have functions other than as sources of energy or building materials. Vitamins are components of some enzyme systems that are vital to the cells. Minerals are required to maintain specific concentrations of certain inorganic ions in the cellular and extracellular fluids. They are also utilized in a variety of other ways—for example, calcium and phosphorus in bones and teeth and iron in hemoglobin. Needed in relatively small amounts, vitamins and minerals are termed **micronutrients.** Water is absolutely essential to every diet, because most biochemical reactions occur in aqueous solution.

micronutrient

Nutrients supply the needs of the body's many different cells. For example, red blood cells must produce hemoglobin and thus need iron. Muscle cells produce muscle fibers and need much protein. As they do their jobs, cells wear out and must be replaced. Red blood cells are renewed on average every 6 weeks, whereas the cells lining the digestive tract must be replaced every 3 days.

The six classes of nutrients provide the basis for a healthy diet. This chapter will briefly describe (1) the relationship between diet and nutrition, (2) some characteristics of each nutrient category, and (3) the processes by which some nutrients are digested.

32.2 Diet

Diet is the food and drink that we consume. The foods we consume determine the nutrients available to our bodies. Thus our health depends directly on our diet.

diet

Unfortunately, some dietary choices are not clear. Controversy surrounds many food selections. Are "fast foods" unhealthy? Should we avoid food additives? What are "natural" foods? To eat intelligently, we must understand the role of the food nutrients in maintaining good health.

Nutritionists have the difficult problem of deciding the kinds of nutrients that should be in a diet. They often establish dietary need by correlating physical well-being with nutrient consumption. James Lind (1716–1794), a physician in the British Navy during the 18th century, was one of the first to use this approach. In a study of scurvy, a disease that afflicted sailors on long voyages, he placed seamen who suffered from scurvy on various diets, some of which contained citrus fruits. By observing changes in the conditions of the seamen, Lind was able to conclude that citrus fruits provide a nutrient that prevents scurvy. This and later work eventually led to the requirement of limes and lemons in the diets of the British Merchant Marine (1765) and the British Navy (1795) (hence the origin of the name "limey" for British sailor). Scurvy is now recognized as a deficiency disease caused by a lack of vitamin C, ascorbic acid.

As scientists like Lind discovered and studied more nutrients, they found that the minimum quantities needed for good health vary from person to person. Thus, rather than establish uniform minimum requirements, nutritionists developed a standard called the recommended dietary allowance. The **recommended dietary allowance (RDA)** for a nutrient is an average value that has been shown to maintain health for large groups of people. Although the dietary needs of individuals vary considerably, RDAs are useful yardsticks for judging healthful diets. Table 32.1 shows the RDAs for a variety of nutrients.

recommended dietary allowance (RDA)

Practice 32.1

Referring to Table 32.1,
(a) What is a young woman's iron RDA in grams?
(b) Which vitamin has the smallest RDA?

TABLE 32.1   **Recommended Daily Dietary Allowances[a]**

	Age (yrs)	Weight		Height		Protein (g)	Fat-soluble vitamins		
		(kg)	(lb)	(cm)	(in.)		Vita-min A (μg RE)[b]	Vita-min D (μg)[c]	Vita-min E (mg α-TE)[d]
Infants	0.0–0.5	6	13	61	24	13	375	5	3
	0.5–1.0	9	20	71	28	14	375	5	4
Children	1–3	13	29	90	35	16	400	5	6
	4–6	20	44	112	44	24	500	5	7
	7–10	28	62	132	52	28	700	5	7
Males	11–14	45	99	157	62	45	1000	5	10
	15–18	66	145	176	69	59	1000	5	10
	19–24	72	160	177	70	58	1000	5	10
	25–50	79	174	176	70	63	1000	5	10
	51+	77	170	173	68	63	1000	10	10
Females	11–14	46	101	157	62	46	800	5	8
	15–18	55	120	163	64	44	800	5	8
	19–24	58	128	164	65	46	800	5	8
	25–50	63	138	163	64	50	800	5	8
	51+	65	143	160	63	50	800	10	8
Pregnant						60	800	10	10
Lactating									
1st 6 mo						65	1300	10	12
Next 3 mo						62	1200	10	11

Reference: *Recommended Dietary Allowances,* Food and Nutrition Board, National Research Council–National Academy of Sciencies.
[a]The allowances are intended to provide for individual variations among most normal persons as they live in the United States under usual environmental stresses. Diets should be based on a variety of common foods in order to provide other nutrients for which human requirements have been less well defined. These values are currently being revised as a joint Canadian–U.S.A. project. New values will be termed Dietary Reference Intakes (DRI).
[b]Retinol equivalents; 1 retinol equivalent = 1 μg retinol or 6 μg β-carotene.
[c]As cholecalciferol; 10 μg cholecalciferol = 400 IU of vitamin D. Given as DRI (1997).
[d]α-Tocopherol equivalents; 1 mg α-tocopherol = 1 α-TE.
[e]Given as DRI (1997).

 32.3   Energy in the Diet

An important component of every diet is the energy allowance, which derives primarily from the energy-containing nutrients, the carbohydrates, lipids, and proteins. These molecules are a rich source of reduced carbons. As we will see in Chapter 33, almost all of the energy for life is derived from reactions in which oxidation of carbon compounds occurs in the cells.

The dietary energy allowance varies with activity, body size, age, and sex. Thus a 65-kg male office worker requires a diet that furnishes 2700 kcal/day, whereas a man of similar weight working as a carpenter requires about 3000 kcal/day. Women generally use less energy than men, and energy use decreases with age. Nutritionists express energy in units of kilocalories (kcal) or its equivalent, the large Calorie (Cal). Remember, there are 4.184 kJ per kilocalorie (or Calorie).

TABLE 32.1	Recommended Daily Dietary Allowances[a] Continued

	Water-soluble vitamins							Minerals						
	Vitamin C (mg)	Thiamin (mg)	Riboflavin (mg)	Niacin (mg)	Vitamin B$_6$ (mg)	Folic acid (μg)	Vitamin B$_{12}$ (μg)	Ca (mg)[e]	P (mg)[e]	Mg (mg)[e]	Fe (mg)	Zn (mg)	I (mg)	Se (μg)
Infants	30	0.3	0.4	5	0.3	25	0.3	210	100	30	6	5	40	10
	35	0.4	0.5	6	0.6	35	0.5	220	275	75	10	5	50	15
Children	40	0.7	0.8	9	1.0	50	0.7	500	460	80	10	10	70	20
	45	0.9	1.1	12	1.1	75	1.0	800	500	130	10	10	90	20
	45	1.0	1.2	13	1.4	100	1.4	800	500	130	10	10	120	30
Males	50	1.3	1.5	17	1.7	150	2.0	1300	1250	240	12	15	150	40
	60	1.5	1.8	20	2.0	200	2.0	1300	1250	410	12	15	150	40
	60	1.5	1.7	19	2.0	200	2.0	1000	700	400	10	15	150	70
	60	1.5	1.7	19	2.0	200	2.0	1000	700	420	10	15	150	70
	60	1.2	1.4	15	2.0	200	2.0	1200	700	420	10	15	150	70
Females	50	1.1	1.3	15	1.4	150	2.0	1300	1250	240	15	12	150	45
	60	1.1	1.3	15	1.5	180	2.0	1300	1250	360	15	12	150	50
	60	1.1	1.3	15	1.6	180	2.0	1000	700	310	15	12	150	55
	60	1.1	1.3	15	1.6	180	2.0	1000	700	320	15	12	150	55
	60	1.0	1.2	13	1.6	180	2.0	1200	700	320	10	12	150	55
Pregnant	70	1.5	1.6	17	2.2	400	2.2	1200	1200	350	30	15	175	65
Lactating														
1st 6 mo	95	1.6	1.8	20	2.1	280	2.6	1200	1200	320	15	19	200	75
Next 3 mo	90	1.6	1.7	20	2.1	260	2.6	1200	1200	240	15	16	200	75

The balance between energy needs and the energy allowance is of vital importance. Calorie deficiency leads to a condition called *marasmus,* which affects many of the world's poor, particularly children. Marasmus is a wasting disease due to starvation. People suffering from this disease have limited diets that often consist of bulky, carbohydrate-containing foods—for example, sweet potatoes (32% carbohydrate, 1% fat, 2% protein). These foods are not energy-rich (about 160 kcal for a medium-size sweet potato). Children, with their small stomachs, have difficulty consuming enough to satisfy their energy requirements. It is estimated that 15–20% of the people in underdeveloped countries suffer from malnutrition due to insufficient calories.

In contrast, many people in developed countries consume far more calories than they need for health and well-being. Food is abundant, much of it rich in energy. For example, a fast-food lunch might consist of a hamburger (560 kcal), french fries (220 kcal), and a chocolate shake (380 kcal), which would supply nearly one-half of the

Food for Thought: For the first time in history, a large proportion of the world's population, mainly in developed countries, have an overabundance of food.

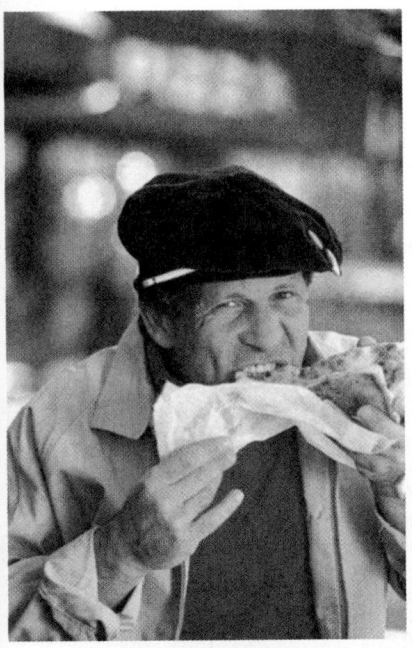

daily energy allowance for an adult male. Excess accumulated calories lead to a condition called *obesity,* which is characterized by an overabundance of fatty tissue and by many attendant health problems.

Each pound of body fat contains about 3500 kcal of energy, or between 1 and 2 days' total RDA for energy. Because fat contains so much energy, it is normally accumulated only slowly. Unfortunately, for the same reason, fat is also very difficult to lose. To lose 1 lb of fat, a 65-kg man would have to swim nonstop for about 10 hours, play tennis continuously for about 8 hours, or run for 4 hours. Exercise alone is generally not sufficient to cure obesity; a change in diet is also necessary.

Thus many people in the affluent, developed countries find it necessary to choose a restricted-calorie diet to lose weight. The difficulties associated with selecting a restricted-calorie diet have led to the creation of numerous fad diets that fail to provide sound nutrition. Nutritionists counsel that successful and nutritionally sound dieting can be accomplished by following these simple guidelines: (1) Reduce calorie intake by only a moderate amount (a 500-kcal/day reduction causes a loss of about 1 lb of fat per week) and be prepared to continue the diet for a long time; (2) carefully select foods so that the diet contains adequate amounts of all nutrients. The primary function of any diet is to maintain good health.

Food for Thought: Carbohydrates are the most versatile of energy nutrients. Humans have more different metabolic options for carbohydrates than for any other energy nutrient.

32.4   Carbohydrates in the Diet

Carbohydrates are the major nutrient in most human diets (see Figure 32.1). They are justly described as macronutrients because they are a major source of usable, reduced carbon atoms and, therefore, of dietary energy. About half of the average daily calorie requirement derives from carbohydrates. These molecules are perhaps the most easily metabolized (see Chapter 34) of the energy-supplying nutrients and can be used for energy under both aerobic and anaerobic conditions. Additionally, carbons from carbohydrates are used to build other cellular molecules—amino acids, nucleic acids, as well as other carbohydrates. Excess dietary carbohydrate is most often converted to fat.

Dietary carbohydrates are primarily the polysaccharides starch and cellulose, the disaccharides lactose and sucrose, and the monosaccharides glucose and fructose. Seeds are the most common source of starch, grains being about 70% by mass starch, while dried peas and beans contain about 40% starch. A second major source of starch is tuber and root crops such as potatoes, yams, and cassava. The disaccharide lactose is an important component of milk, and sucrose is usually consumed as refined sugar (derived from sugar beets or sugar cane). The monosaccharides are often found in fruits.

The polysaccharides, also called *complex carbohydrates,* are difficult to digest because of their complex structures (see Section 27.15). Starch is digested slowly, enabling the body to control distribution of this energy nutrient. Cellulose is not digested by humans; however, it is a major source of *dietary fiber.* As cellulose passes through the digestive tract, it absorbs water and provides dietary bulk. This bulk acts to prevent constipation and diverticulosis, a weakening of the intestinal walls. Many nutritionists recommend a daily fiber intake of 15–30 g, supplied by such foods as whole wheat bread (1.8 g/slice) and bran cereals (7.5 g/0.5 cup).

Although no RDA has been set for carbohydrate, many nutritionists recommend that a minimum of about 1000 kcal/day be derived from this source. In the American

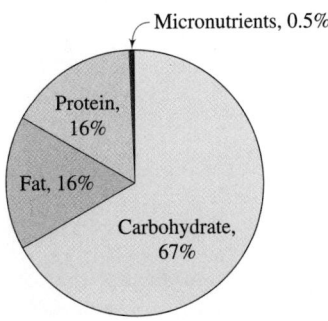

Micronutrients, 0.5%

Protein, 16%

Fat, 16%

Carbohydrate, 67%

▲
**FIGURE 32.1
Relative masses of nutrients in a typical diet.**

diet, most of these calories are derived from starch, although in recent years an increasing amount has come from sucrose.

The high sucrose content of many modern diets is due primarily to the large amounts of sucrose in commercially prepared foods. In 1909, an American consumed an average of about 84 lb of sucrose annually in prepared foods and beverages; by 1994, this figure had risen to about 150 lb annually (almost 40% of all carbohydrate consumed). This large increase was caused mainly by (1) increased consumption of prepared foods and (2) an increased percentage of sucrose added to these foods by the manufacturers in attempts to gain larger shares of the market. (It is well known that many people, especially children, have a preference for sweet foods.) Although the consumption of other sweeteners such as high-fructose corn syrup and aspartame has risen, the consumption of sucrose remains high.

The increase in sucrose consumption troubles nutritionists for several reasons. First, sucrose is a prime factor in the incidence of dental caries. Because it is readily used by oral bacteria, sucrose promotes growth of the microorganisms that cause tooth decay. Second, ingested sucrose is rapidly hydrolyzed to monosaccharides, which are promptly absorbed from the intestine, leading to a rapid increase in blood-sugar levels. Wide variations in the blood-sugar level can stress the body's hormonal system. Finally, sucrose is said to provide "empty calories." This means that sucrose supplies metabolic energy (calories) but lacks other nutrients. Nutritionists usually recommend starch over sucrose as a major source of dietary carbohydrate.

Fats in the Diet

Fat is a more concentrated source of dietary energy than carbohydrate. Not only does fat contain more carbons per unit mass, but the carbons in fats are more reduced. As an energy source, fats provide about 9 kcal/g, whereas carbohydrates provide only about 4 kcal/g. As we will see in Chapter 35, energy can only be obtained from fat when oxygen is present; fat metabolism is strictly aerobic in humans. The carbons from fats can be used to synthesize amino acids, nucleic acids, and other fats, but our bodies cannot achieve a net synthesis of carbohydrate from fat. Thus fats tend not to be as versatile a nutrient as carbohydrates.

Fats contribute much of the dietary energy of many foods. For example, french fried potatoes contain about 18% fat by weight, yet this fat provides about 40% of the calories in this food. A cup of whole milk contains 170 kcal, while a cup of skim (nonfat) milk contains 80 kcal. Many nutritionists counsel that the best way to reduce calorie intake is to eat foods containing less fat. Although fats are macronutrients, it is recommended that fat intake should not exceed 25–30% of the daily energy allowance.

In a diet, both the kind and the amount of fat are important. Fatty acids from meat and dairy products are relatively saturated, whereas those from plant sources are generally more unsaturated. Because there is a probable link between high consumption of saturated fats and atherosclerosis, the U.S. Department of Agriculture and other agencies concerned with nutrition recommend that American diets should contain about equal portions of polyunsaturated and saturated fatty acids. Polyunsaturated fats in the diet can be increased by (1) cooking with vegetable oils such as corn or canola oil and (2) using soft margarine, which usually contains more unsaturated fats than hard or stick margarine or butter.

Food for Thought: Of the three essential fatty acids, linolenic acid has the added benefit of being an ω-3 fatty acid (see Chapter 28).

▲
**Kwashiorkor (protein defi-
ciency) is a serious problem
for children in many poverty-
ridden areas of the world.**

**Food for Thought:
Kwashiorkor, as with many
nutritional deficiencies, can
occur even with more than
enough food to eat.**

complete protein

Polyunsaturated fats generally also contain the three essential fatty acids—linoleic acid, linolenic acid, and arachidonic acid:

$$CH_3(CH_2)_4CH=CHCH_2CH=CH(CH_2)_7CO_2H$$
linoleic acid

$$CH_3CH_2CH=CHCH_2CH=CHCH_2CH=CH(CH_2)_7CO_2H$$
linolenic acid

$$CH_3(CH_2)_4CH=CHCH_2CH=CHCH_2CH=CHCH_2CH=CH(CH_2)_3CO_2H$$
arachidonic acid

These unsaturated fatty acids have been shown to relieve the deleterious physiological changes, such as poor growth, skin lesions, kidney damage, and impaired fertility, that result from a totally fat-free diet. One essential biochemical function of these fatty acids is as precursors for prostaglandin synthesis.

32.6   Proteins in the Diet

The third macronutrient is protein, with an average energy yield of about 4.2 kcal/g. This energy is derived primarily from reduced carbons. However, recall that proteins also contain about 16% nitrogen; in fact, protein is the primary source of nitrogen in our diets. As discussed in Chapter 35, perhaps the most important function of dietary protein is to provide for synthesis of nitrogen-containing molecules such as nucleic acids, enzymes and other proteins, nerve transmitters, and many hormones. Because these materials are critical for human growth, protein malnutrition, or *kwashiorkor,* is a particularly serious problem. Unlike carbohydrates and fats (excluding the essential fatty acids), a specific RDA has been established for dietary protein (see Table 32.1).

Kwashiorkor can occur even when calorie intake is sufficient and thus is an especially insidious form of malnutrition. In many poverty-ridden areas of the world, the only reliable source of protein for children is mother's milk. After weaning, the children eat a protein-poor grain diet. Although the calorie intake is sufficient, these children show the stunted growth, poor disease resistance, and general body wasting that are characteristic of kwashiorkor.

Proteins are obtained from animal sources such as meat, milk, cheese, and eggs, and from plant sources such as cereals, nuts, and legumes (peas, beans, and soybeans). Animal proteins have nutritive values that are generally superior to those of vegetable proteins in that they supply all of the 20 amino acids that the body uses. In contrast, a single vegetable source often lacks several amino acids. This deficiency is critical, because humans cannot synthesize a group of amino acids called the essential amino acids (Table 32.2). These ten *essential amino acids* must be obtained from the diet.

Nutritionists often speak of a dietary source (or sources) of complete protein. A **complete protein** is one that supplies all of the essential amino acids. Animal products are, in general, sources of complete proteins. However, by either choice or necessity, animal protein is seldom consumed by a significant fraction of the world's population. Besides the moral, ethical, and religious reasons given for limiting consumption of animal protein, these proteins are relatively expensive sources of amino

TABLE 32.2	Common Dietary Amino Acids		
Essential		**Nonessential**	
Arginine*	Methionine	Alanine	Glutamine
Histidine*	Phenylalanine	Asparagine	Glycine
Isoleucine	Threonine	Aspartic acid	Proline
Leucine	Tryptophan	Cysteine	Serine
Lysine	Valine	Glutamic acid	Tyrosine

*Essential for growing children

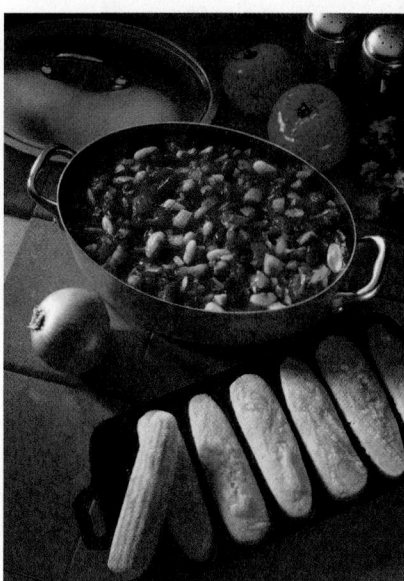

acids. Thus most people in underdeveloped countries subsist primarily on vegetable protein. A constant danger inherent in a vegetarian diet is that the source of vegetable protein may be deficient in several of the essential amino acids; that is, the protein may be incomplete. For this reason nutritionists recommend that several sources of vegetable protein be included with each meal in a vegetarian diet. As an example, soybeans, which are rich in lysine, might supplement wheat, which is lysine deficient.

▲
In a vegetarian diet it is important to include several sources of protein, such as beans and rice, to produce a complete protein.

Vitamins

Vitamins are a group of naturally occurring organic compounds that are essential for good nutrition and must be supplied in the diet. Whereas the energy-supplying nutrients are digested and metabolized extensively, vitamins are often used after only minimal modification. Some of the vitamins necessary for humans are listed in Table 32.3. Note that vitamins are often classified according to their solubility, those that are fat soluble and those that are water soluble. The structural formulas of several vitamins are shown in Figure 32.2.

vitamin

A prolonged lack of vitamins in the diet leads to vitamin deficiency diseases such as beriberi, pellagra, pernicious anemia, rickets, and scurvy. Left uncorrected, a vitamin deficiency ultimately results in death. Even when supplementary amounts of vitamins are provided, impaired growth due to a vitamin deficiency may be irreversible. For example, it is difficult to correct the distorted bone structures resulting from a childhood lack of vitamin D. Thus it is especially important that children receive sufficient vitamins for proper growth and development.

Vitamins are required in only small amounts (see the RDAs in Table 32.1) and are classed as micronutrients. For example, a typical diet might include 250 g of carbohydrate (a macronutrient) and only 2 mg of vitamin B_6. However, the biochemistry of life cannot continue without vitamins. Each vitamin serves at least one specific purpose for an organism. The water-soluble compounds are generally involved in cellular metabolism of the energy-supplying nutrients. For example, thiamin is required to achieve a maximum energy yield from carbohydrates, and pyridoxine is of central importance in protein metabolism (see Chapters 34 and 35). Niacin and riboflavin are key components in almost all cellular redox reactions. The fat-soluble

TABLE 32.3	Some of the Most Important Vitamins	

Vitamin	Important dietary sources	Some deficiency symptoms
FAT SOLUBLE		
Vitamin A (Retinol)	Green and yellow vegetables, butter, eggs, nuts, cheese, fish liver oil	Poor teeth and gums, night blindness
Vitamin D (Ergocalciferol, D_2; cholecalciferol, D_3)	Egg yolk, milk, fish liver oils; formed from provitamin in the skin when exposed to sunlight	Rickets (low blood-calcium level, soft bones, distorted skeletal structure)
Vitamin E (α-Tocopherol)	Meat, egg yolk, wheat germ oil, green vegetables; widely distributed in foods	Not definitely known in humans
Vitamin K (Phylloquinone, K_1; menaquinone, K_2	Eggs, liver, green vegetables; produced in the intestines by bacterial reactions	Blood is slow to clot (antihemorrhagic vitamin)
WATER SOLUBLE		
Vitamin B_1 (Thiamin)	Meat, whole-grain cereals, liver, yeast, nuts	Beriberi (nervous system disorders, heart disease, fatigue)
Vitamin B_2 (Riboflavin)	Meat, cheese, eggs, fish, liver	Sores on the tongue and lips, bloodshot eyes, anemia
Vitamin B_6 (Pyridoxine)	Cereals, liver, meat, fresh vegetables	Skin disorders (dermatitis)
Vitamin B_{12} (Cyanocobalamin)	Meat, eggs, liver, milk	Pernicious anemia
Vitamin C (Ascorbic acid)	Citrus fruits, tomatoes, green vegetables	Scurvy (bleeding gums, loose teeth, swollen joints, slow healing of wounds, weight loss)
Niacin (Nicotinic acid and amide)	Meat, yeast, whole wheat	Pellagra (dermatitis, diarrhea, mental disorders)
Biotin (Vitamin H)	Liver, yeast, egg yolk	Skin disorders (dermatitis)
Folic acid	Liver, wheat germ, yeast, green leaves	Macrocytic anemia, gastrointestinal disorders

vitamins often serve very specialized functions. Vitamin D acts as a regulator of calcium metabolism. One function of vitamin A is to furnish the pigment that makes vision possible, while vitamin K enables blood clotting to occur normally.

Because some vitamin functions are not well understood, miraculous properties have been ascribed to these substances, such as vitamins C and E. In the absence of conclusive scientific studies, it is difficult to judge the merits of some of these claims.

Although vitamins are required only in small quantities, their natural availability is also low. Nutritionists caution that a diet should be balanced to include adequate sources of vitamins. As you can see from Table 32.3, the dietary sources of vitamins are varied. In general, fruits, vegetables, and meats are rich sources of the water-soluble vitamins; and eggs, milk products, and liver are good sources of the fat-soluble vitamins.

A seemingly balanced diet may be deficient in vitamins due to losses incurred in food processing, storing, and cooking. As much as 50–60% of the water-soluble

Food for Thought: There is no difference between "artificial vitamins" and "natural vitamins." Since vitamins are well-characterized organic chemicals, chemists can synthesize vitamins of equal potency to the naturally synthesized vitamins.

vitamin A
(retinol)

vitamin B₁
(thiamin)

vitamin K₁
(phylloquinone)

vitamin C
(ascorbic acid)

vitamin E
(α-tocopherol)

vitamin D
(ergocalciferol, D₂)

▲
FIGURE 32.2
The structure of selected vitamins.

vitamins in vegetables can be lost during cooking. Vitamin C can be destroyed by exposure to air. Removal of the outside hull from grains drastically decreases their B vitamin content. Thus, when polished rice became a dietary staple in the Orient, beriberi (the thiamin-deficiency disease) grew to epidemic proportions.

32.8 Minerals

A number of inorganic elements are needed for good health. Like vitamins, minerals are classified as micronutrients. The body does not synthesize minerals. Those that must be ingested in relatively large amounts, the *major elements,* include sodium, potassium, calcium, magnesium, chloride, and phosphate (phosphorus). Elements in a second group, the *trace elements,* are required in much smaller amounts. As scientific studies of the body's mineral requirements proceed, the list of needed trace elements continually lengthens. A recent compilation of these trace elements is given in Table 32.4.

Mineral nutrients differ from organic nutrients in that the body, in general, uses minerals in the ionic form in which they are absorbed. Although these elements are required for good health, they can also be toxic if ingested in quantities that are too large. Nutritionists have established RDAs for some of these minerals and warn against excess intake.

The major minerals—sodium, potassium, and chloride—are responsible for maintaining the appropriate salt levels in body fluids. Many enzyme reactions require an optimal salt concentration of 0.1–0.3 M. In addition, individual elements serve specific functions; nerve transmission, for example, requires a supply of extracellular sodium and intracellular potassium. Although there is no RDA for these two elements, 2–4 g/day represents an average NaCl consumption. But because high levels of NaCl can contribute to high blood pressure, many nutritionists advise using salt in moderation.

Calcium and magnesium serve many roles in the body, including being required by some enzymes. Fully 90% of all body calcium and an important percentage of the magnesium are found in the bones and teeth. Calcium also is needed for nerve transmission and blood clotting.

Trace elements are similar to vitamins in that (1) they are required in small amounts and (2) food contains only minute quantities of them. Usually a normal diet

TABLE 32.4   Required Trace Elements

Element	Function	Human deficiency signs
Fluorine	Structure of teeth, possibly of bones; possible growth effect	Increased incidence of dental caries; possibly risk factor for osteoporosis
Silicon	Calcification; possible function in connective tissue	Not known
Vanadium	Not known	Not known
Chromium	Efficient use of insulin	Relative insulin resistance, impaired glucose tolerance, elevated serum lipids
Manganese	Mucopolysaccharide metabolism, superoxide dismutase enzyme	Not known
Iron	Oxygen and electron transport	Anemia
Cobalt	Part of vitamin B_{12}	Only as vitamin B_{12} deficiency
Nickel	Interaction with iron absorption	Not known
Copper	Oxidative enzymes; interaction with iron; cross-linking of elastin connective protein	Anemia, changes of ossification; possibly elevated serum cholesterol
Zinc	Numerous enzymes involved in energy metabolism and in transcription and translation	Growth depression, sexual immaturity, skin lesions, depression of immunocompetence, change of taste acuity
Arsenic	Not known	Not known
Selenium	Glutathione peroxidase; interaction with heavy metals	Endemic heart problems conditioned by selenium deficiency
Molybdenum	Xanthine, aldehyde, and sulfide oxidase enzymes	Not known
Iodine	Constituent of thyroid hormones	Goiter, depression of thyroid function, cretinism

contains adequate quantities of all trace elements. The exact function of many trace elements remains unknown.

The most notable exception to the preceding generalizations is iron. This element is part of hemoglobin, the oxygen-binding protein of the blood. (Iron is also a critical component of some enzymes.) Relatively large amounts of iron are required for hemoglobin replenishment. Unfortunately, many foods are not rich enough in iron to provide the necessary RDA. This is especially true for the higher iron RDA for women. Therefore, nutritionists sometimes recommend a daily iron dietary supplement.

Practice 32.2

Referring to Table 32.1, why is the largest RDA for calcium linked to the largest RDA for vitamin D?

 Water

▲
Water is a universal nutrient. It has many functions for keeping our bodies healthy.

Water is the solvent of life. As such it carries nutrients to the cells, allows biochemical reactions to proceed, and carries waste from the cells. Water makes up approximately 55–60% of the total body mass.

Our bodies are constantly losing water via urine (700–1400 mL/day), feces (150 mL/day), sweat (500–900 mL/day), and expired air (400 mL/day). If dehydration is to be prevented, this water output must be offset by water intake. In normal diets, liquids provide about 1200–1500 mL of water per day. The remaining increment is derived from food (700–1000 mL/day) and water formed by biochemical reactions (metabolic water; 200–300 mL/day). Water losses vary and depend on the intake and the activity of the individual.

Food for Thought: Water is an important nutrient that is often overlooked; many Americans drink too little water each day.

A proper water balance is maintained by control of water intake and water excretion. The kidney is the center for control of water output. Water intake is controlled by the thirst sensation. We feel thirsty when one or both of the two following events occur: (1) As the body's water stores are depleted, the salt concentration rises. Specific brain cells monitor this salinity and initiate the thirst feeling. (2) As water is drawn away from the salivary glands, the mouth feels dry; this event also leads to the sensation of thirst.

 Nutrition Content Labeling

Nearly all foods in our supermarkets have been processed to some degree. And loss of nutrients is a problem with some kinds of food processing. In general, the more extensive the processing, the greater the loss. Processing can be quite extensive for such items as luncheon meats, TV dinners, and breakfast cereals.

In an effort to aid the consumer, nutritionists have established procedures for monitoring the nutritional values of processed foods. These values are reported on a package panel termed *Nutrition Facts* (see Figure 32.3). Much can be learned about food nutrients by examining such panels. First, some general observations: (1) Many people's daily energy requirement centers around a 2000-kcal (Cal) or a 2500-kcal

NUTRITION FACTS
Serving Size 1 cup (50g)
Servings Per Container about 7

Amount Per Serving	1 Cup Cereal	Cereal with 1 Cup Vitamins A&D Skim Milk
Calories	180	270
Calories from Fat	15	15
		% Daily Value*
Total Fat 1.5g†	2%	2%
Saturated Fat 0g	0%	0%
Polyunsaturated Fat 1g		
Monounsaturated Fat 0g		
Cholesterol 0mg	0%	1%
Sodium 0mg	0%	5%
Total Carbohydrate 41g	14%	18%
Dietary Fiber 6g	26%	26%
Sugars 0g		
Protein 5g		
Vitamin A	**	10%
Vitamin C	**	4%
Calcium	2%	30%
Iron	8%	10%
Vitamin D	**	25%
Thiamin	6%	10%
Riboflavin	4%	20%
Niacin	15%	15%
Phosphorus	20%	45%
Magnesium	15%	20%

†Amount in 1 Cup Cereal. One cup skim milk contributes an additional 90 calories, less than 5mg cholesterol, 125mg sodium, 12g total carbohydrate (12g sugars), and 8g protein.
*Percent Daily Values are based on a 2,000 calorie diet. Your daily values may be higher or lower depending on your calorie needs.
**Contains less than 2% of the daily value of these nutrients.

	Calories:	2000	2500
Total Fat	Less than	65g	80g
Sat Fat	Less than	20g	25g
Cholesterol	Less than	300mg	300mg
Sodium	Less than	2,400mg	2,400mg
Total Carbohydrate		300g	375g
Dietary Fiber		25g	30g

(a) "Shredded wheat" cereal

NUTRITION FACTS
Serving Size ½ cup (48g)
Servings Per Container about 9

Amount Per Serving

Calories 210	
Calories from Fat 70	
	% Daily Value*
Total Fat 8g	12%
Saturated Fat 3.5g	17%
Polyunsaturated Fat 1g	
Monounsaturated Fat 3g	
Cholesterol 0mg	0%
Sodium 15mg	1%
Potassium 210mg	6%
Total Carbohydrate 33g	11%
Other Carbohydrate 18g	
Dietary Fiber 4g	16%
Sugars 12g	
Protein 5g	
Vitamin A	0%
Vitamin C	0%
Calcium	4%
Iron	6%
Thiamin	10%
Phosphorus	10%
Magnesium	10%
Copper	15%

*Percent Daily Values are based on a 2,000 calorie diet. Your daily values may be higher or lower depending on your calorie needs.

	Calories:	2000	2500
Total Fat	Less than	65g	80g
Sat Fat	Less than	20g	25g
Cholesterol	Less than	300mg	300mg
Sodium	Less than	2,400mg	2,400mg
Total Carbohydrate		300g	375g
Dietary Fiber		25g	30g
Potassium		3,500mg	3,500mg

Calories per gram:
Fat 9 ● Carbohydrate 4 ● Protein 4

(b) "Granola" cereal

▲
FIGURE 32.3
Two examples of Nutrition Facts panels.

(Cal) diet; the Nutrition Facts calculations are based on these approximations. (2) Fats are subdivided to recognize the important health difference between saturated and unsaturated fatty acids; cholesterol is given its own category. (3) Carbohydrates are categorized as sugars and fiber with the remainder being complex carbohydrates such as starch. (4) Total protein is always listed. (5) These nutritional facts are based on a standard serving size (set by the federal government).

Figure 32.3 compares the Nutrition Facts for two breakfast cereals. Notice that the serving size is the same (about 50 g) although the volume of one cereal is smaller. While both cereals are primarily carbohydrates, the "granola" cereal contains about 25% (12 g out of 48 g) sugar. In contrast, the "shredded wheat" cereal contains only complex carbohydrates with a slightly higher Percent Daily Value of fiber. Both cereals are relatively low in fat, but the "granola" cereal contains more. Since common grains contain about the same percentage protein, both cereals contain 5 g of protein. The "shredded wheat" cereal manufacturer has opted to include nutritional information after adding 1 cup of skim milk. An additional food item such as milk can significantly change nutrient levels. Both cereals list vitamins and minerals; the list may vary depending on what is included in the food item. Finally, the recommended nutrient amounts for a healthful daily diet are given at the bottom of these nutritional facts (based on either 2000-Cal or 2500-Cal diets).

The Nutrition Facts panel gives us the amount of nutrients suggested for a 2000-kcal or a 2500-kcal diet. The Percent Daily Values are based on these suggestions. For example, the percent total carbohydrate in 48 g of granola cereal is readily calculated for a 2000-kcal diet (Figure 32.3) thus:

$$\frac{33 \text{ g carbohydrate}}{300 \text{ g carbohydrate}} \times 100 = 11\%$$

The nutrient amounts given at the bottom of the Nutrition Facts panel have been chosen with these important guidelines in mind: (1) Carbohydrate should be the bulk of a diet; (2) saturated fat should be a small proportion of the total fat intake; (3) consumption of both sodium and cholesterol should be minimized. We can estimate our nutrient intake and also learn more about nutrition from the Nutrition Facts panel.

Practice 32.3

Referring to the shredded wheat Nutrition Facts in Figure 32.3a, if 6 g of dietary fiber supplies 26% of the Percent Daily Value, how many grams of dietary fiber should be consumed per day?

 ## 32.11 Food Additives

Various chemicals are often added to foods during processing. In fact, more than 3000 of these *food additives* have been given the "generally recognized as safe" (GRAS) rating by the U.S. Food and Drug Administration.

The purpose of these food additives varies. Some additives enhance the nutritional value (e.g., when a food is vitamin enriched). Other additives serve as preservatives. For example, sodium benzoate is used to inhibit bacterial growth, and BHA (butylated hydroxyanisole) and BHT (butylated hydroxytoluene) are used as antioxidants. Still other additives improve the appearance and flavor of a food—for example, emulsifiers, thickeners, anticaking agents, flavors, flavor enhancers, nonsugar sweeteners, and colors (see Table 32.5).

There has been, and continues to be, a great deal of controversy concerning the use of food additives. Due to the nature and complexity of the subject, this debate will no doubt continue. Controversy centers around the problem of balancing the benefits derived from additives against the risk to consumers. The use of at least some additives are necessary to prepare many foods. Discovering whether a risk exists and assessing the degree of risk requires long, difficult, and expensive research. As a case in point, salting and smoking were used in curing meats for centuries before there was any knowledge that these processes might involve a hazard. Research has shown that both processes involve risks to at least some consumers: Salt aggravates certain cardiovascular conditions, and smoke produces carcinogens in the meat. Yet in the eyes of many consumers, these risks are outweighed by the benefits to be had from salted and smoked meats. On the other hand, few consumers indeed would accept a compound known to be very carcinogenic as an additive in their ice cream, even though that compound could improve the flavor of the ice cream remarkably!

Owing to the technical nature of the task, consumers must rely largely on the judgment and integrity of the professional people who have been assigned the responsibility of protecting our food supply. Consumers should also inform themselves as fully as possible concerning the nature, purpose, and possible hazards of the additives that are used or proposed for use in our foods and, as responsible citizens, make sure that our government provides adequate support to the professionals charged with protecting our food.

TABLE 32.5	Some Common Food Additives	
Food additive	**Purpose**
Sodium benzoate Calcium lactate Sorbic acid	Antimicrobials (prevention of food spoilage)
BHA (Butylated hydroxyanisole) BHT (Butylated hydroxytoluene) EDTA (Ethylenediaminetetraacetic acid)	Antioxidants (prevention of changes in color and flavor)
Calcium silicate Silicon dioxide Sodium silicoaluminate	Anticaking agents (keeps powders and salt free-flowing)
Carrageenan Lecithin	Emulsifiers (aids even distribution of suspended particles)
Pectin Propylene glycol	Thickeners (imparts body and texture)
MSG (Monosodium glutamate) Hydrolyzed vegetable protein	Flavor enhancers (supplements or modifies taste)

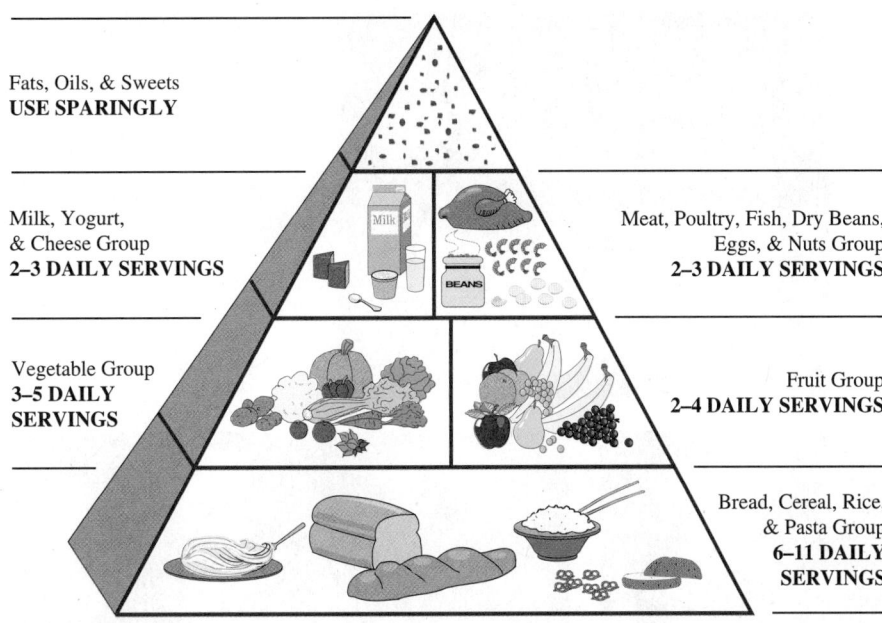

◀ FIGURE 32.4
The food pyramid: a guide
to a balanced diet.

Fats, Oils, & Sweets
USE SPARINGLY

Milk, Yogurt,
& Cheese Group
2–3 DAILY SERVINGS

Meat, Poultry, Fish, Dry Beans,
Eggs, & Nuts Group
2–3 DAILY SERVINGS

Vegetable Group
**3–5 DAILY
SERVINGS**

Fruit Group
2–4 DAILY SERVINGS

Bread, Cereal, Rice,
& Pasta Group
**6–11 DAILY
SERVINGS**

32.12   A Balanced Diet

To summarize the preceding sections, the macronutrients (carbohydrates, fats, and proteins) comprise the majority of our diets. These nutrients supply the molecules we need for energy, growth, and maintenance. A second group of nutrients—vitamins, minerals, and water—is not used for energy but is nevertheless essential to our existence. Vitamins provide organic molecules that cannot be made in the body, and minerals provide the inorganic ions needed for life. Water is the solvent in which most of the chemical reactions essential to life occur. Vitamins and minerals are required only in small amounts, from a few micrograms to a few milligrams per day. But water must be consumed in large quantities, 2–3 L/day.

For health and well-being, each of six groups of nutrients must be in our diet. With the variety of foods available, how can we make sure that our diets contain enough of all the needed nutrients? How can we make sure that our diet is balanced?

To answer these questions, nutritionists have divided foods into five groups; (1) milk products; (2) vegetables; (3) fruits; (4) cereal products; and (5) meats, poultry, fish, beans, eggs, and nuts. Each group is a good source of one or more nutrients. To obtain a balanced mixture of carbohydrate, fat, vitamins, and minerals, a diet should contain food from several classes. Figure 32.4 is a pictorial guide to a balanced diet.

A balanced diet must include several food groups, even though each food may contain many nutrients. For example, compare the three breakfasts in Figure 32.5, each of which supplies about 600 kcal of food energy. By choosing only doughnuts (cereal food groups) and coffee, the consumer would gain some nutrients but only in small quantities. A breakfast of cold cereal (cereal group) and milk (milk group), toast (cereal group) with margarine (fats) and jelly (sweets), and coffee includes

FIGURE 32.5 ▶
A comparison of nutritional
values for three breakfasts.
(No RDA has been established
for carbohydrates and fats.)

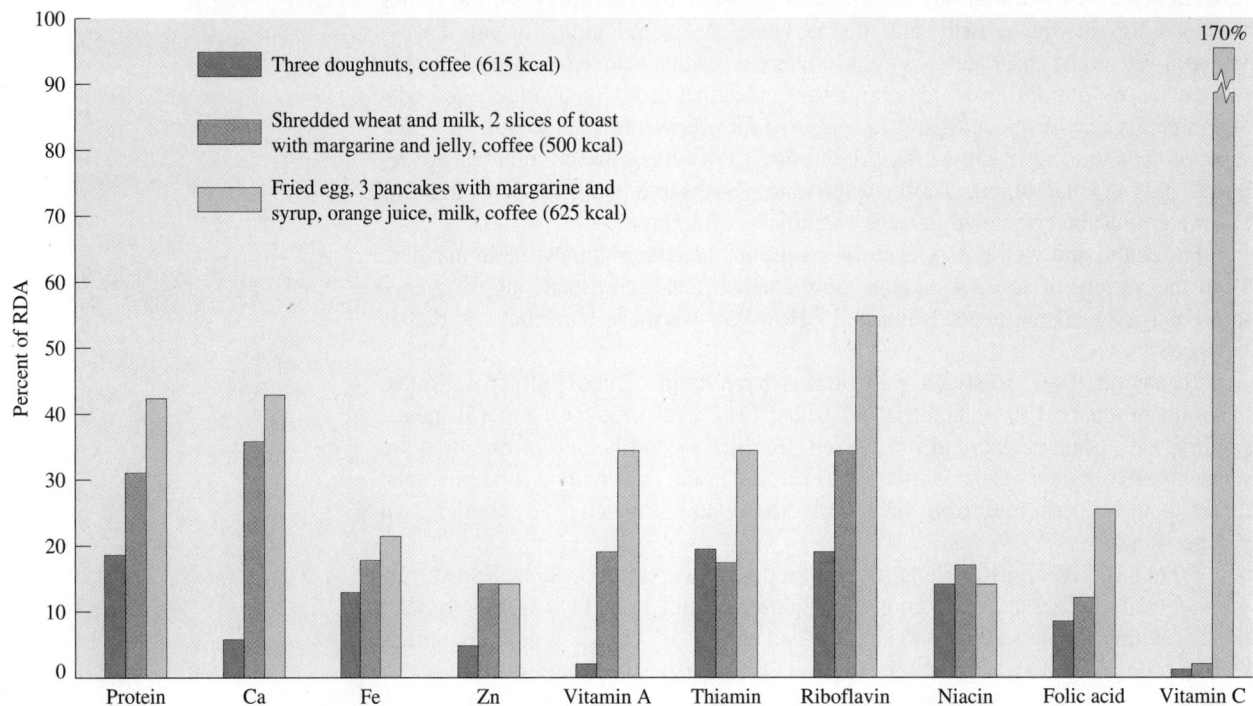

Ice Cream: Food for Thought

For many people there is nothing quite as tasty as a serving of cool, smooth, rich ice cream. The texture is satin smooth and it literally melts in your mouth; you experience a slow increase in richness as one ingredient melds with another. And, although ice cream comes from the freezer, the temperature never seems to be too cold. All of these sensations and more combine to make eating ice cream a simple pleasure.

Like most foods, ice cream is a mixture of nutrients, and some of these nutrients are less healthy than others. For example, scientists recommend that we limit our fat and sugar consumption. This presents a dilemma for ice cream makers: How are they to decrease the fat and sugar content of ice cream while retaining eating enjoyment? This dilemma is shared by many food manufacturers. We can literally take a food apart into its separate nutrients. We can remove unhealthy components. But, can we put it back together again and have it taste as good? In essence, food scientists are being asked to be food designers. Ice cream manufacturing is a good example of this process.

Ice cream is a simple food composed of only a small number of major ingredients:

1. Water (between 60 and 70%) is the largest single ingredient, mostly in the form of small ice crystals that impart the cool sensation. To retain a smooth texture, these crystals must be very small (no larger than 30–45 μm in diameter).
2. Fats are another component of ice cream (the minimum for ice cream is 10% milk fat from cream in the United States). Fats are broken up into small globules (1–10 μm) and provide the creamy texture for the finished product. Fats are important for mouth feel because they reduce the sensation of cold and, as they warm to body temperature, slowly release the flavor of the ice cream.
3. Nonfat milk solids provide the third important basic component of ice cream. Milk proteins (casein and proteins from whey) help stabilize the mixture of fats and water. Milk sugar (lactose) contributes some sweetness and, along with salts, helps to lower the freezing point of the ice cream.
4. To these ice cream components are added emulsifiers, thickeners, sweeteners, and flavorings. Emulsifiers allow the fat to mix thoroughly with the water and also allow more air to be incorporated into the mixture. Chemicals such as mono- and diacylglycerols are often used as emulsifiers. Thickeners, such as carrageenan, guar gum, alginates, and pectins, ensure that ice cream components do not separate in transport or storage. Sweeteners include sucrose and corn syrup and often comprise from 12 to 20% of the ice cream. Finally, there are the flavorings.

Decreasing the sugar content has proven to be the easiest part of ice cream redesign. Nonnutritive sweeteners like aspartame substitute for sucrose or corn syrup. The milk sugar lactose can be broken down to its component monosaccharides to increase ice cream sweetness. Both changes retain sweetness while decreasing the amount of sugars needed.

In contrast, lowering the fat content of ice cream has proved difficult. Fats provide the creamy texture for ice cream. Perhaps more importantly, fats control the release of ice cream flavor so that the consumer can savor each mouthful. At this writing there is no ideal fat substitute. Some manufacturers have opted for nonfat polymers that form a suspension of small globules (several microns in diameter); these globules provide a creamy texture for the ice cream. Both carbohydrates (e.g., cellulose, modified starches, and pectins) and proteins (egg albumen

▲
Can food designers keep the desirable qualities of ice cream while trying to make it a healthier food?

and proteins from whey) have been used as substitutes. Unfortunately, these substitutes don't "melt in your mouth"; they don't release the ice cream flavors slowly, letting them blend as full-fat ice cream does. Other manufacturers have opted for lower-calorie modified fats, "designer fats (e.g., olestra)." These molecules give the character of fat but are harder to digest (and thus provide fewer calories). Again, these substitutes yield a creamy product but lack the proper flavor release.

Ice cream can be made healthier. However, since we often choose food based not only on the nutritional value but also on an enjoyment factor, ice cream redesign is not a simple task.

more food groups. Still more food groups are found in a breakfast of orange juice (fruit group), a fried egg (meat group), pancakes (cereal group) with margarine (fats) and syrup (sweets), milk (milk group), and coffee. Many other breakfasts could be chosen instead, yet an important generalization can be drawn from these three examples. The nutritional value of a meal improves if at least several food groups are included. The third breakfast in this illustration is the most balanced. Although the consumer often does not know the nutrient content of a specific food, overall nutrition can be ensured by choosing a balanced diet.

After the food composing a balanced diet is eaten, it must be digested, absorbed, and transported in order for the proper nutrients to reach the cells. Eating puts food into the alimentary canal (which includes the mouth, esophagus, stomach, small intestine, and large intestine). In a sense this canal is an extension of our external environment; that is, until food is broken down and processed, it cannot enter our internal environment to reach the cells. After digestion and absorption, the nutrients are transported to the cells by the blood and lymph systems. The liver plays a vital role in controlling nutrient levels in the blood and neutralizing toxic substances.

 Human Digestion

digestion

The human digestive tract is shown in Figure 32.6. Although food is broken up mechanically by chewing and by a churning action in the stomach, digestion is a chemical process. **Digestion** is a series of enzyme-catalyzed reactions by which large molecules are hydrolyzed to molecules small enough to be absorbed through the intestinal membranes. Foods are digested to smaller molecules according to this general scheme:

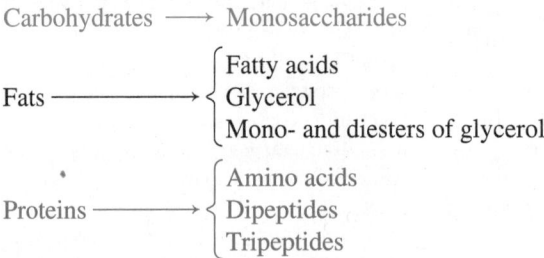

Carbohydrates $\longrightarrow$ Monosaccharides

Fats $\longrightarrow$ $\begin{cases} \text{Fatty acids} \\ \text{Glycerol} \\ \text{Mono- and diesters of glycerol} \end{cases}$

Proteins $\longrightarrow$ $\begin{cases} \text{Amino acids} \\ \text{Dipeptides} \\ \text{Tripeptides} \end{cases}$

Food passes through the human digestive tract (Figure 32.6) in this sequence:

Mouth $\longrightarrow$ esophagus $\longrightarrow$ stomach $\longrightarrow$ small intestine (duodenum, jejunum, and ileum) $\longrightarrow$ large intestine

Five principal digestive juices (or fluids) enter the digestive tract at various points:

1. Saliva from three pairs of salivary glands in the mouth
2. Gastric juice from glands in the walls of the stomach
3. Pancreatic juice secreted by the pancreas enters the duodenum through the pancreatic duct
4. Bile secreted by the liver enters the duodenum via a duct from the gall bladder
5. Intestinal juice from intestinal mucosal cells

◀ **FIGURE 32.6**
The human digestive tract.

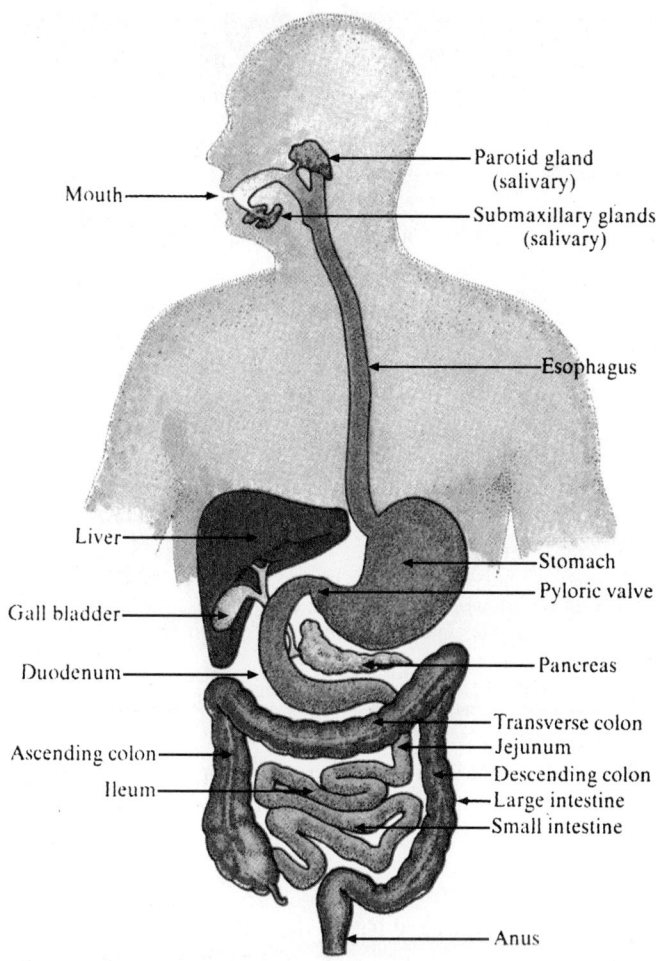

The main functions and principal enzymes found in each of these fluids are summarized in Table 32.6. The important digestive enzymes occur in gastric, pancreatic, and intestinal juices. An outline of the digestive process follows.

Detailed accounts of the various stages of digestion are to be found in biochemistry and physiology textbooks.

Salivary Digestion Food is chewed (masticated) and mixed with saliva in the mouth, and the hydrolysis of starch begins. The composition of saliva depends on many factors—age, diet, condition of teeth, time of day, and so on. Normal saliva is about 99.5% water. Saliva also contains mucin (a glycoprotein); a number of mineral ions such as K^+, Ca^{2+}, Cl^-, PO_4^{3-}, SCN^-; and one enzyme, salivary amylase (ptyalin). The pH of saliva ranges from slightly acidic to slightly basic, with the optimal pH about 6.6–6.8.

Mucin acts as a lubricant and facilitates the chewing and swallowing of food. The enzyme salivary amylase catalyzes the hydrolysis of starch to maltose:

$$\text{Starch + Water} \xrightarrow{\text{salivary amylase}} \text{Maltose}$$

Salivary amylase is inactivated at a pH of 4.0, so it has very little time to act before the food reaches the highly acidic stomach juices.

TABLE **32.6**	**Digestive Fluids**	
Fluid (volume produced daily)	**Source**	**Principal enzymes and/or function**
Saliva (1000–1500 mL)	Salivary glands	Lubricant, aids chewing and swallowing; also contains salivary amylase (ptyalin), which begins the digestion of starch
Gastric juice (2000–3000 mL)	Glands in stomach wall	Pepsin, gastric lipase; pepsin catalyzes partial hydrolysis of proteins to proteoses and peptones in the stomach
Pancreatic juice (500–800 mL)	Pancreas	Trypsinogen, chymotrypsinogen, procarboxypeptidase (converted after secretion to trypsin, chymotrypsin, and carboxypeptidase, respectively, which continue protein digestion); amylopsin (α-amylase, a starch-digestive enzyme), steapsin (a lipase)
Bile (500–1000 mL)	Liver	Contains no enzymes but does contain bile salts, which aid digestion by emulsifying lipids; serves to excrete cholesterol and bile pigments derived from hemoglobin
Intestinal juice	Intestinal mucosal cells	Contains a variety of finishing enzymes: sucrase, maltase, and lactase for carbohydrates; aminopolypeptidase and dipeptidase for final protein breakdown; intestinal lipase; nucleases and phosphatase for hydrolysis of nucleic acids

Saliva is secreted continuously, but the rate of secretion is greatly increased by the sight and odor, or even the thought, of many foods. The mouth-watering effect of the sight or thought of pickles is familiar to many of us. This is an example of a *conditioned reflex.*

Gastric Digestion When food is swallowed, it passes through the esophagus to the stomach. In the stomach, mechanical action continues; food particles are reduced in size and mixed with gastric juices until a material of liquid consistency, known as *chyme,* is obtained.

Gastric juice is a clear, pale yellow, acidic fluid having a pH of about 1.5–2.5. It contains hydrochloric acid; the mineral ions Na^+, K^+, Cl^-; some phosphates; and the digestive enzymes pepsin and lipase. The flow of gastric juice is accelerated by conditioned reflexes and by the presence of food in the stomach. The secretion of the hormone *gastrin* is triggered by food entering the stomach. This hormone, produced by the gastric glands, is absorbed into the bloodstream and returned to the stomach wall where it stimulates the secretion of additional gastric juice. Control of gastric secretion by this hormone is but one of many chemical control systems that exist in the body.

The chief digestive function of the stomach is the partial digestion of protein. The principal enzyme of gastric juice is pepsin, which digests protein. The enzyme is secreted in an inactive form called pepsinogen, which is activated by hydrochloric acid to pepsin. Pepsin catalyzes the hydrolysis of proteins to fragments called proteoses and peptones, which are still fairly large molecules. Pepsin splits the peptide bonds adjacent to only a few amino acid residues, particularly tyrosine and phenylalanine:

$$\text{Protein} + \text{Water} \xrightarrow{\text{pepsin}} \text{Proteoses} + \text{Peptones}$$

The second enzyme in the stomach, gastric lipase, is a fat-digesting enzyme. Its action in the stomach is slight because the acidity is too high for lipase activity. Food can be retained in the stomach for as long as 6 hours. It then passes through the pyloric valve into the duodenum.

Intestinal Digestion The next section of the digestive tract, the small intestine, is where most of the digestion occurs. The stomach contents are first made alkaline by secretions from the pancreatic and bile ducts. The pH of the pancreatic juice is 7.5–8.0, and the pH of bile is 7.1–7.7. The shift in pH is necessary because the enzymes of the pancreatic and intestinal juices are active only in an alkaline medium. Enzymes that digest all three kinds of food—carbohydrates, fats, and proteins—are secreted by the pancreas. Pancreatic secretion is stimulated by hormones secreted into the bloodstream by the duodenum and the jejunum.

The enzymes occurring in the small intestine include pancreatic amylases (diastase) that hydrolyze most of the starch to maltose, and carbohydrases (α-amylase, maltase, sucrase, and lactase) that complete the hydrolysis of disaccharides to monosaccharides. The proteolytic enzymes trypsin and chymotrypsin attack proteins, proteoses, and peptones, hydrolyzing them to dipeptides. Then the peptidases—carboxypeptidase, aminopeptidase, and dipeptidase—complete the hydrolysis of proteins to amino acids. Pancreatic lipases catalyze the hydrolysis of almost all fats. Fats are split into fatty acids, glycerol, and mono- and diesters of glycerol by these enzymes.

The liver is another important organ in the digestive system. A fluid known as bile is produced by the liver and stored in the gall bladder, a small organ located on the surface of the liver. When food enters the duodenum, the gall bladder contracts, and the bile enters the duodenum through a duct that is also used by the pancreatic juice. In addition to water, the major constituents of bile are bile acids (as salts), bile pigments, inorganic salts, and cholesterol. The bile acids are steroid monocarboxylic acids, two of which are shown here:

cholic acid   chenodeoxycholic acid

The bile acids are synthesized in the liver from cholesterol, which is also synthesized in the liver. The presence of bile in the intestine is important for the digestion and absorption of fats. When released into the duodenum, the bile acids emulsify the fats, allowing them to be hydrolyzed by the pancreatic lipases. About 90% of the bile salts are reabsorbed in the lower part of the small intestine and are transported back to the liver and used again.

Most of the digested food is absorbed from the small intestine. Undigested and indigestible material passes from the small intestine to the large intestine, where it is retained for varying periods of time before final elimination as feces. Additional chemical breakdown, sometimes with the production of considerable amounts of gases, is brought about by bacteria (or rather by bacterial enzymes) in the large intestine. For a healthy person this additional breakdown is not important from the standpoint of nutrition because absorption of nutrients does not occur from the large intestine. However, large amounts of water, partly from digestive juices, are absorbed from the large intestine so that the contents become more solid before elimination as feces.

Practice 32.4

Complete digestion of starch yields which monosaccharide?

 Absorption

absorption

For digested food to be utilized in the body, it must pass from the intestine into the blood and lymph systems. The process by which digested foods pass through the membrane linings of the small intestine and enter the blood and lymph is called **absorption.** Absorption is complicated, and we will consider only an overview of it.

After the food you have eaten is digested, your body must absorb billions upon billions of nutrient molecules and ions into the bloodstream. The absorption system is in the membranes of the small intestine, which, upon microscopic inspection, are seen to be wrinkled into hundreds of folds. These folds are covered with thousands of small projections called *villi.* Each projection is itself covered by many minute folds called *microvilli.* The small intestine's wrinkles, folds, and projections increase the surface area available for absorption. The average small intestine is about 4 m (13.3 ft) long and is estimated to contain about 8360 m^2 (90,000 ft^2) of absorbing surface.

The inner surface of the small intestine is composed of mucosal cells that produce many enzymes, such as disaccharidases, aminopeptidases, and dipeptidases, needed to complete the digestive process. As digestion is completed, the resulting nutrient molecules are absorbed by the mucosal cells and transferred to the blood and lymph systems. Water-soluble nutrients such as monosaccharides, glycerol, short-chain fatty acids, amino acids, and minerals enter directly into the bloodstream. Fat-soluble nutrients such as long-chain fatty acids and monoacylglycerols first enter the lymph fluid and then enter the bloodstream where the two fluids come together.

An important factor in the absorption process is that the membranes of the intestine are selectively permeable; that is, they prevent the passage of most large molecules but allow the passage of smaller molecules. For example, polysaccharides, disaccharides, and proteins are not ordinarily absorbed, but generally monosaccharides and amino acids are.

▲ **Absorption of digested food occurs in the small intestine through the villi and microvilli shown here.**

 Liver Function

The liver is the largest organ in the body and performs several vital functions. Two of these functions are the regulation of the concentrations of organic nutrients in the blood and the removal of toxic substances from the blood.

The concentration of blood sugar (glucose) is controlled and maintained by processes that occur in the liver. After absorption, excess glucose and other monosaccharides are removed from the blood and converted to glycogen in the liver. The liver is the principal storage organ for glycogen. As glucose is used in other cells, the stored liver glycogen is gradually hydrolyzed to maintain the appropriate blood-glucose concentration. Liver function is under sensitive hormonal control, as are most vital body functions. This regulation will be discussed further in Chapter 34.

A second major function of the liver is the detoxification of harmful and potentially harmful substances. This function apparently developed as higher vertebrates appeared in the evolutionary time scale. The liver is able to deal with most of the toxic molecules that occur in nature. For example, ethanol is oxidized in the liver, and nitrogenous metabolic waste products are converted to urea for excretion.

Organic chemists have learned to synthesize new substances that have no counterparts in the biological world. They are used as industrial chemicals, insecticides, drugs, and food additives. When these potentially toxic substances are ingested, even in small amounts, the body is faced with the difficult challenge of metabolizing or destroying substances that are unlike any found in nature.

The liver meets this challenge and deals with most of these foreign molecules through an oxidation system located in the endoplasmic reticulum. Bound to these intracellular membranes are enzymes that catalyze reactions between oxygen and the foreign molecules. As these molecules are oxidized, they become more polar and water soluble. Finally, the oxidation products of the potential toxins are excreted in the urine or bile fluid.

For example, most automobile antifreeze solutions contain ethylene glycol, a toxic substance. Even though this compound does not occur naturally, the liver can metabolize ethylene glycol. When small amounts are ingested, the following chemical changes occur:

$$\underset{\substack{\text{ethylene}\\\text{glycol}}}{\text{H}-\overset{\overset{\displaystyle\text{OH}}{|}}{\underset{\underset{\displaystyle\text{H}}{|}}{\text{C}}}-\overset{\overset{\displaystyle\text{OH}}{|}}{\underset{\underset{\displaystyle\text{H}}{|}}{\text{C}}}-\text{H}} \xrightarrow{\text{oxidation}} \underset{\text{glyoxal}}{\text{H}-\overset{\overset{\displaystyle\text{OH}}{|}}{\underset{\underset{\displaystyle\text{H}}{|}}{\text{C}}}-\overset{\overset{\displaystyle\text{O}}{\|}}{\text{C}}-\text{H}} \xrightarrow{\text{oxidation}} \underset{\text{glycolate}}{\text{H}-\overset{\overset{\displaystyle\text{OH}}{|}}{\underset{\underset{\displaystyle\text{H}}{|}}{\text{C}}}-\overset{\overset{\displaystyle\text{O}}{\|}}{\text{C}}-\text{O}^-} \xrightarrow{\text{oxidation}}$$

$$\underset{\text{glyoxalate}}{\text{H}-\overset{\overset{\displaystyle\text{O}}{\|}}{\text{C}}-\overset{\overset{\displaystyle\text{O}}{\|}}{\text{C}}-\text{O}^-} \xrightarrow{\text{oxidation}} \underset{\text{formate}}{\text{H}-\overset{\overset{\displaystyle\text{O}}{\|}}{\text{C}}-\text{O}^-} + \underset{\text{bicarbonate}}{\text{HO}-\overset{\overset{\displaystyle\text{O}}{\|}}{\text{C}}-\text{O}^-}$$

Thus ethylene glycol is converted by oxidation to two more polar (charged) acid anions that are easily eliminated from the body.

Unfortunately, oxidation is not effective for some compounds. Halogenated hydrocarbons, which are particularly inert to oxidation, accumulate in fatty tissue or in the liver itself. Examples of halogenated hydrocarbons are carbon tetrachloride, hexachlorobenzene, DDT, dioxins, and polychlorinated biphenyls (PCBs). Some compounds become more toxic after oxidation. For example, polycyclic hydrocarbons (which can be formed when food is barbecued) become carcinogenic upon partial oxidation. Methanol becomes particularly toxic because it is converted by oxidation to formaldehyde. One of the most serious dangers of environmental pollution lies in the introduction of compounds that the liver cannot detoxify.

In a sense the liver is the final guardian along the pathway by which nutrients pass to the cells. This pathway starts with a balanced diet. Once foods are digested and absorbed, the liver adjusts nutrient levels and removes potential toxins. The blood can then provide nutrients for the cellular biochemistry that constitutes life.

Concepts in Review

1. Distinguish a nutrient from a food.
2. Describe the difference between a minimum dietary requirement and a recommended dietary allowance.
3. Summarize the importance of nutrients in metabolism.
4. Briefly discuss the major functions of dietary protein.
5. Discuss the dangers of kwashiorkor.
6. Explain why fats are a more concentrated source of metabolic energy than carbohydrates.
7. List some of the dangers of a high-sucrose diet.
8. Briefly describe the importance of dietary fiber.
9. List the essential fatty acids and essential amino acids.
10. Discuss a major danger inherent in a vegetarian diet.
11. Summarize some of the functions of the water-soluble vitamins.
12. Discuss some of the functions of the fat-soluble vitamins.
13. State the general results of vitamin deficiencies.
14. List the major mineral elements.
15. Compare the similarities between trace mineral elements and vitamins.
16. Briefly discuss the importance of water in the diet.
17. List four purposes served by food additives.
18. Understand the information given in the Nutrition Facts panel.
19. List the five principal digestive juices and where they originate in the body.
20. List the principal enzymes of the various digestive juices.
21. Give the main digestive functions of each of the five principal digestive juices.
22. List the classes of products formed when carbohydrates, fats, and proteins are digested.
23. Briefly describe the absorption of nutrients from the digestive tract into the lymph and blood.
24. Discuss the role of the liver in maintaining blood-glucose levels.
25. Explain how the liver metabolizes potentially toxic compounds.

Key Terms

The terms listed here have been defined within this chapter. Section numbers are referenced in parentheses for each term.

absorption (32.14)	digestion (32.13)	nutrient (32.1)	vitamin (32.7)
complete protein (32.6)	macronutrient (32.1)	recommended dietary	
diet (32.2)	micronutrient (32.1)	allowance (RDA) (32.2)	

Questions

Questions refer to tables, figures, and key words and concepts defined within the chapter. A particularly challenging question or exercise is indicated with an asterisk.

1. How does a food differ from a nutrient?
2. What is meant by the term *energy allowance?*
3. How does marasmus differ from kwashiorkor?
4. What is meant by the statement that candy provides empty calories?
5. What structural features do the essential fatty acids have in common?

6. What is meant by the term *essential fatty acid*?

7. List the essential amino acids.

8. How do animal proteins differ nutritionally from vegetable proteins?

9. Which vitamins are water soluble? Which are fat soluble?

10. List three functions that can be attributed to vitamins.

11. Distinguish the major elements from the trace elements.

12. List two major biological functions for calcium.

13. What percentage of the average water consumption comes from solid foods? How much from liquids?

14. List five common categories of food additives.

15. What are the five principal digestive juices?

16. Which federal agency is responsible for regulating the use of food additives?

17. What enzymes are in each of the digestive juices?

18. What is chyme?

19. What is the digestive function of the liver?

20. How do the intestinal mucosal cells aid digestion?

21. How does the liver metabolize toxic compounds?

22. What is the meaning of "GRAS"?

Paired Exercises

These exercises are paired. Each odd-numbered exercise is followed by a similar even-numbered exercise. Answers to the even-numbered exercises are given in Appendix VI.

23. A sirloin steak (85 g) has the following nutritional composition:

Protein	23	g
Carbohydrates	0	
Fats	19	g
Sodium	52	mg
Potassium	297	mg
Magnesium	23	mg
Iron	2.5	mg
Zinc	4.7	mg
Calcium	9.0	mg
Vitamin A	15	μg
Vitamin C	0	
Thiamin	92	μg
Riboflavin	218	μg
Niacin	3.2	mg
Vitamin B_6	330	μg
Folic acid	7	μg
Vitamin B_{12}	2	μg

(a) What is the mass of macronutrients contained in this food?

(b) What is the mass of minerals?

(c) What is the mass of vitamins?

24. A battered and fried chicken breast (280 g) has the following nutritional composition:

Protein	70	g
Carbohydrates	25	g
Fats	37	g
Sodium	770	mg
Potassium	564	mg
Magnesium	68	mg
Iron	3.5	mg
Zinc	2.7	mg
Calcium	56	mg
Vitamin A	56.5	μg
Vitamin C	0	
Thiamin	322	μg
Riboflavin	408	μg
Niacin	29.5	mg
Vitamin B_6	1.2	mg
Folic acid	16	μg
Vitamin B_{12}	0.82	μg

(a) What is the mass of macronutrients contained in this food?

(b) What is the mass of minerals?

(c) What is the mass of vitamins?

25. Fat yields about 9 kcal/g.

(a) How much energy is available from fat in sirloin steak from Exercise 23?

(b) If the total energy content is 270 kcal, what percentage of energy is derived from the fat?

26. Fat yields about 9 kcal/g.

(a) How much energy is available from fat in chicken breast from Exercise 24?

(b) If the total energy content is 728 kcal, what percentage of energy is derived from the fat?

27. For 15- to 18-year-old women, the RDA for calcium is 1.3 g. What percentage of the RDA is supplied by the sirloin steak in Exercise 23?

28. For 51+-year-old women, the RDA for calcium is 1.2 g. What percentage of the RDA is supplied by the chicken breast from Exercise 24?

29. The RDA for vitamin B_{12} is 2.0 μg for adults. What percentage of the RDA is supplied by the sirloin steak in Exercise 23?

30. The RDA for vitamin B_{12} is 2.0 μg for adults. What percentage of the RDA is supplied by the chicken breast in Exercise 24?

31. List three classes of nutrients that are considered macronutrients.

32. List two classes of nutrients that are considered micronutrients.

33. List three classes of nutrients that do *not* commonly supply energy for the cell.

34. List three classes of nutrients that commonly supply energy for the cell.

35. Why is starch important in the diet?

36. Why is cellulose important in the diet?

37. Is saliva slightly acidic, slightly basic, or neutral?

38. Is pancreatic juice slightly acidic, slightly basic, or neutral?

39. In what parts of the digestive system are proteins digested?

40. In what parts of the digestive system are carbohydrates digested?

41. Pancreatic amylase is a digestive enzyme found in the small intestine. What nutrient class does this enzyme digest?

42. Pepsin is a digestive enzyme found in the stomach. What nutrient class does this enzyme digest?

Additional Exercises

These exercises are not paired and provide additional practice on concepts covered in this chapter.

43. Milk is a food. List four nutrients that can be obtained from milk.

44. You are told that a new diet will cause you to lose 9 kg (20 lb) of fat in 1 week. Is this reasonable? Briefly explain.

45. In what ways are the trace elements similar to vitamins? How do they differ?

46. Explain why a tablespoon of butter (a fatty food) approximately doubles the calorie content of a medium-size baked potato (a carbohydrate food).

47. Given the Nutrition Facts panel in the adjacent column from cream of chicken soup:
(a) What percentage of total energy is provided by the fat?
(b) How many grams of protein will be provided by two servings of this food?
(c) For a person on a reduced-calorie diet (1000 kcal/day), what percentage of the total energy is supplied by one serving?
(d) What percentage of the total carbohydrates are fiber?

48. Why must the food of higher animals be digested before it can be utilized?

49. Galactosemia is an inherited condition in some babies. It is a deficiency in the enzyme that catalyzes the conversion of galactose to glucose. Symptoms are lack of appetite, weight loss, diarrhea, and jaundice. What staple of a baby's diet must be changed to correct this problem?

NUTRITION FACTS	
Serv. Size 1/2 cup (125g)	
Servings about 2 1/2	
Amount Per Serving	
Calories 100	Fat cal 50
	% Daily Value*
Total Fat 5g	8%
Saturated fat 2g	9%
Cholesterol 10mg	3%
Sodium 860mg	36%
Total carbohydrate 10g	3%
Fiber 2g	9%
Sugars 1g	
Protein 3g	
Vitamin A 10% ● Vitamin C 0%	
Calcium 0% ● Iron 2%	
*Percent Daily Values are based on a 2,000 calorie diet.	

50. One large egg contains 270 mg of cholesterol. How does this compare with the daily recommended amount of cholesterol? (See the Nutrition Facts panel in Figure 32.3.)

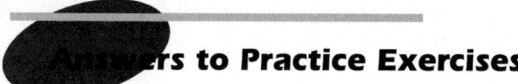

Answers to Practice Exercises

32.1 **(a)** A young woman's iron

$$RDA = (15 \text{ mg}) \left(\frac{1 \text{ g}}{1000 \text{ mg}} \right) = 1.5 \times 10^{-2} \text{g}$$

(b) Vitamin B_{12} has the smallest RDA.

32.2 Vitamin D acts as a regulator for calcium metabolism, and so the Vitamin D RDA is high when the calcium RDA is also high.

32.3 Fiber daily value: $\dfrac{6 \text{ g} \times 100\%}{26\%} = X \text{ g}$

Fiber daily value = 20 g

32.4 Glucose

CHAPTER 33

Bioenergetics

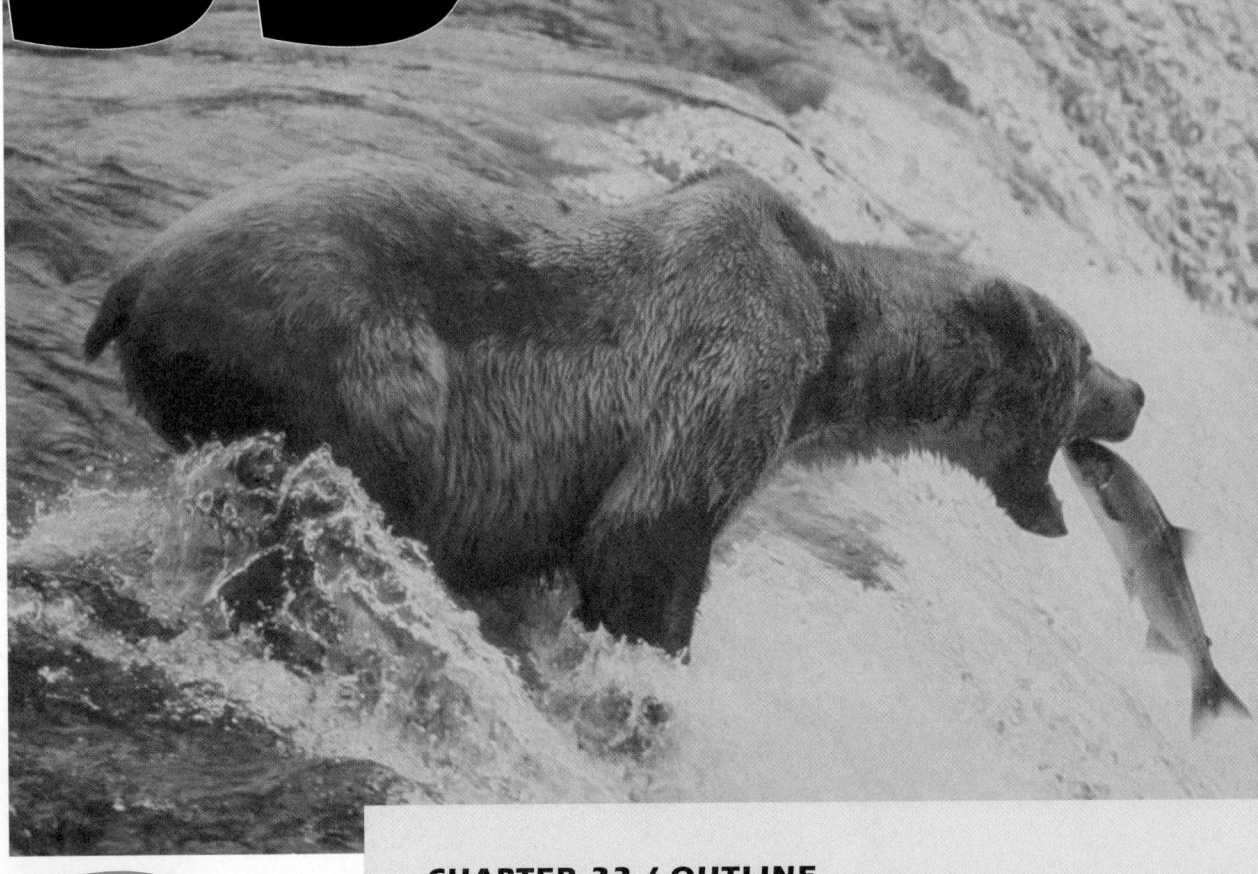

▲ Finding adequate sources of energy is a constant challenge for all living organisms, including this bear.

CHAPTER 33 / OUTLINE

If you travel through the midwestern United States on a warm night in early summer, you may be lucky enough to see small yellow lights dancing through the fields. This beautiful sight is evidence that the fireflies, or lightning bugs, have returned. Their flickering light does not arise from a small fire or electrical discharge but rather from a chemical reaction. Bioenergetics are at work here.

The monarch butterfly is a striking insect with a wing span of several inches. As it flies, it is buffeted by the wind, first one direction and then another. Clearly, the monarch is light and not a powerful flier. Yet each year (in one leg of a complex, multigeneration migration) these small creatures fly from southern Canada to winter in central Mexico, a distance of over 2000 miles. Chemical reactions in the insects' cells make energy available for this arduous trip. Bioenergetics are at work again.

While you have been reading these paragraphs, fully 20% of the oxygen you breathed has been used by your brain. Chemical processes have reacted this oxygen with blood glucose, yielding the energy needed to comprehend your reading. This is another example of bioenergetics—the chemical processes directly related to energy needs and uses in life.

33.1 Energy Changes in Living Organisms

One of the basic requirements for life is a source of energy. **Bioenergetics** is the study of the transformation, distribution, and utilization of energy by living organisms. Bioenergetics includes the radiant energy of sunlight used in photosynthesis, electrical energy of nerve impulses, mechanical energy of muscle contractions, heat energy liberated by chemical reactions within cells, and potential energy stored in "energy-rich" chemical bonds. The major source of biological energy includes chemical reactions occurring inside cells.

bioenergetics

The bioenergetics of a cell can be compared to the energetics of a manufacturing plant (see Figure 33.1). First, energy is delivered to the cell or plant. Second, this energy is commonly converted to a more usable form. Third, work is done (See Table 33.1).

33.2 Metabolism and Cell Structure

The sum of all chemical reactions that occur within a living organism is defined as **metabolism.** Many hundreds of different chemical reactions occur in a typical cell. To make sense of this myriad of reactions, biochemists subdivide metabolism into two contrasting categories—*anabolism* and *catabolism*. **Anabolism** is the process by which simple substances are synthesized (built up) into complex substances. **Catabolism** is the process by which complex substances are broken down into simpler substances. Anabolic reactions usually involve carbon reduction and consume cellular energy, whereas catabolic reactions usually involve carbon oxidation and produce energy for the cell.

metabolism

anabolism

catabolism

FIGURE 33.1 ▶
A typical manufacturing plant converts electrical energy to mechanical energy and uses this energy to change raw materials to finished products. This is illustrated here as robots weld the undercarriage of a car on an assembly line.

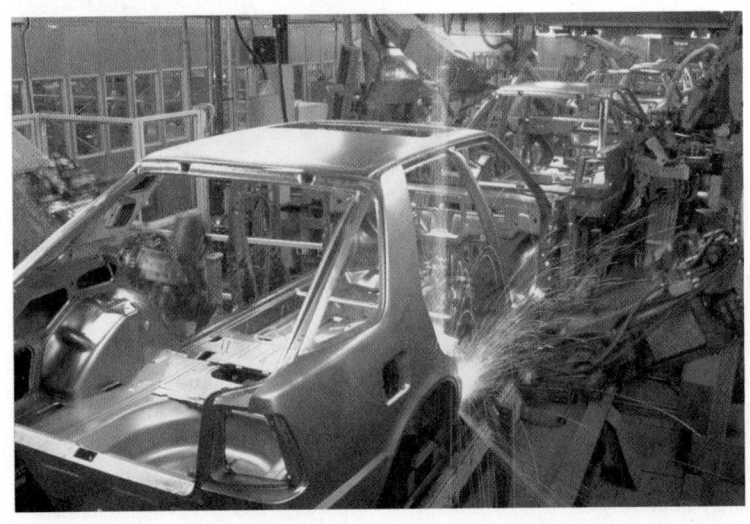

TABLE 33.1	Delivery, Conversion, and Use of Energy in Energetics		
Step	**Manufacturing plant energetics**	**Animal cell bioenergetics**	**Plant cell bioenergetics**
1. Energy delivery	Energy is delivered as electricity, coal, oil, gas, etc.	Energy is delivered as reduced carbon atoms.	Energy is delivered as sunlight.
2. Energy conversion	Electricity, coal, oil, gas, etc., is converted to mechanical energy.	Energy in reduced carbon atoms is converted to high-energy phosphate bonds.	Sunlight is converted to chemical energy.
3. Work done	Mechanical energy converts raw materials into products.	High-energy phosphate bonds are used to do the work of the cell.	High-energy phosphate bonds are used to do the work of the cell.

Cells segregate many of their metabolic reactions into specific, subcellular locations. The simple **procaryotes**—cells without internal membrane-bound bodies—have a minimum amount of spatial organization (see Figure 33.2). The anabolic processes of DNA and RNA synthesis in these cells are localized in the nuclear material, whereas most other metabolic reactions are spread throughout the cytoplasm.

In contrast, metabolic reactions in the cells of higher plants and animals are often segregated into specialized compartments. These cells, the **eucaryotes**, contain internal, membrane-bound bodies called **organelles** (see Figure 33.2). It is within the organelles that many specific metabolic processes occur.

In the eucaryotic cell, most of the DNA and RNA syntheses are localized in the nucleus. Anabolism of proteins takes place in the ribosomes, whereas that of carbohydrates and lipids occurs primarily in the cytoplasm.

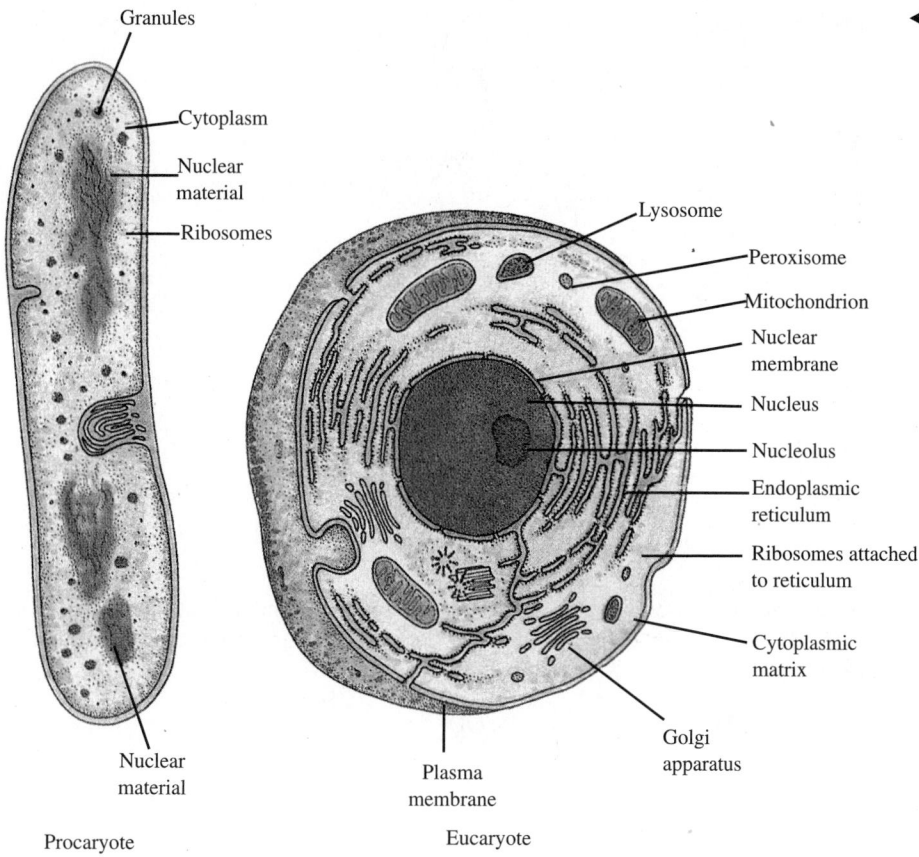

◀ **FIGURE 33.2**
**A procaryotic and a eucary-
otic cell. The procaryote
lacks much of the orga-
nized structure and the numerous organelles found in
the eucaryote.**

There are a variety of specialized catabolic organelles within a eucaryotic cell. The lysosome contains the cell's digestive enzymes, and the peroxisome is the site of oxidative reactions that form hydrogen peroxide. Perhaps the most important catabolic organelle is the mitochondrion (plural, mitochondria). This membrane-bound body provides most of the energy for a typical cell. The energy is released by catabolic processes, which oxidize carbon-containing nutrients. Mitochondria consume most of the O_2 that is inhaled and produce most of the CO_2 that is exhaled by the lungs.

33.3 Biological Oxidation–Reduction: Energy Delivery

The ultimate source of biological energy on earth is sunlight. Plants capture light energy and transform it to chemical energy by a process called *photosynthesis* (see Section 33.6). This chemical energy is stored in the form of reduced carbon atoms in carbohydrate molecules. It is important to understand that the energy contained in carbohydrates, lipids, and proteins originally came from sunlight.

Humans, as well as other animals, draw most of their energy from foodstuffs that contain reduced carbons (see Chapter 32). For example, after we eat a meal,

FIGURE 33.3 ▶
Energy flow through metabolism using the important energy nutrients, carbohydrates. Reduction of carbon stores energy while oxidation of carbon releases energy.

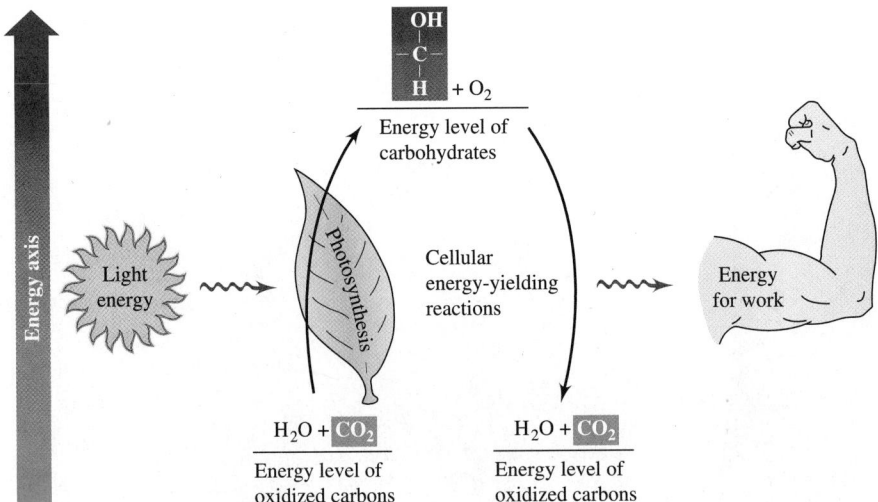

nutrients such as carbohydrates and fats are metabolized and transported to our cells. The carbons in these compounds are in a reduced state and thus contain stored energy.

typical fatty
acid carbon
(oxidation number = −2)

typical carbohydrate
carbon
(oxidation number = 0)

carbon dioxide
carbon
(oxidation number = +4)

Through the cell's metabolism these reduced carbons are oxidized, step by step, and are eventually converted to carbon dioxide.

Practice 33.1

Which molecule is likely to provide more cellular energy? Briefly explain.

Figure 33.3 is an energy diagram summarizing the energy flow through metabolism. Such a diagram will be used several times in this chapter. The black arrows trace progress from one chemical to the next. Because each chemical has its own special energy level, it is presented at a particular height with respect to the energy

(a) Chloroplast

(b) Mitochondrion

▲
FIGURE 33.4
(a) A chloroplast and (b) a mitochondrion.

axis (the higher the level, the more energy the carbon atoms contain). When energy levels change, energy must be released or absorbed as shown by the red arrows. Thus, in Figure 33.3, photosynthesis causes carbons to move to a higher energy level as they are reduced to carbohydrate carbons. The carbohydrate carbons are then oxidized to a lower energy level with the release of energy to do work.

In eucaryotic cells, specific organelles are present that specialize in redox reactions. The *mitochondria* (see Figure 33.4), often called the powerhouses of the cell, are the sites for most of the catabolic redox reactions. *Chloroplasts* (see Figure 33.4) are organelles found in higher plants and contain an electron-transport system that is responsible for the anabolic redox reactions in photosynthesis (see Section 33.6).

To move electrons from one place to another (often outside of the mitochondrion or chloroplast), the cell uses a set of redox coenzymes. The redox coenzymes act as temporary storage places for electrons. The three most common redox coenzymes (nicotinamide adenine dinucleotide, NAD^+; nicotinamide adenine dinucleotide phosphate, $NADP^+$; and flavin adenine dinucleotide, FAD) are shown in Figure 33.5. Humans synthesize NAD^+ and $NADP^+$ from the vitamin niacin while FAD is made from the vitamin riboflavin. In each case, the vitamin provides the reaction center of the coenzyme.

Recall from Section 30.1 that a coenzyme is an organic compound that is used and reused to help an enzyme-catalyzed reaction.

FIGURE 33.5
Structures of the redox co-enzymes (a) nicotinamide adenine dinucleotide (NAD⁺), (b) nicotinamide adenine dinucleotide phosphate (NADP⁺), and (c) flavin adenine dinucleotide (FAD).

(a) **NAD⁺**

(b) **NADP⁺**

(c) **FAD**

The addition or removal of electrons occurs in only one portion of each of these complex molecules. The nicotinamide ring is the reactive component within NAD^+ or $NADP^+$:

NAD⁺ or NADP⁺
(oxidized form)

NADH or NADPH
(reduced form)

(R represents the remainder of each molecule.)

For FAD, the flavin ring is the reactive component:

FAD
(oxidized form)

$+ 2 e^- + 2 H^+ \rightleftharpoons$

FADH$_2$
(reduced form)

(R represents the remainder of each molecule.)

A very important function of these redox coenzymes is to carry electrons to the mitochondrial electron-transport system. As the coenzymes are oxidized, molecular oxygen is reduced:

$$2 \text{ FADH}_2 + O_2 \xrightarrow[\text{electron transport}]{\text{mitochondrial}} 2 \text{ FAD} + 2 \text{ H}_2\text{O}$$

reduced
coenzyme

oxidized
coenzyme

$$2 \text{ NADH} + 2 \text{ H}^+ + O_2 \xrightarrow[\text{electron transport}]{\text{mitochondrial}} 2 \text{ NAD}^+ + 2 \text{ H}_2\text{O}$$

reduced
coenzyme

oxidized
coenzyme

With the movement of electrons, energy is released. In fact, over 85% of a typical cell's energy is derived from this redox process. However, the released energy is not used immediately by the cell but is instead stored, usually in high-energy phosphate bonds such as those in ATP.

Remember, the mitochondria are the powerhouses of the cell!

Practice 33.2

What is wrong with the following equation for biological redox? Briefly explain.

$$\text{CH}_3-\text{OH} + \text{FADH}_2 \longrightarrow \text{H}_2\text{C}=\text{O} + \text{FAD}$$

33.4   *High-Energy Phosphate Bonds*

Cells need an energy delivery system. Most cellular energy is produced in the mitochondria, but this energy must be transported throughout the cell. Such a delivery system is required to carry relatively large amounts of energy and to be easily accessible to cellular reactions. Molecules that contain high-energy phosphate bonds meet this need.

The most common high-energy phosphate bond within the cell is the phosphate anhydride bond (or phosphoanhydride bond) (also see Section 31.4):

$$
\underset{\text{phosphate}}{\overset{\displaystyle O^-}{\underset{\displaystyle O}{^-O-\overset{|}{\underset{||}{P}}-OH}}} + \underset{\text{phosphate}}{\overset{\displaystyle O^-}{\underset{\displaystyle O}{HO-\overset{|}{\underset{||}{P}}-OR}}} + 35\ \text{kJ} \rightleftharpoons \underset{\text{diphosphate}\atop\text{anhydride}}{\overset{\displaystyle O^- \quad O^-}{\underset{\displaystyle O \quad\ O}{^-O-\overset{|}{\underset{||}{P}}\sim O-\overset{|}{\underset{||}{P}}-O-R}}} + H_2O
$$

Phosphate anhydride bond

(R represents the remainder of each molecule.)

A relatively large amount of energy is required to bond the two negatively charged phosphate groups. The repulsion between these phosphates causes the phosphate anhydride bond to behave somewhat like a coiled spring. When the bond is broken, the phosphates separate rapidly and energy is released.

The phosphate anhydride bond is an important component of the nucleotide triphosphates, the most important of which is adenosine triphosphate (ATP):

adenosine triphosphate (ATP)

Adenosine triphosphate plays an important role in all cells, from the simplest to the most complex. It functions by storing and transporting the energy in its high-energy phosphate bonds to the places in the cell where energy is needed. ATP is the common intermediary in energy metabolism.

The cell realizes several advantages by storing energy in ATP. First, the stored energy is easily accessible to the cell; it is readily released by a simple hydrolysis reaction yielding adenosine diphosphate (ADP) and an inorganic phosphate ion (P_i):

adenosine triphosphate (ATP)

adenosine diphosphate (ADP)

Second, ATP serves as the common energy currency for the cell. Energy from catabolism of many different kinds of molecules is stored in ATP. For example, the energy obtained from oxidation of carbohydrates, lipids, and proteins is stored in ATP. This process is analogous to an economic system that values all goods and services in terms of a common currency such as the dollar. Buying and selling within the system is thus greatly simplified. In the cell, energy utilization is greatly simplified by converting stored energy to ATP, the common energy currency.

33.5 Phosphorylation: Energy Conversion

Thus far we have considered two forms for chemical storage of biological energy: reduced carbon atoms and high-energy phosphate bonds. It is vital that the cell be able to convert one form of stored energy to the other.

Figure 33.6 summarizes biological energy conversion. In the first process, a variety of cellular reactions oxidize the carbons of nutrients, converting these molecules from a high energy level to a lower energy state. The energy released (red arrow) is used to create high-energy phosphate bonds. Finally, when a cellular reaction needs energy, or work must be accomplished, molecules with high-energy phosphate bonds are recycled to low-energy molecules plus inorganic phosphate.

Energy is stored in phosphate anhydride bonds through two biological processes—*substrate-level phosphorylation* and *oxidative phosphorylation*. **Substrate-level phosphorylation** is the process whereby energy derived from

substrate-level
phosphorylation

FIGURE 33.6▶
Energy flow from nutrients with reduced carbons (energy-yielding nutrients) to high-energy phosphate bonds that are used to do work.

FIGURE 33.7▶
Some biological molecules that contain high-energy phosphate bonds (phosphorylated substrates).

phosphoenolpyruvate
(from glucose metabolism)

1,3 diphosphoglycerate
(from glucose metabolism)

phosphocreatine
(found in muscle tissue)

oxidation is used to form high-energy phosphate bonds on various biochemical molecules (substrates) (Figure 33.7).

$$\text{R}-\text{OH} + \text{HO}-\overset{\displaystyle O^-}{\underset{\displaystyle O}{\overset{|}{\underset{||}{P}}}}-O^- + \text{Energy} \longrightarrow \text{R}-\text{O}\sim\overset{\displaystyle O^-}{\underset{\displaystyle O}{\overset{|}{\underset{||}{P}}}}-O^- + H_2O$$

substrate   phosphate ion   (derived from redox)   phosphorylated substrate

(R represents the remainder of the substrate molecule.)

In a succeeding reaction, the phosphorylated substrate often transfers the phosphate to ADP and forms ATP.

$$R-O\sim \overset{\overset{\displaystyle O^-}{|}}{\underset{\underset{\displaystyle O}{||}}{P}}-O^- \;+\; {}^-O-\overset{\overset{\displaystyle O^-}{|}}{\underset{\underset{\displaystyle O}{||}}{P}}\sim O-\overset{\overset{\displaystyle O^-}{|}}{\underset{\underset{\displaystyle O}{||}}{P}}-OCH_2 \quad \text{Adenine} \;+\; H^+ \longrightarrow$$

phosphorylated
substrate

OH OH

adenosine diphosphate (ADP)

$$R-OH \;+\; {}^-O-\overset{\overset{\displaystyle O^-}{|}}{\underset{\underset{\displaystyle O}{||}}{P}}\sim O-\overset{\overset{\displaystyle O^-}{|}}{\underset{\underset{\displaystyle O}{||}}{P}}\sim O-\overset{\overset{\displaystyle O^-}{|}}{\underset{\underset{\displaystyle O}{||}}{P}}-OCH_2 \quad \text{Adenine}$$

substrate

OH OH

adenosine triphosphate (ATP)

This process is called substrate-level phosphorylation because ADP gains a phosphate from a cellular substrate.

Substrate-level phosphorylation is found most commonly in the catabolism of carbohydrates—that is, glycolysis (see Chapter 34). This process accounts for only a small amount of a resting cell's total energy production and does not require oxygen. Under anaerobic conditions (e.g., when the bloodstream cannot deliver enough O_2 to hardworking muscles), substrate-level phosphorylation may be the cell's principal means of forming ATP.

Oxidative phosphorylation is a process that directly uses energy from redox reactions to form ATP. This process occurs in the mitochondria and depends on the mitochondrial electron-transport system. The enzyme-catalyzed oxidation and phosphorylation reactions are coupled in such a way that energy released by the oxidation of a coenzyme is used to form ATP. The overall process is shown in Figure 33.8. Note that, for each $FADH_2$ oxidized, two ATPs are formed; for each NADH oxidized, three ATPs are produced. This combination of mitochondrial electron transport and oxidative phosphorylation produces most cellular ATP.

oxidative phosphorylation

Several important reaction sequences depend on electron transport and oxidative phosphorylation to produce ATP. Cells can derive energy from fats (see Chapter 35) only when oxidative phosphorylation is functioning. The citric acid cycle (see Chapter 34), which completes the oxidation of most nutrients, forms ATP using oxidative phosphorylation. Because electron transport and oxidative phosphorylation require oxygen, the processes that depend on this means of producing ATP are aerobic. Oxygen must be available during oxidative phosphorylation. Thus the major energy-producing reaction sequences in the cell function only in the presence of oxygen.

FIGURE 33.8
Mitochondrial electron transport and oxidative phosphorylation. (a) FADH₂ oxidation releases less energy than (b) NADH oxidation.

(a)

(b)

Practice 33.3

How many moles of ATP can be produced from 0.25 mol of $(NADH + H^+)$ and 0.45 mol of $FADH_2$ during mitochondrial electron transport and oxidative phosphorylation?

33.6 Photosynthesis

Light from the sun is the original source of nearly all energy for biological systems. Many kinds of cells can transform chemical energy to a form useful for doing work. However, there are also cells that can transform sunlight into chemical energy. Such cells use **photosynthesis,** a process by which energy from the sun is converted to chemical energy that is stored in chemical bonds.

Photosynthesis is performed by a wide variety of organisms, both eucaryotic and procaryotic. Besides the higher plants, photosynthetic eucaryotes include multicellular green, brown, and red algae and unicellular organisms such as euglena. Pho-

photosynthesis

tosynthetic procaryotes include the green and purple bacteria and the blue-green algae. Although the photosynthetic importance of higher plants is usually emphasized, it has been estimated that more than half of the world's photosynthesis is carried out by unicellular organisms.

Photosynthesis in higher plants is a complex series of reactions in which carbohydrates are synthesized from atmospheric carbon dioxide and water:

$$6\ CO_2 + 6\ H_2O + 2820\ kJ \longrightarrow \underset{\text{glucose}}{C_6H_{12}O_6} + 6\ O_2$$

Sunlight provides the large energy requirement for this process. An important side benefit of photosynthesis is the generation of oxygen, which is crucial to all aerobic metabolism.

The 1961 Nobel prize in chemistry was awarded to the American chemist Melvin Calvin (1911–1992), of the University of California at Berkeley, for his work on photosynthesis. Calvin and his co-workers used radioactive carbon tracer techniques to discover the details of the complicated sequence of chemical reactions that occur in photosynthesis.

Photosynthesis traps light energy by reducing carbons. For eucaryotes the necessary electron-transfer reactions are segregated in the chloroplast (see Figure 33.4). Like the mitochondrion, the chloroplast contains an electron-transport system within its internal membranes. Unlike the mitochondrial system, which oxidizes coenzymes to liberate energy, the chloroplast electron-transport system reduces coenzymes with an input of energy:

$$NADP^+ + H_2O + Energy \xrightarrow[\text{electron transport}]{\text{chloroplast}} NADPH + H^+ + \tfrac{1}{2}O_2$$

The chloroplasts capture light energy and place it in chemical storage.

The photosynthetic mechanism is complex, but it can be divided into two general components—the *dark reactions* and the *light reactions.* The dark reactions produce glucose from carbon dioxide, reduced coenzymes, and ATP. No light is needed and, in nature, these reactions continue during the night.

The light reactions of photosynthesis form the ATP and NADPH needed to produce glucose. The mechanism for capturing light energy is unique to the photosynthetic process. Although much research has been devoted to this topic, not all of the details are clear. In general, light is absorbed by colored compounds (pigments) located in the chloroplasts. The most abundant of these pigments is chlorophyll. Once the light energy is absorbed, it is transferred to specific molecules (special chlorophylls) that lose electrons. These energized electrons travel through the chloroplast electron-transport system, as shown in Figure 33.9. Two events follow in quick succession: First, the electrons lost by these special chlorophylls are moved to higher energy levels until they can reduce molecules of the coenzyme $NADP^+$. Second, the special chlorophylls that lost electrons now regain them. Water is the electron donor, giving up electrons and producing oxygen gas (and hydrogen ions) in the process. The overall redox reaction moves four electrons from two water molecules to produce two molecules of NADPH.

$$2\ H_2O + 2\ NADP^+ \xrightarrow{\text{light}} 2\ NADPH + O_2 + 2\ H^+$$

▲
Green algae shows spiral chloroplasts within the cytoplasm in the cells.

▲
This greenhouse creates ideal conditions for photosynthesis.

Artificial Photosynthesis

We use huge quantities of energy in our modern society and are constantly looking for more. Our petroleum supplies are being depleted, and scientists are searching for a cleaner source of energy. Where are they looking? Toward the sun, the same source of energy used by plants during photosynthesis. Sunlight is a readily available source *if* an inexpensive way to trap this energy can be found.

Solar cells (photovoltaic cells) can be used to power your calculator, but this approach to trapping the sun's energy requires expensive, carefully produced components. An alternative approach copies photosynthesis—that is, uses sunlight to drive a chemical reaction.

Imagine drawing some tap water and generating enough hydrogen and oxygen to power your cooking and heating. The simple artificial photosynthetic cell shown in the photograph does just that. Like natural photosynthesis, artificial photosynthesis requires several components (summarized in the diagram):

1. There must be a means of absorbing the light energy (a photosensitive electrode).
2. Electrons must be energized and moved (electrodes and wires move electrons from one solution to another).
3. The energized electrons must be used to make high-energy chemicals (hydrogen and oxygen gases are often produced).

An artificial photosynthetic cell is much cheaper and easier to build than a photovoltaic cell. In fact, most households have the wires, glass jars, and water necessary to build artificial photosynthetic cells. What is missing is a photosensitive electrode that does not

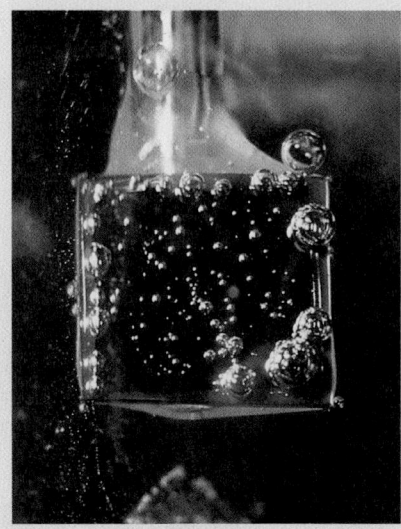

▲ **Photograph of an artificial photosynthetic cell generating hydrogen gas.**

corrode over time. A silicon electrode, for example, is converted to silicon dioxide as it absorbs light in an aqueous solution. Photosynthesis runs for a brief time and then stops because of electrode corrosion.

Scientists appear to be close to providing an inexpensive solution to the electrode corrosion problem. Titanium dioxide (the white pigment in many paints) forms a noncorroding electrode but only absorbs ultraviolet light (a small fraction of the total sunlight). Engineers have borrowed an idea from natural photosynthesis; by adding colored pigments, the amount of absorbed light can be increased. Titanium dioxide can be coated with colored dyes to absorb up to 90% of all sunlight.

Using inexpensive materials, artificial photosynthesis can capture and convert about 20% of the sun's energy. This may seem to be a small percentage, but natural photosynthesis achieves only about a 3% conversion. Furthermore, consider the amount of energy that strikes the earth's surface every day. Even 20% represents a huge amount of usable energy. Today, the biggest problem with artificial photosynthesis lies with the photosensitive electrode. However, a commercial device is already in production in the United States. Within the next 15 to 20 years, artificial photosynthesis may trigger a revolution in the cost of energy.

▲ **Schematic diagram of an artificial photosynthetic cell.**

◀ **FIGURE 33.9**
The movement of electrons from water to NADP$^+$ in the photosynthetic electron transport pathway: Note that light energy causes the electrons to become more energetic so that they can reduce NADP$^+$.

Photosynthesis uses light energy to force electrons to higher energy levels, as shown in Figure 33.9. As noted in the preceding paragraph, these energetic electrons are used to reduce NADP$^+$. But also notice that, when the electron loses energy in the middle of this electron-transport process, the released energy is used to make ATP. Thus the light reactions of photosynthesis supply both the NADPH and ATP needed to make glucose.

Concepts in Review

1. Explain how an anabolic process differs from a catabolic process.

2. List three major classes of biochemical substances that provide metabolic energy.

3. Explain how reduced carbon atoms are important in the production of cellular energy.

4. Explain why fats generally contain more energy than carbohydrates.

5. Explain the difference between a procaryotic and a eucaryotic cell.

6. Define the role of the mitochondria in metabolism.

7. Briefly describe electron transport.

8. Discuss the function of NAD$^+$, NADP$^+$, and FAD in biological processes.

9. Explain the importance of the high-energy phosphate bond in metabolism.

10. Give two advantages for using ATP as a common energy currency for the cell.

11. Contrast substrate-level phosphorylation with oxidative phosphorylation.

12. Compare the role of the mitochondrion with that of the chloroplast.

13. Outline the principal steps in the overall process of photosynthesis.

Key Terms

The terms listed here have been defined within this chapter. Section numbers are referenced in parentheses for each term.

anabolism (33.2)
bioenergetics (33.1)
catabolism (33.2)

eucaryote (33.2)
metabolism (33.2)
organelle (33.2)

oxidative phosphorylation (33.5)
photosynthesis (33.6)

procaryote (33.2)
substrate-level phosphorylation (33.5)

Questions

Questions refer to tables, figures, and key words and concepts defined within the chapter. A particularly challenging question or exercise is indicated with an asterisk.

1. Contrast the type of energy delivered to plants with that delivered to animals.

2. Describe a common energy conversion found in both plants and animals.

3. Why are fats and carbohydrates good sources of cellular energy?

4. Give a general structure for the most common high-energy phosphate bond found in the cell. With what compound is it generally associated?

5. Why is ATP known as the "common energy currency" of the cell?

6. What is an oxidation–reduction coenzyme?

7. Draw the ring structure portion of NAD^+ that becomes reduced during metabolism.

8. How does oxidative phosphorylation differ from substrate-level phosphorylation?

9. In what part of the eucaryotic cell does oxidative phosphorylation occur?

10. Compare the structural similarities between choroplasts and mitochondria.

11. What role do chloroplast pigments serve in photosynthesis?

12. Give the overall reaction for photosynthesis in higher plants.

Paired Exercises

These exercises are paired. Each odd-numbered exercise is followed by a similar even-numbered exercise. Answers to the even-numbered exercises are given in Appendix VI.

13. Which molecule delivers more energy in biological redox?

    CH_3OH or CO_2

    Briefly explain.

14. Which molecule delivers more energy in biological redox?

$$\overset{O}{\overset{\|}{CH_3CH}} \quad \text{or} \quad \overset{O}{\overset{\|}{CH_3C-OH}}$$

    Briefly explain.

15. What is wrong with the following equation for biological redox?

    $NADH + FAD \rightarrow NAD^+ + FADH_2$

    Briefly explain.

16. What is wrong with the following equation for biological redox?

    $NAD^+ + FAD \rightarrow NADH + H^+ + FADH_2$

    Briefly explain.

17. Many cells convert glucose to carbon dioxide. Is this an anabolic or a catabolic process? Briefly explain.

18. Many cells convert acetate to long-chain fatty acids. Is this an anabolic or a catabolic process? Briefly explain.

19. Your friend states, "The chloroplasts in this procaryotic cell are undergoing photosynthesis." What is wrong with this statement?

20. Your friend states, "The chloroplasts in this eucaryotic cell are undergoing oxidative phosphorylation." What is wrong with this statement?

21. Explain why the chemical changes in the mitochondria are said to be catabolic.

22. Explain why the chemical changes in the chloroplasts are said to be anabolic.

23. List three important characteristics of an anabolic process.

24. List three important characteristics of a catabolic process.

25. In mitochondrial electron transport, how many moles of oxygen gas are reduced to water by 2.38 mol of $FADH_2$?

26. In mitochondrial electron transport, how many moles of oxygen gas are reduced to water by 0.67 mol of NADH?

27. In mitochondrial chemical processes, how many moles of electrons will reduce 11.75 mol of NAD^+ to NADH?

28. In mitochondrial chemical processes, how many moles of electrons will reduce 0.092 mol of FAD to $FADH_2$?

29. How many high-energy phosphate anhydride bonds are contained in the following formula?

$$
\begin{array}{c}
\text{CH}_2\text{—O—P—O}^- \\
\text{(structure)}
\end{array}
$$

30. How many high-energy phosphate anhydride bonds are contained in the following formula?

$$
\begin{array}{c}
\text{CH}_2\text{—O—P—O}^- \\
\text{(structure)}
\end{array}
$$

31. The equation for conversion of ADP to ATP shows an energy requirement of 35 kJ. How much energy would be required to convert 0.55 mol of ADP to ATP?

32. The equation for conversion of ATP to ADP shows an energy release of 35 kJ. How much energy would be released from 1.65 mol of ATP?

33. The following compound is formed during glucose catabolism:

$$
\begin{array}{c}
\text{COOH} \quad \text{O}^- \\
\text{(structure)} \\
\text{CH}_2 \quad \text{O}
\end{array}
$$

How many ATPs can be formed from one molecule of this compound during substrate-level phosphorylation? Explain.

34. The following compound is formed during glucose catabolism:

$$
\begin{array}{c}
\text{O} \quad \text{O}^- \\
\text{C—O—P—O}^- \\
\text{(structure)} \\
\text{CHOH} \\
\text{CH}_2\text{—O—P—O}^-
\end{array}
$$

How many ATPs can be formed from one molecule of this compound during substrate-level phosphorylation? Explain.

35. How many ATPs would be formed from 4 NADH and 2 $FADH_2$ using mitochondrial electron transport and oxidative phosphorylation?

36. How many ATPs would be formed from 2 NADH and 3 $FADH_2$ using mitochondrial electron transport and oxidative phosphorylation?

Additional Exercises

These exercises are not paired and provide additional practice on concepts covered in this chapter.

37. One of your student colleagues has just told you that photosynthesis is a process by which carbon atoms are oxidized. Do you agree? Briefly explain.

38. Based on the function of the mitochondria, explain why these organelles are composed of about 90% membrane by mass.

39. Could a higher plant cell survive with chloroplasts but no mitochondria? Explain.

40. What structural similarities, if any, exist between ATP and NAD^+?

Answers to Practice Exercises

33.1 3-Hydroxybutyric acid is likely to provide more energy because the carbons are more reduced than those in D-erythrose.

33.2 The oxidized form (FAD) should react with CH_3OH to yield the reduced form ($FADH_2$) and $H_2C{=}O$.

33.3 1.65 mol of ATP

Carbohydrate Metabolism

The contestants in this bicycle race will metabolize a great amount of carbohydrates before the race is over.

CHAPTER 34 / OUTLINE

Have you ever stopped to think about why carbohydrates are a staple of our world? Carbohydrates are everywhere. Our diets consist mainly of carbohydrates, many building materials in our homes are carbohydrates, and most of the plant kingdom consists of carbohydrates. But *why* is this class of molecules so prevalent? Biochemists might answer this question in a number of ways, but one of the most important answers is in the air around us. Carbohydrates can be made from carbon dioxide, water, and sunlight through *photosynthesis,* so the constituents for carbohydrate metabolism are incredibly abundant.

Of course, a solid carbohydrate like sugar, which is sweet to the taste, has little in common with gaseous carbon dioxide, which is tasteless and can be hazardous in high concentrations. But a series of chemical reactions can convert carbon dioxide and water to sugar. Photosynthesis is an example of carbohydrate metabolism and illustrates the fact that metabolism can achieve remarkable conversions. Our world depends on such metabolic processes as photosynthesis.

34.1 Metabolic Pathways

Biochemists define a **metabolic pathway** as a series of biochemical reactions that serve a specific purpose. Carbohydrates are rich sources of energy for most organisms. As we will see in this chapter, a number of different metabolic pathways (and a large number of chemical reactions) are required to obtain energy.

metabolic pathway

Sunlight provides the energy that is stored in carbohydrates:

$$6\ CO_2 + 6\ H_2O + 2820\ kJ \xrightarrow[\substack{from \\ sunlight}]{\substack{enzymes \\ chlorophyll}} C_6H_{12}O_6 + 6\ O_2 \qquad (1)$$

Equation (1) summarizes the production of glucose (a carbohydrate) and oxygen by the endothermic process called *photosynthesis.* Note that, in the overall transformation, carbon atoms from carbon dioxide are reduced and oxygen atoms from water are oxidized. The light energy is used to cause a net movement of electrons from water to carbon dioxide. As the carbons become more reduced, more energy is stored. Photosynthesis also supplies free oxygen to the atmosphere and is an example of an anabolic pathway.

Equation (2) represents the oxidation of glucose and corresponds to the reversal of the overall photosynthesis reaction. Electrons move from carbohydrate carbons to oxygen. Energy stored in the reduced carbon atoms in glucose is released and can then be used by the cell to do work. This is an example of a catabolic process.

$$C_6H_{12}O_6 + 6\ O_2 \xrightarrow{enzymes} 6\ CO_2 + 6\ H_2O + 2820\ kJ \qquad (2)$$

Glucose can be burned in oxygen in the laboratory to produce carbon dioxide, water, and heat. But in the living cell, the oxidation of glucose does not proceed directly to carbon dioxide and water. Instead, like photosynthesis, the overall process proceeds by a series of enzyme-catalyzed intermediate reactions. These intermediate steps channel some of the liberated energy into uses other than heat production. Specifically, a portion of the energy is stored in the chemical bonds of ATP.

Practice 34.1

Is the following chemical process catabolic or anabolic? Briefly explain.

$$
\begin{array}{c}
\text{COOH} \\
| \\
\text{HOCH} \\
| \\
\text{HC}-\text{COOH} \\
| \\
\text{CH}_2 \\
| \\
\text{COOH}
\end{array}
+ \text{NAD}^+ \longrightarrow
\begin{array}{c}
\text{COOH} \\
| \\
\text{C}=\text{O} \\
| \\
\text{CH}_2 \\
| \\
\text{CH}_2 \\
| \\
\text{COOH}
\end{array}
+ \text{NADH} + \text{H}^+ + \text{CO}_2
$$

34.2   Exercise and Energy Metabolism

Working or playing, studying or watching TV—everything we do requires metabolic energy. Sometimes we need the energy quickly, sometimes it can be delivered more slowly. Metabolism is a complex interplay: Chemical reactions within cells lead to chemical transport between cells.

Carbohydrate catabolism is designed to release energy relatively quickly, so this form of catabolism is activated during strenuous muscular exercise. When a muscle contracts, energy is consumed. Muscle contraction, as with almost all mechanical work, uses ATP. But, just like ready cash, ATP is in short supply. Muscle tissue can contract for no more than several seconds before the supply of high-energy phosphate bonds is depleted.

After the initial contraction, the muscle cells look for other energy sources. Muscle glycogen is the next available source. This polymer breaks down to glucose, which is oxidized to replenish the ATP supply. Because glucose oxidation is an involved process, muscle contraction must proceed at a slower rate. Unfortunately, even this energy supply is only useful for about 2 minutes of work. Muscles rapidly deplete their glycogen stores and rapidly build up lactic acid.

If this were the end of the story, our work schedule would always be very short. In fact, certain animals respond in just this way. For example, an alligator can run quickly, thrash its tail powerfully while capturing a prey, but only for several minutes. The alligator's exertion must be followed by a rest of several hours as the muscles recuperate their energy stores. Humans (and other mammals) are not so limited because their livers replenish the muscle glucose supply and remove lactic acid. Muscle contractions can continue with help from the liver, but the rate of contraction is slowed further as muscle cells wait on glucose transport.

▲
Even relaxing and watching TV requires metabolic energy.

34.3   The Bloodstream: A Metabolic Connection

The bloodstream transports chemicals from one cell to another. Nutrients (e.g., glucose, amino acids, and fatty acids) and oxygen are delivered; metabolic products (e.g., lactate) and carbon dioxide are removed.

Human muscle cells, like many other cells, must have an adequate oxygen supply. About 5 g of oxygen (4 L at 25° C and 1 atm) is required for every minute of strenuous activity by a young adult. The necessary oxygen is supplied by arterial blood (using the oxygen-transport protein, hemoglobin, Section 29.8) and waste carbon dioxide is removed by venous blood.

Second, hydrogen ions are formed as the muscles work. Two products of cellular metabolism—carbonic acid and lactic acid—are the primary sources of this acidity.

$$CO_2 + H_2O \rightleftharpoons H_2CO_3 \rightleftharpoons H^+ + HCO_3^-$$

$$\underset{\text{lactic acid}}{CH_3CH(OH)COOH} \rightleftharpoons H^+ + \underset{\text{lactate}}{CH_3CH(OH)COO^-}$$

When the blood reaches the lungs, these processes are reversed. Oxygen is inhaled and carbon dioxide is exhaled. The blood becomes less acidic and hemoglobin binds more oxygen. The overall relationship between the muscles and lungs can be summarized in the following way.

Energy use at the muscle	Gas balance restored at the lungs
O_2 used	O_2 inhaled
CO_2 formed	CO_2 exhaled
Acidity increases	Acidity decreases
More O_2 released from hemoglobin	More O_2 binds to hemoglobin

Many nutrient levels in the bloodstream are maintained by the liver. For example, the liver sets normal blood-glucose levels (70–90 mg/100 mL of blood for a healthy person before eating) by either releasing glucose to the blood or removing glucose from the blood. When blood concentrations of lactate increase, it is converted back to glucose in the liver. The relationship between the muscles and the liver can be summarized as follows.

Energy use at the muscle	Nutrient balance restored at the liver
Glucose consumed	Glucose released
Lactate produced	Lactate converted to glucose
Acidity increases	Acidity decreases

Figure 34.1 relates the blood levels of glucose and lactate to cellular metabolic processes within the muscle and liver. As muscles or other tissues use more glucose, the liver increases glucose output to the blood. After a meal, the liver removes excess glucose from the blood.

When blood glucose is in excess, it is converted to glycogen in the liver and in muscle tissue. Glycogen is a storage polysaccharide; it quickly hydrolyzes to replace depleted glucose supplies in the blood. The synthesis of glycogen from glucose is

FIGURE 34.1 ▶
Overview of carbohydrate metabolism.

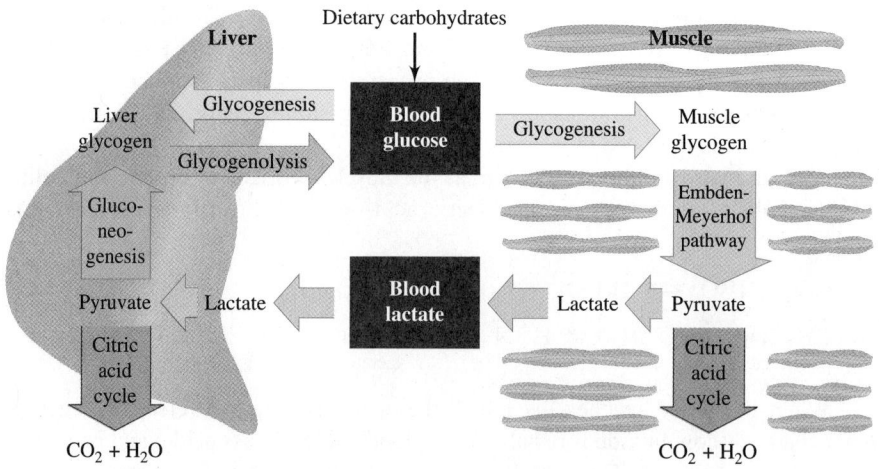

glycogenesis
glycogenolysis

called **glycogenesis;** the hydrolysis, or breakdown, of glycogen to glucose is known as **glycogenolysis** (see Figure 34.1).

Glucose supplies cellular energy via several metabolic pathways. Under anaerobic (oxygen not required) conditions, a general sequence of reactions known as the *Embden–Meyerhof pathway* oxidizes glucose. This reaction sequence is especially important for humans (and other animals) during strenuous activity when muscle tissue generates much lactate. Part of the lactate produced is sent to the liver and is converted to glucose in the *gluconeogenesis* pathway and then to glycogen. The rest is converted to carbon dioxide and water via the *citric acid cycle,* an aerobic (free or respiratory oxygen required) sequence. The *citric acid cycle* is also known as the Krebs cycle or the tricarboxylic acid cycle.

The importance of nutrient transport between cells is illustrated by the disease type II diabetes mellitus (also known as adult-onset or noninsulin-dependent diabetes). This disease affects about 10 million people in the United States and an estimated 120 million people worldwide. Muscle cells of diabetics lose their ability to import glucose because these cells no longer respond normally to the hormone insulin (see Section 34.9). Uncontrolled type II diabetes is characterized by high blood glucose and weak muscles; the muscles become starved for glucose. Often type II diabetes is controlled with exercise, diet, and the sulfonurea drugs (which enhance insulin secretion). Yet, this "breakdown" in nutrient transport is not cured and diabetics face a life of continuing health problems.

Figure 34.1 gives a general overview of carbohydrate metabolism. The entire process requires specific enzymes at each reaction stage and is controlled by chemical regulatory compounds. The net result of these reactions is a conversion of carbohydrate to CO_2 and H_2.

 Anaerobic Sequence

In the absence of oxygen, glucose in living cells is converted to a variety of end products including lactic acid (in muscle) and alcohol (in yeast). A similar conversion occurs in many different cells. At least a dozen reactions, many different

Carbo-loading to Improve Athletic Performance

A typical depletion-taper precompetition program for glycogen-loading		
Day	**Exercise**	**Diet**
1	90-min period at maximum work	Mixed diet (50% carbohydrate)
2–3	Gradual tapering of time and intensity of work	Mixed diet (50% carbohydrate)
4–5	Tapering of time and intensity of work continues	Mixed diet (70% carbohydrate)
6	Complete rest	Mixed diet (70% carbohydrate)
7	Day of competition	High-carbohydrate pre-event diet

The day is at hand. Athletes from around the world arrive at the stadium. As they warm up, they picture the work that lies ahead of them. The marathoners are preparing to run about 26 miles in a little over 2 hours of time. A premier event of the Summer Olympics is about to begin. These athletes have been training for this moment for years. Their bodies show evidence of consistent training—a high percentage of lean body mass and well-developed leg muscles for running. Perhaps as important, these athletes have prepared their body's metabolism to endure the marathon run.

In their pursuit of excellence, endurance athletes are constantly striving for the "winning edge." A new schedule of road work or even a new design for running shoes may provide this winning edge. Diet has always been an important part of training. As scientists have learned more about exercise, they have recognized the importance of carbohydrates in a training diet. Today a carbohydrate-loading diet is an accepted means toward gaining the winning edge. Perhaps you are familiar with athletes who "carbo-load" before events that require endurance. These competitors are trying to increase their body's glycogen stores knowing that carbohydrates provide important metabolic energy.

How effective is a carbohydrate-loading diet? To answer this question, we need to focus on the relationship between diet, metabolism, and exercise. A typical carbohydrate-loading diet is a timed sequence of dietary change coupled with changes in the exercise regimen. An example of such a diet is given in the table.

Diet has an important effect on glycogen stores. A low or carbohydrate-free diet does not provide for the replacement of glycogen stores. After several days on a carbohydrate-free diet (or about 1 day of complete starvation), the liver's glycogen stores are depleted. A 50% carbohydrate diet does not maintain glycogen stores for endurance athletes in training. On the glycogen-loading program, athletes in endurance training deplete their liver glycogen from the first day through the third day.

Glycogen is needed for work—for exercise. So the body attempts to maintain a supply of muscle glycogen. Even after 8 days on a carbohydrate-free diet (or 5 days of total starvation), muscle-glycogen levels have only decreased by about one-third. However, any loss of glycogen decreases the amount of work that can be done. Shifting from a normal diet to carbohydrate-free diet for as little as 3 days can halve the time muscles will work before exhaustion. On this diet, the liver no longer has much glycogen to contribute. After about 1 hour of work, people on carbohydrate-free diets become hypoglycemic. Athletes on the glycogen-loading program must taper their exercise from the second day until competition as their glycogen stores are depleted.

Increased glycogen stores can increase athletic performance. By exercising on a daily basis and by depleting the body's glycogen stores, an athlete prepares for carbo-loading. A switch to a carbohydrate-rich diet for about 3 days (from the fourth day through the seventh day on the above program) increases both the liver and muscle glycogen levels. In fact, these tissues are stimulated to store more than the normal amount of glycogen; they are "loaded" with glycogen. With the extra glycogen comes the ability to work longer. Some studies report an almost tripling of the work time available to the muscles before exhaustion. Carbo-loading before endurance athletics follows sound biochemical principles.

FIGURE 34.2 ▶
Ten-step conversion of glucose to pyruvate via the Embden–Meyerhof pathway (anaerobic sequence). Glycolysis proceeds further, forming lactate. Formation of ATP is denoted in green. The only oxidation–reduction reaction in this pathway is shown in red.

enzymes, ATP, and inorganic phosphate (P_i) are required. Such a sequence of reactions from a particular reactant to end products is called a metabolic pathway.

Embden–Meyerhof pathway

The anaerobic conversion of glucose to pyruvate is known as the **Embden–Meyerhof pathway** (see Figure 34.2). The sequence is a catabolic one in which glucose is oxidatively degraded:

Carbon oxidation commonly provides energy for life (see Section 27.3).

$$C_6H_{12}O_6 \xrightarrow[\text{pathway}]{\text{Embden–Meyerhof}} 2\ CH_3\overset{\displaystyle O}{\overset{\displaystyle \|}{C}}{-}COO^-$$

D-glucose
(sum of oxidation numbers
for six carbons = 0)

pyruvate
(sum of oxidation numbers
for six carbons = +4)

In glucose the sum of the oxidation numbers of the six carbon atoms is zero. After processing one molecule of glucose to two molecules of pyruvate by the pathway, the sum of the oxidation numbers of the six carbon atoms is +4. This makes it evident that an overall oxidation of carbon must occur in the Embden–Meyerhof pathway.

As with most catabolic processes, the Embden–Meyerhof pathway produces energy for the cell. Carbon atoms are oxidized; energy is released and stored in the form of ATP. The pathway uses the process of substrate-level phosphorylation (see Section 33.5)—the energy released from carbon oxidation is used to form a high-energy substrate–phosphate bond, which in turn is used to form ATP. It is interesting to note that there is only one oxidation–reduction reaction in the Embden–Meyerhof pathway—the conversion of glyceraldehyde-3-phosphate to 1,3-bisphosphoglycerate (marked in red on Figure 34.2):

glyceraldehyde-3-phosphate $\qquad$ 1,3 bisphosphoglycerate

Carbon 1 is oxidized from the aldehyde to the carboxylate oxidation state. Simultaneously, the oxidation–reduction coenzyme NAD^+ is reduced. This single oxidation–reduction supplies most of the energy generated by the Embden–Meyerhof pathway, which is used to make these two different high-energy phosphate bonds:

High-energy phosphate bond

1,3-bisphosphoglycerate

High-energy phosphate bond

phosphoenolpyruvate

These high-energy bonds are found on intermediate compounds or pathway substrates. Substrate-level phosphorylation is complete when these substrates transfer their high-energy phosphate bonds to ADP, forming ATP (the ATP formation is shown in green on Figure 34.2).

1,3-bisphosphoglycerate

adenosine diphosphate (ADP)

adenosine triphosphate (ATP)

3-phosphoglycerate

$$H^+ + \underset{\substack{\| \\ CH_2}}{\overset{\substack{COO^- \\ |}}{C}} - O \sim \underset{\substack{\| \\ O}}{\overset{\substack{O^- \\ |}}{P}} - O^- + {}^-O - \underset{\substack{\| \\ O}}{\overset{\substack{O^- \\ |}}{P}} \sim O - \underset{\substack{\| \\ O}}{\overset{\substack{O^- \\ |}}{P}} - OCH_2 \quad O \quad Adenine \longrightarrow$$

OH OH

phosphoenolpyruvate

adenosine diphosphate (ADP)

$${}^-O - \underset{\substack{\| \\ O}}{\overset{\substack{O^- \\ |}}{P}} \sim O - \underset{\substack{\| \\ O}}{\overset{\substack{O^- \\ |}}{P}} \sim O - \underset{\substack{\| \\ O}}{\overset{\substack{O^- \\ |}}{P}} - OCH_2 \quad O \quad Adenine \quad + \quad \underset{\substack{| \\ CH_3}}{\overset{\substack{COO^- \\ |}}{C}}=O$$

OH OH

pyruvate

adenosine triphosphate (ATP)

It is important to note that the Embden–Meyerhof pathway oxidizes carbon and produces ATP in the absence of molecular oxygen. This pathway provides for *anaerobic* energy production. For this reaction sequence to continue, the coenzyme NADH must be recycled back to NAD$^+$; that is, an additional oxidation–reduction reaction is needed to remove electrons from NADH. In the absence of oxygen, pyruvate is used as an electron acceptor. In human muscle cells, pyruvate is reduced directly to lactate:

$$\underset{\substack{\| \\ O}}{CH_3CCOO^-} + NADH + H^+ \rightleftharpoons \underset{\substack{| \\ OH}}{CH_3CHCOO^-} + NAD^+$$

pyruvate   lactate

In yeast cells, pyruvate is converted to acetaldehyde, which is then reduced to ethanol:

$$\underset{\substack{\| \\ O}}{CH_3CCOO^-} + H^+ \xrightarrow[\text{cells}]{\text{in yeast}} \underset{\substack{\| \\ O}}{CH_3C}-H + CO_2$$

pyruvate   acetaldehyde

$$\underset{\substack{\| \\ O}}{CH_3C}-H + NADH + H^+ \rightleftharpoons CH_3CH_2OH + NAD^+$$

acetaldehyde   ethanol

In each case, NADH is reoxidized to NAD$^+$, and a carbon atom from pyruvate is reduced. When lactate is the final product of anaerobic glucose catabolism, the

glycolysis   pathway is termed **glycolysis.** As the Embden–Meyerhof pathway produces equal amounts of pyruvate and NADH, there is just enough pyruvate to recycle all the NADH. This is a good example of an important general characteristic of metabolism: Chemical reactions in the cell are precisely balanced so that there is never a large surplus or a large deficit of any metabolic product. If such a situation does occur, the cell may die.

What glycolysis does for the cell can be summarized with the following net chemical equation:

$$C_6H_{12}O_6 + 2\ ADP + 2\ P_i \longrightarrow 2\ CH_3CH(OH)COO^- + 2\ ATP + 150\ kJ$$

This anaerobic process is not very efficient because only 2 mol of ATP is formed per mole of glucose. In fact, if the energy stored in the two ATPs is added to the energy released as heat during the Embden–Meyerhof pathway, a total of only about 220 kJ/mol is found to have been removed from glucose. Compare this result with the complete oxidation of glucose:

$$C_6H_{12}O_6 + 6\ O_2 \longrightarrow 6\ CO_2 + 6\ H_2O + 2820\ kJ$$

We see that the Embden–Meyerhof pathway releases less than one-tenth of the total energy available in glucose. Thus lactate must still contain much energy in its reduced carbon atoms.

Practice 34.2
_____

If 1.7 mol of glucose is processed via the Embden–Meyerhof pathway, (a) how many moles of pyruvate are produced? (b) how many moles of ATP are produced?

 Citric Acid Cycle (Aerobic Sequence)

As discussed previously, only a small fraction of the energy that is potentially available from glucose is liberated during the anaerobic conversion to lactate (glycolysis). Consequently, lactate remains valuable to the cells. The lactate formed may be (1) circulated back to the liver and converted to glycogen at the expense of some ATP or (2) converted back to pyruvate in order to enter the citric acid cycle.

$$\underset{\text{lactate}}{\overset{\overset{\displaystyle OH}{|}}{CH_3CHCOO^-}} + NAD^+ \rightleftharpoons \underset{\text{pyruvate}}{\overset{\overset{\displaystyle O}{\|}}{CH_3CCOO^-}} + NADH + H^+$$

Pyruvate is the link between the anaerobic sequence (Embden–Meyerhof pathway) and the aerobic sequence (citric acid cycle). Pyruvate itself does not enter into the citric acid cycle. It is first converted to acetyl coenzyme A (acetyl CoA), a complex substance that, like ATP, is of great importance in metabolism:

$$\underset{\text{coenzyme A}}{\overset{\overset{\displaystyle O}{\|}}{CH_3C}-COO^- + CoASH + NAD^+} \longrightarrow \underset{\text{acetyl-CoA}}{\overset{\overset{\displaystyle O}{\|}}{CH_3C}-SCoA} + NADH + CO_2$$

This important reaction depends on the availability of several vitamins. In addition to niacin (needed for synthesis of NAD^+), riboflavin and thiamine are also required.

This is one of the few metabolic reactions that requires thiamine, but because this reaction is key to obtaining large amounts of energy from carbohydrate, thiamine is a vitamin of major importance. Also, our bodies need the vitamin pantothenic acid to synthesize coenzyme A.

Acetyl coenzyme A consists of an acetyl group bonded to a coenzyme A group. Coenzyme A contains the following units: adenine, ribose-3-phosphate, diphosphate, pantothenic acid, and thioethanolamine. Coenzyme A is abbreviated as CoASH or CoA. Acetyl coenzyme A is abbreviated as acetyl-CoA or acetyl-SCoA.

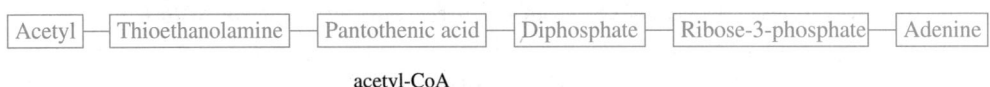

acetyl-CoA

The acetyl group is the group that is actually oxidized in the citric acid cycle. This group is attached to the large carrier molecule as a thioester—that is, by an ester linkage in which oxygen is replaced by sulfur:

$$CH_3-\overset{\overset{\displaystyle O}{\|}}{C}-S-CoA$$

acetyl group

carrier group

thioester linkage

Not only is acetyl-CoA a key component of carbohydrate metabolism, it also (as we will see in Chapter 35), serves to tie the metabolism of fats and that of certain amino acids to the citric acid cycle.

citric acid cycle   The **citric acid cycle** was elucidated by Hans A. Krebs (1900–1981), a British biochemist; thus it is also called the Krebs cycle. For his studies in intermediary metabolism, Krebs shared the 1953 Nobel prize in medicine and physiology with Fritz A. Lipmann (1899–1986), an American biochemist who discovered coenzyme A.

Krebs showed the citric acid cycle to be a series of eight reactions in which the acetyl group of acetyl-CoA is oxidized to carbon dioxide and water, and where many reduced coenzymes (both NADH and $FADH_2$) are formed. These reduced coenzymes then pass electrons through the electron-transport system and, in the process, produce ATP. The citric acid cycle occurs in the mitochondria, which are the primary sites for the production of cellular energy.

The sequence of reactions involved in the citric acid cycle is shown in Figure 34.3. It is important to note that this cycle produces little usable cellular energy directly (only one GTP, or guanosine triphosphate, convertible to ATP). However, many ATPs are produced because of two other processes—electron transport and oxidative phosphorylation (see Section 33.5). Electron transport recycles the large number of reduced coenzymes formed by the citric acid cycle. As a consequence, oxidative phosphorylation produces ATP from ADP and phosphate. Because electron transport requires oxygen and the citric acid cycle depends upon electron transport, the citric acid cycle is an aerobic process.

All three processes—the citric acid cycle (plus the initial conversion of pyruvate to acetyl-CoA), electron transport, and oxidative phosphorylation—team up to produce energy for the cell. All three take place in the mitochondria. Overall, these

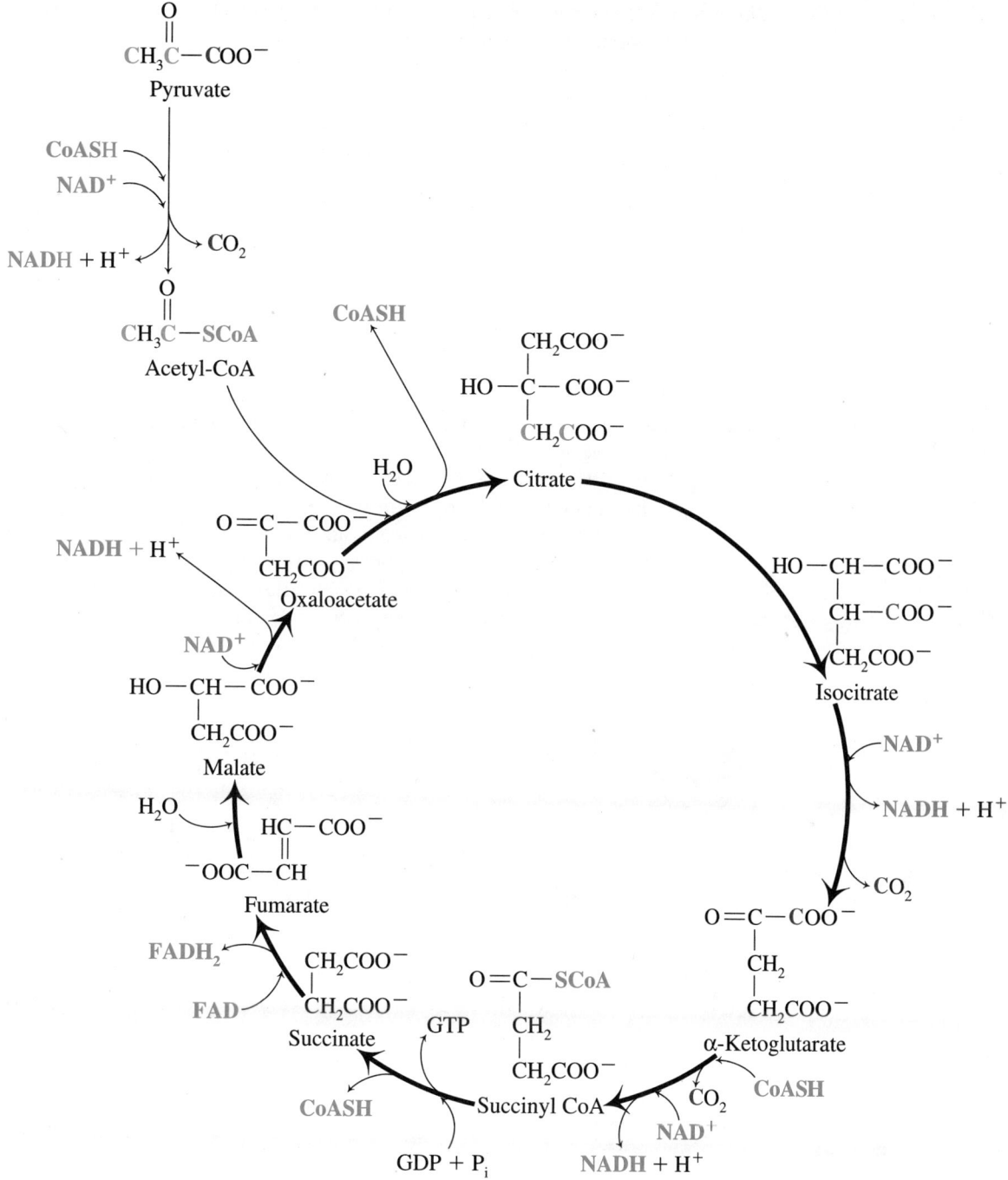

▲
FIGURE 34.3
The citric acid cycle (Krebs cycle). During one cycle, (1) the carbons marked in blue enter the cycle, and (2) the carbons marked in red are lost as CO₂. Coenzymes are marked in green. The cycle is started by the reaction of acetyl-CoA with oxaloacetate to form citrate.

three processes result in the oxidation of 1 pyruvate ion to 3 carbon dioxide molecules and the formation of a maximum of 15 ATP molecules.

$$CH_3\overset{\displaystyle O}{\overset{\displaystyle \|}{C}}-COO^- \longrightarrow 3\ CO_2$$

A large quantity of energy (1260 kJ) becomes available for the cell when these three carbon atoms are fully oxidized. Most of this energy is used to reduce coenzymes:

$$4\ NAD^+ + FAD + 10\ H^+ + 10\ e^- \longrightarrow 4\ NADH + 4\ H^+ + FADH_2$$

These coenzymes, in turn, yield a total of 14 ATP molecules via mitochondrial electron transport and oxidative phosphorylation. Each mole of ATP stores approximately 35 kJ of energy. If we include the single GTP (convertible to ATP) that is formed in the citric acid cycle, the cell obtains about 462 kJ from each mole of pyruvate that is oxidized. By using an aerobic process, the cell produces 30 ATP molecules from the 2 lactates after gaining only 2 ATPs using the anaerobic pathway—that is, glycolysis. The presence of oxygen yields a large energy bonus for the cell.

Myocardial infarction and stroke are two injuries that are especially serious because they deprive rapidly metabolizing tissue of oxygen. When the heart muscle loses at least part of its normal blood supply, myocardial infarction results; a similar occurrence in the brain results in a stroke. In both cases, a lack of blood circulation means at least part of the tissues lose their normal oxygen supply. Suddenly these cells lose most of their capacity to generate ATP. Therefore, the anaerobic process of glycolysis cannot support continued cell viability. Permanent tissue damage results from only minutes of oxygen deprivation.

Energy summation:

Practice 34.3

Using the citric acid cycle, how many moles of (a) NADH and (b) FADH$_2$ are produced from 7.8 mol of acetyl-CoA?

34.6   Gluconeogenesis

A continuous supply of glucose is needed by the body, especially for the brain and the nervous system. But the amount of glucose and glycogen present in the body will only last for about 4 hours at normal metabolic rates. Thus there is a metabolic need for a pathway that produces glucose from noncarbohydrate sources. The formation of glucose from noncarbohydrate sources is called **gluconeogenesis.**

gluconeogenesis

Most of the glucose formed during gluconeogenesis comes from lactate, certain amino acids, and the glycerol of fats. In each case, these molecules are converted to Embden–Meyerhof pathway intermediates (see Figure 34.2). As shown previously, lactate is converted to pyruvate. Amino acids first undergo *deamination* (loss of amino groups) and then are converted to pyruvate or phosphoenolpyruvate. Glycerol is converted to glyceraldehyde-3-phosphate (see Figure 34.4). Most of the steps in the Embden–Meyerhof pathway are reversed to transform these pathway intermediates into glucose.

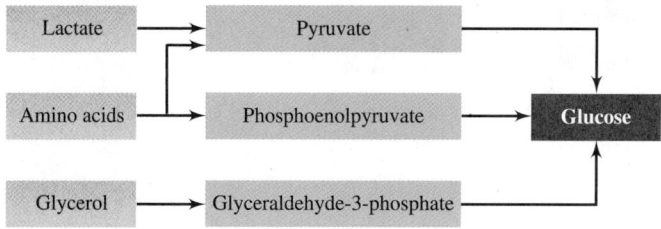

Gluconeogenesis takes place primarily in the liver and also in the kidneys. These organs have the enzymes that catalyze reversal of the Embden–Meyerhof pathway. Because of this capability the liver is the organ primarily responsible for maintaining normal blood-sugar levels.

34.6 Overview of Complex Metabolic Pathways

When we examine a metabolic pathway such as the Embden–Meyerhof pathway (Figure 34.2) or the citric acid cycle (Figure 34.3), the sheer complexity can be overwhelming. We are tempted to ask why a cell uses so many reactions to achieve its goal. Biochemists have asked this question for many years. It appears that a physiological design that includes a number of reactions per pathway achieves several vital objectives for the cell.

First, many of the chemicals that form in the middle of a pathway, *pathway intermediates,* are used in other metabolic processes within the cell.

$A \rightleftarrows B \rightleftarrows C \rightleftarrows D$ Pathway I
                          (B, C = pathway intermediates)

$B \rightleftarrows X \rightleftarrows Y \rightleftarrows Z$ Pathway II

$C \rightleftarrows I \rightleftarrows J \rightleftarrows K$ Pathway III

These intermediates are like interchangeable machine parts. Once such parts are made, they can be used in more than one machine. And once pathway intermediates are formed, they can be used in several pathways and serve a variety of metabolic functions in the cell.

Second, having multiple-step pathways helps the cell to handle metabolic energy efficiently. For many pathways the total energy available is much greater than the cell can handle in a single reaction. As an example, a one-step complete oxidation of glucose yields enough energy to cook a cell, figuratively speaking. To avoid such disasters, the cell extracts only a little energy from glucose at each chemical reaction (Figure 34.5). The quantity of energy released in each step is small enough to be handled by the cell. Thus, for a cell to extract the maximum energy, a metabolic pathway must have a number of steps.

Metabolic reaction sequences seem needlessly complex because we see only a small part of the total cellular function. Although we study the processes separately, it is important to remember that they are all interrelated.

▲
Long-distance runners use gluconeogenesis to provide a source of glucose when more readily available supplies are exhausted.

FIGURE 34.5 ▶
A single-step oxidation process compared with a multiple-step process: In the pathway, A, B, and C represent hypothetical pathway intermediates.

hormones

34.8 Hormones

Hormones are chemical substances that act as control agents in the body, often regulating metabolic pathways. They help to adjust physiological processes such as digestion, metabolism, growth, and reproduction. For example, the concentration of glucose in the blood is maintained within definite limits by hormonal action. Hormones are secreted by the endocrine, or ductless, glands directly into the bloodstream and are transported to various parts of the body where they exert specific control functions. The endocrine glands include the thyroid, parathyroid, pancreas, adrenal, pituitary, ovaries, testes, placenta, and certain portions of the gastrointestinal tract. A hormone produced by one species is usually active in some other species. For example, the insulin used to treat diabetes mellitus in humans can be obtained from the pancreas of animals slaughtered in meat packing plants.

Hormones are often called the chemical messengers of the body. They do not fit into any single chemical structural classification. Many are proteins or polypeptides, some are steroids, others are phenol or amino acid derivatives; examples are shown in Figure 34.6. Because a lack of any hormone often produces serious physiological disorders, many hormones are produced synthetically or are extracted from their natural sources and made available for medical use.

Like vitamins, hormones are generally needed in only minute amounts. Concentrations range from $10^{-6}\,M$ to $10^{-12}\,M$. Unlike vitamins, which must be supplied in the diet, the necessary hormones are produced in the body of a healthy person. A number of hormones and their functions are listed in Table 34.1.

▲
A crystal of thyroxine. This hormone is produced in the thyroid gland and controls metabolic rate.

34.9 Blood Glucose and Hormones

An adequate blood-glucose level must be maintained to ensure good health. To achieve this goal, hormones regulate and coordinate metabolism in specific organs. The hormones control selected enzymes, which, in turn, regulate the rates of reaction

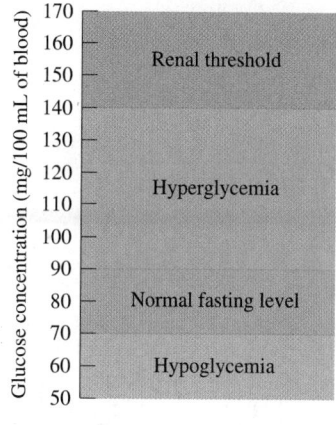

thyroxine

oxytocin

Cys-Tyr-Ile-Gln-Asn-Cys-Pro-Leu-Gly-NH$_2$

$\overset{1}{\text{His}}$-Ser-Gln-Gly-Thr-Phe-Thr-Ser-Asp-Tyr-Ser-Lys-Tyr-Leu-Asp-Ser-Arg-Arg-

$\overset{20}{\text{Ala}}$-Gln-Asp-Phe-Val-Gln-Tyr-Leu-Met-Asn-Thr

glucagon

testosterone

estradiol
(estrogen)

epinephrine
(adrenalin)

◀ **FIGURE 34.6**
Structure of selected hormones. Thyroxine is produced in the thyroid gland; oxytocin is a polypeptide produced in the posterior lobe of the pituitary gland; glucagon is a polypeptide produced in the pancreas; testosterone and estradiol are steroid hormones produced in the testes and the ovaries, respectively; epinephrine is produced in the adrenal glands.

in the appropriate metabolic pathways. Let's examine some of the physiological mechanisms that maintain proper blood-glucose levels.

Glucose concentrations average about 70–90 mg/100 mL of blood under normal fasting conditions—that is, when no nourishment has been taken for several hours. Most people are in a normal fasting condition before eating breakfast. After the ingestion of carbohydrates, the glucose concentration rises above the normal level, and a condition of *hyperglycemia* exists. If the concentration of glucose rises still further, the renal threshold for glucose is eventually reached. The *renal threshold* is the concentration of a substance in the blood above which the kidneys begin to excrete that substance into the urine. The renal threshold for glucose is about 140–170 mg/100 mL of blood. Glucose excreted by the kidneys can be detected in the urine by a test for reducing sugars (e.g., the Benedict test; see Section 27.13). When the glucose concentration of the blood is below the normal fasting level, hypoglycemia exists (see Figure 34.7).

Glucose concentration in the blood is under the control of various hormones. These hormones act as checks on one another and establish an equilibrium condition called *homeostasis*—that is, self-regulated equilibrium. Three hormones—insulin, epinephrine (adrenalin), and glucagon—are of special significance in maintaining glucose concentration within the proper limits. Insulin, secreted by the islets of Langerhans in the pancreas, reduces blood-glucose levels by increasing the rate of glycogen formation. Epinephrine from the adrenal glands and glucagon from the pancreas increase the rate of glycogen breakdown (glycogenolysis) and thereby increase blood-glucose levels. These opposing effects are summarized as follows:

▲ **FIGURE 34.7**
Conditions related to concentration of glucose in the blood.

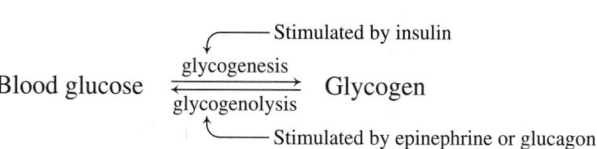

Blood glucose $\underset{\text{glycogenolysis}}{\overset{\text{glycogenesis}}{\rightleftarrows}}$ Glycogen

Stimulated by insulin

Stimulated by epinephrine or glucagon

TABLE 34.1	Selected Hormones and Their Functions	
Hormone	**Source**	**Principal functions**
Insulin	Pancreas	Controls blood-glucose level and storage of glycogen
Glucagon	Pancreas	Stimulates conversion of glycogen to glucose; raises blood-glucose level
Oxytocin	Pituitary gland	Stimulates contraction of the uterine muscles and secretion of milk by the mammary glands
Vasopressin	Pituitary gland	Controls water excretion by the kidneys; stimulates constriction of the blood vessels
Growth hormone	Pituitary gland	Stimulates growth
Adrenocorticotrophic hormone (ACTH)	Pituitary gland	Stimulates the adrenal cortex, which, in turn, releases several steroid hormones
Prolactin	Pituitary gland	Stimulates milk production by mammary glands after birth of a baby
Epinephrine (adrenalin)	Adrenal glands	Stimulates rise in blood pressure, accelerates the heartbeat, decreases secretion of insulin, and increases blood glucose
Cortisone	Adrenal glands	Helps control carbohydrate metabolism, salt and water balance, formation and storage of glycogen
Thyroxine and triiodothyronine	Thyroid gland	Increases the metabolic rate of carbohydrates and proteins
Calcitonin	Thyroid gland	Prevents the rise of calcium in the blood above the required level
Parathyroid hormone	Parathyroid gland	Regulates the metabolism of calcium and phosphate in the body
Gastrin	Stomach	Stimulates secretion of gastric juice
Secretin	Duodenum	Stimulates secretion of pancreatic juice
Estrogen	Ovaries	Stimulates development and maintenance of female sexual characteristics
Progesterone	Ovaries	Stimulates female sexual characteristics and maintains pregnancy
Testosterone	Testes	Stimulates development and maintenance of male sexual characteristics

During the digestion of a meal rich in carbohydrates, the blood-glucose level of a healthy person rises into the hyperglycemic range. This stimulates insulin secretion, and the excess glucose is converted to glycogen, thereby returning the glucose level to normal. A large amount of ingested carbohydrates can overstimulate insulin production and thereby produce a condition of mild hypoglycemia. This in turn triggers the secretion of additional epinephrine and glucagon, and the blood-glucose levels are again restored to normal. The body is able to maintain the normal fasting level of blood glucose for long periods without food by drawing on liver glycogen, muscle glycogen, and finally on body fat as glucose replacement sources. Thus, in a normal person, neither hyperglycemia nor mild hypoglycemia has serious consequences since the body is able to correct these conditions. However, either condition, if not corrected, can have very serious consequences. Since the brain is heavily dependent on blood glucose for energy, hypoglycemia affects the brain and the central nervous system. Mild hypoglycemia can result in impaired vision, dizziness, and fainting spells. Severe hypoglycemia produces convulsions and unconsciousness; if prolonged, it may result in permanent brain damage and death.

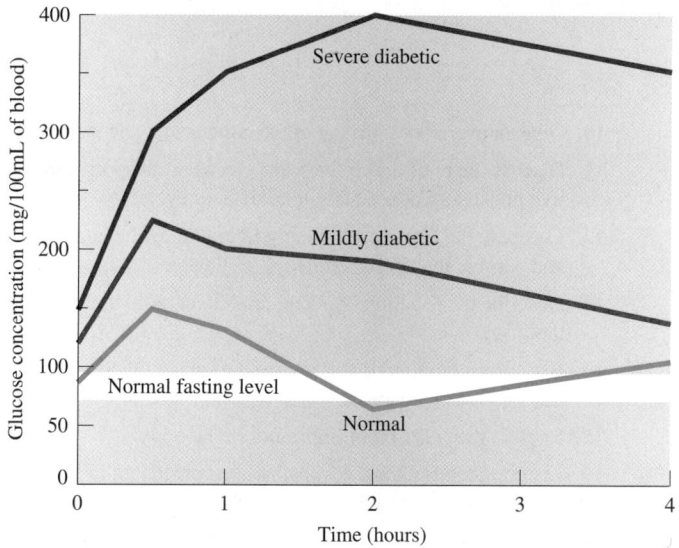

Hyperglycemia can be induced by fear or anger because the rate of epinephrine secretion is increased under emotional stress. Glycogen hydrolysis is thereby speeded up and glucose-concentration levels rise sharply. This whole sequence readies the individual for the strenuous effort of either fighting or fleeing as the situation demands.

Type I diabetes mellitus (also known as juvenile-onset or insulin-dependant diabetes) is a serious metabolic disorder characterized by hyperglycemia, glycosuria (glucose in the urine), frequent urination, thirst, weakness, and loss of weight. Prior to 1921, diabetes often resulted in death. In that year Frederick Banting and Charles Best, working at the University of Toronto, discovered insulin and devised methods for extracting the hormone from animal pancreases. For his work on insulin, Banting, with J. J. MacLeod, received the Nobel prize in medicine and physiology in 1923. Insulin is very effective in controlling diabetes. It must be given by injection, because, like any other protein, it would be hydrolyzed to amino acids in the gastrointestinal tract.

People with mild or borderline diabetes may show normal fasting blood-glucose levels, but they are unable to produce sufficient insulin for prompt control of ingested carbohydrates. As a result, their blood glucose rises to an abnormally high level and does not return to normal for a long period of time. Such a person has a decreased tolerance for glucose, which may be diagnosed by a glucose-tolerance test. After fasting for at least 12 hours, blood and urine specimens are taken to establish a reference level. The person then drinks a solution containing 100 g of glucose (amount for adults). Blood and urine specimens are collected at 0.5-, 1-, 2-, and 3-hour intervals and tested for glucose content. In a normal situation the blood-glucose level returns to normal after about 3 hours. Individuals with mild diabetes show a slower drop in glucose levels, but the glucose level in a severe diabetic remains high for the entire 3 hours. Responses to a glucose tolerance test are shown in Figure 34.8.

In 1978, the synthesis of insulin, identical in structure to that made by the human pancreas, was announced. In 1982, this insulin became the first genetically engineered human protein to receive FDA approval for sale. Today many diabetics choose to use human insulin rather than the animal insulin obtained as a by-product of meat packing.

▲
Insulin monitors are available to assist diabetics in watching their blood-sugar levels.

Concepts in Review

1. Briefly explain why carbohydrates are known as energy-storage molecules.
2. Briefly explain why enzymes are of vital importance in metabolism.
3. Outline the role of the lungs and liver in helping muscle metabolism.
4. State the purpose of the final reactions in anaerobic glucose catabolism.
5. Explain why the Embden–Meyerhof pathway is both catabolic and anaerobic.
6. Explain the role of glycogen in maintenance of blood-glucose levels.
7. Explain the role of epinephrine, insulin, and glucagon in the control of blood-glucose concentration.
8. Briefly describe how substrate-level phosphorylation is involved in the Embden–Meyerhof pathway.
9. Give an overall description of the Embden–Meyerhof pathway (anaerobic) for the metabolism of glucose.
10. Give an overall description of the citric acid cycle (aerobic).
11. Discuss the relationship between electron transport, oxidative phosphorylation, and the citric acid cycle.
12. Compare the amounts of energy liberated by the anaerobic and aerobic metabolic pathways of glucose.
13. Describe the function of acetyl-CoA in carbohydrate metabolism.
14. List three compounds that are used to make glucose during gluconeogenesis.
15. Explain the function of hormones in the body.
16. List the blood-glucose levels that are considered to be normal, hyperglycemic, and hypoglycemic.
17. Explain the renal threshold.
18. Predict what might happen to the blood-glucose level if a large overdose of insulin is taken.
19. Describe the glucose-tolerance test and how it ties in with the condition of diabetes mellitus.

Key Terms

The terms listed here have been defined within this chapter.
Section numbers are referenced in parentheses for each term.

citric acid cycle (34.5)
Embden–Meyerhof pathway (34.4)

gluconeogenesis (34.6)
glycogenesis (34.3)

glycogenolysis (34.3)
glycolysis (34.4)

hormones (34.8)
metabolic pathway (34.1)

Questions

Questions refer to tables, figures, and key words and concepts defined within the chapter. A particularly challenging question or exercise is indicated with an asterisk.

1. Why are carbohydrates considered to be energy-storage molecules?
2. List two general ways in which glucose catabolism differs from photosynthesis.
3. How much energy would be released if 3 mol of glucose is burned to form carbon dioxide and water?
4. Why are enzymes important in metabolism?
5. What form of chemical energy is first used for muscle contraction? What is used next?
6. How is the liver important in muscle contraction?
7. What is the purpose of the final reactions in anaerobic glucose metabolism?
8. How many high-energy phosphate bonds are directly formed in the citric acid cycle? List and identify the reduced coenzymes that are produced in this cycle.
9. What is acetyl-CoA and what is its function in metabolism?
10. What are the general functions of hormones and where are they produced?

11. **(a)** What is the range of glucose concentration in blood under normal fasting conditions?
    (b) What blood-glucose concentrations are considered to be hyperglycemic? Hypoglycemic?

12. What is meant by the renal threshold?

13. How does type II diabetes differ from type I diabetes?

14. Why is gluconeogenesis defined as an anaerobic process?

Paired Exercises

These exercises are paired. Each odd-numbered exercise is followed by a similar even-numbered exercise. Answers to the even-numbered exercises are given in Appendix VI.

15. Draw the structure of the carbohydrate that is produced during glycogenolysis.

16. Draw the structure of the carbohydrate that is produced during glycogenesis.

17. Would you define glycogenesis as anabolic or catabolic? Briefly explain.

18. Would you define glycogenolysis as anabolic or catabolic? Briefly explain.

19. Is the ATP production that is associated with the citric acid cycle substrate-level phosphorylation or oxidative phosphorylation? Briefly explain.

20. Is the ATP production that is associated with the Embden–Meyerhof pathway substrate-level phosphorylation or oxidative phosphorylation? Briefly explain.

21. Circle the high-energy phosphate bond(s) in the following molecule:

$$\begin{array}{c}
\text{O} \qquad\quad \text{O} \\
\| \qquad\quad \| \\
\text{C}-\text{O}-\text{P}-\text{O}^- \\
| \qquad\quad | \\
\qquad\quad \text{O}^- \\
\text{CHOH} \\
| \qquad\quad \text{O} \\
\qquad\quad \| \\
\text{CH}_2-\text{O}-\text{P}-\text{O}^- \\
| \\
\text{O}^-
\end{array}$$

22. Circle the high-energy phosphate bond(s) in the following molecule:

$$\begin{array}{c}
\text{O} \\
\| \\
\text{C}-\text{O}^- \quad \text{O} \\
| \qquad\quad \| \\
\text{C}-\text{O}-\text{P}-\text{O}^- \\
\| \qquad\quad | \\
\text{CH}_2 \qquad \text{O}^-
\end{array}$$

23. Explain why the Embden–Meyerhof pathway is considered to be anaerobic even though the oxidation of glucose occurs.

24. Explain why the citric acid cycle is considered to be aerobic even though no molecular oxygen (O_2) is used in this cycle.

25. How many moles of lactate will be formed from 2.8 mol of glucose through glycolysis?

26. How many moles of ATP will be formed from 0.85 mol of glucose through glycolysis?

27. What are the end products of the anaerobic catabolism of glucose in yeast cells?

28. What are the end products of the anaerobic catabolism of glucose in muscle tissue?

29. Is the Embden–Meyerhof pathway catabolic or anabolic? Briefly explain.

30. Is the gluconeogenesis pathway catabolic or anabolic? Briefly explain.

31. Fill in the coenzyme reactant and product for the following reaction:

$$\begin{array}{c}
\text{CH}_2\text{COO}^- \\
| \qquad\qquad +\underline{\quad} \longrightarrow \\
\text{CH}_2\text{COO}^-
\end{array}
\begin{array}{c}
\text{HC}-\text{COO}^- \\
\| \qquad\qquad +\underline{\quad} \\
\text{CH}-\text{COO}^-
\end{array}$$

32. Fill in the coenzyme reactant and product for the following reaction:

$$\begin{array}{c}
\text{HO}-\text{CHCOO}^- \\
| \qquad\qquad +\underline{\quad} \longrightarrow \\
\text{CH}_2\text{COO}^-
\end{array}
\begin{array}{c}
\text{O}=\text{C}-\text{COO} \\
| \qquad\qquad +\underline{\quad} + \text{H}^+ \\
\text{CH}_2-\text{COO}^-
\end{array}$$

33. Why are electron transport and oxidative phosphorylation needed when the cell uses the citric acid cycle to produce energy?

34. Why are electron transport and oxidative phosphorylation not needed when the cell uses the glycolysis pathway to produce energy?

35. In what way(s) are hormones like vitamins?

36. In what way(s) do hormones differ from vitamins?

37. Exercise and a low-carbohydrate diet can cause a blood-glucose concentration of 65 mg/100 mL of blood. Is this hyperglycemic or hypoglycemic? Briefly explain.

38. After a meal, a diabetic can have a blood-glucose concentration of 350 mg/100 mL of blood. Is this hyperglycemic or hypoglycemic? Briefly explain.

Additional Exercises

These exercises are not paired and provide additional practice on concepts covered in this chapter.

39. Why was the citric acid cycle not involved in the metabolism of life forms before the evolution of photosynthesis?

40. How does the function of hormones in the body differ from that of enzymes?

41. Why is the hormone insulin not effective when taken orally?

42. Predict what might happen to blood-glucose concentrations if a large overdose of insulin is taken by accident.

43. Why is epinephrine sometimes called the emergency or crisis hormone?

44. Write the structure of lactate. Would you predict that this molecule could provide cellular energy? Briefly explain.

45. In our bodies, what would be the most efficient way of using the energy produced: a multireaction system or a single-step reaction? Explain.

46. How does the use of pyruvate differ in aerobic catabolism and anaerobic catabolism?

47. What are the common structural features in NAD, FAD, ADP, ATP, and acetyl-CoA?

48. In animal metabolism, glucose is converted to carbon dioxide. Is this overall process oxidation or reduction? Explain.

49. Define the following terms: (a) glycolysis, (b) gluconeogenesis, (c) glycogenesis, (d) glycogenolysis.

50. Write the net reaction for the formation of two pyruvate from glucose via the Embden–Meyerhoff pathway. (Your answer should include NAD^+ and ADP.)

51. Write structural formulas for (a) lactate, (b) pyruvate, (c) glucose-6-phosphate, (d) fructose-6-phosphate, (e) fructose-1,6-bisphosphate, and (f) adenosine triphosphate.

Answers to Practice Exercises

34.1 The process is catabolic because (a) a six-carbon compound is converted to a five-carbon compound and CO_2, and (b) the top two carbons of the reactant are oxidized.

34.2 (a) 3.4 mol pyruvate (b) 3.4 mol ATP
34.3 (a) 23.4 mol NADH (b) 7.8 mol $FADH_2$

CHAPTER 35 / OUTLINE

Whales have an insulating layer of up to 30 cm in some species. Whale oil, extracted from this blubber, is used to make soap, leather dressing, lubricants, and hydrogenated fats. Whales provide Eskimos with food, clothing, illumination, and cooking oil.

CHAPTER
35

One of the world's most pervasive nutritional problems is protein deficiency, known as kwashiorkor. It especially afflicts children and, when untreated, has a mortality rate of between 30 and 90%. These young people suffer from growth retardation, anemia, liver damage, and often appear bloated because of excess water absorption.

In more affluent societies, nutritional problems are often associated with a high intake of saturated fat—stroke and heart disease are closely correlated to lipid intake.

These disparate nutritional problems point toward an important similarity between protein and lipid metabolism. Our biochemical processes have needs for specific amino acids (proteins) and lipids. No matter how much food is available for our diet, we must also be concerned with meeting requirements for selected nutrients.

35.1 Metabolic Energy Sources

The ability to produce cellular energy is a vital characteristic of every cell's metabolism. As we have seen, carbohydrates are one of the major sources of cellular energy. The other two are lipids and proteins.

Of all the lipids, fatty acids are the most commonly used for cellular energy. Each fatty acid contains a long chain of reduced carbon atoms that can be oxidized to yield energy.

$$CH_3CH_2CH_2CH_2CH_2CH_2CH_2CH_2CH_2CH_2CH_2CH_2CH_2CH_2CH_2COOH$$
palmitic acid

The average oxidation number of the carbon atoms in fatty acids is about -2 compared with 0 in carbohydrates. Thus, when catabolized (oxidized), fatty acids yield more energy per carbon atom than do carbohydrates.

Proteins (amino acids) are also a source of reduced carbon atoms that can be catabolized to provide cellular energy. In addition, amino acids provide the major pool of usable nitrogen for cells. Proteins and amino acids also perform diverse functions. Some of these functions will be considered later in this chapter.

35.2 Fatty Acid Oxidation (Beta Oxidation)

Fats are the most energy-rich class of nutrients. Most of the energy from fats is derived from their constituent fatty acids. Palmitic acid derived from fat yields 39.1 kJ per gram when burned to form carbon dioxide and water. By contrast, glucose yields only 15.6 kJ per gram. Of course, fats are not actually burned in the body simply to produce heat. They are broken down in a series of enzyme-catalyzed reactions that also produce useful potential chemical energy in the form of ATP. In complete biochemical oxidation, the carbon and hydrogen of a fat ultimately are combined with oxygen (from respiration) to form carbon dioxide and water.

In 1904, Franz Knoop, a German biochemist, established that the catabolism of fatty acids involves a process whereby their carbon chains are shortened by two carbon atoms at a time. Knoop knew that animals do not metabolize benzene groups to carbon dioxide and water. Instead, the benzene nucleus remains attached to at least one carbon atom and is eliminated in the urine as a derivative of either benzoic acid or phenylacetic acid.

benzoic acid   phenylacetic acid

Accordingly, Knoop prepared a homologous series of straight-chain fatty acids with a phenyl group at one end and a carboxyl group at the other end. He then fed these benzene-tagged acids to test animals. Phenylaceturic acid was identified in the urine of the animals that ate acids with an even number of carbon atoms; hippuric acid was present in the urine of the animals that consumed acids with an odd number of carbon atoms:

phenylaceturic acid
(metabolic end product when n is even)

hippuric acid
(metabolic end product when n is odd)

These results indicated a metabolic pathway for fatty acids in which the carbon chain is shortened by two carbon atoms at each stage.

Knoop's experiments were remarkable for their time. They involved the use of tagged molecules and served as prototypes for modern research that utilizes isotopes to tag molecules.

Knoop postulated that the carbon chain of a fatty acid is shortened by successive removals of acetic acid units. The process involves the oxidation of the β-carbon atom and cleavage of the chain between the α and β carbons. A six-carbon fatty acid would produce three molecules of acetic acid thus:

First reaction sequence

This C is oxidized.

$$CH_3CH_2CH_2CH_2CH_2COOH \longrightarrow CH_3CH_2CH_2COOH + CH_3COOH$$

Chain is cleaved here.   butyric acid   acetic acid

caproic acid

Second reaction sequence

This C is oxidized.

$$CH_3CH_2CH_2COOH \longrightarrow CH_3COOH + CH_3COOH$$

Chain is cleaved here.

The general validity of Knoop's theory of β-carbon oxidation has been confirmed. However, the detailed pathway for fatty acid oxidation was not established until about 50 years after his original work. Like those of the Embden–Meyerhof and citric acid pathways, the sequence of reactions involved is another fundamental metabolic pathway. **Beta oxidation,** or the *two-carbon chop,* is accomplished in a series of reactions whereby the first two carbon atoms of the fatty acid chain become the acetyl group in a molecule of acetyl-CoA.

beta oxidation

The catabolism proceeds in this manner: A fatty acid reacts with coenzyme A (CoASH) to form an activated thioester. The energy needed for this step of the catabolism is obtained from ATP.

In many ways, a thioester is like the more common ester that is formed from an alcohol and a carboxylic acid.

Step 1. Activation: Formation of thioester with CoA

$$
\underset{\text{fatty acid}}{RCH_2CH_2CH_2\overset{\displaystyle O}{\overset{\|}{C}}-OH} + CoASH + ATP \longrightarrow
$$

$$
\underset{\text{CoA thioester of a fatty acid}}{RCH_2CH_2CH_2\overset{\displaystyle O}{\overset{\|}{C}}-SCoA} + AMP + \underset{\substack{\text{inorganic}\\\text{phosphate}}}{2\ P_i}
$$

The activated thioester next undergoes four more steps in the reaction sequence involving *oxidation, hydration, oxidation,* and *cleavage* to produce acetyl-CoA and an activated thioester shortened by two carbon atoms. The cleavage reaction requires an additional molecule of CoA.

Step 2. Oxidation: Dehydrogenation at carbons 2 and 3 (α- and β-carbons)

$$
RCH_2CH_2CH_2\overset{\displaystyle O}{\overset{\|}{C}}-SCoA + FAD \longrightarrow RCH_2CH{=}CH\overset{\displaystyle O}{\overset{\|}{C}}-SCoA + FADH_2
$$

Hydration of a double bond means adding the components of water to the doubled-bonded carbons.

Step 3. Hydration: Conversion to a secondary alcohol

$$
RCH_2CH{=}CH\overset{\displaystyle O}{\overset{\|}{C}}-SCoA + H_2O \longrightarrow RCH_2\overset{\displaystyle OH}{\overset{|}{C}H}CH_2\overset{\displaystyle O}{\overset{\|}{C}}-SCoA
$$

Step 4. Oxidation: Dehydrogenation of carbon 3 (β-carbon) to a keto group

$$
RCH_2\overset{\displaystyle OH}{\overset{|}{C}H}CH_2\overset{\displaystyle O}{\overset{\|}{C}}-SCoA + NAD^+ \longrightarrow RCH_2\overset{\displaystyle O}{\overset{\|}{C}}CH_2\overset{\displaystyle O}{\overset{\|}{C}}-SCoA + NADH + H^+
$$

Step 5. Carbon-chain cleavage: Reaction with CoA to produce acetyl-CoA and activated thioester of a fatty acid shortened by two carbons

$$
RCH_2\overset{\displaystyle O}{\overset{\|}{C}}CH_2\overset{\displaystyle O}{\overset{\|}{C}}-SCoA + CoASH \longrightarrow RCH_2\overset{\displaystyle O}{\overset{\|}{C}}-SCoA + \underset{\text{acetyl-CoA}}{CH_3\overset{\displaystyle O}{\overset{\|}{C}}-SCoA}
$$

The shortened-chain thioester repeats the reaction sequence of oxidation, hydration, oxidation, and cleavage to shorten the carbon chain further and produce another

◀ **FIGURE 35.1**
**A comparison of the ATPs
produced from 18 carbons
of 1 stearic acid molecule
and 18 carbons of 3
glucose molecules.**

molecule of acetyl-CoA. Thus, for example, eight molecules of acetyl-CoA can be produced from one molecule of palmitic acid.

As in the metabolic pathways for glucose, each reaction in the fatty acid oxidation pathway is enzyme catalyzed. No ATP is directly produced during fatty acid catabolism. Instead, ATP forms when the reduced coenzymes, FADH$_2$ and NADH, are oxidized by the mitochondrial electron-transport system in concert with oxidative phosphorylation. Fatty acid oxidation is aerobic because the products, FADH$_2$ and NADH, can only be reoxidized when oxygen is present.

In general, fatty acid catabolism yields more energy than can be derived from the breakdown of glucose (see Figure 35.1). For example, the reduced coenzymes derived from stearic acid (18 carbons) yield 139 ATPs via electron transport and oxidative phosphorylation. Nine additional ATPs can be obtained from the 9 GTPs formed in the citric acid cycle while 1 ATP is used to start the beta oxidation process. Thus the 18 carbons of stearic acid yield a total of 147 ATPs or about 8.2 ATPs per carbon atom. In contrast, glucose yields between 36 and 38 ATPs as its six carbons are completely oxidized to carbon dioxide. About 6 ATPs per carbon atom are gained from glucose as compared with about 8 ATPs per carbon atom from a fatty acid. Because fatty acid carbons are, in general, more reduced than glucose carbons, fatty acids are a more potent source of energy and yield more ATP molecules during metabolism.

Not surprisingly, the energy-storage molecule of choice in the human body is the fatty acid. On the average, a 70-kg male adult carries about 15 kg of fat (as triacylglycerols) but only about 0.22 kg of carbohydrate (as glycogen). Fat is such an effective way to store energy that obese people could exist for about 1 year without food. Unfortunately, fat is not the best energy-supply molecule for all tissues. For example, the brain normally derives all of its energy needs from glucose, using about 60% of all glucose metabolized by an adult at rest. Thus fatty acids are not a universal energy source although they are our most concentrated supply of energy.

The high energy stored in fats means diets with a high fat content may have too much energy. To decrease calories and because fats are not a universal energy source, many people seek to limit dietary fat intake. Most nutritionists agree that a diet should be limited to no more than 30% calories from fat. Fats, especially saturated fats, are known to increase the risk of heart attack, stroke, and some forms of cancer. Therefore, more and more consumers are shopping for low fat-content foods. However, two important points should be made. First, as shown in Chapter 34 and Section 35.5, carbohydrates and proteins also can be rich energy sources—eating less fat may not be enough to achieve a low-calorie diet. Second, as shown in Section 32.5, some fatty acids are essential for good health—eating too little fat can actually cause health problems. In general, nutritionists advise against trying for a "zero-fat" diet.

The essential fatty acids are linoleic, linolenic, and arachidonic acids.

35.3   Fat Storage and Utilization

Fats (triacylglycerols) are stored primarily in adipose tissue, which is widely distributed in the body. Fat tends to accumulate under the skin (subcutaneous fat), in the abdominal region, and around some internal organs, especially the kidneys. Fat deposited around internal organs acts as a shock absorber, or cushion. Subcutaneous fat acts as an insulating blanket. It is developed to an extreme degree in mammals such as seals, walruses, and whales that live in cold water.

Fat is the major reserve of potential energy. It is metabolized continuously. Stored fat does not remain in the body unchanged; there is a rapid exchange between the triacylglycerols of the plasma lipoproteins and the triacylglycerols in the adipose tissue. The plasma lipoprotein–bound triacylglycerols are broken down by the enzyme lipoprotein lipase that is found on the walls of all capillaries, and the resulting free fatty acids are transported into the adipose cells. The fatty acids are then converted back to triacylglycerols. When the body needs energy from fat, adipose cell enzymes hydrolyze triacylglycerols, and the fatty acids are exported to other body tissues. This vital process is under careful hormonal control. For example, a part of the "fight or flight" response caused by the hormone epinephrine (adrenalin) is an increased fatty acid output from the adipose tissue. Conversely, when there is more energy available in the diet than the body needs, the excess energy is used to make body fat. Continued eating of more food than the body can use results in obesity.

▲
The fat (triacylglycerol) is the large area in the adipose cell.

35.4   Biosynthesis of Fatty Acids (Lipogenesis)

lipogenesis   The biosynthesis of fatty acids from acetyl-CoA is called **lipogenesis.** Acetyl-CoA can be obtained from the catabolism of carbohydrates, fats, or proteins. Fatty acids, after they are synthesized, combine with glycerol to form triacylglycerols, which are stored in adipose tissue. Consequently, lipogenesis is the pathway by which all three of the major classes of nutrients are ultimately converted to body fat.

Is lipogenesis just the reverse of fatty acid oxidation (beta oxidation)? By analogy with carbohydrate metabolism (see Chapter 34), we might expect this to be the case. However, fatty acid biosynthesis is not simply a reversal of fatty acid oxidation. The following are the major differences between the two pathways:

1. Fatty acid catabolism occurs in the mitochondria, but fatty acid anabolism (lipogenesis) occurs in the cytoplasm.
2. Lipogenesis requires a set of enzymes that are different from the enzymes used in the catabolism of fats.
3. In the anabolic pathway (lipogenesis), the growing fatty acid chain bonds to a special acyl carrier protein (ACP—SH). The ACP—SH acts as a "handle" to transfer the growing chain from one enzyme to another through the series of enzyme-catalyzed reactions in the pathway. Coenzyme A is the carrier in fatty acid catabolism.
4. A preliminary set of reactions, involving malonyl-CoA, occurs for each two-carbon addition cycle in the synthesis. Malonyl is a three-carbon group and has no counterpart in the catabolic pathway. Malonyl-CoA is synthesized from acetyl-CoA and carbon dioxide in the presence of the enzyme acetyl-CoA carboxylase, ATP, and the vitamin biotin.

$$
\underset{\text{acetyl-CoA}}{CH_3\overset{\overset{\displaystyle O}{\|}}{C}-SCoA} \; + \; CO_2 \;
\xrightarrow[\substack{\text{acetyl-CoA}\\\text{carboxylase}}]{\text{ATP, biotin}} \;
\underset{\text{malonyl-CoA}}{HO\overset{\overset{\displaystyle O}{\|}}{C}CH_2\overset{\overset{\displaystyle O}{\|}}{C}-SCoA}
$$

The biosynthesis of a fatty acid occurs by addition of successive two-carbon increments to a lengthening chain starting with acetyl-CoA. Each incremental addition follows this five-step pathway or reaction sequence.

Step 1. Acetyl-CoA and malonyl-CoA bonds to separate acyl carrier proteins:

$$
\underset{\text{acetyl-CoA}}{CH_3\overset{\overset{\displaystyle O}{\|}}{C}-SCoA} + \underset{\text{malonyl-CoA}}{HO\overset{\overset{\displaystyle O}{\|}}{C}CH_2\overset{\overset{\displaystyle O}{\|}}{C}-SCoA} + \underset{\substack{\text{acyl protein}\\\text{carrier}}}{2\,ACP-SH} \longrightarrow
$$

$$
\underset{\text{acetyl-ACP}}{CH_3\overset{\overset{\displaystyle O}{\|}}{C}-SACP} + \underset{\text{malonyl-ACP}}{HO\overset{\overset{\displaystyle O}{\|}}{C}CH_2\overset{\overset{\displaystyle O}{\|}}{C}-SACP} + 2\,CoASH
$$

Step 2. Acetyl-ACP and malonyl-ACP condense, with loss of carbon dioxide (decarboxylation):

$$
\underset{\text{acetyl-ACP}}{CH_3\overset{\overset{\displaystyle O}{\|}}{C}-SACP} + \underset{\text{malonyl-ACP}}{HO\overset{\overset{\displaystyle O}{\|}}{C}CH_2\overset{\overset{\displaystyle O}{\|}}{C}-SACP} \longrightarrow
$$

$$
\underset{\text{acetoacetyl-ACP}}{CH_3\overset{\overset{\displaystyle O}{\|}}{C}CH_2\overset{\overset{\displaystyle O}{\|}}{C}-SACP} + CO_2 + ACP-SH
$$

The three steps that follow are approximate reversals of three steps in fatty acid beta oxidation (Section 35.2).

Step 3. Reduction: Hydrogenation of carbon 3 (β-keto group)

$$CH_3 \overset{\overset{O}{\|}}{C} CH_2 \overset{\overset{O}{\|}}{C} - SACP \; + \; NADPH \; + \; H^+ \longrightarrow$$

$$CH_3 \overset{\overset{OH}{|}}{C} H CH_2 \overset{\overset{O}{\|}}{C} - SACP + NADP^+$$

β-hydroxybutyryl-ACP

Step 4. Dehydration: Formation of a double bond between carbons 2 and 3

$$CH_3 \overset{\overset{OH}{|}}{C} H CH_2 \overset{\overset{O}{\|}}{C} - SACP \longrightarrow CH_3 CH = CH \overset{\overset{O}{\|}}{C} - SACP + H_2O$$

crotonyl-ACP

Step 5. Reduction: Hydrogenation of carbons 2 and 3

$$CH_3 CH = CH \overset{\overset{O}{\|}}{C} - SACP + NADPH + H^+ \longrightarrow CH_3 CH_2 CH_2 \overset{\overset{O}{\|}}{C} - SACP + NADP^-$$

butyryl-ACP

This completes the first cycle of the synthesis; the chain has been lengthened by two carbon atoms. Biosynthesis of longer-chain fatty acids proceeds by a series of such cycles, each lengthening the carbon chain by an increment of two carbon atoms. The next cycle begins with the reaction of butyryl-ACP and malonyl-ACP, leading to a six-carbon chain, and so on. This synthesis commonly produces palmitic acid (16 carbons) as its end product. The synthesis of palmitic acid from acetyl-CoA and malonyl-CoA requires cycling through the series of steps seven times. The net equation for the formation of palmitic acid is

$$CH_3 \overset{\overset{O}{\|}}{C} - SCoA + 7 \; HOCCH_2 \overset{\overset{O}{\|} \quad \overset{O}{\|}}{C} - SCoA + 14 \; NADPH + 14 \; H^+ \longrightarrow$$

acetyl-CoA   malonyl-CoA

$$CH_3(CH_2)_{14}COOH + 7 \; CO_2 + 6 \; H_2O + 8 \; CoASH + 14 \; NADP^+$$

palmitic acid

Nearly all naturally occurring fatty acids have even numbers of carbon atoms. A sound reason for this fact is that both the catabolism and the synthesis proceed by two-carbon increments.

In conclusion, it should be noted that the metabolism of fats has some features in common with that of carbohydrates. The acetyl-CoA produced in the catabolism of both carbohydrates and fatty acids can be used as a raw material for making other substances and as an energy source. When acetyl-CoA is oxidized via the citric acid cycle, more potential energy can be trapped in ATP. The ATP in turn serves as the source of energy needed for the production of other substances, including the synthesis of carbohydrates and fats.

Ketone Bodies—A Stress Response

Brain cells require a high and continuous supply of glucose for energy. Extreme circumstances such as long-term, continuous exercise; low-carbohydrate diet; starvation; and untreated diabetes mellitus cause metabolism to change. In such cases the energy obtainable from glucose is severely limited. In response, the body uses fatty acids for fuel molecules and produces small molecules, called *ketone bodies,* as glucose substitutes. These molecules include acetoacetic acid and two of its derivatives, β-hydroxybutyric acid and acetone.

$$
\begin{array}{c}
\text{O} \\
\parallel \\
\text{CH}_3\text{CCH}_2\text{COOH}
\end{array}
$$

acetoacetic acid

$$
\begin{array}{c}
\text{OH} \\
\mid \\
\text{CH}_3\text{CHCH}_2\text{COOH}
\end{array}
$$

β-hydroxybutyric acid

$$
\begin{array}{c}
\text{O} \\
\parallel \\
\text{CH}_3\text{CCH}_3
\end{array}
$$

acetone

Ketone bodies are highly soluble in blood and are able to provide cellular energy via beta oxidation and the citric acid cycle. When blood glucose becomes less available, some of the body's tissues begin to use ketone bodies as substitute energy-supply molecules. For example,

the brain shifts its metabolism to use about 70% ketone bodies after about 40 days of minimal food intake.

An increase in ketone-body concentration is an emergency response to an extreme circumstance. This response can have dangerous side effects. For example, most ketone bodies are carboxylic acids. When their concentration increases in the blood, the acidity rises, causing a condition called *ketosis,* or *ketoacidosis.* A mild state of ketosis (acidosis) is not dangerous, but severe cases can affect respiration; result in dehydration and loss of K^+ and Na^+ due to excessive urination; affect the kidneys; and ultimately lead to coma and death.

35.5 Amino Acid Metabolism

Amino acids serve an important and unique role in cellular metabolism; they are the building blocks of proteins and also provide most of the nitrogen for other nitrogen-containing compounds.

Amino acid metabolism differs markedly from the biochemistry of carbohydrates or fatty acids. Amino acids always contain nitrogen, and the chemistry of this element presents unique problems for the cell. In addition, there is no structure common to all the carbon skeletons of amino acids. The carbohydrates can share a common metabolic pathway, as can fatty acids, because they share common structures. But the carbon skeletons of amino acids vary widely, and the cell must use a different metabolic pathway for almost every amino acid. Thus the metabolism of the carbon structures of amino acids is more complex. In the sections that follow, we present a brief overview of amino acid metabolism together with a more detailed examination of nitrogen metabolism.

35.6 Metabolic Nitrogen Fixation

Nitrogen is an important component of many biochemicals. In addition to being a component of amino acids, nitrogen is found in proteins, nucleic acids, hemoglobin, and many vitamins. Every cell must have a continuous nitrogen supply. This might seem easy to obtain because the atmosphere contains about 78% free (elemental) nitrogen. Unfortunately, most cells cannot use elemental nitrogen.

Elemental nitrogen on earth exists as N_2 molecules, the two atoms being bonded together by a strong and stable triple bond. On an industrial scale, high temperature (400–500°C), high pressure (several hundred atmospheres), and a catalyst are required to react nitrogen gas with the reducing agent, hydrogen, to form ammonia (the Haber process). In the biosphere only a few organisms have the metabolic machinery necessary to use the abundant atmospheric nitrogen. These are procaryotes and include the *Azobacter* species, *Clostridium pasteurianum,* and *Rhizobium* species. By converting elemental nitrogen to compounds, these organisms make nitrogen available to the rest of the biological world.

nitrogen fixation

nitrogen cycle

The conversion of diatomic nitrogen to a biochemically useful form is termed **nitrogen fixation.** The process by which nitrogen is circulated and recirculated from the atmosphere through living organisms and back to the atmosphere is known as the **nitrogen cycle.**

Nitrogen compounds are required by all living organisms. Despite the fact that the atmosphere is about 78% nitrogen, all animals as well as the higher plants are unable to utilize free nitrogen. Higher plants require inorganic nitrogen compounds, and animals must have nitrogen in the form of organic compounds.

Atmospheric nitrogen is fixed—that is, converted into chemical compounds that are useful to higher forms of life—by three general routes:

1. *Bacterial action*   Certain soil bacteria are capable of converting N_2 into nitrates. Most of these bacteria live in the soil in association with legumes (e.g., peas, beans, clover). Some free-living soil bacteria and the blue-green algae, which live in water, are also capable of fixing nitrogen. Of the procaryotes that are able to catalyze nitrogen fixation, *Rhizobium* bacteria deserve special attention. These bacteria flourish on the roots of legumes, in a symbiotic relationship. The bacteria have degenerated essentially into nitrogen-fixing machines that are maintained by the plants in the root nodules. The legumes even provide a special hemoglobin protein to assist the bacteria in their task.

2. *High Temperature*   The high temperature of lightning flashes causes substantial amounts of nitrogen oxide in the atmosphere. This NO is dissolved in rainwater and eventually converts to nitrate ions in the soil. The combustion of fuels also provides temperatures high enough to form NO in the atmosphere. The total amount of nitrogen fixed by combustion is relatively insignificant on a worldwide basis. However, NO produced by combustion, especially in automobile engines, is a serious air pollution problem in some areas.

3. *Chemical fixation*   Chemical processes have been devised for making nitrogen compounds directly from atmospheric nitrogen. By far the most important of these is the Haber process for making ammonia from nitrogen and hydrogen. This process is the major means of production for the millions of tons of nitrogen fertilizers that are produced synthetically each year.

The nitrogen cycle is diagrammed in Figure 35.2. Starting with the atmosphere, the cycle begins with the fixation of nitrogen by any one of the three routes. In the soil, nitrogen or nitrogen compounds are converted to nitrates, taken up by higher plants, and converted to organic compounds. The plants eventually die or are eaten

▲
Legume root nodules containing nitrogen-fixing bacteria.

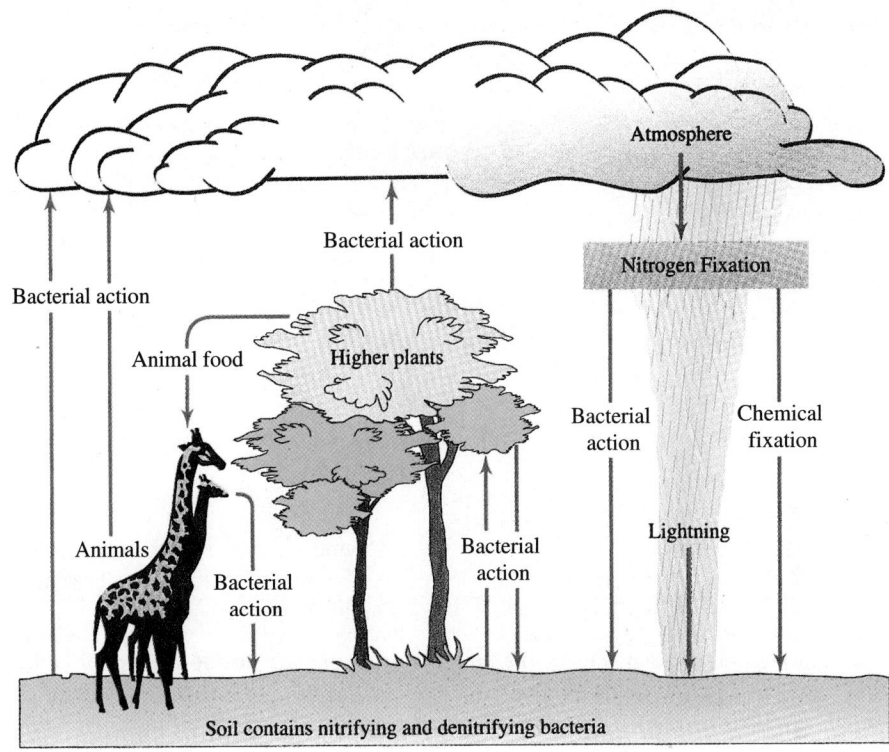

◀ FIGURE 35.2
The nitrogen cycle. Arrows
indicate the movement of
nitrogen through the cycle.

by animals. During the life of the animal, part of the nitrogen from plants in its diet
is returned to the soil in the form of fecal and urinary excreta. Eventually after death,
both plants and animals decompose by bacterial action. Part of the nitrogen from
plant and animal tissues returns to the atmosphere as free nitrogen and part is retained in the soil. The cycle thus continues.

Currently, in many university and industrial genetic-engineering laboratories,
scientists are attempting to transfer the nitrogen-fixation capability of procaryotes to
higher plants. To accomplish this transfer, the appropriate genes must be removed
from procaryotic cells and spliced into the DNA of higher plants. This research is
particularly challenging and, if successful, will have potentially great rewards.
Success will mean that grains and other commercially important crops can be grown
in poor soils using little or no nitrogen fertilizer. This achievement would vastly increase the world's food production capability.

Higher plants use nitrogen compounds primarily to produce proteins, which, in
turn, enter the animal food chain. Both plant and animal proteins are important
human nutrients.

 ## 35.7 Amino Acids and Metabolic
Nitrogen Balance

The total content of nitrogen compounds in the body can be divided into several nitrogen pools (see Figure 35.3). Protein is digested and absorbed to provide the

FIGURE 35.3 ▶
Major biological nitrogen pools related to the central amino acid pool.

amino acid dietary requirements (see Chapter 32). Once absorbed, an amino acid can be

1. incorporated into a protein, or
2. used to synthesize other nitrogenous compounds such as nucleic acids, or
3. deaminated to a keto acid, which can either be used to synthesize other compounds or be oxidized to carbon dioxide and water to provide energy

amino acid pool   Absorbed amino acids enter the amino acid pool. The **amino acid pool** is the total supply of amino acids available for use throughout the body. The amount of amino acids in the pool is maintained in balance with other cellular nitrogen pools (see Figure 35.3).

One particularly important nitrogen pool is composed of the proteins in all the body's tissues. Amino acids continually move back and forth between the amino acid pool and the tissue proteins. In other words, our body proteins are constantly being broken down and resynthesized. The rate of turnover varies with different proteins. In the liver and other active organs, protein molecules may have a half-life of less than 1 week, whereas the half-life of some muscle proteins is about 6 months.

In a healthy, well-nourished adult, the amount of nitrogen excreted is equal to the amount of nitrogen ingested. The nitrogen pools within the body remain constant

nitrogen balance   (see Figure 35.3), and such a person is said to be in **nitrogen balance.**

For a growing child, the amount of nitrogen consumed is more than that excreted. Dietary protein feeds into the amino acid pool, which is used to make many new nitrogen-containing molecules. The contents of the nitrogen pools increase, and

positive nitrogen balance   the child is said to be in **positive nitrogen balance.**

negative nitrogen balance   A fasting or starving person excretes more nitrogen than is ingested; such a person is said to be in **negative nitrogen balance.** Tissue protein breaks down to supply more amino acids to the amino acid pool (see Figure 35.3). These are used as an energy source and nitrogen is excreted. The body's nitrogen pools are depleted during negative nitrogen balance.

Most of the amino acid pool is derived originally from dietary protein. This protein is digested to yield amino acids (see Section 32.13). In contrast, only a very small proportion of the amino acid pool comes directly from dietary amino acids. Recently, it has been suggested that some diets should be supplemented by additional amino acids. Extra amino acid intake reportedly aids muscle building and counters mental depression. Most studies, however, show that the protein in a balanced diet is sufficient to provide all needed amino acids. In fact, sometimes added amino acids are harmful. The food additive, monosodium glutamate (MSG; the sodium salt of the amino acid, glutamic acid), causes allergic reactions in some people. However, more research is certainly needed on the effect of ingesting extra amino acids.

Nitrogen balance depends on the transition between nitrogen pools. If these pathways are disturbed, severe health problems can develop. For example, people suffering from phenylketonuria (PKU) are unable to catabolize phenylalanine. When people with PKU ingest food protein containing this amino acid, it builds up in their body's amino acid pool and eventually causes toxic side reactions. Babies who suffer from PKU develop extensive mental retardation. Fortunately, these symptoms can be avoided by removing most of the phenylalanine from the diet. Products like the artificial sweetener aspartame (Nutrasweet), which contain large amounts of phenylalanine carry a warning label for people with the PKU genetic defect.

As shown in Figure 35.3, amino acids are central to many different metabolic processes. They are the raw material for protein synthesis (see Section 31.11) and can also be used to make other nitrogen-containing molecules. In addition, amino acids can supply metabolic energy if the amino acid nitrogen is first removed (or transferred to another molecule).

35.8 Amino Acids and Nitrogen Transfer

Amino acids are important in metabolism as carriers of usable nitrogen. If an amino acid is not directly incorporated into tissue proteins, its nitrogen may be incorporated into various molecules such as other amino acids, nucleic acid bases, the heme of hemoglobin, and some lipids. In general, when an amino acid is used for some purpose other than protein synthesis, the amino acid carbon skeleton is separated from the amino acid nitrogen.

A process called transamination is responsible for most of the nitrogen transfer to and from amino acids. **Transamination** is the transfer of an amino group from an α-amino acid to an α-keto acid:

transamination

$$
\underset{\substack{| \\ NH_2 \\ \text{\alpha-amino acid}}}{RCH} - COOH \ + \ \underset{\substack{\| \\ O \\ \text{\alpha-keto acid}}}{R'C} - COOH \ \rightleftharpoons \ \underset{\substack{\| \\ O}}{RC} - COOH \ + \ \underset{\substack{| \\ NH_2}}{R'CH} - COOH
$$

Transamination involves many different molecules, with each different transamination requiring a different enzyme (transaminase). For example, one enzyme (glutamic-pyruvic transaminase) catalyzes the conversion of pyruvic acid to L-alanine.

$$
\underset{\text{pyruvic acid}}{\underset{\substack{\| \\ O}}{CH_3C} - COOH} \ + \ \underset{\text{L-glutamic acid}}{\begin{array}{c} COOH \\ | \\ CH_2 \\ | \\ CH_2 \\ | \\ NH_2 - CH - COOH \end{array}} \ \rightleftharpoons \ \underset{\text{L-alanine}}{\underset{\substack{| \\ NH_2}}{CH_3CH} - COOH} \ + \ \underset{\alpha\text{-ketoglutaric acid}}{\begin{array}{c} COOH \\ | \\ CH_2 \\ | \\ CH_2 \\ | \\ O = C - COOH \end{array}}
$$

A different enzyme catalyzes the production of L-aspartic acid from oxaloacetic acid:

$$
\begin{array}{c}
\text{COOH} \\
| \\
\text{CH}_2 \\
| \\
\text{O}=\text{C}-\text{COOH}
\end{array}
\;+\;
\begin{array}{c}
\text{COOH} \\
| \\
\text{CH}_2 \\
| \\
\text{CH}_2 \\
| \\
\text{NH}_2-\text{CHCOOH}
\end{array}
\;\rightleftharpoons\;
\begin{array}{c}
\text{COOH} \\
| \\
\text{CH}_2 \\
| \\
\text{NH}_2-\text{CHCOOH}
\end{array}
\;+\;
\begin{array}{c}
\text{COOH} \\
| \\
\text{CH}_2 \\
| \\
\text{CH}_2 \\
| \\
\text{O}=\text{C}-\text{COOH}
\end{array}
$$

oxaloacetic acid   L-glutamic acid   L-aspartic acid α-ketoglutaric acid

Note that in both of these reactions L-glutamic acid is converted to α-ketoglutaric acid. In fact, most transaminations use L-glutamic acid. Thus this amino acid plays a central role in cellular nitrogen transfer.

Transamination is the first step in the conversion of the carbon skeletons of amino acids to energy-storage compounds. Amino acids that are used to produce glucose are termed **glucogenic amino acids.** Most amino acids are glucogenic (see Table 35.1), but some amino acids are converted to acetyl-CoA. When fed to starving animals, these amino acids cause an increase in the rate of ketone body formation and are therefore called **ketogenic amino acids.** Only leucine and lysine are completely ketogenic, but a few amino acids can be converted to either glucose or acetyl-CoA and are both ketogenic and glucogenic.

Relatively recently, high-protein/low-carbohydrate diets have been touted as a means of losing weight quickly. This diet forces protein to be used for energy. In the liver, glucogenic and ketogenic amino acids are converted to glucose and ketone bodies, respectively. These molecules are transported to other cells via the bloodstream. The high-ketone body concentration can lead to a condition called *ketosis* (see the Chemistry in Action feature Ketone Bodies—A Stress Response). This same situation is experienced during fasting, starvation, or diabetes. Weight can

glucogenic amino acid

ketogenic amino acid

TABLE 35.1   Classification of Amino Acids as Sources of Energy-Storage Molecules

Glucogenic	Ketogenic and glucogenic	Ketogenic
Alanine	Isoleucine	Leucine
Arginine	Phenylalanine	Lysine
Aspartic acid	Tryptophan	
Asparagine	Tyrosine	
Cysteine		
Glutamic acid		
Glutamine		
Glycine		
Histidine		
Methionine		
Proline		
Serine		
Threonine		
Valine		

certainly be lost by a high-protein/low-carbohydrate diet. But unfortunately much of this lost weight is water as the kidneys attempt to remove ketone bodies from the blood. As a result, the body is forced into an emergency situation through this unbalanced diet.

The amino acid pool of Figure 35.3 can now be described in more detail. At the center of this pool is L-glutamic acid, and other amino acids can either add or remove nitrogen from this central compound. L-Glutamic acid can also accept a second nitrogen atom to form L-glutamine:

$$
NH_3 + \quad
\begin{array}{c}
O \\
\parallel \\
C\!-\!OH \\
\mid \\
CH_2 \\
\mid \\
CH_2 \\
\mid \\
NH_2\!-\!CHCOOH
\end{array}
\quad \rightleftharpoons \quad
\begin{array}{c}
O \\
\parallel \\
C\!-\!NH_2 \\
\mid \\
CH_2 \\
\mid \\
CH_2 \\
\mid \\
NH_2\!-\!CHCOOH
\end{array}
\quad + H_2O
$$

L-glutamic acid   L-glutamine

An amide is formed when ammonia or an amine reacts with a carboxylic acid.

Although this reaction can be considered as the simple addition of ammonia to yield an amide, the actual cellular reactions are more complex. The product, L-glutamine, facilitates biological nitrogen transfer, the amide nitrogen being transferable in a number of cellular reactions.

It is worthwhile to examine the synthesis of L-glutamine in more detail. Ammonia is a base and therefore is toxic to the cell. When ammonia forms an amide bond to L-glutamic acid, the nitrogen becomes less basic and also nontoxic. Thus L-glutamine serves as a safe package for transporting nitrogen. In the human body, L-glutamine is the major compound for transferring nitrogen from one cell to another via the bloodstream.

35.9   Nitrogen Excretion and the Urea Cycle

Nitrogen—unlike carbon, hydrogen, and oxygen—is often conserved for reuse by the cell. Nitrogen excretion does occur, however, when an excess of this element exists or when the carbon skeletons of nitrogen-containing compounds are needed for other purposes. Two examples from human nutrition arise: (1) when we consume more protein than is needed (an excess of nitrogen-containing molecules) or (2) when we experience starvation (protein is destroyed to provide energy). Under normal conditions, adult humans excrete 6-18 g of nitrogen per day.

The nitrogen elimination process poses a major problem for the cell. The simplest excretion product is ammonia, but ammonia is basic and therefore toxic. Fish can excrete ammonia through their gills because ammonia is soluble and is swept away by water passing through the gills. Land animals and birds excrete nitrogen in less toxic forms. Birds and reptiles excrete nitrogen as the white solid uric acid, a derivative of the purine bases. Mammals excrete the water-soluble compound urea. Both of these compounds contain a high percentage of nitrogen in a nontoxic form.

uric acid

$$H_2NCNH_2$$

urea

Urea synthesis in mammals follows a pathway called the urea cycle (Figure 35.4), which takes place in the liver. Ammonium ion is first produced from L-glutamic acid in an oxidation–reduction reaction:

L-glutamic acid α-ketoglutaric acid

The ammonium ion quickly reacts with the hydrogen carbonate ion and ATP to form carbamoyl phosphate:

$$NH_4^+ + HCO_3^- + 2\,ATP \longrightarrow NH_2-\overset{\displaystyle O}{\overset{\|}{C}}-O-\overset{\displaystyle O^-}{\underset{\displaystyle O}{\overset{|}{\underset{\|}{P}}}}-O^- + 2\,ADP + P_i$$

carbamoyl phosphate

Finally, carbamoyl phosphate enters the urea cycle. The overall reaction of the urea cycle is

carbamoyl phosphate   L-aspartic acid

urea   fumaric acid

Like the intermediate compounds of the citric acid cycle, the urea cycle intermediates do not appear in the reaction. Note that the urea cycle intermediates are

COOH
|
CH$_2$
|
CH$_2$
|
NH$_2$—CH—COOH

L-glutamic acid

COOH
|
CH$_2$
|
CH$_2$
|
O=C—COOH

α-ketoglutaric acid

NAD$^+$
H$_2$O

H$^+$ + NADH

2 ATP   2 ADP + P$_i$

NH$_3$

HCO$_3^-$

$$NH_2-\overset{\overset{\displaystyle O}{\|}}{C}-O-\overset{\overset{\displaystyle O}{\|}}{\underset{\underset{\displaystyle O^-}{|}}{P}}-O^-$$

carbamoyl phosphate

$$NH_2\overset{\overset{\displaystyle O}{\|}}{C}NH_2$$
urea

NH$_2$
|
CH$_2$
|
CH$_2$
|
CH$_2$
|
NH$_2$—CH—COOH

ornithine

P$_i$

H$_2$O

NH
‖
C—NH$_2$
|
NH
|
CH$_2$
|
CH$_2$
|
CH$_2$
|
NH$_2$—CH—COOH

L-arginine

Urea cycle

NH$_2$
|
C=O
|
NH
|
CH$_2$
|
CH$_2$
|
CH$_2$
|
NH$_2$—CH—COOH

citrulline

COOH
|
CH$_2$
|
NH$_2$—CH—COOH

L-aspartic acid

NH   H
‖ |
C—NH—C—COOH
| |
NH   CH$_2$
| |
CH$_2$ COOH
|
CH$_2$
|
CH$_2$
|
NH$_2$—CH—COOH

argininosuccinic acid

ATP

AMP + PP$_i$

COOH
|
CH
‖
HC
|
COOH

fumaric
acid

◀FIGURE 35.4
Urea cycle. The cycle begins with the reaction between carbamoyl phosphate and ornithine.

FIGURE 35.5 ▶
Simplified diagram showing acetyl-CoA is at the hub of protein, carbohydrate, and fat metabolism.

α-amino acids. These amino acids are rarely used in protein synthesis. Their primary role is in the formation of urea.

Also, it is important to recognize that the cell must expend energy, ATP, to produce urea. Formation of a nontoxic nitrogen excretion product is essential. In fact, one of the major problems of liver cirrhosis, caused by alcoholism, is the impairment of the urea cycle. As liver function is impaired, more nitrogen is excreted as toxic ammonia.

Finally, let's look at the sources of the nitrogen that is excreted as urea. One nitrogen atom in each urea molecule comes from L-glutamic acid. The other nitrogen atom comes from L-aspartic acid, which may have gained its nitrogen from L-glutamic acid by transamination. Thus the amino acid that is central to nitrogen-transfer reactions is also the major contributor to nitrogen excretion.

35.10   Acetyl-CoA, a Central Molecule in Metabolism

Thinking back through metabolism, we can identify some especially important compounds, molecules that are central to this part of biochemistry. For example, glucose is the central compound in carbohydrate metabolism; glutamic acid is central to amino acid metabolism. One compound is at the hub of all common metabolic processes; this compound is acetyl-CoA, which is central to the metabolism of fats, proteins, and carbohydrates (Figure 35.5). This molecule is a critical intermediate in the processes that form and break down both fats and amino acids. In addition, essentially all compounds that enter the citric acid cycle must first be catabolized to acetyl-CoA. In this section we will examine (1) the characteristics of acetyl-CoA that make it a central metabolic compound and (2) the advantages of a centralized metabolism for the cell.

CHEMISTRY IN ACTION An "Obese" Protein and Weight Loss

Obesity is a prevalent problem in the more affluent countries of the world. For example, between one-quarter and one-third of adults in the United States are overweight. With extra weight comes increased health risks and, sometimes social disfavor. We want to feel better and look better. So, in the United States alone, about 30 billion dollars are spent per year on weight loss programs. Unfortunately, most weight lost is regained over time; our bodies act as if they have a natural equilibrium weight.

Recently, a protein has been discovered that seems to establish the normal body weight for each individual. This discovery is the culmination of about 40 years of research. In the mid-1950s, a mutant mouse was discovered that grew to about double the normal weight—an obese mouse. Later it was learned that this obesity resulted from an inactive gene. When this gene is activated, it codes for a protein called the obese protein. Subsequently, this gene has been found to reside in the adipose tissue of humans as well as mice.

The obese protein acts as a hormone; as more fat is accumulated,

more obese protein is released into the bloodstream. The target organ for the obese protein is unknown although there is some evidence that it may be the brain. Wherever its destination, a high concentration of obese protein signals that the body has a large amount of fat reserves. The body's response is twofold. First, appetite is decreased. Second, metabolism is increased to burn fat. These two effects together act as a feedback to control body weight. As fat accumulates, the obese protein serves to stop further accumulation.

Thinner people with low fat reserves will have a correspondingly lower obese protein concentration. This lower concentration leads to a larger appetite and a less active metabolism. The obese protein helps the body stay in a "normal" weight range, getting neither fatter nor thinner.

Doctors who treat obesity are particularly excited about this protein. Imagine injecting a medication that tells the body to decrease eating and to burn up more fat—just the responses for which weight loss clinics have searched. By adjusting the concentration of obese protein, a new

normal body weight can be selected. However, major hurdles must be overcome before this protein can achieve its potential. As scientists point out, human obesity has many causes; it may be that the obese protein will not cure all obesity. Yet, the results are so encouraging that a biotechnology company has purchased the right to use a genetically engineered obese protein gene.

Recall that acetyl-CoA consists of a small two-carbon unit (acetyl group) bonded by thioester linkage to a large organic coenzyme molecule, coenzyme A:

$$CH_3 - \overset{\overset{\textstyle O}{\|}}{C} - S - \boxed{CoA}$$

acetyl group   thioester linkage

This structure makes for an almost ideal central metabolic molecule with the following major advantages:

1. *Potential use in a wide variety of syntheses.* The small size and simple structure of the two-carbon acetyl fragment enables this molecule to be used to build a variety of diverse structures. Complex molecules with very different shapes and functions such as long-chain fatty acids, amino acids, and steroid hormones are synthesized from acetyl-CoA.

2. *Special reactivity of bonds.*   The thioester causes both carbons in the acetyl fragment to be specially reactive. These carbons are "primed" to form new bonds as the acetyl-CoA enters various metabolic pathways.

3. *Structure that is recognizable by a wide variety of enzymes.*   Coenzyme A acts as a kind of "handle" for the various enzymes that catalyze reactions of the acetyl group. Because many enzymes bind tightly to CoA, the acetyl group of acetyl-CoA is involved in a great number of diverse reactions.

Acetyl-CoA is comparable to ATP as a central metabolic molecule. Remember that ATP has (1) potential uses in a wide variety of syntheses, (2) a special reactivity in its phosphate anhydride bonds, and (3) a structure that is recognizable and used by a wide variety of enzymes. For these important reasons, ATP serves as the "common energy currency" for the cell. In analogous fashion, acetyl-CoA might be termed the "common carbon currency" for the cell.

Consideration of these central metabolic compounds raises an important question, "How does centralization aid the cell?" We have seen numerous examples of centralization in biochemistry. Not only are there important central metabolites such as ATP and acetyl-CoA, but there are pathways such as the citric acid cycle that centralize the cellular metabolic machinery. A general answer to this question then is that centralization improves the efficiency of metabolism and ensures that biochemistry is under careful metabolic control.

Greater efficiency results from central metabolic pathways that handle a variety of different nutrients. For example, the citric acid cycle completes carbon oxidation for all energy-supplying nutrients that are first converted to the central metabolite, acetyl-CoA.

Greater control results from the dependence of many "feeder" pathways on a single central process. Control of a central path like the citric acid cycle will affect in a coordinated way the metabolism of a variety of nutrients. As scientists have learned more about metabolism, it has become clear that centralization is an important attribute of the chemistry of life.

Concepts in Review

1. Briefly explain what characteristic of fatty acids allows them to provide large amounts of metabolic energy.

2. Describe Knoop's experiments on fatty acid oxidation and degradation and the conclusions derived from them.

3. Explain what is meant by beta oxidation and beta cleavage in relation to the metabolism of fatty acids.

4. State the purpose of ketone body production and indicate a dangerous side effect.

5. Describe the biosynthesis of fatty acids using palmitic acid as an example.

6. List three major differences between beta oxidation and fatty acid synthesis.

7. Briefly describe how amino acid metabolism differs from carbohydrate and fatty acid metabolism.

8. Briefly describe the nitrogen cycle in nature.

9. List the possible metabolic fates of amino acids in humans.

10. Describe the major purpose of transamination.

11. Explain the importance of L-glutamic acid and L-glutamine in nitrogen metabolism.

12. Explain how the metabolism of proteins (amino acids) is tied into that of carbohydrates and fats.

13. Explain how a lack of essential amino acids in the diet affects the nitrogen balance.

14. Describe the purpose of the urea cycle.

15. Summarize the importance of acetyl-CoA in metabolism.

16. Discuss the major advantages of a centralized metabolism.

Key Terms

The terms listed here have been defined within this chapter.
Section numbers are referenced in parentheses for each term.

amino acid pool (35.7)

beta oxidation (35.2)

glucogenic amino acid (35.8)

ketogenic amino acid (35.8)

lipogenesis (35.4)

negative nitrogen balance (35.7)

nitrogen balance (35.7)

nitrogen cycle (35.6)

nitrogen fixation (35.6)

positive nitrogen balance (35.7)

transamination (35.8)

Questions

Questions refer to tables, figures, and key words and concepts defined within the chapter. A particularly challenging question or exercise is indicated with an asterisk.

1. What major characteristic of a fatty acid allows it to serve as an energy-storage molecule?

2. Briefly describe Knoop's experiment on fatty acid oxidation and catabolism.

3. What is meant by the terms *beta oxidation* and *beta cleavage* in relation to fatty acid catabolism?

4. How are ketone bodies important in energy metabolism?

5. Define *ketosis*.

6. Aside from being a food reserve, what are the two principal functions of body fat?

7. Is fatty acid synthesis simply the reverse of beta oxidation? Briefly explain.

8. Briefly describe the nitrogen cycle in nature.

9. In general terms, how is the citric acid cycle involved in obtaining energy from fats?

10. Briefly describe why soybeans are a crop that enriches the soil.

11. Briefly describe why L-glutamic acid is considered to be the central amino acid of the amino acid pool.

12. In general, what are the metabolic fates of amino acids?

13. Why is L-glutamine a better nitrogen-transport molecule than ammonia?

14. What nitrogen compound is excreted by (a) fish and (b) birds?

15. Give the structures of urea and uric acid.

16. How many ATPs are used to produce urea in the urea cycle?

17. List three reasons why acetyl-CoA makes a good central intermediate in metabolism.

18. Briefly explain why a growing child is in a positive nitrogen balance.

Paired Exercises

These exercises are paired. Each odd-numbered exercise is followed by a similar even-numbered exercise. Answers to the even-numbered exercises are given in Appendix VI.

19. Circle the two carbons that are found in acetyl-CoA after one pass through the beta oxidation pathway by butyric acid, $CH_3CH_2CH_2COOH$.

20. Circle the two carbons that remain after two passes through the beta oxidation pathway by caproic acid, $CH_3CH_2CH_2CH_2CH_2COOH$.

21. Write the structure and mark the β-carbon in palmitic acid.

22. Write the structure and mark the β-carbon in myristic acid.

23. How many molecules of acetyl-CoA can be formed from lauric acid using the beta oxidation pathway?

24. How many molecules of acetyl-CoA can be formed from palmitic acid using the beta oxidation pathway?

25. One $FADH_2$ is produced for each pass through the beta oxidation pathway. How many of these reduced coenzymes will be formed from lauric acid?

26. One NADH is produced for each pass through the beta oxidation pathway. How many of these reduced coenzymes will be formed from palmitic acid?

27. Each $FADH_2$ can yield two ATPs, and each NADH can yield three ATPs. Given that one NADH and one $FADH_2$ is made for each pass through the beta oxidation pathway, how many ATPs will be formed from myristic acid?

28. Each $FADH_2$ can yield two ATPs, and each NADH can yield three ATPs. Given that one NADH and one $FADH_2$ is made for each pass through the beta oxidation pathway, how many ATPs will be formed from lauric acid?

29. How is the acyl carrier protein (ACP) similar to coenzyme A (CoA)?

30. How is the acyl carrier protein (ACP) different from coenzyme A (CoA)?

31. How might a low-protein diet cause a negative nitrogen balance?

32. How might a rapid growth spurt cause a positive nitrogen balance?

33. A diet high in L-leucine was observed to cause an increase in blood levels of

$$CH_3\overset{\overset{\displaystyle O}{\|}}{C}CH_2COOH$$

   acetoacetic acid

Is L-leucine a glucogenic or ketogenic amino acid? Briefly explain.

34. A diet high in L-aspartic acid was observed to cause no change in blood levels of

$$CH_3\overset{\overset{\displaystyle OH}{|}}{C}HCH_2COOH$$

   β-hydroxybutyric acid

Is L-aspartic acid a glucogenic or ketogenic amino acid? Briefly explain.

35. Write the structural formulas of the compounds produced by transamination of the following amino acids:
(a) L-alanine
(b) L-serine

36. Write the structural formulas of the compounds produced by transamination of the following amino acids:
(a) L-aspartic acid
(b) L-phenylalanine

37. What portion of the carbamoyl phosphate molecule

$$H_2N-\overset{\overset{\displaystyle O}{\|}}{C}-O-\overset{\overset{\displaystyle O}{\|}}{\underset{\underset{\displaystyle O^-}{|}}{P}}-O^-$$

is incorporated into urea?

38. What portion of the L-aspartic acid molecule

$$H_2N-\overset{\overset{\displaystyle COOH}{|}\,\overset{\displaystyle CH_2}{|}}{C}H-COOH$$

is incorporated into urea?

Additional Exercises

These exercises are not paired and provide additional practice on concepts covered in this chapter.

39. How is it possible to accumulate fatty tissue (or adipose tissue) even though very little fat is included in the diet?

40. Give the name and structure of a ketone body that does not contain a ketone functional group.

41. Beta oxidation is used by the cell to produce energy, yet no ATPs are formed in this pathway. Briefly explain this seeming contradiction.

42. Based on your knowledge of catabolic pathways, why is the ATP yield from a six-carbon fatty acid greater than the ATP yield from a six-carbon hexose (glucose)?

43. Starting with acetyl-CoA and malonyl-CoA, write the condensed equation for the lipogenesis of myristic acid, $CH_3(CH_2)_{12}COOH$.

44. Starting with L-glutamic acid and

$$CH_3CH_2\overset{\overset{\displaystyle CH_3}{|}}{C}H-\overset{\overset{\displaystyle O}{\|}}{C}-COOH$$

write an equation showing the transamination to isoleucine.

45. You have been invited to a lecture entitled "The Role of Malonyl CoA." Would you expect the lecture to be about anabolism of fats or catabolism of fats or both? Briefly explain.

46. In Step 2 of beta oxidation of a fatty acid, dehydrogenation occurs. Is the fatty acid oxidized or reduced? Briefly explain.

47. The energy content of fats is 9 kcal/g. Calculate the energy content of 1.0 mol of palmitic acid.

PUTTING IT TOGETHER
Review for Chapters 32–35

Multiple Choice: *Choose the correct answer for each of the following.*

1. Milk is a good example of a
   (a) macronutrient   (c) mineral
   (b) food   (d) carbohydrate

2. Which of the following is a micronutrient?
   (a) glucose   (c) iron
   (b) stearic acid   (d) water

3. RDA stands for
   (a) recommended daily allowance
   (c) recommended dietary allowance
   (c) required daily allowance
   (d) required dietary allowance

4. Amylose, a component of starch, is an example of a
   (a) micronutrient   (c) vitamin
   (b) mineral   (d) macronutrient

5. Starvation leads to a condition called
   (a) kwashiorkor   (c) marasmus
   (b) scurvy   (d) no correct answer given

6. Which of the following is <u>not</u> an example of a macronutrient?
   (a) water   (c) fructose
   (b) triacylglycerol (d) soy protein

7. Each pound of fat is equivalent to about
   (a) 3500 kcal   (c) 4.184 kJ
   (b) 500 kcal   (d) 65 kg

8. An example of a complex carbohydrate is
   (a) blood sugar   (c) cellulose
   (b) lactose   (d) sucrose

9. Which of the following is an example of an ω-3 fatty acid?
   (a) linoleic acid   (c) arachidonic acid
   (b) linolenic acid   (d) oleic acid

10. Vitamin C is known as
    (a) a fat-soluble vitamin   (c) ascorbic acid
    (b) citric acid   (d) pantothenic acid

11. A person suffering from beriberi has a deficiency of
    (a) vitamin B_1 (c) vitamin B_2
    (b) vitamin B_{12} (d) niacin

12. Potassium is an example of a
    (a) trace element   (c) major mineral
    (b) macronutrient   (d) no correct answer given

13. A deficiency of iodine commonly causes
    (a) anemia   (c) dental caries
    (b) goiter   (d) impaired glucose tolerance

14. How many grams of calcium are supplied if a serving of cereal (plus milk) gives 30% of a calcium RDA (1200 mg)?
    (a) 0.36 g   (c) 360 g
    (b) 3.6×10^5 g   (d) 0.00036 g

15. If 0.9 mg of zinc is 6% of the daily value, the RDA for zinc is
    (a) 54 mg   (c) 1.5 mg
    (b) 15 mg   (d) 0.15 g

16. Many food additives have
    (a) an RDA
    (b) a GARS rating
    (c) a "generally recognized as sufficient" rating
    (d) no correct answer given

17. Ultimately, carbohydrates are digested to
    (a) disaccharides   (c) lactic acid
    (b) monosaccharides   (d) carbon dioxide

18. A very important enzyme in salivary digestion is
    (a) pepsin   (c) amylase
    (b) lipase   (d) chymotrypsin

19. Bile is important for
    (a) fat digestion   (c) protein digestion
    (b) protein absorption (d) carbohydrate digestion

20. Intestinal microvilli
    (a) secrete bile   (c) synthesize bile
    (b) absorb nutrients   (d) no correct answer given

21. Blood glucose concentration is maintained by the
    (a) liver   (c) kidneys
    (b) gall bladder   (d) gastric juices

22. Hemoglobin binds to carbon dioxide to form
    (a) a heme–carboxyl ion   (c) a bicarbonate ion
    (b) a carbonate ion   (d) a carbamino ion

23. The formula for carbonic acid is
    (a) H_3CO_4 (c) H_2CO_3
    (b) HCO_3^- (d) H_2CO_4

24. Two major waste products of cellular energetics are
    (a) carbonic acid and pyruvic acid
    (b) carbonic acid and lactic acid
    (c) carbolic acid and lactic acid
    (d) carbolic acid and pyruvic acid

25. Which carbon in the following compound provides the most energy upon oxidation?

$$
\begin{array}{c}
\overset{O}{\underset{\|}{}} \quad \overset{O}{\underset{\|}{}} \\
^{1}C - O - P - OH \\
| \qquad | \\
H - ^{2}C - OH \; OH \\
| \\
\qquad O \\
\qquad \| \\
H_2{}^{3}C - O - P - OH \\
| \\
OH
\end{array}
$$

(a) carbon 1   (c) carbon 3
(b) carbon 2   (d) no correct answer given

26. The name of FAD is
(a) flavin adenine dinucleotide
(b) flavin arginine dinucleotide
(c) flavodoxin adenine dinucleotide
(d) flavodoxin arginine dinucleotide

27. NADH is the
(a) oxidized form of nicotinamide adenine dinucleotide
(b) oxidized form of nicotinamide adenine dinucleotide phosphate
(c) reduced form of nitcotinamide adenine dinucleotide
(d) reduced form of nicotinamide adenine dinucleotide phosphate

28. NAD^+ is derived from the vitamin
(a) niacin   (c) pantothenic acid
(b) thiamine   (d) riboflavin

29. Which reaction is <u>not</u> part of the mitochondrial electron transport system?
(a) $2\,NADH + 2\,H^+ + O_2 \longrightarrow 2\,NAD^+ + 2\,H_2O$
(b) $2\,NAD^+ + 2\,H_2O \longrightarrow 2\,NADH + 2\,H^+ + O_2$
(c) $2\,FADH_2 + O_2 \longrightarrow 2\,FAD + 2\,H_2O$
(d) $4\,NADH + 4\,H^+ \longrightarrow 2O_2 \longrightarrow 4\,NAD^+ + 4\,H_2O$

30. ATP is composed of a triphosphate, an adenine, and
(a) a deoxyribose   (c) a glucose
(b) a fructose   (d) a ribose

31. How many high-energy phosphate bonds does the following structure have?

$$
\begin{array}{c}
O \quad\quad O \\
\| \quad\quad \| \\
^-O - P - O - P - O - CH_2 \\
| \quad\quad | \\
O^- \quad\quad O^-
\end{array}
$$

(a) two   (c) one
(b) four   (d) three

32. A process that directly uses redox reaction energy to form ATP is called
(a) mitochondrial electron transport
(b) chloroplast electron transport
(c) oxidative phosphorylation
(d) reductive phosphorylation

33. Photosynthesis uses $6\,CO_2$ to make
(a) one ribose-5-phosphate
(b) glucose
(c) two 3-phosphoglycerates
(d) fructose

34. The chloroplast electron transport produces
(a) NADH   (c) $NADP^+$
(b) $FADH_2$ (d) no correct answer given

35. In photosynthesis, light energy is first absorbed by
(a) $NADP^+$ (c) ATP
(b) chlorophyll   (d) glucose

36. Which of the following is an anabolic pathway?
(a) Embden–Meyerhof pathway   (c) glycogenolysis
(b) glycogenesis   (d) glycolysis

37. How many high-energy phosphate bonds does the following compound (1,3-bisphosphoglycerate) have?

$$
\begin{array}{c}
O \quad\quad O \\
\| \quad\quad \| \\
C - O - P - O^- \\
| \quad\quad | \\
H - C - OH \; O^- \\
| \\
\quad\quad O \\
\quad\quad \| \\
H_2C - O - P - O^- \\
| \\
O^-
\end{array}
$$

(a) one   (c) three
(b) two   (d) four

38. The Embden–Meyerhof pathway converts glucose to two pyruvates. The sum of the oxidation numbers of carbon changes from zero in glucose to what value in two pyruvates?
(a) +6   (c) −2
(b) +4   (d) +2

39. In order to cycle NADH back to NAD^+, muscle tissue converts pyruvate to
(a) acetaldehyde   (c) ethanol
(b) glyceraldehyde   (d) lactate

40. Glycolysis is termed anaerobic because
(a) molecular oxygen is required
(b) only small amounts of ATP are formed
(c) molecular oxygen is not required
(d) NADH is produced

41. Synthesis of coenzyme A requires
(a) pantothenic acid   (c) vitamin D
(b) lactic acid   (d) nicotinic acid

42. The citric acid cycle takes place in the
    (a) cytoplasm (c) chloroplasts
    (b) nuclei   (d) mitochondria

43. The citric acid cycle was elucidated by
    (a) Fritz Lipmann   (c) Hans Krebs
    (b) Franz Meyerhof (d) Melvin Calvin

44. In the citric acid cycle, the acetyl group of acetyl CoA is
    (a) reduced to two carbon dioxides
    (b) oxidized to two carbon dioxides
    (c) oxidized to two NADHs
    (d) reduced to two NADHs

45. The citric acid cycle is aerobic because
    (a) O_2 is used in the cycle
    (b) the cycle depends on chloroplast electron transport
    (c) the cycle depends on mitochondrial electron transport
    (d) the cycle uses O_2 to convert pyruvate to acetyl CoA

46. How many NADHs are produced from 13.2 mol of acetyl CoA by the citric acid cycle?
    (a) 6.6 mol   (c) 39.6 mol
    (b) 13.2 mol   (d) 26.4 mol

47. Which of the following compounds <u>cannot</u> be used to make glucose in gluconeogenesis?
    (a) glycerol   (c) lactate
    (b) acetyl CoA (d) amino acids

48. Which of the following is <u>not</u> an advantage to a multistep process (as compared with a single-step process)?
    (a) Energy can be handled efficiently.
    (b) Fewer enzymes are needed.
    (c) Intermediates can be used in several pathways.
    (d) The energy release per step is small.

49. The glucose renal threshold is reached when
    (a) a condition of hyperglycemia occurs
    (b) blood-glucose levels are too low
    (c) insulin levels are too high
    (d) a fasting state occurs

50. In studying beta oxidation, which of the following scientists designed prototypes of modern equipments that tag molecules with radioactive isotopes?
    (a) Franz Meyerhof   (c) Franz Knoop
    (b) Melvin Calvin   (d) no correct answer given

51. How many ATPs are produced in beta oxidation from one stearic acid molecule?
    (a) 2   (c) 139
    (b) 0   (d) 98

52. The anabolic pathway for fatty acids is called
    (a) "two-carbon chop" (c) glycolysis
    (b) lipogenesis   (d) glycogenesis

53. The vitamin, biotin, is very important in reactions that
    (a) add carbon dioxide to a molecule
    (b) cause hydration of a double bond
    (c) form a thioester
    (d) reduce $NADP^+$

54. Which of the following is <u>not</u> a method for fixing nitrogen?
    (a) lightning   (c) transamination
    (b) Haber process   (d) soil bacteria

55. Acetyl CoA is to the citric acid cycle as one of the following compounds is to the urea cycle.
    (a) glutamic cycle   (c) carbamoyl phosphate
    (b) fumaric acid   (d) glutamine

56. When a person takes in more nitrogen than is excreted, they are said to be in
    (a) negative nitrogen balance
    (b) nitrogen balance
    (c) positive nitrogen balance
    (d) no correct answer given

57. Transamination is the first step in the conversion of glucogenic amino acids to
    (a) ketogenic amino acids   (c) glycogen
    (b) fats   (d) essential amino acids

58. Fish commonly excrete nitrogen in the form of
    (a) urea   (c) uric acid
    (b) ammonia   (d) glutamine

59. How many nitrogen atoms are in one molecule of urea?
    (a) one   (c) three
    (b) two   (d) four

60. The urea cycle occurs in the
    (a) liver   (c) pancreas
    (b) small intestine (d) skeletal muscles

Free Response Questions

1. Is the vitamin below likely to be water-soluble or fat-soluble? Explain.

2. What are the end products of the digestion of polysaccharides, fats and oils, and proteins before they enter the metabolic pathways? Most of these digestive end products are often converted into what common metabolic intermediate?

3. Are the two nutrients shown macronutrients or micronutrients? What class of nutrients do they fall under? Draw the products of digestion (hydrolysis) of these two nutrients and indicate which is more likely to provide more cellular energy. Briefly explain.

$$
\begin{array}{c}
\text{O} \\
\parallel \\
H_2C-O-C-(CH_2)_{16}CH_3
\end{array}
$$

$$
\begin{array}{c}
\text{O} \\
\parallel \\
HC-O-C-(CH_2)_{16}CH_3
\end{array}
$$

$$
\begin{array}{c}
\text{O} \\
\parallel \\
H_2C-O-C-(CH_2)_{16}CH_3
\end{array}
$$

$$
\begin{array}{c}
\text{O} \\
\parallel \\
H_2C-O-C-(CH_2)_7CH=CH(CH_2)_7CH_3
\end{array}
$$

$$
\begin{array}{c}
\text{O} \\
\parallel \\
HC-O-C-(CH_2)_7CH=CH(CH_2)_7CH_3
\end{array}
$$

$$
\begin{array}{c}
\text{O} \\
\parallel \\
H_2C-O-C-(CH_2)_7CH=CH(CH_2)_7CH_3
\end{array}
$$

4. The three macronutrients can be converted into glucose. For example, carbohydrates are digested via salivary and pancreatic juices to give monosaccharides, including glucose directly. *Briefly* outline how proteins and fats/oils can also lead to the production of glucose in humans and indicate the name of the pathway that includes both processes.

5. Complete the following reaction using nicotinamide adenine dinucleotide as the redox coenzyme. Is the coenzyme being reduced or oxidized in this process? Identify coenzymes A and B. Would this coenzyme be shunted to the electron-transport chain to produce ATP? Explain.

$$
\begin{array}{c}
H_2C-OH \\
| \\
C=O \\
| \\
CH_2OPO_3^{2-}
\end{array}
\quad
\xrightarrow{\text{A} \quad \text{B}}
\quad
\begin{array}{c}
H_2C-OH \\
| \\
HO-C-H \\
| \\
CH_2OPO_3^{2-}
\end{array}
$$

6. The complete oxidation of glucose ($C_6H_{12}O_6$) to carbon dioxide and water can be broken down into three main processes, glycolysis, pyruvate oxidation, and the citric acid cycle. The glycolysis of 1 mol of glucose in the skeletal muscle yields 2 mol NADH, 2 ml ATP, and 2 mol of pyruvate. The oxidation of 1 mol of pyruvate yields 1 mol NADH and 1 mol acetyl-CoA. The citric acid cycle produces 3 mol NADH, 1 mol FADH$_2$ and 1 mol ATP (GTP but can be converted to ATP) for every mole of acetyl-CoA entering the cycle. Before a race, an athlete eats 45 g of glucose powder, how many moles of ATP would be produced assuming no ATP is used to transport the glucose to the skeletal muscle, and that all the glucose and coenzymes are oxidized?

7. One mole of which saturated fatty acid was fully catabolized via beta oxidation if the acetyl-CoA from the catabolism produced 96 moles of ATP when used in the citric acid cycle? The citric acid cycle produces 3 moles NADH, 1 mol FADH$_2$ and 1 mol ATP for every mole of acetyl-CoA entering the cycle. Briefly explain.

8. Fill in the missing information in the following metabolic summary.

A (nutrient) $\xrightarrow{\substack{\textbf{B} \text{ (enzyme)} \\ \textbf{C} \text{ (location)}}}$

D (name)

$\textbf{E} \xrightarrow{\textbf{F} \text{ (pathway)}} \text{pyruvate} \longrightarrow \text{acetyl-CoA} \xrightarrow{\textbf{G} \text{ (cycle)}} CO_2$

9. A possible pathway for the nitrogen ingested could be represented as follows. Indicate in which part of the body each of the processes take place.

glu-phe-asp-asn-val-
tyr-trp-ala-val-phe $\longrightarrow$

glu-phe

asp-asn-val-tyr $\longrightarrow$

trp-ala-val-phe

glu
phe
asp
asn
val
tyr
trp
ala
val
phe

$\longrightarrow$ urea

Mathematical Review

Multiplication Multiplication is a process of adding any given number or quantity to itself a certain number of times. Thus, 4 times 2 means 4 added two times, or 2 added together four times, to give the product 8. Various ways of expressing multiplication are

$$ab \qquad a \times b \qquad a \cdot b \qquad a(b) \qquad (a)(b)$$

Each of these expressions means a times b, or a multiplied by b, or b times a.

When $a = 16$ and $b = 24$,   we have $16 \times 24 = 384$.

The expression $°F = (1.8 \times °C) + 32$ means that we are to multiply 1.8 times the Celsius degrees and add 32 to the product. When $°C$ equals 50,

$$°F = (1.8 \times 50) + 32 = 90 + 32 = 122°F$$

The result of multiplying two or more numbers together is known as the *product.*

Division The word *division* has several meanings. As a mathematical expression, it is the process of finding how many times one number or quantity is contained in another. Various ways of expressing division are

$$a \div b \qquad \frac{a}{b} \qquad a/b$$

Each of these expressions means a divided by b.

When $a = 15$ and $b = 3$, $\dfrac{15}{3} = 5$.

The number above the line is called the *numerator;* the number below the line is the *denominator.* Both the horizontal and the slanted (/) division signs also mean "per." For example, in the expression for density, the mass per unit volume:

$$\text{density} = \text{mass/volume} = \frac{\text{mass}}{\text{volume}} = \text{g/mL}$$

The diagonal line still refers to a division of grams by the number of milliliters occupied by that mass. The result of dividing one number into another, is called the *quotient.*

Proper fraction	Decimal fraction	Proper fraction
$\dfrac{1}{8}$	$= 0.125 =$	$\dfrac{125}{1000}$
$\dfrac{1}{10}$	$= 0.1 =$	$\dfrac{1}{10}$
$\dfrac{3}{4}$	$= 0.75 =$	$\dfrac{75}{100}$
$\dfrac{1}{100}$	$= 0.01 =$	$\dfrac{1}{100}$
$\dfrac{1}{4}$	$= 0.25 =$	$\dfrac{25}{100}$

Fractions and Decimals   A fraction is an expression of division, showing that the numerator is divided by the denominator. A *proper fraction* is one in which the numerator is smaller than the denominator. In an *improper fraction,* the numerator is the larger number. A decimal or a decimal fraction is a proper fraction in which the denominator is some power of 10. The decimal fraction is determined by carrying out the division of the proper fraction. Examples of proper fractions and their decimal fraction equivalents are shown in the accompanying table.

Addition of Numbers with Decimals   To add numbers with decimals, we use the same procedure as that used when adding whole numbers, but we always line up the decimal points in the same column. For example, add 8.21 + 143.1 + 0.325:

$$
\begin{array}{r}
8.21 \\
+143.1 \\
+0.325 \\
\hline
151.635
\end{array}
$$

When adding numbers that express units of measurement, we must be certain that the numbers added together all have the same units. For example, what is the total length of three pieces of glass tubing: 10.0 cm, 125 mm, and 8.4 cm? If we simply add the numbers, we obtain a value of 143.4, but we are not certain what the unit of measurement is. To add these lengths correctly, first change 125 mm to 12.5 cm. Now all the lengths are expressed in the same units and can be added:

$$
\begin{array}{r}
10.0 \text{ cm} \\
12.5 \text{ cm} \\
\underline{8.4 \text{ cm}} \\
30.9 \text{ cm}
\end{array}
$$

Subtraction of Numbers with Decimals   To subtract numbers containing decimals, we use the same procedure as for subtracting whole numbers, but we always line up the decimal points in the same column. For example, subtract 20.60 from 182.49:

$$
\begin{array}{r}
182.49 \\
-20.60 \\
\hline
161.89
\end{array}
$$

Multiplication of Numbers with Decimals   To multiply two or more numbers together that contain decimals, we first multiply as if they were whole numbers. Then, to locate the decimal point in the product, we add the number of digits to the right of the decimal in all the numbers multiplied together. The product should have this same number of digits to the right of the decimal point.

Multiply $2.05 \times 2.05 = 4.2025$ (total of four digits to the right of the decimal). Here are more examples:

$14.25 \times 6.01 \times 0.75 = 64.231875$ (six digits to the right of the decimal)
$39.26 \times 60 = 2355.60$ (two digits to the right of the decimal)

If a number is a measurement, the answer must be adjusted to the correct number of significant figures.

Division of Numbers with Decimals   To divide numbers containing decimals, we first relocate the decimal points of the numerator and denominator by moving them to the right as many places as needed to make the denominator a whole number. (Move the decimal of both the numerator and the denominator the same amount and in the same direction.) For example,

$$\frac{136.94}{4.1} = \frac{1369.4}{41}$$

The decimal point adjustment in this example is equivalent to multiplying both numerator and denominator by 10. Now we carry out the division normally, locating the decimal point immediately above its position in the dividend:

$$41 \overline{)1369.4}^{\,33.4} \qquad \frac{0.441}{26.25} = \frac{44.1}{2625} = 2625 \overline{)44.1000}^{\,0.0168}$$

These examples are guides to the principles used in performing the various mathematical operations illustrated. Every student of chemistry should learn to use a calculator for solving mathematical problems (see Appendix II). The use of a calculator will save many hours of doing tedious calculations. After solving a problem, the student should check for errors and evaluate the answer to see if it is logical and consistent with the data given.

Algebraic Equations   Many mathematical problems that are encountered in chemistry fall into the following algebraic forms. Solutions to these problems are simplified by first isolating the desired term on one side of the equation. This rearrangement is accomplished by treating both sides of the equation in an identical manner until the desired term is isolated.

(a) $a = \dfrac{b}{c}$

To solve for b, multiply both sides of the equation by c:

$$a \times c = \frac{b}{\cancel{c}} \times \cancel{c}$$

$$b = a \times c$$

To solve for c, multiply both sides of the equation by $\dfrac{c}{a}$:

$$\cancel{a} \times \frac{c}{\cancel{a}} = \frac{b}{\cancel{c}} \times \frac{\cancel{c}}{a}$$

$$c = \frac{b}{a}$$

(b) $\dfrac{a}{b} = \dfrac{c}{d}$

To solve for a, multiply both sides of the equation by b:

$$\dfrac{a}{\not b} \times \not b = \dfrac{c}{d} \times b$$

$$a = \dfrac{c \times b}{d}$$

To solve for b, multiply both sides of the equation by $\dfrac{b \times d}{c}$:

$$\dfrac{a}{\not b} \times \dfrac{\not b \times d}{c} = \dfrac{\not c}{\not d} \times \dfrac{b \times \not d}{\not c}$$

$$\dfrac{a \times d}{c} = b$$

(c) $a \times b = c \times d$

To solve for a, divide both sides of the equation by b:

$$\dfrac{a \times \not b}{\not b} = \dfrac{c \times d}{b}$$

$$a = \dfrac{c \times d}{b}$$

(d) $\dfrac{b - c}{a} = d$

To solve for b, first multiply both sides of the equation by a:

$$\dfrac{\not a(b - c)}{\not a} = d \times a$$

$$b - c = d \times a$$

Then add c to both sides of the equation:

$$b - \not c + \not c = d \times a + c$$

$$b = (d \times a) + c$$

When $a = 1.8$, $c = 32$, and $d = 35$,

$$b = (35 \times 1.8) + 32 = 63 + 32 = 95$$

Expression of Large and Small Numbers In scientific measurement and calculations, we often encounter very large and very small numbers—for example, 0.00000384 and 602,000,000,000,000,000,000,000. These numbers are troublesome to write and awkward to work with, especially in calculations. A convenient method of expressing these large and small numbers in a simplified form is by means of exponents or powers of 10. This method of expressing numbers is known as **scientific or exponential notation.**

An **exponent** is a number written as a superscript following another number. Exponents are often called *powers* of numbers. The term *power* indicates how many times the number is used as a factor. In the number 10^2, 2 is the exponent, and the number means 10 squared, or 10 to the second power, or $10 \times 10 = 100$. Three other examples are

scientific or exponential
notation

exponent

$$3^2 = 3 \times 3 = 9$$
$$3^4 = 3 \times 3 \times 3 \times 3 = 81$$
$$10^3 = 10 \times 10 \times 10 = 1000$$

For ease of handling, large and small numbers are expressed in powers of 10. Powers of 10 are used because multiplying or dividing by 10 coincides with moving the decimal point in a number by one place. Thus, a number multiplied by 10^1 would move the decimal point one place to the right; 10^2, two places to the right; 10^{-2}, two places to the left. To express a number in powers of 10, we move the decimal point in the original number to a new position, placing it so that the number is a value between 1 and 10. This new decimal number is multiplied by 10 raised to the proper power. For example, to write the number 42,389 in exponential form, the decimal point is placed between the 4 and the 2 (4.2389), and the number is multiplied by 10^4; thus, the number is 4.2389×10^4:

$$42,389 = 4.2389 \times 10^4$$
$$4\,3\,2\,1$$

The exponent of 10 (4) tells us the number of places that the decimal point has been moved from its original position. If the decimal point is moved to the left, the exponent is a positive number; if it is moved to the right, the exponent is a negative number. To express the number 0.00248 in exponential notation (as a power of 10), the decimal point is moved three places to the right; the exponent of 10 is -3, and the number is 2.48×10^{-3}.

$$0.00248 = 2.48 \times 10^{-3}$$
$$1\,2\,3$$

Study the following examples.

$$1237 = 1.237 \times 10^3$$
$$988 = 9.88 \times 10^2$$
$$147.2 = 1.472 \times 10^2$$
$$2,200,000 = 2.2 \times 10^6$$
$$0.0123 = 1.23 \times 10^{-2}$$
$$0.00005 = 5 \times 10^{-5}$$
$$0.000368 = 3.68 \times 10^{-4}$$

Exponents in multiplication and division The use of powers of 10 in multiplication and division greatly simplifies locating the decimal point in the answer. In multiplication, first change all numbers to powers of 10, then multiply the numerical portion in the usual manner, and finally add the exponents of 10 algebraically, expressing them as a power of 10 in the product. In multiplication, the exponents (powers of 10) are added algebraically.

$$10^2 \times 10^3 = 10^{(2+3)} = 10^5$$
$$10^2 \times 10^2 \times 10^{-1} = 10^{(2+2-1)} = 10^3$$

Multiply: (40,000)(4200)
Change to powers of 10: $(4 \times 10^4)(4.2 \times 10^3)$
Rearrange: $(4 \times 4.2)(10^4 \times 10^3)$
                                     $16.8 \times 10^{(4+3)}$
                                     16.8×10^7 or 1.68×10^8 (Answer)

Multiply: (380)(0.00020)
                 $(3.80 \times 10^2)(2.0 \times 10^{-4})$
                 $(3.80 \times 2.0)(10^2 \times 10^{-4})$
                 $7.6 \times 10^{(2-4)}$
                 7.6×10^{-2} or   0.076   (Answer)

Multiply: (125)(284)(0.150)
                 $(1.25 \times 10^2)(2.84 \times 10^2)(1.50 \times 10^{-1})$
                 $(1.25)(2.84)(1.50)(10^2 \times 10^2 \times 10^{-1})$
                 $5.325 \times 10^{(2+2-1)}$
                 5.33×10^3 (Answer)

In division, after changing the numbers to powers of 10, move the 10 and its exponent from the denominator to the numerator, changing the sign of the exponent. Carry out the division in the usual manner and evaluate the power of 10. Change the sign(s) of the exponent(s) of 10 in the denominator and move the 10 and its exponent(s) to the numerator. Then add all the exponents of 10 together. For example,

$$\frac{10^5}{10^3} = 10^5 \times 10^{-3} = 10^{(5-3)} = 10^2$$

$$\frac{10^3 \times 10^4}{10^{-2}} = 10^3 \times 10^4 \times 10^2 = 10^{(3+4+2)} = 10^9$$

Significant Figures in Calculations   The result of a calculation based on experimental measurements cannot be more precise than the measurement that has the greatest uncertainty. (See Section 2.5 for additional discussion.)

Addition and Subtraction   The result of an addition or subtraction should contain no more digits to the right of the decimal point than are contained in the quantity that has the least number of digits to the right of the decimal point.

Perform the operation indicated and then round off the number to the proper number of significant figures:

142.8 g	
18.843 g	93.45 mL
36.42 g	−18.0 mL
198.063 g	75.45 mL
198.1 g (Answer)	75.5 mL (Answer)

Multiplication and Division   In calculations involving multiplication or division, the answer should contain the same number of significant figures as the measurement that has the least number of significant figures. In multiplication or division, the position of the decimal point has nothing to do with the number of significant figures in the answer. Study the following examples:

	Round off to
$(2.05)(2.05) = 4.2025$	4.20
$(18.48)(5.2) = 96.096$	96
$(0.0126)(0.020) = 0.000252$ or	
$(1.26 \times 10^{-2})(2.0 \times 10^{-2}) = 2.520 \times 10^{-4}$	2.5×10^{-4}
$\dfrac{1369.4}{41} = 33.4$	33
$\dfrac{2268}{4.20} = 540.$	540.

Dimensional Analysis   Many problems of chemistry can be readily solved by dimensional analysis using the factor-label or conversion-factor method. Dimensional analysis involves the use of proper units of dimensions for all factors that are multiplied, divided, added, or subtracted in setting up and solving a problem. Dimensions are physical quantities such as length, mass, and time, which are expressed in such units as centimeters, grams, and seconds, respectively. In solving a problem, we treat these units mathematically just as though they were numbers, which gives us an answer that contains the correct dimensional units.

A measurement or quantity given in one kind of unit can be converted to any other kind of unit having the same dimension. To convert from one kind of unit to another, the original quantity or measurement is multiplied or divided by a conversion factor. The key to success lies in choosing the correct conversion factor. This general method of calculation is illustrated in the following examples.

Suppose we want to change 24 ft to inches. We need to multiply 24 ft by a conversion factor containing feet and inches. Two such conversion factors can be written relating inches to feet:

$$\frac{12 \text{ in.}}{1 \text{ ft}} \quad \text{or} \quad \frac{1 \text{ ft}}{12 \text{ in.}}$$

We choose the factor that will mathematically cancel feet and leave the answer in inches. Note that the units are treated in the same way we treat numbers, multiplying or dividing as required. Two possibilities then arise to change 24 ft to inches:

$$(24 \text{ ft})\left(\frac{12 \text{ in.}}{1 \text{ ft}}\right) \quad \text{or} \quad (24 \text{ ft})\left(\frac{1 \text{ ft}}{12 \text{ in.}}\right)$$

In the first case (the correct method), feet in the numerator and the denominator cancel, giving us an answer of 288 in. In the second case, the units of the answer are $ft^2/in.$, the answer being $2.0 \, ft^2/in.$ In the first case, the answer is reasonable because it is expressed in units having the proper dimensions. That is, the dimension of length expressed in feet has been converted to length in inches according to the mathematical expression

$$\text{ft} \times \frac{\text{in.}}{\text{ft}} = \text{in.}$$

In the second case, the answer is not reasonable because the units (ft²/in.) do not correspond to units of length. The answer is therefore incorrect. The units are the guiding factor for the proper conversion.

The reason we can multiply 24 ft times 12 in./ft and not change the value of the measurement is that the conversion factor is derived from two equivalent quantities. Therefore, the conversion factor 12 in./ft is equal to unity. When you multiply any factor by 1, it does not change the value:

$$12 \text{ in.} = 1 \text{ ft} \qquad \text{and} \qquad \frac{12 \text{ in.}}{1 \text{ ft}} = 1$$

Convert 16 kg to milligrams. In this problem it is best to proceed in this fashion:

$$\text{kg} \longrightarrow \text{g} \longrightarrow \text{mg}$$

The possible conversion factors are

$$\frac{1000 \text{ g}}{1 \text{ kg}} \quad \text{or} \quad \frac{1 \text{ kg}}{1000 \text{ g}} \qquad \frac{1000 \text{ mg}}{1 \text{ g}} \quad \text{or} \quad \frac{1 \text{ g}}{1000 \text{ mg}}$$

We use the conversion factor that leaves the proper unit at each step for the next conversion. The calculation is

$$(16 \text{ kg})\left(\frac{1000 \text{ g}}{1 \text{ kg}}\right)\left(\frac{1000 \text{ mg}}{1 \text{ g}}\right) = 1.6 \times 10^7 \text{ mg}$$

Many problems can be solved by a sequence of steps involving unit conversion factors.

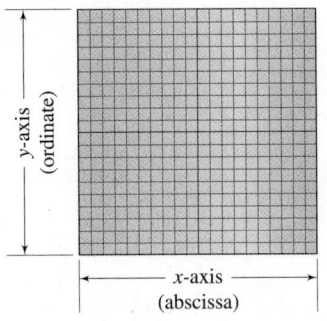

y-axis (ordinate)

x-axis (abscissa)

▲
FIGURE I.1

Graphical Representation of Data   A graph is often the most convenient way to present or display a set of data. Various kinds of graphs have been devised, but the most common type uses a set of horizontal and vertical coordinates to show the relationship of two variables. It is called an x–y graph because the data of one variable are represented on the horizontal or x-axis (abscissa) and the data of the other variable are represented on the vertical or y-axis (ordinate). See Figure I.1.

As a specific example of a simple graph, let us graph the relationship between Celsius and Fahrenheit temperature scales. Assume that initially we have only the information in the table next to Figure I.2.

On a set of horizontal and vertical coordinates (graph paper), scale off at least 100 Celsius degrees on the x-axis and at least 212 Fahrenheit degrees on the y-axis. Locate and mark the three points corresponding to the three temperatures given and draw a line connecting these points (see Figure I.2).

Here is how a point is located on the graph: Using the 50°C–122°F data, trace a vertical line up from 50°C on the x-axis and a horizontal line across from 122°F on the y-axis and mark the point where the two lines intersect. This process is called *plotting*. The other two points are plotted on the graph in the same way. [*Note:* The number of degrees per scale division was chosen to give a graph of convenient size. In this case, there are 5 Fahrenheit degrees per scale division and 2 Celsius degrees per scale division.]

The graph in Figure I.2 shows that the relationship between Celsius and Fahrenheit temperature is that of a straight line. The Fahrenheit temperature corresponding to any given Celsius temperature between 0 and 100° can be determined from the graph. For example, to find the Fahrenheit temperature corresponding to 40°C, trace a perpendicular line from 40°C on the x-axis to the line plotted on the graph. Now trace a horizontal line from this point on the plotted line to the y-axis and read the corresponding Fahrenheit temperature (104°F). See the dashed lines in Figure I.2. In turn, the Celsius temperature corresponding to any Fahrenheit temperature between 32 and 212° can be determined from the graph by tracing a horizontal line from the Fahrenheit temperature to the plotted line and reading the corresponding temperature on the Celsius scale directly below the point of intersection.

The mathematical relationship of Fahrenheit and Celsius temperatures is expressed by the equation $°F = 1.8 \times °C + 32$. Figure I.2 is a graph of this equation. Because the graph is a straight line, it can be extended indefinitely at either end. Any desired Celsius temperature can be plotted against the corresponding Fahrenheit temperature by extending the scales along both axes as necessary.

Figure I.3 is a graph showing the solubility of potassium chlorate in water at various temperatures. The solubility curve on this graph was plotted from the data in the table next to the graph.

In contrast to the Celsius–Fahrenheit temperature relationship, there is no simple mathematical equation that describes the exact relationship between temperature and the solubility of potassium chlorate. The graph in Figure I.3 was constructed from experimentally determined solubilities at the six temperatures shown. These experimentally determined solubilities are all located on the smooth curve traced by the unbroken-line portion of the graph. We are therefore confident that the unbroken line represents a very good approximation of the solubility data for potassium chlorate over the temperature range from 10 to 80°C. All points on

°C	°F
0	32
50	122
100	212

◀ **FIGURE I.2**

Temperature (°C)	Solubility (g KClO$_3$/ 100 g water)
10	5.0
20	7.4
30	10.5
50	19.3
60	24.5
80	38.5

Figure 1.3 ▶

the plotted curve represent the composition of saturated solutions. Any point below the curve represents an unsaturated solution.

The dashed-line portions of the curve are *extrapolations;* that is, they extend the curve above and below the temperature range actually covered by the plotted solubility data. Curves such as this one are often extrapolated a short distance beyond the range of the known data, although the extrapolated portions may not be highly accurate. Extrapolation is justified only in the absence of more reliable information.

The graph in Figure I.3 can be used with confidence to obtain the solubility of KClO$_3$ at any temperature between 10 and 80°C, but the solubilities between 0 and 10°C and between 80 and 100°C are less reliable. For example, what is the solubility of KClO$_3$ at 55°C, at 40°C, and at 100°C?

First draw a perpendicular line from each temperature to the plotted solubility curve. Now trace a horizontal line to the solubility axis from each point on the curve and read the corresponding solubilities. The values that we read from the graph are

40°C	14.2 g KClO$_3$/100 g water
55°C	22.0 g KClO$_3$/100 g water
100°C	59 g KClO$_3$/100 g water

Of these solubilities, the one at 55°C is probably the most reliable because experimental points are plotted at 50°C and at 60°C. The 40°C solubility value is a bit less reliable because the nearest plotted points are at 30°C and 50°C. The 100°C solubility is the least reliable of the three values because it was taken from the extrapolated part of the curve, and the nearest plotted point is 80°C. Actual handbook solubility values are 14.0 and 57.0 g of $KClO_3$/100 g water at 40°C and 100°C, respectively.

The graph in Figure I.3 can also be used to determine whether a solution is saturated or unsaturated. For example, a solution contains 15 g of $KClO_3$/100 g of water and is at a temperature of 55°C. Is the solution saturated or unsaturated? *Answer:* The solution is unsaturated because the point corresponding to 15 g and 55°C on the graph is below the solubility curve; all points below the curve represent unsaturated solutions.

APPENDIX

II

Using a Scientific Calculator

A calculator is useful for most calculations in this book. You should obtain a scientific calculator, that is, one that has at least the following function keys on its keyboard.

Addition $\boxed{+}$ Second function $\boxed{F}$ or $\boxed{INV}$ or $\boxed{Shift}$

Subtraction $\boxed{-}$ Change sign $\boxed{+/-}$

Multiplication $\boxed{\times}$ Exponential number $\boxed{Exp}$

Division $\boxed{\div}$ Lorgarithm $\boxed{Log}$

Equals $\boxed{=}$ Antilogarithm $\boxed{10^x}$

Not all calculators use the same symbolism for these function keys, nor do all calculators work in the same way. The following discussion may not pertain to your particular calculator. Refer to your instruction manual for variations from the function symbols shown above and for the use of other function keys.

The second function key may have a different designation on your calculator.

Some keys have two functions, upper and lower. In order to use the upper (second) function, the second function key $\boxed{F}$ must be pressed in order to activate the desired upper function.

The display area of the calculator shows the numbers entered and often shows more digits in the answer than should be used. Therefore, the final answer should be rounded to reflect the proper number of significant figures of the calculations.

Addition and Subtraction   To add numbers using your calculator,

1. Enter the first number to be added followed by the plus key $\boxed{+}$.
2. Enter the second number to be added followed by the plus key $\boxed{+}$.
3. Repeat Step 2 for each additional number to be added, except the last number.
4. After the last number is entered press the equal key $\boxed{=}$. You should now have the answer in the display area.
5. When a number is to be subtracted, use the minus key $\boxed{-}$ instead of the plus key.

As an example, to add $16.0 + 1.223 + 8.45$, enter 16.0 followed by the $\boxed{+}$ key; then enter 1.223 followed by the $\boxed{+}$ key; then enter 8.45 followed by the $\boxed{=}$ key. Display shows 25.673, which should then be rounded to the answer 25.7.

Examples of Addition and Subtraction			
Calculation	Enter in sequence	Display	Rounded answer
a. 12.0 + 16.2 + 122.3	12.0 $+$ 16.2 $+$ 122.3 $=$	150.5	150.5
b. 132 − 62 + 141	132 $-$ 62 $+$ 141 $=$	211	211
c. 46.23 + 13.2	46.23 $+$ 13.2 $=$	59.43	59.4
d. 129.06 + 49.1 − 18.3	129.06 $+$ 49.1 $-$ 18.3 $=$	159.86	159.9

Multiplication To multiply numbers using your calculator

1. Enter the first number to be multiplied followed by the multiplication key $\times$.
2. Enter the second number to be multiplied followed by the multiplication key $\times$.
3. Repeat Step 2 for all other numbers to be multiplied except the last number.
4. Enter the last number to be multiplied followed by the equal key $=$. You now have the answer in the display area. Round off to the proper number of significant figures.

As an example, to calculate (3.25)(4.184)(22.2) enter 3.25 followed by the $\times$ key; then enter 4.184 followed by the $\times$ key; then enter 22.2 followed by the $=$ key. The display shows 301.8756, which is rounded to the answer 302.

Examples of Multiplication			
Calculation	Enter in sequence	Display	Rounded answer
a. 12 × 14 × 18	12 $\times$ 14 $\times$ 18 $=$	3024	3.0×10^3
b. 122 × 3.4 × 60.	122 $\times$ 3.4 $\times$ 60. $=$	24888	2.5×10^4
c. 0.522 × 49.4 × 6.33	0.522 $\times$ 49.4 $\times$ 6.33 $=$	163.23044	163

Division To divide numbers using your calculator,

1. Enter the numerator followed by the division key $\div$.
2. Enter the denominator followed by the equal key to give the answer.
3. If there is more than one denominator, enter each denominator followed by the division key except for the last number, which is followed by the equal key.

As an example, to calculate $\frac{126}{12}$ enter 126 followed by the $\div$ key; then enter 12 followed by the $=$ key. The display shows 10.5, which should then be rounded to the answer 11.

Examples of Division			
Calculation	Enter in sequence	Display	Rounded answer
a. $\dfrac{142}{25}$	142 $\div$ 25 $=$	5.68	5.7
b. $\dfrac{0.422}{5.00}$	0.422 $\div$ 5.00 $=$	0.0844	0.0844
c. $\dfrac{124}{0.022 \times 3.00}$	124 $\div$ 0.022 $\div$ 3.00 $=$	1878.7878	1.9×10^3

Exponents   In scientific measurements and calculations we often encounter very large and very small numbers. To express these large and small numbers conveniently, we use exponents or powers of 10. A number in exponential form is treated like any other number; that is, it can be added, subtracted, multiplied, or divided.

To enter an exponential number into your calculator first enter the nonexponential part of the number, then press the exponent key $\boxed{\text{Exp}}$, followed by the exponent. For example, to enter 4.9×10^3, enter 4.94, then press $\boxed{\text{Exp}}$, then press 3. When the exponent of 10 is a negative number, press the Change of Sign key $\boxed{+/-}$ after entering the exponent. For example, to enter 4.94×10^{-3}, enter in sequence 4.94 $\boxed{\text{Exp}}$ 3 $\boxed{+/-}$. In most calculators the exponent will appear in the display a couple of spaces after the nonexponent part of the number—for example, 4.94   03 or 4.94   −03.

Examples Using Exponential Numbers			
Calculation	Enter in sequence	Display	Rounded answer
a. $(4.94 \times 10^3)(21.4)$	4.94 $\boxed{\text{Exp}}$ 3 $\times$ 21.4 $=$	105716	1.06×10^5
b. $(1.42 \times 10^4)(2.88 \times 10^{-5})$	1.42 $\boxed{\text{Exp}}$ 4 $\times$ 2.88 $\boxed{\text{Exp}}$ 5 $\boxed{+/-}$ $=$	0.40896	0.409
c. $\dfrac{8.22 \times 10^{-5}}{5.00 \times 10^7}$	8.22 $\boxed{\text{Exp}}$ 5 $\boxed{+/-}$ $\div$ 5.00 $\boxed{\text{Exp}}$ 7 $=$	1.644 −12	1.64×10^{-12}

Logarithms   The logarithm of a number to the base 10 is the power(exponent) to which 10 must be raised to give that number. For example, the log of 100 is 2.0(log $100 = 10^{2.0}$). The log of 200 is 2.3(log $200 = 10^{2.3}$). Logarithms are used in chemistry to calculate the pH of an aqueous acidic solution. The answer (log) should contain the same number of significant figures in the decimal as is in the original number. Thus, the log 100 = 2.0 but the log 100. is 2.000.

The log key on most calculators is a function key. To determine the log using your calculator, enter the number, then press the function ⎡Log⎤ key. For example, to determine the log of 125, enter 125, then press the ⎡Log⎤ key. The display shows 2.097, which is the answer with the correct number of significant figures.

Examples Using Logarithms
Determine the log of the following:

	Enter in sequence	Display	Rounded answer
a. log 42	42 ⎡F⎤ ⎡Log⎤	1.6232492	1.62
b. log 1.62×10^5	1.62 ⎡Exp⎤ 5 ⎡F⎤ ⎡Log⎤	5.209515	5.210
c. log 6.4×10^{-6}	6.4 ⎡Exp⎤ 6 ⎡+/−⎤ ⎡F⎤ ⎡Log⎤	−5.19382	−5.19

Antilogarithms (Inverse Logarithms) An antilogarithm is the number from which the logarithm has been calculated. It is calculated using the ⎡10^x⎤ key on your calculator. For example, to determine the antilogarithm of 2.891, enter 2.891 into your calculator, then press the second function key followed by the ⎡10^x⎤ key, 2.891 ⎡F⎤ ⎡10^x⎤. The display shows 778.0365, which rounds to the answer 778. The answer should contain the same number of figures as there are to the right of the decimal in the number used (2.891) to obtain its antilog.

Examples Using Antilogarithms
Determine the antilogarithm of the following:

	Enter in sequence	Display	Rounded answer
a. antilog 1.628	1.628 ⎡F⎤ ⎡10^x⎤	42.461956	42.5
b. antilog 7.086	7.086 ⎡F⎤ ⎡10^x⎤	12189896	1.22×10^7
c. antilog −6.33	6.33 ⎡+/−⎤ ⎡F⎤ ⎡10^x⎤	4.6773514 −07	4.7×10^{-7}

**Additional Practice Problems. Only the problem,
the display, and the rounded answer are given.**

Problem	Display	Rounded answer
1. $143.5 + 14.02 + 1.202$	158.722	158.7
2. $72.06 - 26.92 - 49.66$	-4.52	-4.52
3. $2.168 + 4.288 - 1.62$	4.836	4.84
4. $(12.3)(22.8)(1.235)$	346.3434	346
5. $(2.42 \times 10^6)(6.08 \times 10^{-4})(0.623)$	916.65728	917
6. $\dfrac{(46.0)(82.3)}{19.2}$	197.17708	197
7. $\dfrac{0.0298}{243}$	1.2263374 -04	1.23×10^{-4}
8. $\dfrac{(5.4)(298)(760)}{(273)(1042)}$	4.2992554	4.3
9. $(6.22 \times 10^6)(1.45 \times 10^3)(9.00)$	8.1171 10	8.12×10^{10}
10. $\dfrac{(1.49 \times 10^6)(1.88 \times 10^6)}{6.02 \times 10^{23}}$	4.6531561 -12	4.65×10^{-12}
11. $\log 245$	2.389166	2.389
12. $\log 6.5 \times 10^{-6}$	-5.1870866	-5.19
13. $\log 24 \times \log 34$	2.1137644	2.11
14. antilog 6.34	2187761.6	2.2×10^6
15. antilog -6.34	4.5708818 -07	4.6×10^{-7}

Vapor Pressure of Water at Various Temperatures

Temperature (°C)	Vapor pressure (torr)	Temperature (°C)	Vapor pressure (torr)
0	4.6	26	25.2
5	6.5	27	26.7
10	9.2	28	28.3
15	12.8	29	30.0
16	13.6	30	31.8
17	14.5	40	55.3
18	15.5	50	92.5
19	16.5	60	149.4
20	17.5	70	233.7
21	18.6	80	355.1
22	19.8	90	525.8
23	21.2	100	760.0
24	22.4	110	1074.6
25	23.8		

Units of Measurement

Physical Constants

Constant	Symbol	Value
Atomic mass unit	amu	1.6606×10^{-27} kg
Avogadro's number	N	6.022×10^{23}/mol
Gas constant	R (at STP)	0.08205 L atm/K mol
Mass of an electron	m_e	9.11×10^{-31} kg
		5.486×10^{-4} amu
Mass of a neutron	m_n	1.675×10^{-27} kg
		1.00866 amu
Mass of a proton	m_p	1.673×10^{-27} kg
		1.00728 amu
Speed of light	c	2.997925×10^8 m/s

SI Units and Conversion Factors

Length

SI unit: meter (m)

1 meter	= 1.0936 yards
	= 100 centimeters
	= 1000 millimeters
1 centimeter	= 0.3937 inch
1 inch	= 2.54 centimeters (exactly)
1 kilometer	= 0.62137 mile
1 mile	= 5280 feet
	= 1.609 kilometers
1 angstrom	= 10^{-10} meter

Mass

SI unit: kilogram (kg)

1 kilogram	= 1000 grams
	= 2.20 pounds
1 gram	= 1000 milligrams
1 pound	= 453.59 grams
	= 0.45359 kilogram
	= 16 ounces
1 ton	= 2000 pounds
	= 907.185 kilograms
1 ounce	= 28.3 g
1 atomic mass unit	= 1.6606×10^{-27} kilogram

Volume

SI unit: cubic meter (m^3)

1 liter	= 10^{-3} m^3
	= 1 dm^3
	= 1.0567 quarts
	= 1000 milliliters
1 gallon	= 4 quarts
	= 8 pints
	= 3.785 liters
1 quart	= 32 fluid ounces
	= 0.946 liter
1 fluid ounce	= 29.6 milliliters

Temperature

SI unit: kelvin (K)

$$0 \text{ K} = -273.15°C$$
$$= -459.67°F$$
$$\text{K} = °C + 273.15$$
$$°C = \frac{°F - 32}{1.8}$$
$$°C = 5/9 \, (°F - 32)$$
$$°F = 1.8(°C) + 32$$
$$°F = 1.8(°C + 40) - 40$$

Energy

SI unit: joule (J)

1 joule	= 1 kg m^2/s^2
	= 0.23901 calorie
1 calorie	= 4.184 joules

Pressure

SI unit: pascal (Pa)

1 pascal	= 1 kg/m s^2
1 atmosphere	= 101.325 kilopascals
	= 760 torr (mm Hg)
	= 14.70 pounds per square inch (psi)

Solubility Table

	F^-	Cl^-	Br^-	I^-	O^{2-}	S^{2-}	OH^-	NO_3^-	CO_3^{2-}	SO_4^{2-}	$C_2H_3O_2^-$
H^+	S	S	S	S	S	I	S	S	I	S	S
Na^+	S	S	S	S	S	S	S	S	S	S	S
K^+	S	S	S	S	S	S	S	S	S	S	S
NH_4^+	S	S	S	S	—	S	S	S	S	S	S
Ag^+	S	I	I	I	I	I	—	S	I	I	I
Mg^{2+}	I	S	S	S	I	d	I	S	I	S	S
Ca^{2+}	I	S	S	S	I	d	I	S	I	I	S
Ba^{2+}	I	S	S	S	ss	d	ss	S	I	I	S
Fe^{2+}	ss	S	S	S	I	I	I	S	ss	S	S
Fe^{3+}	I	S	S	—	I	I	I	S	I	S	I
Co^{2+}	S	S	S	S	I	I	I	S	I	S	S
Ni^{2+}	ss	S	S	S	I	I	I	S	I	S	S
Cu^{2+}	ss	S	S	—	I	I	I	S	I	S	S
Zn^{2+}	ss	S	S	S	I	I	I	S	I	S	S
Hg^{2+}	d	S	I	I	I	I	I	S	I	d	S
Cd^{2+}	ss	S	S	S	I	I	I	S	I	S	S
Sn^{2+}	S	S	S	ss	I	I	I	S	I	S	S
Pb^{2+}	I	I	I	I	I	I	I	S	I	I	S
Mn^{2+}	ss	S	S	S	I	I	I	S	I	S	S
Al^{3+}	I	S	S	S	I	d	I	S	—	S	S

Key: S = soluble in water
ss = slightly soluble in water
I = insoluble in water (less than 1 g/100 g H_2O)
d = decomposes in water

APPENDIX VI

Answers to Even-Numbered Questions and Exercises

Chapter 2

2. 7.6 cm

4. The most dense (mercury) at the bottom and the least dense (glycerin) at the top. In the cylinder the solid magnesium would sink in the glycerin and float on the liquid mercury.

6. 0.789 g/mL < ice < 0.91 g/mL

8. $d = m/V$ specific gravity $= \dfrac{d_{substance}}{d_{water}}$.

10. The number of degrees between the freezing and boiling points of water are

   Fahrenheit   180°F
   Celsius   100°C
   Kelvin   100 K

12. (a) mg   (d) nm
    (b) kg   (e) Å
    (c) m   (f) μL

14. (a) not significant   (d) significant
    (b) significant   (e) significant
    (c) not significant   (f) significant

16. (a) 40.0 (3)   (c) 129,042 (6)
    (b) 0.081 (2)   (d) 4.090×10^{-3} (4)

18. (a) 8.87   (c) 130. (1.30×10^2)
    (b) 21.3   (d) 2.00×10^6

20. (a) 4.56×10^{-2} (c) 4.030×10^1
    (b) 4.0822×10^3 (d) 1.2×10^7

22. (a) 28.1   (d) 2.010×10^3
    (b) 58.5   (e) 2.49×10^{-4}
    (c) 4.0×10^1 (f) 1.79×10^3

24. (a) $\frac{1}{4}$ (c) $1\frac{2}{3}$ or $\frac{5}{3}$
    (b) $\frac{5}{8}$ (d) $\frac{8}{9}$

26. (a) 1.0×10^2
    (b) 4.6 mL
    (c) 22

28. (a) 4.5×10^8 Å   (e) 6.5×10^5 mg
    (b) 1.2×10^{-6} cm (f) 5.5×10^3 g
    (c) 8.0×10^6 mm   (g) 468 mL
    (d) 0.164 g   (h) 9.0×10^{-3} mL

30. (a) 117 ft   (d) 4.3×10^4 g
    (b) 10.3 mi   (e) 75.7 L
    (c) 7.4×10^4 mm³ (f) 1.3×10^3 m³

32. 50. ft/s

34. 0.102 km/s

36. 5.0×10^2 s

38. 3×10^4 mg

40. 3.0×10^3 hummingbirds to equal the mass of a condor

42. $2800

44. 57 L

46. 160 L

48. 4×10^5 m²

50. 5 gal

52. 113°F Summer!

54. (a) 90.°F   (c) 546 K
    (b) −22.6°C   (d) −300 K

56. −11.4°C = 11.4°F

58. 3.12 g/mL

60. 1.28 g/mL

62. 3.40×10^2 g

64. 7.0 lb

66. Yes, 116.5 L additional solution

68. −15°C > 4.5°F

70. *B* is 14 mL larger than *A*.

72. 76.9 g

74. 3.57×10^3 g

76. The container must hold at least 50 mL.

78. The gold bar is not pure gold.

80. 0.841 g/mL

Chapter 3

2. (a) Attractive forces among the ultimate particles of a solid (atoms, ions, or molecules) are strong enough to hold these particles in a fixed position within the solid and thus maintain the solid in a definite shape. Attractive forces among the ultimate particles of a liquid (usually molecules) are sufficiently strong to hold them together (preventing the liquid from rapidly becoming a gas) but are not strong enough to hold the particles in fixed positions (as in a solid).

(b) The ultimate particles in a liquid are quite closely packed (essentially in contact with each other) and thus the volume of the liquid is fixed at a given temperature. But the ultimate particles in a gas are relatively far apart and essentially independent of each other. Consequently, the gas does not have a definite volume.

(c) In a gas the particles are relatively far apart and are easily compressed, but in a solid the particles are closely packed together and are virtually incompressible.

4. mercury and water

6. Three phases are present.

8. A system containing only one substance is not necessarily homogeneous.

10. 30 g Si/1 g H. There are more Si atoms than H atoms.

12. P   Na
Al   N
H   Ni
K   Ag
Mg   Pu

14. sodium   silver
potassium   tungsten
iron   gold
antimony   mercury
tin   lead

16. In an element all atoms are alike, while a compound contains two or more elements that are chemically combined. Compounds may be decomposed into simpler substances while elements cannot.

18. 7 metals   1 metalloid   2 nonmetals

20. aurum

22. A *compound* contains two or more elements chemically combined in a definite proportion by mass. Its properties differ from those of its components. A *mixture* is the physical combining of two or more substances (not necessarily elements). The composition may vary; the substances retain their properties and may be separated by physical means.

24. Compounds are distinguished from one another by their characteristic physical and chemical properties.

26. Cations are positively charged; anions are negatively charged.

28. Homogeneous mixtures contain only one phase; heterogeneous have two or more phases.

30. (a) H_2 **(c)** HCl   **(e)** NO

32. (a) magnesium, bromine
(b) carbon, chlorine
(c) hydrogen, nitrogen, oxygen
(d) barium, sulfur, oxygen
(e) aluminum, phosphorus, oxygen

34. (a) $AlBr_3$ **(c)** $PbCrO_4$
(b) CaF_2 **(d)** C_6H_6

36. (a) 1 atom Al, 3 atoms Br
(b) 1 atom Ni, 2 atoms N, 6 atoms O
(c) 12 atoms C, 22 atoms H, 11 atoms O

38. (a) 2 atoms   **(d)** 5 atoms
(b) 2 atoms   **(e)** 17 atoms
(c) 9 atoms

40. (a) 2 atoms H   **(d)** 4 atoms H
(b) 6 atoms H   **(e)** 8 atoms H
(c) 12 atoms H

42. (a) element   **(c)** element
(b) compound   **(d)** mixture

44. (a) mixture   **(c)** mixture
(b) element   **(d)** compound

46. (a) HO
(b) C_2H_6O
(c) $Na_2Cr_2O_7$

48. No. The only common liquid elements (at room temperature) are mercury and bromine.

50. 72% solids

52. 420 atoms

54. 40 atoms H

56. (a) magnesium, manganese, molybdenum, mendelevium, mercury
(b) carbon, phosphorus, sulfur, selenium, iodine, astatine, boron
(c) sodium, potassium, iron, silver, tin, antimony

58. (a) As temperature decreases, density increases.
(b) 1.28 g/L, 1.19 g/L, 1.09 g/L

Chapter 4

2. solid

4. Water disappears. Gas appears above each electrode and as bubbles in the solution.

6. A new substance is always formed during a chemical change, but never formed during physical changes.

8. (a) 118.0°C + 273 = 391.0 K
   (b) (118.0°C)1.8 + 32 = 244.4°F

10. (a) chemical   (d) chemical
    (b) physical   (e) chemical
    (c) physical   (f) physical

12. The copper wire, like the platinum wire, changed to a glowing red color when heated. Upon cooling, a new substance, black copper(II) oxide, appeared.

14. Reactant: water
    Products: hydrogen, oxygen

16. the transformation of kinetic energy to thermal energy

18. (a) +   (d) +
    (b) −   (e) −
    (c) −

20. 2.2×10^3 J

22. 5.03×10^{-2} J/g °C

24. 5°C

26. 29.1°C

28. 2.929×10^4 J; 44 g coal

30. 654°C

32. 16.7°C

34. 6:06 and 54 s

36. at the same rate

38. The mercury and sulfur react to form a compound since the properties of the product are different from the properties of either reactant. 1.36×10^3 g mercury. This supports the law of conservation of matter since the mass of the product is equal to the mass of the reactants.

Chapter 5

2. A neutron is about 1840 times heavier than an electron.

4. An atom is electrically neutral. An ion has a charge.

6. Isotopes have the same number of protons and electrons. Isotopes have different numbers of neutrons and thus different atomic masses.

8. (a) The nucleus of the atom contains most of the mass.
   (b) The nucleus of the atom is positively charged.
   (c) The atom is mostly empty space.

10. The nucleus of an atom contains nearly all of its mass.

12. Electrons:
    Dalton—Electrons are not part of his model.
    Thomson—Electrons are scattered throughout the positive mass of matter in the atom.
    Rutherford—Electrons are located out in space away from the central positive mass.
    Positive matter:
    Dalton—No positive matter in his model.
    Thomson—Positive matter is distributed throughout the atom.
    Rutherford—Positive matter is concentrated in a small central nucleus.

14. Yes. The isotope of $^{12}_{6}$C with a mass of 12 is an exact number. The mass of other isotopes is not an exact number.

16. Three isotopes of hydrogen have the same number of protons and electrons but differ in the number of neutrons.

18. (a) 80 protons; +80 is the electrical charge on the nucleus.
    (b) Hg

20. The most abundant Ca isotope has a mass number of 40.

22. (a) $^{59}_{27}$Co; 27 protons, 32 neutrons
    (b) $^{31}_{15}$P; 15 protons, 16 neutrons
    (c) $^{184}_{74}$W; 74 protons 110 neutrons
    (d) $^{235}_{92}$U; 92 protons, 143 neutrons

24. 24.31 amu

26. 6.716 amu

28. $1.0 \times 10^5 : 1.0$

30. (a) These atoms are isotopes.
    (b) These atoms are adjacent to each other on the periodic table.

32. $1.9 \times 10^8 : 1.0$

34. ^{210}Bi has the largest number of neutrons (127).

36. 131 g (atomic mass of unknown element)

38.

Element	Protons	Neutrons	Electrons
He	2	2	2
C	6	6	6
N	7	7	7
O	8	8	8
Ne	10	10	10
Mg	12	12	12
Si	14	14	14
S	16	16	16
Ca	20	20	20

40.

Element	Symbol	Atomic No.	Protons	Neutrons	Electrons
(a) platinum	^{195}Pt	78	78	117	78
(b) phosphorus	^{30}P	15	15	15	15
(c) iodine	^{127}I	53	53	74	53
(d) krypton	^{84}Kr	36	36	48	36
(e) selenium	^{79}Se	34	34	45	34
(f) calcium	^{40}Ca	20	20	20	20

Chapter 6

2. Charges on their ions must be equal and opposite in sign.

4. The system for naming binary compounds composed of two nonmetals uses the stem of the second element plus *ide*. A prefix is attached to each element indicating the number of atoms of that element in the compound. Thus N_2O_5 is called dinitrogen pentoxide.

6. Magnesium forms only one series of compounds in which the cation is Mg^{2+}. Thus the name for $MgCl_2$ (magnesium chloride) does not need to be distinguished from any other compound. Copper forms two series of compounds in which the ions are Cu^+ and Cu^{2+}. Thus the name copper chloride does not indicate which compound is in question. Therefore $CuCl_2$ is called copper(II) chloride to indicate that the compound contains the Cu^{2+} ion.

8. (a) BaO (d) $BeBr_2$
   (b) H_2S (e) Li_4Si
   (c) $AlCl_3$ (f) Mg_3P_2

10. Cl^- HSO_4^-
    Br^- HSO_3^-
    F^- CrO_4^{2-}
    I^- CO_3^{2-}
    CN^- HCO_3^-
    O^{2-} $C_2H_3O_2^-$
    OH^- ClO_3^-
    S^{2-} MnO_4^-
    SO_4^{2-} $C_2O_4^{2-}$

12. $(NH_4)_2SO_4$ NH_4Cl $(NH_4)_3AsO_4$ $NH_4C_2H_3O_2$ $(NH_4)_2CrO_4$
    $CaSO_4$ $CaCl_2$ $Ca_3(AsO_4)_2$ $Ca(C_2H_3O_2)_2$ $CaCrO_4$
    $Fe_2(SO_4)_3$ $FeCl_3$ $FeAsO_4$ $Fe(C_2H_3O_2)_3$ $Fe_2(CrO_4)_3$
    Ag_2SO_4 $AgCl$ Ag_3AsO_4 $AgC_2H_3O_2$ Ag_2CrO_4
    $CuSO_4$ $CuCl_2$ $Cu_3(AsO_4)_2$ $Cu(C_2H_3O_2)_2$ $CuCrO_4$

14. (a) carbon dioxide   (f) dinitrogen tetroxide
    (b) dinitrogen oxide   (g) diphosphorus pentoxide
    (c) phosphorus   (h) oxygen difluoride
        pentachloride   (i) nitrogen trifluoride
    (d) carbon tetrachloride   (j) carbon disulfide
    (e) sulfur dioxide

16. (a) potassium oxide   (e) sodium phosphate
    (b) ammonium bromide   (f) aluminum oxide
    (c) calcium iodide   (g) zinc nitrate
    (d) barium carbonate   (h) silver sulfate

18. (a) $SnBr_4$ (d) $Hg(NO_2)_2$
    (b) Cu_2SO_4 (e) TiS_2
    (c) $Fe_2(CO_3)_3$ (f) $Fe(C_2H_3O_2)_2$

20. (a) $HC_2H_3O_2$ (d) H_3BO_3
    (b) HF (e) HNO_2
    (c) $HClO$ (f) H_2S

22. (a) phosphoric acid   (e) hypochlorous acid
    (b) carbonic acid   (f) nitric acid
    (c) iodic acid   (g) hydroiodic acid
    (d) hydrochloric acid   (h) perchloric acid

24. (a) Na_2CrO_4 (h) $Co(HCO_3)_2$
    (b) MgH_2 (i) $NaClO$
    (c) $Ni(C_2H_3O_2)_2$ (j) $As_2(CO_3)_5$
    (d) $Ca(ClO_3)_2$ (k) $Cr_2(SO_3)_3$
    (e) $Pb(NO_3)_2$ (l) $Sb_2(SO_4)_3$
    (f) KH_2PO_4 (m) $Na_2C_2O_4$
    (g) $Mn(OH)_2$ (n) $KSCN$

26. (a) calcium hydrogen sulfate
    (b) arsenic(III) sulfite
    (c) tin(II) nitrite
    (d) iron(III) bromide
    (e) potassium hydrogen carbonate
    (f) bismuth(III) arsenate
    (g) iron(II) bromate
    (h) ammonium monohydrogen phosphate
    (i) sodium hypochlorite
    (j) potassium permanganate

28. (a) FeS_2 (e) $Mg(OH)_2$
    (b) $NaNO_3$ (f) $Na_2CO_3 \cdot 10\,H_2O$
    (c) $CaCO_3$ (g) C_2H_5OH
    (d) $C_{12}H_{22}O_{11}$

30. *-ide:* Suffix is used to indicate a binary compound except for hydroxides, cyanides, and ammonium compounds.
    -ous: Used as a suffix in naming an oxy-acid that has a lower oxygen content than the *ic* acid; also used as a suffix for naming the lower ionic charge of a multivalent metal (e.g., Fe^{2+}, ferrous).
    hypo: Used as a prefix in acids or salts when the polyatomic ion contains less oxygen than that of *-ous* acid or the *-ite* salt.
    per: Used as a prefix in acids or salts when the polyatomic ion contains more oxygen than that of the *-ic* acid or the *-ate* salt.
    -ite: The suffix of a salt derived from an *-ous* acid
    -ate: The suffix of a salt derived from an *-ic* acid.
    Roman numerals indicate the charge on the metal cation.

32. (a) $50e^-$, $50p$ (tin)
    (b) $48e^-$, $50p$ (Sn^{2+})
    (c) $46e^-$, $50p$ (Sn^{4+})

34. $Li_3Fe(CN)_6$
    $AlFe(CN)_6$
    $Zn_3[Fe(CN)_6]_2$

36. ammonium oxide, $(NH_4)_2O$ zinc chloride, $ZnCl_2$
ammonium carbonate,   zinc acetate, $Zn(C_2H_3O_2)_2$
   $(NH_4)_2CO_3$ carbonic acid, H_2CO_3
ammonium chloride, NH_4Cl acetic acid, $HC_2H_3O_2$
ammonium acetate,   hydrochloric acid, HCl
   $NH_4C_2H_3O_2$ water, H_2O
zinc oxide, ZnO
zinc carbonate, $ZnCO_3$

Chapter 7

2. A mole of gold has a higher mass than a mole of potassium.

4. A mole of gold atoms contains more electrons than a mole of potassium atoms.

6. 6.022×10^{23}

8. **(a)** 6.022×10^{23} atoms   **(d)** 16.00 g
(b) 6.022×10^{23} molecules   **(e)** 32.00 g
(c) 1.204×10^{24} atoms

10. Choosing 100.0 g of a compound allows us to simply drop the % sign and use grams for each percent.

12. Molar masses: Units of all answers are in g/mol.
(a) $NaOH$ 40.00 **(f)** C_6H_5COOH 122.1
(b) Ag_2CO_3 275.8 **(g)** $C_6H_{12}O_6$ 180.2
(c) Cr_2O_3 152.0 **(h)** $K_4Fe(CN)_6$ 368.4
(d) $(NH_4)_2CO_3$ 96.09 **(i)** $BaCl_2 \cdot 2\,H_2O$ 244.2
(e) $Mg(HCO_3)_2$ 146.3

14. **(a)** 0.625 mol $NaOH$
(b) 0.275 mol Br_2
(c) 7.18×10^{-3} mol $MgCl_2$
(d) 0.462 mol CH_3OH
(e) 2.03×10^{-2} mol Na_2SO_4
(f) 5.97 mol ZnI_2

16. **(a)** 0.0417 g H_2SO_4 **(c)** 0.122 g Ti
(b) 11 g CCl_4 **(d)** 8.0×10^{-7} g S

18. **(a)** 1.05×10^{24} molecules Cl_2
(b) 1.6×10^{23} molecules C_2H_6O
(c) 1.64×10^{23} molecules CO_2
(d) 3.75×10^{24} molecules CH_4

20. **(a)** 3.271×10^{-22} g Au
(b) 3.952×10^{-22} g U
(c) 2.828×10^{-23} g NH_3
(d) 1.795×10^{-22} g $C_6H_4(NH_2)_2$

22. **(a)** 0.886 mol S
(b) 42.8 mol $NaCl$
(c) 1.05×10^{24} atoms Mg
(d) 9.47 mol Br_2

24. One mole of ammonia contains
(a) 6.022×10^{23} molecules NH_3
(b) 6.022×10^{23} N atoms
(c) 1.807×10^{24} H atoms
(d) 2.409×10^{24} atoms

26. **(a)** 6.0×10^{24} atoms O
(b) 5.46×10^{24} atoms O
(c) 5.0×10^{18} atoms O

28. **(a)** 1.27 g Cl
(b) 9.25×10^{-2} g H
(c) 2.74 g H

30. **(a)** 47.98% Zn   **(d)** 21.20% N
   52.02% Cl 6.100% H
(b) 18.17% N 24.26% S
   9.153% H 48.41% O
   31.16% C   **(e)** 23.09% Fe
   41.51% O 17.37% N
(c) 12.26% Mg 59.53% O
   31.24% P   **(f)** 54.39% I
   56.48% O 45.61% Cl

32. **(a)** KCl 47.55% Cl   **(c)** $SiCl_4$ 83.46% Cl
(b) $BaCl_2$ 34.05% Cl   **(d)** $LiCl$ 83.63% Cl
highest % Cl is $LiCl$
lowest % Cl is $BaCl_2$

34. 24.2% C
4.04% H
71.72% Cl

36. **(a)** $KClO_3$ lower % Cl
(b) $KHSO_4$ higher % S
(c) Na_2CrO_4 lower % Cr

38. Empirical formulas
(a) $CuCl$ **(d)** K_3PO_4
(b) $CuCl_2$ **(e)** $BaCr_2O_7$
(c) Cr_2S_3 **(f)** PBr_8Cl_3

40. V_2O_5

42. The empirical formula is CH_2O. The molecular formula is $C_6H_{12}O_6$.

44. 5.88 g Na

46. 5.54×10^{19} m

48. **(a)** 8×10^{16} drops
(b) 8×10^6 mi^3

50. 10.3 mol H_2SO_4

52. **(a)** H_2O contains the most molecules.
(b) CH_3OH contains the most atoms.

54. 8.66 g Li will combine with 20.0 g S.

56. There is not sufficient S to react with 19.5 g Zn.

58. 4.77 g O in 8.50 g $Al_2(SO_4)_3$

60. Empirical Formula   Molecular Formula
(a) CCl_4 CCl_4
(b) CCl_3 C_2Cl_6
(c) CCl C_6Cl_6
(d) C_3Cl_8 C_3Cl_8

62. 2.4×10^{22} atoms Cu

64. 8.3×10^{-15} mol people

66. 32 g Mg

68. Carbon is the mystery element.

Chapter 8

2. the number of moles of each of the chemical species in the reaction

4. A chemical change that absorbs heat energy is said to be an endothermic reaction. The products are at a higher energy level than the reactants. A chemical change that liberates energy is said to be an exothermic reaction. The products are at a lower energy level than the reactants.

6. (a) $H_2 + Br_2 \longrightarrow 2\,HBr$
   (b) $4\,Al + 3\,C \xrightarrow{\Delta} Al_4C_3$
   (c) $Ba(ClO_3)_2 \xrightarrow{\Delta} BaCl_2 + 3\,O_2$
   (d) $CrCl_3 + 3\,AgNO_3 \longrightarrow Cr(NO_3)_3 + 3\,AgCl$
   (e) $2\,H_2O_2 \longrightarrow 2\,H_2O + O_2$

8. (a) combination
   (b) combination
   (c) decomposition
   (d) double displacement
   (e) decomposition

10. (a) $2\,SO_2 + O_2 \longrightarrow 2\,SO_3$
   (b) $4\,Al + 3\,MnO_2 \xrightarrow{\Delta} 3\,Mn + 2\,Al_2O_3$
   (c) $2\,Na + 2\,H_2O \longrightarrow 2\,NaOH + H_2$
   (d) $2\,AgNO_3 + Ni \longrightarrow Ni(NO_3)_2 + 2\,Ag$
   (e) $Bi_2S_3 + 6\,HCl \longrightarrow 2\,BiCl_3 + 3\,H_2S$
   (f) $2\,PbO_2 \xrightarrow{\Delta} 2\,PbO + O_2$
   (g) $2\,LiAlH_4 \xrightarrow{\Delta} 2\,LiH + 2\,Al + 3\,H_2$
   (h) $2\,KI + Br_2 \longrightarrow 2\,KBr + I_2$
   (i) $2\,K_3PO_4 + 3\,BaCl_2 \longrightarrow 6\,KCl + Ba_3(PO_4)_2$

12. (a) $2\,Cu + S \longrightarrow Cu_2S$
   (b) $2\,H_3PO_4 + 3\,Ca(OH)_2 \xrightarrow{\Delta} Ca_3(PO_4)_2 + 6\,H_2O$
   (c) $2\,Ag_2O \xrightarrow{\Delta} 4\,Ag + O_2$
   (d) $FeCl_3 + 3\,NaOH \longrightarrow Fe(OH)_3 + 3\,NaCl$
   (e) $Ni_3(PO_4)_2 + 3\,H_2SO_4 \longrightarrow 3\,NiSO_4 + 2\,H_3PO_4$
   (f) $ZnCO_3 + 2\,HCl \longrightarrow ZnCl_2 + H_2O + CO_2$
   (g) $3\,AgNO_3 + AlCl_3 \longrightarrow 3\,AgCl + Al(NO_3)_3$

14. (a) $Cu(s) + FeCl_3(aq) \longrightarrow$ no reaction
   (b) $H_2(g) + Al_2O_3(s) \xrightarrow{\Delta}$ no reaction
   (c) $2\,Al(s) + 6\,HBr(aq) \longrightarrow 3\,H_2(g) + 2\,AlBr_3(aq)$
   (d) $I_2(s) + HCl(aq) \longrightarrow$ no reaction

16. (a) $SO_2 + H_2O \longrightarrow H_2SO_3$
   (b) $SO_3 + H_2O \longrightarrow H_2SO_4$
   (c) $Ca + 2\,H_2O \longrightarrow Ca(OH)_2 + H_2$
   (d) $2\,Bi(NO_3)_3 + 3\,H_2S \longrightarrow Bi_2S_3 + 6\,HNO_3$

18. (a) $C + O_2 \xrightarrow{\Delta} CO_2$
   (b) $2\,Al(ClO_3)_3 \xrightarrow{\Delta} 9\,O_2 + 2\,AlCl_3$
   (c) $CuBr_2 + Cl_2 \longrightarrow CuCl_2 + Br_2$
   (d) $2\,SbCl_3 + 3\,(NH_4)_2S \longrightarrow Sb_2S_3 + 6\,NH_4Cl$
   (e) $2\,NaNO_3 \xrightarrow{\Delta} 2\,NaNO_2 + O_2$

20. (a) 2 mol of Na react with 1 mol of Cl_2 to produce 2 mol of NaCl and release 822 kJ of energy. Exothermic
   (b) 1 mol of PCl_5 absorbs 92.9 kJ of energy to produce 1 mol of PCl_3 and 1 mol of Cl_2. Endothermic

22. (a) $2\,Al + 3\,I_2 \longrightarrow 2\,AlI_3 + $ heat
   (b) $4\,CuO + CH_4 + $ heat $\longrightarrow 4\,Cu + CO_2 + 2\,H_2O$
   (c) $Fe_2O_3 + 2\,Al \longrightarrow 2\,Fe + Al_2O_3 + $ heat

24. $P_4O_{10} + 12\,HClO_4 \longrightarrow 6\,Cl_2O_7 + 4\,H_3PO_4$
58 atoms O on each side

26. A balanced equation tells us
   (a) the type of atoms/molecules involved in the reaction.
   (b) the relationship between quantities of the substances in the reaction.
   A balanced equation gives no information about
   (a) the time required for the reaction.
   (b) odor or colors that may result.

28. Zn metal is below Mg on the activity series.

30. (a) $4\,K + O_2 \longrightarrow 2\,K_2O$
   (b) $2\,Al + 3\,Cl_2 \longrightarrow 2\,AlCl_3$
   (c) $CO_2 + H_2O \longrightarrow H_2CO_3$
   (d) $CaO + H_2O \longrightarrow Ca(OH)_2$

32. (a) $Zn + H_2SO_4 \longrightarrow H_2 + ZnSO_4$
   (b) $2\,AlI_3 + 3\,Cl_2 \longrightarrow 2\,AlCl_3 + 3\,I_2$
   (c) $Mg + 2\,AgNO_3 \longrightarrow Mg(NO_3)_2 + 2\,Ag$
   (d) $2\,Al + 3\,CoSO_4 \longrightarrow Al_2(SO_4)_3 + 3\,Co$

34. (a) $AgNO_3(aq) + KCl(aq) \longrightarrow AgCl(s) + KNO_3(aq)$
   (b) $Ba(NO_3)_2(aq) + MgSO_4(aq) \longrightarrow$
$$Mg(NO_3)_2(aq) + BaSO_4(s)$$
   (c) $H_2SO_4(aq) + Mg(OH)_2(aq) \longrightarrow$
$$2\,H_2O(l) + MgSO_4(aq)$$
   (d) $MgO(s) + H_2SO_4(aq) \longrightarrow H_2O(l) + MgSO_4(aq)$
   (e) $Na_2CO_3(aq) + NH_4Cl(aq) \longrightarrow$ no reaction

36. (a) combustion of fossil fuels
   (b) destruction of the rain forests by burning
   (c) increased population

38. The effects of global warming can be reduced by
   (a) developing new energy sources (not dependent on fossil fuels)
   (b) conservation of energy resources
   (c) recycling
   (d) decreased destruction of the rain forests and other forests

Chapter 9

2. **(a)** **(c)** **(e)** are correct; **(b)** **(d)** **(f)** are incorrect
   (b) 10.7 mol HCN, **(d)** 36 mol H_2O, **(f)** 2 mol HCN

4. **(a)** 25.0 mol $NaHCO_3$
   (b) 3.85×10^{-3} mol $ZnCl_2$
   (c) 16 mol CO_2
   (d) 4.3 mol C_2H_5OH

6. **(a)** 1.31 g $NiSO_4$ **(d)** 1.35 g $C_6H_{12}O_6$
   (b) 3.60 g $HC_2H_3O_2$ **(e)** 18 g K_2CrO_4
   (c) 373 g Bi_2S_3

8. HCl

10. **(a)** $\dfrac{3 \text{ mol } CaCl_2}{1 \text{ mol } Ca_3(PO_4)_2}$ **(d)** $\dfrac{1 \text{ mol } Ca_3(PO_4)_2}{2 \text{ mol } H_3PO_4}$

   (b) $\dfrac{6 \text{ mol HCl}}{2 \text{ mol } H_3PO_4}$ **(e)** $\dfrac{6 \text{ mol HCl}}{1 \text{ mol } Ca_3(PO_4)_2}$

   (c) $\dfrac{3 \text{ mol } CaCl_2}{2 \text{ mol } H_3PO_4}$ **(f)** $\dfrac{2 \text{ mol } H_3PO_4}{6 \text{ mol HCl}}$

12. 2.80 mol Cl_2

14. 8.33 mol H_2O

16. 19.7 g $Zn_3(PO_4)_2$

18. 117 g H_2O (steam)
   271 g Fe

20. **(a)** 0.500 mol Fe_2O_3 **(d)** 65.6 g SO_2
   (b) 12.4 mol O_2 **(e)** 0.871 mol O_2
   (c) 6.20 mol SO_2 **(f)** 332 g FeS_2

22. **(a)** H_2S is the limiting reactant and $Bi(NO_3)_3$ is in excess.
   (b) H_2O is the limiting reactant and Fe is in excess.

24. **(a)** 3.0 mol CO_2
   (b) 9.0 mol CO_2
   (c) 1.8 mol CO_2

26. zinc

28. 95.0% yield of Cu

30. 77.8% CaC_2

32. A subscript is used to indicate the number of atoms in a formula. It cannot be changed without changing the identity of the substance. Coefficients are used only to balance atoms in chemical equations. They may be changed as needed to achieve a balanced equation.

34. **(a)** 380 g C_2H_5OH
         370 g CO_2
   (b) 480 mL C_2H_5OH

36. 65 g O_2

38. 1.0 g Ag_2S

40. **(a)** 2.0 mol Cu, 2.0 mol $FeSO_4$, and 1.0 mol $CuSO_4$
   (b) 15.9 g Cu, 38.1 g $FeSO_4$, 6.0 g Fe, and no $CuSO_4$

42. **(a)** 3.2×10^2 g C_2H_5OH
   (b) 1.10×10^3 g $C_6H_{12}O_6$

44. 3.7×10^2 kg Li_2O

46. 13 tablets

Chapter 10

2. A second electron may enter an orbital already occupied if its spin is opposite the electron already in the orbital, and if all other orbitals of the same sublevel contain an electron.

4. Both $1s$ and $2s$ orbitals are spherical in shape and located symmetrically around the nucleus. The radius of the $2s$ orbital is larger than the $1s$. The electrons in the $2s$ orbital are further from the nucleus.

6. $1s, 2s, 2p, 3s, 3p, 4s, 3d, 4p$

8. The Bohr orbit has an electron traveling a specific path, while an orbital is a region of space where the electron is most probably found.

10. s orbital

p orbitals

12. Transition elements are found in the center of the periodic table. The last electrons for these elements are found in the d or f orbitals. Representative elements are located on either side of the periodic table (Groups IA–VIIA). The valence electrons for these elements are found in the s and/or p orbitals.

14. **Atomic number Symbol**

Atomic number	Symbol
8	O
16	S
34	Se
52	Te
84	Po

All these elements have an outermost electron structure of s^2p^4.

16. 32; the 6th period has this number of elements.

18. Ar and K; Co and Ni; Te and I; Th and Pa; U and Np; Lr and Rf; Pu and Am.

20. **(a)** F   9 protons   **(c)** Br   35 protons
   (b) Ag   47 protons   **(d)** Sb   51 protons

22. **(a)** Cl $1s^2 2s^2 2p^6 3s^2 3p^5$
   (b) Ag $1s^2 2s^2 2p^6 3s^2 3p^6 4s^2 3d^{10} 4p^6 5s^1 4d^{10}$
   (c) Li $1s^2 2s^1$
   (d) Fe $1s^2 2s^2 2p^6 3s^2 3p^6 4s^2 3d^6$
   (e) I $1s^2 2s^2 2p^6 3s^2 3p^6 4s^2 3d^{10} 4p^6 5s^2 4d^{10} 5p^5$

24. Bohr said that a number of orbits were available for electrons, each corresponding to an energy level. When an electron falls from a higher energy orbit to a lower orbit, energy is given off as a specific wavelength of light. Only those energies in the visible range are seen in the hydrogen spectrum. Each line corresponds to a change from one orbit to another.

26. 32 electrons in the fourth energy level

28. **(a)** (14p 14n) $2e^-8e^-4e^-$ $^{28}_{14}Si$

   (b) (16p 16n) $2e^-8e^-6e^-$ $^{32}_{16}S$

   (c) (18p 22n) $2e^-8e^-8e^-$ $^{40}_{18}Ar$

   (d) (23p 28n) $2e^-8e^-11e^-2e^-$ $^{51}_{23}V$

   (e) (15p 16n) $2e^-8e^-5e^-$ $^{31}_{15}P$

30. **(a)** Sc
   (b) Zn
   (c) Sn
   (d) Cs

32.

Atomic number	Electron structure
(a) 9	$[He]2s^2 2p^5$
(b) 26	$[Ar]4s^2 3d^6$
(c) 31	$[Ar]4s^2 3d^{10} 4p^1$
(d) 39	$[Kr]5s^2 4d^1$
(e) 52	$[Kr]5s^2 4d^{10} 5p^4$
(f) 10	$[He]2s^2 2p^6$

34. **(a)** (13p 14n) $2e^-8e^-3e^-$ $^{27}_{13}Al$

   (b) (22p 26n) $2e^-8e^-10e^-2e^-$ $^{48}_{22}Ti$

36. K is in the fourth energy level because the $4s$ orbital is at a lower energy level than the $3d$ orbital.

38. Noble gases each have filled s and p orbitals in the outer energy level.

40. All the elements in a group have the same number of outer-shell electrons.

42. All of these elements have an $s^2 d^{10}$ electron configuration in their outermost energy levels.

44. (a) and (f)
   (e) and (h)

46. 7, 33 since they are in the same periodic group.

48. **(a)** nonmetal, I
   (b) metal, W
   (c) metal, Mo
   (d) metalloid, Ge

50. Period 4, Group IIIB

52. Group IIIA contains 3 valence electrons. Group IIIB contains 2 electrons in the outermost level and 1 electron in an inner d orbital. Group A elements are representative while Group B elements are transition elements.

54. Nitrogen has more valence electrons on more energy levels. More varied electron jumps are possible.

56. **(a)** 100%   **(d)** 24%
   (b) 100%   **(e)** 20%
   (c) 19%

58. $\dfrac{1.5 \times 10^8}{1}$

60. The outermost electron structure for both sulfur and oxygen is $s^2 p^4$.

62. In transition elements the last electron added is in a d or f orbital. The last electron added in a representative element is an s or p orbital.

64. The outermost energy level is the 7th. Element 87 contains one electron in the $7s$ orbital.

66. Answers will vary but should include at least a statement about:
   (a) numbering of the elements and their relationship to atomic structure
   (b) division of the elements into periods and groups
   (c) division of the elements into metals, nonmetals, and metalloids
   (d) identification and location of the representative and transition elements

68. **(a)** The two elements are isotopes.
   (b) The two elements are in the same period and are adjacent to each other.

Chapter 11

2. More energy is required for neon because it has a very stable outer electron structure consisting of an octet of electrons in filled orbitals (noble gas electron structure).

4. The first ionization energy decreases from top to bottom because, in successive alkali metals, the outermost electron is farther away from the nucleus and is more shielded from the positive nucleus by additional electron shells.

6. The electron to be removed from Ba is located in an energy level farther away from the nucleus than for Be. Therefore, it requires less energy to remove an electron from Ba than from Be.

8. (a) K > Na   **(d)** I > Br
(b) Na > Mg   **(e)** Zr > Ti
(c) O > F

10. Atomic size increases down the column since each successive element has an additional energy level that contains electrons located farther from the nucleus.

12. Cs· Ba: Tl: ·Pb: ·Po: ·At: :Rn:
Each of these is a representative element and has the same number of electrons in its outer shell as its periodic group number.

14. Valence electrons are the electrons found in the outermost energy level of an atom.

16. An aluminum ion has a +3 charge because it has lost 3 electrons in acquiring a noble gas electron structure.

18. A bromine atom is smaller since it has one less electron than the bromine ion in its outer shell. Also, the bromine ion has 35 protons and 36 electrons resulting in a lessening of the attraction between the electrons and the nucleus.

20.   +   −   +   −
(a) H   Cl   **(d)** I   Br
(b) Li   H   **(e)** Mg   H
(c) C   Cl   **(f)** O   F

22. (a) covalent   **(c)** covalent
(b) ionic   **(d)** ionic

24. (a) $F + 1e^- \rightarrow F^-$ **(b)** $Ca \rightarrow Ca^{2+} + 2e^-$

26. (a) Ca + :Ö: → CaO

(b) Na· + :Br: → NaBr

28. Si (4)   N (5)   P (5)   O (6)   Cl (7)

30. (a) Chloride ion, none
(b) Nitrogen atom, gain $3e^-$ or lose $5e^-$
(c) Potassium atom, lose $1e^-$

32. (a) The potassium atom is larger because it has one more energy level containing an electron than does a potassium ion.
(b) A bromide ion is larger than a bromine atom because it has one more electron than a bromine atom. Both are in the same energy level.

34. (a) SbH_3, Sb_2O_3
(b) H_2Se, SeO_3
(c) HCl, Cl_2O_7
(d) CCl_4, CO_2

36. $BeBr_2$, beryllium bromide
$MgBr_2$, magnesium bromide
$SrBr_2$, strontium bromide
$BaBr_2$, barium bromide
$RaBr_2$, radium bromide

38. (a) Ga: **(b)** $[Ga]^{3+}$ **(c)** $[Ca]^{2+}$

40. (a) covalent   **(c)** covalent
(b) ionic   **(d)** covalent

42. (a) covalent   **(c)** ionic
(b) covalent

44. (a) :Ö::Ö: **(b)** :Br:Br: **(c)** :Ï:Ï:

46. (a) :S̈:H   **(c)** H:N̈:H
        Ḧ   H

(b) :S̈::C::S̈:   **(d)** [H:N̈:H]⁺ [:C̈l:]⁻
                                   H

48. (a) [:Ï:]⁻   **(d)** [:Ö:Cl:Ö:]⁻
                                         :Ö:

(b) [:S̈:]²⁻   **(e)** [:Ö:N::Ö:]⁻
                                    :Ö:

(c) [:Ö:C::Ö:]²⁻
              :Ö:

50. (a) F_2, nonpolar
(b) CO_2, nonpolar
(c) NH_3, polar

52. (a) 2 electron pairs, linear
(b) 4 electron pairs, tetrahedral
(c) 4 electron pairs, tetrahedral

54. (a) tetrahedral
(b) pyramidal
(c) tetrahedral

56. (a) tetrahedral
(b) bent
(c) bent

58. potassium

60. (a) Zn
(b) Be
(c) N

62. Lithium has a +1 charge after the first electron is removed. It takes more energy to overcome that charge and to remove another e^- than to remove an e^- from helium.

64. $SnBr_2$, $GeBr_2$

66. A covalent bond results from the sharing of a pair of electrons between two atoms, while an ionic bond involves the transfer of one or more electrons from one atom to another.

68. F, O, N, Cl.

70. It is possible for a molecule to be nonpolar even though it contains polar bonds.

72. (a) 105°   **(b)** 107°   **(c)** 109.5°   **(d)** 109.5°

74. Fluorine's electronegativity is greater than any other element. Ionic bonds form between atoms of widely different electronegativities. Therefore, CsF, RbF, KF, and FrF would be ionic substances with the greatest electronegativity difference.

76. $S \dfrac{1.40 \text{ g}}{32.07 \dfrac{g}{mol}} = 0.0437 \text{ mol}$ $\dfrac{0.0437}{0.0437} = 1.00$

$O \dfrac{2.10 \text{ g}}{16.00 \dfrac{g}{mol}} = 0.131 \text{ mol}$ $\dfrac{0.131}{0.0437} = 3.00$

Empirical formula is SO_3:

$$\ddot{\text{O}}: $$
$$| $$
$$:\ddot{\text{O}} - \text{S} = \ddot{\text{O}}:$$

Chapter 12

2. The air pressure inside the balloon is greater than the air pressure outside the balloon.

4. 1 torr = 1 mm Hg.

6. 1 atm corresponds to 4 L.

8. The piston would move downward.

10. O_2, H_2S, HCl, F_2, CO_2

12. Rn, F_2, N_2, CH_4, He, H_2, velocities increase as molar masses decrease.

14. (a) pressure   (c) temperature
(b) volume   (d) number of moles

16. A gas is least likely to behave ideally at low temperatures. At very low temperatures, the velocity of the molecules decreases, and attractive forces between the molecules begin to play a significant role.

18. Equal volumes of H_2 and O_2 at the same T and P:
(a) have equal numbers of molecules (Avogadro's law)
(b) mass O_2 = 16 × mass H_2
(c) moles O_2 = moles H_2
(d) average kinetic energies are the same (T same)
(e) rate H_2 = 4 × rate O_2 (Graham's law of effusion)
(f) density O_2 = 16 × density H_2

20. $N_2(g) + O_2(g) \longrightarrow 2\,NO(g)$
1 vol + 1 vol $\longrightarrow$ 2 vol
Therefore, each N_2 and O_2 molecule must consist of two atoms.

22. Conversion of oxygen to ozone is an endothermic reaction. 286 kJ/3 mol O_2 is required to convert O_2 to O_3.

24. Oxygen atom, O; oxygen molecule, O_2; ozone molecule, O_3. An oxygen molecule contains 16 electrons.

26. (a) 715 torr
(b) 953 mbar
(c) 95.3 kPa

28. (a) 0.082 atm
(b) 55.92 atm
(c) 0.296 atm
(d) 0.0066 atm

30. (a) 132 mL
(b) 615 mL

32. 711 mm Hg

34. (a) 6.17 L
(b) 8.35 L

36. 7.8×10^2 mL

38. 33.4 L

40. 681 torr

42. 1.450×10^3 torr

44. 1.19 L C_3H_8

46. 28.0 L N_2

48. 1.33 g NH_3

50. (a) 11 mol H_2S
(b) 14.7 L H_2S
(c) 31.8 L H_2S

52. 2.69×10^{22} molecules CH_4

54. (a) 0.179 g/L He
(b) 2.50 g/L C_4H_8

56. (a) 3.17 g/L Cl_2
(b) 1.46 g/L Cl_2

58. 19 L Kr

60. 72.8 L

62. (a) 5.6 mol NH_3
(b) 0.640 L NO
(c) 1.1×10^2 g O_2

64. 153 L SO_2

66. The can will explode due to buildup of pressure.

68. (a) 5 L Cl_2
(b) 5.0 L NH_3
(c) 5.9 L SO_3

70. (a) CH_3 (b) C_2H_6 (c)

$$H-\underset{\underset{H}{|}}{\overset{\overset{H}{|}}{C}}-\underset{\underset{H}{|}}{\overset{\overset{H}{|}}{C}}-H$$

72. 430 mL CO_2

74. (a) 0.18 mol air
(b) 5.2 g air
(c) 0.3 g air

76. 1.0×10^2 atm

78. 1.5×10^3 torr

80. 65 atm

82. 6.1 L

84. 7.39×10^{21} molecules; 2.22×10^{22} atoms

86. (a) 34 mol H_2
    (b) 121 g H_2

88. 43.2 g/mol (molar mass)

90. 39.9 g/mol (molar mass)

92. 164 g/mol

94. 0.13 mol N_2

96. 279 L H_2 at STP

98. (a) Helium effuses twice as fast as CH_4.
    (b) The gases meet 66.7 cm from the helium end.

100. (a) 10.0 mol CO_2; 3.0 mol O_2; no CO₂
     (b) 29 atm

102. Some ammonia gas dissolves in the water squirted into the flask, lowering the pressure inside the flask. The atmospheric pressure outside is greater than the pressure inside the flask and thus pushes water from the beaker up the tube and into the flask.

104. 0.72 g/L

106. 60.4 atm

Chapter 13

2. H_2S, H_2Se, and H_2Te are gases at 0°C

4.

6 Prefixes preceding the word *hydrate* are used to indicate the number of molecules of water present in the formulas.
mono = 1   penta = 5
di = 2   hexa = 6
tri = 3   hepta = 7
tetra = 4   octa = 8

8. about 70°C

10. case (b), the closed container

12. The vapor pressure would remain unchanged.

14. (a) 88°C
     (b) 78°C
     (c) 16°C

16. melting point, 0°C; boiling point, 100°C (at 1 atm pressure); colorless; odorless; tasteless; heat of fusion, 335 J/g (80 cal/g); heat of vaporization, 2.26 kJ/g (540 cal/g); density = 1.0 g/mL (at 4°C); specific heat = 4.184 J/g °C

18. If you apply heat to an ice-water mixture, the heat energy is absorbed to melt the ice, rather than to warm the water, so the temperature remains constant until all the ice has melted.

20. Ice floats in water because it is less dense than water. Ice sinks in ethyl alcohol because it is more dense than the alcohol.

22. Ethyl alcohol exhibits hydrogen bonding; ethyl ether does not. The higher heat of vaporization indicates stronger intermolecular attraction between the alcohol molecules.

24. Ammonia exhibits hydrogen bonding; methane does not.

26. $H_2NCH_2CH_2NH_2$

28. (a) mercury, acetic acid, water, toluene, benzene, carbon tetrachloride, methyl alcohol, bromine
     (b) Highest boiling point is mercury; lowest is bromine.

30. In a pressure cooker, the temperature at which water boils increases above its normal boiling point, because the water vapor (steam) formed by boiling cannot escape. This results in an increased pressure over the water, and consequently, an increased boiling temperature.

32. As temperature increases, molecular velocities increase, resulting in a higher vapor pressure.

34. Ammonia

36. HF has a higher boiling point because of the strong H-bonding in HF. Neither F_2 nor Cl_2 show hydrogen bonding so F_2, with the lower molar mass, has the lower boiling point.

38. 34.6°C, the boiling point of ether

40. The expected temperature would be 4°C at the bottom of the lake.

42. (a) An anhydride is an oxide that reacts with water to form an acid or a base.
     (b) An acid anhydride is an oxide of a nonmetal.
     (c) A basic anhydride is an oxide of a metal.

44. SO_2, SO_3, N_2O_5

46. [KOH, K_2O] [$Ba(OH)_2$, BaO] [$Ca(OH)_2$, CaO]

48. (a) $Li_2O + H_2O \longrightarrow 2\ LiOH$
     (b) $2\ KOH \xrightarrow{\Delta} K_2O + H_2O$
     (c) $Ba + 2\ H_2O \longrightarrow Ba(OH)_2 + H_2$
     (d) $Cl_2 + H_2O \longrightarrow HCl + HClO$
     (e) $SO_3 + H_2O \longrightarrow H_2SO_4$
     (f) $H_2SO_3 + 2\ KOH \longrightarrow K_2SO_3 + 2\ H_2O$

50. (a) magnesium ammonium phosphate hexahydrate
     (b) iron(II) sulfate heptahydrate
     (c) tin(IV) chloride pentahydrate

52. (a) Distilled water has been vaporized by boiling and recondensed.
     (b) Natural waters are generally not pure, but contain dissolved minerals and suspended matter, and can even contain harmful bacteria.

54. 0.262 mol $FeI_2 \cdot 4\ H_2O$

56. 1.05 mol H_2O

58. 48.67% H_2O

60. $FePO_4 \cdot 4\ H_2O$

62. 5.5×10^4 J

64. 42 g of steam needed; not enough steam (35 g)

66. The system will be at 0°C. It will be a mixture of ice and water.

68. (a) 18.0 g H_2O
(b) 36.0 g H_2O
(c) 18.0 g H_2O

70. Eventually the water will lose enough energy to change from a liquid to a solid (freeze) as the alcohol evaporates.

72. (a) From 0°C to 40°C solid X warms until at 40°C it begins to melt. The temperature remains at 40°C until all of X is melted. After that, liquid X will warm steadily to 65°C where it will begin to boil and remain at 65°C until all the liquid becomes vapor. Beyond 65°C the vapor will warm steadily until 100°C.

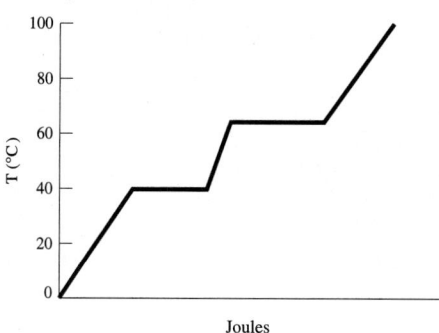

(b) 37,000 J

74. 75°C at 270 torr pressure

76. $MgSO_4 \cdot 7\,H_2O$ $Na_2HPO_4 \cdot 12\,H_2O$

78. chlorine

80. When organic pollutants in water are oxidized by dissolved oxygen, there may not be sufficient dissolved oxygen to sustain marine life.

82. Na_2 zeolite(s) + $Mg^{2+}(aq) \longrightarrow$ Mg zeolite(s) + $2\,Na^+(aq)$

84. 1.6×10^5 cal

86. 3.1 kcal

88. 2.30×10^6 J

90. 40.2 g H_2O

92. Volume of 1.00 mol of liquid water is 18.0 mL. Volume of 1.00 mol of water vapor at STP is 22.4 L.

94. (a) Some O_2 remains unreacted.
(b) 20.0 mL O_2 unreacted

Chapter 14

2. 4.5 g NaF

4. For lithium and sodium halide, solubility is
$$F^- < Cl^- < Br^- < I^-$$
For potassium halides, solubility is
$$Cl^- < Br^- < F^- < I^-$$

6. KNO_3

8. 6×10^2 cm^2

10. The dissolving process involves solvent molecules attaching to the solute ions or molecules. This rate decreases as more of the solvent molecules are already attached to solute molecules. As the solution becomes saturated, the number of unused solvent molecules decreases. Also, the rate of recrystallization increases as the concentration of dissolved solute increases.

12. The solution level in the thistle tube will fall.

14. It is not always apparent which component in a solution is the solute. For example, when mixing equal volumes of two liquids, designation of the solute and solvent is arbitrary.

16. Yes. Metal alloys are metals dissolved in metals.

18. Hexane and benzene are both nonpolar molecules. Therefore, hexane will dissolve in benzene. Benzene, being nonpolar, has no attraction for ionic sodium chloride.

20. Air is considered to be a solution because it is a homogeneous mixture of several gaseous substances and does not have a fixed composition.

22. The solubility of gases in liquids is greatly affected by the pressure of a gas above the liquid; little effect for solids in liquids.

24. In a saturated solution, the net rate of dissolution is zero. The rate of dissolving is equal to that of crystallization.

26. The solution contains 16 mol of HNO_3 per liter of solution.

28. The champagne would spray out of the bottle.

30. Water molecules can pass through a semipermeable membrane in both directions. The net result is that more water molecules pass from the pure water to the sugar solution.

32. A lettuce leaf immersed in salad dressing containing salt and vinegar will become limp and wilted as a result of osmosis.

34. (a) $1\,M$ NaOH   1 L
(b) $0.6\,M$ $Ba(OH)_2$ 0.8 L
(c) $2\,M$ KOH   0.5 L
(d) $1.5\,M$ $Ca(OH)_2$ 0.33 L

36. The vapor pressure of the solution is lower than that of the pure solvent and intersects the vapor pressure curve of the solvent below the freezing point of the pure solvent (see Figure 14.8a).

38. The presence of the methanol lowers the freezing point of the water.

40. The molarity of a 5 molal solution is less than $5\,M$.

42. Reasonably soluble: (c) $CaCl_2$ (d) $Fe(NO_3)_3$
Insoluble: (a) PbI_2 (b) $MgCO_3$ (e) $BaSO_4$

44. (a) 7.41% $Mg(NO_3)_2$
(b) 6.53% $NaNO_3$

46. 544 g solution

48. (a) 8.815% NaOH   (b) 15% $C_6H_{12}O_6$

50. 37.5 g K_2CrO_4

52. 33.6% NaCl

54. 22% C_6H_{14}

56. (a) 2.5 M HCl
(b) 0.59 M $BaCl_2 \cdot 2\ H_2O$
(c) 2.19×10^{-3} M $Al_2(SO_4)_3$
(d) 0.172 M $Ca(NO_3)_2$

58. (a) 3.5×10^{-4} mol NaOH
(b) 16 mol $CoCl_2$

60. (a) 13 g HCl   (b) 8.58 g $Na_2C_2O_4$

62. (a) 3.91×10^4 mL   (b) 7.82×10^3 mL

64. (a) 6.0 M HCl   (b) 0.064 M $ZnSO_4$

66. (a) 2.0×10^2 mL 12 M HCl
(b) 16 mL 16 M HNO_3

68. (a) 1.83 M H_2SO_4 (b) 0.30 M H_2SO_4

70. (a) 33.3 mL of 0.250 M Na_3PO_4
(b) 1.10 g $Mg_3(PO_4)_2$

72. (a) 0.13 mol Cl_2
(b) 16 mol HCl
(c) 1.3×10^2 mL of 6 M HCl
(d) 3.2 L Cl_2

74. (a) 5.5 m $C_6H_{12}O_6$ (b) 0.25 m I_2

76. (a) 10.74 m (b) $-20.0°C$ (c) 105.50°C

78. 163 g/mol

80. 96.5 g 10% NaOH solution

82. $C_8H_4N_2$

84. 6.4 g KNO_3; 0.14 M

86. (a) 4.5 g NaCl
(b) 450. mL H_2O must evaporate.

88. 210. mL solution

90. 6.72 M HNO_3

92. 540. mL water to be added

94. To make 250. mL of 0.625 M KOH, take 31.3 mL of 5.00 M KOH and dilute with water to a volume of 250. mL.

96. 2.08 M HCl

98. 12.0 g $Mg(OH)_2$ is the more effective acid neutralizer.

100. 6.2 m H_2SO_4
5.0 M H_2SO_4

102. (a) 2.9 m (b) 101.5°C

104. (a) 8.04×10^3 g $C_2H_6O_2$
(b) 7.24×10^3 mL $C_2H_6O_2$
(c) $-4.0°F$

106. 0.46 L HCl

108. (a) 7.7 L H_2O must be added
(b) 0.0178 mol
(c) 0.0015 mol

110. Mix together 667 mL 3.00 M HNO_3 and 333 mL 12.0 M HNO_3 to get 1000. mL of 6.00 M HNO_3.

112. 2.84 g $Ba(OH)_2$ is formed.

114. (a) 0.011 mol Li_2CO_3 (c) 3.2×10^2 mL solution
(b) 14 g Li_2CO_3 (d) 1.5%

Chapter 15

2. An electrolyte must be present in the solution for the bulb to glow.

4. First, the orientation of the polar water molecules about the Na^+ and Cl^- ions is different. Second, more water molecules will fit around Cl^-, since it is larger than the Na^+ ion.

6. tomato juice

8. Arrhenius: $HCl + NaOH \longrightarrow NaCl + H_2O$
Brønsted–Lowry: $HCl + KCN \longrightarrow HCN + KCl$
Lewis: $AlCl_3 + NaCl \longrightarrow AlCl_4^- + Na^+$

10. acids, bases, salts

12. Hydrogen chloride reacts with polar water molecules to produce H_3O^+ and Cl^- ions, which conduct an electric current. HCl does not react with or ionize in hexane, hence it does not conduct an electric current.

14. CH_3OH is a nonelectrolyte; NaOH is an electrolyte. This indicates that the OH group in CH_3OH must be covalently bonded to the CH_3 group.

16. Dissociation is the separation of already existing ions in an ionic compound. The dissolving of NaCl is a dissociation process. The dissolving of HCl in water is an ionization process because ions are formed as a result of the reaction of HCl and H_2O.

18. Ions are hydrated in solution because there is an electrical attraction between the charged ions and the polar water molecules.

20. (a) $[H^+] = [OH^-]$
(b) $[H^+] > [OH^-]$
(c) $[OH^-] > [H^+]$

22. HCl is a polar molecule, consequently it is much more soluble in the polar solvent, water, than the nonpolar solvent, hexane.

24. The fundamental difference is in the size of the particles. The particles in true solutions are less than 1 mm in size. The particles in colloids range from 1–1000 mm in size.

26. Dialysis is the process of removing dissolved solutes from colloidal dispersion by the use of a dialyzing membrane. Dialysis is used in artificial kidneys to remove soluble waste products from the blood.

28. (a) $HCl - Cl^-$; $NH_4^+-NH_3$
    (b) $HCO_3^--CO^2{_3}$; H_2O-OH-
    (c) $H_3O^+-H_2O$; $H_2CO_3-HCO_3^-$
    (d) $HC_2H_3O_2-C_2H_3O_2^-$; $H_3O^+-H_2O$

30. (a) $NaOH(aq) + HBr(aq) \longrightarrow NaBr(aq) + H_2O(l)$
    (b) $KOH(aq) + HCl(aq) \longrightarrow KCl(aq) + H_2O(l)$
    (c) $Ca(OH)_2(aq) + 2\,HI(aq) \longrightarrow CaI_2(aq) + 2\,H_2O(l)$
    (d) $Al(OH)_3(s) + 3\,HBr(aq) \longrightarrow AlBr_3(aq) + 3\,H_2O(l)$
    (e) $Na_2O(s) + 2\,HClO_4(aq) \longrightarrow 2\,NaClO_4(aq) + H_2O(l)$
    (f) $3\,LiOH(aq) + FeCl_3(aq) \longrightarrow Fe(OH)_3(s) + 3\,LiCl(aq)$

32. (a) $NaHCO_3$—salt   (e) RbOH—base
    (c) $AgNO_3$—salt   (f) K_2CrO_4—salt
    (d) HCOOH—acid

34. (a) $0.75\,M\ Zn^{2+}$, $1.5\,M\ Br^-$
    (b) $4.95\,M\ SO_4^{2-}$, $3.30\,M\ Al^{3+}$
    (c) $0.680\,M\ NH_4^+$, $0.340\,M\ SO_4^{2-}$
    (d) $0.0628\,M\ Mg^{2+}$, $0.126\,M\ ClO_3^-$

36. (a) 4.9 g Zn^{2+}, 12 g Br^-
    (b) 8.90 g Al^{3+}, 47.6 g SO_4^{2-}
    (c) 1.23 g NH_4^+, 3.28 g SO_4^{2-}
    (d) 0.153 g Mg^{2+}, 1.05 g ClO_3^-

38. (a) $[K^+] = 1.0\,M$, $[Ca^{2+}] = 0.5\,M$, $[Cl^-] = 2.0\,M$
    (b) No ions are present in the solution.
    (c) $0.67\,M\ Na^+$, $0.67\,M\ NO_3^-$

40. (a) $0.147\,M$ NaOH
    (b) $0.964\,M$ NaOH
    (c) $0.4750\,M$ NaOH

42. (a) $H_2S(g) + Cd^{2+}(aq) \longrightarrow CdS(s) + 2\,H^+(aq)$
    (b) $Zn(s) + 2\,H^+(aq) \longrightarrow Zn^{2+}(aq) + H_2(g)$
    (c) $Al^{3+}(aq) + PO_4^{3-}(aq) \longrightarrow AlPO_4(s)$

44. (a) $2\,M$ HCl
    (b) $1\,M\ H_2SO_4$

46. 40.8 mL of $0.245\,M$ HCl

48. 7.0% NaCl in the sample

50. 0.936 L H_2

52. (a) pH = 7.0   (b) pH = 0.30   (c) pH = 4.00

54. (a) pH = 4.30   (b) pH = 10.47

56. 3.0×10^3 mL

58. The ionization of the acetic acid solution increases the particle concentration above that of the alcohol solution. Therefore, the acetic acid solution freezes at a lower temperature.

60. A hydronium ion is a hydrated hydrogen ion. (H_3O^+)

62. (a) 100°C   pH = 6.0
        25°C   pH = 7.0
    (b) H^+ concentration is higher at 100°C.
    (c) The water is neutral at both temperatures.

64. $0.201\,M$ HCl

66. 0.673 g KOH

68. 13.9 L of $18.0\,M\ H_2SO_4$

70. pH = 1.1, acidic

72. (a) $2\,NaOH(aq) + H_2SO_4(aq) \longrightarrow Na_2SO_4(aq) + 2\,H_2O(l)$
    (b) 1.0×10^2 mL NaOH
    (c) 0.71 g Na_2SO_4

74. $12\,M\ HNO_3$

76. Statement (c) is correct.

Chapter 16

2. The reaction is endothermic because the increased temperature increases the concentration of product (NO_2) present at equilibrium.

4. The sum of the pH and the pOH is 14. A solution whose pH is -1 would have a pOH of 15.

6. The order of molar solubilities is: $AgC_2H_3O_2$, $PbSO_4$, $BaSO_4$, $AgCl$, $BaCrO_4$, $AgBr$, AgI, PbS.

8. When 0.010 mol H^+ is added to the buffer solution, the H^+ reacts with excess $C_2H_3O_2^-$ forming un-ionized $HC_2H_3O_2$. The added H^+ is essentially removed from solution, resulting in very little change in the $[H^+]$ and the pH of the solution.

10. The rate increases because the number of collisions increases due to the added reactant.

12. Increasing the temperature causes the velocity of the molecules to increase. Faster moving molecules increase the number of collisions between molecules, resulting in an increased rate of reaction.

14. The equilibrium shifts to the right, yielding a higher percent ionization.

16. The pH of the water can be different at different temperatures, because the degree of ionization varies with temperature.

18. HNO_3 has a higher $[H^+]$ than H_2O. The HNO_3 removes $C_2H_3O_2^-$ ions from $AgC_2H_3O_2$ equilibrium forming un-ionized $HC_2H_3O_2$, which allows more $AgC_2H_3O_2$ to dissolve. Adding HCl to $AgC_2H_3O_2 \rightleftarrows Ag^+ + C_2H_3O_2^-$ forms a precipitate of AgCl resulting in increased solubility of $AgC_2H_3O_2$.

20. A buffer solution contains a weak acid or base plus a salt of that weak acid or base. When a small amount of a strong acid (H^+) is added to this buffer solution, the H^+ reacts with anions of the salt, thus neutralizing the added acid. When a strong base, OH^-, is added, it reacts with un-ionized acid to neutralize the added base. As a result, in both cases, the approximate pH of the solution is maintained.

22. (a) $H_2O(l) \overset{100°C}{\rightleftharpoons} H_2O(g)$
    (b) $SO_2(l) \rightleftharpoons SO_2(g)$

24. (a) $[NH_3]$, $[O_2]$, and $[N_2]$ will be increased. $[H_2O]$ will be decreased. Reaction shifts left.
   (b) The addition of heat will shift the reaction to the left.

26. (a) left   I   I   D (add NH_3)
   (b) left   I   I   D (increase volume)
   (c) no change   N   N   N (add catalyst)
   (d) ?   ?   I   I (add H_2 and N_2)

28.

Reaction	Increase temperature	Increase pressure	Add catalyst
(a)	right	left	no change
(b)	left	left	no change
(c)	left	left	no change

30. Equilibrium shift
   (a) left   **(b)** right   **(c)** right

32. (a) $K_{eq} = \dfrac{[H^+][ClO_2^-]}{[HClO_2]}$

   (b) $K_{eq} = \dfrac{[H^+][C_2H_3O_2^-]}{[HC_2H_3O_2]}$

   (c) $K_{eq} = \dfrac{[NO]^4[H_2O]^6}{[NH_3]^4[O_2]^5}$

34. (a) $K_{sp} = [Fe^{3+}][OH^-]^3$
   (b) $K_{sp} = [Sb^{5+}]^2[S^{2-}]^5$
   (c) $K_{sp} = [Ca^{2+}][F^-]^2$
   (d) $K_{sp} = [Ba^{2+}]^3[PO_4^{3-}]^2$

36. If H^+ is increased,
   (a) pH is decreased   **(c)** OH^- is decreased
   (b) pOH is increased   **(d)** K_W remains unchanged

38. (a) $Ca(CN)_2$, basic   **(c)** $NaNO_2$, basic
   (b) $BaBr_2$, neutral   **(d)** NaF, basic

40. (a) $NH_4^+(aq) + H_2O(l) \rightleftharpoons H_3O^+(aq) + NH_3(aq)$
   (b) $SO_3^{2-}(aq) + H_2O(l) \rightleftharpoons OH^-(aq) + HSO_3^-(aq)$

42. (a) $OCl^-(aq) + H_2O(l) \rightleftharpoons OH^-(aq) + HOCl(aq)$
   (b) $ClO_2^-(aq) + H_2O(l) \rightleftharpoons OH^-(aq) + HClO_2(aq)$

44. When excess base gets into the bloodstream it reacts with H^+ to form water. Then H_2CO_3 ionizes to replace H^+, thus maintaining the approximate pH of the solution.

46. (a) $K_a = 5.7 \times 10^{-6}\, M$
   (b) pH = 5.24
   (c) 2.3×10^{-3}% ionized

48. $K_a = 7 \times 10^{-10}$

50. (a) pH = 3.72   **(b)** pH = 4.23   **(c)** pH = 4.72

52. $K_a = 7.3 \times 10^{-6}$

54. $[OH^-] = 1.0\, M$, pOH = 0.00, pH = 14.00,
   $[H^+] = 1 \times 10^{-14}\, M$

56. (a) pH = 11.4, pOH = 2.60
   (b) pH = 4.73, pOH = 9.8
   (c) pH = 9.1, pOH = 4.92

58. (a) $[OH^-] = 2.5 \times 10^{-6}$ **(b)** $[OH^-] = 1.1 \times 10^{-13}$

60. (a) $[H^+] = 2.2 \times 10^{-9}$ **(b)** $[H^+] = 1.4 \times 10^{-11}$

62. (a) 1.2×10^{-23} ZnS   **(c)** 1.81×10^{-18} Ag_3PO_4
   (b) 2.6×10^{-13} $Pb(IO_3)_2$ **(d)** 5.13×10^{-17} $Zn(OH)_2$

64. (a) $4.5 \times 10^{-5}\, M$
   (b) $7.6 \times 10^{-10}\, M$

66. (a) 8.9×10^{-4} g $BaCO_3$
   (b) 9.3×10^{-9} g $AlPO_4$

68. Precipitation occurs.

70. 2.5×10^{-12} mol AgBr will dissolve.

72. pH = 4.74

74. Change in pH = $4.74 - 4.72 = 0.02$ units in the buffered solution. Initial pH = 4.74.

76. (a) 3.16 mol HI
   (b) 0.57 mol I_2
   (c) $K_{eq} = 57$

78. $K_{eq} = 29$

80. Hypochlorous acid: $K_a = 3.5 \times 10^{-8}$
   Propanoic acid: $K_a = 1.3 \times 10^{-5}$
   Hydrocyanic acid: $K_a = 4.0 \times 10^{-10}$

82. (a) Precipitation occurs.
   (b) Precipitation occurs.
   (c) No precipitation occurs.

84. No precipitate of $PbCl_2$

86. $K_{eq} = 1.1 \times 10^4$

88. $8.0\, M\, NH_3$

90. $K_{sp} = 4.00 \times 10^{-28}$

92. (a) The temperature could have been cooler.
   (b) The humidity in the air could have been higher.
   (c) The air pressure could have been greater.

94. $K_{eq} = 1$, (c) is the correct answer.

96. $K_{eq} = 3$

98. (a) After an initial increase, $[OH^-]$ will be neutralized and equilibrium shifts to the right.
   (b) $[H^+]$ will be reduced (reacts with OH^-). Equilibrium shifts to the right.
   (c) $[NO_2^-]$ increases as equilibrium shifts to the right.
   (d) $[HNO_2]$ decreases as equilibrium shifts to the right.

100. 1.1 g $CaSO_4$

Chapter 17

2. (a) Al
   (b) Ba
   (c) Ni

4. (a) $2\,Al + Fe_2O_3 \longrightarrow Al_2O_3 + 2\,Fe + heat$
   (b) Al is more active than Fe.
   (c) No. Fe is less reactive than Al.
   (d) Yes. Al is more reactive than Cr.

6. (a) Oxidation occurs at the anode.
       $2\,Cl^-(aq) \longrightarrow Cl_2(g) + 2e^-$
   (b) Reduction occurs at the cathode.
       $Ni^{2+}(aq) + 2e^- \longrightarrow Ni(s)$
   (c) The net chemical reaction is

   $$Ni^{2+}(aq) + 2\,Cl^-(aq) \xrightarrow[\text{energy}]{\text{electrical}} Ni(s) + Cl_2(g)$$

8. (a) It would not be possible to monitor the voltage produced, but the reactions in the cell would still occur.
   (b) If the salt bridge were removed, the reaction would stop.

10. $Ca^{2+} + 2\,e^- \longrightarrow Ca$ cathode reaction, reduction
     $2\,Br^- \longrightarrow Br_2 + 2\,e^-$ anode reaction, oxidation

12. Since lead dioxide and lead(II) sulfate are insoluble, it is unnecessary to have salt bridges in the cells of a lead storage battery.

14. Reduction occurs at the cathode.

16. A salt bridge permits movement of ions in the cell. This keeps the solution neutral with respect to the charged particles (ions) in the solution.

18. (a) $K\underline{Mn}O_4$ $+7$ **(d)** $K\underline{Cl}O_3$ $+5$
   (b) $\underline{I}_2$ 0 **(e)** $K_2\underline{Cr}O_4$ $+6$
   (c) $\underline{N}H_3$ -3 **(f)** $K_2\underline{Cr}_2O_7$ $+6$

20. (a) $\underline{O}_2$ 0 **(c)** $Fe(\underline{O}H)_3$ -2
   (b) $\underline{As}O_4{}^{3-}$ $+5$ **(d)** $\underline{I}O_3{}^-$ $+5$

22. (a) $SO_3{}^{2-} + H_2O \longrightarrow SO_4{}^{2-} + 2\,H^+ + 2\,e^-$
       S is oxidized
   (b) $NO_3{}^- + 4\,H^+ + 3e^- \longrightarrow NO + 2\,H_2O$
       N is reduced
   (c) $S_2O_4{}^{2-} + 2\,H_2O \longrightarrow 2\,SO_3{}^{2-} + 4\,H^+ + 2\,e^-$
       S is oxidized
   (d) $Fe^{2+} \longrightarrow Fe^{3+} + 1\,e^-$
       Fe is oxidized

24. Equation (1):
   (a) As is oxidized, Ag^+ is reduced.
   (b) Ag^+ is the oxidizing agent, AsH_3 the reducing agent.
   Equation (2):
   (a) Br is oxidized, Cl is reduced.
   (b) Cl_2 is the oxidizing agent, NaBr the reducing agent.

26. (a) $3\,Cl_2 + 6\,KOH \longrightarrow KClO_3 + 5\,KCl + 3\,H_2O$
   (b) $3\,Ag + 4\,HNO_3 \longrightarrow 3\,AgNO_3 + NO + 2\,H_2O$
   (c) $3\,CuO + 2\,NH_3 \longrightarrow N_2 + 3\,Cu + 3\,H_2O$
   (d) $3\,PbO_2 + 2\,Sb + 2\,NaOH \longrightarrow$
       $3\,PbO + 2\,NaSbO_2 + H_2O$
   (e) $5\,H_2O_2 + 2\,KMnO_4 + 3\,H_2SO_4 \longrightarrow$
       $5\,O_2 + 2\,MnSO_4 + K_2SO_4 + 8\,H_2O$

28. (a) $6\,H^+ + ClO_3{}^- + 6\,I^- \longrightarrow 3\,I_2 + Cl^- + 3\,H_2O$
   (b) $14\,H^+ + Cr_2O_7{}^{2-} + 6\,Fe^{2+} \longrightarrow$
       $2\,Cr^{3+} + 6\,Fe^{3+} + 7\,H_2O$
   (c) $2\,H_2O + 2\,MnO_4{}^- + 5\,SO_2 \longrightarrow$
       $4\,H^+ + 2\,Mn^{2+} + 5\,SO_4{}^{2-}$
   (d) $6\,H^+ + 5\,H_3AsO_4 + 2\,MnO_4{}^- \longrightarrow$
       $5\,H_3AsO_4 + 2\,Mn^{2+} + 3\,H_2O$
   (e) $8\,H^+ + Cr_2O_7{}^{2-} + 3\,H_3AsO_3 \longrightarrow$
       $2\,Cr^{3+} + 3\,H_3AsO_4 + 4\,H_2O$

30. (a) $H_2O + 2\,MnO_4{}^- + 3\,SO_3{}^{2-} \longrightarrow$
       $2\,MnO_2 + 3\,SO_4{}^{2-} + 2\,OH^-$
   (b) $2\,H_2O + 2\,ClO_2 + 2\,OH^- + SbO_2{}^- \longrightarrow$
       $2\,ClO_2{}^- + Sb(OH)_6{}^-$
   (c) $8\,Al + 3\,NO_3{}^- + 18\,H_2O + 5\,OH^- \longrightarrow$
       $3\,NH_3 + 8\,Al(OH)_4{}^-$
   (d) $4\,OH^- + 2\,H_2O + P_4 \longrightarrow 2\,HPO_3{}^{2-} + 2\,PH_3$
   (e) $2\,Al + 6\,H_2O + 2\,OH^- \longrightarrow 2\,Al(OH)_4{}^- + 3\,H_2$

32. (a) The oxidizing agent is $KMnO_4$.
   (b) The reducing agent is HCl.
   (c) $3.011 \times 10^{24} \dfrac{\text{electrons}}{\text{mol } KMnO_4}$

34. $20.2\ L\ Cl_2$

36. $66.2\ mL$ of $0.200\ M\ K_2Cr_2O_7$ solution

38. 91.3% KI

40. $5.560\ mol\ H_2$

42. The electrons lost by the species undergoing oxidation must be gained (or attracted) by another species, which then undergoes reduction.

44. Sn^{4+} can only be an oxidizing agent.
   Sn^0 can only be a reducing agent.
   Sn^{2+} can be both oxidizing and reducing agents.

46. Equations (a) and (b) represent oxidations.

48. (a) $F_2 + 2\,Cl^- \longrightarrow 2\,F^- + Cl_2$
   (b) $Br_2 + Cl^- \longrightarrow NR$
   (c) $I_2 + Cl^- \longrightarrow NR$
   (d) $Br_2 + 2\,I^- \longrightarrow 2\,Br^- + I_2$

50. $4\,Zn + NO_3{}^- + 10\,H^+ \longrightarrow 4\,Zn^{2+} + NH_4{}^+ + 3\,H_2O$

52. (a) Pb is the anode.
   (b) Ag is the cathode.
   (c) Pb (anode)
   (d) Ag (cathode)
   (e) Electrons flow from the lead through the wire to the silver.
   (f) Positive ions flow through the salt toward the negatively charged strip of silver; negative ions flow toward the positively charged strip of lead.

Chapter 18

2. Alpha particles are much heavier.

4. Contributors to the early history of radioactivity include
   (a) Henri Becquerel—discovered radioactivity.
   (b) Marie and Pierre Curie—discovered polonium and radium.
   (c) Wilhelm Roentgen—discovered X rays and developed the technique for producing them.
   (d) Ernest Rutherford—discovered alpha and beta particles, established the link between radioactivity and transmutation, and produced the first successful man-made transmutation.
   (e) Otto Hahn and Fritz Strassmann—were first to produce nuclear fission.

6. *Isotope* is used with reference to atoms of the same element that contain different masses. *Nuclide* infers any isotope of any atom.

8.

	Charge	Mass	Nature of particles	Penetrating power
Alpha	+2	4 amu	He nucleus	low
Beta	−1	$\dfrac{1}{1837}$ amu	electron	moderate
Gamma	0	0	electro-magnetic radiation	high

10. A disintegration series is a series of α and β emissions leading to production of a stable nuclide.

12. $^{232}_{90}\text{Th} \xrightarrow{-\alpha} {}^{228}_{88}\text{Ra} \xrightarrow{-\beta} {}^{228}_{89}\text{Ac} \xrightarrow{-\beta} {}^{228}_{90}\text{Th} \xrightarrow{-\alpha}$

$^{224}_{88}\text{Ra} \xrightarrow{-\alpha} {}^{220}_{86}\text{Rn} \xrightarrow{-\alpha} {}^{216}_{84}\text{Po} \xrightarrow{-\alpha}$

$^{212}_{82}\text{Pb} \xrightarrow{-\beta} {}^{212}_{83}\text{Bi} \xrightarrow{-\beta} {}^{212}_{84}\text{Po} \xrightarrow{-\alpha} {}^{208}_{82}\text{Pb}$

14. $^{211}_{83}\text{Bi} \longrightarrow {}^{4}_{2}\text{He} + {}^{207}_{81}\text{Tl}$ $^{207}_{81}\text{Tl} \longrightarrow {}^{0}_{-1}\text{e} + {}^{207}_{82}\text{Pb}$

16. Natural uranium is 99+% U-238. Commercial nuclear reactors use U-235 enriched uranium as a fuel. Slow neutrons will cause the fission of U-235, but not U-238. Fast neutrons are capable of a nuclear reaction with U-238 to produce fissionable Pu-239. A breeder reactor converts nonfissionable U-238 to fissionable Pu-239, and in the process, manufactures more fuel than it consumes.

18. A certain amount of fissionable material (a critical mass) must be present before a self-supporting chain reaction can occur. Without a critical mass, too many neutrons from fissions will escape, and the reaction cannot reach a chain reaction status, unless at least one neutron is captured for every fission that occurs.

20. When radioactive rays pass through normal matter, they cause that matter to become ionized (usually by knocking out electrons). Therefore, the radioactive rays are classified as ionizing radiation.

22. Strontium-90 has two characteristics that create concern. Its half-life is 28 years, so it remains active for a long period of time (disintegrating by emitting β radiation). The other characteristic is that Sr-90 is chemically similar to calcium, so when it is present in milk, it is deposited in bone tissue along with calcium. Red blood cells are produced in the bone marrow. If the marrow is subjected to beta radiation from strontium-90, the red blood cells will be destroyed, increasing the incidence of leukemia and bone cancer.

24. In living species, the ratio of carbon-14 to carbon-12 is constant due to the constant C-14/C-12 ratio in the atmosphere and food sources. When a species dies, life processes stop. The C-14/C-12 ratio decreases with time because C-14 is radioactive and decays according to its half-life, while the amount of C-12 in the species remains constant. Thus, the age of an archaeological artifact containing carbon can be calculated by comparing the C-14/C-12 ratio in the artifact with the C-14/C-12 ratio in the living species.

26. The half-life of carbon-14 is 5668 years.

$$(4 \times 10^6 \text{ years})\left(\frac{1 \text{ half-life}}{5668 \text{ years}}\right) = 7 \times 10^2 \text{ half-lives}$$

700 half-lives would pass in 4 million years. Not enough C-14 would remain to allow detection with any degree of reliability. C-14 dating would not prove useful in this case.

28.

	Protons	Neutrons	Nucleons
(a) $^{235}_{92}\text{U}$	92	143	235
(b) $^{82}_{35}\text{Br}$	35	47	82

30. Its atomic number increases by one, and its mass number remains unchanged.

32. Equations for alpha decays:
   (a) $^{192}_{78}\text{Pt} \longrightarrow {}^{4}_{2}\text{He} + {}^{188}_{76}\text{Os}$
   (b) $^{210}_{84}\text{Po} \longrightarrow {}^{4}_{2}\text{He} + {}^{206}_{82}\text{Pb}$

34. **(a)** $^{239}_{93}\text{Np} \longrightarrow {}^{0}_{-1}\text{e} + {}^{239}_{94}\text{Pu}$
   (b) $^{90}_{38}\text{Sr} \longrightarrow {}^{0}_{-1}\text{e} + {}^{90}_{39}\text{Y}$

36. $^{30}_{15}\text{P} \longrightarrow {}^{30}_{14}\text{Si} + {}^{0}_{+1}\text{e}$

38. **(a)** $^{66}_{29}\text{Cu} \longrightarrow {}^{66}_{30}\text{Zn} + {}^{0}_{-1}\text{e}$
   (b) $^{0}_{-1}\text{e} + {}^{7}_{4}\text{Be} \longrightarrow {}^{7}_{3}\text{Li}$
   (c) $^{27}_{13}\text{Al} + {}^{4}_{2}\text{He} \longrightarrow {}^{30}_{14}\text{Si} + {}^{1}_{1}\text{H}$
   (d) $^{85}_{37}\text{Rb} + {}^{1}_{0}\text{n} \longrightarrow {}^{82}_{35}\text{Br} + {}^{4}_{2}\text{He}$

40. 3 half-lives; 1/8 remaining

42. **(a)** 5.4×10^{11} J/mol
   (b) 0.199% mass loss

44. 11,340 years old

46. **(a)** 0.0424 g/mol
   (b) 3.8×10^{12} J/mol

48. (a) Geiger counter: Radiation passes through a thin glass window into a chamber filled with argon gas and containing two electrodes. Some of the argon ionizes, sending a momentary electrical impulse between the electrodes to the detector. This signal is amplified electronically and read out on a counter or as a series of clicks.

(b) Scintillation counter: Radiation strikes a scintillator, which is composed of molecules that emit light in the presence of ionizing radiation. A light-sensitive detector counts the flashes and converts them into a digital readout.

(c) Film badge: Radiation penetrates a film holder. The silver grains in the film darken when exposed to radiation. The film is developed at regular intervals.

50. Fission is the process of splitting a large nucleus into two roughly equal mass pieces. Fusion is the process of combining two relatively small nuclei to form a single larger nucleus.

52. (a) $^{235}_{92}\text{U} + ^1_0\text{n} \longrightarrow ^{143}_{54}\text{Xe} + 3^1_0\text{n} + ^{90}_{38}\text{Sr}$
(b) $^{235}_{92}\text{U} + ^1_0\text{n} \longrightarrow ^{102}_{39}\text{Y} + 3^1_0\text{n} + ^{131}_{53}\text{I}$
(c) $^{14}_{7}\text{N} + ^1_0\text{n} \longrightarrow ^1_1\text{H} + ^{14}_{6}\text{C}$

54. $^{236}_{92}\text{U} \longrightarrow ^{90}_{38}\text{Sr} + 3^1_0\text{n} + ^{143}_{54}\text{Xe}$

56. (a) $^{87}_{37}\text{Rb} \longrightarrow ^{0}_{-1}\text{e} + ^{87}_{38}\text{Sr}$
(b) $^{87}_{38}\text{Sr} \longrightarrow ^{0}_{+1}\text{e} + ^{87}_{37}\text{Rb}$

58. 7680 g

60. (a) 0.500 g left
(b) 0.250 g left
(c) 0.0625 g
(d) 9.77×10^{-4} g

62. 2.267×10^4 years

Chapter 19

2. The carbon atom has only two unshared electrons, making two covalent bonds logical, but in CH_4, carbon forms four equivalent bonds. Promoting one $2s$ electron to the empty $2p$ orbital would make four bonds possible, but without hybridization, we could not explain the fact that all four bonds in CH_4 are identical, and the bond angles are equal (109.5°).

4. (a) A molecule of ethane, C_2H_6, contains seven sigma bonds.
(b) A molecule of butane, C_4H_{10}, contains thirteen sigma bonds.
(c) A molecule of 2-methylpropane also contains thirteen sigma bonds.

6. *Iso:* in the word isomers, means compounds having the same molecular formula but different structural formulas.

Iso also refers to a particular isomer with a methyl branch at the end of a carbon chain. For example, isopentane:

$$CH_3 - \underset{\underset{CH_3}{|}}{CH}CH_2CH_3$$

Sec: a secondary carbon atom, one that is bonded to two other carbon atoms: $CH_3CH_2\underset{sec\text{- carbon atom}}{CHCH_3}$ secondary butyl group

Terts or *t-:* a tertiary carbon atom. A carbon atom that is bonded to three other carbon atoms as in *tert*-butyl group:

$$CH_3\underset{\underset{CH_3}{|}}{\overset{\overset{CH_3}{|}}{C}} \leftarrow tert \text{ carbon atom}$$

8. Lewis structures:

(a) CH_4 $H:\overset{\overset{H}{..}}{\underset{\underset{H}{..}}{C}}:H$ **(b)** C_3H_8 $H:\overset{\overset{H}{..}}{\underset{\underset{H}{..}}{C}}:\overset{\overset{H}{..}}{\underset{\underset{H}{..}}{C}}:\overset{\overset{H}{..}}{\underset{\underset{H}{..}}{C}}:H$

(c) C_5H_{12} $H:\overset{\overset{H}{..}}{\underset{\underset{H}{..}}{C}}:\overset{\overset{H}{..}}{\underset{\underset{H}{..}}{C}}:\overset{\overset{H}{..}}{\underset{\underset{H}{..}}{C}}:\overset{\overset{H}{..}}{\underset{\underset{H}{..}}{C}}:\overset{\overset{H}{..}}{\underset{\underset{H}{..}}{C}}:H$

10. Formulas (b), (e), and (f) are identical. The others are different.

12. The formulas in Exercise 10 contain the following numbers of methyl groups:
(a) 4 **(b)** 4 **(c)** 4 **(d)** 4 **(e)** 4 **(f)** 4

14. hexane $CH_3CH_2CH_2CH_2CH_2CH_3$ $CH_3CH_2CH_2\underset{\underset{CH_3}{|}}{CH}CH_3$

$CH_3CH_2\underset{\underset{CH_3}{|}}{CH}CH_2CH_3$ $CH_3CH_2\underset{\underset{CH_3}{|}}{\overset{\overset{CH_3}{|}}{C}}CH_3$ $CH_3\underset{\underset{CH_3}{|}}{\overset{\overset{CH_3}{|}}{CH}}CHCH_3$

16. (a) CH_3Br, one: CH_3 Br   **(b)** C_2H_5Cl, one: C_2H_5Cl
(c) C_4H_9I, four: $CH_3CH_2CH_2CH_2I$ $CH_3CH_2CHICH_3$
    $(CH_3)_2CHCH_2I$ $(CH_3)_3CHI$
(d) C_3H_6BrCl, five: $CH_3CH_2CHBrCl$ $CH_3CHClCH_2Br$
    $CH_3CHBrCH_2Cl$ $CH_2ClCH_2CH_2Br$
    $CH_3CBrClCH_3$

18. IUPAC names
(a) chloroethane
(b) 1-chloro-2-methylpropane
(c) 2-chlorobutane
(d) methylcyclopropane
(e) 2,4-dimethylpentane

20. (a) $CH_3CH_2\underset{\underset{CH_2CH_3}{|}}{CH}CH_2\overset{\overset{CH_3}{|}}{CH}CH_3$

(b) $CH_3CH_2CH_2\underset{\underset{C(CH_3)_3}{|}}{CH}CH_2CH_2CH_3$

(c) $CH_3CHCH_2CCH_2CH_2CHCHCH_2CH_3$

with substituents: CH_3, CH_2CH_3, $CHCH_3$ (top, the $CHCH_3$ bearing CH_3), and CH_3, CH_3 (bottom)

(d) $CH_3-C-CHCH_2CH_2CH_2CH_2CH_3$

with $H_3CCH_2CH_3$ (top) and CH_3 (bottom)

(e) cyclohexane ring with CH_2CH_3 and CH_2CH_3 substituents

22. (a) 3-methyl-5-ethyloctane $CH_3CH_2CHCH_2CHCH_2CH_2CH_3$ with CH_3 and CH_2CH_3 substituents

Ethyl should be named before methyl (alphabetical order): The numbering is correct. The correct name is 5-ethyl-3-methyloctane.

(b) 3,5,5-triethylhexane $CH_3CH_2CHCH_2CCH_3$ with CH_2CH_3, CH_2, CH_2CH_3, CH_3 substituents

The name is not based on the longest carbon chain (7 carbons). The correct name is 3,5-diethyl-3-methylheptane.

(c) 4,4-dimethyl-3-ethylheptane

$CH_3CH_2CH-CCH_2CH_2CH_3$ with CH_3CH_2, CH_3 and CH_3 substituents

Ethyl should be named before dimethyl (alphabetical order). The correct name is 3-ethyl-4,4-dimethylheptane.

24. $CH_3CH_2CH_2CH_2CH_2CH_2Cl$
$CH_3CH_2CH_2CH_2CHClCH_3$
$CH_3CH_2CH_2CHClCH_2CH_3$

26. (a) $CH_3CH_2CH_3 + Br_2 \xrightarrow{hv} CH_3CH_2CH_2Br +$
$CH_3CHBrCH_3 + HBr$
(b) $CH_3CH_2CH_3 + 5 O_2 \xrightarrow{\Delta} 3 CO_2 + 4 H_2O$

28. (a) $CH_3-C(CH_3)(CH_3)$—cyclohexane ring—$C(CH_3)(CH_3)-CH_3$

(b) cyclohexane ring with CH_3 and $CH_2CH_2CH_3$ substituents

(c) cyclohexane ring with Cl, CH_2CH_3, CH_3CH_2, Cl substituents

30. The formula for dodecane is $C_{12}H_{26}$

32. Data: $\dfrac{1 \text{ gal}}{60 \text{ miles}}$; 60 miles traveled; $\dfrac{19 \text{ mol } C_8H_{18}}{\text{gal}}$

$2 C_8H_{18} + 25 O_2 \longrightarrow 16 CO_2 + 18 H_2O$

$\dfrac{1 \text{ gal}}{60 \text{ miles}} \times 60 \text{ miles} = 1 \text{ gal gasoline used}$

$19 \text{ mol } C_8H_{18} \times \dfrac{16 \text{ mol } CO_2}{2 \text{ mol } C_8H_{18}} = 1.5 \times 10^2 \text{ mol } CO_2$

$PV = nRT \quad V = \dfrac{nRT}{P}$

$V = \dfrac{1.5 \times 10^2 \text{ mol } CO_2}{1 \text{ atm}} \times \dfrac{0.0821 \text{ L-atm} \times 293 \text{ K}}{} =$

$3.6 \times 10^3 \text{ L } CO_2$

34. (a) elimination
(b) substitution
(c) addition

36. Propane is the most volatile.
Volatility: propane > butane > hexane

38. (a) sp^3 hybrid orbitals in carbon
(b) Structural isomers are not possible because only one carbon atom is present.
(c) dichlorodifluoromethane
(d) The closest classification for Freon-12 (CF_2Cl_2) in Table 20.1 is alkyl halide.

40. cyclopentane ring with CH_3, H_3C, and CH_3 substituents

1,3,5-trimethylcyclopentane is an incorrect name for this compound. Numbering the ring counterclockwise gives the correct name, 1,2,4-trimethylcyclopentane.

Chapter 20

2.
$Cl\text{-}C=C\text{-}Cl$ with H, H (cis structure)
$H\text{-}C=C\text{-}Cl$ with Cl, H (trans structure)
$H-C-C-H$ with H, Cl (top) and Cl, H (bottom)

cis-1,2-dichloroethene   trans-1,2-dichloroethene   1,2-dichloroethane

We are able to get *cis-trans* isomers of ethene because there is no rotation around the carbon–carbon double bond, so the two structures shown are not the same. With 1,2-dichloroethane, there is free rotation of the carbon–carbon single bond. In the structure shown, exchanging the chlorine on the right with either hydrogen atom on that carbon appears to make a different isomer, but rotation makes any of the three positions equivalent.

4. Acetylene presents two different explosion hazards:
(a) It can form an explosive mixture with oxygen or air.
(b) When highly compressed or liquified, acetylene may decompose violently, either spontaneously or from a slight shock.

6. During the 10-year period from 1935 to 1945, the major source of aromatic hydrocarbons shifted from coal tar to petroleum due to the rapid growth of several industries that used aromatic hydrocarbons as raw material. These industries include drugs, dyes, detergents, explosives, insecticides, plastics, and synthetic rubber. Since the raw material needs far exceeded the aromatics available from coal tar, another source had to be found, and processes were developed to make aromatic compounds from alkanes in petroleum. World War II, which occurred in this period, put high demands on many of these industries, particularly explosives.

8. The 11-*cis*-isomer of retinal combines with a protein (opsin) to form the visual pigment rhodopsin. When light is absorbed, the 11-*cis* double bond is converted to a *trans*-double bond. This process initiates the mechanism of visual excitation, which our brains perceive as light or light forms.

10. (a) propane   **(b)** propene   **(c)** propyne

12. (a) C_3H_5Cl

CH_3 C=C H / Cl *trans*-1-chloropropene

CH_3 C=C Cl / H *cis*-1-chloropropene

$CH_3CCl=CH_2$ 2-chloropropene

$CH_2ClCH=CH_2$ 3-chloropropene

(b) chlorocyclopropane

14. (a) $CH_2=CHCCH_2CH_3$ with CH_2CH_3 and CH_3 substituents   **(d)**

(b) H C=C H with two phenyl groups   **(e)** with CH_3

(c) $CH\equiv CCHCH_3$ with phenyl   **(f)** $CH(CH_3)_2$

16. (a) $CH_2=CHCHCH_3$ with CH_2CH_3 Longest chain contains five carbon atoms. Correct name: 3-methyl-1-pentene

(b) (cyclohexene with Cl)   The C=C bond in cyclohexene is numbered so that substituted groups have the smallest numbers. Correct name: 1-chlorocyclohexene

(c) CH_3 C=C H / $CH_2CH_2CH_3$ Numbering was started from the wrong end of the structure. Correct name: *trans*-2-hexene

18. (a) cis-4-methyl-2-hexene   **(b)** 2,3-dimethyl-2-butene
(c) 3-isopropyl-1-pentene

20. all the pentynes, C_5H_8

1-pentyne   2-pentyne   3-methyl-1-butyne CH_3

$CH_3CH_2CH_2C\equiv CH$ $CH_3CH_2C\equiv CCH_3$ $CH_3CHC\equiv CH$

22. Only structure (c) will show *cis-trans* isomers:
$CH_2ClCH=CHCH_2Cl$

24. (a) $CH_3CH_2CH_2CH=CH_2 + H_2O \xrightarrow{H^+} CH_3CH_2CH_2CHCH_3$ with OH

(b) $CH_3CH_2CH=CHCH_3 + HBr \longrightarrow$
$CH_3CH_2CHBrCH_2CH_3 + CH_3CH_2CH_2CHBrCH_3$

(c) $CH_2=CHCl + Br_2 \longrightarrow CH_2BrCHClBr$

(d) (benzene ring with) $CH=CH_2$ $+ HCl \longrightarrow$ (benzene ring with) $CHClCH_3$

(e) $CH_2=CHCH_2CH_3 + KMnO_4 \xrightarrow[\text{cold}]{H_2O} CH_2CHCH_2CH_3$ with OH OH

26. (a) $CH_3C\equiv CH + H_2(1 \text{ mol}) \xrightarrow[\text{1 atm}]{Pt, 25°C} CH_3CH=CH_2$

(b) $CH_3C\equiv CCH_3 + Br_2(2 \text{ mol}) \longrightarrow CH_3CBr_2CBr_2CH_3$

(c) two-step reaction:
$CH_3C\equiv CH + HCl \longrightarrow CH_3CCl=CH_2 \xrightarrow{HCl}$
$CH_3CCl_2CH_3$

28. When cyclopentene, (cyclopentene ring) reacts with

(a) Cl_2 the product is (cyclopentane ring)—Cl with Cl 1,2-dichlorocyclopentane

(b) HBr the product is (cyclopentane ring)—Br bromocyclopentane

(c) H_2, Pt the product is (cyclopentane ring) cyclopentane

(d) H_2O, H^+ the product is (cyclopentane ring)—OH cyclopentanol

30. (a) $CH_3C\equiv CCH_2CH_3 + Cl_2(1 \text{ mole}) \longrightarrow$
$CH_3CCl=CClCH_2CH_3$
$CH_3CCl=CClCH_2CH_3 + H_2 \xrightarrow[\text{1 atm}]{25°C}$
$CH_3CHClCHClCH_2CH_3$

(b) $CH_3C\equiv CCH_2CH_3 + HCl \longrightarrow$
$CH_3CCl=CHCH_2CH_3$
$CH_3CCl=CHCH_2CH_3 + HCl \longrightarrow$
$CH_3CCl_2CH_2CH_2CH_3$
Can also yield $CH_3CH_2CCl_2CH_2CH_3$ at the same time.

(c) $CH_3C\equiv CCH_2CH_3 + 2\ Cl_2 \longrightarrow$
$CH_3CCl_2CCl_2CH_2CH_3$

32. (a) [benzene ring with CH$_3$]   **(c)** [benzene ring with OH]

(b) [benzene ring with CH$_2$CH$_3$]   **(d)** [anthracene structure]

34. (a) [benzene ring with Cl, Cl, NO$_2$]   **(c)** [diphenyl with CH and CH$_3$]

(b) [benzene ring with NO$_2$, NO$_2$]   **(d)** [benzene ring with CH=CH$_2$]

36. (a) trichlorobenzenes

[three benzene ring structures with Cl substituents]

1, 2, 3-trichlorobenzene   1, 2, 4-trichlorobenzene   1, 3, 5-trichlorobenzene

(b) the benzene derivatives of formula C$_8$H$_{10}$:

[benzene ring with CH$_3$, CH$_3$]   [benzene ring with CH$_3$, CH$_3$]

o-xylene or 1, 2-dimethylbenzene   m-xylene or 1, 3-dimethylbenzene

[benzene ring with CH$_3$ and CH$_3$]   [benzene ring with CH$_2$CH$_3$]

p-xylene or 1, 4-dimethylbenzene   ethylbenzene

38. The isomers that can be written substituting a third chlorine atom on o-dichlorobenzene are:

[benzene ring with Cl, Cl, Cl]   [benzene ring with Cl, Cl, Cl]

1, 2, 3-trichlorobenzene   1, 2, 4-trichlorobenzene

40. (a) styrene   **(d)** isopropylbenzene
(b) m-nitrotoluene   **(e)** 2,4,6-tribromophenol
(c) 2,4-dibromobenzoic acid

42. (a) [benzene] + CH$_3$CHCH$_3$ (with Cl) $\xrightarrow{\text{AlCl}_3}$ [benzene with CHCH$_3$ and CH$_3$] + HCl

isopropylbenzene

(b) [benzene with CH$_3$] + KMnO$_4$ $\xrightarrow{\text{H}_2\text{O}}$ [benzene with COOH]

benzoic acid

44. Two tests can be used:
(a) Baeyer test—hexene will decolorize KMnO$_4$ solution; cyclohexane will not.
(b) In the absence of sunlight, hexene will react with and decolorize bromine; cyclohexane will not.

46. (a) $\overset{+}{\text{C}}\text{H}_3$

(b) $\text{CH}_3\text{CH}_2\overset{+}{\text{C}}\text{H}_2$

(c) CH$_3$ | CH$_3$C$^+$ | CH$_3$

(d) $\text{CH}_3\text{CH}_2\text{CH}_2\text{CH}_2\overset{+}{\text{C}}\text{H}_2$

48. The reaction mechanism by which benzene is brominated in the presence of FeBr$_3$:

$\text{FeBr}_3 + \text{Br}_2 \longrightarrow \text{FeBr}_4^- + \text{Br}^+$

[benzene] + Br$^+$ $\longrightarrow$ [cyclohexadienyl cation with H, Br]

[cyclohexadienyl cation with H, Br] + FeBr$_4^-$ $\longrightarrow$ [benzene with Br] + FeBr$_3$ + HBr

50. (a) $\text{CH}_3\text{CHClCH}_2\text{CH}_3 \xrightarrow{-\text{HCl}} \text{CH}_2{=}\text{CHCH}_2\text{CH}_3 +$ $\text{CH}_3\text{CH}{=}\text{CHCH}_3$

(b) $\text{CH}_2\text{ClCH}_2\text{CH}_2\text{CH}_2\text{CH}_3 \xrightarrow{-\text{HCl}} \text{CH}_2{=}\text{CHCH}_2\text{CH}_2\text{CH}_3$

(c) [cyclohexane with Cl] $\xrightarrow{-\text{HCl}}$ [cyclohexene]

52. Baeyer test: Add KMnO$_4$ solution to each sample. The KMnO$_4$ will lose its purple color with 1-heptene. There will be no reaction (no color change) with 1-heptane.

54. [pentagon]   cyclopentane   Other possibilities are:

cis and trans

Chapter 21

2. Those polymers which soften on reheating are thermoplastic polymers; those which set to an infusible solid and do not soften on reheating are thermosetting polymers.

4. Vulcanization is the process of heating raw rubber with sulfur. Sulfur atoms are introduced as cross links between the polymeric chains.

6. polystyrene: $\left[CH_2-CH\right]_n$ C_8H_8

molar mass of 1 unit:

$8(12.01 \text{ g/mol}) + 8(1.008 \text{ g/mol}) = 104.1 \text{ g/mol}$

molar mass: $(3000 \text{ units})\left(\dfrac{104.1 \text{ g/mol}}{\text{unit}}\right) = 3 \times 10^5 \text{ g/mol}$

= molar mass

8. (a) Saran $\left(CH_2CCl_2\right)_n$ **(d)** Polystyrene $\left[CH_2CH\right]$

  (b) Orlon $\left(\begin{array}{c}CH_2CH \\ | \\ CN\end{array}\right)_n$

  (c) Teflon $\left(CF_2CF_2\right)_n$

10. polyethylene free radical (2 units) starting with RO· free radical $ROCH_2CH_2CH_2CH_2$·

12. (a) ethylene $\left(CH_2-CH_2\right)_n$
  (b) chloroethene $\left(CH_2CHCl\right)$
  (c) 1-butene $\left(\begin{array}{c}CH_2CH \\ | \\ CH_2CH_3\end{array}\right)_n$

14. Two possible ways in which crylonitrile can polymerize to form Orlon are:

$\left(\begin{array}{c}CH_2CHCH_2CHCH_2CHCH_2CH \\ | \qquad | \qquad | \qquad | \\ CN \quad CN \quad CN \quad CN\end{array}\right)_n$

$\left(\begin{array}{c}CH_2CH-CHCH_2CH_2CH-CHCH_2 \\ | \qquad | \qquad\quad | \qquad | \\ CN \quad CN \qquad CN \quad CN\end{array}\right)_n$

16. Gutta percha (all trans)

$\left(\begin{array}{c}CH_2 \\ \diagdown \\ CH_3\end{array}C=C\begin{array}{c}H \\ \diagup \\ CH_2-CH_2\end{array}\begin{array}{c}CH_3 \\ \diagdown \\ \end{array}C=C\begin{array}{c}CH_2-CH_2 \\ \diagup \\ H\end{array}\begin{array}{c}CH_3 \\ \diagdown \\ \end{array}C=C\begin{array}{c}H \\ \diagup \\ CH_2\end{array}\right)_n$

18. Yes, styrene—butadiene rubber contains carbon–carbon double bonds and is attacked by the ozone in smog, causing "age hardening" and cracking.

24. (a) synthetic natural rubber: $CH_2=CCH=CH_2$
$\qquad\qquad\qquad\qquad\qquad\qquad\qquad\quad\;\, |$
$\qquad\qquad\qquad\qquad\qquad\qquad\qquad\;\; CH_3$

  (b) neoprene rubber: $CH_2=CCl-CH=CH_2$

22. (a) Isotactic polypropylene

$\left(\begin{array}{c}CH_2CHCH_2CHCH_2CHCH_2CH \\ | \qquad | \qquad | \qquad | \\ CH_3 \quad CH_3 \quad CH_3 \quad CH_3\end{array}\right)_n$

  (b) Another form of polypropylene

$\left(\begin{array}{c}\qquad CH_3 \qquad\quad CH_3 \\ \qquad | \qquad\qquad | \\ CH_2CHCH_2CHCH_2CHCH_2CH \\ | \qquad\qquad | \\ CH_3 \qquad\quad CH_3\end{array}\right)_n$

(other structures are possible)

24. $\left(\begin{array}{c}CH_2-CH-CH_2-CH-CH_2-CH \\ \qquad | \qquad\qquad | \qquad\qquad | \\ \qquad C=O \qquad\;\; C=O \qquad\;\; C=O \\ \qquad | \qquad\qquad | \qquad\qquad | \\ \qquad O \qquad\qquad O \qquad\qquad O \\ \qquad | \qquad\qquad | \qquad\qquad | \\ \quad CH_2CH_3 \quad CH_2CH_3 \quad CH_2CH_3\end{array}\right)_n$

26. (a) From 2,3-dimethyl-1,3-butadiene,

$$CH_2=C-C=CH_2$$
$$\qquad | \quad\; |$$
$$\qquad H_3C \;\; CH_3$$

this polymer can be made $\left(\begin{array}{c}CH_2C=CCH_2 \\ \quad | \quad\; | \\ \quad H_3C \;\; CH_3\end{array}\right)_n$

  (b) If produced by the free-radical mechanism, a random mixture of cis and trans connections are made. It is possible, using catalysts, for the reaction to proceed by an ionic mechanism which will give a stereochemically controlled polymer.

28. $CH_2=CH_2 \qquad\qquad CF_2=CF_2$
      ethylene   tetrofluoroethylene
$\quad\left(CH_2-CH_2-CF_2-CF_2-CH_2-CH_2-CF_2-CF_2\right)_n$
      (many other structures are possible)

30. The methyl methacrylate polymers result from an addition reaction because no small molecule is split out. The polymer forms in a manner similar to that of polyethylene.

Chapter 22

2. Alkenes are almost never made from alcohols because the alcohols are almost always the higher value material. This is because recovering alkenes from hydrocarbon sources in an oil refinery (primarily catalytic cracking) is a relatively cheap process.

4. Oxidation of primary alcohols yields aldehydes. Further oxidation yields carboxylic acids. Examples are:

$$\qquad\qquad\qquad\qquad\qquad\qquad\;\; O$$
$$\qquad\qquad\qquad\qquad\qquad\qquad\;\; ||$$
$$CH_3CH_2OH + [O] \longrightarrow CH_3C-H + H_2O$$

$$\;\; O \qquad\qquad\qquad\qquad\quad O$$
$$\;\; || \qquad\qquad\qquad\qquad\quad ||$$
$$CH_3C-H + [O] \longrightarrow CH_3C-OH$$

6. (a) Methanol is a poisonous liquid capable of causing blindness or death if taken internally. Exposure to methanol vapors is also very dangerous.
  (b) Ethanol can act as a food, a drug, and a poison. The body can metabolize small amounts of ethanol to produce energy; thus it is a food. It depresses brain functions so that activities requiring skill and judgment are impaired; thus it is a drug. With very large consumption, the depression of brain function can lead to unconsciousness and death; thus it is a poison.

8. Low molar mass ethers present two hazards.
  (a) They are very volatile and their highly flammable vapors form explosive mixtures with air.

(b) They also slowly react with oxygen in the air to form unstable explosive peroxides.

10. Fats are a major source of energy in the body. Infants who die from SIDS, due to a lack of energy, are unable to degrade fats. One of the intermediate compounds in the degradation of fats is an alcohol, which is oxidized to a ketone.

12. (a) $CH_3CHCH_2CH_3$
     |
     OH

   (b) $CH_3CCH_2CH_3$ (with CH_3 above C and OH below C)

   (c) CH_3CHCH_3
       |
       OH

   (d) cyclopentane—OH

   (e) $CH_3CH_2CH_2CH_2CH_2SH$

   (f) $CH_3CH_2CHCHCH_2CH_3$ (with CH_2CH_3 above second C, OH below third C)

14. (a) C_3H_8O

Alcohols	Ethers
$CH_3CH_2CH_2OH$	$CH_3OCH_2CH_3$
$CH_3CH(OH)CH_3$	

   (b) C_4H_8O

Alcohols	Ethers
$CH_3CH_2CH_2CH_2OH$	$CH_3OCH_2CH_2CH_3$
$CH_3CH_2CH(OH)CH_3$	$CH_3OCH(CH_3)_2$
$(CH_3)_2CHCH_2OH$	$CH_3CH_2OCH_2CH_3$
$(CH_3)_3C{-}OH$	

16. The primary alcohols in Question 14 are:
    $CH_3CH_2CH_2OH$ $CH_3CH_2CH_2CH_2OH$
    $(CH_3)_2CHCH_2OH$
    The secondary alcohols are:
    $CH_3CH(OH)CH_3$ $CH_3CH_2CH(OH)CH_3$
    The tertiary alcohol is $(CH_3)_3C{-}OH$

18. (a) ethanol (ethyl alcohol)
    (b) 2-phenylethanol
    (c) 3-methyl-3-pentanol
    (d) 1-methylcyclopentanol
    (e) 3-pentanol
    (f) 1,2-propanediol
    (g) 4-ethyl-2-hexanol

20. (a) 1-methylcyclopentene (cyclopentene with CH_3)
    (b) $CH_3CH{=}CHCH_3$ 2-butene
    (c) $CH_3C{=}CHCH_2CH_3$ (with CH_3 above C)   2-methyl-2-pentene

22. (a) cyclopentane—OH cyclopentanol
    (b) cyclopentane with CH_3 and OH 1-methylcyclopentanol
    (c) $CH_3CHCHCHCH_3$ (with CH_3 and CH_3 above, OH below)   2,4-dimethyl-3-pentanol

24. Oxidation of a secondary alcohol:

$$CH_3CHCH_2CH_3 \longrightarrow CH_3CCHCH_3$$

(with CH_3 above, OH below left; CH_3 above, O below right)   ketone

26. (a) $CH_3CH{=}CHCH_3 + H_2O \xrightarrow{H^+} CH_3CH_2CH_2CH_3$ (with OH below)

    (b) $CH_3CH_2CH_2CH{=}CH_2 + H_2O \xrightarrow{H^+} CH_3CH_2CH_2CHCH_3$ (with OH below)

    (c) $CH_3C{=}CHCH_3 + H_2O \xrightarrow{H^+} CH_3CCH_2CH_3$ (with CH_3 below left; OH above, CH_3 below right)

28. (a) $CH_3CHCH_2CH_3$ 2-butanol
        |
        OH

    (b) $CH_3CHCHCH_2CH_3$ (with CH_2CH_3 above, OH below)   3-ethyl-2-pentanol

    (c) cyclopentane—OH cyclopentanol

30. (a) $2\,CH_3CH_2OH + 2\,Na \longrightarrow 2\,CH_3CH_2O^-Na^+ + H_2$

    (b) $CH_3CH_2CH_2CH_2OH \xrightarrow[H_2SO_4]{K_2Cr_2O_7} CH_3CH_2CH_2C{-}H$ (with O double bond above C)

    (c) cyclopentane—$CH{=}CH_2 \xrightarrow[H_2SO_4]{H_2O}$ cyclopentane—$CHCH_3$ (with OH below)

    (d) $CH_3CH_2C{-}OCH_3 + NaOH \longrightarrow$ (with O double bond above C)
        $CH_3CH_2C{-}O^-Na^+ + CH_3OH$ (with O double bond above C)

32. (a) benzene ring with OH top and NO_2 bottom
    (b) benzene ring with OH top, H_3C left, CH_3 right
    (c) benzene ring with OH and OH adjacent

34. (a) *p*-dihydroxybenzene   (c) 2,4-dinitrophenol
   (hydroquinone)
(b) 2,4-dimethylphenol   (d) *m*-hexylphenol

36. Order of decreasing solubility in water:
[a (highest), d, c, b (lowest)]
(a) $CH_3CH(OH)CH(OH)CH_2OH$
(d) $CH_3CH(OH)CH_2CH_2OH$
(c) $CH_3CH_2CH_2CH_2OH$
(b) $CH_3CH_2OCH_2CH_3$

38. The six isomeric ethers of $C_5H_{12}O$ are:

$CH_3OCH_2CH_2CH_2CH_3$
methyl *n*-butyl ether
(1-methoxybutane)

$CH_3OCHCH_2CH_3$
  |
 CH_3
methyl sec-butyl ether
(2-methoxybutane)

$CH_3OCH_2CHCH_3$
      |
     CH_3
methyl isobutyl ether
(1-methoxy-2-methylpropane)

CH_3OCCH_3
  |
 CH_3
  (with CH_3 above)
methyl *t*-butyl ether
(2-methoxy-2-methylpropane)

$CH_3CH_2OCH_2CH_2CH_3$
ethyl propyl ether
(1-ethoxypropane)

$CH_3CH_2OCHCH_3$
      |
     CH_3
ethyl isopropyl ether
(2-ethoxypropane)

40. $CH_3CH_2OCH_2CH_3 + 6\,O_2 \xrightarrow{\Delta} 4\,CO_2 + 5\,H_2O$

42. These are possible combinations of reactants to make the following ethers by the Williamson syntheses:
(a) $CH_3CH_2CH_2OCH_2CH_2CH_3$
   $CH_3CH_2CH_2ONa + CH_3CH_2CH_2Cl$

(b) $HCOCH_2CH_2CH_3$
   with CH_3 above and CH_3 below
   $HCONa + CH_3CH_2CH_2Cl$
   with CH_3 above and CH_3 below
          or $CH_3CH_2CH_2ONa + CH_3CHCH_3$
                                          |
                                         Cl

(c)

44.

cis-1,2-cyclopentanediol   trans-1,2-cyclopentanediol

46. The common phenolic structure in the catecholamines is catechol, *o*-dihydroxybenzene:

catechol

48. Methyl alcohol is converted to formaldehyde by an oxidation reaction:

$$CH_3OH \xrightarrow{[O]} HCH$$
(with O double bonded above HCH)

50.

$CH_3CH_3 + Cl_2 \xrightarrow{light} CH_3CH_2Cl + HCl$

52. Differentiation between phenols and alcohols:
Phenols are acidic compounds; alcohols are not acidic. Phenols react with NaOH to form salts; alcohols do not react with NaOH.
The OH group is bonded to a benzene ring in phenols and to a nonbenzene carbon atom in alcohols.

54. Only compound (a) will react with NaOH:

Chapter 23

2. Aldehydes and ketones have lower boiling points than alcohols of similar molar mass because they do not form hydrogen bonds, since they contain no —OH groups.

4. The strength of collagen depends on aldol condensations. After collagen is formed, aldehydes add along its length. Collagen fibers adjacent to each other undergo an aldol condensation. The cross-linking bonds between collagen strands form a strong network, giving collagen its strength.

6. The carbonyl group C=O, having a double bond, contains a sigma and a pi bond.

8. (a) ethanal, acetaldehyde   (d) 2-chloro-5-isopropyl-
(b) butanal   benzaldehyde
(c) benzaldehyde   (e) 3-hydroxybutanal

10. (a) 2-butanone, methyl ethyl ketone (MEK)
(b) 3,3-dimethylbutanone, methyl *t*-butyl ketone
(c) 2,5-hexanedione
(d) 1-phenyl-2-propanone, benzyl methyl ketone

12. (a)

$$HOCH_2CH_2\overset{\overset{\displaystyle O}{\|}}{C}-H$$

(b) $CH_3CH_2\overset{\overset{\displaystyle O}{\|}}{C}\underset{\underset{\displaystyle CH_3}{|}}{C}HCH_2CH_3$

(c) cyclohexanone (structure)

(d) $CH_3\underset{\underset{\displaystyle Cl}{|}}{C}HCH_2\underset{\underset{\displaystyle Cl}{|}}{C}HCH_2\underset{\underset{\displaystyle Cl}{|}}{C}H\overset{\overset{\displaystyle O}{\|}}{C}-H$

(e) $CH_3CH=CHCH_2\overset{\overset{\displaystyle O}{\|}}{C}-H$

14. Higher boiling point
  (a) pentanal   **(c)** 2-hexanone
  (b) benzyl alcohol   **(d)** 1-butanol

16. Higher aqueous solubility
  (a) acetaldehyde   **(c)** propanal
  (b) 2,4-pentanediol

18. (a) 1-propanol:

$$CH_3CH_2CH_2OH \xrightarrow[H_2SO_4]{K_2Cr_2O_7} CH_3CH_2\overset{\overset{\displaystyle O}{\|}}{C}-H \text{ or } CH_3CH_2COOH$$

$$CH_3CH_2CH_2OH + O_2 \xrightarrow[\Delta]{Cu} CH_3CH_2\overset{\overset{\displaystyle O}{\|}}{C}-H$$

(b) 2,3-dimethyl-2-butanol:

$$CH_3\underset{\underset{\displaystyle H_3C}{|}}{C}H\underset{\underset{\displaystyle CH_3}{|}}{C}CH_3 + O_2 \xrightarrow{Ag}{\Delta} \text{No reaction (3° alcohol)}$$
with either oxidizing agent

20. (a) An aldehyde group, $-C=O$ must be present to give a positive Fehling test.
  (b) The visible evidence for a positive Fehling test is the formation of brick-red Cu_2O, which precipitates during the reaction.
  (c) $CH_3\overset{\overset{\displaystyle H}{|}}{C}=O + 2 Cu^{2+} \xrightarrow[H_2O]{NaOH}$
  $CH_3COONa + Cu_2O(s)$ (brick red)

22. (a) propanal $\longrightarrow CH_3CH_2COONa + Cu_2O(s)$
  (b) acetone $\longrightarrow$ no reaction
  (c) 3-methylpentanal $\longrightarrow CH_3CH_2\underset{\underset{\displaystyle CH_3}{|}}{C}HCH_2COONa + Cu_2O(s)$

24. (a) 3-pentanone:

$$2 CH_3CH_2\overset{\overset{\displaystyle O}{\|}}{C}CH_2CH_3 \xrightarrow[NaOH]{dilute} CH_3CH_2\underset{\underset{\displaystyle HO}{|}}{C}\underset{\underset{\displaystyle CH_3}{|}}{\overset{\overset{\displaystyle CH_2CH_3}{|}}{C}}H-\overset{\overset{\displaystyle O}{\|}}{C}CH_2CH_3$$

(b) propanal:

$$2 CH_3CH_2\overset{\overset{\displaystyle H}{|}}{C}=O \xrightarrow[NaOH]{dilute} CH_3CH_2\underset{\underset{\displaystyle OH}{|}}{C}H\underset{\underset{\displaystyle H}{|}}{\overset{\overset{\displaystyle CH_3}{|}}{C}}C=O$$

26. (a) $CH_3CH_2\overset{\overset{\displaystyle H}{|}}{C}=O + 2 CH_3CH_2CH_2OH \overset{dry\ HCl}{\rightleftharpoons}$
$CH_3CH_2\underset{\underset{\displaystyle H}{|}}{\overset{\overset{\displaystyle OCH_2CH_2CH_3}{|}}{C}}-OCH_2CH_2CH_3 + H_2O$

(b) cyclohexanone $+ CH_3OH \overset{H^+}{\rightleftharpoons}$ (cyclohexane with OH and OCH_3)

(c) $CH_3CH_2CH_2CH(OCH_3)_2 \underset{H^+}{\overset{H_2O}{\rightleftharpoons}}$
$CH_3CH_2CH_2\overset{\overset{\displaystyle H}{|}}{C}=O + 2 CH_3OH$

28. (a) (benzaldehyde) $\overset{\overset{\displaystyle H}{|}}{C}=O + HCN \xrightarrow{OH^-}$ (benzene)$\underset{}{\overset{\overset{\displaystyle OH}{|}}{C}}H-CN$

(b) (benzene)$\underset{}{\overset{\overset{\displaystyle OH}{|}}{C}}H-CN + H_2O \xrightarrow{H^+}$ (benzene)$\overset{\overset{\displaystyle OH}{|}}{C}HCOOH$

(c) (benzene)$\overset{\overset{\displaystyle OH}{|}}{C}HCOOH \xrightarrow[H_2SO_4]{K_2Cr_2O_7}$ (benzene)$\overset{\overset{\displaystyle O}{\|}}{C}-COOH$

30. $CH_3\overset{\overset{\displaystyle O}{\|}}{C}-H$ ethanal

32. The secondary alcohol in compound I is oxidized to the ketone (compound II).

34. $CH_3\overset{\overset{\displaystyle H}{|}}{C}=O + HCN \longrightarrow CH_3\underset{\underset{\displaystyle OH}{|}}{C}-CN \xrightarrow[H^+]{H_2O} CH_3\underset{\underset{\displaystyle OH}{|}}{C}H-COOH$

36. Pyruvic acid is changed to lactic acid by a reduction reaction.

38. (a) 3-pentanone; 3-pentanol $CH_3CH_2CHCH_2CH_3$
$\qquad\qquad\qquad\qquad\qquad\qquad\qquad\ \ |$
$\qquad\qquad\qquad\qquad\qquad\qquad\qquad\ OH$

(b) methyl ethyl ketone: 2-butanol $CH_3CHCH_2CH_3$
$\qquad\qquad\qquad\qquad\qquad\qquad\qquad\ \ |$
$\qquad\qquad\qquad\qquad\qquad\qquad\qquad OH$

(c) 4-phenyl-2-butanone: 4-phenyl-2-butanol

$\qquad\qquad\qquad\qquad CH_2CH_2CHCH_3$
$\qquad\qquad\qquad\qquad\qquad\qquad\quad\ |$
$\qquad\qquad\qquad\qquad\qquad\qquad\quad OH$

40. Benzaldehyde isomers of formula $C_9H_{10}O$:

42. 4-unit polymer of acetaldehyde:

Chapter 24

2. The butyric acid solution would be expected to have the more objectionable odor because salts normally exhibit little or no odor since they are ionic and therefore have low volatility. For example, a dilute solution of acetic acid (vinegar) has considerable odor, but sodium acetate does not.

4. (a) The major difference between fats and oils is that fats are solids at room temperature, oils are liquids. The fatty acids in the molecules are mostly saturated in fats; more unsaturated in oils.

(b) Soaps are the sodium salts of high molar mass fatty acids. Syndets are synthetic detergents and occur in several different forms. They have cleansing action similar to soaps, but have different structural and solubility characteristics, such as being soluble in hard water. Some syndets also contain long hydrocarbon chains.

(c) Hydrolysis is the breaking apart of an ester in the presence of water to form an alcohol and a carboxylic acid. Mineral acids or digestive enzymes are used to speed the process. Saponification breaks the ester apart using sodium hydroxide to form alcohols and salts of carboxylic acids.

6. The principal advantage of synthetic detergents over soap is that the syndets do not form insoluble precipitates with the ions in hard water (Ca^{2+}, Mg^{2+}, Fe^{3+}).

8. methyl salicylate (pain relieving liniments)
phenyl salicylate (protective coating for pills)

10. In a condensation reaction one of the products formed is a small molecule, such as water. In polyester formation between an alcohol and a carboxylic acid, water is formed in addition to the polyester.

12. (a) *o*-chlorobenzoic acid   **(d)** 2-hydroxybutanoic acid
(b) oleic acid   **(e)** ammonium propanoate
(c) *m*-toluic acid

14. (a) $COOH$
$\qquad\ |$
$\qquad COOH$

(b) $CH_3CH_2CH_2CH_2COOH$

(c)

(d) $CH_3CH_2CH=CHCH_2CH=CHCH_2CH=CH(CH_2)_7COOH$

(e)

(f) $CH_3CH_2COONH_4$

(g) CH_3CHCH_2COOH
$\qquad\ |$
$\qquad OH$

16. Decreasing pH means increasing acidity:

18. (a) methyl methanoate   methyl formate
(b) propyl benzoate   propyl benzoate
(c) ethyl propanoate   ethyl propionate

20. (a)

$$CH_3\overset{\displaystyle O}{\overset{\|}{C}}-OCH_2CH_2CH_3$$

(b)

$$\overset{\displaystyle O}{\overset{\|}{C}}-OCH_3 \text{ (benzene ring)}$$

(c)

$$CH_3CH_2CH_2CH_2CH_2\overset{\displaystyle O}{\overset{\|}{C}}-OCH_2CH_3$$

22. (a) COONa ... COONa (benzene ring)   disodium-*m*-phthalate

(b) $HOOCCH_2CH_2COOH$ succinic acid

(c) $CH_3(CH_2)_7CH=CH(CH_2)_7COONa + H_2O$
sodium oleate

(d) $CH_3CH_2CH_2COOH$ butanoic acid

(e) COONa (benzene ring) + ONa (benzene ring)

sodium benzoate   sodium phenolate

24. (a) methanol and propanoic acid: $CH_3CH_2\overset{\displaystyle O}{\overset{\|}{C}}-OCH_3$

(b) 1-octanol and acetic acid: $CH_3\overset{\displaystyle O}{\overset{\|}{C}}-OCH_2(CH_2)_6CH_3$

(c) ethanol and butanoic acid: $CH_3CH_2CH_2\overset{\displaystyle O}{\overset{\|}{C}}-OCH_2CH_3$

26. (a) isopropyl formate: $CH_3\underset{\underset{OH}{|}}{C}HCH_3 + HCOOH$

(b) diethyl adipate: $CH_3CH_2OH + \underset{\underset{\underset{COOH}{|}}{(CH_2)_4}}{COOH}$

(c) benzyl benzoate: CH_2OH (benzene ring) + COOH (benzene ring)

28. (a)

$\overset{\displaystyle O}{\overset{\|}{C}}-Cl$ (benzene ring) $+ NH_3 \longrightarrow$ $\overset{\displaystyle O}{\overset{\|}{C}}-NH_2$ (benzene ring)

(b) $CH_3CH_2COOH + SOCl_2 \longrightarrow CH_3CH_2\overset{\displaystyle O}{\overset{\|}{C}}-Cl$

(c) $CH_3CH_2\overset{\displaystyle O}{\overset{\|}{C}}-Cl + CH_3OH \longrightarrow CH_3CH_2\overset{\displaystyle O}{\overset{\|}{C}}-OCH_3$

(d) $CH_3CH_2CH_2C\equiv N + H_2O \xrightarrow{H+} CH_3CH_2CH_2COOH$

(e) CH_2CH_2COOH (benzene ring) $+ SOCl_2 \longrightarrow$ $CH_2CH_2\overset{\displaystyle O}{\overset{\|}{C}}-Cl$ (benzene ring)

30. (a) Benzoic acid and ethyl benzoate: benzoic acid is an odorless solid; ethyl benzoate is a fragrant liquid.

(b) Succinic acid and fumaric acid: fumaric acid has a carbon–carbon double bond and will readily add and decolorize bromine; succinic acid will not decolorize bromine.

32. (a)

$$\underset{CH_2-C}{\overset{CH_2-C}{CH_2}}\quad O \quad \text{glutaric anhydride}$$

(b) $\underset{\underset{COOCH_2CH_3}{|}}{\overset{\overset{COOCH_2CH_3}{|}}{CH_2}}$ diethyl malonate

(c) OH (benzene ring) $+ CH_3COOH$

phenol   acetic acid

34. (a) cis-oleic acid:

$$CH_3(CH_2)_7 \quad \quad (CH_2)_7COOH$$
$$C=C$$
$$H \quad \quad H$$

(b) cis,cis-linoleic acid:

$$CH_3(CH_2)_4 \quad CH_2 \quad (CH_2)_7COOH$$
$$C=C \quad C=C$$
$$H \quad H\,H \quad H$$

36. Sodium lauryl sulfate would be more effective as a detergent in hard water than sodium propyl sulfate because the hydrocarbon chain is only three carbons long in the latter, not long enough to dissolve grease well.

38. Only (c), $CH_3(CH_2)_{10}CH_2O(CH_2CH_2O)_7CH_2CH_2OH$, would be a good detergent in water. It is nonionic.

40. (a) $CH_3O-\underset{\underset{OCH_3}{|}}{\overset{\overset{O}{\|}}{P}}-OCH_3 + 3\,H_2O$

(b) $CH_3O-\underset{\underset{OH}{|}}{\overset{\overset{O}{\|}}{P}}-\overset{\overset{CH_3}{|}}{O}CHCH_2CH_3$

42.

COOH
$OC-CH_3$ (benzene ring) $+ CH_3COOH$
O

aspirin   acetic acid
(acetylsalicylic acid)

The reactants are salicylic acid and acetic anhydride.

44. $Cl-\overset{\overset{\displaystyle O}{\|}}{C}-Cl$

The functional groups are acid chlorides.
The acid from which phosgene is derived is carbonic acid,
$HO-\overset{\overset{\displaystyle O}{\|}}{C}-OH$

46.

$$CH_2O-\overset{\overset{\displaystyle O}{\|}}{C}(CH_2)_{10}CH_3$$
$$|\qquad\overset{\displaystyle O}{\|}$$
$$CHO-\overset{}{C}(CH_2)_{14}CH_3$$
$$|\qquad\overset{\displaystyle O}{\|}$$
$$CH_2O-\overset{}{C}(CH_2)_7CH=CH(CH_2)_7CH_3$$

There would be two other triacylglycerols containing all three of these acids. Each of the three acids can be the middle carbon.

48.

$$CH_3O-\overset{\overset{\displaystyle O}{\|}}{C}(CH_2)_{16}CH_3 \qquad\qquad CH_2OH$$
$$|\qquad\overset{\displaystyle O}{\|} \qquad\qquad\qquad\qquad |$$
$$CHO-\overset{}{C}(CH_2)_{16}CH_3 + 3\ KOH \longrightarrow CHOH + 3\ CH_3(CH_2)_{16}COOK$$
$$|\qquad\overset{\displaystyle O}{\|} \qquad\qquad\qquad\qquad\qquad\qquad |$$
$$CH_2O-\overset{}{C}(CH_2)_{16}CH_3 \qquad\qquad CH_2OH$$

The solubility in water will change considerably. Glyceryltristearate is insoluble in water but the products glycerol and potassium stearate (a salt) are soluble in water.

50. $CH_3(CH_2)_{14}\overset{\overset{\displaystyle O}{\|}}{C}-O(CH_2)_{29}CH_3$

52. $CH_3(CH_2)_{12}\overset{\overset{\displaystyle O}{\|}}{C}-OCHCH_3$
$$\qquad\qquad\qquad\qquad |$$
$$\qquad\qquad\qquad\qquad CH_3$$

54. Smell both samples. Butanoic acid has an unpleasant rancid odor; ethyl butanoate has a pleasant odor of pineapple.

56. The statement is false. When methyl propanoate is hydrolyzed, propanoic acid and methanol are formed.

58. (a) Mass percent of oxygen in Dacron:

$$\left(\overset{}{\underset{}{C}}\text{—}\bigcirc\text{—}\overset{}{\underset{}{C}}-OCH_2CH_2-O\right)_n \text{Dacron}$$

Mass of one unit of Dacron: = 192.2 g/mol
Mass of oxygen in one unit of Dacron = 64.00 g/mol
$$\frac{64.00\ \text{g/mol O}}{192.2\ \text{g/mol Dacron}} \times 100 = 33.30\%\ \text{O}$$

(b) $(105\ \text{units})\left(\dfrac{192.2\ \text{g}}{\text{unit}}\right) = 2.02 \times 10^4\ \text{g}$

60. A glyptal polyester would most likely be thermosetting because glycerol is trifunctional and would thus allow cross linking among chains in forming the ester polymer.

Chapter 25

2. Amides: Unsubstituted amides (except formamide) are solids at room temperature. Many are odorless and colorless. Low molar-mass amides are water soluble. Solubility in water decreases as the molar mass increases. Amides are neutral compounds. The NH_2 group is capable of hydrogen bonding. Amines: Low molar-mass amines are flammable gases with an ammonia-like odor. Aliphatic amines up to six carbon atoms are water soluble. Many amines have a "fishy" odor and many have very foul odors. Aromatic amines occur as liquids and solids. Soluble aliphatic amines give basic solutions. Aromatic amines are less soluble in water and less basic than aliphatic amines. The NH_2 group is capable of hydrogen bonding.

4. (a) Heterocyclic compounds are those in which all the atoms in the ring are not alike.
(b) The number of heterocyclic rings in each of the compounds is:
(i) purine, 2; (ii) ampicillin, 2; (iii) methadone, 0; (iv) nicotine, 2.

6. Ammonia is a toxic, basic, water soluble compound that can increase the pH of the blood and the urine and would be painful to pass through bodily tissue. However, ammonia is converted in the liver to the neutral diamide urea, which is water soluble and is excreted in the urine.

8. (a) $\underset{\underset{\displaystyle NH_2}{|}}{CH_3CHCH_2CHClCH_3}$

(c) [benzene ring with NH_2 at top and Cl at lower right]

(b) [benzene ring]$-\overset{\overset{\displaystyle }{\underset{\displaystyle \|}{C}}}{\underset{\displaystyle O}{}}-N\overset{\displaystyle -CH_2CH_3}{\underset{\displaystyle -CH_2CH_3}{}}$

10. (a) 1-amino-3-methylbutane
(b) formamide
(c) N-phenylpropanamide
(d) ethyldiisopropylamine
N-ethyl-N-isopropyl-2-propane

12. Increasing solubility in water

[benzene ring]$-\overset{\overset{\displaystyle O}{\|}}{C}-NH_2 < CH_3(CH_2)_4\overset{\overset{\displaystyle O}{\|}}{C}-NH_2 < CH_3\overset{\overset{\displaystyle O}{\|}}{C}-NH_2$

14. Several possibilities:

16. (a) neither acid nor base   (d) neither acid nor base
    (b) base   (e) neither acid nor base
    (c) base   (f) acid

18. Organic products
    (a)
    $CH_3C-N(CH_2CH_3)_2$
    (b)
    CH_2COOH + $CH_3\overset{+}{N}H_2CH_3$
    (c)
    $CH_3CHCH_2C-NHCH_3$
    CH_3

20. Structures of amines with formula C_3H_9N:
    $CH_3CH_2CH_2NH_2$ CH_3CHCH_3 $CH_3CH_2NHCH_3$ CH_3NCH_3
                          NH_2 CH_3
         1°   1°   2°   3°

22. (a) secondary   (c) secondary
    (b) tertiary   (d) tertiary

24. The isopropylamine solution in 1.0 M KOH would have the more objectionable odor because it would be in the form of the free amine, while in the acid solution the amine would form a salt that will have little or no odor.

26. (a) N-ethylaniline
    (b) 2-aminobutane
    (c) diphenylamine
    (d) ethylammonium bromide
    (e) pyridine
    (f) N-ethylacetamide

28. (a) $CH_3CH_2CH_2CH_2NCH_2CH_2CH_3$
                         $CH_2CH_2CH_2CH_3$
    (b)
    $\overset{+}{N}H_2Cl^-$
    CH_3
    (c) $CH_3CH_2NH_3^+Cl^-$
    (d) $CH_3CH_2CH_2CHCH_2OH$
                       NH_2

(e)
$CH_3CH_2CH_2C-CH-NH-$
CH_3 CH_3
(f) $H_2NCH_2CH_2NH_2$

30. (a) $CH_3CH_2C-NH_2$ (b) $CH_3CH_2CH_2CH_2NH_2$
    (c)   (d)

32. Amphetamines: stimulate the central nervous system; used to treat depression, narcolepsy, and obesity
    Tranquilizers: used to modify psychotic behavior and relieve pressure and anxiety
    Antibacterial agents: used as antibiotics

34.
    $-C-NH_2$ amide
    Some classes of biochemicals that contain an amide are:
    • Antibacterial agents such as ampicillin
    • B-vitamins such as niacin (nicotinamide)
    • Barbiturates such as Nembutal
    • Tranquilizers such as Valium (diazepam)

36.

38. $H_2N-CH_2CH_2CH_2CH_2-NH_2$
    Putrescine has two primary amine bonds.

40.

42.

Chapter 26

2. When the axes of two pieces of polaroid film are parallel, the maximum brightness of light passes through both. When one piece is rotated 90° the polaroid appears black, indicating very little light passing through.

4. Enantiomers are nonsuperimposable mirror-image isomers. Diastereomers are stereoisomers that are not enantiomers (not mirror-image isomers).

6. Most chiral molecules are stereospecific in their biological activity. Therefore, a racemic mixture of a drug provides only half the bioactive material prescribed. By using a single isomer of a compound, the dosage can be cut in half and no possible side effects can be encountered from its enantiomer.

8. Diastereoisomers do not have identical physical properties, so the differences form a basis for chemical or physical separation. Differences of boiling point, freezing point, and solubilities are most commonly used.

10. (a) a wood screw, (c) the letter g, (d) this textbook

12. Number of chiral carbon atoms
(a) 0 (b) 0 (c) 3 (d) 2

14. (a), (b), (c) will show optical activity.

16. Fructose, which has three chiral carbon atoms, will have eight possible stereoisomers. This can be determined from 2^n. $2^3 = 8$.

18. The two projection formulas (A) and (B) are the same compound, for it takes two changes to make (B) identical to (A).

(A) (B) 1st change in (B) (H and CH₃) 2nd change in (B) (F and CH₃)

20. (+)-alanine H₂N—C with COOH, H, CH₃   (−)-alanine H—C with COOH, NH₂, CH₃

(a) CH₃ / H₂N—H / COOH
(b) NH₂ / H—COOH / CH₃
(c) COOH / H—CH₃ / NH₂
(d) CH₃ / H—NH₂ / COOH
(e) COOH / CH₃—H / NH₂
(f) H / CH₃—NH₂ / COOH

22. (a) 2,3-dichlorobutane

CH₃ / H—Cl / Cl—H / CH₃   CH₃ / Cl—H / H—Cl / CH₃   CH₃ / H—Cl / H—Cl / CH₃

enantiomers   meso compound

(b) 2,4-dibromopentane

CH₃ / H—Br / H—H / Br—H / CH₃   CH₃ / Br—H / H—H / H—Br / CH₃   CH₃ / H—Br / H—H / H—Br / CH₃

enantiomers   meso compound

(c) 3-hexanol

CH₂CH₃ / H—OH / CH₂CH₂CH₃   CH₃CH₂ / HO—H / CH₃CH₂CH₂

enantiomers

24. All the stereoisomers of 3,4-dichloro-2-methylpentane:

CH(CH₃)₂ / H—Cl / H—Cl / CH₃   CH(CH₃)₂ / Cl—H / Cl—H / CH₃   CH(CH₃)₂ / H—Cl / Cl—Cl / CH₃   CH(CH₃)₂ / Cl—H / H—Cl / CH₃

A   B   C   D

Compounds A and B and C and D are pairs of enantiomers. There are no meso compounds. Pairs of diastereomers are A and C, A and D, B and C, and B and D.

26. The four stereoisomers of 2-chloro-3-hexene:

CH₃ / H—C—Cl / H—C / H—C‖ / CH₂CH₃   CH₃ / Cl—C—H / H—C / H—C‖ / CH₂CH₃   CH₃ / H—C—Cl / H—C / C—H‖ / CH₂CH₃   CH₃ / Cl—C—H / H—C / C—H‖ / CH₂CH₃

cis   cis   trans   trans

The two cis compounds are enantiomers and the two trans compounds are enantiomers.

28. (a) CH₃CH₂CHBr₂   CH₃CHBrCH₂Br
(i)   (ii)

CH₃CBr₂CH₃   CH₂BrCH₂CH₂Br
(iii)   (iv)

(b) (ii) is chiral

CH₂Br / H—Br / CH₃   CH₂Br / Br—H / CH₃   enantiomers

(i), (iii), and (iv) are achiral; there are no meso compounds.

30. Assume (+)-2-chlorobutane is

CH₃ / CH₂ / H—C—Cl / CH₃

All possible isomers formed when (+)-2-chlorobutane is further chlorinated to dichlorobutane are:

$$\begin{array}{ccccc}
CH_3 & CH_2Cl & CH_3 & CH_3 & CH_3 \\
| & | & | & | & | \\
CH_2 & CH_2 & CH_2 & H-C-Cl & Cl-C-H \\
| & | & | & | & | \\
H-C-Cl & H-C-Cl & Cl-C-Cl & H-C-Cl & H-C-Cl \\
| & | & | & | & | \\
CH_2Cl & CH_3 & CH_3 & CH_3 & CH_3 \\
A & B & C & D & E
\end{array}$$

Compounds A, B, and E would be optically active: C does not have a chiral carbon atom; D is a meso compound.

32. If 1-chlorobutane and 2-chlorobutane were obtained by chlorinating butane, and then distilled, they would be separated into the two fractions, because their boiling points are different. 1-Chlorobutane has no chiral carbon, so would not be optically active. 2-Chlorobutane would exist as a racemic mixture (equal quantities of enantiomers) because substitution of Cl for H on carbon-2 gives equal amounts of the two enantiomers. Distillation would not separate the enantiomers because their boiling points are identical. The optical rotation of the two enantiomers of the 2-chlorobutane fraction would exactly cancel, and thus would not show optical activity.

34. Compound (d) is meso.

$$\begin{array}{cc}
 & H \\
 & | \\
Br- & -CH_3 \\
Cl- & -CH_3 \\
Br- & -CH_3 \\
 & | \\
 & H
\end{array}$$

36. (c) is chiral.

enantiomers

38. $(CH_3)_2CHCH_2$— ⬡ —C=COOH
with CH$_3$ above and H below, — chiral carbon

$$HOOC-\underset{H}{\overset{CH_3}{C}}-⬡-CH_2CH(CH_3)_2$$

40. (a)

chiral carbon

caraway

(b) The spearmint molecule is the optical isomer of the caraway molecule and differs from it in structure at the chiral carbon atom.

42. Ephedrine has two chiral carbons and can have four stereoisomers. This number is calculated using 2^n.
$$2^2 = 4.$$

44. (−) placed in front of a name is used to indicate the rotation of plane-polarized light to the left, and (+) for rotation to the right. Therefore, we can write (−)-methorphan for levomethorphan and (+)-methorphan for dextromethorphan.

There is no obvious correlation between the structures of enantiomers and the direction in which they rotate plane-polarized light.

46. Mesostructures for alkanes:

(a) C_8H_{18}

$$\begin{array}{cc}
 & CH_3 \\
H- & -H \\
H- & -CH_3 \\
H- & -CH_3 \\
H- & -H \\
 & CH_3
\end{array}$$

(b) C_9H_{20}

$$\begin{array}{cc}
 & CH_3 \\
H- & -H \\
H- & -CH_3 \\
H- & -H \\
H- & -CH_3 \\
H- & -H \\
 & CH_3
\end{array}$$

(c) $C_{10}H_{22}$

$$\begin{array}{cccccc}
 & CH_3 & & CH_3 & & CH_3 \\
H- & -H & H- & -H & H- & -H \\
H- & -CH_3 & H- & -CH_3 & H- & -CH_3 \\
H- & -H & H- & -CH_3 & CH_3- & -H \\
H- & -H & H- & -CH_3 & H- & -CH_3 \\
H- & -CH_3 & H- & -H & H- & -H \\
H- & -H & CH_3 & & CH_3 \\
 & CH_3
\end{array}$$

$$\begin{array}{cccc}
 & CH_3 & & CH_3 \\
H- & -CH_3 & & -CH_3 \\
H- & -CH_3 & CH_3- & -H \\
H- & -CH_3 & CH_3- & -H \\
H- & -CH_3 & & -CH_3 \\
 & CH_3 & & CH_3
\end{array}$$

Chapter 27

2. The notations D and L in the name of a carbohydrate specify the configuration on the last chiral carbon atom (from C-1) in the Fischer projection formula. If the —OH is written to the right of that carbon the compound is a D-carbohydrate. If the —OH is written to the left, it is an L-carbohydrate. For example, D-glyceraldehyde and L-glyceraldehyde

4. Galactosemia is the inability of infants to metabolize galactose. The galactose concentration increases markedly in the blood and also appears in the urine. Galactosemia causes vomiting, diarrhea, enlargement of the liver, and often mental retardation. If not recognized a few days after birth it can lead to death.

6. A carbohydrate forms a five-member or six-member heterocyclic ring (one oxygen atom, the rest carbon atoms). If it forms a five-member ring, it is termed a furanose, after the compound furan. If it forms a six-membered ring it is termed a pyranose, after the compound pyran.

8. The cyclic forms of monosaccharides are hemicetals.

10. (a) sucrose: sugar beets and sugar cane
(b) lactose: milk
(c) maltose: sprouting grain and partially hydrolyzed starch

12. Invert sugar is sweeter than sucrose because it is a 50-50 mixture of fructose and glucose. Glucose is somewhat less sweet than sucrose, but fructose is much sweeter, so the mixture is sweeter.

14. In the Benedict test, both concentrated and dilute glucose solutions will give a precipitate. However, during the reaction the dilute solution will appear more greenish-yellow while the concentrated solution will appear more reddish.

16. Some people, as they grow older, stop producing the enzyme lactase and then lose the ability to digest lactose. This leads to lactose intolerance, a gaseous condition that produces intestinal discomfort.

18. (a) D-glyceraldehyde   L-glyceraldehyde

H—C=O   H—C=O
| |
H—C—OH   HO—C—H
| |
CH₂OH   CH₂OH

(b) If D-glyceraldehyde is reacted with hydrogen in the presence of a platinum catalyst, the product will be glycerol.

CH₂OH
|
H—C—OH
|
CH₂OH

20. The epimer could differ at carbon 2 or carbon 3 or carbon 4 or carbon 5. One possible answer is:

CHO
HO——H
HO——H
HO——H
H——OH
CH₂OH

22.

CHO
HO——H
H——OH
HO——H
HO——H
CH₂OH
L-glucose

CH₂OH
C=O
H——OH
HO——H
HO——H
CH₂OH
L-fructose

CHO
H——H
HO——H
HO——H
CH₂OH
L-2-deoxyribose

24. Either the Fischer projection formulas or Haworth formulas are satisfactory.

β-D-glucopyranose   α-D-galactopyranose   β-D-mannopyranose

26. The glucose units in cellulose are connected by β-1,4-glycosidic linkages. The human digestive system does not have the enzymes to catalyze the hydrolysis of cellulose.

28. The Kiliani–Fischer synthesis of D-ribose starts with the proper D-triose, D-glyceraldehyde.

(Answer continued on next page)

The circled hemiacetal structures allow these two disaccharides to be reducing sugars.

40. The systematic name for maltose is α-D-glucopyranosyl-(1,4)-α-D-glucopyranose. The systematic name for lactose is β-D-galactopyranosyl-(1,4)-α-D-glucopyranose.

42. (a) D-ribose and D-2-deoxyribose. The D-2-deoxyribose has no OH group on the number 2 carbon, only 2 hydrogen atoms. Both are five-carbon sugars.

(b) Amylose and amylopectin. Amylose is a straight-chain polysaccharide; amylopectin has branched chains and more monomer units per molecule. Both are large polysaccharides composed of α-D-glucose units.

(c) Lactose and isomaltose. These sugars are disaccharides. Lactose is composed of one galactose unit and one glucose unit while isomaltose is composed of two glucose units. The monosaccharide units in lactose are linked by a β-1,4-glycosidic bond whereas the units in isomaltose are linked by an α-1,6-glycosidic bond. Both disaccharides also contain hemiacetal structures.

44.

```
       COOH
HO ——— H
HO ——— H
H  ——— OH
H  ——— OH
       COOH
```
mannaric acid

46.

```
   HC=O            HC=O            HC=O            HC=O
H ——— OH     HO ——— H      H ——— OH      HO ——— H
H ——— OH      H ——— OH     HO ——— H      HO ——— H
HO ——— H     HO ——— H      HO ——— H      HO ——— H
   CH2OH           CH2OH           CH2OH           CH2OH
     A               B               C               D
```
Epimers are A and B, A and C, B and D, C and D.

48. The Haworth formula for β-D-glucopyranosyl-(1,4)-α-D-galactopyranose is

The Haworth formula for β-D-galactopyranosyl-(1,6)-β-D-mannopyranose is

---

```
                           CN
                    HO ——— H
                    H  ——— OH
                    H  ——— OH
       HC=O              CH2OH
 Na(Hg)  H ——— OH  HCN
         H ——— OH          CN
         CH2OH       H ——— OH
                    H ——— OH
                    H ——— OH
                        CH2OH
```

```
   CN                COOH               HC=O
H ——— OH          H ——— OH    Na(Hg)  H ——— OH
H ——— OH   H2O/H+  H ——— OH   ———————  H ——— OH
H ——— OH          H ——— OH           H ——— OH
   CH2OH             CH2OH             CH2OH
                                      D-ribose
```

30. No, D-2-deoxygalactose is not the same as D-2-deoxyglucose. D-Galactose differs from D-glucose at carbon 4, so replacement of the carbon 2 OH with an H does not make these two sugars identical.

32. The monosaccharide composition of:
(a) lactose: one glucose and one galactose unit
(b) amylopectin: many glucose units
(c) cellulose: many glucose units
(d) sucrose: one glucose and one fructose unit

34. Both maltose and isomaltose are disaccharides composed of two glucose units. The glucose units in maltose are linked by an α-1,4-glycosidic bond while the glucose units of isomaltose are linked by an α-1,6-glycosidic bond.

36. Both maltose and isomaltose will show mutarotation. Both disaccharides contain a hemiacetal structure that will open, allowing mutarotation.

38.

maltose

lactose

50. Aspartame supplies many fewer calories than sucrose. In addition, oral bacteria cannot use aspartame as efficiently as sucrose and will form fewer dental caries.

52. High fructose corn syrup is produced by breaking down some corn starch polymers to D-glucose monomers, which are then converted to D-fructose, a very sweet monosaccharide.

54. Compound A must be a reducing disaccharide because it produces a reddish color in the Benedict test. Sucrose is nonreducing so compound A must be maltose.

56.

```
      HC=O            HC=O
   H ──┼── OH   HO ──┼── H
  HO ──┼── H     H ──┼── OH
      CH₂OH          CH₂OH
```
Enantiomer of I

These compounds are not epimers because they differ at more than one chiral carbon.

58. Cellulose, amylose, and amylopectin are polymers of glucose. Cellulose exists in the form of fibers, is not digestible by humans, and therefore remains in the digestive tract as fibers. Amylose and amylopectin are digested to glucose, which is dissolved into the bloodstream.

60. No, the classmate should not be believed. Although D-glucose and D-mannose are related as epimers, it is pairs of *enantiomers* that have equal and opposite optical rotation.

Chapter 28

2. Although caproic acid, $CH_3(CH_2)_4COOH$, has the same number of polar bonds as stearic acid, $CH_3(CH_2)_{16}COOH$, caproic acid has a shorter, nonpolar hydrocarbon chain and, therefore, is more water soluble.

4. The three essential fatty acids are linoleic, linolenic, and arachidonic acids. Diets lacking these fatty acids lead to impaired growth and reproduction, and skin disorders such as eczema and dermatitis.

6. Aspirin relieves inflammation by blocking the conversion of arachidonic acid to prostaglandins.

8. A membrane lipid must be **(a)** partially hydrophobic to act as a barrier to water and **(b)** partially hydrophilic, so that the membrane can interact with water along its surface.

10. The four classes of eicosanoids are prostaglandins, prostacyclins, thromboxanes, and leukotrienes.

12. Atherosclerosis is the deposition of cholesterol and other lipids on the inner walls of the large arteries. These deposits, called plaque, accumulate, making the arterial passages narrower and narrower. Blood pressure increases as the heart works to pump sufficient blood through the restricted passages. This may lead to a heart attack, or the rough surface can lead to coronary thrombosis.

14. Dietary fish oils provide fatty acids that inhibit the formation of thromboxanes—compounds that participate in blood clotting.

16. Because the interior of a lipid bilayer is very hydrophobic, molecules with hydrophilic character cross the lipid bilayer with difficulty.

18. Both facilitated diffusion and active transport catalyze movement of compounds through membranes. Active transport requires an input of energy whereas facilitated diffusion does not.

20. Up to 80% of the total body cholesterol is metabolically produced. The statin drugs block metabolic production of cholesterol.

22.

```
                  O
                  ‖
  CH₂O─C(CH₂)₁₄CH₃                palmitic acid
   |              O
   |              ‖
  CHO─C(CH₂)₁₄CH₃                 palmitic acid
   |              O
   |              ‖
  CH₂O─C(CH₂)₇CH=CH(CH₂)₇CH₃   oleic acid
```

There is one other possible triacylglycerol with the same components.

24. Yes, a triacylglycerol that contains three units of stearic acid would be more hydrophobic than a triacylglycerol that contains three units of myristic acid. Larger molecules tend to be more hydrophobic than smaller molecules, and a triacylglycerol with three stearic acid units is larger than a triacylglycerol with three myristic acid units.

26.

```
  CH₂OH
   |
  CHOH                               glycerol
   |
  CH₂OH
  CH₃(CH₂)₇CH=CH(CH₂)₇COOH      oleic acid
  CH₃(CH₂)₁₄COOH                 palmitic acid
  CH₃(CH₂)₁₆COOH                 stearic acid
```

28.

```
                O
                ‖
  CH₂O─C(CH₂)₁₆CH₃
   |            O
   |            ‖
  CHO─C(CH₂)₁₆CH₃
   |            O              CH₃
   |            ‖               |
  CH₂O─P─OCH₂CH₂N⁺─CH₃
              |               |
              O⁻              CH₃
```

30.

```
  OH
   |
  CHCH=CH(CH₂)₁₂CH₃
   |        O
   |        ‖
  CHNH─C(CH₂)₁₆CH₃
   |        O
   |        ‖
  CH₂O─P─OCH₂CH₂N⁺H₃
           |
           O⁻
```

32.

OH
|
CHCH=CH(CH₂)₁₂CH₃
|
CHNH—C(CH₂)₇CH=CH(CH₂)₇CH₃   (O double bond)
|
CH₂O

CH₂OH
OH, O
OH
OH

OH
|
CHCH=CH(CH₂)₁₂CH₃
|
CHNH—C(CH₂)₇CH=CH(CH₂)₇CH₃   (O double bond)
|
CH₂O

CH₂OH
OH, O
OH
OH

34.

HO—C(CH₂)₁₂CH₃   (O below)

→ hydrophilic exterior (—COOH)

→ hydrophobic interior (—(CH₂)₁₂CH₃)

36. HDL (high density lipoprotein) differs from LDL (low density lipoprotein) in that
(a) LDL has a lower density than HDL
(b) LDL delivers cholesterol to peripheral tissues whereas HDL scavenges cholesterol and returns it to the liver.

38. (a) Both compounds have two long hydrophobic chains.
(b) For both compounds, the primary alcohol can react further with either acids (to form esters) or sugars (to form acetals).

40. Phosphate ion will move from a region of low concentration to a region of high concentration as it moves from a 0.1 M solution across a membrane to 0.5 M solution. This process requires energy and can be accomplished by active transport.

42. Thromboxanes act as vasoconstrictors and stimulate platelet aggregation. Leukotrienes have been associated with many of the symptoms of an allergy attack (e.g., an asthma attack).

44. Both olestra and natural fats contain fatty acid units. However, in natural fats, the fatty acid units are linked to glycerol while in olestra the fatty acid units are linked to sucrose. Olestra has many of the qualities of a fat. However, it is not shaped like a natural fat and is not attacked by digestive enzymes. Thus, olestra passes through the stomach and the intestines undigested and does not provide any caloric value to the cells.

46. This meal is changed by increasing the fat content from 10 grams to 15 grams, an increase of 5 grams. Since each gram of fat yields an average of 9.5 Cal an increase of 5 grams equates to an increase of 47.5 Cal. Thus the meal will now contain 282 Cal.

48. $CH_3(CH_2)_7CH=CH(CH_2)_7COOH$
oleic acid or 9-octadeceneoic acid

Chapter 29

2. All amino acids and proteins contain carbon, hydrogen oxygen, and nitrogen. Sulfur is contained in some of the amino acids, and thus in most proteins.

4. The amino acids that are essential to humans are isoleucine, leucine, lysine, methionine, phenylalanine, threonine, tryptophan, and valine.

6. At its isoelectric point, a protein molecule must have an equal number of positive and negative charges.

8. The sulfur-containing amino acid, cysteine, has the special role in protein structure of creating disulfide bonding between polypeptide chains, which helps control the shape of the molecule.

10. The α-helix forms a tube composed of a spiraling polypeptide chain whereas the β-pleated sheet forms a plane composed of polypeptide chains aligned roughly parallel to each other.

12. Amino acids containing a benzene ring give a positive xanthoproteic test (formation of yellow reaction products). Among the common amino acids, these would include phenylalanine, tryptophan, and tyrosine.

14. Protein column chromatography uses a column packed with polymer beads (solid phase) through which a protein solution (liquid phase) is passed. Proteins separate based on differences in how they interact with the solid phase. The proteins move through the column at different rates and can be collected separately.

16. In ordinary electrophoresis the rate of movement of a protein depends on its charge and size. In SDS electrophoresis a detergent, sodium dodecyl sulfate, is added to the protein solution, which masks the differences in protein charges, leaving the separation primarily due to the size of the various proteins.

18.   D-serine   L-serine (form commonly found
                                   in proteins)

$$\underset{CH_2OH}{\overset{COOH}{H-\overset{|}{\underset{|}{C}}-NH_2}} \quad \underset{CH_2OH}{\overset{COOH}{H_2N-\overset{|}{\underset{|}{C}}-H}}$$

20. 33 g

22. $\underset{NH_3^+}{\overset{O}{NH_2\overset{\|}{C}-CH_2-\underset{|}{CH}-COO^-}}$

24. For tryptophan:
    (a) Zwitterion formula   **(b)** formula in 0.1 M H_2SO_4

$$\underset{NH_3^+}{CH_2CHCOO^-} \qquad \underset{NH_3^+}{CH_2CHCOOH}$$

(indole ring with N-H)   (indole ring with N-H)

    (c) formula in 0.1 M NaOH

$$\underset{NH_2}{CH_2CHCOO^-}$$

(indole ring with N-H)

26. Ionic equations showing how leucine acts as a buffer toward:
    (a) H^+

$$\underset{NH_3^+}{(CH_3)_2CHCH_2CHCOO^-} + H^+ \longrightarrow \underset{NH_3^+}{(CH_3)_2CHCH_2CHCOOH}$$

    (b) OH^-

$$\underset{NH_3^+}{(CH_3)_2CHCH_2CHCOO^-} + OH^- \longrightarrow$$
$$\underset{NH_2}{(CH_3)_2CHCH_2CHCOO^-} + H_2O$$

28. $\underset{NH_3^+}{(CH_3)_2CHCHCOO^-}$

30.

$$\underset{O}{\overset{CH_3}{\underset{\|}{\overset{|}{NH_2CH_2C-NHCHCOOH}}}} \qquad \underset{O}{\overset{CH_3}{\underset{\|}{\overset{|}{NH_2CHC-NHCH_2COOH}}}}$$
    — peptide bond   — peptide bond
      Gly-Thr   Thr-Gly

32. (a) $\underset{O}{\overset{CH_3 \qquad CH_3}{\underset{\|}{NH_2CHC-NHCHCOOH}}}$

    (b) $\underset{O \qquad\quad O}{\overset{CH_2OH}{\underset{\|\qquad\quad\|}{NH_2CHC-NHCH_2C-NHCH_2COOH}}}$

(c) $\underset{O \qquad\qquad O}{\overset{CH_2OH \qquad\qquad CH_3}{\underset{\|\qquad\qquad\|}{NH_2CHC-NHCH_2C-NHCHCOOH}}}$

34. Tyr-Asp-Ala   Tyr-Ala-Asp   Asp-Tyr-Ala
    Asp-Ala-Tyr   Ala-Tyr-Asp   Ala-Asp-Tyr

36. $\underset{H}{\overset{O \;\; CH_3}{\underset{|}{-CH_2\overset{\|}{C}N\overset{|}{C}H-}}}$ involved in hydrogen bonding

38. $-CH_2COO^-\;\overset{+}{N}H_3CH_2CH_2CH_2CH_2-$
                  — ionic bond

40. The tripeptide, Gly-Ser-Asp, will
    (a) react with $CuSO_4$ (biuret test) to give a violet color. The tripeptide has the required two peptide bonds.
    (b) not react to give a positive xanthoproteic test because there are no benzene rings in this tripeptide.
    (c) react with ninhydrin to give a blue solution.

42.

$$\underset{NH_2CHCOOH,}{\overset{CH_3}{|}} \qquad \underset{NH_2CHCOOH,}{\overset{COOH}{\overset{|}{\overset{CH_2}{\overset{|}{CH_2}}}}} \qquad \underset{NH_2CHCOOH,}{\overset{OH}{\overset{(benzene)}{\overset{|}{CH_2}}}}$$

44. $6.8 \times 10^4 \dfrac{g}{mole}$ (molar mass of hemoglobin)

46. Phe-Ala-Ala-Leu-Phe-Gly-Tyr

48. (a) Arginine will not migrate to either electrode in an electrolytic cell at a pH of 10.8.
    (b) Arginine will migrate toward the positive electrode at pH greater than 10.8, that is, more basic than its isoelectric point.

50. A small protein like ribonuclease (or myoglobin) would also have a small number of protein domains. Small proteins (e.g. myoglobin, ribonuclease) commonly fold into one globular unit (one domain) while larger proteins will fold into more than one globular unit (more than one domain).

52. The immunoglobulin hypervariable regions allow the body to produce millions of different immunoglobulins, each with its distinct amino acid sequence and unique antigen binding site.

54. thirty-nine dipeptides

56. Glutamic acid is the only one of these three amino acids with polar bonds in its side chain. Thus, glutamic acid will be the most polar.

Chapter 30

  2. Enzymes serve as catalysts for chemical reactions in the body.

4. Sucrase catalyzes the hydrolysis of sucrose; lactase catalyzes the hydrolysis of lactose; maltase catalyzes the hydrolysis of maltose.

6. The reaction rate often can be increased (1) by increasing the reactant (substrate) concentration and (2) by raising the reaction temperature.

8. When substrates bind to an active site, they may be converted more easily to products because (1) the substrates are close together (proximity catalysis); (2) the substrates are oriented to best react (productive binding hypothesis); (3) when the substrates bind to the active site, they change shape to be more like products (strain hypothesis).

10. Pectinase is used to remove the peeling from oranges and grapefruits. The pectinase penetrates the peeling and dissolves the white stringy material that attaches the peeling to the fruit. The peeling can then be easily removed from the fruit.

12. $4 \times 10^{-7} M/s$

14. 1×10^3 reactants converted to products by three pepsin molecules

16. The first enzyme would have a narrow active site which is long enough to fit the four carbon carboxylic acid, butanoic acid ($CH_3CH_2CH_2COOH$). The second enzyme would have a narrow active site which would be shorter because it need only fit the two carbon carboxylic acid, acetic acid (CH_3COOH).

18. An enzyme-catalyzed rate decreases at high temperatures because the enzyme loses its natural shape (denatures). As an enzyme denatures it ceases to be an effective catalyst.

20. If the turnover number for an enzyme is increased, the enzyme can convert more reactants to products per unit time; the enzyme is a better catalyst. Any process that increases an enzyme's catalytic abilities is termed activation.

22. Yes. Enzyme B is more effective than enzyme A based on a comparison of turnover numbers (9.8×10^{-1}/s for enzyme B vs. 0.05/s for enzyme A).

24. Glutamine is a starting material (reactant) for the liver metabolic pathway that produces urea. Thus, glutamine control of this pathway must be "feedforward". Since an increase in glutamine concentration causes the metabolic pathway to speed up, this control must be feedforward activation.

26. Blood clots in the brain are the major cause of strokes. Specific proteases are used to dissolve these blood clots to alleviate the cause of strokes.

28. Feedback inhibition means that when a large amount of product has been formed, the beginning of a process will slow down. Thus, feedback inhibition protects against overproduction. "Feedback activation" means that when a large amount of product is formed, the beginning of the process will accelerate. A state of overproduction will worsen and might lead to cell death.

30. The enzyme lactase catalyzes hydrolysis of the disaccharide lactose to produce two monosaccharides, D-glucose and D-galactose.

Chapter 31

2. A nucleoside is a purine or pyrimidine base linked to a sugar molecule, usually ribose or deoxyribose. A nucleotide is a purine or pyrimidine base linked to a ribose or doxyribose sugar which in turn is linked to a phosphate group.

4. The major function of ATP in the body is to store chemical energy, and to release it when called upon to carry out many of the complex reactions that are essential to most of our life processes.

6. Complementary bases are the pairs that hydrogen bond ("fit") to each other between the two helixes of DNA.

8. Since there are only four different bases to make up the code, one nucleotide could only specify four possible amino acids; two nucleotides could specify 16 amino acids; three nucleotides could specify 64 amino acids. Since there are at least 20 amino acids needed, three nucleotides are required.

10. A codon is a triplet of three nucleotides, and each codon specifies one amino acid. The cloverleaf model of transfer RNA has an anticodon loop consisting of seven unpaired nucleotides. Three of these make up the anticodon, which is complementary to, and hydrogen-bonds to the codon on mRNA.

12. From time to time a new trait appears in an individual that is not present in either parents or ancestors. These traits, which are generally the result of genetic or chromosomal changes, are called mutations.

14. With the knowledge of gene structure, genetic therapists can observe DNA with genetic defects and work toward deleting the error and replacing it with a corrected gene. Geneticists will be able to tell in advance who is prone to getting certain diseases. Also they will be able to more accurately detect and predict birth defects, and in general have a better understanding of inherited diseases.

16. (a) G, guanosine
    (b) GMP, guanosine-5′-monophosphate
    (c) dGDP, deoxyguanosine-5′-diphosphate
    (d) CTP, cytidine-5′-triphosphate

18. (a) U

(b) UMP

(c) CTP

(d) dTMP

20. There are several possible sequences for the three-nucleotide single-stranded RNA. One possible structure follows:

22.

(dotted lines are hydrogen bonds)

24. DNA is considered to be the genetic substance of life, because it contains the sequence of bases that carries the code for genetic characteristics.

26. Transcription is the process of making RNA using a DNA template, whereas translation is a process for making protein using an mRNA template.

28. Both mRNA and rRNA can be found in the ribosomes. rRNA serves as part of the ribosome structure, whereas mRNA serves as a template for protein synthesis.

30. $3 \times 573 = 1719$

32. (a) CGAAUCUGGACU
   (b) GCUUAGACCUGA
   (c) Arg-Ile-Trp-Thr

34. Translation makes a polymer of amino acids by forming amide bonds to connect the amino acids to each other. The amide bond combines an amine with a carboxylic acid.

36. Translation initiation occurs when the ribosome reaches a special AUG or GUG codon along the mRNA. Since there is commonly more than one AUG or GUG codon, the ribosome must use other information to choose the special AUG or GUG. This codon is the starting point for protein synthesis and is bound by either a special tRNA carrying *N*-formyl methionine (in procaryotes) or a tRNA carrying methionine (in eucaryotes).

38. (a) GCG   **(c)** CUA
   (b) UGU   **(d)** AAG

40. In RNA the guanine content does not have to be equal to the cytosine content, because RNA is a single strand. Its complements are on the DNA template, not on the RNA. They have to be equal in DNA, because it is a double helix, with each molecule hydrogen-bonded to its complement.

42. (a) Phe-His-Lys
   (b) AAAGTATTC

44. The Human Genome Project is a cooperative effort by leading scientific laboratories to sequence the entire human genome (approximately 3 billion base pairs).

Chapter 32

2. The "energy allowance" represents the recommended dietary energy supply needed to maintain health.

4. Candy contains primarily sucrose, a source of calories, but little or no other nutrients. Thus, candy is said to provide "empty calories."

6. Essential fatty acids are required in the diet for normal growth and development.

8. Animal proteins are generally a source of all twenty common amino acids, whereas vegetable proteins may be deficient in one or several of the amino acids.

10. Vitamin D: functions as a regulator of calcium metabolism; Vitamin K: enables blood clotting to occur normally;

Vitamin A: functions to furnish the pigment that makes vision possible. Many vitamins act as coenzymes.

12. Calcium is a major constituent of bones and teeth and is also important in nerve transmission among other functions.

14. Some common categories of food additives are (1) nutrients; (2) preservatives; (3) anticaking agents; (4) emulsifiers; (5) thickeners; (6) flavor enhancers; (7) colors; (8) nonsugar sweeteners.

16. The U.S. Food and Drug Administration is responsible for regulating the use of food additives.

18. Chyme is the material of liquid consistency found in the stomach and consisting of food particles reduced to a small size resulting from gastric digestion of food.

20. Intestinal mucosal cells produce many enzymes needed to complete digestion (e.g., disaccharidases, aminopeptidases, and dipeptidases).

22. GRAS means Generally Recognized As Safe.

24. (a) 132 g
    (b) 1464 mg
    (c) 31.5 mg

26. (a) 3×10^2 kcal
    (b) 40%

28. 4.7%

30. 41%

32. The two micronutrient classes are minerals and vitamins.

34. The three classes of nutrients that commonly supply energy for the cells are proteins, carbohydrates, and lipids.

36. Cellulose is important as dietary fiber. Although it is not digested, cellulose absorbs water and provides dietary bulk which helps maintain a healthy digestive tract.

38. Pancreatic juice is slightly basic.

40. Carbohydrates are digested in the mouth and small intestine.

42. Pepsin digests proteins.

44. 9 kg of body fat represents about 80,000 kcal of energy. Complete starvation for about 30 days would cause a net loss of 9 kg of fat. Thus, this new diet is unreasonable.

46. Fat (9 kcal/g) contains more than twice as much energy as carbohydrate (4 kcal/g). Thus, a tablespoon of butter will greatly increase the calorie content of a medium size baked potato.

48. The foods of higher animals must be digested before they can be utilized because foods are primarily large molecules. Foods must be broken down to much smaller molecules in order to pass through the intestinal walls into the blood and lymph systems where they can be utilized in the metabolic process of animals.

50. Nutritionists recommend less than 300 mg of dietary cholesterol per day. One large egg (270 mg cholesterol) represents almost the total daily recommended amount (about 90% of the RDA).

Chapter 33

2. Mitochondrial electron transport and oxidative phosphorylation are found both in plants and animals. Electrons are moved from NADH and $FADH_2$ to molecular oxygen to release energy. This energy allows ATP to be formed.

4. High energy phosphate bond

$$^-O-\overset{\overset{\displaystyle O}{\|}}{\underset{\underset{\displaystyle O^-}{|}}{P}}\sim O-\overset{\overset{\displaystyle O}{\|}}{\underset{\underset{\displaystyle O^-}{|}}{P}}-O-R$$

It is generally associated with adenosine triphosphate (ATP).

6. An oxidation-reduction coenzyme is a reusable organic compound which helps an enzyme carry out an oxidation-reduction reaction.

8. Oxidative phosphorylation uses the mitochondrial electron transport system to directly produce ATP from oxidation-reduction reactions. Substrate-level phosphorylation does not use the electron transport system, but, instead, involves transfer of a phosphate group from a substrate to ADP to form ATP.

10. Both chloroplasts and mitochondria are organelles that are bonded by two membranes. In both cases they contain many folded internal membranes (which facilitate oxidation–reduction reactions).

12. The overall photosynthesis reaction in higher energy plants is $6\ CO_2 + 6\ H_2O + 2820\ kJ \longrightarrow C_6H_{12}O_6 + 6\ O_2$. Chloroplast pigments trap light to provide energy for photosynthesis.

14. The average oxidation state of carbon in ethanal in -2. The average oxidation state of carbon in acetic acid is 0. The carbons in ethanal, being in a lower oxidation state, will deliver more energy in biological redox reactions.

16. Both oxidized forms of the coenzymes (NAD^+ and FAD) are on the same side of the equation. For a reaction to occur one reduced coenzyme (NADH or FADH) must be paired with one of the oxidized coenzymes.

18. The conversion of acetate to long-chain fatty acids is anabolic because:
    (a) a smaller molecule (acetate) is converted to a larger molecule (long-chain fatty acids)
    (b) the carbons from acetate become reduced (on the average) as they are converted to long-chain fatty acids

20. Chloroplasts carry out photosynthesis, not oxidative phosphorylation.

22. The chemical changes in the chloroplast are said to be anabolic because:

(a) smaller carbon dioxide molecules are converted to larger glucose molecules

(b) as this transformation takes place, the carbons become progressively more reduced

(c) this process requires an input of energy (from sunlight)

24. Three important characteristics of a catabolic process are (1) complex substances are broken down into simpler substances; (2) carbons are often oxidized; (3) cellular energy is often produced.

26. 0.34 and O_2

28. 0.18 mol e^-

30. two high energy phosphate bonds

32. 58 kJ

34. Since the phosphorylated substrate 1,3-diphosphoglycerate contains only one high energy phosphate bond, substrate-level phosphorylation can produce only one ATP.

36. 12 ATP

38. The mitochondria carry out much of the oxidation-reduction reactions for the cell. Because these reactions involve electron movement, they are more easily controlled in a nonaqueous environment. Thus mitochondria contain 90% membrane by mass.

40. Both ATP and NAD^+ contain a ribose that is linked to an adenine at carbon-1 and to a phosphate at carbon-5.

Chapter 34

2. Glucose catabolism breaks down glucose to the smaller carbon dioxide in an oxidative process. This produces energy for the cell and is catabolic. Photosynthesis produces glucose from carbon dioxide via a reductive path. Light energy is required and this is an anabolic process.

4. Enzymes are needed so that biochemical reactions will proceed fast enough to keep the cell alive.

6. The liver replenishes muscle glucose (via the blood) by (a) releasing glucose from liver glycogen stores and (b) by converting lactic acid back into glucose.

8. One high energy phosphate bond (in GTP) is formed directly in the citric acid cycle. In addition, one $FADH_2$ and three NADHs are produced.

10. Hormones are chemical substances that act as control or regulatory agents in the body. Hormones are secreted by the endocrine, or ductless glands directly into the bloodstream and are transported to various parts of the body to exert specific control functions.

12. The renal threshold is the concentration of a substance in the blood above which the kidneys begin to excrete that substance.

14. In gluconeogenesis, glucose is synthesized from noncarbohydrate sources: lactate, amino acids, and glycerol. These

three substances are converted to Emden–Meyerhof pathway intermediates, which are then converted to glucose. Since the Embden–Meyerhof pathway is anaerobic, gluconeogenesis is considered to be an anaerobic process.

16. Glycogenesis is the process by which glucose is used to produce glycogen.

Glycogen

18. Glycogenolysis is catabolic because it starts with larger precursors (glycogen molecules) and forms smaller products (glucose molecules).

20. The Embden-Meyerhof pathway uses substrate-level phosphorylation to produce ATP. That is, following carbon oxidation a phosphate group is transferred from a substrate to ADP, forming ATP.

22.

24. Although no molecular oxygen (O_2) is used in the citric acid cycle, reduced coenzymes, NADH and $FADH_2$ are produced. These reduced coenzymes must be oxidized by electron transport so that the citric acid cycle can continue. And, it is electron transport that uses molecular oxygen. Thus, the citric acid cycle depends on the presence of molecular oxygen and is considered to be aerobic.

26. 1.7 mol ATP

28. The end product of the anaerobic catabolism of glucose in muscle tissue is lactate.

30. The gluconeogenesis pathway is considered to be anabolic because (a) smaller, noncarbohydrate precursors such as lactate are converted to the larger product, glucose, and (b) as this conversion takes place, the carbons become progressively more reduced.

32.

$$HO-\underset{\underset{CH_2COOH}{|}}{CH}COOH + \underline{NAD^+} \longrightarrow$$

$$O=\underset{\underset{CH_2-COOH}{|}}{C}-COOH + \underline{NADH} + H^+$$

34. The glycolysis pathway uses substrate-level phosphorylation to produce ATP and thus does not need electron transport and oxidative phosphorylation.

36. Unlike vitamins, which must be supplied in the diet, hormones are produced in the body.

38. The normal blood glucose concentration is 70–90 mg/100 mL of blood under fasting conditions. Hyperglycemia occurs when the blood glucose concentration rises above this range. Thus, 350 mg glucose/100 mL blood is hyperglycemic.

40. Hormones function as chemical messengers. They are chemical substances that act as control or regulatory agents in the body. Enzymes are catalysts for specific reactions, allowing these reactions to occur faster and under milder conditions than would otherwise be possible.

42. If a large overdose of insulin was taken by accident, the blood glucose concentration would drop to a low level, probably in the hypoglycemic range. This could result in fainting, convulsions, and unconsciousness.

44.

$$\underset{\underset{OH}{|}}{CH_3CH}COO^-$$

lactate

All three carbons of lactate are more reduced than the carbon in CO_2. Thus, lactate can provide more cellular energy upon further oxidation of the carbons.

46. In aerobic catabolism, pyruvate is converted to carbon dioxide yielding many ATPs while consuming molecular oxygen. In anaerobic catabolism, pyruvate is used to recycle NADH back to NAD^+.

48. Oxidation. The average oxidation number of carbon in glucose is 0. The oxidation number of carbon in carbon dioxide is +4. Thus the conversion of glucose to carbon dioxide is oxidation.

50. $C_6H_{12}O_6 + 2\ NAD + 2\ ADP + 2\ P_i \longrightarrow$
 glucose

$$2\ CH_3\overset{\overset{O}{\|}}{C}-COO^- + 2\ NADH + 2\ ATP$$
pyruvate

Chapter 35

2. Knoop prepared a homologous series of straight-chain fatty acids with a phenyl group at one end and a carboxyl group at the other end. He fed each of these acids to test animals and analyzed the urine of the animals for metabolic products of the acid. Those animals that had been fed acids with an even number of carbons excreted phenylaceturic acid. Those animals that had been fed acids with an odd number of carbon atoms excreted hippuric acid. The experiment showed that the metabolism of the acid must occur by shortening the chain two carbon atoms at a time.

4. Ketone bodies are relatively soluble in the bloodstream and can be transported to energy-deficient cells in time of need. Ketone bodies supply energy via β-oxidation and the citric acid cycle.

6. Fat acts as a cushion or shock absorber around internal organs and as an insulating blanket.

8. The nitrogen cycle is the process by which nitrogen is circulated and recirculated from the atmosphere through living organisms and back to the atmosphere. Beginning with the atmosphere, nitrogen is fixed by bacterial action, combustion, or chemical fixation. In the soil nitrogen is converted to nitrates, which are absorbed by higher plants and animals and converted to organic compounds. Eventually, the nitrogen is returned to the soil in the form of urea and feces or through bacterial decomposition of dead plants and animals. Some nitrogen remains in the soil and the rest returns to the atmosphere as free nitrogen.

10. Soybeans enrich the soil because their roots serve as host to the symbiotic, nitrogen-fixing bacteria.

12. Metabolic fate of amino acids:
 (a) incorporation into a protein
 (b) utilization in the synthesis of other nitrogenous compounds such as nucleic acids
 (c) deamination to a keto acid
 (d) Ultimately, excess nitrogen compounds in the body are converted to urea and excreted in the urine and the feces.

14. Nitrogen is excreted as (a) ammonia by fish and (b) uric acid by birds.

16. three ATPs

18. For a growing child the amount of nitrogen consumed is more than that excreted. Consequently the amount of amino acids from dietary protein increases in the nitrogen pools and the child is said to be in positive nitrogen balance.

20. $\boxed{CH_3CH_2}\ CH_2CH_2CH_2COOH$; the two carbons circled remain.

22. $CH_3(CH_2)_{10}*CH_2CH_2COOH$

24. 8 acetyl-CoA

26. 7 NADH

28. 25 ATP formed from lauric acid

30. (a) ACP is a protein whereas CoA is a coenzyme
 (b) ACP is used in fatty acid synthesis whereas CoA is used in the β-oxidation pathway.

32. A rapid growth spurt commonly means the body is using ingested amino acids to build new proteins. Thus, less nitrogen will be excreted and the nitrogen balance should become more positive.

34. L-aspartic acid is not a ketogenic amino acid because a diet high in L-aspartic acid caused no increase in blood levels of the ketone body β-hydroxybutyric acid. (Other ketone bodies would not increase their blood levels also.) Because L-aspartic acid is not a ketogenic amino acid, it must be a glucogenic amino acid.

36. (a) $\overset{\overset{\text{O}}{\|}}{\text{HOOCCH}_2\text{C}}\text{COOH}$

(b) $\text{C}_6\text{H}_5-\text{CH}_2\overset{\overset{\text{O}}{\|}}{\text{C}}\text{COOH}$

38.
$$\underset{(\text{H}_2\text{N})-\text{CH}-\text{COOH}}{\overset{\text{COOH}}{\underset{|}{\overset{|}{\text{CH}_2}}}}$$

40. β-hydroxybutyric acid:

$$\underset{\qquad\;\;|}{\underset{\text{OH}}{\text{CH}_3\text{CHCH}_2\text{COOH}}}$$

42. The six-carbon fatty acid yields more ATP than the six-carbon hexose during catabolism primarily because more acetyl-CoA molecules are formed and can feed into the citric acid cycle, electron transport, and oxidative phosphorylation.

44.
$$\underset{\text{L-glutamic acid}}{\overset{\text{CH}_3\quad\text{O}}{\text{CH}_3\text{CH}_2\text{CH}-\text{C}-\text{COOH} + \underset{\;\;|}{\underset{\text{NH}_2}{\text{HOOCCH}_2\text{CH}_2\text{CHCOOH}}}}} \xrightarrow{\text{enzyme}} \underset{\text{isoleucine}}{\underset{\;\;|}{\underset{\text{NH}_2}{\overset{\text{CH}_3}{\text{CH}_3\text{CH}_2\text{CHCHCOOH}}}}} + \underset{\text{α-ketoglutaric acid}}{\overset{\text{O}}{\text{HOOCCH}_2\text{CH}_2\text{CCOOH}}}$$

46. In step 2, the fatty acid carbons at the alpha and beta positions change their oxidation numbers from -2 to -1; they are oxidized.

PHOTO CREDITS

Entries in boldface refer to key terms, which are defined in the text.